HANDBOOK OF
VIDEO
DATABASES
DESIGN AND APPLICATIONS

EDITED BY
Borko Furht
Oge Marques

CRC PRESS

Boca Raton   London   New York   Washington, D.C.

Library of Congress Cataloging-in-Publication Data

Furht, Borivoje.
    Handbook of video databases : design and applications / Borko Furht, Oge Marques
         p. cm.
       Includes bibliographical references and index.
       ISBN 0-8493-7006-X (alk. paper)
        1. Optical storage devices. 2. Image processing--Digital techniques. 3. Database
    management. 4. Image processing--Databases. 5. Video recordings--Databases. I.
    Marques, Oge. II. Title.

TA1635.F88 2003
006.7—dc22
                                      2003060762

Visit the CRC Press Web site at www.crcpress.com

To Sandra and Tanya
– BF

To my niece Maria Lucia
– OM

Preface

Technological advances over the last several years have made it possible to construct large image and video libraries comprised of hundreds of terabytes of data. As a consequence, there is a great demand for the capability to provide databases that can effectively support storage, search, retrieval, and transmission of image and video data. The purpose of the Handbook of Video Databases is to provide a comprehensive reference on advanced topics in this field. The Handbook is intended both for researchers and practitioners in the field, and for scientists and engineers involved in the design, development, and applications of video databases. The Handbook can also be used as the textbook for graduate courses in this area.

This Handbook comprises 45 chapters contributed by more than 100 world-recognized experts in the field. The Handbook covers various aspects of video databases, including video modeling and representation, segmentation and summarization, indexing and retrieval, video transmission, design and implementation, and other topics. Section I introduces fundamental concepts and techniques in designing modern video databases. Section II covers concepts and techniques applied for video modeling and representation, while Section III deals with techniques and algorithms for video segmentation and summarization. Section IV describes tools and techniques for designing and interacting with video databases, and Section V discusses audio and video indexing and retrieval techniques. Section VI focuses on video transmission techniques, including video streaming and emerging video compression algorithms. Section VII describes video processing techniques and their relationship to the design and implementation of video databases. Section VIII describes several prototypes and commercial projects in the area of video databases. Finally, Section IX provides a collection of answers from the world-renowned experts in the field to fundamental questions about the state of the art and future research directions in video databases and related topics.

We would like to thank all the authors for their individual contributions of chapters to the Handbook. Without their expertise and effort, this Handbook would never have come to fruition. CRC Press editors and staff also deserve our sincere recognition for their support throughout the project.

Borko Furht and Oge Marques
Boca Raton, Florida

Editors

Borko Furht is a professor and chairman of the Department of Computer Science and Engineering at Florida Atlantic University (FAU) in Boca Raton, Florida. Before joining FAU, he was a vice president of research and a senior director of development at Modcomp, a computer company of Daimler Benz, Germany, a professor at University of Miami in Coral Gables, Florida, and senior scientist at the Institute "Boris Kidric"-Vinca, Belgrade, Yugoslavia. He received Ph.D., MSEE, and B.Sc. (Dipl. Eng.) degrees from the University of Belgrade, Yugoslavia. He is the author of numerous scientific and technical papers, books, and holds two patents. His current research is in multimedia systems, video coding and compression, video databases, wireless multimedia, and Internet computing. He has received several technical and publishing awards, research grants from NSF, NASA, IBM, Xerox, and Racal Datacom, and has consulted for many high-tech companies including IBM, Hewlett-Packard, Xerox, General Electric, JPL, NASA, Honeywell, Cordis, and RCA. He is a founder and editor-in-chief of the *Journal of Multimedia Tools and Applications* (Kluwer). He has given many invited talks, keynote lectures, seminars, and tutorials.

Oge Marques is an assistant professor in the Department of Computer Science and Engineering at Florida Atlantic University (FAU) in Boca Raton, Florida. He received his B.S. degree in Electrical Engineering from *Centro Federal de Educação Tecnológica do Paraná* (CEFET-PR) in Curitiba, Brazil, a Master's degree in Electronic Engineering from *Philips International Institute of Technological Studies* in Eindhoven, The Netherlands, and a Ph.D. degree in Computer Engineering from Florida Atlantic University. During the last five years, he wrote two books, several book chapters and papers in the fields of Digital Image Processing and Visual Information Retrieval. His fields of interest include Visual Information Retrieval, Digital Image Processing, Video Processing and Communications, and Wireless Networks. He is a member of the ACM, IEEE, and the Phi Kappa Phi.

Contributors

Lalitha Agnihotri
Philips Research
Briarcliff Manor, New York, USA

Mohamed Ahmed
National Research Council of Canada
Ottawa, Canada

John G. Apostolopoulos
Hewlett-Packard Laboratories
Palo Alto, California, USA

Edoardo Ardizzone
University of Palermo
Palermo, Italy

Juergen Assfalg
Università di Firenze
Firenze, Italy

Ramazan Savaş Aygün
State University of New York at Buffalo
Buffalo, New York, USA

Bruno Bachimont
INA
Bry-Sur-Marne, France

Prithwish Basu
Boston University
Boston, Massachusetts, USA

Lee Begeja
AT&T Labs – Research
Middletown, New Jersey, USA

Marco Bertini
Università di Firenze
Firenze, Italy

Alan C. Bovik
University of Texas at Austin
Austin, Texas, USA

Lekha Chaisorn
National University of Singapore
Singapore

Shermann S.M. Chan
City University of Hong Kong
Hong Kong

A. Chandrashekhara
National University of Singapore
Singapore

Tsuhan Chen
Carnegie Mellon University
Pittsburgh, Pennsylvania, USA

Yen-Kuang Chen
Intel Corporation
Santa Clara, California, USA

Sen-ching Samson Cheung
University of California
Berkeley, California, USA

N. Chokkareddy
University of Texas at Dallas
Dallas, Texas, USA

Michael G. Christel
Carnegie Mellon University
Pittsburgh, Pennsylvania, USA

Tat-Seng Chua
National University of Singapore
Singapore

Carlo Colombo
Università di Firenze
Firenze, Italy

Alberto Del Bimbo
Università di Firenze
Firenze, Italy

Edward Delp
Purdue University
West Lafayette, Indiana, USA

Nevenka Dimitrova
Philips Research
Briarcliff Manor, New York, USA

John K. Dixon
Michigan State University
East Lansing, Michigan, USA

Chabane Djeraba
Nantes University
Nantes, France

Gwenaël Doërr
Eurécom Institute
Sophia-Antipolis, France

Jean-Luc Dugelay
Eurécom Institute
Sophia-Antipolis, France

Wolfgang Effelsberg
University of Mannheim
Mannheim, Germany

Ahmet Ekin
University of Rochester
Rochester, New York, USA

Alexandros Eleftheriadis
Columbia University
New York, New York, USA

Dirk Farin
University of Mannheim
Mannheim, Germany

HuaMin Feng
National University of Singapore
Singapore

Borko Furht
Florida Atlantic University
Boca Raton, Florida, USA

Ashutosh Garg
University of Illinois at Urbana-
Champaign
Urbana, Illinois, USA

Jérôme Gensel
LSR-IMAG
Grenoble, France

Shahram Ghandeharizadeh
University of Southern California
Los Angeles, California, USA

David Gibbon
AT&T Labs – Research
Middletown, New Jersey, USA

Yihong Gong
NEC USA, Inc.
Cupertino, California, USA

William I. Grosky
University of Michigan-Dearborn
Dearborn, Michigan, USA

Thomas Haenselmann
University of Mannheim
Mannheim, Germany

Younes Hafri
INA
Bry-Sur-Marne, France

Rune Hjelsvold
Gjøvik University College
Gjøvik, Norway

Matthew Holliman
Intel Corporation
Santa Clara, California, USA

Danny Hong
Columbia University
New York, New York, USA

Kien A. Hua
University of Central Florida
Orlando, Florida, USA

Thomas S. Huang
University of Illinois at Urbana-
Champaign
Urbana, Illinois, USA

Z. Huang
National University of Singapore
Singapore

Horace H.S. Ip
City University of Hong Kong
Hong Kong

Radu Jasinschi
Philips Research
Briarcliff Manor, New York, USA

Hari Kalva
Mitsubishi Electric Research Labs
Murray Hill, New Jersey, USA

Ahmed Karmouch
University of Ottawa
Ottawa, Canada

Norio Katayama
National Institute of Informatics
Japan

Wang Ke
Boston University
Boston, Massachusetts, USA

Seon Ho Kim
University of Denver
Denver, Colorado, USA

Stephan Kopf
University of Mannheim
Mannheim, Germany

Igor Kozintsev
Intel Corporation
Santa Clara, California, USA

Rajesh Krishnan
Boston University
Boston, Massachusetts, USA

Gerald Kühne
University of Mannheim
Mannheim, Germany

Marco La Cascia
University of Palermo
Palermo, Italy

Chin-Hui Lee
National University of Singapore
Singapore

Dongge Li
Philips Research
Briarcliff Manor, New York, USA

Qing Li
City University of Hong Kong
Hong Kong

Rainer Lienhart
Intel Corporation
Santa Clara, California, USA

Thomas D.C. Little
Boston University
Boston, Massachusetts, USA

Zhu Liu
AT&T Labs – Research
Middletown, New Jersey, USA

Oge Marques
Florida Atlantic University
Boca Raton, Florida, USA

Hervé Martin
LSR-IMAG
Grenoble, France

Thomas McGee
Philips Research
Briarcliff Manor, New York, USA

Sharad Mehrotra
University of California, Irvine
Irvine, California, USA

Philippe Mulhem
IPAL-CNRS
Singapore

Milind R. Naphade
University of Illinois at Urbana-
Champaign
Urbana, Illinois, USA

Michael Ortega-Binderberger
University of Illinois at Urbana-
Champaign
Urbana, Illinois, USA

Charles B. Owen
Michigan State University
East Lansing, Michigan, USA

A. Picariello
Università di Napoli "Federico II"
Napoli, Italy

B. Prabhakaran
University of Texas at Dallas
Dallas, Texas, USA

Rohit Puri
University of California
Berkeley, California, USA

Bernard Renger
AT&T Labs – Research
Middletown, New Jersey, USA

Simone Santini
University of California, San Diego
La Jolla, California, USA

M. L. Sapino
Università di Torino
Torino, Italy

Shin'ichi Satoh
National Institute of Informatics
Japan

Cyrus Shahabi
University of Southern California
Los Angeles, California, USA

Behzad Shahraray
AT&T Labs – Research
Middletown, New Jersey, USA

Hamid R. Sheikh
University of Texas at Austin
Austin, Texas, USA

Bo Shen
Hewlett-Packard Laboratories
Palo Alto, California, USA

John R. Smith
IBM T. J. Watson Research Center
Hawthorne, New York, USA

Michael A. Smith
AVA Media Systems
Austin, Texas, USA

Yuqing Song
State University of New York at Buffalo
Buffalo, New York, USA

V. S. Subrahmanian
University of Maryland
College Park, Maryland, USA

Wai-tian Tan
Hewlett-Packard Laboratories
Palo Alto, California, USA

H. Lilian Tang
University of Surrey
United Kingdom

Kwok Hung Tang
Michigan State University
East Lansing, Michigan, USA

Mounir Tantaoui
University of Central Florida
Orlando, Florida, USA

A. Murat Tekalp
University of Rochester
Rochester, New York, USA

Luis Torres
Technical University of Catalonia
Barcelona, Spain

Agma Juci M. Traina
University of São Paulo at São Carlos
São Carlos, Brazil

Caetano Traina Jr.
University of São Paulo at São Carlos
São Carlos, Brazil

Nuno Vasconcelos
HP Cambridge Research Laboratory
Cambridge, Massachusetts, USA

Subu Vdaygiri
Siemens Corporate Research, Inc.
Princeton, New Jersey, USA

Anthony Vetro
Mitsubishi Electric Research Labs
Murray Hill, New Jersey, USA

Howard D. Wactlar
Carnegie Mellon University
Pittsburgh, Pennsylvania, USA

Yao Wang
Polytechnic University
Brooklyn, New York, USA

Zhou Wang
University of Texas at Austin
Austin, Texas, USA

Susie J. Wee
Hewlett-Packard Laboratories
Palo Alto, California, USA

Fan Xiao
Michigan State University
East Lansing, Michigan, USA

Minerva Yeung
Intel Corporation
Santa Clara, California, USA

Andre Zaccarin
Université Laval
Quebec City, Canada

Avideh Zakhor
University of California
Berkeley, California, USA

Aidong Zhang
State University of New York at Buffalo
Buffalo, New York, USA

Cha Zhang
Carnegie Mellon University
Pittsburgh, Pennsylvania, USA

Rong Zhao
State University of New York
Stony Brook, New York, USA

Ji Zhou
Michigan State University
East Lansing, Michigan, USA

Xiang Sean Zhou
Siemens Corporate Research, Inc.
Princeton, New Jersey, USA

John Zimmerman
Philips Research
Briarcliff Manor, New York, USA

Roger Zimmermann
University of Southern California
Los Angeles, California, USA

Table of Contents

SECTION IV – Designing and Interacting with Video Databases

SECTION V – Audio and Video Indexing and Retrieval

SECTION VII – Video Processing

SECTION VIII – Case Studies and Applications

SECTION IX – Panel of Experts: The Future of Video Databases

1

INTRODUCTION TO VIDEO DATABASES

Oge Marques and Borko Furht
Department of Computer Science and Engineering
Florida Atlantic University
Boca Raton, FL, USA
`omarques@acm.org, borko@cse.fau.edu`

1. INTRODUCTION

The field of distributed multimedia systems has experienced an extraordinary growth during the last decade. Among the many visible aspects of the increasing interest in this area is the creation of huge digital libraries accessible to users worldwide. These large and complex multimedia databases must store all types of multimedia data, e.g., text, images, animations, graphs, drawings, audio, and video clips. Video information plays a central role in such systems, and consequently, the design and implementation of video database systems has become a major topic of interest.

The amount of video information stored in archives worldwide is huge. Conservative estimates state that there are more than 6 million hours of video already stored and this number grows at a rate of about 10 percent a year [1]. Projections estimate that by the end of 2010, 50 percent of the total digital data stored worldwide will be video and rich media [5]. Significant efforts have been spent in recent years to make the process of video archiving and retrieval faster, safer, more reliable and accessible to users anywhere in the world. Progress in video digitization and compression, together with advances in storage media, have made the task of storing and retrieving raw video data much easier. Evolution of computer networks and the growth and popularity of the Internet have made it possible to access these data from remote locations.

However, raw video data by itself has limited usefulness, since it takes far too long to search for the desired piece of information within a videotape repository or a digital video archive. Attempts to improve the efficiency of the search process by adding extra data (henceforth called *metadata*) to the video contents do little more than transferring the burden of performing inefficient, tedious, and time-consuming tasks to the cataloguing stage. The challenging goal is to devise better ways to automatically store, catalog, and retrieve video information with greater understanding of its contents. Researchers from various disciplines have acknowledged such challenge and provided a vast number of algorithms, systems, and papers on this topic during recent years. In addition to these localized efforts, standardization groups have been working on new standards, such as MPEG-7, which provide a framework for multimedia content description.

The combination of the growing number of applications for video-intensive products and solutions – from personal video recorders to multimedia collaborative systems – with the many technical challenges behind the design of contemporary video database systems yet to be overcome makes the topics discussed in this Handbook of extreme interest to researchers and practitioners in the fields of image and video processing, computer vision, multimedia systems, database systems, information retrieval, data mining, machine learning, and visualization, to name just a few (Figure 1.1).

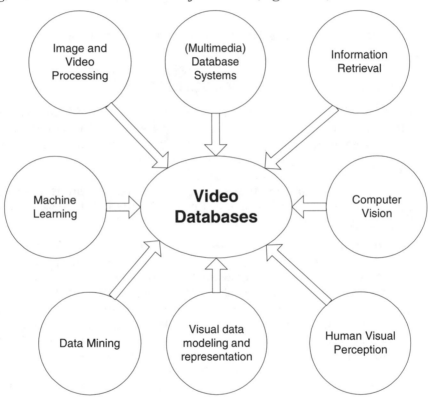

Figure 1.1 Visual Information Retrieval blends together many research disciplines.

In this chapter we present a general overview of the central topic in this Handbook: video databases. Its main goal is to introduce basic concepts behind general-purpose database systems and their extension to multimedia database systems, particularly image and video database systems. Section 2 introduces basic database concepts and terminology. In Section 3 we briefly outline the main steps of the database design process. Section 4 extends the discussion from general databases to multimedia databases and their own particular requirements and characteristics. Section 5 narrows down the field even further, focusing on the aspects that are specific to image and video databases. The main goals this Handbook are outlined in Section 6. Finally, Section 7 provides the reader with an overview of the other chapters of this Handbook and how they have been organized.

2. BASIC DATABASE CONCEPTS

A *database* is a logically coherent collection of related data, where *data* refers to known facts that can be recorded and that have implicit meaning [2]. A *database management system* (DBMS) is a collection of programs that enables users to create and maintain a database. Together, the database and DBMS software are called a *database system*. Contemporary DBMS packages have a modular design and adopt a client-server architecture. A *client* module, running in the user's workstation, handles user interaction and provides friendly GUIs (graphical user interfaces). The *server* module is responsible for data storage, access, search, and other operations.

Database systems should be designed and built in a way as to hide details of data storage from its end users. Such abstraction can be achieved using a *data model*, normally defined as a collection of concepts that can be used to describe the structure of a database, i.e., its data types, relationships, and constraints imposed on data. Most data models also include a set of basic operations for specifying retrievals and updates on the database. An *implementation* of a given data model is a physical realization on a real machine of the components of the abstract machine that together constitute that model [3].

Data models can be categorized in three main groups:

(1) **High-level or conceptual data models**: use concepts such as entitics, attributes, and relationships, which are close to the way many users perceive data.

(2) **Low-level or physical data models**: describe details on how data is stored in a computer, which make them meaningful to computer specialists, not to end users.

(3) **Representational (or implementation) data models**: widely used intermediate category, which aims at concepts that may be understood by end users but that are not far from the way data is organized in a computer. The most widely used representational data models are the *relational data model* and the *object data model*.

The description of a database is called the *database schema*, typically represented in a diagram (usually referred to as *schema diagram*) where each object is called a *schema construct*. The database schema is specified during database design and is expected to remain unchanged unless the requirements of the database applications change.

A typical architecture for a database system, proposed by the ANSI/SPARC Study Group on Database Management Systems and known as the ANSI/SPARC architecture, is shown in Figure 1.2. This architecture is divided into three levels, each of which has its own schema:

(1) The **internal level** (also known as the *physical level*) has an internal schema, which describes the physical storage structure of the database.

(2) The **conceptual level** (also known as the *community logical level*) has a conceptual schema, which describes the structure of the whole database for a community of users.

(3) The **external level** (also known as the *user logical level* or *view level*) has a number of external schemas or user views. Each external schema describes the part of the database that a particular user group is interested in and hides the rest of the database from that user group.

The three-schema architecture involves mappings – represented by the double-pointed arrows – that can be updated any time a partial change in some of the database's schemas take place. The ability of changing the schema at one level of a database system without having to change the schema at the next higher level is known as *data independence*.

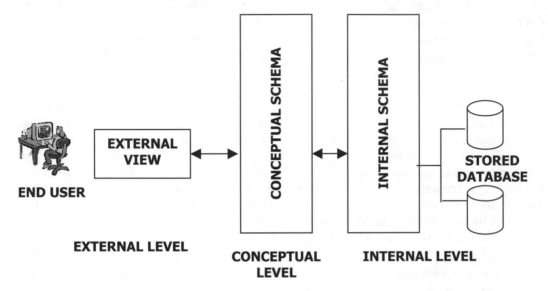

Figure 1.2 The three levels of the ANSI/SPARC architecture.

A DBMS must provide appropriate languages and interfaces for each category of users. The type of language that allows an end user to manipulate, retrieve, insert, delete, and modify data in the database is known as *data manipulation language* (DML). There are two main types of DMLs: high-level DMLs (such as Structured Query Language – SQL) can work on a set of records, while low-level DMLs retrieve and process one record at a time. A high-level DML used in a stand-alone interactive manner is also called a *query language*.

3. THE DATABASE DESIGN PROCESS

The problem of database design can be summarized in a question: Given some body of data to be represented in one or more databases, how do we decide on a suitable logical structure for that data such that the information needs of the users are properly accommodated?

Different authors suggest slightly different procedures for database design. In essence, all procedures contain the following stages:

(1) **Requirements collection and analysis**: the process of collecting and analyzing information about the part of the organization that is to be supported by the database application, and using this information to identify the users' requirements of the new system [4]. Requirement specification techniques include OOA (object-oriented analysis) and DFDs (data flow diagrams),

(2) **Conceptual database design**: this phase involves two parallel activities: (a) *conceptual schema design*, which produces a conceptual database schema based on the requirements outlined in phase 1; and (b)

transaction and application design, which produces high-level specifications for the applications analyzed in phase 1. Complex databases are normally designed using a top-down approach and use the terminology of the Entity-Relationship (ER) model or one of its variations.

(3) **Logical design**: at this stage the internal schemas produced in phase 2(a) are mapped into conceptual and external schemas. Ideally the resulting model should be independent of a particular DBMS or any physical consideration.

(4) **DBMS selection**: the selection of a commercially available DBMS to support the database application is governed by technical, economic, and sometimes even political factors. Some of the most relevant technical factors include: the type of data model used (e.g., relational or object), the supported storage structures and access paths, the types of high-level query languages, availability of development tools, and utilities, among many others.

(5) **Physical design**: the process of choosing specific storage structures and access methods used to achieve efficient access to data. Typical activities included in this phase are: choice of file organization (heap, hash, Indexed Sequential Access Method – ISAM, B+-tree, and so on), choice of indexes and indexing strategies, and estimation of disk requirements (e.g., access time, total capacity, buffering strategies).

(6) **Implementation and testing**: the designed database is finally put to work and many unanticipated problems are fixed and the overall performance of the database system is fine-tuned.

4. MULTIMEDIA DATABASES

The design and implementation of multimedia databases create additional challenges due to the nature of multimedia data and the requirements of possible applications. Multimedia applications can be categorized in three main groups, each of which poses different data management challenges [2]:

(a) **Repository applications**: Large amounts of multimedia data and associated metadata are stored for retrieval purposes. These repositories may be distributed or centralized and can be managed using conventional DBMS. Examples of multimedia data stored in such repositories include medical and satellite images, and engineering drawings.

(b) **Presentation applications**: Applications that involve delivery of multimedia content to a possibly remote location, subject to temporal constraints. In these applications, data is consumed as it is delivered, as opposed to being stored for later processing. A number of new potential problems, such as jitter, latency, and the corresponding need to maintain and guarantee "quality of service" (QoS), come into play. Examples of such applications include audio and video broadcasting over the Internet.

(c) **Collaborative work using multimedia information**: a new breed of multimedia applications in which a geographically dispersed group of professionals (e.g., engineers or medical doctors) work together on a common, multimedia-intensive, task.

Accommodating massive amounts of text, graphics, images, animations, audio, and video streams into a database system is far from trivial and the popularization of multimedia databases has raised a number of complex issues for database designers. Some of these issues are [2]:

- *Modeling*: Multimedia information includes media objects, associated metadata, and the objects' temporal and spatial characteristics. This information is continuously manipulated and modified by applications. Some of the different techniques used for modeling multimedia data are [6]:

    (a) **object-oriented modeling**: inspired by the object-oriented paradigm, it organizes multimedia information into hierarchical structures in which each multimedia object (e.g., text, audio, video, image) has its set of variables, methods, and messages to which it responds. Such a model can enforce concepts such as encapsulation, data hiding, and multiple inheritance, and can handle the metadata as well. Some of its drawbacks include the difficulty of accessing objects in a collective – rather than individual – manner, the need to handle the database schema independently from the class hierarchy, and the impossibility of creating new objects that are based on portions of existing objects and need only to inherit part of their attributes.

    (b) **temporal models**: multimedia objects have associated temporal characteristics, which are particularly important in presentation-type applications. These characteristics specify parameters such as: time instant of an object presentation, duration of presentation, and synchronization among objects in the presentation. Temporal models are called *hard* when the temporal relationships are specified in a precise manner with exact values for time instants and duration of presentations, or *flexible* if they allow a range of values to be specified for each time-related parameter.

    (c) **spatial models**: multimedia applications are constrained by the size of each window and the window layout. These constraints must be taken into account either in a hard way (assigning specific values for the x and y coordinates of each window corner), or in a flexible way, using difference constraints and specifying relative positions among the various windows in a presentation.

- *Design*: The conceptual, logical, and physical design of multimedia databases remains an area of active research. The general design methodology summarized in Section 3 can still be used as a starting point, but performance and fine-tuning issues are more complex than in conventional databases.

- *Storage*: Storage of multimedia information in conventional magnetic media brings new problems, such as representation, compression/decompression, mapping to device hierarchies, archiving, and buffering during the I/O operations.

- *Queries and retrieval*: Efficient query formulation, query execution, and optimization for multimedia data is still an open problem and neither query languages nor keyword-based queries have proven to be completely satisfactory.

- *Performance*: While some multimedia applications can tolerate less strict performance constraints (e.g., the maximum time to perform a content-based query on a remote image repository), others are inherently more critical, such as the minimum acceptable frame rate for video playback.

Recent developments in Multimedia Database Systems are expected to bring together two disciplines that have historically been separate: *database management* and *information retrieval*. The former assumes a rigid structure for data and derives the meaning of a data instance from the database schema, while the latter is more concerned with modeling the data content, without paying much attention to its structure [2].

There are very few commercial multimedia database management solutions currently available, e.g., MediaWay's Chuckwalla Broadband Media Management System. Nonetheless, many well-known DBMSs support multimedia data types; examples include Oracle 8.0, Sybase, Informix, ODB II, and CA-JASMINE. The way multimedia extensions are handled by each of these systems is ad hoc, and does not take into account interoperability with other products and solutions. It is expected that the MPEG-7 standard will promote the necessary standardization.

5. IMAGE AND VIDEO DATABASES

Image and video databases have particular requirements and characteristics, the most important of which will be outlined in this Section and described in much more detail along other chapters in this Handbook. Some of the technical challenges are common to both types of media, such as the realization that raw contents alone are not useful, unless they are indexed for further query and retrieval, which makes the effective indexing of images and video clips an ongoing research topic.

After having been catalogued, it should be possible to query and retrieve images and video clips based on their semantic meaning, using a measure of similarity between the query terms (textual or otherwise) and the database contents. Since in most cases there is no exact matching between the query and its expected result, there is a need to extend the information retrieval approaches to search based on similarity to the case where the mapping between visual and semantic similarity is not straightforward. These issues will be explored within the realm of image databases in Section 5.1. An extension to video and its specific challenges and complexities will be presented in Section 5.2.

5.1 IMAGE DATABASES

The design of image databases brings about the need for new abstractions and techniques to implement these abstractions.

Since raw images alone have very limited usefulness, some technique must be devised to extract and encode the image properties into an alphanumerical format that is amenable to indexing, similarity calculations, and ranking of best results. These properties can be extracted using state-of-the-art computer vision algorithms. Examples of such properties include: shape, color, and texture. These descriptors are recognizably limited and very often fail to capture the semantic meaning of the image, giving rise to the well-known *semantic gap* problem. Moreover, some of these descriptors are only useful if applied to the relevant objects within an image, which calls for some sort of segmentation to occur. Automatic segmentation of a scene into its relevant objects is still an unresolved problem, and many attempts in this field get around this problem by including the user in the loop and performing a semi-automatic segmentation instead.

Storing a set of images so as to support image retrieval operations is usually done with spatial data structures, such as R-trees and a number of variants (e.g., R$^+$ trees, R* trees, SS-trees, TV-trees, X-trees, M-trees) proposed in the literature.

Indexing of images is another open research problem. Raw images must have their contents extracted and described, either manually or automatically. Automatic techniques usually rely on image processing algorithms, such as shape-, color-, and texture-based descriptors. Current techniques for content-based indexing only allow the indexing of simple patterns and images, which hardly match the semantic notion of relevant objects in a scene. The alternative to content-based indexing is to assign index terms and phrases through manual or semi-automatic indexing using textual information (usually referred to as *metadata*), which will then be used whenever the user searches for an image using a query by keyword.

The information-retrieval approach to image indexing is based on one of the three indexing schemes [2]:

1. *Classificatory systems*: images are classified hierarchically according to the category to which they belong.
2. *Keyword-based systems*: images are indexed based on the textual information associated with them.
3. *Entity-attribute-relationship systems*: all objects in the picture and the relationships between objects and the attributes of the objects are identified.

Querying an image database is fundamentally different from querying textual databases. While a query by keyword usually suffices in the case of textual databases, image databases users normally prefer to search for images based on their visual contents, typically under the *query by example* paradigm, through which the user provides an example image to the system and (sometimes implicitly) asks the question: "Can you find and retrieve other pictures that look like this (and the associated database information)?" Satisfying such a query is a much more complex task than its text-based counterpart for a number of reasons, two of which are: the inclusion of a picture as part of a query and the notion of "imprecise match" and how it will be translated into criteria and rules for similarity measurements. The most critical requirement is that a similarity measure must behave well and mimic the human notion of similarity for any pair of images, no matter how different they are, in contrast with what would typically be required from a matching technique, which only has to behave well when there is relatively little difference between a database image and the query [7].

There are two main approaches to similarity-based retrieval of images [8]:

1. *Metric approach*: assumes the existence of a distance metric d that can be used to compare any two image objects. The smaller the distance between the objects, the more similar they are considered to be. Examples of widely used distance metrics include: Euclidean, Manhattan, and Mahalanobis distance.

2. *Transformation approach*: questions the claim that a given body of data (in this case, an image) has a single associated notion of similarity. In this model, there is a set of operators (e.g., translation, rotation, scaling) and cost functions that can be optionally associated with each operator. Since it allows users to personalize the notion of similarity to their needs, it is more flexible than the metric approach; however, it is less computationally efficient and its extension to other similar queries is not straightforward.

There are many (content-based) image retrieval systems currently available, in the form of commercial products and research prototypes. For the interested reader, Veltkamp and Tanase [9] provide a comprehensive review of existing systems, their technical aspects, and intended applications.

5.2 VIDEO DATABASES

The challenges faced by researchers when implementing image databases increase even further when one moves from images to image sequences, or video clips, mostly because of the following factors:

- increased size and complexity of the raw media (video, audio, and text);
- wide variety of video programs, each with its own rules and formats;
- video understanding is very much context-dependent;
- the need to accommodate different users, with varying needs, running different applications on heterogeneous platforms.

As with its image counterpart, raw video data alone has limited usefulness and requires some type of annotation before it is catalogued for later retrieval. Manual annotation of video contents is a tedious, time consuming, subjective, inaccurate, incomplete, and – perhaps more importantly – costly process. Over the past decade, a growing number of researchers have been attempting to fulfill the need for creative algorithms and systems that allow (semi-) automatic ways to describe, organize, and manage video data with greater understanding of its semantic contents.

The primary goal of a Video Database Management System (VDBMS) is to provide pseudo-random access to sequential video data. This goal is normally achieved by dividing a video clip into segments, indexing these segments, and representing the indexes in a way that allows easy browsing and retrieval. Therefore, it can be said that a VDBMS is basically a database of indexes (pointers) to a video recording [10].

Similarly to image databases, much of the research effort in this field has been focused on modeling, indexing, and structuring of raw video data, as well as finding suitable similarity-based retrieval measures. Another extremely important aspect of a VDBMS is the design of its graphical user interface (GUI). These aspects will be explored in a bit more detail below.

5.2.1 The main components of a VDBMS

Figure 1.3 presents a simplified block diagram of a typical VDBMS. Its main blocks are:

- *Digitization and compression*: hardware and software necessary to convert the video information into digital compressed format.

- *Cataloguing*: process of extracting meaningful story units from the raw video data and building the corresponding indexes.
- *Query / search engine*: responsible for searching the database according to the parameters provided by the user.
- *Digital video archive*: repository of digitized, compressed video data.
- *Visual summaries*: representation of video contents in a concise, typically hierarchical, way.
- *Indexes*: pointers to video segments or story units.
- *User interface*: friendly, visually rich interface that allows the user to interactively query the database, browse the results, and view the selected video clips.

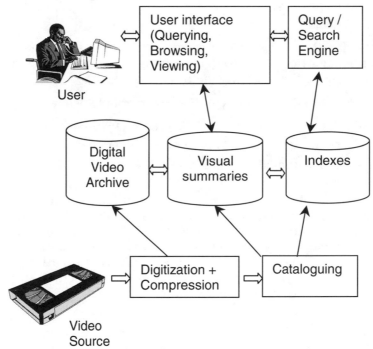

Figure 1.3 Block diagram of a VDBMS.

5.2.2 Organization of video content

Because video is a structured medium in which actions and events in time and space convey stories, a video program must not be viewed as a non-structured sequence of frames, but instead it must be seen as a document. The process of converting raw video into structured units, which can be used to build a visual table of contents (ToC) of a video program, is also referred to as *video abstraction*. We will divide it into two parts:

1. Video modeling and representation
2. Video segmentation (parsing) and summarization

Video modeling and representation

Video modeling can be defined as the process of designing the representation for the video data based on its characteristics, the information content, and the applications it is intended for. Video modeling plays a key role in the design of VDBMSs, because all other functions are more or less dependent on it.

The process of modeling video contents can be a challenging task, because of the following factors:

- video data carry much more information than textual data;
- interpretation is usually ambiguous and depends on the viewer and the application;
- the high dimensionality of video data objects;
- lack of a clear underlying structure;
- massive volume (bytes);
- relationships between video data segments are complex and ill-defined.

When referring to contents of video data, the following distinctions should be made, according to their type and level [11]:

- *Semantic content*: the idea or knowledge that it conveys to the user, which is usually ambiguous, subjective, and context-dependent.
- *Audiovisual content*: low-level information that can be extracted from the raw video program, usually consisting of color, texture, shape, object motion, object relationships, camera operation, audio track, etc.
- *Textual content*: additional information that may be available within the video stream in the form of captions, subtitles, etc.

Some of the requirements for a video data model are [11]:

- Support video data as one of its data types, just like textual or numeric data.
- Integrate content attributes of the video program with its semantic structure.
- Associate audio with visual information.
- Express structural and temporal relationships between segments.
- Automatically extract low-level features (color, texture, shape, motion), and use them as attributes.

Most of the video modeling techniques discussed in the literature adopt a hierarchical video stream abstraction, with the following levels, in decreasing degree of granularity:

- *Key-frame*: most representative frame of a shot.
- *Shot*: sequence of frames recorded contiguously and representing a continuous action in time or space.
- *Group*: intermediate entity between the physical shots and semantic scenes that serves as a bridge between the two.
- *Scene* or *Sequence*: collection of semantically related and temporally adjacent shots, depicting and conveying a high-level concept or story.
- *Video program*: the complete video clip.

A video model should identify physical objects and their relationships in time and space. Temporal relationships should be expressed by: before, during, starts, overlaps, etc., while spatial relationships are based on projecting objects on a 2-D or 3-D coordinate system.

A video model should also support annotation of the video program, in other words the addition of metadata to a video clip. For the sake of this discussion, we consider three categories of metadata:

- *content-dependent metadata* (e.g., facial features of a news anchorperson);
- *content-descriptive metadata* (e.g., the impression of anger or happiness based on facial expression);
- *content-independent metadata* (e.g., name of the cameraman).

Video data models can usually be classified into the following categories [11] (Figure 1.4):

- *Models based on video segmentation*: adopt a two-step approach, first segmenting the video stream into a set of temporally ordered basic units (shots), and then building domain-dependent models (either hierarchy or finite automata) upon the basic units.
- *Models based on annotation layering* (also known as *stratification models*): segment contextual information of the video and approximate the movie editor's perspective on a movie, based on the assumption that if the annotation is performed at the finest grain (by a *data camera*), any coarser grain of information may be reconstructed easily.
- *Video object models*: extend object-oriented data models to video. Their main advantages include the ability to represent and manage complex objects, handle object identities, encapsulate data and associated methods into objects, and inherit attribute structures and methods based on class hierarchy.

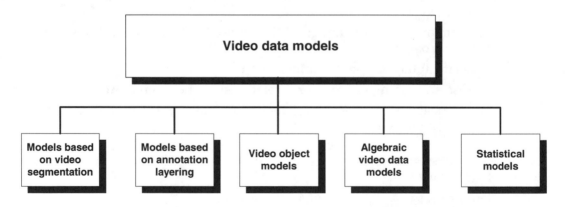

Figure 1.4 Classification of video data models.

- *Algebraic video data models*: define a video stream by recursively applying a set of algebraic operations on the raw video segment. Their fundamental entity is a *presentation* (multi-window, spatial, temporal, and content combination of video segments). Presentations are described by video expressions, constructed from raw segments using video algebraic operations. As an example of the algebraic approach to video manipulation, the reader is referred to Chapter 19 of this Handbook, where Picariello, Sapino, and Subrahmanian introduce AVE! (Algebraic Video Environment), the first algebra for querying video.

- *Statistical models*: exploit knowledge of video structure as a means to enable the principled design of computational models for video semantics, and use machine learning techniques (e.g., Bayesian inference) to learn the semantics from collections of training examples, without having to rely on lower level attributes such as texture, color, or optical flow. Chapter 3 of this Handbook contains an example of statistical modeling of video programs, the Bayesian Modeling of Video Editing and Structure (BMoViES) system for video characterization, developed by Nuno Vasconcelos.

Video segmentation

Video segmentation (also referred to as *video parsing*) is the process of partitioning video sequences into smaller units. Video parsing techniques extract structural information from the video program by detecting temporal boundaries and identifying meaningful segments, usually called *shots*. The shot ("a continuous action on screen resulting from what appears to be a single run of the camera") is usually the smallest object of interest. Shots are detected automatically and typically represented by key-frames.

Video segmentation can occur either at a shot level or at a scene level. The former is more often used and sometimes referred to as shot detection. Shot detection can be defined as the process of detecting transitions between two consecutive shots, so that a sequence of frames belonging to a shot will be grouped together. There are two types of shot transitions: abrupt transitions (or *cuts*) and gradual transitions (e.g., fade-in, fade-out, dissolve). Earlier work on shot detection addressed only the detection of cuts, while more recent research results report successful techniques for detection of gradual transitions as well.

An alternative to shot detection, scene-based video segmentation consists of the automatic detection of semantic boundaries (as opposed to physical boundaries) within a video program. It is a much more challenging task, whose solution requires a higher level of content analysis, and the subject of ongoing research. Three main strategies have been attempted to solve the problem:

- Segmentation based on film production rules (e.g., transition effects, shot repetition, appearance of music in the soundtrack) to detect local (temporal) clues of macroscopic change;

- Time-constrained clustering, which works under the rationale that semantically related contents tend to be localized in time;

- *A priori* model-based algorithms, which rely on specific structural models for programs whose temporal structures are usually very rigid and predictable, such as news and sports.

Video segmentation can occur either in the uncompressed or compressed domain. In the uncompressed domain, the basic idea upon which the first algorithms were conceived involved the definition of a similarity measure between successive images and the comparison of successive frames according to that measure: whenever two frames are found to be sufficiently dissimilar, there may be a cut. Gradual transitions are found by using cumulative difference measures and more sophisticated thresholding schemes. Temporal video segmentation techniques that work directly on the compressed video streams were motivated by the computational savings resulting from not having to perform decoding/re-encoding, and the possibility of exploiting pre-computed features, such as motion vectors (MVs) and block averages (DC coefficients), that are suitable for temporal video segmentation.

Temporal video segmentation has been an active area of research for more than 10 years, which has resulted in a great variety of approaches. Early work focused on cut detection, while more recent techniques deal with gradual transition detection. Despite the great evolution, most current algorithms exhibit the following limitations [12]:

- They process unrealistically short gradual transitions and are unable to recognize the different types of gradual transitions;

- They involve many adjustable thresholds;

- They do not handle false positives due to camera operations.

For a comprehensive review of temporal video segmentation, the interested reader is referred to a recent survey by Koprinska and Carrato [12].

Video summarization

Video summarization is the process by which a pictorial summary of an underlying video sequence is presented in a more compact form, eliminating – or greatly reducing – redundancy. Video summarization focuses on finding a smaller set of images (key-frames) to represent the visual content, and presenting these key-frames to the user.

A still-image abstract, also known as a static storyboard, is a collection of salient still images or key-frames generated from the underlying video. Most summarization research involves extracting key-frames and developing a browser-based interface that best represents the original video. The advantages of a still-image representation include:

- still-image abstracts can be created much faster than moving image abstracts, since no manipulation of the audio or text information is necessary;

- the temporal order of the representative frames can be displayed so that users can grasp concepts more quickly;

- extracted still images are available for printing, if desired.

An alternative to still-image representation is the use of video skims, which can be defined as short video clips consisting of a collection of image sequences and the corresponding audio, extracted from the original longer video sequence. Video skims represent a temporal multimedia abstraction that is played rather than viewed statically. They are comprised of the most relevant phrases, sentences, and image sequences and their goal is to present the original video sequence in an order of magnitude less time. There are two basic types of video skimming:

- Summary sequences: used to provide a user with an impression of the video sequence.

- Highlights: contain only the most interesting parts of a video sequence.

Since the selection of highlights from a video sequence is a subjective process, most existing video-skimming work focuses on the generation of summary sequences.

A very important aspect of video summarization is the development of user interfaces that best represent the original video sequence, which usually translates into a trade-off between the different levels and types of abstractions presented to the user: the more condense the abstraction, the easier it is for a potential user to browse through, but maybe the amount of information is not enough to obtain the overall meaning and understanding of the video; a more detailed abstraction may present the user with enough information to comprehend the video sequence, which may take too long to browse.

Emerging research topics within this field include adaptive segmentation and summarization (see Chapters 11 and 12) and summarization for delivery to mobile users (see Chapter 14).

5.2.3 Video indexing, querying, and retrieval

Video indexing is far more difficult and complex than its text-based counterpart. While on traditional DBMS, data are usually selected based on one or more unique attributes (key fields), it is neither clear nor easy to determine what to index a video data on. Therefore, unlike textual data, generating content-based video data indexes automatically is much harder.

The process of building indexes for video programs can be divided into three main steps:

1. *Parsing*: temporal segmentation of the video contents into smaller units.

2. *Abstraction*: extracting or building a representative subset of video data from the original video.

3. *Content analysis*: extracting visual features from representative video frames.

Existing work on video indexing can be classified in three categories [11]:

1. **Annotation-based indexing**

    Annotation is usually a manual process performed by an experienced user, and subject to problems, such as: time, cost, specificity, ambiguity, and bias, among several others. A commonly used technique consists of assigning keyword(s) to video segments (shots). Annotation-based indexing techniques are primarily concerned with the selection of keywords, data structures, and interfaces, to facilitate the user's effort. But even with additional help, keyword-based annotation is inherently poor, because keywords:

    *   Do not express spatial and temporal relationships;

    *   Cannot fully represent semantic information and do not support inheritance, similarity, or inference between descriptors;

    *   Do not describe relations between descriptions.

    Several alternatives to keyword-based annotation have been proposed in the literature, such as the multi-layer, iconic language, Media Streams [14].

2. **Feature-based indexing**

Feature-based indexing techniques have been extensively researched over the past decade. Their goal is to enable fully automated indexing of a video program based on its contents. They usually rely on image processing techniques to extract key visual features (color, texture, object motion, etc.) from the video data and use these features to build indexes. The main open problem with these techniques is the semantic gap between the extracted features and the human interpretation of the visual scene.

3. Domain-specific indexing

Techniques that use logical (high-level) video structure models (*a priori* knowledge) to further process the results of the low-level video feature extraction and analysis stage. Some of the most prominent examples of using this type of indexing technique have been found in the area of summarization of sports events (e.g., soccer), such as the work described in Chapters 5 and 6 of this Handbook.

The video data retrieval process consists of four main steps [11]:

1. User specifies a query using the GUI resources.

2. Query is processed and evaluated.

3. The value or feature obtained is used to match and retrieve the video data stored in the VDB.

4. The resulting video data is displayed on the user's screen for browsing, viewing, and (optionally) query refining (relevance feedback).

Queries to a VDBMS can be classified in a number of ways, according to their content type, matching type, granularity, behavior, and specification [11], as illustrated in Figure 1.5. The semantic information query is the most difficult type of query, because it requires understanding of the semantic content of the video data. The meta information query relies on metadata that has been produced as a result of the annotation process, and therefore, is similar to conventional database queries. The audiovisual query is based on the low-level properties of the video program and can be further subdivided into: spatial, temporal, and spatio-temporal. In the case of deterministic query, the user has a clear idea of what she expects as a result, whereas in the case of browsing query, the user may be vague about his retrieval needs or unfamiliar with the structures and types of information available in the video database.

Video database queries can be specified using extensions of SQL for video data, such as TSQL2, STL (Spatial Temporal Logic), and VideoSQL. However, query by example, query by sketch, and interactive querying/browsing/viewing (with possible relevance feedback) are more often used than SQL-like queries.

Query processing usually involves four steps [11]:

1. Query parsing: where the query condition or assertion is usually decomposed into the basic unit and then evaluated.

2. Query evaluation: uses pre-extracted (low-level) visual features of the video data.

3. Database index search.

4. Returning of results: the video data are retrieved if the assertion or the similarity measurement is satisfied.

As a final comment, it should be noted that the user interface plays a crucial role in the overall usability of a VDBMS. All interfaces must be graphical and should ideally combine querying, browsing summaries of results, viewing (playing back) individual results, and providing relevance feedback and/or query refinement information to the system. Video browsing tools can be classified in two main types:

- *Time-line display of video frames and/or icons*: video units are organized in chronological sequence.

- *Hierarchical or graph-based story board*: representation that attempts to present the structure of video in an abstract and summarized manner.

This is another very active research area. In a recent survey, Lee, Smeaton, and Furner [13] categorized the user-interfaces of various video browsing tools and identified a superset of features and functions provided by those systems.

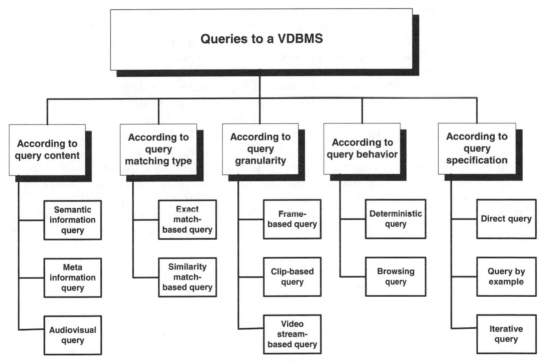

Figure 1.5 Classification of queries to a VDBMS.

6. OBJECTIVES OF THIS HANDBOOK

This Handbook was written to serve the needs of a growing community of researchers and practitioners in the fields of database systems, information retrieval, image and video processing, machine learning, data mining, human-computer interaction, among many others, and provide them with a comprehensive overview of the state of the art in this exciting area of video databases and applications. With contributions from more than 100 recognized world experts in the subject, it is an authoritative reference for most of the relevant research that is being carried out in this field. In addition to chapters that provide in-depth coverage of many topics introduced in this chapter, it showcases many novel applications and provides pointers to hundreds of additional references, becoming a one-stop reference for all the relevant issues related to video databases.

7. ORGANIZATION OF THE HANDBOOK

In the remainder of the Handbook, some of the world leading experts in this field examine the state of the art, ongoing research, and open issues in designing video database systems.

Section II presents concepts and techniques for video modeling and representation. In Chapter 2, Garg, Naphade, and Huang discuss the need for a semantic index associated with a video program and the difficulties in bridging the gap that exists between low-level media features and high-level semantics. They view the problem of semantic video indexing as a multimedia understanding problem, in which there is always a context to the co-occurrence of semantic concepts in a video scene and propose a novel learning architecture and algorithm, describing its application to the specific problem of detecting complex audiovisual events. In Chapter 3, Vasconcelos reviews ongoing efforts for the development of statistical models for characterizing semantically relevant aspects of video and presents a system that relies on such models to achieve the goal of semantic characterization. In Chapter 4, Eleftheriadis and Hong introduce *Flavor* (Formal Language for Audio-Visual Object Representation), an object-oriented language for bitstream-based media representation. The very active field of summarization and understanding of sports videos is the central topic of Chapter 5, where Assfalg, Bertini, Colombo, and Del Bimbo report their work on automatic semantic video annotation of generic sports videos, and Chapter 6, where Ekin and Tekalp propose a generic integrated semantic-syntactic event model to describe sports video, particularly soccer video, for search applications.

Section III describes techniques and algorithms used for video segmentation and summarization. In Chapter 7, Ardizzone and La Cascia review existing shot boundary detection techniques and propose a new neural network-based segmentation technique, which does not require explicit threshold values for detection of both abrupt and gradual transitions. In Chapter 8, Chua, Chandrashekhara, and Feng provide a temporal multi-resolution analysis (TMRA) framework for video shot segmentation. In Chapter 9, Smith, Watclar, and Christel describe the creation of video summaries and visualization systems through multimodal feature analysis, combining multiple forms of image, audio, and language information, and show results of evaluations and user studies under the scope of the Informedia Project at Carnegie Mellon University. In Chapter 10, Gong presents three video content summarization systems developed by multidisciplinary researchers in NEC USA, C&C Research Laboratories. These summarization systems are able to produce three kinds of motion video summaries: (1) audio-centric summary, (2) image-centric summary, and (3) audio-visual content summary. In Chapter 11, Mulhem, Gensel, and Martin present then their work on the VISU model, which allows both to annotate videos with high level semantic descriptions and to query these descriptions for generating video summaries. This discussion is followed by a broad overview of adaptive video segmentation and summarization approaches, presented by Owen and Dixon in Chapter 12. In Chapter 13, Owen, Zhou, Tang, and Xiao describe augmented imagery and its applications as a powerful tool for enhancing or repurposing content in video databases, including a number of interesting case studies. In Chapter 14, Ahmed and Karmouch present a new algorithm for video indexing, segmentation and key framing, called the binary penetration algorithm, and show its extension to a video web service over the World Wide Web for multiple video formats. Concluding the Section, in Chapter 15, Zhao and Grosky revisit the *semantic gap* problem and introduce a novel technique for spatial color indexing, *color anglogram*, and its use in conjunction with a dimension reduction technique, latent semantic indexing (LSI), to uncover the semantic correlation between video frames.

Section IV examines tools and techniques for designing and interacting with video databases. In Chapter 16, Hjelsvold and Vdaygiri present the result of their work while developing two interactive video database applications: HotStreams™ – a system for delivering and managing personalized video content – and TEMA (Telephony Enabled Multimedia Applications) – a platform for developing Internet-based multimedia applications for Next Generation Networks (NGN). In Chapter 17, Huang, Chokkareddy, and Prabhakaran introduce the topic of animation databases and present a toolkit for animation creation and editing. In Chapter 18, Djeraba, Hafri, and Bachimont explore different video exploration strategies adapted to user requirements and profiles, and introduce the notion of probabilistic prediction and path analysis using Markov models. Concluding the Section, in Chapter 19, Picariello, Sapino, and Subrahmanian introduce AVE! (Algebraic Video Environment), the first algebra for querying video.

The challenges behind audio and video indexing and retrieval are discussed in Section V. It starts with a survey of the state of the art in the area of audio content indexing and retrieval by Liu and Wang (Chapter 20). In Chapter 21, Ortega-Binderberger and Mehrotra discuss the important concept of relevance feedback and some of the techniques that have been successfully applied to multimedia search and retrieval. In Chapter 22, Santini proposes a novel approach to structuring and organizing video data using *experience units* and discusses some of its philosophical implications. Farin, Haenselmann, Kopf, Kühne, and Effelsberg describe their work on a system for video object classification in Chapter 23. In Chapter 24, Li, Tang, Ip, and Chan advocate a web-based hybrid approach to video retrieval by integrating the query-based (database) approach with the content-based retrieval paradigm and discuss the main issues involved in developing such a web-based video database management system supporting hybrid retrieval, using their VideoMAP* project as an example. In Chapter 25, Zhang and Chen examine the *semantic gap* problem in depth. The emergence of MPEG-7 and its impact on the design of video databases is the central topic of Chapter 26, where Smith discusses the topic in great technical detail and provides examples of MPEG-7-compatible descriptions. In Chapter 27, Satoh and Katayama discuss issues and approaches for indexing of large-scale (tera- to peta-byte order) video archives, and report their work on Name-It, a system that associate faces and names in news videos in an automated way by integration of image understanding, natural language processing, and artificial intelligence technologies. The next two chapters cover the important problem of similarity measures in video database systems. In Chapter 28, Cheung and Zakhor discuss the problem of video similarity measurement and propose a randomized first-order video summarization technique called the Video Signature (ViSig) method, whereas in Chapter 29, Traina and Traina Jr. discuss techniques for searching multimedia data types by similarity in databases storing large sets of multimedia data and present a flexible architecture to build content-based image retrieval in relational databases. At the end of the Section, in Chapter 30, Zhou and Huang review existing relevance feedback techniques and present a variant of discriminant analysis that is suited for small sample learning problems.

In Section VI we focus on video communications, particularly streaming, and the technological challenges behind the transmission of video across communication networks and the role played by emerging video compression algorithms. In Chapter 31, Hua and Tantaoui present several cost-effective techniques to achieve scalable video streaming, particularly for video-on-demand (VoD) systems. In Chapter 32, Shahabi and Zimmermann report their work designing, implementing, and evaluating a scalable real-time streaming architecture, Yima. In Chapter 33, Zhang, Aygün, and Song present the design strategies of a middleware for client-server distributed multimedia applications, termed *NetMedia*, which provides services to support synchronized presentations of multimedia data to higher level applications. Apostolopoulos, Tan, and Wee examine the challenges that make simultaneous delivery and playback of video difficult, and explore algorithms and systems that enable streaming of pre-encoded or live video over packet networks such as the Internet in Chapter 34. They provide a comprehensive tutorial and overview of video streaming and communication applications, challenges, problems and possible solutions, and protocols. In Chapter 35, Ghandeharizadeh and Kim discuss the continuous display of video objects using heterogeneous disk subsystems and quantify the tradeoff associated with alternative multi-zone techniques when extended to a configuration consisting of heterogeneous disk drives. In Chapter 36, Vetro and Kalva discuss the technologies, standards, and challenges that define and drive universal multimedia access (UMA). In Chapter 37, Basu, Little, Ke, and Krishnan look at the dynamic stream clustering problem, and present the results of simulations of heuristic and approximate algorithms for clustering in interactive VoD systems. In Chapter 38, Lienhart, Kozintsev, Chen, Holliman, Yeung, Zaccarin, and Puri offer an overview of the key questions in distributed video management, storage and retrieval, and delivery and analyze the technical challenges, some current solutions, and future directions. Concluding the Section, Torres and Delp provide a summary of the state of the art and trends in the fields of video coding and compression in Chapter 39.

Section VII provides the reader with the necessary background to understand video processing techniques and how they relate to the design and implementation of video databases. In Chapter 40, Wee, Shen, and Apostolopoulos present several compressed-domain image and video processing algorithms designed with the goal of achieving high performance with computational efficiency, with emphasis on transcoding algorithms for bitstreams that are based on video compression algorithms that rely on the block discrete cosine transform (DCT) and motion-compensated prediction, such as the ones resulting from predominant image and video coding standards in use today. In Chapter 41, Wang, Sheikh, and Bovik discuss the very important, and yet largely unexplored, topic of image and video quality assessment. Concluding the Section, in Chapter 42, Doërr and Dugelay discuss the challenges behind extending digital watermarking, the art of hiding information in a robust and invisible manner, to video data.

In addition to several projects, prototypes, and commercial products mentioned throughout the Handbook, Section VIII presents detailed accounts of three projects in this field, namely: an electronic clipping service under development at At&T Labs, described by Gibbon, Begeja, Liu, Renger, and Shahraray in Chapter 43; a multi-modal two-level classification framework for story segmentation in news videos, presented by Chaisorn, Chua, and Lee in Chapter 44; and the Video Scout system, developed at Philips Research Labs and presented by Dimitrova, Jasinschi, Agnihotri, Zimmerman, McGee, and Li in Chapter 45.

Finally, Section IX assembles the answers from some of the best-known researchers in the field to questions about the state of the art and future research directions in this dynamic and exciting field.

REFERENCES

[1]   R. Hjelsvold, VideoSTAR – A database for video information sharing, Dr. Ing. thesis, Norwegian Institute of Technology, November 1995.

[2]   R. Elmasri and S. B. Navathe, *Fundamentals of Database Systems – 3rd edition*, Addison-Wesley, Reading, MA, 2000.

[3]   C.J. Date, *An Introduction to Database Systems – 7th edition*, Addison-Wesley, Reading, MA, 2000.

[4]   T. Connolly, C. Begg, and A. Strachan, *Database Systems – 2nd ed.*, Addison-Wesley, Harlow, England, 1999.

[5]   K. Brown, A rich diet: Data-rich multimedia has a lot in store for archiving and storage companies, *Broadband Week*, March 5, 2001.

[6]   B. Prabhakaran, *Multimedia Database Management Systems*, Kluwer, Norwell, MA, 1997.

[7]   S. Santini, and R. Jain, Image databases are not databases with images", Proc. 9th International Conference on Image Analysis and Processing (ICIAP '97), Florence, Italy, September 17-19, 1997.

[8]   V.S. Subrahmanian, *Principles of Multimedia Database Systems*, Morgan Kaufmann, San Francisco, CA, 1998.

[9]   R.C. Veltkamp, and M. Tanase, A survey of content-based image retrieval systems, in *Content-Based Image and Video Retrieval*, O. Marques, and B. Furht, Eds., Kluwer, Norwell, MA, 2002.

[10] R. Bryll, A practical video database system, Master Thesis, University of Illinois at Chicago, 1998.

[11] A.K. Elmagarmid, H. Jiang, A.A. Helal, A. Joshi, and M. Ahmed, *Video Database Systems*, Kluwer, Norwell, MA, 1997.

[12] I. Koprinska and S. Carrato. Video segmentation: A survey", *Signal Processing: Image Communication*, 16(5), pp. 477-500, Elsevier Science, 2001.

[13] H. Lee, A.F. Smeaton, and J. Furner, User-interface issues for browsing digital video, in Proc. 21st Annual Colloquium on IR Research (IRSG 99), Glasgow, UK, 19-20 Apr. 1999.

[14] M. Davis, Media Streams: An iconic visual language for video representation, in *Readings in Human-Computer Interaction: Toward the Year 2000*, 2nd ed., R. M. Baecker, J. Grudin, W.A.S. Buxton, and S. Greenberg, Eds., Morgan Kaufmann, San Francisco, CA, 1995.

2

MODELING VIDEO USING INPUT/OUTPUT MARKOV MODELS WITH APPLICATION TO MULTI-MODAL EVENT DETECTION

Ashutosh Garg, Milind R. Naphade, and Thomas S. Huang
Department of Electrical and Computer Engineering
University of Illinois at Urbana-Champaign
Urbana, Illinois, USA
`{ashutosh,milind,huang}@ifp.uiuc.edu`

1. INTRODUCTION

Generation and dissemination of digital media content poses a challenging problem of efficient storage and retrieval. Of particular interest to us are audio and visual content. From sharing of picture albums and home videos to movie advertisement through interactive preview clips, live broadcasts of various shows or multimedia reports of news as it happens, multimedia information has found in the internet and the television powerful media to reach us. With innovations in hand-held and portable computing devices and wired and wireless communication technology (pocket PCs, organizers, cell-phones) on one end and broadband internet devices on the other, supply and dissemination of unclassified multimedia is overwhelming. Humans assimilate content at a semantic level and apply their knowledge to the task of sifting through large volumes of multimodal data. To invent tools that can gain widespread popularity we must try to emulate human assimilation of this content. We are thus faced with the problem of multimedia understanding if we are to bridge the gap between media features and semantics.

Current techniques in content-based retrieval for image sequences support the paradigm of query by example using similarity in low-level media features [1,2,3,4,5,6]. The query must be phrased in terms of a video clip or at least a few key frames extracted from the query clip. Retrieval is based on a matching algorithm, which ranks the database clips according to a heuristic measure of similarity between the query and the target. While effective for browsing and low-level search, this paradigm has limitations. Low-level similarity may not match with the user's perception of similarity. Also, the assumption that clips reflecting desire are available during query is unrealistic. It is also essential to fuse information from multiple modalities, especially the image sequence and audio streams. Most systems use either the image sequence [5,6,4,2,1], or the audio track [7,8,9,10,11,12], while few use both the modalities [13,14,12].

One way of organizing a video for efficient browsing and searching is shown in Figure 2.1. A systematic top-down breakdown of the video into scenes, shots and key frames exists in the form of a table of contents (ToC). To enable access to the video in terms of semantic concepts, there needs to be a semantic index (SI). The links connect entries in the SI to shots/scenes in the ToC and also indicate a measure of confidence.

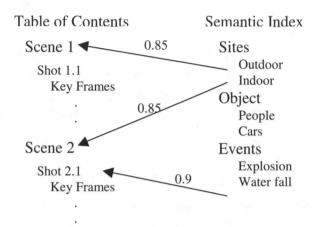

Figure 2.1 Organizing a Video with a Table of Contents (ToC) and a Semantic Index (SI). The ToC gives a top-down break-up in terms of scenes, shots and key frames. The SI lists key-concepts occurring in the video. The links indicate the exact location of these concepts and the confidence measure.

Automatic techniques for generating the ToC exist, though they use low-level features for extracting key frames as well as constructing scenes. The first step in generating the ToC is the segmentation of the video track into smaller units. Shot boundary detection can be performed in compressed domain [15,16,17] as well as uncompressed domain [18]. Shots can be grouped based on continuity, temporal proximity and similarity to form scenes [5]. Most systems support query by image sequence content [2,3,4,5,6] and can be used to group shots and enhance the ability to browse. Naphade et al. [14] presented a scheme that supports query by audiovisual content using dynamic programming. The user may browse a video and then provide one of the clips in the ToC structure as an example to drive the retrieval systems mentioned earlier. Chang et al. [2] allow the user to provide a sketch of a dominant object along with its color shape and motion trajectory. Key frames can be extracted from shots to help efficient browsing.

The need for a semantic index is felt to facilitate search using key words or key concepts. To support such semantics, models of semantic concepts in terms of multimodal representations are needed. For example, a query to find *explosion on a beach* can be supported if models for the concepts *explosion* and *beach* are represented in the system. This is a difficult problem. The difficulty lies in the gap that exists between low-level media features and high-level semantics. Query using semantic concepts has motivated recent research in semantic video indexing [13,19,20,12] and structuring [21,22,23]. We [13] presented novel ideas in semantic indexing by learning probabilistic multimedia representations of semantic events like *explosion* and sites like *waterfall* [13]. Chang et al. [19] introduced the notion of semantic visual templates. Wolf et al. [21] used hidden Markov models to parse video. Ferman et al. [22] attempted to model semantic structures like *dialogues* in video.

The two aspects of mapping low-level features to high-level semantics are the concepts represented by the multiple media and the context, in which they appear. We view the problem of semantic video indexing as a multimedia understanding problem. Semantic concepts do not occur in isolation. There is always a context to the co-occurrence of semantic concepts in a video scene. We presented a probabilistic graphical network to model this context [24,25] and demonstrated that modeling the context explicitly provides a significant improvement in performance. For further details on modeling context, the reader is referred to [24,25]. In this paper we concentrate on the problem of detecting complex audiovisual events. We apply a novel learning architecture and algorithm to fuse information from multiple loosely coupled modalities to detect audiovisual events such as explosion.

Detecting semantic events from audio-visual data with spatio-temporal support is a challenging multimedia understanding problem. The difficulty lies in the gap that exists between low-level features and high-level semantic labels. Often, one needs to depend on multiple modalities to interpret the semantics reliably. This necessitates efficient schemes, which can capture the characteristics of high level semantic events by fusing the information extracted from multiple modalities.

Research in fusing multiple modalities for detection and recognition has attracted considerable attention. Most techniques for fusing features from multiple modalities having temporal support are based on Markov models. Examples include the hidden Markov model (HMM) [26] and several variants of the HMM, like the coupled hidden Markov model [27], factorial hidden Markov model [28], the hierarchical hidden Markov model [13], etc. A characteristic of these models is the stage, at which the features from the different modalities are merged.

We present a novel algorithm, which combines feature with temporal support from multiple modalities. Two main features that distinguish our model from existing schemes are (a) the ability to account for non-exponential duration and (b) the ability to map discrete state input sequences to decision sequences. The standard algorithms modeling the video-events use HMMs, which model the duration of events as an exponentially decaying distribution. However, we argue that the duration is an important characteristic of each event and we demonstrate it by the improved performance over standard HMMs. We test the model on the audio-visual event *explosion*. Using a set of hand-labeled video data, we compare the performance of our model with and without the explicit model for duration. We also compare performance of the proposed model with the traditional HMM and observe that the detection performance can be improved.

2. PRIOR ART

In this section we review prior art in the fields of event detection and information fusion using multimedia features.

2.1 EVENT DETECTION

Recent work in temporal modeling of image sequences includes work in parsing and structuring as well as modeling visual events. Statistical models like the hidden Markov models (HMM) have been used for structuring image sequences [5,21,22]. Yeung et al. introduced dialog detection [5]. Topical classification of image sequences can provide information about the genres of videos like news, sports, etc. Examples include [29]. Extraction of semantics from image-sequences is difficult. Recent work dealing with semantic analysis of image sequences include Naphade et al. [13,30], Chang et al. [19], and Brand et al. [27]. Naphade et al. [13,30] use hidden Markov models to detect events in image sequences. Chang et al. [2] allow user-defined templates of semantics in image sequences. Brand et al. [27] use coupled HMMs to model complex actions in *Tai Chi* movies.

Recent work in segmentation and classification of audio streams includes [7,8,9,10,11,12,31,32]. Naphade and Huang [7] used hidden Markov models (HMMs) for representing the probability density functions of auditory features computed over a time series. Zhang and Kuo [12] used features based on heuristics for audio classification. HMMs have been successfully applied in speech recognition.

Among the state-of-the-art techniques in multimedia retrieval very few techniques use multiple modalities. Most techniques, using audiovisual data, perform temporal segmentation on one medium and then analyze the other medium. For example the image sequence is used for temporal segmentation and the audio is then analyzed for classification. Examples include [33,34], and the Informedia project [35] that uses the visual stream for segmentation and the audio stream for content classification. Such systems also exist for particular video domains like broadcast news [36], sports [12,29,37], meeting videos, etc. Wang et al. [34] survey a few techniques for analysis using a similar approach for similar domains. In case of domain-independent retrieval, while existing techniques attempt to determine what is going on in the speech-audio, most techniques go as far as classifying the genre of the video using audiovisual features. Other techniques for video analysis include the unsupervised clustering of videos [38]. Naphade et al. [14] have presented an algorithm to support query by audiovisual content. Another popular domain is the detection and verification of a speaker using speech and an image sequence obtained by a camera looking at the person [39]. This is particularly applicable to the domain of intelligent collaboration and human-computer interaction. Recent work in semantic video indexing includes Naphade et al. [13,40,41].

2.2 FUSION MODELS

Audio-Visual analysis to detect the semantic concepts in videos poses a challenging problem. One main difficulty arises from the fact that the different sensors are noisy in terms of the information they contain about different semantic concepts. For example, based on pure vision, it is hard to make out between a explosion and normal fire. Similarly, audio alone may give confusing information. On one hand one may be able to filter out the ambiguity arising from one source of information by looking (analysing) other source. While, on the other hand, these different sources may provide complementary information, which may be essential in inference.

Motivated by these difficulties, in the past few years a lot of research has gone into developing algorithms for fusing information from different modalities. Since different modalities may not be sampled at the same temporal rate, it becomes a challenging problem to seamlessly integrate different modalities (e.g., audio is normally sampled at 44KHz whereas video is sampled at 30fps.) At the same time, one may not even have the synchronized streams (sources of information) or the sources of information may have very different characteristics (audio - continuous, inputs to the computer through keyboard - discrete.) If we assume that one can get features from the different streams on a common scale of time, the two main categories of fusion models are those that favor early integration of features versus those that favor late integration. Early integration refers to combining the information at the level of raw features. Simple early integration is often observed in the form of concatenation of weighted features from different streams. More involved models of early integration have been proposed by using some form of Markov models. [27] have proposed the coupled hidden Markov models and used it for detection of human activities. Ghahramani et al. [28] have proposed the factorial hidden Markov models. The main difference in these models arises from the conditional independence assumptions that they make between the states of the different information sources. They assume that the different sources arc tightly coupled and model them using a single generative process.

In many situations, especially when the different sources are providing complementary information, one may prefer late integration. It refers to doing inferencing of each stream independently of the others and then combining the output of the two. This is especially important and shows improved results as how one looks at the essential information contained in the different streams and the sensor depend characteristics do not play any role. It also allows one to learn different models for each source independently of one another and then combine the output. One may simply look at the weighted decisions of different sources or may actually use probabilistic models to model the dependencies. For example [42] have proposed the use of dynamic Bayesian networks over the output of the different streams to solve the problem of speaker detection. Similarly, [13] have proposed the use of hierarchical HMMS.

We observed that in the case of movie, the audio and the visual streams normally carry complementary information. For example, a scene of explosion is not just characterized by a huge thunder but also a visual effect corresponding to bright red and yellow colors. Motivated by this fact we propose the use of late coupling which seems to be better suited for this framework. Fusion of multimodal feature streams (especially audio and visual feature streams) has been applied to problems like Bimodal speech [43], speaker detection [42], summarization of video [36], query by audio-visual content [14] and event detection in movies [13]. Examples of fusion of other streams include fusion of text and image content, motion and image content, etc.

3. PROBABILISTIC MODELING OF MEDIA FEATURES

We presented a probabilistic architecture of multiject models representing sites, objects and events for capturing semantic representations [13,24]. Bayes decision theory [44] and statistical learning form the core of our architecture. We briefly review the characteristics and the assumptions of this architecture.

3.1 PROBABILISTIC MULTIMEDIA OBJECTS (MULTIJECTS)

Semantic concepts (in video) can be informally categorized into *objects*, *sites* and *events*. Any part of the video can be explained as an object or an event occurring at a site or location. Such an informal categorization is also helpful for selecting structures of the models, which are used to represent the concepts. For the automatic detection of such semantic concepts we propose *probabilistic multimedia objects* or *multijects*.

A multiject represents a semantic concept that is supported by multiple media features at various levels (low level, intermediate level, high level) through a structure that is probabilistic [13,30]. Multijects belong to one of the three categories: objects (*car*, *man*, *helicopter*), sites (*outdoor*, *beach*) or events (*explosion*, *man-walking*, *ball-game*). Figure 2.2 illustrates the concepts of a multiject.

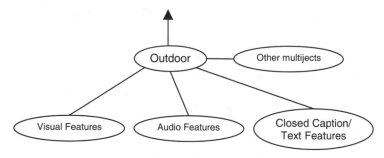

Figure 2.2 A probabilistic multimedia object (multiject).

A multiject is a flexible, open-ended semantic representation. It draws its support from low-level features of multiple media including audio, image, text and closed caption [13]. It can also be supported by intermediate-level features, including semantic templates [2]. It can also use specially developed high-level feature detectors like face detectors. A multiject can be developed for a semantic concept if there is some correlation between low-level multimedia features and high-level semantics. In the absence of such correlation, we may not be able to learn a sufficiently invariant representation. Fortunately many semantic concepts are correlated to some multimedia features, and so the framework has the potential to scale.

Multijects represent semantic concepts that have static as well as temporal support. Examples include sites like *sky*, *water-body*, *snow*, *outdoor*, *explosion*, *flying helicopter*, etc. Multijects can exist locally (with support from regions or blobs) or globally (with support from the entire video frame). The feature representation also corresponds to the extent of the spatiotemporal support of the multiject. We have described techniques for modeling multijects with static support in Naphade and Huang [24]. In this paper we concentrate on multiject models for events with temporal support in media streams.

3.2 ASSUMPTIONS

We assume that features from audiovisual data have been computed and refer to them as X. We assume that the statistical properties of these features can be characteristic signatures of the multijects. For distinct instances of all multijects, we further assume that these features are independent identically distributed random variables drawn from known probability distributions, with unknown deterministic parameters. For the purpose of classification, we assume that the unknown parameters are distinct under different hypotheses and can be estimated. In particular, each semantic concept is represented by a binary random variable. The two hypotheses associated with each such variable are denoted by H_i, $i \in \{0,1\}$, where 0 denotes absence and 1 denotes presence of the concept. Under each hypothesis, we assume that the features are generated by the conditional probability density function $P_i(X)$, $i \in \{0,1\}$. In case of site multijects, the feature patterns are static and represent a single frame. In case of events, with spatiotemporal support, X represents a time series of features over segments of the audiovisual data. We use the *one-zero* loss function [45] to penalize incorrect detection. This is shown in Equation 1:

$$\lambda(\alpha_i \mid w_i) = 0 \qquad i = j$$
$$= 1 \qquad i \neq j$$

(2.1)

The risk corresponding to this loss function is equal to the average probability of error and the conditional risk with action α_i is $1 - P(\omega_i|x)$. To minimize the average probability of error, class ω_i must be chosen, which corresponds to the maximum a posteriori probability $P(\omega_i|x)$. This is the minimum probability of error (MPE) rule.

In the special case of binary classification, this can be expressed as deciding in favor of ω_1 if

$$\frac{p(x \mid \omega_1)}{p(x \mid \omega_2)} > \frac{(\lambda_{12} - \lambda_{22})P(\omega_2)}{(\lambda_{21} - \lambda_{11})P(\omega_1)}$$

(2.2)

The term $p(x|\omega_j)$ is the *likelihood* of ω_j, and the test based on the ratio in Equation (2) is called the *likelihood ratio test* [44,45].

3.3 MULTIJECT MODELS FOR AUDIO AND VISUAL EVENTS

Interesting semantic events in video include *explosion, car chase*, etc. Interesting semantic events in audio include *speech, music, explosion, gunshots*, etc. We propose the use of hidden Markov models for modeling the probability density functions of media features under the positive and negative hypotheses.

We model a temporal event using a set of states with a Markovian state transition and a Gaussian mixture observation density in each state. We use continuous density models in which each observation probability distribution is represented by a mixture density. For state j the probability $b_j(\mathbf{o}_t)$ of generating observation $\mathbf{o}_t$ is given by Equation (3):

$$b_j(\mathbf{o}_t) = \sum_{m=1}^{M_j} c_{jm} N(\mathbf{o}_t; \mu_{jm}, \Sigma_{jm}) \qquad (2.3)$$

where M_j is the number of mixture components in state j, c_{jm} is the weight of the mth component and $N(o; \mu, \Sigma)$ is the multivariate Gaussian with mean μ and covariance matrix Σ as in Equation (4):

$$N(\mathbf{o}; \mu, \Sigma) = \frac{1}{\sqrt{(2\pi)^n |\Sigma|}} e^{-\frac{1}{2}(\mathbf{o}-\mu)'\Sigma^{-1}(\mathbf{o}-\mu)} \qquad (2.4)$$

The parameters of the model to be learned are the transition matrix A, the mixing proportions c, and the observation densities b. With q_t denoting the state at instant t and q_{t+1} the state at t+1, elements of matrix A are given by $a_{ij}=P(q_{t+1}=j|q_t=i)$. The Baum-Welch re-estimation procedure [46,26] is used to train the model and estimate the set of parameters. Once the parameters are estimated using the training data, the trained models can then be used for classification as well as state sequence decoding [46,26]. For each event multiject, a prototype HMM with three states and a mixture of Gaussian components in each state is used to model the temporal characteristics and the emitting densities of the class. For each mixture, a diagonal covariance matrix is assumed.

We have developed multijects for audio events like *human-speech, music* [7] and *flying helicopter* [40], and visual events like *explosion* [13], etc.

4. A HIERARCHICAL FUSION MODEL

In this paper, we discuss the problem of fusing multiple feature streams enjoying spatio-temporal support in different modalities. We present a hierarchical fusion model (see Figure 2.8), which makes use of late integration of intermediate decisions.

To solve the problem we propose a Hierarchical duration dependent input output Markov model. There are four main considerations that have led us to the particular choice of the fusion architecture.

- We argue that the different streams contain information which is correlated to one another only at a high level. This assumption allows us to process output of each source independently of one another. Since these sources may contain information which has highly temporal structure, we propose the use of hidden Markov models. These models are learned from the data and then we decode the hidden state sequence, characterizing the information describing the source at any given time. This assumption leads us to a hierarchical approach.

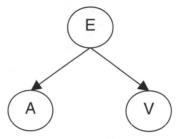

Figure 2.3: Consider the multimodal concept represented by node E and multimodal features represented by nodes A and V in this Bayesian network. The network implies that features A and V are independent given concept E.

- We argue that the different sources contain independent information. At high level, the output of one source is essentially independent of the information contained in the other. However, conditioned upon a particular bi-modal concept, these different sources may be dependent on one another. This suggests the use of an alternative dependence assumption contrary to conventional causal models like HMMs. This is illustrated in Figures 2.3 and 2.4.

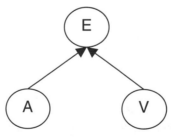

Figure 2.4 Consider the multimodal concept represented by node E and multimodal features represented by nodes A and V in this Bayesian network. The network implies that features A and V are dependent given concept E.

Note the difference between the assumptions implied by Figures 2.3 and 2.4. This idea can be explained with an example event, say an explosion. Suppose the node E represents event explosion and the nodes A and V represent some high characterization of audio and visual features. According to Figure 2.3 given the fact that an explosion is taking place, the audio and visual representations are independent. This means that once we know an explosion event is taking place, there is no information in either channel for the other. On the contrary Figure 2.4 implies that given an explosion event the audiovisual representations are dependent. This means that the audio channel and video channel convey information about each other only if the presence of event explosion is known. These are two very different views to the problem of modeling dependence between multiple modalities.

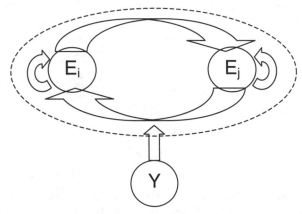

Figure 2.5 This figure illustrates the Markovian transition of the input output Markov model. Random variable E can be present in one of the two states. The model characterizes the transition probabilities and the dependence of these on input sequence y.

- We want to model temporal events. We therefore need to expand the model in Figure 2.4 to handle temporal dependecy in the event E. We make the usual Markovian assumption thus making the variable E at any time dependent only on its value at the previous time instant. This leads to an *input output Markov model* (IOMM). This is shown in Figure 2.5.

- Finally, an important characteristic of semantic concepts in videos is their duration. This points to the important limitation of hidden Markov models. In HMMs the probability of staying in any particular state decays exponentially. This is the direct outcome of the one Markov property of these models. To alleviate this problem, we explicitly model these probabilities. This leads us to what we call duration dependent input output Markov model. Note that because of this explicit modeling of the state duration, these are not really one Markov models but are what have been called in the past semi-Markov models. This leads to the *duration dependent input output Markov model* (DDIOMM) as shown in Figure 2.6.

Figure 2.7 shows the temporally rolled out Hierarchical model. This model takes as input, data from multiple sensors, which in the case of videos reduces to two streams Audio and Video. Figure 2.8 shows these two streams with one referring to audio features or observations $AO_1,...,AO_T$ and the other to video features $VO_1,...,VO_T$. The media information represented by these features is modeled using hidden Markov models which here are referred to as media HMM as these are used to model each media. Each state in the media HMMs represents a stationary distribution and by using the Viterbi decoder over each feature stream, we essentially cluster features spatio-temporally and quantize them through state identities. State-sequence-based processing using trained HMMs can be thought of as a form of guided spatio-temporal vector quantization for reducing the dimensionality of the feature vectors from multiple streams [13].

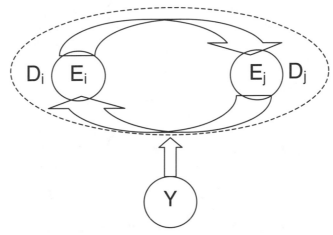

Figure 2.6 This figure illustrates the Markovian transition of the input output Markov model along with duration models for stay in each state. Random variable E can be present in one of the two states. The model characterizes the transition probabilities and the dependence of these on input sequence y and duration of stay in any state. Note that missing self transition arrows.

Once the individual streams have been modeled, inferencing is done to decode the state sequence for each stream of features. More formally, consider S streams of features. For each stream, consider feature vectors $f_1^{(s)},...,f_T^{(s)}$ corresponding to time instants t=1,...,T. Consider two hypotheses H_i, i∈ {0,1} corresponding to the presence and absence of an event E in the feature stream. Under each hypothesis we assume that the feature stream is generated by a hidden Markov model [26] (the parameters of the HMM under each hypothesis are learned using the EM algorithm [26]). Maximum likelihood detection is used to detect the underlying hypothesis for a given sequence of features observed and then the corresponding hidden state sequence is obtained by using the Viterbi decoding (explained later). Once the state sequence for each feature stream is obtained, we can use these intermediate-level decisions from that feature stream [13] in a hierarchical approach.

Hierarchical models use these state sequences to do the inferencing. Figure 7 shows the fusion model, where standard HMM is used for fusing the different modalities. However this model has many limitations - it assumes the different streams are independent for a particular concept and models a exponentially decaying distribution for a particular event, which as discussed is not true in general.

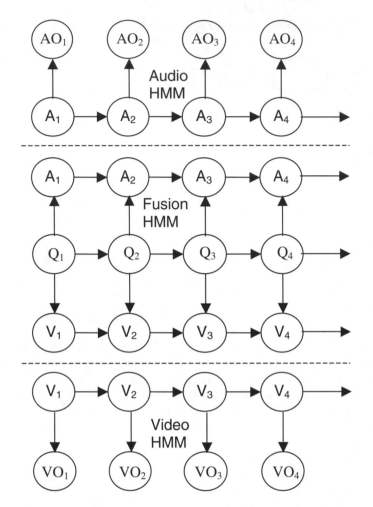

Figure 2.7 Hierarchical multimedia fusion using HMM. The media HMMs are responsible for mapping media observations to state sequences.

Figure 2.8 illustrates in Hierarchical structure that uses DDIOMM. DDIOMM helps us to get around these, providing a more suitable fusion architecture. It is discussed in more detail in the next section.

4.1 THE DURATION DEPENDENT INPUT OUTPUT MARKOV MODEL

Consider a sequence $\mathcal{Y}$ of symbols $y_1,...,y_T$ where $y_i \in \{1,...,N\}$. Consider another sequence Q of symbols $q_1,...,q_T$, where $q_i \in \{1,...,M\}$. The model between the dotted lines in Figure 2.8 illustrates a Bayesian network involving these two sequences. Here $y=\{A,V\}$ (i.e., Cartesian product of the audio and video sequence). This network can be thought of as a mapping of the input sequence $\mathcal{Y}$ to the output sequence Q. We term this network the *Input Output Markov Model*. This network is close in spirit to the input output hidden Markov model [47].

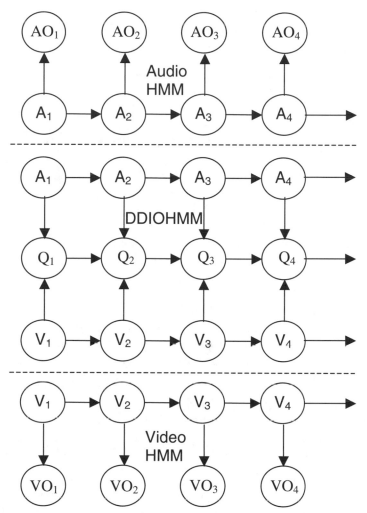

Figure 2.8 Hierarchical multimedia fusion. The media HMMs are responsible for mapping media observations to state sequences. The fusion model, which lies between the dotted lines, uses these state sequences as inputs and maps them to output decision sequences.

The transition in the output sequence is initially assumed to be Markovian. This leads to an exponential duration density model. Let us define $A=\{A_{ijk}\}$, $i,j \in \{1,...,M\}$, $k \in \{1,...,N\}$ as the map. $A_{ijk}=P(q_t=j|q_{t-1}=i,y_t=k)$ is estimated from the training data through frequency counting and tells us the probability of the current decision state given the current input symbol and the previous state. Once A is estimated, we can then predict the output sequence Q given the input sequence $\mathcal{Y}$ using Viterbi decoding.

The algorithm for decoding the decision is presented below.

$$\delta_t(j) = \max_Q P(q_1,...,q_{t-1},q_t = j \mid y_1,...,y_t) \qquad (2.5)$$

Q_t indicates all possible decision sequences until time t-1. Then this can be recursively represented.

$$\delta_t(j) = \max_{i=1:M} \delta_{t-1}(i) A_{ijk}$$

$$\Delta_t(j) = \arg\max_{i=1:M} \delta_{t-1}(i) A_{ijk}$$

$$P^* = \max_{i=1:M} \delta_T(i)$$

P^* is the probability of the best sequence given the input and the parametric mapping A. We can then backtrack the best sequence Q^* as follows

$$q_T^* = \arg\max_{i=1:M} \delta_T(i)$$

Backtracking

$$for\ \ l = 1:T-1$$

$$q_{T-l}^* = \Delta_{T-l+1}(q_{T-l+1}^*)$$

In the above algorithm, we have allowed the density of the duration of each decision state to be exponential. This may not be a valid assumption. To rectify this we now introduce the duration dependent input output Markov model (DDIOMM). Our approach in modeling duration in the DDIOMM is similar to that by Ramesh et al. [48].

This model, in its standard form, assumes that the probability of staying in a particular state decays exponentially with time. However, most audio-visual events have finite duration not conforming to the exponential density. Ramesh et al. [48] have shown that results can improve by specifically modeling the duration of staying in different states. In our current work, we show how one can enforce it in case of models like duration dependent IOMM.

Let us define a new mapping function A={A_{ijkd}}, i,j∈{1,...,M}, k∈{1,...,N}, d∈{1,...,D} where A_{ijkd}=P(q_t=j|q_{t-1}=I,$d_{t-1}(i)$=d,y_t=k). A can again be estimated from the training data by frequency counting. We now propose the algorithm to estimate the best decision sequence given A and the input sequence $\mathcal{Y}$. Let

$$\delta_t(j,d) = \max_Q P(q_1,...,q_{t-1}, q_t = j, d_t(j) = d \mid y_1,...,y_t) \qquad (2.6)$$

where Q_t indicates all possible decision sequences until time t-1 and $d_t(i)$ indicates the duration in terms of discrete time samples for which the path has continued to be in the state q_t. This can be recursively computed as follows

$$\delta_1(j,1) = P(q_1 \mid y_1)$$

$$\delta_1(j,d) = 0,\ \ d > 1$$

$$\delta_{t+1}(j,1) = \max_{i=1:M, i \neq j} \max_{d=1:D} \delta_t(i,d) A_{ijkd}$$

$$\delta_{t+1}(j,d+1) = \delta_t(j,d) A_{jjkd}$$

Let

$$\Delta_t(j,i) = \arg\max_{d=1:D} \delta_{t-1}(i,d) A_{ijkd} \quad 1 \le i \le j, \quad i \ne j$$

$$\Psi_t(j) = \arg\max_{i=1:M} \delta_{t-1}(i,\Delta_t(j,i)) A_{ijk\Delta_t(j,i)}$$

Finally

$$\eta(i) = \arg\max_{d=1:D} \delta_T(i,d) \quad 1 \le i \le M$$

$$P^* = \max_{i=1:M} \delta_T(i,\eta(i))$$

P^* is the probability of the best sequence given the input and the parametric mapping A. We can then backtrack the best sequence Q^* as follows

$$q_T^* = \arg\max_{i=1:M} \delta_T(i,\eta(i))$$

Backtracking

$$x = \eta(q_T^*) \quad t = T \quad z = x$$

while $t > 1$

$$q_{t-x+1}^* = q_t^* \quad l = 1,...,z-1$$

$$q_{t-x}^* = \Psi_{t-x+1}(q_x^*)$$

$$z = \Delta_{t-x+1}(q_t^*, q_{t-x}^*)$$

$$t = t - x \quad x = z$$

Using the above equations, one can decode the hidden state sequence. Each state or the group of states correspond to the state of the environment or the particular semantic concept that is being modeled. In the next section, we compare the performance of these models with the traditional HMMs and show that one can get huge improvements.

5. EXPERIMENTAL SETUP, FEATURES AND RESULTS

We compare the performance of our proposed algorithm with the IOMM as well as with the traditional HMM with its states being interpreted as decisions. We use the domain of movies and the audio-visual event *explosion* for comparison. Data from a movie are digitized. We have over 10000 frames of video data and the corresponding audio data split in 9 clips. The data are labeled manually to construct the ground truth. Figure 2.9 show some typical frames of the video sequence.

From the visual stream we extract several features describing the color (HSV histograms, multiple order moments), structure (edge direction histogram) and texture (statistical measures of gray level co-occurrence matrices at multiple orientations) of the stream [49]. From the audio stream we extract 15 MFCC coefficients, 15 delta coefficients and 2 energy coefficients [7]. As described earlier we train HMMs for the positive as well as the negative hypothesis for the event *explosion*. HMMs for audio streams and video streams are separately trained. Each HMM has 3 states corresponding (intuitively) to the beginning, middle and end state of the event. Using the pair of models for the positive and negative hypothesis we then segment each clip into two types of segments corresponding to the presence or absence of the event. Within each segment the best state sequence decoded by the Viterbi algorithm is available to us.

Figure 2.9 Some typical frames from a video clip.

For the audio and video streams synchronized at the video frame rate, we can now describe each frame by a symbol from a set of distinct symbols. Let v_s^h denote the number of states v corresponding to hypothesis h and feature stream s. Then the total number of distinct symbols needed to describe each audio-visual frame jointly is given by $\prod_{s=1}^{S} \sum_{h=1}^{H} v_s^h$.

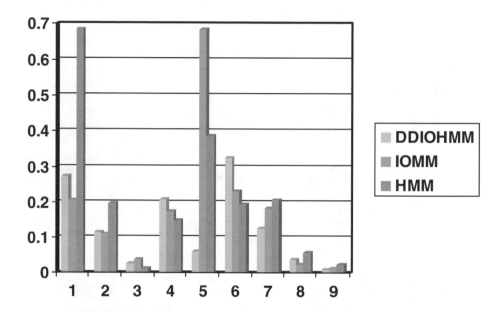

Figure 2.10 Classification error for the nine video clips using the leave-one-clip-out evaluation strategy. The maximum error for the DDIOMM is the least among the maximum error of the three schemes.

This forms the input sequence $\mathcal{Y}$. Similarly, with h^s denoting the number of hypotheses for stream s the total number of distinct symbols needed to describe the decision for each audio-visual frame jointly is given by $\prod_{s=1}^{S} h^s$. This forms the symbols of our decision sequence Q.

Table 2.1: Comparing overall classification error.

	HMM	IOMM	DDIOMM
Classification Error (%)	20.15	18.52	13.23

We report results using a leave-one-clip-out strategy. The quantum of time is a single video frame. To report performance objectively, we compare the prediction of the fusion algorithm for each video frame to our ground truth. Any difference between the two constitutes to a false alarm or misdetection. We also compare the classification error of the three schemes. Figure 2.10 shows error for each clip using the three schemes. Among the three schemes, the maximum error across all clips is least for the DDIOMM. Table 2.1 shows the overall classification error across all the clips.

Clearly the overall classification error is the least for the DDIOMM. We also compare the detection and false alarm rates of the three schemes. Figure 2.11 shows the detection and false alarm for the three schemes. Figures 2.10 and 2.11 and Table 2.1 thus show that the DDIOMM performs better event detection than the IOMM as well as the HMM.

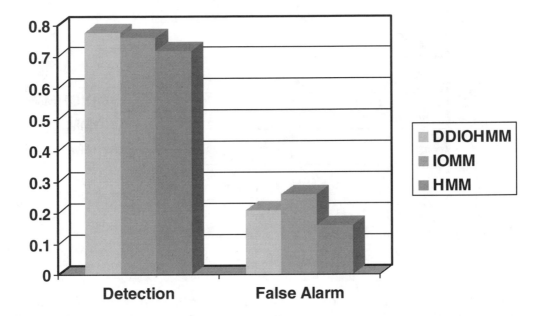

Figure 2.11 Comparing detection and false alarms. DDIOMM results in best detection performance.

6. CONCLUSION

In this paper we have analyzed the problem of detecting temporal events in videos using multiple sources of information. We have argued that some of the main characteristics of this problem are the duration of the events and the dependence between the different concepts. In the past, using standard HMMs for fusion, both of these issues were ignored. We have shown as how one can use the standard probabilistic models, modify them and obtain superior performance.

In particular, we present a new model, the duration dependent input output Markov model (DDIOMM) for performing integration of intermediate decisions from different feature streams to detect events in multimedia. The model provides a hierarchical mechanism to map media features to output decision sequences through intermediate state sequences. It forces the multimodal input streams to be dependent given the target event. It also supports discrete non-exponential duration models for events. By combining these two features in the framework of generative models for inference, we present a simple and efficient decision sequence decoding Viterbi algorithm. We demonstrate the strength of our model by experimenting with audio-visual data from movies and the audio-visual event *explosion*. Experiments comparing the DDIOMM with the IOMM as well as the HMM reveal that the DDIOMM results in lower classification error and improves detection.

REFERENCES

[1]   J. R. Smith and S. F. Chang, Visualseek: A Fully Automated Content-based Image Query System, in *Proceedings of ACM Multimedia*, Boston, MA, Nov. 1996.

[2]   D. Zhong and S. F. Chang, Spatio-Temporal Video Search Using the Object-based Video Representation, in *Proceedings of IEEE International Conference on Image Processing*, Santa Barbara, CA, Oct. 1997, Vol. 1, pp. 21-24.

[3]   Y. Deng and B. S. Manjunath, Content Based Search of Video Using Color, Texture and Motion, in *Proceedings of IEEE International Conference on Image Processing*, Santa Barbara, CA, Oct. 1997, Vol. 2, pp. 534-537.

[4]   H. Zhang, A. Wang, and Y. Altunbasak, Content-based Video Retrieval and Compression: A Unified Solution, in *Proceedings of IEEE International Conference on Image Processing*, Santa Barbara, CA, Oct. 1997, Vol. 1, pp. 13-16.

[5]   M. M. Yeung and B. Liu, Efficient Matching and Clustering of Video Shots," in *Proceedings of IEEE International Conference on Image Processing*, Washington, D.C., Oct. 1995, Vol. 1, pp. 338-341.

[6]   M. R. Naphade, M. M. Yeung, and B. L. Yeo, A Novel Scheme for Fast and Efficient Video Sequence Matching Using Compact Signatures, in *Proceedings of SPIE Storage and Retrieval for Multimedia Databases*, Jan. 2000, Vol. 3972, pp. 564-572.

[7]   M. R. Naphade and T. S. Huang, Stochastic Modeling of Soundtrack for Efficient Segmentation and Indexing of Video, in *Proceedings of SPIE Storage and Retrieval for Multimedia Databases*, Jan. 2000, Vol. 3972, pp. 168-176.

[8]   D. Ellis, Prediction-driven Computational Auditory Scene Analysis, Ph.D. Thesis, MIT, Cambridge, MA, 1996.

[9]   M. Akutsu, A. Hamada, and Y. Tonomura, Video Handling with Music and Speech Detection, *IEEE Multimedia*, Vol. 5, No. 3, pp. 17-25, 1998.

[10]   P. Jang and A. Hauptmann, Learning to Recognize Speech by Watching Television, *IEEE Intelligent Systems Magazine*, Vol. 14, No. 5, pp. 51-58, 1999.

[11]   E. Wold, T. Blum, D. Keislar, and J. Wheaton, Content-based Classification Search and Retrieval of Audio, *IEEE Multimedia*, Vol. 3, No. 3, pp. 27-36, 1996.

[12]   T. Zhang and C. Kuo, An Integrated Approach to Multimodal Media Content Analysis, in *Proceedings of SPIE, IS&T Storage and Retrieval for Media Databases*, Jan. 2000, Vol. 3972, pp. 506-517.

[13]   M. Naphade, T. Kristjansson, B. Frey, and T. S. Huang, Probabilistic Multimedia Objects (Multijects): A Novel Approach to Indexing and Retrieval in Multimedia Systems, in *Proceedings of IEEE International*

Conference on Image Processing, Chicago, IL, Oct. 1998, Vol. 3, pp. 536-540.

[14] M. R. Naphade, R. Wang, and T. S. Huang, Multimodal Pattern Matching for Audio-Visual Query and Retrieval, in *Proceedings of SPIE, Storage and Retrieval for Media Databases*, Jan. 2001, Vol. 4315, pp. 188-195.

[15] B. L. Yeo and B. Liu, Rapid Scene Change Detection on Compressed Video, *IEEE Transactions on Circuits and Systems for Video Technology*, Vol. 5, No. 6, pp. 533-544, Dec. 1995.

[16] J. Meng, Y. Juan, and S. F. Chang, Scene Change Detection in a MPEG Compressed Video Sequence, in *Proceedings of the SPIE Symposium*, San Jose, CA, Feb. 1995, Vol. 2419, pp. 1-11.

[17] H. J. Zhang, C. Y. Low, and S. Smoliar, Video Parsing Using Compressed Data, in *Proceedings of SPIE Conference on Image and Video Processing II*, San Jose, CA, 1994, pp. 142-149.

[18] M. Naphade, R. Mehrotra, A. M. Ferman, J. Warnick, T. S. Huang, and A. M. Tekalp, A High Performance Shot Boundary Detection Algorithm Using Multiple Cues, in *Proceedings of IEEE International Conference on Image Processing*, Chicago, IL, Oct. 1998, Vol. 2, pp. 884-887.

[19] S. F. Chang, W. Chen, and H. Sundaram, Semantic Visual Templates - Linking Features to Semantics, in *Proceedings of IEEE International Conference on Image Processing*, Chicago, IL, Oct. 1998, Vol. 3, pp. 531-535.

[20] R. Qian, N. Hearing, and I. Sezan, A Computational Approach to Semantic Event Detection, in *Proceedings of Computer Vision and Pattern Recognition*, Fort Collins, CO, June 1999, Vol. 1, pp. 200-206.

[21] W. Wolf, Hidden Markov Model Parsing of Video Programs, in *Proceedings of International Conference on Acoustics Signal and Speech Processing*, 1997.

[22] M. Ferman and A. M. Tekalp, Probabilistic Analysis and Extraction of Video Content,' in *Proceedings of IEEE International Conference on Image Processing*, Kobe, Japan, Oct. 1999.

[23] N. Vasconcelos and A. Lippman, Bayesian Modeling of Video Editing and Structure: Semantic Features for Video Summarization and Browsing, in *Proceedings of IEEE International Conference on Image Processing*, Chicago, IL, Oct. 1998, Vol. 2, pp. 550-555.

[24] M. R. Naphade and T. S. Huang, A Probabilistic Framework for Semantic Video Indexing, Filtering and Retrieval, *IEEE Transactions on Multimedia, Special issue on Multimedia over IP*, Vol. 3, No. 1, pp. 141-151, March 2001.

[25] M. R. Naphade, I. Kozintsev, and T. S. Huang, Probabilistic Semantic Video Indexing, NIPS 2000, pp. 967-973.

[26] L. R. Rabiner, A Tutorial on Hidden Markov Models and Selected Applications in Speech Recognition, *Proceedings IEEE*, Vol. 77, No. 2, pp. 257-286, Feb. 1989.

[27] M. Brand, N. Oliver, and A. Pentland, Coupled Hidden Markov Models for Complex Action Recognition, in *Proceedings of Computer Vision and Pattern Recognition*, 1997, pp. 994-999.

[28] Z. Ghahramani and M. Jordan, Factorial Hidden Markov Models, *Machine Learning*, Vol. 29, pp. 245-273, 1997.

[29] V. Kobla, D. DeMenthon, and D. Doermann, Identifying Sports Video Using Replay, Text and Camera Motion Features, in *Proceedings of SPIE Storage and Retrieval for Media Databases*, Jan. 2000, Vol. 3972, pp. 332-343.

[30] M. R. Naphade, Video Analysis for Efficient Segmentation, Indexing and Retrieval, M.S. Thesis, University of Illinois at Urbana-Champaign, 1998.

[31] Z. Liu, Y. Wang, and T. Chen, Audio Feature Extraction and Analysis for Scene Segmentation and Classification, *VLSI Signal Processing Systems for Signal, Image and Video Technology*, Vol. 20, pp. 61-79, Oct. 1998.

[32] E. Scheirer and M. Slaney, Construction and Evaluation of a Robust Multifeatures Speech/Music Discriminator, in *Proceedings of IEEE Intl. Conf. on Acoustic, Speech and Signal Processing*, Munich, Germany, 1997, Vol. 2, pp. 1331-1334.

[33] J. Nam, A.E. Cetin, and A.H. Tewfik, Speaker Identification and Video Analysis for Hierarchical Video Shot Classification, in *Proceedings of IEEE International Conference on Image Processing*, Santa Barbara, CA, Oct. 1997, Vol. 2, pp. 550-555.

[34] Y. Wang, Z. Liu, and J. Huang, Multimedia Content Analysis Using Audio and Visual Information, *IEEE Signal Processing Magazine*, Vol. 17, No. 6, pp. 12-36, Nov. 2000.

[35] H. Wactlar, T. Kanade, M. Smith, and S. Stevens, Intelligent Access to Digital Video: The Informedia Project, *IEEE Computer Digital Library Initiative Special Issue*, No. 5, May 1996.

[36] Y. Nakamura and T. Kanade, Semantic Analysis for Video Contents Extraction - Spotting by Association in News Video, in *Proceedings of ACM International Multimedia Conference*, Nov. 1997.

[37] D. D. Saur, Y. P. Tan, S. R. Kulkarni, and P. J. Ramadge, Automated Analysis and Annotation of Basketball Video, in *Proceedings of SPIE Symposium*, 1997, Vol. 3022, pp. 176-187.

[38] B. Clarkson and A. Pentland, Unsupervised Clustering of Ambulatory Audio and Video, in *Proceedings of IEEE International Conference on Accoustics Speech and Signal Processing*, 1999.

[39] J. Rehg, K. Murphy, and P. Fieguth, Vision-based Speaker Detection Using Bayesian Networks, in *Proceedings of Computer Vision and Pattern Recognition*, Fort Collins, CO, June 1999, Vol. 2, pp. 110-116.

[40] M. R. Naphade and T. S. Huang, Recognizing High-level Audio-Visual Concepts Using Context," submitted to IEEE International Conference on Image Processing, 2001.

[41] M. R. Naphade, A. Garg, and T. S. Huang, Duration Dependent Input Output Markov Models for Audio-Visual Event Detection, submitted to IEEE International Conference on Multimedia and Expo, Tokyo, Japan, 2001.

[42] A. Garg, V. Pavlovic, M. Rehg, and T. S. Huang, Integrated Audio/Visual Speaker Detection Using Dynamic Bayesian Networks, in *Proceedings of IEEE Conference on Automatic Face and Gesture Recognition*, March 2000.

[43] T. Chen and R. Rao, Audio-Visual Integration in Multimodal Communication, *IEEE Proceedings*, Vol. 86, No. 5, pp. 837-852, 1998.

[44] R. O. Duda and P. E. Hart, *Pattern Classification and Scene Analysis*, Wiley Eastern, New York, 1973.

[45]   H. V. Poor, *An Introduction to Signal Detection and Estimation*, Springer-Verlag, New York, 2 edition, 1999.

[46]   L. E. Baum and T. Petrie, Statistical Inference for Probabilistic Functions of Finite State Markov Chains, *Annals of Mathematical Statistics*, Vol. 37, pp. 1559-1563, 1966.

[47]   Y. Bengio and P. Frasconi, Input/Output HMMs for Sequence Processing, *IEEE Transactions on Neural Networks*, Vol. 7, No. 5, pp. 1231-1249, 1996.

[48]   P. Ramesh and J. Wilpon, Modeling State Durations in Hidden Markov Models for Automatic Speech Recognition, in *Proceedings of International Conference on Acoustics, Speech and Signal processing*, Mar. 1992, Vol. 1, pp. 381-384.

[49]   M. R. Naphade and T. S. Huang, Semantic Video Indexing Using a Probabilistic Framework, in *Proceedings of IAPR International Conference on Pattern Recognition*, Barcelona, Spain, Sep. 2000, Vol. 3, pp. 83-88.

3

STATISTICAL MODELS OF VIDEO STRUCTURE AND SEMANTICS

Nuno Vasconcelos
HP Cambridge Research Laboratory
Cambridge, Massachusetts, USA
nuno.vasconcelos@hp.com

1. INTRODUCTION

Given the recent advances on video coding and streaming technology and the pervasiveness of video as a form of communication, there is currently a strong interest in the development of techniques for browsing, categorizing, retrieving, and automatically summarizing video. In this context, two tasks are of particular relevance: the decomposition of a video stream into its component units, and the extraction of features for the automatic characterization of these units. Unfortunately, current video characterization techniques rely on image representations based on low-level visual primitives (such as color, texture, and motion)[1] that, while practical and computationally efficient, fail to capture most of the structure that is relevant for the perceptual decoding of the video. As a result, it is difficult to design systems that are truly useful for naive users. Significant progress can only be attained by a deeper understanding of the relationship between the message conveyed by the video and the patterns of visual structure that it exhibits.

There are various domains where these relationships have been thoroughly studied, albeit not always from a computational standpoint. For example, it is well known by film theorists that the message strongly constrains the stylistic elements of the video [2,3], which are usually grouped into two major categories: the *elements of montage* and the *elements of mise-en-scene*. Montage refers to the temporal structure, namely the aspects of film editing, while mise-en-scene deals with spatial structure, i.e., the composition of each image, and includes variables such as the type of set in which the scene develops, the placement of the actors, aspects of lighting, focus, camera angles, and so on.

Building computational models for these stylistic elements can prove useful in two ways: on one hand it will allow the extraction of *semantic features* enabling video characterization and classification much closer to that which people use than current descriptors based on texture properties or optical flow. On the other hand, it will provide *constraints* for the low-level analysis algorithms required to perform tasks such as video segmentation, key-framing, and so on.

The first point is illustrated by Figure 3.1, where we show how a collection of promotional trailers for commercially released feature films populates a 2-D feature space based on the most elementary characterization of montage and mise-en-scene: average shot duration vs. average shot activity.[1] Despite the coarseness of this characterization, it captures aspects that are important for semantic movie classification: close inspection of the genre assigned to each movie by the *motion picture association of America* reveals that in this space the movies cluster by genre!

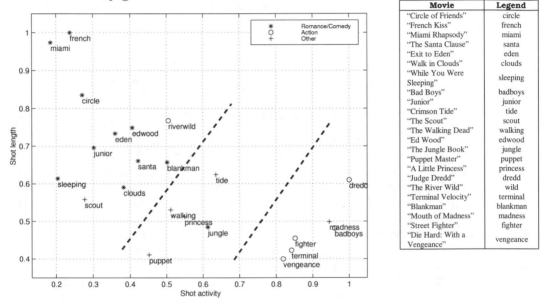

Movie	Legend
"Circle of Friends"	circle
"French Kiss"	french
"Miami Rhapsody"	miami
"The Santa Clause"	santa
"Exit to Eden"	eden
"Walk in Clouds"	clouds
"While You Were Sleeping"	sleeping
"Bad Boys"	badboys
"Junior"	junior
"Crimson Tide"	tide
"The Scout"	scout
"The Walking Dead"	walking
"Ed Wood"	edwood
"The Jungle Book"	jungle
"Puppet Master"	puppet
"A Little Princess"	princess
"Judge Dredd"	dredd
"The River Wild"	wild
"Terminal Velocity"	terminal
"Blankman"	blankman
"Mouth of Madness"	madness
"Street Fighter"	fighter
"Die Hard: With a Vengeance"	vengeance

Figure 3.1 Shot activity vs. duration features. The genre of each movie is identified by the symbol used to represent the movie in the plot (© 2000 IEEE).

The long-term goal of our video understanding research is to exploit knowledge of video structure as a means to enable the principled design of computational models for video semantics. This knowledge can either be derived from existing theories of content production, or learned from collections of training examples. The basic idea is to ground the semantic analysis directly on the high-level patterns of structure rather than on lower level attributes such as texture, colour, or optical flow. Since prior knowledge plays a fundamental role in our strategy, we have been placing significant emphasis on the use of Bayesian inference as the computational foundation of all content characterization work. In this chapter, we 1) review ongoing efforts for the development of statistical models for characterizing semantically relevant aspects of video and 2) present our first attempt at building a system that relies on such models to achieve the goal of semantic characterization. We show that accounting for video structure can both lead significant improvements in low-level tasks such as shot segmentation, and enable surprisingly accurate semantic classification in terms of primitives such as action, dialog, or the type of set.

---

[1] The activity features are described in section 4.

2. THE ROLE OF VIDEO STRUCTURE

The main premise behind our work is that the ability to infer semantic information from a video stream is a monotonic function of the amount of structure exhibited by the video. While newscasts are at the highly structured end of the spectrum and, therefore, constitute one of the simplest categories to analyse [4,5,6], the raw output of a personal camera exhibits almost no structure and is typically too difficult to characterize semantically [7]. Between these two extrema, there are various types of content for which the characterization has a varying level of difficulty. Our interests are mostly in domains that are more generic than newscasts, but still follow enough content production codes to exhibit a significant amount of structure. A good example of such a domain is that of feature films.

We have already mentioned that, in this domain, computational modelling of the elements of montage and mise-en-scene is likely to place a central role for any type of semantics based processing. From the content characterization perspective, the important point is that both montage and mise-en-scene tend to follow some very well established production codes or rules. For example, a director trying to put forth a text deeply rooted in the construction of character (e.g., a drama or a romance) will necessarily have to rely on a fair amount of facial close-ups, as close-ups are the most powerful tool for displaying emotion,[2] an essential requirement to establish a bond between audience and characters. If, on the other hand, the goal is to put forth a text of the action or suspense genres, the elements of mise-en-scene become less relevant than the rhythmic patterns of montage. In action or suspense scenes, it is imperative to rely on fast cutting, and manipulation of the cutting rate is the tool of choice for keeping the audience "at the edge of their seats." Directors who exhibit supreme mastery in the manipulation of the editing patterns are even referred to as *montage directors*.[3]

While there is a fundamental element of montage, the shot duration, it is more difficult to identify a single defining characteristic of mise-en-scene. It is, nevertheless, clear that scene activity is an important one: while action movies contain many active shots, character based stories are better conveyed by scenes of smaller activity (e.g., dialogues). Furthermore, the amount of activity is usually correlated with the amount of violence in the content (at least that of a gratuitous nature) and can provide clues for its detection. It turns out that activity measures also provide strong clues for the most basic forms of video parsing, namely the detection of shot boundaries. For these reasons we start the chapter by analysing a simple model of video structure that reduces montage to shot duration and mise-en-scene to shot activity. It is shown that even such a simple model can lead to shot segmentation algorithms that significantly outperform the current state of the art.

This, of course, does not mean that the semantic characterization problem is solved. In the second part of the chapter, we introduce a generic Bayesian

---

[2] The importance of close-ups is best summarized in the quote from Charles Chaplin: "Tragedy is a close-up, comedy a long shot."
[3] The most popular example in this class is Alfred Hitchcock, who relied intensively on editing to create suspense in movies like "Psycho" or "Birds"[2].

architecture that addresses that more ambitious problem. The basic idea is to rely on 1) a multitude of low-level sensors for visual events that may have semantic relevance (e.g., activity, skin tones, energy in some spatial frequency bands, etc.) and 2) knowledge about how these sensors react to the presence of the semantic stimulae of interest. These two components are integrated through a Bayesian formalism that encodes the knowledge of sensor interaction into a collection of conditional probabilities for sensor measurements given semantic state.

The shot segmentation module and the Bayesian semantic classification architecture form the basis of the Bayesian Modelling of Video Editing and Structure (BMoViES) system for video characterization. An overview of this system is presented in the third part of the chapter, which also illustrates various applications of semantic modelling in video retrieval, summarization, browsing, and classification. More details about the work here discussed can be found in references [7-13].

3. MODELS FOR TEMPORAL VIDEO STRUCTURE

Because shot boundaries can be seen as arrivals over discrete, non-overlapping temporal intervals, a Poisson process seems an appropriate model for shot duration [14]. However, events generated by Poisson processes have inter-arrival times characterized by the exponential density, which is a monotonically decreasing function of time. This is clearly not the case for the shot duration, as can be seen from the histograms in Figure 3.2. In this work, we consider two alternative models, the Erlang and Weibull distributions.

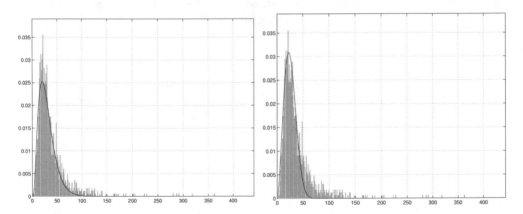

Figure 3.2 Shot duration histogram, and maximum likelihood fit obtained with the Erlang (left) and Weibull (right) distributions (© 2000 IEEE).

3.1 THE ERLANG MODEL

Letting τ be the time since the previous boundary, the Erlang distribution [14] is described by

$$\varepsilon_{r,\lambda}(\tau) = \frac{\lambda^r \tau^{r-1} e^{-\lambda\tau}}{(r-1)!}.$$

It is a generalization of the exponential density, characterized by two parameters: the order r, and the expected inter-arrival time $(1/\lambda)$ of the underlying Poisson process. When *r=1*, the Erlang distribution becomes the exponential distribution. For larger values of *r*, it characterizes the time between the *r*[th] order inter-arrival time of the Poisson process. This leads to an intuitive explanation for the use of the Erlang distribution as a model of shot duration: for a given order *r*, the shot is modelled as a sequence of *r* events which are themselves the outcomes of Poisson processes. Such events may reflect properties of the shot content, such as "setting the context" through a wide-angle view followed by "zooming in on the details" when *r=2*, or "emotional build-up" followed by "action" and "action outcome" when *r=3*.

Figure 3.2 presents a shot duration histogram, obtained from the training set to be described in section 5.2, and its maximum likelihood (ML) Erlang fit.

3.2 THE WEIBULL MODEL

While the Erlang model provides a good fit to the empirical density, it is of limited practical utility due to the constant arrival rate assumption [15] inherent to the underlying Poisson process. Because λ is a constant, the expected rate of occurrence of a new shot boundary is the same if 10 seconds or 1 hour have elapsed since the occurrence of the previous one. An alternative model that does not suffer from this problem is the Weibull distribution [15], which generalizes the exponential distribution by considering an expected rate of arrival of new events that is a function of time τ

$$\lambda(\tau) = \frac{\alpha \tau^{\alpha-1}}{\beta^{\alpha}}$$

and of the parameters α and β; leading to a probability density of the form

$$w_{\alpha,\beta}(\tau) = \frac{\alpha \tau^{\alpha-1}}{\beta^{\alpha}} \exp\left[-\left(\frac{\tau}{\beta}\right)^{\alpha}\right].$$

Figure 3.2 presents the ML Weibull fit to the shot duration histogram. Once again we obtain a good approximation to the empirical density estimate.

4. MODELS OF SHOT ACTIVITY

Given a sequence of images, various features can be used to obtain an estimate of the amount of activity in the underlying scene. In our work we have considered two metrics that can be derived from low-level image measurements: the *colour histogram distance* and the *tangent distance* [12,13] between successive images in the sequence. Since space is limited and the histogram distances have been widely used as a similarity measure for object recognition [16], content-based retrieval [17], and temporal video segmentation [18], we restrict our attention to them in this chapter. In particular we will rely on the well established L_1 norm of the histogram difference,

$$D(\mathbf{a}, \mathbf{b}) = \sum_i |a_i - b_i|$$

where **a** and **b** are histograms of successive frames.

Statistical modelling of the histogram distance features requires the identification of the various states through which the video may progress. For simplicity, we restrict ourselves to a video model composed of two states: "regular frames" (*S = 0*) and "shot transitions" (*S = 1*). The fundamental principles are however applicable to more complex models. As illustrated by Figure 3.3, for regular frames the distribution is asymmetric about the mean, always positive and concentrated near zero. This suggests that a mixture of Erlang distributions is an appropriate model for this state, a suggestion that is confirmed by the fit to the empirical density obtained with the expectation-maximization (EM) algorithm, also depicted in the figure. On the other hand, for shot transitions the fit obtained with a simple Gaussian model is sufficient to achieve a reasonable approximation to the empirical density. In both cases, a uniform mixture component is introduced to account for the tails of the distributions [13].

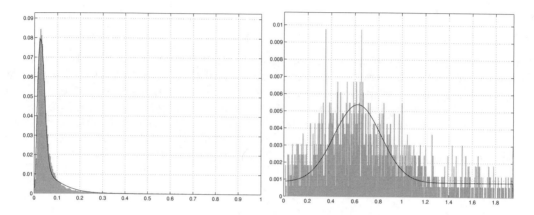

Figure 3.3 Left: Conditional activity histogram for regular frames, and best fit by a mixture with three Erlang and a uniform component. Right: Conditional activity histogram for shot transitions, and best fit by a mixture with a Gaussian and a uniform component (© 2000 IEEE).

5. SHOT SEGMENTATION

Because shot segmentation is a prerequisite for virtually any task involving the understanding, parsing, indexing, characterization, or categorization of video, the grouping of video frames into shots has been an active topic of research in the area of multimedia signal processing. Extensive evaluation of various approaches has shown that simple thresholding of histogram distances performs surprisingly well and is difficult to beat [18].

5.1 BAYESIAN SHOT SEGMENTATION

We have developed an alternative formulation that regards the problem as one of statistical inference between two hypotheses:
- $\mathcal{H}_0$: no shot boundary occurs between the two frames under analysis ($S = 0$),
- $\mathcal{H}_1$: a shot boundary occurs between the two frames ($S = 1$),

for which the minimum probability of error decision is achieved by a likelihood ratio test [19] where $\mathcal{H}_1$ is chosen if

$$L = \log \frac{P(D \mid S = 1)}{P(D \mid S = 0)} > 0 \qquad (3.1)$$

and $\mathcal{H}_0$ is chosen otherwise. It can easily be verified that thresholding is a particular case of this formulation, in which both conditional densities are assumed to be Gaussian with equal covariance. From the discussion in the previous section, it is clear that this does not hold for real video. One further limitation of the thresholding model is that it does not take into account the fact that the likelihood of a new shot transition is dependent on how much time has elapsed since the previous one.

We have, however, shown in [13] that the statistical formulation can be extended to overcome these limitations, by taking into account the shot duration models of section 3. For this, one needs to consider the *posterior odds ratio* between the two hypotheses

$$\frac{P(S_{\tau,\tau+\delta}=1\,|\,S_{0,\tau}=0,D_0,\ldots,D_\tau)}{P(S_{\tau,\tau+\delta}=0\,|\,S_{0,\tau}=0,D_0,\ldots,D_\tau)} \tag{3.2}$$

where $S_{a,b}$ is the state of interval $[a,b]$, δ is the duration of each frame (inverse of frame rate), and D_τ is the activity measurement at time τ. Under the assumption of stationarity for D_τ, and a generic Markov property (D_τ independent of all previous D and S given $S_{\tau,\tau+\delta}$), it is possible to show that the optimal decision (in the minimum probability of error sense) is to declare that a boundary exists in $[\tau,\tau+\delta]$ if

$$\log\frac{P(D_\tau|\,S_{\tau,\tau+\delta}=1)}{P(D_\tau|\,S_{\tau,\tau+\delta}=0)} > \log\frac{\int_{\tau+\delta}^{\infty}p(\alpha)d\alpha}{\int_{\tau}^{\tau+\delta}p(\alpha)d\alpha} = T(\tau) \tag{3.3}$$

and that there is no boundary otherwise. Comparing this with (1), it is clear that the inclusion of the shot duration prior transforms the fixed threshold approach into an adaptive one, where the threshold depends on how much time has elapsed since the previous shot boundary.

It is possible to derive closed-form expressions for the optimal threshold for both the Erlang and Weibull priors [13], namely

$$T(\tau) = \log\frac{\sum_{i=1}^{r}\varepsilon_{i,\lambda}(\tau+\delta)}{\sum_{i=1}^{r}\left[\varepsilon_{i,\lambda}(\tau)-\varepsilon_{i,\lambda}(\tau+\delta)\right]}$$

for the former and

$$T(\tau) = -\log\left\{\exp\left[\frac{(\tau+\delta)^\alpha-\tau^\alpha}{\beta^\alpha}\right]-1\right\}$$

for the latter. The evolution of the thresholds over time for both cases is presented in Figure 3.4. While in the initial segment of the shot the threshold is large and shot changes are unlikely to be accepted, the threshold decreases as the scene progresses increasing the likelihood that shot boundaries will be declared. The intuitive nature of this procedure is a trademark of Bayesian inference [20] where, once the mathematical details are worked out, one typically ends up with a solution that is intuitively correct.

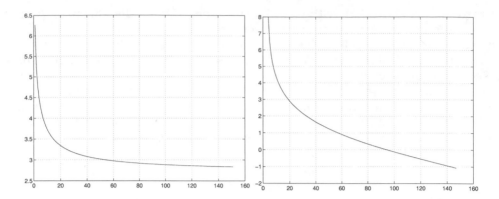

Figure 3.4 Temporal evolution of the Bayesian threshold for the Erlang (left) and Weibull (right) priors (© 2000 IEEE).

While qualitatively similar, the two thresholds exhibit important differences in terms of their steady state behaviour. Ideally, in addition to decreasing monotonically over time, the threshold should not be lower bounded by a positive value as this may lead to situations in which its steady-state value is large enough to miss several consecutive shot boundaries. While the Erlang prior fails to meet this requirement (due to the constant arrival rate assumption discussed in section 3), the Weibull prior does lead to a threshold that decreases to -∞ as τ grows. This guarantees that a new shot boundary will always be found if one waits long enough. In any case, both the Erlang and Weibull priors lead to adaptive thresholds that are more intuitive than the fixed threshold in common use for shot segmentation.

5.2   SHOT SEGMENTATION RESULTS

The performance of Bayesian shot segmentation was evaluated on a database containing the promotional trailers of Figure 3.1. Each trailer consists of 2 to 5 minutes of video and the total number of shots in the database is 1959. In all experiments, performance was evaluated by the *leave-one-out* method. Ground truth was obtained by manual segmentation of all the trailers.

We evaluated the performance of Bayesian models with Erlang, Weibull, and Poisson shot duration priors and compared them against the best possible performance achievable with a fixed threshold. For the latter, the optimal threshold was obtained by brute-force, i.e., testing several values and selecting the one that performed best. The total number of errors, misses (boundaries that were not detected), and false positives (regular frames declared as boundaries) achieved with all priors are shown in Figure 3.5. It is visible that, while the Poisson prior leads to worse accuracy than the static threshold, both the Erlang and the Weibull priors lead to significant improvements. The Weibull prior achieves the overall best performance decreasing the error rate of the static threshold by 20%. A significantly more detailed analysis of the relative performance of the various priors can be found in [13].

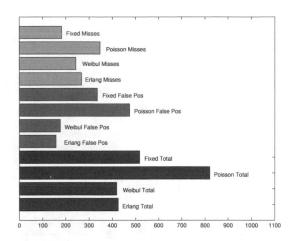

Figure 3.5 Total number of errors, false positives, and missed boundaries achieved with the different shot duration priors (© 2000 IEEE).

The reasons for the improved performance of Bayesian segmentation are illustrated in Figure 3.6, which presents the evolution of the thresholding process for a segment from one of the trailers in the database ("blankman"). Two approaches are depicted: Bayesian with the Weibull prior, and standard fixed thresholding. The adaptive behaviour of the Bayesian threshold significantly increases the robustness against spurious peaks of the activity metric originated by events such as very fast motion, explosions, camera flashes, etc. This is visible, for example, in the shot between frames 772 and 794 which depicts a scene composed of fast moving black and white graphics where figure and ground are frequently reversed. While the Bayesian procedure originates a single false-positive, the fixed threshold produces several. This is despite the fact that we have complemented the plain fixed threshold with commonly used heuristics, such as rejecting sequences of consecutive shot boundaries [18]. The vanilla fixed threshold would originate many more errors.

6. SEMANTIC CHARACTERIZATION

While the results above indicate that shot activity is sufficient for purposes of temporal segmentation, the extraction of full-blown semantic descriptions requires a more sophisticated characterization of the elements of mise-en-scene. In principle, it is possible to train a classifier for each semantic attribute of interest, but such an approach is unlikely to scale as the set of attributes grows. An alternative, which we adopt, is to rely on an intermediate representation, consisting of a *limited* set of sensors tuned to visual attributes that are likely to be semantically relevant, and a sophisticated form of *sensor fusion* to infer semantic attributes from sensor outputs. Computationally, this translates into the use of a *Bayesian network* as the core of the content characterization architecture.

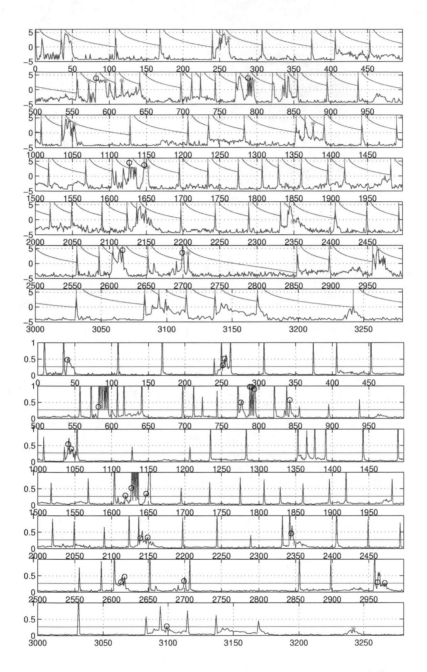

Figure 3.6 Evolution of the thresholding operation for a challenging trailer. Top: Bayesian segmentation. The likelihood ratio and the Weibull threshold are shown. Bottom: Fixed threshold. The histogram distances and the optimal threshold are presented. In both graphs, misses are represented as circles and false positives as stars (© 2000 IEEE).

6.1 BAYESIAN NETWORKS

Given a set of random variables **X**, a fundamental question in probabilistic inference is how to infer the impact on a subset of variables of interest **U** of the observation of another (non-overlapping) subset of variables **O** in the model, i.e., the ability to compute $P(\mathbf{U}|\mathbf{O} = \mathbf{o})$. While this computation is, theoretically, straightforward to perform using

$$P(\mathbf{U}|\mathbf{O}=\mathbf{o})=\frac{\sum_{\mathbf{H}}P(\mathbf{U},\mathbf{H},\mathbf{O}=\mathbf{o})}{\sum_{\mathbf{H},\mathbf{U}}P(\mathbf{U},\mathbf{H},\mathbf{O}=\mathbf{o})}$$

where $\mathbf{H} = \mathbf{X} \setminus \{\mathbf{U} \cup \mathbf{O}\}$, and the summations are over all the possible configurations of the sets $\mathbf{H}$ and $\mathbf{O}$; in practice, the amount of computation involved in the evaluation of these summations makes the solution infeasible even for problems of relatively small size. A better alternative is to explore the relationships between the variables in the model to achieve more efficient inference procedures. This is the essence of Bayesian networks.

A Bayesian network for a set of variables $\mathbf{X} = \{\mathbf{X}_1, \ldots, \mathbf{X}_n\}$ is a probabilistic model composed of 1) a graph $\mathcal{G}$, and 2) a set of local probabilistic relations $\mathcal{P}$. The graph consists of a set of nodes, each node corresponding to one of the variables in $\mathbf{X}$, and a set of links (or edges), each link expressing a probabilistic relationship between the variables in the nodes it connects. Together, the graph $\mathcal{G}$ and the set of probabilities $\mathcal{P}$ define the joint probability distribution for $\mathbf{X}$. Denoting the set of parents of the node associated with $\mathbf{X}_i$ by $\mathbf{pa}_i$, this joint distribution is

$$P(\mathbf{X})=\prod_{i} P(\mathbf{X}_i | \mathbf{pa}_i).$$

The ability to decompose the joint density into a product of local conditional probabilities allows the construction of efficient algorithms where inference takes place by propagation of beliefs across the nodes in the network [21,22].

6.2 BAYESIAN SEMANTIC CONTENT CHARACTERIZATION

Figure 3.7 presents a Bayesian network that naturally encodes the content characterization problem. The set of nodes $\mathbf{X}$ is the union of two disjoint subsets: a set $\mathbf{S}$ of sensors containing all the leaves (nodes that do not have any children) of the graph, and a set $\mathbf{A}$ of semantic content attributes containing the remaining variables. The set of attributes is organized hierarchically, variables in a given layer representing higher level semantic attributes than those in the layers below.

The visual sensors are tuned to visual features deemed relevant for the semantic content characterization. The network infers the presence/absence of the semantic attributes given these sensor measurements, i.e., $P(\mathbf{a}|\mathbf{S})$, where $\mathbf{a}$ is a subset of $\mathbf{A}$. The arrows indicate a causal relationship from higher-level to lower-level attributes and from attributes to sensors and, associated with the set of links converging at each node, is a set of conditional probabilities for the state of the node variable given the configuration of the parent attributes. Variables associated with unconnected nodes are marginally independent.

6.3 SEMANTIC MODELLING

One of the strengths of Bayesian inference is that the sensors of Figure 3.7 do not have to be flawless since 1) the model can account for variable sensor precision, and 2) the network can integrate the sensor information to disambiguate conflicting hypothesis. Consider, for example, the simple task of detecting sky in a sports database containing pictures of both skiing and sailing competitions. One way to achieve such a goal is to rely on a pair of sophisticated water and sky detectors. The underlying strategy is to interpret the images first and then characterize the images according to this interpretation.

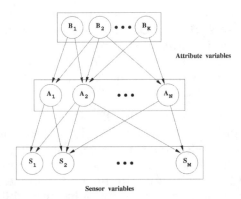

Figure 3.7 A generic Bayesian architecture for content characterization. Even though only three layers of variables are represented in the figure, the network could contain as many as desired (© 1998 IEEE).

While such strategy could be implemented with Bayesian procedures, a more efficient alternative is to rely on the model of Figure 3.8. Here, the network consists of five semantic attributes and two simple sensors for large white and blue image patches. In the absence of any measurements, the variables *sailing* and *skiing* are independent. However, whenever the sensor of blue patches becomes active, they do become dependent (or, in the Bayesian network lingo, *d-connected* [21]) and the knowledge of the output of the sensor of white patches is sufficient to perform the desired inference.

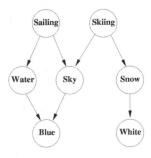

Figure 3.8 A simple Bayesian network for the classification of sports (© 1998 IEEE).

This effect is known as the "explaining away" capability of Bayesian networks [21]. Although there is no direct connection between the *white* sensor and the *sailing* variable, the observation of white reduces the likelihood of the sailing hypothesis and this, in turn, reduces the likelihood of the *water* hypothesis. So, if the *blue* sensor is active, the network will infer that this is a consequence of the presence of sky, even though we have not resorted to any sort of sophisticated sensors for image interpretation. That is, the white sensor explains away the firing of the blue sensor.

This second strategy relies more in modelling the semantics and the relationships between them than in the classification of the image directly from the image observations. In fact, the image measurements are only used to discriminate between the different semantic interpretations. This has two practical advantages. First, a much smaller burden is placed on the sensors which, for example, do not have to know that "water is more textured than sky," and are therefore significantly easier to build. Second, as a side effect of the sky detection process, we obtain a semantic interpretation of the images that, in this case, is sufficient to classify them into one of the two classes in the database.

7. THE BMoViES SYSTEM

The Bayesian shot segmentation module of section 5 and the Bayesian architecture of section 6 form the basis of the BMoViES system for video classification, retrieval summarization, and browsing. This system is our first attempt at evaluating the practical feasibility of extracting semantics in a domain as sophisticated as that of movies.

7.1 THE ATTRIBUTES

In the current implementation, the system recognizes four semantic shot attributes: the presence/absence of a close-up, the presence/absence of a crowd in the scene, the type of set (nature vs. urban), and if the shot contains a significant amount of action or not. These attributes can be seen as a minimalist characterization of mise-en-scene which, nevertheless, provides a basis for categorizing the video into relevant semantic categories such as "action vs dialog," "city vs country side," or combinations of these. Also, as discussed in section 2, it captures the aspects of mise-en-scene that are essential for the inference of higher level semantic attributes such as suspense or drama.

7.2 THE SENSORS

Currently, the sensor set consists of three sensors measuring the following properties: shot activity, texture energy, and amount of skin tones in the scene. Shot activity is measured as discussed in section 4. The texture energy sensor performs a 3-octave wavelet decomposition of each image, and measures the ratio of the total energy in the high-pass horizontal and vertical bands to the total energy in all the bands other than the DC. It produces a low output whenever there is a significant amount of vertical or horizontal structure in the images (as is the case in most man-made environments) and a high output when this is not the case (as is typically the case in natural settings). Finally, the skin tones sensor identifies the regions of each image that contain colours consistent with human skin, measures the area of each of these regions and computes the entropy of the resulting vector (regarding each component as a probability). This sensor outputs a low value when there is a single region of skin and high values otherwise. The situation of complete absence of skin tones is also detected, the output of the sensor being set to one.

Sensor measurements are integrated across each shot by averaging the individual frame outputs. In order to quantize the sensor outputs, their range was thresholded into three equally sized bins. In this way, each sensor provides a ternary output corresponding to the states *no, yes,* or *maybe.* For example, the activity sensor can output one of three states: "there is no significant activity in this shot," "there is a significant amount of activity in this shot," or "maybe there is significant activity, I can't tell."

7.3   THE NETWORK

The Bayesian network implemented in BMoViES is presented in Figure 3.9. The parameters of this model can either be learned from training data [23] or set according to expert knowledge. In the current implementation we followed the latter approach. Both the structure and the probabilities in the model were hand-coded, using common-sense (e.g., the output of the skin tones sensor will be *yes* with probability *0.9* for a scene of a crowd in a man-made set). No effort was made to optimize the overall performance of the system by tweaking the network probabilities.

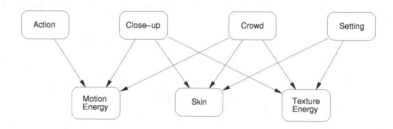

Figure 3.9 Bayesian network implemented in BmoViES (© 1998 IEEE).

To see how explaining away occurs in BMoViES, consider the observation of a significant amount of skin tones. Such observation can be synonymous with either a close-up or a scene of a crowd. However, if a crowd is present there will also be a significant response by the texture sensor, while the opposite will happen if the shot consists of a close-up. Hence, the texture sensor "explains away" the observation of skin tones, and rules out the close-up hypothesis, even though it is not a crowd detector.

8. APPLICATIONS AND EXPERIMENTAL RESULTS

The most obvious application of BMoViES is the classification of movies or movie clips into semantic categories. These categories can then be used as basis for the development of content filtering agents that will screen a media stream for items that comply with a previously established user-profile. However, filtering is not the only problem of interest, since users may also be interested in actively searching for specific content in a given database. In this case, the problem becomes one of information retrieval. As illustrated by Figure 3.10 the classification and retrieval problems are dual.

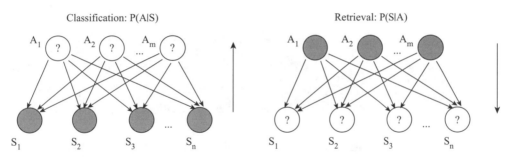

Figure 3.10 The duality between classification and retrieval (© 1998 IEEE).

A classifier is faced with data observations and required to make inferences with respect to the attributes on the basis of which the classification is to be performed. On the other hand, a retrieval system is faced with attribute specifications (e.g., "find all the shots containing people in action") and required to find the data that best satisfies these specifications. Hence, while classification requires the ability to go from sensor observations to attributes, retrieval requires the ability to go from attribute specifications to sensor configurations.

This type of two-way inferences requires a great deal of flexibility from the content characterization architecture that cannot be achieved with most statistical classifiers (such as neural networks [24], decision trees [25], or support vector machines [26]) which require a very precise definition of which variables are inputs and which ones are outputs. Since a Bayesian network does not have inputs or outputs, but only hidden and observed nodes, at a given point in time any node can be an input or an output, and the two dual problems can be handled equally well. Hence, it provides a unified solution to the problems of information filtering and retrieval.

8.1 CLASSIFICATION

To evaluate the accuracy of the semantic classification of BMoViES, we applied the system to a database of about 100 video clips (total of about 3000 frames) from the movie "Circle of friends." The database is a sub-sampling of approximately 25 minutes of film, and contains a wide variety of scenes and high variation of imaging variables such as lighting, camera viewpoints, etc. To establish ground truth, the video clips were also manually classified.

Table 3.1 Classification accuracy of BmoViES (© 1998 IEEE).

Attribute	Action	Close-up	Crowd	Set
Accuracy (%)	90.7	88.2	85.5	86.8

Table 3.1 presents the classification accuracy achieved by BMoViES for each of the semantic attributes in the model. Overall the system achieved an accuracy of 88.7%. Given the simplicity of the sensors this is a very satisfying result. Some of the classification errors, which illustrate the difficulty of the task, are presented in Figure 3.11.

Figure 3.11 Classification errors in BMoViES. People were not detected in the left three clips; crowd was not recognized on the right (© 1998 IEEE).

8.2 RETRIEVAL

As a retrieval system, BMoViES supports standard query by example, where the user provides the system with a video clip and asks it to "find all the clips that look like this". BMoViES then searches for the video clip in the database that maximizes the likelihood $P(S|A)$ of sensor configurations given attribute specifications.

The point that distinguishes BMoViES from most of the current retrieval systems is that the retrieval criterion, *semantic similarity*, is much more meaningful than the standard *visual similarity* criterion. In fact, whenever a user orders the machine to "search for a picture like this," the user is with high likelihood referring to pictures that are semantically similar to the query image (e.g., "pictures which also contain people") but which do not necessarily contain identical patterns of colour and texture.

Figure 3.12 Example based retrieval in BMoViES. The top left image is a key frame of the clip submitted to the retrieval system. The remaining images are key frames of the best seven matches found by the system (© 1998 IEEE).

Figure 3.12 presents an example of retrieval by semantic similarity. The image in the top right is a key-frame of the clip submitted as a query by the user, and the remaining images are key frames of the clips returned by BMoViES. Notice that most of the suggestions made by the system are indeed semantically similar to the query, but very few are similar in terms of colour and texture patterns.

8.3 RELEVANCE FEEDBACK

Because retrieval of visual information is a complex problem, it is unlikely that any retrieval system will be able to always find the desired database entries in response to a user request. Consequently, the retrieval process is usually interactive, and it is important that retrieval systems can learn from user feedback in order to minimize the number of iterations required for each retrieval operation. If it is true that simply combining a relevance feedback mechanism with a low-level image representation is unlikely to solve the retrieval problem, it is also unlikely that the sophistication of the representation can provide a solution by itself. The goal is therefore to build systems that combine sophisticated representations and learning or relevance feedback mechanisms. Once again, the flexibility inherent to the Bayesian formulation enables a unified framework for addressing both issues.

In the particular case of Bayesian architectures, support for relevance feedback follows immediately from the ability to propagate beliefs, and the fact that nodes can be either hidden or observed at any point in time. The situation is illustrated in Figure 3.13, which depicts two steps of an interactive retrieval session. The user starts by specifying a few attributes, e.g., "show me scenes with people." After belief propagation the system finds the sensor configurations that are most likely to satisfy those specifications and retrieves the entries in the database that lead to those configurations. The user inspects the returned sequences and gives feedback to the system, e.g., "not interested in action scenes," beliefs are updated, new data fetched from the database, and so on. Since the attributes known by the system are semantic, this type of feedback is very intuitive, and much more tuned to the way in which the user evaluates the content himself than relevance feedback based on low level features.

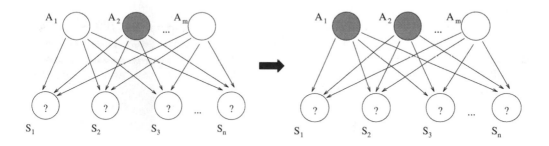

Figure 3.13 Relevance feedback in the Bayesian setting (© 1998 IEEE).

Figure 3.14 illustrates the ability of BMoViES to support meaningful user interaction. The top row presents the video clips retrieved in a response to a query where the action attribute was instantiated with *yes*, and the remaining attributes with *don't care*. The system suggests a shot of ballroom dancing as the most likely to satisfy the query, followed by a clip containing some graphics and a clip of a rugby match. In this example, the user was not interested in clips containing a lot of people. Specifying *no* for the crowd attribute led to the refinement shown in the second row of the figure. The ballroom shot is no longer among the top suggestions, which tend to include at most one or two people. At this point, the user specified that he was looking for scenes shot in a natural set, leading the system to suggest the clips shown in the third row of the figure. The clips that are most likely to satisfy the specification contain scenes of people running in a forest. Finally, the specification of *no* for the close-up attribute led to the suggestion of the bottom row of the figure, where the clips containing close-ups were replaced for clips where the set becomes predominant.

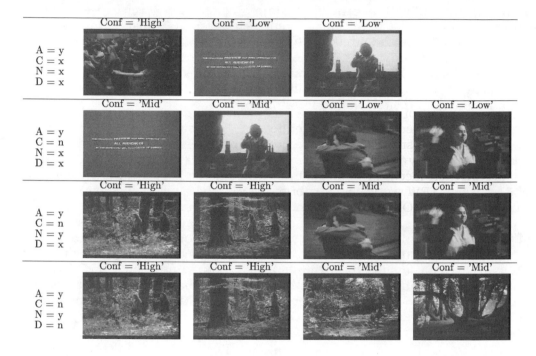

Figure 3.14 Relevance feedback in BMoViES. Each row presents the response of the system to the query on the left. The action (A), crowd (C), natural set (N), and close-up (D) attributes are instantiated with yes (y), no (n), or don't care (x). The confidence of the system of each of the retrieved clips is shown on top of the corresponding key frame (© 1998 IEEE).

8.4   SEMANTIC USER INTERACTION

In addition to intuitive relevance feedback mechanisms, the semantic characterization performed by BMoViES enables powerful modes of user interaction via a combination of summarization, visualization, and browsing. For a system capable of inferring content semantics, summarization is a simple outcome of the characterization process. Because system and user understand the same language, all that is required is that the system can display the inferred semantic attributes in a way that does not overwhelm the user. The user can then use his/her own cognitive resources to extrapolate from these semantic attributes to other attributes, usually of higher semantic level, that may be required for a coarse understanding of the content.

Semantic time-lines

In BMoViES, this graphical summarization is attained in the form of a time-line that displays the evolution of the state of the semantic attributes throughout the movie. Figure 3.15 presents the time-lines resulting from the analysis of the promotional trailers of the movies "Circle of friends" (COF) and "The river wild" (TRW). Each line in the time-line corresponds to the semantic attribute identified by the letter on the left margin – "A" for action, "D" for close-up, "C" for crowd, and "S" for natural set - and each interval between small tick marks displays the state of the attribute in one shot of the trailer - filled (empty) intervals mean that the attribute is active (not present). The shots are represented in the order by which they appear in the trailer.

By simple visual inspection of these time-lines, the user can quickly extract a significant amount of information about the content of the two movies. Namely, it is possible to immediately understand that while COF contains very few action scenes, consists mostly of dialogue, and is for the most part shot in man-made sets, TRW is mostly about action, contains few dialogues, and is shot in the wilderness. When faced with such descriptions, few users looking for a romance would consider TRW worthy of further inspection, and few users looking for a thriller would give COF further consideration.

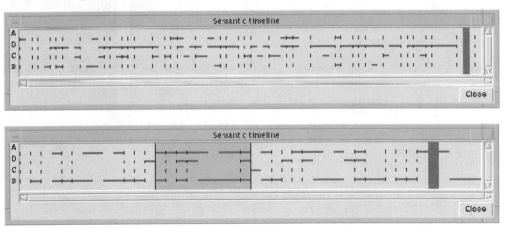

Figure 3.15 Semantic time-lines for the trailers of the movies "Circle of friends" (top) and "The river wild" (bottom) (© 1998 IEEE).

In fact, it can be argued that given a written summary of the two movies few people would have doubts in establishing the correspondence between summaries and movies based on the information provided by the semantic time-lines alone. To verify this the reader is invited to consider the following summaries, extracted from the Internet Movie Database [27].

Circle of Friends:

A story about the lives, loves, and betrayals of three Irish girls, Bennie, Eve, and Nan as they go to Trinity College, Dublin. Bennie soon seems to have found her ideal man in Jack, but events conspire to ruin their happiness.

The River Wild:

Gail, an expert at white water rafting, takes her family on a trip down the river to their family's house. Along the way, the family encounters two men who are inexperienced rafters that need to find their friends down river. Later, the family finds out that the pair of men are armed robbers. The men then physically force the family to take them down the river to meet their accomplices. The rafting trip for the family is definitely ruined, but most importantly, their lives are at stake.

Since visual inspection is significantly easier and faster than reading text descriptions, the significance of this conjecture is that semantic summarization could be a viable replacement for the textual summaries that are now so prevalent. We are currently planning experiments with human subjects that will allow a more objective assessment of the benefits of semantic summarization.

Semantic content-access and browsing

It can obviously be argued that the example above does not fully stretch the capabilities of the semantic characterization, i.e., that the movies belong to such different genres that the roughest of the semantic characterizations would allow a smart user to find the desired movie. What if instead of distinguishing COF from TRW, we would like to differentiate TRW from "Ghost and the Darkness" (GAD)? GAD is summarized as follows:

Ghost and the Darkness:

Set in 1898, this movie is based on the true story of two lions in Africa that killed 130 people over a nine-month period, while a bridge engineer and an experienced old hunter tried to kill them.

From this summary, one would also expect GAD to contain a significant amount of action, little dialogue, and be shot in the wilderness. How would the semantic characterization help here?

There are two answers to this question. The first is that it would not because the characterization is not *fine* enough to distinguish between TRW and GAD. The solution would be to augment the system with finer semantic attributes, e.g., to subdivide the natural set attribute into classes like "river," "forest," "savannah," "desert," etc. The second, significantly simpler, is that while simply looking at the time-lines would not help, interacting with them would.

Consider the action scenes in the two movies. While in TRW we would expect to see a river, a woman, and good and bad guys, in GAD we would expect to see savannah, lions, and hunters. Thus, the action scenes would probably be the place to look first. Consider next the TRW time-line in the bottom of Figure 3.15. The high concentration of action shots in the highlighted area indicates that this is likely to be the best area to look for action. This is confirmed by Figure 3.16, which presents key frames for each of the shots in the area. By actually viewing the shots represented in the figure, it is clear that the action occurs in a river, that there are good and bad guys (the first, and third shots depict a fight), and there are a woman and a child in the boat, i.e., even when the information contained in them is not enough to completely disambiguate the content, the semantic attributes provide a way to quickly *access* the relevant portions of the video stream. Semantic-based access is an important feature on its own for browsing as it allows users to quickly move on to the portions of the video that really interest them.

9. CONCLUSION

Digital video is now becoming prevalent and will likely become the major source of information and entertainment in the next few decades. With decreasing production and storage costs and faster networking, it will soon be possible to access massive video repositories. It is, however, not clear that such repositories will be useful in the absence of powerful search paradigms and intuitive user interfaces. So far, most research on video analysis in the context of large databases has focused on low-level processes (such as colour or texture characterization) that are not likely to solve the problem.

In this chapter, we have presented an alternative view, which is directly targeted at recovering video semantics. We have argued for a principled formulation based on Bayesian inference, and showed that it has great potential for both low-level tasks, such as shot segmentation, and high-level semantic characterization. In particular, it was shown that principled models of video structure can make a significant difference at all levels. These ideas were embodied in the BMoViES system to illustrate the fact that, once a semantic representation is in place, problems such as retrieval, summarization, visualization, and browsing become significantly simpler to address.

Figure 3.16 Key-frames of the shots in the highlighted area of the time-line in the bottom of Figure 3.15. The shot (correctly) classified as not containing action is omitted (© 1998 IEEE).

ACKNOWLEDGMENT

The author gratefully acknowledges Andrew Lippman and Giri Yiengar for many interesting discussions on the topics of this chapter and on the work presented above.

REFERENCES

[1] P. Aigrain, H. Zhang, and D. Petkovic, Content-based Representation and Retrieval of Visual Media: A State-of-the-Art Review, *Multimedia Tools and Applications*, Vol. 3:179 - 202, 1996.

[2] D. Bordwell and K. Thompson, *Film Art: an Introduction*, McGraw-Hill, 1986.

[3] K. Reisz and G. Millar, *The Technique of Film Editing*, Focal Press, 1968.

[4]   Y. Ariki and Y. Saito, Extraction of TV News Articles Based on Scene Cut Detection Using DCT Clustering, Proc. IEEE Int. Conf. on Image Processing, pages 847- 850, Lausanne, 1996.

[5]   A. Hanjalic, R. Lagendijk, and J. Biemond, Template-based Detection of Anchorperson Shots in News Programs, in Proc. IEEE Int. Conf. on Image Processing, pages 148- 152, Chicago, 1998.

[6]   C. Low, Q. Tian, and H. Zhang, An Automatic News Video Parsing, Indexing, and Browsing System, in ACM Multimedia Conference, 1996, Boston.

[7]   A. Lippman, N. Vasconcelos, and G. Iyengar, Humane Interfaces to Video, in Proc. 32nd Asilomar Conference on Signals, Systems, and Computers, Asilomar, California, 1998.

[8]   N. Vasconcelos and A. Lippman, Towards Semantically Meaningful Feature Spaces for the Characterization of Video Content, in Proc. IEEE Int. Conf. on Image Processing, Santa Barbara, 1997.

[9]   N. Vasconcelos, and A. Lippman, Bayesian Modelling of Video Editing and Structure: Semantic Features for Video Summarization and Browsing, in Proc. IEEE Int. Conf. on Image Processing, Chicago, 1998.

[10] N. Vasconcelos, and A. Lippman, A Bayesian Framework for Semantic Content Characterization, in Proc. IEEE Conf. on Computer Vision and Pattern Recognition, Santa Barbara, 1998.

[11] N. Vasconcelos, and A. Lippman, Bayesian Video Shot Segmentation, Proc. Neural Information Processing Systems, Denver, Colorado, 2000.

[12] N. Vasconcelos and A. Lippman, Multi-resolution Tangent Distance for Affine Invariant Classification, in Proc. Neural Information Processing Systems, Denver, Colorado, 1977.

[13] N. Vasconcelos and A. Lippman, Statistical Models of Video Structure for Content Analysis and Characterization, *IEEE Transactions on Image Processing*, 9(1):3-19, January 2000.

[14] A. Drake, *Fundamentals of Applied Probability Theory*, McGraw-Hill, 1987.

[15] R. Hogg and E. Tanis, *Probability and Statistical Inference*, Macmillan, 1993.

[16] M. Swain and D. Ballard, Color Indexing, *International Journal of Computer Vision*, Vol. 7(1):11-32, 1991.

[17] W. Niblack, R. Barber, W. Equitz, M. Flickner, E. Glasman, D. Petkovic, P. Yanker, C. Faloutsos, and G. Taubin, The QBIC Project: Querying Images by Content Using Color, Texture, and Shape, in Proc. SPIE Conf. Storage and Retrieval from Image and Video Databases, pp. 173-181, Feb. 1993, San Jose, CA.

[18] J. Boreczky and L. Rowe, Comparison of Video Shot Boundary Detection Techniques, Proc. SPIE Conf. on Visual Communication and Image Processing, 1996.

[19] H. Van Trees, *Detection, Estimation, and Modulation Theory*, John Wiley & Sons, 1968.

[20] A. Gelman, J. Carlin, H. Stern, and D. Rubin, *Bayesian Data Analysis*, Chapman and Hall, 1995.

[21] J. Pearl, *Probabilistic Reasoning in Intelligent Systems: Networks of Plausible Inference*, Morgan Kaufmann, 1988.

[22] D. Spiegelhalter, A. Dawid, S. Lauritzen, and R. Cowell, Bayesian Analysis in Expert Systems, *Statistical Science*, 8(3):219-283, 1993.

[23] D. Heckerman, A Tutorial on Learning with Bayesian Networks, Technical Report MSR-TR-96-06, Microsoft Research, March 1995.

[24] C. Bishop, *Neural Networks for Pattern Recognition*, Oxford University Press, 1995.

[25] L. Breiman, J. Friedman, R. Olshen, and C. Stone, *Classification and Regression Trees*, Wadsworth, 1984.

[26] V. Vapnik, *The Nature of Statistical Learning Theory*, Springer-Verlag, New York, 1995.

[27] Internet Movie Database, http://us.imdb.com/.

4

FLAVOR: A LANGUAGE FOR MEDIA REPRESENTATION

Alexandros Eleftheriadis and Danny Hong
Department of Electrical Engineering
Columbia University
New York, New York, USA
`{eleft,danny}@ee.columbia.edu`

1. INTRODUCTION

Flavor, which stands for Formal Language for Audio-Visual Object Representation, originated from the need to simplify and speed up the development of software that processes coded audio-visual or general multimedia-information. This includes encoders and decoders as well as applications that manipulate such information. Examples include editing tools, synthetic content creation tools, multimedia indexing and search engines, etc. Such information is invariably encoded in a highly efficient form to minimize the cost of storage and transmission. This source coding [1] operation is almost always performed in a bitstream-oriented fashion: the data to be represented is converted to a sequence of binary values of arbitrary (and typically variable) lengths, according to a specified syntax. The syntax itself can have various degrees of sophistication. One of the simplest forms is the GIF87a format [2], consisting of essentially two headers and blocks of coded image data using the Lempel-Ziv-Welch compression. Much more complex formats include JPEG [3], MPEG-1 [4], MPEG-2 [5, 6] and MPEG-4 [7, 8], among others.

General-purpose programming languages such as C++ [9] and Java [10] do not provide native facilities for coping with such data. Software codec or application developers need to build their own facilities, involving two components. First, they need to develop software that deals with the bitstream-oriented nature of the data, as general-purpose microprocessors are strictly byte-oriented. Second, they need to implement parsing and generation code that complies with the syntax of the format at hand (be it proprietary or standard). These two tasks represent a significant amount of the overall development effort. They also have to be duplicated by everyone who requires access to a particular compressed representation within their application. Furthermore, they can also represent a substantial percentage of the overall execution time of the application.

Flavor addresses these problems in an integrated way. First, it allows the "formal" description of the bitstream syntax. Formal here means that the description is based on a well-defined grammar, and as a result is amenable to software tool manipulation. In the past, such descriptions were using ad hoc conventions involving tabular data or pseudo-code. A second and key aspect of Flavor's architecture is that this description has been designed as an extension of C++ and Java, both heavily used object-oriented languages in multimedia applications development. This ensures seamless integration of Flavor code with both C++ and Java code and the overall architecture of an application.

Flavor was designed as an object-oriented language, anticipating an audio-visual world comprised of audio-visual objects, both synthetic and natural, and combining it with well-established paradigms for software design and implementation. Its object-oriented facilities go beyond the mere duplication of C++ and Java features, and introduce several new concepts that are pertinent for bitstream-based media representation.

In order to validate the expressive power of the language, several existing bitstream formats have already been described in Flavor, including sophisticated structures such as MPEG-2 Systems, Video and Audio. A translator has also been developed for translating Flavor code to C++ or Java code. Since Version 5.0, the translator has been enhanced to support XML features. With the enhanced translator, Flavor code can also be used to generate corresponding XML schema. In addition, the generated C++ or Java code can include the method for producing XML documents that represent the bitstreams described by the Flavor code. Detailed description about the translator and its features are given in Section 4 of this chapter.

Emerging multimedia representation techniques can directly use Flavor to represent the bitstream syntax in their specifications. This will allow immediate use of such specifications in new or existing applications, since the code to access/generate conforming data can be generated directly from the specification and with zero cost. In addition, such code can be automatically optimized; this is particularly important for operations such as Huffman encoding/decoding, a very common tool in media representation.

In the following, we first present a brief background of the language in terms of its history and technical approach. We then describe each of its features, including declarations and constants, expressions and statements, classes, scoping rules and maps. We also describe the translator and its simple run-time API. A brief description of how the translator processes maps to generate entropy encoding/decoding programs is given as well. Then, the XML features offered by the translator are explained. Finally, we conclude with an overview of the benefits of using the Flavor approach for media representation. More detailed information and publicly available software can be found in the Flavor web site at: http://flavor.sourceforge.net.

Note that Flavor is an open source project under Flavor Artistic License as defined in the Flavor web site. As a result, the source code for the translator and the run-time library is available as part of the Flavor package. The complete package can be downloaded from http://www.sourceforge.net /projects/flavor.

2. BRIEF OVERVIEW

Flavor provides a formal way to specify how data is laid out in a serialized bitstream. It is based on a *principle of separation* between bitstream parsing operations and encoding, decoding and other operations. This separation acknowledges the fact that different tools can utilize the same syntax, but also that the same tool can work unchanged with different bitstream syntax. For example, the number of bits used for a specific field can change without modifying any part of the application program.

Past approaches for syntax description utilized a combination of tabular data, pseudo-code, and textual description to describe the format at hand. Taking MPEG as an example, both MPEG-1 and MPEG-2 specifications were described using C-like pseudo-code syntax (originally introduced by Milt Anderson, Bellcore), coupled with explanatory text and tabular data. Several of the lower and most sophisticated layers (e.g., macroblock) could only be handled by explanatory text. The text had to be carefully crafted and tested over time for ambiguities. Other specifications (e.g., GIF and JPEG) use similar bitstream representation schemes, and hence share the same limitations.

Other formal facilities already exist for representing syntax. One example is ASN.1 (ISO International Standards 8824 and 8825). A key difference, however, is that ASN.1 was not designed to address the intricacies of source coding operations, and hence cannot cope with, for example, variable-length coding. In addition, ASN.1 tries to hide the bitstream representation from the developer by using its own set of binary encoding rules, whereas in our case the binary encoding is the actual target of description.

There is also some remote relationship between syntax description and "marshalling," a fundamental operation in distributed systems where consistent exchange of typed data is ensured. Examples in this category include Sun's ONC XDR (External Data Representation) and the rpcgen compiler that automatically generates marshalling code, as well as CORBA IDL, among others. These ensure, for example, that even if the native representation of an integer in two systems is different (big versus little endian), they can still exchange typed data in a consistent way. Marshalling, however, does not constitute bitstream syntax description because: 1) the programmer does not have control over the data representation (the binary representation for each data type is predefined), 2) it is only concerned with the representation of simple serial structures (lists of arguments to functions, etc.). As in ASN.1, the binary representation is "hidden" and is not amenable to customization by the developer. One could parallel Flavor and marshalling by considering the Flavor source as the XDR layer. A better parallelism would be to view Flavor as a parser-generator like yacc [11], but for bitstream representations.

It is interesting to note that all prior approaches to syntactic description were concerned only with the definition of message structures typically found in communication systems. These tend to have a much simpler structure compared with coded representations of audio-visual information (compare the UDP packet header with the baseline JPEG specification, for example).

A new language, Bitstream Syntax Description Language (BSDL) [12, 13], has recently been introduced in MPEG-21 [14] for describing the structure of a bitstream using XML Schema. However, unlike Flavor, BSDL is developed to address only the high-level structure of the bitstream, and it becomes almost impossible to fully describe bitstream syntax on a bit-per-bit basis. For example, BSDL doesn't have a facility to cope with variable-length coding, whereas in Flavor, map (described in Section 3.5) can be used. Also, the BSDL description would be overly verbose, requiring a significant effort to review and modify the description with the human eye. More detailed information about BSDL is given in Section 4.7.2.

Flavor was designed to be an intuitive and natural extension of the typing system of object-oriented languages like C++ and Java. This means that the bitstream representation information is placed together with the data declarations in a single place. In C++ and Java, this place is where a class is defined.

Flavor has been explicitly designed to follow a declarative approach to bitstream syntax specification. In other words, the designer is specifying how the data is laid out on the bitstream, and does not detail a step-by-step procedure that parses it. This latter procedural approach would severely limit both the expressive power as well as the capability for automated processing and optimization, as it would eliminate the necessary level of abstraction. As a result of this declarative approach, Flavor does not have functions or methods.

A related example from traditional programming is the handling of floating point numbers. The programmer does not have to specify how such numbers are represented or how operations are performed; these tasks are automatically taken care of by the compiler in coordination with the underlying hardware or run-time emulation libraries.

An additional feature of combining type declaration and bitstream representation is that the underlying object hierarchy of the base programming language (C++ or Java) becomes quite naturally the object hierarchy for bitstream representation purposes as well. This is an important benefit for ease of application development, and it also allows Flavor to have a very rich typing system itself.

"HelloBits"

HelloBits – Traditionally, programming languages are introduced via a simple "Hello World!" program, which just prints out this simple message on the user's terminal. We will use the same example with Flavor, but here we are concerned about bits, rather than text characters. Figure 4.1 shows a set of trivial examples indicating how the integration of type and bitstream representation information is accomplished. Consider a simple object called HelloBits with just a single value, represented using 8 bits. Using the MPEG-1/2 methodology, this would be described as shown in Figure 4.1(a). A C++ description of this single-value object would include two methods to read and write its value, and have a form similar to the one shown in Figure 4.1(b). Here getuint() is assumed to be a function that reads bits from the bitstream (here 8) and returns them as an unsigned integer (the most significant bit first); the putuint() function has similar functionality but for output purposes. When HelloBits::get() is called, the bitstream is read and the resultant quantity is placed in the data member Bits. The same description in Flavor is shown in Figure 4.1(c).

As we can see, in Flavor the bitstream representation is integrated with the type declaration. The Flavor description should be read as: Bits is an unsigned integer quantity represented using 8 bits in the bitstream. Note that there is no implicit encoding rule as in ASN.1: the rule here is embedded in the type declaration and indicates that, when the system has to parse a HelloBits data type, it will just read the next 8 bits as an unsigned integer and assign them to the variable Bits.

These examples, although trivial, demonstrate the differences between the various approaches. In Figure 4.1(a), we just have a tabulation of the various bitstream entities, grouped into syntactic units. This style is sufficient for straightforward representations, but fails when more complex structures are used (e.g., variable-length codes). In Figure 4.1(b), the syntax is incorporated into hand-written code embedded in get() and put() or an equivalent set of methods. As a result, the syntax becomes an integral part of the encoding/decoding method even though the same encoding/decoding mechanism could be applied to a large variety of similar syntactic constructs. Also, it quickly becomes overly verbose.

Flavor provides a wide range of facilities to define sophisticated bitstreams, including if-else, switch, for and while constructs. In contrast with regular C++ or Java, these are all included in the data declaration part of the class, so they are completely disassociated from code that belongs to class methods. This is in line with the declarative nature of Flavor, where the focus is on defining the structure of the data, not operations on them. As we show later on, a translator can automatically generate C++ and/or Java methods (get() and put()) that can read or write data that complies to the Flavor-described representation.

In the following we describe each of the language features in more detail, emphasizing the differences between C++ and Java. In order to ensure that Flavor semantics are in line with both C++ and Java, whenever there was a conflict a common denominator approach was used.

3. LANGUAGE OVERVIEW

3.1   DECLARATIONS AND CONSTANTS

3.1.1 Literals

All traditional C++ and Java literals are supported by Flavor. This includes integers, floating-point numbers and character constants (e.g., 'a'). Strings are also supported by Flavor. They are converted to arrays of characters, with or without a trailing '\0' (null character).

Additionally, Flavor defines a special binary number notation using the prefix 0b. Numbers represented with such notation are called binary literals (or bit strings) and, in addition to the actual value, also convey their length. For example, one can write 0b011 to denote the number 3 represented using 3 bits. For readability, a bit string can include periods every four digits, e.g., 0b0010.1100.001. Hexadecimal or octal constants used in the context of a bit string also convey their length in addition to their value. Whenever the length of a binary literal is irrelevant, it is treated as a regular integer literal.

3.1.2 Comments

Both multi-line /**/ and single-line // comments are allowed. The multi-line comment delimiters cannot be nested.

3.1.3 Names

Variable names follow the C++ and Java conventions (e.g., variable names cannot start with a number). The keywords that are used in C++ and Java are considered reserved in Flavor.

3.1.4 Types

Flavor supports the common subset of C++ and Java built-in or fundamental types. This includes `char`, `int`, `float` and `double` along with all appropriate modifiers (`short`, `long`, `signed` and `unsigned`). Additionally, Flavor defines a new type called `bit` and a set of new modifiers, `big` and `little`. The type `bit` is used to accommodate bit string variables and the new modifiers are used to indicate the endianess of bytes. The `big` modifier is used to represent the numbers using big-endian byte ordering (the most significant byte first) and the `little` modifier is used for the numbers represented using the little-endian method. By default, big-endian byte ordering is assumed. Note that endianess here refers to the bitstream representation, not the processor on which Flavor software may be running. The latter is irrelevant for the bitstream description.

Flavor also allows declaration of new types in the form of classes (refer to Section 3.3 for more information regarding classes). However, Flavor does not support pointers, references, casts or C++ operators related to pointers. Structures or enumerations are not supported either, since they are not supported by Java.

Syntax	No. of Bits	Mnemonic
HelloBits {		
Bits	8	uimsbf
}		

(a)

```
class HelloBits {
  unsigned int Bits;
  void get() {
    Bits = ::getuint(8);
  }
  void put() {
    ::putuint(8, Bits);
  }
};
```

(b)

```
class HelloBits {
  unsigned int(8) Bits;
};
```

(c)

Figure 4.1 HelloBits. (a) Representation using the MPEG-1/2 methodology. (b) Representation using C++ (A similar construct would also be used for Java). (c) Representation using Flavor.

3.1.5 Declarations

Regular variable declarations can be used in Flavor in the same way as in C++ and Java. As Flavor follows a declarative approach, constant variable declarations with specified values are allowed everywhere (there is no constructor to set the initial values). This means that the declaration 'const int a = 1;' is valid anywhere (not just in global scope). The two major differences are the declaration of parsable variables and arrays.

Parsable Variables

Parsable variables are the core of Flavor's design; it is the proper definition of these variables that defines the bitstream syntax. Parsable variables include a parse length specification immediately after their type declaration, as shown in Figure 4.2. In the figure, the `blength` argument can be an integer constant, a non-constant variable of type compatible to `int` or a map (discussed later on) with the same type as the variable. This means that the parse length of a variable can be controlled by another variable. For example, the parsable variable declaration in Figure 4.3(a) indicates that the variable a has the parse length of 3 bits. In addition to the parse length specification, parsable variables can also have the `aligned` modifier. This signifies that the variable begins at the next integer multiple boundary of the length argument – `alength` – specified within the alignment expression. If this length is omitted, an alignment size of 8 is assumed (byte boundary). Thus, the variable a is byte-aligned and for parsing, any intermediate bits are ignored, while for output bitstream generation the bitstream is padded with zeroes.

```
[aligned(alength)] type(blength) variable [=value];
```

Figure 4.2 Parsable variable declaration syntax.

As we will see later on, parsable variables cannot be assigned to. This ensures that the syntax is preserved regardless if we are performing an input or output operation. However, parsable variables *can be redeclared*, as long as their type remains the same, only the parse size is changed, and the original declaration was not as a `const`. This allows one to select the parse size depending on the context (see Expressions and Statements, Section 3.2). On top of this, they obey special scoping rules as described in Section 3.4.

In general, the parse size expression must be a non-negative value. The special value 0 can be used when, depending on the bitstream context, a variable is not present in the bitstream but obtains a default value. In this case, no bits will be parsed or generated; however, the semantics of the declaration will be preserved. The variables of type `float`, `double` and `long double` are only allowed to have a parse size equal to the fixed size that their standard representation requires (32 and 64 bits).

Look-Ahead Parsing

In several instances, it is desirable to examine the immediately following bits in the bitstream without actually removing the bits from the input stream. To support this behavior, a '*' character can be placed after the parse size parentheses. Note that for bitstream output purposes, this has no effect. An example of a declaration of a variable for look-ahead parsing is given in Figure 4.3(b).

Parsable Variables with Expected Values

Very often, certain parsable variables in the syntax have to have specific values (markers, start codes, reserved bits, etc.). These are specified as initialization values for parsable variables. Figure 4.3(c) shows an example. The example is interpreted as: **a** is an integer represented with 3 bits, and must have the value 2. The keyword `const` may be prepended in the declaration, to indicate that the parsable variable will have this constant value and, as a result, cannot be redeclared.

As both parse size and initial value can be arbitrary expressions, we should note that the order of evaluation is parse expression first, followed by the initializing expression.

```
aligned int(3) a;        aligned int(3)* a;       aligned int(3) a=2;
```
          (a)   (b)   (c)

Figure 4.3 (a) Parsable variable declaration. (b) Look-ahead parsing. (c) Parsable variable declaration with an expected value.

Arrays

Arrays have special behavior in Flavor, due to its declarative nature but also due to the desire for very dynamic type declarations. For example, we want to be able to declare a parsable array with different array sizes depending on the context. In addition, we may need to load the elements of an array one at a time (this is needed when the retrieved value indicates indirectly if further elements of the array should be parsed). These concerns are only relevant for parsable variables. The array size, then, does not have to be a constant expression (as in C++ and Java), but it can be a variable as well. The example in Figure 4.4(a) is allowed in Flavor.

An interesting question is how to handle initialization of arrays, or parsable arrays with expected values. In addition to the usual brace expression initialization (e.g., 'int A[2] = {1, 2};'), Flavor also provides a mechanism that involves the specification of a single expression as the initializer as shown in Figure 4.4(b). This means that all elements of A will be initialized with the value 5. In order to provide more powerful semantics to array initialization, Flavor considers the parse size and initializer expressions as executed per each element of the array. The array size expression, however, is only executed once, before the parse size expression or the initializer expression.

Let's look at a more complicated example in Figure 4.4(c). Here A is declared as an array of 2 integers. The first one is parsed with 3 bits and is expected to have the value 4, while the second is parsed with 5 bits and is expected to have the value 6. After the declaration, a is left with the value 7. This probably represents the largest deviation of Flavor's design from C++ and Java declarations. On the other hand it does provide significant flexibility in constructing sophisticated declarations in a very compact form, and it is also in line with the dynamic nature of variable declarations that Flavor provides.

```
int a = 2;       int A[3] = 5;    int a = 1;                int(2) A[[3]] = 1;
int(2) A[a++];                     int(a++) A[a++] = a++;    int(4) B[[2]][3];
```
          (a)   (b)   (c)   (d)

Figure 4.4 Array. (a) Declaration with dynamic size specification. (b) Declaration with initialization. (c) Declaration with dynamic array and parse sizes. (d) Declaration of partial arrays.

Partial Arrays

An additional refinement of array declaration is partial arrays. These are declarations of parsable arrays in which only a subset of the array needs to be declared (or, equivalently, parsed from or written to a bitstream). Flavor introduces a double bracket notation for this purpose. In Figure 4.4(d), an example is given to demonstrate its use. In the first line, we are declaring the 4-th element of A (array indices start from 0). The array size is unknown at this point, but of course it will be considered at least 4. In the second line, we are declaring a two-dimensional array, and in particular only its third row (assuming the first index corresponds to a row). The array indices can, of course, be expressions themselves. Partial arrays can only appear on the left-hand side of declaration and are not allowed in expressions.

3.2   EXPRESSIONS AND STATEMENTS

Flavor supports all of the C++ and Java arithmetic, logical and assignment operators. However, parsable variables cannot be used as lvalues. This ensures that they always represent the bitstream's content, and allow consistent operations for the translator-generated `get()` and `put()` methods that read and write, respectively, data according to the specified form. Refer to Section 4.1 for detailed information about these methods.

Flavor also supports all the familiar flow control statements: `if-else`, `do-while`, `while`, `for` and `switch`. In contrast to C++ and Java, variable declarations are not allowed within the arguments of these statements (i.e., 'for (int i=0; ;);' is not allowed). This is because in C++ the scope of this variable will be the enclosing one, while in Java it will be the enclosed one. To avoid confusion, we opted for the exclusion of both alternatives at the expense of a slightly more verbose notation. Scoping rules are discussed in detail in Section 3.4. Similarly, Java only allows Boolean expressions as part of the flow control statements, and statements like 'if (1) { ... }' are not allowed in Java. Thus, only the flow control statements with Boolean expressions are valid in Flavor.

Figure 4.5 shows an example of the use of these flow control statements. The variable b is declared with a parse size of 16 if a is equal to 1, and with a parse size of 24 otherwise. Observe that this construct would not be meaningful in C++ or Java as the two declarations would be considered as being in separate scopes. This is the reason why parsable variables need to obey slightly different scoping rules than regular variables. The way to approach this to avoid confusion is to consider that Flavor is designed so that these parsable variables are properly defined at the right time and position. All the rest of the code is there to ensure that this is the case. We can consider the parsable variable declarations as "actions" that our system will perform at the specified times. This difference, then, in the scoping rules becomes a very natural one.

```
if (a == 1) {
    int(16) b; // b is a 16 bit integer
} else {
    int(24) b; // b is a 24 bit integer
}
```

Figure 4.5 An example of a conditional expression.

3.3   CLASSES

Flavor uses the notion of classes in exactly the same way as C++ and Java do. It is the fundamental structure in which object data are organized. Keeping in line with the support of both C++ and Java-style programming, classes in Flavor cannot be nested, and only single inheritance is supported. In addition, due to the declarative nature of Flavor, methods are not allowed (this includes constructors and destructors as well).

Figure 4.6(a) shows an example of a simple class declaration with just two parsable member variables. The trailing ';' character is optional accommodating both C++ and Java-style class declarations. This class defines objects that contain two parsable variables. They will be present in the bitstream in the same order they are declared. After this class is defined, we can declare objects of this type as shown in Figure 4.6(b).

```
class SimpleClass {
    int(3) a;
    unsigned int(4) b;
}; // The trailling ';' is optional
```

(a)

```
class SimpleClass(int i[2]) {
    int(3) a = i[0];
    unsigned int(4) b = i[1];
};
```

(c)

```
SimpleClass a;
```

(b)

```
int(2) v[2];
SimpleClass a(v);
```

(d)

Figure 4.6 Class. (a) A simple class definition. (b) A simple class variable declaration. (c) A simple class definition with parameter types. (d) A simple class variable declaration with parameter types.

A class is considered parsable if it contains at least one variable that is parsable. The `aligned` modifier can prepend declaration of parsable class variables in the same way as parsable variables.

Class member variables in Flavor do not require access modifiers (`public`, `protected`, `private`). In essence, all such variables are considered public.

3.3.1 Parameter Types

As Flavor classes cannot have constructors, it is necessary to have a mechanism to pass external information to a class. This is accomplished using *parameter types*. These act the same way as formal arguments in function or method declarations do. They are placed in parentheses after the name of the class. Figure 4.6(c) gives an example of a simple class declaration with parameter types. When declaring variables of parameter type classes, it is required that the actual arguments are provided in place of the formal ones as displayed in Figure 4.6(d).

Of course the types of the formal and actual parameters must match. For arrays, only their dimensions are relevant; their actual sizes are not significant as they can be dynamically varying. Note that class types are allowed in parameter declarations as well.

3.3.2 Inheritance

As we mentioned earlier, Flavor supports single inheritance so that compatibility with Java is maintained. Although Java can "simulate" multiple inheritances through the use of interfaces, Flavor has no such facility (it would be meaningless since methods do not exist in Flavor). However, for media representation purposes, we have not found any instance where multiple inheritances would be required, or even be desirable. It is interesting to note that all existing representation standards today are not truly object-based. The only exception, to our knowledge, is the MPEG-4 specification that explicitly addresses the representation of audio-visual objects. It is, of course, possible to describe existing structures in an object-oriented way but it does not truly map one-to-one with the notion of objects. For example, the MPEG-2 Video slices can be considered as separate objects of the same type, but of course their semantic interpretation (horizontal stripes of macroblocks) is not very useful. Note that containment formats like MP4 and QuickTime are more object-oriented, as they are composed of object-oriented structures called 'atoms.'

Derivation in C++ and Java is accomplished using a different syntax (`extends` versus ':'). Here we opted for the Java notation (also ':' is used for object identifier declarations as explained below). Unfortunately, it was not possible to satisfy both.

In Figure 4.7(a) we show a simple example of a derived class declaration. Derivation from a bitstream representation point of view means that B is an A with some additional information. In other words, the behavior would be almost identical if we just copied the statements between the braces in the declaration of A into the beginning of B. We say "almost" here because scoping rules of variable declarations also come into play, as discussed in Section 3.4.

Note that if a class is derived from a parsable class, it is also considered parsable.

3.3.3 Polymorphic Parsable Classes

The concept of inheritance in object-oriented programming derives its power from its capability to implement polymorphism. In other words, the capability to use a derived object in place of the base class is expected. Although the mere structural organization is useful as well, it could be accomplished equally well with containment (a variable of type A is the first member of B).

Polymorphism in traditional programming languages is made possible via vtable structures, which allow the resolution of operations during run-time. Such behavior is not pertinent for Flavor, as methods are not allowed.

A more fundamental issue, however, is that Flavor describes the bitstream syntax: the information with which the system can detect which object to select *must be present in the bitstream.* As a result, traditional inheritance as defined in the previous section *does not* allow the representation of polymorphic objects. Considering Figure 4.7(a), there is no way to figure out by reading a bitstream if we should read an object of type A or type B.

Flavor solves this problem by introducing the concept of *object identifiers* or IDs. The concept is rather simple: in order to detect which object we should parse/generate, there must be a parsable variable that will identify it. This variable must have a different expected value for any class derived from the originating base class, so that object resolution can be uniquely performed in a well-defined way (this can be checked by the translator). As a result, object ID values must be constant expressions and they are always considered constant, i.e., they cannot be redeclared within the class.

In order to signify the importance of the ID variables, they are declared immediately after the class name (including any derivation declaration) and before the class body. They are separated from the class name declaration using a colon (':'). We could rewrite the example of Figure 4.7(a) with IDs as shown in Figure 4.7(b). Upon reading the bitstream, if the next 1 bit has the value 0 an object of type A will be parsed; if the value is 1 then an object of type B will be parsed. For output purposes, and as will be discussed in Section 4, it is up to the user to set up the right object type in preparation for output.

```
class A {                        class A:int(1) id=0 {
  int(2) a;                        int(2) a;
}                                }

class B extends A {              class B extends A:int(1) id=1 {
  int(3) b;                        int(3) b;
}                                }
            (a)                               (b)
```

Figure 4.7 Inheritance. (a) Derived class declaration. (b) Derived class declaration with object identifiers.

The name and the type of the ID variable are irrelevant, and can be anything that the user chooses. It cannot, however, be an array or a class variable (only built-in types are allowed). Also, the name, type and parse size must be identical between the base and derived classes. However, object identifiers are not required for all derived classes of a base class that has a declared ID. In this case, only the derived classes with defined IDs can be used wherever the base class can appear. This type of polymorphism is already used in the MPEG-4 Systems specification, and in particular the Binary Format for Scenes (BIFS) [15]. This is a VRML-derived set of nodes that represent objects and operations on them, thus forming a hierarchical description of a scene.

The ID of a class is also possible to have a range of possible values which is specified as `start_id .. end_id`, inclusive of both bounds. See Figure 4.8 for an example.

```
class slice:aligned bit(32) slice_start_code=0x00000101 .. 0x000001AF {
  ...
}
```

Figure 4.8 A class with id range.

3.4   SCOPING RULES

The scoping rules that Flavor uses are identical with C++ and Java with the exception of parsable variables. As in C++ and Java, a new scope is introduced with curly braces ({ }). Since Flavor does not have functions or methods, a scope can either be the global one or a scope within a class declaration. Note that the global scope cannot contain any parsable variable, since it does not belong to any object. As a result, global variables can only be constants.

Within a class, all parsable variables are considered as class member variables, regardless of the scope they are encountered in. This is essential in order to allow conditional declarations of variables, which will almost always require that the actual declarations occur within compound statements (see Figure 4.5). Non-parsable variables that occur in the top-most class scope are also considered class member variables. The rest live within their individual scopes.

This distinction is important in order to understand which variables are accessible to a class variable that is contained in another class. The issues are illustrated in Figure 4.9. Looking at the class A, the initial declaration of i occurs in the top-most class scope; as a result i is a class member. The variable a is declared as a parsable variable, and hence it is automatically a class member variable. The declaration of j occurs in the scope enclosed by the if statement; as this is not the top-level scope, j is not a class member. The following declaration of i is acceptable; the original one is hidden within that scope. Finally, the declaration of the variable a as a non-parsable would hide the parsable version. As parsable variables do not obey scoping rules, this is not allowed (hiding parsable variables of a base class, however, is allowed). Looking now at the declaration of the class B, which contains a variable of type A, it becomes clear which variables are available as class members.

```
class A {                         class B {
   int i = 1;                        A a;
   int(2) a;                         a.j = 1; // Error, j not a class member
   if (a == 2) {                     int j = a.a + 1; // OK
      int j = i;                     j = a.i + 2      // OK
      int i = 2; // Hides i, OK      int(3) b;
      int a;     // Hides a, error }
}
```

Figure 4.9 Scoping rules example.

In summary, the scoping rules have the following two special considerations. Parsable variables do not obey scoping rules and are always considered class members. Non-parsable variables obey the standard scoping rules and are considered class members only if they are at the top-level scope of the class.

Note that parameter type variables are considered as having the top-level scope of the class. Also, they are not allowed to hide the object identifier, if any.

3.5 MAPS

Up to now, we have only considered fixed-length representations, either constant or parametric. A wide variety of representation schemes, however, rely heavily on entropy coding, and in particular Huffman codes [1]. These are variable-length codes (VLCs), which are uniquely decodable (no codeword is the prefix of another). Flavor provides extensive support for variable-length coding through the use of maps. These are declarations of tables in which the correspondence between codewords and values is described.

Figure 4.10 gives a simple example of a map declaration. The map keyword indicates the declaration of a map named A. The declaration also indicates that the map converts from bit string values to values of type int. The type indication can be a fundamental type, a class type, or an array. Map declarations can only occur in global scope. As a result, an array declaration will have to have a constant size (no non-constant variables are visible at this level). After the map is properly declared, we can define parsable variables that use it by indicating the name of the map where we would put the parse size expression as follows: 'int(A) i;'. As we can see the use of variable-length codes is essentially identical to fixed-length variables. All the details are hidden away in the map declaration.

```
map A(int) {
   0b0,  1,
   0b01, 2
}
```

Figure 4.10 A simple map declaration.

The map contains a series of entries. Each entry starts with a bit string that declares the codeword of the entry followed by the value to be assigned to this codeword. If a complex type is used for the mapped value, then the values have to be enclosed in curly braces. Figure 4.11 shows the definition of a VLC table with a user-defined class as output type. The type of the variable has to be identical to the type returned from the map. For example, using the declaration - `YUVblocks(blocks_per_component) chroma_format;` - we can access a particular value of the map using the construct: `chroma_format.Ublocks`.

```
// The output type of a map is defined in a class
class YUVblocks {
  unsigned int Yblocks;
  unsigned int Ublocks;
  unsigned int Vblocks;
}

/* A table that relates the chroma format with
 * the number of blocks per signal component
 */
map blocks_per_component(YUVblocks) {
  0b00, {4,1,1}, // 4:2:0
  0b01, {4,2,2}, // 4:2:2
  0b10, {4,4,4}  // 4:4:4
}
```

Figure 4.11 A map with defined output type.

As Huffman codeword lengths tend to get very large when their number increases, it is typical to specify "escape codes," signifying that the actual value will be subsequently represented using a fixed-length code. To accommodate these as well as more sophisticated constructs, Flavor allows the use of parsable type indications in map values. This means that, using the example in Figure 4.10, we can write the example in Figure 4.12. This indicates that, when the bit string `0b001` is encountered in the bitstream, the actual return value for the map will be obtained by parsing 5 more bits. The parse size for the extension can itself be a map, thus allowing the cascading of maps in sophisticated ways. Although this facility is efficient when parsing, the bitstream generation operation can be costly when complex map structures are designed this way. None of today's specifications that we are aware of require anything beyond a single escape code.

```
map A(int) {
  0b0,   1,
  0b01,  2,
  0b001, int(5)
}
```

Figure 4.12 A map declaration with extension.

The translator can check that the VLC table is uniquely decodable, and also generate optimized code for extremely fast encoding/decoding using our hybrid approach as described in [16].

4. THE FLAVOR TRANSLATOR

Designing a language like Flavor would be an interesting but academic exercise, unless it was accompanied by software that can put its power into full use. We have developed a translator that evolved concurrently with the design of the language. When the language specification became stable, the translator was completely rewritten. The most recent release is publicly available for downloading at http://www.sourceforge.net/projects/flavor.

4.1   RUN-TIME API

The translator reads a Flavor source file (.fl) and, depending on the language selection flag of the code generator, it creates a pair of .h and .cpp files (for C++) or a set of .java files (for Java). In the case of C++, the .h file contains the declarations of all Flavor classes as regular C++ classes and the .cpp file contains the implementations of the corresponding class methods (put() and get()). In the case of Java, each .java file contains the declaration and implementation of a single Flavor class. In both cases, the get() method is responsible for reading a bitstream and loading the class variables with their appropriate values, while the put() method does the reverse. All the members of the classes are declared public, and this allows direct access to desired fields in the bitstream.

The translator makes minimal assumptions about the operating environment for the generated code. For example, it is impossible to anticipate all possible I/O structures that might be needed by applications (network-based, multi-threaded, multiple buffers, etc.). Attempting to provide a universal solution would be futile. Thus, instead of having the translator directly output the required code for bitstream I/O, error reporting, tracing, etc., a run-time library is provided. With this separation, programmers have the flexibility of replacing parts of, or the entire library with their own code. The only requirement is that this custom code provides an identical interface to the one needed by the translator. This interface is defined in a pure virtual class called IBitstream. Additionally, as the source code for the library is included in its entirety, customization can be performed quite easily. The Flavor package also provides information on how to rebuild the library, if needed. Custom library can be built simply by deriving from the IBitstream class. This will ensure compatibility with the translator.

The run-time library includes the Bitstream class that is derived from the IBitstream interface, and provides basic bitstream I/O facilities in terms of reading or writing bits. A Bitstream reference is passed as an argument to the get() and put() methods.

If parameter types are used in a class, then they are also required arguments in the get() and put() methods as well. The translator also requires that a function is available to receive calls when expected values are not available or VLC lookups fail. The function name can be selected by the user; a default implementation (flerror) is included in the run-time library.

For efficiency reasons, Flavor arrays are converted to fixed size arrays in the translated code. This is necessary in order to allow developers to access Flavor arrays without needing special techniques. Whenever possible, the translator automatically detects and sets the maximum array size; it can also be set by the user using a command-line option. Finally, the run-time library (and the translator) only allows parse sizes of up to the native integer size of the host processor (except for double values). This enables fast implementation of bitstream I/O operations.

For parsing operations, the only task required by the programmer is to declare an object of the class type at hand, and then call its get() method with an appropriate bitstream. While the same is also true for put() operation, the application developer must also load all class member variables with their appropriate values before the call is made.

4.2   INCLUDE AND IMPORT DIRECTIVES

In order to simplify the source code organization, Flavor supports %include and %import directives. These are the mechanisms to combine several different source code files into one entity, or to share a given data structure definition across different projects.

4.2.1 Include Directive

The statement – %include "file.fl" – will include the specified .fl file in the current position and will flag all of its content so that no code is generated. Figure 4.13(a) displays a .fl file (other.fl) that is included by another .fl file (main.fl). The other.fl file contains the definition of the constant a. The inclusion makes the declaration of the a variable available to the main.fl file. In terms of the generated output, Figure 4.13(b) outlines the placement of information in different files. In the figure, we see that the main and included files each keep their corresponding implementations. The generated C++ code maintains this partitioning, and makes sure that the main file includes the C++ header file of the included Flavor file.

```
// In the file, other.fl
const int a = 4;

// In the file, main.fl
%include "other.fl"

class Test {
  int(a) t; // The variable 'a' is included from the other.fl file
}
```

(a)

```
// In the file, other.h        // In the file, other.cpp      // In the file, main.h
extern const int a;            #include "other.h"             #include "other.h"
                               const int a = 4;               …
```

(b)

Figure 4.13 The %include directive. (a) The other.fl file is included by the main.fl file. (b) The other.h and other.cpp files are generated from the other.fl file whereas the main.h file is generated from the main.fl file.

The %include directive is useful when data structures need to be shared across modules or projects. It is similar in spirit to the use of the C/C++ preprocessor #include statement in the sense that it is used to make general information available at several different places in a program. Its operation, however, is different as Flavor's %include statement does not involve code generation for the included code. In C/C++, #include is equivalent to copying the included file in the position of the #include statement. This behavior is offered in Flavor by the %import directive.

Similarly, when generating Java code, only the .java files corresponding to the currently processed Flavor file are generated. The data in the included files are allowed to be used, but they are not generated.

4.2.2 Import Directive

The %import directive behaves similarly to the %include directive, except that full code is generated for the imported file by the translator, and no C++

#include statement is used. This behavior is identical to how a C++ preprocessor #include statement would behave in Flavor.

Let's consider the example of the previous section, this time with an %import directive rather than an %include one as shown in Figure 4.14(a). As can be seen from Figure 4.14(b), the generated code includes the C++ code corresponding to the imported .fl file. Therefore, using the %import directive is exactly the same as just copying the code in the imported .fl file and pasting it in the same location as the %import statement is specified. The translator generates the Java code in the same way.

```
%import "other.fl"

class Test {
    int(a) t; // The variable 'a' is imported from the other.fl file
}
```

(a)

```
// In the file, main.h              // In the file, main.cpp
extern const int a;                 const int a = 4;
...                                 ...
```

(b)

Figure 4.14 The %import directive. (a) The main.fl file using the %import directive. (b) The main.h and main.cpp files generated from the main.fl file defined in (a).

Note that the Java import statement behaves more like Flavor's %include statement, in that no code generation takes place for the imported (included) code.

4.3 PRAGMA STATEMENTS

Pragma statements are used as a mechanism for setting translator options from inside a Flavor source file. This allows modification of translation parameters (set by the command-line options) without modifying the makefile that builds the user's program, but more importantly, it allows very fine control on which translation options are applied to each class, or even variable. Almost all command-line options have pragma equivalents. The ones excluded were not considered useful for specification within a source file.

Pragma statements are introduced with the %pragma directive. It can appear wherever a statement or declaration can. It can contain one or more settings, separated by commas, and it cannot span more than one line. After a setting is provided, it will be used for the remainder of the Flavor file, unless overridden by a different pragma setting. In other words, pragma statements do not follow the scope of Flavor code. A pragma that is included in a class will affect not only the class where it is contained but all classes declared after it. An example is provided in Figure 4.15.

In this example, we start off setting the generation of both get() and put() methods, enabling tracing and setting the maximum array size to 128 elements. Inside the Example class, we disable the put() method output. This class reads a chunk of data, which is preceded by its size (length, a 10-bit quantity). This means that the largest possible buffer size is 1024 elements. Hence for the data array that immediately follows, we set the array size to 1024, and then switch it back to the default of 128. Finally, at the end of the class we select a different tracing function name; this function is really a method of a class, but this is irrelevant for the translator. Since this directive is used when the `get()` method code is produced, it will affect the entire class despite the fact that it is declared at its end.

Note that these pragma settings remain in effect even after the end of the `Example` class.

```
// Activate both put and get, generate tracing code, and set array size to 128
%pragma put, get, trace, array=128

class Example {
  %pragma noput                  // No put() method needed

  unsigned int(10) length;
  %pragma array=1024             // Switch array size to 1024
  char(3) data[length];

  %pragma array=128              // Switch array size back to 128
  %pragma trace="Tracer.trace"   // Use custom tracer
}

// The above settings are still active here!
```

Figure 4.15 Some examples of using pragma statements to set the translator options at specific locations.

4.4   VERBATIM CODE

In order to further facilitate integration of Flavor code with C++/Java user code, the translator supports the notion of verbatim code. Using special delimiters, code segments can be inserted in the Flavor source code, and copied verbatim at the correct places in the generated C++/Java file. This allows, for example, the declaration of constructors/destructors, user-specified methods, pointer member variables for C++, etc. Such verbatim code can appear wherever a Flavor statement or declaration is allowed.

The delimiters `%{` and `%}` can be used to introduce code that should go to the class declaration itself (or the global scope). The delimiters `%p{` and `%p}`, and `%g{` and `%g}` can be used to place code at exactly the same position they appear in the `put()` and `get()` methods, respectively. Finally, the delimiters `%*{` and `%*}` can be used to place code in both `put()` and `get()` methods. To place code specific to C++ or Java, `.c` or `.j` can be placed before the braces in the delimiters, respectively. For example, a verbatim code segment to be placed in the `get()` method of the Java code will be delimited with `%g.j{` and `%g.j}`.

The Flavor package includes several samples on how to integrate user code with Flavor-generated code, including a simple GIF parser. Figure 4.16 shows a simple example that reads the header of a GIF87a file and prints its values. The print statement which prints the values of the various elements is inserted as verbatim code in the syntax (within `%g{` and `%g}` markers, since the code should go in the `get()` method). The implementation of the print method for C++ code is declared within `%.c{` and `%.c}`, and for Java, the corresponding implementation is defined within `%.j{` and `%.j}`. The complete sample code can be found in the Flavor package.

```
class GIF87a {
  char(8) GIFsignature[6] = "GIF87a" // GIF signature

  %g{ print(); %g}

  ScreenDescriptor sd;                // A screen descriptor

  // One or more image descriptors
  do {
    unsigned int(8) end;

    if (end == ',') {      // We found an image descriptor
      ImageDescriptor id;
    }
    if (end == '!') {      // We found an extension block
      ExtensionBlock eb;
    }
    // Everything else is ignored
  } while (end != ';');     // ';' is the end-of-data marker

  %.c{
  void print() {…}
  %.c}

  %.j{
  void print() {…}
  %.j}
}
```

Figure 4.16 A simple Flavor example: the GIF87a header. The usage of verbatim code is illustrated.

4.5 TRACING CODE GENERATION

We also included the option to generate bitstream tracing code within the get() method. This allows one to very quickly examine the contents of a bitstream for development and/or debugging purposes by creating a dump of the bitstream's content. With this option, and given the syntax of a bitstream described in Flavor, the translator will automatically generate a complete C++/Java program that can verify if a given bitstream complies with that syntax or not. This can be extremely useful for codec development as well as compliance testing.

4.6 MAP PROCESSING

Map processing is one of the most useful features of Flavor, as hand-coding VLC tables is tedious and error prone. Especially during the development phase of a representation format, when such tables are still under design, full optimization within each design iteration is usually not performed. By using the translator, such optimization is performed at zero cost. Also, note that maps can be used for fixed-length code mappings just by making all codewords have the same length. As a result, one can very easily switch between fixed and variable-length mappings when designing a new representation format.

When processing a map, the translator first checks that it is uniquely decodable, i.e., no codeword is the prefix of another. It then constructs a class declaration for that map, which exposes two methods: getvlc() and putvlc(). These take as arguments a bitstream reference as well as a pointer to the return type of the map. The getvlc() method is responsible for decoding a map entry and returning the decoded value, while the putvlc() method is responsible for the output of the correct codeword. Note that the defined class does not perform any direct bitstream I/O itself, but uses the services of the Bitstream class instead.

This ensures that a user-supplied bitstream I/O library will be seamlessly used for map processing.

According to Fogg's survey on software and hardware VLC architectures [17], fast software decoding methods usually exploit a variable-size look-ahead window (multi-bit lookup) with lookup tables. For optimization, the look-ahead size and corresponding tables can be customized for position dependency in the bitstream. For example, for MPEG video, the look-ahead can be selected based on the picture type (I, P, or B).

One of the fastest decoding methods is comprised of one huge lookup table where every codeword represents an index of the table pointing to the corresponding value. However, this costs too much memory. On the other end of the spectrum, one of the most memory efficient algorithms would be of the form of a binary tree. The tree is traversed one bit at a time, and at each stage one examines if a leaf node is reached; if not, the left or right branch is taken for the input bit value of 0 or 1, respectively. Though efficient in memory, this algorithm is extremely slow, requiring N (the bit length of the longest codeword) stages of bitstream input and lookup.

In [16] we adopted a hybrid approach that maintains the space efficiency of binary tree decoding, and most of the speed associated with lookup tables. In particular, instead of using lookup tables, we use hierarchical, nested `switch` statements. Each time the read-ahead size is determined by the maximum of a fixed step size and the size of the next shortest code. The fixed step size is used to avoid degeneration of the algorithm into binary tree decoding. The benefit of this approach is that only complete matches require `case` statements, while all partial matches can be grouped into a single `default` statement (that, in turn, introduces another `switch` statement).

With the above-mentioned approach, the space requirement consists of storage of the case values and the comparison code generated by the compiler (this code consists of just 2 instructions on typical CISC systems, e.g., a Pentium). While slightly larger than a simple binary tree decoder, this overhead still grows linearly with the number of code entries (rather than exponentially with their length). This is further facilitated by the selection of the step size. When the incremental code size is small, multiple `case` statements may be assigned to the same codeword, thus increasing the space requirements.

In [16] we compared the performance of various techniques, including binary tree parsing, fixed step full lookups with different step sizes and our hybrid switch statement approach. In terms of time, our technique is faster than a hierarchical full-lookup approach with identical step sizes. This is because switching consumes little time compared to fixed-step's function lookups. Furthermore, it is optimized by ordering the `case` statements in terms of the length of their codeword. As shorter lengths correspond to higher probabilities, this minimizes the average number of comparisons per codeword.

With Flavor, developers can be assured of extremely fast decoding with minimal memory requirement due to the optimized code generation. In addition, the development effort and time in creating software to process VLCs is nearly eliminated.

4.7   XFLAVOR

Since Version 5.0, the Flavor translator is also referred to as XFlavor, an extended version that provides XML features. XFlavor has the capability to transform a Flavor description into an XML schema, and it can also produce code for generating XML documents corresponding to the bitstreams described by Flavor. Additionally, as a part of XFlavor, a compression tool (Bitstream Generator) for converting the XML representation of the data back into its original bitstream format is provided. Thus, XFlavor consists of the translator and Bitstream Generator.

The purpose of converting multimedia data in binary format into an equivalent XML document [18] is for easier and more flexible manipulation of the data. In the XML document, the bitstream layer is abstracted from applications and the semantic values of the data (e.g., width and height of an image) are directly available. In the bitstream format, such values must be extracted via bit string manipulations, according to the specified syntax.

Another advantage of using XML representation of data is that software tools are provided with generic access to multimedia data (usually with a generic XML parser). For example, a search engine that uses DC values to search for images/videos can work with all of JPEG, MPEG-1, 2, 4, and H.26x formats if XML is used to represent the data. However, if each individual binary format is used, then different search engines must be created for each format, as a different syntax is used to represent DC values by the different formats. This requires a different parser for each bitstream.

Additionally, as the name suggests, XML documents are extensible. With XML representation of data, extra information can be added to (or deleted from) the given document and a software tool can use both the new and old documents without changing the parser. On the contrary, with a GIF image, for example, adding extra information anywhere in the bitstream (other than at the end of the bitstream) renders the bitstream useless. Additionally, the extra information cannot be easily distinguished from the bitstream.

With the XML document generating code (the translator automatically generates a `putxml()` method for creating XML documents), a bitstream representing multimedia data can be read in and a corresponding XML document can be automatically generated. Unfortunately, XML documents are too verbose and unfavorable for storage or network transmission. For this reason, XFlavor also provides the feature for converting the documents back into their corresponding binary format. When processing a bitstream, it can be converted to an XML file (in whole or in part) or the XML representation can be stored in memory. Then, after processing the data, it can be converted back into its original bitstream format.

Also, in order to make the XML representation of the multimedia data strictly conforming to the original bitstream specification, the corresponding XML schema can be used. The schema [19] defines the syntax of the data and this is used as the guideline for determining the "validity" of XML represented data. Also, with the schema generated from the bitstream specification (Flavor description), the conforming XML documents can be exactly converted back into the original bitstream format. The converted data, then, can be processed in the bitstream domain (which yields faster processing) by the bitstream specification compliant decoders.

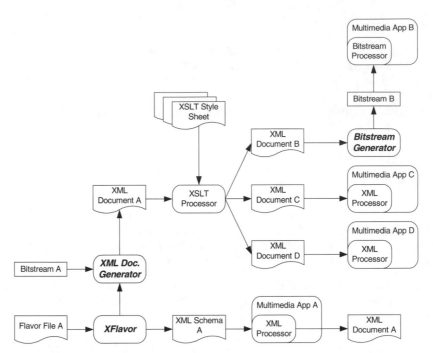

Figure 4.17 An illustration of XML features offered by XFlavor that are used for different applications.

Figure 4.17 illustrates the functions and some of the possible applications of XFlavor. Starting with a Flavor description of a multimedia bitstream – *Flavor File A* – XFlavor can generate an XML schema or XML document generating code. Then, applications can use the generated schema to create new multimedia content in the XML form (e.g., *Multimedia App A*). Additionally, the bitstream representation of multimedia content can be converted into an XML document for more flexible and extensible processing of the data. Once the data is in the XML form, different applications can easily manipulate it (e.g., *Multimedia App C* and *D*) or XML tools can be used to manipulate the data (e.g., *XSLT Processor*). Also, for space and time critical applications, the XML representation of the data can be converted back into the original bitstream representation (e.g., *Multimedia App B*).

In the following, we describe some of the benefits obtained by using the XML features.

4.7.1 Universal Parser

XML representation allows easy access to different elements. It allows an application to modify and add/delete elements of an XML document, and yet allows different applications to still work with the modified document. In other words, even if the structure of the document is changed, the correct semantics of the original elements can still be obtained. Any application can add proprietary information into the document without ever worrying about breaking the semantics of the original data, and thus preserving the interoperability of the data. This is the benefit of a self-describing format.

Additionally, XML based applications offer benefits such as providing support for different bitstream syntax without reprogramming/recompiling. An application built for Syntax A can still work with a document with Syntax B as long as the required elements are still available in the document and the tags remain the same.

4.7.2 Generic Content Manipulation

With XML representation, all the XML related technologies and software tools can be directly used. One very important XML technology is XSLT [20]. XSLT is a language defined using XML for transforming XML documents. With XML alone, interoperability is only obtained among the applications that understand a certain set of predefined tags. However, with XSLT, any application can be enabled to use any XML document as long as the actual data in the document is useful to the application. This "universal interoperability" is achieved by transforming XML documents into different structures usable for different applications.

Using such technology, a transcoder can be easily created by simply declaring a set of rules to transform the source elements. The style sheet (containing XSLT rules) along with the source document can be fed into an XSLT processor to generate a new document with desired structure and information. A set of style sheets can also be used to manipulate the elements of a given document. Similarly, simple text manipulating tools such as Perl can also be used to easily manipulate the data in the XML form.

Such an idea has recently been proposed in MPEG-21 for easy manipulation of scalable content with the introduction of a bitstream syntax description language – BSDL. BSDL is derived from XML Schema and it is used to describe (in high-level) scalable content. Generic software tools are also provided so that, using the BSDL description, scalable content can be converted back and forth between binary and XML representations. Once the content is in the XML form, generic software can be used to parse and manipulate the content. The goal is, depending on the conditions related to the user, terminal, environment, etc., to easily provide modified content that satisfies the conditions.

Though, the software tools in BSDL seem to provide the same functionalities as that of XFlavor, there are some differences between them. Using BSDL, each one of the supported bitstream syntaxes must be hand-described. Due to the nature of the language, the description can get quite verbose and the process of the description can be error-prone and tedious. As a solution, XFlavor can be used to automatically generate XML schema (BSDL description) corresponding to the given Flavor bitstream description.

Additionally, BSDL uses generic software to convert bitstream descriptions into XML documents. The software is very slow, and for real-time applications, there is a need to provide content-specific (syntax-specific) software tools so that fast conversion can take place. For each bitstream syntax description, XFlavor provides code to generate corresponding XML document much faster than the generic software tool.

Finally, the goal of BSDL is to provide high-level description of scalable content and, as described in Section 2, it lacks certain facilities to describe the whole bitstream syntax in a concise manner.

4.7.3 Compression

With the very high usage of XML applications, it is very important to come up with an efficient representation of XML documents. Because an XML document is text based, it is not efficient in a compression sense, and thus, it is not preferable for storage or transmission. Additionally, textual representation of multimedia data requires more processing time than binary representation. To fix these problems, binary representation of the data can be created for the given XML representation.

MPEG-7 [21] has already standardized binary encoding of its XML documents (BiM [22]). For wireless applications, where resources are very scarce, a binary version of XML is being deployed (WBXML [23]). Alternatively, there are a number of XML compression tools such as XMill [24] and XMLZip [25]. Though these can be applied to the XML representation of multimedia data, they do not yield as efficient result as the original compressed bitstream.

XMill is a compression tool based on a novel compression algorithm derived from the "grouping strategy" [24]. This algorithm sometimes yields better compression result than conventional compression (*gzip*) on the original data; however, this is usually the case for text data. For multimedia data originally represented by a compressed bitstream, it is very difficult to reduce the corresponding XML document to a size smaller than the original data. XMLZip is also a compression tool developed by XML Solutions (now a part of Vitria Technology) for compressing XML documents while maintaining the XML structure. As expected, it yields worse compression than XMill, but it offers DOM API [26] in the compressed domain, which ultimately results in faster XML parsing and processing. BiM is a generic XML tool used for MPEG-7 description encoding/decoding. Similar to XMLZip, BiM provides XML APIs for legacy systems. It also provides additional features such as streaming of XML documents and fast skipping of elements.

Table 4.1 lists the compression effectiveness of XMill, XMLZip, BiM and XFlavor (Bitstream Generator). Two GIF87a images (File 1 and File 2) are first converted into corresponding XML documents (using XFlavor), and they are compressed using the XML compression tools. Millau [27] was also examined; however, the encoder and decoder were not available for testing at the time of this writing. As shown in the table, all three compressors (XMill, XMLZip and BiM) have CE greater than 1. XMill produces the best result among the three because its sole purpose is to compress the data, whereas XMLZip and BiM produce much bigger files because they add some useful features in the compressed domain. With XMill, the compressed XML data has to be decompressed in order to process the data. With XFlavor, however, the XML data can be converted to the original bitstream syntax and the resulting data can be processed using existing decoders that conform to the given bitstream specification. Additionally, XFlavor converts the data in XML format into the original (compressed) bitstream, and the size of the bitstream is always smaller than the same XML data compressed using XMill.

In summary, if storage or transmission is the main concern, then the original bitstream should be used. Upon the time of processing, the bitstream can be converted to an XML file. Additionally, it is beneficial to maintain the bitstream representation of the data allowing fast processing for the applications like Multimedia App B in Figure 4.17.

Table 4.1 Compression effectiveness (CE) for different XML compressors. CE = Compressed XML file / Original (bitstream) file.

	File 1 (in Bytes)	CE	File 2 (in Bytes)	CE
Original File Size	*10,048*		*278,583*	
XML File Size	*305,634*		*8,200,278*	
XMill (gzip, CF = 6)	14,348	1.43	362,173	1.30
XMill (gzip, CF = 9)	14,311	1.42	357,335	1.28
XMill (bzip)	11,120	1.11	285,088	1.02
XMLZip	18,561	1.85	471,203	1.69
BiM	45,264	4.50	-	-
XFlavor	10,048	1.00	278,583	1.00

5. CONCLUSION

Flavor's design was motivated by our belief that content creation, access, manipulation and distribution will become increasingly important for end-users and developers alike. New media representation forms will continue to be developed, providing richer features and more functionalities for end-users. In order to facilitate this process, it is essential to bring syntactic description on par with modern software development practices and tools. Flavor can provide significant benefits in the area of media representation and multimedia application development at several levels.

First, it can be used as a media representation document tool, substituting ad hoc ways of describing a bitstream's syntax with a well-defined and concise language. This by itself is a substantial advantage for defining specifications, as a considerable amount of time is spent to ensure that such specifications are unambiguous and bug-free.

Second, a formal media representation language immediately leads to the capability of automatically generating software tools, ranging from bitstream generators and verifiers, as well as a substantial portion of an encoder or decoder.

Third, it allows immediate access to the content by any application developer, for such diverse use as editing, searching, indexing, filtering, etc.

With appropriate translation software, and a bitstream representation written in Flavor, obtaining access to such content is as simple as cutting and pasting the Flavor code from the specification into an ASCII file, and running the translator.

Flavor, however, does not provide facilities to specify how full decoding of data will be performed as it only addresses bitstream syntax description. For example, while the data contained in a GIF file can be fully described by Flavor, obtaining the value of a particular pixel requires the addition of LZW decoding code that must be provided by the programmer. In several instances, such access is not necessary. For example, a number of tools have been developed to do automatic indexing, searching and retrieval of visual content directly in the compressed domain for JPEG and MPEG content (see [28, 29]). Such tools only require parsing of the coded data so that DCT coefficients are available, but do not require full decoding. Also, emerging techniques, such as MPEG-7, will provide a wealth of information about the content without the need to decode it. In all these cases, parsing of the compressed information may be the only need for the application at hand.

Flavor can also be used to redefine the syntax of content in both forward and backward compatible ways. The separation of parsing from the remaining code/decoding operations allows its complete substitution as long as the interface (the semantics of the previously defined parsable variables) remains the same. Old decoding code will simply ignore the new variables, while newly written encoders and decoders will be able to use them. Use of Java in this respect is very useful; its capability to download new class definitions opens the door for such downloadable content descriptions that can accompany the content itself (similar to self-extracting archives). This can eliminate the rigidity of current standards, where even a slight modification of the syntax to accommodate new techniques or functionalities render the content useless in non-flexible but nevertheless compliant decoders.

Additionally, with XFlavor, generic software tools can easily be created that can access multimedia data in different bitstream formats. XFlavor enables the data to be converted from its binary representation to XML representation, and vice versa. Once in the XML form, due to the self-describing nature of XML documents, multimedia data can be easily manipulated without knowing the syntax. The bitstream layer is abstracted and the semantic values are directly available to the software tools. Once the data is manipulated, it can be put back into its binary form for storage, transmission and fast processing.

In [30], Jon Bosak states "XML can do for data what Java has done for programs, which is to make the data both platform-independent and vendor-independent." This is the reason why XML is becoming the standard for data exchange as well as media-independent publishing. XFlavor, by providing the XML features, applies the benefits provided by XML in the Information Technology world into the world of media representation.

More information about Flavor can be obtained from [31, 32, 33], and more detailed information about XFlavor can be found in [34, 35].

ACKNOWLEDGMENTS

The authors gratefully acknowledge Olivier Avaro (France Telecom), Carsten Herpel (Thomson Multimedia) and Jean-Claude Dufourd (ENST) for their contributions in the Flavor specification during the development of the MPEG-4 standard. We would also like to acknowledge Yihan Fang and Yuntai Kyong, who implemented earlier versions of the Flavor translator.

REFERENCES

[1]   T. M. Cover and J. A. Thomas, *Elements of Information Theory*, Wiley, 1991.

[2]   CompuServ Inc., *Graphics Interchange Format*, 1987.

[3]   ISO/IEC 10918 International Standard (JPEG), *Information Technology - Digital Compression and Coding of Continuous-Tone Still Images*, 1994.

[4]   ISO/IEC 11172 International Standard (MPEG-1), *Information Technology - Coding of Moving Pictures and Associated Audio for Digital Storage Media at up to about 1,5 Mbit/s*, 1993.

[5]   ISO/IEC 13818 International Standard (MPEG-2), *Information Technology - Generic Coding of Moving Pictures and Associated Audio Information*, 1996.

[6]   B. G. Haskell, A. Puri, and A. N. Netravali, *Digital Video: An Introduction to MPEG-2*, Chapman and Hall, 1997.

[7]   ISO/IEC 14496 International Standard (MPEG-4), *Information Technology - Coding of Audio-Visual Objects*, 1999.

[8]   Tutorial Issue on the MPEG-4 Standard, *Signal Processing: Image Communication*, Vol. 15, No. 4-5, January 2000.

[9]   B. Stroustrup, *The C++ Programming Language*, Addison-Wesley, 2nd Edition, 1993.

[10]   K. Arnold, J. Gosling, and D. Holmes, *The Java Programming Language*, Addison-Wesley, 3rd Edition, 1996.

[11]   J. Levine, T. Mason, and D. Brown, *lex & yacc*, O'Reilly, 2nd Edition, 1992.

[12]   S. Devillers and M. Caprioglio, Bitstream Syntax Description Language (BSDL), *ISO/IEC JTC1/SC29/WG11 M7433*, Sydney, July 2001.

[13]   S. Devillers, M. Amielh, and T. Planterose, Bitstream Syntax Description Language (BSDL), *ISO/IEC JTC1/SC29/WG11 M8273*, Fairfax, May 2002.

[14]   ISO/IEC JTC1/SC29/WG11 N4801, *MPEG-21 Overview v.4*, Fairfax, May 2002. (http://mpeg.telecomitalialab.com/standards/mpeg-21/mpeg-21.htm).

[15]   ISO/IEC 14496 1 International Standard, *Information Technology - Coding of Audio-Visual Objects - Part 1: Systems*, 2001.

[16]   Y. Fang and A. Eleftheriadis, Automatic Generation of Entropy Coding Programs Using Flavor, in Proceedings of IEEE Workshop on Multimedia Signal Processing, Redondo Beach, CA, December 1998, pp. 341-346.

[17]   C. Fogg, Survey of Software and Hardware VLC Architectures, *SPIE Vol. 2186 Image and Video Compression*, 1994.

[18]   W3C Recommendation, *Extensible Markup Language (XML) 1.0 (Second Edition)*, October 2000.

[19]   W3C Recommendation, *XML Schema Part 0: Primer, XML Schema Part 1: Structures, XML Schema Part 2: Datatypes*, May 2001.

[20]   W3C Recommendation, *XSL Transformations (XSLT) Version 1.0*, November 1999.

[21]   A. Avaro and P. Salembier, MPEG-7 Systems: Overview, *IEEE Transactions on Circuits and Systems for Video Technology*, Vol. 11 No. 6, June 2001, pp. 760-764.

[22]   http://www.expway.tv, *BiM*.

[23]   W3C Note, *WAP Binary XML Content Format*, June 1999.

[24]   H. Liefke and D. Suciu, XMill: An Efficient Compressor for XML Data, in *Proceedings of ACM SIGMOD International Conference on Management of Data*, 2000, pp. 153-164.

[25]   http://www.xmls.com, *XMLZip*.

[26]   W3C Recommendation, *Document Object Model (DOM) Level 2 Specification*, November 2000.

[27]   M. Girardot and N. Sundaresan, Efficient Representation and Streaming of XML Content over the Internet Medium, *in Proceedings of IEEE International Conference on Multimedia and Expo (ICME)*, Vol. 1, New York, NY, July 2000, pp. 67-70.

[28]   J. R. Smith and S. F. Chang, VisualSEEk: A Fully Automated Content-Based Image Query System, in *Proceedings of ACM International Conference Multimedia*, Boston, MA, November 1996.

[29]   S. W. Smoliar and H. Zhang, Content-Based Video Indexing and Retrieval, *IEEE Multimedia Magazine*, Summer 1994.

[30]   J. Bosak, Media-Representation Publishing: Four Myths about XML, *IEEE Computer*, Vol. 31 No. 10, October 1998, pp. 120-122.

[31]   Y. Fang and A. Eleftheriadis, A Syntactic Framework for Bitstream-Level Representation of Audio-Visual Objects, *in Proceedings of IEEE International Conference on Image Processing*, Lausanne, Switzerland, September 1996, pp. II.426-II.432.

[32]   A. Eleftheriadis, Flavor: A Language for Media Representation, in Proceedings of *ACM International Conference on Multimedia*, Seattle, WA, November 1997.

[33]   A. Eleftheriadis, A Syntactic Description Language for MPEG-4, *Contribution ISO/IEC JTC1/SC29/WG11 M546*, Dallas, November 1995

[34]   D. Hong and A. Eleftheriadis, XFlavor: Bridging Bits and Objects in Media Representation, in Proceedings of *IEEE International Conference on Multimedia and Expo*, Lausanne, Switzerland, August 2002.

[35]   D. Hong and A. Eleftheriadis, XFlavor: The Next Generation of Media Representation, *Technical Report ee2002-05-022*, Department of Electrical Engineering, Columbia University, New York, May 2002.

5

INTEGRATING DOMAIN KNOWLEDGE AND VISUAL EVIDENCE TO SUPPORT HIGHLIGHT DETECTION IN SPORTS VIDEOS

J. Assfalg, M. Bertini, C. Colombo, and A. Del Bimbo
Università di Firenze
Dipartimento di Sistemi e Informatica
Via S.Marta 3, 50139 Firenze, Italy
`delbimbo@dsi.unifi.it`

1. INTRODUCTION

The dramatic quantity of videos generated by digital technologies has originated the need for automatic annotation of these videos, and the consequent need for techniques supporting their retrieval. Content-based video annotation and retrieval is therefore an active research topic. While many of the results in content-based image retrieval can be successfully applied to videos, additional techniques have to be developed to address their peculiarity. In fact, videos add the temporal dimension, thus requiring to represent object dynamics. Furthermore, while we often think of a video just as of a sequence of images, it is actually a compound medium, integrating such elementary media as realistic images, graphics, text and audio, each showing characteristic presentation affordances [8]. Finally, application contexts for videos are different than those for images, and therefore call for different approaches in the way in which users may annotate, query for, and exploit archived video data.

This results in video streams to go through a complex processing chain, which comprises a variable number of processing steps. These steps include, but are not limited to, temporal segmentation of the stream into shots [18], detection and recognition of text appearing in captions [11, 9], extraction and interpretation of the audio track (including speech recognition) [1, 19, 36], visual summarization of shot content [10] and semantic annotation [2, 13]. In general, a bottom-up approach, moving from low level perceptual features to high level semantic descriptions, is followed. While this general framework roughly meets the requirements of a variety of application domains, the specificity of each domain has to be also addressed when developing systems that are expected to effectively support users in the accomplishment of their tasks. This specificity affects different stages in the development of content-based video retrieval systems, including selection of relevant low level features, models of specific domain knowledge supporting representation of semantic information [2], querying and visualization interfaces [23].

The huge amount of data delivered by a video stream requires development of techniques supporting an effective description of the content of a video.

This necessarily results in higher levels of abstraction in the annotation of the content, and therefore requires investigation and modeling of video semantics. This further points out that general purpose approaches are likely to fail, as semantics inherently depends on the specific application context. Semantic modeling of content of multimedia databases has been addressed by many researchers. From a theoretical viewpoint, the semantic treatment of a video requires the construction of a hierarchical data model including, at increasing levels of abstraction, four main layers: raw data, feature, object and knowledge [2]. For each layer, the model must specify both the elements of representation *(what)* and the algorithms used to compute them *(how)*. Upper layers are typically constructed by combining the elements of the lower layers according to a set of rules (however they are implemented). Concrete video retrieval applications by high-level semantics have been reported on in specific contexts such as movies, news and commercials [5, 7].

Due to its enormous commercial appeal, sports video represent another important application domain, where most of the research efforts have been devoted so far on the characterization of single, specific sports. Miyamori et al. [12] proposed a method to annotate the videos with human behavior. Ariki et al. [3] proposed a method for classification of TV sports news videos using DCT features. Among sports that have been analyzed so far, we can cite soccer [24, 17, 25, 26], tennis [27], basketball [28, 29, 15], baseball [30], American football [21].

This chapter illustrates an approach to semantic video annotation in the specific context of sports videos. Within scope of the EU ASSAVID[1] (Automatic Segmentation and Semantic Annotation of Sports Videos) project a number of tools supporting automatic annotation of sports videos were developed.

In the following sections methodological aspects of top-down and bottom-up approaches followed within the project will be addressed, considering generic sports videos. Videos are automatically annotated according to elements of visual content at different layers of semantic significance. Unlike previous approaches, videos can include several different sports and can also be interleaved with non-sport shots. In fact, studio/interview shots can be recognized and distinguished from sports action shots; the latter are then further decomposed into their main visual and graphic content elements, including sport type, foreground vs. background, text captions, and so on. Relevant semantic elements are extracted from videos by suitably combining together several low level visual primitives such as image edges, corners, segments, curves, color histograms, etc., according to context-specific aggregation rules. From section 4 to the end, detection of highlights will be

[1] This work was partially supported by the ASSAVID EU Project (The ASSAVID consortium comprises ACS SpA (I), BBC R&D (UK), Institut Dalle Molle D'Intelligence Artificielle Perceptive (CH), Sony BPE (UK), University of Florence (I), University of Surrey (UK).

discussed, analyzing in particular soccer videos. Soccer specific semantic elements are recognized using highlights models, based on low and medium level cues such as playfield zone and motion parameters; thus application to different kind of sports can be easily envisioned.

2. THE APPLICATION CONTEXT

The actual architecture of a system supporting video annotation and retrieval depends on the application context, and in particular on end users and their tasks. While all of these different application contexts demand for a reliable annotation of the video stream to effectively support selection of relevant video segments, it is evident that, for instance, service providers (e.g., broadcasters, editors) or consumers accessing a Video-On-Demand service have different needs [16].

In the field of supporting technologies for the editorial process, for both the old and new media, automatic annotation of video material opens the way to the economic exploitation of valuable assets. In particular, in the specific context of sports videos, two possible scenarios can be devised for the reuse of archived material within broadcasting companies: i) *posterity logging*, which is known as one key method of improving production quality by bringing added depth and historical context to recent events. Posterity logging is typically performed by librarians to make a detailed annotation of the video tapes, according to standard format; ii) *production logging*, where broadcasters use footage recorded few hours before, that may be even provided by a different broadcaster, and thus is not indexed, to annotate it in order to edit and produce a sports news program. Production logging is typically carried out live (or shortly after the event) by an assistant producer to select relevant shots to be edited into a magazine or news program that reports on sports highlights of the day (e.g., BBC's "Match of the day" or Eurosport's "Eurogoal"). An example of posterity logging is the reuse of shots that show the best actions of a famous athlete: they can be reused later to provide an historical context. An example of production logging is the reuse of highlights, such as soccer goals or tennis match points, to produce programs that contain the best sport actions of the day.

In both scenarios, video material, which typically originates "live," should be annotated automatically, as detailed manual annotation is mostly impractical. The level of annotation should be sufficient to enable simple text-based queries. The annotation process includes such activities as segmentation of the material into shots, grouping and classification of the shots into semantic categories (e.g., type of sport), supporting query formulation and retrieval of events that are significant to the particular sport.

In order to achieve an effective annotation, it is important to have a clear insight into the current practice and established standards in the domain of professional sports videos, particularly concerning the nature and structure of their content. Videos comprising the data set used in the experiments reported on in the following include a wide variety of typologies. The sport library department of a broadcaster may collect videos from other departments as well as other broadcasters, e.g., a broadcaster of the country that hosts the Olympic Games.

Videos differ from each other in terms of sports types (outdoor and indoor sports) and number of athletes (single or teams). Also, videos differ in terms of editing, as some of them represent so called *live feeds* of a single camera for a complete event, some include different feeds of a specific event edited into a single stream, and some others only feature highlights of minor sports assembled in a summary. Very few assumptions can be made on the presence of a spoken commentary or super-imposed text, as their availability depends on a number of factors, including technical facilities available on location, and on the agreements between the hosting broadcaster and the other broadcasters. As shown in Figure 5.1, the typical structure of a sports video includes sport sequences interleaved with studio scenes, possibly complemented with superimposed graphics (captions, logos, etc.).

| Studio / interviewee | Graphic objects | Playfield | Audience | Player | Player |

Figure 5.1 Typical sequence of shots in a sports video.

3. THE COMPUTATIONAL APPROACH

Following the above description of a typical sports video, the annotation task is organized into four distinct subtasks: *i)* shot pre-classification (aimed at extracting the actual sports actions from the video stream); *ii)* classification of graphic features (which, in sports videos, are mainly text captions that are not synchronized with shot changes); *iii)* camera and object motion analysis and mosaicing; and *iv)* classification of visual shot features. The first three subtasks are sport independent, while the fourth requires embedding some domain knowledge, e.g., information on a playfield. This subtask can be further specialized for different kinds of sports, such as detecting specific parts of a playfield, e.g., the soccer goal box area. Video shot segmentation will not be described, readers can find a thorough description of several video segmentation algorithms in [6] and [18]. While this general framework meets the requirements of a variety of sport domains, the specificity of each domain has to be addressed when developing systems that are expected to detect highlights specific of different sports. In Section 4 it will be shown how to combine results deriving from these subtasks, to recognize specific highlights.

This section expounds on the contextual analysis of the application domain and on the implementations of modules supporting each of the subtasks: contextual analysis examines specificity of data and provides an overview on the rationale underlying selection of relevant features; the implementation describes how to compute the features, and how the feature combination rules are implemented.

Sports Shots Pre-classification

Contextual analysis. The anchorman/interview shot classification module provides a simple preliminary classification of shot content, which can be further exploited and enhanced by subsequent modules. The need for this type of classification stems from the fact that some video feeds contain interviews and studio scenes featuring anchorman and athletes. An example is that of the Olympic Games, where the material that must be logged is often pre-edited by the hosting broadcaster, and may contain such kinds of shots. The purpose of this module is to roughly separate shots that contain possible sport scenes from shots that do not contain sport scenes. To this end, a statistical approach can be followed to analyze visual content similarity and motion features of the anchorman shots, without requiring any predefined shot content model to be used as a reference. In fact, this latter constraint is required in order to be able to correctly manage interviews, which do not feature a standard studio set-up, as athletes are usually interviewed near the playfield, and each interview has a different background and location. Also the detection of studio scenes requires such independence of a shot content model, since the "style" changes often, and each program has its unique style that would require the creation and maintenance of database of shot content models.

Studio scenes show a well defined syntax: shot location is consistent within the video, the number of cameras and their view field is limited, the sequence of shot content can be represented by a repeating pattern. An example of such a structure is shown in Figure 5.2, where the first frames of the five successive shots comprising a studio scene are shown.

Figure 5.2 Studio scene with alternating anchorman shots.

Implementation. Shots of the studio/interview are repeated at intervals of variable length throughout the sequence. The first step for the classification of these shots stems from this assumption and is based on the computation, for each video shot S_k, of its shot lifetime $L(S_k)$. The shot lifetime measures the shortest temporal interval that includes all the occurrences of shots with similar visual content, within the video. Given a generic shot S_k, its lifetime is computed by considering the set $T_K = \{T_i \mid \sigma(S_k, S_i) < \tau_S\}$, where $\sigma(S_k, S_i)$ is a similarity measure applied to keyframes of shots S_k and S_i, τ_S a similarity threshold and t_i is the value of the time variable corresponding to the occurrence of the keyframe of shot S_i. The lifetime of shot S_k is defined as $L(S_K) = \max(T_K) - \min(T_K)$. Shot classification is based on fitting values of $L(S_k)$ for all the video shots in a bimodal distribution. This allows for the determination of a threshold value t_l that is used to classify shots into the *sport* and *studio/interview* categories. Particularly, all the shots S_k such that $L(S_K) > t_l$ are classified as studio/interview shots, where t_l was determined according to the statistics of the test database, and set to 5 sec. Remaining shots are classified as *sport* shots. Typical videos in the target domain do not contain complete studio shows, and, in feeds produced on location, interviews have a limited time and shot length. This allows for the reduction of false detections caused by the repetition of similar sport scenes (e.g., as in the case of edited magazine programs or summaries) by limiting the search of similar shots to a window of shots. The adopted similarity metric is a histogram intersection of the mean color histogram of shots. Usage of the mean histogram takes into account the dynamics of sport scenes. In fact even if some scenes take place in the same location, and thus the color histogram of their first frame may be similar, the following actions yield a different color histogram. When applied to studio/interview shots, where the dynamics of changes of lighting of the scene are much more compressed, and the reduced camera and objects movement do not introduce new objects, we get a stable histogram.

Although the mean color histogram accounts for minor variations due to camera and objects movement, it does not take into account spatial information. Results of the first classification step are therefore refined by considering motion features of the studio/interview shots. This develops on the assumption that in an anchorman shot, both the camera and the anchorman are almost steady. In contrast, for sport shots, background objects and camera movements—persons, free-hand shots, camera panning and zooming, changes in scene lighting—cause relevant motion components throughout the shot. Classification refinement is performed by computing an index of the quantity of motion Q_M, for each possible anchorman shot. Only those shots whose Q_M doesn't exceed a threshold τ_M are definitely classified as studio/interview shots.

Identification of Graphic Features

Contextual analysis. In sports videos, graphic objects (GO) may appear everywhere within the frame, even if most of the time they are placed in the lower third or quarter of the image. Also the vertical and horizontal ratio of the GO zones varies, e.g., the roster of a team occupies a vertical box, while the name of a single athlete usually occupies a horizontal box (see Figure 5.3). For text graphics, character fonts may vary in size and typeface, and may be super-imposed either on an opaque background or directly on the image captured by the camera. GOs often appear and disappear gradually, through dissolve or fade effects. These properties call for automatic GO localization algorithms with the least amount of heuristics and possibly no training.

Several features such as edges and textures have been used in past research as cues of super-imposed GOs [14] [31]. Such features represent global properties of images, and require the analysis of large frame patches. Moreover, also natural objects such as woods and leaves, or man-made objects such as buildings and cars may present a local combination of such features that can be wrongly classified as a GO [9].

In order both to reduce the visual information to a minimum and to preserve local saliency, we have elected to work with image corners, extracted from luminance information of images. Corners are computed from luminance information only; this is very appealing for the purpose of GO detection and localization in that it prevents many misclassification problems from arising with color-based approaches. This fact is particularly important when considering the characteristics of television standards, which require a spatial sub-sampling of the chromatic information; thus the borders of captions are affected by color aliasing. Therefore, to enhance readability of characters the producers typically exploit luminance contrast, since luminance is not spatially subsampled and human vision is more sensitive to it than to color contrast. Another aspect that must be considered when analyzing GO detection algorithms is they do not require any knowledge or training on super-imposed captions features.

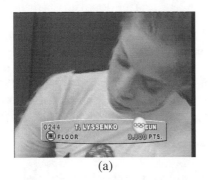

(a)

(b)

Figure 5.3 Examples of superimposed graphic objects.

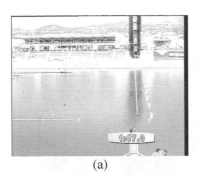

(a)

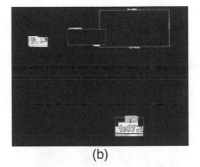

(b)

Figure 5.4 a) Source frame; b) Detected captions with noise removal.

Implementation. The salient points of the frames, which are to be analyzed in the following steps, are extracted using the Harris algorithm, from the luminance map, extracted from each frame. Corner extraction greatly reduces the number of spatial data to be processed by the GO detection and localization system. The most basic property of GOs is the fact that they must remain stable for a certain amount of time, in order to let people read and understand them. This property is used in the first step of GO detection. Each corner is checked to determine if it is still present in the same position in at least 2 more frames within a sliding window of 4 frames.

Each corner that complies with this property is marked as *persistent*, and is kept for further analysis, while all the others are discarded. Every 8th frame is processed to extract its corners, thus further reducing the computational resources needed to process a whole video. This choice develops on the assumption that, in order to be perceived and understood by the viewer, a GO must be stable on the screen for 1 second. The patch surrounding each corner is inspected, and if there are not enough neighboring corners (i.e., corners whose patches do not intersect), the corner is not considered in further processing.

This process, which is repeated a second time in order to eliminate corners that get isolated after the first processing, avoids having isolated high contrast background objects contained within static scenes are recognized as possible GO zones.

An unsupervised clustering is performed on the corners that comply with the temporal and spatial features described above. This is aimed at determining bounding boxes for GOs (Figures 5.4 and 5.5). For each bounding box the percentage of pixels that belong to the corner patches is calculated, and if it is below a predefined threshold the corners are discarded. This strategy reduces the noise due to high contrast background during static scenes, which typically produce small scattered zones of corners that cannot be eliminated by the spatial feature analysis. An example of GO detection is provided in Figure 5.5.

Evaluation of results takes into account GO detection (whether the appearance of a GO is correctly detected) and correct detection of the GO's bounding box. Typical results of GO detection have a precision of 80.6% and recall of 92%. Missed detections are experienced mostly in VHS videos, and only few in DV videos. The GO bounding box miss rate is 5%. Results also included false detections, due to scene text.

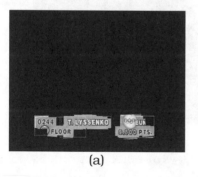

(a)   (b)

Figure 5.5 Detection results for the frames in Figure 5.3.

Camera and object motion analysis and mosaicing

Contextual analysis. Camera and object motion parameters have been used to detect and recognize highlights in sports videos. Use of lack of motion and camera operations has been described in the specific context of soccer by [4], [17]. Usage of camera motion (based on 3 parameter model) for the purpose of video classification has been discussed in [22]. Mosaicing is an intra-shot video summarization technique which is applied to a video shot. Mosaicing is based on the visual tracking of all motions in the shot, either due to camera action or to scene objects moving independently from the camera. Image pixels whose motion is due to camera action are labeled as background, and the others as foreground. A single mosaic image summarizing the whole visual content of the shot is then incrementally obtained from all new background pixels as they enter in the field of view of the camera; camera action is also recorded for every shot frame. Foreground information is stored separately, by tracing frame by frame the image shape and trajectory of each of the objects in individual motion. Figure 5.6 and 5.7 show a mosaic image and pixels associated to the diver in an intermediate frame of the diving shot. Use of mosaic images has been described by [24] and [26] for highlight presentation. Also in [17] mosaic images are used to show soccer highlights.

Two of the main paradigms for motion estimation are: feature-based and correlation-based, each complementing the other [10]. Feature-based techniques compute image motion by first tracing features of interest (edges, corners, etc.) from frame to frame, and then inferring a set of motions compatible with the set of feature correspondences. Feature-based motion computation is known to be robust (since it works with a selected number of salient image locations), to work well even with large displacements, and to be quite fast, thanks to the limited number of image locations being analyzed. Yet, in applications requiring that each image pixel be labeled as belonging to a motion class, the feature-based approach loses much of its appeal, since passing from sparse computations to dense estimates is both an ambiguous and slow task. Correlation-based techniques compute the motions of whole image patches through block matching. Different from feature-based methods, these approaches are intrinsically slow due to the high number of elementary operations to be performed at each pixel. Moreover, block matching works only for small image displacements, thus requiring the use of a multiresolution scheme in the presence of large displacements. Nonetheless, correspondence-based motion computation techniques automatically produce dense estimates, and can be optimized in order to work reasonably fast.

Implementation. In this paragraph a basic feature-based algorithm for mosaicing using corner features is first outlined [32]. The image motion between successive shot frames is evaluated through a three-step analysis: (1) corner detection; (2) corner tracking; (3) motion clustering and segmentation. A fourth step, namely mosaic updating, concludes the processing of a generic shot frame.

Corner Detection. An image location is defined as a corner if the intensity gradient in a patch around it is distributed along two preferred directions. Corner detection is based on the Harris algorithm (see Figure 5.8).

Corner Tracking. To perform intra-shot motion parameters estimation, corners are tracked from frame to frame, according to an algorithm originally proposed in [35] and modified by the authors to enhance tracking robustness. The algorithm optimizes performance according to three distinct criteria, namely:

- Frame similarity: The image content in the neighborhood of a corner is virtually unchanged in two successive frames; hence, the matching score between image points can be measured via a local correlation operator.
- Proximity of Correspondence: As frames go by, corner points follow smooth trajectories in the image plane, thus allowing to reduce the search space for each corner in a small neighborhood of its expected location, as inferred based on previous tracking results.
- Corner Uniqueness: Corner trajectories cannot overlap, i.e., it is not possible that at the same time two corners share the same image location. Should this happen, only the corner point with higher correlation would be maintained, while the other would be discarded.

Figure 5.6 Mosaic image of a dive.

Figure 5.7 Intermediate diver frame.

Since the corner extraction process is heavily affected by image noise (the number and individual location of corner varies significantly in successive frames; also, a corner extracted in one frame, albeit still visible, could be ignored in the next one), the modified algorithm implements three different corner matching strategies, ensuring that the above tracking criteria are fulfilled:

1. strong match, taking place between pairs of locations classified as corners in two consecutive frames;
2. forced match, image correlation within the current frame, in the neighborhood of a previously extracted corner;

3. backward match, image correlation within the previous frame, in the neighborhood of a currently extracted corner.

These matching strategies ensure that a corner trajectory continues to be traced even if, in some instants, the corresponding corner fails to be detected.

Motion clustering and segmentation. After corner correspondences have been established, a motion clustering technique is used to obtain the most relevant motions present in the current frame. Each individual 2D motion of the scene is detected and described by means of the 6-parameter affine motion model. Motion clustering takes place starting from the set of corner correspondences found for each frame. A robust estimation method is adopted, guaranteeing on the one hand an effective motion clustering, and on the other a good rejection of false matches (clustering outliers). The actual motion-based segmentation is performed by introducing spatial constraints to the classes obtained via the previous motion clustering phase. Compact image regions featuring homogenous motion parameters -thus corresponding to single, independently moving objects- are extracted by region growing. The motion segmentation algorithm is based on the computation of an a posteriori error obtained by plain pixel differences between pairs of frames realigned according to the extracted affine transformations. Figure 5. 9 shows the results of the segmentation of the frame of Figure 5. 8 into its independent motions.

Figure 5.8 Image corners.

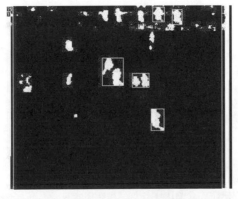

Figure 5.9 Segmented image (all black pixels are considered part of the background).

Mosaic updating. All the black-labeled pixels of Figure 5. 9 are considered as part of the background, and are thus used to create the mosaic image of Figure 5. 10. In this algorithm, to obtain the corresponding mosaic pixel, all background pixels present in multiple shot frames were simply averaged together.

Figure 5.10 Mosaic image.

Classification of Visual Shot Features

Contextual analysis. Generic sports videos feature a number of different scene types, intertwined with each other in a live video feed reporting on a single event, or edited into a magazine summarizing highlights of different events.

A preliminary analysis of videos reveals that 3 types of scenes prevail: *playfield*, *player* and *audience* (see Figure 5.11). Most of the action of a sports game takes place on the playfield. Hence, the relevance of playfield scenes, showing mutual interactions among subjects (e.g.,: players, referees, etc.) and objects (e.g.,: ball, goal, hurdles, etc.). However, along with playfield scenes, a number of scenes appear in videos, featuring close-ups of players, or framing the audience. The former typically show a player that had a relevant role in the most recent action (e.g., the athlete who just failed throwing the javelin, or the player who shot the penalty). The latter occur at the beginning and at the end of an event, when nothing is happening on the playfield, or just after a highlight (e.g.,: in soccer, when a player shoots a goal, audience shots are often shown immediately after). It is thus possible to use these scenes as cues for detection of highlights.

In a sample of 1267 keyframes, extracted randomly from our data set (obtained from the BBC Sports Library), approximately 9% were audience scenes, whereas player scenes represented up to 28% of the video material. To address the specificity of such a variety of content types, we devised a hierarchical classification scheme. The first stage performs a classification in terms of the categories of *playfield*, *player* and *audience*, with a twofold aim: on the one hand, this provides an annotation of video material that is meaningful for users' tasks; on the other hand it is instrumental for further processing, such as identification of sports type and highlight detection.

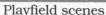
Playfield scenes   Player scenes   Audience scenes

Figure 5.11 Although sports events reported on in a video may vary significantly, distinguishing features are shared across this variety. For example, playfield lines is a concept that is explicitly present in some outdoor and indoor sports (e.g.,: athletics or swimming), but can also be extended to other sports (e.g.,: cycling on public roads). Similarly, player and audience scenes appear in most sports videos.

Inspection of video material reveals that: *i)* *playfield* shots typically feature large homogeneous color regions and distinct long lines; *ii)* in *player* shots, the shape of the player appears distinctly in the foreground, and the background of the image tends to be homogeneous, or blurred (either because of camera motion or lens effects); *iii)* in *audience* shots, individuals in the audience do not always appear clearly, but the audience as a whole appears as a texture. These observations suggest that basic edge and shape features could significantly help in differentiating among *playfield*, *player* and *audience* scenes. It is also worth pointing out that models for these classes do not vary significantly across different sports, events and sources.

We propose here that identification of the type of sport represented in a shot relies on playfield detection. In fact, we can observe that: *i)* most sports events take place in a play field, with each sport having its own playfield; *ii)* each playfield has a number of distinguishing features, the most relevant of which is color; *iii)* the playfield appears in a large number of frames of a video shot, and often covers a large part of the camera frame (i.e., a large area of single images comprising the video). Hence, playfield, and objects that populate it, may effectively support identification of sports types. Therefore, in our approach, sports type identification is applied to playfield shots output by the previous classification stage.

Playfield shape and playfield lines can be used also to perform highlight detection, besides sport classification. For example detection of a tennis player near the net may be considered a cue of a volley; tracking swimmers near the turning wall helps identify the backstroke turn, etc. In Section 4 playfield shape and playfield lines features will be used to detect soccer highlights.

Implementation. A feature vector comprising edge, segment and color features was devised. Some of the selected features are represented in Figure 5.12 for representatives of the three classes (*playfield*, *player* and *audience*, respectively).

Edge detection is first performed, and a successive growing algorithm is applied to edges to identify segments in the image [20]. The distribution of edge intensities is analyzed to evaluate the degree of uniformity. Distributions of lengths and orientations of segments are also analyzed, to extract the maximum length of segments in an image, as well as to detect whether peaks exist or not in the distribution of orientations. This choice was driven by the following observations: playfield lines are characteristic segments in playfield scenes [17], and determine peaks in the orientation histogram, and also feature longer segments than other types of scenes; audience scenes are typically characterized by more or less uniform distributions for edge intensities, segments orientation and hue; player scenes typically feature fewer edges, a uniform segment orientation distribution, and short segments.

Color features were also considered, both to increase robustness to the first classification stage (e.g., audience scenes display more uniform color distributions than playfield or player scenes), and to support sports type identification.

In fact, the playfield each sport usually takes place on typically features a few dominant colors (one or two, in most cases). This is particularly the case in long and mid-range camera takes, where the frame area occupied by players is only a fraction of the whole area. Further, for each sport type, the color of the playfield is fixed, or varies in a very small set of possibilities. For example, for soccer the playfield is always green, while for swimming it is blue. Color content is described through color histograms. We selected the HSI color space, and quantized it into 64 levels for hue, 3 levels for saturation and 3 levels for intensity. Indices describing the distribution (i.e., degree of uniformity, number of peaks) were also derived from these distributions.

Two neural network classifiers are used to perform the classification tasks. To evaluate their performance, over 600 frames were extracted from a wide range of video shots, and were manually annotated to define a ground truth.

Table 5.1 Results for the classification of keyframes in terms of *playfield*, *player* and *audience* classes.

Class	Correct	Missed	False
Playfield	80.4%	19.6%	9.8%
Player	84.8%	15.2%	15.1%
Audience	92.5%	7.5%	9.8%

Frames were then subdivided into three sets to perform training, testing and evaluation of the classifiers. The aforementioned edge, segment and color features were computed for all of the frames. Results for the scene type classification are summarized in Table 5.1. It is worth pointing out that extending this classification scheme to shots, rather than just limiting it to keyframes, will yield an even better performance, as integration of results for keyframes belonging to the same shot reduces error rates. For instance, some keyframes of a playfield shot may not contain playfield lines (e.g.,: because of a zoom-in), but some others will. Hence, the whole shot can be classified as a playfield shot.

Results on sports type identification are shown in Table 5.2. The first column of figures refers to an experiment carried out on a data set including also player and audience scenes, whereas the second column of figures summarizes an experiment carried out on the output of a filtering process keeping only playfield frames. As expected, in the former case we obtained lower success rates. By comparing results in the two columns, we can observe that introduction of the *playfield*, *player* and *audience* classes is instrumental to improve identification rates for sports types. On average, these improve by 16%, with a maximum of 26%. The highest improvement rates are observed for those sports where the playfield is shown only for small time intervals (e.g., high diving), or in sports where only one athlete takes part in the competition, videos of which frequently show close-ups of the athlete (e.g., javelin).

Table 5.2 Sports type identification results. The evaluation set in the first experiment comprised also *player* and *audience* scenes. The second experiment was carried out on *playfield* scenes only.

Sports type	All frames	Playfield only
High diving	56.9%	83.2%
Floor	78.7%	97.4%
Field hockey	85.0%	95.1%
Long horse	53.4%	64.3%
Javelin	37.8%	58.8%
Judo	80.6%	96.9%
Soccer	80.3%	93.2%
Swimming	77.4%	96.1%
Tennis	69.1%	94.5%
Track	88.2%	92.7%

4. A REAL APPLICATION: DETECTION AND RECOGNITION OF SOCCER HIGHLIGHTS

Among the many sports types, soccer is for sure one of the most relevant and worldwide diffused. In the following sections we report on our experience in the classification of soccer highlights, using an approach based on temporal logic models. The method has been tested using several soccer videos containing a wide range of different video editing and camera motion styles, as produced by several different international broadcasters. Considering a variety of styles is of paramount importance in this field, as otherwise the system lacks robustness. In fact, videos produced by different directors display different styles in the length of the shots, in the number of cameras, in the editing effects.

We review hereafter previous work related to soccer videos. The work presented in [24] is limited to detection and tracking of both the ball and the players; the authors do not attempt to identify highlights. In [17], the authors rely on the fact that the playing field is always green for the purpose of extracting it. Successive detection of ball and players is limited to the field, described by a binary mask. To determine position of moving objects (ball and players) within the field, the central circle is first located, and a four-point homographic planar transformation is then performed, to map image points to the model of the playing field. Whenever the central circle is not present in the current frame, a mosaic image is used to extend the search context. In this latter case, the mosaicing transformation is combined with the homographic transformation. This appears to be a fairly expensive approach. In [25] a hierarchical E-R model that captures domain knowledge of soccer has been proposed. This scheme organizes basic actions as well as complex events (both observed and interpreted), and uses a set of (nested) rules to tell whether a certain event takes place or not. The system relies on 3D data of position of players and ball, which are obtained from either microwave sensors or multiple video cameras. Despite the authors' claim that, unlike other systems, their own works on an exhaustive set of events, only little evidence of this is provided, as only a basic action (*deflection*) and a complex event (*save*) are discussed. In [26] has been proposed the usage of panoramic (mosaic) images to present soccer highlights: moving objects and the ball are super-imposed on a background image featuring the playing field. Ball, players and goal posts are detected. However, despite the title, only presentation of highlights is addressed, and no semantic analysis of relevant events is carried out.

5. ANALYSIS OF THE VIDEOS AND EXTRACTED FEATURES

Inspection of tapes showed that producers of videos use a main camera to follow the action of the game; since game action depends on the ball position, there exists a strong correlation between the movement of the ball and camera action. The main camera is positioned along one of the long sides of the playing field. In Figure 5.13 some typical scenes taken with the main camera are shown.

Identification of the part of the playing field currently framed and camera action are among the most significant features that can be extracted from shots taken by the main camera; these features can be used to describe and identify relevant game events. Typical actions featured by the main camera are: *i*) pan, *ii*) tilt and *iii*) zoom. Pan and tilt are used to move from a part of the playing field to another one, while zoom is used to change the framing of the subject.

Highlights that we have elected for thorough investigation are: i) forward launches, *ii*) shoots on goal, *iii*) turnovers, *iv*) penalty kicks, free kicks next to the goal box and corner kicks. These are typical highlights shown in TV news and magazine programs summarizing a match, even if these actions do not lead to the scoring of a goal. Moreover, penalty kicks and corners are often used to calculate statistics for a match. All of the above highlights are part of attack actions taking place in the goal box area, and are therefore strictly related to goal actions. It must also be noted that for each of the above highlights there exist two different versions, one for each playing field side. The system that we present can discriminate each case.

6. PLAYFIELD ZONE CLASSIFICATION

The soccer playfield has been divided in 12 zones, 6 for each side (Figure 5.14). These zones have been chosen so that the change from one to the other indicates a change in the action shown, such as a defense action that changes into a counter-attack, or an attack that enters the goal box. Moreover it must be noted that typical camera views are associated with the selected zones. In some cases the zones can't be recognized from simple image analysis (e.g., zones near the center of the playfield), but the temporal analysis solves this problem.

The features used to recognize the playfield zones are playfield lines and the playfield shape. Figure 5.15 shows an example of extraction of playfield shape and playfield lines. From these features we calculate a five elements vector composed by the following elements:

- playfield shape descriptor F: six different views of the playfield are recognized. They are shown in Figure 5.16 a);
- playfield line orientation descriptor O: playfield lines are extracted, and their directions are quantized in 32 bins. These bins are grouped to calculate the number of horizontal and vertical lines, and the number of lines whose angle is greater or smaller than the specified line orientation descriptor, as shown in Figure 5.16 b);
- playfield size descriptor R: this descriptor signals if the percentage of pixels of a frame, that belong to the playfield, is above a threshold;
- playfield corner position C: this descriptor may assume 4 values: absence of playfield corner in the image, or presence of one of three possible positions of the playfield corner that is farthest from the camera, as shown in Figure 5.17 a);
- midfield line descriptor M: this descriptor is similar to the C descriptor, but it indicates the presence and type of the midfield line (Figure 5.17b).

A Naïve Bayes classifier has been used to classify each playfield zone Zx. This choice is motivated by the fact that: $i)$ we are interested in confidence values for each zone, to handle views that are not easily classified; $ii)$ some descriptors are not useful for some zones (e.g., the C descriptor is not useful for the midfield zone); $iii)$ some zones are mutually exclusive and some are not, thus it is not possible to define them as different values of a single variable. In Figure 5.18 the classifiers for Z1 and Z6 are shown. It must be noted that in some cases some values of the descriptors correspond to a single value of the variable of the classifier, e.g.,:

$$F_{Z1} = \begin{cases} 1 \text{ if } F = F_1 \\ 2 \text{ if } F = F_2 \\ 3 \text{ otherwise} \end{cases} \qquad F_{Z6} = \begin{cases} 1 \text{ if } F = F_3 \\ 2 \text{ otherwise} \end{cases}$$

Playfield zone classification is performed through the following steps: playfield lines and shape are extracted from the image, descriptors are then calculated and the observation values of the classifier variables are selected. The classifier with the highest confidence value (if above a threshold) is selected.

7. CAMERA MOTION ANALYSIS

As noted in section 5, camera parameters are strongly related to ball movement. Pan, tilt and zoom values are calculated for each shot, using techniques shown in section 3. The curves of these values are filtered and quantized, using 5 levels for pan, 3 for tilt and zoom. Analysis of the videos has shown that conditions such as a flying ball rather than a change of direction can be observed from camera motion parameters. Through heuristic analysis of these values, three descriptors of *low*, *medium* and *high* ball motion have been derived. Figure 5.19 shows a typical shot action: at time t_{start} a player kicks the ball toward the goal post. Acceleration and deceleration (at time t_{OK}) can be identified. Often zoom on the goal post can be observed.

8. MODEL CHECKING

Information provided from playfield zone recognition and camera motion analysis is used to create temporal logic models of the highlights. In fact the highlights can be characterized by the playfield zone where they happen, and how the action develops through the playfield. For instance the forward launch requires that the ball moves quickly from midfield toward the goal box. The models of all the actions have been implemented as FSM. Figure 5.20 and 5.21 show two FSM, one for the goal shot and one for the turnover.

The system has been tested on about one 80 sequences, selected from 15 European competitions. Tables 5.3 and 5.4 report the results. From the results it can be noticed that those highlights that can be easily defined (free/penalty kicks and shots on goal) obtain a good recognition rates.

Recognition of shots on goal is very good, especially if we consider that the false detection is due to attack actions near the goal box. Since the ball often moves too fast and is too small, we did not attempt to detect the goal. Anyway this highlight could be recognized combining together the shot on goal and the caption that shows the new goal count. Also the good result of forward launches detection is encouraging, since this highlight is usually very similar to other actions. It must be noticed that while this highlight is not important *per se*, it is linked to other important highlights, such as counterattacks. The recognition of turnovers is critical since its definition is quite fuzzy; probably adding other features such as player position would reduce the false detection. Other improvements may arise considering more cues, such as players trajectories, obtained from the object motion analysis described in section 3.

ACKNOWLEDGMENTS

This work was supported by the ASSAVID EU Project, under contract IST-13082. The consortium comprises ACS SpA (I), BBC R&D (UK), Institut Dalle Molle D'Intelligence Artificielle Perceptive (CH), Sony BPE (UK), University of Florence (I), University of Surrey (UK).

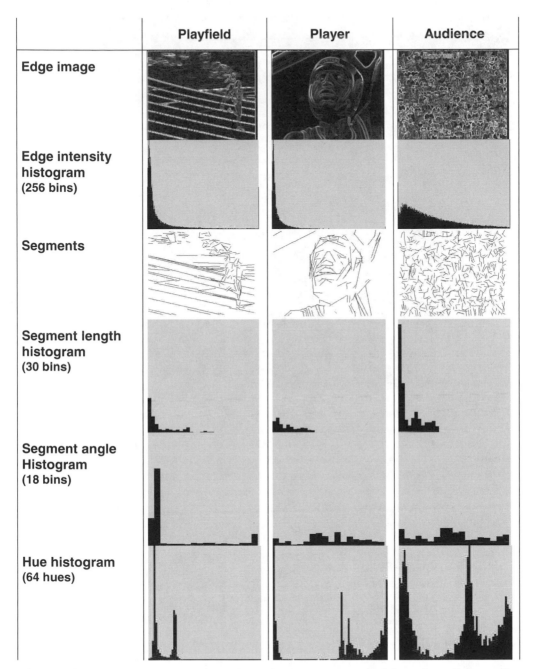

Figure 5.12 Edge, segment length and orientation, and hue distribution for the three representative sample images in the first row of Figure 5.11. Synthetic indices derived from these distributions allow to differentiate among the three classes of *playfield*, *player*, and *audience*. (Please, note that hue histograms are scaled to the maximum value).

Figure 5.13 Example of frames taken from the main camera.

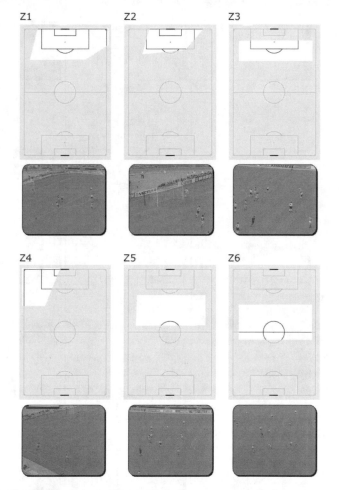

Figure 5.14 Playfield zones; Z7 to Z12 are symmetrical.

Figure 5.15 Original image, playfield shape and lines in soccer.

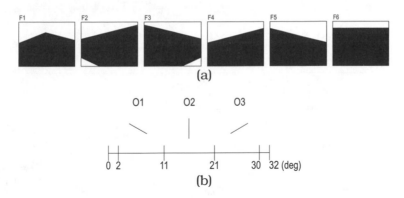

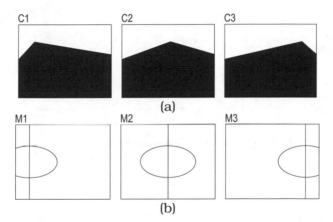

Figure 5.16 (a) Playfield shape descriptor F. (b) Playfield line orientation descriptor O.

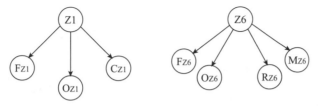

Figure 5.17 (a) Playfield corner position descriptor C. (b) Midfield line descriptor M.

Figure 5.18 Naïve Bayes networks: Z1 and Z6 zone classifiers.

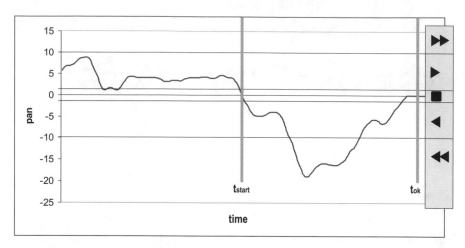

Figure 5.19 Typical shot action: at time t_{start} the ball is kicked toward the goal post. Symbols on the right identify low, medium and high motion.

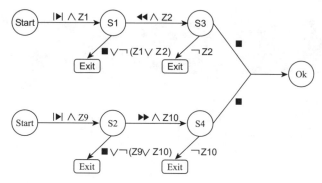

Figure 5.20 Shot model: on the arcs are reported the camera motion and playfield zones needed for the state transition. The upper branch describes a shot in the left goal post, the lower branch the shot in the right goal post. If state *OK* is reached then the highlight is recognized

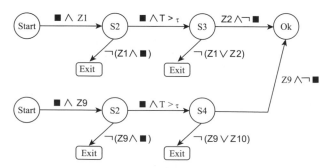

Figure 5.21 Restart model: the final transition requires a minimum time length.

Table 5.3 Highlight classification results. The Corner/Penalty kick class comprises also free kicks near the goal box zone

	Detect.	Correct	Miss.	False
Forward launch	36	32	1	4
Shots on goal	18	14	1	4
Turnover	20	10	3	10
Corner/pen. kick	13	13	2	0

Table 5.4 The *Other* class contains actions that were not modeled. The Corner/Penalty kick class comprises also free kicks near the goal box zone.

Classific. result	Actual highlight				
	Fwd. launch	Sh. On goal	Turnover	C./p. kick	Other
Fwd. launch	32	1	1	0	2
Shots on goal	1	14	0	0	3
Turnover	0	0	10	0	10
Corner/pen. kick	0	0	0	13	2

REFERENCES

[1] J.Ajmera, I.McCowan, and H.Bourlard, Robust HMM-Based Speech/Music Segmentation, in *Proceedings of ICASSP*, 2002.

[2] W. Al-Khatib, Y. F. Day, A. Ghafoor, and P. B. Berra, Semantic Modeling and Knowledge Representation in Multimedia Databases, *IEEE Trans. On Knowledge and Data Engineering* 11(1), 1999.

[3] Y.Ariki, and Y.Sugiyama, Classification of TV Sports News by DCT Features using Multiple Subspace Method, in *Proc. 14th International Conference on Pattern Recognition (ICPR'98)*, pp. 1488–1491, 1998.

[4] A.Bonzanini, R.Leonardi, and P.Migliorati, Exploitation of Temporal Dependencies of Descriptors to Extract Semantic Information, in *Int'l Workshop on Very Low Bitrate Video (VLBV'01)*, Athens(GR), 2001.

[5] C. Colombo, A. Del Bimbo, and P. Pala, Semantics in Visual Information Retrieval, *IEEE MultiMedia* 6(3):38–53, 1999.

[6] A. Del Bimbo, *Visual Information Retrieval*, Academic Press, London 1999.

[7] S. Eickeler and S. Muller, Content-Based Video Indexing of TV Broadcast News Using Hidden Markov Models, in *Proc. IEEE Int. Conf. on Acoustics, Speech, and Signal Processing (ICASSP)*, pp. 2997–3000, 1999.

[8] R. S. Heller and C. D. Martin, A Media Taxonomy, *IEEE Multimedia*, 2(4):36–45, Winter 1995.

[9] H. Li, D. Doermann, Automatic Identification of Text in Digital Video Key Frames, *Proc. Int. Conf. on Pattern Recognition ICPR'98*, 1998.

[10] M. Irani, P. Anandan, J. Bergen, R. Kumar, and S. Hsu. Efficient Representations of Video Sequences and Their Applications, *Signal Processing: Image Communication*, 8(4):327-351, 1996.

[11] R. Lienhart, Indexing and Retrieval of Digital Video Sequences Based On Automatic Text Recognition, in *Proc. of 4-th ACM International Multimedia Conference*, 1996.

[12] H. Miyamori, S.-I. Iisaku, Video annotation for content-based retrieval using human behavior analysis and domain knowledge, in *Proc. Int. Workshop on Automatic Face and Gesture Recognition 2000*, 2000.

[13] J.Kittler, K.Messer, W.J.Christmas, B.Levienaise-Obadia, D.Koubaroulis, Generation of semantic cues for sposrts video annotation, in *Proc. ICIP 2001*, 2001.

[14] T. Sato, T. Kanade, E. K. Hughes, M. A. Smith, Video OCR for Digital News Archive, in *Proc. IEEE Int. Workshop on Content–Based Access of Image and Video Databases CAIVD'98*, pp. 52–60, 1998.

[15] W. Zhou, A. Vellaikal, and C.C.J. Kuo, Rule-based video classification system for basketball video indexing, in *Proc. ACM Multimedia 2000 Workshop*, pp. 213–216, 2000.

[16] N. Dimitrova et al., Entry into the Content Forest: The Role of Multimedia Portals, *IEEE MultiMedia*, Summer 2000.

[17] Y. Gong, L.T. Sin, C.H. Chuan, H. Zhang, and M. Sakauchi, Automatic Parsing of TV Soccer Programs, in *Proc. of the Int'l Conf. on Multimedia Computing and Systems (ICMCS'95)*, Washington, D.C, May 15-18, 1995.

[18] U. Gargi, R. Kasturi, and S.H. Strayer, Performance Characterization of Video-Shot-Change Detection Methods, in *IEEE Transactions on Circuits and Systems for Video Technology*, Vol. 10, No. 1, 2000.

[19] S.Pfeiffer, S.Fischer, and W.Effelsberg, Automatic Audio Content Analysis, *in Proc. ACM Multimedia 96*, pp. 21-30, 1996.

[20] R.C. Nelson, Finding Line Segments by Stick Growing, *IEEE Transactions on PAMI*, 16(5):519-523, May 1994.

[21] S.S.Intille and A.F.Bobick, Recognizing Planned, Multi-person Action, in *Computer Vision and Image Understanding*, (1077-3142) 81(3):414-445, March 2001.

[22] P.Martin-Granel, M.Roach, and J.Mason, Camera Motion Extraction Using Correlation for Motion-Based Video Classification, in *Proc. of IWFV 2001*, Capri(I), 2001.

[23] S. Santini and R. Jain, Integrated Browsing and Querying for Image Databases, *IEEE Multimedia*, July-Sept. 2000.

[24] S.Choi, Y.Seo, H.Kim, K.-S.Hong, Where Are the Ball and Players? : Soccer Game Analysis with Color-based Tracking and Image Mosaick, in *Proc. of Int'l Conf. Image Analysis and Processing (ICIAP'97)*, 1997.

[25] V.Tovinkere and R.J.Qian, Detecting Semantic Events in Soccer Games: Towards a Complete Solution, in *Proc. of Int'l Conf. on Multimedia and Expo (ICME 2001)*, pp. 1040–1043, 2001.

[26] D.Yow, B.-L.Yeo, M.Yeung, and B.Liu, Analysis and Presentation of Soccer Highlights from Digital Video, in *Proc. of 2nd Asian Conf. on Computer Vision (ACCV'95)*, 1995.

[27] G. Sudhir, J.C.M. Lee, and A.K. Jain, Automatic Classification of Tennis Video for High-level Content-based Retrieval, in *Proc. of the Int'l Workshop on Content-Based Access of Image and Video Databases (CAIVD '98)*, 1998.

[28] S.Nepal, U.Srinivasan, G.Reynolds, Automatic Detection of 'Goal' Segments in Basketball Videos, in *Proc. of ACM Multimedia*, pp. 261-269, 2001.

[29] D.D.Saur, Y.-P.Tan, S.R.Kulkami, and P.J.Ramadge, Automatic Analysis and Annotation of Baskctball Vidco, *Storage and Retrieval for Image and Video Databases* V, pp. 176-187, 1997

[30] Y.Rui, A.Gupta, and A.Acero, Automatically Extracting Highlights for TV Baseball Programs, in *Proc. of ACM Multimedia*, 2000

[31] Y. Zhong, H. Zangh, and A. K. Jain, Automatic Caption Localization in Compressed Video, *IEEE Trans. on Pattern Analysis and Machine Intelligence* 22(4):385-392, 2000.

[32] G. Baldi, C. Colombo and A. Del Bimbo, A Compact and Retrieval-oriented Video Representation Using Mosaics, in *Proc. 3rd Int'l Conference on Visual Information Systems VISual99*, pages 171-178, 1999.

[33] T.-J. Cham and R. Cipolla. A Statistical Framework for Long-range Feature Matching in Uncalibrated Image Mosaicing, in *Proc. Int'l Conf. on Computer Vision and Pattern Recognition CVPR'98*, pages 442-447, 1998.

[34] H.S. Sawhney and S. Ayer, Compact Representations of Videos Though Dominant and Multiple Motion Estimation, *IEEE Transactions on Pattern Analysis and Machine Intelligence*, 18:814–830, 1996.

[35] L.S. Shapiro, H. Wang, and J.M. Brady. A Matching and Tracking Strategy for Independently Moving, Non-rigid Object, in *Proc. British Machine Vision Conference*, pages 306-315, 1992.

[36] H.D. Wactlar, A.G. Hauptmann, and M.J. Witbrock, Informedia: News–On–Demand Experiments in Speech Recognition, in *ARPA Speech Recognition Workshop*, 1996.

6

A GENERIC EVENT MODEL AND SPORTS VIDEO PROCESSING FOR SUMMARIZATION AND MODEL-BASED SEARCH

Ahmet Ekin and A. Murat Tekalp

Department of Electrical and Computer Engineering
University of Rochester, Rochester, NY, 14627 USA
`{ekin,tekalp}@ece.rochester.edu`

1. INTRODUCTION

In the last decade, several technological developments enabled the creation, compression, and storage of large amounts of digital information, such as video, images, and audio. With increasing usage of Internet and wireless communications, ubiquitous consumption of such information poses several problems. We address two of those: efficient content-based indexing (description) of video information for fast search, and summarization of video in order to enable delivery of the essence of the content over low-bitrate channels. The description and summarization problems have also been the focus of recently completed ISO MPEG-7 standard [1], formally Multimedia Content Description Interface, which standardizes a set of descriptors and description schemes. However, MPEG-7 does not normatively specify media processing techniques to extract these descriptors or to summarize the content, which are the main goals of this chapter. We also propose a generic integrated semantic-syntactic event model for search applications, which we believe is more powerful than the MPEG-7 Semantic Description Scheme.

Although the proposed semantic-syntactic event model is generic, its automatic instantiation for generic video and automatic generation of summaries that capture the essence for generic video are not simple. Hence, for video analysis, we focus on *sports video;* in particular *soccer* video, since sports video appeals to large audiences and its distribution over various networks should contribute to quick adoption and widespread usage of multimedia services worldwide. The model should provide a mapping between the *low-level video features* and *high-level events and objects*, which can be met for certain objects and events in the domain of sports video, and the processing for summary generation should be *automatic*, and in *real,* or *near real-time.*

Sports video is consumed in various scenarios, where network bandwidth and processing time determine the depth of the analysis and description. To this effect, we classify four types of users:

- **TV user:** The bandwidth is not a concern for the TV user. However, the user may not afford to watch a complete game due to either being away from home or having another game on TV at the same time. A user with software-enhanced set-top box and a personal digital recorder (PDR) can record a customized summary of a game. Furthermore, TV users with web access, which is one of the applications addressed by TV Anytime [2], may initiate searches on remote databases and retrieve customized summaries of past games through their TVs.

- **Mobile user:** The primary concern for the mobile user is insufficient bandwidth. With the advent of the new 3G wireless standards, mobile users will have faster access. However, live streaming of a complete game may still be impractical or unaffordable. Instead, the user may prefer receiving summaries of essential events in the game in near real-time. The value of sports video drops significantly after a relatively short period of time [3]; hence summarization of sports video in near real-time is important. In this case, the service provider should perform the video summarization.

- **Web user:** The web user may or may not share the same bandwidth concerns with the mobile user. A web user with a low bitrate connection is similar to the mobile user above. He/she may receive summaries of essential events that are computed on the server side. On the other hand, a web user with a high bitrate connection is similar to the TV user category. He/she may initiate searches on remote databases and retrieve either complete games or summaries of past games.

- **Professional user:** Professional users include managers, players, and sports analysts. Their motive is to extract team and player statistics for developing game plans, or to assess player performance, or scouting. These users are interested in performing searches on past games based on low-level motion features, such as object trajectories or semantic labels. The time constraint for processing is not very critical, but the accuracy in descriptor extraction is important; hence, semi-automatic video analysis algorithms are more applicable.

The next section briefly explains related works in video modeling, description, and analysis. We present a generic semantic-syntactic model to describe sports video for search applications in Section 3. In Section 4, we introduce a sports video analysis framework for video summarization and instantiation of the proposed model. The summaries can be low-level summaries computed on the server (mobile user) or customized summaries computed at the user side (TV user). The model instantiation enables search applications (web user or professional user). Finally, in Section 5, we describe a graph-based querying framework for video search and browsing with the help of the proposed model.

2.   BACKGROUND

This section presents the prior art in semantic video models, description, and analysis.

2.1   SEMANTIC VIDEO MODELING

Semantic video models are those that capture the structure and semantics of video programs. We can classify them as textual models, which employ only keywords and structured annotations to represent semantics, and integrated models, which employ both textual and content-based low-level features in order to perform mixed-level queries. A mixed-level query is one that includes both high-level semantic concepts and low-level content-based features, e.g.,, "Object 1 is on the left of Object 2 and participates in Event A," or "Object 1 participates in Event A and follows trajectory T."

Among the textual models, Oomoto *et al.* [4] have designed an object-based model for their video database system, OVID. A video object in OVID refers to a meaningful scene in terms of an object identifier, an interval, and a collection of attribute-value pairs. Hjelsvold and Midstraum [5] have developed a generic video model that captures video structure at various levels to enable specification of the video structure, the annotations, and the sharing and reuse in one model. The thematic indexing is achieved by annotations defined for video segments, and by specific annotation entities corresponding to persons, locations, and events. Adali *et al.* [6] introduce AVIS with a formal video model in terms of interesting video objects. A video object in AVIS refers to a semantic entity that attracts attention in the scene. The model includes events, as the instantiation of activity types, and the roles of objects in the events. Al Safadi and Getta [7] have designed a semantic video model to express various human interpretations. Their conceptual model constitutes semantic units, description of semantic units, association between semantic units, and abstraction mechanisms over semantic units. These textual models, in general, do not include low-level features; for example, they lack object motion modeling.

In contrast, Hacid et al. [8] have extended a textual database model to include low-level features. Their model consists of two layers: 1) Feature and content (audiovisual) layer that contains low-level features, 2) Semantic layer that provides conceptual information. The first layer is based on the information obtained from QBIC [9]. This work is a good starting point for a system to handle mixed-level queries, but their framework does not support object-based motion description and mid-level semantics for motion.

As a sign of a certain level of maturity reached in the field of content-based retrieval, the ISO MPEG-7 standard (formally Multimedia Content Description Interface) provides normative tools to describe multimedia content by defining a normative set of descriptors (D) and description schemes (DS). One of these DSs, the SemanticDS, introduces a generic semantic model to enable semantic retrieval, in terms of objects, events, places, and semantic relations [1]. MPEG-7 also provides low-level descriptors, such as color, texture, shape, and motion, under a separate SegmentDS. Thus, in order to perform mixed-level queries with spatio-temporal relations between objects, one needs to instantiate both a SemanticDS and a SegmentDS with two separate graph structures. This unduly increases representation complexity, resulting in inefficiency in query resolution and problems due to the independent manipulations of a single DS. Furthermore, there is no context-dependent classification of object attributes in the MPEG-7 SemanticDS. For example, if an object appears in multiple events, all attributes (relations) of the object related to all events are listed within the same object entity. This aggravates the inefficiency problem.

We propose a new generic integrated semantic-syntactic model to address the above deficiencies in Section 3.

2.2   AUTOMATIC SPORTS VIDEO ANALYSIS

Semantic analysis of sports video involves use of *cinematic features*, to detect shot boundaries, shot types, and replays, as well as *object-based* features to detect and track players and referee, and detect certain game specific events such as goals and plays around the penalty box in soccer videos.

The earliest works on sports video processing have used object color and texture features to generate highlights [10] and to parse TV soccer programs [11]. Object motion trajectories and interactions are used for football play classification [12] and for soccer event detection [13]. Both [12] and [13], however, rely on pre-extracted accurate object trajectories, which are obtained manually in [12]; hence, they are not practical for real-time applications. LucentVision [14] and ESPN K-Zone [15] track only specific objects for tennis and baseball, respectively. The former analyzes trajectory statistics of two tennis players and the ball. The latter tracks the ball during pitches to show, as replays, if the strike and ball decisions are correct. The *real-time* tracking in both systems is achieved by extensive use of a priori knowledge about the system setup, such as camera locations and their coverage, which limits their application. Cinematic descriptors are also commonly employed. The plays and breaks in soccer games are detected by frame view types in [16]. Li and Sezan summarize football video by play/break and slow-motion replay detection using both cinematic and object descriptors [17]. Scene cuts and camera motion parameters are used for soccer event detection in [18] where usage of only few cinematic features prevents reliable detection of multiple events. A mixture of cinematic and object descriptors is employed in [19]. Motion activity features are proposed for golf event detection [20]. *Text* information from closed captions and visual features are integrated in [21] for event-based football video indexing. *Audio* features, alone, are proposed to detect hits and generate baseball highlights [22].

In this chapter, we present a framework to process soccer video using both *cinematic* and *object-based* features. The output of this processing can be used to automatically generate summaries that capture the essential semantics of the game, or to instantiate the proposed new model for mixed-level (semantic and low-level) search applications.

3.   A GENERIC SEMANTIC-SYNTACTIC EVENT MODEL FOR SEARCH

This section proposes an integrated semantic-syntactic model, i.e., employing both high- and low-level features, that allows efficient description of video events and the motion of objects participating in these events. The model is an extension of the well-known entity-relationship (ER) database models [23] with object-oriented concepts. The main entities in the model are *events*, *objects* that participate in these events, and *actors* that describe the roles of the objects in these events (by means of object-event relationships). The actor entity allows separation of event independent attributes of objects, such as the name and age of a soccer player, from event dependent roles throughout a video. For example, a player may be "scorer" in one event and "assist-maker" in another, and so on. These event-dependent roles of an object are grouped together according to the underlying event, and constitute separate actor entities, which all refer to the same player object. Low-level object motion and reactions are also event-specific; hence, they are described as attributes of "actor" entities and actor-actor relations, respectively. To describe low-level features, we define a "segment" descriptor which may contain multiple object motion units (EMUs) and reaction units (ERUs), for the description of object motion information (e.g., trajectories) and interactions (e.g., spatio-temporal relations), respectively. In addition to segment level relations, we also define semantic relations between the entities in the model. Figure 6.1 shows the graphical representation of the proposed integrated model, where a rectangle refers to an entity, a diamond shows a relationship with the name of the relationship written on top and arrow pointing the direction of the relationship, and an oval represents an attribute of an entity or a relationship. For example, an event may have, among others, *causal, temporal,* and *aggregation (composedOf)* relationships with other events.

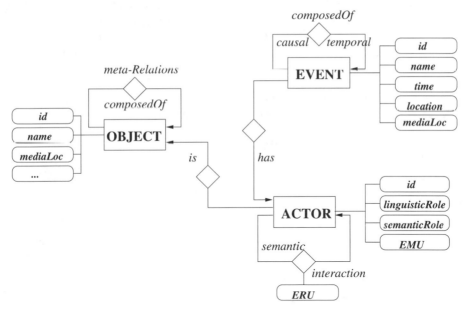

Figure 6.1 The graphical representation of the model.

3.1 MODEL DEFINITION

In the following, we provide formal definitions of all model entities and relationships:

1) **Video Event:** Video events are composed of semantically meaningful object *actions,* e.g., running, and *interactions* among objects, e.g., passing. In order to describe complex events, an event may be considered as the composition of several sub-events, which can be classified as actions and interactions. Actions generally refer to a semantically meaningful motion of a single object; whereas interactions take place among multiple objects. Events and sub-events can be associated with semantic time and location. Formally, a video event is described by *{ID, name, time, location, L}* where *ID* is a

unique id, **L** is one or more media locators of the event life-span, and name, location, and time of the event.

2) **Video Object:** A video object refers to a semantic spatio-temporal entity. Objects have *event-independent,* e.g., name, and *event-dependent* attributes, e.g., low-level object features. Only event-independent attributes are used to describe an object entity. The event-dependent roles of an object are stored in actor entities (defined later). In our model, we allow generalizations and specializations of video objects by the class hierarchy of objects. Formally, a video object can be defined as *{ID, Attr:Val, L}* where *Attr:Val* pairs are multiple event-independent attribute-value pairs and *ID* and *L* are defined in the same way as for events.

3) **Actor:** Video objects play roles in events; hence, they are the actors within events. As such, they assume event-specific semantic and low-level attributes that are stored in an actor entity. That is, the actor entity enables grouping of object roles in the context of a given event. At the semantic level, a video object carries a linguistic role and a semantic role. We adopt the linguistic roles that are classified by SemanticDS of MPEG-7 [24], such as *agentOf* and *patientOf.* Semantic roles also vary with context, for example, a player may assume a *scorer* role in one goal event and an *assist-maker* role in another. At the low-level, we describe object motion by elementary motion units (EMUs) as segment level actor attributes. Formally, an actor entity is described by *{ID, linguistic role, semantic role, E}* where *E* is the list of motion units of a specific object in a single event.

4) **Video Segment:** In general, motion of objects and their interactions within an event may be too complex to describe by a single descriptor at the low-level. Thus, we further subdivide the object motion and interactions into mid-level elementary motion units (EMU) and elementary reaction units (ERU), respectively:

   a) **Elementary Motion Units (EMUs):** Life-span of video objects can be segmented into temporal units, within which their motion is coherent and can be described by a single descriptor. That is, in an event, the motion of an object creates multiple EMUs, which are stored in actor entities. Each EMU is represented with a single motion descriptor, which can be a trajectory descriptor or a parametric motion descriptor [25].

   b) **Elementary Reaction Units (ERUs):** ERUs are spatio-temporal units that correspond to the interactions between two objects. The interactions may be temporal, spatial, and motion reactions. We consider "coexistence" of two objects within an interval and describe their temporal relations by Allen's interval algebra [26]: *equals, before, meets, overlaps, starts, contains, finishes,* and their inverses. Spatial object reactions are divided into two classes: *Directional* and *Topological* relations. Directional relations include *north, south, west, east* as strict directional relations, *north-east, north-west, south-east, south-west* as mixed-directional relations, and *above, below, top, left, right, in front of, behind, near, far* as positional relations. The topological relations include *equal, inside, disjoint, touch, overlap,* and *cover.* Most of the above relations are due to Li et al. [27] and their revision of [28]. Finally, motion reactions include *approach, diverge,* and *stationary.* Each relation can be extended by the application specific attributes, such as velocity and acceleration. As shown in Figure 6.1, ERUs are stored as attributes of actor-actor relationships.

5) **Relations:** We define relations between various entities:

a) **Event-Event Relations:** An event may be composed of other events, called sub-events. Causality relationship may also exist between events; for instance, an object action may cause another action or interaction. Furthermore, the users may also prefer to search the video events from temporal aspect using temporal relations of the events. Therefore, we consider event-event relations in three aspects: *composedOf, causal,* and *temporal. ComposedOf* relation type assumes a single value *composedOf* while the *causality* relation may be one of *resultingIn* and *resultingFrom* values. The *temporal* relations follow Allen's temporal algebra [26].

b) **Object-Object Relations:** Similar to events, an object may be *composedOf* other objects, e.g., a player is a *memberOf* a team. Object meta-relations are defined as those relations that are not visually observable from video content.

c) **Actor-Actor Relations:** A single actor entity contains only one object; therefore, actor-actor relations are defined to keep semantic and low-level object-object relations in the event life-span, such as ERUs.

6) **Semantic Time and Location:** Semantic time and location refer to the world time and location information, respectively. Both may be specified by their widely known name, such as "World Cup" and "Yankee Stadium," or by their calendar and postal attributes, respectively.

7) **Media Locator:** Video objects, events, EMUs, and ERUs contain a set of media locators to indicate their temporal duration and media file location. The media files may be video clips of the same event recorded by different camera settings, still images as keyframes or they may be in other formats, such as document and audio. A media locator contains *{t[start:end], V_i}* corresponding to a temporal interval and *N*, *i=1:N,* media files.

The proposed model is generic in the sense that it can represent any video event, although the examples given in the next subsection are taken from a soccer video. The main differences between our proposed model and the corresponding MPEG-7 Description Schemes (Semantic and Segment DS) are our ability to represent both semantic and low-level descriptors in a single graph, and our representation of event-dependent object attributes in separate actor entities, both of which contribute to increased search efficiency.

3.2   MODEL EXAMPLES

In this section, we describe a video clip of a soccer goal by using the proposed model entities (graph vertices) and relationships (edges). The notation used in this section and in Section 5 relates to the notation in Figure 6.1 as follows: i) An event entity description is labeled as "*name:* **EVENT,**" such as "*Free Kick:* **EVENT**." ii) An actor is shown as "*semantic role:* **ACTOR,**" such as "*Kicker:* **ACTOR.**" iii) A video object is specified either by "*class:* **OBJECT,**" if the described object belongs to a specialized object class, e.g., player, or by "*name:*

OBJECT" if the object has only standard attributes, such as name and media locators.

The *goal* event in the example clip is composed of three sub-events: a *free kick,* a *header,* and a *score* event. We first present the description of the *free kick* and instantiation of its low-level descriptors. Three keyframes of the event are shown in Figure 6.2. In Figure 6.3, the conceptual description of *free kick* sub-event is given, where a *free kick* event *has* two actors with roles, *kicker* and *kicked object.* These actors *interact* during the event and form a set of segment level relations as ERUs. Each actor carries event-specific linguistic roles and low-level object motion attributes as a set of EMUs. The kicker *is* a *player,* and the kicked object is a *ball.* The event-independent attributes of objects, such as *name* and *number* of the player, are stored in video object vertices. Figure 6.3 is the conceptual description of *free kick* event meaning that it does not refer to a specific media segment; therefore, many attributes are not instantiated (shown as "...") except those that stay the same in every *free kick* event, such as *kicker* is always *agentOf* the event.

Figure 6.2 The keyframes of an example free kick event.

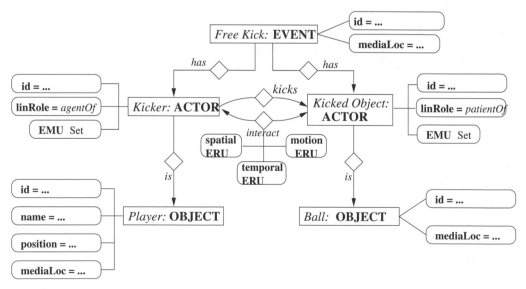

Figure 6.3 The description of the free kick event in the example clip.

In Figure 6.3, low-level spatio-temporal attributes of the player and the ball are represented as EMU and ERU attributes of actor vertices and actor-actor edges, respectively. The detailed description of object motion and object reaction segments for the example scene in Figure 6.2 is presented in Figure 6.4. For simplicity, we assume that the *free kick* sub-event starts at frame #0 of the corresponding video. The player has a media life-span from frame #0 to frame #25 where its motion is described by two EMUs, while the ball appears in the whole event life-span, and its motion attributes create three EMUs. Segment relations between the two objects are valid only in the interval of their coexistence. Temporal interval of player media life-span *starts* the temporal interval of the ball, meaning that their life-span intervals start together, but the player media life-span ends earlier. Motion reactions of the two objects are "approach" before the time point of "kick," and "diverge" after it. The detail of motion descriptions in the model can be adjusted to the requirements by adding attributes to the motion ERU. For instance, the stationarity of the ball during the "approach" relationship is described by the zero velocity of the ball. Topological and directional spatial object relations are also shown in Figure 6.4. The bounding boxes of the player and the ball are *disjoint* starting at frame #0 (Figure 6.2 (left frame)), they *touch* each other from frame#12 to frame#15 (Figure 6.2 (middle frame)) and they are *disjoint* after frame #16. Although two objects always have topological relationships, they may not have a directional relationship for every time instant. That situation is illustrated for the time interval (frame #12, frame #15).

In our second example, we present the complete description of the example video clip that starts with the *free kick* sub-event in the previous example. There are two other sub-events, *header* and *score* (defined as the entering of the ball to the goal), that follow the *free kick* event. Since all of the sub-events have descriptions similar to Figure 6.3 and Figure 6.4, we do not explicitly describe *header* and *score* events. In Figure 6.5, the temporal relationships among the three sub-events are described by using *before* relationship in Allen's interval algebra. Then, the *goal* event is composed of *free kick, header,* and *score*. Three players act in the composite goal event as the *assist maker*, the *scorer*, and the *goalie*.

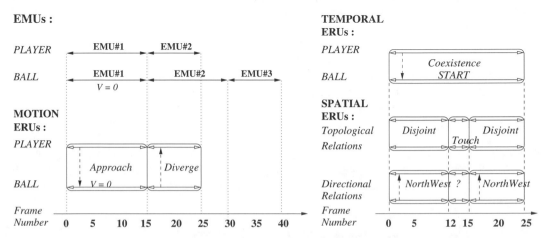

Figure 6.4 The low-level descriptions in the example free kick event

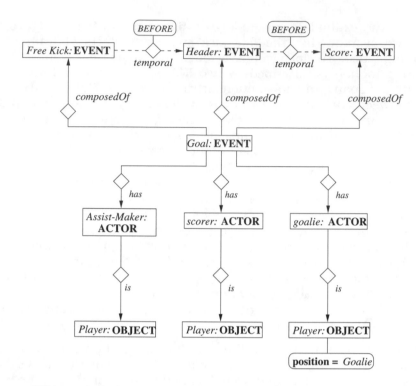

Figure 6.5 The description of the composite goal event

4. SPORTS VIDEO ANALYSIS FOR SUMMARIZATION AND MODEL INSTANTIATION

In this section, we present a sports video analysis framework for video summarization and model instantiation. We discuss the application of the algorithms for soccer video, but the introduced framework is not limited to soccer, and can be extended to other sports. In Figure 6.6, the flowchart of the proposed summarization and analysis framework is shown for soccer video. In the following, we first introduce algorithms using *cinematic* features, such as shot boundary detection, shot classification, and slow-motion replay detection, for low-level processing of soccer video. The output of these algorithms serves for two purposes: 1) Generation of video summaries defined solely by those features, e.g., summaries of all slow-motion replays. 2) Detection of interesting segments for higher level video processing, such as for event and object detection (shown by the segment selection box in Figure 6.6). In Section 4.2, we present automatic algorithms for the detection of soccer events and in Section 4.3, we explain object detection and tracking algorithms for object motion descriptor extraction. Then, in Section 4.4, the generation of summaries is discussed. Finally, we elaborate on the complete instantiation of the model for querying by web and professional users.

4.1 LOW-LEVEL PROCESSING OF SPORTS VIDEO

As mentioned in Section 2.2, semantic analysis of sports video generally involves use of *cinematic* and *object-based* features. Cinematic features refer to those that result from common video composition and production rules, such as shot types and replays. Objects are described by their spatial, e.g., color, texture, and shape, and spatio-temporal features, such as object motions and interactions as explained in Section 3. Object-based features enable *high-level domain analysis*, but their extraction may be *computationally costly* for real-time implementation. Cinematic features, on the other hand, offer *a good trade-off* between the computational requirements and the resulting semantics. Therefore, we first extract *cinematic* features, e.g., shot boundaries, shot classes, and slow-motion replay features, for *automatic real-time* summarization and selection of segments that are employed by higher level event and object detection algorithms.

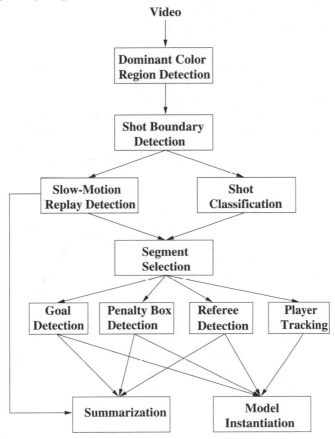

Figure 6.6 The flowchart of the proposed summarization and model instantiation framework for soccer video

In the proposed framework, the first low-level operation is the detection of the dominant color region, i.e., grass region, in each frame. Based on the difference in grass colored pixel ratio between two frames and the color histogram similarity difference, both cut-type shot boundaries and gradual transitions, such as wipes and dissolves, are detected. Then, each shot is checked for the existence of slow-motion replay segment by the algorithm proposed in [29], since replays in sports broadcasts are excellent locators of semantically important segments for high-level video processing. At the same time, the class of the corresponding shot is assigned as one of the following three classes, defined below: 1) long shot, 2) in-field medium shot, 3) close-up or out-of-field shot.

- **Long Shot:** A long shot displays the global view of the field as shown in Figure 6.7 (a); hence, a long shot serves for accurate localization of the events on the field.

- **In-Field Medium Shot:** A medium shot, where a whole human body is usually visible as in Figure 6.7 (b), is a zoomed-in view of a specific part of the field.

- **Close-Up or Out-of-field Shot:** A close-up shot usually shows above-waist view of one person as in Figure 6.7 (c). The audience (Figure 6.7 (d)), coach, and other shots are denoted as out-of-field shots. We analyze both out-of-field and close-up shots in the same category due to their similar semantic meaning.

Shot classes are useful in several aspects. They can be used for segmentation of a soccer video into *plays* and *breaks* [16]. In general, long shots correspond to plays, while close-up and out-of-field shots indicate breaks. Although the occurrence of a single isolated medium shot between long shots corresponds to a play, a group of nearby medium shots usually indicates a break in the game. In addition to play/break segmentation, each shot class has specific features that can be employed for high-level analysis [30]. For example, long shots are appropriate for the localization of the action on the field by field line detection and registration of the current frame onto a standard field. In Figure 6.6, segment selection algorithm uses both shot classes and the existence of slow-motion replays to select interesting segments for higher level processing.

Classification of a shot into one of the above three classes is based on spatial features. Therefore, the class of a shot can be determined from a single keyframe or from a set of keyframes selected according to certain criteria. In order to find the frame view, frame grass colored pixel ratio, **G,** is computed. Intuitively, a low **G** value in a frame corresponds to close-up or out-of-field view, while high **G** value indicates that the frame is of long view type, and in between, medium view is selected. By using only grass colored pixel ratio, medium shots with high **G** value will be mislabeled as long shots. The error rate due to this approach depends on the specific broadcasting style and it usually reaches intolerable levels for the employment of higher level algorithms. We use a compute-easy, yet very efficient, cinematographic measure *for the frames with a high **G** value.* We define regions by using *Golden Section* spatial composition rule [31], which suggests dividing up the screen in 3:5:3 proportions in both directions, and positioning the main subjects on the intersection of these lines. We have revised this rule for soccer video, and divide the *grass region box* instead of the whole frame. *Grass region box* can be defined as the minimum bounding rectangle (MBR), or a scaled version of it, of grass colored pixels. In Figure 6.8, the examples of the regions obtained by *Golden Section* rule are displayed over medium and long views. In the regions, R_1, R_2, and R_3, we define two features to measure the distribution of the grass colored pixels:

- G_{R2} : The grass colored pixel ratio in the second region

- R_{diff}: The average of the sum of the absolute grass color pixel differences between R_1 and R_2, and between R_2 and R_3:

$$R_{diff} = \frac{1}{2}\left\{ \left| G_{R2} - G_{R1} \right| + \left| G_{R2} - G_{R3} \right| \right\}$$ (6.1)

Then, we employ a Bayesian classifier [32] using the above two features for shot classification. The flowchart of the complete shot classification algorithm is shown in Figure 6.9, where the grass colored pixel ratio, G, is compared with a set of thresholds for either shot class decision or the use of Golden Section composition rule. When G has a high value, shot class decision is given by using grass distribution features, G_{R2} and R_{diff}.

4.2   EVENT DETECTION

In the proposed framework, all goal events and the events in and around penalty boxes are detected. Goal events are detected in *real-time* by using only cinematic features. We can further classify each segment as those consisting of events in and around the penalty box. These events may be free kicks, saves, penalties, and so on.

4.2.1   Goal Detection

A goal is scored when the whole of the ball passes over the goal line, between the goal posts and under the crossbar [33]. Unfortunately, it is difficult to verify these conditions automatically and reliably by video processing algorithms. However, the occurrence of a goal is generally followed by a special pattern of cinematic features, which is what we exploit in the proposed goal detection algorithm. A goal event leads to a break in the game. During this break, the producers convey the emotions on the field to the TV audience and show one or more replays for a better visual experience. The emotions are captured by one or more close-up views of the actors of the goal event, such as the scorer and the goalie, and by out-of-field shots of the audience celebrating the goal. For a better visual experience, several slow-motion replays of the goal event from different camera positions are shown. Then, the restart of the game is usually captured by a long shot. Between the long shot resulting in the goal event and the long shot that shows the restart of the game, we define *a cinematic template* that should satisfy the following requirements:

- *Duration of the break:* A break due to a goal lasts no less than 30 and no more than 120 seconds.

- *The occurrence of **at least one** close-up/out-of-field shot:* This shot may either be a close-up of a player or out-of-field view of the audience.

- *The existence of **at least one** slow-motion replay shot:* The goal play is always replayed one or more times.

- *The relative position of the replay shot(s):* The replay shot(s) follow the close-up/out of field shot(s).

In Figure 6.10, the instantiation of the cinematic goal template is given for the first goal in Spain sequence of MPEG-7 test set. The break due to this goal lasts 53 seconds, and three slow-motion replay shots are broadcast during this break. The segment selection for goal event templates starts by detection of the slow-motion replay shots. For every slow-motion replay shot, we find the long shots that define the start and the end of the corresponding break. These long shots must indicate a play that is determined by a simple duration constraint, i.e., long shots of short duration are discarded as breaks. Finally, the conditions of the template are verified to detect goals.

(a)   (b)   (c)   (d)

Figure 6.7 The shot classes in soccer: (a) Long shot, (b) in-field medium shot, (c) close-up shot, and (d) out-of-field shot

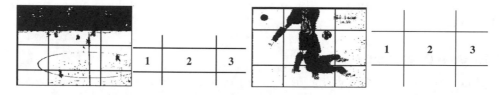

Figure 6.8 Grass/non-grass segmented long and medium views and the regions determined by Golden Section spatial composition rule

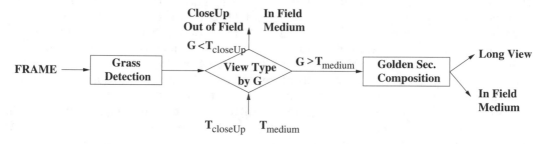

Figure 6.9 The flowchart of the shot classification algorithm

Figure 6.10 The occurrence of a goal and its break: (left to right) goal play as a long shot, close-up of the scorer, out-of-field view of the fans (middle), 3rd slow-motion replay shot, the restart of the game as a long shot

4.2.2   Detection of Events in and around the Penalty Box

The events occurring in and around penalty boxes, such as saves, shots wide, shots on goals, penalties, free kicks, and so on, are important in soccer. To classify a summary segment as consisting of such events, penalty boxes are detected. As explained in Section 4.1, field lines *in a long view* can be used to localize the view and/or register the current frame on the standard field model. In this section, we reduce the penalty box detection problem to the search for three parallel lines. In Figure 6.11, a model of the whole soccer field is shown, and *three parallel field lines*, shown in bold on the right, become visible when the action occurs around one of the penalty boxes.

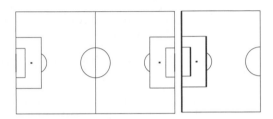

Figure 6.11 Soccer field model (left) and the highlighted three parallel lines of a penalty box

In order to detect three lines, we use the grass detection result in Section 4.1. The edge response of non-grass pixels are used to separate line pixels from other non-grass pixels, where edge response of a pixel is computed by 3x3 Laplacian mask [34]. The pixels with the highest edge response, the threshold of which is *automatically* determined from the histogram of the gradient magnitudes, are defined as line pixels. Then, three parallel lines are detected by Hough transform that employs *size*, *distance*, and *parallelism* constraints. As shown in Figure 6.11, the line in the middle is the *shortest* line and it has a *shorter* distance to the goal line (outer line) than to the penalty line (inner line). A result of the algorithm is shown in Figure 6.12 as an example.

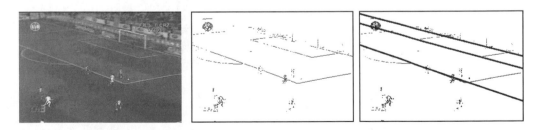

Figure 6.12 Penalty Box detection by three parallel lines

4.3   OBJECT DETECTION AND TRACKING

In this section, we first describe a referee detection and tracking algorithm, since the referee is a significant object in sports video, and the existence of referee in a shot may indicate the presence of an event, such as red/yellow cards and penalties. Then, we present a player tracking algorithm that uses feature point correspondences between frame pairs.

4.3.1 Referee Detection and Tracking

Referees in soccer games wear distinguishable colored uniforms from those of the two teams on the field. Therefore, a variant of our dominant color region detection algorithm is used to detect referee regions. We assume that there is, if any, a single referee in *a medium* or *close-up/out-of-field shot* (we do not search for a referee in a long shot). Then, the horizontal and the vertical projections of the feature pixels can be employed to accurately locate the referee region. The peak of the horizontal and the vertical projections and the spread around the peaks are used to compute the rectangle parameters surrounding the referee region, hereinafter "MBR$_{ref}$." MBR$_{ref}$ coordinates are defined to be the first projection coordinates at both sides of the peak index without enough pixels, which is assumed to be 20% of the peak projection. In Figure 6.13, an example frame, the referee pixels in that frame, the horizontal and vertical projections of the referee region, and the resulting MBR$_{ref}$ are shown.

The decision about the existence of the referee in the current frame is based on the following size-invariant *shape* descriptors:

- *The ratio of the area of the MBR$_{ref}$ to the frame area:* A low value indicates that the current frame does not contain a referee.

- *MBR$_{ref}$ aspect ratio (width/height):* It determines if the MBR$_{ref}$ corresponds to a human region.

- *Feature pixel ratio in the MBR$_{ref}$:* This feature approximates the compactness of the MBR$_{ref}$; higher compactness values are favored.

- *The ratio of the number of feature pixels in the MBR$_{ref}$ to that of the outside:* It measures the correctness of the single referee assumption. When this ratio is low, the single referee assumption does not hold, and the frame is discarded.

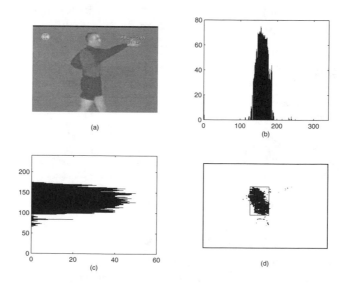

Figure 6.13 Referee Detection by horizontal and vertical projections

Tracking of referee is achieved by region correspondence; that is, the referee template, which is found by the referee detection algorithm, is tracked in the other frames. In Figure 6.14, the output of the tracker is shown for a medium shot of MPEG-7 Spain sequence.

4.3.2 Player Tracking

The players are tracked in the long shot segments that precede slow-motion replay shots of interesting events. These long shots consist of the normal-motion action of the corresponding replay shots. For example, for the replay shots of the goal event in Figure 6.10, we find the long shot whose keyframe is shown as the leftmost frame in Figure 6.10. The aim of *tracking objects* and *registering each frame* onto a standard field is to extract EMU and ERU descriptors by the algorithms in [25] and [35].

Figure 6.14 Referee detection and tracking

The tracking algorithm takes the object position and **object ID** (or **name**), which is a link to high-level object attributes, as its input in the first frame of a long shot. Then, a number of feature points are selected in the object bounding box by Kanade-Lucas-Tomasi (KLT) feature tracker [36], [37]. The motion of the feature points from the previous to the current frame is used to predict the object bounding box location in the current frame. To update the bounding box location, a region-based approach is used, since the use of spatial features allows for the erratic movements of the players. The object regions are found by the dominant color region detection algorithm in Figure 6.6. The object regions extracted by the classifier are fed into the connected component labeling algorithm, and the bounding box for the largest sized region is extracted. The flowchart of the algorithm is given in Figure 6.15.

The bounding box location is corrected by integrating the information about the region bounding box, the initial estimate of the bounding box, and the motion history of the object. If the occlusion is severe, corresponding to a very small or a very large region box, or the loss of many feature points, the region and the estimated bounding boxes will not be reliable. In such cases, the system use the weighted average of the global motion compensated object motion in a set of frames [35]. An example output from the tracking algorithm is presented in Figure 6.16.

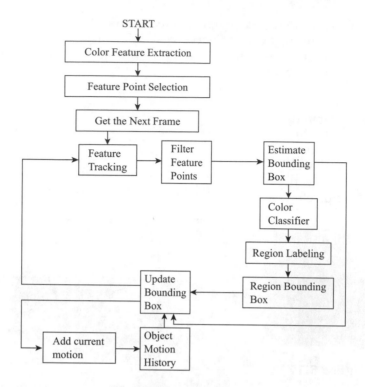

Figure 6.15 The flowchart of the tracking algorithm

Figure 6.16 Example tracking of a player in Spain sequence

The registration of each frame of a long shot involves field line detection. Low-level image processing operations, such as color segmentation, edge detection, and thinning, are applied to each frame before Hough transform. The integration of prior knowledge about field line locations as a set of constraints, such as the number of lines and parallelism, increases the accuracy of the line detector. Unlike the penalty box detection, both horizontal and vertical lines must be detected for accurate registration. Detected lines in several frames are shown in Figure 6.17. After field lines are detected, each frame is registered onto a standard field model, i.e., the model in Figure 6.11, by using any one of global motion models, such as affine and perspective [38]. In this process, a human operator is assumed to label the lines in the first frame of the shot.

4.4   VIDEO SUMMARIZATION

The proposed framework includes three types of summaries: 1) All slow-motion replay shots in a game, 2) all goals in a game, and 3) extensions of both with detected events. The first two types of summaries are based solely on cinematic features, and are generated in real-time; hence they are particularly convenient to TV, mobile, and web users with low bitrate connections. The last type of summary includes events that are detected by also object-based features, such as penalty box and referee detection results; and it is available in near real-time.

Figure 6.17 Line detection examples in Spain sequence.

Slow-motion summaries are generated by shot boundary, shot class, and slow-motion replay features, and consist of slow-motion shots. Depending on the requirements, they may also include all shots in a predefined time window around each replay, or, instead, they can include only the closest long shot before each replay in the summary, since the closest long shot is likely to include the corresponding action in normal motion. As explained in Section 4.2.1, goals are detected in a cinematic template. Therefore, *goal summaries* consist of the shots in the detected template, or in its customized version, for each goal. Finally, summaries extended by referee and penalty box features are generated by concatenating each slow-motion replay shot with the selected segments by the referee and penalty box detection algorithms. The segments are defined as close non-long shots around the corresponding replay for referee detection, and one or more closest long shots *before* the replay for penalty box detection.

4.5   MODEL INSTANTIATION

Model instantiation is necessary for the resolution of model-based queries formed by web and professional users. To instantiate the model, the attributes of the events, objects, and the relationships between them are needed. Some events, such as goals, plays in/around penalty box, and red/yellow cards, can be detected. These events and their actors, i.e., objects, are linked by actor entities, which include both low-level and high-level object roles. The proposed semi-automatic player tracking algorithm, where an annotator specifies the **name,** or **id,** of the player to be tracked, extracts the low-level player motion and interaction features as actor attributes and actor-actor relations, respectively. The high-level roles of the objects, such as *scorer* and *assist-maker,* are already available if the event has an abstract model, as shown in Figure 6.3; otherwise it can be specified by the annotator.

The interactive part of the instantiation process is related to the detection of some specific events, such as header and free kick, which, in the current state of art, is still an unsolved problem involving several disciplines, such as computer vision and artificial intelligence. Furthermore, accurate and reliable extraction of EMU and ERU descriptors involving a soccer ball is usually a very difficult, if not impossible, problem due to the small size of the object and its frequent occlusions with the other objects, which prevent the employment of vision-based tracking algorithms. However, this data may be available when a GPS chip is integrated to soccer balls. Therefore, for the time being, the interactive instantiation of these descriptors for the automatically generated summary segments by an annotator is a feasible solution.

5.   QUERY FORMATION AND RESOLUTION USING THE MODEL

The users may browse through the descriptions, set their preferences by model entities, and form complex queries that require matching a low-level criterion in one part of the scene and high-level constraints in another. We employ a single graph-based query formation and resolution framework for any of the above uses. In the framework, the queries are represented by graph patterns that can be formed by either editing an example description or the database scheme. In order to facilitate query formation by the user, we further define abstract models (model templates) as special graph patterns for certain events in a specific sports type. Then, the user can also form queries by editing an abstract event model from the model database. The relevant sections of the database can be retrieved by matching the query graph pattern with the graphs of the descriptions in the database. The similarity of the query graph pattern to each of the matching sub-graphs in the database is calculated by matching both high- and low-level attributes.

5.1   GRAPH-BASED QUERY FORMATION

The queries are represented by graph patterns, which are defined as instances of the database model with the exception of null or undefined values for some attributes [39]. Graph patterns can be obtained by editing the database scheme (or an example description) or by using abstract models, defined below. Editing the database scheme refers to modifying the model diagram in Figure 6.1 by copying, duplicating, identifying, and deleting some part of the scheme. Therefore, graph patterns conform to the database scheme, and syntactically incorrect queries cannot be formulated. In certain structured domains, such as sports, the number of popular queries may be limited, and model templates, called abstract models as special graph patterns, can be defined for specific events. An abstract model is the abstraction of a specific model instantiation from specific object, location, and time instances; for example, Figure 6.3 is the abstract model of *free kick* event. An abstract model conforms to the database scheme; therefore, the queries based on abstract models are also syntactically correct. A similar concept of abstraction has been included in the MPEG-7 standard, and it is called "formal abstraction" [24]. Browsing is a special case of querying with graph patterns where the query is limited to one or more specific vertices [40]. In Figure 6.18, the graph pattern refers to the query: "find all events where the ball object has a similar trajectory to the example trajectory and the event precedes header and score events." The queried event name is not important; hence, it is not specified (shown by asterisks). The low-level query constraint is specified as an EMU attribute of the actor entity. In addition to enabling the formation of a rich set of queries, the integration of low-level features in the queries removes the dependency to the annotations that may be subjective.

5.2   QUERY RESOLUTION BY GRAPH MATCHING

Queries are resolved by matching the query graph with the sub-graphs of the descriptions in the database. In our application, we use the following model-based constraints to reduce the search space in graph matching: 1) The type of vertices in the description is known to be an event, an object, or an actor with each type having distinguishing features from the others, and 2) The directed edges correspond to different types of relations with type-specific semantic-level and/or low-level attributes. A query as a graph pattern consists of a set of constraints that may be related to vertices, edges, or both. The proposed algorithm starts the search from the vertices and finds a set of matching vertices for each query vertex in each distinct database description. Then, starting with the most constrained query vertex, assumed to be inversely proportional to the number of matching vertices, edge constraints are compared. That is, the algorithm looks for the combinations of the resulting set of vertex matches for the edge constraints. The steps of the recursive search for graph matching are as follows:

Graph Matching Algorithm:
  – *For each query vertex, find the initial set of matching vertices using the query constraints*
  – *Rank the query vertices from the one having the least number of matches to the most*
  – *Check the edge constraints to find the combinations of the vertices that match the query graph. For this purpose, use a recursive search from the most constrained vertex to the least constrained one and return whenever an edge constraint fails*

We use the vertex and edge attributes defined below to find the isomorphism or similarity between two graphs:
  • **Vertex Matching:** When the graph element is a vertex, we have three choices: i) an event, ii) an object, or iii) an actor vertex. Therefore, the most distinguishing attribute for a vertex in finding a match is its type. Then, we evaluate the other specific attributes defined below for each vertex type:
    • **Event:** Event vertices are evaluated by the equivalence of the *name* attribute, such as *goal* in soccer and *fumble* in football.
    • **Actor:** The match between two actor vertices is found by comparing two high-level attributes: *linguistic role* and *semantic role*. Low-level descriptors for object motion can also be used for the same purpose.
    • **Object:** Object vertices are compared by the equivalence of the *object class* and the specified *attribute-value* pairs.
  • **Edge Matching:** The match between two edges is defined by the equivalence of the semantic and segment relations.

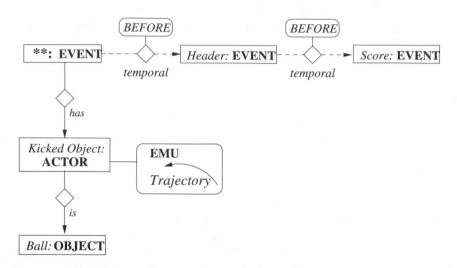

Figure 6.18 The query graph pattern for "find the events where the ball object has a similar trajectory to the example and the event precedes header and score events"

We compute two different cost values to rank the matching graphs:

- **Semantic Cost:** Semantic cost of a match is determined as the cost due to the mismatches in the graph structure, in particular, those of event structure. That is, the graphs that differ in temporal structure of events will be penalized by the cost of the mismatch. The semantic cost value is defined as the sum of the cost of insertions and deletions of event vertices and temporal event edges to match each query event vertex:

$$C_{sem} = \sum_{i=1}^{N} \left(C_{insertion}\left(V_i, E_i\right) + C_{deletion}\left(V_i, E_i\right) \right) \tag{6.2}$$

- **Syntactic Cost:** The syntactic cost is defined as the dissimilarity of two graphs due to their low-level features. As explained in Section 3, in the proposed model, low-level features are defined for actor vertices and actor-actor links as object motion parameters, such as object trajectories and spatio-temporal object interactions, respectively. The metrics to measure the dissimilarity (or similarity) of object motion descriptors are well defined and widely available in the literature. Binary decisions, as matching or not, for spatio-temporal relations are given by the equality of the corresponding reactions in the specified time interval.

ACKNOWLEDGMENT

The authors gratefully acknowledge Dr. Rajiv Mehrotra (Eastman Kodak Company) for his contributions to Sections 3 and 5. This work was supported by National Science Foundation under grant number IIS-9820721 and Eastman Kodak Company.

REFERENCES

[1]   ISO/IEC Committee Draft 15938-5 Information Technology - Multimedia Content Description Interface: Multimedia Description Schemes, ISO/IEC/JTC1/SC29/WG11/N4242, Oct. 2001.

[2]   The TV Anytime Forum: Call for Contributions on Metadata, Content Referencing, and Rights Management, TV014r3, Dec. 1999. http://www.tv-anytime.org/

[3]   S-F. Chang, The holy grail of content-based media analysis, *IEEE Multimedia*, vol. 9, no. 2, pp. 6-10, Apr.-June 2002.

[4]   E. Oomoto and K. Tanaka, OVID: Design and Implementation of a Video-Object Database System, *IEEE Trans. on Knowledge and Data Eng.*, vol. 5, no. 4, pp. 629-643, Aug. 1993.

[5]   R. Hjelsvold and R. Midstraum, Modelling and Querying Video Data, in *Proc. of the 20th VLDB Conf.*, 1994.

[6]   S. Adali, K.S. Candan, S.S. Chen, K. Erol, and V.S. Subrahmanian, The Advanced Video Information System: Data Structures and Query Processing, *Multimedia Systems*, Vol.4, pp. 172-186, 1996.

[7]   L.A.E. Al Safadi and J.R. Getta, Semantic Modeling for Video Content-Based Retrieval Systems, in *23rd AustralAsian Comp. Science Conf.*, pp. 2-9, 2000.

[8]   M. S. Hacid, C. Declcir, and J. Kouloumdjian, A Database Approach for Modeling and Querying Video Data, *IEEE Trans. on Knowledge and Data Eng.*, Vol. 12, No. 5, pp. 729-749, Sept./Oct. 2000.

[9]   M. Flickner, H. Sawhney, W. Niblack, J. Ashley, Q. Huang, B. Dom, M. Gorkani, J. Hafner, D. Lee, D. Petkovic, D.Steele, and P. Yanker, Query by Image and Video Content: The QBIC System, *IEEE Computer*, vol. 28, no. 9, pp. 23-32, Sept. 1995.

[10]   D. Yow, B-L. Yeo, M. Yeung, and B. Liu, Analysis and Presentation of Soccer Highlights from Digital Video, in *Proc. Asian Conf. on Comp. Vision (ACCV)*, 1995.

[11]   Y. Gong, L.T. Sin, C.H. Chuan, H-J. Zhang, and M. Sakauchi, Automatic Parsing of Soccer Programs, in *Proc. IEEE Int'l. Conf. on Mult. Comp. and Sys.*, pp. 167-174, 1995.

[12]   S. Intille and A. Bobick, Recognizing planned, multi-person action, *Computer Vision and Image Understanding*, Vol. 81, No. 3, pp. 414-445, March 2001.

[13]   V. Tovinkere and R. J. Qian, Detecting Semantic Events in Soccer Games: Towards a Complete Solution, in *Proc. IEEE Int'l. Conf. on Mult. and Expo (ICME)*, Aug. 2001.

[14]   G. S. Pingali, Y. Jean, and I. Carlbom, Real Time Tracking for Enhanced Tennis Broadcasts, in *Proc. IEEE Comp. Vision and Patt. Rec. (CVPR)*, pp. 260-265, 1998.

[15]   A. Gueziec, Tracking pitches for broadcast television, *IEEE Computer*, Vol. 35, No. 3, pp. 38-43, March 2002.

[16]   P. Xu, L. Xie, S-F. Chang, A. Divakaran, A. Vetro, and H. Sun, Algorithms and System for Segmentation and Structure Analysis in Soccer Video, in *Proc. IEEE Int'l. Conf. on Mult. and Expo (ICME)*, Aug. 2001.

[17]   B. Li and M. I. Sezan, Event Detection and Summarization in American Football Broadcast Video, in *Proc. of the SPIE Conf. on Storage and Retrieval for Media Databases,* Vol. 4676, pp. 202-213, Jan. 2002.

[18]   R. Leonardi and P. Migliorati, Semantic indexing of multimedia documents, *IEEE Multimedia,* Vol. 9, No. 2, pp. 44-51, Apr.- June 2002.

[19]   D. Zhong and S-F. Chang, Structure Analysis of Sports Video Using Domain Models, in *Proc. IEEE Int'l. Conf. on Mult. and Expo (ICME),* Aug. 2001.

[20]   K. A. Peker, R. Cabasson, and A. Divakaran, Rapid Generation of Sports Video Highlights Using the MPEG-7 Motion Activity Descriptor, in *Proc. of the SPIE conf. on Storage and Retrieval for Media Databases,* Vol. 4676, pp. 318-323, Jan. 2002.

[21]   N. Babaguchi, Y. Kawai, and T. Kitashi, Event based indexing of broadcasted sports video by intermodal collaboration, *IEEE Trans. on Multimedia,* Vol. 4, No. 1, pp. 68-75, March 2002.

[22]   Y. Rui, A. Gupta, and A. Acero, Automatically extracting highlights for TV baseball programs, in *Proc. ACM Multimedia,* 2000.

[23]   P.P. Chen, The Entity-Relationship Model - Toward a Unified View of Data, *ACM Transactions on Database Systems,* Vol. 1, No. 1, pp. 9-36, March 1976.

[24]   ISO/IEC Committee Draft 15938-5 Information Technology - Multimedia Content Description Interface: Multimedia Description Schemes, ISO/IEC/JTC1/SC29/WG11/N3966, March 2001.

[25]   Y. Fu, A. Ekin, A.M. Tekalp, and R. Mehrotra, Temporal Segmentation of Video Objects for Hierarchical Object-based Motion Description' *IEEE Trans. on Image Processing,* Vol. 11, No. 2, pp.135-145, Feb. 2002.

[26]   J.F. Allen, Maintaining knowledge about temporal intervals, *Comm. ACM,* vol. 26, no. 11, pp. 832-843, 1983.

[27]   J.Z. Li, M. T. Ozsu, and D. Szafron, Modeling of Video Spatial Relationships in an Object Database Management System, in *IEEE Proc. Int'l. Workshop on Mult. Dat. Man. Sys.,* 1996.

[28]   M. Egenhofer and R. Franzosa, Point-set topological spatial relations, *Int'l J. of Geographical Information Systems,* vol. 5, no. 2, pp. 161-174, 1991.

[29]   H. Pan, P. van Beek, and M. I. Sezan, Detection of Slow-motion Replay Segments in Sports Video for Highlights Generation, in *Proc. IEEE Int'l. Conf. on Acoustics, Speech, and Signal Processing (ICASSP),* 2001.

[30]   A. Ekin and A.M. Tekalp, A Framework for Analysis and Tracking of Soccer Video, in *Proc. of the IS&T/SPIE Conf. on Visual Com. and Image Proc. (VCIP),* Jan. 2002.

[31]   G. Millerson, *The Technique of Television Production,* 12th Ed., Focal Publishers, March 1990.

[32]   S. Theodoridis and K. Koutroumbas, *Pattern Recognition,* Academic Press, 1999.

[33]   Laws of the Game and Decisions of the International Football Associations Board - 2001, www.fifa.com

[34]   M. Sonka, V. Hlavac, and R. Boyle, *Image Processing, Analysis, and Machine Vision,* 2nd ed., PWS Publishing, 1999.

[35]   A. Ekin, A. M. Tekalp, and R. Mehrotra, Automatic Extraction of Low-Level Object Motion Descriptors, in *Proc. IEEE ICIP,* Thessaloniki, Greece, Oct. 2001.

[36]   C. Tomasi and T. Kanade, Detection and Tracking of Point Features, Carnegie Mellon University Tech. Rep. CMU-CS-91-132, April 1991.

[37] J. Shi and C. Tomasi, Good Features to Track, in *IEEE Conf. on CVPR*, pp. 593-600, 1994.

[38] A. M. Tekalp, *Digital Video Processing*, Prentice Hall, 1995.

[39] M. Gyssens, J. Paredaens, J. van den Bussche, and D. van Gucht, A Graph-oriented Object Database Model, *IEEE Trans. on Knowledge and Data Eng.*, Vol. 6, No. 4, pp. 572-586, Aug. 1994.

[40] M. Andries, M. Gemis, J. Paradeans, I. Thyssens, and J. van den Bussche, Concepts for Graph-Oriented Object Manipulation, *Advances in Database Tech.- EDBT'92*, A. Pirotte et al., Eds., pp. 21-38, 1992, Springer-Verlag.

7

TEMPORAL SEGMENTATION OF VIDEO DATA

Edoardo Ardizzone and Marco La Cascia
Dipartimento di Ingegneria Informatica
University of Palermo
Palermo, ITALY
`{ardizzone,lacascia}@unipa.it`

1. INTRODUCTION

Temporal segmentation of video data is the process aimed at the detection and classification of transitions between subsequent sequences of frames semantically homogeneous and characterized by spatiotemporal continuity. These sequences, generally called camera-shots, constitute the basic units of a video indexing system. In fact, from a functional point of view, temporal segmentation of video data can be considered as the first step of the more general process of content-based automatic video indexing. Although the principal application of temporal segmentation is in the generation of content-based video databases, there are other important fields of application. For example in video browsing, automatic summarization of sports video or automatic trailer generation of movies temporal segmentation is implicitly needed. Another field of application of temporal segmentation is the transcoding of MPEG 1 or 2 digital video to the new MPEG-4 standard. As MPEG-4 is based on video objects that are not coded in MPEG 1 or 2, a temporal segmentation step is in general needed to detect and track the objects across the video.

It is important to point out the conceptual difference between the operation of an automatic tool aimed at the detection and classification of the information present in a sequence and the way a human observer analyzes the same sequence. While the human observer usually performs a semantic segmentation starting from the highest conceptual level and then going to the particular, the automatic tool of video analysis, in a dual manner, starts from the lowest level, i.e., the video bitstream, and tries to reconstruct the semantic content. For this reason, operations that are very simple and intuitive for a human may require the implementation of quite complex decision making schemes. For example consider the process of archiving an episode of a tv show. The human observer probably would start annotating the title of the episode and its number, then a general description of the scenes and finally the detailed description of each scene. On the other side the automatic annotation tool, starting form the bitstream, tries to determine the structure of the video based on the homogeneity of consecutive frames and then can extract semantic information. If we assume a camera-shot is a sequence of semantically homogeneous and spatiotemporally continuous frames then the scene can be considered as an homogeneous sequence of shots and the episode as a homogeneous sequence of scenes. In practice, the first step for the automatic annotation tool is the determination of the structure of the video based on camera-shots, scenes and episodes as depicted in Figure 7.1.

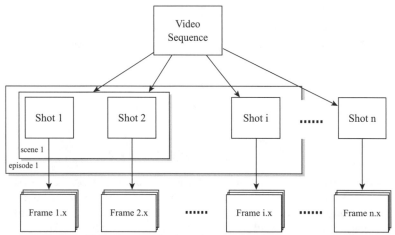

Figure 7.1 General structure of a video sequence.

In a video sequence the transitions between camera-shots can be of different typologies. Even though in many cases the detection of a transition is sufficient, in some cases it is important at least to determine if the transition is abrupt or gradual. Abrupt transitions, also called *cuts*, involve only two frames (one for each shot), while gradual transitions involve several frames belonging to the two shots processed and mixed together using different spatial and intensity variation effects [4].

It is also possible to further classify gradual transitions [28], [5]. Most common gradual transitions are fades, dissolves and wipes. Fade effects consisting of the transition between the shot and a solid background are called fade-out, while opposite effects are called fade-in. Fade-in and fade-out are often used as beginning and end effect of a video. Dissolve is a very common editing effect and consists of a gradual transition between two shots where the first one slowly disappears and, at the same time and at the same place, the second one slowly appears. Finally, wipes are the family of gradual transitions where the first shot is progressively covered with the second one, following a well defined trajectory. While in the dissolve the superposition of the two shots is obtained changing the intensity of the frames belonging to the shots, in the case of the wipes the superposition is obtained changing the spatial position of the frames belonging to the second shot. In Figure 7.2 are reported a few frames from a fade-out, a dissolve and a wipe effect.

The organization of this chapter is as follows. In Sect. 2 we review most relevant shot boundary detection techniques, starting with basic algorithms in uncompressed and compressed domains and then discuss some more articulated methodologies and tools. In Sect. 3 we propose a new segmentation technique, grounded on a neural network architecture, which does not require explicit threshold values for detection of both abrupt and gradual transitions. Finally, we present in Sect. 4 a test bed for experimental evaluation of the quality of temporal segmentation techniques, employing our proposed technique as an example of use.

(a)

(b)

(c)

Figure 7.2 Examples of editing effects. (a) fade-out, (b) dissolve, (c) wipe.

2. TEMPORAL SEGMENTATION TECHNIQUES

It has been pointed out that the aim of temporal segmentation is the decomposition of a video in camera-shots. Therefore, the temporal segmentation must primarily allow to exactly locate transitions between consecutive shots. Secondarily, the classification of the type of transition is of interest. Basically, temporal segmentation algorithms are based on the evaluation of the quantitative differences between successive frames and on some kind of thresholding. In general an effective segmentation technique must combine an interframe metric computationally simple and able to detect video content changes with robust decision criteria.

In subsections 2.1-2.4 we will review some of the most significant approaches proposed in recent years and discuss their strength and limitations. These techniques also constitute the building blocks of more sophisticated segmentation methodologies and tools, such as the ones illustrated in subsection 2.5, and may be used to evaluate new segmentation techniques.

2.1 TECHNIQUES BASED ON PIXELS DIFFERENCE METRICS

Metrics classified as PDM (pixels difference metrics) are based on the intensity variations of pixels in equal position in consecutive frames. Temporal segmentation techniques using interframe differences based on color are conceptually similar, but they are not very popular as they have a greater computational burden and almost the same accuracy of their intensity based counterpart.

A basic PDM is the sum of the absolute differences of intensity of the pixels of two consecutive frames [8]. In particular, indicating with $Y(x, y, j)$ and $Y(x, y, k)$ the intensity of the pixels at position (x, y) and frames j and k, the metric can be expressed in the following way:

$$\Delta f = \sum_x \sum_y |Y(x, y, j) - Y(x, y, k)| \tag{7.1}$$

where the summation is taken all over the frame.

The same authors propose other metrics based on first and second order statistical moments of the distributions of the levels of intensity of the pixels. Indicating with μ_k and σ_k respectively the values of mean and standard deviation of the intensity of the pixels of the frame k, it is possible to define the following interframe metric between the frames j and k:

$$\lambda = \frac{\left[\dfrac{\sigma_j + \sigma_k}{2} + \left(\dfrac{\mu_j - \mu_k}{2} \right)^2 \right]^2}{\sigma_j \sigma_k} \tag{7.2}$$

This metric has been also used in [7], [30], and is usually named *likelihood ratio*, assuming a uniform second order statistic. Other metrics based on statistical properties of pixels are:

$$\eta_1 = \frac{|\mu_j - \mu_k| \, |\sigma_j^2 - \sigma_k^2|}{\sigma_j \sigma_k \left(\dfrac{\mu_j + \mu_k}{2} \right)} \qquad \eta_1 \geq 0 \tag{7.3}$$

$$\eta_2 = \frac{\mu_j \sigma_j^2}{\mu_k \sigma_k^2} \qquad \mu_j > \mu_k, \, \sigma_j > \sigma_k \tag{7.4}$$

$$\eta_3 = \left(\frac{\mu_j \sigma_j}{\mu_k \sigma_k} \right)^2 \qquad \mu_j > \mu_k, \, \sigma_j > \sigma_k \tag{7.5}$$

Associating a threshold to a metric it is possible to detect a transition whenever the metric value exceeds the threshold value. As it was pointed out in [8], PDMs offer the best performance for the detection of abrupt transitions. In general all the PDMs are particularly exposed to the effects of noise, camera/object movements or sudden lighting changes, leading to temporally or spatially localized luminance perturbations, and then to a potentially large number of false transitions. From this point of view, slightly better performances are obtained considering block-based comparisons between successive frames, with matching criteria based on values of mean and variance in corresponding blocks. Several block-based PDMs are reviewed in [18].

In order to limit the effects of local perturbations, some authors have developed top-down methods based on mathematical models of video. For example, in [1] a differential model of motion picture is presented, where three factors concur to the intensity variations of pixels of consecutive frames: a small amplitude additive zero-centered Gaussian noise, essentially modelling the noise effects of camera, film and digitizer; the *intrashot* intensity variations due to camera/object motion and focal length or lightness changes; and finally the *intershot* variations due to the presence of abrupt or gradual transitions. In [34], another approach for gradual transitions detection based on a model of intensity changes during fade out, fade in and dissolve effects is presented.

Another interesting approach based on PDMs has been proposed in [6]. In this paper the authors propose the use of *moment invariants* of the image. Properties such as the scale and rotation invariance make them particularly suited to represent the frame. Denoting by $Y(x, y)$ the intensity at the position (x, y), the generic moment of order pq of the image is defined as:

$$m_{pq} = \sum_x \sum_y x^p y^q Y(x, y) \tag{7.6}$$

Moment invariants are derived from normalized central moments defined as:

$$n_{pq} = \frac{1}{m_{00}^\gamma} \sum_x \sum_y (x - \bar{x})^p (y - \bar{y})^q Y(x, y) \tag{7.7}$$

where:

$$\gamma = 1 + \frac{(p+q)}{2} \qquad \bar{x} = \frac{m_{10}}{m_{00}} \qquad \bar{y} = \frac{m_{01}}{m_{00}} \tag{7.8}$$

Limiting our attention to the first three moment invariants, these are defined as:

$$\phi_1 = n_{20} + n_{02}$$

$$\phi_2 = (n_{20} - n_{02})^2 + 4n_{11}^2 \tag{7.9}$$

$$\phi_3 = (n_{30} - 3n_{12})^2 + (3n_{21} - n_{03})^2$$

These three numbers may be interpreted as the components of a vector, say σ that can be used to represent the image:

$$\sigma = \{\phi_1, \phi_2, \phi_3\} \tag{7.10}$$

The interframe metric adopted in [6] is the Euclidean distance between the vector σ associated to frames j and k:

$$f_{moms}(j, k) = \left| \sigma_j - \sigma_k \right|^2 \tag{7.11}$$

Although the use of moment invariants can lead to robust segmentation algorithms, with respect to noise and other local perturbations, techniques based on statistical properties generally exhibit a computational load not adequate for specific applications, e.g., real time systems.

2.2 TECHNIQUES BASED ON HISTOGRAM DIFFERENCE METRICS

Metrics classified as HDM (histograms difference metrics) are based on the evaluation of the histograms of one or more channels of the adopted color space. As it is well known, the histogram of a digital image is a measure that supplies information on the general appearance of the image. With reference to an image represented by three color components, each quantized with 8 bit/pixel, a three-dimensional histogram or three monodimensional histograms can be defined. Although the histogram does not contain any information on the spatial distribution of intensity, the use of interframe metrics based on image histograms is very diffused because it represents a good compromise between the computational complexity and the ability to represent the image content.

In recent years several histogram-based techniques have been proposed [9], [6], [23], [37]; some of them are based only on the luminance channel, others on the conversion of 3-D or 2-D histograms to linear histograms [2]. An RGB 24 bit/pixel image would generate an histogram with 16.7 millions of bins and this is not usable in practice. To make manageable histogram-based techniques, a coarse color quantization is needed. For example in [2] the authors use an histogram in the HSV color space using 18 levels for hue and 3 for saturation and value leading to a easily tractable 162 bins histogram. Other approaches are based both on PDM and HDM. For example, in [24] two metrics are combined together: a PDM metric is used to account for spatial variation in the images and a HDM metric to account for color variations. Ideally, a high value in both metrics corresponds to a transition; however the choice of thresholds and weights may be critical.

In what follows some of the most popular HDM metrics are reviewed. In all the equations M and N are respectively the width and height (in pixels) of the image, j and k are the frame indices, L is the number of intensity levels and $H[j, i]$ the value of the histogram for the i-th intensity level at frame j. A commonly used metric [9] is the *bin-to-bin* difference, defined as the sum of the absolute differences between histogram values computed for the two frames:

$$f_{db2b}(j,k) = \frac{1}{2MN} \sum_{i=0}^{L-1} |H(j,i) - H(k,i)| \qquad (7.12)$$

The metric can easily be extended to the case of color images, computing the difference separately for every color component and weighting the results. For example, for a RGB representation we have:

$$f_{drgb}(j,k) = \frac{r}{s} f_{db2b}(j,k)^{(red)} + \frac{g}{s} f_{db2b}(j,k)^{(green)} + \frac{b}{s} f_{db2b}(j,k)^{(blue)} \quad (7.13)$$

where r, g and b are the average values of the three channels and s is:

$$s = \frac{(r+g+b)}{3} \qquad (7.14)$$

Another metric, used for example in [9], [6], is called *intersection* difference and is defined in the following way:

$$f_{d\,int}(j,k)=1-\frac{1}{MN}\sum_{i=0}^{L-1}\min[H(j,i),H(k,i)] \qquad (7.15)$$

In other approaches [28], the *chi-square test* has been used, which is generally accepted as a test useful to detect if two binned distributions are generated from the some source:

$$f_{dchi2}(j,k)=\sum_{i=0}^{L-1}\frac{[H(j,i)-H(k,i)]^2}{[H(j,i)+H(k,i)]^2} \qquad (7.16)$$

Also the correlation between histograms is used:

$$f_{dcorr}(j,k)=1-\frac{\mathrm{cov}(j,k)}{\sigma_j\sigma_k} \qquad (7.17)$$

where *cov(i, j)* is the covariance between frame histograms:

$$\mathrm{cov}(j,k)=\frac{1}{L}\sum_{i=0}^{L-1}[H(j,i)-\mu_j][H(k,i)-\mu_k] \qquad (7.18)$$

and μ_j and σ_j represent the mean and the standard deviation, respectively, of the histogram of the frame *j*:

$$\mu_j=\frac{1}{L}\sum_{i=0}^{L-1}H(j,i) \qquad (7.19)$$

$$\sigma_j=\sqrt{\frac{1}{L}\sum_{i=0}^{L-1}[H(j,i)-\mu_j]^2} \qquad (7.20)$$

All the metrics discussed so far are global, i.e., based on the histogram computed over the entire frame. Some authors propose metrics based on histograms computed on subframes. For example, in [5] a rectangular 6x4 frame partitioning is used. Each frame is subdivided into 6 horizontal blocks and 4 vertical blocks. The reason for this asymmetry is that horizontal movements are statistically more frequent than vertical ones. Then an HDM is used to compute difference between corresponding blocks in consecutive frames, after an histogram equalization. The shape of histograms is also taken into account for each block, using histogram moments up to the third order. Region differences are also weighted, using experimentally determined coefficients, to account for different contributions, in correspondence of a shot transition, of low and high intensity levels. The global metric is finally obtained adding up the contribution of all the sub-regions. In despite of its greater complexity, the performance of this technique remains, in many cases, quite sensitive to the choice of thresholds and weights.

Both PDM and HDM techniques are based on the computation of a similarity measure of two subsequent frames and on comparison of this measure with a threshold. The choice of the threshold is critical, as too low threshold values may lead to false detections and too high threshold values may cause the opposite effect of missed transitions.

To limit the problem of threshold selection, several techniques have been proposed. For example, in [5] a short-term analysis of the sequence of interframe differences is proposed to obtain temporally localized thresholds. The analysis is performed within a temporal window of appropriate size, for example 7 frames. This approach is more robust to local variations of brightness and camera/object motion. In detail, naming D_k the difference metric between the frames k and k-1 and W_k the short-term temporal window centered on the frame k, we have:

$$\rho(k) = \frac{D_k}{M_k} \qquad\qquad \text{where} \quad M_k = \max_{k \in W_k} \{D_k\} \qquad\qquad (7.21)$$

If $\rho(k)$ exceeds a threshold, computed experimentally, then an abrupt transition at frame k is detected. In a similar approach based on temporal windows and local thresholds [7], the statistical properties of a bin-to-bin histogram metric are evaluated over a 21-frame temporal window and used to detect transitions.

2.3 TECHNIQUES FOR DETECTION OF EDITING EFFECTS

The techniques reported in the previous sections, based on PDMs or HDMs, are mainly suited for the detection of abrupt transitions. Other techniques have been developed to detect gradual transition like fades or wipes. For example, in [5] the authors propose a technique to detect fading effects based on a linear model of the luminance L in the CIE L*u*v color space. Assuming that chrominance components are approximately constant during the fading, the model for a fade-out is:

$$L(x, y, t) = L(x, y, t_0)\left(1 - \frac{t - t_0}{d}\right); \qquad t \in [t_0, t_0 + d] \qquad\qquad (7.22)$$

where $L(x, y, t)$ is the luminance of the pixel at position (x, y) and time t, t_0 is the time of beginning of the fading effect and d its duration. Similarly the model for a fade-in is:

$$L(x, y, t) = L(x, y, t_0 + d)\left(\frac{t - t_0}{d}\right); \qquad t \in [t_0, t_0 + d] \qquad\qquad (7.23)$$

Even if the behavior of real luminance is not strictly linear during the fading, a technique based on the recognition of pseudo-linearity of L may be used to detect transitions. Even in this case, however, local thresholds have to be considered. Moreover, for some kind of video with very fast dynamic characteristics (for example TV commercials) this technique cannot be used. Other approaches have been proposed to overcome this limitation. For example, in [12] a production-based model of the most common editing effects is used to detect gradual transitions; in [37] a simple and effective two-thresholds technique based on HDM is reported. The two thresholds respectively detect the beginning and the end of the transition. The method, called *twin-comparison*, takes into account the cumulative differences between frames of the gradual transition. In the first pass a high threshold T_h is used to detect cuts; in the second pass a lower threshold T_l is employed to detect the potential starting frame F_s of a gradual transition. F_s is then compared to subsequent frames. An accumulated comparison is performed as during a gradual transition this difference value increases. The end frame F_e of the transition is detected when the difference between consecutive frames decreases to less than T_l, while the accumulated comparison has increased to a value higher than T_h. If the consecutive difference falls below T_l before the accumulated difference exceeds T_h, then the potential start frame F_s is dropped and the search continues for other gradual transitions.

Specific techniques have been aimed at the detection of other gradual transitions. For example, in [32] a two-step technique for the detection of wipe effects is proposed. It is based on statistical and structural properties of the video sequence and operates on partially decompressed MPEG streams. In [26] the authors propose a technique for transition detection and camera motion analysis based on spatiotemporal textures (spatiotemporal images will be further treated in the next subsection 2.5). The analysis of the texture changes can lead to the estimation of shooting conditions and to the detection of some types of wipes.

Finally, techniques based on higher level image features have also been tried. For example, in [35] the analysis of intensity edges between consecutive frames is used. During a cut or a dissolve, new intensity edges appear far from the locations of the old edges. Similarly, old edges disappear far from the location of new edges. Thus, by counting the entering and exiting edge pixels, cuts, fades and dissolves may be detected and classified.

2.4 TECHNIQUES OPERATING ON COMPRESSED VIDEO

Due to the increasingly availability of MPEG [19] compressed digital video, many authors have focused their attention on temporal segmentation techniques operating directly on the compressed domain or on a partially decompressed video sequence. Before discussing some of presented methods, we shortly review the fundamentals of MPEG compression standard.

MPEG uses two basic compression techniques: 16 x 16 macroblock-based motion compensation to reduce temporal redundancy and 8 x 8 Discrete Cosine Transform (DCT) block-based compression to capture spatial redundancy. An MPEG stream consists of three types of pictures, I, P and B, which are combined in a repetitive pattern called group of picture (GOP).

    I (*Intra*) frames provide random access points into the compressed data and are coded using only information present in the picture itself. DCT coefficients of each block are quantized and coded using Run Length Encoding (RLE) and entropy coding. The first DCT coefficient is called DC term and is proportional to the average intensity of the respective block.

    P (*Predicted*) frames are coded with forward motion compensation using the nearest previous reference (I or P) pictures.

    B (*Bi-directional*) pictures are also motion compensated, this time with respect to both past and future reference frames.

Motion compensation is performed finding for each 16 x 16 macroblock of the current frame the best matching block in the respective reference frame(s). The residual error is DCT-encoded and one or two motion vectors are also transmitted.

A well known approach to temporal segmentation in the MPEG compressed domain, useful for detecting both abrupt and gradual transitions, has been proposed in [33] using the DC sequences. A DC sequence is a low resolution version of the video, since it is made up of frames where each pixel is the DC term of a block (see Figure 7.3). Since this technique uses I, P and B frames, a partial decompression of the video is necessary. The DC terms of I frames are directly available in the MPEG stream, while those of B and P frames must be estimated using the motion vectors and the DCT coefficients of previous I frames. This reconstruction process is computationally very expensive.

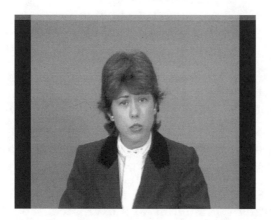

Figure 7.3 Sample frame from a sequence and corresponding DC image.

Differences of DC images are compared and a sliding window is used to set the thresholds for abrupt transitions. Both PDM and HDM metrics are suited as similarity measures, but pixel differences-based metrics give satisfactory results as DC images are already smoothed versions of the corresponding full images. Gradual transitions are detected through a accurate temporal analysis of the metric.

The technique reported in [28] uses only I pictures. It is based on the chi-square test applied to the luminance histogram and to row and column histograms of DC frames. The use of horizontal and vertical projections of the histogram introduces local information that is not available in the global histogram. In particular, for frames of M x N blocks, row and column histograms are defined as:

$$X_i = \frac{1}{M}\sum_{j=1}^{M} b_{0,0}(i,j) \quad ; \qquad i = 1,...,N$$

$$Y_j = \frac{1}{N}\sum_{i=1}^{N} b_{0,0}(i,j) \quad ; \qquad j = 1,...,M \qquad\qquad (7.24)$$

where $b_{0,0}(i,j)$ is the DC coefficient of the block (i,j). The three interframe differences computed on the three histograms are then used in a binned decision scheme to detect abrupt or gradual transitions. As only I frames are used, the DC recovering is eliminated. Note that as in a video there are typically two I frames per second, the analysis based on I frames only is adequate to approximately detect abrupt transitions. For gradual transitions this coarse temporal sub-sampling of the video may introduce more serious problems.

Another technique operating in the compressed domain is presented in [3], which is based on the correlation of DCT coefficients for M-JPEG compressed sequences. Other authors [38] extended this approach to MPEG sequences.

2.5 OTHER TECHNIQUES

Many authors tried to organize one or more of the previously described basic algorithms within more general frameworks, with the aim of defining and implementing more robust techniques. In particular, most of the approaches discussed so far rely on suitable thresholding of similarity measures between successive frames. However, the thresholds are typically highly sensitive to the type of input video.

In [11], the authors try to overcome this drawback by applying an unsupervised clustering algorithm. In particular, the temporal video segmentation is viewed as a 2-class clustering problem (scene change and no scene change) and the well-known K-means algorithm [27] is used to cluster frame dissimilarities. Then the frames from the cluster scene change which are temporary adjacent are labeled as belonging to a gradual transition and the other frames from this cluster are considered as cuts. Both chi-square statistics and HDMs were used to measure frame similarity, both in RGB and YUV color spaces. This approach is not able to recognize the type of the gradual transitions, but it exhibits the advantage that it is a generic technique that not only eliminates the need for threshold setting but also allows multiple features to be used simultaneously to improve the performance. The same limitations and advantages characterize the technique presented at end of this chapter.

Among others, Hanjalick recently proposed a robust statistical shot-boundary detector [14]. The problem of shot-boundary detection is analyzed in detail and a conceptual solution to the shot-boundary detection problem is presented in the form of a statistical detector based on minimization of the average detection-error probability. The idea is that to draw reliable conclusions about the presence or absence of a shot boundary, a clear separation should exist between discontinuity-value ranges for measurements performed within shots and at shot boundaries. Visual-content differences between consecutive frames within the same shot are mainly caused by two factors: object/camera motion and lighting changes. Unfortunately, depending on the magnitude of these factors, the computed discontinuity values within shots vary and sometimes may cause detection mistakes.

An effective way to reduce the influence of motion and lighting changes on the detection performance is to embed additional information in the shot boundary detector. The main characteristic of this information is that it is not based on the range of discontinuity values but on some other measurements performed on a video. For example, this information may result from the comparison between the measured pattern formed by discontinuity values surrounding the interval of frames taken into consideration and a known template pattern of a shot boundary. Various template patterns, specific for different types of shot boundaries (cuts or fades, wipes and dissolves), may be considered.

Another type of useful additional information may result from observation of the characteristic behavior of some visual features for frames surrounding a shot boundary for the case of gradual transitions. For example, since a dissolve is the result of mixing the visual material from two neighboring shots, it can be expected that variance values measured per frame along a dissolve ideally reveal a downwards-parabolic pattern [22]. Hence, the decision about the presence of a dissolve can be supported by investigating the behavior of the intensity variance in the suspected series of frames.

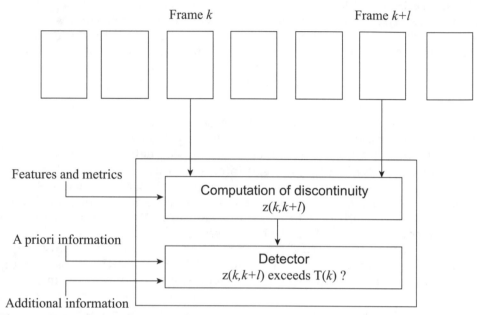

Figure 7.4 Representation of the shot boundary detector proposed in [14].

Further improvement of the detection performance can be obtained by taking into account a priori information about the presence or absence of a shot boundary at a certain time stamp along a video. The difference between additional information and a priori information is that the latter is not based on any measurement performed on the video. An example of a priori information is the dependence of the probability for shot boundary occurrence on the number of elapsed frames since the last detected shot boundary. While it can be assumed zero at the beginning of a shot, this probability grows and converges to the value 0.5 as the number of frames in the shot increases. The main purpose of this probability is to make the detection of one shot boundary immediately after another one practically impossible and so to contribute to a reduction of false detection rate.

Combining measurements of discontinuity values with additional and a priori information may result in a robust shot-boundary detector, e.g., by continuously adapting the detection threshold $T(k)$ for each frame k. Statistical decision theory is employed to obtain $T(k)$ based on the criterion that average probability for detection mistakes is minimized. To this aim, all detection errors (e.g., both missed and false detections) are treated equally; thus the detector performance is determined by the average probability that any of the errors occurs. This average probability is expressed in terms of the probability of missed detection and of the probability of false detection. The necessary likelihood functions are pre-computed using training data. Both additional information (for cuts and dissolves) and a priori information are taken into account in the computation of the average probability, while discontinuity values are computed by using a block-matching, motion compensated procedure [31]. The input sequence is assumed to be a MPEG partially decoded sequence, i.e., a DC sequence. This makes possible to limit the block dimensions to 4 x 4 pixels. The system performance is evaluated by using five test sequences (different from those employed for training) belonging to four video categories: movies, soccer game, news and commercials. The authors report a 100% precision and recall for cuts, 79% precision and 83% recall for dissolve detection.[1]

Another probabilistic approach was presented in [21], where Li et al. propose a temporal segmentation method based on spatial–temporal joint probability image (ST-JPI) analysis. Here, joint probabilities are viewed as a similarity estimate between two images (only luminance images are considered in the paper). Given two images $A(x, y)$ and $B(x, y)$, a joint probability image (JPI) is a matrix whose element value $JPI_{A,B}(i_1, i_2)$ is the probability that luminance i_1 and i_2 appear at the same position in image A and image B, respectively. Each element of a JPI corresponds to an intensity pair in two images. The distribution of the values in a JPI maps the correlation between two images: for two identical images, the JPI shows a diagonal line, while for two independent images the JPI consists of a uniform distribution. Because of the high correlation between video frames within a single shot, a JPI derived from two frames belonging to the same shot usually has a narrow distribution along the diagonal line. On the contrary, since narrative and visual content changes between two consecutive shots, then a uniform distribution is expected in the JPI. Thus the JPI behavior may be used to develop transition detection methods.

In particular, a spatial-temporal joint probability image (ST-JPI) is defined as a series of JPIs in chronological order, with all JPIs sharing the same initial image. The ST-JPI reflects the temporal evolution of video contents. For example, if a ST-JPI is derived between frames 0 and T, and a cut happens within this frame interval, the JPIs before the cut have very limited dispersion from the diagonal line, while after the cut uniform JPIs are usually obtained. The shift from narrow dispersion JPIs to uniform JPIs happens instantaneously at the cut position. By estimating the uniformity of JPIs, cuts can be detected and reported.

Detection of gradual transitions is also obtained with this approach, even if in a more complicated way. In particular, two kinds of gradual transition are considered, *cross* transitions and *dither* transitions. During a cross transition, every pixel value gradually changes from one shot to another, while during a dither transition a small portion of pixels abruptly change from pixels values from the first shot to those of the second shot every moment. With time, more and more pixels change until all of the pixels change into the second video shot. Wipes, fades and various types of dissolve may be described in terms of this scheme.

Template ST-JPIs are derived both for cross transitions and dither transitions. The detection of gradual transitions is performed by analyzing the pattern match between model ST-JPIs and the ST-JPI derived for the frame interval under consideration. Experiments performed on several digital videos of various kind gave the following results: 97% (recall) and 96% (precision) for cuts, 82% and 93% for cross transitions, 75% and 81% for dither transitions.

---

[1] Recall and precision are popular quality factors whose formal definition is given in the following Sect. 4.

The algorithm proposed in [25] is based on the coherence analysis of temporal slices extracted from the digital video. Temporal slices are extracted from the video by slicing through the sequence of video frames and collecting temporal signatures. Each of these slices contains both spatial and temporal information from which coherent regions are indicative of uninterrupted video partitions separated by camera breaks (cuts, wipes and dissolves). Each spatiotemporal slice is a collection of scans, namely horizontal, vertical or diagonal image stripes, as a function of time. The detection of a shot boundary therefore becomes a problem of spatiotemporal slice segmentation into regions each of a coherent rhythm. Properties could further be extracted from the slice for both the detection and classification of camera breaks. For example, cut and wipes are detected by color-texture properties, while dissolves are detected by mean intensity and variance. The analysis is performed on the DC sequence extracted from a MPEG video. The approach has been tested by experiments on news sequences, documentary films, movies, and TV streams, with the following results: 100% (recall) and 99% (precision) for cuts, 75% and 80% for wipes, 76% and 77% for dissolves.

Finally, a fuzzy theoretic approach for temporal segmentation is presented in [16], with a fusion of various syntactic features to obtain more reliable transition detection. Cuts are detected using histogram intersection; gradual changes are detected using a combination of pixel difference and histogram intersection, while fades are detected using a combination of pixel difference, histogram intersection and edge-pixel-count. Frame-to-frame differences of these properties are considered as the input variable of the problem expressed in fuzzy terms. In particular, the linguistic variable inter-frame-difference is fuzzified so that it can be labeled as "negligible," "small," "significant," "large" or "huge." The values of metric differences are represented as these linguistic terms. To this aim, appropriate class boundaries and membership functions must be selected for each category. This is made by modeling the interframe property difference through the Rayleigh distribution. The appropriateness of this model has been tested by fitting Rayleigh distribution (chi-square test) to interframe difference data for about 300 video sequences of various kind, having 500-5000 frames each. Fuzzy rules for each property are derived by taking into account the current interframe difference, the previous interframe difference and the next interframe difference.

3. A MLP-BASED TECHNIQUE

As already stated, one of the critical aspects of most of the techniques discussed in previous section is the determination of thresholds or, in general, the definition of criteria of detection and classification of the transitions. Moreover, most of the techniques present in literature are strongly dependent on the kind of sequences analyzed. To cope with these problems we propose the use of a neural network that analyzes the sequence of interframe metric values and is able to detect shot transitions, also producing a coarse classification of the detected transitions. This approach may be considered a generalization of the MLP-based algorithm already proposed in [2].

3.1 SHORT NOTES ON NEURAL NETWORKS

In the last decades, neural networks [29] have been successfully used in many problems of pattern recognition and classification. Briefly, a neural network is a set of units or nodes connected by links or synapses. A numeric weight is associated to each link; the set of weights represents the memory of the network, where knowledge is stored. The determination of these weights is done during the learning phase. There are three basic classes of learning paradigms [15]: supervised learning (i.e., performed under an external supervision), reinforcement learning (i.e., through a trial-and-error process) and unsupervised learning (i.e., performed in a self-organized manner).

The network interacts with the environment in which it is embedded through a set of input nodes and a set of output nodes, respectively. During the learning process, synaptic weights are modified in an orderly fashion so as input-output pairs fit a desired function. Each processing unit is characterized by a set of connecting links to other units, a current activation level and an activation function used to determine the activation level in the next step, given the input weights.

A multilayer perceptron or MLP exhibits a network architecture of the kind shown in Figure 7.5. It is a multilayer (i.e., the network units are organized in the form of layers) feedforward (i.e., signals propagate through the network in a forward direction) neural network characterized by the presence of one or more hidden layers, whose nodes are correspondingly called hidden units. This network architecture, already proposed in the fifties, has been applied successfully to solve diverse problems after the introduction [17], [29] of the highly popular learning algorithm known as error *back-propagation* algorithm. This supervised learning algorithm is based on the error-correction learning rule.

Basically, the back-propagation process consists of two passes through the different network layers, a *forward* pass and a *backward* pass. In the forward pass, an input vector (training pattern) is applied to the input nodes, and its effect propagates through the network, layer by layer, so as to produce a set of outputs as the actual response of the network. In this phase the synaptic weights are all fixed. During the backward pass, the synaptic weights are all adjusted in accordance with the error-correction rule. Specifically, the actual response of the network is subtracted from the desired response to produce an error signal. This error signal is then propagated backward through the network, and the synaptic weights are adjusted so as to make the actual response of the network move closer to the desired response. The process is then iterated until the synaptic weights stabilize and the error converges to some minimum, or acceptably small, value. In practical applications, learning results from the many presentations of a prescribed set of training examples to the network. One complete presentation of the entire training set is called an *epoch*. It is common practice to randomize the order of presentation of training examples from one epoch to the next.

3.2 USE OF MLPs IN TEMPORAL SEGMENTATION

We propose the use of a MLP with an input layer, an hidden layer and an output layer, whose input vector is a set of interframe metric difference values. The training set is made up by examples extracted from sequences containing abrupt transitions, gradual transitions or no transition at all. We adopted the bin-to-bin luminance histogram difference as an interframe metric. This choice is due to the simpleness of this metric, to its sufficient representativity with respect to both abrupt and gradual transitions, and to its ability to provide a simple interpretation model of the general video content evolution. As an example, Figure 7.6 illustrates the evolution of the bin-to-bin metric within a frame interval of 1000 frames, extracted from a soccer video sequence. Five cuts and four dissolves are present in the sequence, and all these shot boundaries are clearly present in the figure. In the same figure, it is also evident as gradual transitions are sometimes very difficult to distinguish from intrashot sequences (e.g., compare A and B in the figure).

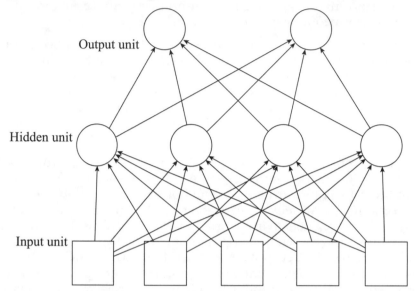

Figure 7.5 An example of multilayer perceptron with one hidden layer.

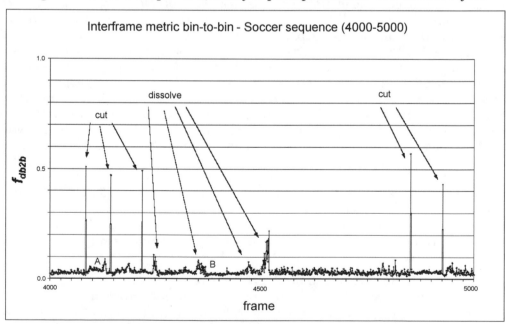

Figure 7.6 Interframe bin-to-bin metric for a soccer sequence.

The MLP's input units are fed with the values of the HDM computed within a temporal window of several frames (the details about the choice of the number of input and hidden units will be provided in the next section). The output units are three, each of them representing one of three distinct possibilities: abrupt transition, gradual transition or no break at all. Output units assume real values in the range 0 – 1. Table 7.1 shows the output values used during the training process, in correspondence to the following input configurations: abrupt transition at the center of the temporal window, gradual transition with its maximum located near the center of the temporal window, no-transition.

Table 7.1 Output values of the MLP during the training phase

	O_1	O_2	O_3
Abrupt transition	0.99	0.01	0.01
Gradual transition	0.01	0.99	0.01
No transition	0.01	0.01	0.99

During the learning phase, the position of each transition in the input video sequence is known, thus defining the correct triple of output values for each position of the input temporal window, i.e., for each input pattern in the training set. The above described back-propagation process therefore allows for the weights adjustment. When the trained network analyzes an unknown input sequence, at each step of the analysis the highest value of the output triple determines the detection of an abrupt or gradual transition, or the absence of transitions, in the analyzed temporal window. As a consequence, explicit threshold values are not required for the decision.

4. PERFORMANCE EVALUATION OF TEMPORAL SEGMENTATION TECHNIQUES

In this section we give some guidelines for the performance evaluation of temporal segmentation techniques. In particular, we define a test bed made of two techniques, respectively named T1 and T2, used as comparison terms and suitable for detection of abrupt and abrupt/gradual transitions, respectively. Both T1 and T2 techniques require a manual or semi-automatic work for adapting thresholds and other parameters to the feature of input sequences, in order to obtain performances near to the optimum for both of them. It is with these optimal or near-to-optimal performances that a new temporal segmentation technique can be compared. As an example, we will evaluate the performance of our MLP-based approach. The design of this evaluation framework is part of a project on video temporal segmentation currently under development at the *Computer Science and Artificial Intelligence* lab of the University of Palermo.

We also describe the dataset we used for the evaluation. We used four video sequences representative of different types of video. In particular, we considered two sport sequences (soccer and F1 world championship), one news sequence and one naturalistic documentary. Details on test sequences are reported in Table 7.2.

Most of the gradual transitions are dissolves. A few wipes are present and they are mainly in the F1 sequence. As previously stated both the T2 technique and the MLP-based one are aimed to the detection of a gradual transition but cannot discriminate between different kinds of gradual transitions.

In order to evaluate the performance of a segmentation algorithm we should primarily consider correctly detected transitions, false detections and missed transitions. Then we should analyze the performance with respect to gradual transition classification accuracy.

Recall and precision factors are normally used to evaluate performance [28] of transition detection techniques. These quality factors are defined as follows:

$$P_c = \frac{n_c}{n_c + n_f} \; ; \qquad R_c = \frac{n_c}{n_c + n_m} \qquad (7.25)$$

where n_c is the number of transitions correctly detected, n_f is the number of false positives and n_m is the number of missed detections. Ideally, $n_f = n_m = 0$ so that both the factors are 1. In our test bed we found useful a new quality factor defined in the following way:

$$Qr = \frac{n_c}{n_c + n_m + n_f} = \frac{n_c}{n_c + n_{tot} - n_c + n_f} = \frac{n_c}{n_{tot} + n_f} \qquad (7.26)$$

where n_{tot} is the total number of transitions and $n_m = n_{tot} - n_c$. This quality factor takes simultaneously into account the ability to avoid both false and missed detections.

To evaluate the performance of transition classification techniques a different quality factor is needed. Assume n_{cc} as the number of abrupt transitions detected and correctly classified, n_{cg} as the number of gradual transitions detected and correctly classified and n_{sw} the number of transitions detected but misclassified. We can define the following quality factor:

$$I_{sw} = \frac{n_{cc} + n_{cg}}{n_{cc} + n_{cg} + n_{sw}} \qquad (7.27)$$

This quality factor is 1 if $n_{sw}=0$. Note that, as previously stated, I_{sw} is only a measure of transition classification accuracy and then cannot replace the quality factor Q_r that is a measure of transition detection performance. The two factors should always be used together to evaluate the performance of a system.

Table 7.2 Characteristics of dataset

Sequence	Frame rate (fps)	# of frames	# of abrupt transitions	# of gradual transitions	Total # of transitions
Soccer	25	19260	115	51	166
F1	25	20083	106	14	120
News	25	20642	122	6	128
Nature	25	25016	150	12	162
	Total	85001	493	83	576

4.1 PERFORMANCE EVALUATION OF THE TECHNIQUE T1

We used the interframe metrics bin-to-bin, chi-square and histogram correlation, and called **T1** the technique whose result coincides, in each analyzed case, with the output of the best performing of the three methods. We evaluated the quality factor Q_r for the different HDM used varying the value of the threshold.

Using the annotated dataset described in the previous section we were able to determine for each sequence the optimal threshold. For example, in Figures 7.7 and 7.8 is reported the quality factor Q_r as a function of the threshold using the bin-to-bin metric, respectively, for soccer and nature videos. Obviously, as the threshold increases the number of false detections decreases but the number of missed detections increases. From Figures 7.7 and 7.8 it is possible to see that Q_r for the nature video is higher, meaning a better performance, than Q_r for the soccer video. This is due to the larger presence of gradual transitions in soccer video that are misdetected.

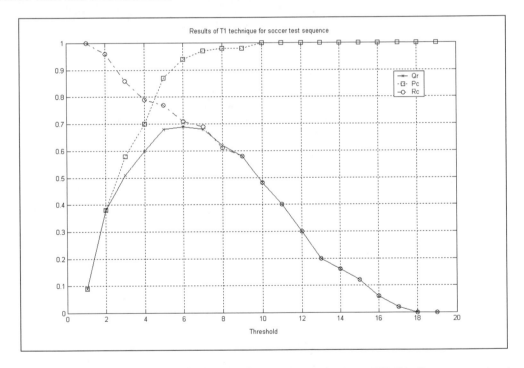

Figure 7.7 Performance of abrupt detection technique T1 for the *soccer* test sequence.

For the other sequences and for the other metrics we obtained very similar shapes of the curves of Figures 7.7 and 7.8. In Tables 7.3-7.5 it is summarized the performance of this technique on the four different test video and using three different histogram metrics in correspondence to the optimal value of the threshold:

Table 7.3 Performance using the b2b metric

Sequence	Optimal threshold	Qr
Soccer	0.30	0.68
News	0.30	0.76
F1	0.30	0.72
Nature	0.35	0.79

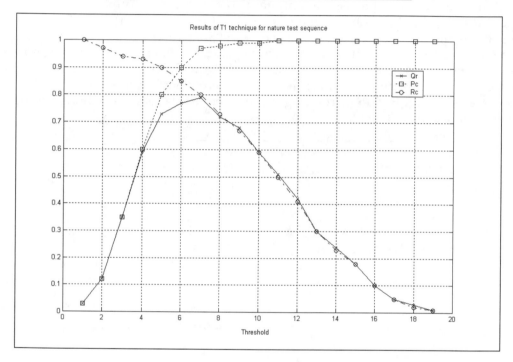

Figure 7.8 Performance of abrupt detection technique T1 for the *nature* test sequence.

Table 7.4 Performance using the chi-square metric

Sequence	Optimal threshold	Qr
Soccer	0.35	0.62
News	0.35	0.74
F1	0.40	0.69
Nature	0.40	0.75

Table 7.5 Performance using the histogram correlation metric.

Sequence	Optimal threshold	Qr
Soccer	0.20	0.64
News	0.25	0.74
F1	0.25	0.71
Nature	0.25	0.77

It should be noted that the bin-to-bin metric, despite of its simpleness, exhibits the best behavior for the proposed input sequences.

4.2 PERFORMANCE EVALUATION OF THE TECHNIQUE T2

To evaluate the performance of a temporal segmentation technique in presence of gradual transitions, as a term of comparison we implemented a multi-threshold technique, called **T2**, inspired by the algorithm presented in [7]. The technique T2, similarly to other techniques presented in literature, is based on the observation that the video properties relevant for the shot transition detection are intrinsically local, i.e., depending on the behavior of the interframe metric within a temporal window of a few decades of frames. Within the window, a shot boundary may be detected and classified analyzing statistical factors like the mean and the standard deviation of the sequence of interframe metric values. In particular, the local standard deviation calculated when an abrupt transition is present within the window is very different from those obtained in case of a gradual transition, due to the more distributed changes of the interframe metric values in the latter case.

Table 7.6 shows the definition and the meaning of thresholds and other factors intervening in the transition detection process.

Table 7.6 Factors intervening in transition detection (technique T2)

Parameter	Description
w	Temporal window of analysis
$T_{hc} = \mu_w + \alpha_c \sigma_w$	High threshold for abrupt transitions
$T_{hg} = \mu_w + \alpha_g \sigma_w$	High threshold for gradual transitions
$T_{lc} = \beta_c \mu_T$	Low threshold for abrupt transitions
$T_{lg} = \beta_g \mu_T$	Low threshold for gradual transitions

In the above definitions, μ_w and σ_w, respectively, are the mean and the standard deviation of the interframe metric values within the temporal window w, while μ_T is the global mean of the interframe metric. The coefficients α_c, α_g, β_c and β_g are determined experimentally, as well as the optimal size of the window w.

The technique T2 operates in the following way. Shot boundary occurrences are detected in correspondence to the central pair of frames within the current temporal window. To detect a cut, the corresponding interframe metric value must be the maximum of values belonging to the current window, and also greater than the local thresholds T_{hc} and T_{hg}. Moreover, it must also be greater than the global threshold T_{lc}, whose value is depending, through the coefficient β_c, on the global mean of the interframe metric. Similarly, a gradual transition is detected in correspondence to the central pair of frames within the current temporal window if the corresponding interframe metric value is a local maximum, and if it is greater than T_{hg} and T_{lg}, but not greater than T_{hc}.

It should be noted that verifying the presence of a local maximum before declaring a transition is a condition necessary to avoid the detection of more than one transition within the same temporal window. Moreover, two different local thresholds, T_{lc} and T_{lg}, have been introduced to overcome the problem of local perturbation effects, which could cause exceeding the threshold T_{hg}, as it has also been experimentally verified. T_{lc} and T_{lg} must be different, because the average metric values related to abrupt transitions are very different from the average metric values related to gradual transitions.

Figure 7.9 shows the behavior of the interframe metric bin-to-bin for the soccer sequence, outlining the variation of local thresholds T_{hc} and T_{hg} (the values $w = 21$, $\alpha_c = 4$ and $\alpha_g = 1.5$ have been assumed). It should be noted that both the thresholds are exceeded in correspondence to abrupt transitions, while T_{hg} (but not T_{hc}) is sometimes exceeded when searching for gradual transitions.

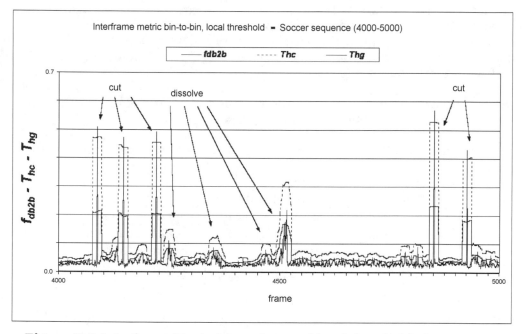

Figure 7.9 Interframe bin-to-bin metric and local thresholds for a soccer sequence.

These aspects are even more evident in Figure 7.10, where the metric bin-to-bin is shown for a shorter sequence of 250 frames extracted from the same video sequence. In this figure, two cuts of different value are reported, along with a gradual transition (dissolve) characterized by prominent fluctuations.

As the technique T2 is largely dependent on various parameters and thresholds, we carefully evaluated its performance by letting the parameters vary in quite large intervals, with small variation steps (intervals and steps were determined on the basis of preliminary experiments), as shown in Table 7.7.

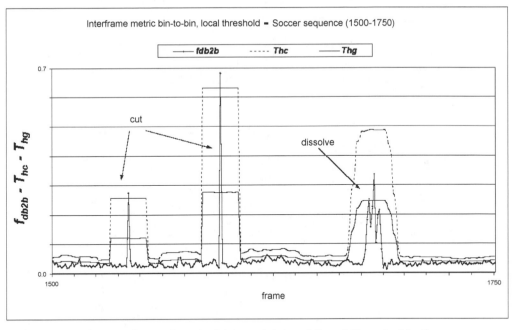

Figure 7.10 Interframe bin-to-bin metric and local thresholds for a soccer sequence.

Table 7.7 Range of parameters used for experiments

Parameter	Range	Step
w	15 – 49	2
α_c	2.0 – 5.0	0.5
α_g	1.0 – 3.0	0.5
β_c	2.0 – 5.0	0.5
β_g	1.0 – 4.0	0.5

In Tables 7.8-7.10 we report the best result obtained for each metric and each sequence and the set of parameters that led to the result:

Table 7.8 Optimal result using bin-to-bin metric

Sequence	Q_r	I_{sw}	W	α_c	α_g	β_c	β_g
Soccer	0.84	0.94	21	4.0	1.5	3.0	2.0
F1	0.82	0.92	25	3.5	1.5	3.0	2.0
News	0.83	0.97	21	3.5	2.0	3.0	2.0
Nature	0.83	0.96	21	3.5	2.0	3.0	2.0

Table 7.9 Optimal result using chi-squares metric

Sequence	Q_r	I_{sw}	W	α_c	α_g	β_c	β_g
Soccer	0.72	0.80	29	3.5	2.0	3.5	3.0
F1	0.75	0.76	27	3.0	2.0	3.0	2.5
News	0.78	0.86	27	3.5	2.5	3.5	2.5
Nature	0.71	0.88	25	3.0	2.5	3.5	3.0

Table 7.10 Optimal result using correlation metric

Sequence	Q_r	I_{sw}	W	α_c	α_g	β_c	β_g
Soccer	0.76	0.90	25	4.0	2.5	2.0	1.5
F1	0.77	0.92	29	4.0	2.5	2.5	1.5
News	0.78	0.86	27	4.0	2.5	2.0	1.5
Nature	0.75	0.79	25	4.5	2.5	2.5	2.0

4.3 PERFORMANCE EVALUATION OF THE MLP-BASED TECHNIQUE

To evaluate the performance of the proposed MLP-based technique we used the soccer and news video as training set and the nature and F1 video as test set. After each training epoch, the MLP performance is evaluated both on the training and test set.

The MLP architecture has been determined experimentally. The number of input units is obviously related to the probable number of frames belonging to a gradual transition. From video data available in our dataset we observed that the duration of gradual transitions is usually between 10 and 20 frames, i.e., below one second at 25 fps, a result in accord with modern electronic editing tools. We then tried several networks with 21 or 31 input units, a choice allowing to capture an entire transition without the risk of enclosing more than one transition within an unique temporal window. A negative consequence of this choice is the impossibility of dealing with very short shot transitions, like those ones typical of some commercial or musical sequences. However, we assumed our dataset more representative of kinds of video sequences of interest for video segmentation.

We tried network architectures with different numbers of hidden units. If the number of input units is intuitively related to the duration of gradual transitions, the choice of the number of hidden units is critical, in the sense that a limited number of hidden units can limit the ability of the network to generalize, while an excessive number of hidden units can induce problems of overtraining. Even this aspect has been investigated experimentally.

In Table 7.11 the performance of different MLP architectures after 50 training epochs is reported. The network architecture is indicated with three numbers indicating, respectively, the number of input, hidden and output units. The performance is expressed in terms of the quality factors previously defined, both on training set and testing set:

Table 7.11 Performance of MLPs of different architecture

	(21,10,3)		(21,40,3)		(21,100,3)		(31,40,3)	
	g_r	I_{sw}	g_r	I_{sw}	g_r	I_{sw}	g_r	I_{sw}
Soccer	0.87	0.92	0.95	0.99	0.86	0.90	0.99	0.99
News	0.84	0.94	0.94	1.00	0.84	0.94	0.94	1.00
F1	0.84	0.90	0.89	0.97	0.82	0.91	0.88	0.97
Nature	0.80	0.91	0.91	0.99	0.77	0.88	0.89	0.96

From inspection of Table 7.11 it is evident the best performance is obtained using a network with 21 input units and 40 hidden units. As expected, the network with 10 or 100 hidden units does not perform well. Since the use of 31 input units does not give significant performance improvement on our dataset, the (21-40-3) network has been selected for its lower computational load and for its better ability to detect temporally close transitions.

4.4 COMPARISON OF DIFFERENT TECHNIQUES

In this section we compare the three techniques we described in the previous sections. In Table 7.12 a summary of most important results is reported. Note that the results of techniques T1 and T2 were obtained using a choice of optimal parameters (for T1 and T2). Similarly the results of the MLP-based technique for the sequences soccer and news are computed on the training set. These results are then to be considered optimal and, in practice, we should expect worse performance.

On the other side, it is important to note that the performance of the MLP-based classifier on the sequences F1 and Nature, which are outside of the training set, is not based on the knowledge of the ground truth and then may be considered representative of the algorithm behavior. If we used the optimal parameters computed for techniques T1 and T2 for the sequences soccer and news to analyze the sequences F1 and Nature, we would obtain a worse performance.

Finally, although there is no theoretical justification, we note that, independently of the decision technique, best results are obtained using bin-to-bin metric.

Table 7.12 Performance comparison among the described techniques

	T1		T2		MLP	
	g_r	I_{sw}	g_r	I_{sw}	g_r	I_{sw}
Soccer	0.68	---	0.84	0.94	0.95	0.99
News	0.76	---	0.82	0.92	0.94	1.00
F1	0.72	---	0.83	0.97	0.89	0.97
Nature	0.79	---	0.83	0,96	0.91	0.99

5. CONCLUSIONS

In the last decade, with the increasing availability of digital video material, advanced applications based on storage and transmission of video data (digital libraries, video communication over the networks, digital TV, video-on-demand, etc.) began to appear. Their full utilization depends on the capability of information providers to organize and structure video data, as well as on the possibility of intermediate and final users to access the information.

From this point of view, automatic video temporal segmentation is a necessary task. As we showed in this chapter, much work has been done to individuate algorithms and methods in this area, but results are often too much depending on the peculiarity of input data. Research effort is still necessary to find generally effective and computationally manageable solutions to the problem.

REFERENCES

[1]   P. Aigrain, and P. Joly, The Automatic Real Time Analysis of Film Editing and Transition Effects and its Applications, *Computer & Graphics*, Vol. 18, No. 1, pp. 93-103, 1994.

[2]   E. Ardizzone, and M. La Cascia, Automatic Video Database Indexing and Retrieval, *Multimedia Tools and Applications*, Vol. 4, pp. 29-56, 1997.

[3]   F. Arman, A. Hsu, and M. Y. Chiu, Feature Management for Large Video Databases, in Proceedings IS&T/SPIE Conf. Storage and Retrieval for Image and Video Databases I, Vol. SPIE 1908, pp. 2-12, 1993.

[4]   Y. A. Aslandogan, and C. T. Yu, Techniques and Systems for Image and Video Retrieval, *IEEE Transactions On Knowledge and Data Engineering*, Vol. 11, No. 1, pp. 56-63, 1999.

[5]   J. M. Corridoni and A. Del Bimbo, Structured Representation and Automatic Indexing of Movie Information Content, *Pattern Recognition*, vol. 31, n. 12, pp. 2027-2045, 1998.

[6]   A. Dailianas, R. B. Allen, and P. England, Comparison of Automatic Video Segmentation Algorithms, *in Proceedings of SPIE Photonics West*, 1995.SPIE, Vol. 2615, pp. 2-16, Philadelphia, 1995.

[7]   R. Dugad, K. Ratakonda, and N. Ahuja, Robust Video Shot Change Detection, IEEE Second Workshop on Multimedia Signal Processing, pp. 376-381, Redondo Beach, California, 1998.

[8]   R. M. Ford, C. Robson, D. Tample, and M. Gerlach, Metrics for Scene Change Detection in Digital Video Sequences, in IEEE International Conference on Multimedia Computing and Systems (ICMCS '97), pp. 610-611, Ottawa, Canada, 1997.

[9]   U. Gargi, R. Kasturi, and S. H. Strayer, Performance Characterization of Video-Shot-Change Detection Methods, *IEEE Transactions on Circuits and Systems for Video Technology*, Vol. 10, No. 1, pp. 1-13, 2000.

[10]   R. C. Gonzalez and R. E. Woods, *Digital Image Processing*, Addison-Wesley, 1992.

[11]   B. Gunsel, A.M. Ferman, and A.M. Tekalp, Temporal video segmentation using unsupervised clustering and semantic object tracking, *Journal of Electronic Imaging*, Vol. 7, No. 3, pp. 592-604, 1998.

[12]   A. Hampapur, R. Jain, and T. Weymouth, Production Model Based in Digital Video Segmentation, *Multimedia Tools and Applications*, Vol. 1, No. 1, pp. 9-46, 1995.

[13]   A. Hampapur, R. Jain, and T. Weymouth, Indexing in Video Databases, *in IS&T SPIE Proceedings: Storage and Retrieval for Image and Video Databases III*, Vol. 2420, pp. 292-306, San Jose, 1995.

[14]	A. Hanjalick, Shot Boundary Detection: Unraveled and Resolved?, *IEEE Transactions on Circuits and Systems for Video Technology*, Vol. 12, No. 2, pp. 90-105, 2002.

[15]	S. Haykin, *Neural Networks – A Comprehensive Foundation*, MacMillan College Publishing Company, 1994.

[16]	R.S. Jadon, S. Chaudhury, and K.K. Biswas, A fuzzy theoretic approach for video segmentation using syntactic features, *Pattern Recognition Letters*, Vol. 22, pp. 1359-1369, 2001.

[17]	T. Khanna, *Foundation of Neural Networks*, Addison-Wesley, 1990.

[18]	I. Koprinska and S. Carrato, Temporal Video Segmentation: A Survey, *Signal Processing: Image Communication*, Vol. 16, pp. 477–500, 2001.

[19]	D. Le Gall, A Video Compression Standard for Multimedia Applications, *Communications of the ACM*, Vol. 34, No. 4, pp. 46-58, 1991.

[20]	M. S. Lee, Y. M. Yang, and S. W. Lee, Automatic Video Parsing Using Shot Boundary Detection and Camera Operation Analysis, *Pattern Recognition*, Vol. 34, pp. 711-719, 2001.

[21]	Z.-N. Li, X. Zhong, and M.S. Drew, Spatial-temporal joint probability images for video segmentation, *Pattern Recognition*, Vol. 35, pp. 1847-1867, 2002.

[22]	R. Lienhart, Reliable transition detection in videos: A survey and practitioner's guide, International Journal Image Graphics, Vol. 1, No.3, pp. 469-486, Aug. 2001.

[23]	A. Nagasaka and Y. Tanaka, Automatic Video Indexing and Full-Video Search for Objects Appearances, *in Visual Databases Systems II*, E. Knuth and L. M. Wegner (Eds.), Elsevier Science, pp. 113-127, 1992.

[24]	M. R. Naphade, R. Mehrotra, A.M. Ferman, J. Warnick, T. S. Huang, and A. M. Tekalp, A High-Performance Shot Boundary Detection Algorithm Using Multiple Cues, *in IEEE International Conference on Image Processing* (ICIP '98), Vol. 1, pp. 884-887, Chicago, 1998.

[25]	C.-W. Ngo, T.-C. Pong, and R.T. Chin, Video Partitioning by Temporal Slice Coherency, *IEEE Transactions on Circuits and Systems for Video Technology*, Vol. 11, No. 8, pp. 941-953, 2001.

[26]	C. W. Ngo, T. C. Pong, and R. T. Chin, Camera Breaks Detection by Partitioning of 2D Spatio-temporal Images in MPEG Domain, in IEEE International Conference on Multimedia Systems (ICMCS '99), vol. 1, pp. 750-755, Italy, 1999.

[27]	T.N. Pappas, An adaptive clustering algorithm for image segmentation, *IEEE Transaction on Signal Processing*, Vol. 40, pp. 901-914, 1992

[28]	N. V. Pathel and I. K. Sethi, Video Shot Detection and Characterization for Video Databases, *Pattern Recognition, Special Issue on Multimedia*, Vol. 30, pp. 583-592, 1997.

[29]	S. J. Russell and P. Norvig, *Artificial Intelligence – A Modern Approach*, Prentice Hall, 1995.

[30]	K. Sethi, and N. V. Patel, A Statistical Approach to Scene Change Detection, *in IS&T SPIE Proceedings: Storage and Retrieval for Image and Video Databases III*, Vol. 2420, pp. 329-339, San Jose, 1995.

[31]	B. Shahraray, Scene Change Detection and Content-based Sampling of Video Sequences, *in Proceedings of IS&T/SPIE*, Vol. 2419, Feb. 1995, pp. 2–13.

[32]   M. Wu, W. Wolf, and B. Liu, An Algorithm for Wipe Detection, in IEEE International Conference on Image Processing (ICIP '98), Vol. 1, pp. 893-897, Chicago, 1998.

[33]   L. Yeo and B. Liu, Rapid Scene Analysis on Compressed Video, *IEEE Transactions on Circuits and Systems for Video Technology*, Vol. 5, No. 6, pp. 533-544, 1995.

[34]   H. Yu, G. Bozdagi, and S. Harrington, Feature-based Hierarchical Video Segmentation, in IEEE International Conference on Image Processing (ICIP'97), Santa Barbara, 1997, pp. 498-501.

[35]   R. Zabih, J. Miller, and K. Mai, A Feature-Based Algorithm for Detecting and Classifying Scene Breaks, in Proceedings of ACM Multimedia '95, pp. 189-200, San Francisco, 1995.

[36]   H. J. Zhang, C. Y. Low, S. W. Smoliar, and J. H. Wu, An Integrated System for Content-Based Video Retrieval and Browsing, *Pattern Recognition*, Vol. 30, No. 4, pp. 643-658, 1997.

[37]   H. J. Zhang, A. Kankanhalli, and S. W. Smoliar, Automatic Partitioning of Full-motion Video, *Multimedia Systems*, Vol. 1, No. 1, pp. 10-28, 1993.

[38]   H. J. Zhang, C. Y. Low, and S. W. Smoliar, Video Parsing Using Compressed Data, *in Proceedings IS&T/SPIE, Image and Video Processing II*, pp. 142-149, 1994.

8

A TEMPORAL MULTI-RESOLUTION APPROACH TO VIDEO SHOT SEGMENTATION

Tat-Seng Chua, A. Chandrashekhara, and HuaMin Feng
Department of Computer Science
National University of Singapore
Singapore
{chuats, chandra, fenghm}@comp.nus.edu.sg

1. INTRODUCTION

The rapid accumulation of huge amounts of digital video data in archives has led to many video applications, which has necessitated the development of many video-processing techniques. These applications should have the ability to represent, index, store and retrieve video efficiently. Since video is a time-based media, it is very important to break the video streams into basic temporal units, called shots [45]. This kind of temporal segmentation is very useful in most video applications. The temporal segmentation is the first and the most basic step in the structured modeling of video [35][2]. When attempts are made to let computers understand videos, these shots serve as the basic blocks to construct the whole story. The understanding of video usually requires the understanding of the relationship between shots [6]. While in an application of video retrieval system, indexing the shots seems to be an inevitable step [13][5]. Therefore a good algorithm for the temporal segmentation of video can be helpful in all of these systems.

The temporal partitioning of video is generally called video segmentation or shot boundary detection [13][45][42]. To fulfill the task of partitioning video, video segmentation needs to detect the joining of two shots in the video stream and locate the position of these joins. These joins are made by the video editing process, which appears to be of two different types based on the technique involved [12]. If the video editor does nothing but directly concatenates the two shots together, the join is termed an abrupt transition, which is named CUT. If the video editor uses some special technique such as fade in/out, wipe, dissolve or morphing to make the joint appear smooth visually, the join will be a gradual transition, which is called GT. Due to the presence of these transitions and the wide varying lengths of GTs, the task of detecting the type and location of the transition of video shot is a complex task. Moreover, GT can be of varying temporal durations and are generated by different special techniques involved in the editing. These have made GT much more difficult to handle than CUT.

Most research uses different techniques to detect different types of GT. However, it can be observed that the various types of video transitions can be modeled as a temporal multi-resolution edge phenomenon. The temporal resolution of a video stream can be high (i.e., the original video stream) or low (i.e., by temporal sub-sampling of video frames), and different types of video transitions only differ in their characteristics in the temporal scale space. For example, longer GT's cannot be observed at a high temporal resolution but are apparent at a lower temporal resolution of the same video stream. Thus we claim that the transition of video shots is a multi-resolution phenomenon. Information across resolutions will be used to help detect as well as locate both the CUT and GT transition points. Since wavelet is well known for its ability to model sharp discontinuities and processing signals according to scales [4], we use Canny-like B-Spline wavelets in this multi-resolution analysis.

While various concepts of video segmentation have been explained and elaborated in other chapters of this book, this chapter is intended to provide a temporal multi-resolution analysis (TMRA) framework for video shot segmentation. In the next section, a short review of shot segmentation techniques is discussed. Section 3 reviews the basic theory behind video representation and multi-resolution analysis. Section 4 discusses the TMRA framework. Section 5 provides a brief account of simulation results. Section 6 concludes with the future discussions of work in this area.

2. REVIEW OF TECHNIQUES

This section presents related research in video modeling, techniques for video segmentation for CUT and GT detection and wavelets theory for video content analysis.

2.1 VIDEO MODELING

There are two approaches proposed to model the temporal logical structure of video, namely the structured modeling approach [3] and stratification [23].

The structured modeling approach is based on the idea of dividing the video sequences into atomic shots. Each shot corresponds to an event and is used as a basic unit for manipulation. The contextual information between video shots is modeled using additional higher-level constructs. The whole video is represented in a hierarchical concept structure. The basic idea of structure-based modeling is illustrated in Figure 8.1. This modeling reveals the basic necessity for video segmentation. It is the first step to extract the lowest level information.

The stratification modeling is more flexible. It is based on logical structure rather than temporal structure. The video sequences are modeled as overlapping chunks called strata. Each stratum represents one single concept that can be easily described. The content information of a subsequence is derived from the union of descriptions of all associated strata. One example of video stratification modeling on news video can be seen in Figure 8.2.

In another modeling approach, [9] builds a quad-tree for the temporal structure to facilitate navigation on video frames. In this model, the video is indexed in multi-temporal resolution. Different temporal resolutions correspond to different layers in the quad-tree. When browsing the video, switching between different temporal resolutions will result in different playback rates.

[9] only separates the sequences in units of frames, which is not as flexible as shots when moving up to higher-level analysis. In addition, the static quad-tree structure construction uses too much redundant data because in most cases, adjacent frames are quite similar in content. However, the idea of studying video in multi-temporal resolution can be applied to video segmentation problems.

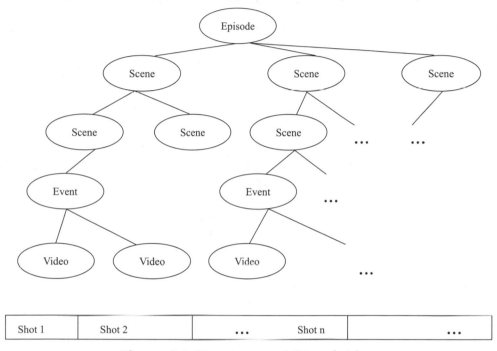

Figure 8.1 Structure modeling of video.

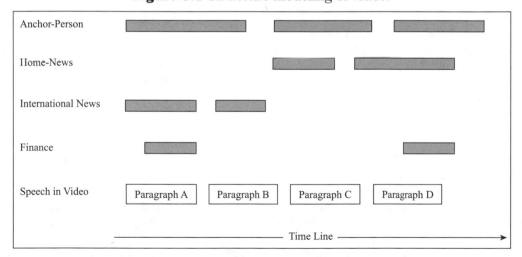

Figure 8.2 Stratification Modeling of Video.

In another multi-resolution approach, [16] models video by generating a trail of points in a low-dimensional space where each point is derived from physical features of a single frame in the video clip. Intuitively, this leads to clusters of points whose frames are similar in this reduced dimension feature space and correspond to parts of the video clip where little or no change in content is present. This modeling has a great advantage in dimensionality reduction and has good visual representation. Since this model maps the video content into another space, the features in the new space can be utilized to analyze video content that may bring in new insights.

Based on the modeling techniques presented, it can be seen that video is actually a multi-level structured medium. This feature makes video suitable to be studied in a multi-resolution view. Instead of studying the video data stream directly, one way is to transform the video stream into a trajectory in another feature space. Some phenomenon that is not observable in the original data space may have features that are observable in the new space. Another advantage is that the complexity of the problem may be reduced when the dimensionality is reduced. Other similar modeling approaches may be found in [33].

2.2 VIDEO SEGMENTATION

Video segmentation is a fundamental step in video processing and has received a lot of attention in recent years. A number of techniques have been suggested for video segmentation for both the raw and compressed video streams. These techniques can be divided broadly into 6 classes: pixel or block comparison, histogram comparison, methods based on DCT coefficients and motion vectors in MPEG encoded video sequences, feature comparison and video editing model-based methods. Some of the approaches will be described in this section.

Pixel or Region Comparison

The change between frames can be detected by comparing the difference in intensity values of corresponding pixels in the two frames. The algorithm counts the number of pixels or regions changed, and the shot boundary is declared if the percentage of the total number of pixels or region changed exceeds a certain threshold [1][8][15][24].

However, large camera and object movements in a shot may result in false shot boundaries being detected using such an algorithm. This is because these movements bring about pixel or region changes that are big enough to exceed the threshold. Fortunately, the effects brought about by the camera and object movements can be reduced to some degree by enlarging the regions to compute the average intensity values.

Histogram Comparison

Nagasaka et al. [24] presented the earliest work that demonstrates the importance of video segmentation. They observed that there is a large difference in histograms between the two frames separated by a CUT. They evaluated several basic functions for computing the difference histogram measurement for CUT detection. Their implementations are based mainly on the differences between the gray level and color histograms. Because the histogram features are more global-based, it is less sensitive to camera motion and object movement. Similar to the pixel or region based approach, when the difference between frames is above some threshold, a shot boundary is declared.

In order to eliminate momentary noise, they assume that its influence is always not more than half a whole frame, and apply sub-frame techniques to remove the effect. But this is not always true in a real case. For example, camera flash always causes the whole picture to be brightened, and large objects moving fast into the view will not follow the half-frame assumption. In this sense, their noise tolerance level is too limited.

Twin-Comparison

Although the pixel or region based algorithms and histogram-based algorithms are not perfect, these techniques are well developed and can achieve a high-level of accuracy in CUT detection. However, the accuracy of these techniques for GT detection is not high. One of the most successful early methods that attempted to handle both CUT and GT is the twin-comparison method [45]. Under the observation that a sharp change in contents between successive frames indicates a CUT, and a moderate change corresponds to a GT, the algorithm sets two thresholds. If the difference is above the upper threshold, a CUT is declared. If the difference is between the upper and a lower threshold, the algorithm starts to accumulate the difference between successive frames. When this accumulated difference exceeds the high threshold, a GT is declared.

The algorithm also analyses motion vectors in video frames to distinguish camera motions like panning and zooming from that of GTs. The algorithm is simple and intuitive, and has been found to be effective on a wide variety of video. The twin-comparison algorithm seems to be the first real approach to solve the GT problem by extending the histogram comparison method from CUT to GT domain.

DCT Coefficients in MPEG

Solving the video segmentation problem by extracting image features from decompressed video stream is obviously not very efficient. A number of researchers have developed techniques that work directly on the MPEG video stream without decompression [46][40][29][26].

In MPEG compression the image is divided into a set of 8 by 8 pixel blocks. The pixels in the blocks are transformed into 64 coefficients using the discrete cosine transform (DCT), which are quantized and Huffman entropy encoded. In this case, the most popular way to work with compressed video stream is to utilize the DCT coefficients since these coefficients in a frequency domain are mathematically related to the spatial domain.

The most well known work in compressed domains is the DC image that was introduced by [40]. The DC images are generated by some coefficients, DC+2AC, from the original frame and the reduced image has similar content representation as the old frames. They also applied template matching and global color statistic comparison (RGB color histogram) in DC images to solve the video segmentation problem [40]. They first computed the inter-frame difference sequence, and then applied a sliding window of fixed size m, using dynamic threshold to declare video transitions. Their experimental results showed that the algorithm works 70 times faster than in raw domains with similar accuracy.

Motion Information in MPEG

In MPEG compressed video, motion information is available readily in the form of motion vectors and the type of coding of the macroblocks. The compressed data consists of I-, P- and B-frames. An I-frame is completely intra-frame coded. A P-frame is predictive coded with motion compensation from past I- or P-frames. Both these frames are used for bi-directional motion compensation of B-frames.

Macroblock type information is a simple video coding feature, which is very helpful for video indexing and analysis. In I-frames, macroblocks are intra-coded. In P-frames, it is forward predicted as well as intra-coded and skipped. In B-frames, macroblock is one of the five possible types: forward predicted, backward predicted, bi-directional predicted, intra-coded and skipped. The count of these types of macroblocks and their ratios are very useful.

Zhang et al. [46] have experimented with motion-based segmentation using the motion vectors as well as DCT coefficients. Similarly Meng et al. [22] use the ratio between the numbers of intra-coded macroblocks to detect scene changes in P- and B- frames. Kobla et al. [17] use macroblock type information for the shot boundary detection. Nang et al. [25] compute macroblock type changes to detect shot boundaries.

Camera motion should be taken into account during shot segmentation. Kobla et al. [17] and Zhang et al. [46] estimate camera pan and tilt using motion vectors.

The motion vectors represent a sparse approximation to real optical flow and the macroblock coding type is encoder specific. Gargi et al. [11] mentioned that the block matching methods do not do well compared to intensity/color-based algorithms.

Video Editing Model-Based Methods

GT is produced during the video editing stage when the video editor is used to merge different video shots using a variety of techniques to produce different types of gradual transitions. Thus, GT detection can be considered as an inverse process of the video editing process and also has a direct relationship with the different editing techniques applied. By modeling techniques to produce different video transitions, researchers can find a way to solve this problem [12].

Hampapur et al. [12] proposed a framework to solve the video segmentation problem by building up different video editing models for different GTs. Their work classifies the GTs based on 3 basic editing models, namely, the chromatic edits, spatial edits and mixed chromatic and spatial edits. When processing the whole video, they calculated the chromatic images and spatial images from the differential images generated from the consecutive frames. By observing the different responses of these two images, not only can they declare the video transition, they can also roughly classify the video transitions. The disadvantage of their technique is that when faced with new types of transitions, a new threshold for the responses must be set for detection.

Other Feature-Based Approaches

There are many other approaches to handle different types of GTs. [29] adopted a statistical approach that analyzes the distribution of difference-histograms and tries to characterize different types of transitions with different distribution patterns. Zabih et al. [43] proposed a feature-based algorithm that uses the patterns of the increasing and decreasing edges in the video frames to detect GTs. In fact, up until now, more than one hundred algorithms or approaches have been proposed to solve the video segmentation problem. Recent research focus is moving from CUT to GT, from simple shot detection to more complex scene detection, and more work is being done in the compressed domain. The amount of research work carried out in this topic demonstrates that it is an important and difficult problem.

2.3 WAVELETS

Wavelets analysis is a mathematical theory that was introduced to the engineering world in recent years. Wavelets are functions that satisfy certain mathematical requirements and are used in representing data or other functions, and wavelet algorithms process data at different scales or resolutions. This enabled the gross feature as well as small detailed features of the data to be noticed at the same time, thus achieving the effects of seeing both the forest and the trees, so to speak. Working with this advantage, wavelets are suitable for multi-resolution analysis and have been used in many fields including data compression, astronomy, acoustics, nuclear engineering, sub-band coding, signal and image processing, neuron-physiology, music, magnetic resonance imaging, speech discrimination, optics, fractals, turbulence, earthquake-prediction, radar, human vision and pure mathematics.

Wavelets have also been applied to solve the video segmentation problem. [42] used wavelets to spatially decompose every frame into the low and high resolution component to extract the edge spectrum average feature to detect fade, and applied double chromatic difference on the low-resolution component to identify dissolve transitions.

2.4 TMRA

The temporal multi-resolution analysis (TMRA) approach used in this chapter is built upon the ideas presented in the early works reviewed. Essentially, TMRA models the video streams as trails in low-dimensional space as is done in [16]. However, TMRA adopts the global or spatial color representation for each video frame, and maps the frames into a fixed low dimensional space. TMRA also performs multi-resolution wavelets analysis on features as is adopted in [9][42]. However, their approaches analysed one video frame at a time and thus they are essentially single temporal resolution approaches. In contrast, TMRA considers varying number of frames, and thus different temporal resolution, in the analysis. By analysing video transitions in the multiple temporal resolution space, we are able to develop a unifying approach to characterize different types of transitions. In contrast, methods that rely on single temporal resolution need different techniques to handle different types of transitions.

The only work that uses the temporal resolution idea is [40]. However, their approach uses a fixed temporal resolution of 20, which can handle only short transitions of about 12-30 frames in length. By the simultaneous analysis of multi-temporal-resolution, TMRA is able to handle transitions of arbitrary length, and the detection is not sensitive to threshold selection. This chapter is based on our published work on TMRA for video segmentation [19] with several modifications and improvements over it.

3. BASIC THEORY

This section discusses theories related to this research. The section first presents a video model built on the content feature space. This is followed by a discussion of the feature spaces that were adopted in the research. Next, the observations on video transitions and the modeling of video transitions are discussed in a multi-temporal-resolution manner. Lastly, wavelets to model the multi-temporal resolution phenomenon are introduced.

3.1 VIDEO MODEL

Video can be mathematically modelled through the content features of the frames in the video data stream. The content of the frames is normally represented by some low-level features such as the color, shape, texture or motion. By considering a specific feature such as the color, a video stream can be represented as a trajectory in a multi-dimensional space, in this case, color space.

Figure 8.3 illustrates the trajectory of a video stream in a RGB 3-D color space. For illustration purposes a very simple mapping is performed, by computing the average RGB color for every frame in the video stream using

$$f = I_{average} = \frac{\sum I_{xy}}{w \times h} \tag{8.1}$$

where $I_{xy} = (R_{xy}, G_{xy}, B_{xy})$ represents the RGB value of every pixel (x, y) of the frame, and w and h are respectively the width and height of the frame. In other words, only one average color is used to represent a frame. In this way, every frame is mapped into a point in the RGB color space and connecting the points in their temporal order of occurrence in the video stream will map the temporal sequence of frames into a spatial trajectory of points in the color space. As a result, every video can be mapped into a spatial trajectory in a preferred feature space.

Different content features extracted from a video stream will lead to different feature spaces where the video will be mapped in a different corresponding trajectory. The dimension of a feature space depends on the dimensionality of the chosen feature. In the illustration of Figure 8.3, RGB forms a 3-D feature space. If a 64-bin color histogram is used to represent each frame, a 64-D feature space will be obtained. If the feature space is well chosen and corresponds closely by the user's perception of video contents, the similarity between any two frames is expected to be modelled closely by the distance between the two corresponding points in the space. Thus, similar frames will be closer in the feature space and dissimilar frames will be farther apart.

Assume that n-color histogram is used to represent the contents of each frame in a video. Every frame in the video can be represented as a frame vector

$$f = (x_1, x_2, ..., x_n) \tag{8.2}$$

where x_i is the ith bin of the color histogram. Then, the temporal video stream can be shown as a feature vector stream

$$v = f(t) = (x_1(t), x_2(t), ..., x_n(t)) \tag{8.3}$$

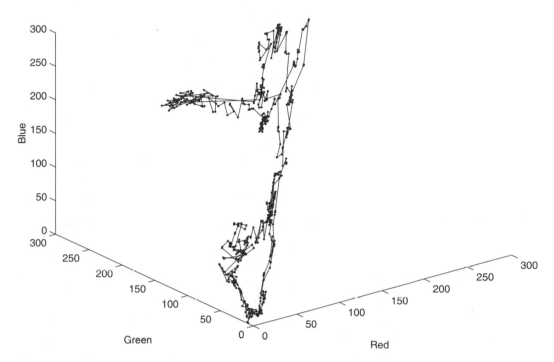

Figure 8.3 Transitions on the video trajectory in RGB color space.

3.2 FEATURE SPACE

To map the video into a trajectory in the feature space, feature descriptions of every frame must be obtained. In the uncompressed video stream where the frames are already obtained after decompression, the feature descriptions can be calculated by using image processing and machine vision techniques. For example, a 64-bin histogram description for the frame is a good feature description in color space and it can be obtained by first quantizing the image into 64 colors and then counting the pixels for every color.

In the compressed domain, image feature description of every frame is not obtained so easily as in uncompressed domain. This is because the data in the compressed video stream are DCT transformed, and the compressed video stream has its special structure that needs special treatment. To extract the image feature description, it is usually necessary to analyze the DCT coefficients for every frame. For each 8x8 blocks of a single DCT-based encoded video frame f, the DC coefficient gives the average luminance value of the whole block. In this case, all the DC coefficients for the whole frame can already be used to construct a meaningful image description about this frame at a lower spatial resolution.

While the DC coefficients can be obtained directly for I frames, it is necessary to apply motion compensation in order to extract them from the B or P frames. This is because in the MPEG coding system, the B or P frames only record the difference from reference frames for better efficiency. A method of extracting DC values from compressed MPEG stream is discussed in [40].

Thus, the image descriptions of every frame of the video stream can be obtained in both the uncompressed and compressed domains. Using the image descriptions, the video stream is mapped into the trajectory in the feature space. The mapping is actually a dimensionality reduction process. Usually, the video data is presented in a high dimensionality vector of the form

$$I_n = (I(1,1), I(1,2), ..., I(x,y), ..., I(w,h)) \; 1 \quad x \leq w, 1 \leq y \leq h \qquad (8.4)$$

where I_n is the nth frame image description, x and y are the pixel location in the image, $I(x,y)$ is the color of the pixel location at (x,y) and w and h are the width and height of the image. The feature description of the video frame is

$$F_n = (F_1, F_2, ..., F_k) \qquad (8.5)$$

where F_n is the n th frame feature description, and k is the dimension of the feature space.

In most cases, k is much smaller than $w \times h$. For an example, using a 64-bin histogram to represent a 320x240 resolution video frame reduces the dimensionality of the frame by $1/1200$. The feature space must be chosen such that it is able to effectively discriminate frames. Thus, it is very important to choose a well-defined feature space.

Besides having a low dimension, other desirable properties of a feature space for video segmentation are as follows:

- It must have a reliable similarity measure and must model the user's perception of video sequence well.

- It must have a fixed dimensionality to form a stable feature space.

- It must be easy to extract the feature vector "automatically" from the video stream.

Feature Space for Global Feature – Histogram Representations

In the raw domain of video where frames are processed after full decompression, color histogram is the most basic and effective representation of image content of every frame. It models the global content of the image and provides a good basis for evaluating different images. Even in the compressed domain, by extracting the information from the DC coefficients, a DC histogram for every frame can also be obtained.

In this research, histogram-based methods for both the uncompressed (to form the 64-bin histogram) and the compressed video streams (to form the 64-bin DC histogram) have been implemented. The cost of computing the histogram is very low. For a good trade-off between efficiency and the quality of representation, 64-bin color histogram has been shown experimentally to be sufficient for most video processing applications, including video segmentation.

Feature Space for Region-Based Feature – DC64

Since the DC coefficients for every frame can be extracted and directly reflect the average luminance value of the block, the image description formed by DC coefficients already formed a good basis for a region-based representation of every frame. The feature vector for every frame is

$$V_f = \{dc_1, dc_2, \ldots\} \tag{8.6}$$

where dc's are the DC coefficients for every block. This description can easily be used to extract the color distribution information of a frame with a 1/64-dimensionality reduction. This is called the DC image of the current frame. To achieve even lower dimensionality, the frame can be equally split into 8 by 8 sub-windows and the DC coefficients are averaged in the sub-window. Composing these 64 values can form a reduced DC image that is 8 by 8 in size. This DC64 image, or the reduced DC image, represents the spatial information that can be used as basis to represent the color-spatial information of the frame.

Feature Space for Motion-Based Feature – MA64

Since the motion vectors and the macroblock type information can be easily extracted from the MPEG compressed stream, it forms a good basis for a motion-based representation of every frame. The major problem in shot boundary detection is the flash and camera/object motion resulting in wrong transition detection. The motion based feature vector is robust in characterizing motion accurately and is used to identify: a) transition boundary points; and b) elimination of wrong transitions due to motion. A motion angle (MA) histogram representing the directions of motion is constructed. The motion vector direction is quantized to 60 values with each bin representing 6-degree increment from 0 to 360 degrees. This feature vector is concatenated with 4 more values representing counts of forward, backward, intra and skipped macroblocks in the frame. This forms the compact representation of a MA64 feature vector.

$$V_f = \{ma_1, ma_2, \ldots ma_{60}, fc, bc, ic, sc\} \tag{8.7}$$

where ma's are the motion angles, fc is the forward predicted motion vector count, bc is the backward predicted motion vector count, ic is the intra-coded macroblock count and sc is the skipped macroblock count. This feature vector characterizes motion continuity and is used to observe the level of motion activity in the video sequence.

3.3 MULTI-RESOLUTION PHENOMENA OF TRANSITIONS

One of the advantages of modeling video as a trajectory in the feature space is that it provides a natural visual representation of video stream that can be observed easily. Similar frames are closer on the trajectory in the space and different frames are farther away from each other. Thus, video transitions can be studied by observing the video trajectory in the feature space.

It is easily shown from Figure 8.3 that video shot transitions correspond to regions where the contents change rapidly. These changes represent the discontinuities in the trajectory and can be considered as the boundaries in a video, similar to the concept of edges in images. Thus, algorithms similar to edge detection in computer vision can be applied since all these points of rapid variations can be identified as local maxima in the first order derivatives of the video signals in a suitable domain.

By empirically observing GTs in many video streams, it is found that different types of GTs exist like fade-in/fade-out, dissolve, wipe and morphs. Moreover, the length of the transitions can vary widely. Sample MPEG-7 data shows that GT length can vary from 2 frames to several hundreds frames (e.g., 487 frames). In this case, it cannot be claimed that all the video transitions will bring in sufficiently "rapid" changes in the video frame content (or in the video trajectory in a proper feature space). Instead, it is true that whatever is the type or the length of the transition there will always be a change big enough that is observable at some temporal resolution. So, the transitions must be defined with respect to different resolutions. By viewing the video at multiple resolutions simultaneously, the detection of both CUTs and GTs can be unified. By employing a framework of temporal multi-resolution analysis, the information across multiple temporal resolutions can also be utilized to solve the problem of video segmentation.

3.4 WAVELETS

Based on the observations stated in the previous section, video transitions can be uniformly modeled as maxima of first order derivatives in some resolutions of the video signal. Thus, wavelets can be applied in solving the problem of video segmentation. [4][7][20[21]

Consider the temporal sequence of video frames as samples from a continuous signal $f(t)$, and the video signal in the ath temporal resolution is denotated as $f_a(t)$. Denoting the Gaussian kernel by

$$\theta(x) = \frac{1}{\sqrt{2\pi}} e^{-\frac{1}{2}x^2}$$ (8.8)

then $f_a(t)$ can be obtained by convolving $f(t)$ with $\theta(t)$

$$f_a(t) \quad f(t) \quad \theta(\frac{t}{a}) = \frac{1}{a}\theta(\frac{x}{a})f(t)dt$$ (8.9)

Here, the temporal resolution means sampling video stream by grouping frames. The higher is the a, the lower the temporal resolution will be, and the coarser the sampling will become. This is because the convolution of $f(t)$ and $\theta(t)$ removes the details (high frequency components) of $f(t)$ in time and spatial domain. Since both CUT and GT all correspond to the maxima of the first order derivatives in low resolution video signal, these transitions can be detected by simply calculating the first order derivative of $f_a(t)$ when a is sufficiently large.

The whole process of computing $f_a(t)$ in different resolutions and calculating the first order derivative is equivalent to applying Canny wavelets transformation on $f(t)$. Canny wavelets are defined from the first order derivative of the Gaussian kernel. The reason for n derivatives of the Gaussian θ

$$\theta^{(n)} = \frac{\partial^n \theta(x)}{\partial x^n} \qquad (8.10)$$

to be wavelets for $n > 0$ can be shown by the lemma that follows:

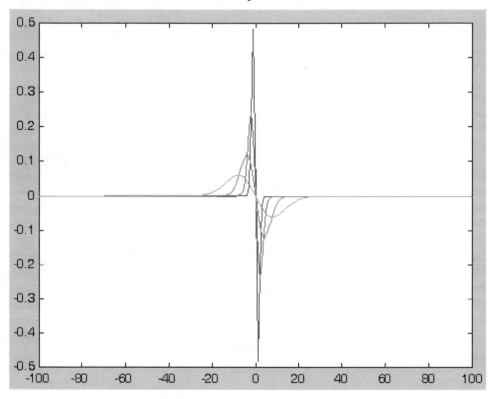

Figure 8.4 Canny wavelets in R0-R3.

Definition: Let $n \in N$. A wavelet ψ has n vanishing moments (i.e., of order n), if for all integers $k < n$:

$$\int_{-\infty}^{\infty} \varphi(x) x^k \, dx = 0 \qquad (8.11)$$

and

$$\int_{-\infty}^{\infty} \varphi(x) x^n \, dx \neq 0 \qquad (8.12)$$

Lemma: Let ϕ be an n-times differentiable function and $\phi^{(n)} \in L^2(R), \varphi^{(n)} \neq 0$. Then, it follows that $\psi = \phi^{(k)}$ is a wavelet.

Denoting $\theta_a = \frac{1}{a}\theta(\frac{x}{a})$ as the Gaussian at resolution a, then the following can be obtained:

$$W_a f(x) \quad f \quad \varphi_a(*) \quad f = (a\frac{d\theta_a}{dx})(x) \quad = a\frac{d}{dx}(f * \theta_a)(x) \quad (8.13)$$

Canny wavelets, which are shown in Figure 8.4, are derived from the first order derivative of the Gaussian kernel as $\frac{\partial \theta(x)}{\partial x}$. Here, it shows that the wavelet transform using Canny wavelets will obtain the first order derivative of the $f(t)$ at various resolutions. Thus, the problem of video transition detection can be simplified to local maxima detection after wavelet transformation.

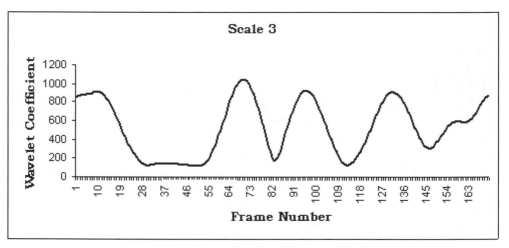

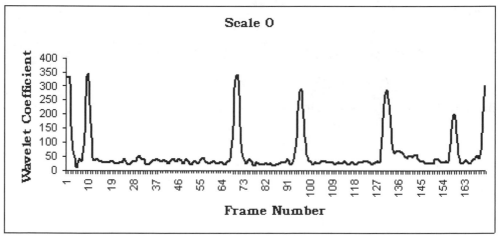

Figure 8.5 Transitions shown at two resolutions.

Figure 8.5 shows the result of using Canny wavelet transformation on a daily video stream in different temporal resolution. By carefully examining the maxima corresponding to video transitions, it can be noticed that a GT, which is not observable at a high resolution (scale 0), can be easily detected at a lower resolution (scale 3).

Although all the transitions are shown in low resolution, different phenomena can be observed at different resolutions. For CUT, the significant maxima are observable both at a high resolution and at a coarser resolution. On the other hand, GTs only show the maxima at coarse temporal resolutions. By using wavelet transform, the analysis of the information across resolutions becomes easy.

Beside the convenience of the correspondence of maxima and the boundary, wavelet also provides a simple way to measure the regularity of the signal by using Lipschitz exponent α. It is proved that

$$\left| W_{2^j} f(x) \right| = O(2^{\alpha j}) \tag{8.14}$$

The scale space can be constructed by dyadic scale 2^j. Thus, the evolution of the amplitudes of the absolute maxima can be traced across the scales. Then, the degree of smoothness and the corresponding scale where the transition exists can be obtained. Using this information, it is possible to distinguish the CUT's from GT's.

4. TMRA SYSTEM

This section discusses the design and the detailed algorithm of the temporal multi-resolution analysis framework that can handle both CUT and GT transitions in a consistent manner. The major problem in shot boundary detection is the noise introduced by camera, object motion and flashes. We discuss the solution to eliminate the wrong boundaries and improve the system precision.

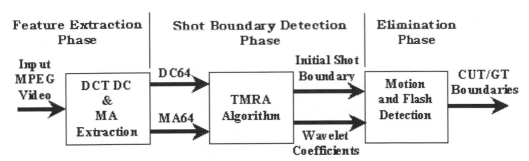

Figure 8.6 Schematic of TMRA system for shot boundary detection.

4.1 SYSTEM OVERVIEW

The TMRA system has 3 phases. In the feature extraction phase, feature vectors suitable for the TMRA are computed. In the shot boundary detection phase, transitions are characterized and TMRA algorithm is applied to obtain the transition boundaries. This result will include many wrong transitions (insertions). In the elimination phase, motion analysis and flash detection are applied to remove insertions due to motion and abrupt flash noises. Figure 8.6 shows the system architecture for shot boundary detection. The following sections discuss in detail each of these three phases.

4.2 FEATURE EXTRACTION

In the current implementations, four popular feature representations for video streams are considered. To show that the framework can fit in uncompressed domain and compressed domain, the 64-bin color histogram and the 64-bin motion angle histogram are extracted in both domains. To show that the framework can fit into any well-defined feature space, the DC64 feature space is also used for the compressed domain.

64-bin Color Histogram in Raw Domain

The color representation in the 64-bin color histogram feature space is in 6-bit coding, 2 bits for each color channel (R, G, B). Consider the color histogram $h(n)$ for frame n denoted as $h(n) = (b_1, b_2, ..., b_n)$ where b_i gives the ith color bin. The distance for frame i and frame j in this feature space (64-bin color histogram) is defined as:

$$Diff(h(i), h(j)) = \sqrt{\sum_k (b_{ik} - b_{jk})^2}$$

(8.15)

64-bin DCT DC Histogram from Compressed Domain – DC64

While it is straightforward to extract DC coefficients from I frame in MPEG video, motion compensation needs to be implemented in the DC extraction in P and B frames. The DC image is obtained after calculating the DC coefficients. In the system implementation, only the DC coefficients are used instead of DC+2AC [40]. This is because processing speed is more important in compressed domain and AC is less important when calculating the global feature such as color histogram.

The DC image obtained in compressed domain can be used to form the DCHistogram feature for videos. In this feature space, the DCHistogram for every frame is denoted as $DH(k)$, where $DH(k) = (dcl_1, dcl_2, ..., dcl_{64})$, and the dcl_j represents the j th level of luminance. The distance between two frames in this feature space (DCHistogram) can be illustrated as

$$Diff(DH(i), DH(j)) = \sqrt{\sum_k (dcl_{ik} - dcl_{jk})^2}$$

(8.16)

Furthermore, we obtain the *reduced DC image* from *DC image* by dividing the original DC image into 8x8 bins, and computing the value of each bin as the average of the DC coefficients of all pixels covered by that bin. In this way, we reduce the original DC image to 64 bins, represented by: $rdc_l = (rp_1, rp_2, ..., rp_{64})$. Here rdc_l denotes the reduced dc image for frame l, and rp_m is the m th bin of the reduced DC image. It should be noted that each bin is attached to a specific location and thus this is a color-spatial representation.

The distance between two frames in the reduced DC feature space is:

$$Diff(rdc_i, rdc_j) = \sqrt{\sum_k (rp_{ik} - rp_{jk})^2}$$

(8.17)

64-bin Motion Angle Histogram from Compressed Domain - MA64

The motion vector represents the motion of a macroblock of 16x16 pixels. It contains information about the type of temporal prediction and corresponding motion vectors used for motion compensation. A continuously strong inter-frame reference will be present in a stream as long as there is motion continuity. The temporal change in motion based feature vector can be used to define and characterize the occurrence of a transition. Significant instability of this feature is used as a sign of shot changes. So we incorporated a motion-based feature vector in our system along with the color based feature vector aiming to: (a) eliminate the falsely detected transitions caused by camera/object motion; and (b) accurately detect transition boundary. In MPEG compressed stream only forward motion vectors are used. In the case of B frames, averaging and approximations are applied to obtain forward motion vectors. Alternatively in the raw domain we compute the motion vectors using block based motion estimation such as spiral search method [44]. Since the motion vectors tend to be sparse, a 3x3 median filtering is applied to the motion vectors. Also boundary blocks are not considered, as they tend to be erroneous. Motion angle is computed as

$$ma(i, j) = \tan^{-1}\left(\frac{mv_y(i, j)}{mv_x(i, j)}\right) \tag{8.18}$$

where i, j represents macroblock row and column value.

The motion vectors obtained can be used to form the Motion Angle Histogram feature for videos. By quantizing the motion angles into intervals of 6 degrees, we can obtain a reduced MAHistogram consisting of 60 bins, represented by: $rma_l = (rp_1, rp_2, ..., rp_{60})$, where rp_m is the m th bin of the reduced motion angle. It should be noted that each bin is attached to a specific location and thus this is a color-spatial representation.

The distance between two frames in the reduced motion feature space is:

$$Diff(rma_i, rma_j) = \sqrt{\sum_k (rp_{ik} - rp_{jk})^2} \tag{8.19}$$

To make our feature vector consistent with the DC64 feature space, we extended our 60-bin motion angle feature to a 64-bin feature space (MA64) by concatenating the counts of forward predicted macroblocks, backward predicted macroblocks, intra-coded macroblocks and skipped macroblocks. In the case of raw domain block based motion estimation only forward prediction is employed in which case we forced $fc \neq bc = 0$ which is not used in any of our computations for shot boundary detection. This feature space is very useful in identifying the transition boundaries and also in observing the motion activity.

4.3 SHOT BOUNDARY DETECTION

In this phase, TMRA algorithm is applied to determine the potential transition point and their type. It performs the wavelet transformation on the color and motion based feature vectors. The choice of feature vector is application specific and some examples are the HSV color based histogram, 256-bin intensity histogram, etc.

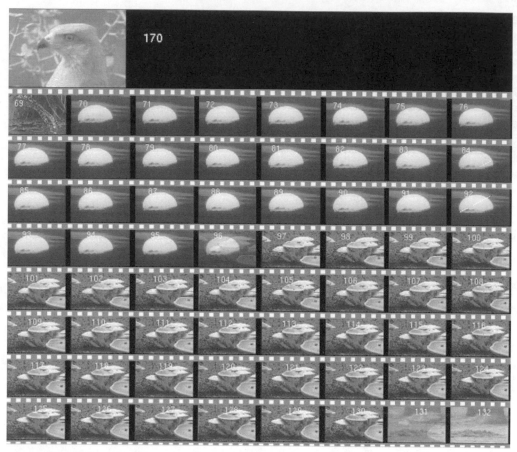

Figure 8.7 Sample frames from the sequence Echosystem.mpg.

Transition Characterization

By making the fundamental observation that a video shot boundary is a multi-resolution phenomenon, video transitions can be characterized by the following features:
- Resolution of the transition;
- Strength of the transition;
- Length of the transition.

These features can be obtained respectively by:
- Counting the scale where the transition has been found; it is also the scaling parameter of the wavelet transform a in Equation 13.
- Computing the coefficients after the wavelet transformation as $\left| W_a f(x_0) \right|$, where x_0 is the frame where the maxima is found in scale a.
- Computing the distance between start and end of transition.

In this way, all the transition points are represented in the whole framework by the three multi-resolution features.

TMRA Algorithm

Using the ideas in transition characterization, a *temporal multi-resolution algorithm* was developed to detect and characterize both CUT and GT shot boundaries. To illustrate the overall algorithm, we use a video named "Echosystem.mpg." This is an MPEG-1 compressed video sequence with 170 frames. The video sequence contains 2 CUT's and 3 GT's. Sample sequence with frames 69-132 is shown in Figure 8.7. Figure 8.8 shows the frames following the 5 transitions. The overall algorithm is described as follows.

Figure 8.8 Post transition frames of the video sequence Echosystem.mpg.

Table 8.1 Transition Boundary for Echosystem.mpg

Transition Type	Start Frame	End Frame
GT	9	11
CUT	70	71
GT	95	97
GT	130	132
CUT	159	160

1. Extract the feature vectors $f(n)$ from the video stream

In the uncompressed domain, two feature vectors (DCHistogram and MA64) are obtained. DCHistogram is composed of the 64 color-histogram of every frame. MA64 is computed from the motion vectors computed from the block based motion estimation algorithm in raw domain.

In the compressed domain, three feature vectors are obtained in the model, *DC64, DCHistogram and MA64*. In the DC cases, DC coefficients are extracted first. The DCHistogram is obtained by counting the 64-bin luminance histogram of the DC image. The *DC64* is obtained by splitting the *DC image* into 8 by 8 sub-windows and averaging every window to form the 64D feature vector. MA64 is computed from the extracted motion vectors from the compressed stream and along with the macroblock type counts.

2. ***Apply wavelet transformation on the video feature vectors***
This is done by computing the *Canny wavelet transform* of $f(x)$ for the temporal resolutions of $a = 2^j$ for j = 0 to 4. We thus obtain a set of coefficients of the form $W_c = |W_a f(x)|$. By applying the transformation on the compressed domain DC64 and MA64 feature vectors, the corresponding wavelet coefficients obtained are W_{DC64} and W_{MA64}, respectively. Figure 8.9 shows the W_{DC64} values of the video "Echosystem.mpg."

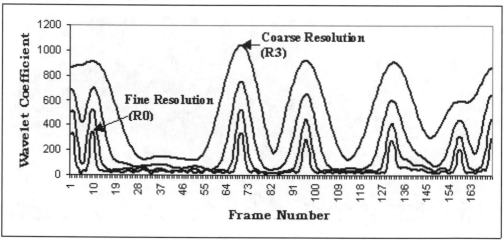

Figure 8.9 Examples of coefficients after the transformation (W_{DC64}).

3. ***Select the potential transition points at coarsest resolutions based on $|W_a f(x)|$ of DC64.***
All *local maxima* are selected as potential *transition points*. The selection of local maxima is quite straightforward by tracing the feature vectors. When the feature vectors are changing from increasing to decreasing, the points are flagged as local maxima. This can be illustrated in Figure 8.10. In the illustration, the arrows indicate the trend of the signal, either increasing or decreasing, and the circles indicate the selected local maxima. Most of the local maxima in the higher temporal resolution space are either false or will be merged or transformed into a smaller number of local maximas in a coarser resolution space. And in the coarsest resolution space, only valid local maxima that correspond to transitions will be likely to remain. Thus for efficiency and effectiveness reasons, this step starts by selecting potential transition points in the coarsest resolution space.

4. ***Find the 'start' and 'end' of each potential transition point using $|W_a f(x)|$ of MA64 at finest resolution.***
The goal of video segmentation is not only to detect the occurrence of a transition, but also to locate the exact positions of the CUT/GT to segment the video. In a GT, both start position and end position need to be detected. Low resolution of W_{DC64} and W_{MA64} helps to detect the occurrence; while high-resolution helps to characterize the start and end of the transition.

W^0_{DC64} represents high resolution R0 and W^3_{DC64} represents coarse resolution R3, since gradual transition will have a slightly larger difference between frames than those within the normal shot. So in the higher resolution, the boundaries would show up as local maxima points. To identify this boundary, we use both the DC64 and MA64 wavelet coefficients. For coarse resolution (Resolution 3) DC64 wavelet coefficients are used and for high resolution (Resolution 0) MA64 wavelet coefficients are used. The reason for this choice is because DC64 feature space fails to characterize the beginning and ending frames of the transitions accurately at the high resolution. This is due to the fact that the rate of change in DC64 feature space is not high and doesn't result in a distinguishable peak. Hence we designed a new feature space based on the direction of motion along with counts representing static (skip) and intra (blocks having significant change) blocks. As a result of this we observe that even a gradual change resulted in an abrupt spike (peak) at the start and end of the transition. This is captured as the boundary of the transition.

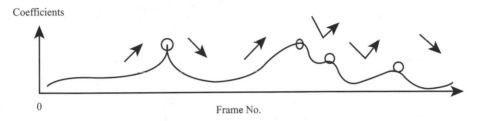

Figure 8.10 Selecting the local maxima.

For each potential transition point identified as local maxima in W^3_{DC64}, we use it as the anchor point to search for the two nearest and strongest local maxima in W^0_{MA64}. Thus for each potential transition we can identify the 'start' and the 'end' of the transition. This is illustrated in Figure 8.11. The two strong local maxima in W^0_{MA64} are determined within the range of valley points enclosing the anchor point in W^3_{DC64}. Two valley points enclose each potential transition point.

5. ***Remove the potential transition points that correspond to noise by adaptive threshold***

The problem of choosing an appropriate threshold is a key issue in applying TMRA for shot boundary detection. Heuristically chosen global threshold is not suitable as the shot content changes from scene to scene. So adaptive thresholds are better than a simple global threshold. The wavelet coefficient for DC64 will have many weak local maxima. These points should be deleted before the rest of the algorithm steps are applied. This will improve the performance of the system.

The adaptive threshold is determined by using a sliding window. The size of sliding window is a very important factor, which affects the final result directly. When the sliding window size is small, the threshold is low and more wrong transition points are selected. When the sliding window size is large, the threshold is higher, and thus may cause some correct transitions to be missed. By heuristics, we chose the sliding window size to be the sum of maximum distance between local maxima points and valley points. The threshold value is the average of local maxima points within the sliding window range. All local maxima below the adaptive threshold value are removed. This is illustrated in Figure 8.12.

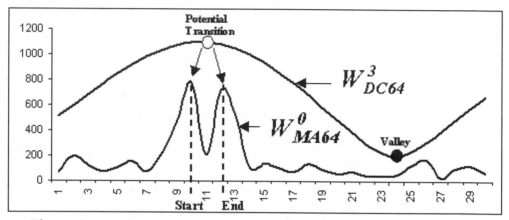

Figure 8.11 Finding the start and end frame number of a transition.

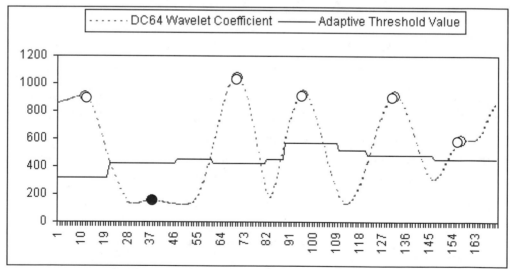

Figure 8.12 Adaptive threshold for wavelet coefficients.

6. *Classify transition type*

This is done by observing that the strength of potential transition points that correspond to noise tends to get severely attenuated at coarser resolutions, while those that correspond to CUT and GT will not be affected much. Thus by starting from coarse resolution space, only those genuine transitions remain. For each potential transition, we classify the transition points that correspond to CUT from those that correspond to GT by determining the distance between the 'start' and 'end' frames. For CUT, the frame difference is less than two and for GT it is more than two frames.

4.4 ELIMINATION

The last phase of the TMRA algorithm is the elimination of wrong transitions due to motion and abrupt flashes in the shot. The following sections give a brief description on the method to characterize such activities in the multi-resolution framework.

Abrupt Flash Detection

Flash occurrences are common in video sequences, which often makes the shot detection method wrongly classify them as cut. In video segmentation system, flash detection is one of the difficult tasks that seriously affect the system performance. With the TMRA, this problem can be easily resolved by comparing W_{MA64} at different resolutions. Figure 8.13 shows an example where the strength of local maximas corresponding to a flash decreases drastically as the resolution decreases. This is because the duration of flash tends to be short. It thus results in large local maxima at higher resolution space, and attenuates quickly in lower resolution space. This can be extended to any other similar effects of flash, for example, the fire and fireworks, etc. By using this characteristic we can detect and remove such abrupt noises.

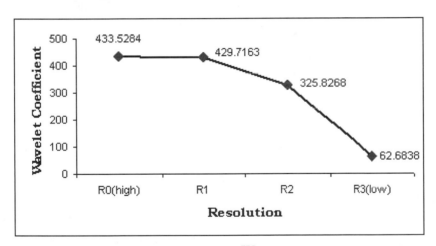

Figure 8.13 Decreasing strength of W_{MA64} with lowering resolution corresponding to a flash.

Camera and Object Motion Detection

Fast camera and object motions account for a large number of wrong detection of transitions. Camera motion and object motion have the same effect as the gradual transition (for slow motion) and cut (fast motion). It is very difficult to distinguish them from correct transitions. With the TMRA framework, the MA64 feature vector is specially designed to capture the motion information.

As we expect the changes during a gradual transition to be relatively smooth, in principle, we expect the mean absolute differences (MAD) of DC64 and MA64 for the correct transitions (CUT/GT) to be consistent. Figure 8.14 illustrates the consistent and non-consistent portions of the mean absolute differences for DC64 and MA64 features. At CUT and GT points, MADs are consistent across

DC64 and MA64. If the MAD changes are not consistent across DC64 and MA64, we conclude that it is a wrong transition caused by camera/object motion.

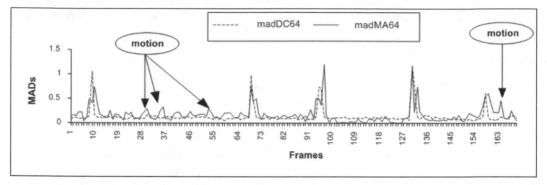

Figure 8.14 Mean absolute differences for DC64 and MA64 features.

For each potential transition, the quadratic distance between mean absolute differences (QMAD) of the feature vector is computed. Three such quadratic MADs are computed between frames selected in the following manner.

$$QMAD_{before} = \left(\sum_{i=start-k}^{start-1} \sum_{j=i+1}^{start} |f_i - f_j| \right) / C_k^2 \qquad (8.20)$$

$$QMAD_{int\ ra} = \left(\sum_{i=start}^{end-1} \sum_{j=i+1}^{end} |f_i - f_j| \right) / C_{end\ start+1}^2 \qquad (8.21)$$

$$QMAD_{after} = \left(\sum_{i=end}^{end+k-1} \sum_{j=i+1}^{end+k} |f_i - f_j| \right) / C_k^2 \qquad (8.22)$$

where *start* and *end* represents the start and end frame number of potential transition. *k* is a scalar chosen to be between 4-9. C_k^r is the normalizing factor in the above equations.

Here, $QMAD_{int\ ra}$ is computed for the range of (*start*-1) to (*end*+1);

$QMAD_{before}$ is computed based on the sequence of frames from (*start*-k) and *start*; and

$QMAD_{after}$ is computed based on the frame sequence *end* and (*end*+k)

The QMADs for the potential transitions (CUTs and GTs) are computed in both the DC64 and MA64 feature spaces. The following condition is checked to decide whether a transition is due to motion or is a real CUT/GT transition.

$$\left(\begin{array}{c} \left(QMAD_{int\ ra}^{(dc)} < \left(QMAD_{before}^{(dc)} + QMAD_{after}^{(dc)} \right)/2 \right) \wedge \\ \left(QMAD_{int\ ra}^{(mv)} < \left(QMAD_{before}^{(mv)} + QMAD_{after}^{(mv)} \right)/2 \right) \end{array} \right) \qquad (8.23)$$

During a transition, the $QMAD^{(dc)}$ value within the transition tends to be more than the QMAD value in the *before* and *after* regions. The motion angle variation within the transition is minimal compared to the neighbouring *before* and *after* regions. By using this approach we can eliminate 70-80% of falsely detected transitions due to camera and object motions.

5. SIMULATION RESULTS

The MPEG-7 data set consists of 10 categories of video in over 13 hours. The 8 categories compose 12.5 hours, which contained about 5256 CUTs and 1085 GTs. The ground truth for the 8 categories of videos was manually established. The ground truth contains the location and the type of all transitions found in the video. Table 8.3 shows the distribution of different transition types in different categories.

Table 8.2 Videos for Evaluation

# of files	Time(hh:mm:ss)	# of Frames	Kbytes
32	12:31:57	1,177,723	10,696,164

Table 8.3 Transition Type Distribution

Category	Cut	Dissolve	Fade	Other	Wipe
Animation	85.3%	8.6%	2.9%	3.1%	0.0%
Commercial	87.0%	4.3%	1.0%	7.7%	0.0%
Documentary	67.4%	19.0%	7.8%	5.8%	0.1%
Home video	100.0%	0.0%	0.0%	0.0%	0.0%
Movie	96.9%	0.3%	2.4%	0.3%	0.1%
MTV&TV	76.2%	17.8%	0.1%	3.5%	2.4%
News	88.2%	9.6%	0.1%	1.4%	0.6%
Sports	68.2%	17.4%	0.0%	10.7%	3.6%
TOTAL:	82.9%	10.4%	2.1%	3.6%	0.9%

5.1 EVALUATION OF RESULTS

The effectiveness of the algorithm was evaluated on the whole MPEG-7 test data set using the *precision* and *recall* measures, which are widely used in the field of information retrieval.

$$Precision = \frac{NumberofCorrectlyDetectedTransitions}{TotalNumberofTransitionsDetected} \times 100\% \qquad (8.24)$$

$$Recall = \frac{correctDetected}{totalTransition} \times 100\% \qquad (8.25)$$

The twin-comparison algorithm, which is the most commonly used technique today to detect GTs, was first run on the data set. Here, the twin-comparison algorithm was implemented using the 64-bin color histogram, but without performing motion detection. It can be noticed in Table 8.4 that the twin-comparison method has a very low recall and precision when detecting GTs. This is because the algorithm is threshold-sensitive, and requires different thresholds for different video types

Table 8.4 Result Using Twin-Comparison

Transition Type	Precision	Recall
CUT	40%	92%
GT	15%	58%

Next, we apply the TMRA algorithm using the same feature space (64-bin color histogram). Table 8.5 summarizes the initial result that demonstrates that TMRA approach outperforms the twin-comparison method especially on GT detection. The TMRA algorithm could achieve very high recall of over 97% for both CUT's and GTs. However, there is too much false detection that made our precision not satisfactory. To demonstrate that the TMRA framework is independent of the content feature used, the framework has also been applied in other feature spaces like DCHistogram and DC64 in the compressed domain. The results are respectively shown in Table 8.6 and Table 8.7. The result in Table 8.6 shows that the TMRA can achieve good results in this feature space. It can also be seen that the precision for GT detection is increased. This is because DC histogram smoothes the noise more effectively. Thus, less false detection was introduced in the compressed domain.

Table 8.5 Result on TMRA using 64-bin Histogram Without Elimination Phase

Transition Type	Precision	Recall
CUT	68%	98%
GT	51%	97%

Table 8.6 Result on TMRA using DC Histogram Without Elimination Phase

Transition Type	Precision	Recall
CUT	75%	97%
GT	68%	97%

Table 8.7 Result on TMRA using DC64 Without Elimination Phase

Transition Type	Precision	Recall
CUT	66%	98%
GT	53%	96%

Table 8.6 and Table 8.7 shows that the use of compressed domain feature 64-bin DC histogram and DC64 is effective. This resulted in the improvement in performance over the use of 64-bin color histogram (uncompressed domain). This is extremely encouraging as it demonstrates the effectiveness of using compressed domain features in video boundary detection. Among the two-feature

space for compressed video, the 64-bin DC histogram gives the best performance. In particular, it achieves high recall of 97% and precision of over 68% for GT's. The low precision of only 68% for GTs is due to the high number of false detections. Most of the false GT detections were due to motion (either camera motion or object motion). Thus, it is necessary to detect motions and to distinguish phenomena of motion change from transitions.

After applying the elimination phase using the motion-based feature MA64, many false transitions were eliminated and resulted in very high precision with little sacrifice in recall values. From the results shown in Table 8.8, it is observed that the use of motion information can increase the precision from 68% to 87% for GT's while maintaining excellent recall value of 90%.

Table 8.8 Result on TMRA Integrated with Elimination Phase Using DC64 and MA64

Transition Type	Precision	Recall
CUT	96%	92%
GT	87%	90%

All the results show that the TMRA framework is effective for video transition detections and can be applied with different frame features in raw domain or in compressed domain.

6. CONCLUSION

In this work, it can be shown that a temporal video sequence can be visualized as a trajectory of points in the multi-dimensional feature space. By studying different types of transitions in different categories of videos, it can be observed that the shot boundary detection is a temporal multi-resolution phenomenon. From this insight, a TMRA algorithm is developed for video segmentation by using Canny wavelet and applying multi-temporal-resolution analysis on the video stream. The solution is a general framework for all kinds of video transitions.

TMRA is implemented and tested on the whole MPEG-7 video data set consisting of over 13 hours of video. The results demonstrate that the method is very effective and has good tolerance to most common types of noise occurring in videos. Also, the results show that this method is independent of the feature space and can work particularly well in compressed domain DC features.

Further enhancements of this work can be carried out as follows. First, we will investigate the use of other features especially those at the semantic level to analyse video data. A well-constructed feature space can help not only to improve the performance of the algorithm, but also to perform structure extraction of video data.

Next, work can also be done to select more appropriate mother wavelets in a suitable feature space to perform TMRA on the video stream. An appropriate choice of wavelet and feature space may result in better techniques to detect and recognize gradual transitions.

Acknowledgment
The authors would like to acknowledge the support of the National Science and Technology Board, and the Ministry of Education of Singapore for supporting this research under research grant RP3989903.

REFERENCES

[1]   Aigrain and Joly, The automatic real-time analysis of film editing and transition effects and its application. *Comput. & Graphics*, 1994. Vol. 18, No. 1, pp. 93-103.

[2]   T.-S. Chua and J. C. McCallum, *Multimedia Modeling*, in Encyclopedia of Computer Science and Technology, A. Kent and J.G. Williams, Eds., Marcel Dekker, Vol. 35, 1996, pp. 253-287.

[3]   T.-S. Chua and L.-Q. Ruan, *A video retrieval and sequencing system. ACM Trans. Information Systems*, Oct. 1995, Vol. 13, No. 4, pp. 373-407.

[4]   A. Cohen and R. D. Ryan, *Wavelets and Multiscale Signal Processing*, Chapman and Hall Publishers, 1995.

[5]   J. M. Corridoni, and A. Del Bimbo, Strucutred Representation and Automatic Indexing of Movie Information Content, *Pattern Recognition*, 1998, Vol. 31, No.12, pp. 2027-2045.

[6]   G. Davenport, T. A. Smith, and N. Pincever, Cinematic Primitives for Multimedia, *IEEE Computer Graphics and Applications*, July 199, Vol. 11, No.4, pp. 67-74

[7]   C. J. G. Evertsz, K. Berkner and W. Berghorn., *A Local Multiscale Characterization of Edges applying the Wavelet Transform*. Proc. Nato A.S.I., Fractal   Image   Encoding   and   Analysis,   Trondheim,   July   1995.

[8]   A. M. Ferman and A. M. Tekalp, Editing Cues for Content-Based Analysis and Summarization of Motion Pictures. In SPIE Conference on Storage and Retrieval for Image and Video Database IV, 1998, Vol. 3312.

[9]   A. Finkelstein, C. E. Jacobs, and D. H. Salesin, *Multiresolution Video*, in Proc. of the 23rd Annual Conference on Computer Graphics, 1996, pp. 281-290.

[10]   R. M. Ford, *A Fuzzy Logic Approach to Digital Video Segmentation*, in SPIE Conference on Storage and Retrieval for Image and Video Database IV, 1998, Vol. 3312.

[11]   U. Gargi, R. Kasturi, and S. H. Strayer, *Performance Characterization of Video Shot Change Detection Methods*, IEEE Transactions on Circuits and Systems for Video Technology, Feb. 2000, Vol. 10, No. 1.

[12]   A. Hampapur, R. Jain, and T. E. Weymouth, *Production Model Based Digital Video Segmentation*, Multimedia Tools and Applications, 1995, Vol. 1, No. 1, pp 9-46.

[13]   F. Idris and S. Panchanathan, *Review of Image and Video Indexing Techniques*, Journal of Visual Communication and Image Representation, June 1997, Vol. 8, No. 2, pp. 146-166.

[14]   H. Jiang, A.Helal, A. K. Elmagarimid and A. Joshi, *Scene Change Detection Techniques for Video Database Systems, ACM Multimedia Systems*, 1998, Vol. 6, pp. 186-195.

[15]   R. Kasturi and R. Jain, *Dynamic Vision*. Computer Vision: Principles, R. Jain and R. Kasturi, Eds., IEEE Computer Society Press, Washington, 1991, pp. 469-480.

[16]   V. Kobla, D.S. Doermann, and C. Faloutsos, *VideoTrails: Representing and Visualizing Structure in Video Sequences.* in Proceedings of ACM Multimedia Conference, 1997: pp. 335-346.

[17]   V. Kobla, D.S. Doermann, and K.-I Lin, *Archiving, indexing and retrieval of video in compressed domain, in Proc. SPIE Conf. Storage and Archiving Systems*, SPIE, 1996] Vol. 2916: pp. 78-89.

[18]   Y. Li, Y. Lin, and T. S. Chua, *Semantic Feature Model for Text Classification and Segmentation*, School of Computing, National University of Singapore: Singapore, 2000.

[19]   Y. Lin, M. S. Kankanhalli, and T.S. Chua, *Temporal Multi-resolution Analysis for video Segmentaton*, in Proc. SPIE Conf. Storage and Retrieval for *Media Database VIII*, SPIE, Jan. 2000, Vol. 3970, pp.494-505.

[20]   S. Mallat, *Multiresolution Approximations and Wavelet Orthonormal Bases of $L^2(R)$*, American Mathematical Society, 1989.

[21]   S. Mallat and S. Zhong, *Signal Characterization from Multiscale Edges*, IEEE, 1990, pp. 132-145.

[22]   J. Meng, Y. Juan, and S.-F. Chang, *Scene Change Detection in an MPEG Compressed Video Sequence*, in IS&T/SPIE Symposium Proceedings, Feb. 1995, Vol. 2419.

[23]   K. S. Mohan and T. S. Chua, *Video modeling using strata-based annotation, IEEE Multimedia*, 2000, Vol. 7, pp. 68-74.

[24]   A. Nagasaka and Y. Tanaka, *Automatic Video Indexing and Full-Video Search for Object Appearances*, *Visual Database Systems II*, L.M. Wegner and E. Knuth, Eds., Elsevier, 1992, IFI, pp. 113-127.

[25]   J. Nang, S. Hong, and Y. Ihm, *An Efficient Video Segmentation Scheme for MPEG Video Stream Using Macroblock Information*, in *Proc. ACM Multimedia 99*, Oct. 1999. Orlando, FL.

[26]   C.W. Ngo, T. C. Pong and R. T. Chin, *A Survey of Video Parsing and Image Indexing Techniques in Compressed Domain*, in *Symposium on Image, Speech, Signal Processing, and Robotics (Workshop on Computer Vision)*, Hong Kong, 1998, Vol. 1, pp. 231-236.

[27]   N. V. Patel and I. K. Sethi, *Compressed Video Processing for Cut Detection. IEEE Proceedings Visual Image Signal Processing*, 1996, Vol. 143, No. 5, pp. 315.

[28]   L. Q. Ruan and T. S. Chua, *A Frame-Based Interactive Video Retrieval System*, in Proc. 2nd International Computer Science Conference, Data and Knowledge Engineering: Theory and Applications, 1992.

[29]   I. K. Sethi and N. Patel, *A Statistical Approach to Scene Change Detection*, in SPIE Conference on Storage and Retrieval for Image and Video Database III, 1995, Vol. 2420, pp. 329-338.

[30]   B. Shahraray, *Scene Change Detection and Content-based Sampling of Video Sequences.* SPIE Conference on Digital Video Compression: Algorithms and Technologies, 1995] Vol.2419: pp. 2-13.

[31]   K. Shen and E. J. Delp, *A Fast Algorithm for Video Parsing Using MPEG Compressed Sequences*, in Proc. IEEE International Conference on Image Processing, Oct. 1995, pp. 252-255.

[32]   M. H. Song, T. H. Kwon, W. M. Kim, H. M. Kim, and B. D. Rhee, *On Detection of Gradual Scene Changes for Parsing of Video Data*, in SPIE Conference on Storage and Retrieval for Image and Video Database IV, 1998, Vol. 3312.

[33]   X. D. Sun, M. S. Kankanhalli, Y. W. Zhu, and J. K. Wu., *Content-Based Representative Frame Extraction for Digital Video,* in Proc. IEEE International Conference on Multimedia Computing and Systems, Jul. 1998,   pp. 190-193.

[34]   M. S. Toller, H. Lewis, and M. S. Nixon, *Video Segmentation using Combined Cues,* in SPIE Conference on Storage and Retrieval for Image and Video Database IV, 1998, Vol. 3312.

[35]   Y. Tonomura, A. Akutsu, Y. Taniguchi, and G. Suzuki, *Structured Video Computing, IEEE Multimedia,* 1994, Vol. 1, No. 3, pp. 34-43.

[36]   Y.-P. Wang and S. L. Lee, *Scale-Space Derived From B-Spline, IEEE Trans. on Pattern Analysis & Machine Intelligence,* October 1998, Vol.20, No.10.

[37]   J. Wei, M. S. Drew, and Z. N. Li, *Illumination Invariant Video Segmentation by Hierarchical Robust Thresholding,* in SPIE Conference on Storage and Retrieval for Image and Video Database IV, 1998, Vol. 3312.

[38]   C. S. Won, D. K. Park and S. J. Yoo, *Extracting Image Features from MPEG-2 Compressed Stream,* in SPIE Conference on Storage and Retrieval for Image and Video Database IV, 1998, Vol. 3312.

[39]   W. Xiong and J. C. M. Lee, *Automatic Dominant Camera Motion Annotation for Video,* in SPIE Conference on Storage and Retrieval for Image and Video Database IV, 1998, Vol. 3312.

[40]   B. L. Yeo and B. Liu, *Rapid Scene Analysis on Compressed Video,* IEEE *Transactions on Circuits and Systems for Video Tech*nology, Dec. 1995, Vol. 5, No. 6.

[41]   M. M. Yeung, B. L. Yeo, W. Wolf, and B. Liu, *Video Browsing using Clustering and Scene Transitions on Compressed Sequences,* in Proc. Multimedia Computing and Networking , San Jose, Feb. 1995.

[42]   H. H. Yu, Wayne, and Wolf, *Multi-resolution Video Segmentation Using Wavelet Transformation,* in SPIE Conference on Storage and Retrieval for Image and Video Database IV, 1998, Vol. 3312, pp. 176-187.

[43]   R. Zabih, J. Miller, and K. Mai, *A Feature-Based Algorithm for Detecting and Classifying Scene Breaks,* in Fourth ACM Multimedia Conference, 1995, pp. 189-200.

[44]   Th. Zahariadis and D. Kalivas, *A Spiral Search Algorithm for fast estimation of block motion vectors,* in EUSIPCO-96, Italy, September 1996.

[45]   H.J. Zhang, A. Kankanhalli, and S. W. Smoliar, *Automatic partitioning of full-motion video, ACM Multimedia Systems,* July 1993, Vol. 1, No. 1, pp. 10-28.

[46]   H.J. Zhang, C. Y. Low, Y. Gong, and S. W. Smoliar, *Video Parsing and Browsing Using Compressed Data, Multimedia Tools and Applications,* 1995, Vol. 1, No. 1, pp. 91-111.

9

VIDEO SUMMARIES THROUGH MULTIMODAL ANALYSIS

Michael A. Smith[1], Howard D. Wactlar[2] and Michael G. Christel[2]

[1]AVA Media Systems, Austin, TX, USA

[2]Department of Computer Science
Carnegie Mellon University
Pittsburgh, PA, USA
msmith@savasystems.com, hdw@cs.cmu.edu,
christel@cs.cmu.edu

1. INTRODUCTION

Automatic video summarization requires characterization and a means to rank or select a small subset of a video as summary output. For accurate characterization, techniques must be used that incorporate audio, image and language features from video. Multimodal analysis uses different features to achieve improved results over a single mode of data. A single modal feature description may be combined or used individually to select video for summarization.

The accuracy of a video summary is often based on subjective matters such as ease-of-use, visual display and user control. User-studies and feedback surveys are usually the preferred means to assess an accuracy or quality measure to the final result. Visualization and display are also important elements in creating video summaries. A summary system may accurately locate the most important areas of a video, but the display of the results must be shown in a manner that is useful to the viewer.

The remainder of the chapter is presented as follows: Section 2, A description of image, audio, language and other video features; Section 3, Video surrogates and the process of combining video features for characterization; Section 4, Methods for evaluation and user-studies; and Section 5, Visualization techniques for summaries.

2. IMAGE, AUDIO AND LANGUAGE FEATURES

The techniques for single modal image, audio and language analysis have been studied in a number of applications. Video may consist of all these modalities and its characterization must include them individually or in combination. This requires integrating these techniques for extraction of significant information, such as specific objects, audio keywords and relevant video structure. In order to characterize video, features may be visual, audible or some text-based interpretation.

A feature is defined as a descriptive parameter that is extracted from a video stream [54]. Features may be used to interpret visual content, or as a measure for similarity in image and video databases. In this section, features are described as 1. Statistical - Features are extracted from an image or video sequence without regard to content, 2. Compressed Domain – Features extracted from compressed data and 3. Content-Based – Features that attempt to describe content.

2.1 STATISTICAL FEATURES

Certain features may be extracted from image or video without regard to content. These features include such analytical features as shot changes, motion flow and video structure in the image domain, and sound discrimination in the audio domain. In this section we describe techniques for image difference and motion analysis as statistical features. For more information on statistical features, see Section V.

2.1.1 Image Difference

A difference measure between images serves as a feature to measure similarity. In the sections below, we describe two fundamental methods for image difference: Absolute difference and Histogram difference. The absolute difference requires less computation, but is generally more susceptible to noise and other imaging artifacts, as described below.

Absolute Difference
This difference is the sum of the absolute difference at each pixel. The first image I_t is analyzed with a second image, I_{t-T}, at a temporal distance T. The difference value is defined as,

$$D(t) = \sum_{i=0}^{M} \left| I_{(t-T)}(i) - I_t(i) \right|$$

where M is the resolution, or number of pixels in the image. This method for image difference is noisy and extremely sensitive to camera motion and image degradation. When applied to sub-regions of the image, $D(t)$ is less noisy and may be used as a more reliable parameter for image difference.

$$D_s(t) = \sum_{j=S}^{\frac{H}{n}} \sum_{i=S}^{\frac{W}{n}} \left| I_{(t-T)}(i, j) - I_t(i, j) \right|$$

$D_s(t)$ is the sum of the absolute difference in a sub-region of the image, where S represents the starting position for a particular region and n represents the number of sub-regions. H and W are image height and width respectively.

We may also apply some form of filtering to eliminate excess noise in the image and subsequent difference. For example, the image on the right in Figure 9.1 represents the output of a Gaussian filter on the original image on the left.

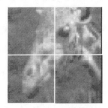 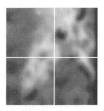

Figure 9.1. Left: original; Right: filtered.

Histogram Difference

A histogram difference is less sensitive to subtle motion, and is an effective measure for detecting similarity in images. By detecting significant changes in the weighted color histogram of two images, we form a more robust measure for image correspondence. The histogram difference may also be used in sub-regions to limit distortion due to noise and motion.

$$D_H(t) = \sum_{v=0}^{N} \left| H_{(t-1)}(v) - H_t(v) \right|$$

The difference value, $D_H(t)$, will rise during shot changes, image noise and camera or object motion. In the equation above, N represents the number of bins in the histogram, typically 256. Two adjacent images may be processed, although this algorithm is less sensitive to error when images are separated by a spacing interval, D_i. D_i is typically on the order of 5 to 10 frames for video encoded at standard 30 fps. An empirical threshold may be set to detect values of $D_H(t)$ that correspond to shot changes. For inputs from multiple categories of video, an adaptive threshold for $D_H(t)$ should be used.

If the histogram is actually three separate sets for RGB, the difference may simply be summed. An alternative to summing the separate histograms is to convert the RGB histograms to a single color band, such as Munsell or LUV color [36].

2.1.2 Video Shot Change

An important application of image difference in video is the separation of visual shots [5]. Zhang *et al.* define a shot as a set of contiguous frames representing a continuous action in time or space [23, 60]. A simple image difference represents one of the more common methods for detecting shot changes. The difference measures, $D(t)$ and $D_H(t)$, may be used to determine the occurrence of a shot change. By monitoring the difference of two images over some time interval, a threshold may be set to detect significant differences or changes in scenery. This method provides a useful tool for detecting shot cuts, but is susceptible to errors during transitions. A block-based approach may be used to reduce errors in difference calculations. This method is still subject to errors when subtle object or camera motion occurs.

The most fundamental shot change is the video cut. For most cuts, the difference between image frames is so distinct that accurate detection is not difficult. Cuts between similar shots, however, may be missed when using only static properties such as image difference. Several research groups have developed working techniques for detecting shot changes through variations in image and histogram difference [5, 10, 11, 12, 23, 35].

A histogram difference is less sensitive to subtle motion, and is an effective measure for detecting shot cuts and gradual transitions. By detecting significant changes in the weighted color histogram of each successive frame, video sequences can be separated into shots. This technique is simple, and yet robust enough to maintain high levels of accuracy.

2.1.3 Video Shot Categories

There are a variety of complex shot changes used in video production, but the basic premise is a change in visual content. Certain shot changes are used to imply different themes and their detection is useful in characterization. Various video cuts, as well as other shot change procedures, are listed below [53].

- **Fast Cut** - A sequence of video cuts, each very short in duration.
- **Distance Cut** - A camera cut from one perspective of a shot to another.
- **Inter-cutting** - Shots that change back and forth from one subject to another.
- **Fades and Dissolves** - A shot that fades over time to a black background and shots that change over time by dissolving into another shot.
- **Wipe** - The actual format may change from one genre to the next.

2.1.4 Motion-based Shot Change Analysis

An analysis of the global motion of a video sequence may also be used to detect changes in scenery [57]. For example, when the error in optical flow is high, this is usually attributed to its inability to track a majority of the motion vectors from one frame to the next. Such errors can be used to identify shot changes. A motion-controlled temporal filter may also be used to detect dissolves and fades, as well as separate video sequences that contain long pans. The use of motion as a statistical feature is discussed in the following section. The methods for shot detection described in this section may be used individually or combined for more robust segmentation.

2.1.5 Motion Analysis

Motion characteristics represent an important feature in video indexing. One aspect is based on interpreting camera motion [2, 53]. Many video shots have dynamic camera effects, but offer little in the description of a particular segment. Static shots, such as interviews and still poses, contain essentially identical video frames. Knowing the precise location of camera motion can also provide a method for video parsing. Rather than simply parse a video by shots, one may also parse a video according to the type of motion.

An analysis of optical flow can be used to detect camera and object motion. Most algorithms for computing optical flow require extensive computation, and more often, researchers are exploring methods to extract optical flow from video compressed with some form of motion compensation. Section 2.2 describes the benefits of using compressed video for optical flow and other image features.

Statistics from optical flow may also be used to detect shot changes. Optical flow is computed from one frame to the next. When the motion vectors for a frame are randomly distributed without coherency, this may suggest the presence of a shot change. In this sense, the quality of the camera motion estimate is used to segment video. Video segmentation algorithms often yield false shot changes in the presence of extreme camera or object motion. An analysis of optical flow quality may also be used to avoid false detection of shot changes.

Optical flow fields may be interpreted in many ways to estimate the characteristics of motion in video. Two such interpretations are the camera motion and object motion.

Camera Motion

An affine model was used in our experiments to approximate the flow patterns consistent with all types of camera motion.

$$u(xi, yi) = axi + byi + c$$
$$v(xi, yi) = dxi + eyi + f$$

Affine parameters a, b, c, d, e and f are calculated by minimizing the least squares error of the motion vectors.

We also compute average flow $\bar{v}$ and $\bar{u}$. Where $\bar{v}$ and $\bar{u}$,

$$\bar{u} = \sum_{i-0}^{N} axi + byi + c$$

$$\bar{v} = \sum_{i-0}^{N} dxi + eyi + f$$

Using the affine flow parameters and average flow, we classify the flow pattern. To determine if a pattern is a zoom, we first check if there is the convergence or divergence point (x0,y0), where: $u(xi, yi) = 0$ and $v(xi, yi) = 0$.

If the above relation is true, and (x0,y0) is located inside the image, then it must represent the focus of expansion. If $\bar{v}$ and $\bar{u}$ are large, then this is the focus of the flow and camera is zooming. If (x0,y0) is outside the image, and or are large, then the camera is panning in the direction of the dominant vector.

If the above determinant is approximately 0, then (x0,y0) does not exist and the camera is panning or static. If $\bar{v}$ or $\bar{u}$ are large, the motion is panning in the direction of the dominant vector. Otherwise, there is no significant motion and the flow is static. We may eliminate fragmented motion by averaging the results in a W frame window over time. An example of the camera motion analysis results is shown in Figure 9.2.

Object Motion

Object motion typically exhibits flow fields in specific regions of an image, while camera motion is characterized by flow throughout the entire image. The global distribution of motion vectors distinguishes between object and camera motion. The flow field is partitioned into a grid as shown in Figure 9.2. If the average velocity for the vectors in a particular grid is high, then that grid is designated as containing motion. When the number of connected motion grids, G_m,

$$G_m(i) = \begin{cases} 0 & (G_m(i-1)=0, G_m(i+1)=0,...M) \\ 1 & \text{otherwise} \end{cases}$$

is high (typically $G_m > 7$), the flow is some form of camera motion. $G_m(i)$ represents the status of motion grid at position i and M represents the number of neighbors. A motion grid should consist of at least a 4x4 array of motion vectors. If G_m is not high, but greater than some small value (typically 2 grids), the motion is isolated in a small region of the image and the flow is probably caused by object motion. This result is averaged over a frame window of width W_A, just as with camera motion, but the number of object motion regions needed is typically on the order of 60%. Examples of the object motion analysis results are shown in Figure 9.2.

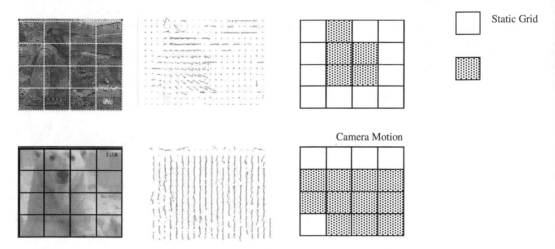

Figure 9.2. Camera and object motion detection.

2.1.6 Texture, Shape and Position

Analysis of image texture is useful in the discrimination of low interest video from video containing complex features [36]. A low interest image may also contain uniform texture, as well as uniform color or low contrast. Perceptual features for individual video frames were computed using common textual features such as coarseness, contrast, directionality and regularity.

The shape and appearance of objects may also be used as a feature for image correspondence. Color and texture properties will often change from one image to the next, making image difference and texture features less useful.

2.1.7 Audio Features

In addition to image features, certain audio features may be extracted from video to assist in the retrieval task [25]. Loud sound, silence and single frequency sound markers may be detected analytically without actual knowledge of the audio content.

Loud sounds imply a heightened state of emotion in video, and are easily detected by measuring a number of audio attributes, such as signal amplitude or power. Silent video may signify an area of less importance, and can also be detected with straightforward analytical estimates. A video producer will often use single frequency sound markers, typically a 1000 Hz. tone, to mark a particular point in the beginning of a video. This tone may be detected to determine the exact point in which a video will start.

2.1.8 Hierarchical Video Structure

Most video is produced with a particular format and structure. This structure may be taken into consideration when analyzing particular video content. News segments are typically 30 minutes in duration and follow a rigid pattern from day to day. Commercials are also of fixed duration, making detection less difficult.

Another key element in video is the use of the black frame. In most broadcast video, a black frame is shown between a transition of two segments. In news broadcast this usually occurs between a story and a commercial. By detecting the location of black frames in video, a hierarchical structure may be created to determine transitions between segments. A black frame or any single intensity image may be detected by summing the total number of pixels in a particular color space.

In the detection of the black frame, I_{high}, the maximum allowable pixel intensity is on the order of 20% of the maximum color resolution (51 for a 256 bit image), and I_{low}, the minimum allowable pixel intensity, is 0. The separation of segments in video is crucial in retrieval systems, where a user will most likely request a small segment of interest and not an entire full-length video. There are a number of ways to detect this feature in video, the simplest being to detect a high number of pixels in an image that are within a given tolerance of being a black pixel.

2.2 Compressed Domain Features

In typical applications of multimedia databases, the materials, especially the images and video, are often in a compressed format [3, 59]. To deal with these materials, a straightforward approach is to decompress all the data, and utilize the same features as mentioned in the previous section. Doing so, however, has some disadvantages. First, the decompression implies extra computation. Secondly, the process of decompression and re-compression, often referred to as recoding, results in further loss of image quality. Finally, since the size of decompressed data is much larger than the compressed form, most hardware and CPU cycles are needed to process and store the data. The solution to these problems is to extract features directly from the compressed data. Below are a number of commonly used compressed-domain features [51].

2.2.1 Motion Compensation

The motion vectors that are available in all video data compressed using standards such H.261, H.263, MPEG-1 and MPEG-2 are very useful. Analysis of motion vectors can be used to detect shot changes and other special effects such as dissolve, fade in and fade out. For example, if the motion vectors for a frame are randomly distributed without coherency that may suggest the presence of a shot change. Motion vectors represent low-resolution optical flow in the video, and can be used to extract all information that can be extracted using the optical flow method.

The percentage of each type of block in a picture is also a good indicator of shot changes, too. For a P frame, a large percentage of intra blocks implies a lot of new information for the current frame that cannot be predicted from the previous frame. Therefore, such a P-frame indicates the beginning of a new shot right after a shot change [59].

2.2.2 DCT Analysis

The DCT (Discrete Cosine Transform) provides a decomposition of the original image in the frequency domain. Therefore, DCT coefficients form a natural representation of texture in the original image. In addition to texture analysis, DCT coefficients can also be used to match images and to detect shot changes. The DC components are a low-resolution representation of the original image, averaged over 8x8 blocks. This implies much less data to manipulate, and for some applications, DC components already contain sufficient information. For color analysis, usually only the DC components are used to estimate the color histogram. For shot change detection, usually only the DC components are used to compare the content in two consecutive frames.

The parameters in the compression process that are not explicitly specified in the bitstream can be very useful as well. One example is the bit rate, i.e., the number of bits used for each picture. For intra coded video (i.e., no motion compensation), the number of bits per picture should remain roughly constant for a shot segment and should change when the shot changes. For example, a shot with simple color variation and texture requires fewer bits per picture compared to a shot that has detailed texture. For intercoding, the number of bits per picture is proportional to the action between the current picture and the previous picture. Therefore, if the number of bits for a certain picture is high, we can often conclude that there is a shot cut.

The compressed-domain approach does not solve all problems, though. To identify useful features from compressed data is typically difficult because each compression technique poses additional constraints, e.g., non-linear processing, rigid data structure syntax, and resolution reduction.

2.2.3 Compression Standards

Another issue is that compressed-domain features depend on the underlying compression standard. For different compression standards, different feature extraction algorithms have to be developed. Ultimately, we would like to have new compression standards with maximal content accessibility. MPEG-4 and MPEG-7 already have considered this aspect. In particular, MPEG-7 is a standard that goes beyond the domain of "compression" and seeks efficient representation of image and video content. The compressed-domain approach provides significant advantages but also brings new challenges. For more details on compressed-domain video processing, please see Chapter 43. For more details on MPEG-7 standard, please see Chapter 28.

2.3 Content-Based Features

Section 2 described a number of features that can be extracted using well-known techniques in image and audio processing. Section 2.3 described how many of these features are computed or approximated using encoded parameters in image and video compression. Although in both cases there is considerable understanding of the structure of the video, the features in no way estimate the actual image or video content.

In this section we describe several methods to approximate the actual content of an image or video. For many users, the query of interest is text based, and therefore, the content is essential. The desired result has less to do with analytical features such as color, or texture, and more with the actual objects within the image or video.

2.3.1 Object and Face Detection

Identifying significant objects that appear in the video frames is one of the key components for video characterization. Several working systems have generated reasonable results for the detection of a particular object, such as human faces, text or automobiles [52]. These limited domain systems have much greater accuracy than do broad domain systems that attempt to identify all objects in the image. Recent work in perceptual grouping [30] and low-level object recognition have yielded systems capable of recognizing small groups such as certain four-legged mammals, flowers, specific terrain, clothing and buildings [18, 20, 29].

Face detection and recognition are necessary elements for video characterization. The "talking head" image is common in interviews and news clips, and illustrates a clear example of video production focusing on an individual of interest. The detection of a human subject is particularly important in the analysis of news footage. An anchorperson will often appear at the start and end of a news broadcast, which is useful for detecting segment boundaries. In sports, anchorpersons will often appear between plays or commercials. A close-up is used in documentaries to introduce a person of significance.

The detection of humans in video is possible using a number of algorithms [28]. The Eigen Faces work from MIT is one of the earliest and widely used face detection algorithms [53]. Figure 9.3 shows examples of faces detected using the Neural Network Arbitration method [47]. Sneiderman and Kanade extended this method to detect faces at 90 degree rotation [52].   Most techniques are dependent on scale, and rely heavily on lighting conditions, limited occlusion and limited facial rotation.   Recent work for biometric and security based face recognition has also contributed to better characterization systems for video summarization [61].

2.3.2 Captions and Graphics

Text and graphics are used in a variety of ways to convey content to the viewer. They are most commonly used in broadcast news, where information must be absorbed in a short time. Examples of text and graphics in video are discussed below.

Video Captions
Text in video provides significant information as to the content of a shot. For example, statistical numbers and titles are not usually spoken but are included in captions for viewer inspection. Moreover, this information does not always appear in closed captions so detection in the image is crucial for identifying potentially important regions.

Figure 9.3. Recognition of video captions and faces.

In news video, captions of the broadcasting company are often shown at low opacity as a watermark in a corner without obstructing the actual video. A ticker tape is widely used in broadcast news to display information such as the weather, sports scores or the stock market. In some broadcast news, graphics such as weather forecasts are displayed in a ticker-tape format with the news logo in the lower right corner at full opacity. Captions that appear in the lower third portion of a frame are almost always used to describe a location, person of interest, title or event in news video. In Figure 9.3, the anchorperson's location is listed.

Video captions are used less frequently in video domains other than broadcast news. In sports, a score or some information about an ensuing play is often shown in a corner or border at low opacity. Captions are sometimes used in documentaries to describe a location, person of interest, title or event. Almost all commercials use some form of captions to describe a product or institution, because their time is limited to less than a minute in most cases.

A producer will seldom use fortuitous text in the actual video unless the wording is noticeable and easy to read in a short time. A typical text region can be characterized as a horizontal rectangular structure of clustered sharp edges, because characters usually form regions of high contrast against the background. By detecting these properties, we can extract potentially important regions from video frames that contain textual information. Most captions are high contrast text such as the black and white chyron commonly found in news video. Consistent detection of the same text region over a period of time is probable since text regions remain at an exact position for many video frames. This may also reduce the number of false detections that occur when text regions move or fade in and out between shots.

Text regions can be characterized as a horizontal rectangular structure of clustered sharp edges. Characters usually form regions of high contrast against the background. By detecting these properties we can extract regions from video frames that contain textual information. We first apply a global horizontal differential filter, F_{HD}, to the image. There are many variations to the differential filter, but the most common format is shown as,

$$F_{HD} = \left[-\frac{1}{2} \quad 1 \quad \frac{1}{2} \right]$$

An appropriate binary threshold should be set for extraction of vertical edge features. A smoothing filter, F_S, is then used to eliminate extraneous fragments, and to connect character sections that may have been detached. Individual regions must be identified through cluster detection. A bounding box, B_B, should be computed for selection of text regions. We now select clusters with bounding regions that satisfy constraints in cluster size, C_S, cluster fill-factor, C_{FF}, and horizontal-vertical aspect ratio.

$$C_{FF}(n) = \frac{C_S(n)}{B_{B\,area}(n)}$$

A cluster's bounding region must have a small vertical-to-horizontal aspect ratio as well as satisfy various limits in height and width. The fill factor of the region should be high to insure dense clusters. The cluster size should also be relatively large to avoid small fragments. Other controlling parameters are listed below.

Finally, we examine the intensity histogram of each region to test for high contrast. This is because certain textures and shapes appear similar to text but exhibit low contrast when examined in a bounded region.

For some fonts a generic optical character recognition (OCR) package may accurately recognize video captions. For most OCR systems, the input is an individual character. This presents a problem in digital video since most of the characters experience some degradation during recording, digitization and compression. For a simple font, we can search for blank spaces between characters and assume a fixed width for each letter [48].

Graphics

A graphic is usually a recognizable symbol, which may contain text. Graphic illustrations or symbolic logos are used to represent many institutions, locations and organizations. They are used extensively in news video, where it is important to describe the subject matter as efficiently as possible. A logo representing the subject is often placed in a corner next to an anchorperson during dialogue. Detection of graphics is a useful method for finding changes in semantic content. In this sense, its appearance may serve as a shot break. Recognition of corner regions for graphics detection may be possible through an extension of the shot detection technology. Histogram difference analysis of isolated image regions instead of the entire image can provide a simple method for detecting corner graphics. In Figure 9.4, a change is detected by the appearance of a graphics logo in the upper corner, although no shot is detected.

Frame *t*   **Frame *t+T***

Figure 9.4. Graphics detection through sub-region histogram difference.

2.3.3 Articulated Objects

A particular object is usually the emphasis of a query in image and video retrieval. Recognition of articulated objects poses a great challenge, and represents a significant step in content-based feature extraction. Many working systems have demonstrated accurate recognition of animal objects, segmented objects and rigid objects such as planes or automobiles.

The recognition of a single object is only one potential use of image based recognition systems. Discrimination of synthetic and natural backgrounds, or an animated or mechanical motion would yield a significant improvement content-based feature extraction.

2.3.4 Audio and Language

An important element in video indexing creation is the audio track. Audio is an enormous source for describing video content. Words specific to the actual content, or Keywords can be extracted using a number of language processing [37, 38]. Keywords may be used to reduce indexing and provide abstraction for video sequences. There are many possibilities for language processing in video, but the audio track must first exist as an ASCII document or speech recognition is necessary.

Audio Segmentation

Audio segmentation is needed to distinguish spoken words from music, noise and silence. Further analysis through speech recognition is necessary to align and translate these words into text. Audio selection is made on a frame-by-frame basis, so it is important to achieve the highest possible accuracy. At a sampling rate of 8 KHz, one frame corresponds to 267 samples of audio. Techniques in language understanding are used for selecting the most significant words and phrases.

Audio segmentation is also used to parse or separate audio into distinct units. The units may be described as phrases or used as input for a speech recognition system. The duration of the phrase can be controlled and modified based on the duration and genre of the summary. Shorter phrases improve speech recognition efficiency when the duration of the audio input is long.

Keyword Spotting

For documentaries, a digital ASCII version of the transcript is usually provided with an analog version of the video. From this we can identify keywords and phrases. Language analysis works on the audio transcript to identify keywords in it. We use the well-known technique of TF-IDF (Term Frequency Inverse Document Frequency) to measure relative importance of words for the video document [38].

A high TF-IDF value signifies relative high importance. Words that appear often in a particular segment, but appear relatively infrequently in the standard corpus, receive the highest weights. Punctuation in the transcript provides a means to identify sentences. With the sentence structure, we can parse smaller important regions, such as noun phrases and conjunction phrases. The link grammar parser [22] developed at Carnegie Mellon was used to parse noun phrases. With audio alignment from speech recognition, we can extract the phrases with the highest TF-IDF values as the audio portion of the summary. We can also look for conjunction phrases, slang words and question words to alter the frequency weighting for the TF-IDF analysis.

Speech Recognition

In order to use the audio track, we must isolate each individual word. To transcribe the content of the video material, we recognize spoken words using a speech recognition system. Speaker independent recognition systems have made great strides as of late and offer promise for application in video indexing [24, 23]. Speech recognition works best when closed-captioned data are available.

Closed-Captions

Captions usually occur in broadcast material, such as sitcoms, sports and news. Documentaries and movies may not necessarily contain captions. Closed-captions have become more common in video material throughout the United States since 1985 and most televisions provide standard caption display. Captions usually occur in United States broadcast material.

2.3.5 Embedded Video Features

A final solution for content-based feature extraction is the use of known procedures for creating video. Video production manuals provide insight into the procedures used during video editing and creation. There are many textbooks and journals that describe the editing and production procedures for creating video segments. Pryluck published one of the most well known works in this area [45, 46].

One of the most common elements in video production is the ability to convey climax or suspense. Producers use a variety of different effects, ranging from camera positioning, lighting and special effects to convey this mood to an audience. Detection of procedures such as these is beyond the realm of present image and language understanding technology. However, many of the important features described in sections 2, 3 and 4 were derived from research in the video production industry.

Structural information as to the content of a video is a useful tool for indexing video. For example, the type of video being used (documentaries, news footage, movies and sports) and its duration may offer suggestions to assist in object recognition. In news footage, the anchorperson will generally appear in the same pose and background at different times. The exact locations of the anchorperson can then be used to delineate story breaks. In documentaries, a person of expertise will appear at various points throughout the story when topical changes take place. There are also many visual effects introduced during video editing and creation that may provide information for video content. For example, in documentaries the shots prior to the introduction of a person usually describe their accomplishments and often precede shots with large views of the person's face.

A producer will often create production notes that describe in detail action and scenery of a video, shot by shot. If a particular feature is needed for an application in image or video databases, the description may have already been documented during video production.

Another source of descriptive information may be embedded in the video stream in the form of timecode and geospatial (GPS/GIS) data. These features are useful in indexing precise segments in video or a particular location in spatial coordinates. Aeronautic and automobile surveillance video will often contain GPS data that may be used as a source for indexing.

3. GENERATING VIDEO SURROGATES AUTOMATICALLY

The concept of summarization has existed for some time in areas such as text abstraction, video editing, image storyboards and other applications. In Firmin's evaluation of automatic text summarization systems, he divides summaries into three categories: 1) Indicative, providing an indication of a central topic, 2) Informative, summaries that serve as substitutes for full documents, and 3) Evaluative, summaries that express the author's point of view [42]. For video, the function of a summary is similar, but there are additional opportunities to express the result as text, imagery, audio, video or some combination. They may also appear very different from one application to the next. Just as summaries can serve different purposes, their composition is greatly determined by the video genre being presented [32]. For sports, visuals contain lots of information; for interviews, shot changes and imagery provide little information; and for classroom lecture, audio is very important.

Summarization is inherently difficult because it requires complete semantic understanding of the entire video (a task difficult for the average human). The best image understanding algorithms can only detect simple characteristics like those discussed in Section 2. In the audio track we have keywords and segment boundaries. In the image track we have shot breaks, camera motion, object motion, text captions, human faces and other static image properties.

3.1 Multimodal Processing for Deriving Video Surrogates
A video can be represented with a single thumbnail image. Such a single image surrogate is not likely to be a generic, indicative summary for the video in the news and documentary genres. Rather, the single image can serve as a query-relevant indicative summary. News and documentary videos have visual richness that would be difficult to capture with a single extracted image. For example, viewers interested in a NASA video about Apollo 11 may be interested in the liftoff, experiences of the astronauts on the moon or an expert retrospective looking back on the significance of that mission. Based on the viewer's query, the thumbnail image could then be a shot of the rocket blasting off, the astronauts walking on the moon or a head shot of the expert discussing the mission, respectively. Figure 9.5 illustrates how a thumbnail could be chosen from a set of shot key frame images.

Given the NASA video, speech recognition breaks the dialogue into time-aligned sequences of words. Image processing breaks the visuals down into a sequence of time-aligned shots. Further image processing then extracts a single image, i.e., frame, from the video to represent each shot. When the user issues a query, e.g., "walk on the moon," language processing can isolate the query terms to emphasize (e.g., walk and moon), derive additional forms (e.g., "walking" for "walk"), and identify matching terms in the dialogue text. In the case illustrated in Figure 9.5, the query walk on the moon matches most often to the shot at time 3:09 when the dialogue includes the phrase Walking on the moon. The shot at time 3:09 is represented by a black and white shot of two astronauts on the moon, and this shot's key frame image is then chosen as the image surrogate for this video given this query context. Speech, image and language processing each contribute to the derivation of a single image to represent a video document.

3.2 Feature Integration
The features described in previous sections may be used with rules that describe a particular type of video shot to create an additional set of content-based features [52]. By experimentation and examples from video production standards, we can identify a small set of heuristic rules for assessing a "summary rank" or priority to a given subset of video. Once these subsets are identified, a summary may be generated based on the application and user specifications. In most cases these rules involve the integration of image processing features with audio and language features. Below is a description of four rule-based features suitable for most types of video.

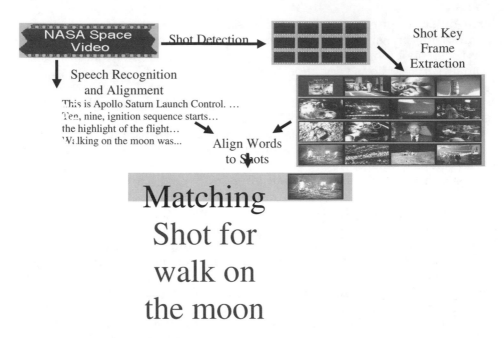

Figure 9.5. Thumbnail selection from shot key frames.

Image and Audio Selection

The first example of a video surrogate is the selection or ranking of audio based on image properties, and vice versa. Audio is parsed and recognized through speech recognition, keyword spotting and other language and audio analysis procedures. A ranking is added to the audio based on its proximity to certain imagery. If an audio phrase with a certain keyword value is within some short duration of a video caption, it will receive a higher summary rank than an audio phrase with a similar keyword value. An example of this is shown in Figure 9.6.

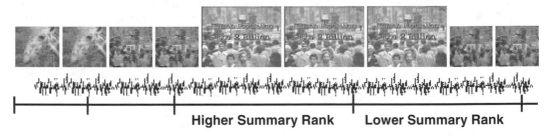

Higher Summary Rank   Lower Summary Rank

Figure 9.6. Audio with unsynchronized imagery.

Introduction Scenes

The shots prior to the introduction of a person usually describe their accomplishments and often precede shots with large views of the person's face. A person's name is generally spoken and then followed by supportive material. Afterwards, the person's actual face is shown. If a shot contains a proper name or an on-screen text title, and a large human face is detected in the shots that follow, we call this an Introduction Scene. Characterization of this type is useful when searching for a particular human subject because identification is more reliable than using the image or audio features separately. This effect is common in documentaries based on human recognition or achievement.

Inter-cut Scenes

In this example, we detect an inter-cutting sequence to rank an adjacent audio phrase. This describes the selection of the inter-cutting shot. The color histogram difference measure gives us a simple routine for detecting similarity between shots. Shots between successive imagery of a human face usually imply illustration of the subject. For example, a video producer will often interleave shots of research between shots of a scientist. Images that appear between two similar scenes that are less than T_{SS} seconds apart are characterized as an adjacent similar scene. An example of this effect is shown in Figure 9.7a. Audio near the inter-cut receives a higher rank than subsequent audio.

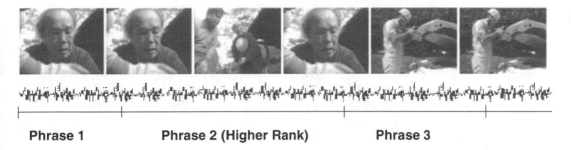

| Phrase 1 | Phrase 2 (Higher Rank) | Phrase 3 |

Figure 9.7a. Inter-cut shot detection for image and audio selection.

Short Successive Effect

Short successive shots often introduce an important topic. By measuring the duration of each scene, S_D, we can detect these regions and identify short successive sequences. In practice, summary creation involves selecting the appropriate keywords or audio and choosing a corresponding set of images. Candidates for the image portion of a summary are chosen by two types of rules: 1) Primitive Rules, independent rules that provide candidates for the selection of image regions for a given keyword, and 2) Meta-Rules, higher order rules that select a single candidate from the primitive rules according to global properties of the video. For some cases, the image selection portion will supersede an audio keyphrase for summary selection. An example of this effect is shown in Figure 9.7b, where each thumbnail represents approximately 10 frames, or 1/3 a second of video. The first image appears at a normal pace. It is followed by three short shots and the normal pace resumes with the image of the temple.

3.3 Summary Presentations

The format for summary presentation is not limited to shortened viewing of a single story. Several alterations to the audio and image selection parameters may result in drastically different summaries during playback. Summary presentations are usually visual and textural in layout. Textual presentations provide more specific information and are useful when presenting large collections of data. Visual, or iconic, presentations are more useful when the content of interest is easily recalled from imagery. This is quite often the case in stock footage video where there is no audio to describe the content. Many of the more common formats for summarizing video are described below. Section 5 describes recent applications that use multimodal features for visualizing video summaries.

Figure 9.7b. Short successive effects.

Titles – Text Abstracts

Titles have a long history of research through text abstraction and language understanding [38]. The basic premise is to represent a document by a single sentence or phrase. With video, the title is derived from the transcript, closed-captions, production notes or on-screen text.

Thumbnails and Storyboards – Static Filmstrips

The characterization analysis used for selecting important image and audio in summaries may be applied to select static poster-frames. These frames may be shown as a single representative image, or thumbnail; or as a sequence of images over time, such as a storyboard.

Duration Control

For various applications, a user will want to vary the compaction level during playback. This flexibility should exist in much the same that videocassette editors allow for speed control with a roll bar. Such a feature is implemented for variable rate summarization as an option for the user interface.

Skims – Dynamic Summaries

Skims provide summarization in the form of moving imagery. In short, they are subsets of an original sequence placed together to form a shorter video. They do not necessarily contain audio [50].

User Controlled Summaries

For indexing, an interactive weighting system is used to allow for personal text queries during summary creation. A text query may be entered and used during the TF-IDF weighting process for keyword extraction or during speech alignment.

Summarizing Multiple Documents

A summary is not limited to coverage of a single document or video. It may encompass several sets of data at an instant or over time. In the case of news, it is necessary to analyze multiple stories to accurately summarize the events of a day.

4. EVALUATION AND USER STUDIES

The Informedia Project at Carnegie Mellon University has created a multi-terabyte digital video library consisting of thousands of hours of video, segmented into tens of thousands of documents. Since Informedia's inception in 1994, numerous interfaces have been developed and tested for accessing this library, including work on multimedia surrogates that represent a video document in an abbreviated manner [11, 56]. The video surrogates build from automatically derived descriptive data, i.e., metadata, such as transcripts and representative thumbnail images derived from speech recognition, image processing and language processing. Through human computer interaction evaluation techniques and formal user studies, the surrogates are tested for their utility as indicative and informative summaries, and iteratively refined to better serve users' needs. This section discusses evaluations for text titles, thumbnail images, storyboards and skims.

Various evaluation methods are employed with the video surrogates, ranging from interview and freeform text user feedback to "discount" usability techniques on prototypes to formal empirical studies conducted with a completed system. Specifically, transaction logs are gathered and analyzed to determine patterns of use. Text messaging and interview feedback allow users to comment directly on their experiences and provide additional anecdotal comments. Formal studies allow facets of surrogate interfaces to be compared for statistically significant differences in dependent measures such as success rate and time on task.

Discount usability techniques, including heuristic evaluation, cognitive walkthrough and think-aloud protocol, allow for quick evaluation and refinement of prototypes [41]. Heuristic evaluation lets usability specialists review an interface and categorize and justify problems based on established usability principles, i.e., heuristics. With cognitive walkthrough, the specialist takes a task, simulates a user's problem-solving process at each step through the interaction and checks if the simulated user's goals and memory content can be assumed to lead to the next correct action. With think-aloud protocol, a user's interaction with the system to accomplish a given task is videotaped and analyzed, with the user instructed to "think aloud" while pursuing the task.

4.1 TEXT TITLES

A great deal of text can be associated with a video document in the news and documentary genres. The spoken narrative can be deciphered with speech recognition and the resulting text time-aligned to the video [56]. Additional text can also be generated through "video OCR" processing, which detects and translates into ASCII format the text overlaid on video frames [48]. From this set of words for a video, a text title can be automatically extracted, allowing for interfaces as shown in Figure 9.8, where the title for the third result is shown.

These text titles act as informative summaries, a label for the video that remains constant across all query and browsing contexts. Hence, a query on OPEC that matches the third result shown in Figure 9.8 would present the same title.

Initially, the most significant words, as determined by the highest TF-IDF value, were extracted from a video's text metadata and used as the title. Feedback from an initial user group of teachers and students at a nearby high school showed that the text title was referred to often, and used as a label in multimedia essays and reports, but that its readability needed to be improved. This feedback took the form of anecdotal reporting through email and a commenting mechanism located within the library interface, shown as the pull-down menu "Comments!" in Figure 9.8. Students and teachers were also interviewed for their suggestions, and timed transaction logs were generated of all activity with the system, including the queries, which surrogates were viewed and what videos were watched [8].

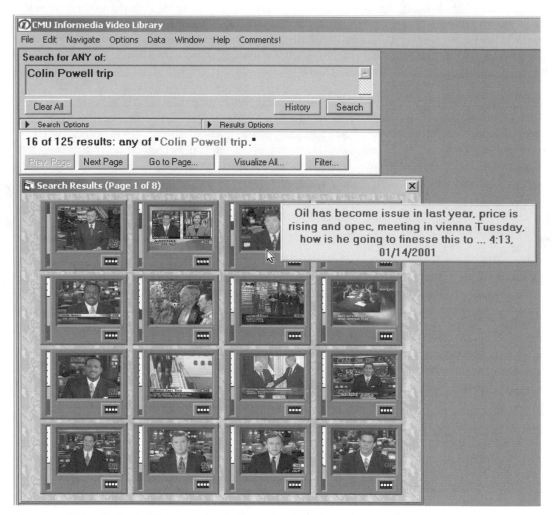

Figure 9.8. Informedia results list, with thumbnail surrogates and title shown for third video result.

The title surrogate was improved by extracting phrases with high TF-IDF score, rather than individual words. Such a title is shown in Figure 9.8. The title starts with the highest scoring TF-IDF phrase. As space permits, the highest remaining TF-IDF phrase is added to the list, and when complete the phrases are ordered by their associated video time (e.g., dialogue phrases are ordered according to when they were spoken). In addition, user feedback noted the importance of reporting the copyright/production date of the material in the title, and the size in standard hours:minutes:seconds format. The modified titles were well-received by the users, and phrase-based titles remain with the Informedia video library today, improved by new work in statistical analysis and named entity extraction.

Figure 9.8 shows how the text title provides a quick summary in readable form. Being automatically generated, it has flaws, e.g., it would be better to resolve pronoun references such as "he" and do better with upper and lower case. As with the surrogates reported later, though, the flaws with any particular surrogate can be compensated with a number of other surrogates or alternate views into the video. For example, in addition to the informative title summary it would be nice to have an indicative summary showing which terms match for any particular video document.

With Figure 9.8, the results are ordered by relevance as determined by the query engine. For each result, there is a vertical relevance thermometer bar that is filled in according to the relevance score for that result. The bar is filled in with colors matching the colors used for the query terms. In this case, with Colin colored red, "Powell" colored violet, and trip colored blue, the bar shows at a glance which terms match a particular document, as well as their relative contribution to that document. The first six results of Figure 9.8 all match on all three terms, with the next ten in the page only matching on "Colin" and "Powell." Hence, the thermometer bar provides the indicative summary showing matching terms at a glance. It would be nice to show not only what terms match, but also match density and distribution within the document, as is done with TileBars [26]. Such a more detailed surrogate is provided along with storyboards and the video player, since those interfaces include a timeline presentation.

4.2 THUMBNAIL IMAGE

The interface shown in Figure 9.8 makes use of thumbnail images, i.e., images reduced in resolution by a quarter in each dimension from their original pixel resolution of 352 by 240 (MPEG-1). A formal empirical study was conducted with high school scholarship students to investigate whether this thumbnail interface offered any advantages over simply using a text menu with all of the text titles [9]. Figure 9.9 illustrates the 3 interfaces under investigation: text menus, "naïve" thumbnails in which the key frame for the first shot of the video document is used to represent the document, and query-based thumbnails. Query-based thumbnails select the key frame for the highest scoring shot for the query, as described earlier in Section 3.

The results of the experiment showed that a version of the thumbnail menu had significant benefits for both performance time and user satisfaction: subjects found the desired information in 36% less time with certain thumbnail menus over text menus. Most interestingly, the manner in which the thumbnail images were chosen was critical. If the thumbnail for a video segment was taken to be the key frame for the first shot of the segment (treatment B), then the resulting pictorial menu of "key frames from first shots in segments" produced no benefit compared to the text menu. Only when the thumbnail was chosen based on usage context, i.e., treatment C's query-based thumbnails, was there an improvement. When the thumbnail was chosen based on the query, by using the key frame for the shot producing the most matches for the query, then pictorial menus produced clear advantages over text-only menus [C9].

This empirical study validated the use of thumbnails for representing video segments in query result sets, as shown in Figure 9.8. It also provided evidence that leveraging from multiple processing techniques leads to digital video library interface improvements. Thumbnails derived from image processing alone, such as choosing the image for the first shot in a segment, produced no improvements over the text menu. However, through speech recognition, natural language processing and image processing, improvements can be realized. Via speech recognition the spoken dialogue words are tightly aligned to the video imagery. Through natural language processing the query is compared to the spoken dialogue, and matching words are identified in the transcript and scored. Via the word alignment, each shot can be scored for a query, and the thumbnail can then be the key frame from the highest scoring shot for a query. Result sets showing such query-based thumbnails produce advantages over text-only result presentations and serve as useful indicative summaries.

	Search Results	? X
[1.00]	INV0641	The Great Dinosaur Hunt in Paris, Gould consid
[0.47]	INV0638	Nanotyrannus resembles Troodon, and many livi
[0.30]	INV071	Environmental destruction causes extinction of
[0.30]	INV061	The study of dinosaurs has been dynamic over
[0.24]	INV069	Paleontologists recreate and display the creatu
[0.21]	INV0637	The nanotyrannus was similar to the Tyrannosa
[0.20]	INV0629	Today scavengers at flood sites are birds, maki
[0.20]	INV066	In the 18th century, extinction was a controvers
[0.17]	INV0636	One can find new species and ideas on the she
[0.17]	PLE0436	A new theory links the extinction of dinosaurs t
[0.16]	PLE0444	The existence of Nemesis may illuminate human
[0.14]	INV0642	Dinosaurs roamed the earth for far longer than

A. Text Menu

B. "Naïve" Thumbnails

3 treatments used in empirical study, each treatment always representing 12 video documents

(views are scaled down here to fit into a single figure)

C. Query-based Thumbnails

Figure 9.9. Snapshots of the 3 treatments used in thumbnail empirical study with 30 subjects.

4.3 STORYBOARDS

Rather than using only a single image to summarize a video, another common approach presents an ordered set of representative thumbnails simultaneously on a computer screen [23, 31 54, 58, 60]. This storyboard interface, referred to in the Informedia library as "filmstrip," is shown for a video clip in Figure 9.10. Storyboards address certain deficiencies of the text title and single image surrogates. Text titles and single thumbnail images (see Figure 9.8) are quick communicators of video segment content, but do not present any temporal details. The storyboard surrogate communicates information about every shot in a video segment. Each shot is represented in the storyboard by a single image, or key frame. Within the Informedia library interface, the shot's middle frame is assigned by default to be the key frame. If camera motion is detected and that motion stops within the shot, then the frame where the camera motion ends is selected instead. Other image processing techniques, such as those that detect and avoid low-intensity images and those favoring images where faces or overlaid text appears, further refine the selection process for key frames [52].

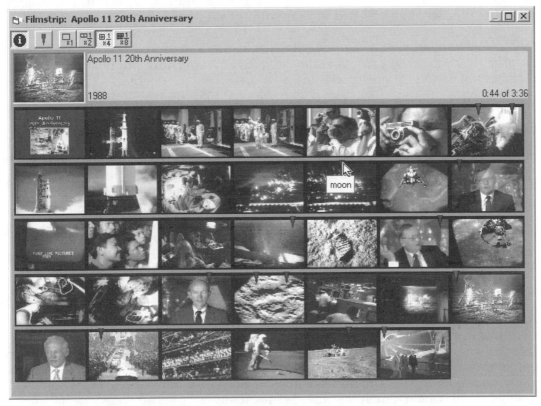

Figure 9.10. Storyboard, with overlaid match "notches" following query on "man walking on the moon."

Figure 9.11. Reduced storyboard display for same video represented by full storyboard in Figure 9.10.

As with the thermometer bar of Figure 9.8, the storyboard can indicate which terms match by drawing color-coded match "notches" at the top of shots. The locations of the notches indicate where the matches occur, showing match density and distribution within the video. Should the user mouse over a notch, as is done in the twelfth shot of Figure 9.10, the matching text for that notch is shown, in this case "moon." As the user mouses over the storyboard, the time corresponding to that storyboard location is shown in the information bar at the top of the storyboard window, e.g., the time corresponding to the mouse location inside of the twelfth shot in Figure 9.10 is 44 seconds into the 3 minute and 36 second video clip. The storyboard summary facilitates quick visual navigation. To jump to a location in the video, e.g., to the mention of "moon" in the twelfth shot, the mouse can be clicked at that point in the storyboard.

One major difficulty with storyboards is that there are often too many shots to display in a single screen. An area of active research attempts to reduce the number of shots represented in a storyboard to decrease screen space requirements [6, 34, 58]. In Video Manga [55, 6], the interface presents thumbnails of varying resolutions, with more screen space given to the shots of greater importance. In the Informedia storyboard interface, the query-based approach is again used: the user's query context can indicate which shots to emphasize in an abbreviated display. Consider the same video document represented by the storyboard in Figure 9.10. By only showing shots containing query match terms, only 11 of the 34 shots need to be kept. By reducing the resolution of each shot image, screen space is further reduced. In order to see visual detail, the top information bar can show the storyboard image currently under the mouse pointer in greater resolution, as illustrated by Figure 9.11. In this figure, the mouse is over the ninth shot at 3:15 into the video, with the storyboard showing only matching shots at 1/8 resolution in each dimension.

4.4 STORYBOARD PLUS TEXT

Another problem with storyboards is that they apply to varying degrees of effectiveness depending on the video genre [32]. For visually rich genres like travelogues and documentaries, they are very useful. For classroom lecture or conference presentations, they are far less important. For a genre such as news video, in which the information is conveyed both through visuals (especially field footage) and audio (such as the script read by the newscaster), a mixed presentation of both synchronized shot images and transcript text extracts may offer benefits over image-only storyboards. Such a "storyboard plus text" surrogate is shown in Figure 9.12.

This storyboard plus text surrogate led to a number of questions:
- Does text improve navigation utility of storyboards?
- Is less text better than complete dialogue transcripts?
- Is interleaved text better than block text?

For example, the same video represented in Figure 9.12 could have a concise storyboard-plus-text interface in which the text corresponding to the video for an image row is collapsed to at most one line of phrases, as shown in Figure 9.13.

Figure 9.12. Scaled-down view of storyboard with full transcript text aligned by image row.

To address these questions, an empirical study was conducted in May 2000 with 25 college students and staff [12]. Five interface treatments were used for the experiment: image-only storyboard (like Figure 9.10), interleaved full text (Figure 9.12), interleaved condensed text (Figure 9.13), block full text (in which image rows are all on top, with text in a single block after the imagery) and block condensed text.

The results from the experiment showed that text clearly improved utility of storyboards for a known item search into news video, with statistically significant results [12]. This was in agreement with prior studies showing that the presentation of captions with pictures can significantly improve both recall and comprehension, compared to presenting either pictures or captions alone [43, 27, 33]. Interleaving text with imagery was not always best: even though interleaving the full text by row like Figure 9.12 received the highest satisfaction measures from subjects, their task performance was relatively low with such an interface.

The fastest task times were accomplished with the interleaved condensed text and the block full text. Hence, reducing text is not always the best surrogate for faster task accomplishment. However, given that subjects preferred the interleaved versions and that condensed text takes less display space than blocking the full transcript text beneath the imagery, the experiment concluded that an ideal storyboard plus text interface is to condense the text by phrases, and interleave such condensed text with the imagery.

Figure 9.13. Storyboard plus concise text surrogate for same video clip represented by Figure 9.12.

4.5 SKIM

While storyboard surrogates represent the temporal dimension of video, they do so in a static way: transitions and pace may not be captured; audio cues are ignored. The idea behind a "video skim" is to capture the essence of a video document in a collapsed snippet of video, e.g., representing a 5 minute video as a 30 second video skim that serves as an informative summary for that longer video. Skims are highly dependent on genre: a skim of a sporting event might include only scoring or crowd-cheering snippets, while a skim of a surveillance video might include only snippets where something new enters the view.

Skims of educational documentaries were studied in detail by Informedia researchers and provided in the digital library interface. Users accessed skims as a comprehension aid to understand quickly what a video was about. They did not use skims for navigation, e.g., to jump to the first point in a space documentary where the moon is discussed. Storyboards serve as much better navigation aids because there is no temporal investment that needs to be made by the user; for skims, the user must play and watch the skim.

For documentaries, the audio narrative contains a great deal of useful information. Early attempts at skims did not preserve this information well.

Snippets of audio for an important word or two were extracted and stitched together in a skim, which was received poorly by users, much like early text titles comprised of the highest TF-IDF words were rejected in favor of more readable concatenated phrases. By extracting audio snippets marked by silence boundaries, the audio portion of the skim became greatly improved, as the skim audio was more comprehensible and less choppy.

A formal study was conducted to investigate the importance of aligning the audio with visuals from the same area of the video, and the utility of different sorts of skims as informative summaries. Specifically, it was believed that skims composed of larger snippets of dialogue would work better than shorter snippets, the equivalent of choosing phrases over words. A new skim was developed that comprised snippets of audio bounded by significant silences, more specifically audio signal power segmentation [10]. The transcript text for the audio snippets was ranked by TF-IDF values and the highest valued audio snippets were included in the skim, with the visual portion for the skim snippets being in the close neighborhood of the audio.

Five treatments were seen by each of 25 college students:
- DFS: a default skim using short 2.5 second components, e.g., comprising seconds 0-2.5 from the full source video, then seconds 18.75-21.25, seconds 37.5-40, etc.
- DFL: a default skim using long 5 second components, e.g., consisting of seconds 0-5, then seconds 37.5-42.5, seconds 75-80, etc.
- NEW: a new skim outlined above and discussed in more detail in [10]
- RND: same audio as NEW but with reordered video to test synchronization effects
- FULL: complete source video, with no information deleted or modified

These treatment derivations are illustrated in Figure 9.14.

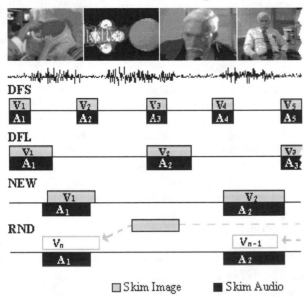

☐ Skim Image   ■ Skim Audio

Figure 9.14. Skim treatments used in empirical study on skim utility as informative summary.

Following a playing of either a skim or the full video, the subject was asked which of a series of images were seen in the video just played, and which of a series of text summaries would make sense as representing the full source video. As expected, the FULL treatment performed best, i.e., watching the full video is an ideal way to determine the information content of that full video. The subjects preferred the full video to any of the skim types. However, subjects favored the NEW skim over the other skim treatments, as indicated by subjective ratings collected as part of the experiment. These results are encouraging, showing that incorporating speech, language and image processing into skim video creation produces skims that are more satisfactory to users.

The RND skim distinguished itself as significantly poorer than NEW on the text-phrase gisting instrument, despite the fact that both RND and NEW use identical audio information. This result shows that the visual content of a video skim does have an impact on its use for gisting. The DFS and DFL skim treatments did not particularly distinguish themselves from one another, leaving open the question of the proper component size for video skims. The larger component size, when used with signal-power audio segmentation, produced the NEW skim that did distinguish itself from the other skims. If the larger component size is used only for subsampling, however, it yields no clear objective or subjective advantage over short component size skims, such as DFS. In fact, both DFS and DFL often rated similarly to RND, indicating perhaps that any mechanistically subsampled skim, regardless of granularity, may not do notably well.

While very early Informedia skim studies found no significant differences between a subsampled skim and a best audio and video skim, this study uncovered numerous statistically significant differences [10]. The primary reasons for the change can be traced to the following characteristics of the audio data in the skim:
- Skim audio is less choppy due to setting phrase boundaries with audio signal-processing rather than noun-phrase detection.
- Synchronization with visuals from the video is better preserved.
- Skim component average size has increased from three seconds to five.

Although the NEW skim established itself as the best design under study, considerable room for improvement remains. It received mediocre scores on most of the subjective questions, and its improvement over the other skims may reflect more on their relatively poor evaluations than on its own strengths. NEW did distinguish itself from RND for the image recognition and text-phrase gisting tasks, but not from the DFS and DFL skims. The NEW skim under study achieved smoother audio transitions but still suffered abrupt visual changes between image components. Transitions between video segments should also be smoothed — through dissolves, fades or other effects — when they are concatenated to form a better skim.

4.6 LESSONS LEARNED FROM SINGLE DOCUMENT VIDEO SURROGATES

In summary, usage data, HCI techniques and formal experiments have led to the refinement of single document video surrogates in the Informedia digital video library over the years. Thumbnail images are useful surrogates for video, especially as indicative summaries chosen based on query-based context. The image selection for thumbnails and storyboards can be improved via camera motion data and corpus-specific rules. For example, in the news genre shots of the anchorperson in the studio and weather reporter in front of a map typically contribute little to the visual understanding of the news story. Such shots can be de-emphasized or eliminated completely from consideration as single image surrogates or for inclusion in storyboards.

Text is an important component of video surrogates. Text titles are used as identifying labels for a video document and serve as a quick identifier. Adding synchronized text to storyboards helps if interlaced with the imagery. Assembling from phrases (longer chunks) works better than assembling from words (shorter chunks).

Showing distribution and density of match terms is useful, and can naturally be added to a storyboard or a video player's play progress bar. The interface representation for the match term can be used to navigate quickly to that point in the video where the match occurs.

Finally, with a temporal summary like skims, transition points between extracted snippets forming the skim are important. When the audio for a skim breaks at silence points, the skim is received much better than skims with abrupt, choppy transitions between audio snippets.

5. VISUALIZATION FOR SUMMARIZATION

As digital video assets grow, so do result sets from queries against those video collections. In a library of a few thousand hours of video comprising tens of thousands of video documents, many queries return hundreds or thousands of results. Paging through those results via interfaces like Figure 9.8 becomes tedious and inefficient, with no easy way to see trends that cut across documents. By summarizing across a number of documents, rather than just a surrogate for a single video, the user can:

- be informed of such trends cutting across video documents
- be shown a quick indicator as to whether the results set, or in general the set of video documents under investigation, satisfies the user's information need
- be given a navigation tool via the summary to facilitate targeted exploration

As automated processing techniques improve, e.g., speech, image and language processing, more metadata is generated with which to build interfaces into the video. For example, all the metadata text for a video derived from speech recognition, overlaid text VOCR processing and other means can be further processed into people's names, locations, organizations and time references via named entity extraction. Named entity extraction from broadcast news speech transcripts has been done by MITRE via Alembic [39], and BBN with Nymble [4, 40]. Similarly, Informedia processing starts with training data where all words are tagged as people, organizations, locations, time references or something else. A tri-gram language model is built from this training data using a statistical language modeling toolkit [17], which alternates between a named-entity tag and a word, i.e., –person- Rusty –person- Dornin –none- reporting –none- for –organization- CNN –none- in –location- Seattle. To label named entities in new text, a lattice is built from the text where each text word can be preceded by any of the named-entity tags. A Viterbi algorithm then finds the best path through the named-entity options and the text words, just like speech recognition hypothesis decoding.

With a greater volume of metadata describing video, e.g., lists of people's names, locations, etc., there needs to be an overview capability to address the information overload. Prior work in information visualization has offered many solutions for providing summaries across documents and handling volumes of metadata, including:

- Visualization by Example (VIBE), developed to emphasize relationships of result documents to query words [44]
- Scatter plots for low dimensionality relationships, e.g., timelines for emphasizing document attributes mapped to production date [11, 14]
- Colored maps, emphasizing geographic distribution of the events covered in video documents [13, 14]

Each technique can be supplemented with dynamic query sliders [1], allowing ranges to be selected for attributes such as document size, date, query relevance, and geographic reference count.

Consider a query on "air crash" against a 2001 CNN news video subset in the Informedia library. This query produces 998 documents, which can be overviewed using the timeline visualization shown in Figure 9.15. The visualizations shown here convey semantics primarily through positioning, but could be enriched to overlay other information dimensions through size, shape and color, as detailed elsewhere for the Informedia library [15, 11]. By dragging a rectangle bounding only some of the green points representing stories, the user can reduce the result set to just those documents for a certain time period and/or relevance range.

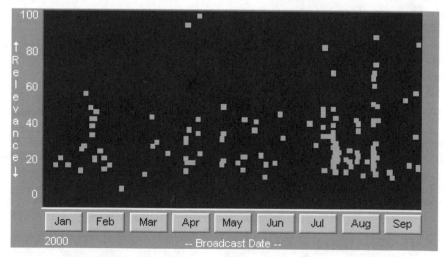

Figure 9.15. Timeline overview of "air crash" query.

For more complex word queries, the VIBE plot of documents to query terms can be used to understand the mapping of results to each term and to navigate perhaps to documents matching 2 words but not a third from the query. VIBE allows users unfamiliar or uncomfortable with Boolean logic to be able to manipulate results based on their query word associations.

For a news corpus, there are other attributes of interest besides keywords, such as time and geography. Figure 9.16 shows both of these attributes in use, as well as the search engine relevance score. The video documents' location references are identified automatically through named entity extraction, and stored with the document metadata for use in creating such maps.

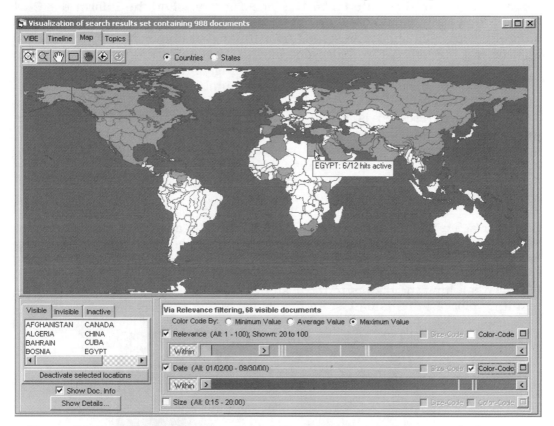

Figure 9.16. Map visualization for results of "air crash" query, with dynamic query sliders for control and feedback.

Figure 9.16 illustrates the use of direct manipulation techniques to reduce the 998 documents in Figure 9.15 to a set of 68 documents under current review. By moving the slider end point icon ▣ with the mouse, only those documents having relevance ranking of 20 or higher are left displayed on the map. As the end point changes, so does the number of documents plotted against the map, e.g., if Brazil only appeared in documents ranked with relevance score 19 or lower, then Brazil would initially be colored on the map but drop out of the visible, colored set with the current state of the slider shown in Figure 9.16. Similarly, the user could adjust the right end point for the slider, or set it to a period of say one month in length in the date slider and then slide that one month active period within January through September 2000 and see immediately how the map animates in accordance with the active month range.

The map is color-coded based on date, and the dynamic sliders show distribution of values based on the country under mouse focus, e.g., the 6 "Egypt" stories with relevance > 20 have the relevance and date distribution shown by the yellow stripes on the relevance bars. Eick has previously reported on the benefits of using sliders as a filtering mechanism, color scale, and to show data distributions to make efficient use of display space [19].

The visualizations shown in Figure 9.15 and 9.16 do not take advantage of the visual richness of the material in the video library. For the Informedia CNN library, over 1 million shots are identified with an average length of 3.38 seconds, with each shot represented by a thumbnail image as shown in earlier Figures 9.8 through 9.13. Video documents, i.e., single news stories, average 110 seconds in length, resulting in an average image count for document storyboards for these stories of 32.6. These thumbnails can be used instead of points or rectangles in VIBE and timeline plots, and can be overlaid on maps as well. Ongoing research is looking into reducing the number of thumbnails intelligently to produce a more effective visualization [16]. For example, by using query-based imagery (keeping only the shots where matches occur), and folding in domain-specific heuristics, the candidate thumbnail set for a visualization can be greatly reduced. With the news genre, a heuristic in use is to remove all studio shots of anchors and weather maps [16].

Consider again the 998 documents returned from the "air crash" query shown in earlier figures and densely plotted in Figure 9.15. Through a VIBE plot mapping stories to query words, the user can limit the active set to only the results matching both "air" and "crash"; through a world map like that of Figure 9.16, the user can limit the results to those stories dealing with regions in Africa. The resulting plot of 11 remaining stories is shown in Figure 9.17.

Figure 9.17. Filtered set of video documents from Figure 9.15, with added annotations.

With the increased display area and reduced number of stories in the active set, more information can be shown for each story cluster. Synchronization information kept during the automatic detection of shots for each video, representative images for each shot and dialogue alignment of spoken words to video can be used to cluster text and images around times within the video stories of interest to the user. For a query result set, the interesting areas are taken to be those sections of the video where query terms ("air crash") are mentioned.

Prior work with surrogates underscores the value of text phrases as well, so text labels for document clusters will likely prove useful. Initial investigations into displaying common text phrases for documents focused by visualization filters, e.g., the 11 documents from the full set of 998 remaining in Figure 9.17, have shown that the text communicates additional facts to the user, supplementing the visualization plots [16]. Of course, more extensive evaluation work like that conducted for Informedia single document surrogates will need to be performed in order to determine the utility of video digests for navigating, exploring and collecting information from news libraries.

Initial interfaces for the Informedia digital video library interface consisted of surrogates for exploring a single video document without the need to download and play the video data itself. As the library grew, visualization techniques such as maps, timelines, VIBE scatter plots and dynamic query sliders were incorporated to allow the interactive exploration of sets of documents. Shahraray notes that "well-designed human-machine interfaces that combine the intelligence of humans with the speed and power of computers will play a major role in creating a practical compromise between fully manual and completely automatic multimedia information retrieval systems" [7]. The power of the interface derives from its providing a view into a video library subset, where the user can easily modify the view to emphasize various features of interest. Summaries across video documents let the user browse the whole result space without having to resort to the time-consuming and frustrating traversal of a large list of documents. The visualization techniques discussed here allow the user efficient, effective direct manipulation to interact with and change the information display.

6. CONCLUSION

This chapter describes the creation of video summaries and visualization systems through multimodal feature analysis. The integration of multiple features provides insights into video characterization not present with a single feature. Multimodal analysis is necessary for combining multiple forms of image, audio and language information, as well as incorporating new forms of meta-data in the future. Surrogates have proven advantages for allowing information from a video document to be found and accessed quickly and accurately. Visualization techniques addressing text corpora and databases have been shown to be applicable to video libraries as well.

Access to video information is greatly improved through advances in video summarization and visualization. The video summary provides an efficient means for abstracting, shortening or simply browsing a large collection. Future digital video library interfaces that summarize sets of video documents and leverage from the library's multiple media and rich visual nature need to be designed and evaluated so that the wealth of material within such libraries can be better understood and efficiently accessed.

ACKNOWLEDGMENTS

A portion of the material in this chapter is based on research at AVA Media Systems. Section 2 contains contributions from earlier work with Tsuhan Chen,

Carnegie Mellon University, Electrical Engineering Department. The Informedia material is based on work supported by the National Science Foundation (NSF) under Cooperative Agreement No. IRI-9817496. Christel and Wactlar are also supported in part by the advanced Research and Development Activity (ARDA) under contract number MDA908-00-C-0037. CNN and WQED Communications in Pittsburgh, PA supplied video to the Informedia library for sole use in research. Their video contributions as well as video from NASA, the U.S. Geological Survey and U.S. Bureau of Reclamation are gratefully acknowledged. More details about Informedia research and AVA Media Systems can be found at http://www.informedia.cs.cmu.edu/ and http://www.savasystems.com.

REFERENCES

[1]   C. Ahlberg, and B. Shneiderman, Visual Information Seeking: Tight Coupling of Dynamic Query Filters with Starfield Displays, in Proc. ACM CHI '94, Boston, MA, April 1994, 313-317.

[2]   A. Akutsu and Y. Tonomura, Video Tomography: An Efficient Method for Camerawork Extraction and Motion Analysis, *in Proc. ACM Multimedia Conference*, 1994, pp. 349--356.

[3]   F. Arman, A. Hsu, and M. Chiu, Image Processing on Compressed Data for Large Video Databases, in *Proceedings of the ACM MultiMedia*, California, June 1993, pp. 267-272.

[4]   D. M. Bikel, S. Miller, R. Schwartz, and R. Weischedel, Nymble: A High-Performance Learning Name-finder, *in* Proc. 5th Conf. on Applied Natural Language Processing (ANLP), Washington DC, April 1997, pp. 194-201.

[5]   J.S. Boreczky and L.A. Rowe, Comparison of Video Shot Boundary Detection Techniques, in SPIE Conf. on Visual Communication and Image Processing, 1996.

[6]   J. Boreczky, A. Girgensohn, G. Golovchinsky, and S. Uchihashi, An Interactive Comic Book Presentation for Exploring Video, in *Proc. CHI '00*, 2000, pp. 185-192.

[7]   Chang, Moderator, Multimedia Access and Retrieval: The State of the Art and Future Directions, in *Proc. ACM Multimedia '99*, Orlando, FL, October 1999, pp. 443-445.

[8]   M.G. Christel and K. Pendyala, Informedia Goes to School: Early Findings from the Digital Video Library Project, *D-Lib Magazine*, September 1996.
         http://www.dlib.org/dlib/september96/informedia/09christel.html.

[9]   M.G. Christel, D.B. Winkler and C.R. Taylor, Improving Access to a Digital Video Library, in *Human-Computer Interaction INTERACT '97: IFIP TC13 International Conference on Human-Computer Interaction*, July 1997, Sydney, Australia, S. Howard, J. Hammond, & G. Lindgaard, Eds. London: Chapman & Hall, 1997, pp. 524-531.

[10]   M.G. Christel, M.A. Smith, C.R Taylor, and D.B. Winkler, Evolving Video Skims into Useful Multimedia Abstractions, in *Proc. of the CHI '98 Conference on Human Factors in Computing Systems*, C. Karat, A. Lund, J. Coutaz, and J. Karat, Eds., Los Angeles, CA, April 1998, pp. 171-178.

[11]   M. Christel, Visual Digests for News Video Libraries, in *Proc. ACM Multimedia '99*, Orlando FL, Nov. 1999, ACM Press, pp. 303-311.

[12]   M.G. Christel and A.S. Warmack, The Effect of Text in Storyboards for Video Navigation, *in Proc. IEEE International Conference on Acoustics,*

Speech, and Signal Processing (ICASSP), Salt Lake City, UT, May 2001, Vol. III, pp. 1409-1412.

[13] M., Christel, A.M, Olligschlaeger, and C. Huang, Interactive Maps for a Digital Video Library, *IEEE MultiMedia* 7(1), 2000, pp. 60-67.

[14] G. Crane, R. Chavez, et al., Drudgery and Deep Thought, *Comm. ACM* 44(5), 2001, pp. 34-40.

[15] M. Christel and D. Martin, Information Visualization within a Digital Video Library, *Journal of Intelligent Information Systems* 11(3), 1998, pp. 235-257.

[16] M. Christel, A. Hauptmann, H. Wactlar, and T. Ng, Collages as Dynamic Summaries for News Video, in *Proc. ACM Multimedia '02*, Juan-les-Pins, France, Dec. 2002, ACM Press.

[17] P. Clarkson and R. Rosenfeld, Statistical Language Modeling Using the CMU-Cambridge Toolkit, in Proc. Eurospeech '97, Rhodes, Greece, Sept. 1997, Int'l Speech Communication Assoc., pp. 2707-2710.

[18] Y. Deng, B.S. Manjunath, C. Kenney, M.S. Moore, and H.Shin, An efficient color representation for image retrieval, *IEEE Transactions on Image Processing*, Vol. 10, No.1, IEEE, Jan. 2001, pp.140-147.

[19] S.G. Eick, Data Visualization Sliders, in *Proc. ACM Symposium on User Interface Software and Technology*, Marina del Rey, CA, Nov. 1994, ACM Press, pp. 119-120.

[20] D.A. Forsyth, J. Haddon, and S. Ioffc, Finding Objects by Grouping Primitives, in *Shape, Contour and Grouping in Computer Vision*, D.A. Forsyth, J.L. Mundy, R. Cipolla, and V. DiGes'u, Eds., Springer-Verlag, 2000, LNCS 1681.

[21] U. Gargi, R. Kasturi, and S. H. Strayer, Performance Characterization of Video-Shot-Change Detection Methods, *IEEE Transaction on Circuits and Systems for Video Technology*, Vol. 10, No. 1, February 2000.

[22] D. Grinberg, J. Lafferty and D. Sleator, A robust parsing algorithm for link grammars, Carnegie Mellon University Computer Science technical report CMU-CS-95-125, and *Proc. Fourth International Workshop on Parsing Technologies*, Prague, September 1995.

[23] A. Hampapur, R. Jain, and T. Weymouth, Production Model Based Digital Video Segmentation, *Multimedia Tools and Applications*, Vol. 1, 1995, pp. 9-46.

[24] A. Hauptmann, M. Witbrock, A. Rudnicky, and S. Reed, Speech for Multimedia Information Retrieval, *User Interface Software and Technology Conference* (UIST'95), Pittsburgh, PA, November, 1995

[25] A. Hauptmann and M. Smith, Text, Speech, and Vision for Video Segmentation: The Informedia Project, in *AAAI Fall 1995 Symposium on Computational Models for Integrating Language and Vision*.

[26] M. A. Hearst, TileBars: Visualization of Term Distribution Information in Full Text Information Access, *Proceedings of the ACM CHI'95 Conference on Human Factors in Computing Systems*, Denver, CO, May, 1995, 59-66.

[27] M. Hegarty and M.A. Just, Constructing mental models of machines from text and diagrams, *Journal of Memory & Language*, Dec. 1993, 32(6), pp. 717 – 742.

[28] E. Hjelmås and B. K. Low, Face Detection: A survey, *Computer Vision and Image Understanding* Vol. 83, No. 3, September 2001.

[29]   Q. Iqbal and J. K. Aggarwal, Retrieval by Classification of Images Containing Large Manmade Objects Using Perceptual Grouping, *Pattern Recognition Journal*. Vol. 35, No. 7, July 2002, pp. 1463-1479.

[30]   Q. Iqbal and J. K. Aggarwal, Perceptual Grouping for Image Retrieval and Classification, in 3rd IEEE Computer Society Workshop on Perceptual Organization in Computer Vision, July 8, 2001, Vancouver, Canada, pp. 19.1-19.4.

[31]   A. Komlodi and L. Slaughter, Visual Video Browsing Interfaces Using Key Frames, in *CHI '98 Summary*, ACM, New York, 1998, pp. 337-338.

[32]   F. Li, A. Gupta, E. Sanocki, L. He, and Y. Rui, Browsing Digital Video, *in Proc. ACM CHI '00*, ACM Press, 2000, pp. 169-176.

[33]   A. Large, J. Beheshti, A. Breuleux, and A. Renaud, Multimedia and comprehension: The relationship among text, animation, and captions. *Journal of American Society for Information Science*, June 1995, 46(5), pp. 340 – 347.

[34]   R. Lienhart et al., Video Abstracting, *Communications of the ACM, 40*, 12, 1997, pp. 54-62.

[35]   R. Lienhart, Comparison of Automatic Shot Boundary Detection Algorithms, in Proc. *Storage and Retrieval for Still Image and Video Databases* VII 1999, SPIE 3656-29, January 1999.

[36]   B. S. Manjunath, J.-R. Ohm, V. V. Vinod, and A. Yamada, Color and Texture Descriptors, *IEEE Transactions Circuits and Systems for Video Technology*, Special Issue on MPEG-7, 2001.

[37]   M. Mauldin, *Conceptual Information Retrieval: A Case Study in Adaptive Partial Parsing*, Kluwer Academic Press, September 1991.

[38]   M. Mauldin, Information Retrieval by Text Skimming, Ph.D. Thesis, Carnegie Mellon University, August, 1989 (also available as CMU Computer Science technical report CMU-CS-89-193).

[39]   A. Merlino, D. Morey, and M. Maybury, Broadcast News Navigation using Story Segmentation, in *Proc. ACM Multimedia '97*, ACM Press, 197, pp. 381-391.

[40]   D. Miller, R. Schwartz, R. Weischedel, and R. Stone, Named Entity Extraction for Broadcast News, in *Proc. DARPA Broadcast News Workshop*, Washington, DC., March 1999.
http://www.nist.gov/speech/publications/darpa99/html/ie20/ie20.htm

[41]   J. Nielsen and R.L. Mack, Eds., *Usability Inspection Methods*, John Wiley & Sons, New York, NY, 1994.

[42]   T. Firmin and M.J. Chrzanowski, An Evaluation of Automatic Text Summarization Systems, in M.T. Maybury, Ed., *Advances in Automatic Text Summarization*, The MIT Press, Cambridge, MA, 1999.

[43]   G.C. Nugent, Deaf students' learning from captioned instruction: The relationship between the visual and caption display, *Journal of Special Education*, 1983, 17(2), pp. 227-234.

[44]   K.A. Olsen, R.R. Korfhage, et al., Visualization of a Document Collection: The VIBE System, *Information Processing & Management*, 29(1), 1993, pp. 69-81.

[45]   C. Pryluck, When Is a Sign Not a Sign, Revised and reprinted, *Journal of Dramatic Theory and Criticism*, 6(Spring 1992)2, pp. 221-231,

[46]   C. Pryluck, Meaning in Film/Video: Order, Time, and Ambiguity, *Journal of Broadcasting*, 26(Summer 1982)3, pp. 685- 695 (with Charles Teddlie, Richard Sands).

[47] H. Rowley, T. Kanade, and S. Baluja, Neural Network-Based Face Detection, *IEEE Transactions on Pattern Analysis and Machine Intelligence*, January 1998.

[48] T. Sato, T. Kanade, E. Hughes, and M. Smith, Video OCR for Digital News Archive, in Proc. Workshop on Content-Based Access of Image and Video Databases, IEEE, Los Alamitos, CA, 1998, pp. 52-60.

[49] M. Smith and T. Kanade, Video Skimming and Characterization through the Combination of Image and Language Understanding Techniques, in Computer Vision and Pattern Recognition Conference, San Juan, Puerto Rico, June 1997, pp. 775-781.

[50] M. Smith, Integration of Image, Audio, and Language Technology for Video Characterization and Variable-Rate Skimming, Ph.D. Thesis, Carnegie Mellon University. January 1998 (also available as text, Kluwer Academic Press, March 2003).

[51] M. Smith and T. Chen, Image and Video Indexing and Retrieval, in *The Handbook of Image and Video Processing*, A.C. Bovik, Ed., Academic Press, New York, 2000.

[52] H. Schneiderman and T. Kanade, Object Detection Using the Statistics of Parts, *International Journal of Computer Vision*, 2002.

[53] K-K. Sung and T. Poggio, Example-Based Learning for View-Based Human Face Detection, *Pattern Analysis and Machine Intelligence*, Vol. 20, No. 1, January 1998.

[54] Y. Taniguchi, A. Akutsu, Y. Tonomura, and H. Hamada, An Intuitive and Efficient Access Interface to Real-Time Incoming Video Based on Automatic Indexing, in *Proc. ACM Multimedia Conf.*, ACM Press, New York, pp. 25-33.

[55] S. Uchihashi, J. Foote, A. Girgensohn, and J. Boreczky, Video Manga: Generating Semantically Meaningful Video Summaries, in *Proc. ACM Multimedia*, ACM Press, 1999, pp. 383-392.

[56] H. Wactlar, M. Christel, Y. Gong, and A. Hauptmann, Lessons Learned from the Creation and Deployment of a Terabyte Digital Video Library, *IEEE Computer*, 32(2), Fcb. 1999, pp. 66-73.

[57] W. Xiong and J. C.-M. Lee, Efficient scene change detection and camera motion annotation for video classification, in *Computer Vision and Image Understanding*, 71, 1998, pp. 166-181.

[58] B.-L. Yeo and M.M. Yeung, Retrieving and Visualizing Video, *Comm. ACM* 40 (12), 1997, pp. 43-52.

[59] H.J. Zhang, C.Y. Low, and S.W Smoliar, Video parsing and browsing using compressed data, *Multimedia Tools and Applications*, 1, 1995, pp. 89-111.

[60] H.J. Zhang, S.W. Smoliar, J.H. Wu, C.Y. Low, and A. Kankanhalli, A Video Database System for Digital Libraries, *Lecture Notes in Computer Science*, 916, 1995, pp. 253-264.

[61] Technical Notes: Biometric Consortium, www.biometrics.org, 2002.

10

AUDIO AND VISUAL CONTENT SUMMARIZATION OF A VIDEO PROGRAM

Yihong Gong
NEC USA, Inc.
C&C Research Laboratories
10080 North Wolfe Road, SW3-350
Cupertino, CA, U.S.A.
ygong@ccrl.sj.nec.com

1. INTRODUCTION

Video is a voluminous, redundant, and time-sequential medium whose overall content can not be captured at a glance. The voluminous and sequential nature of video programs not only creates congestion in computer systems and communication networks, but also causes bottlenecks in human information comprehension because humans have a limited information processing speed. In the past decade, great efforts have been made to relieve the computer and communication congestion problems. However, in the whole video content creation, processing, storage, delivery, and utilization loop, the human bottleneck problem has long been neglected. Without technologies enabling fast and effective content overviews, browsing through video collections and finding desired video programs from a long list of search results will remain arduous and painful tasks.

Automatic video content summarization is one of the promising solutions to the human bottleneck problem. Video summarization is a process intended to create a concise form of the original video in which important content is preserved and redundancy is eliminated. Similar to paper abstracts and book prefaces, video summaries are valuable for accelerating human information comprehension. A concise and informative video summary enables users to quickly perceive the general content of a video and helps them to decide whether the content is of interest or not. In today's fast-paced world with floods of information, such a video summary will remarkably enhance the users' ability to sift through huge volumes of video data, and facilitate their decisions on what to take and what to discard. When accessing videos from remote servers, a compact video summary can be used as a video "thumbnail" to the original video, which requires much less efforts to download and comprehend. For most home users with limited network bandwidths, this type of video thumbnail can well prevent users from spending minutes or tens of minutes to download lengthy video programs, only to find them irrelevant. With the deployment and evolution of the third-generation mobile communication networks, it will soon be possible to access videos using palm computers and cellular phones. For wireless video accesses, video summaries will become a must because mobile devices have limited memory and battery capacities, and are subject to expensive airtime charges. For video content retrieval, a video summary will allow the users to quickly browse through large video libraries and efficiently spot the desired videos from a long list of search results.

There are many possible ways to summarize video content. To date, the most common approach is to extract a set of keyframes from the original video and display them as thumbnails in a storyboard. However, keyframes extracted from a video sequence are a static image set that contains no temporal properties nor audio information. While keyframes are effective in helping the user to identify the desired shots from a video, they are far from sufficient for the user to get a general idea of the video content, and to judge if the content is relevant or not.

In this chapter, we present three video content summarization systems developed by multidisciplinary researchers in NEC USA, C&C Research Laboratories. These summarization systems are able to produce three kinds of motion video summaries: (1) audio-centric summary, (2) image-centric summary, and (3) audio-visual content summary. The audio-centric summary is created by using text summarization techniques to select video segments that contain semantically important spoken sentences; the image-centric summary is composed by eliminating redundant video frames and preserving visually rich video segments; and the audio-visual summary is constructed by summarizing the audio and visual contents of the original video separately, and then integrating the two summaries with a partial alignment. A Bipartite Graph-based audio-visual alignment algorithm was developed to efficiently find the best alignment solution between the audio and the visual summaries that satisfies the predefined alignment requirements. These three types of motion video summaries are intended to provide natural, diverse, and effective audio/visual content overviews for a broad range of video programs.

In the next section, related video summarization work is briefly reviewed. A discussion of the different types of video programs and their appropriate summarization methods is provided in section 3. Sections 4, 5, and 6 describe the audio-centric summarization, the image-centric summarization, and the audio-visual summarization methods, respectively. Section 7 provides a brief account of future video content summarization research. A glossary of important terms can be found at the end of the chapter.

2. BRIEF REVIEW OF RELATED WORK

To date, video content overview has mainly been achieved by using keyframes extracted from original video sequences. Many works focus on breaking video into shots, and then finding a fixed number of keyframes for each detected shot. Tonomura et al. used the first frame from each shot as a keyframe [1]. Ueda et al. represented each shot using its first and last frames [2]. Ferman and Tekalp clustered the frames in each shot, and selected the frame closest to the center of the largest cluster as the keyframe [3].

An obvious disadvantage to the above equal-number keyframe assignments is that long shots in which camera pan and zoom as well as object motion progressively unveil the entire event will not be adequately represented. To address this problem, DeMenthon et al. proposed to assign keyframes of a variable number according to the activity level of each scene shot [4]. Their method represents a video sequence as a trajectory curve in a high dimensional feature space, and uses the recursive binary curve splitting algorithm to find a set of perceptually significant points to approximate the video curve. This approximation is repeated until the approximation error falls below a user specified value. Frames corresponding to the perceptually significant points are then used as keyframes to summarize the video content. As the curve splitting algorithm assigns more points to larger curvatures, this method naturally assigns more keyframes to shots with more variations.

Keyframes extracted from a video sequence may contain duplications and redundancies. In a television program with two talking persons, the video camera usually switches back and forth between the two persons, with the insertion of some overall views of the scene. Applying the above keyframe selection methods to this kind of video sequences will yield many keyframes that are almost identical. To remove redundancies from keyframes, Yeung et al. selected one keyframe from each video shot, performed hierarchical clustering on these keyframes based on their visual similarity and temporal distance, and then retained only one keyframe per cluster [5]. Girgensohn and Boreczky also applied the hierarchical clustering technique to group the keyframes into as many clusters as specified by the user. For each cluster, a keyframe is selected such that the constraints of an even distribution of keyframes over the length of the video and a minimum distance between keyframes are met [6].

Apart from the above methods of keyframe selection, summarizing video content using keyframes has its own inherent limitations. A video program is a continuous audio/visual recording of real-world scenes. A set of static keyframes captures no temporal properties nor audio content of the original video. While keyframes are effective in helping the user to identify the desired frames or shots from a video, they are far from sufficient for the user to get a general idea of the video content, and to judge if the content is relevant or not. Besides, a long video sequence, e.g., one or two hours, is likely to produce thousands of keyframes, and this excessive number of keyframes may well create another information flood rather than serving as an information abstraction.

There have been research efforts that strive to output motion video summaries to accommodate better content overviews. The CueVideo system from IBM provides two summarization functions: moving story board (MSB) and fast video playback. The MSB is composed of a slide show that displays a string of keyframes, one for each shot, together with a synchronized playback of the entire audio track of the original video. The time scale modulation (TSM) technique is employed to achieve a faster audio playback speed that preserves the timbre and quality of the speech [7]. Although the MSB can shorten the video watching time by 10% to 15%, it does not provide a content abstract of the original video. Therefore, rather than considering the MSB as a video summarization tool, it is more appropriate to consider it as a lower bitrate and higher speed video playback method. On the other hand, the fast video playback function from the CueVideo system plays long, static shots with a faster speed (higher frame rate), and plays short, dynamic shots with a slower speed (lower frame rate) [8]. However, this variable frame rate playback causes static shots to look more dynamic, and dynamic shots to look more static, therefore dramatically distorting the temporal characteristics of the video sequence.

The Informedia system from Carnegie Mellon University provides the video skim that strives to summarize the input video by identifying those video segments that contain either semantically important scenes or statistically significant keywords and phrases [9]. The importance of each video segment is measured using a set of heuristic rules that are highly subjective and content specific. This rule-based approach has certainly limited the ability of the system to handle diverse video sequences. I. Yahiaoui et al. also proposed a similar method that summarizes multi-episode videos based on statistics as well as heuristics [10].

3. THREE TYPES OF VIDEO SUMMARIZATION

Video is a medium which contains both audio and image tracks. Based on which track of a video is being analyzed for content summarization, we can classify a video summarization method into one of the following three basic categories: audio-centric summarization, image-centric summarization, and audio-visual summarization. An audio-centric summarization can be achieved by analyzing mainly the audio track of the original video and selecting those video segments that contain either important audio sounds or semantically important speeches. Conversely, an image-centric summarization can be accomplished by focusing on the image track of the video and selecting those video segments whose image contents bear either visual or semantic significances. Finally, we can achieve an audio-visual summarization by decoupling the audio and image track of the input video, summarizing the two tracks separately, and then integrating the two summaries according to certain alignment rules.

Audio-centric summarization is useful for summarizing video programs whose visual contents are less significant than their audio contents. Examples of such video programs include news conferences, political debates, seminars, etc., where shots of talking heads, static scenes, and audiences consist of the most part of their visual contents. In contrast, image-centric summarization is useful for summarizing video programs which do not contain many significant audio contents. Action movies, television broadcast sports games, and surveillance videos are good examples of this kind of video programs.

There are certain video programs that do not have a strong synchronization between their audio and visual contents. Consider a television news program in which an audio segment presents information concerning the number of casualties caused by a recent earthquake. The corresponding image segment could be a close shot of a reporter in the field, of rescue teams working at the scene of a collapsed building, or of a regional map illustrating the epicenter of the earthquake. The audio content is related to, but does not directly refer to the corresponding image content. This kind of video production patterns is very common among such video programs as news, documentaries, etc. For these video programs, since there is no strong synchronization between the audio and visual contents, and a video segment containing significant audio content does not necessarily contain significant visual content at the same time (and vice versa), an audio-visual summarization is the most appropriate summarization type because it can maximize the coverage for both the audio and visual contents of the original video without having to sacrifice either of them.

4. AUDIO-CENTRIC SUMMARIZATION

In this section, we present an audio-centric summarization system that creates a motion video summary of the input video by selecting video segments that contain semantically important speeches. This summarization system is especially effective for summarizing video programs that record meetings, report news events, disseminate knowledge, such as news reports, documentaries, conference/seminar videos, etc.

4.1 SYSTEM OVERVIEW

Figure 10.1 shows the block diagram of the audio-centric summarization system. For the audio track of the input video, speech recognition is first conducted to obtain a speech transcript that includes the recognized sentences along with their time codes within the audio track (Module 1). Next, a text summarization method (described in Section 4.2) is applied to the speech transcript to obtain an importance rank for each of the sentences in the transcript (Module 2). An audio-centric summary is created by selecting the video segments of the sentences in descending order of their importance ranks (until the user specified summary length is reached), and then concatenating the selected video segments in their original time order (Module 3). Because each sentence in the speech transcript has a time code indicating its start and end positions in the original video sequence, the video segment containing a particular sentence can be easily determined using the time codes.

If closed captions for the original video are available, they will be utilized to reduce errors in the speech transcript generated by the speech recognition engine (Module 4). In fact, closed captions often contain many errors such as typing mistakes, the dropping of certain words/phrases, etc. because they are usually generated by human operators during the live broadcast of a video program. Nonetheless, by aligning the speech transcript with the corresponding closed captions, and then taking as many grammatically correct segments as possible, we are able to improve the correctness of the speech transcript by a large percentage.

Obviously, the text summarization module (module 2) is the centerpiece of the whole audio-centric summarization system, and the eventual summarization quality of the system is solely determined by the performance of this module. In the following subsections, we present two text summarization methods developed by us, and describe their systematic performance evaluations in detail.

4.2 TEXT SUMMARIZATION

A document usually consists of several topics. Some topics are described intensively by many sentences, and hence form the major content of the document. Other topics may be just briefly mentioned to supplement the major topics, or to make the story more complete. A good text summary should cover the major topics of the document as much as possible, and at the same time keep redundancy to a minimum.

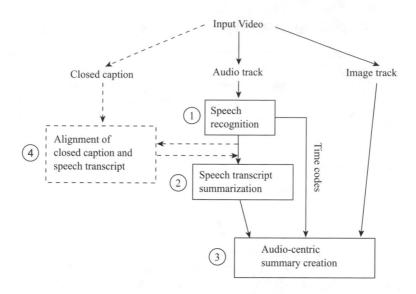

Note: the dash arrows and rectangle denote optional operations

Figure 10.1 Block diagram of the audio-centric summarization system.

In this section, we present two methods that create text summaries by selecting sentences based on the relevance measure and the latent semantic analysis. Both the methods need to first decompose the document into individual sentences, and to create the weighted term-frequency vector for each of the sentences. Let $T_i = [t_{1i}, t_{2i}, \cdots, t_{ni}]^t$ be the term-frequency vector of passage i, where element t_{ji} denotes the frequency in which term j occurs in passage i. Here passage i could be a phrase, a sentence, a paragraph of the document, or could be the whole document itself. The weighted term-frequency vector $A_i = [a_{1i}, a_{2i}, \cdots, a_{ni}]^t$ of passage i is defined as:

$$a_{ji} = L(t_{ji}) \cdot G(t_{ji}) \tag{10.1}$$

where $L(t_{ji})$ is the local weighting for term j in passage i, and $G(t_{ji})$ is the global weighting for term j. When the weighted term-frequency vector A_i is created, we have the further choice of using A_i with its original form, or normalizing it by its length $|A_i|$. There are many possible weighting schemes. In Section 4.3.3, we inspect several major weighting schemes and discuss how these weighting schemes affect the summarization performances.

4.2.1 Summarization by Relevance Measure

After the document has been decomposed into individual sentences, we compute the relevance score of each sentence against the whole document. We then select the sentence k with the highest relevance score, and add it to the summary. Once the sentence k has been added to the summary, it is eliminated from the candidate sentence set, and all the terms contained in k are eliminated from the original document. For the remaining sentences, we repeat the steps of relevance measure, sentence selection, and term elimination until the number of selected sentences has reached the predefined value. The operation flow is as follows:

1. Decompose the document into individual sentences, and use these sentences to form the candidate sentence set S.

2. Create the weighted term-frequency vector A_i for each sentence $i \in S$, and the weighted term-frequency vector D for the whole document.

3. For each sentence $i \in S$, compute the relevance score between A_i and D, which is the inner product between A_i and D.

4. Select sentence k that has the highest relevance score, and add it to the summary.

5. Delete k from S, and eliminate all the terms contained in k from the document. Re-compute the weighted term-frequency vector D for the document.

6. If the number of sentences in the summary reaches a predefined number, terminate the operation; otherwise, go to Step 3.

In Step 4 of the above operations, the sentence k that has the highest relevance score with the document is the one that best represents the major content of the document. Selecting sentences based on their relevance scores ensures that the summary covers the major topics of the document. On the other hand, eliminating all the terms contained in k from the document in Step 5 ensures that the subsequent sentence selection will pick the sentences with the minimum overlap with k. This leads to the creation of a summary that contains little redundancy.

4.2.2 Summarization by Latent Semantic Analysis

Inspired by the latent semantic indexing, we applied singular value decomposition (SVD) to the task of text summarization. The process starts with the creation of a terms-by-sentences matrix $\mathbf{A} = [A_1, A_2, \cdots, A_n]$ with each column vector A_i representing the weighted term-frequency vector of sentence i in the document under consideration. If there are a total of m terms and n sentences in the document, then we will have an $m \times n$ matrix $\mathbf{A}$ for the document. Since every word does not normally appear in each sentence, the matrix $\mathbf{A}$ is usually sparse.

Given an $m \times n$ matrix $\mathbf{A}$, where without loss of generality $m \geq n$, the SVD of $\mathbf{A}$ is defined as [11]:

$$\mathbf{A} = \mathbf{U \Sigma V}^t \tag{10.2}$$

where $\mathbf{U} = [u_{ij}]$ is an $m \times n$ column-orthonormal matrix whose columns are called left singular vectors; $\mathbf{\Sigma} = \text{diag}(\sigma_1, \sigma_2, \cdots, \sigma_n)$ is an $n \times n$ diagonal matrix whose diagonal elements are non-negative singular values sorted in descending order; and $\mathbf{V} = [v_{ij}]$ is an $n \times n$ orthonormal matrix whose columns are called right singular vectors. If rank$(\mathbf{A}) = r$, then Σ satisfies

$$\sigma_1 \geq \sigma_2 \cdots \geq \sigma_r , \sigma_{r+1} = \cdots = \sigma_n = 0. \tag{10.3}$$

The interpretation of applying the SVD to the terms-by-sentences matrix $\mathbf{A}$ can be made from two different viewpoints. From the transformation point of view, the SVD derives a mapping between the m-dimensional space spanned by the weighted term-frequency vectors and the r-dimensional singular vector space with all of its axes being linearly independent. This mapping projects each column vector A_i in matrix $\mathbf{A}$, which represents the weighted term-frequency vector of sentence i, to column vector $\Psi_i = [v_{i1}, v_{i2}, \cdots, v_{ir}]^t$ of matrix $\mathbf{V}^t$, and maps each row vector j in matrix $\mathbf{A}$, which tells the occurrence count of the term j in each of the documents, to row vector $\Phi_j = [u_{j1}, u_{j2}, \cdots, u_{jr}]$ of matrix $\mathbf{U}$. Here each element v_{ix} of Ψ_i, u_{jy} of Φ_j is called the index with the ith, jth singular vectors, respectively.

From the semantic point of view, the SVD derives the latent semantic structure from the document represented by matrix $\mathbf{A}$ [12]. This operation reflects a breakdown of the original document into r linearly-independent base vectors or concepts. A unique SVD feature which is lacking in conventional IR technologies is that the SVD is capable of capturing and modelling the interrelationships among terms so that it can semantically cluster terms and sentences. Consider an extreme document with the following terms-by-sentences matrix:

Sentences

$$\mathbf{A} = \begin{array}{c} \\ \\ \text{Terms} \\ \\ \\ \\ \end{array} \quad \begin{array}{cccccc} S_1 & S_2 & S_3 & S_4 & S_5 & S_6 \end{array}$$

$$\mathbf{A} = \begin{bmatrix} 1 & 2 & 3 & 0 & 0 & 0 \\ 3 & 6 & 9 & 0 & 0 & 0 \\ 4 & 8 & 12 & 0 & 0 & 0 \\ 0 & 0 & 0 & 5 & 10 & 15 \\ 0 & 0 & 0 & 2 & 4 & 6 \\ 0 & 0 & 0 & 7 & 14 & 21 \end{bmatrix} \begin{array}{l} \text{physician} \\ \text{hospital} \\ \text{medicine} \\ \text{car} \\ \text{driver} \\ \text{speeding} \end{array}$$

The document consists of only six words and six sentences. Sentences 1 to 3 consist of only the words *physician, hospital,* and *medicine*; sentences 4 to 6 consist of only the words *car, driver,* and *speeding*. Obviously, the whole document consists of only two topics: Sentences 1 to 3 represent the first topic, which is about hospitals, and sentences 4 to 6 represent the second topic, which is about drivers. Because topic 2 has higher term frequencies than topic 1, it can be considered more important than topic 1. With topic 1, because sentence 3 has the highest term frequencies, it can be considered to best cover the topic. Similarly, sentence 6 can be considered to best cover topic 2.

By performing SVD on the above terms-by-sentences matrix **A**, we get a projection in the singular vector space shown in Figure 10.2.

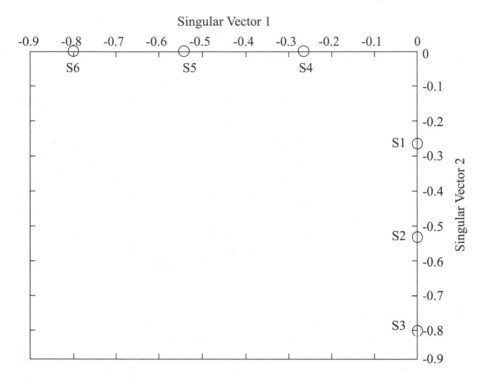

Figure 10.2 The projection of the above terms-by-sentences matrix **A** into the singular vector space.

Amazingly, sentences 4 to 6 all lie on the first singular vector, and sentences 1 to 3 all on the second singular vector. The singular vectors are sorted in descending order of their corresponding singular values (e.g., the singular value of the first singular vector is larger than that of the second singular vector). On the two axes, sentences with the highest term frequencies have the highest index values. This extreme example strongly supports the hypotheses that

- Each singular vector from the SVD represents a salient concept/topic of the document.

- The magnitude of the corresponding singular value of a singular vector represents the degree of importance of the salient concept/topic.

- The sentence having the largest index value with a singular vector X_i is the one that best covers the topic represented by X_i.

Based on the above discussion, we present the following SVD-based document summarization method.

1. Decompose the document D into individual sentences, and use these sentences to form the candidate sentence set S, and set $k = 1$.

2. Construct the terms-by-sentences matrix **A** for the document D.

3. Perform the SVD on $\mathbf{A}$ to obtain the singular value matrix Σ, and the right singular vector matrix $\mathbf{V}^t$. In the singular vector space, each sentence i is represented by the column vector $\Psi_i = [v_{i1}, v_{i2}, \cdots, v_{ir}]^t$ of $\mathbf{V}^t$.

4. Select the kth right singular vector from matrix $\mathbf{V}^t$.

5. Select the sentence which has the largest index value with the kth right singular vector, and include it in the summary.

6. If k reaches the predefined number, terminate the operation; otherwise, increment k by one, and go to Step 4.

In Step 5 of the above operation, finding the sentence that has the largest index value with the kth right singular vector is equivalent to finding the column vector Ψ_i whose k^{th} element v_{ik} is the largest. By the hypothesis, this operation is equivalent to finding the sentence describing the salient concept/topic represented by the kth singular vector. Since the singular vectors are sorted in descending order of their corresponding singular values, the kth singular vector represents the kth most important concept/topic. Because all the singular vectors are independent of each other, the sentences selected by this method contain minimum redundancy.

4.3 PERFORMANCE EVALUATION

In this section, we describe the data corpus constructed for performance evaluations, present various evaluation results, and make in-depth observations on some aspects of the evaluation outcomes.

4.3.1 Data Corpus

Our evaluations on the two text summarization methods have been conducted using a database of two months of the CNN Worldview news program. Excluding commercial advertisements, one day broadcast of the CNN Worldview program lasts for about 22 minutes, and consists of 15 individual news stories on average. The evaluation database consists of the closed captions of 549 news stories whose lengths are in the range of 3 to 105 sentences. As summarizing short articles does not make much sense in real applications, for our evaluations we eliminated all the short stories with less than ten sentences, resulting in 243 documents.

Table 10. 1 provides the particulars of the evaluation database. Three independent human evaluators were employed to conduct manual summarization of the 243 documents contained in the evaluation database. For each document, each evaluator was requested to select exactly five sentences which he/she deemed the most important for summarizing the story.

Table 10.1 Particulars of the Evaluation Database.

Document Attributes	Values
Number of documents	549
Number of documents with more than 10 sentences	243
Average sentences/document	21
Minimum sentences/documents	3
Maximum sentences/documents	105

Because of the disparities in the evaluators' sentence selections, each document can have between 5 to 15 sentences selected by at least one of the evaluators. Table 10.2 shows the statistics of the manual summarization results. As evidenced by the table, the disagreements among the three evaluators are much more than expected: each document has an average of 9.0 sentences selected by at least one evaluator, and among these 9.0 selected sentences, only 1.2 sentences receive a unanimous vote from all three evaluators. Even when the sentence selection is determined by a majority vote, we still get a lower than expected overlapping rate: an average of 2.5 sentences per document. These statistics suggest that for many documents in the database, the manual summarization determined by a majority vote could be very short (2.5 sentence per document), and this summary length is below the best fixed summary length (three to five sentences) suggested in [13]. For this reason, we decided to evaluate our two summarization methods using each of the three manual summarization results, as well as the combined result determined by a majority vote.

Table 10.2 Statistics of the Manual Summarization Results.

Summarization Attributes	Values	Average sentences/doc
Total number of sentences	7053	29.0
Sentences selected by 1 person	1283	5.3
Sentences selected by 2 person	604	2.5
Sentences selected by 3 persons	290	1.2
Total number of selected sentences	2177	9.0

4.3.2 Performance Evaluation

We used the recall (R), precision (P), along with F to measure the performances of the two summarization methods. Let S_{man} and S_{sum} be the set of sentences selected by the human evaluator(s), and the summarizer, respectively. The standard definitions of R, P, and F are defined as follows:

$$R = \frac{|S_{man} \cap S_{sum}|}{|S_{man}|}$$

$$P = \frac{|S_{man} \cap S_{sum}|}{|S_{sum}|}$$

$$F = \frac{2RP}{R+P}$$

For our evaluations, we set the length of the machine generated text summaries to the length of the corresponding manual summaries. When the evaluation is performed using each individual manual summarization result, both $|S_{man}|$ and $|S_{sum}|$ are equal to five. When the evaluation is performed using the combined result determined by a majority vote, $|S_{man}|$ becomes variable, and $|S_{sum}|$ is set to the value of $|S_{man}|$.

The evaluation results are shown in Table 10. 3. These results are generated using the weighting scheme that uses binary local weighting, no global weighting, nor normalization (denoted in short as BNN, see Section 4.3.3 for a detailed description). As evidenced by the results, despite the very different approaches taken by the two summarizers, their performance measures are quite compatible. This fact suggests that the two approaches may interpret each other. The first summarizer (the one using the relevance measure) takes the sentence that has the highest relevance score with the document as the most important sentence, while the second summarizer (the one based on the latent semantic analysis) identifies the most important sentence as the one that has the largest index value with the most important singular vector. On the other hand, the first summarizer eliminates redundancies by removing all the terms contained in the selected sentences from the original document, while the second summarizer suppresses redundancies by using the kth singular vector for the kth round of sentence selection. The first method is straightforward, and it is relatively easy for us to give it a semantic interpretation. In the second method, there has been much debate about the meaning of the singular vectors when a collection of text (which could be sentences, paragraphs, documents, etc.) are projected into the singular vector space. Surprisingly, the two different methods create very similar summaries. This mutual resemblance enhances the belief that each important singular vector does capture a major topic/concept of a document, and two different singular vectors do capture two semantically independent topics/concepts that have the minimum overlap.

Table 10.3 Evaluation Results.

Test Data	First Summarizer			Second Summarizer		
	R	P	F	R	P	F
Assessor 1	0.57	0.60	0.58	0.60	0.62	0.61
Assessor 2	0.48	0.52	0.50	0.49	0.53	0.51
Assessor 3	0.55	0.68	0.61	0.55	0.68	0.61
Majority vote	0.52	0.59	0.55	0.53	0.61	0.57

4.3.3 Weighting Schemes

In our performance evaluations, we studied the influence of different weighting schemes on summarization performances as well. As shown by Equation (1), given a term i, its weighting scheme is defined by two parts: the local weighting $L(i)$ and the global weighting $G(i)$. Local weighting $L(i)$ has the following four possible alternatives:

1. No weight: $L(i) = tf(i)$ where $tf(i)$ is the number of times term i occurs in the sentence.

2. Binary weight: $L(i) = 1$ if term i appears at least once in the sentence; otherwise, $L(i) = 0$.

3. Augmented weight: $L(i) = 0.5 + 0.5 \cdot \dfrac{tf(i)}{tf(\max)}$ where $tf(\max)$ is the frequency of the most frequently occurring term in the sentence.

4. Logarithmic weight: $L(i) = \log(1 + tf(i))$.

Possible global weighting $G(i)$ can be:

1. No weight: $G(i) = 1$ for any term i.

2. Inverse document frequency: $G(i) = \log(N/n(i))$ where N is the total number of sentences in the document, and $n(i)$ is the number of sentences that contain term i.

When the weighted term-frequency vector A_k of a sentence k is created using one of the above local and global weighting schemes, we further have the choice of:

1. Normalization: which normalizes A_k by its length $|A_k|$.

2. No normalization: which uses A_k in its original form.

Therefore, for creating vector A_k of sentence k, we have a total of $4 \times 2 \times 2 = 16$ combinations of the possible weighting schemes. In our experimental evaluations, we have studied nine common weighting schemes, and their performances are shown in Figure 10.3. As seen from the figure, summarizer 1 is less sensitive than summarizer 2 to the changes of weighting schemes. Any of the three local weighting schemes (i.e., Binary, Augmented, logarithm) produces quite similar performance readings. Adding a global weighing and/or the vector normalization deteriorates the performance of summarizer 1 by 2 to 3% in average. In contrast, summarizer 2 reaches the best performance with the binary local weighting, no global weighing, and no normalization for most of the cases, while its performance drops a bit by adding the global weighing, and deteriorates dramatically by adding the normalization into the formula.

4.3.4 Further Observations

Generic text summarization and its evaluation are very challenging. Because no queries or topics are provided to the summarization system, summarization outputs and performance judgments tend to lack consensus. In our experiments, we have seen a large degree of disparities in the sentence selections among the three independent evaluators, resulting in lower than expected scores (F=0.55 for summarizer 1, F=0.57 for summarizer 2) from the performance evaluation by a majority vote.

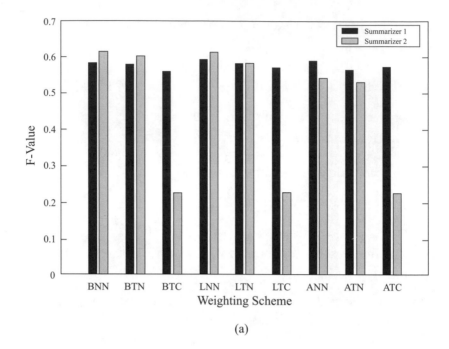

(a)

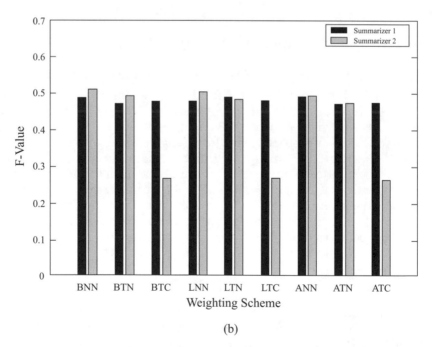

(b)

Figure 10.3 The influence of different weighting schemes on the summarization performances. (a),(b),(c),(d): Evaluation using the manual summarization result from evaluator 1, 2, 3, and the one determined by a majority vote, respectively. The notation of weighting schemes is the same as the one from the SMART system. Each weighting scheme is denoted by three letters. The first, second, and third letters represent the local weighting, the global weighing, and the vector normalization, respectively. The meaning of the letters are as follows: N: No weight, B: Binary, L: Logarithm, A: Augmented, T: Inverse document frequency, C: Vector normalization.

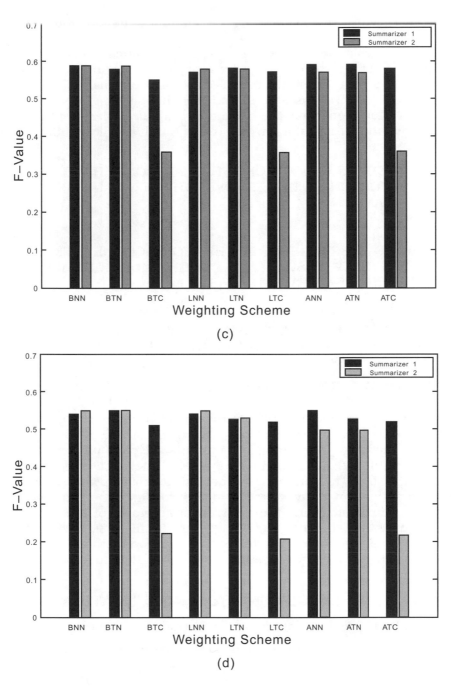

Figure 10.3 (Continued)

It is observed from Table 10.3 that the two text summarizers both receive better scores when they are evaluated using the manual summarization results from evaluator 1 and 3. However, when evaluated using evaluator 2's results, the performance scores drop by 10% in average, dramatically dragging down the performance scores for the evaluation by a majority vote. An in-depth analysis of the cause of this large difference has revealed different manual summarization patterns among the three evaluators. Consider the following passage taken from a CNN news story reporting the recent Israeli-Palestinian conflicts, political efforts for restoring the calm in the region, and hostile sentiments among Palestinian people:

(1) ......IN RECENT VIOLENCE MORE THAN 90 PEOPLE HAVE BEEN KILLED, THOUSANDS MORE INJURED, THE OVERWHELMING MAJORITY OF THOSE ARE PALESTINIANS.

......

(2) NOW AFTER A BRIEF LULL IN THE VIOLENCE, NEW FIGHTING, NEW CLASHES ERUPTED THURSDAY, AND TONIGHT MORE GUNFIRE REPORTED WITH MORE INJURIES OF PALESTINIANS.

......

(3) IN THE NORTHERN WEST BANK TOWN NABLUS, ISRAELI TANKS EXCHANGED FIRE WITH PALESTINIAN GUNMEN, KILLED AT LEAST 3 OF THEM ON WEDNESDAY.

The above three sentences all cover the topic of Israeli-Palestinian conflicts. Our two summarizers both selected sentence (1), and discarded (2) and (3) because of their similarities to sentence (1). On the other hand, both evaluators 1 and 3 selected sentence (1), while evaluator 2 picked all the three sentences for summarizing the topic. This example represents a typical pattern that happens repeatedly in the whole evaluation process. The fact suggested by this phenomenon is that, to summarize a document, some people strive to select sentences that maximize the coverage of the document's main content, while others tend to first determine the most important topic of the document, and then collect only the sentences that are relevant to this topic. Evidently, when it comes to the evaluation of our two text summarization methods, the former type of evaluators generates a higher accuracy score than the latter.

5. IMAGE-CENTRIC SUMMARIZATION

We set the following goals for our image-centric summarization system:

1. The user is able to affect the summarization outcome;

2. Within the user specified time length, the summarization minimizes the redundancy while preserving visually rich content of the original video program.

The first goal aims to meet different users' content overview requirements. The second goal strives to turn the difficult, subjective visual content summarization problem into a feasible and objective one. By common sense, an ideal visual content summary should be the one that retains only visually or semantically important segments of a given video. However, finding such video segments requires an overall understanding of the video content, which is beyond our reach given the state of the art of current computer vision and image understanding techniques. On the other hand, it is relatively easy to measure the level of visual content change, and to identify duplicates and redundancies in a video sequence. For the purpose of visual content overviews, the video watching time will be largely shortened, and the visual content of the original video will not be dramatically lost if we eliminate duplicate shots and curtail lengthy and static shots. Therefore, instead of relying on heuristically picking visually "important" segments for generating summaries, we choose to eliminate duplicates and redundancies while preserving visually rich contents in the given video. The following subsections present the system overview, and describe the image-centric summarization process in more detail.

5.1 SYSTEM OVERVIEW

We enable the user to affect the summarization process by specifying two parameters: the summary length T_{len}, and the minimum time length of each shot T_{min} in the summary. Given T_{len} and T_{min}, the maximum number of scene shots a video summary can incorporate equals $N = T_{len}/T_{min}$. Here the parameter T_{min} can be considered as a control knob for the user to select between depth-oriented and breadth-oriented summaries. A smaller value for T_{min} will produce a breadth-oriented summary that consists of more shots that each is shorter in length, while a larger value for T_{min} will produce a depth-oriented summary that consists of less shots that each is longer in length.

Figure 10.4 shows the block diagram of the image-centric summarization system. For the image track of the input video, shot boundary detection is first conducted to segment the video into individual scene shots (Module 1). Shot clustering is then conducted to group all the shots into $N = T_{len}/T_{min}$ clusters based on their visual similarities (Module 2). The clustering process generates shot clusters that meet the following conditions:

1. All the shots within the same cluster are visually similar.

2. Any pair of shots from two different clusters are remarkably different with respect to their visual contents.

Using this shot cluster set, an image-centric summary is created by selecting the longest shot from each cluster as the representative shot, taking a chunk from each of the representative shots, and then concatenating them in their original time order (Module 3).

Image-centric summaries are particularly useful for summarizing entertainment video programs whose visual contents play a more important role than the corresponding audio contents. Examples of such videos include action and combat movies, cartoons, TV broadcasted sports games, etc.

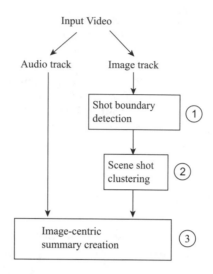

Figure 10.4 Block diagram of the image-centric summarization system.

5.2 SCENE SHOT CLUSTERING

Scene shot clustering is the most challenging and important part of the image-centric summarization system. Here, the challenges are:

- We have to generate shot clusters of the requested number N;

- The generated N shot clusters have to all satisfy the two conditions listed in Section 5.1.

The task is challenging because in many cases, satisfying one requirement will require breaking the other.

In the literature, hierarchical clustering is a commonly used technique for dynamically grouping scene shots into a specified number of clusters. However, this technique does not meet the above requirements for the following reasons:

(a) The computation time is $O(n^3)$ for building up the entire hierarchy, where n is the number of scene shots.

(b) When the clustering hierarchy reaches certain heights, some major, visually different scene shots will be merged together.

(c) At the lower layers of the clustering hierarchy, many visually similar shots have yet to be merged properly, and there are still some shot clusters that are visually similar to one another.

If the summarization process has to take clusters from layers described in either (b) or (c), this summary will either drop some major shots, or contain many duplicates.

To overcome the problems associated with the hierarchical clustering methods, we developed a novel method that performs scene shot clustering using the minimum spanning tree (MST) algorithm together with the upper-bound threshold T_{high} and the lower-bound threshold T_{low}. This method produces shot clusters that meet the two conditions listed in Section 5.1. Under the extreme circumstance where the required number of clusters N is either very high or very low, and can not be generated without breaking the two conditions, we use either T_{high} or T_{low} to generate clusters of number N' that best approaches N. If $N' > N$, we sort all the shot clusters in descending order of their combined time lengths, and discard those clusters at the bottom of the list. If $N' < N$, we define a visual content richness metric, and assign playback time to each shot cluster according to its content richness measure.

5.2.1 Visual Content Richness Metric

For many video programs, the degree of visual change is a good indicator of the amount of visual content conveyed by a video. From the viewpoint of visual content richness, a dynamic video with many changes contains richer visual content than a static video with almost no changes. Based on this observation, a simple visual content richness metric $R(S)$ for a video segment S can be defined as follows:

$$R(S) = \frac{1}{K} \sum_{i=1}^{K-1} d(f_{i+1} - f_i) \tag{10.4}$$

where f_i is the ith frame of the video segment S, K is the total number of frames in S, and $d(f_{i+1} - f_i)$ is the distance between frames $i+1$ and i in the image feature space.

5.2.2 Clustering Scene Shots Using Minimum Spanning Tree

Minimum spanning tree (MST) is a special kind of graph that connects all its vertices using the shortest path [15]. Let $G = (V, E)$ be a connected, undirected graph where V is the set of vertices, and E is the set of possible interconnections between pairs of vertices. For each edge $(u, v) \in E$, we define a weight $w(u, v)$ specifying the cost to connect u and v. The MST of the graph $G = (V, E)$ is defined as $T = (V, P)$ where P is an acyclic subset $P \subseteq E$ that connects all the vertices, and minimizes the total weight

$$w(P) = \sum_{(u,v) \in P} w(u, v) \tag{10.5}$$

Construction of MST T for a given graph G is not unique, and could have multiple solutions.

For our scene shot clustering problem, an MST T is constructed for such a graph $G = (V, E)$ where each vertex $v \in V$ represents a scene shot of the original video, and each edge $(u, v) \in E$ represents the distance between shots u and v in the image feature space. By definition, $T = (V, P)$ connects the shot set V with the shortest edge set P. Given a graph $G = (V, E)$, its MST T can be built in time $O(|E| \cdot \lg |V|)$. The computation time is a dramatic improvement compared to the time $O(|V|^3)$ required for the hierarchical clustering.

Once an MST $T = (V, P)$ is constructed for the shot set V, we sort all the edges in P in descending order of their lengths. As T is a tree structure, which means that any two vertices in T are connected by a unique simple path, a cut at any edge will break the tree into two subtrees. Therefore, if N shot clusters are required to compose the video summary, we can cut T at its top $N-1$ longest edges to obtain N subtrees, and to use each subtree to form a shot cluster. Let L_{N-1} denote the length of the $N-1$'th longest edge of T. All of the N subtrees obtained above have the property that the distance between an arbitrary vertex and its nearest neighbour is less than, or equal to L_{N-1}. Because the $N-1$'th longest edge of T defines the upper-bound edge of the subsequent N subtrees, to highlight its special role in the shot clustering process, we call it the thresholding edge for cutting the MST.

To achieve a shot clustering meeting the two conditions listed in Section 5.1 using MST, we must put certain restrictions on cutting the tree. Our experiments have shown that, using image features with a reasonable discrimination power, one can easily find an upper-bound threshold T_{high} and a lower-bound threshold T_{low} that divide pairs of scene shots into three categories:

1. The two shots are completely different when their distance in the feature space exceeds T_{high}.

2. The two shots are similar when their distance in the feature space is smaller than T_{low}.

3. When the distance is between T_{high} and T_{low}, the two shots could be judged as either similar or different depending on how strict the similarity criterion is.

The upper-bound and the lower-bound thresholds create an ambiguous zone of judgment. This ambiguous zone provides us with a range of thresholds for cutting the MST. To ensure that the clustering process does not separate visually similar shots into different clusters, nor merge completely different shots into the same clusters, the length of the thresholding edge for cutting the MST must be between T_{high} and T_{low}.

5.2.3 Creating Video Summaries

When the required number N of shot clusters is either very high or very low, and can not be generated without breaking the two conditions listed in Section 5.1, we use either T_{high} or T_{low} to generate clusters of number N' that best approaches N, and perform the following procedures for the three different cases:

1. $N' = N$: From each cluster, take the starting portion of the representative shot together with its corresponding audio segment for T_{min} seconds, and concatenate these segments in time order to create the video summary.

2. $N' > N$: Sort the N' clusters in descending order of their combined time lengths. A cluster's combined time length is the sum of time lengths of all the

scene shots contained in the cluster. Select the top N clusters in the sorted list, and perform the operations same to the case $N' = N$ to create the video summary.

3. $N' < N$: Compute the average visual content richness $\overline{R}(\Phi_i)$ for each cluster Φ_i, and assign playback time t_i defined below to Φ_i:

$$
t_i = \min\left(T_{len} \cdot \frac{\overline{R}(\Phi_i)}{\sum_{i=1}^{N'} \overline{R}(\Phi_i)}, \tau_i \right)
\tag{10.6}
$$

where τ_i is the combined time length of Φ_i. If t_i is shorter than T_{min}, set $t_i = T_{min}$. If t_i is longer than the longest shot in Φ_i, and Φ_i consists of multiple scene shots, sort the shots in descending order of their lengths, take the longest shot, the second longest shot, etc., until the total length reaches t_i. Once the video segments have been collected from all the shot clusters, concatenate them in time order to create the video summary.

The above operations accomplish the following two things: when the video summary can incorporate less shots ($N' > N$), those trivial shot clusters with short combined time lengths will be discarded, while redundancies and duplicates will be kept to a minimum; when the video summary can accommodate more shots ($N' < N$), those shot clusters with a richer visual content metric will be given more exposures within the summary.

In summary, the whole video summarization process consists of the following major operations:

1. Prompt the user to specify the video summary length T_{len}, and the minimum time length of visual segments T_{min}.

2. Segment the video sequence into individual scene shots.

3. Perform scene shot clustering using the MST algorithm together with the upper-bound threshold T_{high} and the lower-bound threshold T_{low}. The clustering process strives to generate clusters of the required number N; if impossible, then generate clusters of a number N' which best approaches N within the given threshold range.

4. Create the video summary using the procedures in accordance to the cases $N' = N$, $N' > N$, and $N' < N$.

5.3 EXPERIMENTAL EVALUATION

We have tested the image-centric summarization system using five different video programs. The test video programs consist of news reports, documentary, political debate, and live coverage of the tally dispute for the 2000 presidential election, each of which last from 5 to 22 minutes.

Original video shots

Video summary

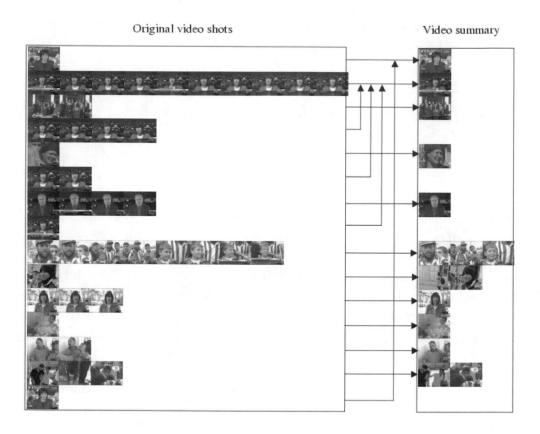

Figure 10.5 An image-centric summarization result.

Figure 10.5 shows the summarization result on a six-minute news report covering recent Israeli-Palestinian conflict (News report 1 in Table 10.4). The sequence consists of 28 shots, and Figure 10.5 displays the 15 major shots. Each row in the left hand rectangle represents a shot in the original video, and the number of frames in each row is proportional to the time length of the corresponding shot. The same row in the right hand rectangle depicts the video segment(s) taken from the corresponding shot cluster. The user has specified the two parameters $T_{len} = 60$ seconds, and $T_{min} = 2$ seconds ($N = 30$ clusters). In this sequence, the field reporter appeared four times, which are at row 2, 4, 6, and 8, respectively. As these four shots are quite static and visually similar, they were clustered together, and were assigned one unit of playback time T_{min} (row 2 in the right hand rectangle). The similar situation occurs for shot 1 and 15 as well. Shot 9 is a long, dynamic shot that contains many visual changes. It was grouped into one cluster, and was assigned approximately three units of playback time ($3 \cdot T_{min}$). Similarly, as shots 10 and 14 contain high degrees of visual change, they were also assigned longer than one unit of playback time.

Table 10.4 shows the detailed evaluation results on five different video sequences. The political debate video contains many shots that display either the same speakers, or the same global view of the studio. As shown in the table, sixteen visually similar shots have been properly merged by the summarization system. However, the merge of the similar shots has yielded less than the requested number of clusters for creating a two-minute summary (case $N' < N$). Therefore, fourteen shots with higher degrees of visual content change have received longer than one unit of playback time. On the other hand, the documentary program consists of no duplicate shots but many long, dynamic shots that gradually unveil the ongoing event in the field. Twelve dynamic shots have been assigned longer than one unit of playback time, while no shots have been merged in generating the summary.

Table 10.4 Experimental Evaluation.

Video Contents	Time Length	Summary Length T_{len}	Total Shots	Shots Merged	Properly Merged	Shots Assigned More time
Documen-tary	5 min.	1 min.	17	0	0	12
Political debase	10 min.	2 min.	62	16	16	14
News report 1	6 min.	1 min.	28	6	6	7
News report 2	13 min.	1.5 min.	48	13	13	10
Live event coverage	22 min.	2 min.	42	13	13	28

6. AUDIO-VISUAL SUMMARIZATION

In developing the audio-visual summarization system, we strive to produce a motion video summary for the original video that:

1. provides a natural, effective audio and visual content overview;

2. maximizes the coverage for both audio and visual contents of the original video without having to sacrifice either of them.

To accomplish these goals, we compose an audio-visual summary by decoupling the audio and the image track of the input video, summarizing the two tracks separately, and then integrating the two summaries with a loose alignment.

6.1 SYSTEM OVERVIEW

Figure 10.6 is the block diagram of the audio-visual summarization system. It is a systematic combination of the audio-centric and the image-centric summarization systems. The summarization process starts by receiving the user's input of the two summarization parameters: the summary length T_{len}, and the minimum time length of each image segment T_{min} in the summary. The meaning and the intended use of these two parameters are the same as described in Section 5.1.

The audio content summarization is accomplished using the same process as described in Section 4. It consists of conducting speech recognition on the audio track of the original video to obtain a speech transcript, applying the text summarization method described in Section 4.2 to obtain an importance rank for each sentence in the transcript, and selecting audio segments of the sentences in descending order of their importance ranks until the user specified summary length is reached.

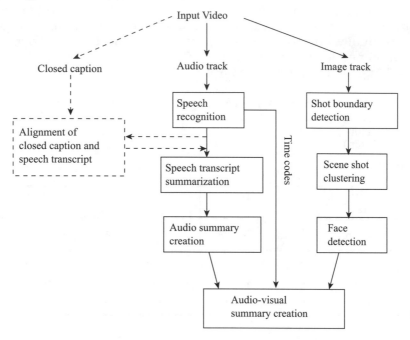

Note: the dash arrows and rectangle denote optional operations

Figure 10.6 The block diagram of the audio-visual summarization system.

The first two steps of the visual content summarization, shot boundary detection and scene shot clustering, are the same as described in Section 5, while the visual summary composition is conducted based on both the audio summarization and the shot clustering results. The whole visual content summarization process consists of the following major steps. First, shot boundary detection is conducted to segment the image track into individual scene shots. Next, shot clustering is performed to group scene shots into the required number $N = T_{len} / T_{min}$ of clusters based on their visual similarities. The summary length T_{len} is then divided into N time slots each of which lasts for T_{min} seconds, and each time slot is assigned to a suitable shot from an appropriate shot cluster. The assignment of a time shot to an appropriate shot cluster is made by the alignment process to fulfill the predefined alignment constraints (see Section 6.2 for detailed descriptions). Once a shot is assigned a time slot, its beginning segment (T_{min} seconds long) is collected, and a visual summary is created by concatenating these collected segments in their original time order. Moreover, face detection is conducted for each scene shot to detect the most salient frontal face that appears steadily in the shot. Such a face is considered as a speaker's face, and will play an important role in the alignment operation.

For the alignment task, to archive the summarization goals listed at the beginning of this section, we partially align the spoken sentences in the audio summary with the associated image segments in the original video. With video programs such as news and documentaries, a sentence spoken by an anchor person or a reporter lasts for ten to fifteen seconds on average. If a full alignment is made between each spoken sentence in the audio summary and its corresponding image segment, what we may get in the worst case is a video summary whose image part consists mostly of anchor persons and reporters. The summary created this way may look natural and smooth, but it is at the great sacrifice of the visual content. To create a content rich audio-visual summary, we developed the following alignment operations: for each spoken sentence in the audio summary, if the corresponding image segment in the original video displays scenes rather than the speaker's face, perform no alignment operations. The visual summary associated with this spoken sentence can be created by selecting image segments from any appropriate shot clusters. If the corresponding image segment does display the speaker's face, align the spoken sentence with its corresponding image segment for the first $T_{\min}$ seconds, and then fill the remaining portion of the associated visual summary with image segments from appropriate shot clusters. The decision of selecting which shot from which cluster is made by the alignment process to fulfill the predefined alignment constraints.

6.2 ALIGNMENT OPERATIONS

Let $A(t_i, \tau_i)$, $I(t_i, \tau_i)$ denote the audio, image segments that start at time instant t_i, and last for τ_i seconds, respectively. The alignment operation consists of the following two main steps.

1. For a spoken sentence $A(t_i, \tau_i)$ in the audio summary, check the content of its corresponding image segment $I(t_i, \tau_i)$ in the original video. If $I(t_i, \tau_i)$ shows a close-up face, and this face has not been aligned with any other component in the audio summary, align $A(t_i, \tau_i)$ with $I(t_i, \tau_i)$ for $T_{\min}$ seconds. Otherwise, do not perform the alignment operation for $A(t_i, \tau_i)$. This $T_{\min}$ seconds alignment between $A(t_i, \tau_i)$ and $I(t_i, \tau_i)$ is called an alignment point.

2. Once all the alignment points are identified, evenly assign the remaining time period of the summary among the shot clusters which have not received any playback time slot. This assignment must ensure the following two constraints:

   - Single assignment constraint: Each shot cluster can receive only one time slot assignment.

   - Time order constraint: All the image segments forming the visual summary must be in original time order.

The following subsections explain our approach to realizing the above alignment requirements.

6.2.1 Alignment Based on Bipartite Graph

Assume that the whole time span T_{len} of the video summary is divided by the alignment points into P partitions, and the time length of partition i is T_i (see Figure 10.7).

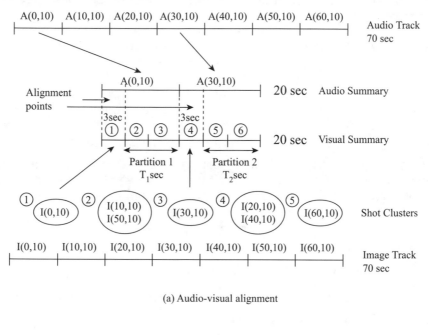

(a) Audio-visual alignment

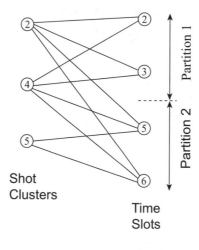

(b) Bipartite graph

Figure 10.7 An example of the audio-visual alignment and the corresponding bipartite graph.

Because each image segment forming the visual summary must be at least T_{min} seconds long (a time duration of T_{min} seconds long is called a time slot), partition i will be able to provide $S_i = \lceil T_i / T_{min} \rceil$ time slots, and hence the total number of available time slots becomes $S_{total} = \sum_{i=1}^{P} S_i$. Here the problem becomes as follows: Given a total of N shot clusters and S_{total} time slots, determine a best matching between the shot clusters and the time slots which satisfies the above two constraints. By some reformulation, this problem can be converted into the following maximum-bipartite-matching (MBM) problem [15]. Let $G = (V, E)$ represent an undirected graph where V is a finite set of vertices and E is an edge set on V. A bipartite graph is an undirected graph $G = (V, E)$ in which V can be partitioned into two sets L and R such that $(u, v) \in E$ implies either $u \in L$ and $v \in R$ or $u \in R$ and $v \in L$. That is, all edges go between the two sets L and R. A matching is a subset of edges $M \subseteq E$ such that for any vertex pair (u, v) where $u \in L$ and $v \in R$, at most one edge of M connects between u and v. A maximum matching is a matching M such that for any matching M', we have $|M| \geq |M'|$.

To apply the MBM algorithm to our alignment problem, we use each vertex $u \in L$ to represent a shot cluster, and each vertex $v \in R$ to represent a time slot. An edge (u, v) exists if a shot cluster u is able to take time slot v without violating the time order constraint. If a shot cluster consists of multiple scene shots, this cluster may have multiple edges that leave from it and enter different vertices in R. A maximum-bipartite-matching solution is a best assignment between all the shot clusters and the time slots. Note that a best assignment is not necessarily unique.

6.2.2 Alignment Process Illustration

Figure 10.7(a) illustrates the alignment process using a simple example. In this figure, the original video program is 70 seconds long, which consists of 7 scene shots and 7 spoken sentences each of which lasts for 10 seconds. The user has set $T_{len} = 20$ seconds, and $T_{min} = 3$ seconds. Assume that the audio summarization has selected two spoken sentences $A(0,10)$ and $A(30,10)$, and that the shot clustering process has generated five shot clusters as shown in the figure. As the audio summary is formed by $A(0,10)$ and $A(30,10)$, we must first examine the contents of the corresponding image segments $I(0,10)$ and $I(30,10)$ to determine whether the alignment operations are required. Suppose that $I(0,10)$ and $I(30,10)$ display the faces of the spoken sentences $A(0,10)$, $A(30,10)$, respectively, and that $I(0,10)$, $I(30,10)$ have not been aligned with other audio segments yet. Then, according to the alignment rules, $I(0,10)$ will be aligned with $A(0,10)$, and $I(30,10)$ with $A(30,10)$ for $T_{min} (= 3)$ seconds. Because $I(0,10)$ and $I(30,10)$ have been used once, they will not be used in other parts of the visual summary. By these two alignment points, the remaining time period of the visual summary is divided into two partitions, with each lasting for 7 seconds that can provide at most 2 time slots. Because there are three shot clusters and four time slots left for the alignment, we have a bipartite graph for the alignment task shown in Figure 10.7(b). Since shot cluster 2 consists of two shots: $I(10,10)$ and $I(50,10)$, it could take a time slot in either partition 1 or partition 2. If $I(10,10)$ is selected from cluster 2, it can take either time slot 2 or 3 in partition 1. On the other hand, if $I(50,10)$ is selected, it can take either time slot 5 or 6 in partition 2. Therefore, we have four edges leaving from cluster 2, each entering time slots 2, 3, 5, and 6, respectively. Similarly, there are four edges leaving from cluster 4, and two edges leaving from shot cluster 5, respectively.

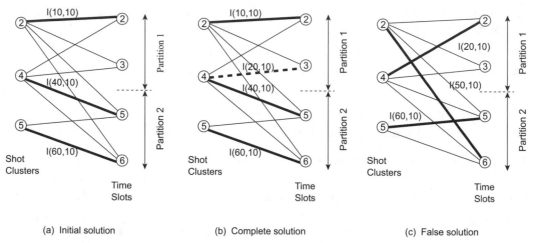

(a) Initial solution   (b) Complete solution   (c) False solution

Figure 10.8 Alignment solutions: the coarse lines represent the assignment of the shot clusters to the time shots; the notation I(j,k) on each coarse line tells which shot from the cluster has been selected, and assigned to the time slot.

There are several possible maximum matching solutions for the bipartite graph in Figure 10.7(b). Figure 10.8(a) shows one solution where the coarse lines represent the assignment of the shots to the time slots. Note that in this solution time slot 3 remains unassigned. This example illustrates a fact that, although the MBM algorithm will find a best matching between the available shot clusters and time slots, it may leave some time slots unassigned, especially when the number of available shot clusters is less than that of available time slots. To fill these unassigned time slots, we loosen the single assignment constraint, examine those clusters with multiple scene shots, and select an appropriate shot that has not been used yet, and that satisfies the time order constraint. In the above example, the blank time slot 3 is filled using the shot $I(20,10)$ in cluster 4 (coarse dashed line in Figure 10.8(b)).

In the case when the number of available shot clusters is more than that of available time shots, some shot clusters will not be assigned time slots within the visual summary. The MBM algorithm determines which cluster to discard and which cluster to take during its process of finding the best matching solution.

It is noticed that the MBM algorithm may generate some false solutions, and Figure 10.8(c) shows such an example. Here, because shot $I(60,10)$ has been placed before shot $I(50,10)$, it has violated the time order constraint. However, this kind of false solution can be easily detected, and can be corrected by sorting the image segments assigned to each partition into their original time order. In the above example, the time order violation can be corrected by exchanging the two image segments assigned to time slots 5 and 6 in Partition 2.

In a summary, Step 2 of the alignment operation (Section 6.2) can be described as follows:

1. After the alignment points have been identified, determine the number of shot clusters and time slots that are left for the assignment, and construct a bipartite graph accordingly.

2. Apply the MBM algorithm to find a solution.

3. Examine the solution with the time order constraint; if necessary, sort the image segments assigned to each partition into their original time order.

4. If there exist unassigned time slots, examine those shot clusters with multiple scene shots, and select an appropriate shot that has not been used yet, and that satisfies the time order constraint.

6.3 SUMMARIZATION PERFORMANCES

Conducting an objective and meaningful evaluation for an audio-visual content summarization method is difficult and challenging, and is a open issue deserving more research. The challenge is mainly from the fact that research for audio-visual content summarization is still at its early stage, and there are no agreed-upon metrics for performance evaluations. This challenge is further compounded by the fact that different people carry different opinions and requirements towards summarizing the audio-visual contents of a given video, making the creation of any agreed-upon performance metrics even more difficult.

Our audio-visual content summarization system has the following characteristics:

- The audio content summarization is achieved by using the latent semantic analysis technique to select representative spoken sentences from the audio track.

- The visual content summarization is performed by eliminating duplicates/redundancies and preserving visually rich contents from the image track.

- The alignment operation ensures that the generated audio-visual summary maximizes the coverage for both audio and visual contents of the original video without having to sacrifice either of them.

In Section 4.3, systematic performance evaluations have been conducted on the text summarization method. The evaluations were carried out by comparing the machine generated summaries with the manual summaries created by three independent human evaluators, and the F-value was used to measure the overlap degrees between the two types of summaries. It has been shown that the text summarization method achieved the F-value in a range of 0.57 and 0.61 for multiple test runs, the performance compatible with the top-ranking state-of-the-art text summarization techniques [13],[15].

With respect to the visual content summarization, as a visual summary is composed by first grouping visually similar shots into the same clusters, and then selecting at most one shot segment from each cluster, this visual summarization method ensures that duplicates and redundancies are diminished and visually distinct contents are preserved within the visual summary.

The alignment operation partially aligns each spoken sentence in the audio summary to the image segment displaying the speaker's face, and fills the remaining period of the visual summary with other image segments. In fact this alignment method is a mimic of news video production technique commonly used by major TV stations. A common pattern for news programs is that an anchor person appears on the screen and reports the news for several seconds. After that, the anchor person continues his/her reports, but the image part of the news video switches to either field scenes or some related interesting scenes. By doing so, visual contents of news broadcast are remarkably enriched, and viewers will not get bored. On the other hand, by mimicking this news video production technique in our summarization process, we get an audio-visual content summary which provides a richer visual content and a more natural audio-visual content overview. Such audio-visual summaries dramatically increase the information intensity and depth, and lead to a more effective video content overview.

Figure 10.9 illustrates the process of summarizing a 12-minute CNN news program reporting the Anthrax threat after the September 11 terrorist attack. The news program consists of 45 scene shots in its image track, and 117 spoken sentences in its audio track. Keyframes of all the shots are displayed at the left hand side of the Figure 10.9 in their original time order. Obviously, this news program contains many duplicated shots which arise from video cameras switching forth and back among anchor persons and field reporters.

The user has set $T_{len} = 60$ seconds, and $T_{min} = 2.5$ seconds. The shot clustering process has generated 25 distinct shot clusters, and the audio summarization process has selected sentences 1, 3, 38, 69, 103 for composing the audio summary. The clustering result is shown by the table in the middle of the figure. It is clear from the clustering result that all the duplicated shots have been properly grouped into the appropriate clusters, and there is no apparent misplacement among the resultant clusters. Because the total time length of these five sentences equals 70 seconds, the actual length of the produced audio-visual summary exceeds the user specified summary length by 10 seconds.

Among the five sentences comprising the audio summary, four sentences have their corresponding image segments containing the speakers' faces. Each of these four audio sentences has been aligned with its corresponding image segment for $T_{min} = 2.5$ seconds. The dashed lines between the audio and the visual summaries denote these alignments.

With the actual $T_{len} = 70$ seconds, and $T_{min} = 2.5$ seconds, the audio-visual summary can accommodate 28 time slots. As there are a total of only 25 distinct shot clusters, some shot clusters were assigned more than one time slot (e.g., clusters 0, 15, 17, 20, 21). To find an alignment solution that fulfills the two alignment constraints listed in Section 6.2, several shot clusters were not assigned any time slots, and were consequently discarded by the alignment algorithm (e.g., clusters 6, 7, 8, 11). The configuration of the audio and the visual summaries are displayed at the right hand side of the figure.

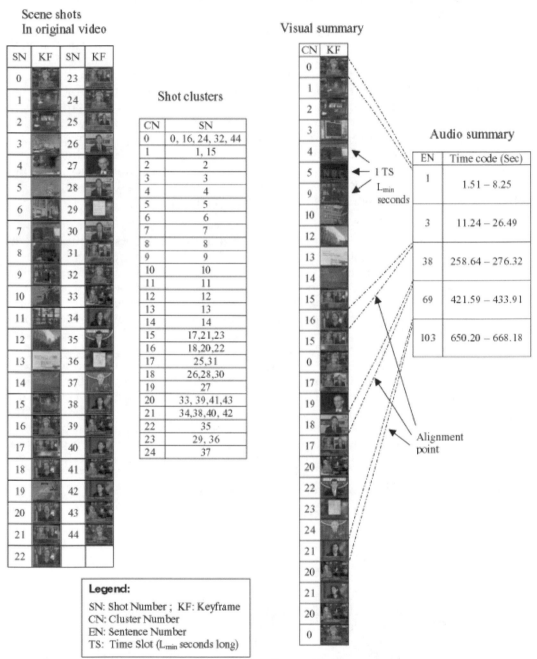

Figure 10.9 Audio-visual summarization of a 12-minute news program.

Video summarization examples can be viewed at http://www.ccrl.com/~ygong/ VSUM/VSUM.html.

7. CONCLUSION

In this chapter, we have presented three video summarization systems that create motion video summaries of the original video from different perspectives.

The audio-centric summarization system summarizes the input video by selecting video segments that contain semantically important spoken sentences; the image-centric summarization system creates a summary by eliminating duplicates and redundancies while preserving visually rich contents in the given video; and the audio-visual summarization system constitutes a summary by decoupling the audio and the image track of the input video, summarizing the two tracks separately, and then integrating the two summaries with a loose alignment. These three summarization systems are capable of covering a variety of video programs and customer needs. The audio-centric summarization is applicable for such video programs that information is mainly conveyed by audio speeches and visual content is a subordinate to the audio; the image-centric summarization is useful when the visual content delivers the main theme of the video program; and the audio-visual summarization is suitable when the user wants to maximize the coverage for both audio and visual contents of the original video without having to sacrifice either of them.

Video summarization is a process of creating a concise form of the original video in which important content is preserved and redundancy is eliminated. The process is very subjective because different people may have different opinions on what constitutes the important content for a given video. Especially for entertainment videos such as movies and dramas, different people watch them and appreciate them from many different perspectives. For example, given an action movie with a love story, some viewers may be touched by the faithful love between the main figures, some may be impressed by the action scenes, some may be interested only in a particular star, etc. These different interests reflect the difference in individuals' ages, personal tastes, cultural backgrounds, social ranks, etc. The three video summarization systems presented in this chapter strive to summarize a given video program from three different aspects. They cover certain information abstraction needs, but are not omnipotent for the whole spectrum of users' requirements. An ultimate video summarization system is the one that is able to learn the viewpoint and the preference from a particular user, and create summaries accordingly. With a such system intelligence, a personalized content abstraction service will become available, and a wide range of information intelligence requirements will be fulfilled.

REFERENCES

[1]   Y. Tonomura, A. Akutsu, K. Otsuji, and T. Sadakata, Videomap and Videospaceicon: Tools for Anatomizing Video Content, in Proc. ACM SIGCHI'93, 1993.

[2]   H. Ueda, T. Miyatake, and S. Yoshizawa, Impact: An Interactive Natural-Motion-Picture Dedicated Multimedia Authoring System, in Proc. ACM SIGCHI'91, New Orleans, Apr. 1991.

[3]   A. Fermain and A. Tekalp, Multiscale Content Extraction and Representation For Video Indexing, in Proc. SPIE on Multimedia Storage and Archiving Systems II, Vol. 3229, 1997.

[4]   D. DeMenthon, V. Kobla, and D. Doermann, Video Summarization by Curve Simplification, in Proc. ACM Multimedia, 1998.

[5]   M. Yeung, B. Yeo, W. Wolf, and B. Liu, Video Browsing Using Clustering and Scene Transitions on Compressed Sequences, in Proc. SPIE on Multimedia Computing and Networking, Vol. 2417, 1995.

[6]   A. Girgensohn and J. Boreczky, Time-Constrained Keyframe Selection Technique, in Proc. IEEE Multimedia Computing and Systems (ICMCS'99), 1999.

[7]   A. Amir, D. Ponceleon, B. Blanchard, D. Petkovic, S. Srinivasan, and G. Cohen, Using Audio Time Scale Modification for Video Browsing, in Proc. 33rd Hawaii International Conference on System Science (HICSS-33), Hawaii, Jan. 2000.

[8]   S. Srinivasan, D. Ponceleon, A. Amir, and D. Petkovic, What Is in that Video Anyway? In Search of Better Browsing, in Proc. IEEE International Conference on Multimedia Computing and Systems, Florence, Italy, June 1999, pp. 388-392.

[9]   M. A. Smith and T. Kanade, Video Skimming and Characterization Through the Combination of Image and Language Understanding Techniques, in Proc. IEEE International Conference on Computer Vision and Pattern Recognition (CVPR'97), Puerto Rico, June 1997, pp. 775-781.

[10]   I. Yahiaoui, B. Merialdo, and B. Huet, Generating Summaries of Multi-episode Video, in Proc. IEEE International Conference on Multimedia and Expo (ICME2001), Tokyo, Japan, Aug. 2001.

[11]   W. Press et al., *Numerical Recipes in C: The Art of Scientific Computing*, 2nd Ed., Cambridge University Press, Cambridge, England, 1992.

[12]   S. Deerwester, S. Dumais, G. Furnas, T. Landauer, and R. Harshman, Indexing by latent semantic analysis," *Journal of the American Society for Information Science*, Vol. 41, 1990, pp. 391-407.

[13]   J. Goldstain, M. Kantrowitz, V. Mittal, and J. Carbonell, Summarizing Text Documents: Sentence Selection and Evaluation Metrics, in Proc. ACM SIGIR'99, Berkeley, California, Aug. 1999.

[14]   E. Hovy and C. Lin, Automated Text Summarization in Summarist, in Proc. of the TIPSTER Workshop, Baltimore, MD, 1998.

[15]   T.H. Cormen, C.E. Leiserson, and R.L. Rivest, *Introduction to Algorithms*, The MIT Press, Cambridge, MA, 1990.

11
ADAPTIVE VIDEO SUMMARIZATION

Philippe Mulhem[1], Jérôme Gensel[2], and Hervé Martin[2]
[1]: IPAL-CNRS, Singapore,
[2]:LSR-IMAG, Grenoble, France
mulhem@lit.a-star.edu.sg, Jerome.Gensel@imag.fr,
Herve.Martin@imag.fr

1. INTRODUCTION

One of the specific characteristics of the video medium is to be a temporal medium: it has an inherent duration and the time spent to find information present in a video depends somehow on its duration. Without any knowledge about the video, it is necessary to use some Video Cassette Recorder facilities in order to decrease the search time. Fortunately, the digitalization of videos overcome former material constraints that forced a sequential reading of the video. Nowadays, the technological progress makes it possible to achieve some treatments on the images and segments which compose the video. For instance, it is very easy now to edit certain images or sequences of images of the video, or to create non-destructive excerpts of videos. It is also possible to modify the order of the images in order to make semantic groupings of images according to a particular center of interest. These treatments are extensively used by film-makers who use numeric benches of assembly such as Final Cut Pro [1] and Adobe Premiere [2].

In such software, users have to intervene in most of the stages of the process in order to point out the sequences of images to be treated and to specify the treatment to be made by choosing appropriate operators. Such treatments mainly correspond to operations of manipulation such as, for instance, cutting a segment and copying it out in another place. Such software offer simple interfaces that fit the requirements of applications like personal video computing and, in a more general way, this technology is fully adapted to the treatment of one video. However, in the case of large video databases used and manipulated at the same time by various kinds of users, this technology is less well suited. In this case, a video software has to offer a set of tools in able to i) handle and manage multiple video documents at the same time, ii) find segments in a collection of videos corresponding to given research criteria, (iii) create dynamically and automatically (i.e., with no intervention of the user) a video comprising the montage of the result of this search.

In order to fulfill these three requirements, video software must eventually model the semantics conveyed by the video, as well as its structure which can be determined from the capturing or editing processes. According to the nature of the semantics, the indexing process in charge of extracting the video semantics can be manual, semi-automated or automated. For instance, some attempts have been made to automatically extract some physical features such as color, shape and texture in a given frame and to extend this description to a video segment. Obviously, it is quite impossible to extract some high-level information such as the name of an actor or the location of a sequence. Thus, automatic indexing is generally associated with manual indexing.

The expected result of this indexing process is a description of the content of video in a formalism that allows accessing, querying, filtering, classifying and reusing the whole or some parts of the video. Offering metadata structures for describing and annotation audiovisual (AV) content is the goal of the MPEG-7 standard [3] which supports a range of descriptions ranging from the low-level signal features (shape, size, texture, color, movement, position...) to the highest semantic level (author, date of creation, format, objects and characters involved, their relationships, spatial and temporal constraints...). MPEG-7 standard descriptions are defined using the XML Schema Language and Description Schemes can then be instantiated as XML documents.

Using such descriptions, it is possible to dynamically create summaries of videos. Briefly, a video summary is an excerpt of a video that is supposed to keep the relevant parts of the video while dropping the less interesting parts of the video. This notion of relevance is subjective, interest-driven, related to the context and therefore it is hard to specify and necessitates some preference criteria to be given. It is our opinion that in order to create such a summary, the semantics previously captured by the semantic data model can be exploited while taking into account the preference criteria.

In this paper, we present then the VISU model which is the result of our ongoing research in this domain. This model capitalizes amount of work made in the field of information retrieval, notably the use of Conceptual Graphs [4]. We show how the VISU model adapts and extends these results in order to satisfy the constraints inherent to the video medium. Our objective is to annotate videos using Conceptual Graphs in order to represent complex descriptions associated with frames or segments of frames of the video. Then, we take advantage of the material implication of Conceptual Graphs on which is based an efficient graph matching algorithm [5] which allows formulation of queries. We propose a query formalism that provides a way to specify retrieval criteria. These criteria are useful to users for adapting the summary to their specific requirements. The principles of the query processing are also presented. Finally, we discuss about time constraints that must be solved to create a summary with a given duration. The paper is organized as follows. In the next section, the different approaches used to capture video semantics are presented. The section 3 proposes an overview of the main works in the field of video summarization. The section 4 presents the video data model, and the query models of the VISU system we are currently developing. Section 4 gives some concluding remarks and perspectives to this work.

2. VIDEO SEMANTICS

2.1. ANNOTATION-BASED SEMANTICS

An annotation represents any symbolic description of a video, or an excerpt of a video. However, when considering annotated symbolic description of videos, it should be noted that the representation may not be always extracted automatically. In contexts where enough knowledge can be used like in sports videos [6] or in news videos [7][8], automatic processes are able to extract abstract semantics, but in general cases systems are only able to provide help for users to describe the content of videos [9] or low level representations of the video content.

Many approaches have been proposed for modeling video semantics. According to the linear and continuous aspect of a video, many models are based on a specification of strata. Strata can be supported by different knowledge or data representation: VSTORM [10] or Carlos *et al.* [11] adopt an object or prototype approach, and recently, the MPEG committee has chosen the family of XML languages for the definition and the extension of the MPEG-7 standard audio-visual (AV) descriptions. We give below the principles of the stratum-based models, and the different representations of the strata.

2.1.1. Strata-based fundamentals

A stratum is a list of still images or frames that share a common semantics. So, a strata can be associated with some descriptions of the video excerpts it encompasses. In some models, it is allowed to specify overlapping between strata.

In [12][13], the authors propose a stratification approach inspired from [14] to represent high level description of videos. A stratum is a list of non-intersected temporal intervals. Each temporal interval is expressed using frame numbers. A video is thus described by a set of strata. Two types of strata arc proposed: *dialog stratum* and *entity stratum.* An entity stratum reflects the occurrence of an object, of a concept, or of a text, etc., and has a Boolean representation. For instance, an entity stratum can be used to store the fact that a person occurs, to express the mood of a sequence in the video (*e.g.,* 'sadness'), and to represent the structure of the video in term of shots or scenes. A dialog stratum describes the dialog content for each interval considered, and speech recognition may also be used in this case to extract such strata. Retrieval is here based on Boolean expressions for entity strata and on vector space retrieval [15] for dialog strata. The automatic extraction of semantic features, like Action/Close-Up/Crowd/Setting [16] allows also the definition of strata which are characterized by these features.

2.1.2. Strata structures and representations

In the stratification work of [13], the strata are not structured and no explicit link between strata exists. In [17], the authors define a video algebra that defines *nested strata*. Here, the structure is defined in a top-down way to refine the content description of video parts. Such nesting ensures consistency in the description of strata, because the nested strata correspond necessarily to nested time intervals. From a user perspective, browsing in a tree structure is certainly easier than viewing a flat structure. In that case, however, the question related to the retrieval of the video parts according to a query is far more complex than with flat strata.

Such tree-based content representation is also proposed in [18]. The approach there consists in finding ways to evaluate queries using database approaches. A tree structure describes when objects occur, and an SQL-like query language (using specific predicates or functions dedicated to the management of the object representation) allows the retrieval of the video excerpts that correspond to the query. Other database approaches have been proposed, like [19], in order to provide database modeling and retrieval on strata-based video representation. An interesting proposal, AI-Strata [20], has been dedicated to formalize strata and relations among strata. The AI-strata formalism is a graph-based AV documentation model. Root elements are *AV units* or strata to which *annotation elements* are attached. These annotations elements derive from a knowledge base into which abstract annotation elements (classes) are described and organized in a specialisation hierarchy. The exploitation of this graph structure is based on a sub-graph matching algorithm, a query being formulated in the shape of a so-called *potential graph*. In this approach, one annotation graph describes how the video document decomposes into scenes, shots and frames, the objects and events involved in the AV units, their relationships.

In V-STORM [10], a video database management system written using 02, we proposed to annotate a video at each level of its decomposition into sequences, scenes, shots and frames. A 02 class represents an annotation and is linked to a sub-network of other 02 classes describing objects or events organized in specialization and composition hierarchies. Using an object-based model to describe the content of a stratum has some advantages: complex objects can be represented; objects are identified; attributes and methods can be inherited. Following a similar approach in the representation formalism, authors in [11] have proposed a video description model based on a prototype-instance model. Prototypes can be seen as objects which can play the roles of both instances and classes. Here, the user describes video stories by creating or adapting prototypes. Queries are formulated in the shape of new prototypes which are classified in the hierarchies of prototypes describing the video in order to determine the existing prototype(s) that match best the query and are delivered as results.

Among the family of MPEG standards dedicated to videos, the MPEG-7 standard addresses specifically the annotation problem [3][21]. The general objective of MEPG-7 is to provide standard descriptions for the indexing, searching and retrieval of AV content. MPEG-7 *Descriptors* can either be in a XML (and then human-readable, searchable, filterable) form or in a binary form when consuming storage, transmission and streaming are required. MPEG-7 Descriptors (which are representations of features) can describe low-level features (such as color, texture, sound, motion, or such as location, duration, format) which can be automatically extracted or determined, but also higher level features (such as regions, segments, their spatial and temporal structure, or such as objects, events, and their interaction, or such as author, copyright, date of creation, or such as users preferences, summaries). MPEG-7 predefines *Description Schemes* which are structures made of descriptors and the relationships between descriptors or other description schemes. Descriptors and

Description Schemes are specified and can be defined using the MPEG-7 Description Definition Language which is based on the XML Schema Language with some extensions concerning vectors, matrices and references. The *Semantic Description Scheme* is dedicated to provide data about objects, concepts, places, time in the narrative world and abstraction. The descriptions can be very complex: it is possible to express, using trees or graphs, actions or relations between objects, states of objects, abstraction relationships, abstract concepts (like 'happiness') in the MPEG-7 Semantic Description Scheme. Such Description Schemes can be related to temporal intervals like in the strata-based approach.

To conclude, annotation-based semantics generally relies on a data structure that allows expression of a relationship between a continuous list of frames and its abstract representation. The difference among approaches is the capabilities offered by the underlying model to formalize strata and links between strata and the structure of the annotations. Our opinion is that this description is a key-point for generating video summarization. Thus, proposing a consistent and sound formalism will help in avoiding inconsistencies and fuzzy interpretation of annotations.

2.2. LOW LEVEL CONTENT-BASED SEMANTICS

We define the low level content-based semantics of videos as the elements that can be extracted automatically from the video flow without considering specific knowledge related to a specific context. It means that the video is processed by an algorithm that captures various signal information about frames and that proposes an interpretation in term of color, shape, structure and object motion of these features.

We can roughly separate the different content-based semantics extracted from videos into two categories: single frame-based and multiple frame-based. Single frame-based extractions consider only one frame at a time, while multiple frame-based extractions use several frames, mostly sequences of frames.

Generally, single frame extraction is performed using segmentation and region feature extractions using colors, textures and shapes (like the one used in still image retrieval, like QBIC [22] and Netra [23] for instance). MPEG-7 proposes, in its visual part, description schemes that supports color descriptions of images, groups of images and/or image regions (different color spaces, using dominant colors or histograms), texture descriptions of image regions (low level based on Gabor filters, high level based on 3 labels, namely regularity, main direction and coarseness), shape of image regions based on curvature scale spaces and histograms of shapes. Retrieval is then based on similarity measures between the query and the features extracted from the images. However, the use of usual still image retrieval systems on each frame of video documents is not adequate in term of processing time and of accuracy: consecutive frames in videos are usually quite similar, and this feature has to be taken into account..

Features related to sequences of frames can be extracted from averaging over the sequence, in order to define, like in [24], the ratio of saturated colors in commercials.

Others approaches propose to define and use motion of visible objects and motion of camera. MPEG-7 defines motion trajectory based either on key points and interpolation techniques, and parametric motion based on parametric estimations using optical flow techniques or on usual MPEG-1, MPEG-2 motion vectors. In VideoQ [25], the authors describe ways to extract object motion from videos and to process queries involving motion of objects. Works in the field of databases also consider the modeling of objects motion [26]. In this case, the concern is *not* how to extract the objects and their motion, but how to represent the objects and their motion in a database for allowing fast retrieval (on an object oriented database system) of a video's part according to SQL-like queries based on object motion.

Indexing formalisms used at this level are too low-level to be straightforwardly used in a query process by consumers for instance but on the other hand such approaches are not tightly linked to specific contexts (for instance motion vectors can be extracted from any video). Thus, a current trend in this domain is the merger of content-based information with other semantic information in order to provide usable information.

2.3. STRUCTURE-BASED SEMANTICS

It is widely accepted that video documents are hierarchically structured into clips, scenes, shots and frames. Such structure usually reflects the creation process of the videos. Shots are usually defined as continuous sequences of frames taken without stopping the camera. Usually, scenes are defined as sequences of contiguous shots that are semantically related, but in [27] [28] the shots of a sequence may not be contiguous. A clip is a list of scenes.

After the seminal work of Nagasaka and Tanaka in 1990 [29], work has been done for detecting shot boundaries in a video flow. Many researchers [30][31][32][33] have focused on trying to detect the different kinds of shot transitions that occur in video. The TREC video track 2001 [34] compared different temporal segmentation approaches, and if the detection of cuts between shots is usually successful, the detection of fades does not achieve very high success rates.

Other approaches focus on scene detection, like [27] using shot information and multiple cues like audio consistency between shots and the close caption of the speeches of the video. Bolle *et al.* [35] use types of shots and predefined rules to define scenes. Authors in [36] extend the previous work of Bolle *et al.* by including vocal emotion changes.

Once the shots and sequences are extracted, it is then possible to describe the content of each structural element [37] in a way to retrieve video excerpts using queries or by navigating in a synthetic video graph.

The video structure provides a sound view of a video document. From our point of view, a video summary must keep this view of the video. Like the table of contents of a book, the video structure provides a direct access to video segments.

3. SUMMARIZATION

Video summarization aims at providing an abstract of a video for shortening the navigation and browsing the original video. The problematic of video summarization is to be able to present in a synthetic way the content of video, while preserving 'the essential message of the original' [38]. According to [39] there are two different types of video abstracts: still-image and moving-image abstracts. Still-image abstracts, like with Video Manga [40][41], are presentations of salient images or key-frames while moving-image abstracts consist of a sequence of image sequences. The former are referred to as *video summary*, the latter as *video skimming*.

Video skimming is a more difficult process since it imposes an audio-visual synchronization of the selected sequences of images in order to restitute a coherent abstract of the entire video content. Video skimming can be achieved by using audio time scale modification [42] which consists in compressing the video and speeding up the audio and the speech while preserving the timbre, the voice quality and the pitch. Another approach consists in highlighting the important scenes (sequences of frames) in order to build a video trailer. In [38][43], scenes with a lot of contrast, scenes in the average coloration of the video, as well as scenes with a lot of different frames are automatically detected and are integrated in the trailer as they are supposed to be important scenes. Action scenes (containing explosion, gun shot, rapid camera movement) are also detected. Close captioning can also be used for selecting audio segments that contain some selected keywords [44] that together with the corresponding image segments put in chronological order will constitute an abstract. Clustering is also often used to gather video frames that share similar color or motion features [45]. Once the set of frame clusters is obtained, the more representative key-frame is extracted from each cluster. The video skimming is built by assembling video shots that contain these keyframes. In the work of [46] sub-parts of shots are processed using a hierarchical algorithm to generate the video summaries.

Video summary can be seen as a more simple task to perform since it consists in extracting from a video sequences of frames or sequences of segments of video as the best abstract of it, without considering audio segments selection and synchronization or close caption. The representation of video summaries can be composed of still images or of moving images, and may use the video cinematographic structure (clip, scenes, shots, frame) as well. The problem is still to define the relevant parts of video (still or moving images) to be kept in the summary.

Many existing approaches consider only signal-based summary generation. For instance Sun and Kankanhalli [47] proposed the Content Based Adaptive Clustering that defines a hierarchical removal of clusters of images according to color differences. This work is conceptually similar to [48] (based on Genetic Algorithms) and [49] (based on singular value decomposition). Other approaches, like [50] and [51], try to use objects and/or background for summary generation, hoping for more meaningful results, at the expense of an increased complexity. Another signal-level feature present on videos is motion; authors in [52] proposed to use such motion and gesture recognition of people in the context of filmed talk with slides. In [53], MPEG-7 content representations are used to generate semantics summaries based on relevant shots that can be subsampled based on the motions/colors in each shot.

In some specific repetitive contexts, like for electrocardiograms [54], the process uses a priori knowledge to extract summaries. Other contexts, like broadcast news [55], help the system to find out the important parts of the original videos.

Approaches propose also to use multiple features to summarize videos: [55] use closed caption extraction and speaker change detection, when [44] extract human faces and significant audio parts.

The process of defining video summaries is complex and prone to errors. We consider that the use of strata-like information in a video database environment is able to produce meaningful summaries by using high level semantic annotations. In order to achieve this goal, we propose to use the powerful formalism of Conceptual Graphs in order to represent complex video content. We are currently developing this approach in a generator of adaptive video summaries, called VISU (which stands for VIdeo SUmmarization). This system is based on two models we present in the following section.

4. THE VISU MODELS

We present in this section the two models of the VISU system, which allows both to annotate videos with high level semantic descriptions and to query these descriptions for generating video summaries. Thus, the VISU System relies on two models: an annotation model which uses strata and conceptual graphs for representing annotations, and a query model, based on SQL, for describing the expected summary. The query processing exploits an efficient graph matching algorithm for extracting the frames or sequences of frames which answer the query before to generate the summary according to some relevance criteria. We describe here the VISU models and the process of summary generation.

4.1. THE ANNOTATION-BASED STRUCTURE MODEL

In this model, objects, events and actions occurring in a video are representation units linked to *strata* similar to those presented in section 2. Moreover, we organize these representation units in Conceptual Graphs wherein nodes correspond to concepts, events or/and actions which are linked together in complex structures. A video or a segment of video can be annotated using several strata. Each stratum is associated with a set of chronologically ordered segments of video. Then, as shown in Figure 11.1, two strata involved in a video may share some common segments of video, meaning that the object, event or action they respectively represent both appear simultaneously in those segments. Provided that, for each segment of video where it occurs, a stratum is associated with the starting time and the ending time of this segment, the overlapping segments can be computed as the intersection of the two sets of segments.

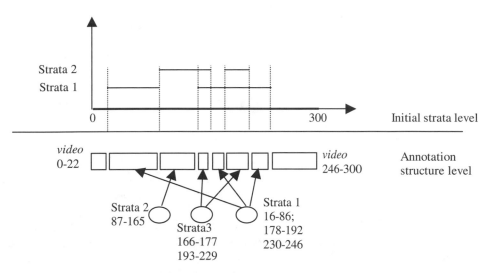

Figure 11.1 An example of annotation-based structure for a video. A video of 300 frames is here annotated using 2 initial strata, namely Strata1 and Strata2. The video segments between frames 87 and 165 are related only to the description of stratum 2; the frame intervals [166, 177] and [193,229] are described by a generated strata Strata3, corresponding to a conjunction of the annotations of the strata 1 and 2, where the frame intervals [178,192] and [230,246] are described only by the stratum 1.

Annotations associated with strata can be seen made of elementary descriptions (representation units) or more complex descriptions organized in Conceptual Graphs [4]. The Conceptual Graph formalism is based on a graphical representation of knowledge. Conceptual Graphs can express complex representations, and are able to be used efficiently for retrieval purposes [5].

More formally, Conceptual graphs are bipartite oriented graphs, composed of two kinds of nodes: concepts and relations.

- A Concept, noted "[T: r]" in an alphanumeric way and T: r in a graphical way, is composed of a concept type T and a referent r. Concept types are organized in a lattice that represents a generic/specific partial order. For the concept to be correct syntactically, the referent has to conform to the concept type, according to the predefined *conformance relationship*. Figure 11.2 shows a simple lattice that describes the concept types *Person*, *Woman* and *Man*; T_c and $\perp_c$ represent respectively the most generic and the most specific concepts of the lattice. A referent r may be individual (i.e., representing one uniquely identified instance of the concept type), or generic (i.e., representing any instance of the concept type, and noted by a star: "*").
- A relation R is noted "(R)" in a alphanumeric way and ⓇR in a graphical way. Relations are also represented in a lattice expressing a generic/specific partial order.

Figure 11.2 An example of concept types lattice.   **Figure 11.3** An example of relations lattice

We call *arch* a triplet (concept, relation, concept) that links three nodes including one relation and two concepts in a conceptual graph. In the following, we refer to a triplet "concept→relation→concept" as an *arch*.

Conceptual graphs can be used to represent complex descriptions. For instance, the Conceptual Graph G1, noted [Man: #John]→(talk_to)→[Woman: #Mary] can describe the semantic content of a frame or segment of frames. In this description, [Man: #John] represents the occurrence of a man, John, [Woman: #Mary] represents the occurrence of a woman, Mary, and the graph (which here, reduces to a simple arch) expresses the fact that John is talking to Mary. Figure 11.4 presents the graphical representation of this simple graph. It can be noticed that the expressive power of Conceptual Graphs is able to represent hierarchies of concepts and relations between objects as proposed by the MPEG-7 committee.

Figure 11.4 An example of conceptual graph.

Syntactically correct Conceptual Graphs are called *Canonical Graphs*. They are built using a set of basic graphs that constitutes the *Canonical Base*, from the following construction operators:
- *joint*: this operator joins graphs that contain an identical (same concept type and referent) concept,
- the *restriction* operator constrains a concept by replacing a generic referent by a specific one,
- the *simplification* operator removes duplicate relations that may occur after a joint operation, for instance,
- the *copy* operator copies a graph.

As shown by Sowa, the advantage of using conceptual graphs is that a translation, noted φ, exists between these graphs and the first order logic, providing conceptual graphs with a sound semantics. We exploit this property to ensure the validity of the summary generation process. This process is based on the material implication of first order logic (see [4]).

According to the conceptual graphs formalism described above, the *joint* operator can be used to achieve the fusion of graphs when possible. Thus, the resulting joint graph G of two graphs G_i and G_j, performed on one concept C_{ik} of G_i and one concept C_{jl} of G_j, where C_{ik} is identical to C_{jl}, is such that:

-   every concept of G_i is in G
-   every concept of G_j except C_{jl} is in G
-   every arch of G_j containing C_{jl} is transformed in G in an arch containing C_{ik}
-   every arch of G_i is in G
-   every arch of G_j that do not contain C_{jl}, is in G

As described above, an annotation of a strata may correspond to a single graph description, or to a conjunction of graphs under the form of one graph or a non-singleton set of graphs. More precisely, consider for instance two annotations s1 and s2 described respectively by the graphs g1, "[Man: John]->(talk_to)->[Woman: Mary]," and g2, "[Man: John]->(playing)->[Piano: piano1]." If the video segments corresponding to the related strata intersect, then the graph describing the generated stratum from the strata related to g1 and g2 is the graph g3, i.e. the graph that join g1 and g2 on their common concept [Man: John], as presented in Figure 11.5. The graph g3 is the best representation for the conjunction of the graphs g1 and g2, since in g3 we explicitly express the fact that the same person (John) is playing and talking at the same time. If we consider now two descriptions g3 "[Man: John]->(talk_to)->[Woman: Mary]" and g4 "[Man: Harry]->(playing)->[Piano: piano1]," then the description that corresponds to the generated strata cannot be the result of a joint operation because the two graphs g3 and g4 do not have a common concept, and then the annotation of the intersection is the set { g3, g4}. Generated strata are necessary to avoid mismatches between query expressions and initial strata graph descriptions.

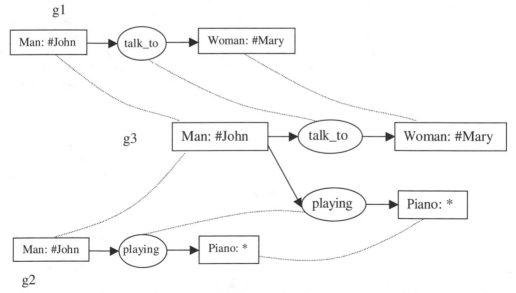

Figure 11.5 Joint of two graphs (g1 and g2) on a concept [Man: #John].

Figure 11.5 describes the joint of two graphs but it should be noticed that in our annotation-based structure, we try to joint all the graphs that correspond to generated strata. Such a task is simply achieved through an iterative process that first finds the identical concepts of strata, and then generates joint graphs when possible. If no joint is possible for some graphs, then these graphs are kept such as they are in the set of graphs describing the generated stratum.

Our objective is to exploit the complex description of Conceptual Graphs involving strata in order to generate meaningful summaries. This goal requires associating a measure of the relative importance of the graph elements: the concepts and the arches. Concepts are important because they express the basic elements present in the description. But arches are important too because they represent relationships between elements. The solution adopted here is based on the weighting scheme of the well known *tf*idf* (term frequency * inverse document frequency) product used for 30 years in the text-based information retrieval domain [56]. The term frequency is related to the importance of a term in a document, while the inverse document frequency is related to the term's power to distinguish documents in a corpus. Historically, these *tf* and *idf* values were defined for words, but here, we use them to evaluate the relevance of concepts and arches in graphs. We describe below how these values are computed, respectively, for a concept and an arch, in a Conceptual Graph:

- The term frequency *tf* of a concept C in a conceptual graph description Gsj is defined as the number of concepts in Gsj that are *specific* concepts of C. A concept X, [*typeX* : *referentX*] is a specific of Y, [*typeY* : *referentY*], if *typeX* is a specific of *typeY* according to the lattice of concept type (i.e., *typeX* is a sub-type of *typeY*) and referentX=referentY (they represent the same individual) or referentX is an individual referent and referentY is the generic referent '*'. In the graph "[*Man: #John*]->(*talk_to*)->[*Man: *]*," representing the fact that John is talking to an unidentified man, *tf*([*Man: #John*]) is equal to 1, and *tf*([*Man: *]) is equal to 2 because both [*Man: #John*] and [*Man: *] are specific of [*Man: *].
- The inverse document frequency *idf* of a concept C is based on the relative duration of the video parts that are described by a specific of C, using a formula inspired from [57]: $idf(c)=log(1+D/d(c))$ with D the duration of the video and $d(C)$ the duration of the occurrence of C or a specific of C. For instance, if John appears on 10% of the video idf([*Man: #John*])=1.04, and if a Man occurs 60% of the video, idf([*Man: *])=0.43.
- For an arch A, the principle is similar to concepts: the term frequency *tf(A)* is based on the number of different arches that are specific of A in a considered graph. An arch *(C1x, Rx, C2x)* is a specific of an arch *(C1y, Ry, C2y)* if and only if the concept *C1x* is a specific of *C1y*, and if the concept *C2x* is a specific of the concept *C2y*, and if the relation *Rx* is a specific of the relation *Ry* according to the relation lattice (the record-type associated to Rx is a sub-record-type of the record-type associated to Ry).
- The idf of an arch A is defined similarly to concepts, and is based on the relative duration of the video parts that are described by specific arches of A.

4.2. CINEMATOGRAPHIC STRUCTURE MODEL

The cinematographic structure Model is dedicated to represent the organization of the video according to what we presented in part 2.3. This structure is a tree that reflects the compositional aspect of the video; the node corresponds to a structural level and a frame number interval. The inclusion of parts is based on the interval inclusion.

As explained previously, one video is structured into scenes, shots and frames. We choose to limit the structure to shots and frames, because according to the state of the art shot boundaries can be extracted with an accuracy greater than 95% (at least for cuts, as described in [58]), while automatic scene detection is still not effective enough for our needs.

```
SUMMARY
FROM video
WHERE graph [WITH PRIORITY {HIGH|MEDIUM|LOW}]
        [ {AND|OR|NOT} graph [WITH PRIORITY {HIGH|MEDIUM|LOW}] ]*
DURATION integer {s|m}
```

Figure 11.6. Query syntax.

4.3. QUERY MODEL

We use a query model in order to describe the expected content of the adaptive summary and the expected duration of the summary through the formulation of a query. For instance, a query Q1 that expresses the fact that we look for video parts where John talks to somebody can be written: "[Man: #John]→(talk_to)→[Human: *]." It is obvious that the graph G1 presented in section 4.1.1 and noted [Man: #John]→(talk_to)→[Woman: #Mary] is an answer for the query Q1. In the context of VISU, this means that a frame or segment of frames (assuming that duration constraints are satisfied) described by the graph G1 should be present in a summary described by means of the query Q1.

In the query model of VISU, we also allow users to assign weights to the different parts of the query. These weights reflect the relative importance of different contents of the videos. A syntax of a query is given in Figure 11.6, using the usual Backus-Naur semantics for the symbols "[", "]", "{", "}" and "*".

In Figure 11.6, *video* denotes the initial video that is to be summarized. *Graph* is represented as a set of arches in an alphanumerical linear form. An arch is represented by "*[Type1: referent1 | id1]->(relation)->[Type2: referent2 | id2]*," where the concept identifiers *id1* and *id2* define uniquely each concept in a way to represent concepts that occur in more than one arch. The concept identifiers are used because a linear form is not able without such identifiers to fully represent graphs, particularly when considering generic referents. For instance, a graph representing that a man, John, is talking to a unidentified woman, and at the same time John is smiling to another unidentified woman, is represented as: "*{[Man: John | 0]->(talk_to)->[Woman: * | 1], [Man: John | 0]->(smile)->[Woman: * | 2]}.*" Without such identifiers, it would not be possible to distinguish between the fact that John smiles to the same woman or to another woman. *integer* after the keyword DURATION corresponds to the expected time of the summary with the unit *s* for seconds and *m* for minutes.

To illustrate, let us consider that we want to obtain a summary:
- taken from the video named "Vid001"

- showing a man, John, is talking to a woman, Mary, and at the same time John is at the right of Mary, with a high importance
- or showing snow falling on houses
- and having a duration of 20 seconds.

The query which corresponds to the expected generation can be formulated as follows:

SUMMARY
FROM Vid001
WHERE {[Man: John|0]->(talk_to)->[Woman: Mary|1],
         [Man: John|0]->(right_of)->[Woman: Mary|1]}WITH PRIORITY HIGH
  OR {{[SNOW:*|0]->(falling)->[BUILDING:*|1]} WITH PRIORITY MEDIUM
DURATION 20s

4.4. SUMMARY GENERATION WITH VISU

The query processing relies on two phases. One is dedicated to the video content and the other is dedicated to the duration of the resulting summary. We describe each of these phases in the following.

4.4.1. Information Retrieval theoretical background

To evaluate the relevance value between a query Qi and a document stratum Dj, we consider the use of the logical model of information retrieval [59]. This meta-model stipulates, in the *Logical Uncertainty Principle*, that the evaluation of the relevance of a document (represented by the sentence y) to a query (represented by a sentence x) is based on "the minimal extend" to be added to the data set in order to assess the certainty of the implication $x \rightarrow y$. In our case the $x \rightarrow y$ is related to the existing knowledge in the Conceptual Graph Canon (composed of the canonical base, the concept type hierarchy, the relation hierarchy and the conformance relationship), and the "$\rightarrow$" symbol is the material implication $\supset$ of first order logic. In our process, once the information of an annotation implies an annotation the relevance value of a stratum Sj according to a query Qi can then be computed as we explain in the following.

4.4.2. Matching of graphs

The conceptual graph formalism allows to search in graphs using the project operator [4]. The project operator is equivalent to the material implication according to the semantics given to conceptual graphs. The *project* operator, a sub-graph search, takes into account the hierarchy of concepts types and of relations. The projection of a query graph Gqi, noted $\pi_{Gsj}(Gqi)$, into a conceptual graph Gsj, only concludes on the existence of a sub-graph in the description that is specific to the query graph. In the work of [5], it has been shown that the projection on graphs can be implemented very efficiently in term of search algorithm complexity.

We quantify the matching between a content query graph Gq and an annotation graph Gs by combining concepts matching and arches matching, inspired by [60] and [61]:

$$F(Gq, Gs) = \Sigma\{tf(c).idf(c) \,|\, c \text{ in concepts of } \pi_{Gs}(Gq)\}$$
$$+ \Sigma\{tf(a).idf(a) \,|\, a \text{ in arches of } \pi_{Gs}(Gq)\} \qquad (1)$$

In formula (1), the *tf* and *idf* concepts and arches are defined according to section 4.1.1.

Since we defined that an annotation may be one graph or a set of conceptual graphs, we now have to express how to match a query graph and a set of graphs corresponding to an annotation. Consider an annotation composed of a set S of graphs Gs, $1 \le k \le N$. The match M of the query graph Gq and S is:

$$M(Gq,S) = \max_{Gs \in S} (F(Gq,Gs))$$

4.4.3. Query expression evaluation

We describe here how the priorities in the sub-query expression "{graph} WITH PRIORITY P," P being "HIGH," "MEDIUM" or "LOW" are taken into account in order to reflect the importance of the sub-expression for the summary generation. The notion of priority reflects the importance of the different sub-parts of the query content, and was defined in the MULTOS system [62] or in the OFFICER system [63]. A query sub-expression is then evaluated for each graph of each annotation of the annotation-based structure, using a simple multiplication rule: if the matching value obtained for a node is v, then the matching value for the query sub-expression is: $v \times p$, with p equal to 0.3 for "LOW," 0.6 for "MEDIUM" and 1.0 for "HIGH."

Complex query expressions composed by Boolean expressions of elementary sub-query expressions are evaluated for strata annotations using Lukasiewicz-like definitions for n-valued logics:

- an expression composed of "A AND B" is evaluated as the minimum of the matching for each sub-queries A and B,
- an expression composed of "A OR B" is evaluated as the maximum of the matching values of each sub-query A and B,
- an expression composed of "NOT B" is evaluated as the negative of the matching value for B.

Then, we are able to give a global matching value for each of the annotations of the annotation-based structure.

The duration part of the query is used to constrain the result to a given duration. Three cases might arise:

- The duration of all the relevant video parts is larger than the duration expected. Then we avoid presenting the video parts that are less relevant, i.e., the video parts that have the lower matching values. We notice, however, that this is only processed for video parts that match with a positive matching value for the summary generation criteria.
- The duration of all the relevant video parts is equal to the duration expected. Then the result is generated with all the relevant video parts.
- The duration of the relevant video parts is smaller than the duration expected. Then the result is generated using all the relevant parts obtained and using the cinematographical structure of the video: consider that the query asks for x seconds and that the duration of the relevant semantic parts is y seconds (we have then y < x by hypothesis). The remaining video to

add to the relevant parts must be x-y seconds long. We include in the result continuous excerpts of (x-y)/n seconds for each of the n shots that do not intersect with any relevant video parts. The excerpts correspond to the shot parts having the more motion activity in each shot, consistently with the work of [47] on signal-based summarization.

In any case, we force the result generated to be monotonic according to the frame sequence: for each frame fr_i and fr_j in the generated result corresponding respectively to the frames fo_k and fo_l in the original video, if the frame fo_k is before (resp. after) the frame fo_l in the original video, then the frame fr_i is before (resp. after) the frame fr_j in the generated result.

5. CONCLUSION AND FUTURE WORK

In this chapter, we presented various requirements for generating video summarization. Existing works on this field rely more or less on each of the following aspects of videos:

- o The video-structure level that expresses how the video is organized in term of sequences, scenes and shots. Such structure also helps to capture temporal properties of a video.
- o The semantic level that expresses the semantics of parts of the video. Various forms of semantics may be either automatically or manually extracted.
- o The signal level that represents features related to the video content in terms of colors/textures and motion.

Many works have been done in these different fields. Nevertheless, we show that a better and powerful formalism has to be proposed in order to facilitate the dynamic creation of adaptive video summaries. It is the reason why we proposed the VISU model. This complex model is based both on a stratum data model and on a conceptual graph formalism in order to represent video content. We also make use of the cinematographic structure of the video and some low-level features in a transparent way for the user during the generation of summaries.

An originality of the VISU model is to allow the expression of rich queries for guiding video summary creation. Such rich queries can only be fulfilled by the system if the representation model is able to support effectively complex annotations; that is why we choose to use the conceptual graph formalism as a basis for the annotation structure in VISU. Our approach also makes use of well known values in Information Retrieval, namely the term frequency and inverse document frequency. Such values are known to be effective for retrieval and we apply them on annotations.

The proposed model will be extended to support temporal relations between annotations in the future. We will also in the future use this work to propose a graphical user interface in order to generate automatically query expressions.

REFERENCES

[1] [Final Cut Pro: http://www.apple.com/fr/finalcutpro/software

[2] Adobe Premiere: http://www.adobe.fr/products/premiere/main.html

[3] MPEG-7 Committee, Overview of the MPRG-7 Standard (version 6.0), Report ISO/IEC JTC1/SC29/WG11 N4509, J. Martinez Editor, 2001.

[4] J. F. Sowa, *Conceptual Structures: Information Processing in Mind and Machines*, Addison-Wesley, Reading (MA), USA, 1984.

[5] I. Ounis and M. Pasça, RELIEF: Combining Expressiveness and Rapidity into a Single System, ACM SIGIR 1998, Melbourne, Australia, 1998, pp. 266-274.

[6] N. Babaguchi, Y. Kawai, and T. Kitahashi, Event Based Video Indexing by Intermodal Collaboration, in Proceedings of First International Workshop on Multimedia Intelligent Storage and Retrieval Management (MISRM'99), Orlando, FL, USA, 1999, pp. 1-9.

[7] B. Merialdo, K. T. Lee, D. Luparello and J. Roudaire, Automatic Construction of Personalized TV News Programs, in Proceedings of the Seventh ACM International Conference on Multimedia, Orlando, FL, USA, 1999, pp. 323-331.

[8] H.-J. Zhang, S. Y. Tan, S. W. Smoliar, and G. Y. Hone, Automatic Parsing and Indexing of News Video, *Multimedia Systems*, Vol.2, No. 66, 1995, pp. 256-266.

[9] M. Kankahalli and P. Mulhem, Digital Albums Handle Information Overload, *Innovation Magazine*, 2(3), National University of Singapore and World Scientific Publishing, 2001, pp. 64-68.

[10] R. Lozano and H. Martin, Querying Virtual Videos Using Path and Temporal Expressions, in Proceedings of the 1998 ACM Symposium on Applied Computing, February 27 - March 1,1998, Atlanta, GA, USA, 1998.

[11] R. P. Carlos, M. Kaji, N. Horiuchi, and K. Uehara, Video Description Model Based on Prototype-Instance Model, in Proceedings of the Sixth International Conference on Database Systems for Advanced Applications (DASFAA), April 19-21, Hsinchu, Taiwan, pp. 109-116.

[12] M. S. Kankahalli and T.-S. Chua, Video Modeling Using Strata-Based Annotation, *IEEE Multimedia*, 7(1), Mar. 2000, pp. 68-74.

[13] T.-S. Chua, L. Chen and J. Wang, Stratification Approach to Modeling Video, *Multimedia Tools and Applications*, 16, 2002, pp. 79-97.

[14] T.G. Aguierre Smith and G. Davenport, The Stratification System: A Design Environment for Random Access Video, in Proc. 3rd International Workshop on Network and Operating System Support for Digital Audio and Video, La Jolla, CA, USA, 1992, pp. 250-261.

[15] G. Salton and M. J. McGill, *Introduction to Modern Information Retrieval*, McGraw-Hill, New-York, 1983.

[16] N. Vasconcelos and A. Lippman, Bayesian Modeling of Video Editing and Structure: Semantic Features for Video Summarization and Browsing, in ICIP'98, pp. 153-157, 1998.

[17] R. Weiss, A. Duda, and D. Gifford, Composition and Search with Video Algebra, *IEEE Multimedia*, 2(1), 1995, pp 12-25.

[18] V. S. Subrahmanian, *Principles of Multimedia Database Systems*, Morgan Kaufmann, San Francisco, 1997.

[19] R. Hjesvold and R. Midtstraum, Modelling and Querying Video Data, VLDB Conference, Chile, 1994, pp.686-694.

[20] E. Egyed-Zsigmond, Y. Prié, A. Mille and J.-M. Pinon, A Graph Based Audio-Visual Annotation and Browsing System, in Proceedings of RIAO'2000, Volume 2, Paris, France, 2000, pp. 1381-1389.

[21] P. Salembier and J. Smith, MPEG-7 Multimedia Description Schemes, *IEEE Transactions on Circuits and Systems for Video Technology*, Vol. 11, No. 6, June 2001, pp. 748-759.

[22] M. Flickner, H. Sawhney, W. Niblack, J. Ashley, Q. Huang, B. Dom, M. Gorkani, J. Hafner, D. Lee, D. Petkovic, D. Steele, and P. Yanker, Query by Image and Video Content: the QBIC System, *IEEE Computer*, 28(9), 1995, pp. 23-30.

[23] W. Y. Ma and B. S. Manjunath, NETRA: A Toolbox for Navigating Large Image Databases, in Proceedings of the IEEE ICIP'97, Santa Barbara, 1997, pp. 568-571.

[24] J. Assfalg, C. Colombo, A. Del Bimbo, and P. Pala, Embodying Visual Cues in Video Retrieval, in IAPR International Workshop on Multimedia Information Analysis and Retrieval, LNCS 1464, Hong-Kong, PRC, 1998, pp. 47-59.

[25] S.-F. Chang, W. Chen, H.J. Horace, H. Sundaram, and D. Zhong, A Fully Automated Content Based Video Search Engine Supporting Spatio-Temporal Queries, *IEEE Transactions CSVT*, 8, (5), 1998, pp. 602-615.

[26] J. Li, T. Özsu, and D. Szafron, Modeling of Moving Objects in a Video Database, in IEEE International Conference on Multimedia Computing and Systems (ICMCS), Ottawa, Canada, 1997, pp. 336 – 343.

[27] Y. Li, W. Ming and C.-C. Jay Kuo, Semantic Video Content Abstraction Based on Multiple Cues, in IEEE International Conference on Multimedia and Expo (ICME) 2001, Tokyo, Japan, 2001.

[28] Y. Li and C.-C. Jay Kuo, Extracting Movie Scenes Based on Multimodal Information, in SPIE Proceedings on Storage and Retrieval for Media Databases 2002 (EI2002), Vol. 4676, San Jose, USA, 2002, pp.383-394.

[29] A. Nagasaka and Y. Tanaka, Automatic Scene-Change Detection Method for Video Works, in Second Working Conference on Visual Database Systems, 1991, pp. 119-133.

[30] R. Zabih, J. Miller, and K. Mai, Feature-Based Algorithms for Detecting and Classifying Scene Breaks, in Proceedings of the Third ACM Conference on Multimedia, San Francisco, CA, November 1995, pp 189-200.

[31] G. Quénot and P. Mulhem, Two Systems for Temporal Video Segmentation, in CBMI'99, Toulouse, France, October, 1999, pp.187-193.

[32] H. Zhang, A. Kankanhalli, S. W. Smoliar, Automatic Partitioning of Full-Motion Video, *Multimedia Systems*, Vol. 1, No. 1, 1993, pp. 10-28.

[33] P. Aigrain and P. Joly, The Automatic Real-Time Analysis of Film Editing and Transition Effects and Its Applications, *Computer and Graphics*, Vol. 18, No. 1, 1994, pp. 93-103.

[34] A. F. Smeaton, P. Over, and R. Taban, The TREC-2001 Video Track Report, in The Tenth Text Retrieval Conference (TREC 2001), NIST Special Publication, 2001, pp. 500-250.
http://trec.nist.gov/pubs/trec10/papers/TREC10Video_Proc_Report.pdf

[35] R. M. Bolle, B.-L. Yeo, and M. M. Leung, Video Query: Beyond the Keywords, IBM Research Report RC 20586 (91224), 1996.

[36] J. Nam, Event-Driven Video Abstraction and Visualization, *Multimedia Tools and Applications*, 16, 2002, pp. 55-77.

[37] B.-L. Yeo amd M. M. Yeung, Retrieving and Visualizing Video, *Communication of the ACM*, 40(12), 1997, pp.43-52.

[38] S. Pfeiffer, R. Lienart, S. Fisher, and W. Effelsberg, Abstracting Digital Movies Automatically, *Journal of Visual Communication and Image Representation*, Vol. 7, No. 4, 1996, pp. 345-353.

[39] Y. Li, T. Zhang, and D. Tretter, An Overview of Video Abstraction Techniques, HP Laboratory Technical Report HPL-2001-191, 2001.

[40] S. Uchihashi, J. Foote, A. Girgensohn, and J. Boreszki, Video Manga: Generating Semantically Meaningful Video Summaries, in ACM Multimedia'99, Orlando (FL), USA, 1999, pp. 383-392.

[41] S. Uchihashi and J. Foote, Summarizing Video Using a Shot Importance Measure and a Frame-Packing Algorithm, in ICASSP'99, Phoenix (AZ),Vol. 6, 1999, pp. 3041-3044.

[42] A. Amir, D. B. Ponceleon, B. Blanchard, D. Petkovic, S. Srinivasan, and G. Cohen, Using Audio Time Scale Modification for Video Browsing, in Hawaii International Conference on System Sciences, Maui, USA, 2000.

[43] R. Lienhart, S. Pfeiffer, and W. Effelsberg, Video Abstracting, *Communication of the ACM*, 40(12), 1997, pp.55-62.

[44] M. Smith and T. Kanade, Video Skimming and Characterization Through the Combination of Image and Language Understanding Techniques, in IEEE Computer Vision and Pattern Recognition (CVPR), Puerto Rico, 1997, pp. 775-781.

[45] A. Hanjalic, R.L. Lagendijk, and J. Biemond, Semi-Automatic News Analysis, Classification and Indexing System based on Topics Preselection, in SPIE/IS&T Electronic Imaging'99, Storage and Retrieval for Image and Video Databases VII, Vol. 3656, San Jose, CA, USA, 1999, pp. 86-97.

[46] R. Lienhart, Dynamic Video Summarization of Home Video, in SPIE 3972: Storage and Retrieval for Media Databases, 2000, pp. 378-389.

[47] X.D. Sun and M.S. Kankanhalli, Video Summarization Using R-Sequences, *Journal of Real-Time Imaging*, Vol. 6, No. 6, 2000, pp. 449-459.

[48] P. Chiu, A. Girgensohn, W. Polak, E. Rieffel, and L. Wilcox A Genetic Algorithm for Video Segmentation and Summarization, in IEEE International Conference on Multimedia and Expo (ICME), 2000, pp. 1329-1332.

[49] Y. Gong and X. Liu, Generating Optimal Video Summaries, in IEEE International Conference on Multimedia and Expo (III), 2000, pp. 1559-1562.

[50] J. Oh and K. A. Hua, An Efficient Technique for Summarizing Videos using Visual Contents, in Proceedings of IEEE International Conference on Multimedia and Expo. July 30 - August 2, 2000, pp. 1167-1170.

[51] D. DeMenthon, V. Kobla, and D. Doermann, Video Summarization by Curve Simplification, in ACM Multimedia 98, Bristol, Great Britain, 1998, pp. 211-218.

[52] S. X. Ju, M. J. Black, S. Minneman, and D. Kimber, Summarization of video-taped presentations: Automatic analysis of motion and gesture, *IEEE Transactions on Circuits and Systems for Video Technology*, Vol. 8, No. 5, 1998, pp. 686-696.

[53] B. L. Tseng, C.-Y. Lin, and J. R. Smith, Video Summarization and Personnalization for Pervasive Mobile Devices, in SPIE Electronic Imaging - Storage and Retrieval for Media Databases, Vol. 4676, San Jose (CA), 2002, pp. 359-370.

[54] S. Ebadollahi, S. F. Chang, H, Wu, and S. Takoma, Echocardiogram Video Summarization, in SPIE Medical Imaging, 2001, pp. 492-501.

[55] M. Maybury and A. Merlino, Multimedia Summaries of Broadcast News, *IEEE Intelligent Information Systems*, 1997, pp. 422-429.

[56] G. Salton, A. Wong, and C. S. Yang, A vector space model for automatic indexing, *Communication of the ACM*, 18, 1975, pp. 613-620.

[57] G. Salton and C. Buckley, Term-weighting approaches in automatic text retrieval, *Information Processing and Management*, Vol. 24, John Wiley and Sons Publisher, 1988, pp. 513-523.

[58] Y-F. Ma, J. Shen, Y. Chen, and H.-J. Zhang, MSR-Asia at TREC-10 Video Track: Shot Boundary Detection Task, in The Tenth Text Retrieval Conference (TREC 2001), NIST Special Publication 500-250, 2001. http://trec.nist.gov/pubs/trec10/papers/MSR_SBD.pdf

[59] C. J. van Rijsbergen, A non-classical logic for information retrieval, Computer Journal, 29, 1986, pp. 481-485.

[60] S. Berretti, A. Del Bimbo, and E. Vicario, Efficient Matching and Indexing of Graph Models in Content-Based Retrieval, *IEEE Transactions on PAMI*, 23(10), 2001, pp.1089-1105.

[61] P. Mulhem, W.-K. Leow, and Y.-K. Lee, Fuzzy Conceptual Graphs for Matching of Natural Images, in International Conference on Artificial Intelligence (IJCAI'01), Seattle, USA, 2001, pp. 1397-1402.

[62] F. Rabitti, Retrieval of Multimedia Documents by Imprecise Query Specification, LNCS 416, in Advances in Databases Technologies, EDBT'90, Venice, Italy, 1990.

[63] B. Croft, R. Krovetz, and H, Turtle, Interactive Retrieval of Complex Documents, *Information Processing and Management*, Vol. 26, No. 5, 1990.

12

ADAPTIVE VIDEO SEGMENTATION AND SUMMARIZATION

Charles B. Owen and John K. Dixon
Department of Computer Science and Engineering
Michigan State University
East Lansing, Michigan, USA
`cbowen,dixonjoh@cse.msu.edu`

1. INTRODUCTION

Many techniques have been developed for segmentation and summarization of digital video. The variety of methods is partially due to the fact that different methods work better on different classes of content. Histogram-based segmentation works best on color video with clean cuts; motion-based summarization works best on video with moving cameras and a minimum of disjoint motion. Recognizing that there is no single, best solution for each of these problems has led to the ideas in this chapter for integrating the variety of existing algorithms into a common framework, creating a composite solution that is sensitive to the class of content under analysis, the performance of the algorithms on the specific content, and the best combination of the results. This chapter presents a survey of the existing techniques used to perform video segmentation and summarization and highlights ways in which a composite solution can be developed that adapts to the underlying video content.

As massive quantities of digital video accumulate in corporate libraries, public archives, and home collections, locating content in that video continues to be a significant problem. A key element of any method that attempts to index digital video is effective segmentation and summarization of the video. Video segmentation involves decomposing a video sequence into shots. Summarization involves developing an abstraction for the video sequence and creating a user-interface that facilitates browsing of the abstraction. Indexing methods require minimization of redundancy for effective performance. Ideally, an indexing method would be given a compact summarization of the content with only salient elements subject to analysis. And, given the limited performance of indexing methods and the questionable ability of humans to pose exact queries, it is essential that results be presented in a way that allows for very fast browsing by the user. The goal of adaptive video segmentation and summarization is to create algorithmic solutions for each area of analysis that are independent of the video class, non-heuristic in nature, and useful for both indexing and browsing purposes. Such tools can go a long ways towards unlocking these massive vaults of digital video.

Great progress has been made on the segmentation and summarization of digital video. However, it is common for this research to focus on specific classes of video. Major projects have analyzed CNN Headline News, music videos, and late night comedians [1-7]. This work answered many questions about how to analyze video where the structure is known in advance. However, any general solution must work for a wide variety of content without the input of manually collected structural knowledge. This fact has been well known in the mature document analysis community for many years. Researchers have attempted to apply ideas gleaned by that community and new approaches derived from analysis of past segmentation and structural analysis results to develop solutions that are adaptive to the content and able to function for video ranging from commercial edited motion pictures, to home movies produced by amateurs, to field content captured in military settings.

The primary goal of adaptive video segmentation is to apply known algorithms based on motion, color, and statistical characteristics in parallel, with instrumentation of the algorithms, content analysis, and results fusion to determine a best segmentation with maximum accuracy and performance. The text document retrieval community has recognized the value of adaptive application of multiple algorithmic search mechanisms. This strategy can also be applied to shot segmentation.

Likewise, results fusion and adaptation techniques can also be applied to video summarization. One of the critical tools of any indexing and browsing environment is effective summarization. Video to be indexed must be presented to an indexing system with a minimum of redundancy so as to avoid redundant retrieval results and to maximize the disparity in the indexing space. Likewise, search results must be presented as compact summaries that allow users to choose the correct result or adapt the search as quickly as possible. Again, many different approaches for video summarization exist. This toolbox of approaches can be utilized as the basis for an adaptive solution for video summarization that will draw on the strengths of the different approaches in different application classes.

2. REVIEW OF VIDEO SEGMENTATION AND SUMMARIZATION

Network constraints or bandwidth limitations do not always allow for the dissemination of high-bandwidth mediums such as video. Moreover, based on its content, it may be inefficient to transmit an entire video. In a wireless environment the amount of available bandwidth does not allow for the transmission of an entire video sequence. In such an environment, there needs to be a mechanism to transmit only the information of relative importance. Often, the content of a video sequence can be captured and conveyed with a few representative images. Imagine recording and capturing the video of a still scene where the content remains fixed or has minimal changes. It would be impractical and inefficient to capture and transmit the entire video because of the massive amounts of redundancy. Little if anything has changed from the first frame to the last. The content and semantic meaning of the video could be captured with a few frames, thus tremendously reducing the amount of needed bandwidth. Eliminating the redundancy is effective as long as it preserves the overall meaning and nature of the video. Much research has been conducted on developing content-based summarizations of digital video. The goal of the research has been to detect temporal boundaries, identify meaningful segments from the video, and extract and construct a subset of the video from the original via keyframes or key-sequences [8]. Thus, summarization can be thought of as a two-step process of video segmentation and abstraction.

Commercial video sequences can be thought of as a hierarchy of a sequence of stories, which are composed of a set of scenes, which are composed of a set of shots, with each shot containing a sequence of frames. The most common segmentation algorithms rely on low-level image features at the shot level to partition the video. This is because partitioning of a video at the scene or story level is difficult because there is no standard or universal definition of scenes or stories. A shot is an unbroken sequence of frames taken from one camera [9]. There are two basic types of shot transitions: abrupt and gradual. Abrupt transitions, called cuts, occur when a frame from a subsequent shot immediately follows a frame from the previous shot. These types of transitions are relatively easy to detect. Gradual transitions consist of slow change between frames from one shot to a frame of a different shot. These types of transitions include cross-dissolves, fade-ins, fade-outs, and other graphical editing effects such as wipes [10]. A fade-in is the gradual increase of intensity starting from one frame to the next. A fade-out is a slow decrease in brightness from one frame to the next. A cross-dissolve is when one frame is superimposed on another, and while one frame gets dimmer, the other frame gets brighter. A dissolve can be considered an overlapping of a fade-in and a fade-out [11]. Gradual transitions are more difficult to detect. This is because camera and object motion can inhibit the accurate detection of gradual transitions, causing false positives.

There have been several research projects comparing and evaluating the performance of shot detection techniques. Koprinska et al. [9] provides a survey of the existing approaches in the compressed and uncompressed domain. Dailianas [11] compared several segmentation algorithms across different types of video. Lienhart [12] evaluated the performance of various existing shot detection algorithms on a diverse set of video sequences with respect to the accuracy of each detection method, and the recognition of cuts, fades, and dissolves. Lupatini et al. [13] compared and evaluated the performance of three classes of shot detection algorithms: histogram-based, motion-based, and contour-based. Boreczsky and Rowe [14] compared various compressed and uncompressed video shot detection algorithms.

2.1   FRAME-BASED TECHNIQUES

The majority of the shot-based segmentation algorithms operate in the uncompressed domain. A similarity measure between successive frames is defined and compared against a predetermined threshold. A cut is determined when the distance value between two images falls below this predetermined threshold. Gradual transitions can be detected by adaptive thresholding techniques [15] or using cumulative difference measures.

One way to detect possible transitions between successive frames is to compare the corresponding pixel values between the two frames and count how many pixels have changed. If the number of changed pixels is above a predetermined threshold, a shot is detected [9]. A potential problem with this implementation is its sensitivity to camera motion. If a camera moves a few pixels between two successive frames, a large number of pixels will be counted as being changed. Zhang [15] attempts to reduce this effect with the use of a smoothing filter. Before each pixel comparison the candidate pixel is replaced with the average value of the pixels within its 3x3 neighborhood. Additionally, this filter is used to reduce noise in the input images.

A likelihood ratio has been utilized to compare successive frames based on the assumption of uniform second-order statistics over a region [15]. In this algorithm each frame is subdivided into blocks and the corresponding blocks are compared based on statistical characteristics of their intensity values. One advantage that this method has over the pixel difference method is that it improves the tolerance against noise associated with camera and object movement. Additionally, the likelihood ratio has a broader dynamic range than does the percentage used with the pixel difference method [15]. This broader dynamic range makes it easier to choose a threshold t to distinguish between changed and unchanged areas. A problem with the likelihood ratio algorithm is that it is possible for two different corresponding blocks to have the same density function, causing no segment to be detected.

Histogram-based algorithm techniques use a distance metric between histograms as a similarity measure. The basic assumption is that content does not change abruptly within shots, but does across shot boundaries [12]. Once an image has been represented as a histogram there are various distance metrics that can be used. Zhang et al. [15] concluded that the χ^2 method of histogram comparison enhances the difference between two frames across a cut; however it also increases the difference between frames with small camera and object movements. Additionally, it was concluded that the overall performance of the χ^2 method is little better than the absolute bin difference method, with χ^2 being more computationally intensive.

Histograms are attractive because they are effective at determining abrupt changes between frames. They are also tolerant to translational and rotational motions about the view axis, and change gradually with the angle of view, change in scale, or occlusion [16]. However, they completely ignore the spatial distribution of intensity values between frames. Consecutive frames that have different spatial distribution, but have similar histograms, are considered similar. Zhang et al. [15] and Nagaska and Tanaka [17] are examples of color-based histogram implementations.

One solution to the problems associated with global histograms is to create local histograms. Local histograms segment a frame into smaller blocks and compute histograms for each region. This method is tolerant to local changes in motion; however it still sensitive to changes in luminance over an entire frame [13]. Nagaska and Tanaka [17] split each frame into 16 uniform blocks and evaluate the difference between histograms in corresponding blocks. The χ^2 method is used to compute a distance metric between frames. The largest difference value is discarded in order to reduce the effects of noise, object, and camera movements.

Gargi et al. [10] experimented with computing histograms for various shot detection methods in different color spaces. The color spaces include RGB, HSV, YIQ, XYZ, L*a*b*, L*u*v*, Munsell, and Opponent. They concluded that the Munshell space produced the best performance results. The Munshell space is used because it is close to the human perception of colors [16]. Additionally, Zhang et al. [16] used a dominant color technique that only examined colors corresponding to the histograms with the most bins. This is based on the assumption that small histogram bins are likely to contain noise, distorting shot detection results.

Lupatini et al. [13] compared the performance of twelve shot detection methods based on histograms, motion, and contours. Although the best performance was achieved with histogram-based algorithms that use color or hue information, these algorithms failed when there was a continuous camera motion due to long camera pans that changed the content of the scene. Boreczky [14] compared the performance of three histogram-based algorithms, a motion-based algorithm, and an algorithm based on DCT coefficients. They concluded that the histogram-based algorithms perform better in general than the motion-based and DCT-based algorithms. However, the histogram-based algorithms did a poor job of identifying gradual transitions. As a result, the authors suggest using an additional edge-based algorithm to accurately identify gradual transitions. Histogram-based algorithms alone cannot accurately detect shot boundaries. Multiple methods need to be utilized in order to accurately characterize the different types of shot transitions.

Ferman et al. [18] introduced a temporal video segmentation technique based on 2-class clustering to eliminate the subjective nature of selecting thresholds. Video segmentation is treated as a 2-class clustering problem, where the two classes are "scene change" and "no scene change." The K-means clustering algorithm [19] is applied to the similarity measure of color histograms between successive frames. The χ^2 method and the histogram difference method are used to compute the similarity metric in the RGB and YUV color spaces. From their experiments, the χ^2 method in the YUV color space detected the larger number of correct transitions; however when the complexity of the distance metric is factored in, the histogram difference method in the YUV color space was the best in terms of overall performance. One major advantage of this technique is that it eliminates the need to select a predetermined threshold. Additionally multiple features can be used to improve performance of the algorithm. In subsequent research, Ferman and Tekalp [18] utilize two features, histograms and pixel differences, for video segmentation via clustering.

Zabith et al. [20] detect cuts, fades, dissolves, and wipes based on the appearance of intensity edges that are distant from edges in the previous frame. A summarization of the edge pixels that appear far from an existing pixel (entering edge pixel) and edge pixels that disappear far from an existing pixel (exiting pixel) are used to detect cuts, fades, and dissolves. The method is further improved to tolerate camera motion by performing motion compensation. The global motion between frames is calculated and used to align frames before detecting the entering and disappearing edge pixels. One disadvantage of this technique is that it is not able to handle multiple moving objects [9]. Another disadvantage is false positives due to the limitations of edge detection. Changes in image brightness, or low quality frames, where edges are harder to detect, may cause false positives. Lienhart [12] experimented with the edge-based segmentation algorithm of Zabith et al. [20] and concluded that false positives can arrive from abrupt entering and existing lines of text. In order to reduce the false positive, the classification of hard cuts was extended. Additionally, Lienhart [12] found that hard cuts from monochrome images were commonly classified as fades.

2.2   OTHER TECHNIQUES

Boreczky and Wilcox [21] segment video using Hidden Markov Models (HMM) [22]. The features used for segmentation are the distance between gray-level histograms, an audio distance based on the acoustic difference in intervals just before and just after the frames, and an estimate of the object motion between frames. States in the HMM are used to model fades, dissolves, cuts, pan, and zoom. The arcs between states model the allowable progression of states. The arc from a state to itself denotes the length of time the video is in that state. Transition probabilities and the means and variances of the Gaussian distribution are learned during the training phase using the Baum-Welch algorithm [22]. Training data consists of features (histograms, audio distance, and object motion) from a collection of video content. After the parameters are trained, segmenting the video into its shots and camera motions is performed using the Viterbi algorithm [22]. One advantage of this technique is that thresholds are not subjectively determined; they are learned automatically based on the training data. Another advantage of this technique is that it allows for the inclusion of multiple features in the training data.

Most video is stored in a compressed format such as MPEG. As a result, there has been considerable interest in performing video segmentation directly on the coded MPEG compressed video [23-27]. One advantage to performing video segmentation in the compressed domain is the decreased computational complexity when decoding is avoided. Additionally, algorithm operations are faster [9]. Additionally, the encoded video inherently contains computed features such as motion vectors and block averages that can be utilized. However, the speed and efficiency of algorithms in the compressed domain comes at the cost of the increased complexity of the algorithm implementation.

2.3   SUMMARIZATION

Video summarization attempts to present a pictorial summary of an underlying video sequence in a more compact form, eliminating redundancy. Video summarization focuses on finding a smaller set of images to represent the visual content, and presenting these keyframes to the user. Abstraction involves mapping an entire segment of video into a smaller number of representative images [16]. A still-image abstract, also known as a static storyboard, is a collection of salient still images or keyframes generated from the underlying video. Most summarization research involves extracting keyframes and developing a browser-based interface that best represents the original video. Li et al. [28] describes three advantages of a still-image representation: 1) still-image abstracts can be created much faster than moving image abstracts, since no manipulation of the audio or text information is necessary, 2) the temporal order of the representative frames can be displayed so that users can grasp concepts more quickly, and 3) extracted still images are available for printing, if desired. There have been numerous research efforts on keyframe extraction [29-31]. Li et al. [28] groups these techniques into the following categories: sampling-based, shot-based, color-based, motion-based, mosaic-based, and segment-based keyframe extraction techniques.

2.3.1 Still Image Representation

In some systems the first and last frame of each shot are selected to represent shot content [26, 28, 32-34]. Although this may reduce the total number of keyframes and provide information about the total number of keyframes a priori, this method is not an accurate and sufficient representation of the shot content. It does not characterize or capture dynamic action or motion within a shot. Ideally, keyframes should be extracted based on the underlying semantic content. But, semantic analysis of a video is a difficult research problem. As a result, most keyframe extraction techniques rely on low-level image features, such as color and motion.

Zhang et al. [16] extract keyframes based on color content. Keyframes are extracted in a sequential manner. The density of the keyframe selection process can be adjusted. However, the default is that the first and last frames of each shot are considered keyframes. Once a keyframe has been selected, a color histogram comparison method is employed on subsequent frames and the previous keyframe. If the distance metric exceeds a predetermined threshold, a keyframe is selected. The Munsell space was chosen to define keyframes, because of its relation to human perception [16]. The color space is quantized into 64 "super-cells" using a standard least squares clustering algorithm. A 64-bin histogram is calculated for each keyframe, with each bin having the normalized count of the number of pixels that fall in the corresponding supercell.

Ferman and Tekalp [35] propose a keyframe extraction method based on the clustering of frames within a shot. Frames within a shot are clustered into a group based on a color histogram similarity measure. The frame closest to the center of the largest cluster is selected as the keyframe for the shot. Zhuang et al. [36] propose a method for keyframe extraction based on unsupervised clustering. The color histogram is used to represent the visual content within a frame. A 16x8 2D histogram is used in the HSV color space. For each cluster, the frame that is closest to the centroid is selected as the keyframe.

Wolf proposes a motion based keyframe selection algorithm based on optical flow [37]. The algorithm computes the flow field for each frame. The sum of the magnitudes of the optical flow components is used as the motion metric. The keyframe selection process selects keyframes that are at the local minima of motion between two local maximas. Dixon and Owen propose a method based on motion analysis and overlap management [38]. This method is optimized for content wherein camera motion dominates the motion field and saliency requires coverage of the viewable area.

One of the main problems with keyframe extraction algorithms is that they produce an excessive amount of keyframes from long video sequences. Additionally, due to redundant scenes and alternating shots, most keyframe extraction algorithms that operate on single shots produce redundant keyframes. In order to combat these problems Girgensohn et al. [29] calculate an importance measure for each segment based on its rarity and duration. A pictorial summary is created based on the segments whose scores are above a threshold. Keyframes are selected near the center of each segment. Uchihashi et al. [39] attempted to reduce the overall number of keyframes by developing pictorial summaries based on a shot importance measure and frame-packing algorithm. Their hierarchical clustering method is used to group similar shots together and a score is given to each cluster. A keyframe is extracted from each cluster and the size of the keyframe is based on the score given to each cluster. A frame-packing algorithm is used to visually display the different sized keyframes to the user in a unified manner.

2.3.2 Moving Image Representation

Video Skims are short video clips consisting of a collection of image sequences and the corresponding audio, extracted from the original longer video sequence. Video skims represent a temporal multimedia abstraction that is played rather than viewed statically. They are comprised of the most relevant phrases, sentences, and image sequences. The goal of video skims is to present the original video sequence in an order of magnitude less time [40]. There are two basic types of video skimming: summary sequence and highlight [28]. Summary sequences are used to provide a user with an impression of the video sequence, while a highlight video skim contains only the most interesting parts of a video sequence. The MoCA project [41] developed automated techniques to extract highlight video skims to produce movie trailers. Scenes containing important objects, events, and people are used to develop the video skims. Selecting highlights from a video sequence is a subjective process; as a result most existing video-skimming work focuses on the generation of summary sequences [28].

The Informedia Project at Carnegie Mellon University [40, 42] utilizes speech, closed caption text, speech processing, and scene detection to automatically segment news and documentary video. They have created a digital library with over a terabyte of video data. One aspect of the summarization method of the Informedia Digital Video Library System is composed of partitioning the video into shots and keyframes. Multi-level video summarization is facilitated through visual icons, which are keyframes with a relevance measure in the form of a thermometer, one-line headlines, static filmstrip views, utilizing one frame per scene change, active video skims, and the transcript following of an audio track. The text keywords are extracted from the transcript and the closed captioning text by using a Term Frequency Inverse Document Frequency (TF-IDF) technique. The audio data is extracted from the segments corresponding to the selected keywords and its neighboring areas to improve audio comprehension. Image extraction is facilitated by selecting frames with faces and text, frames following camera motion, frames with camera motion and face or text, and frames at the beginning of a video sequence. Video Skimming is created by the confluence of the extracted audio and image extraction. Experiments of this skimming approach have shown impressive results on limited types of documentary video that have explicit speech or text contents [28]. It remains unclear whether this technique may produce similar results with video containing more complex audio contents.

2.3.3 User Interfaces

The second aspect of video summarization involves developing user-interfaces that best represent the original video sequence. The user- interface design is the part of a system that combines the video technology with the users requirements. If users cannot use the overall system effectively, then the system fails. The major concerns developing effective user interfaces is the trade-off between the different levels and types of abstractions presented to the user [43]. The more condense the abstraction, the easier it is for a potential user to browse through; however, a condensed abstraction may not provide enough information to obtain the overall meaning and understanding of the video. Contrastingly, a more detailed abstraction may present the user with enough information to comprehend the video sequence; however the amount of information in the abstraction may take a potential user a long time to browse.

Lee and Smeaton [43] surveyed and categorized the user-interfaces of various browsing tools. Some of the criteria that were used in classifying the browsing tools were whether they had a single keyframe display, a keyframe storyboard, options to change the density of the keyframes, interactive hierarchical keyframe browser, Time playback of keyframes, video skims, VCR like playback, and keyframes with playback synchronization. They concluded that in order to develop an effective user-interface, it was necessary to identify the different classes of users and the features and functionality needed for each class.

Researchers have also focused on results from user studies to determine the best possible user-interface for maximum efficiency and user comprehension. Christel et al. [40, 44] performed user studies on secondary school and college students on the effectiveness of the multiple levels of abstraction and summarization techniques utilized in the Informedia Digital Library [45]. Studies were done with respect to the relative value of each abstraction method and its impact on the understanding of the content. The abstraction methods used in the Informedia Digital Library are text titles or headlines, thumbnail keyframe images or poster frames, filmstrips, video skims, and match bars. The Informedia Digital library then provides the user with multiple levels of abstraction and summarization. Although speech and text processing can significantly aid in the creation of a rich summarization technique, these processing genres are beyond the scope of our research.

Komlodi et al. [46] performed user studies on 30 undergraduate psychology majors to determine the effectiveness of three keyframe selection interfaces. Comparisons were made with respect to user performance, interface efficiency, and user satisfaction. The study experimented with using static storyboard displays as well as dynamic slideshow displays. The three displays that were used were static storyboard with 12 keyframes, static storyboard with 4 keyframes, and dynamic slideshow with 12 keyframes. The dependent variables that were measured in the experiment were object identification, action identification, gist comprehension, selection precision, selection recall, examination time, and user satisfaction. The experiment performed by Komlodi et al. [46] concluded that static keyframe displays are better for object identification than dynamic keyframe displays. An adaptive algorithm must be able to determine when object identification is important for a certain type of video and provide a display that maximizes the users understanding and comprehension of the content.

Wei et al. [47] studied the effectiveness of video summaries containing keyframes only, keywords and phrases only, and a combination of keyframes and keywords or phrases with respect to user comprehension and understanding. The study concluded that summaries containing both text and imagery are more effective than either modality alone.

2.4   SUMMARY

This section has presented a broad view of the current research related to video segmentation and summarization. Most video segmentation research focuses on the creation of algorithms for specific classes of content. When the type of video is known a priori, any number of algorithms can be chosen with relative success. However, when the type of video is unknown, an adaptive method is needed to adjust to the type of content for the best possible segmentation result and the best summarization method for maximum user comprehension and efficiency.

3. ADAPTIVE VIDEO SEGMENTATION

Digital video has become ubiquitous in today's society. It comes in many forms including music videos, movies, surveillance video, Unmanned Aerial Vehicle (UAV) video, home movies, and news broadcasts. Although there are many similarities, each type of video has its own distinct and defining characteristics. For example, news broadcasts can be thought of as a series of anchorperson and story shots [2, 4]. UAV video consists of constant and continuous camera motion [38]. As a result of these and other distinct characteristics, there is no catchall segmentation algorithm. If the type of video is not known before analysis, how does one determine the best possible segmentation algorithm to use? The solution to this problem yields developing a composite solution of the existing algorithms that is sensitive to the video under analysis. To date, little research effort has been conducted regarding creating a composite video segmentation solution. This section describes the current research in this area as well as suggestions for developing a composite solution.

3.1   SEGMENTATION

Document retrieval methods have shown increased performance when combining results from various document representations. Katzer et al. [48] compared text document retrieval performance using different document representation methods. Their results discovered that the different document representations retrieved different sets of relevant documents. As a result, performing information retrieval with multiple document representations improved retrieval performance over using a single method. As the document retrieval methods rely on various document representations, the various video segmentation algorithms rely on different characteristics of the underlying video. Color-based methods create histograms of the frame content and compute distance metrics between histograms to search for shot boundaries [10, 12, 13,

15-17]. Model-based methods create Hidden Markov Models (HMM) of each possible state and transition in a video sequence [14, 22]. Edge based methods utilize edge maps of the frame content to search for segmentation boundaries [12, 20]. Motion-based methods rely on velocity and displacement vectors to compute the amount of motion between video frames to determine shot boundaries [38]. The results obtained from the document filtering community with respect to the combination of multiple methods to increase retrieval performance suggests that the various digital video representations can be combined to increase shot boundary segmentation detection performance.

Research efforts, to date, have focused on the implementation of segmentation algorithms independently [5, 8, 16, 49]. Limited research has attempted to combine the shot boundary detection algorithms into a composite system [50]. The process of combining various input sources can be achieved in a variety of ways. Most of the research on combining multiple input sources has come from the area of document retrieval [51-58]. In this area, researchers attempt to combine multiple representations of queries, documents, or multiple retrieval techniques. Research in this field has shown that significant improvements can be achieved by combining multiple evidences [51, 56]. One application that attempts to incorporate results from multiple input sources is the development of metasearch engines. A metasearch engine is a system that provides access to multiple existing search engines. Its primary goal is to collect and reorganize the results of user queries of multiple search engines. There has been considerable research with respect to the effectiveness and performance of metasearch engines [59-66]. The result-merging step of a metasearch engine combines the query results of several search engines into a single result. A metasearch engine usually associates a weight or similarity measure to each of the retrieved documents and returns a ranked list of documents based on this value [62]. These weights can be derived from an adjustment of the document ranks value of the local search engines or defining a global value based on all the retrieved documents. This type of approach only focuses on the results of multiple searches and not a combination of the actual search engines.

Browne [50] experimented with combining three shot boundary algorithms. The shot detection algorithms used for their research are color histograms, edge detections, and encoded macroblocks. It was concluded from the experiments that a dynamic threshold implementation of each algorithm improved shot boundary detection performance. Weighted Boolean logic was used to combine the three shot boundary detection algorithms. The algorithm works as follows: The three shot detection algorithms are executed in parallel using dynamic thresholds for each algorithm. A shot is determined in a hierarchical manner. If the color histogram algorithm is above its adaptive threshold, a shot boundary is detected. If the edge detection algorithm is above its adaptive threshold and the color histogram algorithm is above a minimum threshold then a shot boundary is detected. Lastly, if the encoded macroblock algorithm is above its threshold and the color histogram algorithm is above its minimum threshold, a shot boundary is detected.

One problem with the Brown implementation is that the three algorithms utilized are not really combined. Each algorithm is run independently and the results of the various algorithms are combined for further analyses. As a result, the algorithm does not try to determine the most appropriate algorithm to use based on the content. Moreover, the implementation does not determine how well an algorithm works for certain content. All three algorithms are always run simultaneously. Additionally, the dynamic thresholds are only utilized within the context of each individual algorithm. There is no mechanism for the combined algorithms to adapt to a global threshold.

Figure 12.1 depicts an adaptive video shot boundary detection technique motivated by the metasearch engine approach. Each algorithm is run independently and the results of each algorithm are fused together by a data fusion technique. Data fusion can be facilitated by Boolean logic or a weighting scheme.

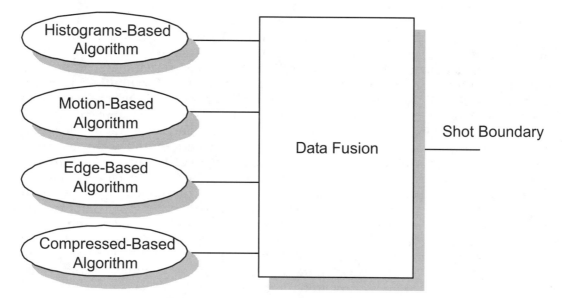

Figure 12.1 Multiple Methods

An improved approach to creating an adaptive system fusing multiple approaches draws on experience gained in document filtering. Document filtering attempts to select documents identified as relevant to one or more query profiles [53]. Documents are accepted or rejected that are independent of previous or subsequent examined documents. Hull et al. developed statistical classifiers of the probability of relevance and combined these relevance measures to improve filtering performance. The solution they chose is based on machine learning. Each classifier utilizes a different learning algorithm. The hypothesis in their experiment was that different learning algorithms used different document representations and optimization strategies; thus the combination of multiple algorithms would perform better than any individual method. This study concluded that the combination of multiple strategies could improve performance under various conditions.

Associating probabilities with each shot detection algorithm could facilitate transitioning this approach to the video shot boundary domain. Different learning algorithms use different document representations; similarly the various shot boundary detection techniques require different representations of the video. Histogram-based techniques require the creation of color histograms for each video frame, whereas motion-based techniques require the creation of motion vectors to determine object and camera motion between frames. These probabilities could determine how certain a particular type of algorithm will perform in the presence of a shot boundary. Figure 12.2 depicts a multiple method video shot boundary detection technique motivated by the combination of multiple probabilities from different algorithms.

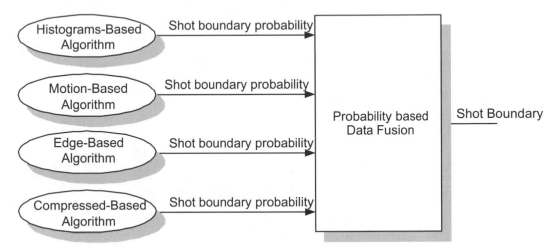

Figure 12.2 Multiple Probability Methods

One technique for combining the results of multiple methods would be simple averaging of the probabilities. If the average probability of the multiple methods is above a threshold, a shot boundary is detected. Hull et al. state that it may be more useful to look at averaging the log-odds ratios instead. If one algorithm determines with high probability that a shot boundary is detected or not detected the average log odds will show this certainty much more directly than will the average probability [53]. One problem with this technique is that each algorithm's probability is given equal weight. A better strategy would be to give more weight to the algorithms based on the content. Certain algorithms perform better on some content than others.

This method is based on the instrumentation of segmentation methods with assignment of a probability or likelihood that a shot boundary is observed. The wide variety of algorithms needs to be benchmarked so as to ascertain performance over a wide class of video content. This content will be utilized to determine a statistical method to assign confidence values to results based on the content analysis. For example, the confidence measure for a threshold-based histogram method may be based on variance properties of the histogram differences. Large variances indicate content that is not well modelled by histogram methods; so confidence will be low.

Katzer et al. [48] compared document retrieval performance using different document representation methods. They demonstrated that performing information retrieval with multiple document representations improved retrieval performance over using a single method. Transitioning this idea to the digital video domain suggests that the various digital video representations can be combined to increase shot boundary detection performance. Figure 12.3 depicts a multiple representation video shot boundary detection technique.

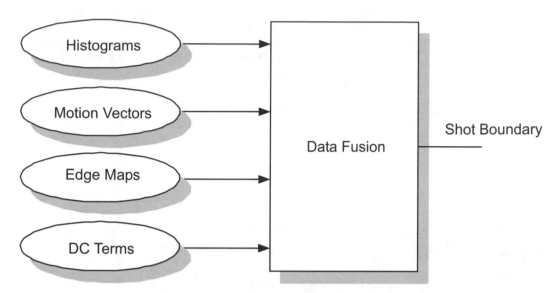

Figure 12.3 Multiple Inputs

In this technique, multiple video representations (histograms, motion vectors, edge maps, DC terms, etc.) can be used together to determine shot boundaries. This technique differs from the previous techniques described above in that multiple algorithms for each type of input source are not used. Instead, all different video representations are sent to a data fusion engine. This data fusion would use the various digital video representations in a unified algorithm to detect shot boundaries. This unified algorithm would be based on translation of existing algorithmic solutions.

4. ADAPTIVE VIDEO SUMMARIZATION

At present, most research efforts focus on developing and implementing the various summarization methods independently. There is little research currently dedicated to the development of an adaptive summarization technique. Some systems provide the user with various summarization methods and allow the user to choose which method is most appropriate. As stated above, there are various types of digital video, i.e., music video, home video, movies, news broadcasts, surveillance video, etc. The most effective summarization method is the one that allows for the user to gain the maximum understanding and competency of the original video. For the various types of digital video, the level of understanding varies. For example, for surveillance video it may be necessary to view a highly detailed summary of a video sequence. News broadcasts may only need a brief keyframe summary of the anchorperson and story shots. The level of understanding is based on the users information need. When the type of video is unknown before analysis, how does one determine the best summarization method to utilize? The solution to this problem involves creating a composite summarization solution that adapts to the different classes of video. This section describes the current research in this area as well as suggestions for further research.

The Físchlár system is an online video management system that allows for the browsing, indexing, and viewing of television programs.   Summarization is facilitated through organizing the video content into shots, groups of shots, scenes, story lines, and objects and events. Browsing in the Físchlár system is facilitated via 6 browser interfaces [1, 30, 49]. A scrollbar browser allows a user to scroll through the available keyframes. The advantage of this browser is that it is easy to use. However, viewing large documents with large numbers of keyframes can be overwhelming to the user. A slideshow browser presents each individual keyframe in approximately two-second intervals. The advantage of this browser is that it preserves the temporal nature of the video. The disadvantage of this browser is   that a slideshow can be too long for large video sequences causing the user to lose track of their location in the video.   A timeline browser allows a user to see a fixed number of keyframes (24) at a time in temporal order. The advantage of this type of browser is that the user can view the keyframes in sets. A ToolTip is also used to display the start and end time of each segment within the set of keyframes. An overview/detail browser displays the number of significant keyframes to the user selected by the scene level segmentation.   The user then has a choice of viewing more detail by selecting a keyframe that displays the timeline browser for that particular segment of video.   A dynamic overview browser allows a user to view a condensed set of significant keyframes.   When the mouse scrolls over the significant keyframes the keyframes flip through the detailed keyframes within the segment. A small timeline appears beneath each keyframe to let the user know where they are in temporal order. This technique allows a user to view the keyframes without major screen changes; however the time that it takes to view each significant keyframe's detailed keyframes may be too long. The hierarchical browser groups keyframes in a hierarchal tree structure that the user can navigate. The top level of the tree structure indicates the entire programs with the subsequent level indicating further segmentations of the previous levels.

The Informedia Digital Video Library System [40, 45] utilizes speech, closed caption text, speech processing, and scene detection to automatically segment news and documentary video. One aspect of the summarization method of the Informedia Digital Video Library System is composed of partitioning the video into shots and keyframes. Browsing is facilitated in this system by filmstrips. Filmstrips consist of thumbnail images along with the video segment metadata. The thumbnail is selected based on the similarity between the text that is associated with the frame and the user query. The user can use the browsing interface to view the metadata and jump to a particular point in the video segment for playback. During playback, the text-based transcript is displayed with highlighted keywords.

We are advocating the development of new video summarization algorithms based on the composition of multiple existing summarization algorithms. Key to the success of such an endeavour is the ascertainment of what constitutes good keyframe choices. Unlike segmentation, it is not obvious what the ground truth is for any given set of data. For this task we propose a controlled user study wherein users are provided with a simple interface for selection of keyframes for a corpus of test content. From the data gained in this study and user survey material, we will create a mean keyframe selection for the corpus and analyze the performance of existing algorithms relative to that corpus.   The key measures of performance are: 1) do the algorithms-under-test produce similar volumes of keyframes over similar regions and 2) how close are the algorithmically selected keyframes to those selected by the test population. From this measure of performance, the content characteristics can be correlated with the internal measures in the algorithm and video content characteristics so as to assess the appropriate classes of content for each summarization method.

5. CONCLUSION

Increased processor capabilities and falling storage costs are making the accumulation of large quantities of digital video not only practical, but also relatively easy. However, storing this vast quantity of data is of little use if users cannot locate content they are interested in. Indexing and browsing digital video effectively requires a variety of technologies. This chapter presented an overview of the existing segmentation and summarization techniques as well as suggestions for future research in developing a composite solution that can adapt to the underlying class of video. When the type of video is known a priori, any number of algorithms can be employed that take advantage of the known characteristics of the video. When the type of video is unknown before analysis, how does one choose the appropriate segmentation and summarization that maximizes performance and user information need. A new method for video analysis is needed that:

- Integrate the variety of shot-based segmentation algorithms into an adaptive framework that is sensitive to varying content types.
- Integrate and expand on existing methods for video summarization so as to create an adaptive summarization method.

Many techniques have been developed for segmentation and abstraction. Recognizing that there is no single, best solution for each of these problems has led to the ideas in this chapter for integrating the variety of existing algorithms into a common framework, creating a composite solution that is sensitive to the class of content under analysis, the performance of the algorithms on the specific content, and the best combination of the results.

This research can have significant impact on future systems for indexing and browsing of digital video content:
- Digital video indexing systems can perform much better if the summarizations provided to the indexing engine are smaller, more salient, and more efficient.
- Human browsing of search results will significantly benefit from a better-structured summarization of the video content, allowing choices to be made much faster and with less effort.

REFERENCES

[1]   Hauptman, A. and M. Witbrock, Story segmentation and detection of commercials in broadcast news video, in *Advances in Digital Libraries Conference*. 1998.
[2]   Hauptmann, A.G. and M.J. Witbrock, Story segmentation and detection of commercials in broadcast news video, in *Advances in Digital Libraries*. 1998. p. 168-179.

[3] Ide, I. et al., News Video Classification based on Semantic Attributes of Captions, in ACM International Conference. 1998.

[4] Maybury, M., A. Merlino, and D. Morey, Broadcast News Navigation using Story Segments, in ACM International Multimedia Conference. November, 1997. p. 381-391.

[5] O'Connor, N. et al., News Story Segmentation in the Fischlar Video Indexing System, in ICIP01. 2001. p. Summarizing Video.

[6] Sato, T. et al., Video OCR for Digital News Archive, in CAIVD. 1998. p. 52-60.

[7] Tang, T., X. Gao, and C. Wong, NewsEye: A News Broadcast Browsing and Retrieval System, in Proceedings of 2001 International Symposium on Intelligent Multimedia Video and Speech Processing. May, 2001. p. 150-153.

[8] Zhang, H., Content-based video browsing and retrieval, in *The Handbook of Multimedia Computing*, 1999, CRC Press, Boca Raton, FL. p. 255-280.

[9] Koprinska, I. and S. Carrato, Temporal Video Segmentation: A Survey.

[10] Gargi, U., R. Kasturi, and S. Antani, Performance Characterization and Comparison of Video Indexing Algorithms, in IEEE Conference on Computer Vision and Pattern Recognition. 1998.

[11] Dailianas, A., R. Allen, and P. England, *Comparison of Automatic Video Segmentation Algorithms*, in *SPIE Photonics West*. 1995. p. 2-16.

[12] Lienhart, R., Comparison of automatic shot boundary detection algorithms, in Proc. Storage and Retrieval for Image and Video Databases VII. January, 1999.

[13] Lupatini, G., C. Saraceno, and R. Leonardi, Scene Break Detection: A Comparison, in Proceedings of the RIDE'98 Eighth International Workshop on Research Issues in Data Engineering. 1998: Orlando, Florida. p. 34-41.

[14] Boreczky, J.S. and L.A. Rowe, Comparison of Video Shot Boundary Detection Techniques, in Storage and Retrieval for Image and Video Databases (SPIE). 1996. p. 170-179.

[15] Zhang, H., A. Kankanhalli, and S.W. Smoliar, Automatic partitioning of full-motion video. Multimedia Systems, 1993(1): p. 10-28.

[16] Zhang, H., et al., Video Parsing, Retrieval and Browsing: An Integrated and Content-Based Solution, in ACM Multimedia. 1995. p. 15-24.

[17] Nagasaka, A. and Y. Tanaka, *Automatic video indexing and full-video search for object appearances, in Visual Database Systems II,*. 1992, Elsevier / North-Holland, 113-127.

[18] Ferman, A.M. and A.M. Tekalp, Efficient filtering and clustering for temporal video segmentation and visual summarization. Journal of Visual Communication and Image Representation, 1998.

[19] Jain, A.K. and R.C. Dubes, *Algorithms for Clustering Data.* 1988. Prentice Hall, New York.

[20] Zabith, R., J. Miller, and K. Mai, A Feature-Based Algorithm for Detecting and Classifying Scene Breaks, in ACM Multimedia Proceedings. 1995. p. 189-200.

[21] Boreczky, J.S. and L.D. Wilcox, A Hidden Markov Model Framework for Video Segmentation using Audio and Image Features, in Proc. Int. Conf. Acoustics, Speech, and Signal Processing. 1998. p. 3741-3744.

[22]   Rabiner, L., A Tutorial on Hidden Markov Models and Selected Applications in Speech Recognition, in Proceedings of the IEEE. February, 1989. p. 257-285.

[23]   Liu, H. and G. Zick, *Scene Decomposition of MPEG Compressed Video*, in *Digital Video Compression: Algorithms and Technologies*. 1995.

[24]   Liu, H. and G. Zick, *Automatic Determination of Scence Changes in MPEG Compressed Video*, in *Proc. ICASS-IEEE Int. Symp. Circuits and Systems*. 1995. p. 764-767.

[25]   Arman, F., A. Hsu, and M. Chiu, *Image Processing on Compressed Data for Large Video Databases*, in *ACM Multimedia*. 1993. p. 267-272.

[26]   Zhang, H.J. *et al.*, *Video Parsing Using Compressed Data*, in *Proc. SPIE Conf. Image and Video Processing II*. 1994. p. 142-149.

[27]   Yeo, B. and B. Liu, *Rapid Scene Analysis on Compressed Video*, in *IEEE Transactions on Circuit and Systems for Video Technology*. 1993.

[28]   Li, Y., T. Zhang, and D. Tretter, *An Overview of Video Abstraction Techniques*. 2001: Hewlett-Packard Labs.

[29]   Girgensohn, A. *et al.*, *Keyframe-Based User Interfaces for Digital Video*, in *IEEE Computer*. 2001. p. 61-67.

[30]   Lee, H. *et al.*, *User Interface Design for Keyframe-Based Browsing of Digital Video*, in *WIAMIS 2001 - Workshop on Image Analysis for Multimedia Interactive Services*. 2001. p. 16-17.

[31]   Zhang, H., C. Low, and S. Smoliar. *Video Parsing and Browsing Using Compressed Data*. in *Multimedia Tools and Applications*. 1995.

[32]   England, P., *et al.*, *I/Browse: The Bellcore Video Library Toolkit*, in *Storage and Retrieval for Image and Video Databases (SPIE)*. 1996. p. 254-264.

[33]   Shahraray, B., *Scene change detection and content-based sampling of video sequences*. Digital Video Compression: Algorithms and Technologies, 1995: p. 2--13.

[34]   Ueda, H., *et al.*, *Automatic Structure Visualization for Video Editing*, in *INTERCHI '93 , ACM Press*. 1993. p. 137--141.

[35]   Ferman, A. and A. Tekalp, *Multiscale content extraction and representation for video indexing*, in *Proc. SPIE Multimedia Storage and Archiving Systems II*. 1997. p. 23--31.

[36]   Zhuang, Y., *et al.*, *Adaptive Key Frame Extraction using Unsupervised Clustering*, in *Proc. of Intl. Conf. on Image Processing*. 1998. p. 886--870.

[37]   Wolf, W., *Key frame selection by motion analysis*, in *1996 IEEE International Conference on Acoustics, Speech, and Signal Processing*. 1996: Atlanta, GA. p. 1228 -1231.

[38]   Dixon, J. and C. Owen, *Fast Client-Server Video Summarization for Continuous Capture (poster session)*, in *ACM Multimedia*. 2001.

[39]   Uchihashi, S. and J. Foote, *Summarizing video using a shot importance measure and frame-packing algorithm*, in *Proceedings of ICASSP'99*. 1999. p. 3041-3044.

[40]   Wactlar, H. and M. Christel, *Lessons Learned from the Creation and Deployment of a Terabyte Digital Video Library*. <u>Computer</u>, February, 1999(2): p. 66--73.

[41]   Lienhart, R., S. Pfeiffer, and W. Effelsberg, *Video abstracting*. <u>Communications of the ACM</u>, 1997(12): p. 54--62.

[42] Christel, M.G., *et al.*, *Evolving Video Skims into Useful Multimedia Abstractions*, in *CHI*. 1998. p. 171-178.

[43] Lee, H. and A. Smeaton. *User-interface issues for browsing digital video.* in *Proc. 21st Annual Colloquium on IR Research (IRSG 99)*. 1999.

[44] Christel, M.G., D.B. Winkler, and C.R. Taylor, *Multimedia Abstractions for a Digital Video Library*, in *ACM DL*. 1997. p. 21-29.

[45] Wactlar, H.D., *et al.*, *Intelligent Access to Digital Video: Informedia Project.* Computer, May, 1996(5).

[46] Komlodi, A. and G. Marchioini, *Keyframe Preview Techniques for Video Browsing*, in *ACM International Conference on Digital Libraries*. 1998. p. 118-125.

[47] Ding, W., G. Marchionini, and D. Soergel, *Multimodal Surrogates for Video Browsing*, in *Proceedings of the Fourth ACM International Conference on Digital Libraries*. 1999.

[48] Katzer, J., *et al.*, *A study of the overlap among document representations*, in *Information Technology: Research and Development*. 1982.

[49] O'Connor, N., et al., *Físchlár: an On-line System for Indexing and Browsing of Broadcast Television Content*, in *Proceedings of ICASSP 2001*.

[50] Browne, P., *Evaluating and Combining Digital Video Shot Boundary Detection Algorithms*, in *Irish Machine Vision and Image Processing Conference (IMVIP'2000)*. 2000.

[51] Bartell, B.T., G.W. Cottrell, and R.K. Belew, *Automatic Combination of Multiple Ranked Retrieval Systems*, in *Research and Development in Information Retrieval*. 1994. p. 173-181.

[52] Freud, Y., *et al.*, *An Efficient Boosting Algorithm for Combining Preferences*, in *International Conference on Machine Learning*. 1998.

[53] Hull, D.A., J.O. Pedersen, and H. Schütze, *Method Combination for Document Filtering*, in *Proceedings of SIGIR-96, 19th ACM International Conference on Research and Development in Information Retrieval*, H.-P. Frei, *et al.*, Editors. 1996, ACM Press, New York, p. 279-288.

[54] Lee, J.-H., *Analyses of Multiple Evidence Combination*, in *Research and Development in Information Retrieval*. 1997. p. 267-276.

[55] Lee, J.H., *Combining Multiple Evidence from Different Properties of Weighting Schemes*, in Research and Development in Information Retrieval. 1995. p. 180-188.

[56] Lee, J.H., *Combining Multiple Evidence from Different Relevant Feedback Networks*, in *Database Systems for Advanced Applications*. 1997. p. 421-430.

[57] Losee, R. Comparing Boolean and probabilistic information retrieval systems across queries and disciplines, Journal of the American Society for Information Science. 1997.

[58] Shaw, J.A. and E.A. Fox, *Combination of Multiple Searches*, in *Text Retrieval Conference*. 1994.

[59] Howe, A.E. and D. Dreilinger, *SAVVYSEARCH: A Metasearch Engine That Learns Which Search Engines to Query*. AI Magazine, 1997(2): p. 19-25.

[60] Glover, E.J., *et al.*, *Architecture of a Metasearch Engine that Supports User Information Needs*, in *Eighth International Conference on Information and*

Knowledge Management (CIKM'99). November, 1999, ACM Press: Kansas City, MO. p. 210-216.

[61]   Gravano, L., *et al.*, *STARTS: Stanford proposal for Internet meta-searching*, in *Proc. of the 1997 ACM SIGMOD International Conference On Management of Data*. 1997. p. 207-218.

[62]   Meng, W., C. Yu, and K.-L. Liu, *Building Efficient and Effective Metasearch Engines*. .

[63]   Lawrence, S. and C.L. Giles, *Context and Page Analysis for Improved Web Search*. IEEE Internet Computing, 1998(4): p. 38-46.

[64]   Dreilinger, D. and A.E. Howe, *Experiences with Selecting Search Engines Using Metasearch*. ACM Transactions on Information Systems, 1997(3): p. 195--222.

[65]   Chang, S.-F., *et al.*, *Visual information retrieval from large distributed online repositories*. Communications of the ACM, 1997(12): p. 63-71.

[66]   Beigi, M., A.B. Benitez, and S.-F. Chang, *MetaSEEk: A Content-Based Metasearch Engine for Images*, in *Storage and Retrieval for Image and Video Databases (SPIE)*. 1998. p. 118-128.

13

AUGMENTED IMAGERY FOR DIGITAL VIDEO APPLICATIONS

Charles B. Owen, Ji Zhou, Kwok Hung Tang, and Fan Xiao
Media and Entertainment Technologies Laboratory
Computer Science and Engineering Department,
Michigan State University
East Lansing, MI 48824
cbowen@cse.msu.edu

1. INTRODUCTION

Augmented Reality (AR) is the blending of computer-generated content with the perceived reality. Traditionally, virtual reality (VR) has existed to provide an alternative to reality. Virtual presentations provide users with a view of the world that may not exist for them elsewhere. However, a criticism of VR has been that it cannot hope to completely supplant reality given the incredible power of human senses. And, disconnecting the user from reality is not appropriate for a wide variety of applications. For this reason, researchers have begun to explore the vast range of possibilities between actual reality and virtual environments. Milgram and Kishino proposed a *virtuality continuum*, illustrated in Figure 13.1, to describe concepts beyond these simple ideas of reality and virtual reality [1, 2]. At one end of the scale lie real environments, the world with no computer augmentation. At the other end lies conventional virtual reality, the world replaced by the synthetic. Augmented reality dwells between the two extremes, typically closer to the real than the virtual, modifying the real world by adding virtual elements.

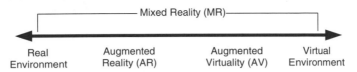

Figure 13.1 The Milgram virtuality continuum (VC)

Augmented reality seeks to blend computer-generated content with reality so as to create a seamless whole, in which the virtual elements have spatial and visual attributes as if they were natural elements of the source. A variety of augmented reality technologies exist. Head mounted displays can augment the vision field and enhance vision. Processing methods can enhance the perception of sound. Haptics can be used to amplify the feeling of touch. This chapter discusses augmented imagery, the enhancement of digital imagery, with specific application to digital video.

Augmented imagery is the modification of images, either still images or video, through the addition of virtual computer-generated elements. Variations on augmented imagery are common in films, video games, TV shows and commercials. Televised coverage of football games now includes on-screen yellow lines indicating the first-down line. Clearly, the referees are not rushing out onto the field with cans of yellow spray paint each time a team achieves a first-down and, in fact, the line is not picked up by the camera at all. The video captured by the camera has been augmented with this new visual feature in real time. Augmented imagery is a powerful tool for film makers, or so-called "movie magicians." From *Star Wars* (1977) to the *Lord of Rings* series (2001), films have been relying on computer-generated imagery (CGI) technologies to create fantastic effects, imaginative figures and non-existent scenes that appear realistic and completely integrated.

Augmenting an image sequence or digital video is, in general, much more complex than simply summing the source images arithmetically. This chapter introduces the concepts, components, and applications of augmented imagery, describing how systems process the real and virtual elements to build a desired result.

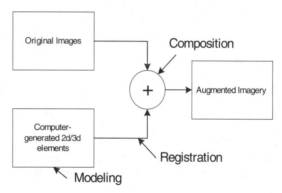

Figure 13.2 Architecture of an augmented imagery system

Augmented imagery systems create imagery by combining 2D or 3D computer-generated elements with original images as illustrated in Figure 13.2. Given this basic system design, an augmented imagery system can be subdivided into three major components: modeling, registration and composition. Modeling is the mathematical description of content, both real and virtual. It is crucial that each environment can be described in a fixed frame of reference. A system adding first-down lines to a football field needs to know the dimensions of the field including any surface curvature. Modeling can include 2D and 3D representations. Registration is the determination of the relationship between the real and virtual models and the computation of appropriate transformations that align the two so as to coincide in the same frame of reference. For many 3D applications, registration consists of determining camera parameters for the virtual model that duplicate those physically realized in the real model. Some applications only require a 2D to 2D transformation that places content in the right location as tracked in the reality image.

Composition is the blending of the real and virtual elements into a single output image. Composition can be as simple as overlaying the virtual elements or as complex as blue-screen matting or segmentation of the football field from the players.

The terms augmented imagery and augmented reality are often used interchangeably. The authors feel that augmented imagery is a sub-category of the more general augmented reality, though the term augmented reality is most used to refer to real-time applications involving head mounted displays or projection systems [2, 3]. Indeed, augmented imagery in real-time using a single camera source is effectively identical to monitor-based augmented reality [4].

Augmented imagery is a powerful tool for enhancing or repurposing content in *video databases*. Virtual sets can be blended with real actors to localize presentations, advertising can be added to existing content to enhance revenue or objectionable elements can be masked or hidden selectively for specific audiences. Advances in processing power and graphics subsystems make it possible for systems to perform many augmented imagery operations in real time on video feeds or while streaming content from video databases.

Research in the field of augmented imagery has been active for a considerable time, with early work primarily focusing on analog and digital composting technologies. Petro Vlahos was one of the early researchers on matting and compositing techniques, and is recognized for his many patents on "Electronic Compositing Systems" [5-12]. Survey papers from Blinn and Smith provide additional reading in this area [13, 14]. Kutulakos and Vallino describe mixing live video with computer-generated graphics objects using affine representations [15]. A number of books, such as *The Art and Science of Digital Compositing* by Ron Brinkmann [16] and *Digital Compositing in Depth* by Doug Kelly [17], give comprehensive descriptions and technical details of digital compositing for augmented imagery.

This chapter is structured as follows: Sections 2, 3 and 4 introduce the three major elements of an augmented imagery system: modeling, registration and composition. Section 5 explores techniques for visual realism, such as anti-aliasing and lighting consistency. Some example applications are described in detail at Section 6, and Section 7 contains conclusions.

2. MODELING

Modeling is the description of virtual and real elements of the environment in mathematical forms and as data structures. Modeling can be as simple as describing the physical dimensions of a planar image in pixels. Many applications, however, require more complex descriptions, such as 3D polygon meshes, scene graphs or constructive solid geometry trees. Modeling must capture characteristics of both the real and virtual elements in the augmentation process. A system might wish to add a billboard advertisement to video of travel on a highway. The billboard would need to be rendered from varying viewpoints as the car passes the location the billboard has been assigned to. Consequently, the billboard model will be represented using 3D modeling techniques from computer graphics [18, 19]. Additionally, characteristics of the road may result in occlusion of the billboard when some physical object, such as a bridge, passes between the viewpoint and the virtual billboard location. In that instance an *occlusion model* of the physical environment is required to render the occlusion area that will block the billboard. This area will vary as the viewpoint changes. Consequently, the geometry of the highway system may be modelled using 3D techniques as well.

The modeling process can be broken into two primary categories: 2D and 3D. 2D representations are usually trivial, consisting only of polygonal boundaries in source images or rectangular region description. A virtual advertisement that is to be placed into a scene need only be described as a simple image and the destination location as a quadrilateral. 3D object representations are more complex, creating a description of an object sufficient to support rendering by a graphics package and display on the computer screen or integration into the projection of a real 3D scene as captured by a camera. A realistic 3D visualization requires object representations that accurately describe all the attributes required for realistic rendering including geometry, surface characteristics and lighting. This section introduces typical methods used for modeling in augmented imagery, such as polygonal meshes, parametric descriptions and fractal construction.

For the rendered image to look "real," the augmented content should maintain the same or similar geometric and spatial characteristics as that of the original video. An augmented imagery designer's tools will often include tape measures and callipers. Surface and lighting characteristics in the virtual environment must match those in the real world. The original video often must be analyzed to acquire data required for modeling. This process is called scene-dependent modeling, which is introduced later in this section.

2.1 3D OBJECT REPRESENTATION

Ideally, augmented imagery integrates images from different sources into a seamless whole. The most common combination is of a real camera image and a virtual image rendered by a graphics system such as OpenGL or Direct3D. Therefore, 3D object representation is the first step for many augmented imagery applications. 3D object representations are used both to render the augmented elements of an image and to indicate where the pixels of the augmented content are visible. The output of the rendering process is often both an image and an alpha map, a matrix of values corresponding to pixels and indicating the transparency of each pixel. Describing an object in most graphics systems requires the specification of geometry and material. Geometric data describes the structure of the object, indicating surfaces, boundaries, and interiors; material data include appearance attributes such as surface reflection characteristics and textures. This section focuses on boundary representations for 3-D object modelling, methods that describe objects by describing the visible surfaces. Alternatives such as constructive solid geometry and volume descriptions are also used, but are beyond the scope of this chapter.

The most commonly used 3-D boundary representation is a set of surface polygons that enclose the object interior and form the "skin" of it. This collection of polygons is called a polygon mesh, a standard means of describing 3D surfaces. A polygon mesh consists of vertices and polygons and can be organized in many ways. A common organization is as a face list of polygons, with each polygon described as a vertex list and a normal list. The vertex list contains coordinate values, and the normal list contains surface orientation information for the face or vertices of the face, information useful in the shading process.

Polygonal mesh representations are simple and fast, since all information is stored and processed with linear equations. Complex curved surfaces can be simulated to a high degree of accuracy through the use of shading techniques that modify the surface normals to blend shading between adjacent surfaces. Most high-performance graphics systems incorporate fast hardware-implemented polygon renderers which can display thousands to millions of shaded polygons (often triangulated) per second. Figure 13.3 illustrates a simple pyramid represented by a polygonal mesh and the corresponding vertex, normal and face list.

	Vertex list (x, y, z)	Normal list (n_x, n_y, n_z)	Face list (vertices, and associated normal)
0	(0, 0, 0)	(-0.897, 0.447, 0)	(0, 1, 4) (0, 0, 0)
1	(0, 0, 1)	(0, 0.447, 0.897)	(1, 2, 4) (1, 1, 1)
2	(1, 0, 1)	(0.897, 0.447, 0)	(2, 3, 4) (2, 2, 2)
3	(1, 0, 0)	(0, 0.447, -0.897)	(3, 0, 4) (3, 3, 3)
4	(0.5, 1, 0.5)	(0, -1, 0)	(0, 1, 2, 3) (4, 4, 4)

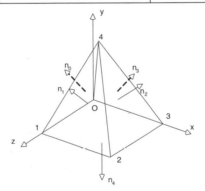

Figure 13.3 A simple polygon mesh description of a pyramid with vertex, normal and face list

Objects that are naturally composed of flat faces can be precisely represented by a polygon mesh. Objects with curved surfaces, such as spheres, cylinders, cones, etc., must be approximated using small polygons, a process referred to as tessellation. The object surface is divided into facets in order to achieve a piecewise approximation. The error due to tessellation is referred to as *geometric aliasing*. Decreasing geometric aliasing requires increasing the number of polygons, which increases the resolution of the description, but also boosts the storage requirements and processing time. If the mesh represents a smoothly curved surface, rendering techniques such as Gouraud shading eliminate or reduce the presence of polygon edge boundaries, simulating smooth surface curves with smaller sets of polygons. Figure 13.4 is a tessellated 3D object; Figure 13.5 is the same object represented with the same level of tessellation, but using Gouraud shading to improve the appearance.

Figure 13.4 A ball represented as a tessellated polygon mesh

Figure 13.5 A ball rendered using Gouraud shading

Parametric representations describe curves and smooth surfaces at a higher level of abstraction, either using parameterizations of mathematical functions (for common objects such as spheres or ellipsoids) or through parameterized sets of equations such as quadratics or cubics. Common basic geometries are often described mathematically because the representation is much smaller than a polygon mesh and is not an approximation. It should be noted that the rendering process of many graphic environments, including OpenGL, converts parameterized representations to polygon-mesh approximations, anyway due to the design of rendering hardware. Quadric surfaces (second-degree equations), such as spheres and ellipsoids, are examples of functional representations, a class of parametric representations.

For complex curves and surfaces that can't be described with a single mathematical function, spline representations can be applied to model objects with piecewise cubic polynomial functions. This method is a staple of computer-aided-design (CAD) and interactive modeling tools used to describe 3D objects with smoothly varying surfaces, such as human faces or bodies. A spline representation specifies a curve or curved surface using a set of *control points*, which are fitted with piecewise continuous parametric polynomial functions using interpolation or approximation. Moving the control points modifies the resulting curve, so control points can be adjusted to arrive at the expected shape. There are several common spline specifications used in 3D object modeling, include Bezier, B-splines and NonUniform Rational B-Splines (NURBS). As an example, to model a Bezier surface, an input mesh of control points is specified so as to connect two sets of orthogonal Bezier curves. A two-dimensional Bezier curve generated from four control points and a Bezier surface are shown in Figure 13.6 and Figure 13.7.

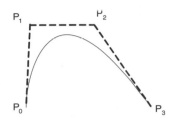

Figure 13.6 A Bezier curve with 4 control points

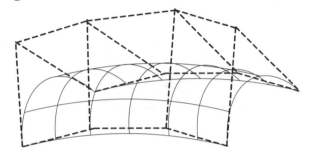

Figure 13.7 A Bezier surface

Fractal representations are commonly used to model naturally chaotic objects such as clouds, snowflakes, terrain, rock, leaves, etc., which are difficult to be described as equations of Euclidean-geometry methods because of their irregular shape or fragmented features. These objects have two common characteristics: infinite detail at every point and self-similarity between the object parts and whole features. Based on these properties, Fractal representations can generate representations of this class of objects by repeatedly applying a specified transformation function to points within a region of space. Greater level of detail with a fractal object can be obtained when more steps in the iterated function system are performed.

2.2 SCENE GRAPH REPRESENTATIONS

Simple polygon mesh representations are often organized hierarchically. Indeed, most complex graphic objects are built by composing other component graphics objects, with only the most primitive elements directly built from polygon meshes. Figure 13.8 is an example of how a highway billboard might be represented hierarchically in a *scene graph*. The billboard is built from a support structure that mounts it to the ground and a structure that serves as a surface for the advertising. Each of these elements can be designed independently, and then placed relative to each other by associating a geometric transformation (specification of orientation and relative location) with each graph edge. Likewise, the subcomponents are, in turn, built from other subcomponents.

The transformation matrix necessary to render a scene graph element in the proper location is the composition of transformations along the edge path to the graph node. Typically, only the leaf nodes will be described using primitives. Scene graphs greatly simplify the representation of complex environments.

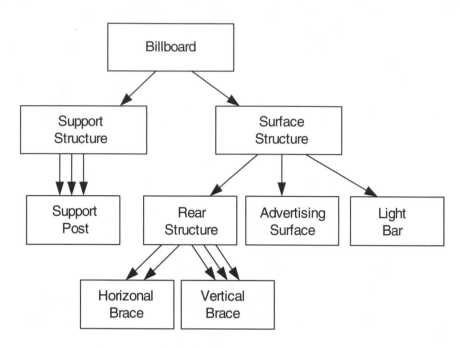

Figure 13.8 Screen graph of a highway billboard

2.3 REPRESENTATION FOR OCCLUSION

The billboard example describes the representation of a graphical object that will be rendered as a virtual element in an augmented image. It is also common that models will be used to describe elements of the real image that may occlude virtual elements. As an example, a virtual insect may fly behind objects in a room. An occlusion model is a graphical model of the real environment that facilitates occlusion of virtual objects.

When the virtual object is rendered, the occlusion model is also rendered. However, the occlusion model is not subject to lighting or surface characteristics, but instead is rendered in such a way as to ensure transparency in the composition process. The most basic solution is to render the occlusion model with a transparent alpha value when generating an alpha map, an image indicating transparent and opaque regions of the result. Alpha maps are described in Section 4. Other examples include the use of a fixed background color, such as blue, that is easily segmented and replaced.

2.4 SCENE-DEPENDENT MODELING

There are often more considerations that must be taken into account than just geometrical and visual properties. Augmented imagery differs from simple image composition in that the augmented elements have geometric consistency with the real image, which makes the resulting combination look "real" rather than artificial. It is often required that the augmented imagery system analyze the original images to acquire information required for modeling, particularly when the patches will "replace" some parts in the background. Scene-dependent modeling, or image-based modeling, attempts to build three-dimensional models based on visual cues represented in two-dimensional images or image sequences. This process can be very complex, requiring computer vision methods to reconstruct underlying image content.

3. REGISTRATION

3.1 INTRODUCTION

One of the most important elements in any augmented reality system is registration. Virtual objects that are inserted into images should appear to really exist in the environment. To achieve such an illusion, virtual objects and real objects must be properly aligned with respect to each other. Registration is a general AR problem; many interactive AR applications, such as surgery systems, require accurate registration. Augmented imagery systems must contend with the incredible sensitivity of human users to registration errors [20].

Registration is the determination of transformations between the real image model and the virtual content model. A transformation is a function from points in one model to points in another. A function that translates pixel coordinates in a virtual advertisement image to pixel coordinates in a destination image is an example transformation. This 2D to 2D example describes the warping of the original image to an arbitrary placement on the screen. This placement may have to contend with perspective projection of the advertisement patch; so the transformation might convert an upright rectangle to an arbitrary quadrilateral.

Transformations in augmented imagery applications can be 2D to 2D or 3D to 2D. 2D to 2D registration assumes a 2D model and is typically used to replace planer regions in images. 3D to 2D registration assumes a 3D graphical model will be aligned with a 3D representation of the environment as projected to a 2D image by the camera system. Both applications are common in systems.

Registration to a camera image requires knowledge of the location of the camera relative to the scene. Many trackers and sensors, such as ultrasonic trackers, GPS receivers, mechanical gyroscopes and accelerometers, have been used to mechanically assist in achieving accurate registration. Tracking is the determination of location and/or orientation of physical objects such as cameras. As an example, a pan/tilt/zoom (PTZ) head measures the physical parameters of a camera. Tracking of just position *or* orientation (but not both) is called 3 degree-of-freedom (3DOF) tracking. Tracking of position *and* orientation is called 6 degree-of-freedom (6DOF) tracking. In addition, other system parameters such as zoom may be tracked. Tracking can be used to determine position and orientation of additional objects in an environment in addition to the camera. Instrumenting a portable, handheld advertisement in a sporting event with tracking would allow for replacement of the advertisement at a later date.

Tracking can be achieved optically from the video image or through instrumentation of the camera or objects. Optical tracking can utilize natural features (unprepared spaces) [21] or placed fiducial marks (prepared spaces) [22]. Rolland, Davis, and Baillot present an overview of tracking systems [23]. Hybrid tracking techniques combine multiple tracking technologies and are employed to compensate for the weakness of individual tracking technologies. Vision-based tracking techniques have received more attention in recent years. Bajura and Neumann [24] point out that vision-based techniques have an advantage in that the digitized image provides a mechanism for bringing feedback into the system. It is possible to aid the registration with that feedback.

3.2 2D-2D REGISTRATION

A simple augmented imagery system might replace a blank planar region in a video sequence with a commercial advertisement. This is a typical 2D-2D registration problem. All that need be done is to track the blank planar correctly and replace pixels in the planar with corresponding pixels in the advertisement. The replacement advertisement is warped to the proper shape and rendered into the space using a transformation that converts locations in the ad to locations in the final image. Region tracking techniques differ in the geometric transformation model chosen and the optimization criteria. Rigid, affine, homographic and deformable transformation models have been used. The optimization criteria include texture correlation [25], mutual information [26] optical flow measurement, etc. Three simple transformation models are introduced here. A review of image registration methods can be found in Brown [27].

3.2.1 Geometric Transformation Models

A geometric transformation model transforms points in one frame of reference to points in another. Transformations are used to translate, rotate, scale and warp 2D and 3D content. This section describes several common 2D geometric transformations and their application in augmented imagery.

Rigid Transformation

A *rigid* transformation maps a point (x, y) to a point (x', y') via rotation and translation only:

$$\begin{pmatrix} x' \\ y' \end{pmatrix} = R_\theta \begin{pmatrix} x \\ y \end{pmatrix} + T \tag{13.1}$$

In this equation, T is a translation vector and R_θ a rotation matrix:

$$R_\theta = \begin{pmatrix} \cos\theta & -\sin\theta \\ \sin\theta & \cos\theta \end{pmatrix} \tag{13.2}$$

It is a convenient notation to use homogeneous coordinates for points. The homogeneous coordinates of a 2D point (x, y) are (sx, sy, s), where s is an arbitrary scale factor, commonly 1.0. Thus Equation 13.1 can be written as:

$$\begin{pmatrix} x' \\ y' \\ 1 \end{pmatrix} = \begin{pmatrix} \cos\theta & -\sin\theta & t_x \\ \sin\theta & \cos\theta & t_y \\ 0 & 0 & 1 \end{pmatrix} \begin{pmatrix} x \\ y \\ 1 \end{pmatrix} \tag{13.3}$$

This transformation is easily modified to include scaling. Rigid transformations are very basic and generally only used for overlay information. As an example, a rigid transformation with scaling is used to place the bug (a small station or network logo) in the corner of a television picture. However, for most applications the rigid transformation is too restrictive.

Affine Transformation

Affine motion describes translation, rotation, scaling and skew. These are the most common objects motions. The mathematic model for the affine transformation is:

$$\begin{pmatrix} x' \\ y' \end{pmatrix} = A \begin{pmatrix} x \\ y \end{pmatrix} + T \tag{13.4}$$

A is an arbitrary 2×2 matrix and T is a vector of dimension two. Using homogeneous coordinates, an affine transformation can be written as:

$$\begin{pmatrix} x' \\ y' \\ 1 \end{pmatrix} = \begin{pmatrix} a_{11} & a_{12} & t_1 \\ a_{21} & a_{22} & t_2 \\ 0 & 0 & 1 \end{pmatrix} \begin{pmatrix} x \\ y \\ 1 \end{pmatrix} \tag{13.5}$$

This model has the property that parallel lines remain parallel under transformation. The affine transformation is simple, but applies well in many applications. V. Ferrari et al. [28] implemented an augmented reality system using an affine region tracker. First, the affine transformation mapping the region from the reference frame to the current frame is computed. Then an augmentation image is attached to that region using an affine transformation.

General Homographic Transformation

The general homographic transformation maps a point (x, y) to a point (x', y') as follows:

$$\begin{pmatrix} sx' \\ sy' \\ s \end{pmatrix} = H \begin{pmatrix} x \\ y \\ 1 \end{pmatrix} \tag{13.6}$$

In this equation, H is an arbitrary 3×3 matrix and s is an arbitrary value other than 0. H has 9 parameters, 8 of which are independent since s can have an arbitrary value. Because H can be arbitrarily scaled, it is common to set the lower-right corner to 1. Unlike rigid and affine motion model, the general homographic transformation can be used to model perspective effects.

Other nonlinear motion models are also used. For example, B. Bascle and R. Deriche [25] applied a deformable motion model to track a region.

3.2.2 Computation of Transformation Matrices

In most cases the required transformation is not known and must be computed from image content or tracking results. In the planer region example, the transformation might be computed from the corner points of the destination region. Let p_1-p_4 be 2D coordinates of the corners of the region that the advertisement is to be written into. A transformation is required that will warp the advertisement such that the corners of the ad, c_1-c_4, will be warped to p_1-p_4. If p_1-p_4 represent a region with the same width and height as the image (unlikely), a rigid transformation can be computed by determining the rotation that will rotate $c_1 c_2$ to the same angle as $p_1 p_2$, then composing this rotation with a translation of c_1 to p_1 (assuming c1 is (0,0)).

A similar solution exists for computing an affine transform. Only three point correspondences are required to compute an affine transform. The problem is easily modelled as:

$$[p_1 \quad p_2 \quad p_3] = T[c_1 \quad c_2 \quad c_3] \tag{13.7}$$

It is assumed in this example that p_1-p_4 and c_1-c_4 are represented as homogeneous coordinates (three element column vectors with the third element set to 1). In this example, T can be easily computed as:

$$T = [p_1 \quad p_2 \quad p_3][c_1 \quad c_2 \quad c_3]^{-1} \tag{13.8}$$

The most general solution for a four point correspondence is a general homographic transform and, indeed, this is the solution required for most applications. Let the values of p_i be x_i, y_i and z_i and the coordinates of c_i be x_i', y_i' and z_i'. The general homographic transformation, as described in equation 13.6, can be written as:

$$x' = \frac{a_{11}x + a_{12}y + a_{13}}{a_{31}x + a_{32}y + 1} \tag{13.9}$$

$$y' = \frac{a_{21}x + a_{22}y + a_{23}}{a_{31}x + a_{32}y + 1} \tag{13.10}$$

Equations 13.9 and 13.10 can be restructured as:

$$x' = a_{11}x + a_{12}y + a_{13} - a_{31}xx' - a_{32}yx' \tag{13.11}$$

$$y' = a_{21}x + a_{22}y + a_{23} - a_{31}xy' - a_{32}yy' \tag{13.12}$$

Equations 13.11 and 13.12 can be used to construct a matrix solution for the values of T, as illustrated in equation 13.13. Equation 13.13 represents the problem as the product of a 8 by 8 matrix of knowns multiplied by an 8 element vector of unknowns equal to an 8 element vector of knowns. The problem is in the familiar Ax=b form for simultaneous equations and can be solved using a variety of conventional matrix methods.

$$
\begin{bmatrix}
x_1 & y_1 & 1 & 0 & 0 & 0 & x_1 x_1{}' & y_1 x_1{}' \\
0 & 0 & 0 & x_1 & y_1 & 1 & x_1 y_1{}' & y_1 y_1{}' \\
& & & & \cdots & & & \\
x_4 & y_4 & 1 & 0 & 0 & 0 & x_4 x_4{}' & y_4 x_4{}' \\
0 & 0 & 0 & x_4 & y_4 & 1 & x_4 y_4{}' & y_4 y_4{}'
\end{bmatrix}
\begin{bmatrix}
a_{11} \\ a_{12} \\ a_{13} \\ a_{21} \\ a_{22} \\ a_{23} \\ a_{31} \\ a_{32}
\end{bmatrix}
=
\begin{bmatrix}
x_1{}' \\ y_1{}' \\ \cdots \\ x_4{}' \\ y_4{}'
\end{bmatrix}
\qquad (13.13)
$$

Similar computations can be used to determine the appropriate affine or general homographic transformation that will correspond to the points. The general homographic transformation is the most common solution for this problem, since it can model the effects of perspective projection.

In general, control points must be determined in the images that are to be corresponded. These control points may be defined in source content or located using tracking, manual selection or image recognition. Fiducial images can be placed in the physical environment to assist in the location of control points. For example, fiducial chemical markers are widely used in medical imaging. Control points can be features of a region. For example corners, intersection of lines can be selected. Obviously, selected features must be robust in the image sequences, meaning they undergo small changes from frame to frame and are easy to detect. Shi and Tomasi [29] proposed a method to select "good" features to track. To match features in each image, template matching is often applied. After n pairs of points are found, a transformation can be computed.

For many applications, more points will be determined than are absolutely needed for the transformation computation. In these cases, least-squares estimation of the transformation allows the additional points to help decrease errors. A least-squares error criterion function for an affine transformation is defined as follows:

$$
E(A,T) = \sum_{i=1}^{n} \left((a_{11} x_i + a_{12} y_i + t_1 - x_i{}')^2 + (a_{21} x_i + a_{22} y_i + t_2 - y_i{}')^2 \right) \quad (13.14)
$$

Least-squares estimation minimizes this error function.

3.3 2D-3D REGISTRATION

2D to 3D registration is typically used to match a virtual camera in a 3D graphics environment to a real camera in the physical world. The starting point for class of registration is nearly always a set of points in the real world and the corresponding points in the camera image. These points may be determined automatically or manually.

3.3.1 The Camera Model.

For convenience homogeneous coordinates will be used to describe the model for a camera. For a 3D point X and a corresponding 2D point x, the homogeneous coordinates can be written as $X = (x, y, z, 1)$ and $x = (u, v, 1)$, respectively. The 2D-3D correspondence is expressed as:

$$
\begin{pmatrix} su \\ sv \\ s \end{pmatrix} = P \begin{pmatrix} x \\ y \\ z \\ 1 \end{pmatrix}
\tag{13.15}
$$

In this equation, P is a 3×4 camera projection matrix, which can be decomposed into:

$$
P = K[R \,|\, t]
\tag{13.16}
$$

R is a 3×3 rotation matrix and t is a translation vector of dimension 3. K is the calibration matrix of the camera. The | symbol indicates concatenation. The equation for K is:

$$
K = \begin{pmatrix} f & s & c_x \\ & af & c_y \\ & & 1 \end{pmatrix}
\tag{13.17}
$$

In this equation, f is the camera focal length, (c_x, c_y) is a principal point, the center of projection in the camera image, s is skew and a is the image aspect ratio. These intrinsic parameters may be calibrated in advance.

In a video-based augmented reality system, the world coordinates of 3D virtual objects are often known, and, if the camera projection matrix is also known, 3D virtual objects can be inserted into the real scene using equation 13.14 to determine the 2D coordinate that any 3D virtual point will project to.

3.3.2 Estimation of Projection Matrix.

Model-based Method.

The most common approach for computing the projection matrix is the model-based method. The projections of landmarks (artificial or natural) in an image are identified and the 3D-2D correspondences are established based on knowledge of the real-world coordinates of those landmarks. Like estimation of transformation matrices in the 2D-2D registration problem, the projection matrix can also be computed by using a least-squares method. The error function is:

$$
E(R, t) = \sum_i d^2(x_i, PX_i)
\tag{13.18}
$$

This function is also referred to as the re-projection error. Computer vision texts describe methods to compute P given this error function [30].

Natural landmarks include street lamps and 3D curves [31, 32]. Multi-ring color fiducials (artificial landmarks) [33] and black squares with different patterns inside [34] have been proposed as fiducials. The primary advantage of model-based registration is high accuracy and absence of drift, as long as enough landmarks are in the view. For outdoor applications, it is hard to prepare the fiducials or measure natural landmarks. For this reason, several extendible tracking methods have been proposed. A tracking method that extends the tracking range to unprepared environments by line auto-calibration is introduced in [35].

Structure From Motion.

A technology which appears to provide general solutions for outdoor applications is *structure-from-motion estimation*. Two-dimensional image motion is the projection of the three-dimensional motion of objects, relative to a camera, onto the image plane. Sequences of time-ordered images allow the estimation of projected two-dimensional image motion as optical flow. Provided that optical flow is a reliable approximation to image motion, the flow data can be used to recover the three-dimensional motion of the camera [21, 36]. A review of computation of optical flow can be found in [37]. However, this method is time-consuming and not suitable for real-time applications in the immediate future.

4. COMPOSITION

Composition is the blending of registered images to produce a final, augmented image. The television industry refers to this process as *keying*. A variety of methods exist for image composition, depending on the source of the real and virtual content. This section describes common composition methods.

Composing two image sources can be thought of as an image "mixing" process, wherein each pixel of the result image is a linear combination of corresponding source pixels, typically with an "alpha" weighting:

$$o(x, y) = \alpha(x, y)i_1(x, y) + (1 - \alpha(x, y))i_2(x, y) \qquad (13.19)$$

In this equation, i_1 and i_2 represent input images and o is the composited output image. α is an alpha map or matte image, a monochrome image with each pixel defined in the range [0,1]. Alpha maps are common in computer graphics applications. Alpha maps, called mattes in the entertainment industry, can be traced back to early photographic methods that combined images using black and white matte negatives. This process is equivalent to determining which portion of an overlay image is transparent, and which portion is opaque. For more detail on composition, see Blinn [13, 14], Brinkmann [16] and Kelly [17].

4.1 STATIC MATTES

If the image being overlaid has a static position and size, a static fixed matte can be defined. A static matte is simply a fixed alpha map. Computer graphics systems such as OpenGL can directly generate alpha maps as an additional color plane for explicit determination of transparency. Systems can also generate depth maps, wherein each pixel indicates the distance to the lit pixel. Depth maps can be rescaled into alpha maps to allow blending based on depth.

4.2 LUMA-KEYING

Beyond applications of the simple static matte, the goal of compositing systems is the automatic generation of an alpha map. The most basic approach for alpha map generation is luma-keying, wherein the alpha map is linearly determined by the luminance of the source image. This is the basic equation for luma-keying:

$$\alpha(x, y) = a_1 i_1(x, y) \qquad (13.20)$$

In this equation, α is the computed map and i_1 is the source image used to determine the alpha map, typically the foreground image. a_1 is a parameter that controls the intensity of the alpha map. a_1 works similarly to a threshold, though the range allows for smooth transitions between the foreground and background regions. It is a convention in alpha map generation that the map is clamped to the range $[0, 1]$.

It should be noted that most source-dependent compositing systems depend on human operators to adjust parameters such as the a_1 value in this example. Smith and Blinn point out that automatic generation of alpha maps is a hard problem (particularly color-based matte generation), and human determination of parameters for best appearance is nearly always required [38]. These parameters can be easily stored with the source video.

4.3 DIFFERENCE MATTES

Given an image i_1 and an image of just the elements of i_1 to be removed (i_0), a difference matte can be generated. As an example, i_0 might be a fixed image of the background of a scene, while i_1 is the same image with a foreground that needs to be extracted. This is a common problem in augmented imagery where existing content needs to be replaced, but may be subject to occlusion. The generated alpha map should have values of 1.0 for pixels in the occluded regions and 0.0 where the source image matches the background image. In the generated image, the source content in the occluded area (someone standing in front of a sign for example) should be used intact, while the content matching the background is replaced. Section 6.3 describes an example of this process in virtual advertising. The equation for a difference matte is:

$$a(x, y) = abs(a_1(i_1(x, y) - i_0(x, y))) - a_2 \qquad (13.21)$$

The result is, of course, clamped to the range $[0, 1]$. a_1 and a_2 are parameterizations of the matte generation process.

Difference mattes are difficult to use. The alignment between the source image and the background must be very exact. Dynamic differences in lighting can confuse the generation process. Most applications for difference mattes assume automatic registration.

4.4 CONSTANT COLOR MATTING

One of the classic methods for combining two video images into one is the constant color matting method (also known as "blue-screen compositing" or "chroma-keying"). This technique is widely used in the film and television industry to generate special effects, where a foreground layer of video is filmed with a special constant color background (usually chroma-key blue or chroma-key green), and then composited with the new background video. On the upper layer, every pixel in the image within a range of predefined brightness level for one color channel is defined as transparent. Vlahos developed the original methods for constant color keying and has a wide variety of patents in the area [5-12]. Smith and Blinn analyze the Vlahos methods and describe problems with color-based matting and propose some solutions [38].

The basic Vlahos equation is:

$$a(x, y) = 1 - a_1 (B(x, y) - a_2 G(x, y))$$ (13.22)

B and G are the blue and green planes of the source image and a_1 and a_2 are parameterizations of the matte generation. Background pixels are indicated by high blue intensity relative to green. In most Vlahos-derived systems, parameters are assigned to physical knobs that are adjusted to achieve the best result.

A common problem in constant color matting is pixels in the foreground layer that match the color range being assigned to the background. This is sometimes evidenced by transparent regions in newscasters who have worn the wrong tie, for example. The Vlahos equation can be swapped to key on green rather than blue, though neither is a complete solution and color problems often exist. Objects on the upper layer need to be carefully selected so that their color does not conflict with the backing color.

Photographing foreground content in front of a blue screen can also result in blue tinges on the edges of the foreground image or blue shading of the image. This phenomenon is referred to as *blue spill* and is quite common and very annoying. A simple solution for blue spill is to replace the blue component of the foreground pixels with $\min(B(x, y), a_2 G(x, y))$. Often this function is parameterized independently.

4.5 COMPOSITION IN AUGMENTED IMAGERY

Most augmented imagery applications can be roughly classified as either adding foreground or background elements to an image. Applications that add foreground elements are often uncomplicated because the alpha map can be directly generated by the graphics system. Adding or replacing background content, or any content subject to potential occlusion by real objects, requires segmentation of the occluding content in an alpha map. Often an alpha map is built from multiple alpha maps generated by different stages of the augmentation process. Section 6 describes case studies of augmented reality applications, two of which have complex alpha map generation processes.

As an example, a virtual first-down line in a football game is composited with the real camera content using an alpha map generated by combining three computed alpha maps: a map generated by the graphics system, an inclusion map of color regions where the line is valid and an exclusion map of color regions where the line is not valid. Inclusion regions are determined by the colors of the grass field. Exclusion regions are determined by player uniform and other field colors that are close to the inclusion region and must be subtracted from the alpha map to ensure the line does not overwrite a player.

5. VISUAL REALISM

The ultimate goal of augmented imagery is to produce seamless and realistic image compositions, which means the elements in the resulting images are consistent in two aspects: spatial alignment and visual representation. This section discusses techniques for visual realism, which attempt to generate images "indistinguishable from photographs" [39].

Most of the source images or videos used in augmented imagery represent scenes of the real world. The expected final composition should not only be consistent and well integrated, but also match or comply with the surrounding environment shown in the scene. It is hard to achieve visual realism in computer-generated images, due to the richness and complexity of the real environment. There are so many surface textures, subtle color gradations, shadows, reflections and slight irregularities that complicate the process of creating a "real" visual experience. Meanwhile, computer graphics itself has inherent weaknesses in realistic representation. This problem is aggravated by the merging of graphics and real content into integrated images, where they are contrasted with each other in the most brutal comparison environment possible. Techniques have been developed for improving visual realism, including anti-aliased rendering and realistic lighting.

5.1 ANTI-ALIASING

Aliasing is an inherent property of raster displays. Aliasing is evidenced by jagged edges and disappearing small objects or details. According to Crow, aliasing originates from inadequate sampling, when a high frequency signal appears to be replaced by its lower "alias" frequency [40]. Computer-generated images are stored and displayed in a rectangular array of pixels, and each pixel is set to "painted" or not in aliased content, based on whether the graphics element covers one particular point of the pixel area, say, its center. This manner of sampling is inadequate in that it causes prominent aliasing, as shown in Figure 13.9. Therefore, if computer-synthesized images are to achieve visual realism, it's necessary to apply anti-aliasing techniques to attenuate the potentially jagged edges. This is particularly true for augmented imagery applications, where the graphics are combined with imagery that is naturally anti-aliased.

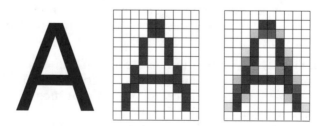

Figure 13.9 Aliased and anti-aliased version of the character "A"

Anti-aliasing is implemented by combined processing of smooth filtering and sampling, in which filtering is used for reducing high frequencies of the original signal and sampling is applied for reconstructing signals in a discrete domain for raster displays. There are commonly two approaches for anti-aliasing: prefiltering and postfiltering. Prefiltering is filtering at object precision before calculating the value of each pixel sample. The color of each pixel is based on the fraction of the pixel area covered by the graphic object, rather than determined by the coverage of an infinitesimal spot in the pixel area. This computation can be very time consuming, as with Bresenham's algorithm [41]. Pitteway and Watkinson [42] developed an algorithm that is more efficient to find the coverage for each pixel, and Catmull's algorithm [43] is another example of a prefiltering method using unweighted area sampling. Postfiltering method implements anti-aliasing by increasing the frequency of the sampling grid and then averaging the results down. Each display pixel is computed as a weighted average of N neighbouring samples. This weighted average works as a lowpass filter to reduce the high frequencies that are subject to aliasing when resampled to a lower resolution. Many different kinds of square masks or window functions can be used in practice, some of which are mentioned in Crow's paper [44]. Postfiltering with a box filter is also called super-sampling, in which a rectangular neighbourhood of pixels is equally weighted. OpenGL supports postfiltering through use of the accumulation buffer.

5.2 IMPLEMENTING REALISTIC LIGHTING EFFECTS

Graphics packages like Direct3D or OpenGL are designed for interactive or real-time application, displaying 3-D objects using constant-intensity shading or intensity-interpolation (Gouraud) shading. These rendering algorithms are simple and fast, but do not duplicate the lighting effects experienced in the real world, such as shadows, transparency and multiple light-source illumination. Realistic lighting algorithms, including ray tracing and radiosity, provide powerful rendering techniques for obtaining global reflection and transmission effects, but are much more complex computationally.

Ray tracing casts a ray (a vector with an associated origin) from the eye through the pixels of the frame buffer into the scene, and collects the lighting effects at the surface-intersection point that is first hit by the ray. Recursive rays are then cast from the intersection point in light, reflection and transmission directions. In Figure 13.10, when a ray hits the first object having reflective and transparent material, these secondary rays in the specular and refractive paths are sent out to collect corresponding illuminations. Meanwhile, a shadow ray pointing to the light source is also sent out, so that any shadow casting can be computed if there are any objects that block the light source from reaching the point. This procedure is repeated until some preset maximum recursion depth or the contribution of additional rays is negligible. The resulting color assigned to the pixel is determined by accumulation of intensity generated from all of these rays. Ray trace rendering reflects the richness of the lighting effects so as to generate realistic images, but the trade-off is considerable computation workload, due to the procedures to find the intersections between a ray and an object. F. S. Hill, Jr. gives detailed description of related contents in his computer graphics text [45].

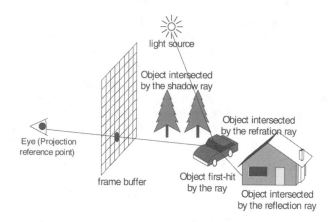

Figure 13.10 A ray tracing model

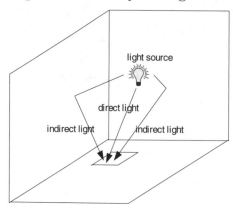

Figure 13.11 Direct and indirect lighting in radiosity

Radiosity is an additional realistic lighting model. Radiosity methods compute shading using the intensity of radiant energy arriving at the surface directly from the light sources or diffuse reflection from other surfaces. This method is based on a detailed analysis of light reflections off surfaces as an energy transfer theory. The rendering algorithm calculates the radiosity of a surface as the sum of the energy inherent in the surface and that received from the emitted energy of other surfaces. Accuracy in the resulting intensity solution requires preprocessing the environment, which subdivides large surfaces into a set of small meshes. A data-driven approach to light calculation is used instead of traditional demand-driven mode. Certain surfaces driving the lighting processing are given initial intensities, and the effects they have on other surfaces are computed in an iterative manner. A prominent attribute of radiosity rendering is viewpoint independence, because the solution only depends on the attributes of source lights and surfaces in the scene. The output images can be highly realistic, featuring soft shadows and color blending effects, which differ from the hard shadows and mirror-like reflections in ray tracing. The disadvantages of radiosity are similar as those of the ray tracing method, in that large computational and storage costs are required. However, if lighting is static in an environment, radiosity can be precomputed and applied in real-time. Radiosity is, in general, the most realistic lighting model currently available.

5.3 VISUAL CONSISTENCY

Using the realistic rendering techniques introduced earlier, augmented imagery systems can generate graphic elements that approximate photography. However, in augmented imagery, the goal is to create realistic integration and mixing of source images with obviously different lighting effects can yield unsatisfying results. Therefore, visual consistency is an important consideration when preparing graphic patches used in augmented imagery. Key to visual consistency is the duplication of conditions from the real environment to the virtual environment. This duplication can be done statically or dynamically. An obvious static approach is to ensure the graphics environment defines lights in locations identical to those in the real world and with the same illumination characteristics. A dynamic approach might include the analysis of shading of patches that will be replaced in order to determine shadows and varying lighting conditions on the patch that can be duplicated on the replacement patch.

Systems may acquire lighting information through analyzing the original scene, as in scene-dependent modeling. But, compared with geometric attributes, complete rebuilding of source lighting is a difficult mission, in that a system must deduce multiple elements from a single input. The problem is due to the complexity of the illumination in the real world, in which the color and intensity of one certain point is the comprehensive results of the lights, object material and environments. A common approach is to adjust lighting interactively when rendering images or graphics elements to be augmented with another source; this depends on the experience and skills of the operator.

6. CASE STUDIES

This section describes three common augmented imagery applications that have enjoyed commercial success. In particular, it is useful to identify how the elements of modeling, registration and composition are accomplished in each application.

6.1 VIRTUAL STUDIOS

Evolving from blue-screen compositing techniques, virtual studios have been widely used in the film and television industries, replacing traditional sets with computer-generated "virtual" sets, which are integrated seamlessly with live video in real time. Virtual studios are a very basic example of augmented imagery, registering a computer graphics model to matching real-world camera images and compositing the result as a foreground/background image. Virtual studios are distinguished from traditional digital composition in that virtual sets are simultaneously updated according to the movements of real cameras, maintaining correct perspective view angles and spatial positions. Virtual set technologies can be utilized in non-real time applications as well (post-production). This section discusses the real-time virtual studio. To achieve proper registration of the virtual set graphics with the real studio camera images, cameras are tracked so as to acquire position, orientation and lens focal length, allowing for adjustment of the corresponding projection transformation required for image rendering. The registration element of the system is also known as "camera matching," the alignment of the virtual camera in computer graphics environment with the real camera. A system diagram for a virtual studio is shown in Figure 13.12. Although the system architecture is simple, there are many technical challenges to practical implementation, due to the combined requirements of real-time performance and realistic image quality.

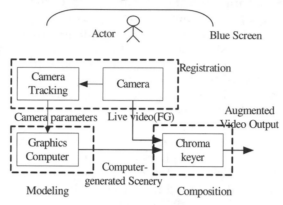

Figure 13.12 Diagram of a virtual studio system

Figure 13.13 An augmented video frame

The first challenge in virtual studios is modeling the generation of virtual environments that will substitute as a real set. A simple approach is to build the graphical scene using 3D modeling software such as Softimage, 3D Studio Max, Maya, etc., in 1:1 scale. But, in practice, many considerations must be taken into account during this task. Graphics hardware cannot render excessively large and complex scenes in real time beyond the hardware's capabilities. Wojdala [46] describes the two main bottlenecks for real-time rendering: the number of polygons and the texture space. It is hard to pose definite limits on these two parameters, because they depend on many factors. But, a number of simplification methods are used to achieve real-time performance, such as replacing flat complex shapes by textures with alpha channels. Another concern is to give the sets realistic looks, such as lighting simulation. Ray tracing or radiosity rendering cannot be used in real-time applications, but can be used to render patches with illuminations and shadows, and map them onto objects in the scene as "faked" lighting effects, adding more credibility to the rendered scene. More discussion on modeling in virtual studio and virtual set design can be found in [47]. Many virtual set applications forego conventional lighting, preferring to precompute lighting and apply it to textures.

Registration in a virtual studio utilizes camera tracking and virtual camera synchronization. The video cameras used in a virtual studio are free to move but must be tracked so that the virtual set can be updated with correct view angles. There are two types of camera tracking methods in common use: pan/tilt/zoom (PTZ) and pattern recognition. In PTZ systems, optical or mechanical encoders are mounted on the tripod or pedestal, which can detect the camera movement relative to a stationary base. If the camera is to be moved around, distance detectors such as infrared or ultrasonic devices are used to measure the camera tripod position. Some commercial tracking systems are now available for this application [48]. Another tracking technology has been developed by Orad Company, which extracts the camera parameters based on the video signal itself. With visible reference points or grid lines in the blue screen, the intrinsic (focus length) and extrinsic (position and orientation) parameters of the camera can be computed using camera calibration algorithms. These two approaches to tracking have advantages or drawbacks, but both of them can provide real-time update of camera spatial information, which can then be used in virtual set rendering.

Composition in virtual studios is generally accomplished using color-based matting or chroma-key. In practice, video compositing is implemented using chroma keyer, a hardware device based on the matting technologies introduced in Section 4. One of the famous keyer manufacturers is Ultimatte. Ultimatte can create seamless composites and preserve fine details such as hair, smoke, mist and shadows.

Virtual studios make it cheaper and easier to produce sets in a computer graphical environment than to build them physically. Continued improvements in rendering quality are narrowing the difference between the real and virtual. Using virtual studios, designers can create scenes previously too difficult, dangerous or expensive. Special effects and animations can be easily imported, adding flexibility to film and television production.

6.2 VIRTUAL FIRST-DOWN LINES

In the late 1990's, television coverage of professional football games began to include an additional marking on the field indicating the first-down line. The line is added electronically in real-time during the broadcast using technology developed by Sportvision, Inc [49]. Adding virtual elements to the presentation of sporting events was not new at the time. The SailTrack system provided virtual presentations of the race course for the America's Cup yacht races in 1992. The presentation could be expanded or panned to angles not possible with aerial cameras and could provide views of an entire large race course. In the tradition of virtual reality systems, SailTrack provided a virtual representation of the course only. The Sportvision system was the first virtual technology to directly augment live broadcast video.

The process of providing a virtual first-down line is complicated by several factors. The video content is sourced by live cameras in real-time. The cameras can pan, tilt and zoom. The first-down line moves throughout the game, and can be anywhere within the bounds of the field. The line must appear to be a physical part of the field. It must not "swim around," but rather appear to be solidly fixed even as the camera moves. The line must also appear on the field, but not on occluding content such as regular field markings, players or equipment.

Figure 13.14 is a simplified block diagram of the virtual first-down system [49]. Cameras provide the video source in the system. Each camera is equipped with pan/tilt/zoom instrumentation (PTZ), sensors that determine the current orientation and zoom characteristics of the camera. These sensors are typically implemented using mechanical attachments to shaft encoders. A shaft encoder measures rotation in discrete steps, typically providing 40,000 steps for each revolution in this application.

Augmentation is always applied to the camera currently generating content on-air. A video switch selects the current on-air camera feed. A computer referred to as the *PTZ concentrator* serves as a switch for PTZ data. The PTZ data is fed to a registration computer, which computes the appropriate transformation for the computer graphics that will allow the virtual elements of the system to register with the real elements. The computed transformation matrix is forwarded to a graphics computer, which renders the virtual first-down line and an alpha map.

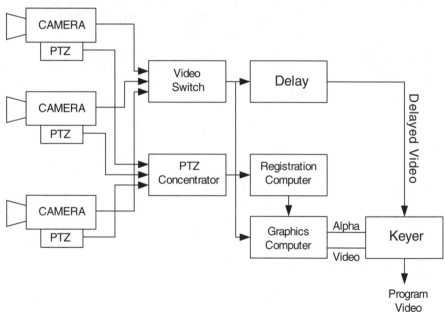

Figure 13.14 Sportvision virtual first-down line system

The graphics computer provides both the registration transformation and the current camera feed. An additional responsibility of the graphics computer is the computation of an alpha map. Alpha maps are described in Section 4. The alpha map is constructed from the location and blending of the first-down line combined with inclusion/exclusion information derived from the current camera image.

The system utilizes an inclusion/exclusion masking mechanism for computation of alpha maps. The initial map is simply the location of the virtual first-down line and is generated using conventional alpha map techniques in computer graphics systems. An inclusion mask is an image that is white where a first-down line can be placed. The line is always placed on the field, so a chrominance value for the grass or turf is input into the system indicating colors that can be replaced with virtual elements. The input image is scanned for chrominance values in the inclusion range, producing the inclusion mask. The inverse of the inclusion map is subtracted from the initial map, with floor values of zero. Exclusion chrominance ranges can also be defined in the system and are used to construct an exclusion mask. Exclusion ranges are useful for specifying colors that are close to the field color, but are not to be overwritten, such as uniform colors. The exclusion mask is subtracted from the generated mask. Finally, the computed alpha map is filtered to soften any sharp edges. Viewers are less sensitive to registration of soft edges than sharp edges.

The computed graphics image and alpha mask are routed to a keyer, a video composition device that blends inputs based on alpha map values at each pixel. The camera tracking and graphics generation process necessarily introduce several frames of latency into the system. A frame delay ensures the camera feed is delayed by an identical latency period, so that no latency is evident in the result.

Modeling in this application is relatively simple. The only rendered content is the first-down line, which is constrained as perpendicular to the field length. The field is 100 yards long and 50 yards wide. The line is rendered as a 3D line in the graphics system with the virtual camera duplicating the position of the real camera. Because occlusion is handled by the inclusion and exclusion maps, an occlusion model is generally not used.

Registration in this application requires the characterization of the camera system using conventional calibration algorithms [30]. In practice, points are located using poles on the field, laser range finders and a laser plane to determine known x,y,z locations and the corresponding u,v image locations. The transformation necessary for registration is then computed as the composition of the results of the PTZ sensors with the camera calibration. The net effect is that the virtual camera mimics the real, physical camera.

6.3 VIRTUAL ADVERTISING

Virtual advertising is a natural application for augmented imagery. The goal of a virtual advertising system is to add commercial content to existing images in such a seamless way that they appear to have been physically placed in the real environment. The ads may be placed where no advertising previously existed. Examples include placing advertising on the surface of a tennis court or the bottom of a swimming pool. New ads can also replace existing ads. A television broadcaster can sell the same stadium location that the local team has already sold or update advertising in older content. Figure 13.15 is an example of virtual advertising. The existing banner in a stadium has been replaced with a new, virtual banner. Virtual advertising is commonly used to replace the stadium-wall ads behind the batter in baseball games.

The development of virtual advertising can be traced to the seminal 1993 patent by Rosser and Leach [50]. They proposed virtual advertising as the placement of a 2D planer image in an existing scene. Later work has added effects such as a ripple to simulate placement in water. A user selects a region of a sample image. The crosshairs in Figure 13.15 illustrate this process. This region is warped to a rectangle of the same pixel dimensions as the substitute image. The warped sample region becomes a template for tracking and alpha map computation.

Figure 13.15 Virtual advertising

Rosser and Leach proposed tracking of the location of the original content using pyramidal pattern matching techniques such as the Burt Pyramid Algorithm. Recent systems have supplemented this vision-based tracking with pan/tilt/zoom camera tracking. The sensor-based tracking allows the location of the patch region to be roughly tracked and improves the performance of the matching algorithm. In either case, a matching algorithm provides for sub-pixel resolution determination of the region boundaries as four region corner points. The matching algorithm must be robust to partial occlusion of the region. A batter may step in front of an ad, particularly corners of the ad, at any time.

Figure 13.16 illustrates the process of computing an alpha map. This is an application of difference mattes as described in Section 4. The template region was captured during the system initialization process. The captured and registered image is the warping of a matching region from a captured frame. The difference between these two images is computed as:

$$\alpha(x, y) = a_1 \left(|t(x, y) - c(x, y)| - a_2 \right) \qquad (13.23)$$

In this equation, t and c are the template and captured image frames. An offset a_2 is subtracted from the absolute value of the difference so as to decrease noise due to small local differences. The result is multiplied by a factor a_1 and then bounded to the range [0,1]. Where the difference is large, the image is likely occluded and the alpha map instructs the keyer to pass the occluding content rather than replacing the advertisement. For all other pixels within the boundaries of the recognized ad, the substitute advertisement replaces the pixel. Some simple refinements have been suggested for this process including low-pass filtering of the difference image to decrease noise and low-pass filtering of the alpha image to smooth edges and better blend the edges into the scene.

Closely related to virtual advertising is *virtual product placement*, the addition or replacement of products in video sequences. Virtual product placement is allowing motion pictures to defer marking of product placement in film until after production and sell supplemental contracts for television appearances. Virtual product placement requires the 3D registration of the replacement product with the original video sequence and typically cannot count on tracking data availability. Vision-based tracking techniques are used after an initial manual placement.

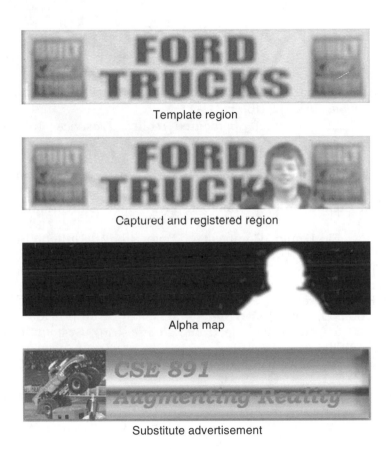

Template region

Captured and registered region

Alpha map

Substitute advertisement

Figure 13.16 Example of computation of an alpha map

7. CONCLUSION

Augmented imagery adds virtual content to existing real video content. This chapter has described basic techniques for augmenting video content and three major example applications. Rapidly advancing processor capabilities are now allowing limited augmentation of content in streams sourced by video database systems. These augmentations could add or change advertisements in order to generate revenue, or they could allow content to be repurposed by changing the locale or background in the sequence. Virtual set technologies allow content to be created with no background at all, leaving the background settings choice as a run-time decision. Augmented imagery merges AR technologies, computer graphics and tracking/identification into systems useful in a wide variety of settings.

ACKNOWLEDGMENT

The authors gratefully acknowledge the National Science Foundation, the Michigan State University intramural research grants program and the MSU manufacturing research consortium, all of whom have contributed to AR programs in our research group.

REFERENCES

[1] Milgram, P. and F. Kishino, A Taxonomy of Mixed Reality Visual Displays. *IEICE Transactions on Information Systems*, 1994. E77-D(12).

[2] Azuma, R.T., Recent Advances in Augmented Reality. *IEEE Computer Science and Applications*, 2001. 21(6): p. 34-47.

[3] Azuma, R.T., A survey of augmented reality. *Presence: Teleoperator and Virtual Environments*, 1977. 6(4): p. 355-385.

[4] Tuceryan, M., et al., Calibration Requirements and Procedures for a Monitor-Based Augmented Reality System. *Transactions on Visualization and Computer Graphics*, 1995. 1(3): p. 255-273.

[5] Vlahos, P., Composite Color Photography, United States patent 3,158,477, Nov. 24, 1964.

[6] Vlahos, P., Electronic Composite Photography, United States patent 3,595,987, Jul. 21, 1971.

[7] Vlahos, P., Electronic Composite Photography with Color Control, United States patent 4,007,487, Feb. 8, 1977.

[8] Vlahos, P., Comprehensive Electronic Compositing System, United States patent 4,100,569, Jul. 11, 1978.

[9] Vlahos, P., Comprehensive Electronic Compositing System, United States patent 4,344,085, Aug. 10, 1982.

[10] Vlahos, P. et al., Encoded Signal Color Image Compositing, United States patent 4,409,611, Oct. 11, 1983.

[11] Vlahos, P. et al., Automated Encoded Signal Color Image Compositing, United States patent 4,589,013, May. 13, 1986.

[12] Vlahos, P. et al., Comprehensive Electronic Compositing System, United States patent 4,625,231, Nov. 25, 1986.

[13] Blinn, J.F., Compositing, Part 2: Practice. *IEEE Computer Graphics and Applications*, 1994. 14(6): p. 78-82.

[14] Blinn, J.F., Compositing, Part 1: Theory. *IEEE Computer Graphics and Applications*, 1994. 14(5): p. 83-87.

[15] Kutulakos, K.N. and J.R. Vallino, Calibration-Free Augmented Reality. *IEEE Transactions on Visualization and Computer Graphics*, 1998. 4(1): p. 1-20.

[16] Brinkmann, R., *The Art and Science of Digital Compositing*. 1994: Morgan Kaufmann.

[17] Kelly, D., *Digital Compositing in Depth: The Only Guide to Post Production for Visual Effects in Film*. 2002: Paraglyph Publishing.

[18] Foley, J.D., A.V. Dam et al., *Computer Graphics: Principles and Practice(2nd Ed. in C)*. 1995: Addison-Wesley Publishing Company.

[19] Foley, J.D., A.V. Dam et al., *Fundamentals of Interactive Computer Graphics*. 1982: Addison-Wesley Publishing Company.

[20] MacIntyre, B., E.M. Coelho, and S.J. Julier. Estimating and Adapting to Registration Errors in Augmented Reality Systems. In *IEEE Virtual Reality Conference 2002*. 2002. Orlando, Florida.

[21] Neumann, U. et al., *Natural feature tracking for augmented reality. IEEE Transactions on Multimedia*, 1999. 1(1): p. 53-64.

[22] Owen, C.B., F. Xiao, and P. Middlin. *What is the best fiducial*. in *The First IEEE International Augmented Reality Toolkit Workshop*. 2002. Darmstadt, Germany.

[23] Rolland, J.P., L.D. Davis, and Y. Baillot, *A survey of tracking technologies for virtual environments*, in *Fundamentals of wearable computers and augmented reality*, W. Barfield and T. Caudell, Editors. 2001, Lawrence Erlbaum, Mahwah: Mahwah, NJ. p. 67-112.

[24] Bajura, M. and U. Neumann, *Dynamic registration correction in video-based augmented reality systems. IEEE Computer Graphics and Applications*, 1995. 15(5): p. 52-60.

[25] Bascle, B. and R. Deriche. *Region tracking through image sequences*. in *Int. Conf. on Computer Vision*. 1995.

[26] Viola, P. and W.M.W. III, *Alignment by maximization of mutual information. International Journal of Computer Vision*, 1997. 24(2): p. 137-154.

[27] Brown, L.G., *A survey of image registration techniques. ACM Computing Surveys*, 1992. 24(4): p. 325-376.

[28] Ferrari, V., T. Tuytelaars, and L.V. Gool. *Markerless augmented reality with a real-time affine region tracker*. in *IEEE and ACM International Symposium on Augmented Reality (ISAR'01)*. 2001. New York, NY.

[29] Shi, J. and C. Tomasi. *Good features to track*. in *IEEE Conference on Computer Vision and Pattern Recognition*. 1994.

[30] Shapiro, G. and G.C. Stockman, *Computer Vision*. 2001, Prentice Hall.

[31] Simon, G. and M.O. Berger. *A two-stage robust statistical method for temporal registration from features of various types*, in *ICCV'98*. 1998.

[32] Berger, M.O., et al., *Mixing synthetic and video images of an outdoor urban environment. Machine Vision and Applications*, 1999. 11(3).

[33] Cho, Y., J. Lee, and U. Neumann. *Multi-ring Color Fiducial Systems and An Intensity-Invariant Detection Method for Scalable Fiducial Tracking Augmented Reality*, in *IEEE International Workshop on Augmented Reality*. 1998.

[34] ARToolKit, http://www.hitl.washington.edu/research/shared_space/.

[35] Jiang, B. and U. Neumann. *Extendible tracking by line auto calibration*. in *IEEE and ACM International Symposium on Augmented Reality*. 2001. New York, NY.

[36] Azuma, R., et al., *Tracking in unprepared environments for augmented reality systems. IEEE Transactions on Computer Graphics*, 1999.

[37] Beauchemin, S.S. and J.L. Barron, *The computation of optical flow. ACM Computing Surveys*, 1995. 27(3).

[38] Smith, A.R. and J.F. Blinn. *Blue Screen Matting*. in *SIGGRAPH '96*. 1996.

[39] Rademacher, P. et al. *Measuring the perception of visual realism in images*. in *12th Eurographics Workshop on Rendering*. 2001. London.

[40] Crow, F.C., *The Aliasing Problem in Computer-Generated Shaded Images. Communications of ACM*, 1977. 20(11): p. 799-905.

[41] Bresenham, J.E., *Algorithm for computer control of a digital plotter. IBM Systems Journal*, 1965. 4(1): p. 25-30.

[42] Pitteway, M.L.V. and D.J. Watkison, *Bresenham's Algorithm with Gray Scale.* *Communications of the ACM*, 1981. 23: p. 625-626.

[43] Catmull, E. *A Hidden-Surface Algorithm With Anti-Aliasing.* in *Proceedings of the 5th annual conference on Computer Graphics and Interactive Techniques.* 1978.

[44] Crow, F.C., *A Comparison of Antialiasing Techniques.* *IEEE Computer Graphics & Applications*, 1981. 1(1): p. 40-48.

[45] Francis. S. Hill, J., *Computer Graphics Using Open GL.* 2nd ed. 2000: Prentice Hall.

[46] Wojdala, A., *Challenges of Virtual Set Technology.* *IEEE Multimedia*, 1998. 5(1): p. 50-57.

[47] Kogler, R.A., *Virtual Set Design.* *IEEE Multimedia*, 1998. 5(1): p. 92-96.

[48] Gibbs, S., C. Arapis, and C. Breiteneder, *Virtual Studio: An Overview.* *IEEE Multimedia*, 1998. 5(1): p. 18-35.

[49] Gloudemans, J.R., et al. Blending a Graphics, United States patent 6,229,550.

[50] Rosser, R.J. and M. Leach. Television displays having selected inserted indicia, United States patent 5,264,933, November 23, 1993.

14

VIDEO INDEXING AND SUMMARIZATION SERVICE FOR MOBILE USERS

Mohamed Ahmed

National Research Council of Canada,
1200 Montreal Road, M50, HPC group,
Ottawa, ON, Canada, K1A 0R6
Mohamed.Ahmed@nrc.ca

Ahmed Karmouch

University of Ottawa,
School of Information Technology & Engineering (SITE),
University of Ottawa, 161 Louis-Pasteur,
Ottawa, ON, Canada, K1N 6N5
Karmouch@site.uottawa.ca

INTRODUCTION

Video information, image processing and computer vision techniques are developing rapidly nowadays because of the availability of acquisition, processing and editing tools which use current hardware and software systems. However, problems still remain in conveying this video data from enterprise video databases or emails to their end users. As shown in Figure 14.1, limiting factors are the resource capabilities in distributed architectures, enterprise policies and the features of the users' terminals. The efficient use of image processing, video indexing and analysis techniques can provide users with solutions or alternatives. In this chapter, we see the video stream as a sequence of correlated images containing in its structure temporal events such as camera editing effects. The chapter will include a new algorithm for achieving video segmentation, indexing and key framing tasks. The algorithm is based on color histograms and uses a binary penetration technique. Although a lot has been done in this area, most work does not adequately consider the optimization of timing performance and processing storage. This is especially the case if the techniques are designed for use in run-time distributed environments. The main contribution is to blend high performance and storage criteria with the need to achieve effective results.

The algorithm exploits the temporal heuristic characteristic of the visual information within a video stream. It takes into consideration the issues of detecting false cuts and missing true cuts due to the movement of the camera, the optical flow of large objects or both. Another issue is the heterogeneous run-time conditions. Thus, we designed a platform-independent XML schema of our video service. We will present our implemented prototype to realize a video web service over the World Wide Web.

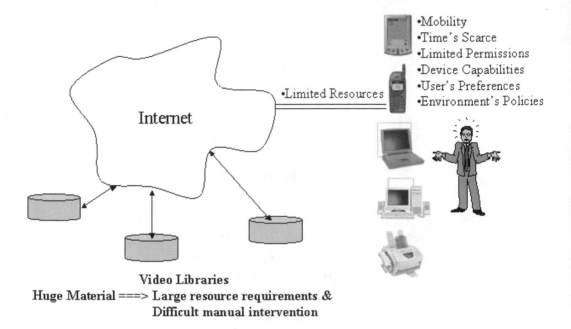

Figure 14.1 Current problems to access video libraries.

As depicted in Figure 14.2, the need for services of content-based video processing and analysis is becoming more useful and sometimes urgent nowadays due to the evolving number, size and complexity of the multimedia information and user's devices. Video data became a very useful way of conveying messages or sharing ideas among end-users. It is used in different aspects of life such as entertainment, digital broadcasting, digital libraries, interactive-TV, video-on-demand, computer-based education, video-conferencing, etc.

This chapter is organized as follows: First, we will provide some related work and literature in section 2. In section 3, we will start by describing the evolving algorithms that we implemented to perform the video segmentation process. We will present our binary penetration approach as a way to achieve this task with higher performance and lower requirements for computational and storage resources. In section 4, we will provide a generalization algorithm for the binary penetration algorithm that handles different camera angles and shot sequences. In section 5, we will describe our video cut detection software system with the use of the binary penetration algorithm. In section 6, we will show how we integrated the video indexing and segmentation algorithm into a web service to expose their functions using XML service interface. Samples of the developed video indexing and key framing service Interface will be presented along with some discussions in section 7. We will draw some conclusions of our work and future steps in section 8.

RELATED WORK

In the past decade, researchers have made considerable contributions in stand-alone image analysis and recognition to allow the partitioning of a video source into segments. In addition, they have developed algorithms to index videos in multimedia databases, allowing users to query and navigate through image or video databases by content. One approach to video segmentation is the use of simple pixel-pair wise comparison [1] to detect a quantitative change between a pair of images. By comparing the corresponding pixels in the two frames, it is easy to determine how many pixels have changed, and the percentage of change. If this percentage exceeds some preset threshold, the algorithm decides that a frame change has been detected. The main value of this method is its simplicity. However, its disadvantages outweigh the advantages. A large processing overhead is required to compare all consecutive frames. Some situations are not accommodated such as when large objects move within one video shot.

Methods using spatial and temporal skips and histogram analysis [2,3,4,5] recognize that, spatially or temporally, many video frames are very similar and that is redundant to analyze them all. These approaches confirmed that the use of color histograms better accommodates the flow motion of the objects and the camera within the same shot. Significant savings in processing time and resources can be achieved during the analysis, either temporally, by sampling every defined number of frames (instead of all frames consecutively), or spatially, by comparing the number of changed pixels to the total frame size. Color histogram distributions of the frames can also be used. The histogram comparison algorithm is less sensitive to object motion than the pixel-pair wise comparison algorithm using various histogram difference measures. Then using a preset threshold, a frame change is determined and so a shot cut operation is detected. Hirzalla [6] has described the detailed design of a key frame detection algorithm using the HVC color space. First, for every two consecutive frames, the system converts the original RBG coloring space of each pixel into the equivalent HVC Histogram coloring representation because the HVC space mirrors human color perception. Instead of intensity distribution, the system uses the hue histogram distribution to perform the comparison. This to some extent reduces the variations in intensity values due to light changes and flashes. Pass et al. [7] proposed a histogram refinement algorithm using color coherence vectors based on local spatial coherence. Wolf [8] provided an algorithm to detect the key frames in MPEG video file format using optical flow motion analysis. Sethi et al. [9] used a similarity metric, which matches both the hue and saturation components of the HSI color space between two images. Pass et al. [10] defined a notion of Joint Histograms. They use many local features in comparison, including color, edge density, texture, gradient magnitude and the rank of the pixels.

After surveying the current approaches for video indexing and segmentation, we found that most need extensive processing power, memory and storage requirements. In addition, most of them are stand-alone applications and filtering processes or algorithms. These algorithms and the current efforts in video analysis and summarization developed mainly to solve specific video processing problems. They were not designed for use in high volume requests and required run-time customization procedures. They do not necessarily perform optimally in terms of processing time besides the terms of accuracy and efficiency. We therefore developed a new algorithm called the binary penetration algorithm for video indexing, segmentation and key framing. We designed the algorithm to be evaluated by processing time as well as by the accuracy of its results. In addition, we believe that our approach could be used as an orthogonal processing mechanism within the other functions for video indexing such as the detection of gradual transitions and the classification of camera motions.

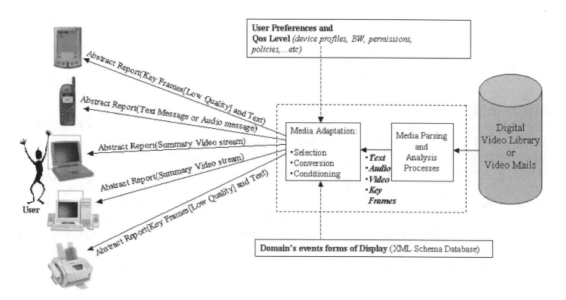

Figure 14.2 The goal: delivering the video message in an adapted form.

Meanwhile as we said, the proliferation and variety of end user devices that can be used by mobile users made it necessary to adapt multimedia services presentation accordingly. There are ongoing research and standardization efforts to enable location-based mobile access of multimedia content. One example is the MPEG-7 working document on mobile requirements and applications [11] from the Moving Picture Experts Group (MPEG). They are currently working on enabling mobile access to specific MPEG-7 information using GPS, context-aware and location-dependent technologies. A user would be able to watch the trailers of current movies wirelessly from the nearest cinema while he is walking or driving his car. Another related ongoing standardization effort is the web services architecture [12], which is a framework to build a distributed computing platform over the web and to enable service requesters to find and remotely invoke published services at service providers. This web service framework uses specific standards such as XML, Web Services Description Language (WSDL) for defining the service and its interfaces and Simple Object Access Protocol (SOAP), which is an XML-based protocol for accessing services, objects and components in a platform-independent manner. Web services architecture also utilizes Universal Description, Discovery and Integration (UDDI) directory and API's for registering, publishing and finding web services similar to the phone's white and yellow pages. It also uses Web Services Inspection Language (WSIL) for locating service providers and for retrieving their description documents (WSDL).

We will provide a mechanism to utilize our algorithms as a video web service over the World Wide Web for multiple video formats. An XML interface for this video service has been designed and then implemented. Thus, users (either mobile or static) or even peer service within the enterprise could utilize the video indexing service seamlessly while hiding the details of the video processing algorithms themselves.

VIDEO CUT DETECTION PROCESS

Usually, within video stream data, camera cut operations represent the most common transition between consecutive shots, much more than other gradual transitions such as dissolving, fading and wiping. Thus, if the timing performance of video cut detection algorithms can be enhanced without reducing their accuracy, the overall performance of video segmentation and indexing procedures will be greatly improved.

For the video cut detection process, our system extracts the consecutive frames needed to detect the camera cut events. The system can define various configurations for the extracted frames. They can be color or gray frames, different image qualities and different image formats (either JPG or BMP). In addition, there is a temporal skip parameter, removing our need to extract and analyze all consecutive frames within the required segment. For cut detection, the system already implements a spatial skip parameter. This improves the performance without sacrificing the accuracy of the detection, capitalizing on the redundant information within the frames. We will describe first the original algorithm, which we call the "six most significant RGB bits with the use of blocks intensity difference." Then, we will present the new recommended algorithm, which we call the "binary penetration" algorithm. This new approach will be seen to further improve the performance of the cut detection procedure by taking into account the temporal heuristics of the video information.

3.1 THE "SIX MOST SIGNIFICANT RGB BITS WITH THE USE OF BLOCKS INTENSITY DIFFERENCE" ALGORITHM

First, the system makes use of the 24-bit RGB color space components of each of the compared pixels (each component has 8-bit representation). However, to speed the performance considerably, the system uses a masking operation to exploit only the two most significant bits (MSBs) of each color, meaning that we actually define only 2^6 (64) ranges of color degrees for the entire RGB color space. The system evaluates the histogram of the corresponding two frames, taking the temporal skip into account. Then, the following formula is used to represent the difference between two frames, f_1 and f_2:

Histogram Difference $(f_1, f_2) = \frac{1}{2} \Sigma \mid H_2(i) - H_1(i) \mid \ / \ \Sigma H_1(i)$

$$\text{for } i=1 \text{ To } N \quad (1)$$

Where,   -$H_1(i)$ is the RGB Histogram distribution for frame M,
                    -$H_2(i)$ is the RGB Histogram distribution for frame (M +
                     Temporal Skip),
                    -$N = 64 =$ the possible 6 bits RGB values

If this histogram difference exceeds some defined threshold, the system decides that the two frames represent an abrupt camera cut. The ½ factor is used for normalization purposes, so that the histogram difference could range from 0% to 100%.

Originally, following the results of the initial approach, the system gave poor results in a few cases. One problem was that the previous algorithm makes use only of the global color distribution information. This meant that the system did not detect video cuts even when there was an actual cut in the compared frames. The two examples shown in Figure 14.3 illustrate this problem. The previous algorithm ignores the locality information of the colors' distribution. This is especially true in the case of two frames of different shots within the same scene, where the background color information is normally similar.

Therefore, we extended the algorithm to suit these circumstances, partitioning each frame into a number of disjoint blocks of equal size. This allowed us to make quicker and better use of the locality information of the histogram distribution. Selecting the number of the blocks used therefore became a system design issue. Our testing results in [13] show that the algorithm behaves consistently with various numbers of blocks.

Nevertheless, using 25 for the number of blocks (5 horizontal blocks * 5 vertical blocks) has shown slightly better overall accuracy than other numbers. It is important to mention that the number of blocks should not be increased very much, for two reasons. First, this will slow the performance of the cut detection process. Second, the algorithm will tend to simulate the pixel pair-wise histogram algorithm. The possibility of detecting false camera cuts is therefore increasing and efficiency is decreasing in the case of motion resulting from quick camera operation or large object optical flow. However, the number of blocks should not be too small as well in order to avoid the overall distribution problem of missing true visual cuts, as shown in Figure 14.3.

Using equation (2), the system evaluates the mean of every corresponding two-block histogram difference between the compared frames to represent the overall two-frame histogram difference (f_1, f_2). Figure 14.4 shows the solution to the previous problem.

Histogram Difference $(f_1, f_2) = \Sigma \Sigma$ Histogram Difference $(b_{1ij}, b_{2ij}) / L^2$
$$\text{for } i, j = 1 \text{ To } L \quad (2)$$

Where,   -Histogram Difference (b_{1ij}, b_{2ij}) is the histogram difference of each two corresponding blocks (sub-images) b_{1ij}, b_{2ij} of the two frames f_1, f_2, evaluated similar to equation (1),
                      -L = number of horizontal blocks = number of vertical blocks

Another problem to be handled is false detection. It occurs mainly because of the use of the temporal skip during processing. If the temporal skip is significantly high, and the change in the same continuous shot is sufficiently quick (because of motion from object flow or camera operations like tilting, panning or zooming), the algorithm will mistakenly recognize the frames as significantly different. We therefore use an additional step. After the first cut detection process, we analyze the changed frames more comprehensively [1], provided that the temporal skip is already greater than one. We do this in order to compensate for the possible velocity of object and camera movements, recognizing that the use of block differences magnifies the effect of these movements. We therefore re-analyze these specific frames but with the temporal skip equal to one, asking the system to extract all the frames in the area being analyzed. As a result, the algorithm became more accurate in maintaining true camera cuts while still rejecting the false cuts obtained from the first phase.

3.2 THE "BINARY PENETRATION" ALGORITHM

Although the use of the "six most significant RGB bits, with the use of blocks intensity difference" algorithm provided us with efficient and reliable results, it lacks the performance required in these kinds of systems, especially if it is to be used within distributed architectures. For this reason, we updated the algorithm with a new approach using the temporal correlation within the visual information of the selected video segment. The result was the design and implementation of the "binary penetration" algorithm.

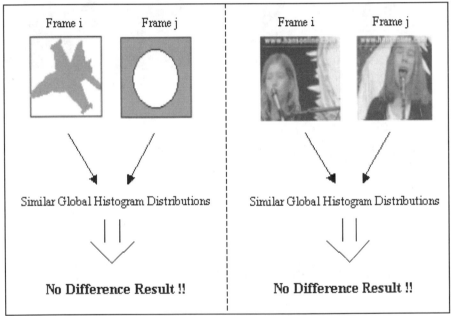

Figure 14.3 Incorrect Frames - Change Decision.

The idea behind this algorithm is to delay the step of analyzing all the consecutive frames of a certain region until the algorithm suggests whether or not they may contain a potential cut. The previous algorithm extracts and analyzes all the frames when the histogram difference exceeds the threshold.

In our new algorithm, we extract and analyze certain frames in a binary penetration manner as shown in Figure 14.5. Initially, the algorithm (Figure 14.6) compares frames 'm' and 'k', which are separated by a temporal skip number of frames. Then, if the difference in the two frames exceeds the threshold, we extract the middle frame and compare it to both ends. If both differences are less than the threshold, we conclude that there is actually no video cut in this region. Processing then continues for the regions that follow. However, if one of the two differences exceeds the threshold, or even if both of them exceed it, we take the greater difference to indicate the possibility of a cut within this half region. We must stress that the temporal skip should be a moderate value that indicates the possibility of finding a maximum of one camera cut operation within each region. We continue the same procedure with the selected half until we find two consecutive frames exceeding the given threshold. This represents a true camera cut operation. The procedure may stop as well when the difference in both halves, at a certain level, is lower than the threshold. That means that there is actually no camera cut operation in this whole region, but merely the movement of a large object or the camera. The use of the temporal skip provides a difference that will pass this cut detection test within upper levels while recognizing false cuts at lower levels.

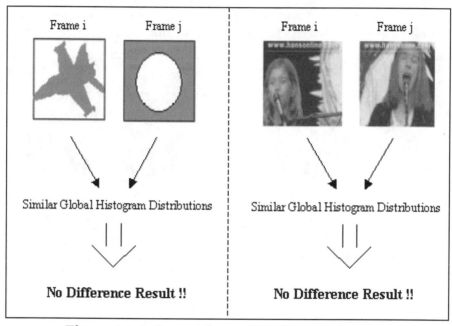

Figure 14.4 Correct frames: unchanged decision.

A GENERALIZED BINARY PENETRATION ALGORITHM

The previous algorithm description handles the most likely cases for detecting video cuts of consecutive shots. However, the use of the temporal skip parameter makes other scenarios possible. Figure 14.7 shows the most important cases of camera and video editing within a video segment. We could define these cases as follows:

Case 1: the entire temporal skip region of processing is included within one continuous shot.
Case 2: one true camera cut exists within the temporal skip period.
Case 3: more than one camera cut exists within the temporal skip period. The editing operation includes the transition between at least three different camera angles.
Case 4: again, more than one camera cut is found within the temporal skip period. However, in this case, the transition returns to the first camera angle

after one or more other camera shots. An example is the interview-like scenario in news or documentary videos.

Case 5: the start and/or end frame of a processing region coincides with the camera cut effect.

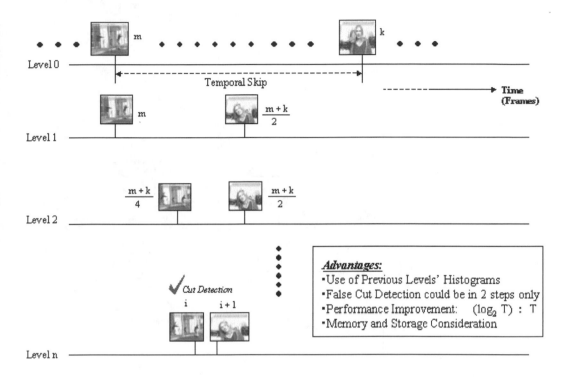

Figure 14.5 A Camera Cut Detection Scenario Using the Binary Penetration.

The described "binary penetration" algorithm will recognize cases 1 and 2 easily. In case 5, the algorithm still works smoothly, using any temporal skip value, because the last frame of the previous processed region is itself the starting frame of the following processing region. So, even if we cannot detect the shot cut effect from the previous region, the algorithm will still detect the effect in the following region. For cases 3 and 4, care is needed; as a result, we made a simple generalization to the algorithm.

This generalization simply just changes one step. Instead of choosing only the higher of the two evaluated differences in any level, which exceeds the defined threshold, we need to continue the penetration in each half of the level. This is done if both differences exceed the threshold, not necessarily only the higher of them. These modifications allow us to discover all the cuts within the temporal skip frames, even if there is more than one such cut.

In case 4, it more likely that camera cuts will be missed, thereby reducing the recall accuracy. We could decrease this effect by reducing the threshold value to pass the first level and discover the actual cuts in the next levels. However, we still need to remember that, in both cases 3 and 4, the problem is avoided if the temporal skip parameter has a moderate value, not large enough to include more than one cut effect.

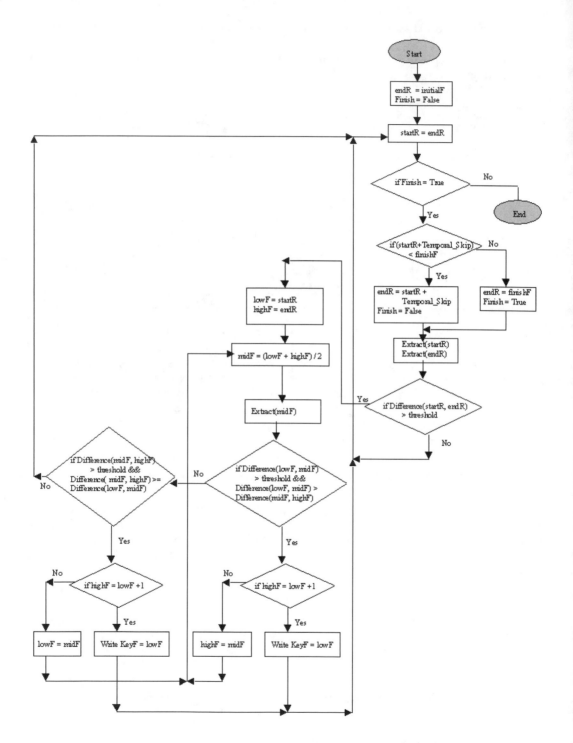

Figure 14.6 The Binary Penetration Algorithm.

VIDEO KEY FRAMING SYSTEM PROCESS

We have developed a prototype system addressing multi-format video visualization and analysis. The system aims to discover the different cut changes within a video file with high accuracy and better performance. Figure 14.8 depicts the architecture of the video segmentation and key framing system. The system has four main components. They are:

1)   The *Media Preparation module* recognizes and verifies the input file format, size, resolution, etc. We also use this module to handle different video formats seamlessly. For instance, it handles compressed video formats through low-level function calls using the software driver for the corresponding format. Therefore, we implemented this function to unify the next processing functions irrespective of the input format. A further responsibility of this module is browsing with different VCR capabilities such as play, stop, pause, resume, rewind, forward, go to certain frame, etc.

2)   The *Media Analysis module* implements two different key framing algorithms, namely the "six most significant RGB bits with the use of blocks intensity difference" both through blind search and through binary penetration, as explained. This allows their relative merits to be studied. This module also updates the key framing index list of the required input segment and provides the updated index list to the *Key Frames Handling module.*

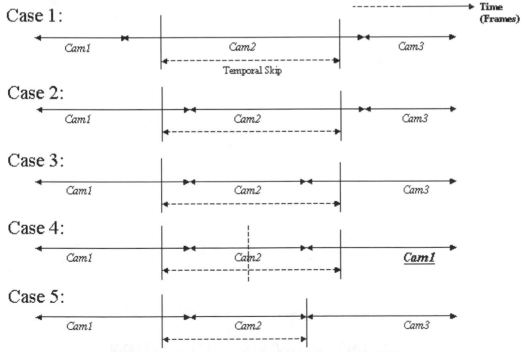

Figure 14.7 Consecutive shots cut possibilities.

3)   The *Frames Management module* uses the media format segment for the analysis. The system uses this module to extract certain frames from the input segment. The *Media Analysis module* requests certain frames within a defined

segment and specifies the region of the requested frames, as well as the temporal skip, frames quality, color/gray condition and frames format. The *Frames Management module* executes the frame extraction function itself. It then replies to the *Frames Analysis module* request with the extracted frame references (i.e., their locations on the local storage). To compensate for the storage requirements, we partition the file into discrete regions for separate analysis.

4)   The *Key Frames Handling module* could use the *Frames Management module* services again to extract the actual frames from the video file by checking their index, received in the key frames list from the *Media Analysis module*. The *Key Frames Handling module* is responsible for providing the user with the key frames result report, including the total processing time and the configuration of the parameters used.

We tried to separate the different modules according to their functions in order to facilitate the future update of any module without affecting the other ones. This is done by a module requesting services from the others through consistent and coordinated interfaces. For example, we could process new media formats without the need to change the current video analysis algorithm. Another example is the possibility of changing the video analysis algorithm itself for the current media formats supported.

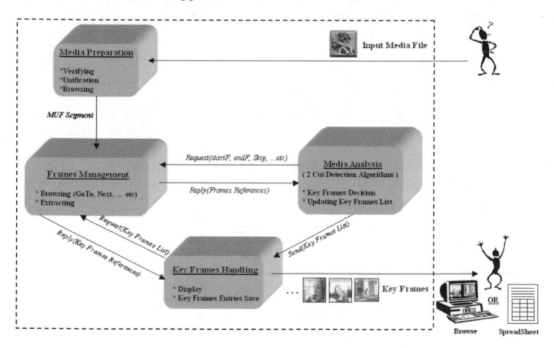

Figure 14.8 Video segmentation process.

MULTIMEDIA SERVICE INTEGRATION

Trying to utilize the developed subsystem of the media key framing algorithm to generate low-level video indexing and summary reports, a batch service program is developed. Figure 14.9 represents the described subsystem of media key framing as a black box and its interface to the corresponding service agent. The process accepts an input request file, which is submitted from a service representative agent. This input mission file contains the parameters of the applied request and it generates a result report that contains the detected key frames along with the confidence percentage of each key frame. Then the service agent utilizes the results and reformats it, using XML as we will show later, before passing this report to the user or storing it for later retrieval requests.

Listing 14.1 depicts an example of an input XML file prepared by the service agent before submitting this mission request to the service process. There are only two mandatory parameters while the other ones are optional. These mandatory parameters are the reference number of the request and the video part itself. The other optional parameters contain some other input parameters about the request. They include as well some parameters that could affect the behaviour of the media key framing algorithm itself such as the threshold for cut detection, the temporal and spatial skips used, start and end of the requested segment of the media file, etc. However, these optional parameters have default values that are included in the algorithm itself, which could be overridden by the input mission request if necessary.

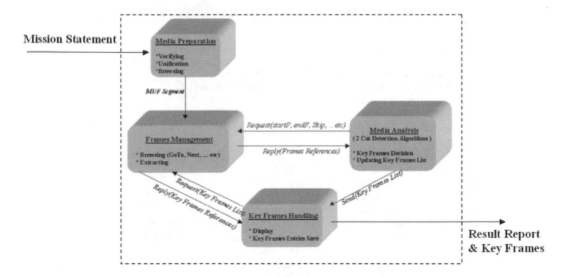

Figure 14.9 The use of the media key framing as a batch process.

Upon a successful completion of the key framing process, the system generates a result report in XML format. Figure 14.10 shows the adopted report structure of our video indexing service. Each report has the four shown main components. First, we will provide the high-level description of these components. Then, a detailed XML result file will be explained later.

The components of the service report are:

-Request Component:
It has the details of the submitted request. It should uniquely describe a certain request. This is guaranteed through a unique serial number that is incremented sequentially each time a new request is submitted to the system.

-User Component:
The details of the user of the system who submitted the request are stored within this part. This could be useful for accounting and billing reasons. It could be useful as well for future access to this report and other submitted queries by this user after authentication.

-Service Component:
This part describes the properties of the selected video service for the mentioned request. The system elects the appropriate service according to accepted level of service for this request after the proposal counter proposal procedure among our system agents [15].

-Result Component:
The results themselves of the video service reside within this component. The detailed description of this component will differ according to the elected service described within the service component. This component should preserve some indication about the processing time or power that have been undergone to process this request for billing and accounting reasons.

Listing 14.1 An example of video service mission request in XML format.

```
<?xml version="1.0" ?>
<mission>
        <ReferenceNumber>624</ReferenceNumber>
        <InputMediaFile>video122.avi</InputMediaFile>
        <UserID>xyz@sol.genie.uottawa.ca</UserID>
        <iDate>17 Aug 1999</iDate>
        <iTime>14:26:47 ET</iTime>
        <StartF>1/4</StartF>
        <EndF>3/4</EndF>
        <TemporalSkip>5</TemporalSkip>
        <SpatialSkip>5</SpatialSkip>
        <Threshold>25</Threshold>
        <Operation>KeyFraming</Operation>
        <KeyframingMethod>6MSB_Blocks</KeyframingMethod>
        <PerformanceMethod>Binary_Penetration</PerformanceMethod>
        <NumberOfBlocks>9</NumberOfBlocks>
        <WorkingDirectory>vb5\for_integration\outputs</WorkingDirectory>
        <FramesFormat>JPG</FramesFormat>
        <ColorGrey>Color</ColorGrey>
        <CompressionRatio>30</CompressionRatio>
        <InitialFormSize>Normal</InitialFormSize>
</mission>
```

Listing 14.2 shows a result record in a well-formed and validated XML format generated by the service agent for the corresponding output results, which are shown in Figure 14.11-b. We chose XML (eXtensible Markup Language) as a result format language because first XML is extensible language while the tags used to markup HTML documents and the structure of HTML documents are predefined. The author of HTML documents can only use tags that are defined in the HTML standard. In contrast, XML allows the author to define his own tags and his own document structure. Second, it is strongly and widely believed that XML will be as important to the future of the Web as HTML has been to the foundation of the Web. XML is regarded as the future for all data transmission and data manipulation over the Web especially among heterogeneous environments and devices. In summary, XML provides an application-independent way to share information. With a DTD (Document Type Definition), independent groups of people or organizations can agree to use a common certain DTD for interchanging and understanding the meaning of the shared data. Thus, any application could use a standard DTD file to verify that data that you receive from the outside world is valid. In addition, you can also use a DTD to verify your own data before delivering it to the user or another organization.

The following XML result report is well formed and validated against the associated DTD structure, presented later, using a third-party software tool called XMLwriter [14].

As mentioned, the XML result file contains four main parts, namely: user, service, request and result as illustrated in Figure 14.10.

The elements used within the *user* part are:
Line 5:   the identity that uniquely describes the user who submitted the request.
Line 6:   title or affiliation of the user.
Line 7:   visiting site or server of the user where he submitted the request.
Line 8:   original home site or server of the user.

The elements used within the *service* part are:
Line 11:   name of the service.
Line 12:   location of the main service.
Line 14-15:   cost attribute name and value of the request.
Line 18-20:   the accepted level of service of the request.

The elements used within the *request* part are:
Line 23:   the request unique reference number.
Line 24:   the name of the media file that has been processed.
Line 25-26:   the start and end points of the media segment within the video.
Line 27:   the name of the operation or function used to process this request.
Line28-29:   the date and time of the request.

The elements used within the *result* part are:
Line 32:   the end status of the request.
Line 33:   the CPU processing time of serving the request.
Line 34:   the initial size of the media file.
Line 35:   total number of video segment frames.
Line 36:   number of processed frames.
Line 38:   count of key frames for this request.
Line 39-42:   key frames numbers or references within the media file.
Line 44:   format of the result key frames.
Line 45:   width and height of the key frames in pixel units.
Line 48:   the protocol that could be used to retrieve and browse this report.
Line 49:   the server address of the XML and frames files.
Line 50:   the path of the XML and frames files within the server given in line 39-42.

Listing 14.2 An example of a result report for the video service in XML format.

```
(1)     <?xml version="1.0" ?>
(2)     <!DOCTYPE report SYSTEM "KF_report.dtd">
[3]     <report>
(4)             <user>
(5)                     <id>xyz@sol.genie.uottawa.ca</id>
(6)                     <title>PhD student</title>
(7)                     <vlocation>Mitel</vlocation>
(8)                     <hlocation>UoOttawa</hlocation>
(9)             </user>
(10)            <service>
(11)                    <name>MediABS</name>
(12)                    <location>UoOttawa</location>
[13]                    <parameter>
[14]                            <key>cost</key>
[15]                            <value>15</value>
(16)                    </parameter>
(17)                    <parameter>
(18)                            <key>quality-of-service</key>
(19)                                    <value>colorKeyFrames</value>
(20)                    </parameter>
(21)            </service>
(22)            <request>
(23)                            <reference>624</reference>
(24)                            <input>video122.avi</input>
(25)                            <startframe>1/4</startframe>
(26)                            <endframe>3/4</endframe>
[27]                    <operation>key framing</operation>
(28)                    <submitdate>17 Aug 1999</submitdate>
(29)                    <submittime>14:26:47 ET</submittime>
(30)            </request>
(31)            <result>
(32)                            <processing>ok</processing>
(33)                    <time>8.19 s</time>
[34]                    <initial_size>2.310 Mb</initial_size>
(35)                    <totalframes>295</totalframes>
(36)                    <processedframes>28</processedframes>
(37)                    <frames>
(38)                                    <number>4</number>
(39)                                    <fvalue>73</fvalue>
(40)                            <fvalue>120</fvalue>
(41)                            <fvalue>185</fvalue>
(42)                            <fvalue>221</fvalue>
(43)                            <total_size>18.428 kb</total_size>
(44)                                    <format>JPG</format>
(45)                                    <dimension>160x120</dimension>
(46)                    </frames>
(47)                    <dlocation>
(48)                                    <protocol>http</protocol>
(49)                                    <host>altair.genie.uottawa.ca</host>
(50)
        <path>\mediabs\outputs\624_result\</path>
(51)                    </dlocation>
(52)                    </result>
[53]            </report>
```

In addition, we defined a Document Type Definition (i.e., DTD) description file associated with all the XML result documents. The main purpose of a DTD description is to define and validate the structure of an XML document when an XML parser within a browser parses the XML document. It defines the XML document structure with a list of approved elements. A DTD can be declared inline within the XML document itself, or as an external reference file. Then, we can use an associated XSLT transformation/ filtering template or a cascade style sheet (CSS) to present the report document on the user's web browser.

In our system, the external DTD file, KF_report.dtd, associated with the provided XML result file is shown in Listing 14.3.

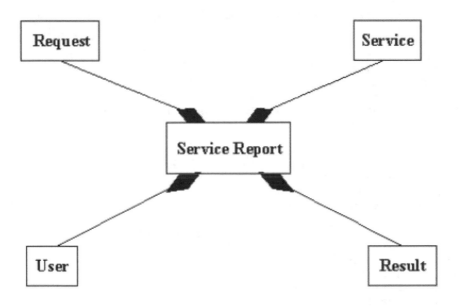

Figure 14.10 Adopted video indexing service structure.

VIDEO INDEXING AND KEY FRAMING SERVICE INTERFACE

Regarding the implementation of our video processing service, we have implemented the video cut detection, media key framing and text-caption detection subsystems as a given service over the distributed environment. This service is developed and run on a PII personal computer that has 266 MHz CPU speed, 128 MB RAM and runs over a Windows NT 4.0 operating system. We integrated those subsystems within our agent-based testbed (please refer to [15] for more details about our agent-based infrastructure, components, security, policies and protocols of negotiations). It is used as a video web service for authorized mobile users of an enterprise over WWW.

Generally, the user can select a portion of the video file he/she is interested in. In addition, we handle different video formats such as AVI, MOV and MPEG file formats seamlessly. The service is structured upon modular components so that we could adopt new video analysis algorithms or handle new video formats with minimum changes. As an example of overall scenario conclusion, Figure 14.11 presents the output user interface to an end-user. The figures present an adapted result that corresponds to different circumstances as follows:

Listing 14.3 The DTD file used to validate the XML service's result report.

```
<?xml version="1.0"?>
<!ELEMENT report (user,service,request,result)>
<!ELEMENT user (id,title,vlocation,hlocation)>
<!ELEMENT id (#PCDATA)>
<!ELEMENT title (#PCDATA)>
<!ELEMENT vlocation (#PCDATA)>
<!ELEMENT hlocation (#PCDATA)>
<!ELEMENT service (name,location,parameter+)>
<!ELEMENT name (#PCDATA)>
<!ELEMENT location (#PCDATA)>
<!ELEMENT parameter (key,value)>
<!ELEMENT key (#PCDATA)>
<!ELEMENT value (#PCDATA)>
<!ELEMENT request
(reference,input,startframe,endframe,operation,submitdate, submittime)>
<!ELEMENT reference (#PCDATA)>
<!ELEMENT input (#PCDATA)>
<!ELEMENT startframe (#PCDATA)>
<!ELEMENT endframe (#PCDATA)>
<!ELEMENT operation (#PCDATA)>
<!ELEMENT submitdate (#PCDATA)>
<!ELEMENT submittime (#PCDATA)>
<!ELEMENT result
(processing,time,initial_size,totalframes,processedframes,frames,dlocati
on)>
<!ELEMENT processing (#PCDATA)>
<!ELEMENT time (#PCDATA)>
<!ELEMENT initial_size (#PCDATA)>
<!ELEMENT totalframes (#PCDATA)>
<!ELEMENT processedframes (#PCDATA)>
<!ELEMENT frames (number,fvalue*,total_size,format,dimension)>
<!ELEMENT number (#PCDATA)>
<!ELEMENT fvalue (#PCDATA)>
<!ELEMENT total_size (#PCDATA)>
<!ELEMENT format (#PCDATA)>
<!ELEMENT dimension (#PCDATA)>
<!ELEMENT dlocation (protocol,host,path)>
<!ELEMENT protocol (#PCDATA)>
<!ELEMENT host (#PCDATA)>
<!ELEMENT path (#PCDATA)>
```

The service could render a normal video streaming result to the user given the availability of communication resources such as the link's bandwidth, CPU power, etc. Also, the system provides the complete stream only if the user has the device capability to browse video contents.

Otherwise, in the case of lack of resources or less device capabilities, the distributed architecture could choose after a negotiation process to furnish automatically only few key frames of the selected video segment to the user through some negotiations between the resources management, available service agents and device handling modules. The system supplies this summary in two defined quality levels.

The first option is to use color and high quality JPG image files. The second possibility is to provide only a gray scale version of these key frames with lower JPG quality. We select the key frame as the 3/4th frame of selected recognized shots. We select this frame to represent the focus of the corresponding shot. In addition, we left 1/4th of the shot because of the possibility of having a gradual transition video editing between two consecutive shots. The result report, delivered to the user, shows the processing time to select and extract these key frames along with the size of the original video segment and the total key frames size for possible corresponding billing procedures.

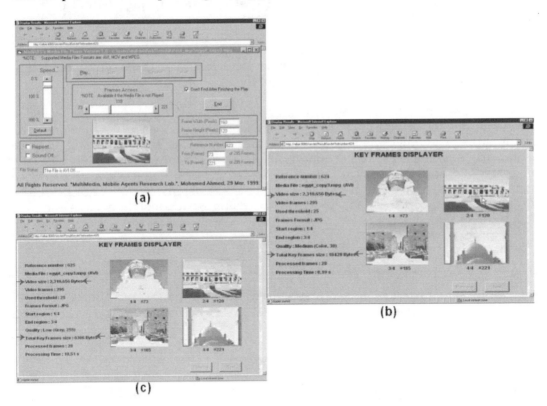

Figure 14.11 Samples of user interface to browse tourist destinations in Egypt.
(a) In the case of highest quality level: Video Streaming;
(b) In the case of medium quality level: Color and High Resolution Key Frames;
(c) In the case of lowest quality level: Gray and Low Resolution Key Frames.

Various examples of this service output interface are shown in Figure 14.11. In this case, a video documentary about different tourist destinations in Egypt is used, that has no text-caption or audio content. Thus in this case, we utilize only the scene change feature using the binary penetration algorithm. Here, similarly for the same given request but within different environments (resources and device capabilities), in Figure 14.11-a, the system provides the whole stream of about 2.3 MB of video content. However, in the case of medium quality level in

Figure 14.11-b, only 18 KB in total size of color and good quality key frames is transferred over the network. In the lowest quality level adopted by the system, Figure 14.11-c represents a total size of just 6.3 KB (of gray scale images and with lower image quality) of key information for the same request.

CONCLUSION

We described the original algorithm initially used in our system for video segmentation, indexing and generating key frames within a video segment. Our approach partitions the selected frames into exclusive blocks in order to resolve the problem of missed cuts. The system uses initially spatial and temporal skip parameters to improve the performance of the analysis.

However, to compensate for the use of the temporal skip, the original algorithm re-analyzes all frames in the region if the first test shows the potential to find a cut. We then explained the updated "binary penetration" algorithm. The algorithm, based on a dichotomy ad hoc image processing, makes use of the temporal correlation heuristics of the visual information within a video stream. The new approach resulted in superior performance over the original algorithm while accuracy was not sacrificed in the analysis.

We presented an extendable system that we developed for the purpose of video browsing and for detecting camera cuts to represent the visual key frames. The architecture, the functions and the uses of each module within the system were explained, together with the scenarios of data and information flow within the modules. We implemented the use of the video key framing system as a service using any java-enabled web browser. In our next steps, we will augment the prototype to support low capacity and processing power mobile devices such as PDA and cellular phones.

ACKNOWLEDGMENT

We would like to thank Roger Impey from NRC Canada and Tom Gray and Ramiro Liscano from Mitel Corporation for their helpful discussions and suggestions.

REFERENCES

[1]   H. Zhang, A. Kankanhalli and S. Smoliar, Automatic Partitioning of Full-Motion Video, Multimedia Systems, Vol. 1, No. 1, 1993, pp. 10-28.
[2]   M. Swain and D. Ballard, Color Indexing, International Journal of Computer Vision, Vol. 7, No. 1, 1991, pp. 11-32.
[3]   J. Boreczky and L. Rowe, A Comparison of Video Shot Boundary Detection Techniques, Journal of Electronic Imaging, Vol. 5, No. 2, 1996, pp. 122-128.
[4]   M. Ahmed and A. Karmouch, Improving Video Processing Performance using Temporal Reasoning, Proceedings of SPIE-Applications of Digital Image Processing XXII, Denver,CO, Vol. 3808, 1999, pp. 645-656.
[5]   M. Ahmed, S. Abu-Hakima and A. Karmouch, Key Frame Extraction and Indexing for Multimedia Databases, Proceedings of Visual Interface 99 Conference, Québec Canada, 1999, pp. 506-511.

[6]   N. Hirzalla, Media Processing and Retrieval Model for Multimedia Documents, Ph.D. Thesis, Ottawa University, January 1997.

[7]   G. Pass and R. Zabih, Histogram Refinement for Content-Based Image Retrieval, Proceedings of IEEE Workshop on Applications of Computer Vision, Sarasota FL USA, 1996, pp. 96-102.

[8]   W. Wolf, Key Frame Selection by Motion Analysis, Proceedings of IEEE International Conference on Acoustics, Speech, and Signal Processing, Atlanta GA USA, Vol. 2, 1996, pp. 1228-1231.

[9]   I. Sethi, I. Coman, B. Day, F. Jiang, D. Li, J. Segovia-Juarez, G. Wei and B. You, Color-WISE: A System for Image Similarity Retrieval Using Color, Proceedings of SPIE, Storage and Retrieval for Image and Video Databases VI, San Jose CA USA, Vol. 3312, 1998, pp. 140-149.

[10]   G. Pass and R. Zabih, Comparing Images Using Joint Histograms, ACM Journal of Multimedia Systems, Vol. 7, No. 3, 1999, pp. 234-240.

[11]   ISO/IEC JTCI/SC29/WG11, Study of MPEG-7 Mobile Requirements & Applications (ver.2), Singapore, 2001.

[12]   S. Graham et al., Building Web Services with Java: Making Sense of XML, SOAP, WSDL and UDDI, Sams Publishing, ISBN 0672321815, 2002.

[13]   M. Ahmed and A. Karmouch, Video Indexing Using a High-Performance and Low-Computation Color-Based Opportunistic Technique, Journal of SPIE Optical Engineering, Vol. 41, No. 2, 2002, pp. 505-517.

[14]   XMLwriter version 1.21, website,
        http://www.xmlwriter.net/index.shtml

[15]   H. Harroud, A. Karmouch, T. Gray and S. Mankovski, An Agent-based Architecture for Inter-Sites Personal Mobility Management System, Proceedings of Mobile Agents for Telecommunication Applications Workshop, MATA'99, Ottawa Canada, 1999, pp. 345-357.

15

VIDEO SHOT DETECTION USING COLOR ANGLOGRAM AND LATENT SEMANTIC INDEXING:
FROM CONTENTS TO SEMANTICS

Rong Zhao
Department of Computer Science
State University of New York
Stony Brook, NY 11794-4400, USA
`rzhao@cs.sunysb.edu`

William I. Grosky
Department of Computer and Information Science
University of Michigan-Dearborn
Dearborn, MI 48128-1491, USA
`wgrosky@umich.edu`

1. INTRODUCTION

The emergence of multimedia technology coupled with the rapidly expanding image and video collections on the World Wide Web have attracted significant research efforts in providing tools for effective retrieval and management of visual information. Video data is available and used in many different application domains such as security, digital library, distance learning, advertising, electronic publishing, broadcasting, interactive TV, video-on-demand entertainment, and so on. As in the old saying, "a picture is worth a thousand words." If each video document is considered a set of still images, and the number of images in such a set might be in hundreds or thousands or even more, it's not so hard to imagine how difficult it could be if we try to find certain information in video documents. The sheer volume of video data available nowadays presents a daunting challenge in front of researchers – How can we organize and make use of all these video documents both effectively and efficiently? How can we represent and locate meaningful information and extract knowledge from video documents? Needless to say, there's an urgent need for tools that can help us index, annotate, browse, and search video documents. Video retrieval is based on the availability of a representation scheme of video contents and how to define such a scheme mostly depends on the indexing mechanism that we apply to the data. Apparently, it is totally impractical to index video documents manually due to the fact that it is too time consuming.

However, state-of-the-art computer science hasn't been mature enough to provide us with a method that is both automatic and able to cope with these problems with the intelligence comparable to that of human beings. Therefore, existing video management techniques for video collections and their users are typically at cross-purposes. While they normally retrieve video documents based on low-level visual features, users usually have a more abstract and conceptual notion of what they are looking for. Using low-level features to correspond to high-level abstractions is one aspect of the *semantic gap* [22] between content-based analysis and retrieval methods and the concept-based users [9, 28, 38, 39, 40].

In this chapter, we attempt to find a solution to negotiating this semantic gap in content-based video retrieval, with a special focus on video shot detection. We will introduce a novel technique for spatial color indexing, *color anglogram*, which is invariant to rotation, scaling, and translation. We will present the results of our study that seeks to transform low-level visual features to a higher level of meaning when we apply this technique to video shot detection. This chapter also concerns another technique to further our exploration in bridging the semantic gap, *latent semantic indexing (LSI)*, which has been used for textual information retrieval for many years. In this environment, LSI is used to determine clusters of co-occurring keywords, sometimes, called *concepts*, so that a query which uses a particular keyword can then retrieve documents perhaps not containing this keyword, but containing other keywords from the same concept cluster. In this chapter, we examine the use of this technique for video shot detection, hoping to uncover the semantic correlation between video frames. Experimental results show that LSI, together with color anglogram, is able to extract the underlying semantic structure of video contents, thus helping to improve the shot detection performance significantly.

The remainder of this chapter is organized as follows. In Section 2, related works on visual feature indexing and their application to video shot detection are briefly reviewed. Section 3 describes the color anglogram technique. Section 4 introduces the theoretical background of latent semantic indexing. Comparison and evaluation of various experimental results are presented in Section 5. Section 6 contains the conclusions, along with proposed future work.

2. RELATED WORKS

The development of video shot detection techniques has a fairly long history already and has become one of the most important research areas in content-based video analysis and retrieval. The detection of boundaries between video shots provides a basis for almost all of the existing video segmentation and abstraction methods [24]. However, it is quite difficult to give a precise definition of a video shot transition since many factors such as camera motions may change the video content significantly. Usually, a shot is defined to be a sequence of frames that was (or appears to be) continuously captured by the same camera [16]. Ideally, a shot can encompass camera motions such as pans, tilts, or zooms, and video editing effects such as fades, dissolves, wipes, and mattes [8, 24]. Basically, video shot transitions can be categorized into two types: *abrupt/sharp shot transitions*, which is also called *cuts*, where a frame from one shot is followed by a frame from a different shot, and *gradual shot transitions*, such as cross dissolves, fade-ins, and fade-outs, and various other editing effects. Methods to cope with these two types of shot transitions have been proposed by many researchers. These methods fall into one of two domains, either uncompressed or compressed, depending on whether it is applied to raw video stream or compressed video data. According to [8], methods working on uncompressed video are, in general, more reliable but require higher storage and computational resources, compared with techniques in the compressed domain. In this chapter, our discussion will focus only on methods in the uncompressed domain. For more details of compressed domain shot detection algorithms and evaluation of their performance, please refer to [8, 16, 17, 24, 26].

Usually, a similarity measure between successive video frames is defined based on various visual features. When one frame and the following frame are sufficiently dissimilar, an abrupt transition (cut) may be determined. Gradual transitions are found by using measures of cumulative differences and more sophisticated thresholding mechanisms.

In [37], *pair-wise pixel comparison*, which is also called *template matching*, was introduced to evaluate the differences in intensity or color values of corresponding pixels in two successive frames. The simplest way is to calculate the absolute sum of pixel differences and compare it against a threshold. The main drawback of this method is that both the feature representation and the similarity comparison are closely related to the pixel position. Therefore, methods based on simple pixel comparison are very sensitive to object and camera movements and noises.

In contrast to pair-wise comparison which is based on global visual features, *block-based* approaches use local characteristics to increase the robustness to object and camera movements. Each frame is divided into a number of blocks that are compared against their counterparts in the successive frame. Typically, the similarity or dissimilarity between two frames can be measured by using a likelihood ratio, as proposed in [25, 37]. A shot transition is identified if the number of changed blocks is above the given threshold. Obviously, this approach provides a better tolerance to slow and small motions between frames.

To further reduce the sensitivity to object and camera movement and thus provide a more robust shot detection technique, histogram comparison was introduced to measure the similarity between successive frames. In fact, histogram-based approaches have been widely used in content-based image analysis and retrieval. Beyond the basic histogram comparison algorithm, several researchers have proposed various approaches to improve it's performance, such as histogram equalization [1], histogram intersection [32], histogram on group of frames [14], and normalized x^2 test [27]. However, experimental results show that approaches which enhance the difference between two frames across a cut may also magnify the difference due to object and camera movements [37]. Due to such a trade-off, for instance, the overall performance of applying x^2 test is not necessarily better than that of the linear histogram comparison, even though it is more time consuming.

Another interesting issue is which color space to use when we consider color-based techniques such as color histogram comparison. As we know, the HSV color space reflects human perception of color patterns. In [18] the performance of several color histogram based methods using different color spaces, including RGB, HSV, YIQ, etc., were evaluated. Experimental results showed that HSV performs quite well with regard to classification accuracy and it is one of those that are the least expensive in terms of computational cost of conversion from the RGB color space. Therefore, the HSV color space will be used in the experimental study in the following sections of this chapter.

The reasoning behind any of these approaches is that two images (or frames) with unchanging background and unchanging (although moving) objects will have minor difference in their histogram [26]. In addition, histograms are invariant to rotation and can minimize the sensitivity to camera movements such as panning and zooming. Besides, they are not sensibly affected by histogram dimensionality [16]. Performance of precision is quite impressive, as is shown in [4, 6]. Finally, histogram comparison doesn't require intensive computation. Although it has these attractive characteristics, theoretically, histogram-based similarity measures may lead to incorrect classifications, since the whole process depends on the distribution and does not take spatial properties into account. Therefore, the overall distribution of features, and thus its histogram, may remain mostly unchanged, even if pixel positions have been changed significantly.

There are many other shot detection approaches, such as clustering-based [13, 21, 31], feature-based [36], and model-driven [1, 5, 23, 35] techniques. For details of these methods please refer to surveys such as [8, 16, 26, 30].

3. COLOR ANGLOGRAM

In this section, we will introduce our *color anglogram* approach, which is a spatial color indexing technique based on Delaunay triangulation. We will first give some Delaunay triangulation-related concepts in computational geometry, and then present the geometric triangulation-based *anglogram* representation for encoding spatial correlation, which is translation, scale, and rotation invariant.

Let $P = \{ p_1, p_2, ..., p_n \}$ be a set of points in the two-dimensional Euclidean plane, namely the *sites*. Partition the plane by labeling each point in the plane to its nearest site. All those points labeled as p_i form the *Voronoi region* $V(p_i)$. $V(p_i)$ consists of all the points x's at least as close to p_i as to any other site:

$$V(p_i) = \{ x: |p_i - x| \le |p_j - x|, \forall j \ne i \}$$

Some points x's do not have a unique nearest site. The set of all points that have more than one nearest site form the *Voronoi diagram* $V(P)$ for the set of sites.

Construct the *dual* graph G for a Voronoi Diagram $V(P)$ as follows: the nodes of G are the sites of $V(P)$, and two nodes are connected by an arc if their corresponding Voronoi polygons share a Voronoi edge. In 1934, Delaunay proved that when the dual graph is drawn with straight lines, it produces a planar triangulation of the Voronoi sites P, so called the *Delaunay triangulation $D(P)$*. Each face of $D(P)$ is a triangle, so called the *Delaunay triangle*.

For example, Figure 15.1(a) shows the Voronoi diagram for a number of 18 sites. Figure 15.1(b) shows the corresponding Delaunay triangulation for the sites shown in Figure 15.1(a), and Figure 15.1(c) shows the Voronoi diagram in Figure 15.1(a) superimposed on the corresponding Delaunay triangulation in Figure 15.1(b). We note that it is not immediately obvious that using straight lines in the dual would avoid crossings in the dual. The dual segment between two sites does not necessarily cross the Voronoi edge shared between their Voronoi regions, as illustrated in Figure 15.1(c).

The proof of Delaunay's theorems and properties is beyond the scope of this chapter, but can be found in [29]. Among various algorithms for constructing the Delaunay triangulation of a set of N points, we note that there are O(*NlogN*) algorithms [11, 15] for solving this problem.

Spatial layout of a set of points can be coded through such an *anglogram* that is computed by discretizing and counting the angles produced by the Delaunay triangulation of a set of unique feature points in the context, given the selection criteria of what the bin size will be, and of which angles will contribute to the final angle histogram. An important property of our proposed anglogram for encoding spatial correlation is its invariance to translation, scale, and rotation. An O(*max(N, #bins)*) algorithm is necessary to compute the anglogram corresponding to the Delaunay triangulation of a set of N points.

The *color anglogram* technique is based on the Delaunay triangulation computed on visual features of images. To construct color anglograms, color features and their spatial relationship are extracted and then coded into the Delaunay triangulation. Each image is decomposed into a number of non-overlapping blocks. Each individual block is abstracted as a unique feature point labeled with its spatial location and feature values. The feature values in our experiment are dominant or average hue and saturation in the corresponding block. Then, all the normalized feature points form a point feature map for the corresponding image. For each set of feature points labeled with a particular feature value, the Delaunay triangulation is constructed and then the feature point histogram is computed by discretizing and counting the number of either the two large angles or the two small angles in the Delaunay triangles. Finally, the image will be indexed by using the concatenated feature point histogram for each feature value. Figure 15.2(a) shows a pyramid image of size 192×128. By dividing the image into 256 blocks, Figure 15.2(b) and Figure 15.2(c) show the image approximation using dominant hue and saturation values to represent each block, respectively. Figure 15.2(d) presents the corresponding point feature map perceptually. Figure 15.2(e) is the Delaunay triangulation of the set of feature points labeled with saturation value 5, and Figure 15.2(f) shows the corresponding anglogram obtained by counting the two largest angles of each triangle. A sample query with *color anglogram* is shown in Figure 15.3.

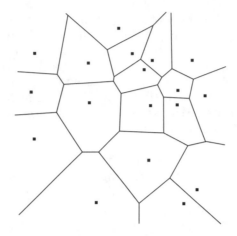

(a) Voronoi Diagram (n=18)

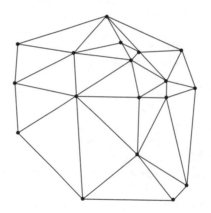

(b) Delaunay Triangulation

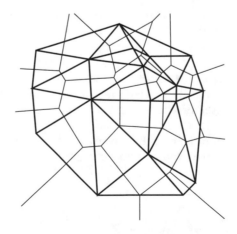

(c) Delaunay Triangulation and Voronoi Diagram

Figure 15.1 A Delaunay Triangulation Example

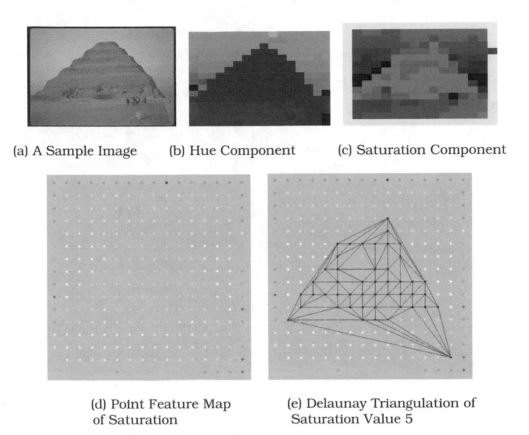

(a) A Sample Image   (b) Hue Component   (c) Saturation Component

(d) Point Feature Map
of Saturation

(e) Delaunay Triangulation of
Saturation Value 5

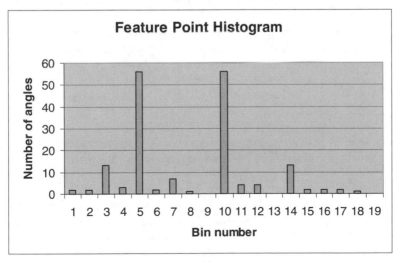

(f) Anglogram of Saturation Value 5

Figure 15.2 A Color Anglogram Example

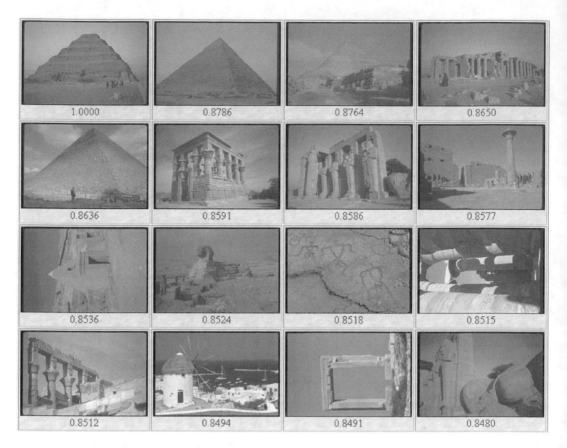

Figure 15.3 A Sample Query Result of Color Anglogram

4. LATENT SEMANTIC INDEXING

Latent Semantic Indexing (LSI) was introduced to overcome a fundamental problem that plagues existing textual retrieval techniques. The problem is that users want to retrieve documents on the basis of conceptual content, while individual keywords provide unreliable evidence about the conceptual meaning of a document. There are usually many ways to express a given concept. Therefore, the literal terms used in a user query may not match those of a relevant document. In addition, most words have multiple meanings and are used in different contexts. Hence, the terms in a user query may literally match the terms in documents that are not of any interest to the user at all.

In information retrieval these two problems are addressed as *synonymy* and *polysemy*. The concept *synonymy* is used to describe the fact that there are many ways to refer to the same object. Users in different contexts, or with different needs, knowledge, or linguistic habits will describe the same concept using different terms. The prevalence of synonyms tends to decrease the *recall* performance of the retrieval. By *polysemy* we refer to the fact that most words have more than one distinct meaning. In different contexts or when used by different people, the same term takes on a varying referential significance. Thus, the use of a term in a query may not necessarily mean that a document containing the same term is relevant at all. Polysemy is one factor underlying poor *precision* performance of the retrieval [7].

Latent semantic indexing tries to overcome the deficiencies of term-matching retrieval. It is assumed that there exists some underlying latent semantic structure in the data that is partially obscured by the randomness of word choice. Statistical techniques are used to estimate this latent semantic structure, and to get rid of the obscuring noise.

The LSI technique makes use of the *Singular Value Decomposition (SVD)*. We take a large matrix of term-document association and construct a semantic space wherein terms and documents that are closely associated are placed near to each other. The singular value decomposition allows the arrangement of the space to reflect the major associative patterns in the data, and ignore the smaller, less important influences. As a result, terms that did not actually appear in a document may still end up close to the document, if that is consistent with the major patterns of association in the data. Position in the transformed space then serves as a new kind of semantic indexing. Retrieval proceeds by using the terms in a query to identify a point in the semantic space, and documents in its neighborhood are returned as relevant results to the query.

Latent semantic indexing is based on the fact that the term-document association can be formulated by using the vector space model, in which each document is represented as a vector, where each vector component reflects the importance of a particular term in representing the semantics of that document. The vectors for all the documents in a database are stored as the columns of a single matrix. Latent semantic indexing is a variant of the vector space model in which a low-rank approximation to the vector space representation of the database is employed. That is, we replace the original matrix by another matrix that is as close as possible to the original matrix but whose column space is only a subspace of the column space of the original matrix. Reducing the rank of the matrix is a means of removing extraneous information or noise from the database it represents. According to [2], latent semantic indexing has achieved average or above average performance in several experiments with the TREC collections.

In the vector space model, a vector is used to represent each item or *document* in a collection. Each component of the vector reflects a particular keyword associated with the given document. The value assigned to that component reflects the importance of the term in representing the semantics of the document.

A database containing a total of d documents described by t terms is represented as a $t \times d$ *term-document matrix A*. The d vectors representing the d documents form the columns of the matrix. Thus, the matrix element a_{ij} is the weighted frequency at which term i occurs in document j. The columns of A are called the *document vectors*, and the rows of A are the *term vectors*. The semantic content of the database is contained in the column space of A, meaning that the document vectors span that content. We can exploit geometric relationships between document vectors to model similarity and differences in content. Meanwhile, we can also compare term vectors geometrically in order to identify similarity and differences in term usage.

A variety of schemes are available for weighting the matrix elements. The element a_{ij} of the term-document matrix A is often assigned such values as $a_{ij} = l_{ij}g_i$. The factor g_i is called the *global weight*, reflecting the overall value of term i as an indexing term for the entire collection. Global weighting schemes range from simple normalization to advanced statistics-based approaches [10]. The factor l_{ij} is a local weight that reflects the importance of term i within document j itself. Local weights range in complexity from simple binary values to functions involving logarithms of term frequencies. The latter functions have a smoothing effect in that high-frequency terms having limited discriminatory value are assigned low weights.

The *Singular Value Decomposition (SVD)* is a dimension reduction technique which gives us reduced-rank approximations to both the column space and row space of the vector space model. SVD also allows us to find a rank-k approximation to a matrix A with minimal change to that matrix for a given value of k [2]. The decomposition is defined as follows,

$$A = U \Sigma V^T$$

where U is the $t \times t$ orthogonal matrix having the left singular vectors of A as its columns, V is the $d \times d$ orthogonal matrix having the right singular vectors of A as its columns, and Σ is the $t \times d$ diagonal matrix having the singular values $\sigma_1 \geq \sigma_2 \geq ... \geq \sigma_r$ of the matrix A in order along its diagonal, where $r \leq min(t, d)$. This decomposition exists for any given matrix A [19].

The rank r_A of the matrix A is equal to the number of nonzero singular values. It follows directly from the orthogonal invariance of the *Frobenius* norm that $|| A ||_F$ is defined in terms of those values,

$$\left\| A \right\|_F = \left\| U\Sigma V^T \right\|_F = \left\| \Sigma V^T \right\|_F = \left\| \Sigma \right\|_F = \sqrt{\sum_{j=1}^{r_A} \sigma_j^2}$$

The first r_A columns of matrix U are a basis for the column space of matrix A, while the first r_A rows of matrix V^T are a basis for the row space of matrix A. To create a rank-k approximation A_k to the matrix A, where $k \leq r_A$, we can set all but the k largest singular values of A to be zero. A classic theorem about the singular value decomposition by Eckart and Young [12] states that the distance between the original matrix A and its rank-k approximation is minimized by the approximation A_k. The theorem further shows how the norm of that distance is related to singular values of matrix A. It is described as

$$\left\| A - A_k \right\|_F = \min_{rank(X) \leq k} \left\| A - X \right\|_F = \sqrt{\sigma_{k+1}^2 + ... + \sigma_{r_A}^2}$$

Here $A_k = U_k \Sigma_k V_k^T$, where U_k is the $t \times k$ matrix whose columns are the first k columns of matrix U, V_k is the $d \times k$ matrix whose columns are the first k columns of matrix V, and Σ_k is the $k \times k$ diagonal matrix whose diagonal elements are the k largest singular values of matrix A. Using the SVD to find the approximation A_k guarantees that the approximation is the best that can be achieved for any given choice of k.

In the vector space model, a user queries the database to find relevant documents, using the vector space representation of those documents. The query is also a set of terms, with or without weights, represented by using a vector just like the documents. The matching process is to find the documents most similar to the query in the use and weighting of terms. In the vector space model, the documents selected are those geometrically closest to the query in the transformed semantic space.

One common measure of similarity is the cosine of the angle between the query and document vectors. If the term-document matrix A has columns $a_j, j = 1, 2, ..., d$, those d cosines are computed according to the following formula

$$\cos\theta_j = \frac{a_j^T q}{\|a_j\|_2 \|q\|_2} = \frac{\sum_{i=1}^{t} a_{ij}q_i}{\sqrt{\sum_{i=1}^{t} a_{ij}^2}\sqrt{\sum_{i=1}^{t} q_i^2}}$$

for $j = 1, 2, ..., d$, where the Euclidean vector norm $||x||_2$ is defined by

$$\|x\|_2 = \sqrt{x^T x} = \sqrt{\sum_{i=1}^{t} x_i^2}$$

for any t-dimensional vector x.

The latent semantic indexing technique has been successfully applied to textual information retrieval, in which it shows distinctive power of finding the latent correlation between terms and documents [2, 3, 7]. This inspired us to apply LSI to content-based image retrieval. In a previous study [38, 39], we made use of the power of LSI to reveal the underlying semantic nature of image contents, and thus to find the correlation between image features and the semantics of the image or its objects. Then in [40] we further extended the power of LSI to the domain of Web document retrieval by applying it to both textual and visual contents of Web documents and the experimental results verified that latent semantic indexing helps improve the retrieval performance by uncovering the semantic structure of Web documents.

5. EXPERIMENTAL RESULTS

In this section we are going to discuss and evaluate the experimental results of applying color anglogram and latent semantic indexing to video shot detection. We will also compare the performance of these methods with that of some existing shot detection techniques. Our data set consisted of 8 video clips of which the total length is 496 seconds. A total of 255 abrupt shot transitions (cuts) and 60 gradual shot transitions were identified in these clips. Almost all of the possible editing effects, such as cuts, fades, wipes, and dissolves, can be found in these clips. These clips contain a variety of categories ranging from outdoor scenes and news story to TV commercials and movie trailers. All these video clips were converted into AVI format using a software decoder/encoder. A sample clip (outdoor scene) is presented in Figures 15.4 and 15.5. Figure 15.4 shows the 4 abrupt transitions and Figure 15.5 shows the gradual transition.

(a) Cut #1, Frame #26 and Frame #27

(b) Cut #2, Frame #64 and Frame #65

(c) Cut #3, Frame #83 and Frame #84

(d) Cut #4, Frame #103 and Frame #104

Figure 15.4 Abrupt Shot Transitions of a Sample Video Clip

Our shot detection evaluation platform is shown in Figure 15.6. This system supports video playback and frame-by-frame browsing and provides a friendly interface. It allows us to compare the performance of various shot detection techniques, such as pair-wise comparison, global and local color histogram, color histogram with x^2, color anglogram, and latent semantic indexing, and a combination of these techniques. Several parameters, such as number of blocks, number of frames, and various thresholds used in feature extraction and similarity comparison, can be adjusted. Statistical results of both abrupt transitions and gradual transitions, together with their locations and strength, are presented in both chart and list format. The results of using various shot detection methods on a sample video clip are shown in Figure 15.7. Our evaluation platform also measures the computational cost of each shot detection process in terms of processing time.

Two thresholds, T_1 and T_2, are involved in the similarity comparison process, which are similar to those in [20, 37]. If the distance between two consecutive frames f_i and f_{i+1} is above T_1, a shot transition is identified between frames f_i and f_{i+1}. If the distance between f_i and f_{i+1} is below T_2, the two frames are considered to be within the same shot. If the distance falls in the range between these two thresholds, further examination will be necessary to determine if the distance results from a gradual transition or not. A certain number of frames, N, can be specified by the user, which allows the system to analyze the accumulative similarity of frames f_{i-N}, f_{i-N+1}, ..., and f_i. This measure will be compared with frame f_{i+1} and thus to determine if a transition exists between f_i and f_{i+1}.

| (a) Frame #36 | (b) Frame #37 | (c) Frame #38 |
| (d) Frame #39 | (e) Frame #40 | (f) Frame #41 |

Figure 15.5 Gradual Shot Transition of a Sample Video Clip

In our experiment, we compared the performance of our color anglogram approach with that of the color histogram method, due to the fact that color histogram provides one of the best performance among existing techniques. Then, we applied the latent semantic indexing technique to both color histogram and color anglogram, and evaluated their shot detection performance. The complete process of visual feature extraction and similarity comparison is outlined as follows.

Each frame is converted into the *HSV* color space. For each pixel of the frame, hue and saturation are extracted and each quantized into a 10-bin histogram. Then, the two histograms h and s are combined into one $h \times s$ histogram with

100 bins, which is the representing feature vector of each frame. This is a vector of 100 elements, $\mathbf{F} = [f_1, f_2, f_3, \ldots f_{100}]^T$.

To apply the latent semantic indexing technique, a feature-frame matrix, $\mathbf{A} = [\mathbf{F_1},\ldots,\mathbf{F_n}]$, where n is the total number of frames of a video clip, is constructed using the feature vector of each frame. Each row corresponds to one of the feature elements and each column is the entire feature vector of the corresponding frame.

Singular Value Decomposition is performed on the feature-frame matrix. The result comprises three matrices, $\mathbf{U}$, Σ, and $\mathbf{V}$, where $\mathbf{A} = \mathbf{U}\Sigma\mathbf{V^T}$. The dimensions of $\mathbf{U}$, Σ, and $\mathbf{V}$ are 100×100, $100 \times n$, and $n \times n$, respectively. For our data set, the total number of frames of each clip, n, is greater than 100. To reduce the dimensionality of the transformed space, we use a rank-k approximation, $\mathbf{A}_k$, of the matrix $\mathbf{A}$, where $k = 12$. This is defined by $\mathbf{A_k} = \mathbf{U_k}\Sigma_k\mathbf{V_k^T}$. The dimension of $\mathbf{A}_k$ is the same as $\mathbf{A}$, 100 by n. The dimensions of $\mathbf{U_k}$, Σ_k, and $\mathbf{V_k}$ are 100×12, 12×12, and $n \times 12$, respectively.

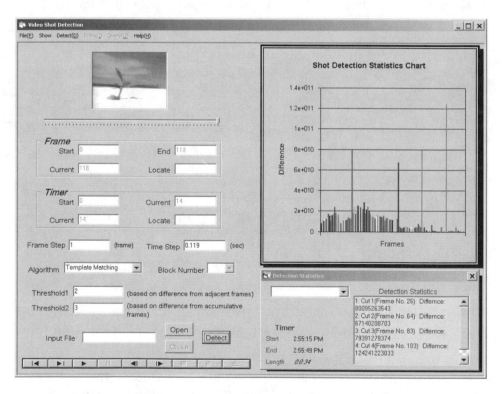

Figure 15.6 Video Shot Detection Evaluation System

The following *normalization* process will assign equal emphasis to each frame of the feature vector. Different components within the vector may be of totally different physical quantities. Therefore, their magnitudes may vary drastically and thus bias the similarity measurement significantly. One component may overshadow the others just because its magnitude is relatively too large. For the feature-frame matrix $\mathbf{A}=[\mathbf{V_1},\mathbf{V_2}, \ldots, \mathbf{V_n}]$, we have $\mathbf{A_{i,j}}$ which is the ith component in vector $\mathbf{V_j}$. Assuming a Gaussian distribution, we can obtain the mean, μ_i, and standard deviation, σ_i, for the ith component of the feature vector across all the frames. Then we normalize the original feature-frame matrix into the range of [-1,1] as follows,

$$A_{i,j} = \frac{A_{i,j} - \mu_i}{\sigma_i}.$$

It can easily be shown that the probability of an entry falling into the range of [-1, 1] is 68%. In practice, we map all the entries into the range of [-1, 1] by forcing the out-of-range values to be either −1 or 1. We then shift the entries into the range of [0, 1] by using the following formula

$$A_{i,j} = \frac{A_{i,j} + 1}{2}.$$

After this normalization process, each component of the feature-frame matrix is a value between 0 and 1, and thus will not bias the importance of any component in the computation of similarity.

One of the common and effective methods for improving full-text retrieval performance is to apply different weights to different components [10]. We apply these techniques to our experiment. The raw frequency in each component of the feature-frame matrix, with or without normalization, can be weighted in a variety of ways. Both global weight and local weight are considered in our approach. A *global weight* indicates the overall importance of that component in the feature vector across all the frames. Therefore, the same global weighting is applied to an entire row of the matrix. A *local weight* is applied to each element indicating the relative importance of the component within its vector. The value for any component $\mathbf{A}_{i,j}$ is thus $L(i, j)G(i)$, where $L(i, j)$ is the local weighting for feature component i in frame j, and $G(i)$ is the global weighting for that component.

Common local weighting techniques include *term frequency*, *binary*, and *log of term frequency*, whereas common global weighting methods include *normal*, *gfidf*, *idf*, and *entropy*. Based on previous research, it has been found that *log (1 + term frequency)* helps to dampen effects of large differences in frequency and thus has the best performance as a local weight, whereas *entropy* is the appropriate method for global weighting [10].

The entropy method is defined by having a component global weight of

$$1 + \sum_j \frac{p_{ij} \log(p_{ij})}{\log(number\_of\_documents)}$$

where

$$p_{ij} = \frac{tf_{ij}}{gf_i}$$

is the probability of that component, tf_{ij} is the raw frequency of component $\mathbf{A}_{i,j}$, and gf_i is the global frequency, i.e., the total number of times that component i occurs in all the frames.

The global weights give less emphasis to those components that occur frequently or in many frames. Theoretically, the entropy method is the most sophisticated weighting scheme, taking the distribution property of feature components over the set of all the frames into account.

We applied color histogram to shot detection and evaluated the results of using it with and without latent semantic indexing. The experimental results are presented in Table 15.1. The measures of *recall* and *precision* are used in evaluating the shot detection performance. Consider an information request *I* and its set *R* of relevant documents. Let $|R|$ be the number of documents in this set. Assume that a given retrieval method generates a document answer set *A* and let $|A|$ be the number of documents in this set. Also, let $|R_a|$ be the number of documents in the intersection of the sets *R* and *A*. Then *recall* is defined as

$$\text{Recall} = |R_a| \ / \ |R|$$

which is the fraction of the relevant documents that has been retrieved, and *precision* is defined as

$$\text{Precision} = |R_a| \ / \ |A|$$

which is the fraction of the retrieved documents that are considered as relevant. It can be noticed that better performance is achieved by integrating color histogram with latent semantic indexing. This validates our beliefs that LSI can help discover the correlation between visual features and higher level concepts, and thus help uncover the semantic correlation between frames within the same shot.

For our experiments with color anglogram, we still use the hue and saturation values in the HSV color space, as what we did in the color histogram experiments. We divide each frame into 64 blocks and compute the average hue value and average saturation value of each block. The average hue values are quantized into 10 bins; so are the average saturation values. Therefore, for each quantized hue (saturation) value, we can apply Delaunay triangulation on the point feature map. We count the two largest angles of each triangle in the triangulation, and categorize them into a number of anglogram bins each of which is 5°. Our vector representation of a frame thus has 720 elements: 36 bins for each of the 10 hue values and 36 bins for each of the 10 saturation values. In this case, for each video clip the dimension of its feature-frame matrix is $720 \times n$, where *n* is the total number of frames. As is discussed above, we reduce the dimensionality of the feature-frame matrix to $k = 12$. Based on the experimental results of our previous studies in [38, 39], we notice that normalization and weighting has a negative impact on the performance of similarity comparison using color anglogram. Therefore, we do not apply normalization and weighting on the elements in the feature-frame matrix.

Table 15.1 Evaluations of Experimental Results

	Abrupt Shot Transition		Gradual Shot Transition	
	Precision	Recall	Precision	Recall
Color Histogram	70.6%	82.7%	62.5%	75.0%
Color Histogram with LSI	74.9%	83.1%	65.8%	80.0%
Color Anglogram	76.5%	88.2%	69.0%	81.7%
Color Anglogram with LSI	82.9%	91.4%	72.6%	88.3%

We compare the shot detection performance of using color anglogram with or without latent semantic indexing, and the results are shown in Table 15.1. From the results we notice that the color anglogram method achieves better performance than color histogram in capturing meaningful visual features. This result is consistent with those of our previous studies in [33, 34, 38, 39, 40]. One also notices that the best performance of both recall and precision is provided by integrating color anglogram with latent semantic indexing. Once again our experiments validated that using latent semantic indexing to uncover the semantic correlations is a promising approach to improve content-based retrieval and classification of image/video documents.

6. CONCLUSIONS

In this chapter, we have presented the results of our work that seeks to negotiate the gap between low-level features and high-level concepts in the domain of video shot detection. We introduce a novel technique for spatial color indexing, color anglogram, which is invariant to rotation, scaling, and translation. This work also concerns a dimension reduction technique, latent semantic indexing (LSI), which has been used for textual information retrieval for many years. In this environment, LSI is used to determine clusters of co-occurring keywords, sometimes, called concepts, so that a query which uses a particular keyword can then retrieve documents perhaps not containing this keyword, but containing other keywords from the same cluster. In this chapter, we examine the use of this technique to uncover the semantic correlation between video frames.

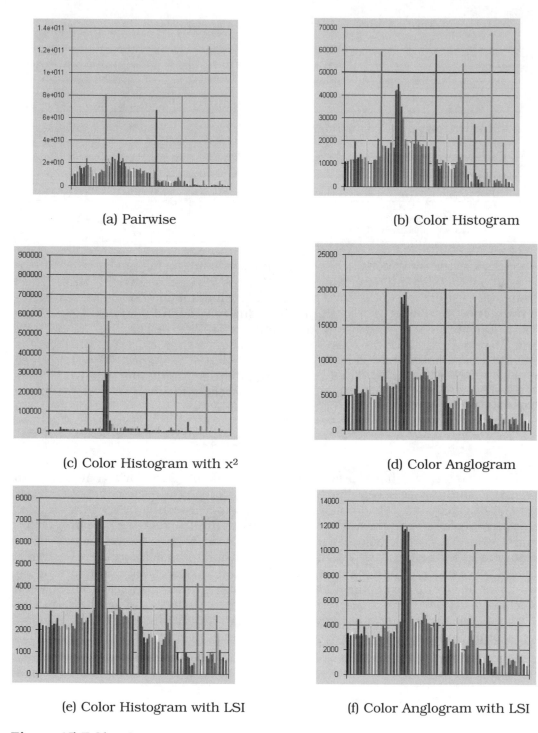

(a) Pairwise

(b) Color Histogram

(c) Color Histogram with x^2

(d) Color Anglogram

(e) Color Histogram with LSI

(f) Color Anglogram with LSI

Figure 15.7 Shot Detection Result of a Sample Video Clip

First of all, experimental results show that latent semantic indexing is able to correlate the semantically similar visual features, either color or spatial color, to construct higher-level concept clusters. Using LSI to discover the underlying semantic structure of video contents is a promising approach to bringing content-based video analysis and retrieval systems to understand the video contents on a more meaningful level. Since the semantic gap is narrowed by using LSI, the retrieval process can better reflect human perception. Secondly, the results proved that color anglogram, our spatial color indexing technique, is more accurate in capturing and emphasizing meaningful features in the video contents than color histogram. Its invariance to rotation, scaling, and translation also provides a better tolerance to object and camera movements, thus helps improve the performance in situations when more complex shot transitions, especially gradual transitions, are involved. Finally, by comparing the experimental results, we validated that the integration of color anglogram and LSI provides a fairly reliable and effective shot detection technique which can help improve the performance of video shot detection. Considering that these results are consistent to those obtained from our previous studies in other application areas [38, 39, 40], we believe that combining their power of bridging the semantic gap can help to bring content-based image/video analysis and retrieval onto a new level.

To further improve the performance of our video shot detection techniques, a more in-depth study of threshold selection is necessary. Even though it is unlikely to totally eliminate user's manual selection of thresholds, how to minimize these interactions plays a crucial role in improving the effectiveness and efficiency in analyzing very large video databases. Besides, it is also interesting to explore the similarity among multiple frames to tackle the problems with complex gradual transitions.

To extend our study on video analysis and retrieval, we propose to use the anglogram technique to represent shape features, and then, to integrate these features into the framework of our shot detection techniques. One of the strengths of latent semantic indexing is that we can easily integrate different features into one feature vector and treat them just as similar components. Hence, ostensibly, we can expand the feature vector by adding more features without any concern. We are also planning to apply various clustering techniques, along with our shot detection methods, to develop a hierarchical classification scheme.

ACKNOWLEDGMENTS

The authors would like to thank Bin Xu for his effort in the implementation of the video shot detection evaluation platform presented in this chapter. The authors are also grateful to Dr. Yi Tao for the figures of Delaunay triangulation and color anglogram examples.

REFERENCES

[1]   P. Aigrain and P. Joly, The Automatic Real-Time Analysis of File Editing and Transition Effects and Its Applications, Computer and Graphics, Volume 18, Number 1, 1994, pp. 93-103.

[2]   M. Berry, Z. Drmac, and E. Jessup, Matrices, Vector Spaces, and Information Retrieval, SIAM Review, Vol. 41, No. 2, 1999, pp. 335-362.

[3]   M. Berry, S. T. Dumais, and G. W. O'Brien, Using Linear Algebra for Intelligent Information Retrieval, SIAM Review, 1995, pp. 573-595.

[4]   J. Boreczky and L. Rowe, Comparison of Video Shot Boundary Detection Techniques, Proceedings of SPIE Conference on Storage and Retrieval for Video Databases IV, San Jose, CA, February 1995.

[5]   J. Boreczky and L. D. Wilcox, A Hidden Markov Model Framework for Video Segmentation Using Audio and Image Features, International Conference on Acoustics, Speech, and Signal Processing, Seattle, WA, 1998, pp. 3741-3744.

[6]   A. Dailianas, R. B. Allen, and P. England, Comparison of Automatic Video Segmentation Algorithms, Proceedings of SPIE Photonics West, Philadelphia, October 1995.

[7]   S. Deerwester, S.T. Dumais, G.W. Furnas, T.K. Landauer, and R. Harshman, Indexing by Latent Semantic Analysis, *Journal of the American Society for Information Science*, Volume 41, Number 6 (1990), pp. 391-407.

[8]   A. Del Bimbo, Visual Information Retrieval, Morgan Kaufmann, San Francisco, CA, 1999.

[9]   N. Dimitrova, H. Zhang, B. Shahraray, I. Sezan, T. Huang, and A. Zakhor, Applications of Video-Content Analysis and Retrieval, *IEEE Multimedia*, July-September 2002.

[10]   S. Dumais, Improving the Retrieval of Information from External Sources, Behavior Research Methods, Instruments, and Computers, Vol. 23, Number 2 (1991), pp. 229-236.

[11]   R. A. Dwyer, A Faster Divide-and-Conquer Algorithm for Constructing Delaunay Triangulations, Algorithmic, Volume 2, Number 2, 1987, pp. 127-151.

[12]   C. Eckart and G. Young, The Approximation of One Matrix by Another of Lower Rank, Psychometrika, 1936, pp. 211-218.

[13]   A. M. Ferman and A. M. Tekalp, Efficient Filtering and Clustering for Temporal Video Segmentation and Visual Summarization, Journal of Visual Communication and Image Representation, Volume 9, Number 4, 1998.

[14]   A. M. Ferman, A. M. Tekalp, and R. Mehrotra, Robust Color Histogram Descriptors for Video Segment Retrieval and Identification, *IEEE Transactions on Image Processing*, Volume 11, Number 5, 2002, pp. 497-507.

[15]   S. Fortune, A Sweepline Algorithm for Voronoi Diagrams, Algorithmic, Volume 2, Number 2, 1987, pp. 153-174.

[16]   U Gargi, R. Kasturi, and S. H. Strayer, Performance Characterization of Video-Shot-Change Detection Methods, *IEEE Transactions on Circuits and Systems for Video Technology*, Volume 10, Number 1, 2000, pp. 1-13.

[17]   U Gargi and R. Kasturi, An Evaluation of Color Histogram Based Methods in Video Indexing, International Workshop on Image Databases and Multimedia Search, Amsterdam, August 1996, pp. 75-82.

[18]   U. Gargi, S. Oswald, D. Kosiba, S. Devadiga, and R. Kasturi, Evaluation of Video Sequence Indexing and Hierarchical Video Indexing, Proceedings of SPIE Conference on Storage and Retrieval in Image and Video Databases, 1995, pp. 1522-1530.

[19]   G. H. Golub and C. Van Loan, Matrix Computation, Johns Hopkins Univ. Press, Baltimore, MD, 1996.

[20]   Y. Gong and X. Liu, Video Shot Segmentation and Classification, International Conference on Pattern Recognition, September 2000.

[21]   B. Gunsel, A. M. Ferman, and A. M. Tekalp, Temporal Video Segmentation Using Unsupervised Clustering and Semantic Object Tracking, Journal of Electronic Imaging, Voulme 7, Number 3, 1998, pp. 592-604.

[22]   V. N. Gudivada and V. V. Raghavan, Content-Based Image Retrieval Systems, *IEEE Computer*, Volume 28, September 1995, pp. 18-22.

[23]   A. Hampapur, R. Jain, and T. E. Weymouth, Production Model Based Digital Video Segmentation, Multimedia Tools and Applications, Volume 1, Number 1, 1995, pp. 9-46.

[24]   A. Hanjalic, Shot-Boundary Detection: Unraveled and Resolved?, *IEEE Transactions on Circuits and Systems for Video Technology*, Volume 12, Number 2, 2002, pp. 90-105.

[25]   R. Kasturi and R. Jain, Dynamic Vision, Computer Vision: Principles, R. Kasturi and R. Jain (Eds.), IEEE Computer Society Press, Washington DC, 1991, pp. 469-480.

[26]   I. Koprinska and S. Carrato, Temporal Video Segmentation: A Survey, Signal Processing: Image Communication, Volume 16, 2001, pp. 477-500.

[27]   A. Nagasaka and Y. Tanaka, Automatic Video Indexing and Full Video Search for Object Appearances, IFIP Transactions on Visual Database Systems II, E. Knuth and L. M. Wegner (Eds.), Elsevier, 1992, pp. 113-127.

[28]   M. R. Naphade and T. S. Huang, Extracting Semantics From Audiovisual Content: The Final Frontier in Multimedia Retrieval, *IEEE Transactions on Neural Networks*, Volume 13, Number 4, 2002, pp. 793-810.

[29]   J. O'Rourke, Computational Geometry in C, Cambridge University Press, Cambridge, England, 1994.

[30]   C. O'Toole, A. Smeaton, N. Murphy, and S. Marlow, Evaluation of Automatic Shot Boundary Detection on a Large Video Test Suite, Challenge of Image Retrieval, Newcastle, England, 1999.

[31]   T. N. Pappas, An Adaptive Clustering Algorithm for Image Segmentation, *IEEE Transactions on Signal Processing*, Volume 40, 1992, pp. 901-914.

[32]   M. J. Swain and D. H. Ballard, Color Indexing, International Journal of Computer Vision, Volume 7, Number 1, 1991, pp. 11-32.

[33]   Y. Tao and W. I. Grosky, Delaunay Triangulation for Image Object Indexing: A Novel Method for Shape Representation, Proceedings of IS&T/SPIE Symposium on Storage and Retrieval for Image and Video Databases VII, San Jose, California, January 23-29, 1999, pp. 631-642.

[34]   Y. Tao and W. I. Grosky, Spatial Color Indexing Using Rotation, Translation, and Scale Invariant Anglograms, Multimedia Tools and Applications, 15, pp. 247-268, 2001.

[35]   H. Yu, G. Bozdagi, and S. Harrington, Feature-Based Hierarchical Video Segmentation, International Conference on Image Processing, Santa Barbara, CA, 1997, pp. 498-501.

[36]   R. Zabih, J. Miller, and K. Mai, A Feature-Based Algorithm or Detecting and Classifying Production Effects, Multimedia Systems, Volume 7, 1999, pp. 119-128.

[37]   H. Zhang, A. Kankanhalli, and S. Smoliar, Automatic Partitioning of Video, Multimedia Systems, Volume 1, Number 1, 1993, pp. 10-28.

[38]   R. Zhao and W. I. Grosky, Bridging the Semantic Gap in Image Retrieval, Distributed Multimedia Databases: Techniques and Applications, T. K. Shih (Ed.), Idea Group Publishing, Hershey, PA, 2001, pp. 14-36.

[39]   R. Zhao and W. I. Grosky, Negotiating the Semantic Gap: From Feature Maps to Semantic Landscapes, Pattern Recognition, Volume 35, Number 3, 2002, pp. 593-600.

[40]   R. Zhao and W. I. Grosky, Narrowing the Semantic Gap – Improved Text-Based Web Document Retrieval Using Visual Features, IEEE Transactions on Multimedia, Volume 4, Number 2, 2002, pp. 189-200.

16

TOOLS AND TECHNOLOGIES FOR PROVIDING INTERACTIVE VIDEO DATABASE APPLICATIONS

Rune Hjelsvold

Gjøvik University College[1]

Gjøvik, Norway

`runehj@hig.no`

Subu Vdaygiri

Siemens Corporate Research, Inc.

Princeton, New Jersey, USA

`subuv@scr.siemens.com`

1. INTRODUCTION

Video databases and applications were traditionally only to be available to users having high-end devices connected to special analog video networks or high-speed local-area networks. This is no longer the case due to technical progress being made in several areas: The steady increase in computational power being built into standard computers has removed the need for high-end computers to be able to enjoy video applications. High network bandwidths are also becoming available in public networks through cable modems and digital subscriber lines [5] enabling customer access to digital video applications through public networks. The most important change that has occurred, however, is the new video compression methods and streaming video players, such as the Microsoft Windows Media player [3] and the RealNetworks player [4], which makes it possible for most Internet users to access digital video services using ordinary modems.

These changes have created new challenges and opportunities for video database application developers. Traditional video database applications were digital

---

[1] The work described in this chapter was performed while the author was affiliated with Siemens Corporate Research, Inc.

library types of applications used to archive, index/catalog, and retrieve elements of a video production [1]. Today, the majority of video database application users are ordinary users on the Internet. These users bring a new set of expectations and requirements to the developers of video database applications. The most important ones are related to (video) quality, availability, and interactivity. This chapter will mostly address the two latter issues and is based on experiences made by the authors while developing two different interactive video database applications: *HotStreams*™ – a system for delivering and managing personalized video content – and *TEMA* (Telephony Enabled Multimedia Applications) – a platform for developing Internet-based multimedia applications for Next Generation Networks (NGN). This chapter is organized as follows:

Section 2 discusses how the World Wide Web and its set of technologies and standards have changed the environment for digital video applications.

Section 3 introduces the interactive video database applications being developed at Siemens Corporate Research.

Section 4 discusses tools and technologies for adding interactivity to video database applications.

The Internet is a heterogeneous network that offers a wide range of bandwidths and that can be accessed by a diverse set of devices having different capabilities, such as screen size, color depth, processing capabilities, etc. *Section 5* discusses how this diversity affects video database applications and describes some tools and technologies to handle differences in network and device capabilities.

Many Internet sites allow their users to create their interest profile. This way, end-users have come to appreciate localized and personalized content and services. *Section 6* shows some technologies that can be used to generate localized and personalized video content.

The Internet has also become a medium for people to share information and experiences through the use of electronic mail, instant messaging, chat rooms, etc. *Section 7* explores tools and technologies that enable video information sharing among end-users.

2. BACKGROUND: VIDEO AND THE WORLD WIDE WEB

World Wide Web browsers and the Internet are becoming the universal tool for information access and management. End-users can read news, manage their bank accounts, search for recipes or instructions, put together an itinerary for their vacation, and perform a large number of information related tasks using the World Wide Web. The web was, in its infancy, providing access to dominantly textual information but as the web has matured, so has multimedia and streaming technologies and the web is now also a tool for end-users to consume digital media. This section discusses some of the characteristics of the World Wide Web that will have the largest impact on video applications and on the databases that support them.

2.1 NETWORK AND DEVICE DIVERSITY ON THE WEB

Diversity is a key characteristic of the World Wide Web. The web serves a wide variety of contents to a diverse group of end-users. There is also a significant variety among the computers and networks that constitute the web. Some users access the web using powerful desktops with broadband network access. Other users access web content using a small-screen device having restricted computational power – such as a PDA – using a low bandwidth wireless connection to the Internet. Video database applications on the web need to address these diversities to ensure that video content can be delivered in ways that are beneficial to the end-users.

2.2 WEB APPLICATIONS – INTERACTIVE AND DYNAMIC

The World Wide Web is based on the hypermedia paradigm and is truly interactive in nature. Web content is usually structured in information nodes that are connected by hyperlinks. End-users have the freedom to select which hyperlinks to follow and which to ignore. Many hyperlinks give direct access to specific parts of the information content when followed and assist the end-user in locating pieces of information of interest quickly. It is likely that end-users will expect video applications provided on the web to offer similar kinds of interactivity.

World Wide Web content is represented in the form of HTML (HyperText Markup Language) [6]. Many documents are statically represented as HTML documents produced by an HTML authoring tool or converted from other document formats. Web content is, however, increasingly being stored in databases where an N-tier web server retrieves the data from the database and generates the corresponding HTML as a response to end-user requests. Templates for these documents can be created as Java Server Pages [7] on web servers supporting Java 2 or as Active Server Pages [8] on Microsoft servers. This simplifies the task of updating information; the author can focus the attention on the content only and does not have to be concerned about layout issues.

2.3 PERSONALIZING AND SHARING INFORMATION AND EXPERIENCES

The diversity in content and user interest makes personalization and localization important on the web. Many sites deal with user profiles – created by the end-users or captured automatically by observing user actions – and present a selected set of content based on the user profile and/or on the end-users' physical location. The web and the Internet are also becoming vehicles for end-users to share experiences through hosted web sites, instant messaging, and technologies that help end-users create content as well as consume it. Video and multimedia will likely play an important role in enabling shared experiences [2].

2.4 MOBILITY AND UBIQUITY

Professor Kleinrock observes that people are getting increasingly nomadic [9]. Cellular phones, PDAs, and other communication gadgets offer ubiquitous connectivity. This means, on one side, that the World Wide Web can be accessed from almost anywhere by a multitude of different devices. Even more, as Professor Kleinrock noted, this also means that an end-user may initiate a task at one location but would like to be able to continue working on the task as the location is changing.

3. SAMPLE INTERACTIVE VIDEO DATABASE APPLICATIONS

This section gives a brief description of two projects at Siemens Corporate Research aimed at developing interactive video database applications. The technical issues encountered during these projects will be discussed in more details later in this chapter.

3.1 HotStreams™ – A SYSTEM FOR PERSONALIZED VIDEO DELIVERY

HotStreams™ is a prototype system for providing personalized hypervideo information services over the World Wide Web. It can be accessed by an end-user using a regular web browser on a desktop or on a PDA with proper plug-ins for receiving streaming content from a streaming media server. Figure 16.1 shows the prototype user interface on a desktop.

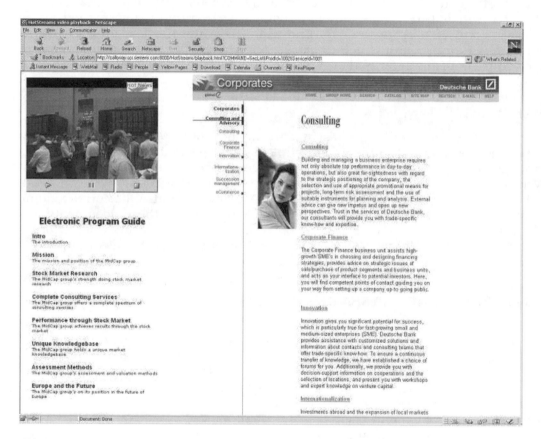

Figure 16.1 A Sample Interactive Video Application Interface (Content Courtesy of the Deutsche Bank)

The user interface consists of three main parts (frames in the HTML terminology). The upper left part contains an embedded video player that can be used to control the playback of the personalized video content. HotStreams™ supports the use of the RealPlayer or Windows Media Player on desktops and the Windows Media Player or the PacketVideo Player on a PDA. The lower left part shows the table of contents for the personalized video. The part to the right is used to display web content that is linked to hotspots in the video.

3.1.1 Interactive Hypervideo

HotStreams™ deliver interactive videos that may contain hyperlinks. These allow the user to retrieve additional information to be displayed in the right part of the interface. Hyperlinks may be visualized by hotspots, which are images overlaying the video window (when the content is streamed to a RealPlayer) or in a separate banner underneath the video (when the Windows Media Player is used for playback). Figure 16.1 shows an image overlaying the video that – when activated by the user – loaded the *Consulting* web page shown in the leftmost frame of the interface.

The user may choose to let the video play back from beginning to end as a regular video. The system also allows the user to control the playback by selecting any of the items from the table of contents shown in the lower right part of the interface. The application will instruct the video server to start streaming the video from the beginning of the selected item in response to this end-user request.

3.1.2 Personalized Content

The HotStreams™ system allows the end-user to customize the contents according to his or her interests and preferences. A user selects one of several predefined versions of the content when connecting to a HotStreams™ server. The financial news prototype shown in Figure 16.1, for instance, defines two such versions; one, called the *headline version*, is intended for casual users and one, called the *extended version*, is intended for investors or customers with a need and an interest to study the financial market in more detail.

3.2 TEMA – A PLATFORM FOR INTEGRATING VIDEO AND TELEPHONY SERVICES

Long distance technical and business communications conducted using traditional media such as voice, fax, and mail are often awkward and time consuming. State-of-the-art, software-only, telephone switches (softswitches) [20] such as the Surpass system by Siemens [19] offer open APIs that facilitate the integration of Internet technologies with traditional telephony systems. The TEMA initiative at Siemens Corporate Research was aimed at improving technical and business communications by integrating interactive multimedia and video with telephony services.

Customer support centers play a significant role in business communications. A customer support agent may often find it difficult to understand an end-user's technical problem or to explain a complex technical solution using only a telephone as the means for communicating. The Surpass-Enabled Multimedia Call center Application (SEMCA) is an application under the TEMA initiative that demonstrates one of many ways in which a business can leverage interactive multimedia in softswitch infrastructures to share and communicate thoughts and ideas through a seamless combination of interactive video, multimedia, and telephony. Figure 16.2 illustrates how TEMA facilitates the transform from single media call centers to multimedia call centers (SEMCAs).

3.2.1 Integrating Interactive Video in Customer Support Centers

A typical SEMCA scenario is described in the following: A customer calls the support center to seek help to fix a problem. She will be connected with a call center agent who assists her in identifying the product and the specifics of the problem she is experiencing. The call center agent is using SEMCA to retrieve multimedia objects from an interactive video database to create a multimedia presentation explaining possible solutions to the customer's problem. Once the multimedia presentation is assembled, SEMCA sends a regular text e-mail message to the end-user containing the link (URL) to the presentation.

The customer, upon receiving the e-mail, can click on the multimedia message URL link, which recreates the multimedia presentation adapted for the customer's device, network, and display capabilities. SEMCA gives the customer a more user-friendly help than searching and receiving multimedia product manuals that can be very large and which in most cases will contain lots of information that is not relevant to the customer's specific problem.

As the customer might wish to be mobile in order to fix her problem, SEMCA allows the customer to transfer the multimedia session to a different device, such as a Pocket PC, for instance.

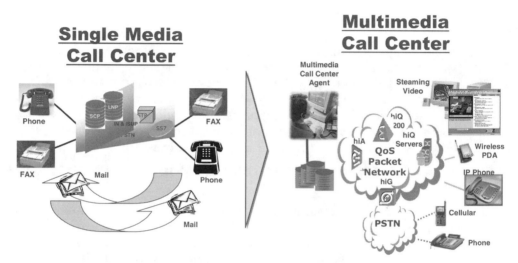

Figure 16.2 TEMA used to Transform Single Media Call Centers

3.2.2 Linking Video Content to Telephone URLs

SEMCA not only customizes the multimedia presentation for different devices but also allows the customer to initiate voice calls on IP based wireless networks using the Surpass platform. The support center agent is able to insert telephony links into the e-mail and the multimedia presentation pertinent to the content being assembled. The customer may establish a telephony connection to a call center just by clicking on one of these telephony links.

4. INTERACTIVITY IN VIDEO DATABASE APPLICATIONS

The two systems developed at SCR are both interactive video database applications that use a video-driven composition technique, similar to the one described by Auffret et al. [17]. The content model underlying this video-driven approach is illustrated in Figure 16.3.

4.1 THE VIDEO-DRIVEN COMPOSITION TECHNIQUE

The content model being used by the video-driven technique is comprised of the following elements (see also [18]):

❏ **Video sequence element (clip).** The video consists of a sequence of video clips called sequence elements. Each such sequence element is a contiguous sequence of frames from a video recording. A sequence element may contain multiple shots if the corresponding video recording contains multiple shots in sequence. Each sequence element keeps the reference to the video recording in which it is defined. Hence the model supports the concatenation of clips from potentially different video recordings. It should be noted that this simple form for composition is not considered as a replacement for a video authoring tool. The objective is for the model to define an assembly of pre-edited video sequences that allows for easy repurposing of such video content.

❑ **Video section.** Contiguous sequence elements are grouped into sections. A section typically represents a meaningful unit of content – e.g., a news story. Sections are used as targets for direct access, i.e., the user may use the table of contents (see below) to start the playback at the very beginning of any given section.

❑ **Video hyperlink.** There may be hotspots within the video. Hotspots have a spatial *and* a temporal range. Two hyperlinks are shown in the example model in Figure 16.3. The temporal range is illustrated by the grey shading. The broken arrows indicate that the target object for the hyperlink will be loaded in one of the application's content frames if the end-user activates the corresponding hotspot.

❑ **Synchronization point.** Synchronization points are temporal locations within the video that defines some transition or change of content in one of the application's content frames. Synchronization points are shown as small dots on the figure. Synchronizations points are useful, for instance, in cases where the interactive video was recorded during a lecture or a business presentation and where the synchronizations points are used to synchronize the "playback" of the slides being used during the presentations with the video playback.

❑ **Table of contents.** The content of each interactive video is listed in a table of contents. Each entry in the table of content is composed of the following fields:

- *Section title.* The title of the section, which will appear in the table of content.

- *Section abstract.* A textual description of the content of the section, which will also be shown in the table of content.

- *Section URL.* A URL that identifies the beginning of the section within the video production. The arrows in Figure 16.3 pointing from the entries in the table of content to the interactive video section indicate the presence of section URLs.

4.2 INTERACTIVE VIDEO TOOLS

The first part of this section discussed the data stored in the video meta-database for generating interactive video applications. Tools are needed for creating the presentations from the data stored in the database and for creating and managing these data. The remaining part of this section discusses such tools.

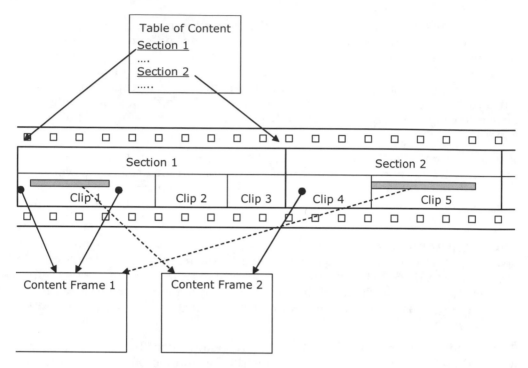

Figure 16.3 The Interactive Video Content Model

4.2.1 Presentation Generation

Generating the presentation is a process consisting of three logical steps, as shown in Figure 16.4. The first step consists of selecting what parts of the interactive video will be delivered. The personalization process used in HotStreams™, for instance, selects the parts of the interactive video that fits the end-user's interest profile. The ultimate goal of the process is to generate scripts that will drive the interactive video presentation. Unfortunately, there exist several script languages that might be used to drive an interactive video production. Hence, it is beneficial to generate an internal, language independent representation of the production first. Support for multiple script languages can then easily be incorporated by implementing a script generator for each of the languages that convert the video from the internal representation to the specific script language.

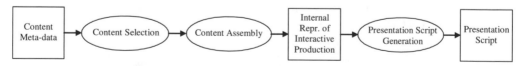

Figure 16.4 The Presentation Generation Process

4.2.2 Content Assembly Tools

HotStreams™ offers a web-based tool for managing the contents of the meta-database [18]. The tool is implemented as a Java applet that runs inside a regular web browser. The tool consists of a number of panels where each panel implements a group of closely related management functions. The *video panel* shown in Figure 16.5, for instance, provides the means needed to create and manage interactive videos, their composition, and the text fields that constitute the table of contents. Similarly, the *hyperlink panel* provides the means needed to create hyperlinks and define their appearance and destination. These panels also provide means to generate meta-data that will be used to create personalized content.

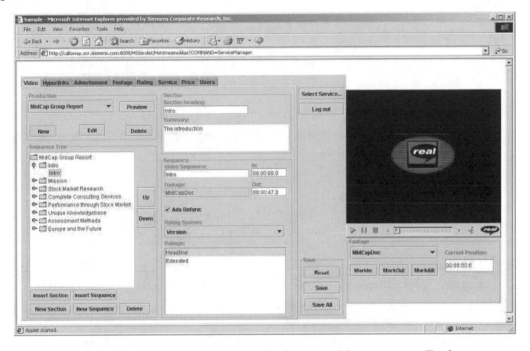

Figure 16.5 The HotStreams™ Content Management Tool

5. NETWORK AND DEVICE DIFFERENCES AND MOBILITY

HotStreams™ was originally designed for a desktop environment and was therefore designed to utilize high-resolution displays. The system was also targeted at customers of broadband networks expecting high quality content. The video content was encoded at high data rates (typically 2-5 mbps). The wireless computing and network technology was changing significantly as the project developed: Powerful PDAs came to market, wireless LANs and other "high" bandwidth wireless networks were installed extensively, and streaming media players became available on PDAs. As a result, wireless devices had become potential end-user devices for the system. The system, however, had to be modified to be able to manage the differences in network and device capabilities.

Any system serving a variety of devices connected to heterogeneous networks needs to find a way to deliver the content in a comprehendible form that will satisfy the end-user. This challenge can be approached in different ways – as illustrated in Figure 16.6. (The term *document* is used in a generic form in the figure and might, for instance, represent a complete video or multimedia production.)

The simplest solution is to prepare different versions of the same document during the authoring phase where each version is targeting a given network and device configuration. The process of creating the right form of the content is then reduced to selecting the pre-authored version that best fits the given network and device configuration. This solution, however, puts an additional burden on the process and does give a recipe for how to handle combinations of network and device configurations that do not exist currently but that might be used in the future.

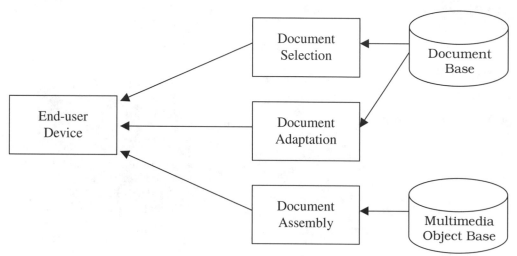

Figure 16.6 Alternative Ways to Deliver Content in the Right Form

The second alternative, *document adaptation*, involves converting and possibly transcoding the contents of the original document to adapt it to the end-users current environment. Ma et al. [10] propose a framework for adaptive content delivery designed to handle different media types and a variety of user preferences and device capabilities. The framework combines several means, such as information abstraction, modality transform, data transcoding, etc., for content adaptation. Mohan et al. [11] propose a multi-modal, multi-resolution scheme called the InfoPyramid that uses transcoding to accomplish multimedia content adaptation. The adaptation may happen at the server or in adaptation proxies in the network, as Fox et al. [12] argues.

One advantage of the adaptation approach is that it stores only one version of the content and thereby keeps the storage requirements to a minimum. The other main benefit is that this approach provides an easy path to incorporate support for new clients. A major disadvantage is the computational resources required to implement a large-scale system that can convert many parallel streams on the fly. HotStreams™ and TEMA are taking the third approach shown in the figure: The document assembly alternative where a unique document is put together on the fly in response to each end-user request. The server will select the best-fit version of each multimedia object to be used in the document at the time of the assembly. The rest of this section discusses how network and device differences and mobility can be addressed in this case.

5.1 MANAGING DISPLAY VARIATIONS

Displays come in a great variety ranging from small, black-and-white, text-only mobile phone displays to true-color, high-resolution computer displays. The screen size is one of the most challenging factors when adapting media-rich presentations. A large-screen interface usually cannot be mapped onto a small-screen device in a satisfactory way without major changes to the layout and without resizing the rendered objects. Most of the existing research has been working on solutions for single-frame interfaces. HotStreams™ and TEMA offer multi-frame interfaces as seen on Figure 16.1. Multi-frame interfaces such as these cannot be nicely displayed on the small screens that PDAs use. The interface needs to be divided over multiple small-screen windows that are displayed one at the time. Figure 16.7 illustrates how a document assembly process will use the meta-data (the *model*) to generate different application interfaces (*views*).

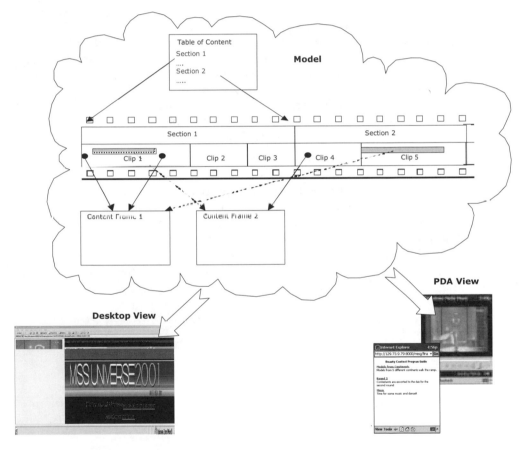

Figure 16.7 Device-dependent View of Interactive Content

The figure shows the model of a video consisting of sections that further consist of clips. Some clips contain hyperlinks that when activated can retrieve additional information related to the given clip. The desktop view shows the original HotStreams™ interface that defines three frames: the video frame located in the upper left corner, the electronic table of content located in the lower left corner, and the more information frame located on the right-hand side. The video frame contains the RealPlayer playing back an MPEG-1 version of the video.

The PDA view shown on the right-hand side of the figure illustrates how the interface has been changed from a single-page, multi-frame view to a multi-page single-frame view. The screen size of the PDA is too small for the simultaneous display of table of content and the video. The figure clearly illustrates that adaptation of multi-frame interfaces for interactive content is not only a page-by-page conversion. In some cases the workflow may need to be modified during the adaptation as well. The desktop workflow, for instance, generates and presents the table of content and the video content as one single task in the workflow. In the PDA workflow the user will go back and forth between the generation and display of the table of contents and the video playback.

5.2 MANAGING MEDIA ENCODING AND NETWORK VARIATIONS

High-quality video streams such as MPEG-1 and MPEG-2 require network and data transport bandwidths in the order of megabits per seconds and require specialized hardware or processing resources that are normally not available on PDAs and mobile devices. Some video file formats, such as the RealNetworks SureStream format [13], for instance, can store data for multiple bandwidths in one file. A video may be encoded in the SureStream format at 56 kbps, 64 kbps, 128 kbps, 256 kbps, and 750 kbps, for example, to accommodate different devices and network bandwidths. Such content could not, however, be delivered to a PDA or a desktop not capable of playing back SureStream encoded videos.

The systems developed at SCR followed a document assembly approach where a single video would be represented by a set of files – each file targeting a given network and device configuration. Figure 16.8 illustrates how the relationships between elements within the interactive video and the corresponding media files are represented in the system. The figure also shows that the systems support multiple streaming servers.

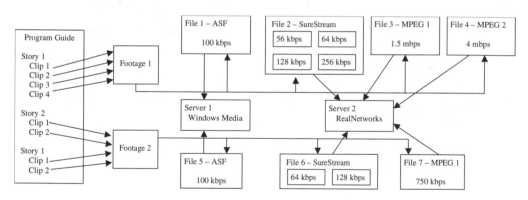

Figure 16.8 Multi-format Video Content Adaptation

Multiple client devices are supported in the meta-database by the presence of *footage* objects. Footage is an abstract representation of one specific video that will have one or more physical instantiations in the form of media files. Footage 1 in the figure, for instance, exists in four different files and in four different encoding formats, and can be played back at 7 different data rates. Clips of an interactive presentation are bound to the abstract footage object and not directly to the media files. New media files can then easily be added to the system whenever there is a need to support additional bit rates or encoding formats without changing any of the productions.

5.3 MANAGING SYSTEM AND SOFTWARE VARIATIONS

The difference in the devices' system and software capabilities is also a challenge for adapting rich, interactive content delivery. The HotStreams™ system, for instance, is generating personalized presentations in the form of SMIL and is utilizing the RealPlayer API to integrate the video player tightly into the web browser. No SMIL-enabled player was available on Microsoft Windows CE-based PDAs when the system was developed. The Microsoft Windows Media player, on the other hand, was available for several Microsoft Windows CE devices. The Windows Media Player defines an API similar to the one offered by the RealPlayer and is interpreting a scripting language similar to SMIL.

Figure 16.9 illustrates the adaptive HotStreams™ content personalization workflow. The first task, *Content Personalization*, retrieves content meta-data from the database and matches these with the end-user election and user profile. The personalized production is represented as a composite JAVA object and passed to the next step, *Advertisement Insertion*. In this step, the system retrieves video commercials according to the end-user location and content type and matches these against the end-user's interest profile and corresponding content clips and hyperlinks are inserted into the production. The production is then passed on to the final step, *Presentation Script Generation*. This step binds the footage objects to media files according to the device profile. Similarly, this step selects the best-suited image, text, and audio files to use if the presentation includes such content. Lastly, this step determines how hyperlinks are to be added to the presentation depending on the device capabilities. A SMIL enabled device may, for instance, have multiple hotspots located at different areas of the presentation simultaneously, while the Windows Media Player can only deal with one MOREINFO hyperlink bound to the BANNER.

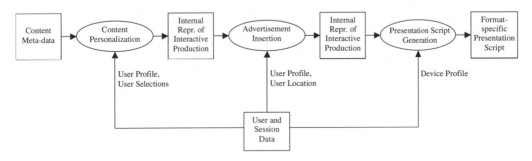

Figure 16.9 Adaptive Personalization Workflow

The Presentation Script Generation step also determines in what scripting language the personalized production should be generated (SMIL or ASX). A SMIL script is generated if the end-user device can handle SMIL, and ASX if the end-user device has the windows media player installed. Figure 16.10 shows typical SMIL and ASX scripts generated by HotStreams™.

Knowing the client's capabilities is a prerequisite to offering content adaptation. The systems discussed in this chapter run on standard World Wide Web components and are consequently limited to base content adaptation on device capability data being exchanged within the web. CC/PP [16] was in its initial stage at the time of these projects. Hence, content adaptation was based on knowledge of the client's operating system: A client running Windows CE was receiving a one-frame/ASX version with videos encoded in Windows Media format while other clients would receive a multi-frame/SMIL version with videos encoded in MPEG-1 or RealMedia formats.

```
<?xml version="1.0"?>
<smil>
 <head>
  <layout>
   <root-layout height="288" width="352" />
   <region id="vcr" top="0" left="0" height="100%"
        width="100%" fit="fill" z-index="1" />
   <region id="aiu_0" top="250" left="287"
        height="17" width="55" z-index="6" />
   <region id="aiu_1" top="10" left="276"
        height="24" width="66" z-index="6" />
  </layout>
  <meta name="title" content="SMIL File" />
  <meta name="abstract"
  <meta name="author" content="HotStreams" />
       content="HotStreams Test Video" />
  <meta name="copyright" content="Siemens" />
 </head>

 <body>
  <par id="section_0" title="SMIL File"
       abstract="HotStreams Test Video">
   <video src="rtsp://HotStreams/Siemens1_1.5mbps.mpg"
       region="vcr" clip-begin="0ms" clip-end="30000ms" />
   <img region="aiu_0"
        src="http://HotStreams/images/BuyNow.jpg">
    <anchor href=http://HotStreams/Siemens.html
            show="new" />
   </img>
  </par>
  <par>
   <video src="rtsp://HotStreams/Video1_1.5mbps.mpg"
       region="vcr" clip-begin="0ms" clip-end="24000ms" />
   <img region="aiu_1"
        src="http://HotStreams/images/MoreInfo.jpg"
        begin="7500ms" end="24000ms">
    <anchor href=http://HotStreams/More1.html
            show="new" />
   </img>
  </par>
```

```
<ASX version="3.0">
  <TITLE>ASX File</TITLE>
  <ABSTRACT>HotStreams Test Video</ABSTRACT>
  <AUTHOR>HotStreams</AUTHOR>
  <COPYRIGHT>Siemens</COPYRIGHT>

  <ENTRY CLIENTSKIP="NO">
   <TITLE>Intro</TITLE>
   <ABSTRACT>The introduction</ABSTRACT>
   <BANNER HREF="images/BuyNow.jpg">
    <MOREINFO HREF="http://HotStreams/Siemens.html" />
   </BANNER>
   <REF HREF="mms://HotStreams/Siemens1_100kbps.asf" />
   <STARTTIME VALUE="0:0:0.0" />
   <DURATION VALUE="0:0:30.0" />
  </ENTRY>
  <ENTRY CLIENTSKIP="NO">
   <TITLE>Intro</TITLE>
   <ABSTRACT>The introduction</ABSTRACT>
   <BANNER HREF="images/MoreInfo.jpg">
    <MOREINFO HREF="http://HotStreams/More1.html" />
   </BANNER>
   <REF HREF="mms://HotStreams/Video_100kbps.asf" />
   <STARTTIME VALUE="0:0:0.0" />
   <DURATION VALUE="0:0:24.0" />
  </ENTRY>
  <ENTRY CLIENTSKIP="NO">
   <TITLE>Intro</TITLE>
   <ABSTRACT>The introduction</ABSTRACT>
   <REF HREF="mms://HotStreams/Video_100kbps.asf" />
   <STARTTIME VALUE="0:0:24.0" />
   <DURATION VALUE="0:0:24.0" />
  </ENTRY>
</ASX>
```

Figure 16.10 SMIL and ASX Scripts for Personalized Video Content

5.4 MANAGING INTERACTION STYLES

Section 5.1 discussed the challenges related to variations in device displays. Devices also differ in what kind of input devices they offer. This further affects the way in which the user may interact with the system. The keyboard and mouse, for instance, encourages a desktop user to be highly interactive, while the remote control interface provided by a set-top box suits better with a laidback interaction style where the user interacts less with the content. The stylus discourages the PDA user from typing in text. An adaptive system will need to take the differences in interaction style into account.

5.5 MANAGING MOBILITY AND UBIQUITY

The original version of HotStreams™ could not accommodate nomadic users. A user might, for instance, follow an interactive training class in the office using her high-resolution/high bandwidth desktop. She might want to continue watching the interactive presentation on her small-screen/low bandwidth PDA on the bus back home. Finally, she might want to complete the class at home in front of her TV connected to a medium-resolution/high bandwidth set-top box.

Figure 16.11 shows the workflow for the scenario described earlier in which the end-user can carry along her personalized presentation from one location and device to another. The end-user is ordering her interactive training class using her desktop that is connected to a high bandwidth network. The desktop has a powerful processor that can decode MPEG-1 data in real time. Hence, the user can view the multi-frame/SMIL based interface to HotStreams™ in the office. The user may decide to stop the interactive video production when she is ready to leave the office. In the bus on her way home, she can connect to the system again – this time from her PDA. The PDA has lower bandwidths available and has Microsoft Media Player installed. Hence the system will generate a different presentation script. The user can now continue viewing her presentation but the content is delivered in the single-frame/ASX view this time. At home she would pick up the presentation from her TV/set-top box. This time the content could be delivered in a SMIL-based view optimized for TV with content delivered in high-quality MPEG-2, depending on the capabilities of her set-top box.

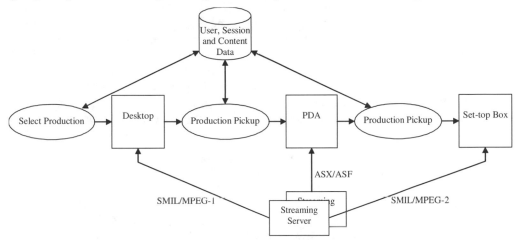

Figure 16.11 Carry-along Workflow

5.6 TOOLS

The first part of this section discussed techniques for handling differences in network and device capabilities and for supporting ubiquity. The remaining part discusses tools needed for offering such functionality in interactive video database applications.

5.6.1 Content Adaptation Tools

Within the framework discussed in Section 4, content adaptation is the responsibility of the *Presentation Script Generation* step shown in Figure 16.4. At this point decisions are made regarding the layout (e.g., one-frame vs. multi-frame layout), media format and encoding, and script file format. Ubiquity, as discussed in Section 5.5, is supported by storing the profiling and selection parameters being used as input to the *Content Selection* step shown in Figure 16.4 in the database for later "pick up." A new script adapted to the current client is generated on the fly every time the end-user "picks up" the stored production.

5.6.2 Content Preparation Tools

A content assembly tool similar to the one described in Section 4.2.2 that allows the content manager to create multimedia presentations without worrying about network and device capabilities is of large value in serving heterogeneous users. There is, in addition, a need for tools to manage the content stored on the various media servers being used and to extract and manage meta-data describing the characteristics of individual media files. The management tool being used in HotStreams™ contains a Media File panel that the content manager can use to view the content on various media servers. In addition, daemon processes run on each media server to monitor the change in availability of various media files and report changes (such as addition and deletion of media files) to the application server.

6. LOCALIZING/PERSONALIZING VIDEO CONTENT

This section discusses a personalization/localization technique that can be used to generate personalized interactive video content. This technique is implemented by the HotStreams™ system. The creation of a personalized video production can be viewed as a filtering process. A complete version of the interactive video production is defined in the meta-database. Each element of the video, such as video scenes and sequences, and hyperlink are classified according to the profiling scheme used for the given type of content. These attributes are matched with the customer's preferences during personalization. The elements that do not match the customer preferences will be filtered out of the personalized version of the video. The filtering process will be described in more detail in the next few subsections.

Personalization is especially important in a streaming video service offered to mobile users. The user may be subscribing to an interactive news service, for instance, and may have to pay for the actual value of a specific news program. At home, the user may want to watch a half-an-hour news presentation on her TV or computer. It is very unlikely, however, that she would be interested in watching the complete 30 minutes' video on her PDA when she is out traveling. Hence she might prefer to receive and pay for only a 3-5 minutes long headline version covering topics that interest her.

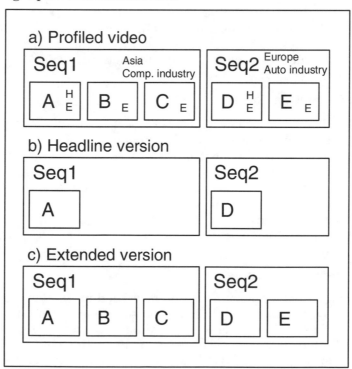

Figure 16.12 Video Node Filtering Examples

6.1 VIDEO NODE PERSONALIZATION

Figure 16.12 shows a simple personalization scenario for financial news. The sample financial news video (a) contains two sequence nodes (Seq1 and Seq2) corresponding to two news stories. The first story in the sample video is about the computer industry in Asia while the second is a story about the European automobile industry. Each news story further consists of several scene nodes (nodes A through E). The meta-data describing each of the scenes contains one attribute that indicates whether the scene is to be included in the headline and/or the extended version of the news. As indicated by a capital H in the figure, scenes A and D are the only ones to be included in the headline version. The extended version, however, will include all five scenes – as can be seen from the presences of a capital E.

Part b) of the figure illustrates what parts of the video are to be included when the user orders a headline version. Scene A from Seq1 and scene D from Seq2 are the only scene nodes to be included in this example. Part c) shows that all parts of the video are included in the extended version.

6.2 HYPERLINK PERSONALIZATION

A hypervideo may also contain hyperlinks to ordinary web pages. Figure 16.13 illustrates how the version attribute of the hyperlink can be used to determine what web page would be loaded in the right part of the HotStreams™ user interface shown in Figure 16.1. Scene node A in the figure contains a hyperlink to web page P1 that will be filtered out in the headline version, because it does not match the headline preference. The hyperlink from scene node D in the figure has two possible terminations, web page P2 or P3, respectively. As shown in part b) of the figure, the hyperlink will link to web page P2 when included in the headline version and to web page P3 when included in the extended version to match the preference selected by the end-user.

6.3 COMMERCIAL INSERTION

HotStreams™ will insert personalized commercials into the video stream – if the service provider decides to. The content manager will specify the locations within the video that can be target locations for inserting commercials using the video panel shown in Figure 16.5. Commercials are selected based on their type attributes. Each video is associated with a set of such types indicating what commercials should be accepted in the video.

6.4 TOOLS

The first part of this section discussed techniques personalizing interactive video content. The remaining part discusses tools needed for offering such functionality in interactive video database applications.

6.4.1 Personalization Tools

Within the framework discussed in Section 4, personalization is one alternative to the *Content Selection* step shown in Figure 16.4. The input to this step is the end-user profile. The end-user may enter this information during the ongoing session or stored from previous sections.

6.4.2 Content Preparation Tools

The *video* and *hyperlink panel*s discussed in Section 4.2.2 provide the means for the content manager to associate profiling attributes values with specific video sequences and hyperlinks. The HotStreams™ content management tool also contains an *advertisement panel* that is used to register and classify commercials that might be selected for insertion into the video streams.

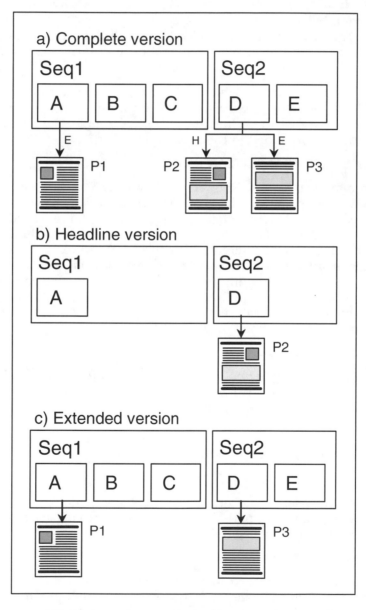

Figure 16.13 Hyperlink Filtering Example

7. SHARING INTERACTIVE EXPERIENCES

Unlike text-based messages, multimedia messaging brings unique challenges. Multimedia messages can originate from large display terminals such as PCs and be sent to devices with limited capabilities like handheld phones. In the case of multimedia messages, the sender cannot have the complete confidence that the recipient will be able to watch the multimedia message for any one of the following reasons:

❑ Display constraints

❑ Processing constraints

❑ Bandwidth constraints

❑ Streaming media player platform

In the reverse case, the sender might have limited device or network capabilities while the recipient(s) might have better display devices and might be able to experience a richer multimedia experience, which the sender could not. For instance, the sender might not have the capabilities to view hyperlinks embedded into the video on his handheld PDA, but there is no reason why the recipient should not be able to experience rich multimedia on his PC when viewing the same content. This shows that there is a need for content adaptation based on device capabilities at both the sender and receiver terminals. The service provider or the content provider should have software which would recognize the capabilities of the client terminal making the request and adapt the multimedia presentation for that terminal.

User mobility creates a demand for location-based services. In most messaging scenarios, the sender and the recipient are at different locations. This provides an opportunity for the content provider to provide location-based advertisements, for instance, along with the actual message content but where the advertisement is based on the location of the receiver of the message rather than the sender.

7.1 SHARING PERSONALIZED CONTENT

The Personalized Multimedia Messaging Service (PMMS) initiative at Siemens Corporate Research adds experience-sharing features to the HotStreams™ architecture by allowing end-users to share particular video stories from customized productions. The messaging module allows an end-user to send a customized production to various recipients. The personalization and adaptation module helps in adapting the multimedia presentation to the recipient's device, streaming platform, bandwidth, network, and location. Many of the features being used by PMMS were also used in the SEMCA application to facilitate experience sharing between the customer and the customer support agent. Figure 16.14 shows the composition workflow and the delivery workflow involved in the experience sharing.

For the recipient to experience the multimedia message as the sender had meant to, PMMS has to adapt the content based on several factors:

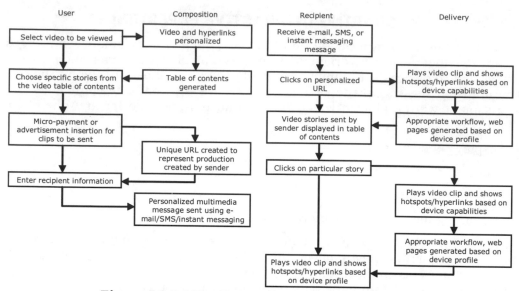

Figure 16.14 The Experience Sharing Workflows

Streaming Platform

PMMS utilizes the HotStreams™ content management tools described earlier to manage media files available for streaming. On various mobile devices, the multimedia capabilities are usually limited to one of the popular streaming platforms. Moreover, most the devices come bundled with a particular media player installed. In the mobile phone video streaming, a new crop of media players based on MPEG4 format, such as PacketVideo, have made their entry. If the sender and recipient are using different video streaming platforms, the message sent by the sender may not be rendered on the recipient's device. PMMS tries to identify the device or the operating system and serves the video file in appropriate format (e.g., Windows Media streaming on the Windows CE platform).

Device and Network Constraints

PMMS utilizes the workflow described earlier in this chapter to adapt the content to the device and network capabilities of the *recipient*.

Location-based Services

PMMS utilizes location information in providing location related information within the message. PMMS can, for instance, insert location-based advertisements in the recipient's video production in cases where the stories are free to the sender. Assume, for instance, that the sender in Italy wishes to share a couple of video clips to a mobile phone user in the USA. PMMS may insert USA related advertisements into the video production before delivery to the recipient. The hotspots included in the video advertisements would similarly point to appropriate websites for that location.

7.2 TOOLS

The workflows shown in Figure 16.14 needed to be supported by tools for the sender to compose the multimedia message used to share experience and tools for delivering the shared experience to the recipient.

7.2.1 Composition Tools

PMMS and SEMCA provide webpages from which the sender can select the video sections that will be part of the content to be shared and for typing the recipients' addresses and a short text message to be included in the e-mail. The application server interfaces a mail server transferring the generated e-mail message to the recipients.

7.2.2 Delivery Tools

The PMMS and SEMCA application servers also implement the tools needed to respond to recipients' requests for shared content and for adapting the content to the recipients' capabilities.

8. CONCLUSION

This chapter has discussed how the World Wide Web age defines new opportunities and challenges for video database applications. The prevalent characteristics of the World Wide Web age were summarized as:

❑   Heterogeneity and large network/device diversity

❑   Interactivity and Dynamic Content Generation

❑   Personalization and Sharing of Information and Experiences

❑   Mobility and Ubiquity

The chapter further discussed some tools and techniques being used to implement two interactive video database applications to deal with these challenges. A content assembly technique was described for dealing with network and device diversity; an interactive video model and a personalization technique was described; and a technique for sharing interactive video content was described. This chapter also explained how these techniques were used in two interactive video database applications being developed at Siemens Corporate Research.

Interactive video database applications are still very much in their infancy. Developers of such applications are facing a situation where a number of tools and techniques exist but where there is lack of a solid usability data that can act as guidelines as to when and how to use the various techniques. Hence, more R&D work is needed to evaluate the usability of the techniques discussed in this chapter and to refine the techniques and tools presented.

ACKNOWLEDGMENT

The authors are greatly thankful for the input from Yves Léauté and Thomas Kühborth during the development of the HotStreams™ system. The authors are also greatly thankful for Amogh Chitnis' and Xavier Léauté's assistance in implementing many of the techniques and tools discussed in this chapter.

REFERENCES

[1]   Hjelsvold, R., Liou, S.-P., and Depommier, R., Multimedia Archiving, Logging, and Retrieval, in Furht, B. (ed), Handbook of Multimedia Computing, CRC Press, 1999.

[2]   Jain, R., TeleExperience: Communicating Compelling Experience, in Proceedings of the 9th ACM International Conference on Multimedia, pp. 1-1, Ottawa, Canada, 2001.

[3]   Microsoft Corporation, Media Players - Windows Media Technologies, Internet URL: http://www.microsoft.com/windows/windowsmedia/players.asp

[4]   RealNetworks, Inc., Products & Services > Media Players, Internet URL: http://www.realnetworks.com/products/media_players.html.

[5]   Starr, T., Cioffi, J. M., and Silverman, P., Understanding Digital Subscriber Line Technology, Prentice Hall PTR, 1998.

[6]   World Wide Web Consortium, HyperText Markup Language (HTML) Home Page. Internet URL: http://www.w3.org/MarkUp/

[7]   Sun Microsystems, Inc., JavaServer Pages™: Dynamically Generated Web Content. Internet URL: http://java.sun.com/products/jsp/

[8]   Microsoft,   Inc.,   Active   Server   Pages.   Internet   URL: http://msdn.microsoft.com/library/default.asp?url=/library/en-us/dnasp/html/msdn_aspfaq.asp

[9]   Kleinrock, L., Nomadicity: Anytime, Anywhere in a Disconnected World, in Mobile Networks and Applications 1, 1996, pp. 351-357.

[10]   Ma, W-Y., Bedner, I., Chang, G., Kuchinsky, A., and Zhang, HJ, Framework for Adaptive Content Delivery in Heterogeneous Network Environments, in Proceedings, SPIE Vol. 3969, Multimedia Computing and Networking 2000, pp. 86-100, San Jose, CA, Jan. 22-28, 2000.

[11]   Mohan, R., Smith, J.R., and Li, C.-S., Adapting Multimedia Internet Content for Universal Access, in IEEE Transactions on Multimedia, 1(1), 1999, pp. 104-114.

[12]   Fox, A., Gribble, S.D., Chawathe, Y., and Brewer, E.A., Adapting to Network and Client Variation Using Infrastructural Proxies: Lessons and Perspectives, in Special Issue on IEEE Personal Communications on Adapation, pp. 10-19, August 1998.

[13]   RealNetworks, Inc., RealSystem Production Guide: RealSystem Release 8 with   RealProducer   8.5,   RealNetworks,   2000.   Internet   URL: http://docs.real.com/docs/realsystem8productionguide.pdf

[14]   The World Wide Web Consortium, Synchronized Multimedia. Internet URL: http://www.w3c.org/AudioVideo/

[15]   Microsoft Corporation, Windows Media Metafile Elements Reference. Internet   URL: http://msdn.microsoft.com/library/psdk/wm_media/wmplay/mmp_sdk/windowsmediametafileelementsreference.htm

[16]   Klyne, G., Franklin, R., Woodrow, C., and Ohto, H., (Eds), Composite Capability/Preference Profiles (CC/PP): Structure and Vocabularies, W3C

Working Draft, March 15, 2001. Internet URL: http://www.w3.org/TR/2001/WD-CCPP-struct-vocab-20010315/

[17]   Auffret, G., Carrive, J., Chevet, O., Dechilly, T., Ronfard, R, and Bachimont, B, Audiovisual-based Hypermedia Authoring: using Structured Representations for Efficient Access to AV Documents. In Proceedings from Hypertext'99, pp. 169-178, Darmstadt, Germany, Feb., 1999.

[18]   Hjelsvold, R., Vdaygiri, S., and Léauté, Y., Web-based Personalization and Management of Interactive Video. In Proceedings of the Tenth International World-Wide Web Conference, pp. 129-139, May 1-2, 2001, Hong Kong.

[19]   Siemens AG, SURPASS by Siemens – IP-based voice-data convergence solutions for building the Next Generation. Internet URL: http://www.siemens.com/surpass

[20]   CommWorks, Softswitch Model Drives New Age of Customized Communication. In International Softswitch Consortium, ISC Reference Materials. Internet URL: http://www.softswitch.org/attachments/CommWorksPositionPaper.pdf

17

ANIMATION DATABASES

Z. Huang
School of Computing
National University of Singapore
Singapore
huangzy@comp.nus.edu.sg

N.Chokkareddy
Department of Computer Science
University of Texas at Dallas
USA
nxc018000@utdallas.edu

B.Prabhakaran
Department of Computer Science
University of Texas at Dallas
USA
praba@utdallas.edu

1. INTRODUCTION

Animations of 3D models are an interesting type of media that are increasingly used in multimedia presentation such as streaming of 3D worlds in advertisements as well as in education material. The most popular way to generate animations of 3D models is motion capture through the use of motion-capture devices and 3D digitizers. In this method, 3D models are first digitized using 3D scanners. Then, applying automatic process and user-computer interactions, the moving structures of the models (e.g., skeletons and joints) and their motion parameters (e.g., translation, rotation, sliding, and deformation) are defined. Third, motion sequences are captured using the optical or magnetic motion capture devices. Finally, animations, the motion sequences and 3D models, are stored in animation databases for further use.

For reusing models and motion sequences, the captured motion sequences may need to be transformed (e.g., changing the speed, time) or may need to be applied to other 3D models. Reusing animations is a research area to devise the systems and methods to transform the existing animations for the new requirements. Few techniques have been proposed for transforming animations. In [2], motions are treated as signals so that the traditional signals processing method can be applied, while preserving the flavor of the original motion. Similar techniques have been described in [11, 4, 5].

1.1 DATABASE APPROACH TO REUSING ANIMATIONS

In this chapter, we consider the issue of generating new animation sequences based on existing animation models and motion sequences. We use an augmented scene graph based database approach for this purpose of reusing models and motions. For this augmented scene graph based animation database, we provide a set of spatial, temporal, and motion adjustment operations. These operations include reusing existing models in a new scene graph, applying motion of a model to another one, and retargeting motion sequence of a model to meet new constraints. Using the augmented scene graph model and the proposed operations, we have developed an animation toolkit that helps users to generate new animation sequences based on existing models and motion sequences. For instance, we may have two different animations where a man is walking and another where (the same or a different) man is waving his hands. We can reuse the walking and waving motion, and generate a new animation of a walking man with a waving hand. The SQL-like operations on the databases and the generated animation using the toolkit are shown in Figure 17.1.

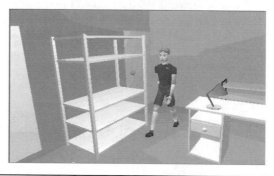

> **INSERT** *Andy* TO *Scene* PARENT *room* WHEN *[6,12]*
> **SAVE AS** *Andy_in_room*
> GET *walking* FROM *Nancy* SAVE AS *walking1*
> USE *walking1* TO *Andy_in_room.Andy*
> CROP *Andy.walking1* BY *50*
> JOIN Nancy.walking WITH translation (0,0,20)

Figure 17.1 An animation reuse example applying a walking sequence of a woman to a man

This toolkit provides an exhaustive set of Graphic User Interfaces (GUIs) that help users to manipulate scene graph structures, apply motion sequences to models, and to edit motion as well as model features. Hence, users of the toolkit do not have to remember the syntax of the SQL-like database operations. We use existing solutions in the literature for applying motion sequences to different models and for retargeting motion sequences to meet different constraints. The current implementation of our animation toolkit uses the motion mapping technique proposed in [8] for applying motion of a model to a different model. It (the toolkit) uses the inverse kinematics technique proposed in [13] for retargeting motion sequences to meet new constraints.

The toolkit can handle animations represented in Virtual Reality Modeling Language (VRML). Apart from VRML, animations are also represented in different standard formats such as MPEG-4 (Moving Pictures Experts Group), SMIL (Synchronized Multimedia Integration Language), and in proprietary formats such as Microsoft PowerPoint. MPEG-4 includes a standard for the description of 3D scenes called BInary Format for Scenes (BIFS). BIFS is similar to VRML with some additional features for compressed representation and streaming of the animated data. SMIL, a recommendation of the World Wide Web Consortium (W3C), incorporates features for describing animations in a multimedia presentation. Microsoft PowerPoint also supports animations. Reusing animations is more general and challenging when animations are represented in different formats, stored in different databases, and the resulting new animation needs to be in a specific format for the presentation. We use a *mediator* type of approach for handling multiple databases of animation formats. This mediator is an eXtensible Markup Language (XML-) based Document Type Definition (DTD) that describes the features of models and motions in a uniform manner. The animations in VRML, MPEG-4 BIFS, SMIL, and Microsoft PowerPoint are mapped onto this XML DTD. The XML DTD is then mapped onto a relational database for querying and reusing the animation models and motion sequences.

1.2 MOTION MANIPULATION

Database approach for reusing motions and models is to adjust the query results from animation databases for new situations while, at the same time, keeping the desired properties of the original models and motions. Manipulation of motion sequences can also be done by using a technique called *inverse kinematics*, originally developed for robotics control [20]. Inverse kinematics is a motion editing technique especially for *articulated figures*. An articulated figure is a structure consisting of multiple components connected by joints, e.g., human and robot arms and legs. An *end effecter* is the last piece of a branch of the articulated figures, e.g., a hand for an arm. Its location is defined in Cartesian space, three parameters for position and another three for orientation. At each joint, there are a number of degrees of freedom (*DOFs*). All DOFs form a *joint (or configuration) space* of the articulated figure (Figure 17.2). Given all the values of DOFs in joint space, the kinematics method to compute the position and orientation of end effecter in Cartesian space is called *direct kinematics*. Inverse kinematics is its opposite. Section 3.2 discusses this issue in broader detail.

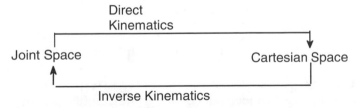

Figure 17.2 Direct and inverse kinematics

2. XML DTD FOR ANIMATION DATABASES

An XML DTD specifies the allowed sets of elements, the attributes of each element, and the valid content of each element. By defining a DTD that incorporates the superset of the nodes specified in VRML, MPEG-4, SMIL, and PowerPoint, XML can adeptly handle multiple databases of animations, as shown in Figure 17.3.

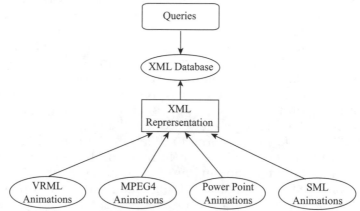

Figure 17.3 Proposed XML mediator for databases of animations

The DTD for the XML mediator is defined based on *augmented scene graphs* proposed in [1]. The *scene graph* is a hierarchical graph structure for 3D models and its generalization, to represent animations [12]. In a scene graph, complete information is defined for all 3D models and other entities that affect the display: lights, colors, sounds, background, and views. Typically, leaf nodes represent geometry of 3D models (shape, structure, size, position, connection, etc.) and interior nodes represent other information such as transformations applied to its child nodes. A sub-tree in the scene graph may represent a complex 3D model in the scene with multiple sub-models. If multiple copies of a complex model are going to populate the scene, then it is efficient to refer (link) the one sub-tree in different parts of the graph. Also, any change in that sub-tree will be reflected in all references. One example of a scene graph is illustrated in Figure 17.4.

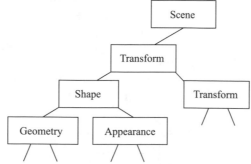

Figure 17.4 An example of scene graph illustrating the model

The basic scene graph was *augmented* in [1] to represent animations of 3D models by introducing a node, called *Interpolator node*, as an internal node to represent motion sequences of 3D models, e.g., a walking sequence for a human body model. An Interpolator node is associated with a 3D model that can be a complex model such as an articulated figure linked by joints of multiple degrees of freedom (DOFs). Each joint can have 6 DOFs, 3 for translation and 3 for rotation. An articulated figure such as a human body may have hundreds of DOFs. In an Interpolator node, key frames are used to represent the motion for all DOFs: *key* $[k_1, k_2, ..., k_i, ..., k_m]$, *keyvalue* $[v_1, v_2, ..., v_i,..., v_m]$, where k_i is the *key frame number* or *key frame time*. The key frame number can be mapped to the time according to the different frame rate standards such as PAL, SECAM, and NTSC. v_i is a vector in the motion configuration space: $v_i=[u_{i,1}, u_{i,2}, ..., u_{i,j}, ..., u_{i,n}]$, where $u_{i,j}$ is a *key value* of DOF j (in displacement for translational DOFs and angle value for rotational DOFs) at k_i. The m and n are the numbers of key frames and DOFs of the model respectively. The *key* and *keyvalue* of an Interpolator node define animation of a 3D model.

Most animation formats are defined based on the above scene graph structure. VRML uses nearly 54 types of scene graph nodes to describe the animation models and sequences [7]. These nodes include *SFNode* (the single field node for scalars), *MFnode* (the multifieldnode for vectors) with field statements that, in turn, contain node (or USE) statements. Each arc in the graph from A to B means that node A has an SFNode or MFNode field whose value directly contains node B. Prototypes and route nodes are some of the other nodes in VRML. MPEG-4 BIFS can be considered as a compressed version of VRML apart from having nodes for body and facial animations. MPEG-4 has 100 nodes defined for animation sequences [7, 9]. Synchronized Multimedia Integration Language (SMIL), a standard proposed by World Wide Web Consortium (W3C), also describes formats for representing animation sequences. Apart from providing constructs for describing spatial and temporal relationships, SMIL also defines animations as a time-based function of a *target element* (or more specifically of some *attribute* of the target element, the *target attribute*). PowerPoint, the popular software from Microsoft for presentations, provides facilities for including sound and video files in a presentation. It also provides support for animations in the presentation. The PowerPoint design is based upon the Component Object Model, which can be easily mapped on to the XML DTD.

2.1 MAPPING XML ANIMATION TO DATABASE

We adopt a relational database approach for representing and storing the DTD of the XML mediator. As shown in the ER diagram of Figure 17.5, the geometric models and motions of each object are stored in two separate tables ("Motion" and "Model") of the database and are linked by the SceneID, the primary key. The model table and the motion table are not the only tables in the database. There are separate tables for storing the interpolator, sensor, and route nodes, which are the main nodes to be considered while looking for a motion. To handle the case of model comparison, information regarding the nodes representing the model is stored in the Model_Nodes table. The scene also has its table in the database, which is linked to the XML_text table by a Transformation relationship and also to the VRML_MPEG4 and PowerPoint tables. The Model_Nodes and PowerPoint database tables have an entry for metadata that provide descriptions of objects and presentations. These descriptions include metadata of the type of object (man, woman, cow, etc.), nature of object (color, size, shape), as well as descriptions of content in presentations such as including text/MPEG-4/VRML/etc. other elements' content, and the order in which they are represented.

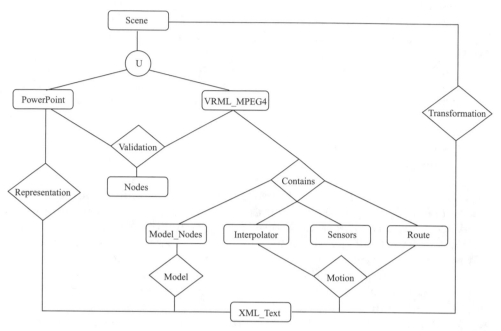

Figure 17.5 The ER diagram of the XML Mediator DTD

The VRML_Text/MPEG4_Text/PwerPt_Text are mapped on to the XML_Text by parsing the content according to the DTD specified. This forms a scene and hence is related to the Scene table from where the mapping on to the respective nodes is done. Names for different models and motions are assigned based on the content in the nodes, i.e., the metadata associated with the nodes. In the case of motion table, identifiers of the interpolator, sensor, and route nodes are used for identifying the metadata (and hence the name). For the model table, identifiers of the nodes that contribute towards the generation of the model are used for metadata (and hence the name) identification. Metadata identification is discussed in more detail, in Section 3.1.2. The motion/model identifiers and the motion/model name form the primary key to identify respective models/motions. These fields characterize a general animation scene in any format. (SMIL being an XML-based language can easily be incorporated into the above ER model. We do not mention it here explicitly, since the current implementation of the animation toolkit does not handle SMIL animations.)

3. ANIMATION TOOLKIT

Access 97 was used as the primary database for the storage of the animations represented in the XML mediator DTD. The animation toolkit uses Visual Basic as well as Java applications for operations on the stored animation models and motion sequences. The toolkit uses existing solutions in the literature for applying motion sequences to different models and for retargeting motion sequences to meet different constraints. The current implementation of our animation toolkit uses the motion mapping technique proposed in [8] for applying motion of a model to a different model. It (the toolkit) uses the inverse kinematics technique proposed in [13] for retargeting motion sequences to meet new constraints. The implementation of the XML mediator animation database toolkit has two phases:

- ***Pre-processing phase***: that helps to generate the XML representation from the various animation formats such as VRML or MPEG-4. It also helps to push the XML representation into Microsoft Access. In the current implementation, we do not handle SMIL animations.

- ***Animation generation phase***: that helps to reuse the models and motion sequences in the animation database, using the spatial, temporal, and motion adjustment operations. Depending on the format in which the animations are to be generated, the second phase applies appropriate mapping techniques for models and motion sequences.

3.1 PRE-PROCESSING PHASE

In the pre-processing phase, animations described in different formats are first mapped on to the general XML representation, as shown in Figure 17.6. XML documents, when parsed, are represented as a hierarchical tree structure in memory that allows operations such as adding data or querying for data. The W3C (World Wide Web Consortium) provides a standard recommendation for building this tree structure for XML documents, called the XML Document Object Model (DOM). Any parser that adheres to this recommendation is called a DOM-based parser. A DOM-based parser provides a programmable library called the DOM Application Programming Interface (DOM API) that allows data in an XML element to be accessed and modified by manipulating the nodes in a DOM tree. In the current implementation, we have used the DOM-based parser available from Xj3D (eXtensible java 3D, developed by Web 3D Consortium)[14]. Once the XML DOM tree is built, it can be stored in the form of a database. In the current implementation, we use Microsoft Access for storing and manipulating the XML DOM tree. This animation database is used in the second phase for reusing models and motion sequences.

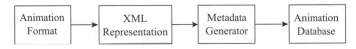

Figure 17.6 Block diagram for the Pre-processing phase

3.1.1 Metadata Generation

Querying the animation databases relies on the metadata associated with animation models and motions. While it is difficult to generate the metadata fully automatically, it is possible to provide semi-automatic techniques. In the first stage, users of this animation toolkit should manually create *reference templates* for motions and models. These reference templates are used for comparison with the XML animation representation in order to generate the metadata.

3.2 ANIMATION GENERATION PHASE

In this second phase, users can query the animation database for models and motion sequences. The resulting models and/or motion sequences can be reused to generate new animation sequences. The authoring of a new animation starts by collecting the necessary objects and motion in the scene. This is done by invoking the Query Module. As shown in Figure 17.9, the Query Module directly interacts with the database to retrieve the animation, object, or motion the user has requested. The user interacts with the system through a Graphical User Interface (GUI). The user can manipulate the query results using the authoring operations. If the result is an object, it can be edited by the spatial operations before it is added to the new scene graph. If the result is a motion, it could be passed to the temporal operations to change time constraints. It can then be either sent to the motion adjustment operations for further editing or added directly to the scene graph. The user can manipulate the nodes of the scene graph by using the authoring operations repeatedly.

Motion adjustment operations use the technique of inverse kinematics. For ease of understanding, we illustrate inverse kinematics in a 2D case. In Figure 17.7, it shows a movement of the articulated figure to have the end effecter to reach the target. This motion sequence represents the original motion.

Now we move the target point to a new position (Figure 17.8(a)) and show how inverse kinematics can be applied to adjust the original motion, i.e., retargeting, generate a new motion (Figure 17.8(b)). Motion in Figure 17.8(b) is achieved by applying the following equation for inverse kinematics:

$$\Delta\theta = J^+\Delta x + (I - J^+J)\Delta z \qquad\qquad (17.1)$$

where:

$\Delta\theta$	is the unknown vector in the joint variation space, of dimension n.
Δx	describes the *main task* as a variation of the end effecter position and orientation in Cartesian space. For example in Figure 17.8(b), the main task assigned to the end of the chain is to follow a curve or a line in the plane under the small movements hypothesis. The dimension m of the main task is usually less than or equal to the dimension n of the joint space.
J	is the Jacobian matrix of the linear transformation, representing the differential behavior of the controlled system over the dimensions specified by the *main task*.
J^+	is the unique pseudo-inverse of J providing the minimum norm solution which realizes the *main task* (Figure 17.8 (a) and (b)).
I	is the identity matrix of the joint variation space n x n.
$(I-J^+J)$	is a projection operator on the *null space* of the linear transformation J. Any element belonging to this joint variation sub-space is mapped by J into the null vector in the Cartesian variation space.
Δz	describes a *secondary task* in the joint variation space. This task is partially realized via the projection on the *null space*. In other words, the second part of the equation does not modify the achievement of the main task for any value of Δz. Usually Δz is calculated so as to minimize a cost function.

The inverse kinematics solver applies the above equation (1) (no secondary task, thus, $\Delta z = 0$) to Figure 17.8(a). Using (1) the offset vector Δx of the target and the end effecter position for each frame, and (2) the configuration at each frame as initial posture, it (the solver) automatically generates a new configuration in joint space to reach the new target. Thus, motion retargeting is handled.

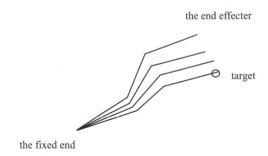

Figure 17.7 The original motion sequence of a 2d articulated figure with 3 DOFs in joint space

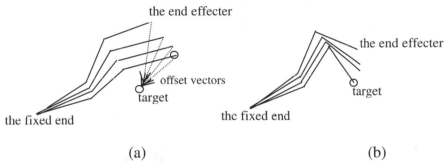

(a)   (b)

Figure 17.8 Retargeting to generate a new motion

Other operations include manipulating the scene graph structure. The toolkit uses VB's Tree View Control Component to implement the scene graph structure, catering to functions such as delete, add, and search. After editing the objects and the motion, the scene graph now reflects the desired animation of the user. The resulting models and motion sequences can be converted to the required animation format. The animation mapper module carries out this conversion. Figure 17.9 describes the different modules and their interactions in the animation toolkit.

3.2.1 Animation Mapper

The *Animation Mapper* module helps in integrating responses to a query (or a set of queries) that are in different animation formats. For instance, the resolution of a user query may involve the following actions. Motion in a VRML animation might need to be applied on a MPEG-4 model and a new MPEG-4 animation is generated. To carry out these actions, information regarding the motion is extracted from the motion and the model tables. This information is then integrated with the metadata content along with key values, timing, and sensor information present in the respective nodes to build the required new animation.

While mapping a scene from one format to another format, care must be taken to handle unsupported types (e.g., the nodes supported by MPEG-4 but not by VRML). For this purpose, the node table definition includes the *File Type* that maps the different nodes (for VRML and MPEG-4) and elements (for PowerPoint) according to the corresponding file type. While incorporating a node in a new animation sequence, it is first checked whether the node type is supported in the desired output format. Node types that are not supported in the output format have to be eliminated.

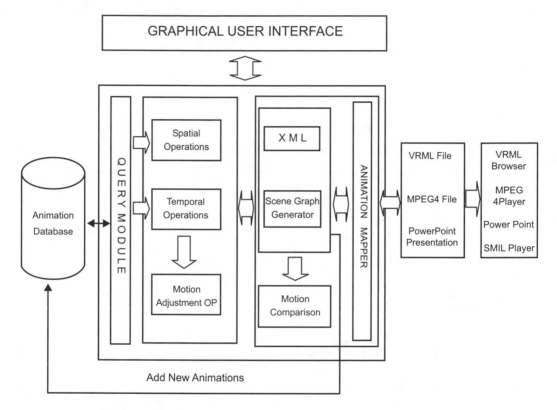

Figure 17.9 XML Mediator-based Animation Toolkit

3.3 QUERY PROCESSING

As discussed above, users can query the XML-based animation database for available animation models and motion sequences. The resulting models and motions can be combined using the spatial, temporal, and motion adjustment operations to generate new animation sequences. The current implementation of the XML-based animation database handles two different types of queries:

- Query on Metadata
- Query by Example

Query on Metadata: Normally, a user's query is resolved based on the metadata associated with a model and/or a motion. For instance, a query can be to retrieve *Walking* motion. This query is applied to the metadata associated with the motion table, and retrieves all the motions with the motion name being *Walking*. The motion sequences are extracted along with the corresponding model, as stored in the model table. The result is displayed as per the user's requested format. If the user is satisfied with the motion, s/he can retain it else the user has the option to choose a different motion. Also, the user can make changes on the displayed motion. For instance, user can modify the speed of the motion or modify the duration of animation (i.e., extend/reduce the animation time). These operations are supported as part of the motion adjustment operations using a set of Graphical User Interfaces (GUIs). The modified animation can be saved as a new entry in the database with the same motion name to process future requests.

Query By Examples: Users can also provide an example animation model and/or motion and query the database to find similar models and/or motions. The example animation model can be provided as a VRML/MPEG-4/PowerPoint file. This file is parsed and the tables in the database are updated appropriately. Also, a new XML file is created and new sceneId's and ObjectId's are assigned. This XML file of the model and/or animation is compared with the reference templates used for models and/or motions in the metadata generation phase. The resolution of queries by examples uses the same set of comparison algorithms discussed in Section 3.1.2 for metadata generation.

3.3.1 Performance of Query Resolution

The comparison of the example (in the query) and the reference motion/model can be realized in real time since the comparison is carried out on only a fixed number of reference templates. For example, given 10000 animation models and motions in the database, there might be only 10-15-reference templates for different motion sequences. (The numbers may be slightly more for models.) Hence, the example query is compared only with these 10-15 reference templates. Once a matching template is identified, models/motions in the database can be retrieved based on the metadata associated with the identified reference template. In the current implementation, the XML database has animation models and motion sequences whose sizes are of the order of 100-200 Kbytes. For these file sizes, the time for one reference template comparison is of the order of microseconds. It should be observed here that during the pre-processing stage we store motions and models separately in different tables. Hence, 100-200 Kbytes size range for simple motions and models is quite reasonable.

The animation model/motion size and hence the timings might increase when complex animations are considered (e.g., continuous sequence of different animation motions such as walking, running, and jumping). We are in the process of populating the database with such complex animation models and motion sequences to measure the performance of queries on complex examples of motions or models.

4. VIEWS AND GUI'S

After the development of the object model, the views with more detailed object models, which include the views, are developed. The model with the motion adjustment view and the query view is given in Figure 17.10.

Views are identified to accommodate key groupings of tasks and to embody key relationships between objects such that each view performs tasks leading to a specific outcome. For this system, seven views are identified: *Query View, Scene Graph View, Motion Adjustment View, Timeline View, Retarget Motion View, Motion Mapping View,* and *Model Metadata View.*

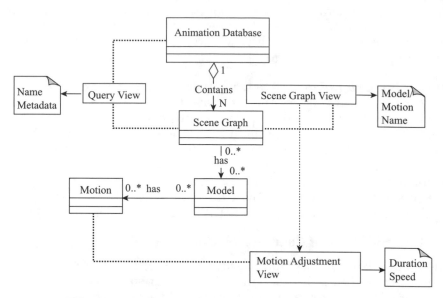

Figure 17.10 Detailed object model with views

4.1 QUERY VIEW

The query view, shown in Figure 17.11, is an integral part of the system. Through this view, the user is able to interact with the database in order to get required models and motions. Within the query view there are two views: one for model query and the other for motion query.

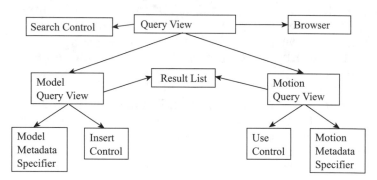

Figure 17.11 Query view

The Query GUI: This interface is the starting point of all animation reuse processes. This GUI contains an ActiveX object Web Browser that invokes the default web browser of the system with the installed VRML browser, such as Cosmo Player or MS VRML viewer. The interface also displays a ranked list of query results. Users can query for models or motions by specifying one or more of the metadata requirements. The search for the animation objects is automatically converted to an SQL-like statement to retrieve the best matching objects from the database. Retrieved objects are shown as a list. By clicking on an object the user selects it and can view it in the browser. Figure 17.13 (a) depicts the selection of a rolling motion after the search. Once the user finds an object of his/her liking s/he can INSERT (for models) or USE (for motion) it in the scene graph. For instance, Insert operation on an animation scene will be carried out using the pseudo-code listed in Figure 17.13 (b).

4.2 SCENE GRAPH VIEW

The motions and models inserted from the query view are displayed in the scene graph view (Figure 17.12). From the scene graph view, the properties of the motions and models can be viewed and changed using other views. VRML text can also be generated through this view.

The Scene Graph GUI: The scene graph can be activated directly by selecting a new file from the menu or via the query GUI upon INSERTing or Using animation objects. The scene graph is the anchoring point of reuse. All operations other than USE and INSERT are carried out, directly or indirectly, via this interface. Depending on the object and type of interaction with this interface, different views will pop up; these views cater to the various modification requirements. Users can DELETE or EXTRACT models and DISREGARD or GET motions from the scene graph. These interactions will lead to the generation of the respective operation command. The user can also explicitly tell the VRML Text Generator to generate the VRML Text using this interface. Figure 17.14 (a) shows this window in detail and Figure 17.14 (b) lists the pseudo-code of generate the VRML text from the scene graph.

4.3 MODEL METADATA VIEW

The model metadata view, Figure 17.15, provides users with the capability to change the properties of the models. The changes can be previewed in a browser before they are applied.

Model Metadata GUI: This GUI is activated when the user double clicks on a model icon in the scene graph. It generates the EDIT operation of the spatial operation set. Users can modify the spatial properties, size, orientation, and position of the models using this. Figure 17.16 shows this window in detail. For the example given, the EDIT statement generated after making the modifications and accepting them will be similar to the following:
EDIT *Dog_Wagging* POSITION *(0, 0, 0)* SIZE *(1, 1, 1)* ORIENTATION *(0, 0, 0, 0)* OF *Scene*

One or more of POSITION, , and ORIENTATION options will be used depending on the parameters changed by the user. Changes made by the user will then be reflected in the VRML file. Users can preview the changes in the browser. Figure 17.17 also shows the *VRML Text* GUI which pops up when the generate VRML button is clicked. This GUI shows the VRML text that is generated for the scene.

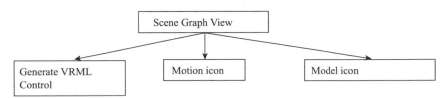

Figure 17.12 Scene Graph View

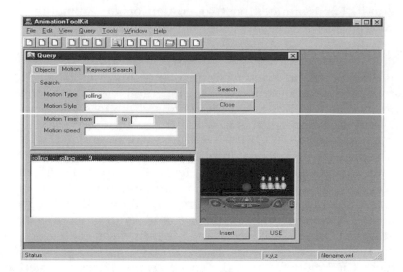

(a)

```
Procedure INSERT(ObjectID as Integer) {
    Open Database for reading
    Query the database table Object Inner Joined with
VRMLTEXT
        through SQL for the ObjectID

    Create a temporary scene graph node
    Read from the database record the necessary record
fields
        To fill in the attributes and metadata required by the
node:
        ObjectID, Category, ID, Name, Type, Size, Position,
        Color and VRMLText
    Parse through the current scene graph

    If no duplicates are found Then
        Insert the temporary node into the scene graph
        Insert in scene graph array
    Else
        Change the unique key of the temporary node and
declare
            As a new instance
        Insert the modified temporary node.
        Insert in scene graph array
    End if
    Close Database
    }
```

(b)

Figure 17.13. Query (motion) GUI and pseudo-code for inserting an object from
the database into the scene graph

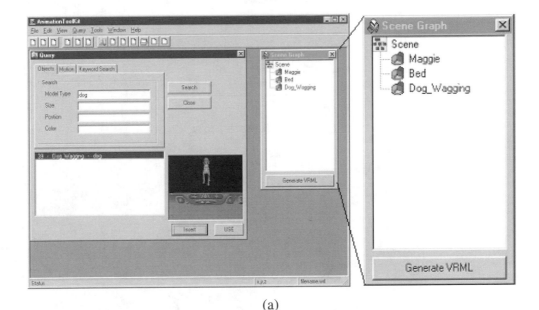

(a)

```
Procedure SceneGraph2File(Root as Node) {
    Declare PrototypeText, NodesText
            RouteText as String
    Declare AnimationEngine as String
     'javascript node that controls motion times

     Traverse all the nodes of the scene graph by DFS
       For each node {
         If node.type = object then
             Append node.prototypetext to PrototypeText
             Append node.text including new translation,
                    scaling an orientation to NodesText
         Else    'if it is a motion
                 Append node.text NodesText
                 Append node.routetext to RouteText
                 Get temporal information
                     and update AnimationEngine
         End if
       }
    Open File
    Save to File ( "#VRML V2.0 utf8" )
    Save to File ( Prototypetext + NodesText +
                   RouteText + AnimationEngine )
    Close File
}
```

(b)

Figure 17.14 The scene graph GUI and the pseudo-code to convert the scene to file

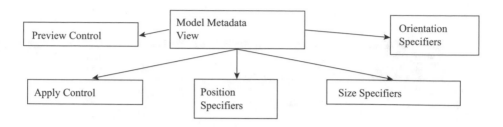

Figure 17.15 Model metadata view

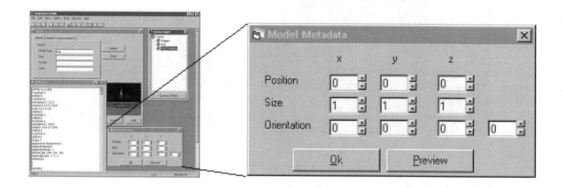

Figure 17.16 Model metadata GUI

4.4 MOTION ADJUSTMENT VIEW

Similar to the model metadata view, the motion adjustment view allows users to alter the temporal properties of the motion, i.e., the duration and the speed. This view is shown in Figure 17.17.

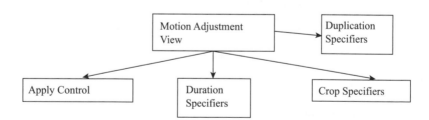

Figure 17.17 Motion Adjustment View

Motion Adjustment GUI: Similar to the Model Metadata GUI, this window is activated when the user double clicks on the motion icons. As the name suggests, this GUI caters to the operations in the Motion Adjustment operation set. This GUI supports the CROP, DUPLICATE, and CHANGE SPEED operations. One or more SQL-like statements will be generated depending on the parameters to which changes have been made. These statements will modify the timing values associated with a motion in the VRML file. A detailed view of this interface can be seen in Figure 17.18. The statements generated for this example will be:

CHANGE SPEED green_*ball.rolling* BY *2*
DUPLICATE green_*ball.rolling* BY *2*

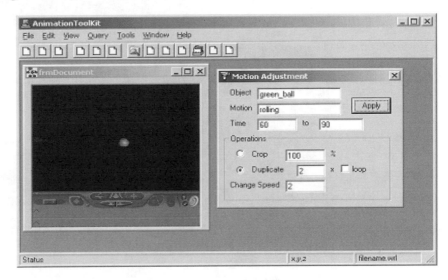

Figure 17.18 Motion adjustment GUI

4.5 MOTION RETARGET VIEW

The retarget motion view, Figure 17.19, facilitates the retargeting of a motion to another scenario, in terms of position and/or model. The effect of the retargeting can be previewed in a browser before they are applied.

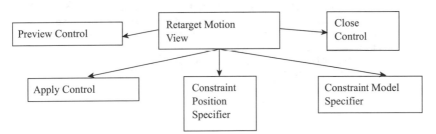

Figure 17.19 Retarget Motion View

For interactive using inverse kinematics for motion retargeting, it is necessary to develop GUI with visualization of 3D models and the animation, including the original and resulting motion sequences. Using GUI the user can specify the articulated figure and new position/orientation for which the motion is retargeted. There are three major components of this GUI, visualization window, scene graph, and control panel. One example is shown in Figure 17.20 (a), where we retarget the kicking motion to the ball. From the control panel, the model's (ball) position can be specified and used as the new target of the end effecter of the front left leg. The 3D model and animation are displayed in the visualization window, being able to be viewed from different viewpoint and zooming factors. In this way, users can intuitively adjust the motion sequences for reuse purpose. Once the result is satisfactory, it will be recorded and stored in database. In Figure 17.20 (b), the pseudo-code is listed for motion mapping.

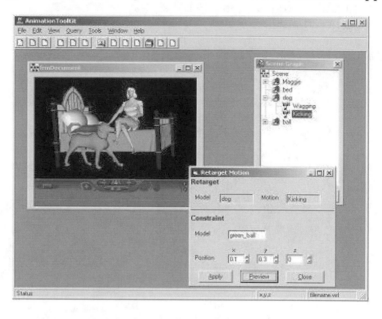

(a)

```
Procedure Use_Motion (MotionNode as SceneGraphNode,
                      ObjectNode as SceneGraphNode ) {
       Parse through the VRML Text Description of the Motion Node
             and extract the defined interpolator nodes and their
             type (Orientation or Position)
       Parse through the VRML Text Description of the Object Node
             and extract the defined joints/segments/nodes
       Using the GUI, a user can assign the correspondence
       between a joint and an interpolator node; otherwise, the
       default one will be applied
       Extract the Timer used in the MotionNode
       Clear the MotionNode.RouteText

       For every Joint in the Object Node {
           If Joint has an assigned Interpolator node Then
                 MotionNode.RouteText= MotionNode.RouteText +
               "ROUTE " + TimerName +".fraction_changed TO" +
               Interpolator +".set_fraction
               If  Interpolator.Type = Orientation Then
                  MotionNode.RouteText = MotionNode.RouteText +
                 "ROUTE " + Interpolator +".value_changed TO" +
                 JointName +".set_rotation
               Else 'Type = Position
                  MotionNode.RouteText = MotionNode.RouteText +
                 "ROUTE " + Interpolator +".value_changed TO" +
                 JointName +".set_translation
               End if  }
       Update Scene Graph Nodes }
```

(b)

Figure 17.20 Motion retargeting GUI and pseudo-code for motion mapping in VRML

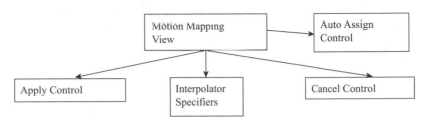

Figure 17.21 Motion mapping view

4.6 MOTION MAPPING VIEW

Using the motion mapping view, Figure 17.21, the interpolators of one model-motion combination can be mapped to another combination of model and motion. The mapping can also be auto assigned.

Motion Mapping GUI: This GUI is activated when a motion is used on a model other than the original model. Using this interface the interpolator nodes of one model can be linked to that of the other model. A sample *Motion Mapping* GUI is shown in Figure 17.22. The pseudo-code for motion mapping is similar to Figure 17.20(b). The USE statement generated after using and mapping the motion will be similar to the following.

Use *wiping* To *Scene_Maggie_on_Bed_with_dog_ball.Maggie*

Map *Barmaid*.r_shoulderXOI.OrientationInterpolator To

Maggie.r_shoulderXOI.OrientationInterpolator

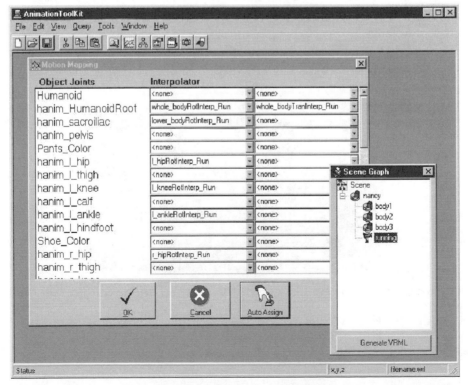

Figure 17.22 Motion mapping GUI

5. ANIMATION GENERATION: AN EXAMPLE

Let us consider a presentation on the solar system using Microsoft PowerPoint. Text may be used in slides to introduce information of each planet such as the name, the size, average temperature, distance to the sun and the earth, its orbit and moving speed. For viewing the moving solar system, the existing solution is to make a scaled-down model of the solar system containing the sun and all planets, with the complicated mechanical configuration and step motors to control their movement. This way is very expensive but works well. For a large audience, one may take the video of the model. Assume that we have the video in MPEG format. We can easily insert it into the PowerPoint. However, the problem is that the MPEG video is shot with a fixed camera. It is impossible for users to see from a new viewpoint with a specific viewing direction; and zooming parameters, furthermore, interact with the system. The XML mediator-based animation toolkit can be applied for this purpose.

> 1.INSERT *SOLAR_SYSTEM (MPEG-4)* TO *Scene*
>   PARENT *STARS_BACKGROUND* SAVE AS
>   *SOLAR_SYSTEM_MOTION*
> 2. GET *rotation, revolution* FROM *CROSSES (VRML)* SAVE AS
>   *rotation1, revolution1*
> 3. USE *rotation1, revolution1* TO
>   *SOLAR_SYSTEM_MOTION. SOLAR_SYSTEM*
> 4. CROP *SOLAR_SYSTEM.rotation1*, *SOLAR_SYSTEM.revolution1*
>   BY *3 0,70.*
> 5. GENERATE **user specified file (VRML/MPEG-4)**
> 6. USE **IN PowerPoint.**

Figure 17.23. Operation sequences for solar systems animation

Figure 17.24. Crosses scene with the rotation and revolution motions

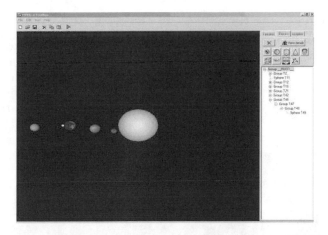

Figure 17.25. Solar system model as viewed in toolkit

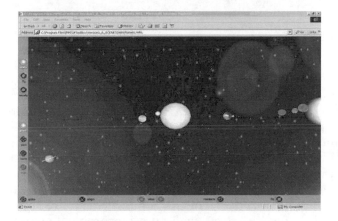

Figure 17.26. Generated VRML file for above queries

It is clear such a task is a complex animation that requires some interaction with more than one object and the combination of objects and motion in one scene. We can interactively generate this animation using our toolkit. The elaborate operation sequences on the XML database are outlined in Figure 17.23. First, we insert a static solar system model to a scene having stars background, by applying a query on the model table. The query results of 3D models are available in MPEG-4 format, in the database (Figure 17.25). Next, the rotation and revolution motions are selected by applying queries on the motion table. In the current animation database, the required rotation and revolution motions are available in VRML format (Figure 17.24). The selected rotation and revolution motions are applied to the solar system model, using the motion adjustment operations. Last, the resulting animation is generated in the required format (Microsoft PowerPoint in this example). The results of the operation sequences on the animation databases are shown in Figure 17.26 and Figure 17.27.

Thus, the animation result will look the same as the video but allow the viewers to interactively see it from any viewpoint. Users can interact with the animation and see the configuration of the solar system at any specific time and simulation of any period by inputting the timing. In the next subsection, we can see that the extensive operation sequences are applied on the XML database through the Graphical User Interfaces (GUIs) and hence a user need not even remember the syntax of the operations to be carried out.

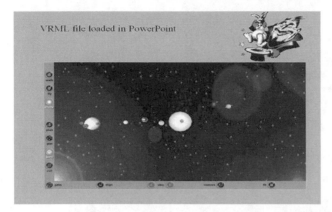

Figure 17.27. Above generated VRML file inserted in PPT

6. CONCLUSION

Animations help in adding an expressive flavor to multimedia presentations. Though creating new animations generally takes lots of effort, a database approach can be utilized for reusing models and motions to create new animations. Since animations are represented in different formats, we have proposed a XML (eXtensible Markup Language) based mediator for handling the multiple formats. The XML mediator can handle animation formats such as VRML, MPEG-4, SMIL, and Microsoft PowerPoint. Other animation formats can easily be incorporated, if needed. The XML based mediator approach is natural since most animation formats are based on a hierarchical structure called *scene graphs*. It also has the advantage that most databases (such as Oracle or Microsoft Access) support XML representations and manipulations.

It should be noted here that the XML mediator based animation toolkit presented in this paper does not try to solve motion transition problems or resolve object collision problems. Similar to other animation tools, users have to manually tweak the animations for proper object interactions.

The current implementation of the animation toolkit using XML mediator supports reuse of models and motions among formats such as VRML, MPEG-4, and Microsoft PowerPoint. We are working on incorporating animations represented in SMIL format. We are also developing more efficient techniques for (semi-) automatic metadata generation. These techniques will also help in efficient resolution of query by examples on complex animation models and motion sequences. The XML mediator approach can be extended to handle other techniques for animation generation. For instance, JPEG/BMP/GIF/TIF images can be used as texture maps in VRML file and then used in the proposed XML-based animation toolkit. In a similar manner, MPEG-1 and MPEG-2 can be converted to MPEG-4 and used in our model.

REFERENCES

[1]   Akanksha, Huang, Z., Prabhakaran, B. and Ruiz, Jr. C. R. Reusing Motions and Models in Animations. Proc. of EGMM 2001, 11-22. Also appears in Jorge, J.A., Correia, N. M., Jones H. and Kamegai M. B. (Eds.) Multimedia 2001, Springer-Verlag/Wien. (2002) 21-32.

[2]   Bruderlin, A. and Williams, L. Motion Signal Processing. Proc. ACM SIGGRAPH '95. (1995) 97-104.

[3]   Elad, M., Tal, A., and Ar, S., Content Based Retrieval of VRML Objects-An Iterative and Interactive Approach, Proc EGMM 2001, pp. 107-118. Also appears in Jorge, J.A., Correia, N. M., Jones H. and Kamegai M. B. (Eds.) Multimedia 2001, Springer-Verlag/Wien, ISBN 3-211-83769-8. (2002).

[4]   Gleicher, M. Retargeting Motion for New Characters. Proc. ACM SIGGRAPH '98. (1998) 33-42.

[5]   Hodgins, J. and Pollard, N. Adapting Simulated Behaviors For New Characters. Proc. ACM SIGGRAPH '97. Los Angeles, CA. (1997) 153-162.

[6]   International Organization for standardization organization, International Demoralization, ISO/IEC JTC1/SC29/WG11, coding of moving pictures and Audio, Overview of the MPEG-4 Standard.

[7]   Close, J.J. et al. VRML Next-Generation (VRML-NG) Requirements Document, Presented to the *VRMLConsortium*, Board of Directors, Draft 1.0, Dec. 1998.

[8]   Lee, W.M. and Lee M.G. An Animation Toolkit Based on Motion Mapping. IEEE Computer Graphics International. (2000) 11-17.

[9]   Hosseini, M. and Georganas, N.D., Suitability of MPEG4's BIFS for Development of Collaborative Virtual Environments, Technical Report, Multimedia Communications Research Laboratory School of Information Technology and Engineering, University of Ottawa, Ottawa, Canada.

[10]   3DMPEG-4 ToolBox: http://uranus.ee.auth.gr/pened99

[11]   Popovic, Z. and Witkin, A. Physically Based Motion Transformation. Proc. ACM SIGGRAPH '99. (1999) 11-19.

[12]   Rohlf, J. and Helman, J., IRIS Performer: A High Performance Multiprocessing Toolkit for Real-Time 3D Graphics, in Proc. ACM SIGGRAPH '95. 550-557.

[13]   Tolani, D., Goswami, A., and Badler, N., Real-time inverse kinematics Techniques for anthropomorphic limbs. *Graphical Models* 62(5), Sept. 2000, 353-388.

[14]   http://www.web3d.org/TaskGroups/source/xj3d.htm

[15]   Ankerst, M., Breunig, M.M., Kriegel, H.P., and Sander, J., Optics: Ordering points to identify the clustering structure. in Proc. of the 1999 ACM SIGMOD Int. Conf. on Management Data, Philadelphia, PA, p. 49-60. (1999).

[16]   Arvo, J. and Novins, K. Fluid sketches: continuous recognition and morphing of simple hand-drawn shapes, in Proc. ACM UIST '00. (2000) 73-80.

[17]   Ayadin, Y., Takahashi, H. and Nakajima, M. Database Guided Animation of Grasp Movement for Virtual Actors, in Proc. Multimedia Modeling '97. (1997) 213-225.

[18]   Ballreich, C.N. - 3D Model. 3Name3D. http://www.ballreich.net/vrml/h-anim/nancy_h-anim.wrl. (1997).

[19]   Boulic R., Huang Z., Magnenat-Thalmann N., and Thalmann D., Goal-Oriented Design and Correction of Articulated Figure 17.Motion with the Track System, *Journal of Comput. & Graphics*, 18 (4), p. 443-452. 1994.

[20]   Fu, K.S., Gonzalez, R.C., and Lee C.S.G., Robotics, Control, Sensing, Vision, and Intelligence, McGraw-Hill, (1987) p. 52-76, 84 -102, and 111-112.

[21]   Geroimenko, V. and Phillips, M. Multi-user VRML Environment for Teaching VRML: Immersive Collaborative Learning. Proc. Information Visualization. (1999).

[22]   Igarashi T. and Hughes J. F. A suggestive interface for 3D drawing. Proc. ACM UIST '01. (2001) 173-181.

[23]   Kakizaki, K. Generating the Animation of a 3D Agent from Explanatory Text. Proc. ACM MM '98. (1998) 139-144.

[24]   Reitemeyer, A. Barmaid Bot. http://www.geometrek.com/web3d/objects.html.

[25]   Roberts, D., Berry, D., Isensee, S., and Mullaly, J. Designing for the User with Ovid: Bridging User Interface Design and Software Engineering. London: Macmillan Technical Publishing. (1998).

[26]   Schroeder W. J., Martin K. M., and Lorenson W. E. The Design and Implementation of an Object-Oriented Toolkit for 3d Graphics and Visualization. IEEE Visualization '96. (1996), 93-100.

[27]   Sommerville, I., Software Engineering, Addison-Wesley, Fifth Edition, (1996)

[28]   Thalmann D., Farenc N., and Boulic R. Virtual Human Life Simulation and Database: Why and How. Proc. International Symposium on Database Applications in Non-Traditional Environments (DANTE'99). IEEE CS Press. (1999).

[29]   The VRML Consortium Incorporated. The Virtual Reality Modeling Language. http://www.vrml.org/Specifications/VRML97/. International Standard ISO/IEC 14772-1: (1997).

[30]   Vcom3D, Inc. - Seamless Solutions, Andy - H-Anim Working Group. http://www.seamless-solutions.com/html/animation/humanoid_animation.htm. (1998).

[31]   Watson J., Taylor D., Lockwood S., Visualization of Data from an Intranet Training Systems using Virtual Reality Modeling Language (VRML). Proc. Information Visualization. (1999).

18

VIDEO INTELLIGENT EXPLORATION

Chabane Djeraba

IRIN, Polytech'Nantes, Nantes University, France

djeraba@irin.univ-nantes.fr

Younes Hafri and Bruno Bachimont

INA, Bry-Sur-Marne, France

yhafri@ina.fr, bbachimont@ina.fr

1. INTRODUCTION

Content-based video exploration (retrieval, navigation, browsing) deals with many applications such as news broadcasting, video archiving, video clip management, advertising, etc.

The different applications mean different exploration parameters. In advertising applications, for example, scenes are generally short. An example of exploration may concern the fact of retrieving all document related to a style of shooting. In video clips, an example of retrieving documents may be based on some dance step described by, e.g., a sketch. In distant learning applications, we distinguish two types of educational documents. In the first one, audio stream presents the main content of tutorials. In the second one, both visual contents and audio stream highlight practical courses. Finally, in broadcasting applications, real-time interactive detection and identifications are of importance.

Different applications mean, too, different user behaviors. Any exploration system should help final users to retrieve scenes within large video bases. Three different user behaviors may be distinguished. In the first one, the user knows that the targeted scene is in the video base; the user will keep exploration until he finds the document. He should be able to describe the scene he looks for, and would be able to see at the first glance whether a suggested scene corresponds to his query. In the second one, the objective is to provide an accurate exploration tool, so that the user may decide as soon as possible whether the target is in the video base or not. In this case, the user does not know if the target scene exists in the video base. In the third one, the user has not a specific scene in mind and the search is fuzzy. The user simply explores video bases based on topics or some event occurring within it. In this case, the exploration should be highly structured, in order to drive the user through the big amounts of video bases. Since the user is not supposed to know exactly what he looks for, he should be able to scroll all responses quickly for selecting the relevant or irrelevant ones. Relevance feedback may be interesting in this case, since it permits the user to ameliorate interactively the quality of explorations. It is therefore essential that exploration systems have to represent retrieval units by compressed, complete and comprehensive descriptions. For all cases, textual annotations or visual characteristics, stored on video bases, should support the content exploration of video documents.

Based on these different applications, exploration parameters and user behaviors, it is evident that there are strong dependencies between content-based video exploration and different domains and usages. The fundamental requirements that ensure the usability of such systems are: obtaining compressed and exhaustive scene representations, and providing different exploration strategies adapted to user requirements.

The scope of the paper deals with the second requirement by investigating a new form of exploration based on historical exploration of video bases and user profiles. This new form of exploration seems to be more intelligent than traditional content-based exploration. The notion of intelligent exploration depends strongly on user-adaptive (profiling) exploration notion.

This new form of exploration induces the answer to difficult problems. An exploration system should maintain over time the inference schema for user profiling and user-adapted video retrieval. There are two reasons for this. Firstly, the information itself may change. Secondly, the user group is largely unknown from the start, and may change during the usage of exploration processes. To address these problems, the approach, presented in this paper, models profile structures. The video server should automatically extract and represent in Markov models user profiles in order to consider the dynamic aspect of user behaviors.

Thus, the main technical contribution of the paper is the notion of probabilistic prediction and path analysis using Markov models. The paper provides solutions, which efficiently accomplish such profiling. These solutions should enhance the day-to-day video exploration in terms of information filtering and searching.

The agreement calls for the participating video bases to track their users so the advertisements can be precisely aimed at the most likely prospects for scenes and shots. For example, a user who looks up a tourist scene about Paris on video bases in the video base might be fed ads for images or scenes of hotels in Paris.

The paper contains the following sections. Section 2 situates our contribution among state-of-the-art approaches. Section 3 describes user-profiling based video exploration. Section 4 highlights the general framework of the system. Finally, section 5 concludes the paper.

2. CONTRIBUTION

There are few intelligent video exploration tools based on user profiles. The majority of state-of-the-art works concerned web applications. More particularly, they concerned user profiles, adaptive web sites [14], web log mining, OLAP [15], [23], [19], intelligent agents that detects user web topics [5], [8], [20], extraction of the most interesting pages [21], study of web of performance of various caching strategies [2] and continuous Markov models to influence caching priorities between primary, secondary and tertiary storages [18].

The particularity of our approach consists of:

-   Applying probabilistic exploration using Markov models on video bases. Each state of the Markov model corresponds to a scene displayed by the user. In addition, each transition corresponds to the probability to explore a scene after another one.

-   Avoiding the problem of Markov model high-dimensionality and sparsity. The Markov model matrix is typically very large ($\aleph*\aleph$ for $\aleph$ scenes). Moreover, the Markov matrix is usually very sparse for high-dimensional matrices. To avoid such problems, the approach consists of clustering video scenes, based on their content, before applying the Markov analysis. Therefore, we propose the grouping of similar states into clusters. The transition matrix lines and columns will concern the gravity centers of clusters. Newest dimension = $\log \aleph$. $\aleph$ is the number of scenes before clustering. The solution is clearly scalable for very large video bases.

To our knowledge, there are no solutions based on video exploration prediction. We believe that applying probabilistic approaches such as Markov models to video exploration inaugurates a new form of video exploitation. Furthermore, when we consider web applications, which are different than video explorations, not all the state-of-the-art works extract the sequence of hyper-link based on probabilistic information.

3. USER PROFILING AND VIDEO EXPLORATION

3.1 DEFINITION

Profiling is a business concept from the marketing community, with the aim of building databases that contain the preferences, activities and characteristics of clients and customers. It has long been part of the commercial sector, but which has developed significantly with the growth of e-commerce, the Internet and information retrieval.

The goal of profiling video exploration is to have the most complete picture of the video users we can.

Mining the videos server, which is also achieved using cookies, carries out profiling scenes off-line. The long-term objective is to create associations between potential commercial video sites and commercial marketing companies. These sites make use of specific cookies to monitor a client's explorations at the video site and record data that users may have provided to the exploration server.

User likes, dislikes, browsing patterns and buying choices are stored as a profile in a database without user knowledge or consent.

3.2 SCENE FEATURES

We consider a robot named Access that extracts, invisibly, features from video user explorations. Access is synchronized with the playing of a scene that resulted from user browsing or queries.

To enable Access to compile meaningful reports for its client video sites and advertisers and to better target advertising to user, Access collects the following type of non-personally identifiable feature about users who are served via Access technology, User IP address, a unique number assigned to every computer on the Internet. Information, which Access can infer from the IP address, includes the user's geographic location, company and type and size of organization (user domain type, i.e., .com, .net, or .edu.), a standard feature included with every communication sent on the Internet. Information, which Access can infer from this standard feature, includes user browser version and type (i.e., Netscape or Internet Explorer), operating system, browser language, service provider (i.e., Wanado, Club internet or AOL), local time, etc., the manner of using the scene or shot visit within an Access client's site. For example, the features represent instant of start and instant of end of video scene the user views. Affiliated advertisers or video publishers may provide access with non-personally identifiable demographic features so that user may receive ads that more closely match his interests. This feature focuses on privacy protections as the access-collected non-personally identifiable data. Access believes that its use of the non-personally identifiable data benefits the user because it eliminates needless repetition and enhances the functionality and effectiveness of the advertisements the users view which are delivered through the Access technology.

The feature collected is non-personally identifiable. However, where a user provides personal data to a video server, e.g., name and address, it is, in principle, possible to correlate the data with IP addresses, to create a far more personalized profile.

3.3 MATHEMATICAL MODELING

Given the main problem "profiling of video exploration," the next step is the selection of an appropriate mathematical model. Numerous time-series prediction problems, such as in [24], supported successfully probabilistic models. In particular, Markov models and Hidden Markov Models have been enormously successful in sequence generation. In this paper, we present the utility of applying such techniques to prediction of video scene explorations.

A Markov model has many interesting properties. Any real world implementation may statistically estimate it easily. Since the Markov model is also generative, its implementation may derive automatically the exploration predictions. The Markov model can also be adapted on the fly with an additional user exploration feature. When used in conjunction with a video server, this later may use the model to predict the probability of seeing a scene in the future given a history of accessed scenes. The Markov state-transition matrix represents, basically, "user profile" of the video scene space. In addition, the generation of predicted sequences of states necessitates vector decomposition techniques.

Markov model creations depend on an initial table of sessions in which each tuple corresponds to a user access.

```
INSERT INTO [Session] ( id, idsceneInput )
SELECT[InitialTable].[Session_ID], [InitialTable].[Session_First_Content]
FROM OriginalTable
GROUP BY [InitialTable].[Session_ID], [InitialTable].[Session_First_Content]
ORDER BY [InitialTable].[Session_ID];
```

3.4 MARKOV MODELS

A set of three elements defines a discrete Markov model:

$$\{\alpha, \beta, \lambda\}$$

α corresponds to the state space. β is a matrix representing transition probabilities from one state to another. λ is the initial probability distribution of the states in α.

Each transition contains the identification of the session, the source scene, the target scene and the dates of accesses. This is an example of transition table creation.

```
SELECT       regroupe.Session_ID,       regroupe.Content_ID      AS      idPageSource,
regroupe.datenormale    AS     date_source,    regroupe2.Content_ID    AS    idPageDest,
regroupe2.datenormale AS date_dest INTO transitions
FROM [2] regroupe sessions from call4] AS regroupe, [2] regroupe sessions from
call4] AS regroupe2
WHERE                     (((regroupe2.Session_ID)=regroupe!Session_ID)                And
((regroupe2.Request_Sequence)=regroupe!Request_Sequence+1));
```

The fundamental property of Markov model is the dependencies of the previous states. If the vector α [ι] denotes the probability vector for all the states at time 'ι', then:

$$\alpha (\iota) = \alpha (\iota-1). \beta$$

If there are '$\aleph$' states in the Markov model, then the matrix of transition probabilities β is of size $\aleph$ x $\aleph$. Scene sequence modeling supports Markov model. In this formulation, a Markov state corresponds to a scene presentation, after a query or a browsing.

Many methods estimate the matrix β. Without loss of generality, the maximum likelihood principle is applied in this paper to estimate β and λ. The estimation of each element of the matrix λ [v, v'] respect the following formula:

$$\alpha (v, v') = \varphi (v, v') / \sum_{v''} \varphi (v, v'). \ \partial (v) = \varphi (v) / \sum_{v'} \varphi (v')$$

φ (v, v') is the count of the number of times v' follows v in the training data. We utilize the transition matrix to estimate short-term exploration predictions. An element of the matrix state, say λ [v, v'], can be interpreted as the probability of transitioning from state v to v' in one step. Similarly, an element of $\alpha^*\alpha$ will denote the probability of transitioning from one state to another in two steps, and so on.

Given the "exploration history" of the user ξ (ι-κ), ξ (ι-κ+1).... ξ (ι-1), we can represent each exploration step as a vector with a probability 1 at that state for that time (denoted by τ (ι-κ), τ (ι-κ+1)... τ (ι-1)). The Markov model represents estimation of the probability of being in a state at time 'ι' is shown in:

$$\alpha\ (\iota) = \tau\ (\iota\text{-}1).\ \alpha$$

The Markovian assumption varies in different ways. In our problem of exploration prediction, we have the user's history available. Answering to which of the previous explorations are "good predictors" for the next exploration creates the probability distribution. Therefore, we propose variants of the Markov process to accommodate weighting of more than one history state. So, each of the previous explorations are used to predict the future explorations, combined in different ways. It is worth noting that rather than compute $\lambda^*\lambda$ and higher powers of the transition matrix, these may be directly estimated using the training data. In practice, the state probability vector $\alpha\ (\iota)$ can be normalized and compared against a threshold in order to select a list of "probable states" that the user will choose.

3.5 PREDICTIVE ANALYSIS

The implementation of Markov models into a video server makes possible four operations directly linked to predictive analysis.

In the first one, the server supports Markov models in a predictive mode. Therefore, when the user sends an exploration request to the video server, this later predicts the probabilities of the next exploration requests of the user. This prediction depends on the history of user requests. The server can also support Markov models in an adaptive mode. Therefore, it updates the transition matrix using the sequence of requests that arrive at the video server.

In the second one, prediction relationship, aided by Markov models and statistics of previous visits, suggests to the user a list of possible scenes, of the same or different video bases, that would be of interest to him, and then the user can go to the next. The prediction probability influences the order of scenes. In the current framework, the predicted relationship does not strictly have to be a scene present in the current video base. This is because the predicted relationships represent user traversal scenes that could include explicit user jumps between disjoint video bases.

In the third one, there is generation of a sequence of states (scenes) using Markov models that predict the sequence of states to visit next. The result returned and displayed to the user consists of a sequence of states. The sequence of states starts at the current scene the user is browsing. We consider default cases, such as, if the sequence of states contains cyclic state, they are marked as "explored" or "unexplored." If multiple states have the same transition probability, a suitable technique chooses the next state. This technique considers the scene with the shortest duration. Finally, when the transition probabilities of all states from the current state are too weak, then the server suggests to the user the come back to the first state.

In the fourth one, we refer to video bases that are often good starting points to find documents, and we refer to video bases that contain many useful documents on a particular topic. The notion of profiled information focuses on specific categories of users, video bases and scenes. The video server iteratively estimates the weights of profiled information based on the Markovian transition matrix.

3.6 PATH ANALYSIS AND CLUSTERING

To reduce the dimensionality of the Markov transition matrix β, a clustering approach is used. It reduces considerably the number of states by clustering similar states into "similar groups." The originality of the approach is to discover automatically the number of clusters. The reduction obtained is about log $\aleph$, where $\aleph$ is the number of scenes before clustering.

3.6.1 How to discover the number of clusters: k

A critical question in the partition clustering method is the number of clusters (k). In the majority of real world experimented methods, the number of cluster k is manually estimated. The estimation needs a minimum knowledge on both data databases and applications, and this requires the study of data. Bad values of k can lead to very bad clustering. Therefore, automatic computation of k is one of the most difficult problems in cluster analysis.

To deal with this problem, we consider a method based on the confidence measures of the clustering. To compute the confidence of the clustering, we consider together two-levels of measures: the global level and the local level. Global Variance Ratio Criterion measure (GVRC) inspired of [4] underlines the global level, and Local Variance Ratio Criterion (LVRC) inspired of [9] measure underlines the local level. GVRC has been never used in partition clustering. It computes the confidence of the whole clusters, and LVRC computes the confidence of individual clusters, known also by silhouette. Therefore, we combine the two levels of measures to correct the computation of the clustering confidence.

Our choices of GVRC and LVRC confidence measures are based on the following arguments. Previous experiments, in [13], in the context of hierarchical clustering, examined thirty measures in order to find the best number of clusters. They apply all of them to a test data set, and compute how many times an index gives the right k. GVRC measure presented the best results in all cases, and GVRCN normalizes GVRC measure (GVRCN value is between 0 and 1). The legitimate questions may be: how is this measure (GVRC) good for partition methods? GVRC is a good measure for hierarchical methods. Is GVRC also good for partition methods? Are there any direct dependencies between data sets (in our case image descriptors) and the confidence of clustering measures? What are relationships between the initial medoids and the measures of the confidence of clustering?

It is too hard to answer all these questions in the same time without theory improvements and exhaustive real world experiments. Therefore, our approach objective is not to answer accurately to all these important questions, but to contribute to these answers by presenting results of a real world experiment. We extend a measure of clustering confidence, by considering the best one presented in the literature [13]. Then, we experimented this measure in our state space.

As said a few lines ago, we combine the two levels of measures to accurate the computation of the clustering confidence. The combination of two levels of measures is an interesting particularity of our method. It avoids the clustering state where the max value of average confidence of clusters does not mean the best confidence of clustering. The method is improved by identifying the best cluster qualities with the lowest variance.

GVRC is described by the following formula.

$$VRC\ (k) = \frac{\dfrac{trace\ (B)}{k-1}}{\dfrac{trace\ (W)}{n-k}}$$

n is the cardinality of the data set. K is the number of clusters. *Trace(B)/(k-1)* is the dispersion (variance) between clusters. *Trace (W)/n-k* is the dispersion within clusters. The expanded formula is:

$$GVRC\ (k) = \frac{n-k}{k-1} \frac{\left(\sum_{i=1}^{n}\left\|x_i - \overline{x}\right\|^2\right) - \left(\sum_{l=1}^{k}\left(\sum_{x_j \in Cl}\left\|x_j - \overline{x_l}\right\|^2\right)\right)}{\sum_{l=1}^{k}\left(\sum_{x_j \in Cl}\left\|x_j - \overline{x_l}\right\|^2\right)}$$

The best clustering coincides with the maximum of *GVRC*. To normalize the *GVRC* value (*0≤GVRCN≤1*), we compute GVRCNormalized (GVRCN), where *GVRC_max = GVRC (k')*, and $\forall\ k$, $0 < k \le k\_max$, $k \ne k'$, *GVRC (k) < GVRC (k')*. *K_max* is the maximum of cluster considered. *K_max* ≤ the cardinal of the data items.

$$GVRCN(k) = \frac{GVRC(k)}{GVRC\_\max}$$

Local Ratio Criterion (LVRC) measures the confidence of the *clusterj (Cj)*. The formula is:

$$LVRC(c_j) = \frac{\sum_{i=1}^{cardinality(cj)} lrc_{xi}}{cardinality\ (c_j)}\ , \text{ with } lrc_{xi} = \frac{b_{xi} - a_{xi}}{\max(a_{xi}, b_{xi})}$$

lvrc measures the probability of x_i to belong to the cluster C_{xi}, where a_{xi} is the average dissimilarity of object x_i to all other objects of the cluster C_{xi} and b_{xi} is the average dissimilarity of object x_i to all objects of the closest cluster C'_{xi} (neighbor of object x_i). Note that the neighbor cluster is like the second-best choice for object x_i. When cluster C_{xi} contains only one object x_i, the s_{xi} is set to zero ($lvrc_{xi} = 0$).

3.6.2 K-automatic discovery algorithm

The algorithm is run several times with different starting states according to different k values, and the best configuration of k obtained from all runs is used as the output clustering.

We consider the sequence variable *sorted_clustering* initialized to empty sequence (< >). *sorted_clustering* contains a sorted, on basis of *confidencei*, sequence of elements in the form of *<confidencei,ki>* where *ki* is the cluster number at *i* iteration and *confidencei* is the global variance ratio criterion and all local variance ratio criterion associated to k_i clusters. *sorted_clustering* = <....., <*confidence*$_{i-1}$,k_{i-1}>, <*confidence*$_i$,k_i>>, where *confidence*$_i$ = <GVRC(k), <LVRC(c$_1$), LVRC(c$_2$), ..., LVRC(c$_{ki}$)> >.

The algorithm follows five steps:

The first one initializes the maximum number of k authorized (max_k), by default $max\_k = n/2$. n is the number of data items.

The second step applies the clustering algorithm at each iteration k_i (for $k=1$ to max_k), computes GVRC(k) and, for each k value, computes LVRC(c$_j$), for $j = 1, ..,$ k. confidence$_i$ = < GVRC(k),<LVRC(c$_1$), LVRC(c$_2$), ..., LVRC(c$_k$)> >.

```
K-discovery()
{
Set max_k //By default max_k = n/2
sorted_clustering ← < > /* empty sequence */
sorted_clustering contains a sequence of sorted qualities of clustering and
associated k. sorted_clustering = <...., <confidence_{i-1},k_{i-1}>, <confidence_i,k_i>>, where
confidence_{i-1} < confidence_{i-1}
<confidence_i,k_i> = <GVRC(k_i), k_i>
// The clustering confidence is often the best when the value is high.
for k=1 to max_k do
  Apply chosen clustering algorithm.
Current ← GVRC(k),<LVRC(c_1), LVRC(c_2), ...., LVRC(c_k) >
/* GVRC(k) = confidence of actual clustering */
sorted_clustering ← insert <k, current> in best_clustering, by considering sorted
sequence of sorted_clustering.
 end for
GVRC_max ←•max_confidence (sorted_clustering) /* GVRC_max = best_clustering */
correct_sorted_clustering ←•< >
for k=1 to max_k do
```

if $VRCN(k) = \dfrac{VRC(k)}{VRC\_max} \geq 1\%$ and $\forall j \in [1,k] LRC(c_k) \geq 1\%$

```
then /* C_k is a correct cluster */
    correct_sorted_clustering        ←        insert        <k,        confidence_k>        in
correct_sorted_clustering>
 end if
end for
if correct_sorted_clustering = <>
then
{
for i=1 to cardinal(correct_sorted_clustering) do
{
```
moving false or weak data items of C_{ij} for which $LRC(c_{ikj}) < 1\%$ to *Noisy_cluster*
```
if cardinal(Noisy_cluster) < 10% then k-discovery() without considering the noisy
data items in Noisy Class
}
Return correct_sorted_clustering.
}
```

The third step normalizes the global variance ratio criterion (GVRC), by computing global variance ratio criterion normalized (GVRCN). *GVRC_max* = *max_GVRC(k)* where *GVRC_max* corresponds to the best_clustering.

The fourth step considers only k clustering for which GVRCN(k) = GVRC(k)/GVRC_max is the minimum value and LVRC(k) is the maximum. So, $\dfrac{\sum_{i=1}^{k} LVRC\ (C_i)}{k} + (1 - GVRCN\ (k))$ is the maximum, and $\forall j \in [1,k] LVRC(c_k) \geq 1\%$. The results are sorted in *correct_sorted_clustering*. If the final *correct_sorted_clustering* is not empty, therefore, we have at least one solution and the algorithm is stopped. If not, the current step is followed by the fifth step.

The fifth step looks for false or weak clusters (LVRC < 1 %). All false or weak data items of these false or weak clusters are moved in a specific cluster, called "Noisy cluster." If the cardinality of the Noisy cluster is less than 10% then we compute the k–discovery without considering the false or weak data items.

However, if the cardinal of the Noisy cluster is greater than 10%, then we consider that there are too many noises and the data item features of the initial databases should be reconsidered before applying again the *k*-discovery algorithm. We deduce that the data item descriptors are not suited to clustering analysis.

3.6.3 Clustering algorithm

The clustering algorithm is a variant of k-medoids, inspired of [16]. The particularity of the algorithm is the replacement of sampling by heuristics. Sampling consists of finding better clustering by changing one medoid. However, finding the best pair (medoid, item) to swap is very costly ($O(k(n-k)^2)$).

```
Clustering()
{
Initialize num_tries and num_pairs
min_cost ← big_number
for k=1 to num_tries do
   current ← k randomly selected items in the entire data set.
   l ← 1
   repeat
     xi ← a randomly selected item in current
     xh ← a randomly selected item in {entire data set current} .
     if TCih <0 then
        current ← currentxi+xh
     else
        j ← j+1
     end if
   until j• num_pairs
   if min_cost < cost(current) then
     best ← current .
   end if
end for
Return best.
}
```

That is why heuristics have been introduced in [16] to improve the confidence of swap (medoid, data item). To speed up the choice of a pair (medoid, data item), the algorithm sets a maximum number of pairs to test (*num_pairs*), then chooses randomly a pair and compares the dissimilarity (the comparison is done by evaluating TC_{ih}).

If this dissimilarity is greater than the actual dissimilarity, just go on choosing pairs until the number of pairs chosen reaches the maximum fixed. The medoids found are very dependent of the *k* first medoids selected. So the approach selects *k* other items and restarts *num_tries* times (*num_tries* is fixed by user). The best clustering is kept after the *num_tries* tries.

4. GENERAL FRAMEWORK

The general framework is composed of three tasks: temporal structure extraction, meta-data generation and Probabilistic prediction.

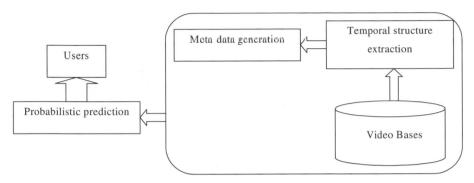

Figure 18.1 General framework

4.1 TEMPORAL STRUCTURE EXTRACTION

The first task "temporal structure extraction" collects temporal data and structures from video bases. The temporal structure of a video document induces its partitioning into basic parts that are defined at four different levels.

Frame level: A frame is the low-level unit of composition. There is insignificant temporal structure at this level.

Shot-level: A shot is a sequence of frames acquired through a continuous camera recording. The partitioning of the video into shots generally does not refer to any semantic analysis. It is the first event of temporal structure.

Scene-level: A scene is a sequence of shots having a common semantic significance. Our approach considers this level as a unit of exploration.

Video-level: The video level represents the whole document.

The first key-level that includes temporal structure is the shot-level, and specific operations (cut, dissolve and wipe) characterize shot boundaries. Cut is a formal boundary between contiguous shots. This generally implies a peak in the difference between color or motion histograms corresponding to the two frames surrounding the cut. Cut detection may therefore simply consist in detecting such peaks. Adding any form of temporal smoothing will also improve the robustness of the detection process. Dissolve is a fuzzy boundary between contiguous shots. The content of last images of the first shot continuously overlaps the first images of the second shot. The major issue here is to distinguish between dissolve effects and changes induced by global motion. Fade-in and fade-out effects are special cases of dissolve transitions where the first or the second scene, respectively, is a dark frame [17]. Wipe is another fuzzy boundary between contiguous shots. The images of the second shot continuously cover or push out of the display of the first shot.

While shot extraction, through cut detection, is relatively easy due to the abrupt nature of transitions, dissolves and wipes are more difficult to detect. Some efficient solutions ([3], [12], [22]) exploit the compressed structure of MPEG files, based on global motion estimation and segmentation. Dissolve effects at high scale consider elaborated operations such as mosaics and whirls. However, depending on the particular type of frame mixing technique, dissolve detectors may be misled by the apparent motion induced by such effects. We are not aware of any technique specialized in detecting elaborated gradual effects.

A deep understanding of the contents of the shots defines a scene. Automated scene annotation relies on a high-level clustering of shots where the indexing data derived from shots composes feature vectors. Depending on the video, the segmentation of shots may lead to a small and manageable set of objects (shot representations). In this case, a human operator can reliably solve the definition of scenes. It is important to note that a shot segmentation performed by a human operator may not be fully reliable, because this task is tedious and calls for a constant concentration. The development of semi-automated segmentation and annotation tools is therefore important in this context [11]. In our framework, the system extracts semi-automatically the scenes that represent Markov states.

4.2 META-DATA GENERATION

The second task is "meta-data generation." The end of video temporal segmentation triggers this task. The objective is to support quick reference [7].

The content of the meta-data varies, depending on the application towards which the video base is oriented. For generic video documents, this data generally includes video scenes and shots boundaries along with some characteristic and visual representation.

One common representation is the choice of one or more key frames within the shot or the scene. The assumptions of the temporal segmentation means normally consistency of all frames within a basic shot. Therefore, the system may derive heuristics for choosing one or more key-frames. The simplest relies on the global position of the frames within the shot. The system may also use some other characteristics such as the corresponding audio stream, for efficient key frame detection [1] [6]. Video micro-segmentation [10] refers to the process of re-segmenting the shots with respect to some heuristics reflecting a comprehension of the video content.

4.3 PROBABILISTIC PREDICTION

The third task "Probabilistic prediction" is composed of four major actions. The first one is Markov model, which consists of a sparse matrix, with a suitable low-level representation, of state transition probabilities, and the initial state probability vector. The second one is user access memory. All user's requests are temporarily saved into the user memory, and flushed once a minimum sample threshold is exceeded, or the session times out. The system assigns to each user a separate memory, and the sequence of user requests temporarily saved in the memory. The third one is an up-date action. This action updates the Markov model with available user path trace. The system typically updates the Markov model, by smoothing the current count matrix with the counts derived from the additional path sequences available. The fourth one is exploration path generator. Given a start scene identifier, the system outputs a sequence of states. The output predicts a chain of scenes the user may follow.

5. CONCLUSION

The intelligent video exploration is based on the fact that the large video bases agree to feed information about their user's reading, shopping and entertainment habits into systems that are already tracking the moves of millions of video users, recording where they go and what they read, often without a user knowledge.

The future video bases, particularly video web services, will amass detailed records of who uses their video bases and how they use them. However, this "new" future industry highlights the most ambitious effort yet to gather disparate bits of personal information into central databases containing electronic information on potentially every person who surfs the video bases. Profiling explorations are in the interest of the user, providing more customized and directed services through the video bases.

Many users may see the practice of video profiling as a violation of basic privacy rights. The system will extract, represent and distribute the data without user consent. In the context of web profiling, many marketing agencies discuss mechanisms for 'opt out,' whereby a user can decide when and where they want their data to be collected and processed. The proliferation of video bases will certainly highlight discussions on such mechanisms. However, in the context of web profiling, many privacy organizations are demanding an 'opt in' policy should be the norm, whereby users are given the freedom to choose anonymity over profiling. When we consider video bases over the Internet, the future of an unrestrained online and industry is clear: hundred of millions of personal documents will contain vast amounts of information about every aspect of people's behaviors. Moreover, this requires the achievement of two objectives that seem to be contradictory: the development of efficient customized-exploration tools with respect for privacy rights.

REFERENCES

[1]   Y. S. Avrithis, A. D. Doulamis, N. D. Doulamis, and S. D. Kollias, A stochastic framework for optimal key frame extraction from MPEG video databases. Computer Vision and Image Understanding, special issue on content-based access for image and video libraries, 75(1/2):3-24, July/August 1999.

[2]   P. Barford, A. Bestavros, A. Bradley, and M. Crovella, Changes in web client access patterns: Characteristics and caching implications, World Wide Web, special issue on Characterization and Performance Evaluation, 1998.

[3]   P. Bouthemy, M. Gelgon and F. Ganansia, A unified approach to shot change detection and camera motion characterisation, Technical Report 1148, IRISA, Rennes, France, 1997.

[4]   T. Calinski and J. Harabasz, A dendrite method for cluster analysis, Communications in Statistics, (3):1_27, 1974.

[5]   D. W. Chueng, B. Kao, and J. W. Lee, Discovering user Access patterns on the World-Wide Web, Proc. Of First Pacific-Asia Conference on Knowledge Discovery and Data Mining (PAKDD-97), 1997.

[6]   M. G. Christel, M. A. Smith, C. R. Taylor, and D. B. Winkler, Evolving video skims into useful multimedia abstractions, In ACM CHI'98 Conference on Human Factors in Computing Systems, Los Angeles, CA, April 1998.

[7]   F. Idris and S. Panchanathan, Review of image and video indexing techniques, Journal of Visual Communication and Image Representation, 8:146-166, 1997.

[8]   T. Joachims, D. Freitag, T. Mitchell, WebWatcher: A Tour Guide for the World Wide Web, in proc. of IJCAI'97, 1997.

[9]   L. Kaufman and P. J. Rousseeuw, Finding Groups in Data: An Introduction to Cluster Analysis, John Wiley & Sons, March 1990.

[10] S. Marchand-Maillet and B. Mérialdo, Stochastic models for face image analysis, In European Workshop on Content-Based Multimedia Indexing, CBMI'99, Toulouse, France, October 25-27 1999.

[11] J. Meng and S.-F. Chang, CVEPS - a compressed video editing and parsing system, In Proceedings of ACM Multimedia 96, Boston, MA, 1996.

[12] R. Milanese, F. Deguillaume and A. Jacot-Descombes, Video segmentation and camera motion characterization using compressed data. In C.-C. J. Kuo, S.-F. Chang and V. N. Gudivada, editors, In SPIE proc. Multimedia Storage and Archiving Systems II, volume 3229, Dallas, TX, 1997.

[13] G. W. Milligan and M. C. Cooper, An examination of procedures for determining the number of clusters in a data set, Psychometrika, 50(1):159_179, 1985.

[14] M. Perkowitz and O. Etzioni, Towards Adaptive Web Sites: Conceptual Framework and Case Study, WWW8, Toronto, 1999.

[15] J. McQueen, Some methods for classification and analysis of multivariate observations. In Proc. of the Fifth Berkeley Symposium on Mathematical Statistics and Probability, pages 281-297, 1967.

[16] R.T. Ng and J. Han, Efficient and effective clustering methods for spatial data mining, Technical Report TR-94-13, Department of Computer Science, University of British Columbia, May 1994.

[17] R. Ruiloba, P. Joly, S. Marchand-Maillet, and G. Quenot, Towards a standard protocol for the evaluation of temporal video segmentation algorithms, In European Workshop on Content-Based Multimedia Indexing, CBMI'99, Toulouse, France, October 25-27 1999.

[18] S. Schechter, M. Krishnan and M. D. Smith, Using path profiles to predict HTTP requests, WWW7, 1998.

[19] C. Shahabi, A. M. Zarkesh, J. Adibi, and V. Shah, Knowledge Discovery from Users Web-Page navigation, IEEE RIDE, 1997.

[20] A. Wexelblat and P. Maes, Footprints: History-Rich Tools for information Foraging, CHI'99, 1999.

[21] T. W. Yan, M. Jacobsen, H. Garcia-Molina and U. Dayal, From user Access Patterns to Dynamic Hypertext linking, Fifth Intl. World Wide Web Conference, May 1996.

[22] H. H. Yu and W. Wolf, A hierarchical multiresolution video shot transition detection scheme, Computer Vision and Image Understanding, 75(1/2):196-213, 1999.

[23] O. R. Zaiane, M. Xin, and J. Han, Discovering Web Access Patterns and Trends by Applying OLAP and data mining Technology on Web logs, Proc. Advances in Digital Libraries Conf. (ADL'98), pages19-29, Santa Barbara, CA, 1998.

[24] Z. Ni, Normal orthant probabilities in the equicorrelated case, Jour. Math. Analysis and Applications, n° 246, 280-295, 2000.

19

A VIDEO DATABASE ALGEBRA

A. Picariello
Dipartimento di Informatica e Sistemistica
Università di Napoli "Federico II"
Via Claudio, 21 – 80125 Napoli, Italy
`antonio.picariello@unina.it`

M.L. Sapino
Dipartimento di Informatica
Università di Torino
Corso Svizzera 185 – 10149 Torino, Italy
`mlsapino@di.unito.it`

V.S. Subrahmanian
Department of Computer Science
University of Maryland
A.V. Williams Building
College Park, MD 20742, USA
`vs@cs.umd.edu`

1. INTRODUCTION

Over the past few years, there has been a spectacular increase in the number of applications that need to organize, store and retrieve video data. Such applications largely fall into three categories.

Annotated video databases: News organizations such as CNN and the BBC have vast archives of video that they need to store, query and retrieve. Much of this data is stored using textual annotations that are often painstakingly hand created. In other cases, such textual annotations are created using free text data that accompanies the video (e.g., transcripts of what is being said in the news program). This body of textual annotations about the video is used to store the video and to retrieve part or all of it.

Image processed video databases: There are numerous applications where the amount of video data being produced is very high, and where the time-frame within which such video must be queryable is very small. Such applications include surveillance video (e.g., in banks, airports, etc.). In such cases, users must be rapidly able to take actions – for instance, when a known terrorist is spotted at an airport, this fact must be immediately made available to appropriate law enforcement authorities before it is too late to act. If the terrorist has previously been seen at various banks, the ability to correlate the airport surveillance video with the bank surveillance videos would greatly enhance the ability to track financial aspects of such crimes.

Military applications also fall into this category. For instance, the use of Predator video for airborne surveillance in military operations represents information that must be acted upon expeditiously. The luxury of waiting till some person looks at the video and creates textual annotations is not an option in such cases, as decisions must be taken in soft real time based on the data contained in such videos.

	Visual	**Textual**
Entities	Color Histograms Texture Maps Shape Descriptors Spatial Position Descriptors	Objects Annotations
Activities	Dancing Gesture Walking Motion Vehicle Tracking	Activity Annotation

Figure 19.1 Visual and Textual Categories

Hybrid video databases: As the name implies, databases of this kind contain both types of data. For example, a news organization with a video of a terrorist activity may want to correlate the terrorist strike with its existing video archives which are processed using textual annotations.

Clearly, the first two types of databases are special cases of hybrid video databases. Hence, in this paper, we focus on hybrid video databases as they cover both the above possibilities.

In this paper, we describe the ***AVE!*** video database system developed by us. ***AVE!*** stands for Algebraic Video Environment. Classical relational databases [22,23] take queries in a declarative query language (SQL or relational calculus are examples) and convert them to a relational algebra query. The relational algebra query is then *optimised*. This is done by using some rewrite rules that hold in the algebra – such rewrite rules make statements of the form "algebraic query expression 1 equals (or returns a subset of) algebraic query expression 2." Without the formal definition of a relational algebra style algebra for video databases, there is little hope to build effective video databases that scale to large numbers of users and data.

This paper represents a first step towards this much broader goal. It describes an algebra that may be used to query video data when either human-created annotations of the video are present, or when video analysis and image processing programs are present, or both. As such, our algebra is general and applies to most kinds of video database applications. To our knowledge, it is the first algebra for querying video (the only prior algebra for video is not for querying video, but for composing videos [24]). In addition, we have implemented a prototype system called *AVE!* that demonstrates our theoretical model.

2. PRELIMINARIES: VIDEO DATA MODEL

Throughout this paper, we will assume that a video v is divided up into a sequence $b_1,...,b_{len(v)}$ of *blocks*. The video database administrator can choose what a block is – he could, for instance, choose a block to be a single frame, or to be the set of frames between two consecutive I-frames (in the case of MPEG video) or something else. The number $len(v)$ is called the *length* of video v. If $1 \leq l \leq u \leq len(v)$, then we use the expression *block sequence* to refer to the closed interval $[l,u]$ which denotes the set of all blocks b such that $l \leq b \leq u$. Associated with any block sequence $[l,u]$ is a set of objects. These objects fall into four categories as shown in Figure 19.1.

- **Visual Entities of Interest** (Visual EOIs for short): An entity of interest is a region of interest in a block sequence (usually when identifying entities of interest, a single block, i.e., a block sequence of length one, is considered). Visual EOIs can be identified using appropriate image processing algorithms. For example, Figure 19. 2 shows a photograph of a stork and identifies three regions of interest in the picture using active vision techniques [1], [2]. This image may have various attributes associated with the above rectangles such as an *id*, a *color histogram*, a *texture map*, etc.
- **Visual Activities of Interest** (Visual AOIs): An activity of interest is a motion of interest in a video segment. For example, a dance motion is an example of an activity of interest. Likewise, a flying bird might be an activity of interest ("flight"). There are numerous techniques available in the image processing literature to extract visual AOIs. These include dancing [3], gestures [4] and many other motions [5].

- **Textual Entities of Interest**: Many videos are *annotated* with information about the video. For example, news videos often have textual streams associated with the video stream. There are also numerous projects [25,26,27] that allow textual *annotations* of video - such annotations may explicitly annotate a video with objects in a block sequence or may use a text stream from which such objects can be derived. A recent commercial product to do this is IBM's AlphaWorks system where a user can annotate a video while watching it.

- **Textual Activities of Interest**: The main difference between Textual EOIs and Textual AOIs is that the latter pertains to activities, while the former pertains to entities. Both are textually marked up.

Figure 19.2(a) A stork

Figure 19.2(b) Region of interests detected by a human eye

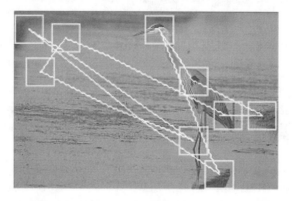

Figure 19.2(c) Region of interests and eye movement with active vision techniques

The algebra and implementation we propose to query and otherwise manipulate video includes elements of all the above. It is the first extension of the relational algebra to handle video databases that includes generic capabilities to handle both annotated video *and* image processing algorithms.

Throughout the rest of this paper, we assume the existence of some set $\mathcal{A}$ whose elements are called attribute names.

Definition (Textual object type) *A textual object type (TOT for short) is inductively defined as follows:*

- **real,int,string** *are TOTs.*
- *If $A_1,...,A_n$ are attribute names and $\tau_1,...,\tau_n$ are TOTs, then $[A_1 : \tau_1,...,A_n : \tau_n]$ is a TOT.*
- *If τ is a TOT, then $\{\tau\}$ is a TOT.*

As usual, every type τ has a *domain*, $dom(\tau)$. The domains of the basic types (**real**, **int**, **string**) are defined in the usual way. The domain of the type $[A_1 : \tau_1,..., A_n : \tau_n]$ is $dom(\tau_1) \times ... \times dom(\tau_n)$. The domain of $\{\tau\}$ is the power-set $2^{dom(\tau)}$ of τ's domain. We return to our Stork example to illustrate various TOTs and their domains.

Example. Consider the stork shown in Figure 19.2. Some textual object type of Figure 19.2(c) might be:

- *Feature:{string} which may specify a set of features.*

Definition (Textual object) *A textual object of type τ is any member of $dom(\tau)$.*

Note that we will often abuse notation and write a textual object of type $[A_1 : \tau_1,...,A_n : \tau_n]$ as $[A_1 : v_1,...,A_n : v_n]$ instead of $[v_1,...,v_n]$ so that the association of values v_i with attributes A_i is explicit (here, the v_i's are all in $dom(\tau_i)$).

Example. Continuing the preceding example, consider the image shown in Figure 19.2(c). Here, the textual objects involved might be:

- Feature:{stork,water}: This may be a human-created annotation reflecting the fact that the image shows a stork and water.

While the concept of a textual object is well suited to human-created annotations, some additional information may be required for image processing algorithms. We now define the concept of a visual type.

Definition (Visual type) *A visual type is any of the following*

- **real,int are Visual Types** *(VTs for short).*

- *If τ is a VT, then $\{\tau\}$ is a VT.*
- *If $A_1,...,A_n$ are strings (attribute names) and $\tau_1,...,\tau_n$ are VT, then $[A_1 : \tau_1,...,A_n : \tau_n]$ is a VT.*

Remark: The reader will note that visual types are a special case of textual types. The reason is that after image processing algorithms are applied to an image or video, the result may be stored in one or more image files which are then referenced by their file name (which is a string). Similarly, other outputs of image matching algorithms are textual (e.g., the name of a person seen in an image, etc.). The key reason we distinguish between visual and textual types is that the former are obtained using image processing algorithms while the latter are obtained using human annotations.

Definition (Visual object) *A visual object is a special kind of textual object over the type*

$$[LLx : real, LLy : real, URx : real, URy : real, A_1 : \tau_1,...,A_n : \tau_n]$$

where $\tau_1,...,\tau_n$ are visual types.

Example. Intuitively, a visual type could be:

- *Red:int* specifying a number from 0-255 denoting the level of redness. *We could have types Green:int and Blue:int in a similar way.*
- *[red:int, green:int, blue:int] is also a type. This may specify average RGB values for a given region of Figure 19.2(c).*

An example visual object could be: [LLx:50,Lly:0,Urx:60,URY:10, red:0,green:6,blue:10]. If we look at the lowest rectangle in Figure 19.2(c), this might describe the average color values associated with that rectangle.

The preceding definitions describe objects of interest, but not activities. In the following, we describe activities of interest in a video.

Definition (textual activity object) *A textual activity object is a special kind of textual object over the type* [**ActName: string, Roles:{[RoleName:string, Player: string]}**]

Definition (visual activity object) *A visual activity object w.r.t video v is a special kind of textual object over the type* [**ActName: string, Start: int, End: int**]

Example. Let us return to the stork of Figure 19.2. Here, the textual activity of the stork is just [ActName:standing,Roles:{ }] describing the fact that the stork is standing. It has no roles nor any players associated with those roles because the stork by itself is not playing a role.

The following example shows a lecture where there are roles.

Example. Figure 19.3 below shows selected frames from a lecture video.

Frame = 5

Frame= 6

Frame = 7

Frame=200

Figure 19.3 A video block showing a lecture

In this case, the visual activity object would be:

[ActName: lecturing,Start:5,End:200]

describing the fact that the activity in question is a lecturing activity and that it starts in video frame 5 and ends in video frame 200.

On the other hand, if a human were to do the annotation, rather than a program, then we would have a textual activity object which may look like this:

[ActName:lecturing, Roles:{ [RoleName:speaker, Player: vs]}].

Again, we emphasize that *visual activities and objects* are identified using image processing algorithms, while human annotations are used to identify non-visual textual objects and activities.

Throughout the rest of this paper, we assume the existence of some arbitrary but fixed set **T** of types including all the types needed for activities and visual objects. We assume that **T** is closed under subtypes. We do not give a formal definition of subtype (as this is fairly obvious) and instead appeal to the reader's intuition.

Definition (video association map) *Suppose* **T** *is some set of types (closed under subtypes). An* association map ρ *is a mapping from block sequences to objects in* $\bigcup_{\tau \in T} d \operatorname{om}(\tau)$ *such that if* $bs \subseteq bs$ *are block sequences, then* $\rho(bs) \subseteq \rho(bs')$.

Intuitively, $\rho(bs)$ specifies objects that occur somewhere in the block sequence bs. Note that the video association map ρ can be specified through the use of image processing algorithms. In our **AVE!** system, ρ can be specified using a variety of image processing algorithms including   adaptive filtering, entropy based detection, Gabor filtering and active vision techniques [2][21], as well as through human-created video annotations.

Example.   Suppose we have a 900 frame video clip of our favorite stork. The video clip may show the stork walking and perhaps catching a fish. In this case, the set **T** of types used may include all the types listed in the preceding examples. The association map ρ may say that:

- ρ ([0,900]) = { Feature:{{stork,water},[red:10,green:15,blue:20]}
- ρ([0,101] = { [ActName:standing,Roles:{ }]}
- ρ([101,250])={[ActName:fishing,Roles:{[RoleName:fisher,Player:stork]} denoting the fact that the stork is playing the role of the fisher (entity catching the fish).
- ρ ([0,900]) may consist of all the above as well as { Feature:{{stork,water},[red:10,green:15,blue:20]}.

This does not mean that the stork is continuously fishing during frames 101-250. It merely means that at some time in this interval, it is fishing. If the user wants to specify that the stork is fishing at each time instance in the [101,250] interval, it can be explicitly stated as $\rho([101,101])$ = { [ActName:fishing,Roles:{ [RoleName:fisher,Player:stork] }, $\rho([102,102])$ =   {   [ActName:fishing,Roles:{ [RoleName:fisher,Player:stork] }, and so on.

As the AND-interpretation of the association map ρ is clearly expressible in terms of the OR-interpretation we have used above, we will assume throughout the rest of this paper that the OR-interpretation is used. Some syntactic sugar can easily be created to make the AND-interpretation explicit for use in a deployed system.

3. THE VIDEO ALGEBRA

Having defined a video formally, we are now ready to define the video algebra. The video algebra consists of extensions of extensions of relational algebra operations (like select, project, Cartesian product, join, union, intersection, and difference).   In the rest of this section, we define these operations one by one. Our first operation is selection.

3.1 SELECTION
Our approach to defining selection consists of three major parts. The first part involves specifying the syntax of a selection condition. The second part involves

specifying when a block sequence satisfies a given selection condition. Once these two parts are defined, it is a simple matter to say that the result of applying a selection (using selection condition *C*) to a video *v* consists of those block sequences in *v* that satisfy selection condition *C*. It is important to note that we will be reasoning about *block sequences* rather than individual blocks. This is necessary because some conditions in a video may need to be evaluated not over a single block, but over a *block sequence*.

3.1.1 Selection Conditions: Syntax

Throughout this paper, we assume the existence of some arbitrary, but fixed set *VP* of *visual predicates*. A *visual predicate* takes a visual object or activity as input (perhaps with other nonvisual inputs as well) and evaluates to either true or false. Examples of visual predicates include:

- ***color(rect1,rect2,dist,d)***: This predicate succeeds if the color histograms of two rectangles (sub-images of an image) are within distance *dist* of each other when using a distance metric *d*.
- **texture(rect1,rect2,dist,d)**: This predicate succeeds if the texture histograms of two rectangles (sub-images of an image) are within distance *dist* of each other when using a distance metric *d*.
- **shape(rect1,rect2,dist,d)**: This predicate succeeds if the shapes of two rectangles are within a distance *dist* of each other when using distance metric *d*.

It is important to note that the above three visual predicates can actually be implemented in many different ways. Section 3.1.3 gives several examples of how to implement these visual predicates using, for instance, color histograms, wavelets for implementing texture distances and wavelet-based shape descriptors.

As our algebra can work with any visual predicates whatsoever, we continue with the algebraic treatment here, and leave the example visual predicate descriptions of Section 3.1.4. We start with the concept of a *path*.

Definition (path) *Suppose o is an object.*

- *If o's type is in {int,real,string} then $\varepsilon_{real}, \varepsilon_{int}, \varepsilon_{string}$ are all paths for o of types **int,real,string**, respectively. When clear from context, we will just write ε (without the subscript).*
- *If o's type is $[A_1 : \tau_1,..., A_n : \tau_n]$ and p_i is a path for an object of type τ, then $A_i \cdot p_i$ is a path for o whose type is the type of p_i.*
- *If o's type is $\{\tau\}$, and p is a path for objects of type τ, then p is a path for o also with the same type τ.*

Intuitively, paths are used to access the interior components of a complex type. For example, if we have an attribute called color of type [red:int,green:int,blue:int] and we want to find all very "bluish" objects, we may use the path expression color.blue to reference the blue component of the color attribute of an object.

We are now ready to define selection conditions.

Definition (selection condition) *Suppose **T** is a set of types VP, p_1 and p_2 are paths over the same type $\tau \in$ **T,** $v \in d$ om(τ), and VP is a set of visual predicates. Then:*

> 1. $p_1 \; \theta \; p_2$, and $p_1 \theta \; v$, and $v\theta \; p_2$ are selection conditions. Here, θ is one of the following:

> > • *If τ is a domain with an associated partial ordering≤ , then $\theta \in \{=, \geq, \leq, >, <\}$.*
> > • *Otherwise θ is the equality symbol "=".*

> 2.   *Every visual predicate is a selection condition.*

> 3.   *If C_1 , C_2 are selection conditions and $k \geq 0$ is an integer, then so are $(C_1 \wedge C_2)$, $(C_1 \vee C_2)$, before(k, C_1), after(k, C_1).*

Example. Here are some simple selection conditions.

- O.color.blue > 200 is satisfied by objects whose "blue" field has a value over 200.
- O1.color.blue > O2.color.blue is satisfied by pairs of objects, with the first one more "blue" than the other one.
- O1.color.blue > O2.color.red is satisfied by pairs of objects such that the blue component of the first object is greater than the red component of the second one.
- O.color.blue > 200 AND O.color.red <20 AND O.color.green < 20 may be used in our stork example to find water (as this is quite blue and the red and green components are very small).
- Color(rect1,rect2,10,L1) is a visual predicate which succeeds if rectangle rect1 is within 10 units of distance of rectangle rect2 using a distance metric called L1. In section 3.1.2, we do introduce one such distance metric.

It is important to note that visual predicates are satisfied by regions, while path conditions such as those in item (1) of the definition of a selection condition are satisfied by objects.

3.1.2 Selection Conditions: Semantics

We are now ready to define the semantics of selection conditions. Suppose v is a video and ρ is its association map. *Throughout this paper, we assume that there is some oracle **O** that can automatically check if a visual predicate is true or not.*

In practice, this oracle is nothing more than the code that implements the visual predicates (e.g., the various matching algorithms). We now define what it means for a block sequence to satisfy a selection condition.

Definition (satisfaction) *Suppose O is an oracle that evaluates visual predicates, b is a block and C is a selection condition. We say that b satisfies C w.r.t. O, denoted $b \models C$, iff:*

1. *C is a visual predicate and $O(C)=$ true.*
2. *C is of the form $(p \theta v)$ and there exists $o \in \rho(b)$, $o.p$ is well defined and $(o.p = v')$ and $(v' \theta v)$ holds.*
3. *C is of the form $(p_1 \theta p_2)$, there exist $o_1, o_2 \in \rho(b)$, not necessarily distinct, $o_1.p_1$ and $o_2.p_2$ are well defined, $(o_1.p_1 = v')$, $(o_2.p_2 = v'')$, and $(v' \theta v'')$ holds.*
4. *C is of the form $(C_1 \wedge C_2)$ and $b \models C_1$ and $b \models C_2$*
5. *C is of the form $(C_1 \vee C_2)$ and $b \models C_1$ and $b \models C_2$*
6. *C is of the form $before(k, C_1)$ and $b' \models C_1$ where $b' = b - k$ and $b - k >= 0$.*
7. *C is of the form $after(k, C_1)$ and $b' \models C_1$ where $b' = b + k$ and $b + k <= len(v)$.*

We are now in a position to define selection.

Definition (selection) *Suppose v is a video and C is a selection condition. The selection on C of video v, denoted $\sigma_C(v)$, is given by:*

$$\sigma_C(v) = \{b \in v \text{ such that } b \models C\}$$

We assume that all blocks in $\sigma_C(v)$ are ordered in the same way as they were in v, i.e., selection preserves order.

If V is a set of videos, $\sigma_C(V) = \{\sigma_C(v) | v \in V\}$.

3.1.3 Example Selection Conditions

In this section, we present examples of some of the different visual predicates informally described in the preceding section. All these visual predicates have been implemented in the **AVE!** system.

3.1.4.a Color Histograms

A **color histogram** is formed by discretizing the colors within an image. In other words, for each distinct color, we associate a bin. Each bin contains those pixels in the image that have that color.

Suppose we have two images and their associated histograms. We measure the similarity of the images based on the similarity of their color histograms.

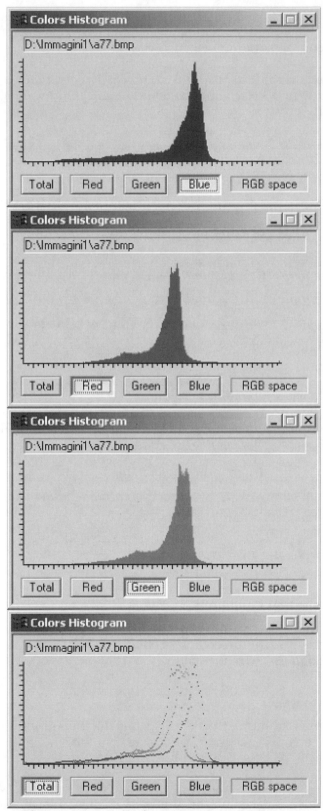

Figure 19.4 Color histograms in RGB space

Several distance functions have been defined in histogram spaces, as proposed by [7], [8], [9], [10], [11]. We now describe a few sample color histogram distance functions.

Let **histogram(*I*)** be a function that returns the color histogram H_I of a given frame *I*. **histogram(*I*)** operates as in the following:

- Map the colors into a discrete color space containing a reduced number of colors n. A good perceptually uniform space is *Hue Saturation Value, HSV* space or alternatively *Opponent Colors* space.

- Build an n-dimensional vector $H_I = h_1, h_2, ..., h_n$, where each $H_I[i] = h_i$ represents the number of pixels with color i in the frame **I**.

In the following we show several color histogram-based distance functions. Figure 19.4 shows an example of histogram for the image in Figure 19.2(a).

We first define the concept of an L1-distance on color histograms due to [7].

Definition (L$_1$ distance) *The **L$_1$-norm** distance function is*

$$L_1(I_1, I_2) = \frac{\sum_{i=1}^{n} \min(H_{I_1}[i], H_{I_2}[i])}{\sum_{i=1}^{n} H_{I_1}[i]}$$

For example, suppose we consider images having 512 colors. In this case, there would be 512 bins for each of the two images I_1, I_2. For each color, we compute the minimum of the number of pixels having that color in image I_1 and I_2. We add all these minima up. The denominator is just the number of pixels in the first image (when both images are of the same size, the denominator is the same irrespective of whether I_1 is considered or I_2 is considered). This function measures nicely the distance between two images. When there is wide variation between the color schemes of the two images, the numerator will be relatively small because for each color, the smaller of the number of pixels in images I_1, I_2 is counted.

Another notion of image distances based on color histograms is the L$_2$-distance given in [11].

Definition (L$_2$ distance) *The **L$_2$-norm** distance function is* given by:

$$L_2 = \sum_k \sum_t (H_{I_1}[t] - H_{I_2}[t]) \times a[k,t] \times (H_{I_1}[k] - H_{I_2}[k])$$

A=a[k,t] is an expression that denotes the cross-correlation between histogram bins.

3.1.4.b Texture Distances

It is well known in the literature [12] that color histograms are not sufficient to describe spatial correlations in images. In this section, we describe how textures may be used to define a distance between two images. Texture and shape have been widely analyzed and several features have been proposed [12], [13], [14], [15]. However, all the proposed visual descriptors exploit spatial interactions between a number of pixels in a certain region. In our **AVE!** system, we have used Wavelet Transform based descriptors in order to provide a multiresolution characterization of the images [16].

Let I be an image defined on a support $\Omega \subseteq \Re^2$. Here, Ω is any arbitrary subset of $\Re^2$. The continuous wavelet transform of I is the functional

$$W_a I(\vec{r}) = \frac{1}{\sqrt{a}} \int_\Omega \overline{\psi}(\frac{\vec{b}-\vec{r}}{a}) I(\vec{b}) d\vec{b}$$

where a is a scaling term and $\vec{r}, \vec{b} \in \Omega$. ψ is the 'mother' wavelet satisfying certain regularity constraints. $\overline{\psi}$ represents the conjugate complex of ψ.

A widely used scheme for image processing applications is Mallat's algorithm [17]. This multiresolution approach is based on the well known subband decomposition technique where an input signal is recursively filtered by perfect reconstruction QMF low-pass and high-pass filters, followed by critical sub-sampling, e.g., subsampling by 2 in the bi-dimensional case [18], [2].

Definition (Wavelet Covariance Signature) *A Wavelet Covariance Signature of a given image is the set of real number:*

$$\{C_{ni}^{X_j,X_i}\}_{n=0,...,d-1;i=1,2,3}^{j,k=1,2,3;j\leq k}$$

where each $C_{n,i}^{X_i,X_j}$ is the covariance of an image I:

$$COV^{n,i}_{wt}(I) = C_{ni}^{X_j,X_i} = \int I_{D_{ni}}^{X_j}(\vec{b}) I_{D_{ni}}^{X_i}(\vec{b}) d\vec{b}$$

I_D^X *being the detail subbands of the X component in a given Color Space of an Image I.*

We are now ready to define the texture distance between two images.

Definition (Texture Distance L_text) *Given two images I_1 and I_2, the texture distance L_text is defined as*

$$L_{text} = 1 - \frac{\sum_{n=0,...,d-1} \sum_{i=1,2,3} | COV^{n,i}_{wt}(I_1) - COV^{n,i}_{wt}(I_2) |}{\min_{\substack{n=0,...,d; \\ i=1,2,3}} (| COV^{n,i}_{wt}(I_1) |, | COV^{n,i}_{wt}(I_2) |)}$$

3.1.4.c Shape Distances

Thus far, we have focused on color distances and texture distances. A third type of distance between two images is based on the shapes of the images involved.

It is well known in the computer vision literature that the wavelet transform provides useful information for curvature analysis [19]. The idea is to find corner points and to use them as high curvature points in a polygonal approximation curve of the image. This task may be performed by means of Local Maxima of Wavelet Transform points. In our **AVE!** system, WT local maxima are calculated by using those points having a WT modulus which is longer than a certain threshold [20].

Definition (Shape Descriptor) *A wavelet based shape descriptor is the sequence:*

$$\left\{ (w_{p_i}, p_i), w_{p_i} \quad \|W_{\frac{=}{a}} I(p_i)\| \mid \|W_a I(p_i)\| \quad T, \; p_i \quad (x_i =\!\!\neq y_i) \quad \in \Re^2 \right\}$$

Definition (Shape Distance) *Given two images I' and I", and two associated wavelet shape descriptors,* $I' = \left\{ (w'_{p'_1}, p'_1),...,(w'_{p'_n}, p'_n) \right\}$ *and* $I'' = \left\{ (w''_{p''_\cdot}, p''_1),...,(w''_{p''_n}, p''_n) \right\}$, *the wavelet shape distance between I' and I" is given by*

$$L_{shape} = \sum_{i=1}^{n} \min_{p''_j \in I(p'_i)} \left\| w'_{p'_i} - w''_{p''_j} \right\|$$

where $I(p'_i) = \left[p'_i - \Delta p'_i, \; p'_i + \Delta p'_i \right]$.

3.2 PROJECTION

In the selection operator defined above, visual objects and regions play a role similar to attributes in relational algebra, while blocks play a role similar to tuples in relational algebra. Object properties, accessible via paths, are tested to evaluate selection conditions on blocks in the same way as attribute values are used to select tuples in relational algebra.

The analogy is further enhanced by the Projection operator that we define below. Given a video, whose blocks contain a number of different objects, projection takes as input a specification of the objects that the user is interested in, and deletes from the input video all objects not mentioned in the object specification list. Of course, pixels that in the original video correspond to the eliminated objects must be set to some value. We will introduce the concept of a *recoloring strategy* that automatically recolors appropriate parts of a video.

To define Projection, we first introduce some notation necessary for the two steps mentioned above.

Object extraction: given a block *b*, and a list *obj-list* of objects, object extraction returns a specification of the pixels associated with the objects mentioned in *obj-list* within block b.

Recoloring: given a block b, a coloring strategy and a specification of a set of pixels, it applies the coloring strategy to all the specified pixels in the block, and returns the newly colored block.

We are now ready to give the definition of projection.

Definition (projection) *Suppose v is a video, obj-list is a list of objects from v and Color is a recolor strategy. The* projection of obj-list from video *v, under recoloring strategy Color, denoted* $\pi_{obj-list,Color}(v)$, *is given by:*

$$\pi_{obj-list,Color}(v) = \{Color\ (b,\ (pixels(b) - Extract\ (b,\ obj\text{-}list)))\ |\ b \in v\ \}$$

We assume that all blocks in $\pi_{obj-list,Color}(v)$ are ordered in the same way as they were in v, i.e., projection preserves order.

In the above, pixels(b) denotes all pixels in block b, and Extract(b,obj-list) returns all pixels in b that are in an object in obj-list. Thus, pixels(b)-Extract(b,obj-list) refers to all pixels in an image that do not involve any of the objects being projected out. These are the pixels that need to be recolored. This is what the Color function does.

If **V** is a set of videos, $\pi_{obj-list,Color}(V) = \left\{\pi_{obj-list,Color}(v) \middle| v \in V\right\}$

Example. Assume we have a couple of videos, v1 and v2, whose visual objects are cats and dogs. Assume also that we are interested in a new video production, only about dogs. If WHITENING is the coloring strategy that sets to white the specified pixels, $\pi_{<dog>,WHITENING}(\{v1,v2\})$ returns the set containing two videos v1' and v2', each one containing the dogs appearing in the corresponding given video, acting in a completely white environment.

3.3.CARTESIAN PRODUCT

In this section, we define the Cartesian Product of two videos. In classical relational databases, the Cartesian product of two relations R,S simply takes each tuple t in R and t' in S and concatenates them together and places the result in the Cartesian product.

In order to successfully define Cartesian Product of two videos, we need to explain what it means to "concatenate" two blocks together. This is done via the concept of a *block merge function.*

Definition (block merge function) *A* block merge function *is any function from pairs of blocks to blocks.*

There are numerous examples of block merge functions. Three simple examples are given below. We emphasize that these are just three examples of such functions – many more exist.

Example. Suppose f1,f2 are any two frames (from videos v1, v2, respectively). The *split-merge, sm(f1,f2),* of f1 and f2 is defined as any frame f that shows both f1 and f2. For example,

- **[left-right split merge]** f's left side may show f1 (in a suitably compressed form) and its right side may show f2.

- **[top-down split merge]** Alternatively, f's top may show f1, and f's bottom may show f2.

- **[embedded split merge]** Alternatively, f may most contain frame f1, but a small window at its top left corner may contain a compressed version of f2.

Suppose $b = f_1,...,f_m$ and $b' = g_1,....,g_n$ are two blocks (continuous sequence of frames). Let $h_i = sm(f_i,g_i)$ if $i <= min(m,n)$. If $m < n$ and $i > =m$, then $h_i = g_i$. If $n < m$ and $i >= n$, then $h_i = f_i$. The *split-merge sm(b1,b2)* of two blocks is then defined to be the sequence $h_1,....,h_{max(m,n)}$.

Note that the above are just three examples of how to merge two blocks. Literally hundreds (if not thousands of variants) are possible.

Definition (cartesian product) *The Cartesian product, $(v_1 \ X_{BM} \ v_2)$ of two videos $v_1=b_1,...,b_m$, $v_2=c_1,...,c_n$, under the block merge function BM consists of the set* $\{ BM(bi,cj) \ | \ i <= m, \ j<=n \}$.

This set is totally ordered under the following ordering. $BM(b_i,c_j) <= BM(b_h,c_l)$ iff either $i < h$ or $(i=h$ AND $j <= l)$.

Suppose $v1 = b1,b2$ and $v2 = c1,c2,c3$ are two videos. Suppose each block consists of just a single frame. Figure 19.5 shows an example of Cartesian product under left-right, top-down and embedded split merge.

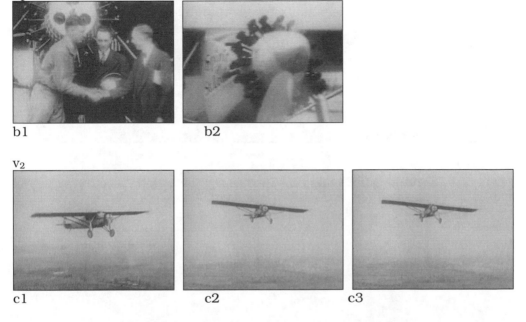

Figure 19.5 Example of Cartesian product

v_1 $\mathbf{X}_{BM}$ v_2 with *left right split merge*

v_1 $\mathbf{X}_{BM}$ v_2 with *top-down split merge*

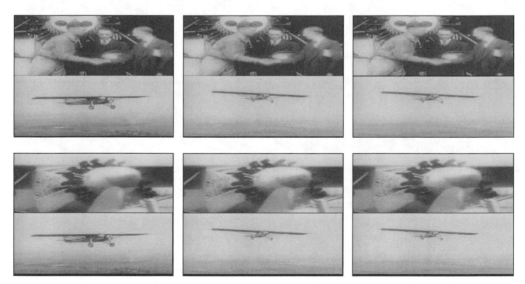

Figure 19.5 Example of Cartesian product (cont'd)

v_1 **X**$_{BM}$ v_2 with *embedded split merge*

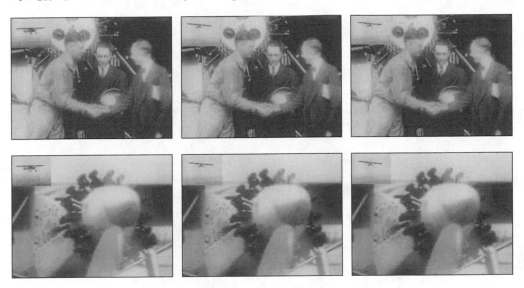

Figure 19.5 Example of Cartesian product (cont'd)

3.4 JOIN

In classical relational databases, the join of two relations under some join condition C is simply the set of all tuples in the Cartesian product of the two relations that satisfy C. Fortunately, the same is true here.

Definition (join) *Suppose v_1, v_2 are two videos, BM is a block merge functio, and C is a selection condition. The join of v_1, v_2 under BM w.r.t. selection condition C, denoted $v_1 \bowtie_C v_2$ consists of the set:*

$$v_1 \bowtie_C v_2 = \left\{ BM(b_1, b_2) \mid b_1 \in v_1, b_2 \in v_2 \wedge BM(b_1, b_2) \models C \right\}$$

This set is totally ordered in the same way as defined in the definition of the Cartesian product.

Intuitively, when we join two videos together, we merge the first block in v_1 with all blocks in v_2 when the merge satisfies C, then continue with the second block in v_1 and so on.

v1=b1,b2,b3,b4,b5,b6

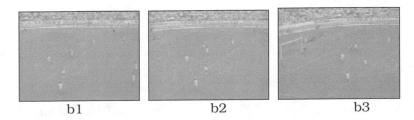

b1   b2   b3

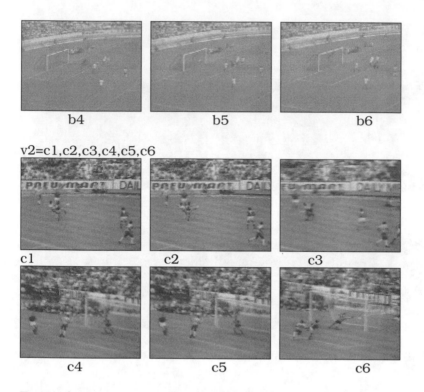

b4   b5   b6

v2=c1,c2,c3,c4,c5,c6

c1   c2   c3

c4   c5   c6

Example. v1 join v2 under Embedded Split Merge w.r.t. selection condition: O.ActName: "goal", i.e., $v_1 \bowtie_{O.ActName="goal"} v_2$

The selection condition is satisfied by v_5 and c_6; v_6 and c_6.

Example. v1 join v2 under Left Right Split Merge w.r.t. selection condition: O.ActName: "goal", i.e., $v_1 \bowtie_{O.ActName="goal"} v_2$

3.5 UNION, INTERSECTION AND DIFFERENCE

The only complication involved in defining these operations is that we must specify the order of the blocks involved.

Definition (Union, Intersection, Difference) *Suppose* $v_1 = b_1, ..., b_m$, $v_2 = c_1, ..., c_n$ *are two videos.*

- *The* union *of* v_1, v_2 *is* $b_1, ..., b_m$, $c_1, ..., c_n$
- *The* intersection *of* v_1, v_2 *consists of all blocks in* v_1 *that also occur in* v_2. *The order of the blocks in* v_1 *is preserved.*
- *The* difference *of* v_1, v_2 *consists of all blocks in* v_1 *that are not present in* v_2. *The order of blocks in* v_1 *is preserved.*

Note that all these operations preserve the order of v_1. Hence, these operations are not symmetric (as is true of standard set theory).

4. IMPLEMENTATION

We have implemented a prototype system called **AVE!** that encodes the algebraic operations described in this paper as well as a large number of image processing algorithms used by various operators in **AVE!**. The system contains approximately 5000 lines of C++ code running under Windows 2000.

Figure 19.6 shows the key parts of the **AVE!** implementation. The implementation consists of the following major parts:

- A graphical user interface through which the user may express his or her query.
- A video query processing engine that contains algorithms to implement the various video algebra operations described here.
- A set of image processing algorithms
- A relational database system in which information about objects, roles, activities, etc. is stored. Queries expressed in the video algebra are converted into queries over the relational database as well as invocations to the image processing algorithms.

The image processing algorithms currently included in **AVE!** are:

- Entropy based detection of objects in video frames (cf.[21])
- Active vision algorithms to detect salient regions in an image (cf. [2])
- A set of morphological functions for common image processing tasks
- Diffusion filtering algorithms.
- Fading algorithms are used to ensure continuity within an answer. In **AVE!**, fading effects are obtained via the use of a Gabor filter applied several times on the last frame in a block.

Figure 19.7 provides a screen dump of the **AVE!** system.

5. RELATED WORK

Oomoto and Tanaka [25] have defined a video-based object oriented data model, OVID. They take pieces of video, identify meaningful features in them and link these features. They also outline a language called VideoSQL for querying such data.

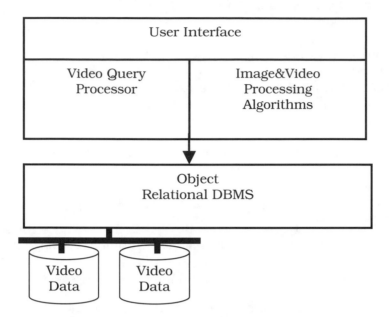

Figure 19.6 *AVE!* key parts

Figure 19.7 The ***AVE!*** system screendump

Adali et al. [26] developed the AVIS video database system that introduced the importance of objects, activities, roles and players used here. They developed index structures to query such videos and algorithms to traverse such index structures to answer queries. Hwang et al. [27] developed a version of SQL to query video databases. However neither of these efforts provided an extension of the relational model of data to process video queries.

Gibbs et al. [28] study how stream-based temporal multimedia data may be modeled using object based methods. However, concepts such as roles and players, the distinction between activities and events and the integration of such video systems with other traditional database systems are not addressed.

Hjelsvold and Midtstraum [29] develop a "generic" data model for capturing video content and structure. Their idea is that video should be included as a data type in relational databases, i.e., systems such as PARADOX, INGRES, etc. should be augmented to handle video data. In particular, they study temporal queries.

Arman et al. [30] develop algorithms that can operate on compressed video directly they can identify scene changes by performing certain computations on DCT coefficients in JPEG and MPEG encoded video. Their effort complements ours neatly in the following way: their algorithms can identify, from compressed video, frame sequences that are of interest, and the objects/roles/events of these frame sequences can be queried using our algebra.

6. CONCLUSIONS

Video databases are an area of growing importance to government and industry. Despite the growing use of video data in commercial and business decision making, and the existence of vast corpuses of video archives, few attempts have been made to place video databases on the same solid theoretical foundations that relational databases are based on.

Relational databases have two key theoretical underpinnings – the relational algebra and the relational calculus. In this paper, we have developed a *video algebra* which is in the same spirit as the relational calculus. To our knowledge, this is the first algebra to query video databases.

Video algebras are important because a deep understanding of equivalences between queries expressed in the algebra can be used to optimise queries as well as to effectively process multiple simultaneous queries. For example, in the relational algebra, equivalence results between algebraic expressions can be used to take a query posed by the user and transform it into an equivalent query whose expected execution cost is smaller. The same holds true of video algebra queries.

In future work, we plan to develop a set of equivalence results for the video algebra and to build a video query optimiser based on the algebra and such equivalence results so that video database systems can scale well to handle large numbers of concurrent queries.

Acknowledgments

Parts of this research were supported by Army Research Lab contract DAAL0197K0135, Army Research Office grant DAAD190010484, DARPA/Rome Labs grant F306029910552 and by the ARL CTA on advanced decision architectures.

REFERENCES

[1] D.H. Ballard, Animate vision. *Artificial Intelligence*, (48):57–86, 1991.

[2] G. Boccignone, A. Picariello, G. Moscato and M. Albanese, Image similarity based on Animate Vision: the Information Path algorithm, submitted to MIS2002, Tempe, October 2002.

[3] K.Koijima, M. Hironaga, S. Nagae and Y. Kawamoto, Human motion analysis using the Rhythm – a reproducing method of human motion, *Journal of Geometry and Graphics*, vol. 5, n. 1, pp. 45-49.

[4] L. Bretzner, I. Laptev, T. Lindeberg, Hand Gesture Recognition using Multi-scale colour features, hierarchical and particle filtering, Proc. Of the fifth IEEE International Conference on Automatic Face and Gesture Recognition (FGR'02), IEEE Computer Society Press, pp. 1-6, 2002.

[5] A. F. Bobick, J.W. Davis, The recognition of human movement using temporal templates, *IEEE Transactions on Pattern Analysis and Machine Intelligence*, vol. 23, n. 3, pp. 257-267, march 2001.

[6] C-L Huang and B-Y. Liao, A Robust Scene-Change Detection Method for Video Segmentation, *IEEE Transactions on Circuits and Systems for Video Technology*, vol. 11, n. 12, pp. 1281-1288., December 2001.

[7] M.J. Swain and D.H. Ballard, Color indexing, *International Journal of Computer Vision*, vol. 7, n. 1, pp. 11-32, 1991

[8] B. V. Funt and G. D. Finlayson, Color Constant Color Indexing, *IEEE Transactions on Pattern Analysis and Machine Intelligence*, vol. 17, n. 5, pp.522-529, may 1995.

[9] W. Niblak et al., The QBIC project : querying images by content using color, texture and shape, *SPIE Storage and Retrieval for Image and Video Databases*, Feb. 1993

[10] M. Flickner et al., Query By Image and Video Content: the QBIC system, *IEEE Computer*, vol. 28, n. 9, 1995, pp. 23-32

[11] J. Hafnerm H. Sawhney, W. Equitz, M. Flickner, W. Niblack, Efficient Color Histogram Indexing for Quadratic Form Distance Functions, *IEEE Transactions on Pattern Analysis and Machine Intelligence*, vol. 17, n. 7, pp. 729-736, july 1995.

[12] T. Caelli and D. Reye, On the classification of image regions by colour, texture and shape, *Pattern Recognition*, vol. 26, n. 4, pp. 461-470, 1993

[13] S. Aksoy and R. M. Haralick, Textural features for image database retrieval, in Proc. IEEE Workshop on Content Based Access of Image and Video Libraries, 1998, pp. 45-49

[14] A. Del Bimbo, *Visual Information Retrieval*, Morgan Kaufmann: San Francisco, CA, 1999

[15] T. Gevers and A.W.M. Smeulders, PicToSeek: combining color and shape invariant features for image retrieval, *IEEE Transactions on Image Processing*, vol. 9, n. 1, pp. 102-119, 2000.

[16] G. Boccignone, A. Chianese and A. Picariello, Wavelet Transform and Image Retrieval: Preliminary Experiments Proc. of International Conference on Computer Vision Pattern Recognition and Image Processing, vol. 2, pp. 164-167, 2000.

[17] S. Mallat, A theory for multiresolution signal decomposition: the wavelet representation, *IEEE Transactions on Pattern Analysis and Machine Intelligence*, vol. 11, n.7, pp. 674-693, 1989.

[18] G. Van de Wouwer et al., Wavelet correlation signatures for color texture characterization, *Pattern Recognition*, vol. 32, pp. 443-451, 1999.

[19] S. Mallat and S. Zhong, Characterization of Signals from Multiscale Edges, *IEEE Transactions on Pattern Analysis and Machine Intelligence*, vol. 14, n.7, pp. 710-732, july 1992.

[20] A. Cheikh et al., MUVIS: a system for content-based indexing and retrieval in large image databases, Proc. SPIE/EI99 Storage and Retrieval for Image and Video Databases, vol 3656, pp. 98-106, 1999.

[21] G. Boccignone, A. Chianese and A. Picariello, Multiresolution spot detection using entropy thresholding, *Journal of Optical Society of America*, A, Vo. 17. No. 7, July 2000, pp. 1160-1171.

[22] J.D. Ullman. *Introduction of Database Systems*, Addison Wesley, 1989.

[23] J. Gehrke and R. Ramakrishnan. Principles of Database Management Systems, McGraw Hill.

[24] R. Weiss, A. Duda and D.K. Gifford. (1994) Content-Based Access to Algebraic Video, Proc. 1994 International Conference on Multimedia Computing and Systems, pps 140--151, IEEE Press.

[25] E. Oomoto and K. Tanaka. (1993) OVID: Design and Implementation of a Video-Object Database System, *IEEE Transactions on Knowledge and Data Engineering*, 5, 4, pps 629--643.

[26] S. Adali , K.S. Candan, S.-S. Chen, K. Erol and V.S. Subrahmanian. (1995) AVIS: Advanced Video Information Systems, *ACM Multimedia Systems Journal*, 4, pps 172-186, 1996.

[27] E. Hwang and V.S. Subrahmanian, Querying Video Libraries, *Journal of Visual Communication and Image Representation*, Vol. 7, Nr. 1, pps 44-60, 1996.

[28] S. Gibbs, C. Breiteneder and D. Tsichritzis. (1994) Data Modeling of Time-Based Media, Proc. 1994 ACM SIGMOD Conference on Management of Data, pps 91--102.

[29] R. Hjelsvold and R. Midtstraum. (1994) Modeling and Querying Video Data, Proc. 1994 International Conference on Very Large Databases, pps 686--694, Santiago, Chile.

[30] F. Arman, A. Hsu and M. Chiu. (1993) Image Processing on Compressed Data for Large Video Databases, First ACM International Conference on Multimedia, pps 267--272.

20

AUDIO INDEXING AND RETRIEVAL

Zhu Liu

AT&T Labs – Research
200 Laurel Avenue South
Middletown, NJ 07748
`zliu@research.att.com`

Yao Wang

Dept. of Electrical Engineering
Polytechnic University
333 Jay Street
Brooklyn, NY 11201
`yao@vision.poly.edu`

INTRODUCTION

With the booming of Internet and the rapid growth of digital storage capability, we are exposed to virtually unlimited volume of multimedia information, including movies, video clips, image animation, still images, music, speech, text, etc. Manually searching useful or interesting pieces of information is exactly the same as finding a needle in a haystack. Automatically generating semantically meaningful index for a large volume of documents is critical for efficient and effective multimedia information retrieval. During the last decade, multimedia content index and retrieval, as a new yet fast growing field, has attracted tremendous interest from researchers worldwide. MPEG-7, formally known as Multimedia Content Description Interface, is an internationally collaborated effort that addresses such a challenge. This chapter will focus on the recent progress of audio indexing and retrieval, a major component within the broad scope. Everyday, we are immersed in ubiquitous audio. Very often, we want to search a piece of audio that either we know or heard before, for example, a specific song or a conference recording, or we are not aware of, for example, a piece of symphony or one speech of President Kennedy. With the support of audio indexing and retrieval services, such a task is an enjoyable experience, otherwise, really a headache. In this chapter, we will survey the state of the art in the area of audio content indexing and retrieval, address audio indexing algorithms and query methods, illustrate several representative audio retrieval systems, and briefly introduce MPEG-7 audio standard.

483

BRIEF REVIEW OF AUDIO INDEXING AND RETRIEVAL

The volume of information accessible over the Internet or within personal digital multimedia collections has exceeded the users' capability to efficiently sift through and find relevant information. The last decade has seen an explosion of automatic multimedia information retrieval (IR) research and development with the booming of Internet and the improvement of digital storage capability. A series of powerful Internet search engines, from Alta Vista, Lycos, to Google, have brought the new information retrieval technologies into reality. Their capabilities of searching billions of documents within a few seconds have brought huge benefits for human beings and totally changed our life and working styles. Google and other popular search engines began to replace the libraries as the starting point in our information exploring and collecting task. While the main focus of the classic IR has been the retrieval of text, recent efforts addressing content-based retrieval of other media, including audio and video, are starting to show promise.

A video sequence is a rich multimedia information source, containing speech, music, text (if closed caption is available), image sequences, animation, etc. Although the human being can quickly and effectively interpret the semantic content, including recognizing the imaged objects, differentiating the types of audio, and understanding the linguistic meanings of speech, computer understanding of a video sequence is still in a primitive stage. There is a strong demand for efficient tools that enable easier dissemination of audiovisual information for the human being. Computer understanding of a video scene is a crucial step in building such tools. Other applications requiring scene understanding include spotting and tracing of special events in a surveillance video, active tracking of special objects in unmanned vision systems, video editing and composition, etc.

Research in video content analysis has focused on the use of speech and image information. Recently, researchers started to investigate the potential of analyzing audio signal. This is feasible because audio conveys discriminative information to identify different video content. For example, the audio characteristics in a football game are easily distinguishable from those in a news report. Obviously, audio information alone may not be sufficient for understanding the scene content, and in general, both audio and visual information should be analyzed. However, because audio-based analysis requires significantly less computation, it can be applied in a pre-processing stage before more comprehensive analysis involving visual information. The audio signal is very dynamic in terms of both waveform amplitude and spectrum distribution, which makes audio content analysis a challenge. Following, we will briefly introduce several sub-areas in this field.

1.1 AUDIO CONTENT SEGMENTATION

The fundamental approach for audio content analysis is breaking audio stream into smaller segments, whose contents are homogenous, and then processing each segment individually by feasible methods. The reason is that audio processing is a domain specific task, and there is no single wide applicable algorithm that suits for all kinds of audio content. For example, different approaches are adopted to deal with speech signal and music signal. Therefore, audio content segmentation is normally the first step in audio analysis system, and different systems may choose different methods depending on real applications. Siegler et al. [34] proposed to use symmetric Kullback-Leibler (KL) Distance as an effective metric for segmenting broadcast news into pieces based on speaker or channel changes. Chen et al. detected changes in speaker identity, and environment or channel conditions in audio stream based on Bayesian Information Criterion (BIC) [4]. Nam and Tewfik [26] proposed to detect sharp temporal variations in the power of the subband signals to segment visual streams. Liu et al. investigated how to segment broadcast news at scene level such that each segment contains a single type of TV program, for example, news reporting, commercial, or games [19].

1.2 AUDIO CONTENT CATEGORIZATION

Audio content categorization is to assign each audio segment to one predefined category. Normally, it is formulated as a pattern recognition problem, where a wide range of pattern classification algorithms can be applied on different sets of acoustic features. Depending on the specified audio categories, various combinations of audio features and classification methods are utilized. Saunders [32] presented a method to separate speech from music by tracking the change of the zero crossing rate. Wold et al. [38] proposed to use nearest neighbour classifier based on weighed Euclidean distance to classify audio into ten categories. Liu *et al.* [16][17] studied the effectiveness of artificial neural network, and Gaussian mixture model (GMM) classifiers on a task of differentiating five types of news broadcast. Tzanetakis et al. [36] classified music into ten categories based on k-nearest neighbor (k-NN) and GMM classifiers. Under certain situations, it is more feasible to segment and categorize audio content jointly. Huang et al. [10] proposed a hidden Markov model (HMM) approach that simultaneously segment and classify video content.

Audio content segmentation and categorization usually serve as the beginning steps in audio content analysis. Then, specific analysis methods can be utilized to process different types of audio; for example, speech and speaker recognition algorithms can be applied on speech signal, and note/pitch detection algorithms can be applied on music signal.

1.3 SPEECH SIGNAL PROCESSING

Although speech signal is normally narrow band, with energy centralizing within 4KHz in frequency domain, it carries rich semantics and wealthy side information. The side information includes language, speaker identification, age, gender, and emotion, etc. Such kind of side information is important for audio indexing, and it will significantly expand audio indexing and query capabilities.

Automatic speech recognition (ASR) transcribes the speech signal into text stream. With decades of hard work in ASR, this technique has matured from research labs to commercial markets. Almost all the state-of-the-art large vocabulary recognition systems are built based on hidden Markov model framework, along with sophisticated methods to improve both the accuracy and the speed. The last decade has witnessed substantial progresses in ASR technology. The ASR systems are able to achieve almost perfect results in restricted domains, e.g., recognize numbers in phone conversation and reasonable results in unrestricted domains, e.g., transcribe news broadcast. Even with an accuracy of about 50% for noisy audio input, ASR still produces half of the right words for index and it is reasonable easy to understand the content. Gauvain et al. [7] provided an overview of recent advances in large vocabulary speech recognition and explored feasible application domains.

How to extract the useful side information has also attracted many research efforts for a long time. Besides the benefit of further improving the ASR performance by adapting acoustic models, side information also provides additional audio query functionalities. For example, we can retrieve the speech of Kennedy if the speaker identification is available. Reynolds presented an overview on various speaker recognition technologies in [29]. Backfried et al. studied automatic identification of four languages in broadcast news based on GMM [1]. Dellaert et al. [6] explored several statistical pattern recognition techniques for emotion recognition in speech. Parris and Carey described a new technique for language independent gender identification based on HMM [27]. The average error rate on a testing database of 11 languages is 2.0%.

1.4 MUSIC SIGNAL PROCESSING

Contrary to narrow band speech signal, music signal is the wide band component of audio signal. Musical information access is a crucial stake regarding the huge quantity available and worldwide interest. As the amount of musical content increases and the Web becomes an important mechanism for distributing music, we expect to see a surging demand for music search services. Many currently available music search engines rely on file names, song titles, and the names of composers or performers. These systems do not make use of the music content directly. To index music based on content, we need to address at least three issues: music categorization, music transcription, and music melody matching.

Similar approaches as those introduced in section 2.2 may be applied to music genre classification. Parallel to speech recognition, music transcription is to convert music signal into a sequence of notes. Musical instrument digital interface (MIDI) is a powerful tool for composers and performers, and it serves as a communication protocol that allows electronic musical instruments to interact with each other. Transforming music into MIDI format is an ideal choice for indexing music archive. Querying by example and humming are natural ways to query music pieces from a large music database. To realize such query capabilities, music melody matching is a key issue which determines the effectiveness and efficiency of music searching. Recent work on query by humming can be found in [8][20].

1.5 AUDIO DATA MINING

Audio data mining is to look for hidden patterns in a group of audio data. For example, in a telemarketing conversation, certain words, sentences, or speech styles indicate the customer's preferences, personality, and shopping patterns, which help a human salesman to quickly decide the best selling strategy. With large collections of thousands of such conversations, audio data mining techniques can be applied to automatically discover the association between speech styles and the shopping patterns. The discoveries can be integrated into an automatic customer service system to serve future customers with human-like intelligence and flexibility. Audio data mining technology combines speech recognition, language processing, knowledge discovery, and indexing and search algorithms to transform the audio data into useful intelligence. Promising application areas for audio mining include customer care call centers, knowledge gathering, law enforcement, and security operations.

Audio mining is not equal to audio query, but they share some fundamental techniques, including audio segmentation and categorization, speech transcription, audio searching, etc. The major difference is that audio query finds something we know that exists, yet audio mining discovers new patterns from the audio archive based on statistical analysis. As technology matures, even greater volumes of data in the form of audio will be captured from television, radio, telephone calls, meetings, conferences, and presentations. Audio mining techniques turn all these audio archives into valuable intelligent knowledge as easily as text mining.

1.6 MUTLI-MODALITY APPROACH

Audio often coexists with other modalities, which include text and video streams. While audio content analysis itself is very useful for video indexing, it is more desirable to combine the information from other modalities, since semantics are embedded in multiple forms that are usually complementary to each other. For example, live coverage on TV about an earthquake conveys information that is far beyond what we hear from the reporter. We can see and feel the effects of the earthquake, while listening to the reporter talking about the statistics. Many efforts have been involved in this field. Wang et al. reported their recent work in combining audio and visual information for multimedia analysis and surveyed related works in [39]. The *Informedia* project at Carnegie Mellon University combined speech recognition, natural language understating, image processing, and text processing for content analysis in creating a terabyte digital video library [37].

With worldwide interests in multimedia indexing and retrieval, a single standard which can provide a simple, flexible, interoperable solution to multimedia indexing and retrieving problems will be extremely valuable. MPEG-7 standard [25] serves such a need by providing a rich set of standardized tools to describe multimedia content. MPEG-7 standardizes the content descriptions and the way for structuring them, but leaves the extraction and usage of them open. Consequently, instead of hurdling the evolution of multimedia content indexing and retrieval technologies, the standard stimulates and benefits from new progresses.

The organization of this chapter is as follows. In Section 3, we describe audio indexing algorithms, including audio feature extraction, audio segmentation, and classification. In Section 4, we show different types of audio query methods. Three representative audio retrieval systems are presented in Section 5. Given the high relevance of MPEG-7 audio with the content of this chapter, we briefly introduce it in Section 6. And finally in Section 7, we indicate some future research directions to conclude this chapter.

AUDIO INDEXING ALGORITHMS

The first step in audio content analysis task is to parse an audio stream into segments, such that the content within each clip is homogeneous. The segmentation criteria are determined by specific domains. In phone conversation, the segment boundaries correspond to speaker turns, and in TV broadcast, these segments may be in-studio reporting, live reporting, and commercials. Depending on applications, different tasks follow the segmentation stage. One important task is the classification of a segment into some predefined category, which can be high level (an opera performance in the Metropolitan Opera House), middle level (a music performance), or low level (a clip in which audio is dominated by music). Such semantic level classification is key to generating audio indexes. Beyond such labelled indexes, some audio descriptors may also be useful as low-level indexes, so that a user can retrieve an audio clip that is aurally similar to an example clip. Finally, audio summarization is essential in building an audio retrieval system to enable a user to quickly skim through a large set of retrieved items in response to a query.

In this section, we first introduce some effective audio features that well represent the audio characteristics, and then present the audio segmentation and categorization methods. We will briefly mention the audio summarization task before we finish this section.

1.7 AUDIO FEATURE EXTRACTION

There are many features that can be used to characterize audio signals. Usually audio features are extracted in two levels: short-term frame-level and long-term clip-level. Here a frame is defined as a group of neighboring samples which last about 10 to 40 milliseconds (ms), within which we can assume that the audio signal is stationary and short-term features such as volume and Fourier transform coefficients can be extracted. The concept of audio frame comes from traditional speech signal processing, where analysis over a very short time interval has been found to be most appropriate.

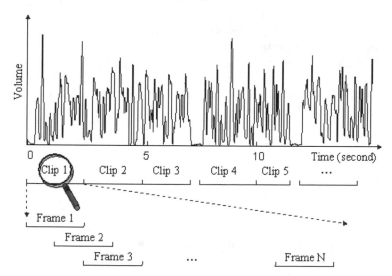

Figure 20.1 Decomposition of an audio signal into clips and frames.

For a feature to reveal the semantic meaning of an audio signal, analysis over a much longer period is necessary, usually from one second to several tens of seconds. Here we call such an interval an audio clip.[1] A clip consists of a

---

[1] In the literature, the term "window" is sometimes used.

sequence of frames and clip-level features usually characterize how frame-level features change over a clip. The clip boundaries may be the result of audio segmentation such that the content within each clip is similar. Alternatively, fixed length clips, usually 2 to 3 seconds (s), may be used. Both frames and clips may overlap with their previous ones, and the overlapping lengths depend on the underlying application. Figure 20.1 illustrates the relation of frame and clip. In the following, we first describe frame-level features, and then move onto clip-level features.

1.7.1 Frame-level features

Most of the frame-level features are inherited from traditional speech signal processing. Generally they can be separated into two categories: time-domain features, which are computed from the audio waveforms directly, and frequency-domain features, which are derived from the Fourier transform of samples over a frame. In the following, we use N to denote the frame length, and $s_n(i)$ to denote the i-th sample in the n-th audio frame.

Volume

The most widely used and easy-to-compute frame feature is volume.[2] Volume is a reliable indicator for silence detection, which may help to segment an audio sequence and to determine clip boundaries. Normally volume is approximated by the root mean square (RMS)[3] of the signal magnitude within each frame. Specifically, the volume of frame n is calculated by

$$v(n) = \sqrt{\frac{1}{N}\sum_{i=0}^{N-1}s_n^2(i)}$$

Note that the volume of an audio signal depends on the gain value of the recording and digitizing device. To eliminate the influence of such device-dependent conditions, we may normalize the volume for a frame by the maximum volume of some previous frames.

Zero crossing rate

Besides the volume, zero crossing rate (ZCR) is another widely used temporal feature. To compute the ZCR of a frame, we count the number of times that the audio waveform crosses the zero axis. Formally,

$$Z(n) = \frac{1}{2}\left(\sum_{i=1}^{N-1}|sign(s_n(i)) - sign(s_n(i-1))|\right)\frac{f_s}{N}$$

where f_s represents the sampling rate. ZCR is one of the most indicative and robust measures to discern unvoiced speech. Typically, unvoiced speech has a low volume but a high ZCR. By using ZCR and volume together, one can prevent low energy unvoiced speech frames from being classified as silent.

---

[2] Volume is also referred to as loudness, although strictly speaking, loudness is a subjective measure that depends on the frequency response of the human listener.
[3] The RMS volume is also referred to as energy.

Pitch

Pitch is the fundamental frequency of an audio waveform, and is an important parameter in the analysis and synthesis of speech and music. Normally only voiced speech and harmonic music have well-defined pitch. But we can still use pitch as a low-level feature to characterize the fundamental frequency of any audio signals. The typical pitch frequency for a human being is between 50Hz to 450Hz, whereas the pitch range for music is much wider. It is difficult to robustly and reliably estimate the pitch value for an audio signal. Depending on the required accuracy and complexity constraint, different methods for pitch estimation can be applied.

One can extract pitch information by using either temporal or frequency analysis. Temporal estimation methods rely on computation of the short time auto-correlation function $R_n(l)$ or average magnitude difference function (AMDF) $A_n(l)$, where

$$R_n(l) = \sum_{i=0}^{N-l-1} s_n(i)s_n(i+l), \quad A_n(l) = \sum_{i=0}^{N-l-1} |s_n(i+l) - s_n(i)|$$

For typical voiced speech, there exist periodic peaks in the auto-correlation function. Similarly, there are periodic valleys in the AMDF. Here peaks and valleys are defined as local extremes that satisfy additional constraints in terms of their values relative to the global extreme and their curvatures. Such peaks or valleys exist in voiced and music frames and they vanish in noise or unvoiced frames.

In frequency-based approaches, pitch is determined from the periodic structure in the magnitude of the Fourier transform or cepstral coefficients of a frame. For example, we can determine the pitch by finding the maximum common divider for all the local peaks in the magnitude spectrum. When the required accuracy is high, a large size Fourier transform needs to be computed, which is time consuming.

Spectral features

The spectrum of an audio frame refers to the Fourier transform of the samples in this frame. The difficulty of using the spectrum itself as a frame-level feature lies in its very high dimension. For practical applications, it is necessary to find a more succinct description. Let $S_n(\omega)$ denote the power spectrum (i.e., magnitude square of the spectrum) of frame n. If we think of ω as a random variable, and $S_n(\omega)$ normalized by the total power as the probability density function of ω, we can define mean and standard deviation of ω. It is easy to see that the mean measures the frequency centroid (FC), whereas the standard deviation measures the bandwidth (BW) of the signal. They are defined as

$$FC(n) = \frac{\int_0^\infty \omega S_n(\omega)d\omega}{\int_0^\infty S_n(\omega)d\omega}, \quad BW^2(n) = \frac{\int_0^\infty (\omega - FC(n))^2 S_n(\omega)d\omega}{\int_0^\infty S_n(\omega)d\omega}$$

It has been found that FC is related to the human sensation of the brightness of a sound we hear.

In addition to FC and BW, Liu et al. proposed to use the ratio of the energy in a frequency subband to the total energy as a frequency domain feature [19], which is referred to as *energy ratio of subband* (ERSB). Considering the perceptual property of human ears, the entire frequency band is divided into four subbands, each consisting of the same number of critical bands, where the critical bands correspond to cochlear filters in the human auditory model [12].

Specifically, when the sampling rate is 22050Hz, the frequency ranges for the four subbands are 0-630Hz, 630-1720Hz, 1720-4400Hz, and 4400-11025Hz. Because the summation of the four ERSB's is equal to one, only first three ratios were used as audio features, referred to as ERSB1, ERSB2, ERSB3, respectively.

Scheirer et al. used spectral rolloff point as a frequency domain feature [33], which is defined as the 95th percentile of the power spectrum. This is useful to distinguish voiced from unvoiced speech. It is a measure of the "skewness" of the spectral shape, with a right-skewed distribution having a higher value. Lu et al. [21] used spectrum flux, which is the average variation value of spectrum between the adjacent two frames in an audio clip.

Mel-frequency cepstral coefficients (MFCC) or cepstral coefficients (CC) are widely used for speech recognition and speaker recognition. While both of them provide a smoothed representation of the original spectrum of an audio signal, MFCC further considers the non-linear property of the human hearing system with respect to different frequencies. Based on the temporal change of MFCC, an audio sequence can be segmented into different segments, so that each segment contains music of the same style, or speech from one person.

1.7.2 Clip-level features

As described before, frame-level features are designed to capture the short-term characteristics of an audio signal. To extract the semantic content, we need to observe the temporal variation of frame features on a longer time scale. This consideration leads to the development of various clip-level features, which characterize how frame-level features change over a clip. Therefore, clip-level features can be grouped by the type of frame-level features that they are based-on.

Volume-based

Considering the difference of gain values in audio digitization systems, the mean volume of a clip does not necessarily reflect the scene content, but the temporal variation of the volume in a clip does. To measure the variation of volume, Liu et al. proposed several clip-level features [19]. The *volume standard deviation* (VSTD) is the standard deviation of the volume over a clip, normalized by the maximum volume in the clip. The *volume dynamic range* (VDR) is defined as (max(v) - min(v))/max(v), where min(v) and max(v) are the minimum and maximum volume within an audio clip. Obviously these two features are correlated, but they do carry some independent information about the audio scene content.

Another feature is *volume undulation* (VU), which is the accumulation of the difference of neighboring peaks and valleys of the volume contour within a clip. Scheirer proposed to use *percentage of "low-energy" frame* [33], which is the proportion of frames with RMS volume less than 50% of the mean volume within one clip. Liu et al. used *non-silence-ratio* (NSR), the ratio of the number of non-silent frames to the total number of frames in a clip, where silence detection is based on both volume and ZCR [19].

The volume contour of a speech waveform typically peaks at 4Hz. To discriminate speech from music, Scheirer et al. proposed a feature called 4Hz modulation energy (4ME) [33], which is calculated based on the energy distribution in 40 subbands. Liu et al. proposed a different definition that can be directly computed from the volume contour. Specifically, it is defined as [19]

$$4ME = \frac{\int_0^\infty W(\omega)|C(\omega)|^2 d\omega}{\int_0^\infty |C(\omega)|^2 d\omega}$$

where $C(\omega)$ is the Fourier transform of the volume contour of a given clip and $W(\omega)$ is a triangular window function centered at 4Hz. Speech clips usually have higher values of 4ME than music or noise clips.

ZCR-based

ZCR contours of different types of audio signal are different. For a speech signal, low and high ZCR periods are interlaced. This is because voiced and unvoiced sounds often occur alternatively in a speech. On the contrary, the mild music has a relatively smooth contour.

Liu et al. used the *standard deviation of ZCR* (ZSTD) within a clip to classify different audio contents [19]. Saunders proposed to use four statistics of the ZCR as features [32]. They are 1) standard deviation of first order difference, 2) third central moment about the mean, 3) total number of zero crossing exceeding a threshold, and 4) difference between the number of zero crossings above and below the mean values. Combined with the volume information, the proposed algorithm can discriminate speech and music at a high accuracy of 98%.

Pitch-based

The patterns of pitch tracks of different audio contents vary a lot. For speech clip, voiced segments have smoothly changed pitch values, while no pitch information is detected in silent or unvoiced segments. For audio with prominent noisy background, no pitch information is detected either. For gentle music clip, since there are always dominant tones within a short period of time, many of the pitch tracks are flat with constant values. The pitch frequency in a speech signal is primarily influenced by the speaker (male or female), whereas the pitch of a music signal is dominated by the strongest note that is being played. It is not easy to derive the scene content directly from the pitch level of isolated frames; but the dynamics of the pitch contour over successive frames appear to reveal the scene content more.

Liu et al. utilized three clip-level features to capture the variation of pitch [19]: *standard deviation of pitch* (PSTD), *smooth pitch ratio* (SPR), and *non-pitch ratio* (NPR). SPR is the percentage of frames in a clip that have similar pitch as the previous frames. This feature is used to measure the percentage of voiced or music frames within a clip, since only voiced and music have smooth pitch. On the other hand, NPR is the percentage of frames without pitch. This feature can measure how many frames are unvoiced speech or noise within a clip.

Frequency-based

Given frame-level features that reflect frequency distribution, such as FC, BW, and ERSB, one can compute their mean values over a clip to derive corresponding clip-level features. Since the frame with a high energy has more influence on the perceived sound by the human ear, Liu et al. proposed to use a weighted average of corresponding frame-level features, where the weighting for a frame is proportional to the energy of the frame [19]. This is especially useful when there are many silent frames in a clip because the frequency features in silent frames are almost random. By using energy-based weighting, their detrimental effects can be removed.

Zhang and Kuo used spectral peak tracks (SPT's) in a spectrogram to classify audio signals [42]. First, SPT is used to detect music segments. If there are tracks which stay at about the same frequency level for a certain period of time, this period is considered a music segment. Then, SPT is used to further classify music segments into three subclasses: song, speech with music, and environmental sound with music background. Song segments have one of three features: ripple-shaped harmonic peak tracks due to voice sound, tracks with longer duration than speech, and tracks with fundamental frequency higher than 300 Hz. Speech with music background segment has SPT's concentrating in the lower to middle frequency bands and has lengths within a certain range. Those segments without certain characteristics are classified as environmental sound with music background.

There are other clip features that are very useful. Some researchers studied the audio feature in compressed domain. Due to the space limit, we cannot include all of them here. Interested readers are referred to [2][3][15][23][28].

1.8 AUDIO SEGMENTATION

Audio segmentation is finding the abrupt change locations along the audio stream. As we indicated before, this task is domain specific, and needs different approaches for different requirements. In this section, we present two segmentation tasks we investigated at two different levels. One is to segment speaker boundaries at the frame level, and the other one is to segment audio scenes, for example, commercials and news reporting in broadcast programs at the clip level.

1.8.1 Speaker segmentation

In our study, we employed 13 MFCC and their first order derivatives as audio features. The segmentation algorithm consists of two steps: splitting and merging. During splitting, we identify possible speaker change boundaries. During merging, neighboring scenes are merged if their contents are similar. In the first step, low energy frames, which are local minimum points on the volume contour, are located as boundary candidates. Figure 20.2 shows the volume contour of an audio file, where all low energy frames are indicated by a circle. For each boundary candidates, the difference between its neighbors (both left and right) is computed. The definition of neighbors is illustrated in the figure, where for frame X, two dotted rectangular windows W1 and W2 are the neighbors of X and each with length of L seconds. If the distance is higher than a certain threshold and it is the maximum in surrounding range, we declare that the corresponding frame is a scene boundary.

Divergence [13] is adopted to measure the difference between two windows. Assume the features in W1 follow Gaussian distribution N(μ_1, Σ_1), where μ_1 is the mean vector, and Σ_1 is the covariance matrix. Similarly, the features in W2 follow N(μ_2, Σ_2). Then the divergence can be simplified as

$$J(W1, W2) = \frac{1}{2} tr(\Sigma_1 - \Sigma_2)(\Sigma_2^{-1} - \Sigma_1^{-1}) + \frac{1}{2} tr(\Sigma_1^{-1} + \Sigma_2^{-1})(\mu_1 - \mu_2)(\mu_1 - \mu_2)'$$

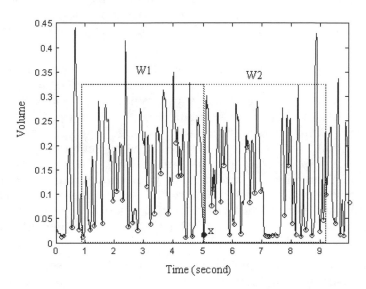

Figure 20.2 Illustration of speaker segmentation.

Such splitting process yields, in general, over segmentation. A merging step is necessary to group similar neighboring segments together to reduce the false speaker boundaries. This is done by comparing the statistical properties of adjacent segments. The same difference can be used based on longer segments (compared to fixed windows in splitting stage), and a lower threshold is applied.

Such an algorithm actually detects the change of acoustic channel property; for example, if the same speaker changed to a different environment, a segment boundary will be declared, although it is not a real speaker change. Testing on two half hour news sequences, 93% true speaker boundaries are detected with a false alarm rate of 22%.

1.8.2 Audio scene segmentation

Here, the audio scenes we considered are different types of TV programs, including news reporting, commercial, basketball, football, and weather forecast. To detect audio scene boundaries, a 14 dimension audio feature vector is computed over each audio clip. The audio features consist of VSTD, VCR, VU, ZSTD, NSR, 4ME, PSTD, SPR, NPR, FC, BW, ERSB1, ERSB2, ERSB3. For a clip to be declared as a scene change, it must be similar to all the neighboring future clips, and different from all the neighboring previous clips. Based on this criterion, we propose using the following measure:

$$Scene-change-index = \frac{\left\|\frac{1}{N}\sum_{i=-N}^{-1}f(i)-\frac{1}{N}\sum_{i=0}^{N-1}f(i)\right\|^2}{\sqrt{(c+\text{var}(f(-N),...,f(-1)))(c+\text{var}(f(0),...,f(N-1)))}},$$

where $f(i)$ is the feature vector of the i-th clip, with $i=0$ representing the current clip, $i>0$ a future clip, and $i<0$ a previous clip, $\|*\|$ is the *L-2* norm, var(...) is the average of the squared Euclidean distances between each feature vector and the mean vector of the N clips considered, and c is a small constant to prevent division by zero. When the feature vectors are similar within previous *N* clips and following *N* clips, respectively, but differ significantly between the two groups, a scene break is declared. If two breaks are closer than N clips away, the one with smaller scene-change-index value is removed. The selection of the window length *N* is critical: If *N* is too large, this strategy may fail to detect scene changes between short audio shots. It will also add unnecessary delay to the processing. Through trials-and-errors, we have found that *N=6* give satisfactory results.

Figure 20.3 (a) shows the content of one testing audio sequence used in segmentation. This sequence is digitized from a TV program that contains seven different semantic segments. The first and the last segments are both football games, between which are TV station's logo shot and four different commercials. The duration of each segment is also shown in the graph. Figure 20.3 (b) shows the scene-change-index computed for this sequence. Scene changes are detected by identifying those clips for which the scene-change-indices are higher than a threshold, D_{min}. We used $D_{min}=3$, which have been found to yield good results through trial-and-error. In these graphs, mark "o" indicates real scene changes and "*" detected scene changes. All the real scene changes are detected using this algorithm. Note that there are two falsely detected scene changes in the first segment of the sequence. They correspond to the sudden appearance of the commentator's voice and the audience's cheering.

1.9 AUDIO CONTENT CLASSFICATION

After audio segmentation, we need to classify each segment into predefined categories. The categories are normally semantically meaningful high level labels that are determined from low level features. The pattern recognition mechanism fits in this gap, and maps the distribution of low level features to high level semantic concepts. In the section, we will present three different audio classification situations: speaker identification, speech/nonspeech classification, and music genre classification.

1.9.1 Speaker recognition

Besides message via words, speaker identities are additional information conveyed in speech signal. Speaker identities are useful in audio content indexing and retrieval. For example, occurrences of the anchorpersons in broadcast news often indicate semantically meaningful boundaries for reported news stories. Speaker recognition aims to detect the speaker identities, and it generally encompasses two fundamental tasks: Speaker identification is the task to determine who is talking from a set of known voices or speakers, and speaker verification is the task of determining whether a person is who he/she claims to be.

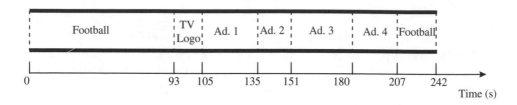

(a)   Semantic contents of the sequence

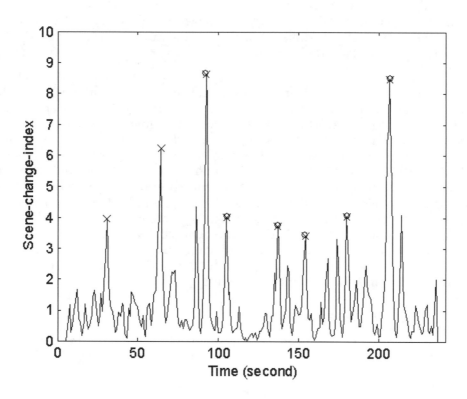

(b) Scene-change-index

Figure 20.3 Content and scene-change-index for one audio stream.

Acoustic features for speaker recognition should have high speaker discrimination power, which means high inter-speaker variability and low intra-speaker variability. Adopted features include linear prediction coefficients (LPC), cepstrum coefficients, log-area ratio (LAR), MFCC, etc., within which MFCC gains more prevalence due to its effectiveness [29]. Depending on the specific applications, different speaker models and corresponding pattern matching methods can be applied. Popular speaker models are dynamic time warping (DTW), hidden Markov model, artificial neural network, and vector quantization (VQ).

Huang et al. studied anchorperson detection, which can be categorized as a speaker verification problem [10]. Detection of anchorperson segments is carried out using text independent speaker recognition techniques. The target speaker (anchorperson) and background speakers are represented by a 64-component Gaussian mixture model with diagonal covariance matrices. The utilized audio features are 13 MFCC coefficients and their first and second order derivatives. A maximum likelihood classifier is applied to detect the target speaker segments. Testing on a dataset of 4 half hour news sequences, this approach successfully detects 91.3% of real anchorperson speech, and the false alarm rate is 1%.

1.9.2 Audio scene detection

Audio scenes are segments with homogeneous content in an audio stream. For example, broadcast news program generally consists of two different audio scenes: news reporting and commercials. Discriminating them is very useful for indexing news content. One obvious usage is to create a summary of news program, where commercial segments are removed.

Depending on the application, different categories of audio scenes and different approaches are adopted. In [32], Saunders considered the discrimination of speech from music. Saraceno and Leondardi further classified audio into four groups: silence, speech, music, and noise [31]. The addition of the silence and noise categories is appropriate, since a large silence interval can be used as segment boundaries, and the characteristic of noise is much different from that of speech or music.

A more elaborate audio content categorization was proposed by Wold et al. [38], which divides audio content into ten groups: animal, bells, crowds, laughter, machine, instrument, male speech, female speech, telephone, and water. To characterize the difference among these audio groups, Wold et al. used mean, variance, and auto correlation of loudness, pitch, brightness (i.e., frequency centroid) and bandwidth as audio features. A nearest neighbor classifier based on weighed Euclidean distance measure was employed. The classification accuracy is about 81% over an audio database with 400 sound files.

Liu et al. [17][19] studied the problem of classifying TV broadcast into five different categories: news reporting, commercial, weather forecast, basketball game, and football game. Based on a set of 14 audio features extracted from audio energy, zero crossing rate, pitch, and spectrogram, a 3 layer feed forward neural network classifier achieves 72.5% accuracy. A classifier based on a hidden Markov model further increases the accuracy by 12%.

Another interesting work related to general audio content classification is by Zhang and Kuo [41]. They explored five kinds of audio features: energy, ZCR, fundamental frequency, timber, and rhythm. Based on these features, a hierarchical system for audio classification and retrieval was built. In the first step, audio data is classified into speech, music, environmental sounds, and silence using a rule-based heuristic procedure. In the second step, environmental sounds are further classified into applause, rain, birds' sound, etc., using an HMM classifier. These two steps provide the so-called coarse-level and fine-level classification. The coarse-level classification achieves 90% accuracy and the fine-level classification achieves 80% accuracy in a test involving ten sound classes.

1.9.3   Music genre classification

Digital music, in all kinds of formats including MPEG Layer 3 (MP3), Microsoft media format, RealAudio, MIDI, etc., is a very popular type of traffic on the Internet. When music pieces are created, they are normally assigned with related metadata by producers or distributors, for example, title, music category, author name, and date. Unfortunately, most of the metadata is not available or is lost in the stages of music manipulation and format conversion. Music genre, as a specific metadata, is important and indispensable for music archiving and querying. For example, a simple query to find all pop music in a digital music database requires the category information. Since manually re-labelling is time consuming and inconsistent, we need an automatic way to classify music genre.

Music genre classification has attracted a lot of research effort in recent years. Tzanetakis et al. [36] explored the automatic classification of audio signals into a hierarchy of music genres. On the first level, there are ten categories: classical, country, disco, hiphop, jazz, rock, blues, reggae, pop, and metal. On the second level, classical music is further separated into choir, orchestra, piano, and string quartet, and jazz is further split into bigband, cool, fusion, piano, quartet, and swing. Three sets of audio features are proposed, which reflect the timbral texture, rhythmic content and pitch content of audio signal, respectively. Timbral texture features include spectral centroid, spectral rolloff, spectral flux, zero crossing rate, and MFCC. Rhythmic content features are calculated based on wavelet transform, where the information of main beat, sub-beats and their periods and strengths are extracted. Pitch content features are extracted based on multiple pitch detection techniques. Utilized pitch features include the amplitude and period of the maximum peaks of pitch histogram, pitch interval between the two most prominent peaks of the pitch histogram, and the sum of the histogram. Tzanetakis et al. tested different classifiers, including simple Gaussian classifier, Gaussian mixture model classifier, and K-nearest neighbor classifier. Among them, GMM with 4 mixtures achieves best classification accuracy, which is 61%. Considering that human beings make 20% to 30% errors on classifying musical genre in a similar task, the performance of automatic music genre classification is reasonably good.

Lambrou et al. [14] investigated the task of classifying audio signal into three different music styles: rock, piano, and jazz. They used zero crossing rate and statistical signal features in wavelet transform domains as acoustic features. Overall, seven statistics are computed, which are first order statistics: mean, variance, skewness, and kurtosis, and second order statistics: angular second moment, correlation, and entropy. Lambrou et al. benchmarked four different classifiers: minimum distance classifier, K-nearest neighbors classifier, least squares minimum distance classifier (LSMDC), and quadrature classifier. Simulation results show that LSMDC gives the best performance with an accuracy of 91.67%.

1.10   AUDIO SUMMARIZATION

The goal of audio summarization is to provide a compact version of the original audio signal in a way that most significant information is kept within a minimum duration. Effective audio summarization can save a tremendous amount of time for the user to digest the audio content without missing any important information. To further reduce the time that users need to skim audio content, the original audio data or the summary can be played back at a faster speed in a way that pitch information is kept. Normally when the speedup is less than two times real time, human beings do not lose much listening comprehension capability.

Huang et al. [11] studied how to summarize the news broadcast in different levels of details. The first level of summary filters out commercials, and second level of summary is composed of all anchorperson speeches, which cover the introductions and summaries of all news stories. The top level summary consists of a reduced set of anchorperson speech such that they cover all reported content, and are least redundant. For more information on audio summarization, please refer to the survey paper [40] of Zechner on spoken language summarization.

AUDIO QUERY METHODS

Besides audio content indexing, query mechanism is the other important component in an audio retrieval system. Here, we briefly introduce the three most commonly used query methods. Specifically, they are query by keywords, query by examples, and query by humming. We also describe the relevance feedback technique, which further improves the audio retrieval performance by user interaction.

1.11 QUERY BY KEYWORDS

Query by keywords or key phrases for audio content basically follows the same approach used in traditional text based information retrieval. The user provides several keywords, or key phrases as searching terms, and the query engine compares them with the textual information attached with audio files in the database to determine the list of returns. Query by text demands low computation resource, yet it requires that the content of each audio file is labelled with semantically meaningful metadata. For example, music songs may have titles, authors, players, and genres, and speech signals have transcriptions, etc. Some of these metadata can only be created manually, yet some of them can be produced by automatic approach; for example, the speech signal can be transcribed by automatic speech recognition technologies.

Query by keywords for audio takes the advantages of classical information retrieval technologies. For example, word stemming can be applied, such that words in different inflections of the same word can also be retrieved. Sometimes, straightforward keyword matching is not adequate, especially when the query is short. In such a situation, the keywords in query may not be the words used in the archive, although they share the same semantic meaning. Automatic query expansion [5][24] is to add more useful words in the query to improve the recall rate. For example, if the query is "AT&T," possible expansion are "AT and T," "telecommunication giant," and "Mike Armstrong (the CEO)". Simple expansion methods include adding corresponding acronyms of query keywords, or use full names of the acronyms in query. More sophisticated approaches utilize statistical relationship among words and relevance feedbacks by users.

Query by keywords for audio has its own shortage. The transcripts produced by ASR may be inaccurate due to the recognition errors or out of vocabulary words. If the query terms in audio samples are not correctly recognized, above mentioned information retrieval techniques no longer work. A promising alternative is to perform retrieval on sub-word acoustic units (e.g., phones or syllables) [5]. For example, a user query is first translated into a set of phone trigrams, and query is conducted by searching all possible three-phone sequences in the lattice output by the recognizer.

1.12 QUERY BY EXAMPLES

Very often, audio query is hard to formulate explicitly in words. For example, it is difficult to explain in text what audio we are looking for if we don't know its title. Query by example is a more natural way for retrieving audio content. Suppose we are looking for a music masterpiece. We have no clue of the title, but we have a short portion of it, for example, a 10 second long clip. Then we can use this piece of audio sample, normally in the format of a file, as a query object. The search engine analyzes the content of query example, computes acoustic features, compares the audio content with audio files in the database, and then generates returns accordingly.

Considering that even the same audio clips may exist in different formats, for example, different sampling rates, different compression methods, etc., it requires the search engine to extract robust acoustic features that survive various kinds of distortions. Another issue is how to speed up the search procedure, which is especially important for large size audio database. Besides optimizing the structure of the database, a more fundamental issue is how to make the comparison between two audio files faster. Liu et al. addressed this problem in [18]. The proposed idea is that instead of comparing the difference of two sets of audio features extracted from audio files, search engine first builds a Gaussian mixture model for each set of audio features, and then compares the distance between two models. Models of archived audio can be built offline, where processing time is not a major concern. The cost to train a GMM for a short query audio sample is neglectable. Another advantage is that the comparison cost between GMMs is independent of the duration of corresponding audio files. As a trade off of the speed and model accuracy, different numbers of mixtures for GMM can be chosen. Liu et al. also proposed a new parametric distance metric for determining the distance between two GMMs efficiently, which makes this approach promising for query by examples in large size audio database.

1.13 QUERY BY HUMMING

One step further than query by examples, query by humming is the most effective way to search audio content. Without forming any query keywords or having a piece of audio sample, the user hums a short clip of audio as a query. The challenges are twofold: robustly extracting the melodic information in query and efficiently matching melody.

Ghias et al. proposed their solution on how to specify a humming query and how to implement fast query execution in music database [8]. The hummed signal is first digitized, and pitch information is tracked. Melodic contour, which is the sequence of relative differences in pitch between successive notes, is then extracted. The query is transformed into a string of three letter alphabet (U, D, S), where U means a note is higher than previous note, D means it is lower than the previous note, and S means they are the same. Similarly, all songs in a music database are pre-processed to convert the melody into a stream of (U, D,

S) characters. A fast approximate string matching algorithm which allows certain number of mistakes is adopted to search similar songs in the database.

Recently, Lu et al. proposed a new method to query by humming in music retrieval [20]. They used a triplet: pitch contour, pitch interval, and duration to represent melody information. A hierarchical matching method was used to make the matching fast and accurate. First, approximate string matching and dynamic programming are applied to align the pitch contours between query and candidate music segments, Then, the similarities of pitch interval and duration according to the matched path are computed. The final rank of candidate music is a weighted summation of the two similarities. Simulations show that 74% of 42 testing queries retrieve correct songs among the top three matches from a database of 1000 MIDI songs. The performance is encouraging for commercial applications of query by humming.

1.14 RELEVANCE FEEDBACK

The previous three query mechanisms accept user's query in different formats, yet they do not involve further user inputs in the retrieval process. Two reasons suggest that user's feedback in the loop of retrieval is desirable. First, the internal low level features, including textual or acoustic features, that are used for audio retrieval normally do not have clear mapping to high level query concepts. It is difficult to accurately catch user's intent in feature space based on one short query. Second, there is variety in human perception. Even with the same query and the same returns, different users may have different opinions on the results. With the user's relevance feedback on a set of retrieved results by labelling them either positive (wanted ones) or negative (others), the search engine has a better chance to capture user's searching goal by analyzing these additional audio samples. The internal query term or search mechanism can be adapted based on the feedback to refine the query results in the next iteration.

Rui et al. [30] investigated relevance feedback techniques in content-based image retrieval. Their approach is to dynamically assign weights, based on user's relevance feedback, to different sets of visual features, which determine the difference of two images. In such a way, there is no longer a burden for users to precisely specify their queries; it is the computer who intelligently finds out user's need and accordingly adjusts the searching scheme. Similar approach is also applicable for audio query system.

AUDIO RETRIEVAL SYSTEMS

There are many audio retrieval systems available in either research laboratories or commercial vendors. Most of them utilize various audio processing algorithms to enhance the retrieving performance. Here, we briefly introduce three representative audio retrieval systems to show the state of the art in this area.

1.15 BBN ROUGH 'N' READY SYSTEM [22]

BBN has conducted a long term research on speech and language processing technologies, and developed an audio indexing system called Rough 'n' Ready system, which provides a *rough* transcription of the speech that is *ready* for browsing. The Rough 'n' Ready system focuses entirely on the linguistic content contained in the audio signal. Overall, the system is composed of three subsystems: the indexer, the server, and the browser. The indexer subsystem takes audio waveform as input and generates a compact structural summarization encoded in XML format. The server subsystem collects and manages the archive, as well as interacts with the browser. The browser handles the user interaction. Its main purpose is to send user queries to the server and present the results in a meaningful way.

Indexer subsystem is the technical core of the entire system, and it is built based on a series of spoken language processing technologies, including speech recognition, speaker recognition, name spotting, topic classification, and story segmentation. ASR module is realized by BBN Byblos, a large vocabulary speaker independent speech recognition system. Byblos employs continuous density hidden Markov models as acoustic models, and n-gram as language models. With multipass search strategy, unsupervised adaptation, and speedup algorithms, including fast Gaussian computation, grammar spreading, and N-Best tree rescoring, Byblos achieves an error rate of 21.4% on DARPA broadcast news test data with a sixty thousand word vocabulary at 3 times real-time performed on a 450M PII processor. Speaker recognition module recognizes a sequence of speakers in speech. It consists of three components: speaker segmentation, speaker clustering, and speaker identification. Name spotting uses IdentifFinder, a hidden Markov model-based name extraction system, to extract important terms from the speech and collect them into a database. OnTopic, a probabilistic HMM-based topic classification module, produces a rank-ordered list of all possible topics with corresponding scores for a given document. Story segmentation detects story boundaries such that each segment covers a coherent set of topics. All these structural features are used to construct highly selective search queries for retrieving specific content from large audio archives.

The technologies developed in Rough 'n' Ready system has reached the stage that commercialization is possible in very near future. There is no doubt that such an advanced audio indexing system will benefit wide ranges of customers in audio browsing and querying.

1.16   HP SPEECHBOT SYSTEM [35]

SpeechBot is claimed to be the first Internet search site for spoken audio on the web. It is built by Compaq research laboratory, which now belongs to Hewlett-Packard. Similar to AltaVista or Google, Speechbot system is a public search site that indexes transcribed audio documents by automatic speech recognition technology, and provides indexing and search functions based on audio content. The system does not serve audio content, but provide links to the relevant segments of the original documents. Currently the system is capable of indexing more than 100 hours of audio data every day, which includes popular talk radio, technical and financial news shows, and video conference recordings. Most of these audio data have no transcript, and when transcript does exist, it will be used to improve the indexing process. Although current recognition error rates are high, yet the system provides acceptable retrieval accuracy, such that the user can effectively find the interesting pieces of audio content easily.

SpeechBot consists of five key components. The *transcoder* module downloads various types of audio files from public web sites, and converts them into raw file format. Relevant metadata, including title, is saved for identifying them in the database. With eight workstations, each handling four streams in real time, the overall throughput of the transcoder module is 768 hours per day. The raw audio files are segmented and passed to the *speech recognizer* module, where they are transcribed in parallel. The speech recognizer is built based on Gaussian mixture, triphone-based hidden Markov model technology. The vocabulary of the recognizer is 64,000 words. With different combinations of model complexity and recognition searching beam, the recognizer provides different accuracies. For 6.5kb/s RealAudio data, the word error rate is in the range of 50%. A farm of 60 processors can process about 110 hours of audio per day, assuming that the recognizer is running at an average speed of 13 times real time. The *librarian* module manages the entire workflow on all modules, and stores metadata and necessary information for user interface. It is built on an Oracle relational database, and contains indices of 15550 hours of audio content at the time when this chapter was written. The *indexer* module basically is a query engine. It provides an efficient catalogue of audio documents based on their transcriptions and metadata. The retrieved document is sorted by relevance based on term frequency inverse document frequency metric. The last module is the *user interface*. The Web server collects users' interactions, passes their queries to Indexer, and generates the retrieval results. The user interface also highlights matching transcripts and expands and normalizes acronyms.

With the help of Speechbot, users are exposed to a huge amount of audio content, which is searchable as easy as text document. The extension of searching capability from text to non-text media is revolutionary.

1.17 AT&T SCANMAIL SYSTEM [9]

SCANMail (Spoken Content-based Audio Navigator Mail) is a system that employs automatic speech recognition, information retrieval, information extraction, and human computer interaction technology to permit users to browse and search their voicemail messages by content through a graphical user interface. Voicemail users receive many voice messages everyday; SCANMail lets users manage their important message in a better way. The system has several useful features: 1) Read transcriptions of messages instead of listening to them, 2) Access messages randomly through a GUI, playing and reading just what's important, 3) Extract telephone numbers automatically from messages, 4) Label messages with caller names even when caller ID is not available, and 5) Use information retrieval to search messages by content.

Figure 20.4 shows the architecture of the SCANMail system. Voicemail messages are retrieved from a messaging system, Audix, via a POP3 server. The *automatic speech recognition* module transcribes voice messages, and the text information is indexed by the *information retrieval* module. The *email server* sends an email with original voice message plus its transcription to the user. Key information in the transcript including phone number, name, date, and time is extracted by the *information extraction* module. The *speaker ID* module compares the audio message against a list of speaker models to propose a caller identification. Users can provide feedback on the speaker ID results such that speaker models will be refined. SCANMail system provides a friendly GUI interface, which allows users to browse and query their voice messages by content. Experiments show that SCANMail offers some increase in efficiency and a significant increase in perceived utility over regular voicemail access by phones. With a variety of random access play and search capabilities, the system makes the daily voice message access an easy and pleasant experience.

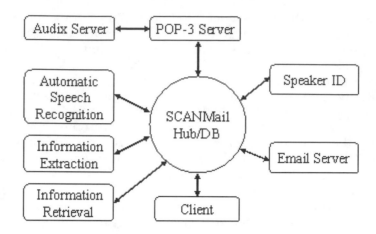

Figure 20.4 The SCANMail architecture.

BRIEF INTRODUCTION ON MPEG-7 AUDIO

MPEG-7 is an ISO/IEC standard developed by MPEG (Moving Picture Experts Group) [25], the committee that developed well known MPEG-1, MPEG-2, and MPEG-4 standards. MPEG-7, formally named "Multimedia Content Description Interface," is a standard for describing the multimedia content data that supports some degree of interpretation of information's meaning, which can be passed onto, or accessed by a device or computer code. MPEG-7 standard provides support to a broad range of applications (for example, multimedia digital libraries, broadcast media selection, multimedia editing, home entertainment devices, etc.). MPEG-7 will also make the web as searchable for multimedia content as it is searchable for text today.

The main elements of MPEG-7 standard are: 1) *Description tools* are composed of Descriptors, which define the syntax and the semantics of each feature (metadata element), and Description Schemes (DS), that specify the structure and semantics of the relationships between their components: Descriptors and Description Schemes. 2) A *Description Definition Language* (DDL) defines the syntax of the MPEG-7 Description tools and allows the creation of new DSs and Ds and the extension and modification of existing DSs. 3) *System tools* support binary coded representation for efficient storage and transmission, transmission mechanisms, multiplexing of descriptions, synchronization of descriptions with content, management, and protection of intellectual property in Ds.

MPEG-7 consists of eight parts: 1) Systems, 2) Description Definition Language, 3) Visual, 4) Audio, 5) Multimedia Description Schemes, 6) Reference Software, 7) Conformance Testing, and 8) Extraction and use of descriptions. MPEG-7 audio provides structures, in conjunction with the Multimedia Description Schemes part of the standard, for describing audio content.

The MPEG-7 audio standard comprises a set of descriptors that can be divided roughly into two classes: low-level or generic tools and high-level or application specific tools. Figure 20.5 shows an overview of these audio descriptors.

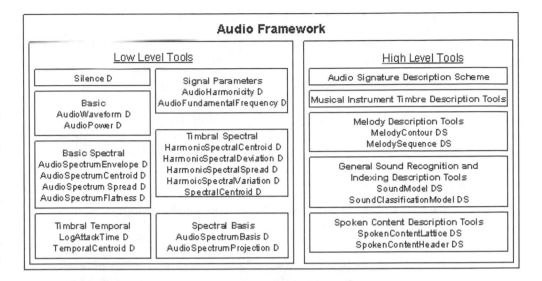

Figure 20.5 Overview of audio framework in MPEG-7 audio.

1.18 LOW LEVEL TOOLS

Low level tools may be used in a variety of applications. Besides silence descriptor, there are six groups of low level audio descriptors.

Basic Audio Descriptors include the AudioWaveform Descriptor, which describes the audio waveform envelope, and the AudioPower Descriptor, which depicts the temporally smoothed instantaneous power.

Basic Spectral Descriptors are derived from a single time-frequency analysis of audio signal. Among this group are the AudioSpectrumEnvelope Descriptor, which is a log-frequency spectrum; the AudioSpectrumCentroid Descriptor, which describes the center of gravity of the log-frequency power spectrum; the AudioSpectrumSpread Descriptor, which represents the second moment of the log-frequency power spectrum; and the AudioSpectrumFlatness Descriptor, which indicates the flatness of the spectrum within a number of frequency bands.

Signal Parameter Descriptors consist of two descriptors: The AudioFundamentalFrequency descriptor describes the fundamental frequency of an audio signal, and the AudioHarmonicity Descriptor represents the harmonicity of a signal.

There are two Descriptors in *Timbral Temporal Descriptors* group. The LogAttackTime Descriptor characterizes the attack of a sound, and the TemporalCentroid Descriptor represents where in time the energy of a signal is focused.

Timbral Spectral Descriptors have five components. The SpectralCentroid Descriptor is the power-weighted average of the frequency of the bins in the linear power spectrum, the HarmonicSpectralCentroid Descriptor is the amplitude-weighted mean of the harmonic peaks of the spectrum, the HarmonicSpectralDeviation Descriptor indicates the spectral deviation of log-amplitude components from a global spectral envelope, the

HarmonicSpectralSpread describes the amplitude-weighted standard deviation of the harmonic peaks of the spectrum, and finally the HarmonicSpectralVariation Descriptor is the normalized correlation between the amplitude of the harmonic peaks between two subsequent time-slices of the signal.

The last group of low level descriptors is *Spectral Basis Descriptors*. It includes the AudioSpectrumBasis Descriptor, which is a series of basis functions that are derived from the singular value decomposition of a normalized power spectrum, and the AudioSpectrumProjection Descriptor, which represents low-dimensional features of a spectrum after projection upon a reduced rank basis.

1.19   HIGH LEVEL TOOLS

High level tools are specialized for domain specific applications. There are five sets of high level tools that roughly correspond to the application areas that are interested in the standard.

Audio Signature Description Scheme statistically summarizes the spectral flatness Descriptor as a condensed representation of an audio signal. It provides a unique content identifier for robust automatic identification of audio signals.

Musical Instrument Timbre Description Tools describe perceptual features of instrument sounds with a reduced set of Descriptors. They relate to the notion such as "attack," "brightness," and "richness" of a sound.

Melody Description Tools include a rich representation for monophonic melodic information to facilitate efficient, robust, and expressive melodic similarity matching. The scheme includes a MelodyContour Description Scheme for extremely terse, efficient melody contour representation, and a MelodySequence Description Scheme for a more verbose, complete, expressive melody representation.

General Sound Recognition and Indexing Description Tools are a collection of tools for indexing and categorization of general sounds, with immediate application to sound effects.

Spoken Content Description Tools allow detailed description of words spoken within an audio stream.

MPEG-7 aims at making the multimedia content more searchable than it is today. The set of descriptors standardized in MPEG-7 audio makes it possible to develop content retrieval tools, systems that are able to access a lot of different audio archives in the same way. They are also useful for the content creator to edit the content, and for content distributors to select and filter the content. Some typical applications of MPEG-7 audio include large scale audio content (radio, TV broadcast, movie, music) archives and retrieval, audio content distribution, education, and surveillance.

CONCLUSIONS

This chapter reviewed the recent progress of audio indexing and retrieval. Audio content indexing and retrieval plays an important role in the field of multimedia information retrieval. Here, we outline several possible future directions in this field to conclude this chapter. First, it remains a challenge to find an application independent approach in unrestricted domains. Audio content analysis in specific domains has been intensively studied, and most of the problems have found feasible solutions. Although different applications share many low level audio processing technologies, there is not a general framework that universally fits most applications. We believe that this topic deserves more research efforts since such a desire is growing rapidly with more and more audio retrieval systems as various domains are developed. Second, personalization demands more attention in future audio retrieval systems. Most of currently available audio retrieval systems ignore the individual characteristics of end users. Personalized profile for each user, which logs the history of user queries and activities, user preferences, etc., is useful for the audio retrieval system to generate better query results tuned for a particular user. Third, commercialization is possible with the current stage of audio analysis capability. After many research efforts invested in recent years, audio retrieval systems in certain fields achieve acceptable performance, and they provide substantial help for the users to find audio content fast and easily. Computation complexity of audio indexing and query continues to be the bottleneck. More efficient technologies are indispensable to increase system scalability and reduce infrastructure cost for commercial systems.

ACKNOWLEDGMENTS

The authors would like to thank Lee Begeja, David Gibbon, Bernard Renger, Aaron Rosenberg, Murat Saraclar, and Behzad Shahraray of AT&T Labs – Research for enlightening technical discussions, and Chun Jin of Carnegie Mellon University for reviewing part of this chapter.

REFERENCES

[1] G. Backfried, R. Rainoldi, and J. Riedler, Automatic Language Identification in Broadcast News, IJCNN-2002, Honolulu, HI, Vol. 2, pp. 1406-1410, 2002.

[2] J.S. Boreczky and L. D. Wilcox, A Hidden Markov Model Framework for Video Segmentation Using Audio and Image Features, ICASSP-1998, Vol. 6, pp. 3741-3744, May 12-15, 1998.

[3] Y. Chang, W. Zeng, I. Kamel, and R. Alonso, Integrated Image and Speech Analysis for Content-based Video Indexing, Proc. 3rd IEEE Int. Conf. Multimedia Computing and Systems, Hiroshima, Japan, pp. 306-313, June 17-23, 1996.

[4] S. Chen and P. Gopalakrishnan, Speaker, Environment and Channel Change Detection and Clustering via the Bayesian Information Criterion, DARPA Speech Recognition Workshop, 1998.

[5] J. Choi, D. Hindle, J. Hirschberg, I. Chagnolleau, C. Nakatani, F. Pereira, A. Singhal, and S. Whittacker, An Overview of The AT&T Spoken Document Retrieval System, DARPA Broadcast News Transcription and Understanding Workshop, 1998.

[6] F. Dellaert, T. Polzin, and A. Waibel, Recognizing Emotion in Speech, ICSLP 1996, Oct. 1996.

[7] J.-L. Gauvain and L. Lamel, Large-vocabulary Continuous Speech Recognition: Advances and Applications, Proc. of the IEEE, Vol. 88, No. 8, pp. 1181-1200, Aug. 2000.

[8]   A. Ghias, J. Logan, D. Chamberlin, B. Smith, Query by Humming, ACM Multimedia 95, 1995.

[9]   J. Hirschberg, M. Bacchiani, D. Hindle, P. Isenhour, A. Rosenberg, L. Stark, L. Stead, S. Whittarker, and G. Zamchick, SCANMail: Browsing and Searching Speech Data by Content, Proc. European Conf. On Speech Communication and Technology, Aalborg, Denmark, Sept. 2001.

[10]   J. Huang, Z. Liu, and Y. Wang, Joint Video Scene Segmentation and Classification Based on Hidden Markov Model, ICME-2000, New York, NY, Aug. 2000.

[11]   Q. Huang, Z. Liu, and A. Rosenberg, Automated Semantic Structure Reconstruction and Representation Generation for Broadcast News, Proc. Of SPIE, Jan. 1999.

[12]   N. Jayant, J. Johnson, and S. Safranek, Signal Compression Based on Models of Human Perception, Proc. of the IEEE, Vol. 81, pp. 1385-1422, Oct. 1993.

[13]   S. Kullback, Information Theory and Statistics, Dover Publications, Inc. 1968.

[14]   T. Lambrou, P. Kudumakis, R. Speller, M. Sandler, and A. Linney, Classification of Audio Signals Using Statistical Features on Time and Wavelet Transform Domains, ICASSP-1998, Vol. 6, pp. 3621-3624, 1998.

[15]   R. Lienhart, S. Pfeiffer, and W. Effelsberg, Scene Determination Based on Video and Audio Features, Proc. IEEE Int. Conf. Multimedia Computing and Systems, Vol. 1, Florence, Italy, pp. 685-690, June 7-11, 1999.

[16]   Z. Liu, J. Huang, Y. Wang, and T. Chen, Audio Feature Extraction & Analysis for Scene Classification, MMSP-1997, pp. 343-348, 1997.

[17]   Z. Liu, J. Huang, and Y. Wang, Classification of TV Programs Based on Audio Information Using Hidden Markov Model, MMSP-1998, pp. 27-32, 1998.

[18]   Z. Liu and Q. Huang, Content-based Indexing and Retrieval-by-example in Audio, ICME-2000, 2000.

[19]   Z. Liu, Y. Wang, and T. Chen, Audio Feature Extraction and Analysis for Scene Segmentation and Classification, J. VLSI Signal Processing Sys. Signal, Image, Video Technol., Vol. 20, pp. 61-79, Oct. 1998.

[20]   L. Lu, H. You, H.-J. Zhang, A New Approach to Query by Humming in Music Retrieval, ICME-2001, 2001.

[21]   L. Lu, H. Jiang, and H. Zhang, A Robust Audio Classification and Segmentation Method, ACM MM-2001, 2001.

[22]   J. Makhoul, F. Kubala, T. Leek, D. Liu, L. Nguyen, R. Schwartz, and A. Srivastava, Speech and Language Technologies for Audio Indexing and Retrieval, Proc. Of the IEEE, Vol. 88, No. 8, pp. 1338-1353, Aug. 2000.

[23]   K. Minami, A. Akutsu, H. Hamada, and Y. Tonomura, Video Handling with Music and Speech Detection, IEEE Multimedia Magazine, Vol. 5, pp. 17-25, July-Sept. 1998.

[24]   M. Mitra, A. Singhal, and C. Buckley, Improving Automatic Query Expansion, ACM SIGIR'98, pp. 206-214, 1998.

[25]   Overview of the MPEG-7 Standard, ISO/IEC JTC1/SC29/WG11, N4509, Dec. 2001.

[26]   J. Nam and A. H. Tewfik, Combined Audio and Visual Streams Analysis for Video Sequence Segmentation, ICASSP-1997, Vol. 3, pp. 2665-2668, 1997.

[27]   E. Parris and M. J. Carey, Language Independent Gender Identification, ICASSP-1996, Vol. 2, pp. 685-688, 1996.

[28]   S. Pfeiffer, S. Fischer, and W. Effelsberg, Automatic Audio Content Analysis, Proc. 4th ACM Int. Conf. Multimedia, Boston, MA, Nov. 18-22, pp. 21-30, 1996.

[29]   D. A. Reynolds, An Overview of Automatic Speaker Recognition Technology, ICASSP-2002, pp. 4072-4075, 2002.

[30]   Y. Rui, T. S. Huang, and S. Mehrotra, Relevance Feedback Techniques in Interactive Content-based Image Retrieval, ICMCS-1999, 1999.

[31]   C. Saraceno and R. Leonardi, Audio As a Support to Scene Change Detection and Characterization of Video Sequences, ICASSP-1997, Vol. 4, pp. 2597-2600, 1997.

[32]   J. Saunders, Real-time Discrimination of Broadcast Speech/Music, in ICASSP-1996, Vol. 2, pp. 993-996, 1996.

[33]   E. Scheirer and M. Slaney, Construction and Evaluation of a Robust Multifeatures Speech/Music Discrimination, ICASSP-1997, Vol. 2, pp. 1331-1334, Apr. 21-24, 1997.

[34]   M. Siegler, U. Jain, B. Raj, R. Stern, Automatic Segmentation, Classification and Clustering of Broadcast News Audio, Proc. DARPA Speech Recognition Workshop, Chantilly, VA pp. 97-99, Feb. 1997.

[35]   J. V. Thong, P. Moreno, B. Logan, B. Fidler, K. Maffey, and M. Moores, Speechbot: An Experimental Speech-Based Search Engine for Multimedia Content on the Web, IEEE Trans. on Multimedia, Vol. 4, No. 1, pp. 88 - 96, March 2002.

[36]   G. Tzanetakis and P. Cook, Musical Genre Classification of Audio Signals, IEEE Trans. On Speech and Audio Processing, Vol. 10, Issue 5, pp. 293-302, July 2002.

[37]   H. D. Wactlar, M. G. Christel, Y. Gong, and A. G. Haupmann, Lessons Learned from Building a Terabyte Digital Video Library, IEEE Computer Magazine, Vol. 32, pp. 66-73, Feb. 1999.

[38]   E. Wold, T. Blum, D. Keislar, and J. Wheaton, Content-based Classification, Search, and Retrieval of Audio, IEEE Multimedia, Vol. 3, No. 2, pp. 27-36, 1996.

[39]   Y. Wang, Z. Liu, and J. Huang, Multimedia Content Analysis, IEEE Signal Processing Magazine, Vol. 17, No. 6, pp. 12-36, Nov. 2000.

[40]   K. Zechner, Summarization of Spoken Language – Challenges, Methods, and Prospects, Speech Technology Expert eZine, Issue 6, Jan. 2002.

[41]   T. Zhang and C.-C.J. Kuo, Hierarchical Classification of Audio Data for Archiving and Retrieving, ICASSP-1999, Vol. 6, pp. 3001-3004, 1999.

[42]   T. Zhang and C.-C.J. Kuo, Video Content Parsing Based on Combined Audio and Visual Information, SPIE's Conference on Multimedia Storage and Archiving Systems IV, Boston, MA, pp. 78-89, Sept., 1999.

21

RELEVANCE FEEDBACK IN MULTIMEDIA DATABASES

Michael Ortega-Binderberger
Department of Computer Science
University of Illinois at Urbana-Champaign
Urbana, IL, USA
`miki@acm.org`

Sharad Mehrotra
Information and Computer Science
University of California, Irvine
Irvine, CA, USA
`sharad@ics.uci.edu`

1. INTRODUCTION

The popularity of web search engines has familiarized countless users with the similarity search paradigm. In this paradigm a user provides an example or simple sketch of desired information to a system and receives a list of items that "best" match the information provided. These results are typically sorted by a system-generated estimate of how closely they match the sketch/requirement provided by users. Consider a typical web search engine. The users' sketch takes the form of keywords and the search engine finds the web pages that best match those keywords.

User expectations have grown to demand powerful and flexible search capabilities for multimedia data such as images and video in addition to the traditional unstructured web pages. Consider a user searching for pictures depicting a "sunset by the sea" in an image database. One possibility is to attach a text description to each image and use standard text search engine techniques to find the results. The problem with this approach is that "a picture is worth a thousand words": it is time consuming to describe each image in sufficient detail to be useful for searching. An alternate possibility is to make the content of the image itself searchable by abstracting some of its properties into a form that can be easily searched. Many image retrieval systems have adopted this approach [7][13][18][27] by using image-processing techniques to extract features that attempt to capture the user's perception of images.

When searching, the results of a user's first search attempt rarely satisfy her information need [26]. This can be due to many reasons. Any search system must first abstract the content of the searchable documents, images or videos into a form that is both searchable and effectively captures some aspect of the user's perception. There may be a gap between these abstractions, also called features, and the way humans really perceive the content. A further problem is the difficulty a user faces in constructing an appropriate example or sketch to submit for search due to interface limitations imposed on her, unfamiliarity with the database content, etc.

To cope with these limitations, users initiate an information discovery cycle [29] whereby they repeatedly modify their original sketch or example in hopes of improving the results. Typically, the sketch changes minimally between search iterations. For example, a user searching for information on image retrieval systems would use the terms "image databases," "image retrieval," or "content-based retrieval" in a web search engine. As a result, there is ample potential for the system to observe the user behaviour and aid her in enhancing her search criteria.

Relevance Feedback is a technique to offload from the user to the search engine the task of discovering a better search query formulation. When users see results, they instantly recognize how relevant, that is, how good or bad, they match their information need. Relevance Feedback refers to the ability of users to communicate, or *feed back*, to the search engine this notion of relevance to their information need. The search engine then uses this relevance information to construct a better sketch and uses it to retrieve improved results to the user. Relevance Feedback is of special interest for multimedia search as compared to textual search. The features derived from multimedia objects are typically more obscure than the sets of keywords used for textual searches. As a result of this feature complexity, there is no counterpart in multimedia retrieval to the ease with which users can manually modify their query formulation in a text search engine, making relevance feedback much more important.

In this chapter, we discuss several relevance feedback techniques that have been successfully applied to multimedia search. Section 2 presents some background on multimedia retrieval. Section 3 discusses the basic relevance feedback concepts we will use during the remainder of the chapter. Section 4 discusses retrieval with only one feature while section 5 discusses retrieval with multiple features. Section 6 describes techniques to reduce the number of relevance feedback iterations to quickly find the optimal results. Section 7 discusses how to evaluate the performance of relevance feedback. Finally, section 8 presents some conclusions and current trends in relevance feedback.

2. BACKGROUND

To enable multimedia retrieval, retrieval systems extract a set of properties from the multimedia objects that capture some aspects of its content. These properties, also called *features*, are what a retrieval system understands of the multimedia objects and thus limit the systems capabilities. In the broadest sense, features are of a textual or visual nature. Textual features include manually or automatically assigned annotations or keywords. Visual features capture properties such as color, texture, shape, faces, etc. and are typically extracted automatically using image processing techniques, although manual extraction (e.g., for segmentation) can also be used. Textual features and associated retrieval have been extensively studied in the field of Information Retrieval [26][2] and we will not discuss them any further.

Visual features are an active research area with many exciting results and a rich literature. Visual features can be broadly classified into general and specific features. General features deal with aspects common to most objects, such as color, texture, shape, etc. Specific features on the other hand focus on properties such as human fingerprints, faces, or gestures.

Many techniques have been developed for both general and specific features. For example, there are many different ways to represent the color content of a multimedia object including color histogram, color moments, color sets [17], etc. This variety corresponds to the subjectivity with which humans perceive the content of multimedia objects, and each of these feature representations capture the feature from a different perspective. Extensive descriptions of feature representations appear in other chapters of this book. In this chapter, we will assume a multimedia retrieval system contains for each multimedia object, a set of features F. Each feature F_i may itself contain a set of feature representations $f_{i,j}$. For example, F_1 can be the color feature with color histogram ($f_{1,1}$) and color moments ($f_{1,2}$) representations, while F_2 can be the texture feature with wavelet ($f_{2,1}$) and Tamura ($f_{2,2}$) representations. Associated with each feature representation $f_{i,j}$ is a set of comparison functions $d_{i,j,k}$ that determine how good two feature representations match each other. Retrieval systems have adopted two interpretations for these functions. Under the distance interpretation, a value of 0 means a perfect match with higher values indicating progressively worse matches. For example, the Euclidean distance metric can be used as the distance function for a feature representation. Under the similarity interpretation, values are in the range [0,1] where a value of 1 means a perfect match and 0 means no match. These two interpretations are generally interchangeable and can easily be converted into each other. We will focus on a distance interpretation in the remainder of the chapter.

Regardless of the details of feature representations, most of them are represented as an array of real values. We can then easily view a feature value as a vector in a multidimensional space. The distance functions for each feature representation can be viewed as determining the distance between two objects from this space.

To explore relevance feedback, we will assume the presence of two features each with two representations as described above. To simplify our discussion, we will assume each of these feature representations to be two-dimensional, with the understanding that the same principles we discuss also apply in higher dimensions. For example, we can interpret the color histogram representation to be the average hue and saturation of an image.

Once the sets of features, representations, and distance functions are established, we must turn to the problem of determining the overall distance between a multimedia object and the query. Because the query model heavily influences how relevance feedback works, we present the different query models in conjunction with the feedback models they support.

3. RELEVANCE FEEDBACK CONCEPTS

Relevance feedback is a technique to offload the work a user performs to improve a search by iteratively reformulating her query. As described above, an initial query formulated by a user may not fully capture her information need due to the complexity of formulating the query, unfamiliarity with the data collection, or inadequacy of the available features. Users then typically manually change the query and re-execute the search until they are satisfied. By using relevance

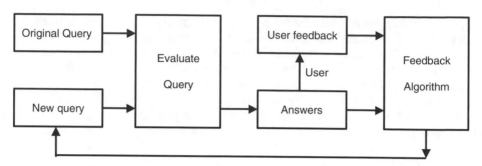

Figure 21.1 Relevance Feedback Cycle

feedback to criticise the answers, the system learns a new query that better captures the user's information need, and therefore relieves the user from reformulating the query herself.

Figure 21.1 shows the overall feedback process. The user formulates an initial query to the retrieval system, which generates a set of answers. The user then examines the answers and provides a judgement as to the quality or *relevance* of the answers. The system uses the original answers and the user supplied feedback, and builds a new query.

There are three main ways for the user to supply relevance feedback:

1. *Goodness / badness of results.* The user looks at individual results and determines if the result is a good or bad instance of her information need. She can provide relevance feedback at varying granularities. Most retrieval systems support a binary approach to relevance: a result is either relevant or not. Typically, the system considers all items to be non-relevant (or neutral) as the user marks only a few items she considers relevant. This binary notion of relevance can be generalized to multiple levels of relevance, as well as non-relevance. Some systems have experimented with varying levels of relevance trading user convenience for a more accurate picture of what the user wants [13]. Empirical studies however have shown that users typically give very little feedback and that the flexibility of multiple levels of relevance is too burdensome [12].

2. *Ranking.* In this approach, the user considers a subset of results at a time and "sorts" them in the order she thinks they should appear. In a sense, the user is performing the task of the retrieval system: to let it imitate her preferred ranking. This approach can be considered an extension of the multiple-relevance-levels approach where there are as many levels of relevance as the user gives relevance judgements, and no items share the same relevance level. The ranking approach gives excellent feedback to the retrieval system, but tends to be burdensome to the user.

3. *Explicit.* For explicit feedback, the retrieval system exposes to the user a visualization of its internal query structure and lets the user interactively manipulate it to improve the query. Examples of such systems are [28], and several text search engines that employ term suggestion. To employ this technique, the user must have some familiarity with the domain, and can quickly become too burdensome for multimedia data; therefore we will not discuss this approach in this chapter.

In this chapter we will concentrate on the first approach to providing feedback, and give also a brief discussion of the second approach, namely ranking. We will denote the answers by a_i, where i indicates the rank of that answer, that is, answers are ordered based on i: $<a_1, a_2, ...>$. We denote the relevance feedback for answer a_i by rf_i. This is a numeric value with the following interpretation:

$$rf_i = \begin{cases} 0 & \text{means no} - \text{information} \\ 1 & \text{means relevant} \\ -1 & \text{means non} - \text{relevant} \end{cases}$$

In general, when a finer gradation of relevance is needed, we use arbitrary positive values to denote relevant answers, and arbitrary negative values to denote non-relevant answers.

The specific query model in use determines how to derive a new query using relevance feedback. The techniques we will discuss assume certain properties of the query model; therefore we describe the query model assumptions together with the relevance feedback techniques.

4. SINGLE FEATURE REPRESENTATION FEEDBACK

A straightforward approach to multimedia retrieval is to use a single feature representation. A single feature representation is easy to extract, manage and search. In fact, text retrieval systems tend to follow this approach using a set of carefully extracted and weighted keywords along with a complex distance function. For the present discussion, we can also consider several concatenated feature representations as a single (longer) feature representation with an integrated distance function. For example, we can concatenate the *color moments* and *wavelet texture* feature representations into a larger feature and still use the Euclidean distance on the combined feature. The ImageRover [27] system, among others, follows this approach.

There are two main query reformulation approaches for single feature representations: single-point and multi-point. We discuss both of them in this section.

4.1 Single-Point Approaches

A simple query model is to use an instance of a single feature representation $f_{i,j}$ as a query point p in a multidimensional space, together with the corresponding set of distance functions D_{ij} for the feature representation. One distance function $d_{ijk} \in D_{ij}$ is the "default" used by the retrieval system in the initial query. We denote the ith value (dimension) of p by $p[i]$. The query model is thus formed by a multidimensional point p and a distance function d_{ijk}: $<p, d_{ijk}>$.

Based on this model, we discuss three relevance feedback-based query formulation approaches: query point movement, affecting the shape of the distance function, and selection of the distance function itself.

4.1.1 Query Point Movement

The most straightforward technique to improve a query is to change the query point itself. Selecting a different query point mimics the way a human user would go about reformulating a query herself by tinkering with the query value without modifying parameters of the distance function itself.

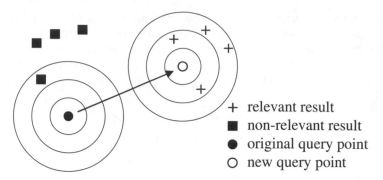

+ relevant result
■ non-relevant result
● original query point
○ new query point

Figure 21.2. Query Point Movemet

The objective of the query point movement approach is to construct a new query point that is "close" to relevant results, and "far" from non-relevant results. Figure 21.2 shows how this approach works for our hypothetical two-dimensional feature representation using a Euclidean distance function. The original and new query points are shown with a set of concentric circles representing iso-distance curves, that is, any point on a curve has the same distance from the query point. The query point moves to an area that is close to relevant results and away from non-relevant results.

Ideally we would know, for the entire database, those results (points) that are deemed relevant, those that are deemed non-relevant, and those deemed neutral (or don't care). Under these assumptions, it can be shown that the optimal query is formed by the centroid among the relevant points minus the centroid among the remaining points. Let $S_{rel} = a_i|rf_i > 0$ be the set of relevant points and let N be the number of points in the database, then the optimal query point p_{opt} is given by:

$$p_{opt} = \frac{1}{|S_{rel}|} \sum_{a_j \in S_{rel}} a_j - \frac{1}{N - |S_{rel}|} \sum_{a_j \notin S_{rel}} a_j$$

In a realistic setting, however, the set of relevant points are not known a priori and must be estimated. A natural way to estimate the relevant and remaining neutral and non-relevant sets of results is to use the relevance information provided by the user.

The best-known approach to achieve query point movement is based on a formula initially developed by Rocchio [23] in the context of textual Information Retrieval. Let S_{rel} be as defined above, and $S_{non-rel} = a_i|rf_i < 0$ be the set of points the user explicitly marked as non-relevant. The new query point is an incremental change over the original query point, which is moved towards the relevant points and away from the non-relevant points:

$$p_{new} = \alpha\, p_{old} + \frac{\beta}{|S_{rel}|} \sum_{a_j \in S_{rel}} a_j - \frac{\gamma}{|S_{non-rel}|} \sum_{a_j \in S_{non-rel}} a_j$$

The speed at which the old query point is moved towards relevant points and away from non-relevant ones is determined by the parameters α, β, and γ, where $\alpha+\beta+\gamma=1$. The purpose of retaining part of the original query point is to avoid "overshooting" and to preserve some part of the user supplied sketch in the hopes it contributes important information to guide the query. In a high dimensional space it is usually difficult to determine in which direction to "go away" from non-relevant points. Although in this formula the old and relevant query points dictate this direction, it is in general advisable to make the parameter γ smaller than β to reduce any unwanted behaviour such as overshooting.

A further enhancement to this formula is to capture multiple levels of relevance by incorporating the user supplied relevance levels rf_j:

$$p_{new} = \alpha\, p_{old} + \frac{\beta}{\sum\limits_{j|rf_j>0} rf_j} \sum\limits_{j|rf_j>0} rf_j\, a_j - \frac{\gamma}{\sum\limits_{j|rf_j<0} rf_j} \sum\limits_{j|rf_j<0} rf_j\, a_j$$

As before, α, β, and γ ($\alpha+\beta+\gamma=1$) control the aggressiveness with which feedback is incorporated.

The main advantages of this approach are its simplicity and generally good results. It is simple and intuitive to understand and closely mimics what a human user would do to improve a result, that is, restating her query with a different query value or sketch.

4.1.2 Reshaping Distance Functions

While we used a standard Euclidean distance function throughout our discussion of the query point movement technique, there are many ways in which its shape can be influenced. Indeed, there is no restriction on the kind of distance function we can use, and its "shape" can be distorted in any arbitrary way that makes sense for that function.

The techniques we will discuss in this section focus on general L_p metrics, of which Euclidian distance is only one instance. L_p metrics compute the distance between two points and are of the form:

$$L_p(x, y) = \sqrt[p]{\sum_i (x[i] - y[i])^p}$$

The well-known Euclidean distance is thus the L_2 metric, while the Manhattan distance is the L_1 metric.

One approach to changing the shape of the distance function is to provide a weight for each dimension in the L_p metric. The interpretation of this is to give more importance to certain elements of the feature representation, for example, the percentage of green pixels in an image may be more important to a user than the percentage of red pixels in the image. The L_p metric becomes:

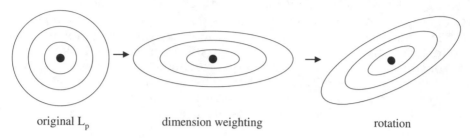

original L_p dimension weighting   rotation

Figure 21.3 Distance Function Shape

$$L_p(x, y) = \sqrt[p]{\sum_i w_i (x[i] - y[i])^p}$$

where $\sum_i w_i = 1$, that is, all the weights add up to 1. Figure 21.3 shows how a standard Euclidean distance function changes when a weight is given for each dimension. Note that the distance function is "stretched" only along the coordinate axis.

To derive a new query for the weighted L_p technique, we must change the weights w_i to new values that better capture the user's information need. To do this, the MARS system [13] suggests choosing weights w_i proportional to the inverse of the standard deviation among the relevant values of dimension i, while Mindreader [11] suggests that using weights w_i proportional to the inverse of the variance among the relevant values of dimension i provides slightly better results. The intuition behind using standard deviation or variance is that a large variation among the values of good results in a dimension means that dimension poorly captures the user's information need; conversely a small variation indicates that the dimension is important to the user's information need and should carry a higher weight.

We derive a new distance function shape through the following steps. First we estimate the new weights:

MARS $w_i^{est} \approx 1/\sigma(a_j[i] \mid a_j \in S_{rel})$

Mindreader $w_i^{est} \approx 1/\sigma^2(a_j[i] \mid a_j \in S_{rel})$

Next, we normalize these weights to ensure they add up to 1 and are compatible with the original weights $\sum_i w_i^{est} = 1$. The final step is to combine the estimated weights with the original weights:

$$w_i^{new} = \alpha\, w_i + \beta\, w_i^{est}$$

As for Rocchio's formula in the previous section, parameters α and β ($\alpha+\beta=1$) control the speed or aggressiveness of the relevance feedback, that is, how much of the original query weight is preserved for the new iteration.

While, as we pointed out earlier, giving a weight to each dimension "stretches" the distance function aligned with the coordinate axis, changes to the distance function can be further generalized by "rotating" the already stretched space as proposed by Mindreader [11]. Figure 21.3 shows the effect of such a rotation in our hypothetical two-dimensional feature representation. The distance function must change from merely weighting individual dimensions to a more general form:

$$d\ (x,y) = (x - y)^T M (x - y)$$

where M is a symmetric square matrix that defines a *generalized ellipsoid distance* and has a dimensionality equal to that of the feature representation. This rotated distance function is also known as *Mahalanobis* distance. We can also write this distance function as:

$$d\ (x, y) = \sum_i \sum_j m_{ij} (x[i] - y[i])(x[j] - y[j])$$

where the weights m_{ij} are the coefficients of the matrix M.
To derive a new query for this distance function we must change the matrix M such that:

$$\min_M \sum_{i|rf_i > 0} rf_i (a_i - p)^T M (a_i - p)$$

And the determinant of M is 1 (*det(M)=1*). Mindreader [11] solves this with *Lagrange multipliers* and constructs an intermediate weighted covariance matrix $C=[c_{jk}]$:

$$c_{jk} = \sum_{i|rf_i > 0} rf_i (a_i[k] - p[k])(a_i[j] - p[j])$$

If C^{-1} exists, then the M is given by:

$$M = (\det(C))^{1/\lambda} C^{-1}$$

where λ denotes the dimensionality of the feature representation.
To compare the dimension weighting and rotation approaches, we note that the rotation approach can capture the case where the region formed by relevant points is diagonally aligned with the coordinate axis. The dimension weighting approach is a special case of the more general rotation approach. A trivial disadvantage of the rotation model as presented in [11] is that it only supports a stretched and rotated version of the L_2 metric; this can however easily be corrected. A more important disadvantage of the rotation approach is that it needs at least as many points to be marked relevant as the dimensionality of the feature representation to prevent the new matrix from being under-specified. This is not a problem if the feature representation has a low number of dimensions (e.g., two) but becomes unworkable if there are many dimensions

(e.g., 64). The dimension weighting approach does not suffer from this limitation.

4.1.3 Distance Function Selection

Another approach to constructing a new query besides selecting a new query point and changing the shape of the distance function is to find a new distance function. It may be entirely possible that something other than an L_2 function or even an L_p metric is what the user has in mind for a particular feature representation. The task of *distance function selection* is to select "the best" distance function d_{ijk} from the set D_{ij} of distance functions that apply to the feature representation f_{ij} (as described in section 4.1).

To find the best matching distance function, we discuss an algorithm [24] that uses a ranked list as feedback, instead of individual relevance values *rf*. The user poses a query and is interested in the top m results that match her query. The objective of the retrieval system is to return to the user a ranked list $<a_1, a_2, ..., a_m>$ based on the best distance function d_{ijk}. To this end, the retrieval system performs the following steps:

1. For each distance function $d_{ijk} \in D_{ij}$, the retrieval system computes an answer ranked list:

$$l_k = <a_{1k}, a_{2k}, ..., a_{\tau mk}>$$

where τ is a small positive integer greater than 1. That is, for each distance function it computes an intermediate answer ranked list l_k larger than the user requested (typically $\tau=2$).

2. Define a *rank* of operator:

$$rank(a_s, l_k) = s \quad \text{(rank of } a_s \text{ in } l_k) \qquad \text{if } a_s \in l_k$$
$$rank(a_s, l_k) = \tau m + 1 \qquad\qquad\qquad\qquad \text{if } a_s \notin l_k$$

The *rank* operator returns the position of the point a_s in the list l_k. If a_s is not found in the list, we set its rank to one beyond the end of the list.

3. For each point in the l_k lists, compute an overall rank *rankAll(a)* by combining the ranks for that point in all the l_k lists. There are at most $m^*|D_{ij}|$ different points to consider, although generally there will be many points that appear in multiple lists. We compute *rankAll(a)* as follows:

$$rankAll(a) = \sum_{k=1}^{|Dij|} rank\ (a, l_k)$$

4. Construct a new combined list l with the top m points based on *rankAll*:

$$l = <a_1, a_2, ..., a_m>$$

This list is presented to the user as the query result.

5. The user returns a new feedback ranked list *rlf* of $r \leq m$ points that is ranked on her perceived information need. This list contains a subset of the answer list l and is arbitrarily ordered:

$$rlf = <a^r_1, a^r_2, ..., a^r_r>$$

We denote by a^r points on which the user gave relevance feedback.

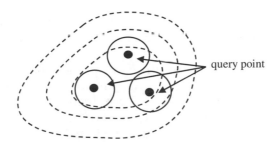

Figure 21.4 Multi-point Query

6. The user-supplied feedback ranked list *rlf* is compared to each answer list l_k corresponding to a distance function in order to determine how closely they mimic the user's preferences, that is, the overall distance between their rank differences:

$$dist(rlf, l_k) = \sum_{s=1}^{m} abs(rank(a_s, rlf) - rank(a_s, l_k))$$

7. Choose *k* such that the distance function d_{ijk} minimizes the overall rank differences between the user supplied *rlf* list and the result list l_k:

$$k = \min_{k=1}^{|D_{ij}|} dist(rlf, l_k)$$

In [24] the authors point out that this process is typically done only once. When the best distance function has been found, it remains in use during later iterations. Nothing, however, prevents an implementation from repeating this process for each relevance feedback iteration.

4.2 Multi-Point Approaches

In a similarity based retrieval paradigm of multimedia objects, as opposed to an exact search paradigm, it makes sense to ask for objects similar to more than one example. The user intuitively says that there is something common among the examples or sketches she supplied and wants the retrieval system to find the best results based on all the examples.

Therefore, an alternative to using a single query point *p* and relying on the distance function alone to "shape" the results is to use multiple query points and aggregate their individual distances to data points into an overall distance. In the single point approach, the distance function alone is responsible for inducing a "shape" in the feature space, which can be visualized with iso-distance curves. The query point merely indicates the position where this "shape" is overlaid on the feature space.

By using multiple query points and an aggregate distance function, the amount and location of query points also influences the "shape" of the query in the feature space. Using multiple query points provides enormous flexibility in the query. For example, it can overcome the limitations of the dimension weighting approach regarding diagonal queries making rotation unnecessary, and can even make dimension weighting unnecessary. Figure 21.4 shows an example of using multiple query points.

To support multiple query points, we extend the query model from section 4.1 to include multiple points and a distance aggregation function. We denote the *n* query points by p_1, p_2, ..., p_n, and the aggregate distance function for feature representation f_{ij} by A_{ij}. The overall query model thus becomes: $< (p_1, p_2, ..., p_n)$, d_{ijk}, $A_{ij}>$.

We will discuss two approaches to multi-point queries: one technique is based on the Query Expansion approach from the MARS [13] system, and the other one is known as FALCON [33]. While both techniques allow the distance function d_{ijk} to be modified as discussed in section 4.1, changes to d_{ijk} are strictly independent of the present discussion and we will therefore restrict ourselves to a simple L_2 metric for d_{ijk}.

4.2.1 Multi-Point Query Expansion

The first approach proposes an aggregation function based on a weighted summation of the distances to the query points. Let there be a weight w_t for each query point p_t with $1 \leq t \leq n$, $0 \leq w_t \leq 1$, and $\sum_{t=1}^{n} w_t = 1$, that is, there is a weight between 0 and 1 for each query point, and all weights add up to 1. The distance aggregation function $A_{ij}(x)$ for computing the distance of a feature representation value (i.e., of an object) x to the query is:

$$A_{ij}(x) = \sum_{t=1}^{n} w_t \, d_{ijk}(x, p_t)$$

Initially, all the weights are initialised to $w_t = 1/n$, that is, they are all equal. To compute a new query using relevance feedback, two things change: (1) the set of query points p_t, and (2) the weights w_t that correspond to each query point p_t. We discuss changing the query points next and defer a discussion of weight updating techniques to section 5, which discusses the techniques that apply here in the context of multiple features.

We present two different heuristic techniques to change the set of query points: *near* and *distant* expansion.

Near expansion: Relevant points (objects) will be added to the query if they are near objects that the user marked relevant. The rationale is that those objects are good candidates for inclusion since they are similar to other relevant objects, and thus represent other relevant objects. After the user marks all relevant objects, the system computes the similarity between all pairs of relevant and query points in order to determine the right candidates to add. For example, suppose that an initial query contains two sample points p_1 and p_2. The system returns a ranked list of objects $<a_1, a_2, ...a_m>$, since p_1 and p_2 were in the query, $a_1=p_1$ and $a_2=p_2$. The user marks a_1, a_2, and a_4 as relevant and a_3 as very relevant. Next, the system creates a distance table of relevant objects:

Point	rf	a_1	a_2	a_3	a_4
a_1	1 (relevant)	0	.4	.1	.5
a_2	1 (relevant)	.4	0	.4	.4
a_3	2 (very relevant)	.05	.2	0	.2
a_4	1 (relevant)	.5	.4	.4	0
Σ		.95	1	.9	1.1

The value in column r row s in the table is computed as $d_{ijk}(a_r, a_s)/rf_s$. Objects with a low distance are added to the query and the objects with a high distance are dropped from the query. In this example, a_3 is added to the query since it has the lowest distance among points not in the query, and a_2 is dropped from the query since it has the highest distance among the query points.

Distant expansion: Relevant points (objects) that are more distant to other relevant objects will be added to the query. The rationale is that if an object is relevant while different from other relevant objects, it may have some interesting characteristics that have not been captured so far. Adding it to the query will give the user an opportunity to include new information not included at the start of the query process. If the point is not useful, it will be eliminated in the next relevance feedback iteration. This approach is implemented as above, except that instead of adding objects with the lowest distance values, those with high distance values are added.

In the near and distant expansion approaches, the number of objects added during an iteration is limited to a constant number of objects (e.g., four objects) to reduce the computation cost impact.

4.2.2 FALCON

The FALCON [33] approach to multipoint queries was specifically aimed at handling disjoint queries. Its aggregate distance function for a data point x with respect to the query is defined as:

$$
A_{ij}(x) = \begin{cases} 0 & \text{if } (\alpha < 0) \wedge \exists i \; d_{ijk}(x, p_i) = 0 \\ \sqrt[\alpha]{\dfrac{1}{n} \sum_{t=1}^{n} d_{ijk}(x, p_t)^{\alpha}} & \text{otherwise} \end{cases}
$$

This aggregate distance function is sensitive to the value of the parameter α and the location of the query points p_i. When α is 1, it behaves very much like the weighted summation approach discussed above. It is when α is negative that it captures with ease disjoint queries; typical values for α are between -2 and -5.

Computing a new query under this approach is as simple as adding all the results (data points) that the user marked as relevant to the query points. That is, the new query becomes $<(q_1, q_2, \ldots, q_n \cup a_i \,|\, rf_i > 0), d_{ijk}, A_{ij}>$. Computing a new query is therefore extremely easy to do, but comes at the cost of increased computation since potentially many query points need to be explored and evaluated. Nevertheless, FALCON is able to easily learn complex queries in the feature space, even "ring" shaped queries.

5. MULTIPLE FEATURE REPRESENTATION FEEDBACK

An alternative to treating multiple multimedia features, as forming a single multidimensional space as discussed in section 4, is to treat each feature representation on its own. Each multimedia object o is represented by a collection of (multidimensional) feature values $f_{ij}(o)$ each based on a different feature representation.

The query is represented using a two-level approach: a collection of individual feature queries and an aggregation function to combine the individual query distances into an overall distance.

The bottom level is formed by a collection C of single-feature representation queries q_{ij} where the query and feedback models are as described in section 4. When doing a search, the user can select which feature representation she is interested in including in the query, possibly even selecting different query and relevance feedback modes for different feature representations. Recall the example described in section 2 where there are two features: color and texture. For the color feature there are two representations: color histogram and color moments. For the texture feature, there are also two feature representations: wavelet and Tamura. As an example, a user can select for her query the color histogram feature representation using a single-point query model and query point movement for relevance feedback, and the wavelet feature representation using a multi-point based query model and multi-point query expansion for relevance feedback.[1] Given a multimedia object, the retrieval system computes the distance between the corresponding feature representation values in the multimedia object and each single feature representation query. In the above example, an object o can have a distance of 0.4 with respect to the color histogram query and a distance of 0.8 with respect to the wavelet texture query.

Once the individual feature query distances have been computed for an object, they are aggregated together into an overall distance for that object. There are many ways to aggregate the distances with regard to individual feature representations; in this section we discuss an approach based on linear weighted summation. In this approach, we assign a weight w_{ij} to each single-feature representation query and use it to scale its distances. The overall distance is the sum of the scaled distances for all single-feature representation queries:

$$d_{overall} = \sum_{\forall ij \in C} w_{ij} q_{ij}(f_{ij}(o))$$

and where $1 = \sum w_{ij}$. By $q_{ij}(f_{ij}(o))$ we mean the distance d_{ijk} based on the single-feature representation f_{ij} value from the multimedia object o, based on its single-feature query representation q_{ij}.[2] The user query for the above example becomes:

$$w_1 d_{color\ histogram}(f_{color\ histogram}(o), q_{color\ histogram}) + w_2 d_{wavelet}(f_{wavelet}(o), q_{wavelet})$$

Suppose in our example $w_1=0.25$ and $w_2=0.75$ and the distance values shown above. The overall distance for object o is $d_{overall}=0.25*0.4 + 0.75*0.8 = 0.1+0.6 = 0.7$.

---

[1] While this demonstrates the flexibility of this approach, in a realistic setting these choices may be hidden from the user and pre-arranged by the system developer.

[2] Remember that this query representation may be any of those discussed in section 4.

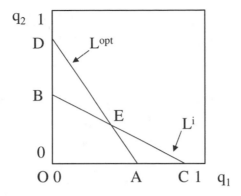

Figure 21.5 Single-feature Distance Graph

As in section 4, the retrieval system returns to the user a ranked list $l=<a_1, a_2, ...>$ with the top results based on distance to the multi-feature query, and obtains relevance judgements rf_s for the result objects a_s.

A query based on the multiple feature query model can change in two ways: by adding/deleting single-feature representation queries, and by changing the weights for each single-feature query called query re-weighting. First we discuss re-weighting based on the work in [21] and later how to add/delete from the query based on the work in [16].

5.1 Query Re-Weighting

Query re-weighting assumes that for the η independent single-feature representation queries q_i, each with a corresponding weight w_i, there exist a set of optimal weights w_i^{opt} that capture the user's information need. The initial query typically starts with default values for the w_i's; typically they all start with the same value: $w_i=1/\eta$.

To describe the re-weighting approach let us assume without loss of generality that there are only two single-feature queries: q_1 and q_2 and which return distances in the range $[0,1]$. Figure 21.5 shows a two dimensional space that shows the distances of q_1 vs. q_2. Since the overall distance function is a weighted summation of the component distances, we can represent the query as a line L. The distance of the line to the origin is determined by how many objects are returned; as more and more objects are returned to the user, the line moves away from the origin. The line L^i represents the query at iteration i of relevance feedback, based on the weights w_1^i and w_2^i. Objects between this line and the origin have been returned and shown to the user, that is, the objects in the region OCB. Assuming that an optimal query with weights w_1^{opt} and w_2^{opt} exists, we can represent it with the line L^{opt}. Objects between this line and the origin are deemed relevant according to the user's information need, that is, objects in the region OAB are relevant. At this point, the user marks as relevant the points that she has seen and are relevant, that is, the points in the area OAEB. Assuming the points A and B are known, we can construct a new line L^{i+1} with weights w_1^{i+1} and w_2^{i+1} for iteration $i+1$ that crosses points A and B for a given set of returned objects. In [21] the authors show that the slope of L^{i+1} lies between the slopes of L^i and L^{opt}. As the user goes through more relevance feedback iterations, the line L^{i+1} converges to the optimal line (weights) L^{opt}. What remains now is how to estimate the points A and B based on the results

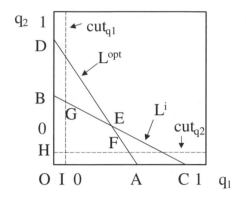

Figure 21.6 Single-feature Distance Graph - Counting

shown to the user and her relevance feedback. We explore two strategies for deriving L^{i+1}: (1) maximum distance and (2) counting.

Maximum distance strategy: Let rl be the set of objects marked relevant by the user (i.e., $rf \geq 1$). Let

$$\Delta q_1 = \max_{o_j \in rl}\{d_{q_1} \mid d_{q_1} \text{ is the distance between } o_j \text{ and the query } q_1\}, \text{ and}$$

$$\Delta q_2 = \max_{o_j \in rl}\{d_{q_2} \mid d_{q_2} \text{ is the distance between } o_j \text{ and the query } q_2\}$$

We use Δq_1 and Δq_2 to estimate the points A and B in Figure 21.5. That is, point A corresponds to $(\Delta q_1, 0)$ and point B corresponds to $(0, \Delta q_2)$. As a result, the slope of the line AB can be estimated as $-\Delta q_2/\Delta q_1$. Since the slope of the line L^{i+1} is the slope of the line AB, and given the constraint $w_1^{i+1} + w_2^{i+1} = 1$, the weights are updated as follows:

$$w_1^{i+1} = \frac{\Delta q_2}{\Delta q_1 + \Delta q_2}$$

$$w_2^{i+1} = \frac{\Delta q_1}{\Delta q_1 + \Delta q_2}$$

Counting strategy: This approach attempts to measure the length of the lines OA and OB by counting the number of objects relevant to the user, which are close to the query based on individual feature queries q_1 and q_2. To see the motivation, consider Figure 21.6. All points between the origin O and the lines labelled cut_{q1} and cut_{q2} are very close to the query based on the single-feature queries q_1 and q_2 respectively. If we assume that objects are uniformly distributed in the similarity space of the user's query, the number of objects close to the query based on single-feature queries q_1 and q_2 that are also relevant to the overall query are proportional to the size of the regions OIGB and OAFH respectively. Since the lines cut_{q1} and cut_{q2} are introduced at equal distance thresholds, the size of the regions OAFH and OIGB is proportional to the lengths of the lines OA and OB. Let Λq_1=cardinality of $a_j \mid d_{q1}(a_j) < cut_{q1}$, and Λq_2=cardinality of $a_j \mid d_{q2}(a_j) < cut_{q2}$. As a result, the ratio $\Lambda q_1/\Lambda q_2$ gives the estimate of the slope of line DC. Given the estimate of the slope, and the constraint that $w_1{}^{i+1} + w_2{}^{i+1} = 1$, a weight update policy is straightforward:

$$w_1^{i+1} = \frac{\Lambda q_1}{\Lambda q_1 + \Lambda q_2}$$

$$w_2^{i+1} = \frac{\Lambda q_2}{\Lambda q_1 + \Lambda q_2}$$

5.2 Single-feature Query Addition/Deletion

A step beyond re-weighting is to add or delete single-feature representation queries to the overall query. Deletion occurs naturally when re-weighting causes a weight to fall to 0. This effectively nullifies any contribution by that feature and thus is eliminated from the overall query. Adding new single-feature queries is more interesting. We present the approach described in [16] which is more general than multimedia retrieval.

To find single-feature query candidates to add to the overall query, a set of candidate single-feature queries is formed. Each single-feature query is composed of a query point and distance function. The distance function is the default for the feature, and the query point is selected by choosing the corresponding feature representation value from the top ranked result a_j with $rf_j > 0$, that is, the query value comes from the best relevant result. The candidate set can clearly be very large depending on the number of feature representations in the retrieval system. While restrictions can be placed on how many candidates to include (e.g., limited to those that can be evaluated in a given amount of time), in practice the number of feature representations is small enough to avoid serious problems. Using the candidate set, we construct a table that incorporates the final results $<a_1, a_2, ..., a_m>$ shown to the user, their corresponding feedback values $<rf_1, rf_2, ..., rf_m>$, and the distance between result a_j and each single-feature candidate query c_k. For example, if $m=2$ and there are two single-feature query candidates, the table is:

Result a_j	Feedback rf_j	$c_1 (a_j)$	$c_2 (a_j)$
a_1	1 (relevant)	0.2	0.1
a_2	0 (neutral)	0.5	0.7

We populate the column for each single-feature candidate query c_k with the distances from c_k to every result a_j. For each column that represents a candidate query we compute the average and standard deviation between relevant and non-relevant results:

$$avg_{rel}^k = average(c_k(a_j) \,|\, rf_j > 0)$$
$$avg_{non-rel}^k = average(c_k(a_j) \,|\, rf_j \leq 0)$$
$$\sigma_{rel}^k = \text{standard deviation}(c_k(a_j) \,|\, rf_j > 0)$$
$$\sigma_{non-rel}^k = \text{standard deviation}(c_k(a_j) \,|\, rf_j \leq 0)$$

Using this information, we compute the "separation" between relevant and non-relevant answers for each c_k:

$$sep^k = (avg^k_{non-rel} - \sigma^k_{non-rel}) - (avg^k_{rel} + \sigma^k_{rel})$$

We add c_k to the overall query iff: $sep^k > 0 \wedge \forall (v \neq k)\ sep^k > sep^v$, that is, we add c_k to the query if there is at least one relevant and non-relevant standard deviation of separation between relevant and non-relevant values for c_k, and choose the c_k with the largest such separation. Experimental results in [16] show that this is a promising approach to adding component single-feature queries to the overall query.

5.3 Distance Normalization

In this discussion we have largely assumed that the distances between different feature representations are of comparable magnitudes. In general, this is not the case. For example, the distance values returned by one feature representation may lie in the range [0,1] while for another feature representation they may lie in the range [0,1000]. Weighting can help alleviate this disparity but is preferably not used for this purpose. It thus may become necessary to normalize the distance ranges between different feature representations and distance functions to make them of comparable magnitudes.

Among the many techniques to alleviate this problem are the min-max and Gaussian normalization approaches. In the min-max approach, the retrieval system keeps track of the minimum and maximum distances observed for each distance function and feature representation and normalizes the final distance by the distance range (maximum-minimum distance). The Gaussian approach advocated in [17], among others, proposes to treat all distances as a Gaussian sequence. To normalize distance values, they are divided by a multiple of the standard deviation to ensure most distances (99%) lie in the range [0,1]. The drawbacks of this technique are a high up-front normalization cost, and that it assumes a specific distribution of distance values which may not hold. In practice, as long as the distances for different feature representations and distance functions are normalized to a comparable range, few unwanted problems appear. Both approaches roughly perform equally well.

6. ACCELERATING RELEVANCE FEEDBACK

While the process of relevance feedback generally kicks in once the user has seen some results, it is desirable to provide the benefits of relevance feedback from the very beginning of the user's search process. A basic limitation of relevance feedback is that it is generally not available for the initial query since it needs the results and user judgements to act. Indeed, all the techniques discussed so far suffer from this limitation.

It is possible to enhance the initial user query with the information gained from relevance feedback in queries made by other users. The field of *Collaborative Filtering* [10] has a rich literature based on the premise that a textual document should be returned to a query not based on its content, but on a measure of popularity based on how many users selected it based on their queries. More broad than collaborative filtering are *recommender* systems that exploit the information acquired and created by one user in her interaction with the system to assist other users [22].

FeedbackBypass [4] is a prominent example of applying collaborative filtering techniques to multimedia retrieval. The objective is to sidestep many feedback iterations and directly go to an optimal query representation given the initial user specification. It focuses on a single feature space approach as presented in section 4. It represents queries as point p in a multidimensional space, together with a distance function d for the feature, that is, the pair $<p,d>$ specifies a query. *FeedbackBypass* supports the weighted Euclidean and *Mahalanobis* distance functions as described in section 4, that is, we can represent the distance function d as a set of weights w for either distance function. Since we can capture d with its corresponding set of weights w, we can represent the query as $<p,w>$. There are two steps in its operation:

1. Build a mapping from initial to optimal query specifications. The objective is to build the mapping:

   $$p^0 \rightarrow (p^{opt}, w^{opt})$$

   A user starts with an initial query point p^0. She provides relevance feedback and eventually arrives at an optimal query representation: $<p^{opt}, w^{opt}>$. This end result of several feedback iterations is encoded as an optimal offset $\Delta p = p^{opt} - p^0$ and the weights w^{opt}, that is, the difference between the initial and final query points and the optimal set of weights. The initial query point p^0, is used as the key into a multidimensional data structure with $(\Delta p, w^{opt})$ as the values. To avoid storing all possible mappings, *FeedbackBypass* uses a simplex tree to merge "close" initial points together. Over time, this mapping is populated with the results of the collective feedback of many users.

2. Given a mapping $p^0 \rightarrow (p^{opt}, w^{opt})$ and a user's initial query point, find the corresponding optimal query specification. To circumvent many feedback iterations, the initial query point p^0 is used as a key into the multidimensional index and the corresponding pair $(\Delta p, w^{opt})$ is retrieved. An optimal query representation is computed as $<p^0 + \Delta p, w^{opt})$ and used to execute the user's query.

In essence, *FeedbackBypass* keeps a history of how query point movement and re-weighting evolved and then applies the learned information to new queries. Over time, as users continue to give relevance feedback, the system keeps learning its mapping, and therefore improving its results.

Another system that uses collaborative filtering for browsing images in the context of a museum web site is described in [14]. It combines multiple visual features (content based filtering) and user recommendations (collaborative filtering) to dynamically guide the user in a virtual museum. This approach predicts the interest rating of a user in an image by matching the visual features of the image to other images and in turn using those images' recommendation rating to form a final estimate. The collaborative component produces a distance value $\rho \in [0,1]$ based on Pearson correlation [14]. The content-based component produces distances for two features, color histogram and texture. The final distance for an image is computed as a weighted summation of the results from visual features and the correlation coefficient:

$$d^{\,combined} = \mu^{\,collaboration} \rho + \mu^{\,texture} d^{\,texture} + \mu^{\,color} d^{\,color} \quad \text{with} \quad \sum \mu^i = 1$$

The weights μ^i are set experimentally.

The overall results for collaborative filtering approaches indicate that the results of a retrieval system improve substantially when taking into consideration the aggregated feedback from all users over the lifetime of the system.

7. EVALUATING RELEVANCE FEEDBACK

There are two aspects to evaluating how good a retrieval system performs its tasks: quantitatively and qualitatively. The quantitative approach to evaluating a system focuses on execution speed: the time it takes the retrieval system to perform a search and return the results to the user. Performance in this sense is of chief importance for an interactive retrieval system and has been extensively studied for many types of search technology including Databases [8] and Information Retrieval systems [32] and is also discussed in other chapters of this book. But when relevance feedback changes the interpretation of a query and a new set of answers is computed, it is often possible to utilize the work already performed for the old query to answer the new query. Work in this area has been done for multimedia retrieval [3] as well as for database search [13]. We do not further discuss techniques for efficient evaluation of new queries; instead we discuss the relevant issues in evaluating the quality of retrieval.

With conventional exact search tools as varied as the UNIX grep utility and SQL databases there is no need to evaluate the quality of the answers since a value either satisfies or does not satisfy the search criteria. A system that returns wrong answers is clearly identifiable and undesirable. But under the similarity search paradigm, answers are given based on how closely they match a user's query and preferences; there is no clear-cut way to distinguish correct answers from wrong answers. Therefore the issue of quality in the results is of prime importance when evaluating a retrieval system.

Since relevance feedback seeks to improve the quality of retrieval, it becomes important to quantify the gains made from iteration to iteration.

We discuss two approaches to evaluate the quality of retrieval results: (1) precision and recall, and (2) normalized recall.

Text retrieval systems typically follow the very popular precision and recall [26] metrics to measure the retrieval performance. Precision and recall are based on the notion that for each query, the collection can be partitioned into two subsets of documents. One subset is the set of *relevant* documents and is based on the user's criteria for relevance to the query. The second is the set of documents actually *retrieved* by the system as the result of the query. Precision and recall are defined as follows:

- *Precision* is the ratio of the number of relevant images retrieved to the total number of images retrieved. Perfect precision (100%) means that all retrieved images are relevant. Precision is defined as:

$$precision = \frac{|\, relevant \cap retrieved \,|}{|\, retrieved \,|}$$

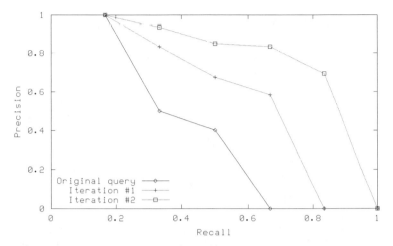

Figure 21.7 Sample Precision-Recall Graph

- *Recall* is the ratio of the number of relevant images retrieved to the total number of relevant images. Perfect recall (100%) can be obtained by retrieving the entire collection, but at a cost in precision. Recall is defined as:

$$recall = \frac{|\, relevant \cap retrieved \,|}{|\, relevant \,|}$$

The retrieval system is characterized by constructing a precision-recall graph for each relevance feedback iteration in a query. The precision-recall graph is constructed by incrementally increasing the size of the retrieved set, that is, by iteratively retrieving the next best result and measuring the precision every time a relevant answer is found (i.e., at different values of recall). Usually, the larger the retrieved set, the higher the recall and the lower the precision become. Figure 21.7 shows an example precision recall graph for the original query and 2 iterations of relevance feedback.

A problem with the precision and recall approach is that it ignores the ordering among relevant results. If we interpret the improvement derived from relevance feedback to include the ordering among relevant results, then we must find a measure that captures this ordering. The query result after each iteration is a list of objects ranked on how closely each object matches the query. The *normalized recall metric* [26] is designed to compare ranked lists with special consideration for the ordering.

The *normalized recall metric* compares two ranked lists, one of *relevant* items (the ground truth) and one of *retrieved* items (the result of a query iteration):

$$recall_{normalized} = 1 - \frac{\sum_{o \in relevant} |\, rank(retrieved,o) - rank(relevant,o) \,|}{(N - |\, relevant \,|) * |\, relevant \,|}$$

where *rank(List,o)* is the rank of object *o* in *List*. The metric computes the rank difference of the two lists and normalizes the result by the highest possible rank difference, thus producing a value in the range 0 (worst) to 1 (best). The metric is sensitive to the position of objects in the ranked list. This sensitivity to rank position is suitable for measuring the effectiveness of the relevance feedback process by comparing the relevant list to the result across feedback iterations. As the lists converge, the metric results in a better value. The normalized recall metric is meant to compare a list of relevant items to the fully ranked list of all objects in a database. The problems with this metric are the following:

- A few poorly ranked, but relevant, items have a great impact because a poorly ranked item contributes a large value of rank difference. In practice, the results seen by the user are much less than the database size.
- The database size has a great impact on the metric. In a large database, only a small part of it is actually explored by the user; the rest is of no use. Therefore, using the database size as an important part of the denominator in the equation obscures the difference of two good retrieved lists because the large denominator dilutes the rank difference significantly.
- Each ranked item is equally important. That is, the best relevant item is as important as the worst relevant, but still relevant, item.

8. CURRENT TRENDS AND FUTURE DIRECTIONS

Relevance feedback is a powerful tool for users to interactively improve their search results. There is extensive evidence in the fields of Information Retrieval and Multimedia Retrieval to support continued efforts in improving techniques to learn from relevance feedback. Several promising directions for using relevance feedback are being explored, including techniques to speed up relevance feedback iterations and finding better ways to identify relevant objects.

8.1 Kernel Approaches

A promising research direction is to explore features in multidimensional spaces of much higher dimensionality than that of the original feature space. For example, [6][9] use kernels distances to effectively map a feature representation space into a much higher dimensional space where they search for suitable separators that cannot be easily represented in the original feature space. [6] uses a support vector machine approach to find linear separators in a very high dimensional space to better characterize relevant results. Another support vector machine approach described in [30] does not return the closest results to the user's query at each iteration; instead it shows different images in hopes to maximize the learning at each iteration.

8.2 Feature Selection

Another approach to using a high dimensional search space is MEGA [1]. MEGA is a query refinement method that starts a general initial query for all users that is composed of a k-CNF Query Concept Space (QCS) and a k-DNF Candidate Concept Space (CCS). This method is based on Valiant's PAC learning model [31] along with a bounded sampling technique that aims at removing the maximum expected number of disjunctive terms from QCS. The refinement involves removing irrelevant conjunctive terms (clauses) from the CCS based on negative examples and removing of irrelevant disjunctive terms from the QCS based on positive examples. The MEGA approach is applicable only in situations where user-specific initial queries are not supplied. MEGA does not provide methods to learn query values and weights. The approach also requires handling an exponential number of terms in queries that results in high learning complexity (the authors suggest a dimensionality reduction method to tackle this). Moreover, it is not clear how CCS and QCS are formed for continuous valued attributes.

Beyond applications in multimedia retrieval, relevance feedback has general applicability to database searches. [28] explores an interactive browser for databases that uses relevance feedback. In contrast, [16] describes a general way to incorporate relevance feedback on arbitrary data types in a database server.

Relevance feedback for multimedia search is a valuable tool that every retrieval system should include. The opportunities for a retrieval system to help the user find her information are tremendous. The retrieval performance improvement that is achievable through relevance feedback frequently improves results so much that users barely believe that the system was able to "outsmart" them and construct better queries. Current developments in relevance feedback techniques will improve retrieval results even more.

ACKNOWLEDGMENTS

The authors gratefully acknowledge the fruitful discussions with Yong Rui, Kriengkrai Porkaew, Kaushik Chakrabarti, Dawid Yimam Sied, and Yiming Ma. This work was supported by NSF grants 9734300, 0086124, 0083489, and 0220069.

REFERENCES

[1]   E. Chang and B. Li, MEGA - The Maximizing Expected Generalization Algorithm for Learning Complex Query Concepts, ACM Transaction on Information Systems (final revision), 2002.

[2]   R. Baeza-Yates and B. Ribeiro-Neto, Modern Information Retrieval, Addison Wesley, May 1999.

[3]   K. Chakrabarti, K. Porkaew and S. Mehrotra, Efficient Query Refinement in Multimedia Databases, Proc. of the 16th Int. Conf. on Data Engineering (ICDE), February 2000.

[4]   I. Bartolini, P. Ciaccia, and F. Waas, FeedbackBypass: A New Approach to Interactive Similarity Query Processing, Proc. 27th VLDB Conference, 2001.

[5]   C. Carson, S. Belongie, H. Greenspan, and J. Malik. Region-based image querying, In Proc. IEEE Workshop on Content-based access of Image and Video Libraries, in conjunction with IEEE CVPR, 1997.

[6]   Y. Chen, X. Zhou, and T. S. Huang, One-class SVM for Learning in Image Retrieval, Proc. IEEE Int. Conf. on Image Processing, pp. 440-447, February 1999.

[7]   M. Flickner, H. Sawhney, W. Niblack, J. Ashley, Q. Huang, B. Dom, M. Gorkani, J. Hafine, D. Lee, D. Petkovic, D. Steele and P. Yanker, Query by Image and Video Content: The QBIC System, IEEE Computer, September 1995.

[8]   J. Gray and A. Reuter, Transaction Processing: Concepts and Techniques, Morgan Kaufman, 1993.

[9]   D. Heisterkamp, J. Peng, H. K. Dai, Adaptive Quasiconformal Kernel Metric for Image Retrieval, Proc. of IEEE Conf. on Computer Vision and Pattern Recognition, pp. 236-243, 2001.

[10]   J. Herlocker, J. Konstan, A. Borchers, J. Riedl, An Algorithmic Framework for Performing Collaborative Filtering. Proceedings of the 1999 Conference on Research and Development in Information Retrieval, August 1999.

[11]   Y. Ishikawa, R. Subramanya and C. Faloutsos, MindReader: Querying Databases Through Multiple Examples, Proc. 24th Int. Conf. Very Large Data Bases (VLDB) 1998.

[12]   B. J. Jansen, A. Spink and T. Saracevic, Real life, real users, and real needs: a study and analysis of user queries on the web, Information Processing and Management, Elsevier Science," vol. 36, no. 2, 2000.

[13]   M. L. Kersten and M. F. N. de Boer, Query Optimisation Strategies for Browsing Sessions, Workshop on Foundations of Models and Languages for Data and Objects, 1993.

[14]   A. Kohrs and B. Merialdo, Improving Collaborative Filtering with Multimedia Indexing Techniques to create User-Adaptive Web Sites, Proc. ACM Multimedia Conf., October 1999.

[15]   S. Mehrotra, Y. Rui, M. Ortega, and T. S. Huang, Supporting Content-based Queries over Images in MARS, Proc. of the 4th IEEE Int. Conf. Multimedia Computing and Systems, Chateau Laurier, Ottawa, Ontario, Canada, Pages 632-633, June 3-6, 1997.

[16]   M. Ortega-Binderberger, K. Chakrabarti, and S. Mehrotra, An Approach to Integrating Query Refinement in SQL, 2002 Conference on Extending Database Systems (EDBT), pages 15-33, March, 2002.

[17]   M. Ortega, Y. Rui, K. Chakrabarti, K. Porkaew, S. Mehrotra, and T. S. Huang, Supporting Ranked Boolean Similarity Queries in MARS, Appeared in IEEE Transaction on Knowledge and Data Engineering, Vol. 10, No. 6, Pages 905-925, December 1998.

[18]   A. Pentland, R. W. Picard and S. Sclarof, Photobook: Tools for Content-Based Manipulation of Image Databases, SPIE vol. l2185, 1994.

[19]   R. W. Picard, Computer Learning of Subjectivity, Proc. ACM Computing Surveys, vol. 27, no. 4, December 1995.

[20]   K. Porkaew, S. Mehrotra, M. Ortega, and K. Chakrabarti, Similarity Search Using Multiple Examples in MARS, Proceedings of the International Conf. on Visual Information Systems (Visual'99), Lecture Notes in Computer Science, Springer Verlag, LNCS Vol. 1614, Pages 68-75, June 1999.

[21]   K. Porkaew, S. Mehrotra, and M. Ortega, Query Reformulation for Content-Based Multimedia Retrieval in MARS, Proc. IEEE Int. Conf. on Multimedia Computing and Systems (ICMCS), vol. 2, June, 1999.

[22]   P. Resnick and H. R. Varian. Recommender systems. Communications of the ACM, 40(3):56--58, 1997.

[23]   J. J. Rocchio. Relevance Feedback in Information Retrieval. In G. Salton, editor, The SMART Retrieval System – Experiments in Automatic Document Processing. Prentice Hall Inc., Englewood Cliffs, NJ, 1971.

[24]   Y. Rui, T. S. Huang, S. Mehrotra, and M. Ortega, Automatic Matching Tool Selection via Relevance Feedback in MARS, Proceedings of the 2nd Int. Conf. on Visual Information Systems, San Diego, California, December, 1997,

[25]   Y. Rui, T. S. Huang, M. Ortega, and S. Mehrotra, Relevance Feedback: A Power Tool for Interactive Content-Based Image Retrieval, IEEE Tran on Circuits   and   Systems   for   Video   Technology, Special Issue on Segmentation, Description, and Retrieval of Video Content,
Vol 8, No. 5, Pages 644-655, September, 1998

[26]   G. Salton and M. J. McGill, Introduction to Modern Information Retrieval McGraw-Hill, March 1984.

[27]   S. Sclaroff, L. Taycher and M. La Cascia, ImageRover: A Content-Based image Browser for the World Wide Web, Proc. IEEE Workshop on Content-Based Access of Image and Video Libraries, 1997.

[28]   J. C. Shafer and R. Agrawal, Continuous Querying in Database-Centric Web Applications, WWW9 conference, May 2000.

[29]   A. Spink and T. Saracevic, Human---computer interaction in information retrieval: nature and manifestations of feedback, Interacting with Computers, Elsevier, vol. 10, no. 3, 1998.

[30]   S. Tong and E. Chang, Support Vector Machine Active Learning for Image Retrieval, ACM International Conference on Multimedia, pp.107-118, Ottawa, October 2001

[31]   L. G. Valiant, A theory of the learnable, Comm. of the ACM, 27(11), November 1984.

[32]   I. Witten, A. Moffat and T. Bell, Managing Gigabytes: Compressing and Indexing Documents and Images, Morgan Kaufman, 2nd ed., May 1999.

[33]   L. Wu, C. Faloutsos, K. Sycara and T. Payne, FALCON: Feedback Adaptive Loop for CONtent-based retrieval, Proc. Int. Conf. Very Large Data Bases (VLDB), 2000.

22

ORGANIZATIONAL PRINCIPLES OF VIDEO DATA

Simone Santini
University of California, San Diego
La Jolla, California, USA
ssantini@ncmir.ucsd.edu

1. INTRODUCTION

Extending his right hand on the book opened in front of him and pointing his left hand toward the majestic silhouette of Notre Dame cathedral visible through the window, the archdeacon Dom Claude pronounced the words that would give Victor Hugo a permanent place in all citation manuals:

Hélas ! ceci tuera cela.[1]

With these words, Dom Claude who (we are, of course, in the novel *Notre Dame de Paris*) spoke at the end of the XV century expressed his preoccupation that the newly invented printing press would destroy the pedagogical function of the cathedral, at the time the greatest effort of the Catholic church to encode in a visible form not only the histories of the Bible but also, subtly encoded, the essence of Thomas Aquinas's scholastic philosophy [1].

The worries of Dom Claude may have been excessive. Just to stay in the neighbourhood of Quasimodo's home, one can notice that Viollet-le-Duc, who presided over the great restoration of Notre Dame in the XIX century, was something of an architecture theoretician, well known for his many books. His professional figure, his culture, and the philosophy of art history upon which the restoration was based would not have been possible without the printing press. Dom Claude's cry, however, is particularly poignant for anybody who might want to analyse the complex relation between multimediality,[2] the Internet, and pre-existing forms of communication like video, images, text, diagrams, and so on.

Of particular interest in our context is the relation between the new channels of information dissemination and video—relations that, if on one hand seem to

---

[1] Alas! This will kill that.

[2] The term multimedia itself is quite infelicitous, at least in the high-tech sense in which it is customarily used: both the media that the archdeacon was indicating, the book and the cathedral, are, according to any reasonable definition, multimedia devices. I will try, whenever possible, to avoid using the term *multimedia* and derivates.

promise new openings and new avenues for the diffusion of video artefacts, on the other hand embody a crisis of the seriality (and, on a different plane, of historicity), so characteristic of video.

The symbiotic/antithetic relationship between the "New Media" [2] and video is, on one hand, at the origin of the frustration that many researchers experiment when trying to merge the two while, on the other hand, can give us some useful information on how such a symbiosis can be made to work.

This chapter will start with a few theoretical statements about video, the Internet, and the relations between the two. One of the results of this analysis will be the division of the general video universe into two parts, which I will call *semiotic* and *phenœstetic*, characterized by very different relations with the Internet medium. I will argue that the two types of videos deserve two very different treatments when it comes to Internet distribution and, in the following sections, I will introduce some of the techniques already developed or hypothesizable for distributing and experiencing these types of video on the Internet, referring mostly to phenœstetic video.

2. AN ARCHÆOLOGY OF INTERNET VIDEO

Video is being hailed, especially in circles prone to an easy technophilia, as the next frontier of the Internet. The premise on which this assessment rests is not unreasonable: the network bandwidth available to the most advanced users is increasing to the point in which the possibility of transmitting videos is no longer a long-term goal. The promises are also rather substantial: as I am writing these words, I am sitting in my office in San Diego watching the news on the third channel of Italian television, albeit in a format just a little larger than a large postal stamp [3].

Yet, there are many problems to be solved before the symbiosis of video and the Internet can be successful, many of them connected to the different, almost incompatible characteristics of the two carriers: where video is a more passive background process, which we can absorb for long periods of time without interruption, the Internet is characterized by a more rapid interaction cycle, and by shorter times of passive absorption. In the case of my Italian news, for instance, I find that almost invariably after a few minutes I put the video window in the background and start doing something else while listening to the news: looking at other Internet sites or, as is the case now, working on an overdue book chapter.

As [2] puts it, one source of incompatibility is that the video screen is *transparent* while the computer screen is *opaque*. That is, the video screen opens a virtual window that reveals a world behind it: one doesn't look *at* the screen, but at what lies behind it. The computer screen, on the other hand, is the location where the tools of interactivity (buttons, menus, etc.) are located and, as such, a user operates *on* the screen and looks *at* it. This opacity makes the computer screen (and, therefore, the Internet, on which the computer screen is displayed) an unlikely site for the lengthy contemplation of a prolonged stream.

In addition to this, a source of contrast is the clash between the *syntagmatic* relations at work in video and the *paradigmatic* relations at work in an interface.

A syntagmatic [4] relation holds between units that are placed next to another in a meaningful structure, for example the words *cat* and *mat* in the expression "the cat is on the mat." On the other hand, one could also construct sentences by choosing possible alternatives to the word "cat," as "the dog is on the mat," or "the table is on the mat." The relation existing between the word "cat" and "dog" or "table" is paradigmatic.

In video, the signification units are related by a syntagmatic relation: they are given meaning by the fact that they follow one another, and this is the reason that the only way of completely understanding video is to sit down and look at it. Interfaces, on the other hand, are paradigmatic: they present a set of alternatives through which one can navigate and each act will result in the presentation of another series of alternative. It is true that the actual act of navigation will create a temporal (therefore syntagmatic) sequence, but this sequence is implicit: what is made explicit in an interface is the paradigmatic relations between the possible alternatives.

In order to facilitate the access to video in a highly interactive medium, the generally accepted solution is that of exposing its *structure*, allowing the user to navigate it, receiving information on the general content of the video and, upon demand, watching short video segments on topics that he or she deems interesting. This is the road followed, for instance, by the French station TV5 for its Internet news [5]: the video window contains a list of the topics covered in the news broadcast together with a brief synopsis, and the user can select which one to watch.

Observations like these have spurred a great interest in the means of defining, describing, and displaying the structure of a video, and in integrating a suitable display of the video structure with the display of part of the video stream. The structure is used, in most cases, to facilitate access to specific parts of video, much in the same way in which a directory structure is used to facilitate access to an otherwise undifferentiated set of files. In order to do this the structure must rely on some *organizational principles* of video.

The principles around which the structure of the video is encoded are taken in a double bind between contrasting requirements. On one hand, they must be semantically meaningful: it would be useless to organize a video along lines that don't make sense to the user, much like it would be quite useless to divide a group of files in directories based, say, on their length or on the value of their fifth byte. On the other hand, if the structure is so complex, or the video library so large and mutable that an automatic structuring mechanism is necessary, it is necessary that the organizational principles be derivable from the video data, that is, that they have an easily identifiable syntactic referent in the video data.

Because of their semantic connection, the organizational principles of video depend on the general structure, purpose, cultural location of the video that is being analyzed. As a start, one can define two broad categories of video: *semiotic* and *phenœstetic*, to which I will now turn.

Semiotic Video

Semiotic video is, loosely speaking, video specifically designed, constructed, or assembled to deliver a message. Produced video, from films to TV programs, news programs, and music videos are examples of semiotic video: they carry an explicit message (or multiple messages), which is encoded not only in their content, but also in the syntax of the video itself.

The syntax of the most common forms of video expression derives from that of cinema. Cinema has been, from many points of view, the form of expression most characteristic of the XX century. Not only did cinema provide a new form for artistic expression, the particular version of narrative it created became the main modality of communication of the century.

The language of cinema was developed in a span of less than 20 years, roughly from the turn of the century to the end of the 1920s. At the beginning of the century, cinema was essentially recorded theatre, while at the end of the 1920s its language was so sophisticated that many people saw the introduction of the spoken word as a useless trick, which added little or nothing to the expressive power of the medium.

The main contribution to the development of the cinematic language came from the Russian constructivists, most notably people like Eisenstein and Vertov. Vertov's *The Man with the Movie Camera* (1929) can be seen as the ultimate catalogue of the expressive possibilities of the film language and of its principal syntactic construct: the *montage*.

The language of montage has extended beyond cinema through its adoption by other cultural media like TV or music videos. Its relative importance as a carrier of the semiosis of a video depends, of course, on the general characteristics of the form of expression in which it is employed, or even on the personal characteristics of the creator of a given message: we have at one extreme film directors who make a very spartan use of montage and, at the other extreme, music videos in which the near totality of the visual message is expressed by montage.

Attempts to an automatic analysis of video have long relied on the structure imposed by montage, due in part to the relative ease of detection of montage artefacts.

Montage is not the only symbolic system at work in semiotic video, although it is the most common and the easiest to detect. Specific genres of video use other means that a long use has promoted to the status of symbols in order to express certain ideas. In cartoons, for instance, characters running very fast always leave a cloud behind them, and, whenever they are hit on the head, they vibrate. These messages are, to an extent, iconic (certain materials do vibrate when they are hit) but their use is now largely symbolic and is being extended especially rather crudely to other genres (such as action movies) that rely on visuals more than on dialog to describe an action; thanks to the possibilities offered by sophisticated computer techniques, the language of action movies is moving away from realism and getting closer and closer to a cartoonish symbolism.

It is quite likely that in the near future, as this type of symbolism becomes even more common, analysis techniques will follow suit and, just as today they try to extract part of the meaning of a film through an analysis of the montage, they will begin analyzing the visual symbols with which directors express their messages.

In addition to these, which we might call *intramedial* messages, there are in many videos *extramedial* messages: shared cultural messages that provide meaning not because of the presence of a certain object in a video, but because of common connotations attached to the object. So, the presence of Adolf Hitler in a video may evoke a meaning of xenophobia, violence, or war, not because of the specific characteristics of the video, but because of the general connotations attached to the image of the German dictator. Directors use these connotations in concert with more strictly cinematic techniques, sometimes to create agreement, sometimes to create contrast both between media (e.g., traditional Jewish chants or peace anthems in the background of a Nazi gathering), or through contrast in time with the use of techniques like collage or pastiche.

All these semantic modalities, although not yet used to the fullest extent of their possibilities, can provide an invaluable aid to the task of identifying the meaning of a piece of semiotic video, at least, within the boundaries of a culture in which certain codes and connotations are shared.

Phenœstetic Video

By contrast, the kind of video that I have called phenœstetic is characterized by the almost absolute absence of cultural references or expressive possibilities that rest on a shared cultural background. It is, to use an evocative, albeit simplistic, image, video that just happens to be. Typical examples of phenœstetic video are given by security cameras or by the more and more common phenomenon of web cameras [6,7].

In this type of video, there is in general no conscious effort to express a meaning except, possibly, through the actions of the people in the video. Expressive means with a simple syntax, like montage, are absent and, in most cases, the identities of the people present in the video will afford no special connotation. In a few words, phenœstetic video is a stream of undifferentiated, uninterrupted narrative, a sort of visual *stream of consciousness* of a particular situation.

It should be apparent that imposing a structure to this kind of video for the purpose of Internet access is much more problematic than with semiotic video, since all the syntactic structures to which the semantic structure is anchored are absent. It would be futile, for instance, to try to infer the character of a phenœstetic video by looking at the average length of the shots [8] or the semiotic characteristics of its color distribution [9], for there are no shots to be detected and colors are not purposefully selected.

The position that I will try to carry forward in this chapter is that the proper way to organize phenœstetic video is around meaningful *events*, that is, that the structure of phenœstetic video is given by the interesting events that take place in it and by the structural relations between these events.

The "catch" of this definition is that there is no absolute or syntactic way to define an event: an event of interest is a purely semantic entity whose nature and location depend on the context in which the video is analyzed. So, for phenœstetic video, there will not be in general *one* structure around which the exploration of the video can proceed, but as many structures as there are contexts in which the video is used.

Experience Units

At this level, video is, in general, no longer alone, but the events of interest are associated to information coming from other media, which completes and complements the video information. It is important, in fact, to realize that on a generalist and interactive carrier like the Internet, media do not, by and large, stand by themselves, but are collected and associated in wider *experience* units. Informally, an experience unit is a complete view of an event of interest, regardless of the specific medium in which the experience is carried, and complete with all the connections between the different sources of information and with other experience units.

It should be stated from the outset that the concept of experience unit is a practical approximation and that, as a principle, it is indefensible. There are,

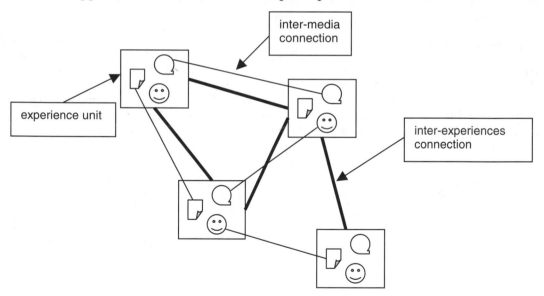

Figure 22.1 Semantic view of a graph.

inexperience, no Leibnitzian *windowless monads*: an experience unit derives its expressivity through the interaction with other experience units. A better way to regard an aggregation of media in a highly interactive carrier is as a graph of units connected by semantically structured edges. Each unit is, in itself, a collection of fragments derived from different media that can, individually, be connected to other fragments belonging to other units. A schematic view of such a graph is shown in Figure 22.1.

Architecturally, the dependence of events on context entails a general scheme like that of Figure 22.2. The video stream is first analyzed by a syntactic detector module, which detects syntactically determined events. These occurrences form then the input of a number of context-dependent semantic detectors, each one of which encodes a particular application domain.

In semantic detectors, events are seen within the context of a particular domain, the data and the relations relevant to that domain are determined and, possibly, sought in other media available to the system.

Consider, as an example, a set of cameras installed in the different rooms of a museum and let us assume that a suitable video processing unit is capable of determining video regions corresponding to the people who walk around and, by some suitable features (the color of the dresses, possibly), is capable of assigning an identifier to each region, and that the region will maintain its identifier across the different rooms in the museum.

Syntactic events in this model can include the following:

1. A visitor (that is, a moving region) goes near a painting.
2. A visitor goes from one room to another.
3. A large assembly of visitors is formed somewhere.
4. A visitor goes into the souvenir shop and buys something (detected, of course, by noticing that the visitor spends some time near the cash register).

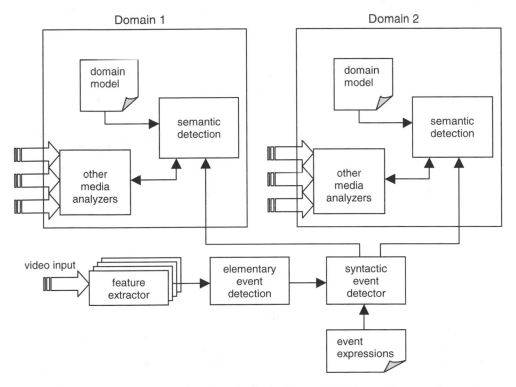

Figure 22.2 A general scheme that shows the dependence of events.

Several people may be interested in determining meaningful events based on these syntactic occurrences. For instance:

1. A security guard will be interested in generically "suspicious" behavior: visitors coming in at unusual times (the security guard will certainly be interested in a "visitor" arriving in the middle of the night!), or, of course, in passing by a small artifact just before that artifact disappears.
2. A museum curator might be interested in people who spend unusual amounts of time in different rooms: she might want to look at the artifact that these visitors look at, and possibly plan a different arrangement of the rooms or of the artifacts inside a room.
3. The person in charge of the gift shop might be interested in people who spend a lot of money (the "buy" event would be, in this case, connected to the cash register to see how much people spend) and, possibly, in tracking the same people during their visit to the museum to see whether the art they are interested in influences their buying and, possibly, to plan for a better offering in the gift shop.

One could continue with a wealth of other professional figures involved in the administration of a museum. The point is that the same syntactic events detected in the same video of people visiting the museum will offer radically different ways of organising the video, depending on the peculiar interests of the people who are looking at the video: it is the role of the viewer that determines what constitutes an event, how events should be connected, and what relevant information should be attached to them. The video data only provides the raw material around which these events are built.

3. STRUCTURING PHENOESTETIC VIDEO

The video that I called *phenoestetic* is constituted by pure video streams coming from a camera placed and, one would say, forgotten, in a place, as is the case with surveillance cameras, web cameras, or, in general, any instance of voyeurism in which the imperative to look at "reality as it is"[3] is more categorical than that of constructing a structured cinematic speech.

The problems of the clash between the syntagm of the video and the paradigm of the interface, and of that between transparent and opaque screens, are, in these cases, even more pronounced than in semiotic video for, in that case, one had a sequence of anchors (the scenes and cuts) which could be presented in a paradigmatic way on an opaque screen and, at the user's whim, explored syntagmatically on a transparent screen.

In phenoestetic one is in the presence of an infinite, unstructured flow of images that, if no suitable organizing principle is found, can only be experienced by endless and continuous observation. This is, of course, unsatisfactory, not only because the endless observation of a video stream clashes with the more interactive attitude of multimedia but because, more simply, in these video most

---

[3] I am well aware, of course, of the fact that there is no observing "reality as it is" through a camera. The very fact of the presence of a camera or, in the case in which the camera is concealed, its framing and de-contextualizing function, are sufficient to alter reality in a significant way. The "real reality" of which I am talking here should therefore be taken with a grain of salt, remembering always that what one observes through a camera is never the same as what one would have observed by being there.

of the time there is absolutely nothing of the remotest interest to observe: the points of interest are few and apart.

There is—one would say—a generalized consensus that in the absence of a syntax like that of the montage, the organization of this type of video should proceed along semantic lines, that is, by identifying nuclei of meaning in the video sequence, and by organizing the video into syntagmatic units around the (paradigmatic) collection of such units.

It should be obvious, however, that nuclei of meaning can only be defined in relation of a certain user or, at least, to a certain prototypical situation. An empty street, with no car and no pedestrian traffic, and only a person idly sitting at a café across the street would appear reassuringly event-less to the security guard of a bank. Yet, to the person sitting at the café—let's imagine him a man who gave his lover a fatal (and final) appointment: we'll run away together today at 5, or we shall never meet again—that same prolonged absence of movement in the street would be excruciatingly eventful.

Approaching video organization from a semantic point of view requires two orders of considerations, the first purely technical, the second related to the observations above:

1. The basis of any kind of video organization is an analysis of the data that the video sequence provides, and this analysis is syntactic in nature. In other words, it is in any case necessary to define a syntactic substratum upon which any semantic construction must rest.
2. The *semeion* of a video will have to be derived from a model of the context in which the video is called to signify. It is painfully well known that such a model is theoretically impossible,[4] and that every formal definition of context will be incomplete in important ways. Nevertheless, it is important that, at least, a framework will be defined to allow interpretation in the most important cases. This aspect also implies that the same syntax of events can have very different interpretations depending on the context under consideration.

---

[4] The impossibility of formally defining context comes, in ultimate analysis, from the consideration that any formalism is, by itself, a meaningless play of symbols, and needs a context to be analyzed and to signify. The formalism in which a context would be defined, therefore, would require an additional context in order to be characterized, this context would require another one, and so on *ad infinitum*. There are, as far as I can see, only two ways of coming out of this impasse. The first is the aprioristic renounce to formally characterize a context. It is quite evident that this position, while philosophically the most satisfactory, is not sustainable in an engineering enterprise, in which the goal is to derive working systems, and an imperfect contextualization of the data is (with some caveats) better than no characterization at all. The second way, coherent with the engineering stance, is the acceptance of the limitations of any formalization of context. We know that our formalization will not be complete, nor do we expect it to be, and we know that this incompleteness will generate ambiguities and misinterpretations. The essential thing, in this case, is not to design systems *as if* the characterization of the context were complete, but to introduce safeguards, in terms of user interaction and interface devices, so that the ambiguities of the context will cause a graceful degradation of the usefulness of the system and not a dramatic breakdown.

I postulate that suitable nuclei of meaning around which a paradigmatic organisation of video data can be built is given by *events*. It would be futile to try to define what an event is in this section. As will be quite evident in the following, and event is whatever the context model says it is, and no more precise definition is possible.

In general terms, however, one can say that if a model is done properly, events will be circumstances bounded in time that cause changes in the interpretation of a situation.

Events arise from the interpretation, done by a context model, of certain syntactic occurrences that can be observed in the video data. I call these syntactic occurrences *hypomena.*[5] A hypomenon is a quantity that is possible to detect in the sequence of video frame and which constitutes the syntactic counterpart to an event which, in this model, is a semantic entity.

I will consider the syntax of video data and its semantics separately, starting with the definition and identification of hypomena, and then proceeding with the definition of events based on these. Since the neologism "hypomenon" is quite unusual and its repeated use might result cumbersome, I will not make use of it, except in circumstances in which the distinction between hypomena and events is crucial. It should be always remembered, however, that in this section whenever I talk about events, I really mean hypomena.

Syntax

The topic of this section is the description of a video sequence in terms of events (hypomena), and their detection in a video sequence given such description. I will start with some epistemological assumptions. Like all such assumptions they constitute a framework (some might say a straitjacket) that, by its very presence delimits the range of phenomena to which the ensuing model will apply, selecting some at the exclusion of others. The pragmatic justification of such a framework is, of course, that many of the phenomena that one intends to study fall within it. I believe this to be true for the model of events that I am going to present but I make, of course, no claims of theoretical completeness.

The assumptions of the model are the following:

1. Events are related to some change or discontinuity in the video sequence. These include discontinuities in a suitable transformed space of the individual frames (e.g., a sudden change in the chromatic distribution of the images) or changes that do not entail discontinuity (an object moves from A to B). Note that this statement does not mean that "events are changes." As the example of the unfortunate lover above reminds us, depending on the particular semantic system employed, the absence of a specific change at a certain time or during a certain time span can constitute an event.

2. Hypomena form a syntactic system at which base are simple (i.e., *atomic*) occurrences. This assumption derives from the requirement that the syntactic formalism should make the structure of a composite event as

---

[5] The word *hypomenon* is, admittedly, a neologism. Its origin is in the Greek words *hypó* (under) and *eimí* (I am).

explicit as possible. The simple events themselves derive from primitive detectors which are outside of the syntactic formalism and are, as such, indivisible inside the formalism, (since the formalism expresses every aspect of the structure we might), they are atomic.

3. Atomic events take place in a single step of a discrete time sequence. This is mainly a technical assumption (discrete time interval make it easier to compose events, as will be seen in the following). The requirement that events take place in a single time step (that is, that they have zero time length) doesn't impact the expressivity of the model: for every event which lasts from t_1 to $t_2 \geq t_1$ it is always possible to define the two events α (the beginning) and ω (the end) which take place at times t_1 and t_2, respectively.

Several systems for the specification of events have been studied in the literature. The formalism introduced here will follow to a large extent that of Snoop [10] and of EPL [11].

Simple and Absolute Events

Let T be the ordered countable set of time instants. Three special constants are defined: the constant $\lozenge$ (now), the constant 0 (beginning of times), and the constant ∞ (infinite time).

The constant $\lozenge$ represents the current system time, 0 is the constant such that, for all times t, $t \geq 0$ is true, and ∞ is the value such that, for all times t, $t \leq \infty$ is true.

An *absolute* event $\langle [t_1; t_2] \rangle$ is the specification of an absolute time t_0 or of a time interval $[t_1, t_2]$, with $t_2 \geq t_1$.

A *simple* event is an event that is detected outside the system using a suitable video processing subsystem.[6] Several video processing systems have been designed to detect specific simple events, and it is beyond the scope of this chapter to analyse them beyond a few bibliographic references. The system in [12] uses colour, texture, and motion to extract motion blobs, and a neural network to determine whether the blob belongs to an object of interest. The system in [13] uses similar features in context of a rule-based system for detecting events of interest in sports videos. Other systems and algorithms useful in this context can be found in [14,15,16,17,18]. In this chapter, I will simply assume that a suitable source of simple events, endowed with suitable attributes, will be available.

Event operators

---

[6] The term *simple* is used here in reference to the event composition system presented here, of course, and it simply means that, from the point of view of the framework presented here, such events are to be regarded as indivisible and devoid of internal structure. Inside the system that detects them, on the other hand, the same event might very well have a structure and further components from which they are derived. These components are, however, assumed to be invisible in the current framework.

I will distinguish between *event types* and *event instances*. An event type is an expression that specifies a class of events using an expression E derived from the language described below. An event instance for E is a specific sequence of events that satisfies the expression E.

I will make a form of closed world assumption in the sense that I will assume that there is a finite database of expressions $\Xi = \{E_1, \dots E_N\}$ upon which the constructs operate.

Let $E_1, \dots E_n$ be types, then the following are event types:

1. (E_1, E_2): sequence of events. An instance of (E_1, E_2) consists of an instance of E_1, immediately followed by an instance of E_2. The meaning of *immediately followed* will be clarified in the following section but, roughly, it means that if E_1 takes place at time t, E_2 takes place at time $t+1$.

2. $(E_1; E_2)$: relaxed sequence of events. An instance of $(E_1; E_2)$ consists of an instance of E_1, eventually followed by an instance of E_2.

3. $(E_1 \vee E_2)$: disjunction. An instance of $(E_1 \vee E_2)$ consists in an instance of either E_1, or E_2.

4. $(E_1 \wedge E_2)$: conjunction. An instance of $(E_1 \wedge E_2)$ consists of an instance of E_1 and of E_2 occurring at the same time.

5. $\neg E$: negation. An instance of $\neg E$ occurs at all times at which no instance of E occurs.

6. $\forall (E_1, E_2)$: universal quantification. An instance of $\forall (E_1, E_2)$ occurs whenever instances of both E_1 and E_2 have taken place, regardless of their order. That is, the hypomenon $\forall (E_1, E_2)$ occurs at the occurrence of the last of the hypomena E_1 and E_2.

Other operations can be defined based on these basic ones. Some examples are:

- The macro *any* ($\aleph$), defined as the disjunction of all atomic events (including the simple events detected outside the system).
- The precedence operator $before(E_1, E_2) \equiv (E_1, (\aleph \vee (\aleph; \aleph))) \wedge E_2$. An instance of this event occurs whenever an instance of E_1 occurs prior to an instance of E_2.
- The first occurrence $first(E) \equiv E \wedge \neg (E; \aleph)$. An instance of this event occurs whenever the fist instance of E occurs.
- The limited universal quantifier $\forall_m (E_1, E_2, \dots E_n)$ with $m \leq n$ is true whenever m of the events $E_1, E_2, \dots E_n$ are verified. The limited universal quantifier can be defined as a disjunction of all the universal quantifiers that give rise to the desired combination, for instance:

$$\forall_2 (E_1, E_2, E_3) = \forall (E_1, E_2) \vee \forall (E_2, E_3) \vee \forall (E_1, E_3) \vee \forall (E_1, E_2, E_3)$$

It is easy to show that all binary operators defined above are associative so that, in most cases, I will use the abbreviated form $(L_1 * L_2 * \cdots * L_n)$ where $*$ is any of the binary operators above in lieu of the more cumbersome $(H_1 * (H_2 * (\cdots * H_n)))$.

Parameters

Event expressions have parameter expressions associated with them, and event instances have associated the corresponding parameter values. Let $P = (\alpha_1 : D_1, \ldots \alpha_n : D_n)$ be a space of named parameters, and let $D = D_1 \times \cdots \times D_n$ be the corresponding data type. Given an event expression E, the expression $E.\alpha_i$ is the expression of the ith parameter of E, and given an instance e, the expression $e.\alpha_i$ is the value of the parameter. I will assume that for each occurrence of an event, the values of its parameters will not change.

Denotational Semantics

This section considers the semantics of events in terms, essentially, of denotational semantics [19]. That is, in this section I am not yet concerned with the *event semantics* that derives from the context model, but only with the clarification of the formal semantics of the event definition language discussed so far.

In order to define the denotational semantics of events, it is first necessary to define the notions of system state and that of valid event occurrences. Informally, each event type (either a primitive event or an event expression) has associated two sets: a set of all event occurrences since the beginning of time, and the set of *valid* event occurrences, that is, of those occurrences that can be used to compute an expression. The need to specify a set of valid occurrences should be clear from the following example. Assume that in the system there are two primitive events, a and b, as well as the composite event $c = a;b$. The events a and b are instantiated as in Figure 22.3.

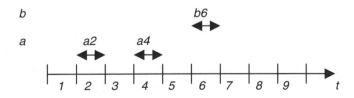

Figure 22.3 The system with two primitive events.

When the event b6 is detected, the conditions are satisfied for the instantiation of the expression $a; b$, but how many such expression will be instantiated? There are two distinct sequences of events that could instantiate c, namely the sequence a2; b6 and the sequence a4; b6. Whether both these two sequences should be instantiated when b6 appears, only a2; b6 should be instantiated, or only a4; b6 rests on the definition of the semantics of the event expressions.

I will consider the semantics in terms of event histories. Each expression, once instantiated in a particular system, determines a function from the set of time instances to a co-domain in which it is possible to see whether at a specific time an event of that instance is occurring and what are the values of the parameters for that occurrence. The occurrence of an event is simply marked by a Boolean value, and the parameters are values of type D, as defined in the previous sub-section. The whole status of an event at a given time is a value in $2 \times D_1 \times \cdots \times D_n = 2 \times D$, where 2 is the data type of Boolean values. The fictitious parameter ω will be assumed to correspond to this component of the data type, so that the function $e.\omega$ is true when the event takes place.

Each primitive event and each event expression in the system determines an event type and, for each event type, a function $[[E]] : T \rightarrow 2 \times D$ is defined. This function is the *semantics* of that event type. Primitive event types come with their semantics, given by the valid occurrences of the events and by the values of the parameters of these occurrences. Note that only the *valid* occurrences of events contribute to their semantics at a given time.

The set of semantics of all the events and event expressions in the system is the *state* of the event detection system. At any time t at which a new primitive event is instantiated, the new state of the system is determined in the following manner:

1. Determine the semantics of all the expressions in the system that may be affected by the new instantiation of the primitive event, computing a new function $[[E]]$ for each of them.
2. Apply a *state transition function*, which depends on the specific semantics that is being used (see below) to the state thus obtained.

In order to define these state transition functions, it is necessary first to define another concept: that of *minimally viable number of instance* of an event a $\theta(a)$. This can be defined as the maximum number of consecutive instances of the event that needs to be preserved in order to be able to compute all the expressions in the system. Consider again the system above, with two sources of primitive events, a and b, and with a single expression: $c = a;b$. In this system, obviously, maintaining a single occurrence of a and a single occurrence of b will, in principle, allow one to compute the expression c. Let us, however, add a third expression $d = (a \vee b);a$. One of the possible ways to instantiate this expression is by the sequence $a2; a4$. If we maintain a single occurrence of the event a in the state, we lose the possibility of detecting this sequence. So, in this case, it is $\theta(a) = 2$.

With this concept in mind, I will define the following three state transition frameworks (these frameworks correspond, mutatis mutandis, to the *Parameter contexts* defined in [10]).

Recent: In this framework, only the most recent instances of the events necessary for the computation of an expression are used. Whenever an event of type E is instantiated, the state computation function will erase all the occurrences of the event except for the last $\theta(E)$. In the case of Figure 22.2, when the event $a2$ occurs, it will become part of the state. When $a4$ arrives, the instance $a2$ will be eliminated, and the event history of a will consist of the only instance $a4$. When $b6$ occurs, the event expression c will be evaluated on the current history consisting, at this time, of the instances $a4$ and $b6$, so that the instances that generates c will be $a4$; $b6$.

Chronological: In this framework, the oldest occurrence of an event that hasn't yet been used contributes to the computation of the expression that need it. The state transition function is composed of two parts: first, when a new primitive event is instantiated, the instance is added to the history of that event. Then, expressions are computed, and each one will use the oldest instances of the events that it needs, and mark them as used. Finally, after all the expressions have been computed, the state computation function will remove all marked instances, unless such removal would leave less than $\theta(E)$ instances.

In the example of Figure 22.2, the events $a2$ and $a4$ will be received and added to the history of the event type a, and similarly the event $b6$ will be added to the history of the event type b. At this point, the expression c will be evaluated, and it will make use of the instances $a2$ and $b6$, marking them as used. Then the state computation function will remove the instance $a2$ and $b6$, since they have already been used.

Continuous: In this framework, all the possible combinations that satisfy an expression are evaluated, and primitive events are discarded only after they have been used at least once. In the previous example, the construction of the state until the arrival of $b6$ proceeds as in the chronological framework but, when $b6$ arrives, both sequences $a2$; $b6$ and $a4$; $b6$ are used to instantiate two instances of the composite event c. After the instantiation, all the used instances that can be removed will be removed. In this case, all the events will be removed.

Absolute events derive their semantics from the interval in which they are defined, so, the semantics of the event $\langle [t_1; t_2] \rangle$ is the function

$$[[\langle [t_1; t_2] \rangle]] = \lambda t.(t \geq t_1 \wedge t \leq t_2)$$

Note that $\langle \lozenge \rangle.\omega$ is true at any time the event is evaluated.

Operators, on the other hand, are maps from semantics to semantics. For instance, the conjunction operator $\wedge$ takes the semantics of two events (i.e., the functions determining their occurrences and the values of their parameters) and maps those into a function that is the semantics of the conjunction of the two events. In other words, the semantics of the conjunction operator is a map:

$$[[\wedge]] : (T \to 2 \times D_a) \times (T \to 2 \times D_b) \to (T \to 2 \times D)$$

The semantics of the events used in the evaluation of a given operator will include, of course, only the instances that are valid at that particular time, the validity being determined by the framework that is being used.

I am making the usual compositionality assumption: the semantics of an expression involving an operator depends only on the semantics of the operator and on those of the events to which the operator is applied; in particular, the semantics of an expression is obtained by applying the operator semantics to the semantics of the events to which it is applied:

$$[[E_1 \Diamond E_2]] = F([[\Diamond]], [[E_1]], [[E_2]]) = [[\Diamond]]([[E_1]], [[E_2]])$$

Because of the definition of the parameter space each event has, roughly speaking, $n+1$ semantics, corresponding to the $n+1$ components of the data type $2 \times D$: there is a function $[[E]].\omega : T \to 2$ that determines the *occurrence semantics* of a given event expression (that is, it determines, given a certain event history, when the events specified by the event expression are taking place), and there are functions $[[E]].\alpha_i : T \to D_i$ that determine the parameter semantics of the event (that is, that compute, at any instant in which the event occurs, the pertinent value of the parameters). These functions are undefined when the event does not take place.

I will begin first by considering the occurrence semantics of the different expression, leaving the parameter semantics for the next section. In order to simplify the notation, I will omit the indication of the component from the semantic functions, writing $[[E]]$ in lieu of $[[E]].\omega$.

Finally, in order to facilitate the expression of the operator semantics, I will introduce two quantities: the *last false* time, and the *last true* time. The first quantity is defined for a time t only if, at t, the event expression that one is considering is true. It is, basically, the last time at which the event expression was false, and is defined as:

$$E^\uparrow(t) = t' : \neg[[E]](t') \wedge \forall t'', t' < t'' \le t \Rightarrow [[E]](t'')$$

The last true time is the dual quantity: it is defined for a time t only if the expression under consideration is false, and is defined as the last time at which the expression was true:

$$E^\downarrow(t) = t' : [[E]](t') \wedge \forall t'', t' < t'' \le t \Rightarrow \neg[[E]](t'')$$

As usual, only valid instances of the events according to the current state transition framework should be considered.

The semantics of the operators defined above is the following.

Sequence:

$$[[E_1, E_2]] = \lambda t.([[E_2]](t) \wedge \neg[[E_1]](t) \wedge E_2^\uparrow(t) = E_1^\downarrow(t-1))$$

The interpretation of this expression is the following: the sequence is true at the current time if the event E_2 is true, the event E_1 is false, and E_2 became true immediately after E_1 became false.

Relaxed sequence:

$$[[E_1, E_2]] = \lambda t.\left([[E_2]](t) \wedge E_1^{\uparrow}(t) \neq \perp \wedge E_1^{\uparrow}(t) \leq E_2^{\uparrow}(t)\right)$$

In this case, the sequence is true at the current time if the event E_2 is true and the event E_2 became true after the event E_1 became true. Note that the relaxed sequence event is true regardless of whether the event E_1 is still happening when the event E_2 happens or not. The diagram of Figure 22.4 shows some examples of the verification of a relaxed sequence (the dashed arrow points from an instance of E_1 to the instance of the relaxed sequence that it enables).

Disjunction:

$$[[E_1 \vee E_2]] = \lambda t.\left([[E_1]](t) \vee [[E_2]](t)\right)$$

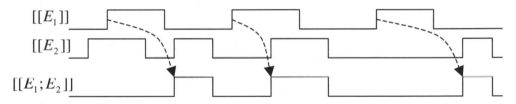

Figure 22.4 Examples of the verification of a relaxed sequence.

Conjunction:

$$[[E_1 \wedge E_2]] = \lambda t.\left([[E_1]](t) \wedge [[E_2]](t)\right)$$

Negation:

$$[[\neg E]](t) = \lambda t.\left(\neg[[E]](t)\right)$$

Universal Quantification:

$$[[\forall(E_1, E_2)]] = \lambda t.\left(\exists t', t'' \leq t : [[E_1]](t') \wedge [[E_2]](t'')\right)$$

The definition of a formal semantics allows one to avoid confusion in the determination of the characteristics of an event expression. Using denotational semantics it is possible, for instance, to verify that the expression *before* is similar to the relaxed sequence but, unlike the latter, it occurs only when the event E_1 finishes before E_2 begins. That is, the second occurrence of the event $[[E_1; E_2]]$ in Figure 22.3 would not take place if $[[E_1; E_2]]$ were replaced by $[[before(E_1, E_2)]]$.

Parameter Semantics

Parameter semantics is strictly connected to the framework in use, because the framework will determine what instances of the events that form a given expression will be used to compute the parameters of the expression. The rules for which instances of an event will be used to compute the parameters of an expression are the same illustrated in the introduction to denotational semantics.

Unlike the occurrence semantics, however, the event operators do not induce any parameter semantics: the function that specifies the parameter of a complex event will have to be specified when the complex event is declared. This computation, however, must satisfy certain conditions, because it must be meaningful regardless the specific chain of events that leads to the instantiation of a complex event. For instance, a parameter of the expression $(E_1 \vee E_2)$ can't depend on a parameter that appears only in E_1, because the event $(E_1 \vee E_2)$ can be instantiated even if E_1 doesn't occur.

For an event E, let $[[E]].P$ be the set of all its parameters. Parameters of two events E_1 and E_2 are considered equal if they have the same name and the same type. Then, the parameters of the event $(E_1 \vee E_2)$ will be computable only if they depend on the parameters in $[[E_1]].P \cap [[E_2]].P$. That is, every parameter α of the disjunction of the two events must be defined as a function

$$[[E_1 \vee E_2]].\alpha = f_\alpha([[E_1]].P \cap [[E_2]].P)$$

The set of parameters $[[E_1]].P \cap [[E_2]].P$ is called the *computability set* for the composite event $(E_1 \vee E_2)$. The computability set depends on the characteristics of the operators applied as indicated in Table 22.1.

Table 22.1

Expression	Computability set
(E_1, E_2)	$[[E_1]].P \cup [[E_2]].P$
$(E_1; E_2)$	$[[E_1]].P \cup [[E_2]].P$
$(E_1 \vee E_2)$	$[[E_1]].P \cap [[E_2]].P$
$(E_1 \wedge E_2)$	$[[E_1]].P \cup [[E_2]].P$
$\neg E$	$\varnothing$
$\forall(E_1, E_2)$	$[[E_1]].P \cup [[E_2]].P$

Event Detection

There are two ways to proceed to event detection: one can either start from the earliest event that can appear in the instantiation of an expression, or from the last. Consider, for example, the expression $E_1; E_2$, and assume that E_1 and E_2 are primitive events.

One way of proceeding is to wait for an event of type E_1 to occur and, when this event occurs, to put the system in a state that waits for E_2 and whenever E_2 arrives, detects the event $E_1; E_2$. This is tantamount to the definition of the state machine of Figure 22.5.

The complete state machine is more complicated, since it must take into account the different frameworks in which the computation is done. The state machine of Figure 22.5, since every new arrival of E_1 causes the re-instantiation of the parameters, works in the recent framework.

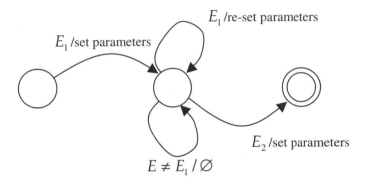

Figure 22.5 State machine.

A state machine like this can be easily defined for each operation, and state machines devoted to the detection of different complex events can be put together in a single push/pop machine to detect events of arbitrary complexity.

A second way of detecting complex events is to wait for the events that can *conclude* an expression. In the case of the expression $E_1; E_2$ this means waiting for the event E_2. After this event has arrived, the system will look in the past history of events for a combination that will satisfy the whole expression. This way reduces event detection, essentially, to a regular expression search, but has the additional complication that a possibly unbounded history of past events must be kept.

4. CONTEXT AND EVENTS

The specification and detection mechanism introduced in the previous section constitutes the syntactic portion of a video organisation system. The semantic part depends on the particular context in which the summarisation is required. The crucial difference between the syntactic and the semantic specification is that, in the syntactic specification, one is only interested in isolating and categorising occurrences within a particular medium (in the present case: video); in the semantics specification, one is interested in describing as precisely as possible a whole domain and the role played by events in relation to the other elements of the domain.

Figure 22.6 shows a diagram of a possible domain specification. Events are created through the satisfaction of a generating condition, which is syntactic in nature. A number of necessary conditions can be specified, together with time relations with the generating conditions, so that an event will actually be instantiated only when the generating condition, as well as all necessary conditions will be satisfied. Notice that the necessary conditions are not necessarily drawn from the same medium as the generating condition, nor are they required to be events. To make a simple example, an event like "people near the Picasso painting" is a syntactic event that will generate the event "possible theft threat" only if the condition "night-time" is verified. The condition "night-time" per se is not, obviously, an event.

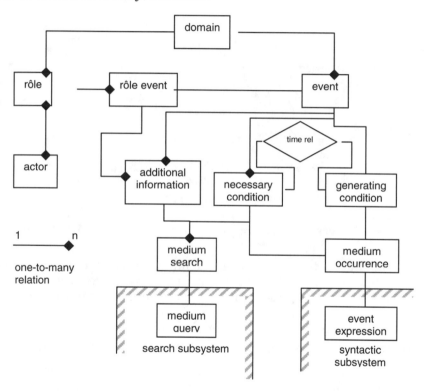

Figure 22.6 Diagram of domain specification.

Users of the system are called *actors* in the diagram, and they participate in certain *roles*. The domain specification will contain additional information about actors and their role, such as relations between different roles, or additional information about actors, which is irrelevant in the present context. Events become such only through association to a role (role events), which incorporate them into the activities and requirements of that role. Role events can specify additional conditions that an event must satisfy in order to be recognised and additional information that needs to be aggregated around that event.

The general concept of role event goes, obviously, beyond the organisation of video: in this scenario role events are focal points for the aggregation of information that can be contained in a variety of media and derived from a variety of sources.

The conditions that lead to the instantiation of an event are, in this case, simpler since in general it is convenient to implement the more complex event composition rules in the syntactic layer.

The semantic level is, however, the ideal location for more "fuzzy" specifications of the time relations between events. For instance, the specification that event E_2 should take place "shortly after" event E_1 should be made into this semantic layer and translated into a suitable time bound for the syntactic layer.

In order to exemplify the matter, I will assume here that all the syntactic events, once detected, are stored in a database that the semantic layer can query. In the example above, then, part of the role of the gift shop manager could be specified as follows:[7]

```
<role name="gift-shop-manager"/>
    <event_track>
        <event type="purchase" name="p"/>
        <event type="cash-register-transaction" name="r"/>
        <event type="painting-visit" name="v">
    </event-track>

    <event name="big-buy/>
        <detection>
            r.amount > $100 and
            coincide(p.time, t.time)
        </detection>
        <associations>
            <association name="painting-visits" type="video">
                <detection>
                    select t.video_link
                    from    VideoEvents t
                    where   t.type = "visit" and t.id = p.id
                </detection>
            <association>
        </associations>
</role>
```

The detection rule looks for purchase videos that coincide with high value transactions (the predicate "coincide" matches two times within a certain error). One association of this event is shown: it looks for all the video fragments of the visit that the same person who bought the souvenirs made to the gallery.

5. CONCLUSIONS

It seems fitting, in order to conclude the discussion of the previous pages, to go beyond the purely technical exposition to look at the organisation of video (and

---

[7] I am using a notation based on XML and interspersed with some SQL query only because these two formalisms are very well known and, I believe, self-explanatory. They are by no means the only or even the best choice.

other data) from a broader perspective. Cameras are becoming cheaper and smaller, and are being placed in many places in which, only a few years ago, their presence would have been unthinkable. This is not an entirely new phenomenon but what is new today is the possibility of inserting these video data into a world-wide network in which they can be analysed, associated to other information, and accessed in a much more convenient way than it was possible before.

The possibility of organising information along semantic lines, and of associating different sources of information around meaningful events, transforms the quantitative increase in the amount of information available into a qualitative leap of the modalities of access of information, a leap fraught with possibilities but also with dangers.

Issues like the people's right to privacy (indeed the very concept of privacy), and the control of information about one's life are already at the centre of the public's attention. Large scale data organisation and association will bring forward other issues, more subtle and complex, like the control of the *inferences* that can be made based on publicly available data: it is well known that even apparently harmless pieces of information can, if accumulated in suitable amounts and analysed with sufficient sophistication, be used to infer information that a reasonable person would consider an intrusion of privacy.

These issues are not eminently technical in nature, but should generate a social, cultural, and political debate in all strata of society. Hopefully, computer professionals will be an active and socially conscious part of this debate. Too often in the past we have seen the role of computer professional reduced to that of technical developers without a voice or, worse, of an amplifier for the voice of the commercial interests of the companies for which they work. As the social issues related to computer technology become more momentous and more pressing, it is essential that the developers of such technology become an active and informed part of the social debate on those issues.

REFERENCES

[1]   E. Panofsky, *Gothic Architecture and Scholasticism,* Latrobe, PA, Archabbey Press, 1951.
[2]   L. Manovich, *The Language of New Media* MIT Press, Cambridge, MA, 2001.
[3]   http://www.rai.it/portale.
[4]   U. Eco, *A Theory of Semiotics*, Indiana University Press, 1979.
[5]   http://www.tv5.org/info/index.php.
[6]   B. R. Farzin, K. Goldberg, and A. Jacobs, A Minimalist Telerobotic Installation on the Internet, in 1st Workshop on Web Robots, International Conference on Robots and Intelligent Systems, September 1998.
[7]   S. Santini, Analysis of Traffic Flow in Urban Areas Using Web Cameras, in Fifth IEEE Workshop on Applications of Computer Vision (WACV 2000), Palms Spring, CA (USA), December 2000.
[8]   N. Vasconcelos and A. Lippman, Towards Semantically Meaningful Feature Spaces for the Characterization of Video Content, in Proc. Int. Conf. Image Processing, Santa Barbara, California, 1997.

[9]   A. Del Bimbo, M. Mugnaini, P. Pala, and F. Turco, F., Visual querying by color perceptive regions, *Pattern Recognition*, 31(9):1241-1253, 1998.

[10] S. Chakravarthy and M. Deepak, *Snoop: An Expressive Event Specification Language for Active Databases*, Techical Report UF-CIS-TR-93-007, Computer and Information Sciences Department, University of Florida, March 1993.

[11] I. Motakis and C. Zaniolo, A formal semantics for Composite Temporal Events in Active Database Rules.

[12] R. Qian, N. Hearing, and I. Sezan, A Computational Approach to Semantic Event Detection, in Proc. Computer Vision and Pattern Recognition, Vol. 1, pp. 200-206, Fort Collins, CO, June 1999.

[13] Wensheng Zhou, Asha Vellaikal and C. C. Jay Kuo, Rule-based Video Classification System for Basketball Video Indexing, in Proceedings on ACM Multimedia Workshop, 2000, pp. 213-216.

[14] S. Chang, J.R. Smith, M. Beigi, and A. Benitez., Visual Information Retrieval from Large Distributed Online Repositories, *Communications of the ACM*, December 1997, pp. 63-71.

[15] Collins, Lipton, Kanade, Fujiyoshi, Duggins, Tsin, Tolliver, Enomoto, and Hasegawa, *A System for Video Surveillance and Monitoring: VSAM Final Report*, Technical report CMU-RI-TR-00-12, Robotics Institute, Carnegie Mellon University, May 2000.

[16] T. Huang, D. Koller, J. Malik, G. Ogasawara, B. Rao, S. Russel, and J. Weber, Automatic Symbolic Traffic Scene Analysis Using Belief Networks, in Proceedings 12th National Conference in AI, 1994, pp. 966-972.

[17] S. K. Bhonsle, A. Gupta, S. Santini, M. Worring, and R. Jain, Complex Visual Activity Recognition Using a Temporally Ordered Database, in 3rd Int. Conf. Visual Information Management, Amsterdam, June 1999.

[18] I. Mikic, S. Santini, and R. Jain, Video Processing and Integration from Multiple Cameras, in Proceedings of the 1998 Image Understanding Workshop, Monterey, CA, November 1998

[19] Lloyd Allison, *A Practical Introduction to Denotational Semantics* Cambridge University Press, 2001.

23

SEGMENTATION AND CLASSIFICATION OF MOVING VIDEO OBJECTS

**Dirk Farin, Thomas Haenselmann, Stephan Kopf,
Gerald Kühne, and Wolfgang Effelsberg**
Praktische Informatik IV
University of Mannheim
L15, 16, 68131 Mannheim, Germany
`effelsberg@informatik.uni-mannheim.de`

1. INTRODUCTION

Despite very optimistic predictions in the early days of Artificial Intelligence research, a computer vision system that interprets image sequences acquired from arbitrary real-world scenes still remains out of reach. Nevertheless, there has been great progress in the field since then and a number of applications emerged within different areas. Of particular interest for several applications are capabilities for *object segmentation* and *object recognition*. Algorithms from the former category support the segmentation of the observed world into semantic entities and thus allow a transition from signal processing towards an object-oriented view. Object recognition approaches allow the classification of objects into categories and enable conceptual representations of still images or videos.

The goal of this chapter is the development of a classification system for objects that appear in videos. This information can be used to index or categorize videos and it thus supports object-based video retrieval. In order to keep the subject manageable, the system is embedded into a set of constraints: The segmentation module relies on motion information; thus it can only detect moving objects. Furthermore, the classification module only considers the two-dimensional shape of the segmented objects. Therefore, just a coarse classification of the objects into generic classes (e.g., cars, people) is possible.

The remainder of the chapter is organized as follows: First of all, in summarizing our approach, Section 2 serves as a guideline through the subsequent sections. In Section 3, camera models and the estimation of their parameters are described. Next, we discuss our approach to object segmentation in Section 4. Section 5 introduces the video object classification system and Section 6 concludes the chapter with experimental results.

2. SYSTEM ARCHITECTURE

Our system for video object classification consists of two components, namely a *segmentation module* and a *classification module* (cf. Figure 23.1).

Based on motion cues the camera motion within the scene is determined (*motion estimation*) and a background image for the entire sequence is constructed (*background mosaic*). During the construction process, parts belonging to foreground objects are removed by temporal filtering. Then, object segmentation is performed by evaluating differences between the current frame and the reconstructed background mosaic (*segmentation*).

The object masks determined by the segmentation algorithm are fed forward to

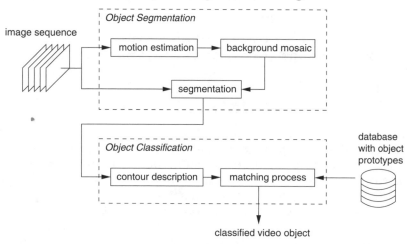

Figure 23.1 Architecture of the video object classification system.

the classification module. For each mask, an efficient shape-based representation is calculated (*contour description*). Then, this description is matched to pre-calculated object descriptions stored in a database (*matching*). The final classification of the object is achieved by integrating the matching results for a number of successive frames. This adds reliability to the approach since unrecognizable single object views occurring in the video are insignificant with respect to the whole sequence. Moreover, it allows an automatic description of object behavior.

3. CAMERA MOTION COMPENSATION

If videos are recorded with a moving camera, not only the foreground objects are moving, but also the background. The first step of our segmentation algorithm determines the motion due to changes in the camera parameters. This allows to stabilize the background such that only the foreground objects are moving relative to the coordinate system of the reconstructed background. It is usually assumed that the background motion is the dominant motion in the sequence, i.e., its area of support is much larger than the foreground objects.

In order to differentiate between foreground and background motion, one has to introduce a regularization model for the motion field. This model should be general enough to describe all types of motion that can occur for a single object, but on the other hand, it should be sufficiently restrictive that two motions that we consider "different" can not be described by the same model. The motion model also allows to determine motion in areas in which the texture content is not sufficient to estimate the correct motion.

3.1 CAMERA MOTION MODELS

We use a world model in which the image background is planar and non-deformable. This assumption, which is valid for most practical sequences, allows us to use a much simpler motion model as would be needed for the general case of a full three-dimensional structure.

Using homogeneous coordinates, the projection of a 3D scene to an image plane can be formulated in the most general case by $(x' \quad y' \quad w)^T = P \cdot (x \quad y \quad z \quad 1)^T$, where P is a 3×4 matrix (see [3,6,7]). As we are only interested in the transformation of one projected image to another projected image at a different camera position (c.f. Fig. 23. 2), we can arbitrarily change the world coordinate system such that the background plane is located at $z = 0$. In this case, the projection equation reduces to

$$\begin{pmatrix} x' \\ y' \\ w \end{pmatrix} = \begin{pmatrix} p_{11} & p_{12} & p_{13} & p_{14} \\ p_{21} & p_{22} & p_{23} & p_{24} \\ p_{31} & p_{32} & p_{33} & p_{34} \end{pmatrix} \begin{pmatrix} x \\ y \\ 0 \\ 1 \end{pmatrix} = \begin{pmatrix} p_{11} & p_{12} & p_{14} \\ p_{21} & p_{22} & p_{24} \\ p_{31} & p_{32} & p_{34} \end{pmatrix} \begin{pmatrix} x \\ y \\ 1 \end{pmatrix}. \tag{23.1}$$

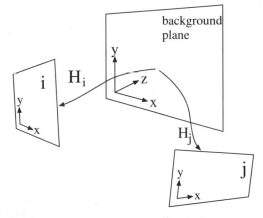

Figure 23.2 Projection of background plane in world coordinates to image coordinate systems of images i and j.

The 3×3 matrix on the right denotes a plane-to-plane mapping (*homography*). Let H_i be the homography to project the background plane onto the image plane of frame i. Then, we can determine the transformation from image plane i to j as

$$H_{ij} = H_j H_i^{-1} = \begin{pmatrix} h_{11} & h_{12} & h_{13} \\ h_{21} & h_{22} & h_{23} \\ h_{31} & h_{32} & h_{33} \end{pmatrix}.$$
(23.2)

Since homogeneous coordinates are scaling invariant, we can set $h'_{ij} = h_{ij} / h_{33}$ and get with a renaming of matrix elements

$$H'_{ij} = \begin{pmatrix} h'_{11} & h'_{12} & h'_{13} \\ h'_{21} & h'_{22} & h'_{23} \\ h'_{31} & h'_{32} & h'_{33} \end{pmatrix} = \begin{pmatrix} a_{11} & a_{12} & t_x \\ a_{21} & a_{22} & t_y \\ p_x & p_y & 1 \end{pmatrix}.$$
(23.3)

Hence, the transformation between image frames can be written as

$$x' = \frac{a_{11}x + a_{12}y + t_x}{p_x x + p_y y + 1}, \qquad y' = \frac{a_{21}x + a_{22}y + t_y}{p_x x + p_y y + 1}.$$
(23.4)

This model is called the *perspective camera motion model*. In this formulation, it is easy to see that the a_{ij} correspond to an affine transformation, t_x, t_y are the translatorial components, and p_x, p_y are the perspective parameters. A disadvantage of the perspective motion model, which will become apparent in the next section, is that the model is non-linear. If the viewing direction does not change much between successive frames, the perspective parameters p_x, p_y can be neglected and can be set to zero. This results in the *affine camera motion model*

$$\begin{pmatrix} x' \\ y' \end{pmatrix} = \begin{pmatrix} a_{11} & a_{12} \\ a_{21} & a_{22} \end{pmatrix} \begin{pmatrix} x \\ y \end{pmatrix} + \begin{pmatrix} t_x \\ t_y \end{pmatrix}.$$
(23.5)

Since the affine model is linear, the parameters can be estimated easily. The selection of the appropriate camera model depends on the application area. While it is possible to use the most general model in all situations, it may be advantageous to restrict to a simpler model. A simple motion model is not only easier to implement, but the estimation also converges faster and is more robust than a model with more parameters. In some applications, it may even be possible to restrict the affine model further to the *translatorial model*. Here, (a_{ij}) equals the identity matrix and only the translatorial components t_x, t_y remain:

$$\begin{pmatrix} x' & y' \end{pmatrix}^T = \begin{pmatrix} x & y \end{pmatrix}^T + \begin{pmatrix} t_x & t_y \end{pmatrix}^T.$$
(23.6)

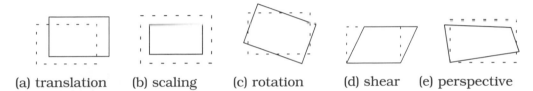

(a) translation   (b) scaling   (c) rotation   (d) shear   (e) perspective

Figure 23.3 Different plane transformations. While transformations (a)-(d) are affine, perspective deformations (e) can only be modeled by the perspective motion model.

3.2 MODEL PARAMETER ESTIMATION

In parameter estimation, we search for the camera model parameters that best describe the measured local motion. Algorithms for camera model parameter estimation can be coarsely divided into two classes: *feature-based estimation* [22] and *direct* (or *gradient-based*) *estimation* [8]. The idea of model estimation based on feature correspondences is to identify a set of positions in the image that can be tracked through the sequence. The camera model is then calculated as the best fit model to these correspondences. In direct matching, the best model parameters are defined as those resulting in the difference frame with minimum energy. This approach is usually solved by a gradient descent algorithm. Hence, it is important to have a good initial estimate of the camera model to prevent getting trapped in a local minimum. As the probability for running into a local minimum increases with large displacements, a pyramid approach is often used. The image is scaled down several levels and the estimation begins at the lowest resolution level. After convergence, the estimation continues at the next higher resolution level until the parameters for the original resolution are found.

Since direct methods provide a higher estimation accuracy than feature-based approaches, but require a good initialization to assure convergence, we are using a two-step process. First, feature-based estimation which can cope with large displacements is used to obtain an initial estimate. Based on this model, a direct method is used to increase the accuracy.

3.3 FEATURE-BASED ESTIMATION

Feature-based estimation is based on a set of features in the image that can be tracked reliably through the sequence. If features can be well localized, image motion can be estimated with high confidence. On the other hand, for pixels inside a uniformly colored region, we cannot determine the correct object motion. Even for pixels that lie on object edges, only the motion component perpendicular to the edge can be determined (see Figure 23.4a). To be able to track a feature reliably, it is required that the neighborhood of the feature shows a structure that is truly two-dimensional. This is the case at corners of regions, or points where several regions overlap (see Figure 23.4b-d).

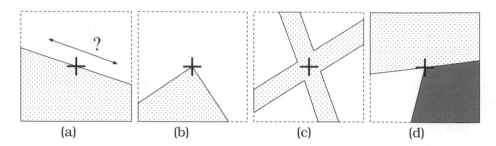

Figure 23.4 Feature points on edges (a) cannot be tracked reliably because a high uncertainty about the position along the edge remains. Feature points at corners (b), crossings (c), or points where several regions overlap (d) can be tracked very reliably.

3.3.1 Feature Point Selection

For the selection of feature points, we employ the *Harris* (or Plessey) corner detector [5] which is described in the following. Let $P = \{p_i\}$ be the set of pixels in an image with associated brightness function $I(p)$. To analyze the structure at pixel $p = (x_p \quad y_p) \in I$, a small neighborhood $N(p) \subset I$ around p is considered.

We denote the image gradient at p as $\nabla I(p) = (g_x(p) \quad g_y(p))^T$. Let us examine how the distribution of gradients has to look like for feature point candidates. Figure 23.5 depicts scatter-plots of the gradient vector components for all pixels inside the neighborhood of some selected image positions. We can see in Figure 23.5c that for neighborhoods that only exhibit one-dimensional structure, the gradients are mainly oriented into the same direction. Consequently, the variance is large perpendicular to the edge and very small along the edge. This small variance indicates that the feature cannot be well localized. Favorably, features should expose a neighborhood where the gradient components are well scattered over the plane and, thus, the variance in both directions is high (cf. Figure 23.5d,e).

Approximating a bivariate Gaussian distribution, we determine the principal axes of the distribution by using a principal component decomposition of the correlation matrix

$$C = \begin{pmatrix} \sum\limits_{i \in N(p)} g_x(i) g_x(i) & \sum\limits_{i \in N(p)} g_x(i) g_y(i) \\ \sum\limits_{i \in N(p)} g_x(i) g_y(i) & \sum\limits_{i \in N(p)} g_y(i) g_y(i) \end{pmatrix}. \tag{23.7}$$

The length of the principal axes corresponds to the eigenvalues $\lambda_1 \le \lambda_2$ of C. Based on these eigenvalues, we can introduce a classification of the pixel p. We differentiate between the classes *flat* for low λ_1, λ_2, *edge* for $\lambda_1 \ll \lambda_2$, or *corner*, *textured* for large λ_1, λ_2. Since the computation of eigenvalues is computationally expensive (note that the computation has to be performed for every pixel in the image), Harris and Stephens proposed to set the classification boundaries such that an explicit computation of the eigenvalues is not required.

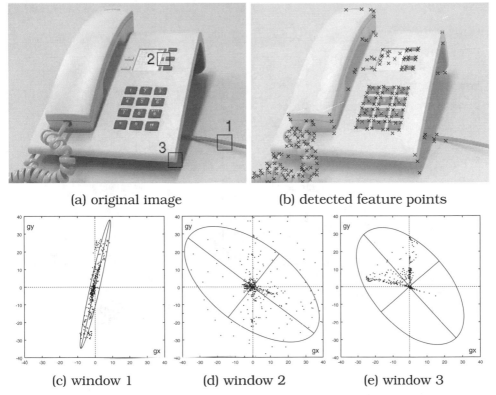

(a) original image

(b) detected feature points

(c) window 1

(d) window 2

(e) window 3

Figure 23.5 Scatter plots of gradient components for a selected set of windows.

Exploiting the fact that $\lambda_1 + \lambda_2 = Tr(C)$ and $\lambda_1\lambda_2 = Det(C)$, they defined a corner response value as

$$r = \lambda_1\lambda_2 - k(\lambda_1 + \lambda_2)^2 = Det(C) - k \cdot Tr(C)^2 \tag{23.8}$$

where k is usually set to 0.06. The class boundaries are chosen as shown in Figure 23.6. After $r(x,y)$ has been computed for each pixel, feature points are obtained from the local maxima of $r(x,y)$ where $\lambda_1 + \lambda_2 = Tr(C) > t_{low}$ (i.e., the pixel is not classified as a *flat* pixel).

To improve the localization of the feature points, Equation 23.7 is modified to a weighted correlation matrix where the gradients are weighted with a Gaussian kernel $w(p)$ as

$$C = \begin{pmatrix} \sum_{i \in N(p)} w(i) g_x(i) g_x(i) & \sum_{i \in N(p)} w(i) g_x(i) g_y(i) \\ \sum_{i \in N(p)} w(i) g_x(i) g_y(i) & \sum_{i \in N(p)} w(i) g_y(i) g_y(i) \end{pmatrix}. \tag{23.9}$$

This increases the weight of central pixels and the feature point is moved to the position of maximum gradient variance. Without this weighting, the best position to place the feature point is not unique. The detector response is equal as long as the corner is completely contained in the neighborhood window. A sample result of automatic feature point detection is shown in Figure 23.5b.

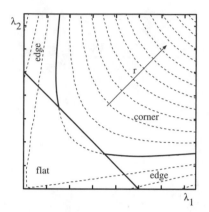

Figure 23.6 Pixel classification based on Harris corner detector response function. The dashed lines are the isolines of r.

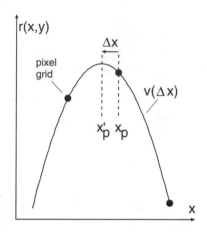

Figure 23.7 Sub-pel feature point localization by fitting a quadratic function through the feature point at x_p and its two neighbors.

3.3.2 Refinement to Sub-Pixel Accuracy

The corner detector described so far locates feature points only up to integer position accuracy. If the true feature point is located near the middle between pixels, jitter may occur. This can be reduced by estimating the sub-pixel position of the feature point.

The refinement is computed independently for the x and y coordinate. In the following, we concentrate on the x direction. The y direction is handled similarly. For each feature point, we match a parabola

$$v(\Delta x) = a \cdot (\Delta x)^2 + b \cdot \Delta x + c$$

through the Harris response surface $r(x, y)$, centered at the considered feature point. The fitted parabola is defined by the values of r at the feature point position and its two neighbors. By setting $\Delta x = x - x_p$, where x_p is the feature point position (cf. Figure 23.7), we get

$$
\begin{aligned}
v(-1) \quad &= r(x_p - 1, y_p) \quad = a - b + c \\
v(0) \quad &= r(x_p, y_p) \quad\quad = c \\
v(1) \quad &= r(x_p + 1, y_p) \quad = a + b + c
\end{aligned}
\tag{23.10}
$$

After setting $dv/d\Delta x = 0$, this leads to

$$\Delta x = \frac{1}{2} \cdot \frac{r(x_p - 1, y_p) - r(x_p + 1, y_p)}{r(x_p + 1, y_p) + r(x_p - 1, y_p) - 2r(x_p, y_p)}. \tag{23.11}$$

Since $r(x_p, y_p)$ is a local maximum, it is guaranteed that $|\Delta x| < 1$. The new feature point position is set to the maximum of v, i.e., $x'_p = x_p + \Delta x$.

3.3.3 Determining Feature Correspondences

After appropriate feature points have been identified, we have to establish correspondences between feature points in successive frames. There are two main problems in establishing the correspondences. First, not every feature point has a corresponding feature point in the other frame. Because of image noise or object motion, new feature points may appear or disappear. Fortunately, the Harris corner detector is very stable so that most feature points in one frame will also appear in the next [19]. The second problem is that the matching can be ambiguous if there are several feature points surrounded by a comparable texture. This may happen, e.g., when there are objects with a regular texture or several identical objects in the image.

Let F_1, F_2 be the set of feature points of two successive images I_1, I_2. Our feature matching algorithm works as follows:

1. For each pair of features $i \in F_1, j \in F_2$ at positions $(x_i; y_i), (x_j; y_j)$, calculate the matching error
   $$d_{i,j} = \sum_{-8 \leq \Delta x < 8} \sum_{-8 \leq \Delta y < 8} |I_1(x_i + \Delta x, y_i + \Delta y) - I_2(x_j + \Delta x, y_j + \Delta y)|.$$
   If the Euclidean distance between the feature points exceeds a threshold $t_{d\,max}$, which is set to about 1/3 of the image width, $d_{i,j}$ is set to infinity. The rationale for this threshold will be given shortly.

2. Sort all matching errors obtained in the last step in ascending order.

3. Discard all matches whose matching error exceeds a threshold $t_{e\,max}$.

4. Iterate through all pairs of feature points with increasing matching error. If neither of the two feature points has been assigned yet, establish a correspondence between the two.

Consequently, the matching process is a greedy algorithm, where best fits are assigned first. If there are single features without a counterpart, the probability that they will be assigned erroneously is low since all features that have correct correspondences have been assigned before and, thus, are not available for assignment any more. Moreover, the matching error will be high so that it will usually exceed $t_{d\,max}$.

There is one special case that justifies the introduction of $t_{e\,max}$. Consider a camera pan. Many feature points will disappear at one side of the image and new feature points will appear at the opposite side. After all feature points that appear in both frames are assigned, only those features at the image border remain. Thus, if the matching error is low, correspondences will be established between just to disappear and just appeared features across the complete image, which is obviously not correct. As we know that there will always be a large overlap between successive frames, we also know that the maximum motion cannot be faster than, say, 1/3 of the image width between frames. Hence, we can circumvent the problem by introducing the maximum distance limit $t_{e\,\mathrm{max}}$.

3.3.4 Model Parameter Estimation by Least Squares Regression

Let $\hat{x}_i, \hat{y}_i$ be the measured position of feature i, which had position x_i, y_i in the last frame. The best parameter set θ should minimize the squared Euclidean distance between the measured feature location and the position according to the camera model $e_i^2 = (x'_i - \hat{x}_i)^2 + (y'_i - \hat{y}_i)^2$. Hence, we minimize the sum of errors

$$E = \sum_i (x'_i - \hat{x}_i)^2 + (y'_i - \hat{y}_i)^2 . \tag{23.12}$$

Using the affine motion model with $\theta = \begin{pmatrix} a_{11} & a_{12} & t_x & a_{21} & a_{22} & t_y \end{pmatrix}$, we obtain the solution by setting the partial derivatives $\partial E / \partial \theta_i$ to zero, which leads to the following linear equation system:

$$\begin{pmatrix} \sum_i x_i^2 & \sum_i x_i y_i & \sum_i x_i & 0 & 0 & 0 \\ \sum_i x_i y_i & \sum_i y_i^2 & \sum_i y_i & 0 & 0 & 0 \\ \sum_i x_i & \sum_i y_i & \sum_i 1 & 0 & 0 & 0 \\ 0 & 0 & 0 & \sum_i x_i^2 & \sum_i x_i y_i & \sum_i x_i \\ 0 & 0 & 0 & \sum_i x_i y_i & \sum_i y_i^2 & \sum_i y_i \\ 0 & 0 & 0 & \sum_i x_i & \sum_i y_i & \sum_i 1 \end{pmatrix} \begin{pmatrix} a_{11} \\ a_{12} \\ t_x \\ a_{21} \\ a_{22} \\ t_y \end{pmatrix} = \begin{pmatrix} \sum_i \hat{x}_i x_i \\ \sum_i \hat{x}_i y_i \\ \sum_i \hat{x}_i \\ \sum_i \hat{y}_i x_i \\ \sum_i \hat{y}_i y_i \\ \sum_i \hat{y}_i \end{pmatrix} \tag{23.13}$$

Clearly, this equation system can be solved efficiently by splitting it up into two independent 3×3 systems.

While this direct approach works for the affine motion model, it is not applicable to the perspective model because the perspective model is non-linear. One solution to this problem is that instead of using Euclidean distances

$$\begin{aligned} e_i^2 &= (x'_i - \hat{x}_i)^2 + (y'_i - \hat{y}_i)^2 \\ &= \left(\frac{a_{11}x_i + a_{12}y_i + t_x}{p_x x_i + p_y y_i + 1} - \hat{x}_i \right)^2 + \left(\frac{a_{21}x_i + a_{22}y_i + t_x}{p_x x_i + p_y y_i + 1} - \hat{y}_i \right)^2 \end{aligned} \tag{23.14}$$

to use algebraic distance minimization, where we try to minimize the residuals

$$r_i^2 = (p_x x_i + p_y y_i + 1)^2 \cdot e_i^2$$

$$= (p_x x_i + p_y y_i + 1)^2 \cdot \left((x_i' - \hat{x}_i)^2 + (y_i' - \hat{y}_i)\right)$$

$$= (a_{11} x_i + a_{12} y_i + t_x - p_x x_i \hat{x}_i - p_y y_i \hat{x}_i - \hat{x}_i)^2 \tag{23.15}$$

$$+ (a_{21} x_i + a_{22} y_i + t_y - p_x x_i \hat{y}_i - p_y y_i \hat{y}_i - \hat{y}_i)^2$$

Note that because of the multiplication with $(p_x x_i + p_y y_i + 1)^2$, minimization of r_i^2 is biased, i.e., feature points with larger x_i, y_i have a greater influence on the estimation. This undesirable effect can be slightly reduced by shifting the origin of the coordinate system into the center of the image.

The sum of residuals r_i^2 can be minimized by a least squares (LS) solution of the overdetermined equation system

$$\begin{pmatrix} x_1 & y_1 & 1 & 0 & 0 & 0 & -x_1\hat{x}_1 & -y_1\hat{x}_1 \\ 0 & 0 & 0 & x_1 & y_1 & 1 & -x_1\hat{y}_1 & -y_1\hat{y}_1 \\ x_2 & y_2 & 1 & 0 & 0 & 0 & -x_2\hat{x}_2 & -y_2\hat{x}_2 \\ 0 & 0 & 0 & x_2 & y_2 & 1 & -x_2\hat{y}_2 & -y_2\hat{y}_2 \\ \vdots & \vdots & \vdots & \vdots & \vdots & \vdots & \vdots & \vdots \\ x_n & y_n & 1 & 0 & 0 & 0 & -x_n\hat{x}_n & -y_n\hat{x}_n \\ 0 & 0 & 0 & x_n & y_n & 1 & -x_n\hat{y}_n & -y_n\hat{y}_n \end{pmatrix} \begin{pmatrix} a_{11} \\ a_{12} \\ t_x \\ a_{21} \\ a_{22} \\ t_y \\ p_x \\ p_y \end{pmatrix} = \begin{pmatrix} \hat{x}_1 \\ \hat{y}_1 \\ \hat{x}_2 \\ \hat{y}_2 \\ \vdots \\ \hat{x}_n \\ \hat{y}_n \end{pmatrix}. \tag{23.16}$$

A number of efficient numerical algorithms (e.g., based on singular value decomposition) can be found for this standard problem in [15].

3.3.5 Robust Estimation

The main drawback of the LS method as described above is that *outliers*, i.e., observations that deviate strongly from the expected model, can totally offset the estimation result. Figure 23.8 illustrates this problem. For this purpose, we calculated motion estimates from two frames of a real-world sequence recorded by a panning camera. The motion vectors due to the pan are clearly visible in the background. In addition, the walking person in the foreground causes motion estimates that deviate from those induced by camera motion. Figure 23.8b displays the global motion field estimated by the LS method. Obviously, the mixture of background and object motion did not lead to satisfactory results. Instead, the camera parameters were optimized to approximate both, background and object motion.

In order to estimate camera parameters reliably from a mixture of background and object motion, one has to distinguish between observations belonging to the global motion model (*inliers*) and those resulting from object motion (*outliers*).

Figure 23.8c displays the same motion estimates as before. This time, however, only a subset of vectors (drawn in black) is used for the LS estimation. The obtained camera motion model depicted in Figure 23.8d is estimated exclusively from inliers and, thus, clearly approximates the panning operation.

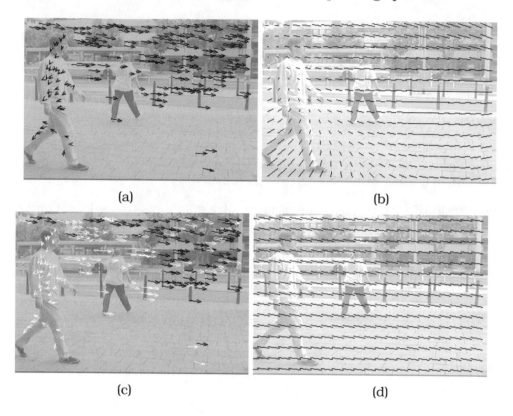

(a)

(b)

(c)

(d)

Figure 23.8 Estimation of camera parameters. (a) motion estimates, (b) estimation by least squares, (c) motion estimates (white vectors are excluded from the estimation), (d) estimation by least-trimmed squares regression.

To eliminate outliers from the estimation process, one can apply *robust regression methods*. Widely used robust regression methods comprise *random sample concensus* (RANSAC) [4], *least median of squares* (LMedS), and *least-trimmed squares* (LTS) regression [17]. Those methods are able to calculate the parameters of our regression problem even when a large fraction of the data consists of outliers.

The basic steps are similar for all of those methods. First of all, a random subset of the data set is drawn and the model parameters are calculated from this subset. The subset size p equals the number of parameters to be estimated. For instance, calculating the 8 parameters of the perspective model requires 4 feature correspondences, each introducing two constraints. A randomly drawn subset containing outliers results in a poor parameter estimation. As a remedy, N subsets are drawn and the parameters are calculated for each of them. By choosing N sufficiently high, one can assure up to a certain probability that at least one "good" subset, i.e., a sample without outliers, was drawn. The probability for drawing at least one subset without outliers is given by $P(\varepsilon, p, N) := 1 - (1 - (1 - \varepsilon)^p)^N$ where ε denotes the fraction of outliers. Conversely, the required number of samples to ensure a high confidence can be determined from this equation.

For each parameter set calculated from the subsets, the model error is measured. Finally, the model which fits the given observation best is retained. The main difference between the three methods RANSAC, LMedS, and LTS is the measure for the model error. We have chosen to use the LTS estimator because it does not require a selection threshold as RANSAC and it is computationally more efficient than LmedS [18]. The LTS estimator can be written as

$$\min_{\hat{\theta}} \sum_{i=1}^{h} (r^2)_{i:n} \tag{23.17}$$

where n is the input data size and $(r^2)_{1:n} \leq \cdots \leq (r^2)_{n:n}$ denote the squared residuals given in ascending order. Similar to the LS method, the sum of squared residuals is minimized. However, only the h smallest squared residuals are taken into account. Setting h approximately to $n/2$ eliminates half of the cases from the summation; thus, the method can cope with about 50% outliers.

Since the computation of the LTS regression coefficients is not straightforward, the basic steps are summarized in the following. For a detailed treatment and a fast implementation called FAST-LTS, we refer to [17,18].

To evaluate Equation 23.17, an initial estimate of the regression parameters is required. For this purpose, a subset S_0 of size p is drawn randomly from the data set. Solving a linear system created by S_0 yields an initial estimate of the parameters denoted by $\hat{\theta}_0$. Using $\hat{\theta}_0$ we can calculate the residuals $r_{0i}, i = 1, \ldots, n$ for all n cases in the data set.

The estimation accuracy is increased by applying several *compaction steps*. Sorting r_{0i} by absolute value yields a subset H_0 containing the h cases with the least absolute residuals. Furthermore, a first quality of fit measure can be calculated from this subset as $Q_0 := \sum_{i \in H_0} (r_0^2)_{i:n}$. Then, based on H_0, a least squares fit is calculated yielding a new parameter estimate $\hat{\theta}_1$. Again, the residuals r_{1i} are calculated and sorted by absolute value in ascending order. This yields a subset H_1 which contains the h cases that possess the least absolute residuals with respect to the parameter set $\hat{\theta}_1$. The quality of fit for H_1 is given by $Q_1 := \sum_{i \in H_1} (r_1^2)_{i:n}$. Due to the properties of LS regression, it can be assured that $Q_1 \leq Q_0$. Thus, by iterating the above procedure until $Q_k = Q_{k-1}$, the optimal parameter set can be determined for a given initial subset.

3.4 DIRECT METHODS FOR MOTION ESTIMATION

Although motion estimation based on feature correspondences is robust and can handle large motions, it is not accurate enough to achieve sub-pixel registration. Particularly in the process of building the background mosaic, small errors quickly sum up and result in clear misalignments. Given a good initial motion estimate from the feature matching approach, a very accurate registration can be calculated using *direct methods*. These techniques minimize the motion compensated image difference given as

$$\min_{\theta} E_{\theta} = \min_{\theta} \sum_{i=(x,y)} \gamma(e_i) = \min_{\theta} \sum_{i=(x,y)} \gamma(I'(x',y') - I(x,y)). \qquad (23.18)$$

$I(x,y)$ denotes the image brightness at position (x,y) and $I'(x',y')$ denotes the brightness at the corresponding pixel (according to the selected motion model) in the other image. For the moment, we assume that $\gamma(e) = e^2$, i.e., we use the sum of squared differences as difference measure. Later, we will replace this by a robust M-estimator.

Minimization of E with respect to the motion model parameter vector θ is a difficult problem and can only be tackled with gradient descent techniques. We are using a Levenberg-Marquardt [9,11,15] minimizer because of its stability and speed of convergence. The algorithm is a combination of a pure gradient descent and a multi-dimensional Newton algorithm.

3.4.1 Levenberg-Marquardt-Minimization

Starting with an estimation $\theta^{(i)}$, the gradient descent process determines the next estimation $\theta^{(i+1)}$ by taking a small step down the gradient

$$\theta^{(i+1)} = \theta^{(i)} - \alpha \cdot \nabla E_{\theta^{(i)}}, \qquad (23.19)$$

where α is a small constant, determining the *step size*. One problem of pure gradient descent is the choice of a good α since small values result in a slow convergence rate, while large values may lead far away from the minimum.

The second method is the Newton algorithm. This algorithm assumes that if $\theta^{(i)}$ is near a minimum, the function to be minimized can often be approximated by a quadratic form Q as

$$E_{\theta} \approx Q_{\theta} = E_{\theta^{(i)}} + \theta \cdot \nabla E_{\theta^{(i)}} + \frac{1}{2} \theta \cdot \nabla^2 E_{\theta^{(i)}} \cdot \theta^T, \qquad (23.20)$$

where $\nabla^2 E$ denotes the Hessian matrix of E.

If the Hessian matrix is positive definite,

$$\theta^{(\min)} = \arg \min_{\theta} Q_{\theta} \qquad \text{iff} \quad \nabla Q_{\theta^{(\min)}} = 0. \qquad (23.21)$$

Hence, because of

$$\nabla Q_{\theta^{(i+1)}} = \nabla E_{\theta^{(i)}} + \nabla^2 E_{\theta^{(i)}} \cdot (\theta^{(i+1)} - \theta^{(i)}) \qquad (23.22)$$

we can directly jump to the minimum of Q by setting

$$\theta^{(i+1)} = \theta^{(i)} - \nabla^2 E_{\theta^{(i)}}^{-1} \cdot \nabla E_{\theta^{(i)}} . \qquad (23.23)$$

Note the similar structure of this equation compared to Equation 23.19. Instead of taking the inverse of the Hessian, the step $\delta\theta^{(i)} = \theta^{(i+1)} - \theta^{(i)}$ can also be computed by solving the linear equation system

$$\nabla^2 E \cdot \delta\theta^{(i)} = -\nabla E . \qquad (23.24)$$

The Levenberg-Marquardt algorithm solves two problems at once. First, the factor α in Equation 23.19 is chosen automatically, and second, the algorithm combines steepest descent and Newton minimization into a unified framework.

If we are comparing the units of $\delta\theta^{(i)}$ with those in the Hessian, we can see that only the diagonal entries of the Hessian provide some information about scale. So, we set the step size (independently for each component θ_k) as

$$\alpha_k = \frac{1}{\lambda (\nabla^2 E)_{kk}} . \qquad (23.25)$$

λ is a new scaling factor which is controlled by the algorithm. To combine steepest descent with the Newton algorithm, [9,11] propose to define a new matrix D with

$$d_{jk} = (\nabla^2 E)_{jk} \qquad \text{for } j \neq k$$

$$d_{kk} = (\nabla^2 E)_{kk} \cdot (1 + \lambda) \qquad \text{(elements on the diagonal)}$$

Replacing the Hessian of Equation 23.24 with D, we get

$$D \cdot \delta\theta^{(i)} = -\nabla E . \qquad (23.26)$$

Note that for $\lambda = 0$ Equation 23.26 reduces to Equation 23.24 (i.e., the Newton algorithm), while for large λ the matrix D becomes diagonal dominant and the algorithm thus behaves like a steepest descent algorithm. The Levenberg-Marquardt minimization algorithm uses λ to control the minimization process and works as follows:

1. Choose an initial λ (e.g., $\lambda = 0.001$).

2. Solve Equation 23.26 to get the parameter update vector $\delta\theta^{(i)}$.

3. If $E_{\theta^{(i)} + \delta\theta^{(i)}} \geq E_{\theta^{(i)}}$, the update does not improve the solution. Hence, we increase λ by a factor of 10 (to reduce the step size) and go back to Step 2.

4. If $E_{\theta^{(i)} + \delta\theta^{(i)}} < E_{\theta^{(i)}}$, the update improves the solution. Hence, we set $\theta^{(i+1)} = \theta^{(i)} + \delta\theta^{(i)}$ and decrease λ by a factor of 10.

5. When λ exceeds a high threshold, even the last small steps did not improve the solution and we stop.

Adapting this technique to our motion estimation problem, we must determine the Hessian matrix and gradient vector for a given parameter estimate θ. In the following, we assume the perspective motion model from Equation 23.4. The gradient vector can be determined easily from

$$\frac{\partial e_i}{\partial \theta_1} = \frac{\partial e_i}{\partial a_{11}} = \frac{\partial I'}{\partial x'} \frac{x_i}{Z_i}$$

$$\frac{\partial e_i}{\partial \theta_2} = \frac{\partial e_i}{\partial a_{12}} = \frac{\partial I'}{\partial x'} \frac{y_i}{Z_i}$$

$$\frac{\partial e_i}{\partial \theta_3} = \frac{\partial e_i}{\partial t_x} = \frac{\partial I'}{\partial x'} Z_i^{-1} \qquad\qquad (23.27)$$

$$\vdots \qquad\qquad \vdots$$

$$\frac{\partial e_i}{\partial \theta_7} = \frac{\partial e_i}{\partial p_x} = -\frac{y_i}{Z_i}\left(x'_i \frac{\partial I'}{\partial x'} + y'_i \frac{\partial I'}{\partial y'} \right)$$

$$\vdots \qquad\qquad \vdots$$

with the shortcut $Z_i = p_x x_i + p_y y_i + 1$. Hence, with $\gamma(e) = e^2$ the gradient vector is simply

$$\nabla E = 2 \cdot \sum_i e_i \cdot \left(\frac{\partial e_i}{\partial \theta_1} \quad \cdots \quad \frac{\partial e_i}{\partial \theta_8} \right). \qquad\qquad (23.28)$$

We simplify the computation of the Hessian by ignoring the second order derivative terms:

$$\frac{\partial^2 e^2_i}{\partial \theta_j \partial \theta_k} = 2 \cdot \left(\frac{\partial e_i}{\partial \theta_j} \cdot \frac{\partial e_i}{\partial \theta_k} + e_i \frac{\partial^2 e_i}{\partial \theta_j \partial \theta_k} \right) \approx 2 \cdot \left(\frac{\partial e_i}{\partial \theta_j} \cdot \frac{\partial e_i}{\partial \theta_k} \right). \qquad\qquad (23.29)$$

Hence, we get

$$(\nabla^2 E)_{jk} = 2 \cdot \sum_i \frac{\partial e_i}{\partial \theta_j} \cdot \frac{\partial e_i}{\partial \theta_k}. \qquad\qquad (23.30)$$

3.4.2 Applying an M-estimator

The algorithm described above assumes that the whole image moves according to the estimated motion model. However, in our application, this is not the case since foreground objects generally move differently. This introduces large matching errors in areas of the foreground objects. The consequence is that this mismatch distorts the estimation because the algorithm also tries to minimize the matching error in the foreground region. As the true object position is not known yet, it is not possible to exclude the foreground regions from the estimation process. This problem can be alleviated by using a limited error function for $\gamma(x)$ instead of the squared error. For simplicity, we use

$$\gamma(e) = \begin{cases} e^2 & \text{for } |e| < t \\ t^2 & \text{else.} \end{cases}$$

(23.31)

Introducing this error function into Equation 23.18 and computing the gradient and Hessian is particularly simple as

$$\frac{\partial \gamma(e_i)}{\partial \theta_i} = \begin{cases} \dfrac{\partial e_i^2}{\partial \theta_i} & \text{if } |e_i| < t \\ 0 & \text{else .} \end{cases}$$

(23.32)

Figure 23.9 shows two difference frames, the first using squared error as matching function and the second using the robust distance function. It can be seen that the registration error in the background region is smaller for the robust distance function.

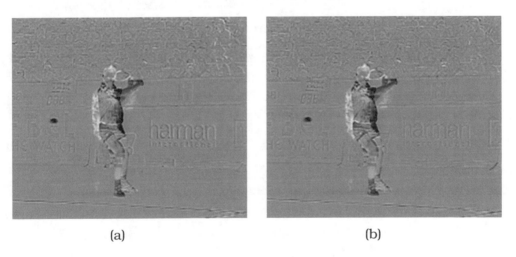

(a)   (b)

Figure 23.9: Difference frame after Levenberg-Marquardt minimization. (a) shows residuals using squared differences as error function, (b) shows residuals with saturated squared differences. It is visible that the robust estimation achieves a better compensation. Especially note the text in the right part of the image.

4. DETERMINING OBJECT MASKS

The principle of our segmentation algorithm is to compute the difference of the current frame to a scene background image which does not contain any foreground objects. The background image is automatically constructed from the sequence such that the background adapts itself to changes or varying illumination. Even if the background is never visible without any foreground objects, the algorithm is capable to artificially recreate it.

Since the difference between input image and background contains much error due to camera noise or small parts of the object having the same color as the background, a regularization of the object shape is applied to the difference frame.

Figure 23.10 Reconstruction of background based on compensation of camera motion between video frames. The original video frames are indicated with borders.

4.1 BACKGROUND RECONSTRUCTION

The motion estimation step provides the motion model $\theta_{j,j+1}$ between consecutive frames j and $j+1$. By considering the transitive closure as the concatenation of motion transformations, we can define all $\theta_{j,k}$ between arbitrary frames j, k. If we fix the first frame as the reference coordinate system for the background reconstruction, we can add frame j to the background by applying the transformation $\theta_{1,j}$. To prevent the drift from slight errors in the motion estimation step, the direct estimation step is not applied to successive frames but to the input frame with respect to the current background mosaic. Figure 23.10 shows how input frames are assembled into a combined mosaic.

In general, the input video will contain foreground objects in most of the frames. However, it is important that the reconstructed background does not contain these objects. As it is not *apriori* clear which parts are foreground and which are background, we define everything as background that is stable for at least b frames. The reconstruction algorithm stores the last $2b$ background mosaics obtained so far. The reconstructed background image is then determined by applying a temporal median filter [12,21] over these pictures (cf. Figure 23.11). Clearly, if at least b pictures have nearly the same color at a pixel, this value will be set in the background reconstruction.

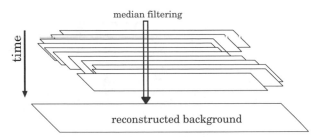

Figure 23.11 Aligned input frames are stacked and a pixel-wise median filter is applied in the temporal direction to remove foreground objects.

This approach works well if the objects are moving in the scene. If they stay too long at the same position, they will eventually become background. A sample reconstructed background from the "stefan" sequence can be seen in Figure 23.20.

4.2 CHANGE DETECTION MASKS

The principle of our segmentation algorithm is to calculate the *change detection mask* (CDM) between the background image and the input frames. In the area where the foreground object is located, the difference between background and input frame will be high. Note that the approach of taking the difference to a reconstructed background has several advantages over taking differences between successive frames:

1. The segmentation boundaries are more exact. If differences are computed between successive frames, not only the new position of an object will have large differences, but also the uncovered background areas. This results in annoying artifacts because fast moving objects are visible twice.
2. Objects that do not move for some time or that are only moving slowly can not be segmented. Moreover, a slowly moving region with almost uniform color would only show differences at the edges in successive frames.
3. The reconstructed background can be used for object-based video coding algorithms like MPEG-4 where the background can be transmitted independently (as a so called "background-sprite"), which reduces the required bit-rate as only the foreground objects have to be transmitted.

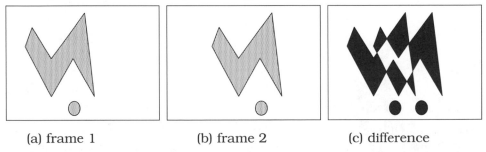

(a) frame 1   (b) frame 2   (c) difference

Figure 23.12 Computing the difference between successive frames results in unwanted artifacts. The first two pictures show two input frames with foreground objects. The right picture shows the difference. Two kinds of artifacts can be observed. First, the circle appears twice since the algorithm cannot distinguish between appearing and disappearing. Second, part of the inner area of the polygon is not filled because the pixels in this area do not change their brightness.

4.3 IMPROVED CHANGE DETECTION BASED ON THE SSD-3 MEASURE

With standard change detection based on squared or absolute differences, a typical artifact can be observed. If the images contain sharp edges or fine texture, these structures usually can not be cancelled completely because of improper filtering and aliasing in the image acquisition process. Hence, fine texture and sharp edges are often accidentally detected as moving objects.

One technique to reduce this effect is to use the *sum of spatial distances* (SSD-3) measure [2] to compute the difference frame. The principle of this measure is to calculate the distance that a pixel has to be moved to reach a pixel of similar brightness in the reference frame. In the one dimensional case, it is defined as

$$d_{SSD-3} = \min(d_{-1}, d_0, d_1) \qquad\qquad (23.33)$$

with

$$d_i = \left| \frac{I(x) - I'(\lfloor x' \rfloor + i)}{I'(\lfloor x' \rfloor + i + 1) - I'(\lfloor x' \rfloor + i)} - (x' - (\lfloor x' \rfloor + i)) \right|. \qquad (23.34)$$

For the two-dimensional case, this measure is computed independently for the horizontal and vertical direction and the minimum is taken. For an in-depth explanation of this measure, see [2]. The difference frames obtained with this measure compared to standard squared error is depicted in Fig. 23.13.

(a) squared error   (b) SSD-3

Figure 23.13 Difference frames using squared error and SSD-3. Note that SSD-3 shows considerably less errors at edges caused by aliasing in the sub-sampling process.

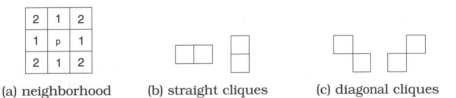

2	1	2
1	p	1
2	1	2

(a) neighborhood   (b) straight cliques   (c) diagonal cliques

Figure 23.14 Definition of pixel neighborhood. Picture (a) shows the two classes of pixel neighbors; straight (1) and diagonal (2). These two classes are used to define the second order cliques. Straight cliques (b) and diagonal cliques (c).

4.4 SHAPE REGULARIZATION USING MARKOV RANDOM FIELDS

If the generation of a binary object mask from the difference frame is done pixel by pixel with a fixed threshold, we have to face a lot of wrongly classified pixels (cf. Figure 23.15a). Since most real objects have smooth boundaries, we improve the segmentation by a shape regularization, which is done using a Markov random field (MRF) model.

The formal definition of our segmentation problem is that we want to assign a label out of the label set $L=\{$ background, foreground $\}$ to each pixel position p.

For each pixel position there is a random variable F_p with values $f_p \in L$. The probability that pixel p is assigned label f_p is denoted as $P(F_p = f_p)$. A random field is Markovian if the probability of a label assignment for a pixel is only dependent on the neighborhood of this pixel:

$$P(F_p = f_p \,|\, F_{I-\{p\}}) = P(F_p = f_p \,|\, F_{N(p)}),\qquad(23.35)$$

where $F_{I-\{p\}}$ denotes the label configuration of the whole image except pixel p and $F_{N(p)}$ the configuration in a neighborhood of pixel p. We define the neighborhood of p as the 8-neighborhood, while differentiating between straight and diagonal neighbors (see Figure 23.14).

Since the probabilities $P(F_p = f_p \,|\, F_{N(p)})$ are usually hard to define, Markov random fields are often modelled as Gibbs random fields (GRF). It can be shown that both descriptions are equivalent [10]. A GRF is defined through the total label configuration probability $P(f)$ as

$$P(f) = Z^{-1} \cdot e^{-\frac{1}{T}U(f)}\qquad(23.36)$$

where Z is a normalization constant to ensure that $\sum_f P(f) = 1$. In the following, we will always set the temperature parameter $T = 1$. $U(f)$ is the *energy function*, which is defined as

$$U(f) = \sum_{c \in C} V_c(f),\qquad(23.37)$$

in which the sum is over all cliques in the image and $V_c(f)$ is a clique potential. Higher clique potentials result in lower probabilities for this clique configuration. Cliques are subsets of related pixel positions in the image. In our application, we are using an Auto-Logistic model, which only uses cliques of single order (the pixels themselves) and of second order (cf. Figure 23.14). Thus, the energy function can be written as

$$U(f) = \sum_p V_1(f_p) + \sum_p \sum_{p' \in N(p)} V_2(f_p, f_{p'}) . \tag{23.38}$$

First order clique potentials $V_1(p)$ are set according to the difference frame information, i.e., how probable a pixel p belongs to foreground objects given its difference frame value $d(p)$.

$$V_1(f_p) = \begin{cases} \beta \cdot e^{-d(p)^2} & \text{for } f_p = \text{foreground, and} \\ \beta \cdot (1.0 - e^{-d(p)^2}) & \text{for } f_p = \text{background.} \end{cases} \tag{23.39}$$

The second order clique potentials are set such that smooth regions are preferred, i.e., cliques that contain different labels are assigned more energy. More specifically, we use

$$V_2(f_p, f_{p'}) = \begin{cases} -\mu & \text{if } f_p = f_{p'} \\ \mu & \text{if } f_p \neq f_{p'} \end{cases} . \tag{23.40}$$

The parameter μ is set differently for the two types of cliques with lower values for diagonal cliques as the corresponding pixels are farther away. The label configuration that maximizes the total field probability (Eq. 36) is obtained through an iterative Gibbs sampling algorithm [10]. Figure 23.15b shows the segmentation mask obtained with our MRF model compared to a pixel based classification. Applying MRF-based classification to each difference frame yields binary object masks for the entire video.

(a) per-pixel classification   (b) MRF-based classification

Figure 23.15 Segmentation results for per-pixel decision between foreground and background object and MRF based segmentation.

5. VIDEO OBJECT CLASSIFICATION

The automatic segmentation procedure as described above provides object masks for each frame of the video. Based on these masks, further high-level processing steps are possible. To enable semantic scene analysis, it is required to assign *object classes* (e.g., persons, animals, cars) to the different masks. Furthermore, *object behavior* (e.g., a person is sitting, stands up, walks away) can be described by observing the object over time.

In our approach, object classification is based on comparing silhouettes of automatically segmented objects to prototypical objects stored in a database. Each real-world object is represented by a collection of two-dimensional projections (or object views). The silhouette of each projection is analyzed by the curvature scale space (CSS) technique. This technique provides a compact representation that is well suited for indexing and retrieval.

Object behavior is derived by observing the transitions between object classes over time and selecting the most probable transition sequence. Since frequent changes of the object class are unlikely, occasional false classifications resulting from errors which occurred in previous processing steps are removed.

5.1 REPRESENTING SILHOUETTES USING CSS

The curvature scale space technique [1,13,14] is based on the idea of curve evolution, i.e., basically the deformation of a curve over time. A CSS image provides a multi-scale representation of the curvature zero crossings of a closed planar contour.

Consider a closed planar curve $\Gamma(u)$ representing an object view,

$$\Gamma(u) = \{(x(u), y(u)) | u \in [0,1]\},$$
(23.41)

with the normalized arc length parameter u. The curve is smoothed by a one-dimensional Gaussian kernel $g(u,\sigma)$ of width σ. The deformation of the closed planar curve is represented by

$$\Gamma(u,\sigma) = \{(X(u,\sigma), Y(u,\sigma)) | u \in [0,1]\},$$
(23.42)

where $X(u,\sigma)$ and $Y(u,\sigma)$ denote the components $x(u)$ and $y(u)$ after convolution with $g(u,\sigma)$. Varying σ is equivalent to choosing a fixed σ' and applying the convolution iteratively.

The curvature $\kappa(u,\sigma)$ of an evolved curve can be computed using the derivatives $X_u(u,\sigma)$, $X_{uu}(u,\sigma)$, $Y_u(u,\sigma)$, and $Y_{uu}(u,\sigma)$ as

$$\kappa(u,\sigma) = \frac{X_u(u,\sigma) \cdot Y_{uu}(u,\sigma) - X_{uu}(u,\sigma) \cdot Y_u(u,\sigma)}{(X_u(u,\sigma)^2 + Y_u(u,\sigma)^2)^{3/2}}.$$
(23.43)

A CSS image $I(u,\sigma)$ is defined by

$$I(u,\sigma) = \{(u,\sigma) | \kappa(u,\sigma) = 0\}.$$
(23.44)

It contains the zero crossings of the curvature with respect to their position on the contour and the width of the Gaussian kernel (or the number of iterations, see Figure 23.16). During the deformation process, zero crossings vanish as transitions between contour segments of different curvature are smoothed out. Consequently, after a certain number of iterations, inflection points disappear and the shape of the closed curve becomes convex. Note that due to the dependence on curvature zero crossings, convex object views cannot be distinguished by the CSS technique.

Significant contour properties that stay intact for a large number of iterations result in high peaks in the CSS image. Segments with rapidly changing curvatures caused by noise produce only small local maxima. In many cases, the peaks in the CSS image provide a robust and compact representation of an object view's contour. Note that a rotation of an object view on the image plane can be accomplished by shifting the CSS image left or right in a horizontal direction. Furthermore, a mirrored object view can be represented by mirroring the CSS image.

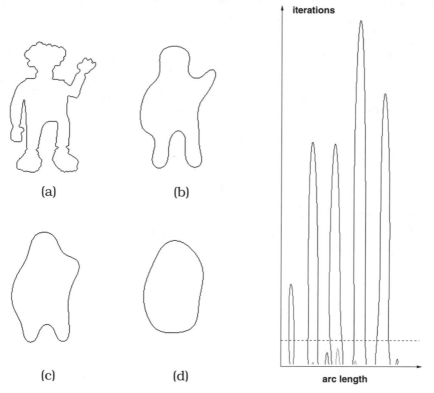

Figure 23.16 Construction of the CSS image. Left: Object view (a) and iteratively smoothed contour (b)-(d). Right: Resulting CSS image.

Each peak in the CSS image is represented by three values, the position and height of the peak and the width at the bottom of the arc-shaped contour. The width specifies the normalized arc length distance of the two curvature zero crossings enframing the contour segment represented by the peak in the CSS image [16].

It is sufficient to extract the significant maxima (above a certain noise level) from the CSS image. For instance, in the example depicted in Figure 23.16, only five data triples remain and have to be stored after a small number of iterations. The database described in the following section stores up to 10 significant data triples for each silhouette.

5.2 BUILDING THE DATABASE

The orientation of an object and the position of the camera have great impact on the silhouette of the object. Therefore, to enable reliable recognition, different views of an object have to be stored in the database.

Rigid objects can be represented by a small number of different views, e.g., for a car, the most relevant views are frontal views, side views, and views where frontal and side parts of the car are visible.

For non-rigid objects, more views have to be stored in the database. For instance, the contour of a walking person in a video changes significantly from frame to frame.

Similar views of one type of object are aggregated to one object class. Our database stores 275 silhouettes collected from a clip art library and from real-world videos. The largest number of images show people (124 images), animals (67 images), and cars (48 images). Based on behavior, the object class *people* is subdivided into the following object classes: standing, sitting, standing up, sitting down, walking, and turning around.

5.3 OBJECT MATCHING

Each automatically segmented object view is compared to all object views in the database. In a first step, the aspect ratio, defined as quotient of object width and height, is calculated. For two object views with significantly different aspect ratios, no matching is carried out. If both aspect ratios are similar, the peaks in the CSS images of the two object views are compared. The basic steps of the matching procedure are summarized in the following [16]:

- First, both CSS representations have to be aligned. For this purpose it might be necessary to rotate or mirror one of them. As mentioned above, shifting the CSS image corresponds to a rotation of the original object view. To align both representations, one of the CSS images is shifted so that the highest peak in both CSS images is at the same position.

- A *matching peak* is determined for each peak in a given CSS representation. Two peaks match if their height, position, and width are within a certain range.

- If a matching peak is found, the Euclidean distance of the peaks in the CSS image is calculated and added to a distance measure. If no matching peak can be determined, the height of the peak is multiplied by a penalty factor and added to the total difference.

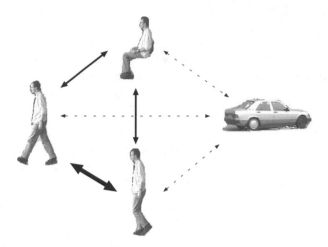

Figure 23.17 Weights for transitions between object classes. Thicker arrows represent more probable transitions.

5.4 CLASSIFYING OBJECT BEHAVIOR

The matching technique described above calculates the best database match for each automatically segmented object view. Since the database entries are labeled with an appropriate class name, the video object can be classified accordingly. The object class assigned to a segmented object mask can change over time because of object deformations or matching errors. Since the probability of changing from one object class to another depends on the respective classes, we assign additionally matching costs for each class change.

Let $d_k(i)$ denote the CSS distance between an input object mask at frame i and object class k from the database. Furthermore, let $w_{k,l}$ denote the transition cost from class k to l (cf. Figure 23.17). Then, we seek the classification vector c, assigning an object class to each input object mask, which minimizes

$$\min_c \sum_i d_{c_i}(i) + w_{c_i, c_{i-1}} \ . \tag{23.45}$$

This optimization problem can be solved by a shortest path search as depicted in Figure 23.18. With respect to the figure, $d_k(i)$ corresponds to costs assigned to the nodes and $w_{k,l}$ corresponds to costs at the edges. The optimal path can be computed efficiently using a dynamic programming algorithm. The object behavior can be extracted easily from the nodes along the minimum cost path.

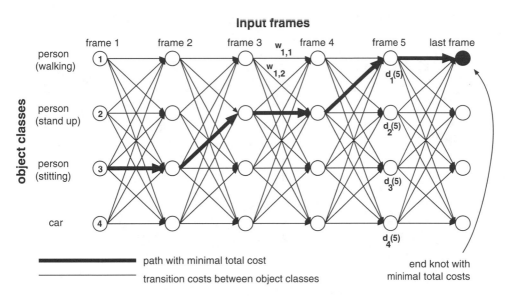

Figure 23.18 Extraction of object behavior.

6. SYSTEM IMPLEMENTATION AND RESULTS

As outlined above, our object classification system consists of a motion-based segmentation module and a shape-based classification module. We start by discussing experimental results achieved by our segmentation module. For this purpose, the segmentation algorithm was applied to two real-world sequences, namely the "stefan" sequence used throughout this chapter and a "road" sequence recorded by a handheld camera. Figures 23.19 and 23.21 depict some of the results. In Figure 23.20, the reconstructed background of the "stefan" sequence is displayed. We observe that the segmentation module separates the moving objects from the background very well. In the case of the "stefan" sequence, some moving parts in the audience are detected. In the "road" sequence the cars (and a pedestrian) are extracted very precisely.

In order to classify the objects resulting from the segmentation process, the classification algorithms are applied to their shapes. Figure 23.22 displays some segmentation masks obtained by our automatic segmentation and the respective best match calculated from the shape database. In three cases the classification is successful and yields a reasonable match. In addition, we observe one mismatch which is due to a segmentation error and the fact that the database did not contain an appropriate representation of the running tennis player.

Finally, let us consider the extraction of object behavior. In Figure 23.23, the left image displays a training sequence used to define object prototypes for a specific object behavior in the database. The right image depicts a test sequence with the automatically assigned class labels. The different stages of the object behavior were determined from the shortest path calculated from the CSS matching results and the object frames were selected from the middle between class transitions.

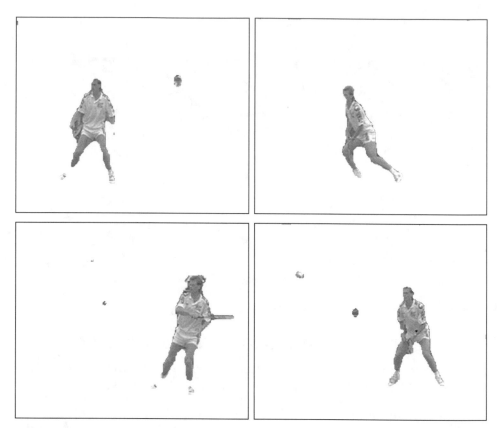

Figure 23.19 Segmentation results of "stefan" test sequence (frames 40, 80, 120, 160).

Figure 23.20 Reconstructed background from "stefan" sequence. Note that the player is not visible even though there is no input video frame without the player.

Figure 23.21 Segmentation results of "road" test sequence (frames 20, 40, 70, 75).

Figure 23.22 CSS matching results for selected frames of the "stefan" sequence.

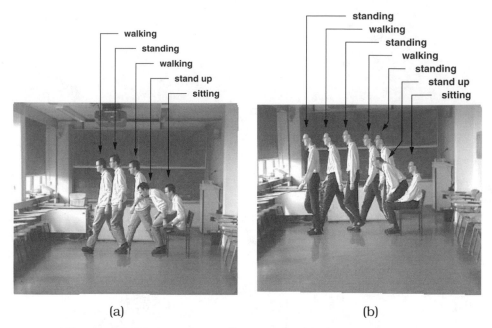

(a)   (b)

Figure 23.23 Automatically extracted behavior description.

REFERENCES

[1]   S. Abbasi and F. Mokhtarian, Shape Similarity Retrieval Under Affine Transform: Application to Multi-view Object Representation and Recognition, in Proc. International Conference on Computer Vision, pp. 450-455, IEEE, 1999.

[2]   D. Farin and P. H. N. de With, A New Similarity Measure for Sub-pixel Accurate Motion Analysis in Object-based Coding, in Proc. of the 5th World Multi-Conference on Systemics, Cybernetics and Informatics (SCI), pp. 244-249, July 2001.

[3]   O. D. Faugeras, *Three-dimensional Computer Vision: A Geometric Viewpoint*, MIT Press, Cambridge, MA, 1999.

[4]   M. Fischler and R. Bolles, Random sample concensus: A paradigm for model fitting with applications to image analysis and automated cartography, *Communications ACM*, 24(6):381-395, 1981.

[5]   C. Harris and M. Stephens, A Combined Corner and Edge Detector, in Proc. Alvey Vision Conference, pp. 147-151, 1988.

[6]   R. Hartley and A. Zisserman, *Multiple View Geometry in Computer Vision*, Cambridge University Press, Cambridge, 2001.

[7]   B. K. P. Horn, *Robot Vision*, MIT Press, Cambridge, MA, 1986.

[8]   M. Irani and P. Anandan, About Direct Methods, in Vision Algorithms: Theory and Practice, International Workshop on Vision Algorithms, pp. 267-277, 1999.

[9]   K. Levenberg, A method for the solution of certain problems in least squares, *Quart. Appl. Math.*, 2:164-168, 1944.

[10] S. Z. Li, *Markov Random Field Modeling in Computer Vision. Artificial Intelligence*, Springer-Verlag, Tokyo, 1995.

[11] D. Marquardt, An algorithm for least-squares estimation of nonlinear parameters, *SIAM J. Appl. Math.*, 11:431-441, 1963.

[12] M. Massey and W. Bender, Salient stills: Process and practice, *IBM Systems Journal*, 35(3/4):557-573, 1996.

[13] F. Mokhtarian, S. Abbasi, and J. Kittler, Efficient and Robust Retrieval by Shape Content Through Curvature Scale Space, in Proc. International Workshop on Image DataBases and MultiMedia Search, pp. 35-42, 1996.

[14] F. Mokhtarian, S. Abbasi, and J. Kittler, Robust and Efficient Shape Indexing Through Curvature Scale Space, in British Machine Vision Conference, 1996.

[15] W. H. Press, S. A. Teukolsky, W. T. Vetterling, and B. P. Flannery, *Numerical Recipes in C: The Art of Scientific Computing*, Cambridge University Press, New York, 1992.

[16] S. Richter, G. Kühne, and O. Schuster, Contour-bascd Classification of Video Objects, in Proc. of SPIE, Storage and Retrieval for Media Databases, Vol. 4315, pp. 608-618, 2001.

[17] P. J. Roussccuw and A. M. Lcroy, *Robust Regression and Outlier Detection*, John Wiley, New York, 1987.

[18] P. J. Rousseeuw and K. Van Driesen, Computing LTS regression for Large Data Sets, *Institute of Mathematical Statistics Bulletin*, 27(6), November/December 1998.

[19] C. Schmid, R. Mohr, and C. Bauckhage, Evaluation of interest point detectors, *International Journal of Computer Vision*, 37(2):151-172, June 2000.

[20] R. Szeliski, Image mosaicing for tele-reality applications, Technical Report 94/2, Digital Equipment Corporation, Cambridge Research, June 1994.

[21] L. Teodosio and W. Bender, Salient video stills: content and context preserved, *ACM Multimedia*, 1993.

[22] P. H. S. Torr and A. Zisserman, Feature Based Methods for Structure and Motion Estimation, Vision Algorithms: Theory and Practice, International Workshop on Vision Algorithms, 278-294, 1999.

24

A WEB-BASED VIDEO RETRIEVAL SYSTEM: ARCHITECTURE, SEMANTIC EXTRACTION, AND EXPERIMENTAL DEVELOPMENT[1]

Qing Li

Department of Computer Engineering and Information Technology
City University of Hong Kong
Tat Chee Avenue, Kowloon, Hong Kong
`itqli@cityu.edu.hk`

H. Lilian Tang

Department of Computing
University of Surrey, UK
`h.tang@surrey.ac.uk`

Horace H. S. Ip

Centre for Innovative Application of Internet and Multimedia
*Technologies (**AIM**tech)*
City University of Hong Kong
Tat Chee Avenue, Kowloon, Hong Kong
`cship@cityu.edu.hk`

Shermann S. M. Chan

Department of Computer Science
City University of Hong Kong
Tat Chee Avenue, Kowloon, Hong Kong
`shermann@cs.cityu.edu.hk`

1. INTRODUCTION

A current important trend in multimedia information management is towards web-based/enabled multimedia search and management systems. Video is a rich and colorful media widely used in many of our daily life applications such as

---

[1] This research has been supported, in part, by a Strategic Research Grant from CityU (Project no. 7001073) and RGC grant No: CityU1150/01E.

education, entertainment, news spreading, etc. Digital videos have diverse sources of origin such as cassette recorder, tape recorder, home video camera, DVD and Internet. Expressiveness of video documents decides their dominative position in the next-generation multimedia information systems. Unlike traditional/static types of data, digital video can provide more effective dissemination of information for its rich content. Collectively, a (digital) video can have several information descriptors: (1) meta data - the actual video frame stream, including its encoding scheme and frame rate; (2) media data - the information about the characteristics of video content, such as visual feature, scene structure and spatio-temporal features; (3) semantic data - the mapping between media data and their high level meanings such as text annotation or concepts relevant to the content of the video, obtained by manual or automatic understanding.

Video meta data is created independently from how its contents are described and how its database structure is organized later. It is thus natural to define "video" and other meaningful constructs such as "scene," "frame" as objects corresponding to their respective inherent semantic and visual contents. Meaningful video scenes are identified and associated with their description data incrementally. But the semantic gap between the user understanding and video content remains a big problem. Depending on the user's viewpoint, the same video/scene may be given different descriptions or interpretations. It is therefore extremely difficult (if not impossible) to describe the whole contents of a video, due to the difficulties in automatic detecting salient features from and interpreting the visual content.

In order to develop an effective web-based video retrieval system, one should go beyond the traditional query-based or purely content-based retrieval (CBR) paradigm. Our standpoint is that videos are multi-faceted data objects, and an effective retrieval system should be able to accommodate all of the above complementary information descriptions and bridge the gap between different elements for retrieving video. As such, in this chapter we advocate a web-based hybrid approach to video retrieval by integrating the query-based (database) approach with the CBR paradigm.

1.1 RELATED WORK

Text-based information retrieval and content-based image retrieval have been two separate topics historically. The former one uses annotation/specification of high-level semantics of images manually. The latter one needs accurate extraction and recognition of low-level image features. Seldom research focuses on integrating these two classes of information; however actual application demonstrates its importance. However, this aspect of work is fast becoming an area of intensive research. Recently, Tang, Hanka and Ip [17] have provided an overall architectural framework for combining iconic and semantic content for image retrieval and developed techniques for extracting semantic information from images for the medical domain. Simone and Ramesh [13] have attempted to integrate both text and visual features for image retrieval. They used labels attached in the HTML files as the textual information, and developed several search engines which analyze low-level image features separately. As the labels extracted from the HTML files may not be relevant to the images, information derived from these two sources may not necessarily be consistent. To reduce the potential semantic inconsistency and the time consuming task in manual textual annotations of images, and to support semantic retrieval of media content, recent works have focused on automatic extraction of semantic information directly from images [15,19].

In the context of videos, previous research focus was on video structuring, such as shot detection and key frame extraction based on visual features in the video. Its application areas have been extended from commercial video databases to home videos (e.g., [9]). But most of the practical systems annotate videos by textual information manually. For example, in OVID [12], each video object has a set of manually annotated attributes and attribute values describing its content. In [1], both the video objects and their spatial relationships are manually annotated in order to support complex spatial queries.

In order to index video automatically, not only the visual contents but also the audio content should be used (see [10,18,21]). Recent work also proposed the use of closed caption as well as other collateral textual information. Ip and Chan have combined lecture scripts and video text analysis to achieve automatic shot segmentation and indexing of lecture videos [7,8]. The approach assumed there is a correspondence between text-script segment and video segment and can be applied to other types of structured video with collateral text such as news videos. An obvious trend in video indexing and retrieval is to use media features derived from a variety of sources such as audio, closed captions, textual scripts and so on instead of simply visual features. Another trend is towards web-enabled/based video management system.

Over the last few years we have been working on developing a generic Video Management and Application Processing (VideoMAP) framework [2,3,20]. A central component of VideoMAP is a query-based video retrieval mechanism called CAROL/ST, which supports spatio-temporal queries [2,3]. While the original CAROL/ST has been designed to work with video semantic data based on an extended object oriented approach, little support has been provided to support video retrieval using visual features. To support a more effective video retrieval system through a hybrid approach, we have been making extensions to the VideoMAP framework, and particularly the CAROL/ST mechanism [5]. At the same time, we have also developed a framework for extracting and combining iconic and semantic features for image retrieval [15,17]. The underlying approach for semantic extraction and automatic image annotation can be extended to video data.

In this chapter, we present our web-based version of VideoMAP (termed as VideoMAP*), and discuss the main issues involved in developing such a web-based video database management system supporting hybrid retrieval. Furthermore, an approach to extracting and representing semantic information from video data in support of hybrid retrieval will be presented.

1.2 CHAPTER ORGANIZATION

The rest of our chapter is organized as follows. We first discuss the philosophical and technical issues on defining and extracting semantics from video key frames and collateral texts such as video scripts and present a representation scheme for video semantics that can be applied to support semantic query processing. This is followed by a review of the main extensions we have made to VideoMAP, the result of which is a comprehensive video database management system which supports CBR and query-based retrieval methods. In section 4 we detail the VideoMAP* framework and its web-based hybrid approach to video retrieval. The specific user interface facilities are described in section 5; example queries are also given to illustrate the expressive power of this mechanism. Finally, we conclude the chapter and offer further research directions in section 6.

2. SEMANTICS IN TEXT AND IMAGES EXTRACTED FROM VIDEO DATA

To locate and represent the semantic meanings in video data is the key to enable intelligent query. This section presents the methodology for knowledge elicitation and, in particular, introduces a preliminary semantic representation scheme that bridges the information gap not only between different media but also between different levels of contents in the media.

Semantics is always desirable for intelligent information systems. However, it is not just difficult to extract semantics from video data, only to define an appropriate set of semantics already presents a significant challenge. Semantics is closely related to contextual information. There are several levels of semantics bearing units in a video system. Roughly, these units can be divided into two levels, static semantics and dynamic semantics. Static semantics are defined in a possible data source that may appear in the application while dynamic semantics is generated according to the application context. In fact, the generation of dynamic semantics is based on static semantics and normally contains high-level information about a wider range of data.

In our work, we focus on two specific elements in the video content: textual script that accompany a video and image content in key frames. Text itself is not the semantics of video content, rather it is a possible way pointing to the video semantics. Text analysis opens windows to approach the video semantics and the information contained in text should be integrated with the visual contents. Generally, the static semantics in textual form are represented in the following contents:

- Meanings in words and phrases.
- Collocations between words and phrases.
- The relationships between words and phrases.
- The concept space.

Traditionally, most of the above information is represented in dictionaries and thesaurus, or contained within a statistical corpus. Dynamic semantics is generated when the information analysis system is making use of the static semantics within the current context in analyzing text. Disambiguation processing paves the path to the rest of the analysis. Examples of dynamic semantics are information about sentences, paragraphs, indexing contents for full text documents, abstracts/summaries of full texts, etc. At the representation level, for general purpose, a semantics network can be deployed.

Compared with the textual semantics extraction, visual semantics poses even more challenges due to the complexity of the data and yet the techniques of computational visual perception are far from mature. In the past years, many researchers mainly focused on the techniques at the perceptual level, in other words, the primitive iconic features that can be extracted from key frames or frame sequence, and developed techniques for detecting texture, colour, shapes, spatial relationships, motion field, etc. However, high-level semantic analysis and interpretation is more desirable for complicated queries. In fact, traditional content-based method and semantic analysis are indispensable and complementary to each other; they should and can be combined to represent the meaning that the underlying visual data depicts. This requires efforts to bridge the gap between low-level syntactic/iconic features that can be automatically detected by conventional image processing tools and high-level semantics that captures the meanings of visual content in different conceptual levels.

2.1 PROCESSES OF SEMANTICS EXTRACTION

2.1.1 Textual Semantic Feature Extraction

Text in video can be found in the form of the video script or video segment annotations or textual description of the video. The basis of the text semantics extraction is done by the creation of a large-scale comprehensive dictionary, a complete control rule system and a concept space. The dictionary stores static syntax and semantics about the words and phrases, which is the basic information for text segmentation and the rest of the text analysis. Each lexical entry contains the information about its category, its classification point in the concept space, the inflection, the frequency and its possible visual attributes. For example, a word "roundish" implies a shape of an object, so "shape" is taken as the visual information in the entry. The rule system provides the possible collocation situations of different words and phrases in a text and which information should be obtained under such situations; in the meantime, it suggests possible solutions for the processing. The collocation situations are described based on collocations statistics that capture the relationships between concepts, words and phrases or even sentences and pieces of texts.

A semantic parser takes control of all this information and generates dynamic semantics for input unknown text. It serves to execute the whole analyzing strategy, to control the pace of analysis, to decide when to execute what kind of analysis, to call which rule base, to invoke different algorithms that are suitable for different analysis. In short, it decides how to approach the analysis for a piece of text. The analysis procedure includes automatic segmentation, ambiguity processing, morphological analysis and syntactic analysis, semantic analysis and complex text context processing. Within our framework, the result is a semantic network Papillon, which will be discussed later in section 2.3.

2.1.2 Semantics Extraction and Semantic Definition for Key Frame Images

The semantics of a video segment can be defined via the key frame images by associating/mapping semantic meanings to low-level image properties of the key frames. The two levels of image contents should be used to complement each other and be integrated to support intelligent query as well as the tasks in an indexing cycle. The first step is to locate the semantics in the images by defining the visual appearance using semantic labels. Then different feature extraction techniques and classifiers were trained to extract and classify low-level image features that have been already associated with semantic labels. This is followed by a process of semantic analysis which aims to ensure the consistency among the semantic labels on different parts of the image by exploiting contextual and high-level domain knowledge. As a summary, the video semantics are analyzed through:

- Defining the semantic label set, which is associated with a set of typical visual appearance that can be found in the video segment

- Building up domain knowledge for interpreting the label set

- Designing the visual detectors to recognize and classify the visual appearances that have been associated with the semantic label set

- Designing a semantic analyzer that can exploit the domain knowledge and dynamic information that is produced by the visual detectors to

  - rectify the results from the visual detectors which are normally not able to provide 100% accuracy for mapping visual content to the set of semantic labels, and

  - generate high level semantics for the whole image and represent them in a semantic representation called Papillon that is introduced later.

To this end, a knowledge elicitation subsystem was developed to support the acquisition of domain knowledge.

Semantic label definition

In order to associate semantic meaning to different visual appearances found in a video, we partition a key frame image into a number of sub-images. The size and the shape of sub-images can be varied according to their suitability to capture the image features. The sets of sub-images serve two purposes: (a) they form the basic units for image analysis such as texture and colour analyses, as well as for semantic analysis; and (b) they form the basic unit for static semantic definition.

Semantic Interpretation of Visual Content and the Semantic Label Set

Generally speaking, the key frame images as well as their object contents under different scales are likely to have different semantics if we classify the semantics in great details. Image content under different scales will show different prominent visual features. Even within the same scale, there are still different semantic (descriptive or interpretive) levels. The choice of semantic level depends on what kind of information the research aims to obtain from the images and how the information is to be categorized. Several levels of interpretations can be defined correspondingly for the features that may appear in the key frame image. The features in each level have their respective knowledge held in different knowledge bases, where such knowledge serves the purposes of reasoning, semantic description and generating annotation.

We have implemented a subsystem for knowledge elicitation to allow a user to interactively assign semantic labels to a large subset of the objects in key frame images obtained from video database. These associations of semantic labels to sub-images depicting the various visual characteristics of the labels were taken as the ground truth and formed the initial set of training samples for designing the feature extraction and recognition processes. The approach was tested, by first giving the visual feature detector a set of training samples containing semantic labels in the form of a mapping between the labels and the corresponding sub-image. The visual feature detector then executes related feature extraction routines and assigns semantic labels to other unknown key frame images.

For a large-scale database that contains a wide range of videos, there is no single feature detector that is capable of extracting all the salient image features. In our research, a set of visual feature detectors were developed for texture, colour and shape measurements that have been associated with the semantic labels.

Incorporated with contextual information in the knowledge base, the semantic analyzer analyzes the labelling results from the visual detectors for the images, and confirms or refutes, as well as explains any unknown regions according to the domain knowledge.

2.2 CONTENT OF THE KNOWLEDGE BASE FOR VISUAL CONTENTS

Depending on the application domain, the defined set of the semantic labels should be described further in the knowledge base, forming the static image semantics. This includes:

- Domain attributes, such as its logical or expected location, logical or expected neighbours, etc.

- Visual attributes, such as colour, shape, size, quantity, as well as the similarity with any other features, and the various relationships including spatial relationship among them.

- Measurement attributes, indicating which detector is the best one that suits this feature.

- Contextual attributes, e.g., the special attributes a video may have when the semantic labels are combined with some situations or other semantic labels.

2.3 PAPILLON – AN INTERMEDIATE SEMANTIC REPRESENTATION

We have devised a semantic representation scheme to represent the semantics for both full text and key frame images. We call this scheme Papillon. The content of Papillon comes from either text analysis or image analysis including its semantic analysis. It contains the static and generated dynamic semantics about the text or images. More importantly, Papillon focuses on fusing the high level information inherent in the different media (image and textual information), and serves as semantic or conceptual representation for them. In a sense, it can be considered as an intermediate semantic representation for the two types of video elements. Furthermore, in our application, textual annotation for unknown key frame images can be automatically generated through Papillon.

Papillon consists of a semantic network, which is a graph, where the nodes in the graph represent concepts, and the arcs represent binary relationships between concepts. The nodes or concepts are objects or features and their descriptions in the image or text, which contains all the information and attributes about the entities including semantic code and concept number in the concept space. The semantic relationships express the inherent semantic relationships between concepts that have been derived from the analysis for text and key frame images. Examples of the semantic relationships are agent, analogue, compose of, degree, direction, duration, focus, frequency, goal, instrument, location, manner, modification, object, origin, parallel, possession, quantity, colour, size, shape, produce, reason, equal, reference, result, scope, attribute, time, value, capacity, condition, comparison, consequence. Detail of this representation scheme can be found in [16].

Semantic information extracted from the key frame and represented within the Papillon can then be incorporated into the object framework for video retrieval to be described in the following section.

3. A "STAND-ALONE" VIDEO RETRIEVAL SYSTEM

As mentioned earlier, we have been developing an extended version of VideoMAP, a "stand-alone" video retrieval system (VRS) for supporting hybrid retrieval of videos through query-based and content-based accesses [5]. The architecture of it is depicted in Fig. 24.1, which is divided into several components: Video acquisition, video analysis, video index and video retrieval.

3.1 PROCESSING FLOWS

The main processing flows of supporting hybrid video retrieval in a stand-alone VRS involve the following:

- Video acquisition: There exists an extensive distribution network of multimedia data globally. And videos, as an important element of multimedia, can be found in a variety of sources.

- Video analysis: The obtained video meta data is only a sequence of frames, without structure and annotation initially, so a series of processes of video indexing and retrieval must be carried out based on content analysis. There are three modules:

  - Segmentation and Clustering Component (SCC): Detect the camera movements in video, cluster the semantically relevant segments into scene and determine which module is necessary for structuring video objects.

  - Feature Extracting Component (FEC): Extract the visual feature vector from the video and other object features, such as the color, texture, shape and so on.

  - Semantic Definition and Extraction Component (SDEC): Define key-frame semantic and extract and associate/map semantic meanings to low-level image properties.

  - Semantic Modeling Component (SMC): Organize video segments into semantically meaningful hierarchies with annotated feature indices.

- Video index: The process of video analysis typically yielded a video hierarchical structure. Indexes are generated for the scene hierarchy, and the video visual features and video segment hierarchies. These indexes are not absolutely self-existent. For example, in the "Segment" and "Keyframe" level of video hierarchy, they also contain the indexes for visual and semantic features.

- Video retrieval: Video retrieval contains two kinds of retrieval formats: CAROL/ST Retrieval - the original retrieval format which mainly uses the semantic annotation and spatio-temporal relationships of video; Content-based Retrieval - the prevailing retrieval format which mainly uses the visual information inherent in the video content.

3.1.1 Foundation Classes

Our "stand-alone" VRS extends a conventional OODB to define video objects through a specific hierarchy (video→scene→segment→keyframe). In addition, it includes a mechanism of CBR to build index on visual features of these objects. Their class attributes, methods and corresponding relations form a complex network (or, a "map"). More details are described in [5].

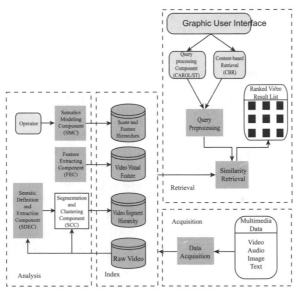

Figure 24.1 A "Stand-alone" VRS architecture.

3.2 A QUERY LANGUAGE (CAROL/ST) WITH CBR

In the hybrid approach, there are three kinds of queries possible for the query-processing component: Text-based Search, Content-based Search and the Hybrid Search. Search paths of the three kinds of queries are described in [5]. In our approach, objects, attributes and methods are stored in separated classes. Therefore, the query processor can search the object database with different entry points. After integrating it with CBR, the search paths become applicable to our hybrid approach. Here, we show some query examples (cf. Figure 24.5 and Figure 24.6). Detailed query syntax and additional features for query processing are described in [6].

4. DEVELOPMENT OF A WEB-BASED VRS

In this section, we describe a web-based version of VideoMAP that we have been developing, which we term as VideoMAP*. In this section, we present a reference architecture for it, and discuss the design issues involved in developing such a web-based video retrieval system.

4.1 REFERENCE ARCHITECTURE

A reference architecture of VideoMAP* is shown in Figure 24.2. It is being built on top of a network and the components can be classified into two parts: server-components and client-components. For the client-components, they are grouped inside the gray boxes; for the server-components, they are shown below the network backbone. The main server-components providing services to the clients include Video Production Database (VPDB) Processing, Profile Processing and Global Query Processing. Theoretically, there is no limit on the number of users working at the same time. There are four kinds of users who can utilize and work with the system: Video Administrator, Video Producer, Video Editor and Video Query Client. Each type of user is assigned with a priority level for the VPDB Processing component to handle the processing requests from the queue.

A request from a user who has the highest priority level is processed first, with the requests from the users of the same priority level being processed in FIFO. The user priority levels of our system, in descending order, are from Video Administrator, Video Producer, Video Editor to Video Query Client.

4.1.1 Main Process Flows

For the four main components of VideoMAP*, their main process flows are described immediately below.

VPDB Processing
The VPDB Processing component is responsible for accepting clients' requests and providing basic database operations of the Global VPDB through CCM/ST (Conceptual Clustering Mechanism supporting Spatio-Temporal semantics) [3]. It cooperates with a Concurrency Control mechanism to ensure data integrity and data consistency among Global VPDB and Local VPDBs. Global VPDB and Local VPDBs store the raw video data together with the semantic information. Global VPDB is the permanent storage of the system and Local VPDBs are the temporary storage for the clients.

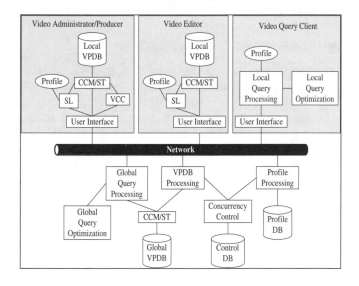

Figure 24.2 Architecture of a VideoMAP*: a web-based video retrieval system.

Concurrency Control
The Concurrency Control mechanism is used to handle multiple-user processing of the system. As a small change to the ToC (Table of Content) of the video program may affect the final display order of the video program [2], data locks cannot be used in the virtual editing of a video program; otherwise, the whole sequence of the video program should be locked entirely. In order to ensure data integrity and data consistency, a complemented mechanism is introduced here. When a user wants to do virtual editing of a video program, he retrieves the video data from the Global VPDB to his Local VPDB. Then the modified video data is saved as a new view of the video program. A collection of views is stored with the original video program. Therefore, Video Query Clients can choose to display the video program from different perspectives. In order to reduce the size

of the database, it is necessary to limit the number of versions of views of each user stored in the database. The Video Administrator can also remove some obsolete videos.

Profile Processing

Profile Processing is used to provide basic database operations of the ProfileDB. It ensures that all the clients can retrieve and update their corresponding user profiles (which are stored in the ProfileDB). There are three types of profiles: common profile, personal profile and the specific video profile. Each profile stores the common knowledge and the Activity Model [3] of which the semantic meanings can be different to different users. Common profile is the standard profile provided to every user, thereby supporting the desired "subjectivity" for virtual editing and access activities. Personal profile is the customized common profile according to the need of the individual user. The Video Producers create the specific video profiles based on the video programs. The personal profile can only be modified by its owner; and the specific video profile can only be modified by the Video Administrator. The profiles are represented in the form of rules for effective reasoning and can be converted into a universal format (e.g., XML) before transmitting to the clients. Therefore, the information can be easily interpreted by the web browsers.

Semantic Query Processing (CAROL/ST) with CBR

The query processing mechanism of VideoMAP* is inherited from, and thus exactly the same as that of the "stand-alone" VRS [5]. We therefore omit its description here for the sake of conciseness.

4.2 DISTRIBUTION DESIGN

In a web-based environment, the cost of data transfer and communication is an important issue worth great attention. In order to reduce this overhead, databases can be fragmented and distributed into different sites. There are many research efforts on data fragmentation of relational databases. On the contrary, there is only limited recent work on data (object) fragmentation in object databases [11,22,23]. As it is time-consuming (and quite often unnecessary) to transmit an entire object database to the client side, especially for a large video object database, it is advantageous to fragment and cluster objects in order to minimizing the overhead of locking and transferring objects. After objects are fragmented, the next step is to allocate the fragments to various sites, which is another open research issue for object databases (see, e.g., [14]).

4.2.1 Global VPDB

In VideoMAP*, different types of users such as Video Producers (VPs), Video Editors (VEs) and the Video Administration (VA) can retrieve the Global VPDB into their Local VPDBs for processing, and their work may affect and be propagated to the Global VPDB. In particular, each type of user may work on a set of video segments and create dynamic objects out of the video segments. It therefore makes sense to cluster the video objects in the same manner as shown in Figure 24.3. Specifically, each raw video is assumed to have a specific domain and relevant domains are grouped together first. Since multiple video segments (may belong to different domains) may be included by one application, raw videos linked by the same application are therefore grouped together. The other objects which have links with these video segments are then grouped into clusters. Further fragmentation based on by user views [4] can be processed if necessary. The clusters can be stored in more than one database; therefore the size of object locking can be minimized.

In order to maintain the fragmentation transparency, a processing component is needed to locate objects. A Global Schema is also needed to contain the meta-data of objects in several levels (by domain, application, link, etc). VP, VE and VA users can easily retrieve the objects, and also easily insert/update the databases through the global schema. Consequently, the global schema also needs to be carefully shared and maintained.

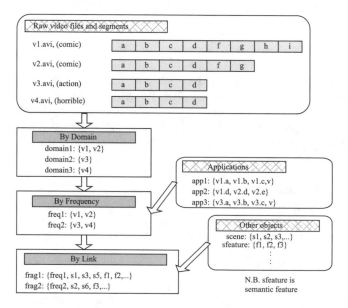

Figure 24.3 Data distribution of Global VPDB.

4.2.2 Global Video Query Database (Global VQDB)

In VideoMAP*, the Global VPDB is replicated into a mirror database (i.e., Global VQDB) for Video Query Client processing. As the Global VQDB is a read-only database and is independent of other processes, it can be fragmented according to the interests of the query clients thoroughly. Since an Activity Model is composed of user-interested activities, and a User Profile is constructed by the user's activity model and query histories, the Global VQDB can be fragmented based on the semantics in multiple levels.

As shown in Figure 24.4, semantic objects (i.e., SemanticFeature, and VisualObject) can be grouped by two ways: by Semantics, and by Activity Model. In the former case, the objects are grouped together by a thesaurus and then they are linked with other associated objects (such as scene, visual feature, etc.). In the latter case, semantic objects are first clustered from the object level to the activity level, and then linked with other associated objects (e.g., scene, visual feature, etc.).

In order to maintain the fragmentation transparency, a processing component is needed to serve as a directory for locating the objects. As in the case of the Global VPDB, it works with the Global Schema in which multiple levels of meta data information are stored.

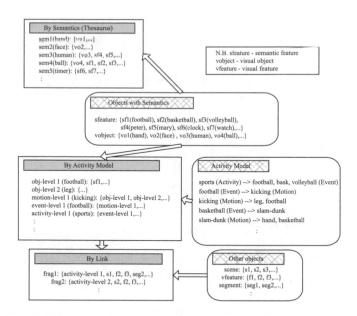

Figure 24.4 Data distribution of Global VQDB.

4.3 DISTRIBUTED QUERY PROCESSING AND OPTIMIZATION

When the query client submits a query, the Local Query Processing component may rewrite and transform it into a number of logical query plans. Since the Global VQDB is fragmented according to the semantics and the Activity Model, the logical query plans can be generated based on User Profiles which contain relevant semantic information. For example, if a user requests for a "ball," the query plans can include "ball," "football" etc. based on his/her profile. Each query plan is then decomposed into sub-queries. After the decomposition of all plans, the sub-queries are sent to the Local Query Optimization component to determine the order of importance. Next, the query plans are sent to the Global Query Processing component and the Global Query Optimization component to search the results by using the information from the Global Schema. As the original query is rewritten into a number of similar queries with different priority levels, the probability of obtaining the right results can be greatly increased.

5. EXPERIMENTAL PROTOTYPING WORK

In this section, we briefly describe our prototype work in terms of the implementation environment, the actual user interface and language facilities, and sample results.

5.1 IMPLEMENTATION ENVIRONMENT

Starting from the first prototype of VideoMAP, our experimental prototyping has been conducted based on Microsoft Visual C++ (as the host programming language), and NeoAccess OODB toolkit (as the underlying database engine). Here we illustrate the main user facilities supported by VideoMAP which is also the kernel for our web-based prototype, i.e., VideoMAP*. It currently runs on Windows and offers a user-friendly graphical user interface supporting two main kinds of activities: Video editing, and Video retrieval.

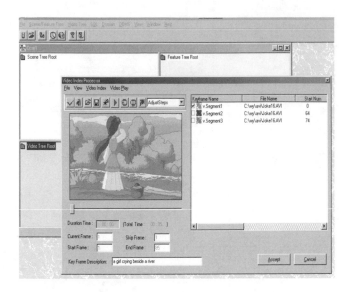

Figure 24.5 Annotating the segments after video segmentation.

5.1.1 Video Editing

When a user invokes the function of video editing, a screen comes up for uploading a new video into the system and annotating its semantics (cf. Figure 24.5). For example, s/he can name the video object and assign some basic descriptions to get started.

A sub-module of Video segmentation is devised to help decompose the whole video stream into segments and to identify the keyframes. Further, the Feature Extraction module is to calculate the visual features of the media object. By reviewing the video abstraction structure composed by the segments and keyframes, the user can annotate the semantics according to his/her understanding and preference (cf. Figure 24.5).

5.1.2 Video Retrieval

Our prototype also provides an interface for the user to issue queries using its query language (i.e. CAROL/ST with CBR). All kinds of video objects such as "Scene," "Segment," "Keyframe" can be retrieved by specifying their semantics or visual information. Figure 24.6 shows a sample query issued by the user, which is validated by a compiler sub-module before execution. The retrieved video objects are then returned to the user in the form of a tree (cf. Figure 24.6), whose node not only can be played out, but also can be used subsequently for formulating new queries in an iterative manner.

5.2 SAMPLE RESULTS

As described earlier, our system supports three primary types of queries, namely: (1) query by semantic information, (2) query by visual information and (3) query by both semantic and visual information (the "hybrid" type). For type (1), VideoMAP supports retrieving all kinds of video objects (Video, Scene, Segment, Keyframe, Feature) based on the semantic annotations input by the editor/operator earlier (cf. section 5.1.1). Figure 24.7 shows the interface for users to specify semantic query, and for displaying the query result video.

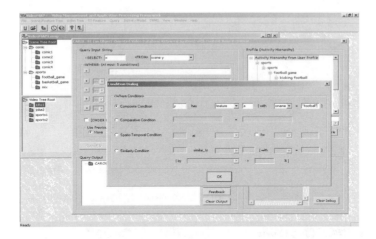

Figure 24.6 Query facilities.

For type (2), users can specify queries which involve visual features and their similarity measurement. Visual similarity considers the feature of color, texture and so on. Users can specify the query by using either individual feature or its combination [5]. Figure 24.8 illustrates the interaction sessions for the user to specify a visual feature-based query (using "similar-to" operator), and the resulting video being displayed to the user. Finally, for type (3), users can issue queries which involve both semantic and visual features of the targeted video objects. These "heterogeneous" features call for different similarity measurement functions, and their integration is being devised in the context of VideoMAP* prototype system.

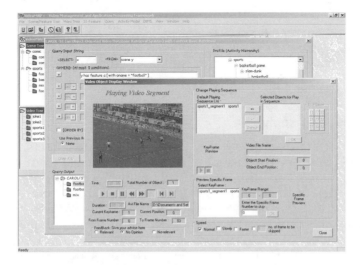

Figure 24.7 Query by semantic information.

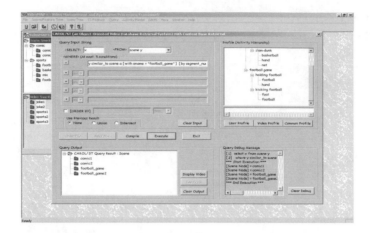

Figure 24.8 Query by visual information.

5.3 THE EXPERIMENTAL USER INTERFACE OF VIDEOMAP*

The experimental prototype of our web-based video query system (VideoMAP*) is built on top of Microsoft Windows 2000. The Query Client Applet is developed using Borland's JBuilder 6 Personal, while the Query Server is implemented using Microsoft Visual C++ 6.0. The kernel of the query server is based on its original standalone system. The client ORB is developed using IONA's ORBacus 4.1 for Java, and the server ORB is developed using IONA's ORBacus 4.1 for C++. The client uses Java Media Framework to display video query result, while the server still uses the same object-oriented database engine, NeoAccess, to store object data.

5.3.1 GUI Overview

Figure 24.9 shows the GUI of the query system. There are two processes: Query Client Applet, and Query Server. The client (left-side) provides several functions, such as Login, DBList, DBCluster, Schema, Query Input, Query Result and Video Display. Some important server messages and query debug messages are shown in the Server/Debug MessageBox. The server console (right-side) shows a list of connection messages.

5.3.2 Sample Query Display

In Figure 24.10, it shows a query of extracting all scene objects that are created in a cluster of databases. The format of the result returned from the query server is XML-like, which contains the type of the result object, the name of the object, the brief description of the object, the video link and the temporal information of the video. Owing to its expressiveness and flexibility, XML is very suitable for Web-based data presentation, in addition to being the standard for multimedia data description (e.g., MPEG-7). More specifically, video data can be separated into XML data description and raw videos. The data description can be easily distributed over the Web but the server keeps its own raw video files. So a user can display the video result by making connection to the web server. This would reduce the loading of the query server when there are multiple user connections.

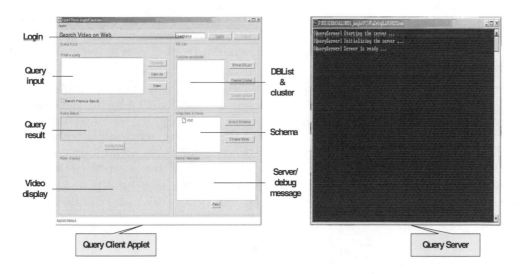

Figure 24.9 GUI of the Web-based video query system (VideoMAP*).

Figure 24.10 A client's query processed by the server (VideoMAP*).

6. CONCLUSION

We have presented a comprehensive web-based video retrieval system, viz., VideoMAP*, and presented in this chapter its central mechanisms which offer a number of unique features. Particularly, we have discussed a methodology for identifying and extracting semantic information from media content which serves towards bridging the semantic gap between media content and its interpretation. The VideoMAP* system provides a hybrid approach to video retrieval over the web, by combining query-based method with content-based retrieval (CBR) functions. In addition, it supports an intermediate semantic scheme (viz., Papillon) for representing both full text and key frame semantics. VideoMAP* also provides concurrency control for virtual editing of video data among different users who are assigned with priority levels, and can accommodate various kinds of profiles containing knowledge information and activity models. These profiles are useful for annotating abstracted information into the video data and for processing queries of the video data. We have also shown the user interface facilities of our system from the perspective of sample user interactions.

There are a number of issues for us to further work on. Among others, we are extending our prototype system to make it as MPEG standard compatible, by allowing input videos to be in either AVI or MPEG format. We have also started to examine the possible performance issues, such as replication technique for the Global VPDB to be distributed into mirror databases, so as to accommodate larger number of Video Query Clients. Finally, the issues of optimizing CAROL/ST with CBR queries, and the possibility of combining local query processing with global query processing in light of overall query optimization are interesting for investigation in the context of VideoMAP*.

REFERENCES

[1]   S. Chang and E. Jungert, Pictorial Data Management Based Upon the Theory of Symbolic Projections, *Journal of Visual Languages and Computations*, Vol.2, No.3, September 1991, pp. 195-215.

[2]   S. S. M. Chan, and Q. Li, Facilitating Spatio-Temporal Operations in a Versatile Video Database System, in Proceedings of the 5th International Workshop on Multimedia Information Systems (MIS'99), USA, 1999, pp. 56-63.

[3]   S. S. M. Chan, and Q. Li, Developing an Object-Oriented Video Database System with Spatio-Temporal Reasoning Capabilities, in Proceedings of the 18th International Conference on Conceptual Modeling (ER'99), France, 1999, pp. 47-61.

[4]   S. S. M. Chan and Q. Li, Architecture and Mechanisms of a Web-based Video Data Management System, in Proceedings of the 1st IEEE International Conference on Multimedia and Expo (ICME'00), USA, 2000.

[5]   S. S. M. Chan, Y. Wu, Q. Li, and Y. Zhuang, A Hybrid Approach to Video Retrieval in a Generic Video Management and Application Processing Framework, in Proceedings of the IEEE International Conference on Multimedia and Expo (ICME'01), Japan, 2001.

[6]   S. S. M. Chan, Y. Wu, Q. Li, and Y. Zhuang, Development of a Web-based Video Management and Application Processing System, in Proceedings of the SPIE International Symposium on Convergence of IT and Communications (ITCom2001), USA, 2001.

[7]   H. H. S. Ip and S. L. Chan, Hypertext-Assisted Video Indexing and Content-based Retrieval, in Proceedings of the 8th ACM Hypertext Conference, UK, 1997, pp. 232-233.

[8]   H. H. S. Ip and S. L. Chan, Automatic Segmentation and Index Construction for Lecture Video, *Journal of Educational Multimedia and Hypermedia*, Vol. 7, No. 1, pp. 91-104, 1998.

[9]   W.-Y. Ma and H.J. Zhang, An Indexing and Browsing System for Home Video, Invited paper, in 10th European Signal Processing Conference (EUSIPCO'2000), Finland, September 2000.

[10]   M. R. Naphade, T. Kristjansson, B. J. Frey, and T. S. Huang, Probabilistic Multimedia Objects Multijects: A novel Approach to Indexing and Retrieval in Multimedia Systems, in Proceedings of the IEEE

International Conference on Image Processing, Vol. 3, pp. 536-540, Oct 1998, Chicago, IL.

[11] M. T. Ozsu, U. Dayal and P. Valduriez, *Distributed Object Management*, Morgan Kaufmann Publishers, San Mateo, California, 1994.

[12] E. Oomoto and K. Tanaka, OVID: Design and Implementation of a Video-Object Database System, *IEEE TKDE*, Vol. 5, No. 4, pp. 629-634, 1993.

[13] S. Santini and R. Jain, Integrated Browsing and Querying for Image Databases, *IEEE Multimedia*, Vol. 7, No. 3, pp. 26-39, 2000.

[14] S-K So, I. Ahamad and K. Karlapalem, Response Time Driven Multimedia Data Objects Allocation for Browsing Documents in Distributed Environments, *IEEE TKDE*, Vol. 11, No. 3, pp. 386-405, 1999.

[15] L. H. Tang, H. H. S. Ip, R. Hanka, K. K. T. Cheung and R. Lam, Semantic Query Processing and Annotation Generation for Content-based retrieval of Histological Images, in Proceedings of the SPIE Medical Imaging, USA, February 2000.

[16] L. H. Tang, *Semantic Analysis of Image content for Intelligent Retrieval and Automatic Annotation of Medical Images*, PhD Dissertation, University of Cambridge, England, 2000.

[17] L. H. Tang, R. Hanka, and H. H. S. Ip, A System Architecture for Integrating Semantic and Iconic Content for Intelligent Browsing of Medical Images, in Medical Imaging 1998: PACS Design and Evaluation: Engineering and Clinical Issues, Steven C. Horii, M.D., G. James Blaine, Editors, Proceedings of the SPIE *3339*, pp. 572-580, San Diego, California, USA, 21-27 February 1998.

[18] F. Wu, Y. Zhuang, Y. Zhang, and Y. Pan, Hidden Markovia-based Audio Semantic Retrieval, *Pattern Recognition and Artificial Intelligence [in Chinese]*, Vol. 14, No. 1, 2001.

[19] A. T. Zhao, L. H. Tang, H. H. S. Ip, and F. Qi, Visual Keyword Image Retrieval Based on Synergetic Neural Network for Web-Based Image Search, *Real Time Systems*, Vol. 21, pp. 127-142, 2001.

[20] R. W. H. Lau, Q. Li, and A. Si, VideoMAP: a Generic Framework for Video Management and Application Processing, in Proceedings of the 33rd International Conference on System Sciences (HICSS-33): Minitrack on New Trends in Multimedia Systems, IEEE Computer Society Press, 2000.

[21] J. Nam, A. E. Cetin, and A. H. Tewfik, Speaker Identification and Video Analysis for Hierarchical Video Shot Classification, in ICIP'97, Vol. 2, pp. 550-555.

[22] K. Karlapalem and Q. Li, A Framework for Class Partitioning in Object Oriented Databases, *Distributed and Parallel Databases*, 8(3):317-350, Kluwer Academic Publishers, 2000.

[23] M. T. Ozsu and P. Valduriez, *Principles of Distributed Database System*, Prentice Hall, Upper Saddle River, New Jersey, 1999.

[24] S. S. M. Chan and Q. Li, Architecture and Mechanisms of a Multi-paradigm Video Querying System over the Web, in Proceedings of the 2nd International Workshop on Cooperative Internet Computing (CIC 2002), Hong Kong, August 2002.

25

FROM LOW LEVEL FEATURES TO HIGH LEVEL SEMANTICS

Cha Zhang and Tsuhan Chen

Department of Electrical and Computer Engineering

Carnegie Mellon University

Pittsburgh, Pennsylvania, USA

{czhang,tsuhan}@andrew.cmu.edu

1. INTRODUCTION

A typical content-based information retrieval (CBIR) system, e.g., an image or video retrieval system, includes three major aspects: feature extraction, high dimensional indexing and system design [1]. Among the three aspects, high dimensional indexing is important for speed performance; system design is critical for appearance performance; and feature extraction is the key to accuracy performance. In this chapter, we will discuss various ways people have tried to increase the accuracy of retrieval systems.

If we think over what "accuracy" means for a retrieval system, we may find it very subjective and user-dependent. The similarity between objects can be very high-level, or *semantic*. This requires the system to measure the similarity in a way human beings would perceive or recognize. Moreover, even given exactly the same inputs, different users probably have different feeling about their similarity. Therefore, a retrieval system also needs to adapt to different users quickly through on-line user interaction and learning.

However, features we can extract from objects are often low-level features. We call these low-level features because most of them are extracted directly from digital representations of objects in the database and have little or nothing to do with human perception. Although many features have been designed for general or specific CBIR systems with high level concepts in mind, and some of them showed good retrieval performance, the gap between low-level features and high-level semantic meanings of the objects has been the major obstacle to better retrieval performance.

Various approaches have been proposed to improve the accuracy performance of CBIR systems. Essentially, these approaches fall into two main categories: to improve the features and to improve the similarity measures. Researchers have tried many features that are believed to be related with human perception, and they are still working on finding more. On the other hand, when the feature set is fixed, many algorithms have been proposed to measure the similarity in a way human beings might take. This includes off-line learning based on some training data, and on-line learning based on the user's feedback.

The chapter is organized as follows. Section 2 overviews some feature extraction algorithms that emphasize the high level semantics. Section 3 discusses the similarity measure. Section 4 presents some off-line learning methods for finding better similarity measures. Section 5 examines algorithms that learn on-line based on the user's feedback. Section 6 concludes the chapter.

2. EXTRACTING SEMANTIC FEATURES

Many features have been proposed for image/video retrieval. For images, often used features are color, shape, texture, color layout, etc. A comprehensive review can be found in [1]. Traditional video retrieval systems employ the same feature set on each frame, in addition to some temporal analysis, e.g., key shot detection [2][3][4][5]. Recently, a lot of new approaches have been introduced to improve the features. Some of them are based on temporal or spatial-temporal analysis, i.e., better ways to group the frames and select better key frames. This includes integrating with other media, e.g., audio, text, etc. Another hot research topic is motion-based features and object-based features. Compared to color, shape and texture, motion-based and object-based features are more natural to human beings, and therefore at a higher level.

Traditional video analysis methods are often shot based. Shot detection methods can be classified into many categories, e.g., pixel based, statistics based, transform based, feature based and histogram based [6]. After the shot detection, key frames can be extracted in various ways [7]. Although key frames can be used directly for retrieval [3][8], many researchers are studying better organization of the video structures. In [9], Yeung *et al.* developed scene transition graphs (STG) to illustrate the scene flow of movies. Aigrain *et al.* proposed to use explicit models of video documents or rules related to editing techniques and film theory [10]. Statistical approaches such as Hidden Markov Model (HMM) [13], unsupervised clustering [14][15] were also proposed. When audio, text and some other accompanying contents are available, grouping can be done jointly [11][12]. There were also a lot of researches on extracting captions from video clips and they can also be used to help retrieval [16][17].

Motion is one of the most significant differences between video and images. Motion analysis has also been very popular for video retrieval. On one hand, motion can help find interesting objects in the video, such as the work by Courtney [18], and Ferman *et al.* [19], Gelgon and Bouthemy [20], Ma and Zhang [21], etc. On the other hand, motion can be directly used as a feature, named by Nelson and Polana [22] as "temporal texture." The work was extended by Otsuka *et al.* [23], Bouthemy and Fablet [24], Szummer and Picard [25], etc.

If objects can be segmented easily, object-based analysis of video sequence is definitely one of the most attractive methods to try. With the improvement on computer vision technologies, many object-based approaches were proposed recently. To name a few, in [18], Courtney developed a system, which allows for detecting moving objects in a closed environment based on motion detection.

Zhong and Chang [26] applied color segmentation to separate images into homogeneous regions, and tracked them along time for content-based video query. Deng and Manjunath [27] proposed a new spatio-temporal segmentation and region-tracking scheme for video representation. Chang et al. proposed to use *Semantic Visual Templates (SVT)*, which is a personalized view of concepts composed of interactively created templates /objects.

This chapter is not intended to cover in depth for features used in the state-of-the-art video retrieval systems. Readers are referred Sections II, III and V for more detailed information.

3. THE SIMILARITY MEASURE

Although new features are being discovered everyday, it is hard to imagine that we can find one set of features that can kill all the applications. Moreover, finding new features requires a lot of trials and errors, which in many cases has few clues to follow. If the feature set has been fixed for a certain retrieval system, another place that researchers can improve is the similarity measure.

It is said that a feature is good if and only if similar objects are close to each other in the feature space, and dissimilar objects are far apart. Obviously, to decide "close" or "far," the similarity measure plays an equally important role as the original feature space. For example, Euclidian distance or other Minkowski-type distances are widely used as similarity measures. A feature space is considered as "bad," probably because it does not satisfy the above criterion under Euclidian distance. However, if the feature space can be "good" by first (nonlinearly) "warping" it and then applying the Euclidian distance or by employing some new distance metrics, it is still fine for our purpose. Here the "warping" can also be considered as a preprocessing step for the feature space. However, in this chapter, such kind of "warping" is included in similarity measure, thus giving the latter a fairly general definition.

The difficult problem is how to get the "warping" function, or, in general, find the right similarity measure. Santini and Jain suggested to do it based on psychological experiments [29]. They analysed some similarity measure proposed in the psychological literature to model human similarity perception, and showed that all of them actually challenge the Euclidean distance assumption in non-trivial ways. Their research implies that a new distance metric is not only a preprocessing method, but also a necessity given the property of human perception. They developed a similarity measure, named *Fuzzy Feature Contrast*, based on fuzzy logic and Tverky's famous *Feature Contrast* [30].

One problem with the psychological view is that it does not have sound mathematical or computational models. A much more popular approach to find the "warping" function is through *learning*. By learning we mean given a small amount of training data, the system can automatically find the right "warping" function to make the feature space better.

Before learning, one must decide what to use as the ground truth data, in other words, what one wants the "warping" function to be optimised for. It turns out that there are two major forms that are most often used: similarity/dissimilarity (SD) and keyword annotations (KA). If we know that some objects are similar to each other while others are not, the feature space should be "warped" so that similar objects get closer, and dissimilar objects get farther. If the training data has some keyword annotated for each object, we want objects that share the same keywords to get closer while otherwise get farther. Both SD and KA have their advantages and good applications. SD is convenient for an end-user, and it does not require any explanation why two objects are similar or not (sometimes the reason is hard to be presented to the system as an end-user). Therefore, SD is suitable for end-user optimised learning, e.g., to learn what the end-user really means by giving some examples. SD is almost exclusively used by relevance feedback – a very hot research topic today. KA is good for system maintainers to improve the general performance of the retrieval system. Recent work in video retrieval has shown an interesting shift from query by example (QBE) [31][1] to query by keywords (QBK). The reason is that it allows the end users to specify queries with keywords, as they have been used to in text retrieval. Moreover, KA allows the knowledge learned to be accumulated by simply adding more annotations, which is often not obvious when using SD. Therefore, by adding more and more annotations, the system maintainer can let the end-user feel that the system works better and better. Both SD and KA have their constraints, too. For example, SD is often too user-dependent and the knowledge obtained is hard to accumulate, while KA is often limited by a predefined small Lexicon.

The learning process is determined not only by the form of the training data, but also by their availability. Sometimes all the training data are available before the learning, and the process can be done in one batch. We often call this *off-line learning*. If the training data are obtained gradually and the learning is progressively refined, we call it *on-line learning*. Both cases were widely studied in the literature. In the following sections, we will focus on applying these learning algorithms on retrieval systems to improve the similarity measure. Off-line learning is discussed in Section 4 and on-line learning is in Section 5.

4. OFF-LINE LEARNING

If all the training data is available at the very beginning, learning can be done in one step. This kind of off-line learning is often applied before the system is open to end-users. Although it might take quite some time, the speed is not a concern, as the end-user would not feel that.

Most off-line learning systems handle keyword annotations (KA). The keywords are often given as a predetermined set, organized in different ways. For example, Basu *et al.* [32] defined a Lexicon as relatively independent keywords describing events, scenes and objects. Many authors prefer the tree structure [34][35][33], as it is clean and easy to understand. Naphade *et al.* [36] and Lee *et al.* [37] used graph structure, which is appropriate if the relationship between keywords is very complex.

Once the training data is given, a couple of learning algorithms, parametric or non-parametric, can be used to learn the concepts behind the keywords. As far as the authors know, at least Gaussian Mixture Model (GMM) [32][35], Support Vector Machine (SVM) [38], Hybrid Neural Network [39], Multi-nets [36], Distance Learning Network [40] and Kernel Regression [33] have been studied in the literature. A common characteristic of these algorithms is that all of them can model potentially any distribution of the data. This is expected because we do not know how the objects that share the same concept are distributed in the low-level feature space. One assumption we can probably make is that in the low-level feature space, if two objects are very close to each other, they should be semantically similar, or be able to infer some knowledge to each other. On the other hand, if two objects are far from each other, the semantic link between them should be weak. Notice that because of the locality of the semantic inference, this assumption allows objects with the same semantic meaning to lie in different places in the feature space, which cannot be handled by simple methods such as linear feature reweighing. If the above assumption does not hold, probably none of the above learning algorithms will help improve the retrieval performance too much. The only solution to this circumstance might be to find better low-level features for the objects.

Different learning algorithms have different properties and are good for different circumstances. Take the Gaussian Mixture Model as an example. It assumes that the objects having the same semantic meaning are clustered into groups. The groups can lie at different places in the feature space, but each of them follows a Gaussian distribution. If the above assumptions are true, GMM is the best way to model the data: it is simple, elegant, easy to solve with algorithms such as EM [41][42] and sound in theoretical point of view. However, the above assumptions are very fragile: we do not know how many clusters the GMM will have, and no real case will happen that each cluster is a Gaussian Distribution. Despite the constraints, GMM is still very popular for its many advantages. Kernel regression (KR) is another popular machine learning technique. Instead of using a global model like GMM, KR assumes some local inference (kernel function) around each training sample. From the unannotated object's point of view, to predict its semantic meaning, an annotated object that is closer will have a higher influence, and a farther one will have less. Therefore, it will have similar semantic meanings to its close-by neighbours. KR can model any distribution naturally, and also has sound theory behind it [43]. The limitation of KR is that the kernel function is hard to select, and the number of samples needed to achieve a reasonable prediction is often high. Support Vector Machine (SVM) [44][45] is a recent addition to the toolbox of machine learning algorithms that has shown improved performance over standard techniques in many domains. It has been one of the most favourite methods among researchers today. The basic idea is to find the hyperplane that has the maximum *margin* towards the sample objects. *Margin* here means the distance the hyperplane can move along its normal before hitting any sample object. Intuitively, the greater the margin, the less the possibility that any sample points will be misclassified. For the same reason, if a sample object is far from the hyperplane, it is less likely to be misclassified. If the reader agrees with the reasoning above, he/she will easily understand the SVM Active Learning approaches introduced in Section 5. For detailed information on SVM, please refer to [44][45].

Although after applying the learning algorithm, the semantic model can be used to tell the similarity between any two objects already, most systems require a fusion step [33][32][35]. The reason is that the performance of the statistically learned models is largely determined by the size of the training data set. Since often the training data is manually made, very expensive and thus small, it is risky to believe that the semantic model is good enough. In [33], semantic distance is combined with low-level feature distance through a weighting mechanism to give the final output, and the weight is determined by the confidence of the semantic distance. In [32], several GMM models are trained for each feature type, and the final result is generated by fusing the outputs of all the GMM models. In [35] where audio retrieval was studied, the semantic space and the feature space are designed symmetrically, i.e., each node in the semantic model is linked to equivalent sound documents in the acoustic space with a GMM, and each audio file/document is linked with a probability model in the semantic space. The spaces themselves are organized with hierarchical models. Given a new query in any space, the system can first search in that space to find the best node, and then apply the link model to get the retrieval results in the other space.

Keyword annotation is very expensive because it requires a lot of manual work. Chang and Li [46] proposed to employ another way of getting the ground truth data. They used 60,000 images as the original set and synthesized another set by 24 transforms such as rotation, scaling, cropping, etc. Obviously, images after the transforms should be similar to the one before the transform. They discovered a perceptual function called *dynamic partial distance function* (DPF). Synthesizing new images by transforms and using them as training data is not new. For example, people play this trick in face recognition systems when the training image set has very few images (e.g., only one). Despite the fact that transforms may not be complete as a model of similarity, this is a very convenient way of getting a lot of training data, and DPF seems to have reasonable performance as reported in [46].

5. ON-LINE LEARNING

Compared to off-line learning, on-line learning does not have the whole set of training data beforehand. The data are often obtained during the process, which makes the learning process a best effort one and highly dependent on the input training data, even the order they come in. However, on-line learning involves the interaction between the system and the user. The system can then quickly modify its internal model in order to output good results for each specific user. As discussed in Section 1, similarity measure in information retrieval systems is highly user-dependent. On-line learning's adaptive property makes it very suitable for such applications.

In retrieval systems, on-line learning is used in three scenarios: relevance feedback, finding the query seed and enhancing the annotation efficiency. They are discussed respectively in the following three subsections.

5.1 RELEVANCE FEEDBACK

Widely used in text retrieval [47][48], relevance feedback was first proposed by Rui *et al.* as an interactive tool in content-based image retrieval [49]. Since then it has been proven to be a powerful tool and has become a major focus of research in this area [50][51][52][53][54][55]. Chapter 23 and Chapter 33 have detailed explanations on this topic.

Relevance feedback often does not accumulate the knowledge the system learned. That's because the end-user's feedback is often unpredictable, and inconsistent from user to user, or even query to query. If the user who gives the feedback is trustworthy and consistent, feedback can be accumulated and added to the knowledge of the system, as was suggested by Lee *et al.* [37].

5.2 QUERY CONCEPT LEARNER

In a query by example [31][1] system, it is often hard to initialise the first query, because the user may not have a good example to begin with. Having got used to text retrieval engines such as Google [56], users may prefer to query the database by keyword. Many systems with keyword annotations can provide such kind of service [32][33][35]. Chang *et al.* recently proposed the SVM Active Learning system [58] and MEGA system [57], which can be an alternate solution.

SVM Active Learning and MEGA have similar ideas but with different tools. They both want to find a query-concept learner that learns query criteria through an intelligent sampling process. No example is needed as the initial query. Instead of browsing the database completely randomly, these two systems ask the user to provide some feedback and try to quickly capture the concept in the user's mind. The key to success is to maximally utilize the user's feedback and quickly reduce the size of the space that the user's concept lies in. Active learning is THE answer.

Active learning is an interesting idea in the machine learning literature. While in traditional machine learning research, the learner typically works as a passive recipient of the data, active learning enables the learner to use its own ability to respond to collect data and to influence the world it is trying to understand. A standard passive learner can be thought of as a student that sits and listens to a teacher, while an active learner is a student that asks the teacher questions, listens to the answers and asks further questions based on the answer. In the literature, active learning has shown very promising results in reducing the number of samples required to finish a certain task [59][60][61].

In practice, the idea of active learning can be translated into a simple rule: if the system is allowed to propose samples and get feedback, always propose those samples that the system is most confused about, or that can bring the greatest information gain.

Following the rule, SVM Active Learning becomes very straightforward. In SVM, objects far away from the separating hyperplane are easy to classify. The most confused objects are those that are close to the boundary. Therefore, during the feedback loop, the system will always propose the images closest to the SVM boundary for the user to annotate.

MEGA system models the query concept space (QCS) in k-CNF and the candidate concept space (CCS) in k-DNF [62]. Here k-CNF and k-DNF are Boolean formulae sets that can virtually model any practical query concepts. The CCS is initialised to be larger than the real concept space, while the QCS is initialised smaller. During the learning process, the QCS keeps being refined by the positive feedbacks, while the CCS keeps shrinking due to the negative samples. The spatial difference between QCS and CCS is the interesting area where most images are undetermined. Based on the idea of active learning, they should be shown to the user for more feedback. Some interesting trade-offs have to be made in selecting these samples [57].

5.3 EFFICIENT ANNOTATION THROUGH ACTIVE LEARNING

Keyword annotation is a very expensive work, as it can only be done manually. It is natural to look for methods that can improve the annotation efficiency. Active learning turns out to be also suitable for this job.

In [33], Zhang and Chen proposed a framework for active learning during the annotation. For each object in the database, they maintain a list of probabilities, each indicating the probability of this object having one of the attributes. During training, the learning algorithm samples objects in the database and presents them to the annotator to assign attributes to. For each sampled object, each probability is set to be one or zero depending on whether or not the corresponding attribute is assigned by the annotator. For objects that have not been annotated, the learning algorithm estimates their probabilities with biased kernel regression. Knowledge gain is then defined to determine, among the objects that have not been annotated, which one the system is the most uncertain of. The system then presents it as the next sample to the annotator to assign attributes to.

Naphade *et al.* proposed a very similar work in [38]. However, they used a support vector machine to learn the semantics. They have essentially the same method as Chang *et al.*'s SVM Active Learning [58] to choose new samples for the annotator to annotate.

6. CONCLUSION

In this chapter, we overviewed the various ways people use to improve the accuracy performance of the content-based information retrieval system. The gap between low-level features and high level semantics has been the main obstacle for developing more successful retrieval systems. It can be expected that it will still remain as one of the most challenging research topics in this field.

REFERENCES

[1]   Y. Rui and T. S. Huang, Image Retrieval: Current Techniques, Promising, Directions and Open Issues, *Journal of Visual Communication and Image Representation*, Vol. 10, No. 4, April 1999.

[2]   A. Nagasaka and Y. Tanaka, Automatic Video Indexing and Full-Video Search for Object Appearance, in Proc. of IFIP 2nd Working Conf. On Visual Database Systems, pp. 113-127, 1992.

[3]   H. J. Zhang, J. H. Wu, D. Zhong, and S. W. Smoliar, Video Parsing, Retrieval and Browsing: An Integrated and Content-Based Solution, *Pattern Recognition*, pp. 643-658, Vol. 30, No. 4, 1997.

[4]   Y. Deng and B. S. Manjunath, Content-Based Search of Video Using Color, Texture and Motion, in Proc. of IEEE Intl. Conf. On Image Processing, pp. 534-537, Vol. 2, 1997.

[5]   P. Aigrain, H. J. Zhang, and D. Petkovic, Content-Based Representation and Retrieval of Visual Media: A State-of-the-Art Review, *International Journal of Multimedia Tools Applications*, pp.179-202, Vol. 3, November 1996.

[6]   J. S. Boresczky and L. A. Rowe, A Comparison of Video Shot Boundary Detection Techniques, in Storage & Retrieval for Image and Video Databases IV, Proc. SPIE 2670, pp. 170-179, 1996.

[7]   W. Wolf, Key frame selection by motion analysis, in IEEE ICASSP, pp. 1228-1231, Vol. 2, 1996.

[8]   M. Flickner *et al.*, Query by Image and Video Content, *IEEE Computer*, pp. 23-32, September 1995.

[9]   M. Yeung, B. L. Yeo, and B. Liu, Extracting Story Units from Long Programs for Video Browsing and Navigation, in Proc. IEEE Conf. on Multimedia Computing and Systems, pp. 296-305, 1996.

[10]   P. Aigrain, P. Joly, and V. Longueville, Medium Knowledge Based Macro-segmentation of Video into Sequences, in *Intelligent Multimedia Information Retrieval*, M. T. Maybury, Ed., pp.159–173. AAAI/MIT Press, 1997.

[11]   A. G. Hauptmann and M. A. Smith, Text, Speech, and Vision for Video Segmentation: The Informedia Project, in AAAI Fall Symposium, Computational Models for Integrating Language and Vision, Boston, 1995.

[12]   R. Lienhart, S. Pfeiffer, and W. Effelsberg, Scene Determination Based on Video and Audio Features, Technical report, University of Mannheim, November 1998.

[13]   J. S. Boreczky and L.D. Wilcox, A Hidden Markov Model Framework for Video Segmentation Using Audio and Image Features, in IEEE ICASSP, pp. 3741-3744, Vol. 6, Seattle, 1998.

[14]   Y. Rui, T. S. Hunag, and S. Mehrotra, Constructing Table-of-Content for Videos, *ACM Multimedia Systems Journal, Special Issue Multimedia Systems on Video Libraries*, pp. 359-368, Vol. 7, No. 5, Sep. 1999.

[15]   D. Zhong, H. Zhang, and S.-F. Chang, Clustering Methods for Video Browsing and Annotation, in SPIE Conference on Storage and Retrieval for Image and Video Databases, pp. 239-246, Vol. 2670, 1996.

[16]   U. Gargi, S. Antani, and R. Kasturi, Indexing Text Events in Digital Video Databases, in Proc. 14th Int'l Conf. Pattern Recognition, pp. 916-918,1998.

[17]   J.C. Shim, C. Dorai, and R. Bolle, Automatic Text Extraction from Video for Content-Based Annotation and Retrieval, in Proc. 14th Int'l Conf. Pattern Recognition, pp. 618-620, 1998.

[18]   J. D. Courtney, Automatic Video Indexing via Object Motion Analysis, Pattern Recognition, pp. 607-625, Vol. 30, No. 4, 1997.

[19]   A. M. Ferman, A. M. Tekalp, and R. Mehrotra, Effective Content Representation for Video, in Proc. of ICIP'98, pp. 521-525, Vol. 3, 1998.

[20]   M. Gelgon and P. Bouthemy, Determining a Structured Spatio-Temporal Representation of Video Content for Efficient Visualization and Indexing, in Proc. 5th Eur. Conf. on Computer Vision, ECCV'98, Freiburg, June 1998.

[21]   Y. F. Ma and H. J. Zhang, Detecting Motion Object by Spatio-Temporal Entropy, in IEEE Int. Conf. on Multimedia and Expo, Tokyo, Japan, August 22-25, 2001.

[22]   R. C. Nelson and R. Polana, Qualitative Recognition of Motion Using Temporal Texture, in Proc. DARPA Image Understanding Workshop, San Diego, CA, pp.555-559, Jan. 1992.

[23]   K. Otsuka, T. Horikoshi, S. Suzuki, and M. Fujii, Feature Extraction of Temporal Texture Based on Spatio-Temporal Motion Trajectory, in Proc. 14th Int. Conf. On Pattern Recognition, ICPR'98, pp.1047-1051, Aug. 1998.

[24]   P. Bouthemy and R. Fablet, Motion Characterization from Temporal Concurrences of Local Motion-Based Measures for Video Indexing, Int. Conf. on Pattern Recognition, ICPR'98, pp. 905-908, Vol. 1, Australia, Aug. 1998.

[25]   M. Szummer and R. W. Picard, Temporal Texture Modeling, in IEEE ICIP'96, pp. 823-826, Sep. 1996.

[26]   D. Zhong and S.F. Chang, Video Object Model and Segmentation for Content-Based Video Indexing, in IEEE Int. Symp. on Circuits and Systems, Hong Kong, June 1997.

[27]   Y. Deng and B. S. Manjunath, Netra-V: Toward an Object-Based Video Representation, *IEEE Transactions CSVT*, pp.616-627, Vol. 8, No. 3, 1998.

[28]   S. F. Chang, W. Chen, and H. Sundaram, Semantic Visual Templates: Linking Visual Features to Semantics, in IEEE ICIP, 1998.

[29]   S. Santini and R. Jain, Similarity Measures, *IEEE Transactions of Pattern Analysis and Machine Intelligence*, pp.871-883, Vol. 21, No. 9, Sep. 1999.

[30]   A. Tversky, Features of Similarity, *Psychological Review*, pp.327-352, Vol. 84, No. 4, July 1977.

[31]   M. Flicker, H. Sawhney, W. Niblack, J. Ashley, Q. Huang, B. Dom, M. Gorkani, J. Hafner, D. Lee, D. Petkovic, D. Steele, and P. Yanker, Query by Image and Video Content: The QBIC System, *IEEE Computer*, pp. 23-32, Vol. 28, No. 9, 1995.

[32]   S. Basu, M. Naphade and J. R. Smith, A Statical Modeling Approach to Content Based Retrieval, in IEEE ICASSP, 2002.

[33]   C. Zhang and T. Chen, An Active Learning Framework for Content Based Information Retrieval, *IEEE Transactions on Multimedia, Special Issue on Multimedia Database*, pp. 260-268, Vol. 4, No. 2, June 2002.

[34]   Y. Park, Efficient Tools for Power Annotation of Visual Contents: A Lexicographical Approach, *ACM Multimedia*, pp. 426-428, 2000.

[35]   M. Slaney, Semantic-Audio Retrieval, in IEEE ICASSP, 2002.

[36]   M. R. Naphade, I. Kozintsev, T. S. Huang, and K. Ramchandran, A Factor Graph Framework for Semantic Indexing and Retrieval in Video, in Proc. IEEE Workshop on Content-based Access of Image and Video Libraries, 2000.

[37]   C. S. Lee, W.-Y. Ma, and H. J. Zhang, Information Embedding Based on User's Relevance Feedback for Image Retrieval, Invited paper, in SPIE Int. Conf. Multimedia Storage and Archiving Systems IV, Boston, pp.19-22, Sep. 1999.

[38]   M. R. Naphade, C. Y. Lin, J. R. Smith, B. Tseng, and S. Basu, Learning to Annotate Video Databases, in SPIE Conference on Storage and Retrieval on Media databases, 2002.

[39]   W. Y. Ma and B. S. Manjunath, Texture Features and Learning Similarity, in IEEE Proceedings CVPR '96, pp. 425-430, 1996.

[40]   D. McG. Squire, Learning a Similarity-Based Distance Measure for Image Database Organization from Human Partitionings of an Image Set, in IEEE Workshop on Applications of Computer Vision (WACV'98), pp.88-93, 1998.

[41]   A. P. Dempster, N. M. Laird, and D. B. Rubin, Maximum-likelihood from Incomplete Data via the EM Algorithm, *Journal of Royal Statistical Society, Ser. B*, pp. 1-38, Vol. 39, No. 1, 1977.

[42]   G. McLachlan and T. Krishnan, *The EM Algorithm and Extensions*, Wiley Series in Probability and Statistics, John Wiley & Sons, New York, 1997.

[43]   R. O. Duda, P. E. Hart, and D. G. Stork, *Pattern Classification (2nd Edition)*, John Wiley & Sons, New York, 2000.

[44] C. Burges, A Tutorial on Support Vector Machines for Pattern Recognition, *Data Mining and Knowledge Discovery*, pp.121-167, Vol. 2, No. 2, 1998.

[45] N. Cristianini and J. Shawe-Taylor, *An Introduction to Support Vector Machines and Other Kernel-Based Learning Methods*, Cambridge University Press, 2000.

[46] E. Chang and B. Li, On Learning Perceptual Distance Function for Image Retrieval, in IEEE ICASSP, 2002.

[47] D. Harman, Relevance Feedback Revisited, in Proc. of the Fifteenth Annual International ACM SIGIR Conf. on Research and Development in Information Retrieval, pp. 1-10, 1992.

[48] G. Salton and C. Buckley, Improving Retrieval Performance by Relevance Feedback, *Journal of the American Society for Information Science*, pp. 288-297, Vol. 41, No. 4, 1990.

[49] Y. Rui, T. S. Huang, M. Ortega, and S. Mehrotra, Relevance Feedback: A Power Tool for Interactive Content-based Image Retrieval, *IEEE Transactions on Circuits and Systems for Video Technology*, pp. 644-655, Vol. 8, No. 5, Sep. 1998.

[50] Y. Ishikawa, R. Subramanya, and C. Faloutsos, Mindreader: Query Database Through Multiple Examples, in Proc. of the 24th VLDB Conf., New York, 1998.

[51] Y. Rui and T. S. Huang, Optimizing Learning in Image Retrieval, in Proc. of IEEE int. Conf. On Computer Vision and Pattern Recognition, Jun. 2000.

[52] Q. Tian, P. Hong, T. S. Huang, Update Relevant Image Weights for Content-based Image Retrieval Using Support Vector Machines, in Proc. Multimedia and Expo IEEE Int. Conf., pp. 1199-1202, Vol. 2 , 2000.

[53] S. Sull, J. Oh, S. Oh, S. M.-H. Song, and S. W. Lee, Relevance Graph-based Image Retrieval, in Proc. Multimedia and Expo IEEE Int. Conf., pp. 713-716, Vol. 2, 2000.

[54] N. D. Doulamis, A. D. Doulamis, and S. D. Kollias, Non-linear Relevance Feedback: Improving the Performance of Content-based Retrieval Systems, in Proc. Multimedia and Expo IEEE Int. Conf., pp. 331-334, Vol. 1, 2000.

[55] T. P. Minka and R. W. Picard, Interactive Learning Using a 'Society of Models', M.I.T Media Laboratory Perceptual Computing Section, Technical Report, No. 349.

[56] www.google.com

[57] E. Chang and B. Li, MEGA – The Maximizing Expected Generalization Algorithm for Learning Complex Query Concepts, *UCSB Technical Report*, August 2001.

[58] S. Tong and E. Chang, Support Vector Machine Active Learning for Image Retrieval, *ACM Multimedia*, 2001.

[59] D. A. Cohn, Z. Ghahramani, and M. I. Jordan, Active Learning with Statistical Models, *Journal of Artificial Intelligence Research*, pp. 129-145, 4, 1996.

[60] A. Krogh and J. Vedelsby, Neural Network Ensembles, Cross Validation, and Active Learning, in *Advances in Neural Information Processing Systems* G. Tesauro, D. Touretzky, and T. Leen, Eds., Vol. 7, MIT Press, Cambridge, MA, 1995.

[61]   D. D. Lewis and W. A. Gale, A Sequential Algorithm for Training Text
        Classifiers, in ACM-SIGIR 94, pp. 3-12, Springer-Verlag, London, 1994.
[62]   M. Kearns, M. Li, and L. Valiant, Learning Boolean Formulae, *Journal of
        ACM*, pp. 1298-1328, Vol. 41, No. 6, 1994.

26

VIDEO INDEXING AND RETRIEVAL USING MPEG-7

John R. Smith
IBM T. J. Watson Research Center
30 Saw Mill River Road
Hawthorne, NY 10532 USA
jrsmith@watson.ibm.com

1. INTRODUCTION

Numerous requirements are driving the need for more effective methods for scarching and retrieving video based on content. Some of the factors include the growing amount of unstructured multimedia data, the broad penetration of the Internet, and new on-line multimedia applications related to distance learning, entertainment, e-commerce, consumer digital photos, music download, mobile media, and interactive digital television. The MPEG-7 multimedia content description standard addresses some aspects of these problems by standardizing a rich set of tools for describing multimedia content in XML [31],[32]. However, MPEG-7 does not standardize methods for extracting descriptions nor for matching and searching. The extraction and use of MPEG-7 descriptions remains a challenge for future research, innovation, and industry competition. As a result, techniques need to be developed for analyzing, extracting, and indexing information from video based using MPEG-7 standard [22][36][37][44]

1.1. VIDEO INDEXING AND RETRIEVAL

The problem of video indexing and retrieval using MPEG-7 involves two processes: (1) producing or extracting MPEG-7 descriptions from video content, and (2) searching for video content based on the MPEG-7 descriptions. In general, in MPEG-7 pull applications the user is actively seeking multimedia content or information. The benefit of MPEG-7 for pull applications is that the queries can be based on standardized descriptions. While content-based video retrieval is useful for many applications such as multimedia databases, intelligent media services, and personalization, many applications require an interface at the semantic level. Ideally, video retrieval involves, for example, description of scenes, objects, events, people, places, and so forth, along with description of features, speech transcripts, closed captions, and so on. MPEG-7 provides rich metadata for describing many aspects of video content including the semantics of real-world scenes related to the video content.

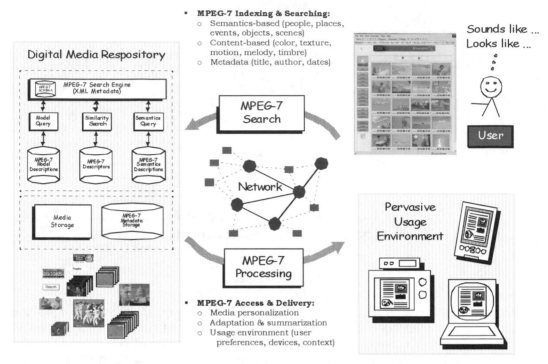

Figure 26.1 MPEG-7 applications include pull-type applications such as multimedia database searching, push-type applications such as multimedia content filtering, and universal multimedia access.

The overall application environment for MPEG-7 video indexing and retrieval is illustrated in Figure 26.1. The environment involves a user seeking information from a digital media repository. The repository stores video content along with corresponding MPEG-7 metadata descriptions. The MPEG-7 metadata potentially gives descriptions of semantics (i.e., people, places, events, objects, scenes, and so on), features (color, texture, motion, melody, timbre, and so on), and other immutable attributes of the digital media (i.e., titles, authors, dates, and so on). The user may be provided with different means for searching the digital media repository, such as by issuing text or key-word queries, by selecting examples of content being sought, by selecting models or illustrating through features, and so on. Another aspect of the environment involves the access and delivery of video from the digital media repository. Given the rich MPEG-7 descriptions, the digital media content can potentially be adapted to the user environment. For example, the video content can be summarized to produce a personalized presentation according to user's preferences, device capabilities, usage context, and so on.

1.1.1. MPEG-7 Standard Elements

MPEG-7 is a standard developed by International Standards Organization (ISO) and International Electrotechnical Commission (IEC), which specifies a "Multimedia Content Description Interface." MPEG-7 provides a standardized representation of multimedia metadata in XML. MPEG-7 describes multimedia content at a number of levels, including features, structure, semantics, models, collections, and other immutable metadata related to multimedia description. The objective of MPEG-7 is to provide an interoperable metadata system that is also designed to allow fast and efficient indexing, searching, and filtering of multimedia based on content. The MPEG-7 standard specifies an industry standard schema using XML Schema Language. The schema is comprised of Description Schemes (DS) and Descriptors. Overall, the MPEG-7 schema defines over 450 simple and complex types. MPEG-7 produces XML descriptions but also provides a binary compression system for MPEG-7 descriptions. The binary compression system allows MPEG-7 descriptions to be more efficiently stored and transmitted. The MPEG-7 descriptions can be stored as files or within databases independent of the multimedia data, or can embedded within the multimedia streams, or broadcast along with multimedia data.

The constructs are defined as follows:

- The Description Definition Language (DDL) is the language specified in MPEG-7 for defining the syntax of Description Schemes and Descriptors. The DDL is based on the XML Schema Language.

- Description Schemes (DS) are description tools defined using DDL that describe entities or relationships pertaining to multimedia content. Description Schemes specify the structure and semantics of their components, which may be Description Schemes, Descriptors, or datatypes. Examples of Description Schemes include: `MovingRegion DS`, `CreationInformation DS`, and `Object DS`.

- Descriptors (D) are description tools defined using DDL that describe features, attributes, or groups of attributes of multimedia content. Example Descriptors include: ScalableColor D, SpatioTemporalLocator D, and AudioSpectrumFlatness D.

- Features are defined as a distinctive characteristic of multimedia content that signifies something to a human observer, such as the "color" or "texture" of an image. This distinguishes Descriptions from Features as follows: consider color to be a feature of an image, then the `ScalableColor D` can be used to describe the color feature.

- Data (Essence, Multimedia Data) is defined as a representation of multimedia in a formalized manner suitable for communication, interpretation, or processing by automatic means. For example, the data can correspond to an image or video.

The MPEG-7 standard specifies the Description Definition Language (DDL) and the set of Description Schemes (DS) and Descriptors that comprise the MPEG-7 schema. However, MPEG-7 is also extensible in that the DDL can be used to define new DSs and Descriptors and extend the MPEG-7 standard DSs and Descriptors. The MPEG-7 schema is defined in such a way that would allow a customer Descriptor to be used together with the standardized MPEG-7 DSs and Descriptors, for example, to include a medical image texture Descriptor within an MPEG-7 image description.

1.1.2. Outline

In this chapter, we examine the application of MPEG-7 for video indexing and retrieval. The chapter is organized as follows: in Section 2, we introduce the

MPEG-7 standard and give examples of the description tools and review the elements of the MPEG-7 standard including description tools and classification schemes. In Section 3, we identify MPEG-7 description tools for video indexing and retrieval and give example descriptions. In Section 4, we discuss video searching. Finally, in Section 5, we examine future directions and make conclusions.

2. METADATA

Metadata has an essential role in multimedia content management and is critical for describing essential aspects of multimedia content, including main topics, author, language, events, scenes, objects, times, places, rights, packaging, access control, content adaptation, and so forth. Conformity with open metadata standards will be vital for multimedia content management systems to allow faster design and implementation, interoperability with broad field of competitive standards-based tools and systems, and leveraging of rich set of standards-based technologies for critical functions such as content extraction, advanced search, and personalization

2.1. MPEG-7 STANDARD SCOPE

The scope of the MPEG-7 standard is shown in Figure 26.2. The normative scope of MPEG-7 includes Description Schemes (DSs), Descriptors (Ds), the Description Definition Language (DDL), and Coding Schemes (CS). MPEG-7 standardizes the syntax and semantics of each DS and D to allow interoperability. The DDL is based on XML Schema Language. The DDL is used to define the syntax of the MPEG-7 DSs and Ds. The DLL allows the standard MPEG-7 schema to be extended for customized applications.

The MPEG-7 standard is "open" on two sides of the standard in that the methods for extraction and use of MPEG-7 descriptions are not defined by the standard. As a result, methods, algorithms, and systems for content analysis, feature extraction, annotation, and authoring of MPEG-7 descriptions are open for industry competition and future innovation. Likewise, methods, algorithms, and systems for searching and filtering, classification, complex querying, indexing, and personalization are also open for industry competition and future innovation.

2.2. MULTIMEDIA DESCRIPTION SCHEMES

The MPEG-7 Multimedia Content Description Interface standard specifies generic description tools pertaining to multimedia including audio and visual content. The MDS description tools are categorized as (1) basic elements, (2) tools for describing content and related metadata, (3) tools for describing content organization, navigation and access, and user interaction, and (4) classification schemes [28][32].

2.2.1. Basic Elements

The basic elements form the building blocks for the higher-description tools. The following basic elements are defined:

- Schema tools. Specifies the base type hierarchy of the description tools, the root element and top-level tools, the multimedia content entity tools, and the package and description metadata tools.

- Basic datatypes. Specifies the basic datatypes such as integers, reals, vectors, and matrices, which are used by description tools.

- Linking and media localization tools. Specifies the basic datatypes that are used for referencing within descriptions and linking of descriptions to multimedia content, such as spatial and temporal localization.

- Basic description tools. Specifies basic tools that are used as components for building other description tools such as language, text, and classification schemes.

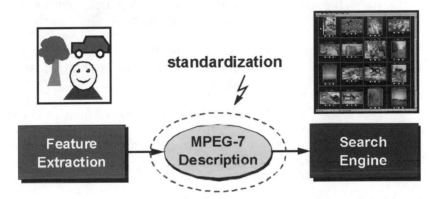

Extraction	MPEG-7 Scope	Use
Content analysis (D, DS)	Description Schemes (DSs)	Searching & filtering
Feature extraction (D, DS)	Descriptors (Ds)	Classification
Annotation tools (DS)	Language (DDL)	Complex querying
Authoring (DS)	Coding Schemes (CS)	Indexing

Figure 26.2 Overview of the normative scope of MPEG-7 standard. The methods for extraction and use of MPEG-7 descriptions are not standardized.

2.2.2. Content Description Tools

The content description tools describe the features of the multimedia content and the immutable metadata related to the multimedia content. The following description tools for content metadata are defined:

- Media description. Describes the storage of the multimedia data. The media features include the format, encoding, storage media. The tools allow multiple media description instances for the same multimedia content.

- Creation & production. Describes the creation and production of the multimedia content. The creation and production features include title, creator, classification, purpose of the creation, and so forth. The creation and production information is typically not extracted from the content but corresponds to metadata related to the content.

- Usage. Describes the usage of the multimedia content. The usage features include access rights, publication, and financial information. The usage information may change during the lifetime of the multimedia content.

The following description tools for content description are defined:

- Structure description tools. Describes the structure of the multimedia content. The structural features include spatial, temporal, or spatio-temporal segments of the multimedia content.

- Semantic description tools. Describes the "real-world" semantics related to or captured by the multimedia content. The semantic features include objects, events, concepts, and so forth.

The content description and metadata tools are related in the sense that the content description tools use the content metadata tools. For example, a description of creation and production or media information can be attached to an individual video or video segment in order to describe the structure and creation and production of the multimedia content.

2.2.3. Content Organization, Navigation, and User Interaction

The tools for organization, navigation and access, and user interaction are defined as follows:

- Content organization. Describes the organization and modeling of multimedia content. The content organization tools include collections, probability models, analytic models, cluster models, and classification models.

- Navigation and Access. Describes the navigation and access of multimedia such as multimedia summaries and abstracts; partitions, views and decompositions of image, video, and audio signals in space, time and frequency; and relationships between different variations of multimedia content.

- User Interaction. Describes user preferences pertaining to multimedia content and usage history of users of multimedia content.

2.2.4. Classification Schemes

A classification scheme is a list of defined terms and their meanings. The MPEG-7 classification schemes organize terms that are used by the description tools. Applications need not use the classification schemes defined in the MPEG-7 standard. They can use proprietary or third party ones. However, if they choose to use the MPEG-7 standard classification schemes defined, no modifications or extensions are allowed. Furthermore, MPEG-7 has defined requirements for a registration authority for MPEG-7 classification schemes, which allows third parties to define and register classification schemes for use by others. All of the MPEG-7 classification schemes are specified using the `ClassificationScheme DS`, that is, they are themselves MPEG-7 descriptions.

2.2.5. MPEG-7 Schema and Values

While MPEG-7 fully specifies the syntax and semantics of the multimedia content description metadata, and allows for extensibility, another important aspect of metadata is the domain of values that populate the structures. We view this as the distinction of the syntax (schema definition) and terms (values). As shown in Figure 26.3, MPEG-7 defines a core set of Descriptors and Description Schemes that form the syntax of the standard schema. This core set is also extensible in that third parties can define new Descriptors and Description Schemes. On the other side, values and terms are required for instantiating the Descriptors and Description Schemes. It is possible for the values to be restricted or managed through MPEG-7 Classification Schemes or Controlled Terms. MPEG-7 specifies a core set of Classification Schemes. However, this core set is also extensible. A registration authority allows for the registration of classification schemes and controlled terms.

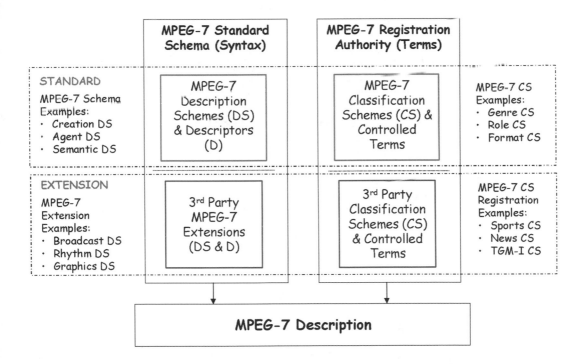

Figure 26.3 The MPEG-7 classification schemes organize terms that are used by the description tools.

2.2.6. Classification Scheme Examples

The following example gives an MPEG-7 description that uses controlled terms from MPEG-7 classification schemes. The example describes the creation information and semantics associated with a video. The creation information describes a creator who has the role of "publisher." The term "publisher" is a controlled term from an MPEG-7 classification scheme. The semantic information describes a sport depicted in the video. The sport "Baseball" is a controlled term that is referenced as term 1.3.4 in the sports classification scheme referenced by the corresponding URN.

```
<Mpeg7>
  <Description xsi:type="CreationDescriptionType">
    <CreationInformation>
      <Creation>
        <Title type="popular">
          All Star Game</Title>
        <Creator>
          <Role href="urn:mpeg:mpeg7:cs:RoleCS:2002:PUBLISHER"/>
          <Agent xsi:type="OrganizationType">
            <Name>
              The Baseball Channel</Name>
          </Agent>
        </Creator>
      </Creation>
    </CreationInformation>
  </Description>
  <Description xsi:type="SemanticDescriptionType">
    <Semantics>
      <Label href="urn:sports:usa:2002:Sports:1.3.4">
        <Name xml:lang="en">
          Baseball</Name>
      </Label>
      <MediaOccurrence>
        <MediaLocator>
          <MediaUri>
            video.mpg </MediaUri>
        </MediaLocator>
      </MediaOccurrence>
    </Semantics>
  </Description>
</Mpeg7>
```

2.2.7. Example MDS Descriptions

The following examples illustrate the use of different MPEG-7 Multimedia Description Schemes in describing multimedia content.

2.2.7.1.   Creation Information

The following example gives an MPEG-7 description of the creation information for a sports video.

```
<Mpeg7>
  <Description xsi:type="CreationDescriptionType">
    <CreationInformation>
      <Creation>
        <Title type="popular">Subway series</Title>
        <Abstract>
          <FreeTextAnnotation>
            Game among city rivals</FreeTextAnnotation>
        </Abstract>
        <Creator>
          <Role href="urn:mpeg:mpeg7:cs:RoleCS:2001:PUBLISHER"/>
          <Agent xsi:type="OrganizationType">
            <Name>Sports Channel</Name>
          </Agent>
        </Creator>
      </Creation>
    </CreationInformation>
  </Description>
</Mpeg7>
```

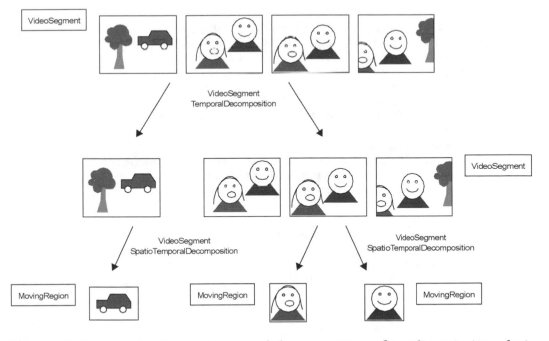

Figure 26.4 Example showing temporal decomposition of a video into two shots and the spatio-temporal of each shot into moving regions.

2.2.7.2.   Free Text Annotation

The following example gives an MPEG-7 description of a car that is depicted in an image.

```
<Mpeg7>
  <Description xsi:type="SemanticDescriptionType">
    <Semantics>
      <Label>
        <Name>Car </Name>
      </Label>
      <Definition>
        <FreeTextAnnotation>
          Four wheel motorized vehicle</FreeTextAnnotation>
      </Definition>
      <MediaOccurrence>
        <MediaLocator>
          <MediaUri>image.jpg </MediaUri>
        </MediaLocator>
      </MediaOccurrence>
    </Semantics>
  </Description>
</Mpeg7>
```

2.2.7.3.   Collection Model

The following example gives an MPEG-7 description of a collection model of "sunsets" that contains two images depicting sunset scenes.

```
<Mpeg7>
  <Description xsi:type="ModelDescriptionType">
    <Model xsi:type="CollectionModelType" confidence="0.75"
      reliability="0.5" function="described">
      <Label>
        <Name>Sunsets</Name>
      </Label>
      <Collection xsi:type="ContentCollectionType">
        <Content xsi:type="ImageType">
          <Image>
            <MediaLocator xsi:type="ImageLocatorType">
              <MediaUri>sunset1.jpg</MediaUri>
            </MediaLocator>
          </Image>
        </Content>
        <Content xsi:type="ImageType">
          <Image>
            <MediaLocator xsi:type="ImageLocatorType">
              <MediaUri>sunset2.jpg</MediaUri>
            </MediaLocator>
          </Image>
        </Content>
      </Collection>
    </Model>
  </Description>
</Mpeg7>
```

2.2.7.4.   Video Segment

The following example gives an MPEG-7 description of the decomposition of a video segment. The video segment is first decomposed temporally into two video segments. The first video segment is decomposed into a single moving region. The second video segment is decomposed into two moving regions.

```xml
<Mpeg7>
  <Description xsi:type="ContentEntityType">
    <MultimediaContent xsi:type="VideoType">
      <Video>
        <TemporalDecomposition gap="false" overlap="false">
          <VideoSegment id="shot1">
            <SpatioTemporalDecomposition>
              <MovingRegion id="car">
                <!-- more elements -->
              </MovingRegion>
            </SpatioTemporalDecomposition>
          </VideoSegment>
          <VideoSegment id="shot2">
            <SpatioTemporalDecomposition>
              <MovingRegion id="person1">
                <!-- more elements -->
              </MovingRegion>
            </SpatioTemporalDecomposition>
            <SpatioTemporalDecomposition>
              <MovingRegion id="person2">
                <!-- more elements -->
              </MovingRegion>
            </SpatioTemporalDecomposition>
          </VideoSegment>
        </TemporalDecomposition>
      </Video>
    </MultimediaContent>
  </Description>
</Mpeg7>
```

Figure 26.5. Example image showing two people shaking hands.

2.2.7.5.   Semantic Event

The following example gives an MPEG-7 description of the event of a handshake between people. Person A is described as the agent or initiator of the handshake event and Person B is described as the accompanier or joint agent of the handshake.

```
<Mpeg7>
  <Description xsi:type="SemanticDescriptionType">
    <Semantics>
      <Label>
        <Name>
          Shake hands </Name>
      </Label>
      <SemanticBase xsi:type="AgentObjectType" id="A">
        <Label href="urn:example:acs">
          <Name>Person A </Name>
        </Label>
      </SemanticBase>
      <SemanticBase xsi:type="AgentObjectType" id="B">
        <Label href="urn:example:acs">
          <Name>Person B </Name>
        </Label>
      </SemanticBase>
      <SemanticBase xsi:type="EventType">
        <Label>
          <Name>Handshake </Name>
        </Label>
        <Definition>
          <FreeTextAnnotation>
            Clasping of right hands by two people</FreeTextAnnotation>
        </Definition>
        <Relation
            type="urn:mpeg:mpeg7:cs:SemanticRelationCS:2001:agent"
            target="#A"/>
        <Relation
            type="urn:mpeg:mpeg7:cs:SemanticRelationCS:2001:accompanier"
            target="#B"/>
      </SemanticBase>
    </Semantics>
  </Description>
</Mpeg7>
```

2.3.   AUDIO DESCRIPTION TOOLS

The MPEG-7 Audio description tools describe audio data [25]. The audio description tools are categorized as low-level and high-level. The low-level tools describe features of audio segments [24]. The high-level tools describe structure of audio content or provide application-level descriptions of audio.

2.3.1. Low-level Audio Tools

The following low-level audio tools are defined in MPEG-7:

- Audio Waveform: Describes the audio waveform envelope for display purposes.

- Audio Power: Describes temporally-smoothed instantaneous power, which is equivalent to square of waveform values.

- Audio Spectrum: Describes features such as audio spectrum envelope (spectrum of the audio according to a logarithmic frequency scale), spectrum centroid (center of gravity of the log-frequency power spectrum), spectrum spread (second moment of the log-frequency power spectrum), spectrum flatness (flatness properties of the spectrum of an audio signal within a given number of frequency bands), spectrum basis (basis functions that are used to project high-dimensional spectrum descriptions into a low-dimensional representation),

- Harmonicity: Describes the degree of harmonicity of an audio signal.

- Silence: Describes a perceptual feature of a sound track capturing the fact that no significant sound is occurring.

2.3.2. High-level Audio Tools

The following high-level audio tools are defined in MPEG-7:

- Audio Signature: Describes a signature extracted from the audio signal that is designed to provide a unique content identifier for purposes of robust identification of the audio signal.

- Timbre: Describes the perceptual feature of an instrument that makes two sounds having the same pitch and loudness sound different. The Timbre Descriptors relate to notions such as "attack," "brightness," or "richness" of a sound.

- Sound Recognition and Indexing: Supports applications that involve audio classification and indexing. The tools include Description Schemes for Sound Model and Sound Classification Model, and allow description of finite state models using Description Schemes for Sound Model State Path and Sound Model State Histogram.

- Spoken Content: Describes the output of an automatic speech recognition (ASR) engine including the lattice and speaker information.

- Melody: describes monophonic melodic information that facilitates efficient, robust, and expressive similarity matching of melodies.

2.3.3. Example Audio Descriptions

The following example describes a melody contour of a song:

```
<Mpeg7>
  <Description xsi:type="ContentEntityType">
    <MultimediaContent xsi:type="AudioType">
      <Audio>
        <AudioDescriptionScheme xsi:type="MelodyType">
          <Meter>
            <Numerator>3</Numerator>
            <Denominator>4</Denominator>
          </Meter>
          <MelodyContour>
```

```
            <Contour>2 -1 -2 1 -1 1 -1</Contour>
            <Beat>1 4 5 7 8 9 9 10</Beat>
          </MelodyContour>
        </AudioDescriptionScheme>
      </Audio>
    </MultimediaContent>
  </Description>
</Mpeg7>
```

The following example describes a continuous hidden Markov model of audio sound effects. Each continuous hidden Markov model has five states and represents a sound effect class. The parameters of the continuous density state model can be estimated via training, for example, using the Baum-Welch algorithm. After training, the continuous HMM model consists of a 3x3 state transition matrix, a 3x1 initial state density matrix, and 3 multi-dimensional Gaussian distributions defined in terms of the mean and variance parameters. Each multi-dimensional Gaussian distribution has six dimensions corresponding to audio features described by the `AudioSpectrumFlatness` D.

```
<Mpeg7>
  <Description xsi:type="ModelDescriptionType">
    <Model xsi:type="ContinuousHiddenMarkovModelType" numOfStates="3">
      <Initial mpeg7:dim="5">
        0.1 0.2 0.1 </Initial>
      <Transitions mpeg7:dim="3 3">
        0.2 0.2 0.6 0.1 0.2 0.1 0.4 0.2 0.1 </Transitions>
      <State>
        <Label>
          <Name>State 1 </Name>
        </Label>
      </State>
      <State>
        <Label>
          <Name>State 2 </Name>
        </Label>
      </State>
      <State>
        <Label>
          <Name>State 3 </Name>
        </Label>
      </State>
      <DescriptorModel>
        <Descriptor xsi:type="AudioSpectrumFlatnessType"
          loEdge="250" highEdge="1600">
          <Vector>1 2 3 4 5 6 </Vector>
        </Descriptor>
        <Field>
          Vector</Field>
      </DescriptorModel>
      <ObservationDistribution xsi:type="GaussianDistributionType"
        dim="6">
        <Mean mpeg7:dim="6">0.5 0.5 0.25 0.3 0.5 0.3 </Mean>
        <Variance mpeg7:dim="6">
          0.25 0.75 0.5 0.45 0.75 0.3</Variance>
      </ObservationDistribution>
      <ObservationDistribution xsi:type="GaussianDistributionType"
        dim="6">
```

```
      <Mean mpeg7:dim="6">0.25 0.4 0.25 0.3 0.2 0.1</Mean>
      <Variance mpeg7:dim="6">
        0.5 0.25 0.5 0.45 0.5 0.2</Variance>
    </ObservationDistribution>
    <ObservationDistribution xsi:type="GaussianDistributionType"
      dim="6">
      <Mean mpeg7:dim="6">0.2 0.5 0.35 0.3 0.5 0.5</Mean>
      <Variance mpeg7:dim="6">
        0.5 0.5 0.5 0.5 0.75 0.5</Variance>
    </ObservationDistribution>
  </Model>
</Description>
</Mpeg7>
```

2.4. MPEG-7 VISUAL DESCRIPTION TOOLS

The MPEG-7 Visual description tools describe visual data such as images and video. The tools describe feature such as color, texture, shape, motion, localization, and faces [5][6][7][8][11][15][16][17][18][21][22][46].

2.4.1. Color

The color description tools describe color information including color spaces and quantization of color spaces. Different color descriptors are provided to describe different features of visual data. The DominantColor D describes a set of dominant colors of an arbitrarily shaped region of an image. The ScalableColor D describes the histogram of colors of an image in HSV color space. The ColorLayout D describes the spatial distribution of colors in an image. The ColorStructure D describes local color structure in an image by means of a structuring element. The GoFGoPColor D describes the color histogram aggregated over multiple images or frames of video.

2.4.2. Texture

The HomogeneousTexture D describes texture features of images or regions based on energy of spatial-frequency channels computed using Gabor filters. The TextureBrowsing D describes texture features in terms of regularity, coarseness, and directionality. The EdgeHistogram D describes the spatial distribution of five types of edges in image regions.

2.4.3. Shape

The RegionShape D describes the region-based shape of an object using Angular Radial Transform (ART). The ContourShape D describes a closed contour of a 2D object or region in an image or video based on Curvature Scale Space (CSS) representation. The 3DShape D describes an intrinsic shape description for 3D mesh models based on a shape index value.

2.4.4. Motion

The CameraMotion D describes 3-D camera motion parameters, which includes camera track, boom, and dolly motion modes; and camera pan, tilt and roll motion modes. The MotionTrajectory D describes motion trajectory of a moving object based on spatio-temporal localization of representative trajectory points. The `ParametricMotion` D describes motion in video sequences including global motion and object motion by describing the evolution of arbitrarily shaped regions over time in terms of a 2-D geometric transform. The `MotionActivity` D describes the intensity of motion in a video segment.

2.4.5. Localization

The Localization Descriptors describe the location of regions of interest in the space and jointly in space and time. The `RegionLocator` describes the localization of regions using a box or polygon. The `SpatioTemporalLocator` describes the localization of spatio-temporal regions in a video sequence using a set of reference regions and their motions.

2.4.6. Face

The `FaceRecognition` D describes the projection of a face vector onto a set of 48 basis vectors that span the space of possible face vectors.

2.4.7. Example Video Descriptions

The following example uses the `ScalableColor` D to describe a photographic image depicting a sunset.

```
<Mpeg7>
  <Description xsi:type="ContentEntityType">
    <MultimediaContent xsi:type="ImageType">
      <Image>
        <MediaLocator>
          <MediaUri>
            image.jpg</MediaUri>
        </MediaLocator>
        <TextAnnotation>
          <FreeTextAnnotation>
            Sunset scene </FreeTextAnnotation>
        </TextAnnotation>
        <VisualDescriptor xsi:type="ScalableColorType" numOfCoeff="16"
          numOfBitplanesDiscarded="0">
          <Coeff>
            1 2 3 4 5 6 7 8 9 0 1 2 3 4 5 6 </Coeff>
        </VisualDescriptor>
      </Image>
    </MultimediaContent>
  </Description>
</Mpeg7>
```

The following example uses the `GoFGoPColor` D to describe a video segment.

```
<VideoSegment>
  <VisualDescriptor xsi:type="GoFGoPColorType" aggregation="Average">
    <ScalableColor numOfCoeff="16" numOfBitplanesDiscarded="0">
      <Coeff>
        1 2 3 4 5 6 7 8 9 0 1 2 3 4 5 6 </Coeff>
    </ScalableColor>
  </VisualDescriptor>
</VideoSegment>
```

The following example uses the RegionShape D to describe a probability model that characterizes oval shapes.

```
<Mpeg7>
  <Description xsi:type="ModelDescriptionType">
    <Model xsi:type="ProbabilityModelClassType" confidence="0.75"
      reliability="0.5">
      <Label relevance="0.75">
        <Name>
          Ovals </Name>
      </Label>
      <DescriptorModel>
        <Descriptor xsi:type="RegionShapeType">
          <MagnitudeOfART>
            3 5 2 5 . . . 6 </MagnitudeOfART>
        </Descriptor>
        <Field>
          MagnitudeOfART</Field>
      </DescriptorModel>
      <ProbabilityModel xsi:type="ProbabilityDistributionType"
        confidence="1.0" dim="35">
        <Mean dim="35">
          4 8 6 9 . . . 5 </Mean>
        <Variance dim="35">
          1.3 2.5 5.0 4.5 . . . 3.2 </Variance>
      </ProbabilityModel>
    </Model>
  </Description>
</Mpeg7>
```

3. VIDEO INDEXING

Rich description of video content is essential for effective video indexing and retrieval applications [42]. Video indexing typically involves the following processing: (1) shot boundary detection – in which the video is partitioned into temporal units, (2) key-frame selection – in which representative frames are selected from each temporal unit, (3) textual annotation or speech transcription – in which textual information is associated with each temporal unit based on human description or speech transcription, (4) feature description – in which feature descriptors are extracted for each temporal unit, possibly from selected key-frame images (i.e., [35]), (5) semantic annotation – in which labels are assigned to describe the semantic content of each temporal unit (i.e., [36]). MPEG-7 provides a rich set of description tools for describing these aspects of video content. The following examples show the use of MPEG-7 for video indexing and retrieval.

3.1.   SHOT BOUNDARY DESCRIPTION

The following example describes the shot boundary segmentation results for video using MPEG-7. In this example, the video has been partitioned into five video segments. The MPEG-7 description indicates the media time for each segment in terms of media time point and duration. The description indicates that there are no gaps or overlaps among the segments.

```
<Mpeg7>
  <Description xsi:type="ContentEntityType">
    <MultimediaContent xsi:type="VideoType">
      <Video id="10">
        <MediaTime>
          <MediaTimePoint>
            T00:00:00:0F30000</MediaTimePoint>
          <MediaDuration>
            PT10M13S22394N30000F</MediaDuration>
        </MediaTime>
        <TemporalDecomposition gap="false" overlap="false">
          <VideoSegment id="shot10_1">
            <MediaTime>
              <MediaTimePoint>
                T00:00:00:0F30000</MediaTimePoint>
              <MediaDuration>
                PT2S19079N30000F</MediaDuration>
            </MediaTime>
          </VideoSegment>
          <VideoSegment id="shot10_2">
            <MediaTime>
              <MediaTimePoint>
                T00:00:02:19079F30000</MediaTimePoint>
              <MediaDuration>
                PT3S2092N30000F</MediaDuration>
            </MediaTime>
          </VideoSegment>
          <VideoSegment id="shot10_3">
            <MediaTime>
              <MediaTimePoint>
                T00:00:05:21171F30000</MediaTimePoint>
              <MediaDuration>
                PT4S1121N30000F</MediaDuration>
            </MediaTime>
          </VideoSegment>
          <VideoSegment id="shot10_4">
            <MediaTime>
              <MediaTimePoint>
                T00:00:09:22292F30000</MediaTimePoint>
              <MediaDuration>
                PT2S26086N30000F</MediaDuration>
            </MediaTime>
          </VideoSegment>
          <VideoSegment id="shot10_5">
            <MediaTime>
              <MediaTimePoint>
                T00:00:12:18378F30000</MediaTimePoint>
              <MediaDuration>
                PT9S24294N30000F</MediaDuration>
            </MediaTime>
          </VideoSegment>
        </TemporalDecomposition>
      </Video>
    </MultimediaContent>
  </Description>
</Mpeg7>
```

3.2.   TEXTUAL ANNOTATION AND SPEECH TRANSCRIPTION

The following example describes transcription of speech for temporal segments of video. In this example, the video has been partitioned into three video segments. The MPEG-7 description indicates a free text annotation for each segment in which the type of text annotation is "transcript" to indicate that it results from speech transcription of the video.

```
<Mpeg7>
  <Description xsi:type="ContentEntityType">
    <MultimediaContent xsi:type="VideoType">
      <Video id="videonnn">
        <TemporalDecomposition gap="false" overlap="false">
          <VideoSegment id="shotxxxxxx">
            <TextAnnotation type="transcript">
              <FreeTextAnnotation> Once upon a time.
              </FreeTextAnnotation>
            </TextAnnotation>
          </VideoSegment>
          <VideoSegment id="shotyyyyyy">
            <TextAnnotation type="transcript">
              <FreeTextAnnotation> There were three bears.
              </FreeTextAnnotation>
            </TextAnnotation>
            <MediaTime>
              <MediaTimePoint>T00:00:10</MediaTimePoint>
              <MediaDuration>PT1M20S</MediaDuration>
            </MediaTime>
          </VideoSegment>
          <VideoSegment id="shotzzzzzz">
            <TextAnnotation type="transcript">
              <FreeTextAnnotation> They were hungry bears.
              </FreeTextAnnotation>
            </TextAnnotation>
          </VideoSegment>
        </TemporalDecomposition>
      </Video>
    </MultimediaContent>
  </Description>
</Mpeg7>
```

3.3.   FEATURE DESCRIPTION

The following example uses the ScalableColor D to describe the color distribution of a key-frame image from a video segment. In this example, the key-frame image depicts a sunset.

```
<Mpeg7>
  <Description xsi:type="ContentEntityType">
    <MultimediaContent xsi:type="VideoType">
      <Video id="videonnn">
        <TemporalDecomposition gap="false" overlap="false">
          <VideoSegment id="shotxxxxxx">
            <SpatioTemporalDecomposition>
              <StillRegion>
                <MediaIncrTimePoint timeUnit="PT1001N30000F">
```

```
                   0 </MediaIncrTimePoint>
                  <SpatialDecomposition>
                    <StillRegion>
                      <TextAnnotation>
                        <FreeTextAnnotation>
                          Sunset scene </FreeTextAnnotation>
                      </TextAnnotation>
                      <VisualDescriptor xsi:type="ScalableColorType"
                        numOfCoeff="16" numOfBitplanesDiscarded="0">
                        <Coeff>
                           1 2 3 4 5 6 7 8 9 0 1 2 3 4 5 6 </Coeff>
                      </VisualDescriptor>
                    </StillRegion>
                  </SpatialDecomposition>
                </StillRegion>
              </SpatioTemporalDecomposition>
            </VideoSegment>
          </TemporalDecomposition>
        </Video>
      </MultimediaContent>
    </Description>
</Mpeg7>
```

The following example uses the GoFGoPColor D to describe the overall color content of a video segment.

```
<Mpeg7>
  <Description xsi:type="ContentEntityType">
    <MultimediaContent xsi:type="VideoType">
      <Video id="videonnn">
        <TemporalDecomposition gap="false" overlap="false">
          <VideoSegment id="shotxxxxxx">
            <VisualDescriptor xsi:type="GoFGoPColorType"
              aggregation="Average">
              <ScalableColor numOfCoeff="16"
                numOfBitplanesDiscarded="0">
                <Coeff>1 2 3 4 5 6 7 8 9 0 1 2 3 4 5 6 </Coeff>
              </ScalableColor>
            </VisualDescriptor>
          </VideoSegment>
        </TemporalDecomposition>
      </Video>
    </MultimediaContent>
  </Description>
</Mpeg7>
```

3.4.   SEMANTIC DESCRIPTION

The following example describes the semantics of temporal segments of video using MPEG-7. This example shows a single temporal segment. The MPEG-7 description indicates a free text annotation for the segment in which the type of text annotation is "scene description" to indicate that it describes the video scene contents.

```
<Mpeg7>
  <ContentDescription xsi:type="ContentEntityType">
    <MultimediaContent xsi:type="VideoType">
      <Video>
        <TemporalDecomposition>
          <VideoSegment>
            <TextAnnotation type="scene description" relevance="1"
              confidence="1">
              <FreeTextAnnotation>
                Sky
              </FreeTextAnnotation>
              <FreeTextAnnotation>
                Water_Body
              </FreeTextAnnotation>
              <FreeTextAnnotation>
                Boat
              </FreeTextAnnotation>
            </TextAnnotation>
            <MediaTime>
              <MediaTimePoint> T00:00:00:0F30000 </MediaTimePoint>
              <MediaIncrDuration timeUnit="PT1001N30000F"> 486
              </MediaIncrDuration>
            </MediaTime>
          </VideoSegment>
        </TemporalDecomposition>
      </Video>
    </MultimediaContent>
  </ContentDescription>
</Mpeg7>
```

The following example describes the semantics of temporal segments of video using MPEG-7 in which there is some uncertainty in the assigned semantic labels. This uncertainty may result from the use of automatic methods for labeling the content [36]. This example shows a single temporal segment. The MPEG-7 description indicates a keyword annotation for each segment. The keywords may belong to a set of controlled terms defined by an MPEG-7 classification scheme. In this example, the first video segment has been labeled "face" with a confidence of 0.75. The second segment has been labeled "outdoors" with a confidence of 0.9.

```
<Mpeg7>
  <Description xsi:type="ContentEntityType">
    <MultimediaContent xsi:type="VideoType">
      <Video id="videonnn">
        <MediaTime>
          <MediaTimePoint>T00:00:00</MediaTimePoint>
          <MediaDuration>PT1M45S</MediaDuration>
        </MediaTime>
        <TemporalDecomposition gap="false" overlap="false">
          <VideoSegment id="shotxxxxxx">
            <TextAnnotation confidence="0.5" relevance="0.75"
              type="label">
            <KeywordAnnotation>
              <Keyword>Face</Keyword>
            </KeywordAnnotation>
            </TextAnnotation>
            <MediaTime>
              <MediaTimePoint>T00:00:00</MediaTimePoint>
```

```
            <MediaDuration>PT0M10S</MediaDuration>
          </MediaTime>
        </VideoSegment>
        <VideoSegment id="shotyyyyyy">
          <TextAnnotation confidence="0.2" relevance="0.5"
            type="label">
            <KeywordAnnotation>
              <Keyword>Outdoors</Keyword>
            </KeywordAnnotation>
          </TextAnnotation>
          <MediaTime>
            <MediaTimePoint>T00:00:10</MediaTimePoint>
            <MediaDuration>PT1M20S</MediaDuration>
          </MediaTime>
        </VideoSegment>
        <VideoSegment id="shotzzzzzz">
          <TextAnnotation confidence="0.9" relevance="1.0"
            type="label">
            <KeywordAnnotation>
              <Keyword>Indoors</Keyword>
            </KeywordAnnotation>
          </TextAnnotation>
          <MediaTime>
            <MediaTimePoint>T00:00:30</MediaTimePoint>
            <MediaDuration>PT1M15S</MediaDuration>
          </MediaTime>
        </VideoSegment>
        <!-- insert more video shots and labels -->
      </TemporalDecomposition>
    </Video>
  </MultimediaContent>
 </Description>
</Mpeg7>
```

4. VIDEO SEARCH

Given these rich MPEG-7 descriptions, it is possible to search for video content at a number of different levels, including features, models, and semantics [35]. The shot boundary description defines the basic unit of matching and retrieval as a video shot or segment. Users may issue queries to the system in several ways: (1) feature-based – by selecting example key-frame images and video segments in which matches are found based on the MPEG-7 feature descriptions, (2) text-based – by issuing text query which is matched against the MPEG-7 textual annotations or speech transcriptions, (3) semantics-based – by issuing text queries or selecting from key-words that are part of MPEG-7 classification scheme, and (4) model-based – by selecting key-words. The distinction between model-based and semantics-based is that the confidence score for model-based search can be used for ranking and fusing results during the video searching process. An example user-interface for MPEG-7 video search engine is shown in Figure 26.6.

Figure 26.6 The MPEG-7 video search engine allows searching based on features, models, and semantics of the video content.

5. CONCLUSIONS

MPEG is continuing to work on MPEG-7 to define new Descriptors and Description Schemes and to complete the specification of MPEG-7 conformance. MPEG is also working to amend the MPEG-2 standard to allow carriage of MPEG-7 descriptions within MPEG-2 streams. An MPEG-7 industry focus group has also been formed to bring together organizations interested in development and deployment of MPEG-7 systems and solutions.

With the growing interest in digital video, efficient and effective systems are needed for video indexing and retrieval. The recently developed MPEG-7 standard provides a standardized metadata system for describing video content using XML. In this paper, we investigated the application of MPEG-7 for indexing and retrieval of video and discussed example instances of MPEG-7 description tools. We showed that MPEG-7 tools can be used to provide rich description of video content in terms of shot boundaries & key-frames, textual annotations and transcriptions, features, semantics, and models. These aspects allow for video searching based on rich description of content and can be used together in an integrated fashion for effective indexing and retrieval of video.

ACKNOWLEDGMENTS

The author gratefully acknowledges members of the Moving Picture Experts Group and the following co-Project Editors of MPEG-7 Multimedia Description

Schemes: Peter van Beek, Ana Benitez, Joerg Heuer, Jose Martinez, Philippe Salembier, Yoshi Shibata, and Toby Walker.

REFERENCES

[1] O. Avaro and P. Salembier, MPEG-7 Systems: Overview, *IEEE Transactions on Circuits and Systems for Video Technology*, 11(6):760-764, June 2001.

[2] A. B. Benitez, et al., Object-based multimedia content description schemes and applications for MPEG-7, *Journal of Signal Processing: Image Communication*, 1-2, pp.235-269, Sept. 2000.

[3] A. B. Benitez, D. Zhong, S.-F. Chang, and J. R. Smith, MPEG-7 MDS Content Description Tools and Applications, in Proc. Computer Analysis of Images and Patterns, CAIP 2001, W. Skarbek Ed., Sept. 5-7, 2001, Lecture Notes in Computer Science 2124 Springer 2001.

[4] A. B. Benitez and J. R. Smith, MediaNet: A Multimedia Information Network for Knowledge Representation, in Proc. SPIE Photonics East, Internet Multimedia Management Systems, November, 2000.

[5] M. Bober, MPEG-7: Evolution or Revolution?, in Proc. Computer Analysis of Images and Patterns, CAIP 2001, W. Skarbek Ed., Warsaw, Poland, Sept. 5-7, 2001, Lecture Notes in Computer Science 2124 Springer 2001.

[6] M. Bober, MPEG-7 visual shape descriptors, *IEEE Transactions on Circuits & Systems for Video Technology (CSVT)*, 6, pp.716-719, July 2001.

[7] M. Bober, W. Price, and J. Atkinson, The Contour Shape Descriptor for MPEG-7 and Its Application, in Proc. IEEE Intl. Conf on Consumer Electronics (ICCE), pp.13-8, June 2000.

[8] L. Cieplinski, MPEG-7 Color Descriptors and Their Applications, in Proc. Computer Analysis of Images and Patterns, CAIP 2001, W. Skarbek Ed., Sept. 5-7, 2001, Lecture Notes in Computer Science 2124 Springer 2001.

[9] S.-F. Chang, T. Sikora, and A. Puri, Overview of the MPEG-7 Standard, *IEEE Transactions on Circuits & Systems for Video Technology (CSVT)*, 6, pp.688-695, July 2001.

[10] N. Dimitrova, L. Agnihotri, C. Dorai, and R. M. Bolle, MPEG-7 Videotext Description Scheme for Superimposed Text in Images and Video, *Journal of Signal Processing: Image Communication*, 1-2, pp.137-155, Sept. 2000.

[11] A. Divakaran, An Overview of MPEG-7 Motion Descriptors and Their Applications, in Proc. Computer Analysis of Images and Patterns, CAIP 2001, Sept. 5-7, 2001, W. Skarbek Ed., Lecture Notes in Computer Science 2124 Springer 2001.

[12] T. Ebrahimi, Y. Abdeljaoued, R. Figueras, I. Ventura and O. D. Escoda, MPEG-7 Camera, in Proc. IEEE Intl. Conf. Image Processing (ICIP), Vol. III, pp. 600-603, Oct.. 2001.

[13] J. Hunter, An overview of the MPEG-7 description definition language (DDL), *IEEE Transactions on Circuits & Systems for Video Technology (CSVT)*, 6, pp.765-772, July 2001.

[14] J. Hunter and F. Nack, An Overview of the MPEG-7 Description Definition Language (DDL) Proposals, *Journal of Signal Processing: Image Communication*, 1-2, pp. 271-293, Sept. 2000.

[15] S. Jeannin, A. Divakaran, "MPEG-7 visual motion descriptors," *IEEE Transactions on Circuits & Systems for Video Technology (CSVT)*, 6, pp.720-724, July 2001.

[16] E. Kasutani, and A. Yamada, The MPEG-7 Color Layout Descriptor: A Compact Image Feature Description for High-Speed Image/Video Segment Retrieval, in Proc. on IEEE Intl. Conf. Image Processing (ICIP), Vol. I, pp. 674-677, Oct. 2001.

[17] B.S. Manjunath, J. R. Ohm, V. V. Vasudevan, and A. Yamada., Color and texture descriptors," *IEEE Transactions on Circuits & Systems for Video Technology (CSVT)*, 6, pp.703-715, July.. 2001.

[18] D. Messing, P. van Beek, and J. Erric, The MPEG-7 Color Structure Descriptor: Image Description Using Color and Local Spatial Information, Proc. on IEEE Intl. Conf. Image Processing (ICIP), Vol. I, pp. 670-673, Oct. 2001.

[19] M. Naphade, C.-Y. Lin, J. R. Smith, B. Tseng, and S. Basu, Learning to Annotate Video Databases, in IS&T/SPIE Symposium on Electronic Imaging: Science and Technology - Storage & Retrieval for Image and Video Databases X, San Jose, CA, January 2002.

[20] A. Natsev, J. R. Smith, Y.-C. Chang, C.-S. Li, and J. S. Vitter, Constrained Querying of Multimedia Databases: Issues and Approaches, in IS&T/SPIE Symposium on Electronic Imaging: Science and Technology - Storage & Retrieval for Image and Video Databases IX, San Jose, CA, January 2001.

[21] J.-R. Ohm, The MPEG-7 Visual Description Framework - Concepts, Accuracy, and Applications, Proc. Computer Analysis of Images and Patterns, CAIP 2001, Sept. 5-7, 2001, W. Skarbek Ed., Lecture Notes in Computer Science 2124, Springer 2001.

[22] J-R. Ohm, Flexible Solutions for Low-Level Visual Feature Descriptors in MPEG-7, Proc. IEEE Intl. Conf on Consumer Electronics (ICCE), 13-6, June 2000.

[23] F. Pereira and R. Kocncn, MPEG-7: A Standard for Multimedia Content Description, *International Journal of Image and Graphics*, pp. 527-546, Vol. 1, No. 3, July 2001.

[24] P. Philippe, Low-level musical descriptors for MPEG-7, *Journal of Signal Processing: Image Communication*, 1-2, pp.181-191, Sept. 2000.

[25] S. Quackenbush and A. Lindsay, Overview of MPEG-7 audio, *IEEE Transactions on Circuits & Systems for Video Technology (CSVT)*, 6, pp.725-729, July 2001.

[26] Y. Rui, T. S. Huang, and S. Mehrotra, Constructing Table-of-Content for Videos, *ACM Multimedia Systems Journal*, Special Issue Multimedia Systems on Video Libraries, Vol. 7, No. 5, Sept. 1999, pp. 359-368.

[27] P. Salembier, R. Qian, N. O'Connor, P. Correia, I. Sezan, and P. van Beek, Description Schemes for Video Programs, Users and Devices, *Signal Processing: Image Communication*, Vol.16(1-2):211-234, September 2000.

[28] P. Salembier and J. R. Smith, MPEG-7 Multimedia Description Schemes, *IEEE Transactions Circuits and Systems for Video Technology*, Vol. 11, No. 6, June 2001.

[29] I. Sezan and P. van Beek, MPEG-7 Standard and Its Expected Role in Development of New Information Appliances, in Proc. IEEE Intl. Conf. on Consumer Electronics (ICCE), June 2000.

[30] T. Sikora, The MPEG-7 visual standard for content description-an overview, *IEEE Transactions on Circuits & Systems for Video Technology (CSVT)*, 6, pp. 696-702, July 2001.

[31] J. R. Smith, B. S. Manjunath, and N. Day, MPEG-7 Multimedia Content Description Interface Standard, in IEEE Intl. Conf. On Consumer Electronics (ICCE), June 2001.

[32] J. R. Smith, MPEG-7 Standard for Multimedia Databases, in ACM Intl. Conf. on Data Management (ACM SIGMOD), May, 2001.

[33] J. R. Smith, Content-based Access of Image and Video Libraries, in Encyclopedia of Library and Information Science, Marcel Dekker, Inc., A. Kent Ed., 2001.

[34] J. R. Smith, Content-based Query by Color in Image Databases, in *Image Databases*, L. D. Bergman and V. Castelli, Eds., John Wiley & Sons, Inc., 2001.

[35] J. R. Smith, S. Srinivasan, A. Amir, S. Basu, G. Iyengar, C.-Y. Lin, M. Naphade, D. Ponceleon, and B. Tseng, Integrating Features, Models, and Semantics for TREC Video Retrieval, in Proc. NIST Text Retrieval Conference (TREC-10), November, 2001.

[36] J. R. Smith, S. Basu, C.-Y. Lin, M. Naphade, and B. Tseng, Integrating Features, Models, and Semantics for Content-based Retrieval, in Proc. NSF Workshop in Multimedia Content-Based Indexing and Retrieval, September, 2001.

[37] J. R. Smith, Y.-C. Chang, and C.-S. Li, Multi-object, Multi-feature Search using MPEG-7, in Proc. IEEE Intl. Conf. On Image Processing (ICIP), Special session on Multimedia Indexing, Browsing, and Retrieval, October, 2001. Invited paper.

[38] J. R. Smith and V. Reddy, An Application-based Perspective on Universal Multimedia Access using MPEG-7, in Proc. SPIE Multimedia Networking Systems IV, August, 2001.

[39] J. R. Smith, MPEG-7's Path for an Intelligent Multimedia Future, in Proc. IEEE Intl. Conf. On Information Technology for Communications and Coding. (ITCC), April, 2001.

[40] J. R. Smith and B. Lugeon, A Visual Annotation Tool for Multimedia Content Description, in Proc. SPIE Photonics East, Internet Multimedia Management Systems, November, 2000.

[41] J. R. Smith and A. B. Benitez, Conceptual Modeling of Audio-Visual Content, in Proc. IEEE Intl. Conf. On Multimedia and Expo (ICME-2000), New York, NY, July, 2000.

[42] J. R. Smith, Video Indexing and Retrieval Using MPEG-7, in Proc. SPIE Multimedia Networking Systems V, August, 2002.

[43] N. Takahashi, M. Iwasaki, T. Kunieda, Y. Wakita, and N. Day, Image Retrieval Using Spatial Intensity Features, in Proc. on Signal Processing: Image Communication, 16 (2000) pp. 45-57, June 2000.

[44] B. Tseng, C.-Y. Lin, and J. R. Smith, Video Summarization and Personalization for Pervasive Mobile Devices, in Proc. IS&T/SPIE Symposium on Electronic Imaging: Science and Technology - Storage & Retrieval for Image and Video Databases X, San Jose, CA, January, 2002.

[45] P. Wu, Y. Choi, Y.M. Ro, and C.S. Won, MPEG-7: Texture Descriptors, *International Journal of Image and Graphics*, pp. 547-563, Vol. 1, No. 3, July 2001.

[46] S. J. Yoon, D. K. Park, C. S. Won, and S-J. Park, Image Retrieval Using a Novel Relevance Feedback for Edge Histogram Descriptor of MPEG-7, in Proc. IEEE Intl. Conf. on Consumer Electronics (ICCE), 20-3, June 2001.

27

INDEXING VIDEO ARCHIVES: ANALYZING, ORGANIZING, AND SEARCHING VIDEO INFORMATION

Shin'ichi Satoh and Norio Katayama

National Institute of Informatics, Japan

{satoh,katayama}@nii.ac.jp

1.   INTRODUCTION

This chapter covers issues and approaches on video indexing technologies for video archives, especially for large-scale archives, for instance, tera- to peta-byte order archives, which are expected to become widespread in a few years. We will discuss the issues from three viewpoints, i.e., analyzing, organizing, and searching video information.

Recent years have seen technical innovation enabling the huge number of broadcast video streams, such as digital TV broadcasting through broadcast satellites (BS), communication satellites (CS), cable television, and broadband networks. The number of accessible channels by end users is getting larger, and may reach several thousands in the near future. Types of video programs in the broadcast video streams cover wide varieties of interest of audience, e.g., news, entertainment, travel, culture, education, etc. In addition, a large amount of cost as well as labor and brand-new technologies make broadcast videos be of extreme high quality, and thus they comprise archival and cultural importance. Once archived, the broadcast video archives may contain any information which ordinary users may want in any situations, although the size of the archives will become huge. If sufficiently intelligent and flexible access to the huge broadcast video archives becomes available, users may obtain almost all information upon their needs from the video stream space.

As the key technology to realize this, video indexing has been intensively studied by many researchers, e.g., content-based annotation using video analysis, efficient browsing, access, and management of video archives, etc. Based on this idea, we have been intensively studying two aspects of video indexing: one is to delineate meaningful content information by video analysis. This approach is important because manual annotation is obviously not feasible for the huge video archives. In doing this, we take advantage of image understanding, natural language processing, and artificial intelligence technologies in an integrated way. The other is to realize efficient organization of multimedia information. For vast video archives, fast access to the archives is indispensable. Since video information is thought to be comprised of high-dimensional features such as color histograms and DCT/Wavelet coefficients for images, keyword vectors for text, and so on, we developed an efficient high-dimensional index structure for the efficient access to video archives. We describe these two activities in the following sections.

We first thought that these two approaches are independent. However, our recent research results revealed that these two are tightly coupled to each other. In this way we extract meaningful contents from videos by associating image and text information as one approach: in other words, correlating skew of data distribution in the text space and in the image space. As for the other approach, the experimental results show that the proposed method achieves efficient search when sufficiently unevenly distributed data is given, whereas only poor search performance is achieved when uniformly distributed data is handled. Based on these observations, we assume that local skew of data distribution has relation to meaning of the data. We studied on searching for meaningful information in video archives and developed a detection method of local skew in the feature space with its application to search technique.

We then conclude this chapter by presenting discussions and future directions of the video indexing research from our viewpoint.

2.   VIDEO ANALYSIS FOR METADATA EXTRACTION

2.1   CONTENT-BASED ANNOTATION FOR VIDEO METADATA

Since information in videos is quite "raw" and dispersed, it is almost impossible to achieve content-based access to videos unless some additional information is available. In order to enable flexible and intelligent access to videos, we somehow need to extract "keywords" which describe contents of videos. Typical "keywords" useful for content-based access to videos include information on:

- what/who appears in the video,
- when the video is broadcasted/recorded,
- where the video is recorded,
- what the video is about, etc.

Video metadata provide these kinds of additional information. MPEG-7 has been standardized (in part as of 2002) to be used as the video metadata standard, which can describe a wide variety of content information for videos. Sufficiently in-depth metadata for large volume of videos are indispensable for content-based access to video archives.

The problem here is how we produce necessary video metadata for large volume of videos. A conservative and unfailing solution to this is manual annotation, since a person can easily perceive contents of videos. On the other hand, some types of information for video metadata are available from production stage of videos, such as shot boundaries, object contours in shots using chroma key, video captions, etc. Even so, automatic extraction of video metadata by using video analysis is still important, especially in the following cases:

- providing metadata which are not available in production stage,

- providing metadata which may be extremely expensive with manual annotation,

- providing metadata of live events (almost) instantaneously,

- whenever possible to save manual labor especially in annotating large volume of video archives.

We focused on the importance of face information in news videos. News' primary role is to provide information on news topics in terms of 5W1H (who, what, when, where, why, and how). Among them "who" information provides the most part of information in news topics. Our goal is to generate annotation to faces appearing in news videos by their corresponding names. This task incurs intensive labor when manual annotation is imposed, i.e., locating every occurrence of faces in video segments, identifying the faces, and somehow naming them. Instead, we developed Name-It, a system that associates faces and names in news videos in an automated way by integration of image understanding, natural language processing, and artificial intelligence technologies. As an example of automated extraction of video metadata, we briefly describe the Name-It system in this section.

2.2 FACE AND NAME ASSOCIATION IN NEWS VIDEOS

The purpose of Name-It is to associate names and faces in news videos [1]. Several potential applications of Name-It might include: (1) News video viewer which can interactively provide text description of the displayed face, (2) News text browser which can provide facial information of names, (3) Automated video annotation generation by naming faces.

To achieve Name-It system, we employ the architecture shown in Figure 27.1. Since we use closed-captioned CNN Headline News for our target, given news are composed of a video portion along with a transcript portion as closed-caption text. From video images, the system extracts faces of persons who might be mentioned in transcripts. Meanwhile, from transcripts, the system extracts words corresponding to persons who might appear in videos. Then, the system evaluates the association of the extracted names and faces. Since names and faces are both extracted from videos, they furnish additional timing information, i.e., at what time in videos they appear. The association of names and faces is evaluated with a "co-occurrence" factor using their timing information. Co-occurrence of a name and a face expresses how often and how well the name coincides with the face in given news video archives. In addition, the system also extracts video captions from video images. Extracted video captions are recognized to obtain text information, and then used to enhance the quality of face-name association.

Component technologies to obtain face, name, and video caption information employ state-of-the-art technologies, which will be briefed here, and related approaches are described elsewhere including other chapters in this handbook. They do not necessarily achieve perfect analysis results, though, properly integrating these results may obtain useful and meaningful contents information. We laid emphasis on the integration technique of imperfect analysis results to reveal its effectiveness in video analysis. We then depict relation between Name-It's integration method and mining the multimedia feature space.

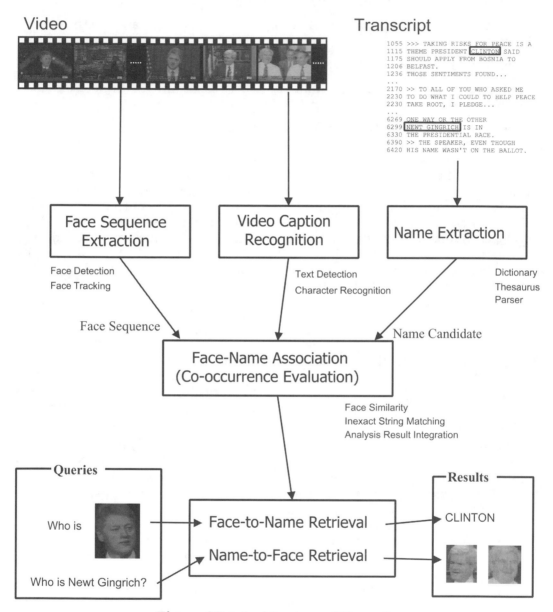

Figure 27.1 Architecture of Name-It.

2.3 IMAGE PROCESSING

The image processing portion of Name-It is necessary for extracting faces of persons who might be mentioned in transcripts. Those faces are typically shown under the following conditions: (a) frontal, (b) close-up, (c) centered, (d) long duration, (e) frequently. Given a video as input, the system outputs a two-tuple list for each occurrence of faces: timing information (start ~ end frame), and face identification information. Some of the conditions above will be used to generate the list; others will be evaluated later using information provided by that list. The image processing portion also contributes for video caption recognition, which provides rich information for face-name association.

Face Tracking

To extract face sequences from image sequences, Name-It applies face tracking to videos. Face tracking consists of 3 components: face detection, skin color model extraction, and skin color region tracking.

First, Name-It applies face detection to every frame within a certain interval of frames, e.g., 10 frames. The system uses the neural network-based face detector [2] which detects mostly frontal faces at various sizes and locations. The face detector can also detect eyes; we use only faces in which eyes are successfully detected to ensure that the faces are frontal and close-up.

Once a face is detected, the system extracts a skin color model of the face. Once a face region is detected in a frame, the skin color model of the face region is captured as the Gaussian model in (R,G,B) space. The model is applied to the subsequent frames to detect skin candidate regions. Face region tracking is continued until a scene change is encountered or until no succeeding face region is found.

Face Identification

To infer the "frequent" occurrence of a face, face identification is necessary. Namely, we need to determine whether one face sequence is identical to another.

To make face identification work effectively, we need to use frontal faces. The best frontal view of a face will be chosen from each face sequence. We first apply the face skin region clustering method to all detected faces. Then, the center of gravity of the face skin region is calculated and compared with the eye locations to evaluate a frontal factor. The system then chooses the face having the largest frontal factor as the most frontal face in the face sequence.

We choose the eigenface-based method to evaluate face identification [3]. Each of the most frontal faces is converted into a point in the 16-dimensional eigenface space. Face identification can be evaluated as the face distance, i.e., the Euclidean distance between two corresponding points in the eigenface space.

Video Caption Recognition

Video captions are directly attached to image sequences, and give text information. In many cases, they are attached to faces, and usually represent persons' names. Thus video caption recognition provides rich information for face-name association, though they do not necessarily appear for all faces of persons of interest.

To achieve video caption recognition [4], the system first detects text regions in video frames. Several filters including differential filters and smoothing filters are employed to achieve this task. Clusters with bounding regions that satisfy several size constraints are selected as text regions. The detected text regions are preprocessed to enhance image quality. First, the filter that minimizes intensities among consecutive frames is applied. This filter suppresses complicated and moving background, yet enhances characters because they are placed at the exact position for a sequence of frames. Next, the linear interpolation filter is applied to quadruple the resolution. Then template-based character recognition is applied. The current system can recognize only upper-case letters, but it achieved 76% character recognition rate.

Since character recognition results are not perfect, inexact matching between the results and character strings is essential to utilize imperfect results for face-name association. We extended the edit distance method [5] to cope with this problem. Assume that C is the character recognition result, and N is a word. The similarity $S_c(C,N)$ is defined using the edit distance to represent that C approximately equals to N.

2.4   NATURAL LANGUAGE PROCESSING

The system extracts name candidates from transcripts using natural language processing technologies. The system is expected not only to extract name candidates, but also to associate them with scores. The score represents the likelihood that the associated name candidate might appear in the video. To achieve this task, combination of lexical and grammatical analysis and the knowledge of the news video structure is employed.

First, the dictionary and parser are used to extract proper nouns as name candidates. The agent of an act such as speech or attending meeting obtains a higher score. In doing this, the parser and thesaurus are essential. In a typical news video, an anchor person appears first, talks about an overview of the news, and mentions the name of the person of interest. The system also uses news structure knowledge like this. Several such conditions are employed for score evaluation. The system evaluates these conditions for each word in transcripts by using a dictionary (the Oxford Advanced Learner's Dictionary [4]), thesaurus (WordNet [7]), and parser (Link Parser [8]). Then, the system outputs a three-tuple list: a word, timing information (frame), and a normalized score.

2.5   INTEGRATION OF PROCESSING RESULTS

In this section, the algorithm for retrieving face candidates by a given name is described. We use the co-occurrence factor to integrate image and natural language analysis. Let N and F be a name and face, respectively. The co-occurrence factor $C(N,F)$ is expected to have a degree that represents the fact that the face F is likely to have the name N. Think of the faces F_a, F_b, ... and the names N_p, N_q, ..., where F_a corresponds to N_p. Then $C(N_p,F_a)$ should have the largest value among co-occurrence factors of any combinations of F_a and the other names (e.g., $C(N_q,F_a)$, etc.), or of the other faces and N_p (e.g., $C(N_p,F_b)$, etc.). Retrieval of face candidates by a given name is realized as follows using the co-occurrence factor:

1.   Calculate co-occurrences of combinations of all face candidates and the given name.

2.   Sort co-occurrences.

3.   Output faces that correspond to the N largest co-occurrences.

Retrieval of name candidates by a face is realized as well.

Integration by Co-occurrence Calculation

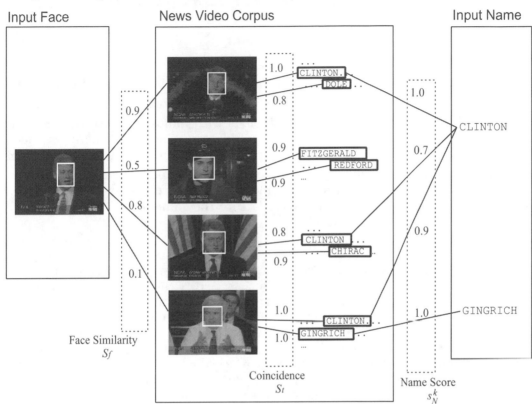

Figure 27.2 Co-occurrence factor calculation.

In this section, the co-occurrence factor $C(N,F)$ of a face F and a name N is defined. Figure 27.2 depicts this process. Assume that we have the two-tuple list of face sequences (timing, face identification): $\{(t_{F_i}, F_i)\}=\{(t_{F_i}, F_1), (t_{F_i}, F_2),$ ...}, the three-tuple list of name candidates (word, timing, score): $\{(N_j, t_{N_j}^k,$ $s_{N_j}^k)\}=\{(N_1, t_{N_1}^1, s_{N_1}^1), (N_1, t_{N_1}^2, s_{N_1}^2), ...,(N_2, t_{N_2}^1, s_{N_2}^1), ...\}$, and the two-tuple list of video captions (timing, recognition result): $\{(t_{C_i}, C_i)\}=\{(t_{C_1}, C_1), (t_{C_2}, C_2), ...\}$. Note that t_{F_i} and t_{C_i} have duration, e.g., $(t_{F_i}^{start} \sim t_{F_i}^{end})$; so we can then define the duration function as $dur(t_{F_i})=t_{F_i}^{end}-t_{F_i}^{start}$. Also note that a name N_j may occur several times in video archives, so each occurrence is indexed by k. We define the face similarity between faces F_i and F_j as $S_f(F_i,F_j)$ using the Euclidean distance in the eigenface space. The caption similarity between a video caption recognition result C and a word N, $S_c(C,N)$, and the timing similarity between times t_i and t_j, $S_t(t_i,t_j)$, are also defined. The caption similarity is defined using the edit distance, and the timing similarity represents coincidence of events. Then the co-occurrence factor $C(N,F)$ of the face F and the name candidate N is defined as follows:

$$C(N,F) = \frac{\sum_i S_f(F_i,F)(\sum_k s_N^k S_t(t_{F_i},t_N^k) + w_c \sum_j S_t(t_{C_j},t_{F_i})S_c(C_j,N))}{\sqrt{\sum_i S_f^2(F_i,F)dur(t_{F_i})\sum_k s_N^{k^2}}}. \qquad (27.1)$$

Intuitively, the numerator of $C(N,F)$ becomes larger if F is identical to F_i while at the same time F_i coincides with N with the large score. To prevent "anchor person problem", (an anchor person coincides with almost any name; a face or name coincides with any name, or face should correspond to no name or face), $C(N,F)$ is normalized with the denominator. w_c is the weight factor for caption recognition results. Roughly speaking, when a name and a caption match and the caption and a face match at the same time, the face equivalently coincides with w_c occurrences of that name. We use 1 for the value of w_c.

2.6 EXPERIMENTS AND DISCUSSIONS

The Name-It system was first developed on an SGI workstation, and now it's working on Windows PC as well. We processed 10 CNN Headline News videos (30 minutes each), i.e., a total of 5 hours of video. The system extracted 556 face sequences from videos. Name-It performs name candidate retrieval by a given face, and face candidate retrieval by a given name from the 5-hours news video archives. In face-to-name retrieval, the system is given a face, then outputs name candidates with co-occurrence factors in descending order. Likewise, in name-to-face retrieval, the system outputs face candidates of a given name with co-occurrence factors in descending order.

Figure 27.3 (a) through (d) show the results of face-to-name retrieval. In each result, an image of a given face and ranked name candidates associated with co-occurrence factors are shown. A correct answer is shown with a circled ranking number. Figure 27.3 (e) through (h) show the results of name-to-face retrieval. The top-4 face candidates are shown in the order from left to right with corresponding co-occurrence factors. These results demonstrate that Name-It achieves effective face-to-name and name-to-face retrieval with actual news videos.

We have to note that there are some faces not being mentioned in the transcripts, but described only in video captions. These faces can be named only by incorporating video caption recognition (e.g., Figure 27.3 (d) and (h)). Although these faces are not always the most important in terms of news topics, namely, "the next to the most important," video caption recognition surely enhances performance of Name-It. The overall accuracy that the correct answer is involved in top-5 candidates is 33% in face-to-name retrieval, and 46% in name-to-face retrieval.

1 MILLER   0.145916
2 VISIONARY   0.114433
3 WISCONSIN   0.1039
4 RESERVATION 0.103132

(a) Bill Miller, singer

1 WARREN   0.177633
2 CHRISTOPHER 0.032785
3 BEGINNING   0.0232368
4 CONGRESS   0.0220912

(b) Warren Christopher, the former U.S. Secretary of State

1 FITZGERALD   0.164901
2 INDIE   0.0528382
3 CHAMPION   0.0457184
4 KID   0.0351232

(c) Jon Fitzgerald, actor

1 EDWARD   0.0687685
2 THEAGE   0.0550148
3 ATHLETES   0.0522885
4 BOWL   0.0508147

(d) Edward Foote, University of Miami President

coocr 0.0298365   coocr 0.0292515   coocr 0.0262954   coocr 0.0249047

(e) given "CLINTON"   (Bill Clinton)

coocr 0.0990763   coocr 0.0374899   coocr 0.0280812   coocr 0.0204594

(f) given "GINGRICH"   (Newt Gingrich, 1st and 2nd candidates)

coocr 0.0777665   coocr 0.041456   coocr 0.0253743   coocr 0.0214603

(g) given "NOMO"   (Hideo Nomo, pitcher of L.A. Dodgers, 2nd candidate)

coocr 0.0225189   coocr 0.0218807   coocr 0.011158   coocr 0.0108854

(h) given "LEWIS"   (Lewis Schiliro, FBI, 2nd candidate)

Figure 27.3 Face and name association results.

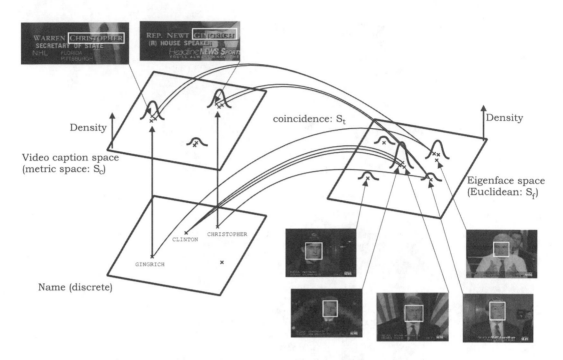

Figure 27.4 Relating Feature Spaces by co-occurrence.

2.7   NAME-IT AS MINING IN FEATURE SPACE

In Name-It process, we assume that a frequently coincident face-name pair may represent a corresponding face-name pair. A face-name pair is frequently coincident if the face (and faces identical to this face) frequently appears, while at the same time the name frequently occurs, and the face and the name coincide in many segments of news video archives. Frequency of occurrence, e.g., of faces, can be regarded as density in the feature space. For example, each face is converted to a point in the eigenface space. By converting all faces in news video archives to points, the faces can be represented as scatter in the eigenface space, and frequently appearing faces correspond to high-density regions in the eigenface space. The same observation is possible for names and video captions. For faces, the feature space is the eigenface space, that is the Euclidean space, and similarity between faces is evaluated by S_f. For names, the feature space is discrete, and thus similarity is evaluated by identity. For video captions, the feature space is the metric space where similarity is evaluated by S_c, which is based on the edit distance. Even though these feature spaces have the different metric systems, density of points can be evaluated based on similarity (i.e., distance) defined in each feature space. The co-occurrence factor of a face-name pair corresponding to coincident high-density regions in the face space, name space, and video caption space will result in larger value (See Equation (27.1).). This observation is depicted in Figure 27.4. From this viewpoint, we can say that Name-It process is closely related to mining in the multimedia feature space.

3. EFFICIENT ACCESS TO LARGE-SCALE VIDEO ARCHIVES

3.1 CONTENT-BASED RETRIEVAL OF MULTIMEDIA INFORMATION

The content-based information retrieval (CBIR) of multimedia information, e.g., images, sounds, videos, etc., is one of the most important components in large-scale video archive systems. In order to achieve the content-based information retrieval, it is necessary to understand the contents of multimedia information. The most typical case is to express the contents by some feature vectors. In this case, the feature is an approximation of contents and information is retrieved based on the similarity of features assuming that the contents having similar features should be similar to each other. The dissimilarity between two feature vectors is often measured by the Euclidean distance. Once a query vector is specified, such a vector that is the most similar to the query vector is found by the nearest-neighbor search. With this approach, various kinds of systems have been developed. For example, QBIC [9] is well known as a typical example of the content-based image retrieval system. The practical effectiveness of this approach has been demonstrated by such prototype systems. However, this approach has two difficulties. Firstly, the feature vector is just an approximation of contents. Therefore, the content is retrieved based on the nearest-neighbor search of feature vectors. Secondly, the dimensionality of feature vectors tends to be quite high (e.g., 64 and 256 dimensional vectors are used by QBIC for representing the hue of colors). Thus, fast nearest-neighbor search of high-dimensional vectors is essential to CBIR. In order to alleviate these difficulties, multidimensional indexing methods have been studied by database system researchers. These methods are supposed to take an important role in building large-scale video archive systems which need to process a huge amount of data in a practical processing time.

In this section, we demonstrate the efficiency of multidimensional indexing methods. First, we describe an example of such indexing methods, the SR-tree [10], which we designed for the efficient similarity retrieval of high dimensional feature vectors. Then, we demonstrate the efficiency of the SR-tree with applying it to face sequence matching.

3.2 MULTIDIMENSIONAL INDEX STRUCTURES

R-Tree and Its Variants

In the database system literature, quite a few index structures have been proposed for high-dimensional indexing. The R-tree [11] and its variant R*-tree [12], which were originally designed as a spatial index structure for geography and CAD databases, are used as an index structure for the feature space of the similarity image retrieval [9]. More recently, some index structures, e.g., the SS-tree [13] and the SR-tree [10], etc., have been proposed especially for the acceleration of the nearest-neighbor search [14].

These index structures divide the search space into a hierarchy of regions (Figure 27.5). In addition to the R-tree, the k-d tree and the quadtree are also well-known hierarchical data structures. These structures differ in the way of dividing the search space. The k-d tree and the quadtree divide the search space into disjoint regions. The k-d tree locates the split plane adaptively based on the data distribution, while the quadtree locates the split plane regularly. On the other hand, the R-tree divides the search space by minimum bounding rectangles with allowing the overlap between them. Among these structures, the R-tree is the most efficient for dynamic updates. With allowing overlap between minimum bounding rectangles, the R-tree is kept balanced even after dynamic updates. Therefore, the R-tree is regarded as one of the most suitable index structures for the online database applications.

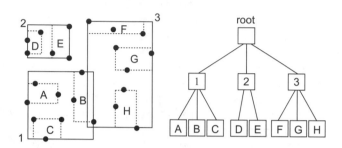

Figure 27.5 Structure of the R-tree.

SR-Tree: An Index Structure for Nearest Neighbor Queries

The way of the region split has great influence on the performance of index structures. Both the SS-tree and the SR-tree are derived from the R-tree; thus they have the similar structure with the R-tree. However, the SS-tree and the SR-tree have better performance than the R-tree with respect to the nearest-neighbor search. This performance improvement comes from the modification of the region split algorithm. The tree construction algorithm is common to the SS-tree and the SR-tree. It is originally designed for the SS-tree; the SR-tree borrows it from the SS-tree. This algorithm splits regions based on the variance of coordinates. When a node is to be split, the coordinate variance on each dimension is calculated and the dimension with the largest variance is chosen for the split dimension. According to the performance evaluation, it is shown that this algorithm generates regions with shorter diameters than the algorithm of the R*-tree does [10].

The difference between the SS-tree and the SR-tree is entries of internal nodes. An internal node of the SS-tree has the following entries:

$$InternalNode_{SS-tree} := \langle\, \langle \vec{c}_1, w_1, r_1, offset_1 \rangle, \ldots, \langle \vec{c}_n, w_n, r_n, offset_n \rangle \,\rangle$$

where $\vec{c}_i$ $(1 \leq i \leq n)$ denotes the centroid of the i-th child node, i.e., the centroid of the points contained in the subtree whose root node is the i-th child node, w_i $(1 \leq i \leq n)$ denotes the weight of the i-th child node, i.e., the number of the points in the subtree, and r_i $(1 \leq i \leq n)$ denotes the radius of the minimum bounding sphere that covers all points in the subtree and whose center is $\vec{c}_i$. Thus, the SS-tree employs minimum bounding spheres instead of minimum bounding rectangles. The tree structure of the SS-tree corresponds to the hierarchy of spherical regions (Figure 27.6). The centroids $\vec{c}_1, \ldots, \vec{c}_n$ and the weights $w_1, \ldots, w_n$ play an important role in the variance-based region split algorithm. They make it possible to estimate the variance of the data distribution at each internal node. The SS-tree outperforms the R-tree by virtue of the variance-based algorithm.

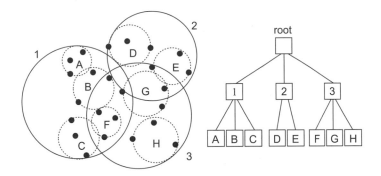

Figure 27.6 Structure of the SS-tree.

The SR-tree is an enhancement of the SS-tree. It employs both the bounding sphere and the bounding rectangle as follows:

$$InternalNode_{SR-tree} := \langle \langle \vec{c}_1, w_1, r_1, MBR_1, offset_1 \rangle, \ldots, \langle \vec{c}_n, w_n, r_n, MBR_n, offset_n \rangle \rangle$$

where $\vec{c}_i$, w_i, and r_i $(1 \le i \le n)$ are the same with those of the SS-tree, and MBR_i $(1 \le i \le n)$ denotes the minimum bounding rectangle of the i-th child node. Thus, the difference between the SS-tree and the SR-tree is the addition of MBR_i. Each sub-region of the SR-tree is determined by the intersection of the bounding sphere (given by $\vec{c}_i$ and r_i) and the bounding rectangle (given by MBR_i). The tree structure of the SR-tree corresponds to the hierarchy of such regions (Figure 27.7). The experimental evaluation has proven that the incorporation of minimum bounding rectangles improves the performance of nearest-neighbor search because minimum bounding rectangles reduce the region size of each node [10].

Nearest-Neighbor Search with Multidimensional Index Structures

The multidimensional index structures can accelerate the nearest-neighbor search since the search can be accomplished with investigating a limited number of nodes (i.e., regions). This could reduce the search space significantly and would be much faster than investigating every feature vector one by one. The search starts from the root node and visits regions in ascending order of the distance from the query point. On visiting each region, the candidate of the nearest neighbor is determined by choosing the nearest point encountered so far. The search continues as long as there remains such a region that is not visited so far and that is closer than the candidate. The search terminates when no region remains to be visited. The final candidate is the answer of the query. The number of nodes to be visited is dependent on various circumstances, e.g., distribution of the data set, location of the query, and the characteristics of the index structure, etc. In the ideal case, the nearest neighbor can be found with traversing the tree   from the root node to a leaf straightly. In this case, the search cost is logarithmic to the size of the database (i.e., the number of feature vectors). The results on the performance evaluation of the SR-tree can be found in [10].

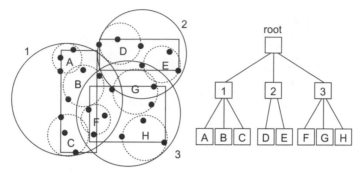

Each region is the intersection of a bounding sphere and a bounding rectangle.

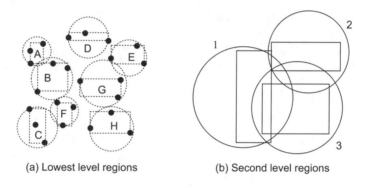

(a) Lowest level regions   (b) Second level regions

Figure 27.7 Structure of the SR-tree.

3.3   EFFICIENT IMPLEMENTATION OF FACE SEQUENCE MATCHING

Similarity Retrieval with Colored Nearest-Neighbor Search

In the case of the similarity retrieval of images, one feature vector is extracted from each image and nearest-neighbor search is conducted with comparing one vector to another. On the other hand, some multimedia information, e.g., videos, permits us to extract multiple feature vectors from each piece of multimedia information. In this case, the colored nearest-neighbor search can be employed for improving the accuracy of the similarity retrieval.

For example, in Figure 27.8, multiple feature vectors which are obtained from each sequence of video frames compose a colored set of feature vectors. The similarity of two vector sets can be measured by the distance of their closest pair as shown in Figure 27.8. In the case of the similarity retrieval of video sequences, the closest pair of two feature vector sets corresponds to the pair of the most similar video frames that are shared by two video sequences. This type of the similarity measurement has been used by the face sequence matching described below and the appearance matching by Nayar et al. [15].

However, this similarity measurement is rather computation intensive. The naive method of finding the closest pair of two vector sets is to compare every combination of vectors. In this case, the computational cost is a serious problem.

Suppose that we have a database containing N video sequences and retrieve such a sequence that is the most similar to a given query. If one vector is extracted from each video sequence, the most similar one can be found by N comparisons of feature vectors. On the other hand, if M feature vectors are extracted from each sequence, $M \times M \times N$ comparisons are required to find the closest pair having the shortest distance. Of course, this estimation based on the naive implementation is like the linear scan but it is obvious that the computational cost is a serious problem for the colored nearest-neighbor search. Therefore, it is necessary to reduce the search cost in order to apply the colored nearest-neighbor search to large-scale video databases. From this perspective, we applied the SR-tree to the colored nearest-neighbor search.

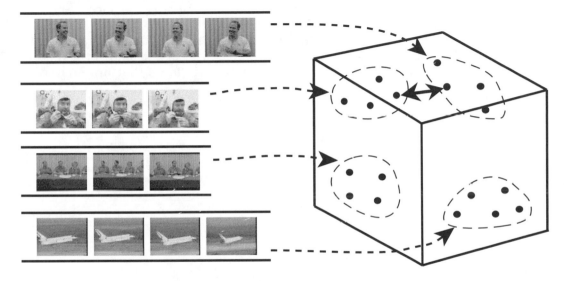

Figure 27.8 Similarity retrieval of video frame sequences based on the closest pair.

Colored NN Search with Multidimensional Index Structures

The colored nearest-neighbor search can be easily conducted with multidimensional index structures by the extension of the non-colored nearest-neighbor search described above. The fundamental flow is similar to the non-colored nearest-neighbor search: nodes of a tree are sorted in ascending order of the distance to the given query and the closest node is visited before the others. For the colored nearest-neighbor search, a query is given by the set of vectors. Therefore, when computing the distance from the query to a node entry (i.e., a child node or a vector), we must choose such a query vector that is closest to the entry and then compute the distance from the vector to the entry. The search consists of two steps as the non-colored nearest-neighbor search. Firstly, the leaf node that is the closest to the query vectors is visited to obtain the initial candidate of the nearest neighbor. Secondly, the search continues with visiting such a node that is closer to the query than the candidate. Every time a node is visited, the candidate is updated. The search terminates when there is no node that is closer to the query than the candidate is. The final candidate is the answer of the query.

In the case of finding k nearest neighbors, k candidates are kept during the search operation in the same way as the non-colored nearest-neighbor search. However, in the colored nearest-neighbor search, at most one vector should be selected for a candidate from each vector set. Therefore, when a new candidate is found, we must test whether the current candidate set contains such a vector that belongs to the same set with the one of the new candidate. If not, we can add the new candidate to the candidate set; if it does, we need to compare the new candidate with the vector belonging to the same set and then only the closer vector should be included in the candidate set.

Performance Evaluation with Face Sequence Matching

In order to evaluate the effectiveness of the multidimensional index structures, we conducted an experiment of face sequence matching. A face sequence consists of the face images extracted from videos. Even if two face sequences are of the same person, they involve the variation in diverse aspects, e.g., facial expression, lighting conditions, etc. Therefore, the closest pair of two face sequences corresponds to the pair of the most similar faces that are shared by those sequences.

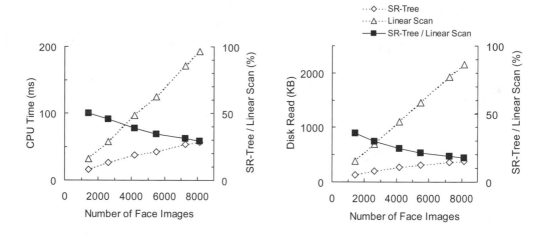

Figure 27.9 Performance of the SR-tree with face sequence matching.

The experiment was conducted with CNN Headline News video. Six data sets are composed by sampling 100, 200, 300, 400, 500, and 566 sequences at random from 566 sequences extracted from the video. The numbers of face images contained in the data sets are 1414, 2599, 4174, 5504, 7257, and 8134 respectively. The SR-tree is used as a multidimensional index structure.

We measured the CPU time and the disk access (the amount of disk read) of finding such a sequence that is closest to a given query sequence. Every sequence in a data set is chosen for a query and the average of them is taken as the experiment result. Programs are implemented in C++. Tests are conducted on a Sun Microsystems workstation, Ultra 60 (CPU: UltraSPARC-II 360MHz, main memory: 512Mbytes, OS: Solaris 2.6). Figure 27.9 shows the measured performance. For comparison, the result of the implementation with the linear scan is also plotted. The horizontal axis indicates the number of face images in a data set. This number corresponds to the size of a database. The vertical axis on the left indicates the CPU time and the one on the right the disk access. The effectiveness of the SR-tree is more significant in the disk access. When the number of face images is 8134, the disk access of the SR-tree is only 18 % of that of the linear scan. This demonstrates that the SR-tree successfully reduces the search space with splitting the data space into the hierarchy of regions. This clearly shows the advantage of multidimensional index structures. The CPU time of the SR-tree is also much smaller than that of the linear scan; the former is 29 % of the latter. In addition, the ratio of the SR-tree to the linear scan, i.e., "SR-tree / Linear Scan" in the figure, decreases as the database size increases. This demonstrates the scalability of the proposed method and implies its effectiveness for large-scale databases.

4. ANALYZING THE FEATURE SPACE FOR FURTHER IMPROVEMENT OF SEARCH EFFICIENCY

4.1 DISTINCTIVENESS OF NEAREST NEIGHBORS IN FEATURE SPACE

Indistinctive Nearest Neighbors

When we use the nearest-neighbor search (NN search), we expect that found neighbors are much closer than the others. However, this intuition is sometimes incorrect in high dimensional space. For example, when points are uniformly distributed in a unit hypercube, the distance between two points is almost the same for any combination of two points. Figure 27.10 shows the minimum, the average, and the maximum of the distances among 100,000 points generated at random in a unit hypercube. As shown in the figure, the minimum of the distances grows drastically as the dimensionality increases and the ratio of the minimum to the maximum increases up to 24 % in 16 dimensions, 40 % in 32 dimensions, and 53 % in 64 dimensions. Thus, the distance to the nearest neighbor is 53 % or more of the distance to the farthest point in 64-dimensional space. In this case, we can consider the nearest neighbor to be *indistinctive*, because the difference between the nearest neighbor and the others is negligible, i.e., the other points are as close to the query point as the nearest neighbor is. From the perspective of the similarity retrieval, when the nearest neighbor is indistinctive, the nearest neighbor has almost the same similarity with the others and does not have distinctive similarity to the query.

As Figure 27.10 shows, indistinctive nearest neighbors are more likely to occur as dimensionality increases. This characteristic can be verified by estimating the distance to k-th nearest neighbor. When N points are distributed uniformly within the hypersphere whose center is the query point, the expected distance to k-th nearest neighbor d_{kNN} is obtained as follows [16]:

$$E\{d_{kNN}\} \approx \frac{\Gamma(k+1/n\}}{\Gamma(k)} \frac{\Gamma(N+1)}{\Gamma(N+1+1/n)} r \qquad (27.2)$$

where n is the dimensionality of the space and r is the radius of the hypersphere. Then, the ratio of the $(k+1)$-NN distance to the k-NN distance is obtained as follows [16]:

$$\frac{E\{d_{(k+1)NN}\}}{E\{d_{kNN}\}} \approx 1 + \frac{1}{k\,n}\,. \tag{27.3}$$

Thus, when points are distributed uniformly around the query point, it is expected that the difference between the k-th and $(k+1)$-th nearest neighbors decreases as the dimensionality increases. This implies that indistinctive nearest neighbors are more likely to occur in high dimensional space than in low dimensional space.

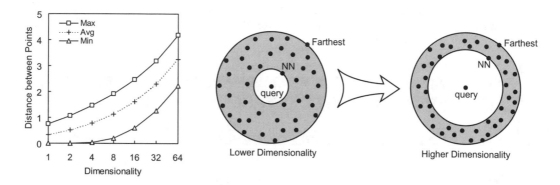

Figure 27.10 Distances among 100,000 points
generated at random in a unit hypercube.

Equation (27.3) also indicates that the ratio of $d_{(k+1)NN}$ to d_{kNN} decreases monotonically as k increases. Therefore, the maximum of the ratio between two nearest neighbors is obtained for the first and the second nearest neighbors. From Equation (27.3), we can estimate the relative difference between the first and the second nearest neighbors when the dimensionality is n:

$$\frac{E\{d_{2NN}\} - E\{d_{1NN}\}}{E\{d_{1NN}\}} \approx \frac{1}{n}\,. \tag{27.4}$$

This equation shows that the relative difference between the first and the second nearest neighbors decreases monotonically as the dimensionality increases (Figure 27.11). Only 20 % difference is expected at 5 dimensions, and only 10 % at 10 dimensions. This equation clearly shows that we could not expect strong distinctiveness of nearest neighbors for high dimensional uniform distribution.

Intrinsic Dimensionality of Feature Space

As shown above, indistinctive nearest neighbors are more likely to appear in high dimensional space, and the expected relative difference between the first and the second nearest neighbors is inversely proportional to the dimensionality of the distribution. However, this does not mean that high dimensional feature space is useless. We should note that the discussion above is based on the uniform distribution. If the data distribution over the entire feature space is uniform, we can say that the feature space is useless, but in real applications, the data distribution is not uniform at all.

Instead of assuming the uniform distribution to the entire space, we should employ the intrinsic dimensionality (or effective dimensionality) which is determined by local characteristics of the data distribution [16]. For example, when the data distribution is governed by a number of dominant dimensions, the intrinsic dimensionality is given by the number of such dominant dimensions. In addition, the intrinsic dimensionality may not be consistent over the data set but vary from one local region to another.

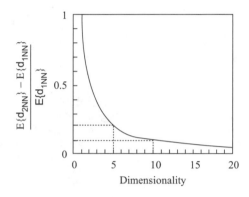

Figure 27.11 Relative difference between the first and the second nearest neighbors.

In real applications, we can expect that the data distribution is so skewed that the intrinsic dimensionality could be much smaller than the dimensionality of the feature space. Therefore, we might have indistinctive nearest neighbors in one region but could have distinctive ones in another. In a region with low intrinsic dimensionality, we can expect distinctive nearest neighbors, while we cannot expect distinctive ones in a region with high intrinsic dimensionality. Thus, the intrinsic dimensionality is the important clue for estimating the distinctiveness of the nearest neighbor.

Harmful effect of indistinctive NNs

Indistinctive nearest neighbors have a harmful effect on the similarity retrieval with the following respects:

(1) NN search performance is degraded.

When the nearest neighbor is indistinctive, there exist many points that have almost the same similarity with the nearest neighbor. Since these points are very strong candidates for the nearest neighbor, NN search operation is forced to examine many points before determining the true nearest neighbor. This degrades the performance of NN search operation.

(2) Less informative result is returned.

When the nearest neighbor is indistinctive, NN search operation returns the closest point among many strong candidates that have almost the same similarity with the nearest neighbor. This means that all of the candidates have slight differences with each other. It is not informative for users to choose the nearest neighbor from plenty of similar candidates.

These effects are extremely harmful to the retrieval systems with human-computer interaction. When the nearest neighbor is indistinctive, the system forces users to wait until a less informative result is answered with slow response. Thus, it is necessary to handle indistinctive nearest neighbors appropriately in order to achieve efficient similarity retrieval of multimedia information.

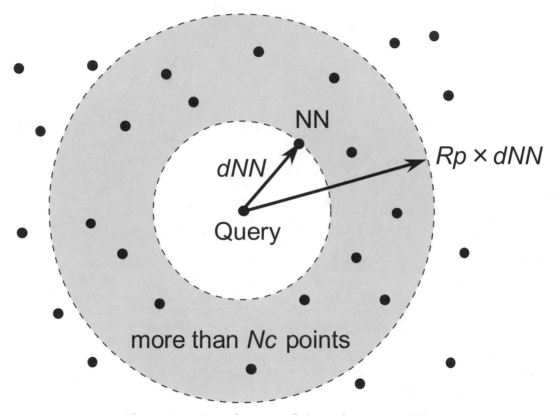

Figure 27.12 Definition of the indistinctive NN.

4.2   ESTIMATING THE DISTINCTIVENESS OF NEAREST NEIGHBORS

Determination of indistinctive NNs

As mentioned above, the intrinsic dimensionality plays an important role in determining the distinctiveness of the nearest neighbor. Therefore, we devised a probabilistic method which determines the distinctiveness of the nearest neighbor based on the intrinsic dimensionality. In the first place, we coined a definition of indistinctive nearest neighbors as follows:

Definition 1: Let d_{NN} be the distance from the query point to a nearest neighbor. Then, the nearest neighbor is indistinctive if more than or equal to N_c points exist within the range of d_{NN} to $R_p \times d_{NN}$ from the query point.

Here, $R_p\,(>1)$ and $N_c\,(>1)$ are controllable parameters (Figure 27.12). R_p determines the proximity of the query point and N_c determines the congestion of the proximity. For example, we set 1.84 to R_p and 48 to N_c at the experiment described in the following section.

As dimensionality increases, we are more likely to have plenty of points in the similar distance with the nearest neighbor. This causes congestion in the proximity of the query point. The above definition determines indistinctive nearest neighbors by detecting the congestion in the proximity. This characteristic can be clearly stated by estimating the probability of congestion. We call this probability the *rejection probability*. When points are distributed uniformly in a local region with the intrinsic dimensionality n, the probability that N_c points exist in the proximity specified by R_p is obtained as follows:

$$Pr\{N_c \text{ or more points in } R_p\} = \left(1-(1/R_p)^n\right)^{N_c} \tag{27.5}$$

According to the definition above, the nearest neighbor is indistinctive when N_c points exist in the proximity specified by R_p. Therefore, Equation (27.5) corresponds to the probability that the nearest neighbor is regarded as being indistinctive when the intrinsic dimensionality is n. This probability increases monotonically as the intrinsic dimensionality increases. Figure 27.13 shows the rejection probability when R_p is 1.84471 and N_c is 48.

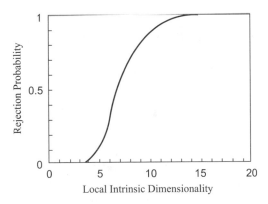

Figure 27.13 Rejection probability when R_p is 1.84471 and N_c is 48.

Distinctiveness-Sensitive NN Search

In order to circumvent the harmful effect of indistinctive nearest neighbors, we developed a new NN search algorithm for multidimensional index structures with applying the distinctiveness estimation method described above [17]. The algorithm tests the distinctiveness of the nearest neighbor in the course of search operation. Then, when it finds the nearest neighbor to be indistinctive, it quits the search and returns a partial result. We call this NN search the *Distinctiveness-Sensitive Nearest-Neighbor Search* since it is sensitive to the distinctiveness of the nearest neighbor. This algorithm not only enables us to determine the distinctiveness of the nearest neighbor but also enables us to cut down the search cost as shown in the following section.

The distinctiveness-sensitive NN search algorithm is an extension of the basic NN search algorithm. The main idea of the extension is counting the number of points that are located within the proximity of the query point in the course of search operation. When it finds the nearest neighbor to be indistinctive, it quits the search and returns the partial result as shown in Figure 27.14. When j-th nearest neighbor is found to be indistinctive during k-nearest neighbor search, the search is terminated and candidates on the termination are returned for j-th to k-th nearest neighbors. Thus, the mixture of the exact nearest neighbors and the nearest neighbor candidates is returned when the indistinctive nearest neighbor is found during the search operation. This algorithm not only enables us to determine the distinctiveness of nearest neighbors but also enables us to cut down the search cost since this algorithm avoids pursuing exact answers for indistinctive nearest neighbors.

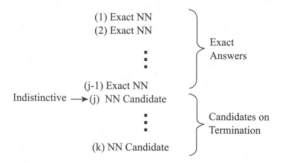

Figure 27.14 Search result of the distinctiveness-sensitive NN search.

4.3   EXPERIMENTAL EVALUATION

We evaluated the performance of the distinctiveness-sensitive NN search with applying it to the similarity retrieval of images. The data set is 60,195 images of Corel Photo Collection contained in the product called Corel Gallery 1,000,000. The similarity of images is measured in terms of the color histogram. Munsell color system is used for the color space. It is divided into nine subspaces: black, gray, white, and six colors. For each image, histograms of four sub-regions, i.e., upper-left, upper-right, lower-left, and lower-right, are calculated in order to take account of the composition of the image. Four histograms are concatenated to compose a 36-dimensional feature vector. Similarity of images is measured by Euclidean distance among these 36-dimensional vectors. We measured the performance of finding 100 nearest neighbors with employing every image as a query. Therefore, one of the found nearest neighbors is the query image itself. Experiments are conducted on Sun Microsystems Ultra 60 (CPU: UltraSPARC-II 360MHz, main memory: 512 Mbytes, OS: Solaris 2.6). Programs are implemented in C++. The VAMSplit R-tree [18] is employed as the index structure, since it has an efficient static construction algorithm suitable for the NN search in high dimensional space. The parameters R_p and N_c are set to 1.84471 and 48 respectively.

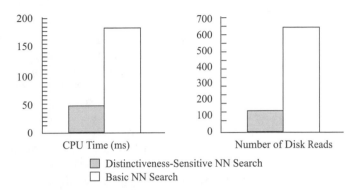

Figure 27.15 Performance of the distinctiveness-sensitive NN search.

We compared the cost of the distinctiveness-sensitive NN search with that of the basic NN search. As shown in Figure 27.15, both the CPU time and the number of disk reads of the distinctiveness-sensitive NN search are remarkably reduced compared with those of the basic NN search. The CPU time is reduced by 75 % and the number of disk reads is reduced by 72 %. This result demonstrates that the distinctiveness-sensitivity NN search enables us to cut down the search cost. Although the reduction rate depends on the data set, this cost reduction capability of the distinctiveness-sensitive NN search should be advantageous to the interactive similarity retrieval systems that need quick response to users.

Table 27.1 Number of distinctive NNs.

# of Distinctive NNs	# of Occurrence	
100	0	(0%)
80 – 99	46	(0.08%)
60 – 79	1	(0.001%)
40 – 59	66	(0.1%)
20 – 39	124	(0.2%)
2 – 19	13918	(23.1%)
1	46040	(76.5%)
Total	60195	(100%)

Table 27.1 shows the number of distinctive nearest neighbors found by the distinctiveness-sensitive NN search. Because we employed every image as a query, the total of the occurrence is equal to the number of images in the data set. Table 27.2 shows what kind of images are retrieved as distinctive nearest neighbors. We examined such search results that have relatively many distinctive nearest neighbors and then determined by hand what kind of images are retrieved as distinctive nearest neighbors. Figure 27.16 shows examples of search results when the number of distinctive nearest neighbors is relatively large. Due to the limitation of space, top 5 images are shown. The numerical value under each image is the distance from the query to the image. We can see that similar images are successfully retrieved by the distinctiveness-sensitive NN search.

The amazing result of Table 27.1 is that we obtained only one distinctive nearest neighbor for 46,040 images. Since each query image is chosen from the images in the data set, the obtained distinctive nearest neighbor is the query itself.

Therefore, we obtained no distinctive nearest neighbor except for the query image. In this case, the query image is surrounded by plenty of neighbors that have almost the same similarity to the query. Since Corel Photo Collection collects a wide variety of photos, it is not strange that an image has no similar one in the collection. In addition, the collection contains some texture photos. In this case, we have plenty of images with small difference. Figure 27.17 shows examples of search results when we obtained no distinctive nearest neighbors except for the query image. Due to the space limitation, top 5 images are shown. Figure 27.17 (a) is the case that the query image is surrounded by plenty of dissimilar images, while Figure 27.17 (b) is the case that the query image is surrounded by plenty of images with small difference. These examples illustrate that the distinctiveness-sensitive NN search allows us to see how retrieved images are significantly close to the query image. This capability should be advantageous to the interactive information retrieval systems.

Table 27.2 Categories of the results.

Category (Observed by Hand)	# of Distinctive NNs (at most)
Texture	92
Sky / Sea	62
Portrait	35
Card	23
Firework	23
Sunset	19
Kung-Fu	18
Steamship	17
Desert	16

In order to validate the results of the distinctiveness-sensitive NN search, we plotted the distribution of nearest neighbors in two-dimensional form. Figure 27.18 shows the examples of the distribution of nearest neighbors; (a) is the case that no distinctive nearest neighbor is found (Figure 27.17 (a)), while (b) is the case that 62 distinctive nearest neighbors are found (Figure 27.16 (b)). The query image is located at the center. The images of 99 nearest neighbors are placed so that the distance from the query image to each image in the figure is proportional to the distance between their feature vectors in the feature space. The inner dashed circle indicates the distance to the 1st nearest neighbor, while the outer solid circle the 99th nearest neighbor. Figure 27.18(a) and (b) are scaled so that their outer solid circles are equal in size. The orientation of each nearest neighbor is not important in this illustration; it is determined by the first and second principal components of the feature vectors of nearest neighbors. As the figure shows, the distribution of nearest neighbors differs significantly from one query point to another. The broader the distribution is, the more the distinctive nearest neighbors are obtained. This tendency meets the aim of the distinctiveness-sensitive NN search, i.e., detecting the congestion of nearest neighbors.

In this experiment, we used still photo images as an example of multimedia information but the similar characteristics of the feature space, i.e., the skewness of distribution and the variety of the distinctiveness of nearest neighbors, can be observed in other types of multimedia information, e.g., video frames, face images, object appearances, etc.

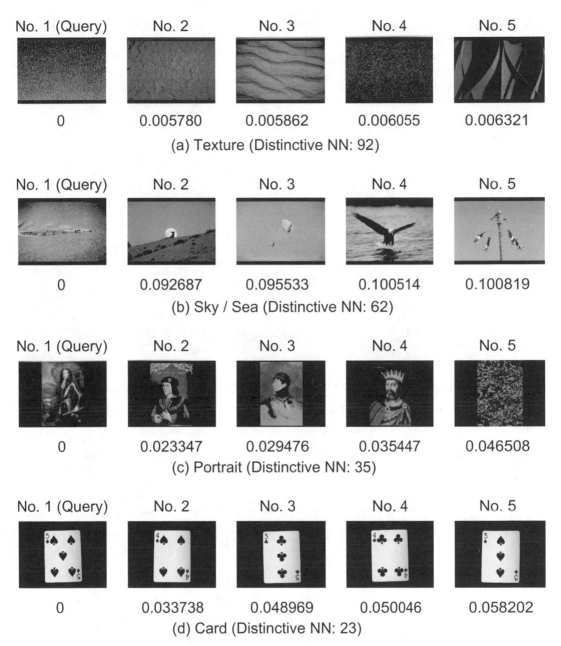

Figure 27.16 Examples of search results.

5. CONCLUSIONS

In this chapter, we emphasize two important aspects in video indexing, namely, analysing and organizing video information. For research effort in video analysis, we described Name-It, which automatically associates faces and names in news videos by properly integrating state-of-the-art technologies of image processing, natural language processing, and artificial intelligence. For video information organization, we developed SR-tree for efficient nearest-neighbor search of multidimensional information. We then revealed that these two aspects are closely related: Name-It extracts meaningful face-name association information by finding coincident high-density regions among the different feature spaces. On the other hand, the similarity retrieval with nearest-neighbor search returns meaningful results when the feature space has skewed distribution. Based on these facts, we studied searching meaningful information in video archives, and developed the distinctiveness detection method and its application to nearest-neighbor search.

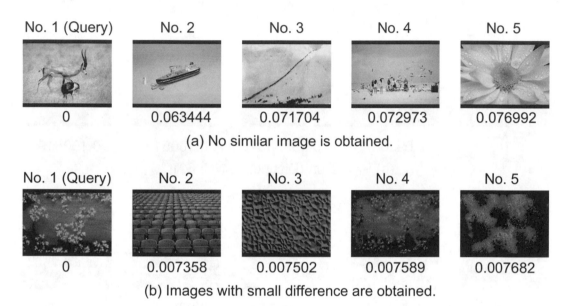

(a) No similar image is obtained.

(b) Images with small difference are obtained.

Figure 27.17 Examples of results when no distinctive NN is found except for the query.

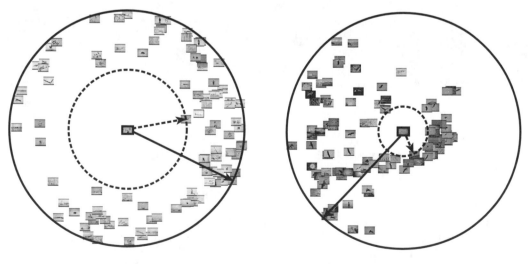

(a) no distinctive NN is found.   (b) 62 distinctive NNs are found.

Figure 27.18 Examples of the distribution of 99 nearest neighbors.

One of the promising future directions toward video indexing might be to integrate the above three approaches to further enhance mining meaningful information from realistic-scale video archives. The presented approaches in this chapter are especially important when the archives are huge and have skewed distribution. As our observation implied that real video archives have skewed distribution in the corresponding feature space, yet at the same time, the size of the video archives should be large enough to be practical. Based on this idea, we are currently developing a broadcast video archive system. The system can acquire digital video archives from seven terrestrial channels available in the Tokyo area, 24 hours a day, in a fully automated way. Thus the system can provide quite large (currently 7000 hours of videos in 10TB disk array) video archives including diverse types of programs. We are testing our technologies with the broadcast video archive system to further enhance the technologies and to realize mining meaningful information from realistic-scale video archives.

REFERENCES

[1] S. Satoh, Y. Nakamura, and T. Kanade, Name-It: Naming and Detecting Faces in News Videos, *IEEE MultiMedia*, Vol. 6, No. 1, January-March 1999, pp. 22-35.

[2] H. A. Rowley, S. Baluja, and T. Kanade, Neural Network-based Face Detection, *IEEE Trans. on PAMI*, Vol. 20, No. 1, 1998, pp. 23-38.

[3] M. Turk and A. Pentland, Eigenfaces for Recognition, *J. Cognitive Neuroscience*, Vol. 3, No. 1, 1991, pp. 71-86.

[4] T. Sato, T. Kanade, E. K. Hughes, M. A. Smith, and S. Satoh, Video OCR: Indexing Digital News Libraries by Recognition of Superimposed Captions, *Multimedia Systems*, Vol. 7, No. 5, 1999, pp. 385-395.

[5] P. A. V. Hall and G. R. Dowling, Approximate String Matching, *ACM Computing Surveys*, Vol. 12, No. 4, 1980, pp. 381-402.

[6] *Oxford Advenced Learner's Dictionary of Current English* (computer usable version), R. Mitton, Ed., 1992, http://ota.ox.ac.uk/.

[7] G. Miller et al., Introduction to WordNet: An Online Lexical Database, *Int'l J. Lexicography*, Vol.3, No. 4, 1990, pp. 235-244.

[8] D. Sleator, Parsing English with a Link Grammar, in Proc. Third Int'l Workshop on Parsing Technologies, 1993.

[9] M. Flickner, H. Sawhney, W. Niblack, J. Ashley, Q. Huang, B. Dom, M. Gorkani, J. Hafner, D. Lee, D. Petkovic, D. Steele, and P. Yanker, Query by Image and Video Content: the QBIC System, *IEEE Computer*, Vol.28, No.9, (Sep. 1995), pp.23-32.

[10] N. Katayama and S. Satoh, The SR-tree: An Index Structure for High-Dimensional Nearest Neighbor Queries, in Proc. of the 1997 ACM SIGMOD, Tucson, USA, (May 1997), pp.369-380.

[11] A. Guttman, R-trees: A Dynamic Index Structure for Spatial Searching, in Proc. ACM SIGMOD, Boston, USA, (Jun. 1984), pp.47-57.

[12] N. Beckmann, H.-P. Kriegel, R. Schneider, and B. Seeger, The R*-tree: an Efficient and Robust Access Method for Points and Rectangles, in Proc. ACM SIGMOD, Atlantic City, USA, (May 1990), pp.322-331.

[13] D. A. White and R. Jain, Similarity Indexing with the SS-tree, in Proc. of the 12th Int. Conf. on Data Engineering (ICDE), New Orleans, USA, (Feb. 1996), pp.516-523.

[14] C. Böhm, S. Berchtold, and D. A. Keim, Searching in high-dimensional spaces: Index structures for improving the performance of multimedia databases, *ACM Computing Surveys*, Vol.33, No.3, (Sep. 2001), pp.322-373.

[15] S. A. Nene and S. K. Nayar, A Simple Algorithm for Nearest Neighbor Search in High Dimensions, IEEE Trans. PAMI, Vol.19, No.9, (Sep. 1997), pp.989-1003.

[16] K. Fukunaga, *Introduction to Statistical Pattern Recognition* (2nd ed.), Academic Press, 1990.

[17] N. Katayama and S. Satoh, Distinctiveness-Sensitive Nearest-Neighbor Search for Efficient Similarity Retrieval of Multimedia Information, in IEEE 17th Int. Conf. on Data Engineering (ICDE), Heidelberg, Germany, (Apr. 2001), pp.493-502.

[18] D. A. White and R. Jain, Similarity Indexing: Algorithms and Performance, in Proc. SPIE Vol.2670, San Diego, USA, (Jan. 1996), pp.62-73.

28

EFFICIENT VIDEO SIMILARITY MEASUREMENT USING VIDEO SIGNATURES[1]

Sen-ching Samson Cheung and Avideh Zakhor
Department of Electrical Engineering and Computer Sciences
University of California
Berkeley, California USA
`sccheung@ieee.org, avz@eecs.berkeley.edu`

1.   INTRODUCTION

The amount of information on the World Wide Web has grown enormously since its creation in 1990. By February 2000, the web had over one billion uniquely indexed pages and 30 million audio, video and image links [1]. Since there is no central management on the web, duplication of content is inevitable. A study done in 1998 estimated that about 46% of all the text documents on the web have at least one "near-duplicate" - document which is identical except for low level details such as formatting [2]. The problem is likely to be more severe for web video clips as they are often stored in multiple locations, compressed with different algorithms and bitrates to facilitate downloading and streaming. Similar versions, in part or as a whole, of the same video can also be found on the web when some web users modify and combine original content with their own productions. Identifying these similar contents is beneficial to many web video applications:

1. As users typically do not view beyond the first result screen from a search engine, it is detrimental to have all "near-duplicate" entries cluttering the top retrievals. Rather, it is advantageous to group together similar entries before presenting the retrieval results to users.

2. When a particular web video becomes unavailable or suffers from slow network transmission, users can opt for a more accessible version among similar video content identified by the video search engine.

3. Similarity detection algorithms can also be used for content identification when conventional techniques such as watermarking are not applicable. For example, multimedia content brokers may use

---

similarity detection to check for copyright violation as they have no right to insert watermarks into original material.

One definition of a video similarity measure suitable for these applications is in terms of the percentage of visually similar frames between two video sequences. This is the video similarity measure used in this chapter. A similar measure, called the Tanimoto measure, is commonly used in comparing text documents where the similarity is defined as the percentage of words common to the two documents [2][3]. There are other similarity measures proposed in the literature and some of them are reviewed in this chapter. No matter which measure is used, the major challenge is how to efficiently perform the measurement. As video is a complex data type, many of the proposed similarity measures are computationally intensive. On the other hand, for every new video added to the database or a query video presented by a user, similarity measurements need to be performed with possibly millions of entries in the database. Thus, it is imperative to develop fast methods in computing similarity measurements for database applications.

Finding visually similar content is the central theme in the area of Content-Based Information Retrieval (CBIR). In the past decade, numerous algorithms have been proposed to identify visual content similar in color, texture, shape, motion and many other attributes. These algorithms typically identify a particular attribute of an image or a video shot as a high-dimensional feature vector. Visual similarity between two feature vectors can then be measured by a metric defined on the feature vector space. The basic premise of this process is that the visual content being analyzed is homogeneous in the attribute of interest. Since a full-length video is typically a collection of video shots with vastly different contents, it must be modeled as a set or a time-series of feature vectors, usually with one vector per video shot. When extending the similarity measurement to video, the first challenge is to define a single measurement to gauge the similarity between two video sequences. Multiple proposals can be found in the literature: in [4], [5] and [6], warping distance is used to measure the temporal edit differences between video sequences. Hausdorff distance is proposed in [7] to measure the maximal dissimilarity between shots. Template matching of shot change duration is used by Indyk et al. [8] to identify the overlap between video sequences. A common step shared by all the above schemes is to match similar feature vectors between two video sequences. This usually requires searching through part of or the entire video. The full computations of these measurements thus require the storage of the entire video, and time complexity that is at least linear in the length of the video. Applying such computations in finding similar content within a database of millions of video sequences may be too complex in practice.

On the other hand, the precise value of a similarity measurement is typically not required. As feature vectors are idealistic models and do not entirely capture the process of how similarity is judged in the human visual system [9], many CBIR applications only require an approximation of the underlying similarity value. As such, it is unnecessary to maintain full fidelity of the feature vector representations, and approximation schemes can be used to alleviate high computational complexity. For example, in a video similarity search system, each video in the database can be first summarized into a compact fixed-size representation. Then, the similarity between two video sequences can be approximated by comparing their corresponding representations.

There are two types of summarization techniques for similarity approximation: the higher-order and the first order techniques. The higher-order techniques summarize all feature vectors in a video as a statistical distribution. These techniques are useful in classification and semantic retrieval as they are highly adaptive and robust against small perturbation. Nonetheless, they typically assume a restricted form of density models such as Gaussian, or mixtures of Gaussian, and require a computationally intensive method like Expectation-Maximization for parameter estimation[10][11][12]. As a result, they may not be applicable for matching the enormous amount of and extremely diverse video content on the web. The first-order techniques summarize a video by a small set of representative feature vectors. One approach is to compute the "optimal" representative vectors by minimizing the distance between the original video and its representation. If the metric is finite-dimensional Euclidean and the distance is the sum of squared metric, the well-known k-means method can be used [13]. For general metric spaces, we can use the k-medoid method which identifies k feature vectors within the video to minimize the distance [14][15]. Both of these algorithms are iterative with each iteration running at $O(l)$ time for k-means, and $O(l^2)$ for k-medoids, where l represents the length of the video. To summarize long video sequences such as feature-length movies or documentaries, such methods are clearly too complex to be used in large databases.

In this chapter, we propose a randomized first-order video summarization technique called the Video Signature (ViSig) method. The ViSig method can be applied to any metric space. Unlike the k-means or k-medoid methods, it is a single-pass $O(l)$ algorithm in which each video is represented by a set of "randomly" selected frames called the ViSig. In this chapter, we show analytically and experimentally that we can obtain a reliable estimate of the underlying video similarity by using a very small ViSig to represent each video in the database. Based on a ground-truth set extracted from a large database of web video clips, we show that the ViSig method is able to achieve almost the same retrieval performance as the k-medoid of the same size, without the $O(l^2)$ complexity of k-medoid.

This chapter is organized as follows: we describe the ViSig method and show a number of analytical results in Section 2. Experimental results on a large dataset of web video and a set of MPEG-7 test sequences with simulated similar versions are used in Section 3 to demonstrate the retrieval performance of our proposed algorithms. We conclude this chapter in Section 4 by discussing related research. The proofs for all the propositions can be found in the appendices. The following is a list of acronyms and notations used in this chapter:

Acronyms:

NVS	Naïve Video Similarity
IVS	Ideal Video Similarity
VVS	Voronoi Video Similarity
ViSig	Video Signature
SV	Seed Vector

VSS$_b$	Basic ViSig Similarity
PDF	Probability Density Function
VG	Voronoi Gap
VSS$_r$	Ranked ViSig Similarity

Notations:

$(F, d(\cdot, \cdot))$	Feature vector space F with metric $d(\cdot, \cdot)$
ε	Frame Similarity Threshold
1_X	Indicator function
$\lvert X \rvert$	Cardinality of set X
$\mathrm{nvs}(X, Y; \varepsilon)$	NVS between X and Y
$[X]_\varepsilon$	Collection of clusters in X
$\mathrm{ivs}(X, Y; \varepsilon)$	IVS between X and Y
$V(X)$	Voronoi Diagram of video X
$V_X(x)$	Voronoi Cell of $x \in X$
$V_X(C)$	Voronoi Cell of a cluster $C \in [X]_\varepsilon$
$g_X(s)$	The frame in X closest to s
$R(X, Y; \varepsilon)$	Similar Voronoi Region
$\mathrm{Vol}(A)$	Volume of a region A
$\mathrm{Prob}(A)$	Probability of event A
$\mathrm{vvs}(X, Y; \varepsilon)$	VVS between X and Y
$\overrightarrow{X_s}$	ViSig of X with respect to the SV set S
$\mathrm{vss}_b(\overrightarrow{X_s}, \overrightarrow{Y_s}; \varepsilon, m)$	VSS$_b$ between $\overrightarrow{X_s}$ and $\overrightarrow{Y_s}$
$f(u; X \cup Y)$	PDF that assigns equal probability to the Voronoi Cell of each cluster in $[X \cup Y]_\varepsilon$
$G(X, Y; \varepsilon)$	VG between X and Y
$Q(g_X(s))$	Ranking function for the ViSig frame $g_X(s)$
$\mathrm{vss}_r(\overrightarrow{X_s}, \overrightarrow{Y_s}; \varepsilon, m)$	VSS$_r$ between $\overrightarrow{X_s}$ and $\overrightarrow{Y_s}$
$d_q(x, y), d_q'(x, y)$	l_1 and modified l_1 color histogram distances
$d(x, y), d'(x, y)$	Quadrant $d_q(x, y), d_q'(x, y)$
ρ	Dominant Color Threshold

rel(X)	The set of video sequences that are subjectively similar (relevant) to video X
ret(X,ε)	The set of video sequences that are declared to be similar to X by the ViSig method at ε level
Recall(ε)	The recall in retrieving the ground-truth by the ViSig method at ε level
Precision(ε)	The precision in retrieving the ground-truth by the ViSig method at ε level

2.   MEASURING VIDEO SIMILARITY

This section defines the *video similarity* models used in this chapter, and describes how they can be efficiently estimated by our proposed algorithms. We assume that individual frames in a video are represented by high-dimensional *feature vectors* from a *metric space* $(F, d(\cdot))^2$. In order to be robust against editing changes in temporal domain, we define a *video X* as a finite set of feature vectors and ignore any temporal ordering. For the remainder of this chapter, we make no distinction between a video *frame* and its corresponding feature vector. The metric $d(x, y)$ measures the visual dissimilarity between frames x and y. We assume that frames x and y are *visually similar* to each other if and only if $d(x, y) \leq \varepsilon$ for an $\varepsilon > 0$ independent of x and y. We call ε the *Frame Similarity Threshold*.

In Section 2.1, we define our target measure, called the Ideal Video Similarity (IVS), used in this chapter to gauge the visual similarity between two video sequences. As we explain in the section, this similarity measure is complex to compute exactly, and requires a significant number of frames to represent each video. To reduce the computational complexity and the representation size, we propose the ViSig method in Section 2.3 to estimate the IVS. We provide an analytical bound on the number of frames required to represent each video in our proposed algorithm. From Sections 2.4 through 2.7, we analyze the scenarios where IVS cannot be reliably estimated by our proposed algorithm, and propose a number of heuristics to rectify the problems.

2.1   IDEAL VIDEO SIMILARITY (IVS)

As mentioned in Section 1, we are interested in defining a video similarity measure that is based on the percentage of visually similar frames between two sequences. A naive way to compute such a measure is to first find the total number of frames from each video sequence that have at least one visually similar frame in the other sequence, and then compute the ratio of this number to the overall total number of frames. We call this measure the *Naïve Video Similarity* (NVS):

---

[2]For all x, y in a metric space F, the function $d(x, y)$ is a metric if a) $d(x, y) \geq 0$; b) $d(x, y) = 0 \Leftrightarrow x = y$; c) $d(x, y) = d(y, x)$; d) $d(x, y) \leq d(x, z) + d(z, y)$, for all z

Definition 1. Naïve Video Similarity

Let X and Y be two video sequences. The number of frames in video X that have at least one visually similar frame in Y is represented by $\sum_{x \in X} 1_{\{y \in Y : d(x,y) \le \varepsilon\}}$, where 1_A is the indicator function with $1_A = 1$ if A is not empty, and zero otherwise. The Naïve Video Similarity between X and Y, $\mathrm{nvs}(X,Y;\varepsilon)$, can thus be defined as follows:

$$\mathrm{nvs}(X,Y;\varepsilon) := \frac{\sum_{x \in X} 1_{\{y \in Y : d(x,y) \le \varepsilon\}} + \sum_{y \in Y} 1_{\{x \in X : d(x,y) \le \varepsilon\}}}{|X| + |Y|} \tag{28.1}$$

If every frame in video X has a similar match in Y and vice versa, $\mathrm{nvs}(X,Y;\varepsilon) = 1$. If X and Y share no similar frames at all, $\mathrm{nvs}(X,Y;\varepsilon) = 0$. Unfortunately, NVS does not always reflect our intuition of video similarity. Most real-life video sequences can be temporally separated into video shots, within which the frames are visually similar. Among all possible versions of the same video, the number of frames in the same shot can be quite different. For instance, different coding schemes modify the frame rates for different playback capabilities, and video summarization algorithms use a single key-frame to represent an entire shot. As NVS is based solely on frame counts, its value is highly sensitive to these kinds of manipulations. To illustrate this with a pathological example, consider the following: given a video X, create a video Y by repeating one single frame in X for a great many times. If $|Y| \gg |X|$, $\mathrm{nvs}(X,Y;\varepsilon) \approx 1$ even though X and Y share one common frame. It is possible to rectify the problem by using shots as the fundamental unit for similarity measurement. Since we model a video as a set and ignore all temporal ordering, we instead group all visually similar frames in a video together into non-intersecting units called clusters. A cluster should ideally contain only similar frames, and no other frames similar to the frames in a cluster should be found in the rest of the video. Mathematically, we can express these two properties as follows: for all pairs of frames x_i and x_j in X, $d(x_i, x_j) \le \varepsilon$ if and only if x_i and x_j belong to the same cluster. Unfortunately, such a clustering structure may not exist for an arbitrary video X. Specifically, if $d(x_i, x_j) \le \varepsilon$ and $d(x_j, x_k) \le \varepsilon$, there is no guarantee that $d(x_i, x_k) \le \varepsilon$. If $d(x_i, x_k) > \varepsilon$, there is no consistent way to group all the three frames into clusters.

In order to arrive at a general framework for video similarity, we adopt a relatively relaxed clustering structure by only requiring the forward condition, i.e., $d(x_i, x_j) \le \varepsilon$ implies that x_i and x_j are in the same cluster. A *cluster* is simply one of the connected components [15] of a graph in which each node represents a frame in the video, and every pair of frames within ε from each other are connected by an edge. We denote the collection of all clusters in video X as $[X]_\varepsilon$. It is possible for such a definition to produce chain-like clusters where one end of a cluster is very far from the other end. Nonetheless, given an appropriate feature vector and a reasonably small ε, we have empirically found most clusters in real video sequences to be compact, i.e., all frames in a cluster are similar to each other. We call a cluster ε-*compact* if all its frames are within ε from each other. The clustering structure of a video can be computed by a simple hierarchical clustering algorithm called the *single-link algorithm* [16].

In order to define a similarity measure based on the visually similar portion shared between two video sequences X and Y, we consider the clustered union $[X \cup Y]_\varepsilon$. If a cluster in $[X \cup Y]_\varepsilon$ contains frames from both sequences, these frames are likely to be visually similar to each other. Thus, we call such a cluster *Similar Cluster* and consider it as part of the visually similar portion. The ratio between the number of Similar Clusters and the total number of clusters in $[X \cup Y]_\varepsilon$ forms a reasonable similarity measure between X and Y. We call this measure the *Ideal Video Similarity* (IVS):

Definition 2. Ideal Video Similarity
Let X and Y be two video sequences. The IVS between X and Y, or ivs$(X,Y;\varepsilon)$, is defined to be the fraction of clusters in $[X \cup Y]_\varepsilon$ that contain frames from sequences, i.e., $C \in [X \cup Y]_\varepsilon$ with $1_{C \cap X} \cdot 1_{C \cap Y} = 1$. Specifically ivs$(X,Y;\varepsilon)$ can be expressed by the following equation:

$$\text{ivs}(X,Y;\varepsilon) := \frac{\sum_{C \in [X \cup Y]_\varepsilon} 1_{C \cap X} \cdot 1_{C \cap Y}}{\left| [X \cup Y]_\varepsilon \right|} \tag{28.2}$$

The main theme of this chapter is to develop efficient algorithms to estimate the IVS between a pair of video sequences. A simple pictorial example to demonstrate the use of IVS is shown in Figure 28.1(a). The feature space is represented as a 2D square. Dots and crosses signify frames from two different video sequences, and frames closer than ε are connected by dotted lines. There are altogether three clusters in the clustered union, and only one cluster has frames from both sequences. The IVS is thus 1/3.

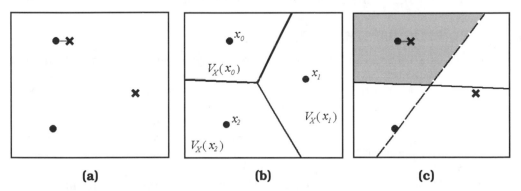

(a) **(b)** **(c)**

Figure 28.1 (a) Two video sequences with IVS equal to 1/3. (b) The Voronoi Diagram of a 3-frame video X. (c) The shaded area, normalized by the area of the entire space, is equal to the VVS between the two sequences shown.

It is complex to precisely compute the IVS. The clustering used in IVS depends on the distances between frames from the two sequences. This implies that for two l-frame video sequences, one needs to first compute the distance between l^2 pairs of frames before running the clustering algorithm and computing the IVS. In addition, the computation requires the entire video to be stored. The complex computation and large storage requirements are clearly undesirable for large database applications. As the exact similarity value is often not required in many applications, sampling techniques can be used to estimate IVS. Consider the following simple sampling scheme: let each video sequence in the database be represented by m randomly selected frames. We estimate the IVS between two sequences by counting the number of similar pairs of frames W_m between their respective sets of sampled frames. As long as the desired level of precision is satisfied, m should be chosen as small as possible to achieve low complexity. Nonetheless, even in the case when the IVS is as large as one, we show in the following proposition that we need a large m to find even one pair of similar frames among the sampled frames.

Proposition 1

Let X and Y be two l-frame video sequences. Assume for every frame x in X, Y has exactly one frame y similar to it, i.e., $d(x,y) \le \varepsilon$. We also assume the same for every frame in Y. Clearly, $\mathrm{ivs}(X,Y;\varepsilon)=1$. The expectation of the number of similar frame pairs W_m found between m randomly selected frames from X and from Y is given below:

$$E(W_m) = \frac{m^2}{l} \qquad (28.3)$$

Despite the fact that the IVS between the video sequences is one, Equation (28.3) shows that we need, on average, $m = \sqrt{l}$ sample frames from each video to find just one similar pair. Furthermore, comparing two sets of $\sqrt{l}$ frames requires l high-dimensional metric computations. A better random sampling scheme should use a fixed-size record to represent each video, and require far fewer frames to identify highly similar video sequences. Our proposed ViSig method is precisely such a scheme and is the topic of the following section.

2.2 VORONOI VIDEO SIMILARITY

As described in the previous section, the simple sampling scheme requires a large number of frames sampled from each video to estimate IVS. The problem lies in the fact that since we sample frames from two video sequences independently, the probability that we simultaneously sample a pair of similar frames from them is rather small. Rather than independent sampling, the ViSig method introduces dependence by selecting frames in each video that are similar to a set of predefined random feature vectors common to all video sequences. As a result, the ViSig method takes far fewer sampled frames to find a pair of similar frames from two video sequences. The number of pairs of similar frames found by the ViSig method depends strongly on the IVS, but does not have a one-to-one relationship with it. We call the form of similarity estimated by the ViSig method the Voronoi Video Similarity (VVS). In this section, we explain VVS and, in Section 2.3, we discuss how it is estimated by the ViSig method. The discrepancies between VVS and IVS, and how they can be rectified by modifying the ViSig method are addressed in Sections 2.4 through 2.7.

The term "Voronoi" in VVS is borrowed from a geometrical concept called the Voronoi Diagram. Given a l-frame video $X = \{x_t : t = 1, \ldots, l\}$, the *Voronoi Diagram* $V(X)$ of X is a partition of the feature space F into l Voronoi Cells $V_X(x_t)$. By definition, the *Voronoi Cell* $V_X(x_t)$ contains all the vectors in F closer to $x_t \in X$ than to any other frames in X, i.e., $V_X(x_t) := \{s \in F : g_X(s) = x_t\}$, where $g_X(s)$ denotes the frame in X closest[3] to s. A simple Voronoi Diagram of a 3-frame video is shown in Figure 28.1(b). We can extend the idea of the Voronoi Diagram to video clusters by merging Voronoi Cells of all the frames belonging to the same cluster. In other words, for $C \in [X]_e$, $V_X(C) := \bigcup_{x \in C} V_X(x)$. Given two video sequences X and Y and their corresponding Voronoi Diagrams, we define the *Similar Voronoi Region* $R(X,Y;\varepsilon)$ as the union of all the intersections between the Voronoi Cells of those $x \in X$ and $y \in Y$ where $d(x,y) \le \varepsilon$:

$$R(X,Y;\varepsilon) := \bigcup_{d(x,y) \le \varepsilon} V_X(x) \cap V_Y(y) \qquad (28.4)$$

It is easy to see that if x and y are close to each other, their corresponding Voronoi Cells are very likely to intersect in the neighborhood of x and y. The larger number of frames from X and from Y that are close to each other, the larger the resulting $R(X,Y;\varepsilon)$ becomes. A simple pictorial example of two video sequences with their Voronoi Diagrams is shown in Figure 28.1(c): dots and crosses represent the frames of the two sequences; the solid and broken lines are the boundary between the two Voronoi Cells of the two sequences represented by dots and crosses respectively. The shaded region shows the Similar Voronoi Region between these two sequences. Similar Voronoi Region is the target region whose volume defines VVS. Before providing a definition of VVS, we need to first clarify what we mean by the volume of a region in the feature space.

We define *volume* function $\mathrm{Vol}: \Omega \to \mathfrak{R}$ to be the Lebesgue measure over the set, Ω, of all the measurable subsets in feature space F [17]. For example, if F is the real line and the subset is an interval, the volume function of the subset is just the length of the interval. We assume all the Voronoi Cells considered in our examples to be measurable. We further assume that F is *compact* in the sense that $\mathrm{Vol}(F)$ is finite. As we are going to normalize all volume measurements by $\mathrm{Vol}(F)$, we simply assume that $\mathrm{Vol}(F) = 1$. To compute the volume of the Similar Voronoi Region $R(X,Y;c)$ between two video sequences X and Y, we first notice that individual terms inside the union in Equation (28.4) are disjoint from each other. By the basic properties of Lebesgue measure, we have

$$\mathrm{Vol}(R(X,Y;\varepsilon)) = \mathrm{Vol}\left(\bigcup_{d(x,y) \le \varepsilon} V_X(x) \cap V_Y(y) \right) = \sum_{d(x,y) \le \varepsilon} \mathrm{Vol}(V_X(x) \cap V_Y(y)) \qquad (28.5)$$

---

[3] If there are multiple x's in X that are equidistant to s, we choose $g_X(s)$ to be the one closest to a predefined vector in the feature space such as the origin. If there are still multiple candidates, more predefined vectors can be used until a unique $g_X(s)$ is obtained. Such an assignment strategy ensures that $g_X(s)$ depends only on X and s but not some arbitrary random choices. This is important to the ViSig method which uses $g_X(s)$ as part of a summary of X with respect to a randomly selected s. Since $g_X(s)$ depends only on X and s, sequences identical to X produce the same summary frame with respect to s.

Thus, we define the *Voronoi Video Similarity* (VVS), $vvs(X,Y;\varepsilon)$, between two video sequences X and Y as,

$$vvs(X,Y;\varepsilon) := \sum_{d(x,y)\le\varepsilon} \text{Vol}\big(V_X(x)\cap V_Y(y)\big) \qquad (28.6)$$

The VVS of the two sequences shown in Figure 28.1(c) is the area of the shaded region, which is about $1/3$ of the area of the entire feature space. Notice that for this example, the IVS is also $1/3$. VVS and IVS are close to each other because the Voronoi Cell for each cluster in the cluster union has roughly the same volume (area).

In general, when the clusters are not uniformly distributed over the feature space, there can be a large variation among the volumes of the corresponding Voronoi Cells. Consequently, VVS can be quite different from IVS. In the following section, we ignore the differences between VVS and IVS and introduce the Basic ViSig method to estimate either similarity measure.

2.3 VIDEO SIGNATURE METHOD

It is straightforward to estimate $vvs(X,Y;\varepsilon)$ by random sampling: First, generate a set S of m independent uniformly distributed random vectors $s_1, s_2, \ldots, s_m$, which we call *Seed Vectors* (SV). By uniform distribution, we mean for every measurable subset A in F, the probability of generating a vector from A is $\text{Vol}(A)$. Second, for each random vector $s\in S$, determine if s is inside $R(X,Y;\varepsilon)$. By definition, s is inside $R(X,Y;\varepsilon)$ if and only if s belongs to some Voronoi Cells $V_X(x)$ and $V_Y(y)$ with $d(x,y)\le\varepsilon$. Since s must be inside the Voronoi Cell of the frame closest to s in the entire video sequence, i.e., $g_X(s)$ in X and $g_Y(s)$ in Y, an equivalent condition for $s\in R(X,Y;\varepsilon)$ is $d\big(g_X(s),g_Y(s)\big)\le\varepsilon$. Since we only require $g_X(s)$ and $g_Y(s)$ to determine if each SV s belongs to $R(X,Y;\varepsilon)$, we can summarize video X by the m-tuple $\overrightarrow{X_s} := \big(g_X(s_1), g_X(s_2), \ldots, g_X(s_m)\big)$ and Y by $\overrightarrow{Y_s} := \big(g_Y(s_1), g_Y(s_2), \ldots, g_Y(s_m)\big)$. We call $\overrightarrow{X_s}$ and $\overrightarrow{Y_s}$ the *Video Signature* (ViSig) with respect to S of video sequences X and Y respectively. In the final step, we compute the percentage of ViSig frame pairs $g_X(s)$ and $g_Y(s)$ with distances less than or equal to ε to obtain:

$$vss_b\big(\overrightarrow{X_s},\overrightarrow{Y_s};\varepsilon,m\big) := \frac{1}{m}\sum_{i=1}^{m} 1_{\{d(g_X(s_i),g_Y(s_i))\le\varepsilon\}} \qquad (28.7)$$

We call $vss_b(\overrightarrow{X_s},\overrightarrow{Y_s};\varepsilon,m)$ the *Basic ViSig Similarity* (VSS$_b$) between ViSig's $\overrightarrow{X_s}$ and $\overrightarrow{Y_s}$. As every SV $s\in S$ in the above algorithm is chosen to be uniformly distributed, the probability of s being inside $R(X,Y;\varepsilon)$ is $\text{Vol}\big(R(X,Y;\varepsilon)\big) = vvs(X,Y;\varepsilon)$. Thus, $vss_b\big(\overrightarrow{X_s},\overrightarrow{Y_s};\varepsilon,m\big)$ forms an unbiased estimator of the VVS between X and Y. We refer to this approach of generating ViSig and computing VSS$_b$ the Basic ViSig method. In order to apply the Basic ViSig method to a large number of video sequences, we must use the same SV set S to generate all the ViSig's in order to compute VSS$_b$ between an arbitrary pair of video sequences.

The number of SV's in S, m, is an important parameter. On one hand, m represents the number of samples used to estimate the underlying VVS and thus, the larger m is, the more accurate the estimation becomes. On the other hand, the complexity of the Basic ViSig method directly depends on m. If a video has l frames, it takes l metric computations to generate a single ViSig frame. The number of metric computations required to compute the entire ViSig is thus $m \cdot l$. Also, computing the VSS_b between two ViSig's requires m metric computations. It is, therefore, important to determine an appropriate value of m that can satisfy both the desired fidelity of estimation and the computational resource of a particular application. The following proposition provides an analytical bound on m in terms of the maximum error in estimating the VVS between any pair of video sequences in a database:

Proposition 2

Assume we are given a database Λ with n video sequences and a set S of m random SV's. Define the error probability $P_{err}(m)$ to be the probability that any pair of video sequences in Λ has their m-frame VSS_b different from the true VVS value by more than a given $\gamma \in (0,1]$, i.e.,

$$P_{err}(m) := \text{Prob}\left(\bigcup_{X,Y \in \Lambda} \left\{ \left| \text{vvs}(X,Y;\varepsilon) - \text{vss}_b(\overrightarrow{X_S}, \overrightarrow{Y_S}; \varepsilon, m) \right| > \gamma \right\} \right) \quad (28.8)$$

A sufficient condition to achieve $P_{err}(m) \leq \delta$ for a given $\delta \in (0,1]$ is as follows:

$$m \geq \frac{2\ln n - \ln \delta}{2\gamma^2} \quad (28.9)$$

It should be noted that the bound (28.9) in Proposition 2 only provides a sufficient condition and does not necessarily represent the tightest bound possible. Nonetheless, we can use such a bound to understand the dependencies of m on various factors. First, unlike the random sampling described in Section 5, m does not depend on the length of individual video sequences. This implies that it takes far fewer frames for the ViSig method to estimate the similarity between long video sequences than random frame sampling. Second, we notice that the bound on m increases with the natural logarithm of n, the size of the database. The ViSig size depends on n because it has to be large enough to simultaneously minimize the error of all possible pairs of comparisons, which is a function of n. Fortunately, the slow-growing logarithm makes the ViSig size rather insensitive to the database size, making it suitable for very large databases. The contribution of the term $\ln \delta$ is also quite insignificant. Comparatively, m is most sensitive to the choice of γ. A small γ means an accurate approximation of the similarity, but usually at the expense of a large number of sample frames m to represent each video. The choice of γ should depend on the particular application at hand.

2.4 SEED VECTOR GENERATION

We have shown in the previous section that the VVS between two video sequences can be efficiently estimated by the Basic ViSig method. Unfortunately, the estimated VVS does not necessarily reflect the target measure of IVS as defined in Equation (28.2). For example, consider the two pairs of sequences in

Figure 28.2(a) and Figure 28.2(b). As in Figure 28.1(c), dots and crosses are frames from the two sequences, whose Voronoi Diagrams are indicated by solid and broken lines respectively. The IVS's in both cases are 1/3. Nonetheless, the VVS in Figure 28.2(a) is much smaller than 1/3, while that of Figure 28.2(b) is much larger. Intuitively, as mentioned in Section 2.2, IVS and VVS are the same if clusters in the clustered union are uniformly distributed in the feature space. In the above examples, all the clusters are clumped in one small area of the feature space, making one Voronoi Cell significantly larger than the other. If the Similar Cluster happens to reside in the smaller Voronoi Cells, as in the case of Figure 28.2(a), the VVS is smaller than the IVS. On the other hand, if the Similar Cluster is in the larger Voronoi Cell, the VVS becomes larger. This discrepancy between IVS and VVS implies that VSS_b, which is an unbiased estimator of VVS, can only be used as an estimator of IVS when IVS and VVS is close. Our goal in this section and the next is to modify the Basic ViSig method so that we can still use this method to estimate IVS even in the case when VVS and IVS are different.

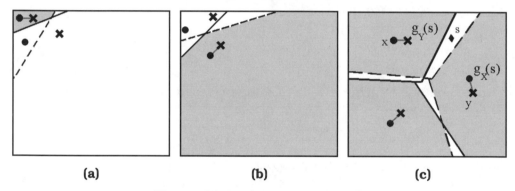

(a)   **(b)**   **(c)**

Figure 28.2 Three examples of VVS.

As the Basic ViSig method estimates IVS based on uniformly-distributed SV's, the variation in the sizes of Voronoi Cells affects the accuracy of the estimation. One possible method to amend the Basic ViSig method is to generate SV's based on a probability distribution such that the probability of a SV being in a Voronoi Cell is independent of the size of the Cell. Specifically, for two video sequences X and Y, we can define the *Probability Density Function* (PDF) based on the distribution of Voronoi Cells in $[X \cup Y]_\varepsilon$ at an arbitrary feature vector u as follows:

$$f(u; X \cup Y) := \frac{1}{\left|[X \cup Y]_\varepsilon\right|} \cdot \frac{1}{\mathrm{Vol}\left(\mathrm{V}_{X \cup Y}(C)\right)} \qquad (28.10)$$

where C is the cluster in $[X \cup Y]_\varepsilon$ with $u \in V_{X \cup Y}(C)$. $f(u; X \cup Y)$ is constant within the Voronoi Cell of each cluster, with the value inversely proportional to the volume of that Cell. Under this PDF, the probability of a random vector u inside the Voronoi Cell $V_{X \cup Y}(C)$ for an arbitrary cluster $C \in [X \cup Y]_\varepsilon$ is given by $\int_{V_{X \cup Y}(C)} f(u; X \cup Y)\, du = 1/\left|[X \cup Y]_\varepsilon\right|$. This probability does not depend on C, and thus, it is equally likely for u to be inside the Voronoi Cell of any cluster in $[X \cup Y]_\varepsilon$.

Recall that if we use uniform distribution to generate random SV's, VSS_b forms an unbiased estimate of the VVS defined in Equation (28.6). If we use $f(u; X \cup Y)$ to generate SV's instead, VSS_b now becomes an estimate of the following general form of VVS:

$$\sum_{d(x,y) \leq \varepsilon} \int_{V_X(x) \cap V_Y(y)} f(u; X \cup Y) \, du \qquad (28.11)$$

Equation (28.11) reduces to Equation (28.6) when $f(u; X \cup Y)$ is replaced by the uniform distribution, i.e., $f(u; X \cup Y) = 1$. It turns out, as shown by the following proposition, that this general form of VVS is equivalent to the IVS under certain conditions.

Proposition 3
Assume we are given two video sequences X and Y. Assume clusters in $[X]_\varepsilon$ and clusters in $[Y]_\varepsilon$ either are identical, or share no frames that are within ε from each other. Then, the following relation holds:

$$\mathrm{ivs}(X, Y; \varepsilon) = \sum_{d(x,y) \leq \varepsilon} \int_{V_X(x) \cap V_Y(y)} f(u; X \cup Y) \, du \qquad (28.12)$$

The significance of this proposition is that if we can generate SV's with $f(u; X \cup Y)$, it is possible to estimate IVS using VSS_b. The condition that all clusters in X and Y are either identical or far away from each other is to avoid the formation of a special region in the feature space called a Voronoi Gap (VG). The concept of VG is expounded in Section 2.5.

In practice, it is impossible to use $f(u; X \cup Y)$ to estimate the IVS between X and Y. This is because $f(u; X \cup Y)$ is specific to the two video sequences being compared, while the Basic ViSig method requires the same set of SV's to be used by all video sequences in the database. A heuristic approach for SV generation is to first select a set Ψ of training video sequences that resemble video sequences in the target database. Denote $T := \bigcup_{Z \in \Psi} Z$. We can then generate SV based on the PDF $f(u; T)$, which ideally resembles the target $f(u; X \cup Y)$ for an arbitrary pair of X and Y in the database.

To generate a random SV s based on $f(u; T)$, we follow a four-step algorithm, called the *SV Generation method*, as follows:

1. Given a particular value of ε_{SV}, identify all the clusters in $[T]_{\varepsilon_{SV}}$ using the single-link algorithm.

2. As $f(u; T)$ assigns equal probability to the Voronoi Cell of each cluster in $[T]_{\varepsilon_{SV}}$, randomly select a cluster C' from $[T]_{\varepsilon_{SV}}$ so that we can generate the SV s within $V_T(C')$.

3. As $f(u; T)$ is constant over $V_T(C')$, we should ideally generate s as a uniformly-distributed random vector over $V_T(C')$. Unless $V_T(C')$ can be easily parameterized, the only way to achieve this is to repeatedly generate uniform sample vectors over the entire feature space until a vector is found inside $V_T(C')$. This procedure may take an exceedingly long time if $V_T(C')$ is small.

To simplify the generation, we select one of the frames in C' at random and output it as the next SV s.

4. Repeat the above process until the required number of SV's has been selected.

In Section 3, we compare performance of this algorithm against uniformly distributed SV generation in retrieving real video sequences.

2.5 VORONOI GAP

We show in Proposition 3 that the general form of VVS using an appropriate PDF is identical to the IVS, provided that all clusters between the two sequences are either identical or far away from each other. As feature vectors are not perfect in modeling a human visual system, visually similar clusters may result in feature vectors that are close but not identical to each other. Let us consider the example in Figure 28.2(c) where frames in Similar Clusters between two video sequences are not identical but within ε from each other. Clearly, the IVS is one. Consider the Voronoi Diagrams of the two sequences. Since the boundaries of the two Voronoi Diagrams do not exactly coincide with each other, the Similar Voronoi Region, as indicated by the shaded area, does not occupy the entire feature space. As the general form of VVS defined in Equation (28.11) is the weighted volume of the Similar Voronoi Region, it is strictly less than the IVS. The difference between the two similarities is due to the unshaded region in Figure 28.2(c). The larger the unshaded region is, the larger the difference between the two similarities. If a SV s falls within the unshaded region in Figure 28.2(c), we can make two observations about the corresponding ViSig frames $g_X(s)$ and $g_Y(s)$ from the two sequences X and Y: (1) they are far apart from each other, i.e., $d(g_X(s), g_Y(s)) > \varepsilon$; (2) they both have similar frames in the other video, i.e., there exists $x \in X$ and $y \in Y$ such that $d(x, g_X(s)) \le \varepsilon$ and $d(y, g_Y(s)) \le \varepsilon$.

These observations define a unique characteristic of a particular region, which we refer to as Voronoi Gap (VG). Intuitively, any SV in VG between two sequences produces a pair of dissimilar ViSig frames, even though both ViSig frames have a similar match in the other video. More formally, we define VG as follows:

Definition 3. Voronoi Gap (VG)

Let X and Y be two video sequences. The VG $G(X,Y;\varepsilon)$ of X and Y is defined by all $s \in F$ that satisfy the following criteria

1. $d(g_X(s), g_Y(s)) > \varepsilon$,

2. *there exists $x \in X$ such that $d(x, g_Y(s)) \le \varepsilon$,*

3. *there exists $y \in Y$ such that $d(y, g_X(s)) \le \varepsilon$.*

In section 2.6, we show that the volume of VG is non-trivial by computing its volume for a particular feature space, namely the Hamming Cube. This implies that it is quite possible to find some of the SV's used in the Basic ViSig method to fall inside the VG. As a result, the performance of the Basic ViSig method in estimating IVS may be adversely affected. In Section 2.7, we introduce the Ranked ViSig method, which mitigates this problem by avoiding SV's that are likely to be inside a VG.

2.6   VORONOI GAP IN HAMMING CUBE

The example in Figure 28.2(c) seems to suggest that VG is small if ε is small. An important question is how small ε should be before we can ignore the contribution of the VG. It is obvious that the precise value of the volume of a VG depends on the frame distributions of the two video sequences, and the geometry of the feature space. It is, in general, difficult to compute even a bound on this volume without assuming a particular feature space geometry and frame distributions. In order to get a rough idea of how large VG is, we compute a simple example using a h-dimensional Hamming cube. A *Hamming cube* is the set containing all the h-bit binary numbers. The distance between two vectors is simply the number of bit-flips to change the binary representation of one vector to the other. Since it is a finite space, the volume function is simply the cardinality of the subset divided by 2^h. We choose the Hamming cube because it is easy to analyze, and some commonly used metrics such as l_1 and l_2 can be embedded inside the Hamming cube with low distortion [18].

To simplify the calculations, we only consider two-frame video sequences in the h-dimensional Hamming cube H. Let $X = \{x_1, x_2\}$ be a video in H. Let the distance between x_1 and x_2 be a positive integer k. We assume the two frames in X are not similar, i.e., the distance between them is much larger than ε. In particular, we assume that $k > 2\varepsilon$. We want to compute the "gap volume", i.e., the probability of choosing a SV s that is inside the VG formed between X and some video sequence in H. Based on the definition of VG, if a two-frame video sequence Y has a non-empty VG with X, Y must have a frame similar to each frame in X. In other words, the IVS between X and Y must be one. Let Γ be the set of all two-frame sequences whose IVS with X is one. The gap volume is thus the volume of the union of the VG formed between X and each video in Γ. As shown by the following proposition, this gap probability can be calculated using the binomial distribution.

Proposition 4

Let $X = \{x_1, x_2\}$ be a two-frame video in the Hamming cube H, and Γ be the set of all two-frame sequences whose IVS with X is one. Define A to be the union of the VG formed between X and every video in Γ, i.e.,

$$A = \bigcup\nolimits_{Y \in \Gamma} G(X, Y; \varepsilon)$$

Then, if $k = d(x_1, x_2)$ is an even number larger than 2ε, the volume of A can be computed as follows:

$$
\begin{aligned}
\mathrm{Vol}(A) \quad &= \quad \mathrm{Prob}(k/2 - \varepsilon \le R < k/2 + \varepsilon) \\
&= \quad \frac{1}{2^k} \sum_{r=k/2-\varepsilon}^{k/2+\varepsilon} \binom{k}{r}
\end{aligned}
\tag{28.13}
$$

where R is a random variable that follows a binomial distribution with parameters k and $1/2$.

We compute $\mathrm{Vol}(A)$ numerically by using the right hand side of Equation (28.13). The resulting plot of $\mathrm{Vol}(A)$ versus the distance k between the frames in X for $\varepsilon = 1, 5, 10$ is shown in Figure 28.3(a). $\mathrm{Vol}(A)$ decreases as k increases and as ε decreases, but it is hardly insignificant even when k is substantially larger than ε. For example, at $k = 500$ and $\varepsilon = 5$, Vol(A) ≈ 0.34. It is unclear whether the same phenomenon occurs for other feature spaces. Nonetheless, rather than assuming that all VG's are insignificant and using any random SV, intuitively it makes sense to identify those SV's that are inside the VG, and discard them when we estimate the corresponding video similarity.

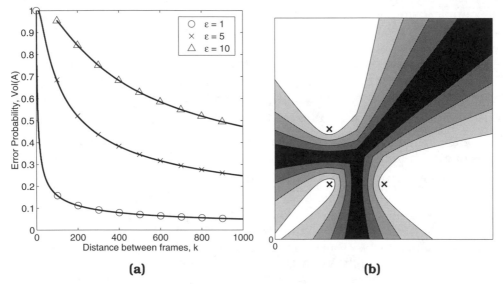

Figure 28.3 (a) The error probability for the Hamming cube at different values of ε and distances k between the frames in the video. (b) Values of ranking function $Q()$ for a three-frame video sequence. Lighter colors correspond to larger values.

2.7   RANKED VISIG METHOD

Consider again the example in Figure 28.2(c). Assume that we generate m random SV's to compute VSS$_b$. If n out of m SV's are inside the unshaded VG, we can reject these n SV's and use the remaining $(m-n)$ SV's for the computation. The resulting VSS$_b$ becomes one which exactly matches the IVS in this example. The only caveat in this approach is that we need an efficient algorithm to determine whether a SV is inside the VG. Direct application of Definition 3 is not very practical, because conditions (2) and (3) in the definition require computing the distances between a ViSig frame of one video and all the frames in the other video. Not only is the time complexity of comparing two ViSig's significantly larger than the VSS$_b$, it defeats the very purpose of using a compact ViSig to represent a video. A more efficient algorithm is thus needed to identify if a SV is inside the VG.

In this section, we propose an algorithm, applied after generating the ViSig, that can identify those SV's which are more likely to be inside the VG. In Figure 28.2(c), we observe that the two sequences have a pair of dissimilar frames that are roughly equidistant from an arbitrary vector s in the VG: x and $g_X(s)$ in the "dot" sequence, and y and $g_Y(s)$ in the "cross" sequence. They are not similar as both $d(x, g_X(s))$ and $d(y, g_Y(s))$ are clearly larger than ε. Intuitively, since vectors such as s inside the VG are close to the boundaries between Voronoi Cells in both sequences, it is not surprising to find dissimilar frames such as x and $g_X(s)$ that are on either side of the boundaries to be roughly equidistant to s. This "equidistant" condition is refined in the following proposition to upper-bound the difference between distance of s and x, and distance of s and $g_X(s)$ by 2ε:

Proposition 5

Let X and Y be two video sequences. Assume all clusters in $[X \bigcup Y]_\varepsilon$ are ε-compact. If a SV $s \in G(X, Y; \varepsilon)$, there exists a frame $x \in X$ such that

1. *x is not similar to $g_X(s)$, the ViSig frame in X with respect to s, i.e.,*
   $d(g_x(s), x) > \varepsilon$.

2. *x and $g_X(s)$ are roughly equidistant to s. Specifically, $d(x, s) - d(g_x(s), s) \le 2\varepsilon$.*

Similarly, we can find a $y \in Y$ that share the same properties with $g_Y(s)$.

The significance of Proposition 5 is that it provides a test for determining whether a SV s can ever be inside the VG between a particular video X and any other arbitrary sequence. Specifically, if there does not exist a frame x in X such that x is dissimilar to $g_X(s)$ and $d(x, s)$ is within 2ε from $d(s, g_Y(s))$, we can guarantee that s will never be inside the VG formed between X and any other sequence. The condition that all Similar Clusters must be ε-compact is to avoid pathological chain-like clusters as discussed in Section 2.1. Such an assumption is not unrealistic for real-life video sequences.

To apply Proposition 5 in practice, we first define a *Ranking Function $Q(\cdot)$* for the ViSig frame $g_X(s)$,

$$Q(g_X(s)) := \min_{x \in X, d(x, g_X(s)) > \varepsilon} d(x, s) - d(g_X(s), s) \qquad (28.14)$$

An example of $Q(\cdot)$ as a function of a 2-D SV s is shown as a contour plot in Figure 28.3(b). The three crosses represent the frames of a video. Lighter color regions correspond to the area with larger $Q(\cdot)$ values, and thus farther away from the boundaries between Voronoi Cells. By Proposition 5, if $Q(g_X(s)) > 2\varepsilon$, s cannot be inside the VG formed between X and any other sequence. In practice, however, this condition might be too restrictive in that it might not allow us to find any SV. Recall that Proposition 5 only provides a sufficient and not a necessary condition for a SV to be inside VG. Thus, even if $Q(g_x(s)) \le 2\varepsilon$, it does not necessarily imply that s will be inside the VG between X and any particular sequence.

Intuitively, in order to minimize the chances of being inside any VG, it makes sense to use a SV s with as large of a $Q(g_X(s))$ value as possible. As a result, rather than using only the ViSig frames with $Q(g_X(s)) > 2\varepsilon$, we generate a large number of ViSig frames for each ViSig, and use the few ViSig frames with the largest $Q(g_X(s))$ for similarity measurements. Let $m' > m$ be the number of frames in each ViSig. After we generate the ViSig $\overrightarrow{X_S}$ by using a set S of m' SV's, we compute and rank $Q(g_X(s))$ for all $g_X(s)$ in $\overrightarrow{X_S}$. Analogous to VSS$_b$ defined in Equation (28.7), we define the *Ranked ViSig Similarity* (VSS$_r$) between two ViSig's $\overrightarrow{X_S}$ and $\overrightarrow{Y_S}$ based on their top-ranked ViSig frames:

$$\text{vss}_r(\overrightarrow{X_S}, \overrightarrow{Y_S}; \varepsilon, m) = \frac{1}{m}\left(\sum_{i=1}^{m/2} 1_{\{d(g_X(s_{j[i]}), g_Y(s_{j[i]})) \le \varepsilon\}} + \sum_{i=1}^{m/2} 1_{\{d(g_X(s_{k[i]}), g_Y(s_{k[i]})) \le \varepsilon\}} \right) \quad (28.15)$$

$j[1],\ldots, j[m']$ and $k[1],\ldots, k[m']$'s denote the rankings of the ViSig frames in $\overrightarrow{X_S}$ and $\overrightarrow{Y_S}$ respectively, i.e., $Q(g_X(s_{j[1]})) \ge Q(g_X(s_{j[2]})) \ge \ldots \ge Q(g_X(s_{j[m']}))$ and $Q(g_Y(s_{k[1]})) \ge Q(g_Y(s_{k[2]})) \ge \ldots \ge Q(g_Y(s_{k[m']}))$. We call this method of generating ViSig and computing VSS$_r$ the Ranked ViSig method. Notice that in the right hand side of Equation (28.15), the first term uses the top-ranked $m/2$ ViSig frames from $\overrightarrow{X_S}$ to compare with the corresponding ViSig frames in $\overrightarrow{Y_S}$, and the second term uses the top-ranked $m/2$ frames from $\overrightarrow{Y_S}$. Computing VSS$_r$ thus requires m metric computations, the same as computing VSS$_b$. This provides an equal footing in complexity to compare the retrieval performances between these two methods in Section 3.

3.   EXPERIMENTAL RESULTS

In this section, we present experimental results to demonstrate the performance of the ViSig method. All experiments use color histograms as video frame features described in more detail in Section 3.1. Two sets of experiments are performed. Results of a number of controlled simulations are presented in Section 3.2 to demonstrate the heuristics proposed in Section 2. In Section 3.3, we apply the ViSig method to a large set of real-life web video sequences.

3.1   IMAGE FEATURE

In our experiments, we use four 178-bin color histograms on the *Hue-Saturation-Value* (HSV) color space to represent each individual frame in a video. *Color histogram* is one of the most commonly used image features in content-based retrieval system. The quantization of the color space used in the histogram is shown in Figure 28.4. This quantization is similar to the one used in [19]. The saturation (radial) dimension is uniformly quantized into 3.5 bins, with the half bin at the origin. The hue (angular) dimension is uniformly quantized at $20°$-step size, resulting in 18 sectors. The quantization for the value dimension depends on the saturation value. For those colors with the saturation values near zero, a finer quantizer of 16 bins is used to better differentiate between gray-scale colors. For the rest of the color space, the value dimension is uniformly quantized into three bins. The histogram is normalized such that the sum of all the bins equals one. In order to incorporate spatial information into the image feature, the image is partitioned into four quadrants, with each quadrant having its own color histogram. As a result, the total dimension of a single feature vector becomes 712.

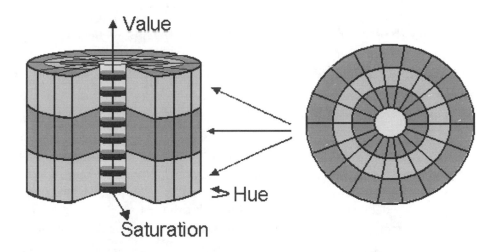

Figure 28.4 Quantization of the HSV color space.

We use two distance measurements in comparing color histograms: the l_1 metric and a modified version of the l_1 distance with dominant color first removed. The l_1 metric on color histogram was first used in [20] for image retrieval. It is defined by the sum of the absolute difference between each bin of the two histograms. We denote the l_1 *metric* between two feature vectors x and y as $d(x, y)$, with its precise definition stated below:

$$d(x, y) := \sum_{i=1}^{4} d_q(x_i, y_i) \text{ where } d_q(x_i, y_i) := \sum_{j=1}^{178} |x_i[j] - y_i[j]| \qquad (28.16)$$

where x_i and y_i for $i \in \{1, 2, 3, 4\}$ represent the quadrant color histograms from the two image feature vectors. A small $d()$ value usually indicates visual similarity, except when two images share the same background color. In those cases, the metric $d()$ does not correctly reflect the differences in the foreground as it is overwhelmed by the dominant background color. Such scenarios are quite common among the videos found on the web. Examples include those video sequences composed of presentation slides or graphical plots in scientific experiments. To mitigate this problem, we develop a new distance measurement which first removes the dominant color, then computes the l_1 metric for the rest of the color bins, and finally re-normalizes the result to the proper dynamic range. Specifically, this new distance measurement $d'(x, y)$ between two feature vectors x and y can be defined as follows:

$$d'(x, y) := \sum_{i=1}^{4} d'_q(x_i, y_i)$$

$$\text{where } d'_q(x_i, y_i) := \begin{cases} \dfrac{2}{2 - x_i[c] - y_i[c]} \displaystyle\sum_{j=1, j \ne c}^{178} |x_i[j] - y_i[j]| & \text{if } x_i[c] > \rho \text{ and } y_i[c] > \rho \\[4mm] \displaystyle\sum_{j=1}^{178} |x_i[j] - y_i[j]| & \text{otherwise.} \end{cases} \qquad (28.17)$$

In Equation (28.17), the dominant color is defined to be the color c with bin value exceeding the Dominant Color Threshold ρ. ρ has to be larger than or equal to 0.5 to guarantee a single dominant color. We set $\rho = 0.5$ in our experiments. When the two feature vectors share no common dominant color, $d'()$ reduces to $d()$.

Notice that the *modified l_1 distance* $d'()$ is not a metric. Specifically, it does not satisfy the triangle inequality. Thus, it cannot be directly applied to measuring video similarity. The l_1 metric $d()$ is used in most of the experiments described in Sections 3.2 and 3.3. We only use $d'()$ as a post-processing step to improve the retrieval performance in Section 3.3. The process is as follows: first, given a query video X, we declare a video Y in a large database to be similar to X if the VSS, either the Basic or Ranked version, between the corresponding ViSig $\overrightarrow{X_s}$ and $\overrightarrow{Y_s}$ exceeds a certain threshold λ. The l_1 metric $d()$ is first used in computing all the VSS's. Denote the set of all similar video sequences Y in the database as $\text{ret}(X, \varepsilon)$. Due to the limitation of l_1 metric, it is possible that some of the video sequences in Π may share the same background as the query video but are visually very different. In the second step, we compute the $d'()$-based VSS between $\overrightarrow{X_s}$ and all the ViSig's in $\text{ret}(X, \varepsilon)$. Only those ViSig's whose VSS with $\overrightarrow{X_s}$ larger than λ are retained in $\text{ret}(X, \varepsilon)$, and returned to the user as the final retrieval results.

3.2   SIMULATION EXPERIMENTS

In this section, we present experimental results to verify the heuristics proposed in Section 2. In the first experiment, we demonstrate the effect of the choice of SV's on approximating the IVS by the ViSig method. We perform the experiment on a set of 15 video sequences selected from the MPEG-7 video data set [21].[4] This set includes a wide variety of video content including documentaries, cartoons, and television drama, etc. The average length of the test sequences is 30 minutes. We randomly drop frames from each sequence to artificially create similar versions at different levels of IVS. ViSig's with respect to two different sets of SV's are created for all the sequences and their similar versions. The first set of SV's are independent random vectors, uniformly distributed on the high-dimensional histogram space. To generate such random vectors, we follow the algorithm described in [22]. The second set of SV's are randomly selected from a set of images in the Corel Stock Photo Collection. These images represent a diverse set of real-life images, and thus provide a reasonably good approximation to the feature vector distribution of the test sequences. We randomly choose around 4000 images from the Corel collection, and generate the required SV's using the SV Generation Method, with ε_{sv} set to 2.0, as described in Section 2.4. Table 28. 1 shows the VSS$_b$ with $m = 100$ SV's per ViSig at IVS levels of 0.8, 0.6, 0.4 and 0.2. VSS$_b$ based on Corel images are closer to the underlying IVS than those based on random vectors. In addition, the fluctuations in the estimates, as

---

[4] The test set includes video sequences from MPEG-7 video CD's v1, v3, v4, v5, v6, v7, v8 and v9. We denote each test sequence by the CD they are in, followed by a number such as v8_1 if there are multiple sequences in the same CD.

indicated by the standard deviations, are far smaller with the Corel images. The experiment thus shows that it is advantageous to use SV's that approximate the feature vector distribution of the target data.

Table 28.1 Comparison between using uniform random and Corel image SV's. The second through fifth columns are the results of using uniform random SV's and the rest are the Corel image SV's. Each row contains the results of a specific test video at IVS levels 0.8, 0.6, 0.4 and 0.2. The last two rows are the averages and standard deviations over all the test sequences.

Seed Vectors	Uniform Random				Corel Images			
IVS	0.8	0.6	0.4	0.2	0.8	0.6	0.4	0.2
v1_1	0.59	0.37	0.49	0.20	0.85	0.50	0.49	0.23
v1_2	0.56	0.38	0.31	0.05	0.82	0.63	0.41	0.18
v3	0.96	0.09	0.06	0.02	0.82	0.52	0.40	0.21
v4	0.82	0.75	0.55	0.24	0.92	0.44	0.48	0.25
v5_1	0.99	0.71	0.28	0.18	0.76	0.66	0.39	0.12
v5_2	0.84	0.35	0.17	0.29	0.81	0.68	0.36	0.10
v5_3	0.97	0.36	0.74	0.07	0.76	0.59	0.51	0.15
v6	1.00	0.00	0.00	0.00	0.79	0.61	0.46	0.25
v7	0.95	0.89	0.95	0.60	0.86	0.60	0.49	0.16
v8_1	0.72	0.70	0.47	0.17	0.88	0.69	0.38	0.20
v8_2	1.00	0.15	0.91	0.01	0.86	0.53	0.35	0.21
v9_1	0.95	0.85	0.54	0.15	0.93	0.56	0.44	0.18
v9_2	0.85	0.70	0.67	0.41	0.86	0.56	0.39	0.17
v9_3	0.90	0.51	0.30	0.10	0.78	0.70	0.45	0.15
v9_4	1.00	0.67	0.00	0.00	0.72	0.45	0.42	0.24
Average	0.873	0.499	0.429	0.166	0.828	0.581	0.428	0.187
Stddev	0.146	0.281	0.306	0.169	0.060	0.083	0.051	0.046

In the second experiment, we compare the Basic ViSig method with the Ranked ViSig method in identifying sequences with IVS one under the presence of small feature vector displacements. As described in Section 2.5, when two frames from two video sequences are separated by a small ε, the Basic ViSig method may underestimate the IVS because of the presence of VG. To combat such a problem, we propose the Ranked ViSig method in Section 2.5. In this experiment, we create similar video by adding noise to individual frames. Most of the real-life noise processes such as compression are highly video dependent, and cannot provide a wide-range of controlled noise levels for our experiment. As such, we introduce artificial noise that directly corresponds to the different noise levels as measured by our frame metric. As shown in [20], the l_1 metric used in histograms is equal to twice the percentage of the pixels between two images that are of different colors. For example, if the l_1 metric between two histograms is 0.4, it implies that 20% of the pixels in the images have different color. Thus, to inject a particular ε noise level to a frame, we determine the fraction of the pixels that need to have different colors and randomly assign colors to them. The color assignment is performed in such a way that ε noise level is achieved exactly. Five ε levels are tested in our experiments: 0.2, 0.4, 0.8, 1.2 and 1.6. In our definition of a feature vector, four histograms are used per frame. This means that an ε of, for example, 1.6 results in an average noise level of 0.4 for each histogram. After injecting noise to create the similar video, a 100-frame Basic ViSig ($m = 100$) and a 500-frame Ranked ViSig ($m' = 500$) are generated for each video. All SV's are randomly sampled from the Corel dataset. To ensure the same computational complexity between the two methods, the top $m/2 = 50$ Ranked ViSig frames are used in computing VSS$_r$. The results are shown in Table 28. 2. The averages and standard deviations over the entire set are shown in the last two rows. Since the IVS is fixed at one, the closer the measured similarity is to one, the better the approximation is. Even though both methods show a general trend of increasing error as the noise level increase, as expected, VSS$_r$ measurements are much closer to one than VSS$_b$.

Table 28.2 Comparison between VSS$_b$ and VSS$_r$ under different level of perturbation. The table follows the same format as in Table 28.1. The perturbation levels ε tested are 0.2, 0.4. 0.8, 1.2 and 1.6.

Algo.	VSS$_b$					VSS$_r$				
ε	0.2	0.4	0.8	1.2	1.6	0.2	0.4	0.8	1.2	1.6
v1_1	0.89	0.76	0.62	0.54	0.36	1.00	1.00	0.90	0.87	0.74
v1_2	0.81	0.73	0.55	0.47	0.34	1.00	0.98	0.83	0.73	0.62
v3	0.90	0.76	0.70	0.42	0.36	1.00	1.00	0.96	0.87	0.72
v4	0.86	0.74	0.64	0.48	0.38	1.00	1.00	0.96	0.83	0.74
v5_1	0.90	0.77	0.64	0.45	0.41	1.00	1.00	0.98	0.79	0.86
v5_2	0.96	0.81	0.52	0.66	0.56	1.00	1.00	1.00	0.86	0.78
v5_3	0.88	0.83	0.59	0.42	0.39	1.00	1.00	0.90	0.83	0.74
v6	0.88	0.72	0.64	0.49	0.49	1.00	1.00	0.98	0.92	0.78
v7	0.89	0.84	0.68	0.46	0.43	1.00	1.00	1.00	0.91	0.78
v8_1	0.85	0.67	0.58	0.52	0.30	1.00	1.00	0.87	0.79	0.73
v8_2	0.90	0.80	0.72	0.59	0.56	1.00	1.00	0.99	0.95	0.86
v9_1	0.87	0.77	0.62	0.67	0.48	1.00	0.99	0.89	0.84	0.82
v9_2	0.82	0.70	0.55	0.50	0.37	1.00	1.00	0.90	0.78	0.59
v9_3	0.86	0.65	0.66	0.49	0.40	1.00	1.00	0.91	0.70	0.58
v9_4	0.92	0.86	0.71	0.61	0.53	1.00	1.00	0.93	0.89	0.82
Avg.	0.879	0.761	0.628	0.518	0.424	1.000	0.998	0.933	0.837	0.744
Stdev	0.038	0.061	0.061	0.080	0.082	0.000	0.006	0.052	0.070	0.088

3.3 WEB VIDEO EXPERIMENTS

To further demonstrate how the ViSig method can be applied to a realistic application, we test our algorithms on a large dataset of web video and measure their retrieval performance using a ground-truth set. Its retrieval performance is further compared with another summarization technique called the k-medoid. k-medoid was first proposed in [14] as a clustering method used in general metric spaces. One version of *k-medoid* was used by Chang et al. for extracting a predefined number of representative frames from a video [7]. Given a l-frame video $X = \{x_t : t = 1, 2, \ldots, l\}$, the k-medoid of X is defined to be a set of k frames $x_{t_1}, x_{t_2}, \ldots, x_{t_k}$ in X that minimize the following cost function:

$$\sum_{t=1}^{l} \min_{j=1,\ldots,k} d(x_t, x_{t_j}) \qquad (28.18)$$

Due to the large number of possible choices in selecting k frames from a set, it is computationally impractical to precisely solve this minimization problem. In our experiments, we use the PAM algorithm proposed in [14] to compute an approximation to the k-medoid. This is an iterative algorithm and the time complexity for each iteration is on the order of l^2. After computing the k-medoid for each video, we declare two video sequences to be similar if the minimum distance between frames in their corresponding k-medoids is less than or equal to ε.

The dataset for the experiments is a collection of 46,356 video sequences, crawled from the web between August and December, 1999. The URL addresses of these video sequences are obtained by sending dictionary entries as text queries to the AltaVista video search engine [23]. Details about our data collection process can be found in [24]. The statistics of the four most abundant formats of video collected are shown in Table 28.3.

Table 28.3 Statistics of collected web video sequences

Video Type	% over all clips	Duration ($\pm$ stddev) minutes
MPEG	31	0.36 ± 0.7
QuickTime	30	0.51 ± 0.6
RealVideo	22	9.57 ± 18.5
AVI	16	0.16 ± 0.3

The *ground-truth* is a set of manually identified clusters of almost-identical video sequences. We adopt a best-effort approach to obtain such a ground-truth. This approach is similar to the pooling method [25] commonly used in text retrieval systems. The basic idea of pooling is to send the same queries to different automatic retrieval systems, whose top-ranked results are pooled together and examined by human experts to identify the truly relevant ones. For our system, the first step is to use meta-data terms to identify the initial ground-truth clusters. Meta-data terms are extracted from the URL addresses and other auxiliary information for each video in the dataset [26]. All video sequences containing at least one of the top 1000 most frequently used meta-data terms are manually examined and grouped into clusters of similar video. Clusters which are significantly larger than others are removed to prevent bias. We obtain 107 clusters which form the initial ground-truth clusters. This method, however, may not be able to identify all the video clips in the dataset that are similar to those already in the ground-truth clusters. We further examine those video sequences in the dataset that share at least one meta-data term with the ground-truth video, and add any similar video to the corresponding clusters. In addition to meta-data, k-medoid is also used as an alternative visual similarity scheme to enlarge the ground-truth. A 7-frame k-medoid is generated for each video. For each video X in the ground-truth, we identify 100 video sequences in the dataset that are closest to X in terms of the minimum distance between their k-medoids, and manually examine them to search for any sequence that is visually similar to X. As a result, we obtain a ground-truth set consisting of 443 video sequences in 107 clusters. The cluster size ranges from 2 to 20, with average size equal to 4.1. The ground-truth clusters serve as the subjective truth for comparison against those video sequences identified as similar by the ViSig method.

When using the ViSig method to identify similar video sequences, we declare two sequences to be similar if their VSS_b or VSS_r is larger than a certain threshold $\lambda \in [0,1]$. In the experiments, we fix λ at 0.5 and report the retrieval results for different numbers of ViSig frames, m, and the Frame Similarity Threshold, ε. Our choice of fixing λ at 0.5 can be rationalized as follows: as the dataset consists of extremely heterogeneous contents, it is rare to find partially similar video sequences. We notice that most video sequences in our dataset are either very similar to each other, or not similar at all. If ε is appropriately chosen to match subjective similarity, and m is large enough to keep sampling error small, we would expect the VSS for an arbitrary pair of ViSig's to be close to either one or zero, corresponding to either similar or dissimilar video sequences in the dataset. We thus fix λ at 0.5 to balance the possible false-positive and false-negative errors, and vary ε to trace the whole spectrum of retrieval performance. To accommodate such a testing strategy, we make a minor modification in the Ranked ViSig method: recall that we use the Ranking Function $Q()$ as defined in Equation (28.14) to rank all frames in a ViSig. Since $Q()$ depends on ε and its computation requires the entire video sequence, it is cumbersome to recompute it whenever a different ε is used. ε is used in the $Q()$ function to identify the clustering structure within a single video. Since most video sequences are compactly clustered, we notice that their $Q()$ values remain roughly constant for a large range of ε. As a result, we a priori fix ε to be 2.0 to compute $Q()$, and do not recompute them even when we modify ε to obtain different retrieval results.

The performance measurements used in our experiments are recall and precision as defined below. Let Λ be the web video dataset and Φ be the ground-truth set. For a video $X \in \Phi$, we define the *Relevant Set* to X, $rel(X)$, to be the ground-truth cluster that contains X, minus X itself. Also recall the definition of the *Return Set* to X, $ret(X,\varepsilon)$, from Section 3.1, as the set of video sequences in the database which are declared to be similar to X by the ViSig method, i.e., $ret(X,\varepsilon) := \{Y \in \Lambda : vss(\overrightarrow{X_s}, \overrightarrow{Y_s}) \le \lambda\} \setminus \{X\}$. vss can be either VSS_b or

VSS$_r$. By comparing the Return and Relevant Sets of the entire ground-truth, we can define the *Recall* and *Precision* as follows:

$$\text{Recall}(\varepsilon) := \frac{\sum_{X \in \Lambda} \left| \text{rel}(X) \cap \text{ret}(X,\varepsilon) \right|}{\sum_{X \in \Lambda} \left| \text{rel}(X) \right|} \quad \text{and} \quad \text{Precision}(\varepsilon) := \frac{\sum_{X \in \Lambda} \left| \text{rel}(X) \cap \text{ret}(X,\varepsilon) \right|}{\sum_{X \in \Lambda} \left| \text{ret}(X,\varepsilon) \right|}$$

Thus, recall computes the fraction of all ground-truth video sequences that can be retrieved by the algorithm. Precision measures the fraction retrieved by the algorithm that is relevant. By varying ε, we can measure the retrieval performance of the ViSig method for a wide range of recall values.

The goal of the first experiment is to compare the retrieval performance between the Basic and the Ranked ViSig methods at different ViSig sizes. The modified l_1 distance on the color histogram is used in this experiment. SV's are randomly selected by the SV Generation Method, with ε_{SV} set to 2.0, from a set of keyframes representing the video sequences in the dataset. These keyframes are extracted by the AltaVista search engine and captured during data collection process; each video is represented by a single keyframe. For the Ranked ViSig method, $m' = 100$ keyframes are randomly selected from the keyframe set to produce the SV set which is used for all ViSig sizes, m. For each ViSig size in the Basic ViSig method, we average the results of four independent sets of randomly selected keyframes in order to smooth out the statistical variation due to the limited ViSig sizes. The plots in Figure 28.5(a) show the precision versus recall curves for four different ViSig sizes: m = 2, 6, 10 and 14. The Ranked ViSig method outperforms the Basic ViSig method in all four cases. Figure 28.5(b) shows the Ranked method's results across different ViSig sizes. There is a substantial gain in performance when the ViSig size is increased from two to six. Further increase in ViSig size does not produce much significant gain. The precision-recall curves all decline sharply once they reach beyond 75% for recall and 90% for precision. Thus, six frames per ViSig is adequate in retrieving ground-truth from the dataset.

In the second experiment, we test the difference between using the modified l_1 distance and the l_1 metric on the color histogram. The same Ranked ViSig method with six ViSig frames is used. Figure 28.5(c) shows that the modified l_1 distance significantly outperforms the straightforward l_1 metric. Finally, we compare the retrieval performance between k-medoid and the Ranked ViSig method. Each video in the database is represented by seven medoids generated by the PAM algorithm. We plot the precision-recall curves for k-medoid and the six-frame Ranked ViSig method in Figure 28.5(d). The k-medoid technique provides a slightly better retrieval performance. The advantages seem to be small considering the complexity advantage of the ViSig method over the PAM algorithm. First, computing VSS$_r$ needs six metric computations but comparing two 7-medoid representations requires 49. Second, the ViSig method generates ViSig's in $O(l)$ time with l being the number of frames in a video, while the PAM algorithm is an iterative $O(l^2)$ algorithm.

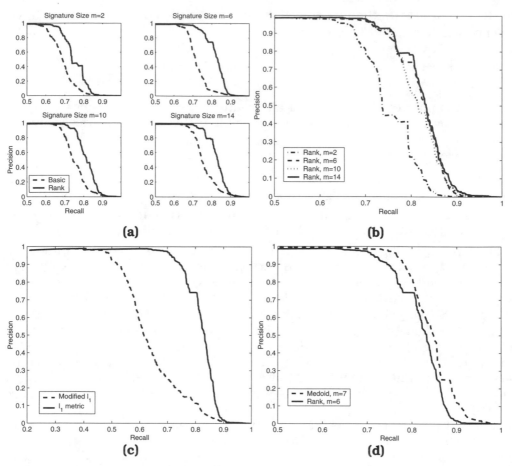

Figure 28.5 Precision-recall plots for web video experiments: (a) Comparisons between the Basic (broken-line) and Ranked (solid) ViSig methods for four different ViSig sizes: m=2, 6, 10, 14; (b) Ranked ViSig methods for the same set of ViSig sizes; (c) Ranked ViSig methods with m=6 based on l_1 metric (broken) and modified l_1 distance (solid) on color histograms; (d) Comparison between the Ranked ViSig method with m=6 (solid) and k-medoid with 7 representative frames (broken).

4.   CONCLUDING REMARKS

In this chapter, we have proposed a class of techniques named ViSig which are based on summarizing a video sequence by extracting the frames closest to a set of randomly selected SV's. By comparing the ViSig frames between two video sequences, we have obtained an unbiased estimate of their VVS. In applying the ViSig method to a large database, we have shown that the size of a ViSig depends on the desired fidelity of the measurements and the logarithm of the database size. In order to reconcile the difference between VVS and IVS, the SV's used must resemble the frame statistics of video in the target database. In addition, ViSig frames whose SV's are inside the VG should be avoided when comparing two ViSig's. We have proposed a ranking method to identify those SV's that are least likely to be inside the VG. By experimenting with a set of MPEG-7 test sequences and their artificially generated similar versions, we have demonstrated that IVS can be better approximated by using (a) SV's based on real images than uniformly random generation, and (b) the ranking method than the basic method. We have further characterized the retrieval performance of different ViSig methods based on a ground-truth set from a large set of web video.

The basic premise of our work is on the importance of IVS as a similarity measurement. IVS defines a general similarity measurement between two sets of objects endowed with a metric function. By using the ViSig method, we have demonstrated one particular application of IVS, which is to identify highly similar video sequences found on the World Wide Web. As such, we are currently investigating the use of IVS on other types of pattern matching and retrieval problems. We have also considered other aspects of the ViSig method in conjunction with its use on large databases. In our recent work [27], we have proposed a novel dimension reduction technique on signature data for fast similarity search, and a clustering algorithm on a database of signatures for improving retrieval performance.

APPENDIX

A   Proof of Proposition 1

Without loss of generality, let $X = \{x_1, x_2, \ldots, x_l\}$ and $Y = \{y_1, y_2, \ldots, y_l\}$ with $d(x_i, y_i) \leq \varepsilon$ for $i = 1, 2, \ldots, l$. Let Z_i be a binary random variable such that $Z_i = 1$ if both x_i and y_i are chosen as sampled frames, and 0 otherwise. Since W_m is the total number of similar pairs between the two sets of sampled frames, it can be computed by summing all the Z_i's:

$$W_m \;=\; \sum_{i=1}^{l} Z_i$$

$$E(W_m) \;=\; \sum_{i=1}^{l} E(Z_i) \;=\; \sum_{i=1}^{l} \mathrm{Prob}(Z_i = 1)$$

Since we independently sample m frames from each sequence, the probability that $Z_i = 1$ for any i is $(m/l)^2$. This implies that $E(W) = m^2 / l$.

B   Proof of Proposition 2

To simplify the notation, let $\rho(X, Y) = \mathrm{vvs}(X, Y; \varepsilon)$ and $\hat{\rho}(X, Y) = \mathrm{vss}_b(\overrightarrow{X_S}, \overrightarrow{Y_S}; \varepsilon, m)$. For an arbitrary pair of X and Y, we can bound the probability of the event $\left| \rho(X, Y) - \hat{\rho}(X, Y) \right| \gg$ by the Hoeffding Inequality [28]:

$$\mathrm{Prob}\left(\left| \rho(x, Y) - \hat{\rho}(X, Y) \right| \right) \leq 2 \exp(\not{} 2^{\,2} m) \tag{28.19}$$

To find an upper bound for $P_{err}(m)$, we can combine (0.1) and the union bound as follows:

$$P_{err}(m) = \text{Prob}\left(\bigcup_{X,Y\in\Lambda} \left|\rho(X,Y)-\hat{\rho}(X,Y)\right| > \right)$$

$$\leq \sum_{X,Y\in\Lambda} \text{Prob}\left(\left|\rho(X,Y)-\hat{\rho}(X,Y)\right| > \right)$$

$$\leq \frac{n^2}{2}\cdot 2\exp(-2\gamma^2 m)$$

A sufficient condition for $P_{err}(m)\leq\delta$ is thus

$$\frac{n^2}{2}\cdot 2\exp(-2\gamma^2 m) \leq \delta$$

$$m \geq \frac{2\ln n - \ln\delta}{2\gamma^2}$$

C   Proof of Proposition 3

For each term inside the summation on the right hand side of Equation (28.12), $d(x,y)$ must be smaller than or equal to ε. If $d(x,y)\leq\varepsilon$, our assumption implies that both x and y must be in the same cluster C belonging to both $[X]_\varepsilon$ and $[Y]_\varepsilon$. As a result, we can rewrite the right hand side of Equation (28.12) based only on clusters in $[X]_\varepsilon \cap [Y]_\varepsilon$:

$$\sum_{d(x,y)\leq\varepsilon}\int_{V_X(x)\cap V_Y(y)} f(u;X\cup Y)\,du = \sum_{C\in[X]_\varepsilon\cap[Y]_\varepsilon}\sum_{x\in C}\int_{V_X(x)\cap V_Y(y)} f(u;X\cup Y)\,du \qquad (28.20)$$

Based on the definition of a Voronoi Cell, it is easy to see that $V_X(z)\cap V_Y(z)=V_{X\cup Y}(z)$ for all $z\in C$ with $C\in[X]_\varepsilon\cap[Y]_\varepsilon$. Substituting this relationship into Equation (28.12), we obtain:

$$\sum_{d(x,y)\leq\varepsilon}\int_{V_X(x)\cap V_Y(x)} f(u;X\cup Y)\,du = \sum_{C\in[X]_\varepsilon\cap[Y]_\varepsilon}\int_{V_{X\cup Y}(C)} f(u;X\cup Y)\,du$$

$$= \sum_{C\in[X]_\varepsilon\cap[Y]_\varepsilon}\int_{V_{X\cup Y}(C)}\frac{1}{\left|[X\cup Y]_\varepsilon\right|\cdot \text{Vol}(V_{X\cup Y}(C))}\,du$$

$$= \frac{1}{\left|[X\cup Y]_\varepsilon\right|}\sum_{C\in[X]_\varepsilon\cap[Y]_\varepsilon}\frac{\int_{V_{X\cup Y}(C)} du}{\text{Vol}(V_{X\cup Y}(C))}$$

$$= \left|[X]_\varepsilon\cap[Y]_\varepsilon\right|/\left|[X\cup Y]_\varepsilon\right|$$

Finally, we note that $[X]_\varepsilon\cap[Y]_\varepsilon$ is in fact the set of all Similar Clusters in $[X\cup Y]_\varepsilon$, and thus the last expression equals to the IVS. The reason is that for any Similar Cluster C in $[X\cup Y]_\varepsilon$, C must have at least one $x\in X$ and one $y\in Y$ such that $d(x,y)\leq\varepsilon$. By our assumption, C must be in both $[X]_\varepsilon$ and $[Y]_\varepsilon$.

D Proof of Proposition 4

Without loss of generality, we assume that x_1 is at the origin with all zeros, and x_2 has k 1's in the rightmost positions. Clearly, $d(x_1, x_2) = k$. Throughout this proof, when we mention a particular sequence $Y \in \Gamma$, we adopt the convention that $Y = \{y_1, y_2\}$ with $d(x_1, y_1) \le \varepsilon$ and $d(x_2, y_2) \le \varepsilon$.

We first divide the region A into two partitions based on the proximity to the frames in X:

$$A_1 := \{s \in A : g_X(s) = x_1\} \text{ and } A_2 := \{s \in A : g_X(s) = x_2\}$$

We adopt the convention that if there are multiple frames in a video Z that are equidistant to a random vector s, $g_Z(s)$ is defined to be the frame furthest away from the origin. This implies that all vectors equidistant to both frames in X are elements of A_2. Let s be an arbitrary vector in H, and R be the random variable that denotes the number of 1's in the rightmost k bit positions of s. The probability that R equals to a particular r with $r \le k$ is as follows:

$$\text{Prob}(R = r) = \frac{1}{2^k} \binom{k}{r}$$

Thus, R follows a binomial distribution of parameters k and $1/2$. In this proof, we show the following relationship between A_2 and R:

$$\text{Vol}(A_2) - \text{Prob}(k/2 \le R < k/2 \mid \varepsilon) \tag{28.21}$$

With an almost identical argument, we can show the following:

$$\text{Vol}(A_1) = \text{Prob}(k/2 - \varepsilon \le R < k/2) \tag{28.22}$$

Since $\text{Vol}(A) = \text{Vol}(A_1) + \text{Vol}(A_2)$, the desired result follows.

To prove Equation (28.21), we first show if $k/2 \le R < k/2 + \varepsilon$, then $s \in A_2$. Assuming the definitions for A and A_2, we need to show two things: (1) $g_X(s) = x_2$; (2) there exists a $Y \in \Gamma$ such that $s \in G(X, Y; \varepsilon)$, or equivalently, $g_Y(s) = y_1$. To show (1), we rewrite $R = k/2 + N$ where $0 \le N < \varepsilon$ and let the number of 1's in s be L. Consider the distances between s and x_1, and between s and x_2. Since x_1 is all zeros, $d(s, x_1) = L$. As x_2 has all its 1's in the rightmost k position, $d(s, x_2) = (L - R) + (k - R) = L + k - 2R$. Thus,

$$
\begin{aligned}
d(s, x_1) - d(s, x_2) &= L - L(+k - 2R) \\
&= 2R - k \\
&= 2N \ge 0,
\end{aligned}
$$

which implies that $g_X(s) = x_2$. To show (2), we define y_1 to be a h-bit binary number with all zeros, except for ε 1's in the positions which are randomly chosen from the R 1's in the rightmost k bits of s. We can do that because $R \ge k/2 \ge \varepsilon$. Clearly, $d(x_1, y_1) = \varepsilon$ and $d(s, y_1) = L - \varepsilon$. Next, we define y_2 by toggling ε out of k 1's in x_2. The positions we toggle are randomly chosen from the same

R 1's bits in s. As a result, $d(x_2, y_2) = \varepsilon$ and $d(s, y_2) = (L - R) + (k - R + \varepsilon) = L + \varepsilon - 2N$. Clearly, $Y := \{y_1, y_2\}$ belongs to Γ. Since

$$
\begin{aligned}
d(s, y_2) - d(s, y_1) &= (L + \varepsilon - 2N) - (L - \varepsilon) \\
&= 2(\varepsilon - N) > 0
\end{aligned}
$$

$g_Y(s) = y_1$ and, consequently, $s \in G(X, Y; \varepsilon)$.

Now we show the other direction: if $s \in A_2$, then $k/2 \le R < k/2 + \varepsilon$. Since $s \in A_2$, we have $g_X(s) = x_2$ which implies that $L = d(s, x_1) \ge d(s, x_2) = L + k - 2R$ or $k/2 \le R$. Also, there exists a $Y \in \Gamma$ with $s \in G(X, Y; \varepsilon)$. This implies $g_Y(s) = y_2$, or equivalently, $d(s, y_1) < d(s, y_2)$. This inequality is strict as equality will force $g_Y(s) = y_2$ by the convention we adopt for $g_Y(\cdot)$. The terms on both sides of the inequality can be bounded using the triangle inequality: $d(s, y_1) \ge d(s, x_1) - d(x_1, y_1) = L - \varepsilon$ and $d(s, y_2) \le d(s, x_2) + d(x_2, y_2) = L + k - 2R + \varepsilon$. Combining both bounds, we have

$$
L - \varepsilon < L + k - 2R + \varepsilon \Rightarrow R < k/2 + \varepsilon
$$

This completes the proof for Equation (28.21). The proof of Equation (28.22) follows the same argument with the roles of x_1 and x_2 reversed. Combining the two equations, we obtain the desired result.

E   Proof of Proposition 5

We prove the case for video X and the proof is identical for Y. Since $s \in G(X, Y; \varepsilon)$, we have $d(g_X(s), g_Y(s)) > \varepsilon$ and there exists $x \in X$ with $d(x, g_Y(s)) \le \varepsilon$. Since all Similar Clusters in $[X \cup Y]_\varepsilon$ are ε-compact, $g_X(s)$ cannot be in the same cluster with x and $g_Y(s)$. Thus, we have $d(g_X(s), x) > \varepsilon$. It remains to show that $d(x, s) - d(g_X(s), s) \le 2\varepsilon$. Using the triangle inequality, we have

$$
\begin{aligned}
d(x, s) - d(g_X(s), s) &\le d(x, g_Y(s)) + d(g_Y(s), s) - d(g_X(s), s) \\
&\le \varepsilon + d(g_Y(s), s) - d(g_X(s), s)
\end{aligned}
\tag{28.23}
$$

$s \in G(X, Y; \varepsilon)$ also implies that there exists $y \in Y$ such that $d(y, g_Y(s)) \le \varepsilon$. By the definition of $g_Y(s)$, $d(g_Y(s), s) \le d(y, s)$. Thus, we can replace $g_Y(s)$ with y in Equation (28.23) and combine with the triangle inequality to obtain:

$$
\begin{aligned}
d(x, s) - d(g_X(s), s) &\le \varepsilon + d(y, s) - d(g_X(s), s) \\
&\le \varepsilon + d(y, g_X(s)) \\
&\le 2\varepsilon
\end{aligned}
$$

REFERENCES

[1] Inktomi Corp., Inktomi webmap, http://www2.inktomi.com/webmap, January 2000.

[2] N. Shivakumar and H. Garcia-Molina, Finding Near-replicas of Documents on the Web, in World Wide Web and Databases, International Workshop WebDB'98, Valencia, Spain, Mar. 1998, pp. 204-212.

[3] A.Z. Broder, S.C. Glassman, M.S. Manasse, and G. Zweig, Syntactic clustering of the Web, in Sixth International World Wide Web Conference, Sept. 1997, Vol. 29, No. 8-13 of Computer Networks and ISDN Systems, pp. 1157-1166.

[4] M.R. Naphade, R. Wang, and T.S. Huang, Multimodal Pattern Matching for Audio-visual Query and Retrieval, in Proceedings of the Storage and Retrieval for Media Datbases 2001, San Jose, USA, Jan 2001, Vol. 4315, pp. 188-195.

[5] D. Adjeroh, I. King, and M.C. Lee, A Distance Measure for Video Sequence Similarity Matching, in Proceedings International Workshop on Multi-Media Database Management Systems, Dayton, OH, USA, Aug. 1998, pp. 72-79.

[6] R. Lienhart, W. Effelsberg, and R. Jain, VisualGREP: A Systematic Method to Compare and Retrieve Video Sequences, in Proceedings of Storage and Retrieval for Image and Video Databases VI, SPIE, Jan. 1998, Vol. 3312, pp. 271-282.

[7] H.S. Chang, S. Sull, and S.U. Lee, Efficient video indexing scheme for content-based retrieval, *IEEE Transactions on Circuits and Systems for Video Technology*, Vol. 9, No. 8, pp. 1269-79, Dec 1999.

[8] P. Indyk, G. Iyengar, and N. Shivakumar, Finding Pirated Video Sequences on the Internet, Tech. Rep., Stanford Infolab, Feb. 1999.

[9] S. Santini and R. Jain, Similarity measures, *IEEE Transactions on Pattern Analysis and Machine Intelligence*, Vol. 21, No. 9, pp. 871-83, Sept 1999.

[10] H. Greenspan, J. Goldberger, and A. Mayer, A Probabilistic Framework for Spatio-temporal Video Representation, in IEEE Conference on Computer Vision and Pattern Recognition, 2001.

[11] G. Iyengar and A.B. Lippman, Distributional Clustering for Efficient Content-based Retrieval of Images and Video, in Proceedings 1998 International Conference on Image Processing, Vancouver, B.C., Canada, 2000, Vol. III, pp. 81-84.

[12] N. Vasconcelos, On the Complexity of Probabilistic Image Retrieval, in Proceedings Eighth IEEE International Conference on Computer Vision, Vancouver, B.C., Canada, 2001, Vol. 2, pp. 400-407.

[13] J. MacQueen, Some Methods for Classification and Analysis of Multivariate Observations, in 5th Berkeley Symposium on Mathematical Statistics, 1967, Vol. 1, pp. 281-297.

[14] L. Kaufman and P.J. Rousseeuw, *Finding Groups in Data*, John Wiley & Sons, New York, 1990.

[15] T. Cormen, C. Leiserson, and R. Riverst, *Introduction to Algorithms*, The MIT Press, Cambridge, Massachusetts, 1992.

[16] R. Sibson, Slink: An optimally efficient algorithm for the single-link cluster method, *The Computer Journal*, Vol. 16, No. 1, pp. 30-34, 1973.

[17] H.L. Royden, *Real Analysis*, Macmillan, New York, 1988.

[18] P. Indyk, High-dimensional Computational Geometry, Ph.D. Thesis, Stanford University, 2000.

[19] J.R. Smith, Integrated Spatial and Feature Image Systems: Retrieval, Analysis and Compression, Ph.D. Thesis, Columbia University, 1997.

[20] M.J. Swain and D.H. Ballard, Color indexing, *International Journal of Computer Vision*, Vol. 7, No. 1, pp. 11-32, November 1991.

[21] MPEG-7 Requirements Group, Description of MPEG-7 Content Set, Tech. Rep. N2467, ISO/IEC JTC1/SC29/WG11, 1998.

[22] A. Woronow, Generating random numbers on a simplex, *Computers and Geosciences*, Vol. 19, No. 1, pp. 81-88, 1993.

[23] AltaVista, http://www.altavista.com, AltaVista Image, Audio and Video search.

[24] S.-C. Cheung and A. Zakhor, Estimation of Web Video Multiplicity, in Proceedings of the SPIE - Internet Imaging, San Jose, California, Jan. 2000, Vol. 3964, pp. 34-36.

[25] K. Sparck Jones and C. van Rijsbergen, Report on the Need For and Provision of an "Ideal" Information Retrieval Test Collection, Tech. Rep. British Library Research and Development Report 5266, Computer Laboratory, University of Cambridge, 1975.

[26] S.-C. Cheung and A. Zakhor, Efficient Video Similarity Measurement and Search, in Proceedings of 7th IEEE International Conference on Image Processing, Vancouver, British Columbia, Sept. 2000, Vol. 1, pp. 85-88.

[27] S.-C. Cheung and A. Zakhor, Towards building a similar video search engine for the World-Wide-Web, submitted to *IEEE Transactions* on Multimedia.

[28] G.R. Grimmett and D.R. Stirzaker, *Probability and Random Processes*, Oxford Science Publications, 1992.

29

SIMILARITY SEARCH IN MULTIMEDIA DATABASES

Agma Juci M. Traina and Caetano Traina Jr.
Department of Computer Science
University of São Paulo at São Carlos
São Carlos, São Paulo, Brazil
`[agma|caetano]@icmc.usp.br`

1. INTRODUCTION AND MOTIVATION

A noticeable characteristic of the current information systems is the ubiquitous use of multimedia data, such as image, audio, video, time series and hypertext. The processing power demanded by the increasing volume of information generated by business and scientific applications has motivated researchers of the data base, artificial intelligence, statistics and visualization fields, who have been working closely to build tools that allow to understand and to efficiently manipulate this kind of data. Besides the multimedia data being complex, they are voluminous as well. Thus, to deal with the multimedia data embedded in the current systems, much more powerful and elaborate apparatuses are required as compared to the needs of the simple data handled by the old systems.

The main challenges concerning complex data are twofold. The first one is how to retrieve the relevant information embedded inside them, allowing the systems to process the data in a straightforward manner. The second challenge refers to how to present the data to the users, allowing them to interact with the system and to take advantage of the data semantics. Therefore, a system dealing with multimedia data has to tackle these two issues: feature extraction and data presentation and browsing. In this chapter, we discuss more deeply the first issue and give directions to the readers for the second issue.

A common problem regarding multimedia or complex data is how to compare and group them using only their embedded information, that is, how to compare and group video, images, audio, DNA strings, fingerprints, etc., based on how similar their contents are. Thus, content-based retrieval systems have been extensively pursued by the multimedia community during the last few years. Initially the specialists in content-based retrieval started working on images. Hence there are many works on content-based image retrieval tackling different approaches [1] [2], such as those based on color distribution [3] [4] [5], shape [6] [7] and texture [8] [9].

With the growing amount of video data available in multimedia systems, such as digital libraries, educational and entertaining systems among others, it became important to manage this kind of data automatically, using the intrinsic characteristics of video. Consequently, content-based video retrieval systems use different views or parts of video data, that is, emphasizing on the spatial aspect of video and annotations [10] or on the audio portion [11] of them.

Obtaining the main characteristics or properties of multimedia data is not a simple task. There are many aspects in each kind of data to be taken in account. For example, when dealing with images, it is well known that the main characteristics or *features* to be extracted are color distribution, shape and texture [12] [4]. The features extracted are usually grouped into vectors, the so-called *feature vectors*. Through the features, the content-based image retrieval system can deal with the meaningful extracted characteristics instead of the images themselves. Feature extraction processes correspond in some way to data reduction processes, as well as consist of a manner to obtain a parametric representation of the data that is invariant to some transformations. Recalling the previous example of images, there are techniques that allow comparison on images represented in different sizes, positioning, brightness during acquisition and even different shapes [13] [3]. However, a common problem, when reducing the information through feature vectors, is that some ambiguity is introduced in the process. That is, two or more distinct objects can be represented by the same feature values. When this happens, other steps of comparisons, using other features of the data must be done to retrieve the desired data.

This chapter discusses techniques for searching multimedia data types by similarity in databases storing large sets of multimedia data. The remainder of this chapter is organized as follows. Section 2 discusses the main concepts involved in supporting search operations by similarity on multimedia data. Section 3 presents a flexible architecture to build content-based image retrieval in relational databases. Section 4 illustrates the application of such techniques in a practical environment allowing similarity search operations. Finally, section 5 presents conclusions regarding future developments in the field, tackling the wide spread of content-based similarity searching in multimedia databases.

2. MAIN CONCEPTS

The main concepts involved in similarity search on multimedia data regarding information representation are:
- the *feature spaces* where the data under analysis is described,
- the *dissimilarity metric* or *distance function* used to compare two instances of data,
- the *searching spaces* and
- the *access methods* specially tailored to quickly process the similarity searching.

These concepts and their relationship are detailed.

2.1 FEATURE SPACES

The first and crucial step in content-based indexing and retrieval of multimedia data is the definition of the proper features to be extracted from the data, aiming to represent these complex data as closely as possible, providing adequate support for the intended use of the data. Based on this representation, access methods must be able to exploit the feature vectors, allowing efficient retrieval when answering similarity queries. It is important to notice that, depending on the intended application, different features would be computed from the same data, and different ways of comparisons would be used to improve the retrieval capability.

As the volume of information in multimedia databases is getting larger every day, it is important to look for feature extraction techniques that do not demand human interaction. There are basically two main approaches for that: extraction of features directly from the raw data, and extraction of features from transformed data, where the information usually comes from a compressed transform domain, such as wavelet coefficients [14]. For example, in image databases it is usual to extract features from raw data, that is, directly from the image pixels. The color histograms represent this kind of feature. Other feature spaces are constructed in a more elaborated way over raw data, thus being in a higher abstraction level. These include the detection of contours, edges and surfaces. Statistical techniques such as determining centroids, moment invariants, principal axis, among others, extract global information from the data or from part of the data. Such statistical results have been used to define feature spaces as well [15].

Among the more abstract feature spaces are the techniques that deal with the relationship between the information already obtained by other feature extractors. Structural features built over pattern graphs [16] or grammars composed from patterns [17] are examples of high-level feature spaces.

The extraction of features based on the transform-domain-based techniques takes advantage of the assumption that transformations are also implemented to compress the data [15]. Thus, extracting features from compressed data requires less memory space and usually less computational effort, with increasing the overall performance of the multimedia system. In the literature there are many transformations used to achieve this intent, such as Short Window Fourier Transform [18], Discrete Cosine Transform (DCT) [18] and Wavelet Transform [19] among others.

The previous discussion about feature spaces leads to an inherent hierarchy of the properties extracted from the multimedia data. Each multimedia data type has specific properties which are used to represent the data. To discuss such properties it is necessary to focus on a specific multimedia data type. In this chapter we frequently based the discussion on images as an exemplifying instrument, because the presented concepts can be applied on multimedia or complex data types with straightforward adaptations. However, video data is composed by other multimedia data, such as sound, images and temporal sequences. Therefore, a special consideration is made with regard to this data type.

Regarding images, we can create four abstraction levels to classify the feature spaces, as shown in Figure 29.1. The level 1 of this figure holds the features that are directly extracted from the raw data or from the pixels. On level 2 are the features based on contours, edges, lines, regions, surfaces and salient regions (sharp borders, corners, crossings). The features represented in level 2 are combined, producing the more structured features that pertain to level 3. Level 4 is composed by the interpretation of the relationship of the features from level 3 and corresponds to the features with the highest semantic properties embedded.   Features from levels 1 and 2 can be processed automatically; however in level 3 a semi-automatic approach would be expected.   Feature spaces handled in level 4 are highly dependent on human interaction, and they basically comprise annotations and descriptions of the images made by humans in a structured language, which can be interpreted by the system.

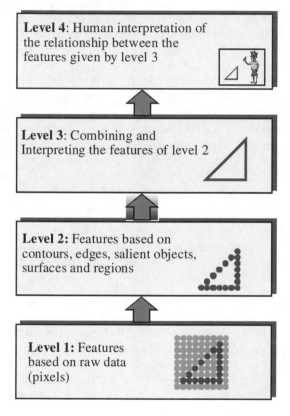

Figure 29.1 Abstraction levels describing the feature spaces for image databases.

2.1.1 Video Databases

Dealing with video databases brings more challenges than is usual with other multimedia data types. The key characteristics of video data are their intrinsic temporal nature as well as the combination of spatial and temporal information. Therefore, to answer similarity queries on video data it is necessary to work on these two aspects. A digital video is frequently represented as a sequence of scenes. Each scene consists of a collection of objects (e.g., a kid playing with a dog in a garden) with their spatial and temporal relationships (the position of the kid and the dog in the scene and the time of movements in the scene). Besides the relationships between the objects in the scene, it is necessary to manage and to process the features of these objects such as color, shape and texture. Therefore, to extract the relevant features from video data, it is necessary to address the spatial and the temporal aspects of the data.

Regarding the spatio-temporal nature of video data, the first step to index video data should be the temporal segmentation of the video elements, which are *episodes, scenes, shots* and *frames* [4]. The elementary video unit is the frame, which actually is a still image. To represent a video by its frames leads to a lot of redundant data, because sequential frames usually have very similar properties. Thus, it is more common to use other techniques based on video segments and *shots*. A shot is a set of contiguous frames representing a continuous action regarding time and space. A scene is composed by a sequence of shots focusing on the same point of location of interest. A series of related scenes forms an episode.

Nonetheless, a shot encompasses a lot of data. Thus, it leads to further summarization through its extracted properties. A condensed representation of a shot can be given by the *key frame* of the shot. The choice of the key frame can be done based on the evaluation of the object's dynamics for every frame inside the shot or even based on simple heuristics. For example, in the Informedia Project [10] the key frame is set as the one in the middle of the shot. A digital video is reduced to a sequence of its key frames.

After establishing a key frame, which is basically a still image, it will be processed by image processing techniques and will have the main features extracted. Such features can be organized hierarchically, and will be used to generate indexes over the video and to allow similarity searching. This approach tackles only the spatial aspect of the video data. Thus, continuing on this aspect, the hierarchy of levels presented in Figure 29.1 should be extended to include video data. Two more levels would appear before the level 1, that is:

> Level 0.0 : partition of each video segment into shots;

> Level 0.1 : choice of a key frame (representative frame) for each shot.

The temporal reference of the key frames should be kept and used to help in further refinements needed.

2.1.2 Searching Complex Data

The general idea behind any content-based multimedia retrieval system is this: For every multimedia data to be indexed, such as image, video or time series, a feature vector representing the data is stored in the database along with the original information. To answer a query, the feature vector of the query object is computed and the database is scanned looking for the corresponding feature vectors that match the searching criteria. When the matching is obtained, the corresponding multimedia data is returned as a query result.

The choice of the meaningful characteristics or properties of multimedia information is made by the specialist on the data domain. Based on the proposal given by such experts, the features corresponding to those characteristics are extracted from every multimedia object stored in the database. Thus, complex data will be "seen" through their features. Let us say that there are n features extracted from a given data. Usually the features are placed in a n-dimensional vector and can be manipulated as an n-dimensional object. After getting the feature vectors of the data it is necessary to compare them, which is done by the use of dissimilarity or distance functions.

Considering images, a usual descriptor, or feature, is the histogram computed from the image. Histograms are used to give a global color distribution of images, as each bin of the histogram keeps the counting of the pixels in the image presented in the corresponding color. That is, a histogram indicates the number of image pixels per each color (or grey level) used in the image quantization. The number of elements of the feature vector which holds the histogram is equal to the number of colors of the image. For example, images acquired in 8 bits or 256 grey levels will produce histograms with 256 elements or dimensions. So the corresponding feature vector is 256-dimensional. Besides given a drastic reduction of the data volume, the extraction of a feature aims to spot some intrinsic and meaningful property of the data, which is usually considered to discriminate the data under analysis. The color distribution is one of the main properties of images, and it is simple to calculate histograms, so it is natural that they are one of the first features to be extracted from images. Figure 29.2 presents three images from a video sequence and their histogram. Instead of comparing the original images, a content-based image retrieval system will deal with the image histograms.

Image and Histogram A

Image and Histogram B

Image and Histogram C

Figure 29.2 An example of feature extraction:
grey scale histograms with 256 bins.

Section 2.4 discusses the applicability of distance functions and metric spaces to speed up searching operations through the use of metric access methods.

2.2 DISTANCE FUNCTION AND METRIC DOMAINS

In this section we present the basic definitions necessary to delimit the scope of this work. The first definition aims to establish a way to compare multimedia and complex data through their features.

> **Definition 1 (metric distance function):** Given a set of objects $S = \{s_1, s_2, ..., s_n\}$ a distance function $d()$ that has the following properties:
> 1. Symmetry: $d(s_1, s_2) = d(s_2, s_1)$
> 2. Non-negativity: $0 < d(s_1, s_2) < 4$, s_1 s_2 and $d(s_1, s_1) = 0$
> 3. Triangle inequality: $d(s_1, s_3)$ # $d(s_1, s_2) + d(s_2, s_3)$
> for any s_1, s_2, s_3 0 S is called a metric distance function, or a metric for short.

These properties are fundamental to build indexing access methods that allow fast and effective searching and retrieval of multimedia data. The triangle inequality property helps diminish the searching space offering a lower bound comparison.

The objects s_i stated in Definition 1 are the feature vectors extracted from the multimedia data. This comparison intends to work on the main characteristics of the multimedia data, aiming to preserve the semantic of the data. Note that the distance function actually measures the dissimilarity between the feature vectors. However, dissimilarity and similarity are easily interchangeable, as one is the opposite of the other. Thus, from now on we will refer to a distance function as a manner to measure the similarity between multimedia data. The domain of the data is stated in the following definition.

> **Definition 2 (metric domain):** Given a set of objects S, and a metric distance function as stated in definition 1, the pair $M = \{S, d()\}$ is a metric domain.

For example, a set of two-dimensional points representing latitude and longitude of cities in a Geographical Information System (GIS) and the Euclidean distance function, which is a metric, constitutes a metric domain. In this example the data are in a spatial domain as well as in a metric domain. In fact, spatial datasets following a metric[1] distance function (such as the L_2 or the Euclidean distance) are special cases of metric domains.[2] Thus, multidimensional feature vectors and a metric measuring the dissimilarity between data are a special case of a metric domain as well.

---

[1]   A L_p-*norm* between two arrays $X(x_1, x_2, ...x_n)$ and $Y(y_1, y_2, ...y_n)$ is expressed as

$$L_p(X,Y) = \sqrt[p]{\sum_{i=1}^{n} w_i(x_i - y_i)^p}$$, where w_i is the weight of each attribute or dimension. The weight is usually equal to one. When $p = 2$, it is the Euclidean norm.

[2] Some authors call metric domains as metric spaces.

An example of a pure metric space is the set of words of a language dictionary, and the L_{Edit}[3] distance function. Clearly the words are not in a spatial domain, but as L_{Edit} gives a way to measure how similar the words are, we have a metric space.

2.3 SEARCHING SPACES AND SIMILARITY QUERIES

As already highlighted in the introductory section of this chapter, multimedia data are usually compared by similarity, because there is no sense in searching for the specific data that the user already has under analysis. What the user customarily desires is to find other data in the database which resembles the given data. It is widely accepted that there are mainly two classes of similarity queries: range queries and nearest neighbors queries. Both of them are defined as follows:

> **Definition 3 (*Range query*):** Given a query object $q \, 0 \, s$, a subset $S \, 0s$ and a maximum search distance r_q, the answer is the subset of S such that the distance between the objects to the query object is less or equal r_q, that is: $Rquery(q, r_q) = \{ s_i \mid s_i \, 0S \, \varpi \, d(s_i, q) \# r_q \}$.

An example of a range query on a word dataset with the L_{Eedit} distance function is: "*Find the words that are within distance 3 from the word 'book'.*"

> **Definition 4 (*Nearest Neighbor query*):** Given a query object $q \, 0 \, s$, and a subset $S0s$, the nearest neighbor of q is the unitary subset of S such that $Nnquery(q) = \{ s_n 0S \mid s_i 0S \, \varpi \, d(s_n, q) \# d(s_i, q) \}$.

A usual variation is the k-nearest neighbor query. It finds the $k > 0$ closest objects to q in the dataset S. An example of a k-nearest neighbor query with $k=5$ is: "*Select the 5 words nearest to 'book'.*" It is possible to have ties when asking for k-nearest neighbor queries. Thus, the system must be able to handle a list of objects that has more than k objects. The number of ties usually increases if the distance function cannot separate the objects well. For example, the L_{Edit} metric always gives integer results from a small range of values. Therefore, many words have the same distance among them, bringing the occurrence of a tie list.

2.4 METRIC ACCESS METHODS

Many approaches to accelerate the processing of similarity queries have been proposed. The majority of them led to the development of metric access methods (MAM). The intent of a MAM is to minimize the number of comparisons (distance calculations) and the number of disk accesses during the query processing.

The study of metric datasets, which are the great majority of multimedia data, including DNA strings, fingerprints, video and images, has attracted the attention of researchers in the database area. Recall that dealing with this kind of data, only the objects (the extracted features) and the distances between them

---

[3] The distance L_{Edit} (or Levenshtein) between two strings X and Y is a metric which counts the minimal number of symbols that have to be inserted, deleted or substituted, to transform X into Y. For example, L_{Edit} ("head", "hobby") = 4, i.e., three substitutions and one insertion.

are available. Usually there is no additional information, as it is given in spatial domains. Therefore, just the objects and the distances between them are all the information that metric access methods can use to organize the data.

The work of Burkhard and Keller [20] is considered the landmark of the MAM area. They presented interesting techniques for partitioning a metric dataset recursively, where the recursive process is materialized as a tree. The first technique partitions a dataset by choosing a representative from the set and grouping the elements with respect to their distance from the representative. The second technique partitions the original set into a fixed number of subsets and chooses a representative from each of the subsets. The representative and the maximum distance from the representative to a point of the corresponding subset are also maintained. The metric tree of Uhlmann [21] and the Vantage-point tree (vp-tree) of Yanilos [22] are somewhat similar to the first technique of [20], as they partition the elements into two groups according to a representative, called a "vantage point." In [22] the vp-tree has also been generalized to a multi-way tree, and in [23] it was further improved through a better algorithm to choose the vantage points and to answer nearest neighbor queries. In order to reduce the number of distance calculations, Baeza-Yates et al. [24] suggested using the same vantage point in all nodes that belong to the same level. Then, a binary tree degenerates into a simple list of vantage points. Another method of Uhlmann [21] is the generalized hyper-plane tree (gh-tree). The gh-tree partitions the dataset into two subsets, by picking two objects as representatives and assigning the remaining to the closest representative. Bozkaya and Ozsoyoglu [25] proposed an extension of the vp-tree called the "multi-vantage-point tree" (mvp-tree), which carefully chooses m vantage points for a node which has a fanout of m^2. The Geometric Near Access Tree (GNAT) of Brin [26] can be viewed as a refinement of the second technique presented in [20]. In addition to the representative and the maximum distance, it is suggested that the distances between pairs of representatives be stored. These distances can be used to prune the searching space using the triangle inequality. Also, an approach to estimate distances between objects using pre-computed distances on selected objects of the dataset is proposed by Shasha and Wang in [27].

All methods aforementioned are static in the sense that they do not support insertions and deletions after the creation of the tree. The M-tree of Ciaccia, Patella and Zezulla [28] was the first method to overcome this deficiency, allowing further insertions. The M-tree is a height-balanced tree where the data elements are stored in the leaves. Note that, the M-tree supports insertions similar to R-trees [29].

For all the metric access methods, the searching space is divided in "regions" or nodes, each region having one or more representatives or centers, which are strategically chosen by the MAM. At the same time, the nodes are hierarchically organized in a tree way. So, each node of a tree has basically the representative, the covering radii and the objects of the dataset which are in the node region. Figure 29.3(a) shows the organization of a node which has maximum capacity equal to 8. The representative (or center) object and the covering radius are depicted along with the remaining objects. For the sake of the clearness, we are drawing the node in a two-dimensional space using the Euclidean distance (L_2) to depict the shape of the node, which results in a circle. Each distance defines a specific shape for the node. Thus, for the Manhattan (L_1) distance the shape of the node is a diamond and for the L_0 or *L-infinity* distance the shape becomes a square. As the shape of the nodes changes depending on the distance used, so does the object composition of the nodes.

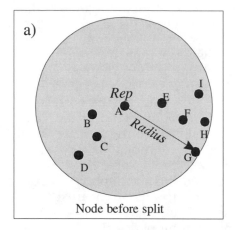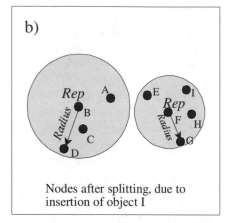

Figure 29.3 Nodes in a two-dimensional space defined by the Euclidean distance. The nodes have the maximum capacity of 8 objects. (a) The representative of the node is the object **A**. At inserting the object **I**, the node surpasses its capacity and needs to be split. (b) After the node splits, there are two nodes, the first one with four objects and its representative is **B**, and the second node with five objects having as representative the object **C**.

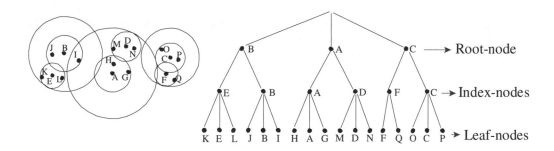

Figure 29.4 A metric tree indexing 17 objects. The objects are organized in three levels. The node capacity is 3.

As the objects are being inserted in the tree, and the nodes achieve their capacity of holding objects, new nodes must be created. This is done by the splitting algorithms presented in the MAM. Figure 29.3(b) illustrates this procedure. At the end of the insertion of the objects the tree is built. Figure 29.4 presents a metric tree built over 17 objects. The capacity, or maximum number of objects in the nodes, of this tree is three. This tree also has three levels, where the root node is on level 1, and it is the only node without a representative. On the second level are the index-nodes, which keep the routing information of the tree. On the third level are the leaf-nodes, which store the indexed objects. The leaf-nodes always will be only on the highest level of the tree. Observe that there is overlap between the regions stated by the nodes of the tree.

As there are many interesting MAMs in the literature, it is important to highlight their particularities. We can say that the difference among the metric access methods presented before is either on the way that the representatives of the nodes are chosen, or in the strategy for splitting the nodes and building the tree, or in the techniques for insertion and data organization. In [30] is presented a survey on metric access methods, comparing these aspects between the MAMs.

A common problem with all metric access methods presented in the literature is the overlap between the nodes imposed by their structures. As the overlap increases, the efficiency of the structure decreases, because all the nodes covering a query region need to be processed during the search operation. The Slim-tree proposed in [31] [32] was developed aiming to minimize the overlap between nodes. This metric access method allows to decrease the amount of overlap between nodes of a tree using the Slim-down algorithm, which achieved a gain in performance for answering similarity queries of up to 50%.

Two aspects involved in measuring the efficiency of a MAM for answering similarity queries are the number of disk access and the number of distance calculations. It was proposed in [33] the Omni-family, which is a set of metric access methods that, taking advantage of global elements called *foci,* can extraordinarily decrease the number of distance calculations when processing similarity queries. Another advantage of such methods is that they can be built on commercial data managers, and improve the performance of these managers without disturbing the execution of previously developed applications.

The improved response in query time already achieved in MAMs had led the performance of similarity queries asking for objects in metric spaces to acceptable levels, that is, compared to the times obtained for their counterparts in spatial domains. Therefore, albeit more research is yet required in techniques to support queries of objects in metric spaces, other demands need to be fulfilled now. Among them, there is the need to represent the structure of the metric objects through its feature vectors, the distance functions and the similarity queries. The next section addresses a generic architecture to store multimedia data following the basic concepts of the relational data model, as well as an extension of the standard relational query language, the SQL.

3. A FLEXIBLE ARCHITECTURE TO STORE MULTIMEDIA DATA

When applications deal with sets of video and images, the usual approach is to store all of them in a single repository. Common examples, both on commercial and research-based systems, are the QBIC (http://wwwqbic. almaden.ibm.com/) [34], VisualSEEk (http://www.ctr.columbia.edu/ VisualSEEk/) [35], Virage (http://www.virage.com/) [36], CueVideo [37] (http://www.almaden.ibm.com /cs/cuevideo/), Chabot [38], etc. Another class of image repositories are the Picture Archiving and Communications Systems (PACS) based on the DICOM image format, widely used in hospital equipments and medical applications [39]. Images in PACS are tagged in the DICOM format [40], and stored in a centralized repository. Images obtained from different kinds of exams and/or equipments are stored together, and classified by tags set individually at each image.

The aforementioned systems store all images in a single dataset. In other words, the logical separations recognized over the full set of images, such as classes or groupings, are recorded through tags set individually at each image. This approach is not suitable to process image retrieval operations in large datasets, because each request for a subset of images tagged in a given way requires processing the full image set, filtering the required images looking for the tag values at each image. A more effective way to store images considering their natural, semantic partitioning is adopting the architecture of relational databases.

Relational Database Management Systems (RDBMS) store large amount of information in tables (also called relations), which are composed by attributes of various types. The use of RDBMS or database management systems (DBMS) for short leads to flexible, expansible and highly maintainable information systems. Each line of a table is called a *tuple* in the relational jargon, and corresponds to a real world entity. Each column of a table is called an attribute, which can have at most one value in each tuple. Each column has its values chosen from a specific data domain, although many attributes can have their values picked from the same domain. For example, the set of grad students in a school could be stored in a `students` table, having as attributes their names, ages, the course they are enrolled in and their advisor. The `name` attribute gets its value from the domain of people names, its `age` from the set of integer numbers that are valid ages for humans, the `course` from the courses offered by the school and the `advisor` attribute from the domain of people names. The attributes `name` and `advisor` are taken from the same domain, because both correspond to people names. They can be compared for identification of the same person, and they can indistinctly be used to find personal data in other tables.

Using these relational concepts, one could want to add two attributes in the `Students` table to store their mugshots, called `FrontView` and `Profile`. As frontal images are completely different from side images of a human face, there is no meaning in comparing a pair of them, and if they are to be used to find data in other tables, the proper image should be selected. Therefore, it is worth to consider `FrontView` and `Profile` as two non-intersecting sets of images, and consequently as elements from distinct domains. Consequently, both sets can be stored as separated sets of images, and whenever an operation is executed to select some images from one set, it would not retrieve images from the other set. This approach is much more effective than considering the whole set of images, as accounted by the traditional image repositories.

Unfortunately, the common RDBMS do not support images adequately. They only enable to store images as *Binary Large OBjects - BLOB* data. Images are retrieved by identifying the tuple where they are stored using other textual or numeric attributes, which are stored together in the same tuple.

In the next section we show how an RDBMS can be extended to support content-based image retrieval in a relational database. Despite enabling the construction of more efficient image storage systems, the relational approach makes it easy to extend existing applications, which already use relational databases, to support images. The usual image repositories are monolithic systems, whose integration with other systems requires cumbersome, and typically inefficient, software interfaces. On the other hand, the relational approach, as described next, affords a modular system that is simple, efficient, expansive and easy to be maintained.

3.1 EXTENDING RDBMS TABLE INFRASTRUCTURE TO SUPPORT IMAGES

The standard relational database access language is the *Structured Query Language -SQL*. It is a declarative retrieval language; that is, it allows the user to declare what he/she wants, and the system decides how to retrieve the required data. This property allows the system to optimize the query execution process, taking advantage of any resource the database manager has, in order to accelerate the retrieval process. To be able to automatically design the query retrieval plan, the RDBMS must have a description of the application data structure, which is called the application schema. The schema includes the structure of the tables and the attribute domains. The SQL commands can be divided in at least two subsets:

- the Data Definition sub-Language (DDL),
- the Data Manipulation sub-Language (DML).

The DML commands allow to query and to update the application data. They are the most used commands during the normal operation of the information system. The DDL commands allow the description of the schema, including the description of the table's structures, attribute's domains, user's access authorizations, etc. Typical RDBMS store the application schema in system-controlled tables, usually called the database dictionary. These tables are defined following the modeling of the relational model itself, so the relational model is called a meta-model. An important consequence is that the modeling of the relational model is independent of the modeling of the application data, and therefore both can be maintained in the same database, without interfering with each other. Since both modelings are maintained in the same database, the DML commands can be seen as those that query and update the data stored in the application tables, and the DDL commands can be seen as those that query and update the data stored in the meta tables.

Considering that the same relational database stores tables for more than one purpose, Figure 29.5 shows the operations executed over each set of tables, to execute DDL and DML commands. In this figure, the schema tables are identified as RDB (the *meta tables*), following the common practice of using the prefix RDB$ preceding the names of these tables. The application tables are identified as the ADB (*application tables*) in Figure 29.5. The different sets of tables are accessed and maintained independently from each other, so they can be seen as conceptually separated databases inside the same physical database. Hence, in the rest of this chapter we call them *logical databases*.

Traditional RDBMS deal with simple data types, letting the attributes be defined only as numbers, small character strings, dates and BLOBs. Therefore, those RDBMS do not support the retrieval of images based on their content. However, a software layer can be developed atop the database system, which monitors the communication stream between the applications and the database server. We describe here the main ideas that enable us to build a generic *Content-based Image Retrieval Core Engine* - CIRCE. This is a software layer that recognizes an extended version of SQL, which supports not only the storage of images as a new data type, but also the indexing and retrieval of images based on their content. Such operations are the basis to support similarity searching. The language extensions were designed aiming to minimize the modifications in the language and at the same time maximizing its power to express image retrieval conditions.

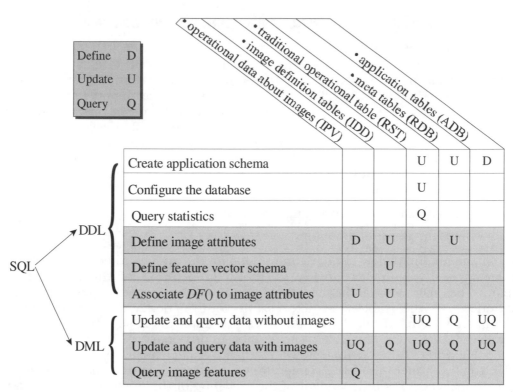

		operational data about images (IPV)	image definition tables (IDD)	traditional operational table (RST)	meta tables (RDB)	application tables (ADB)
DDL	Create application schema			U	U	D
	Configure the database			U		
	Query statistics			Q		
	Define image attributes	D	U		U	
	Define feature vector schema		U			
	Associate $DF()$ to image attributes	U	U			
DML	Update and query data without images			UQ	Q	UQ
	Update and query data with images	UQ	Q	UQ	Q	UQ
	Query image features	Q				

Legend: Define D, Update U, Query Q. Left grouping: SQL → DDL, DML.

Figure 29.5 Operations executed over each logical database by each command in SQL. The definition (D), update (U) and query (Q) operations are performed to process the issued commands. The grey rows of the table refer to image related operations.

As a software layer that intercepts every command between the applications and the database server, CIRCE is able to extend any RDBMS that can store BLOB data types to support images and, with few modifications, also to support the storage and content-based retrieval over any kind of complex data, including long texts, audio and video. Therefore, although in this section we are describing a generic module to extend SQL to support image storage and content-based retrieval, the same concepts can be applied to support the storage and retrieval by content on any kind of multimedia data.

The extensions in SQL to support images are made both in DML and in DDL, which is presented in the grey rows of Figure 29.5. The extensions for update commands in DML correspond to new syntaxes to allow insertion, substitution and deletion of images in attributes of the type image in the application tables. Also, the extension in the query command in DML corresponds to the creation of new search conditions, to allow expressing range and nearest neighbors conditions in the query command. The extensions in the DDL commands require the modeling of the image support, the creation of a new schema to support the description of image definitions, as well as the definition of how the attributes should compose the application tables. The modeling of images is independent both from the relational meta-modeling and from the application modeling. Thus, the image modeling does not interfere with the existing ones. Moreover, as any other modeling, the image modeling also requires specific tables to be created in the database, to store the image definitions and parameters issued by the user through the new image-related DDL commands.

Therefore, two new logical databases are included in a RDBMS, identified as IPV and IDD in Figure 29.5. They will be explained in the next sub-section.

Besides the meta and the application logical databases, a traditional RDBMS stores operational data that helps to improve query processing. Although usually considered as part of the dictionary (and in some cases sharing the same physical tables), operational data in fact correspond to different information from that in the meta schema. The meta schema describes the application's data, like the tables and the attributes that constitutes each table, the domain of each attribute, etc. The operational data usually consist of statistics about the size of the tables, frequency of access, and so on. Therefore, they are stored in another logical database, identified in Figure 29.5 as the traditional operational tables (RST - relational statistics). The main purpose of this logical database is to provide data to create good query plans to execute the DML commands. As the operational data are gathered internally by the system and are used to optimize the query plans, there is no need of commands in SQL to update or retrieve the operational data (although high-end RDBMS usually provide commands for the database administrator to monitor and to fine-tune the database behavior through the operational data).

When dealing with data much more complex than numbers and small character strings, such as images, a more extensive set of operational data is required. Therefore, when an RDBMS is extended to support images, a new logical database is required to store operational data about the images. Differently from the traditional operational data, the end user may be interested in some operational data about the images. Consequently, it is interesting to provide constructs in the DML query command that allow users to query the operational data about images.

Therefore, tables managed by an RDBMS extended to support content-based image retrieval must comprise five logical datasets, as shown in Figure 29.6. The meta tables and the traditional operational data constitute the data dictionary. These five sets of tables are accessed and maintained independently from each other; consequently they are five conceptually separated logical databases inside the same physical database. The next section explains in more detail the two new logical databases IPV and IDD intended to support images, and how they are used in CIRCE to implement the "image data type" in a RDBMS.

3.2 A RELATIONAL DATABASE ARCHITECTURE TO SUPPORT IMAGES

A content-based image retrieval (CBIR) system is built taking advantage of image processing algorithms that we call *extractors*, each one getting values from a set of one or more *features* from each image. The extracted features, also called "ground features" [12], can be applied not only to compose the *feature vectors* used in comparison operations, but also as parameters to control the execution of other extractors. The extracted features can be arranged to create one or more feature vectors for each image, each one following a predefined *schema*, that is, the structure of the feature vector. Different feature vector schemas aim to explore different ways to compare images. For example, images can be compared by their histograms, or by their texture, or by the combination of both of them. For each kind of comparison a different feature vector schema must be used, although some schemas can share the same features already extracted.

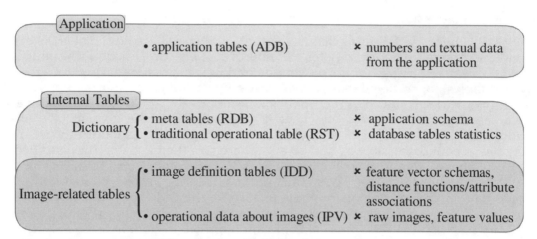

Figure 29.6 The five logical databases required to support the application data and content-based image retrieval in an RDBMS.

Each feature vector schema must be associated to one distance function $DF()$, in order to allow image comparison operations. The main concept in the feature extraction technique is that comparisons between pairs of real images are approximated by comparisons between their feature vectors. Feature vectors create a base to define distance functions that assign a dissimilarity value to each pair of images in a much more efficient way than processing the real images. A metric $DF()$ must obey the non-negativity, symmetry and the triangular inequality properties, as it was stated in section 2.2. Maintaining the images in a metric domain authorizes the use of metric access methods accelerating greatly the image retrieval operations.

Figure 29.7 depicts the steps involved in the feature extraction technique in a data flow diagram. The whole process can be described as follows. Given a set of images from the same domain, they will have their features obtained by extractors. As the extractors are general-purpose algorithms, a given $DF()$ may only use some of the features extracted by each extractor, discarding those not necessary in a particular situation. Moreover, some of the features extracted can be used as parameters to control other extractors executed in sequence. The most interesting features from a set of extractors are merged to create a feature vector. As the same sequence of extractors is used to process every image, we call this sequence and the resulting set of features is a *feature vector schema*. Two feature vectors can be compared only if they have the same feature vector schema. The feature arrays conform to a feature vector schema and from the same image domain can be indexed by a MAM, e.g., the Slim-tree, creating an index structure. At the querying time, the query image given by the user will have its features extracted and the search will be done in the corresponding index structure. The results will return the indication of the images that match the searching criteria specified by the user.

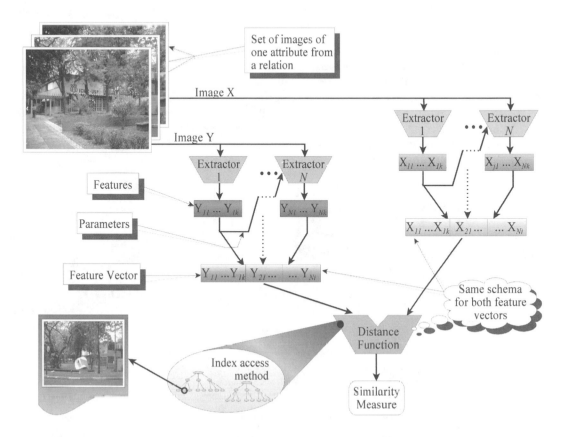

Figure 29.7 Structure of the operations needed to support content-based retrieval operations in a set of images.

To support similarity queries in CBIR, the extensions of SQL must enable the definition of the following concepts:

- the definition of images as a native data type;

- the definition of the feature vectors and the corresponding $DF()$;

- the similarity conditions based on the k-nearest neighbor and range operators.

The main idea is to consider images as another data type supported natively by the DBMS. Therefore, images are stored as attributes in any table that requires them, and each table can have any number of image attributes.

The general architecture of CIRCE, a generic module to enable RDBMS to support CBIR, is shown in Figure 29.8. Three logical databases are represented in this figure: the ADB, IPV and IDD. The ADB corresponds to the existing application database, the IPV corresponds to the operational data about images and IDD corresponds to the image definition database. Applications that do not support images keep the attributes of the tables stored in the ADB as numbers and/or texts. The ADB can be queried using either the standard SQL or the extended SQL. However, when an application is developed using image support, or when an existing one is expanded to support images, the application must use the extended SQL through CIRCE.

Each image attribute defined in a table establishes a set of images disjointed from the set established by the other attributes. That is, each image attribute in each tuple stores only one image. If a query command uses only non-image attributes, the command is sent untouched to the ADB. Whenever an image attribute is mentioned in a query command, CIRCE uses the IPV and IDD databases, together with the metric access methods and the parameter extractors (XP), to modify the query command attending images. This approach effectively implements image retrieval by content.

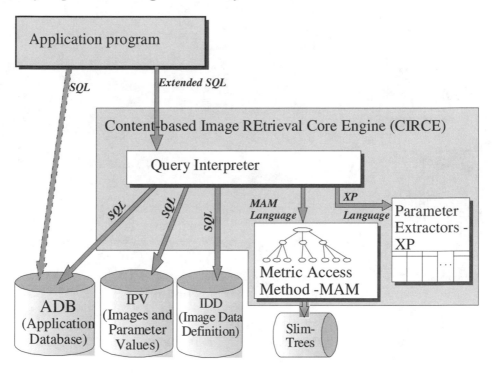

Figure 29.8 Architecture of the image support core engine.

The IPV database (that stores operational data about images) has one table for each attribute of type image defined in the application database. Each table holds at least two attributes: a blob attribute storing the actual image, and an Image Identifier (*ImId*) created by the system. The identifier is a code number, unique for every image in the database, regardless of the table and particular attribute where it is stored. Each "`create table`" command referencing image attributes is modified, so a numeric data type replaces the image data type in the modified command sent to the ADB. The corresponding table in the IPV database is named after the concatenation of the table and attribute names of the original image attribute. Consequently, occurrences of this attribute in further query commands are intercepted by CIRCE and translated accordingly. For example, the `students` table with two image attributes generates two tables in the IPV database named `Students_FrontView` and `Students_Profile`, each one with two attributes: the `image` and the *ImId*.

The comparison of two images requires a metric distance function definition. This is done through the new command in the extended SQL: the "create metric" command. It is part of the Data Definition Language (DDL) of SQL, and enables the specification of the *DF()* by the domain specialist. Each metric *DF()* is associated with at least one image attribute, and an image attribute can have any number of metrics *DF()*. Image attributes that do not have a *DF()* cannot be used in searching conditions; so these attributes are just stored and retrieved in the traditional way. However, if one or more metric *DF()* is defined, it can be used both in content-based search conditions and in metric access methods. Image attributes associated with more than one *DF()* can be used both in searching conditions and in indexing structures. If more than one *DF()* is associated to an indexed image attribute, then each *DF()* will create an independent indexing structure. Comparison operations always use just one *DF()*. Therefore, comparisons involving attributes associated with more than one *DF()* must state which one is intended to be used (although a default one can be defined for each attribute).

A *DF()* definition enrolls one or more extractors and a subset of the features retrieved by each extractor. Here we consider as features only numbers or vectors of numbers. If the extractors used in a *DF()* return only numbers, the feature vectors of every image stored in the associated attribute have the same quantity n of elements. Thus, each feature vector is a point in a n-dimensional space. However, extractors returning vectors may generate feature vectors with different quantity of elements, resulting in a non-spatial domain. This domain can be metric if the *DF()* is metric. To assure this property, we consider the definition of L_p-norm *DF()* over the elements of feature vectors, as defined in section 2.2. The power p of the norm can be specified as zero (Chebychev), one (Manhattan) or two (Euclidean). Each element i in the feature vector has its individual weight w_i. Those elements corresponding to vectors have the difference calculated as the double integral of the curve defined by the vectors [3].

Whenever a *DF()* is defined for an image attribute, the corresponding table in the IPV database is modified to include the elements of the feature vector as numeric attributes. Therefore, whenever an image is stored in the database as the value for an image attribute, every extractor enrolled by each *DF()* associated to the corresponding image attribute is executed, and the extracted features are stored in the corresponding table in the IPV database. The resulting feature vectors used by each *DF()* are inserted in each corresponding indexing structure. This enables the retrieval of images based on more than one similarity criterion, allowing the user to choose which criterion is required to answer each query. Each index structure (built upon a metric access method) stores the *ImId* and the corresponding feature vector extracted from each image. The set of index structures allows the execution of the k-nearest neighbors and range search procedures using the feature vectors as input, and returning the set of *ImId* that answer the query.

The IDD database is the image definition database, that is, the schema for the image attributes. It stores information about each extractor and its parameters. Therefore, whereas the IPV database store the actual image and the values of its features, it is the IDD database that guides the system to store and retrieve the attributes in the IPV database. Whenever a tuple containing image attributes is stored, each image is processed by the set of extractors in the *Parameter Extractor module* (XP), following the definitions retrieved from the IDD database. Subsequently, its *ImId* is created and stored, together with the image and the feature vector, in the IPV database. The original tuple from the application is stored in the ADB, exchanging the images with the corresponding *ImId* identifiers.

Whenever a query has to be answered, the feature vectors of the involved images are used in place of the real image. If the query refers to images not yet stored in the database (images given directly inside the query), the same feature vectors associated to the image attribute being compared are extracted from this image and used to search the database. This feature vector is sent to the corresponding index structure, which retrieves the *ImId* of the images that answer the query. Using these *ImId*, the IPV database is sought to retrieve the actual images that, in turn, are passed as the answer to the requester process.

4. A CASE STUDY: INDEXING AND RETRIEVAL OF IMAGES BY SIMILARITY

When a relational table is composed only by numbers or small character strings, the comparison operations between elements of these domains are always clearly defined: the meaning of `=`, `<`, `like()`, etc. is well known and is executed in the same manner in every DBMS. However, when non-conventional data types are included, the comparison operations between them are not so completely understood, and needs to be declared as part of the description of the data itself. In this section we present an example of how a similarity query can be posed on a database stored in a relational database extended by the CIRCE architecture.

4.1 A CONTENT-BASED IMAGE RETRIEVAL EXAMPLE

To store data in an RDBMS, first of all it is needed to create the tables, using the DDL command "`create table`." Let us resume the example of the previous section. To store the `students` table example, having as attributes their names, ages, the course where they are enrolled, their advisor, their `FrontView` and `Profile` mugshot, the following command should be issued:

```
CREATE TABLE Students      ( Name CHAR(50),
                             Age INTEGER,
                             Course CHAR(15),
                             Advisor CHAR(50),
                             FrontView STILLIMAGE,
                             Profile STILLIMAGE );
```

Notice that the last two lines of this command use the CBIR-extended SQL, as the `STILLIMAGE` keyword is part of that extension. Therefore, CIRCE replaces these two lines, changing the `STILLIMAGE` keyword to `Numeric(10)`, where the image identifier will be stored in place of the actual image. We are using the `STILLIMAGE` keyword to declare image attributes, following the ISO/IEC standard proposal aiming to specify packages of abstract data types in SQL. This proposal, known as the SQL/MM (SQL for MultiMedia applications), includes packages for Still Image, Still Graphic, Animation, Full Motion Video, Audio, Full-Text, Spatial, Seismic, and Music [41, 42]. Although this proposal aims to support images in relational databases, it is intended to be used through the resources for user-defined type provided in SQL-99. Therefore, the image storage, manipulation and retrieval operations addressed by the proposal must be supplied by application-specific procedures. As there is no provision for CBIR in that proposal, the CIRCE approach presented here supplements SQL/MM to allow similarity queries in content-based image retrieval operations.

Attributes other than images can be compared in subsequent DML commands, but the image attributes must have their comparison operations defined in the beginning of the process, because there is no standard way to compare images. In the previous example, this is done through the definition of one or more *DF()*, using the "`create metric`" CIRCE-extended DDL command. This command specifies the schema of a feature vector, selecting the extractors and the features used in each metric. Considering, for example, that `FrontView` images should be compared by their color histogram and by the total area occupied on the photo, the Histogram and TraceObjects extractors can be used, as is presented in the following command:

```
CREATE METRIC FrontViewHistoArea on Students (FrontView) AS (
    HISTOGRAM (HISTOGRAM FrontHistogram INTEGER[256],
               MIN FrontHistoMin INTEGER,
               MAX FrontHistoMax INTEGER)
    TRACEOBJECTS (TOTALAREA FrontViewArea FLOAT)  );
```

Each extractor is an image processing algorithm defined using the resources for the user defined type provided in SQL-99. The extractors are designed to obtain the general and the meaningful properties from images. For example, the `TraceObjects` is a closed border detection algorithm that returns the number of objects found in an image, their total area (in pixels), the minimum bounding rectangles coordinates and area of the five larger objects found in the image. From all these features extracted by the `TraceObjects` extractor, the `FrontViewHistoArea` *DF()* in the example uses only the total areal, referred to in the *DF()* schema as `FrontViewArea`.

After receiving those two commands, CIRCE generates a table in the IPV database named `Students_FrontView`, with the following structure corresponding to this *DF()* schema:

```
CREATE TABLE Students_FrontView ( ImID NUMERIC(10),
                                  FrontView BLOB,
                                  FrontHistogram
                                  INTEGER[256],
                                  FrontHistoMin INTEGER,
                                  FrontHistoMax INTEGER,
                                  FrontViewArea FLOAT  );
```

Thereafter, every tuple inserted in the `Students` table with a non-null value in the `FrontView` attribute will have a corresponding tuple inserted in the `Students_FrontView` table, where the image is stored together with the values of the extracted features and its corresponding image identifier. The identifier is the one stored in the `FrontView` attribute of the `Students` table. The association of each feature attribute in the *DF()* schema with the corresponding feature retrieved by each extractor, as is declared in the "`Create Metric`" command, is stored in tables in the IDD database.

As each *DF*() is uniquely identified (like Students_FrontView in this example), more than a *DF*() can be associated to the same image attribute. This allows that more than one kind of comparison can be made with each image attribute.

In this example, at least another "Create Metric" command must be issued for the Profile attribute, in order to create a *DF*() for it. Thus, the Students_Profile table will also be enlarged with the features used in its corresponding distance functions.

The extracted features are stored in the IPV database tables; therefore its data types must be the traditional ones, like numbers or small character strings. Extractors that return string values are adequate to support some kind of semantic interpretation of the images.

The "Create Metric" command syntax also permits to supply parameters for each extractor used in a *DF*(), through its "PARAMETERS()" clause. Parameters can be either a constant value or the result of extractors declared beforehand in the "Create Metric" command. For example, the Students_FrontView *DF*() could be declared to use histograms with a smaller number of bins, as is shown following:

```
CREATE METRIC FrontViewHistoArea on Students (FrontView) AS
        (      HISTOGRAM PARAMETERS (NBINS=64)
               (HISTOGRAM FrontHistogram INTEGER[64], ...
        );
```

With the Students table populated, it is possible to ask similarity queries over its image attributes. To support this kind of queries, a specific search condition is added to SQL, to be used in the "where" clause of the DML commands. The general syntax of a similarity condition enables both *k*-nearest neighbors and range queries, as shown as follows:

```
SELECT <attributes>
FROM <tables >
WHERE <old_conditions> | <similarity_conditions>
<similarity_conditions> ::= <attribute> NEAR <val>
                           [BY <MetricName>]
                           [RANGE <val_range>]
                           [STOP_AFTER <k_nearest>]
```

where: <attribute> is the name of the image attribute in the table, where the search will be performed;

<val> is the reference image, used as the query center;

<MetricName> is the name of the *DF*() to be used in the comparison operations, required if more than one *DF*() was defined for the involved <attribute>;

<val_range> is the range value for range queries; and

<k_nearest> is the *k* value for *k*-nearest neighbors queries.

For example, the following command retrieves the ten images stored in the FrontView attribute of the Students table that are the most similar to the photo stored in the file PhotoFile.jpg, with the corresponding student's name. As the *DF*() is not specified, it is assumed as the one created by the first "Create Metric" command issued over the FrontView attribute.

```
SELECT Name, FrontView
FROM Students
WHERE FrontView NEAR PhotoFile.jpg STOP AFTER 10;
```

The full power representation of SQL can be exploited. For example, the following command returns both the front view, profile, and the name of the students whose front view image is dissimilar from the front view image of the student called "John Doe" by at most 50 units.

```
SELECT Name, FrontView, Profile
FROM Students
WHERE FrontView NEAR (
            SELECT FrontView
            FROM Students
            WHERE Name='John Doe'
    ) RANGE 50;
```

4.2 EXTENDING CIRCE TO SUPPORT CONTENT-BASED VIDEO RETRIEVAL

The BLOB data type supported by traditional RDBMS is used to store large amounts of data. BLOBs are not analyzed by the system, so they can be used to store video data as well as images. The CIRCE architecture can be used to enable content-based retrieval using similarity queries regarding not only images, but also audio, video and other types of complex data, provided the proper set of feature extractors are defined. Therefore, each complex data type should be supported by a set of specific feature extractors. Moreover, with the video data type, the "PARAMETERS()" clause of the "Create Metric" command must be improved.

The improvement required to support video is due to the fact that image feature vectors are composed only by numeric or short character string values. Regarding audio and video data, the extractors should work on specific key-frames or audio parts that are worth to identify in the original data. The standard SQL is not an adequate platform to support an image or audio processing language; therefore the extractors that process these data types are elementary building blocks, aiming only the declaration of queries over corresponding datasets. Therefore, the ability to support feature vectors that return complex data itself is essential not only for images but also for all the multimedia data. However, when video data are considered, this ability turns out to be quite useful. For example, consider that the similarity of two video sequences should be measured using some image $DF()$ applied over the key-frames of each sequence. The video $DF()$ could be defined using a video processing algorithm that extracts the key-frame, followed by the image feature extractors that receives that frame as an image to have its parameters extracted. Although a two-level architecture (video $\propto$ image) could be built, a more general architecture dealing with other aspects of the video data would be necessary.

5. CONCLUSIONS AND AFTERWORDS

With the increasing use of multimedia data in the information systems nowadays, it is mandatory the DBMS can handle these data properly. Besides to store multimedia data, the database manager has to allow fast and effective data retrieval. As multimedia data usually holds complex information, it is necessary to extract the essence of the data, which is given by the features obtained from the data. Afterwards the multimedia data can be dealt through their features instead of the data themselves. From this time on, the DBMS has to organize and manage the multimedia data and the features extracted in a transparent way. That is, the DBMS must know when to use one or the other, because the operations are done over the features but the results presented to the user are over the original multimedia information.

The relational technology has been the primary resource to store and retrieve large amounts of data represented in numerical or categorical format. Two key foundations in this technology are the ability to develop applications that are independent from the low level considerations of how the data are stored and retrieved, and the ability to perform these operations using algorithms that scale well with the increase of the data volume. The first foundation is achieved through the relational algebra and the SQL language. SQL is a declarative language, where the query commands are written specifying **what** data is required. The system uses the power of the relational algebra to define and optimize **how** to execute retrieval operations. The second ability has been achieved by pre-processing the data as it is being stored, anticipating the operations that would be necessary to answer the query commands. Effective technologies empowering this ability are the use of access methods (which are built based on the comparison properties of the data domains), and the gathering of statistical data about the stored data to drive the execution of the retrieval operations.

This technology depends on precise definition of rules to compare pairs of objects, so the retrieval operations can be executed in a timely manner. Regarding multimedia data, which include image and video data, this does not occur. Multimedia data cannot be ordered, so the relational operators <, #, etc. cannot be used. Even the identity operator is not useful, because two images or videos of the same subject hardly are "equal." Therefore, the first step to entitle supporting content-based multimedia retrieval inside relational database management systems is the definition of rules that enable the comparisons between pairs of multimedia objects. Regarding video and image objects, this has been pursued through the definition of video and image processing algorithms that extracts features from the data, which are used in place of the real objects when executing the comparisons. However, in spite of the advances already achieved, this is an open research area, mainly due to two factors. The first one is that most of the video and image processing algorithms already developed extract relatively low level features that are hard to combine in queries that demand a semantic appreciation of the data. The second factor is that the comparison algorithms are often dependent on the application domain, or even on the particular objective of a query. Both factors indicate that the development of content-based video and image retrieval systems should allow to add new comparison algorithms as they are being built up. Also, it is important to allow the user to choose the algorithm used to answer each query command. These considerations are valid for complex and multimedia data as well.

This chapter also described CIRCE, a generic approach to extend relational database management systems to support content-based video and image retrieval based on similarity queries. It showed how SQL can be extended with a few but powerful new syntax, which enabled to support images as a new data type seamless integrated to the traditional data. The new SQL syntax allowed the definition of image attributes, the definition of the feature vectors and the corresponding *DF*(), as well as the similarity conditions based on the *k*-nearest neighbor and range operators. Furthermore, they enable the creation of access methods based on the feature vectors and distance functions.

Considering the multimedia data as belonging to metric domains, SQL can use the strong formalism of the theory of the metric domains and group theory to maintain strict adherence to the relational algebra, so the powerful query optimization techniques of SQL can be extended to the image-extended commands. Moreover, the client-server architecture of relational system enables to distribute the weight of processing large volumes of data into specific, well-tailored systems to perform the video/image processing and feature extraction, performing indexing operations, and physical storing of the multimedia data. The relational approach also enables the partition of large sets of multimedia objects into disjoint sets, further improving the execution of query operations.

A prototype implementation of CIRCE is under way in the Computer Science Department of the University of Sao Paulo at Sao Carlos. It is being developed aiming to provide content-based image retrieval of medical exams, like tomographies, X-ray, ultrasound, etc. This application explores the full potential of the presented architecture, as it provides an effective way to separate images from different body parts, yet it must deal with different representations (different kind of exams) of them. The prototype is being written in C++ using Oracle 8i as the database server running on a Solaris platform. The client workstations run Windows 2000 on the Radiology Center of the Clinics Hospital of Ribeirvo Preto.

Although the use of CBIR techniques in medical systems is yet exploratory from the medical staff point of view, from the computational view the results already obtained are encouraging. For example, to demonstrate the performance gain of the CIRCE architecture over the traditional one, we modified an existing application that searches a set of X-ray head tomographies looking for a specific disease. In the original version, this application, running at the client station, retrieved the full set of tomographies in the Oracle server using standard SQL. The application pre-processed every image using a histogram-based filter. Using a set of 5049 X-ray head tomographies (a total of approximately 2GBytes of data), the total time to obtain the answer was approximately four hours. In the new version, a distance function was created in CIRCE, using a metric histogram extractor [3] to characterize the images. The pre-processing phase of the application was removed, and the retrieval of the image was performed through a modified SQL command using a range query to retrieve only highly probable images, using the metric histogram as the initial filter. Using the same set of tomographies, the total time to obtain the answer dropped to ten minutes, yet selecting the same images as before. The time to extract the metric histograms and to create the index structure (in a Slim-tree), which must be performed just once, was 25 minutes. Moreover, the resulting architecture of the software improved, as it is more modular and maintainable, and can now be easily configured to embody new metrics and features as they will become available, without even recompiling its source code.

ACKNOWLEDGMENTS

This research has been partially supported by CNPq (Conselho Nacional de Desenvolvimento Científico e Tecnológico, Brazil) under grants 35.0852/94-4, 52.1685/98-6 and 860.068/00-7.

REFERENCES

[1]   A. W. M. Smeulders, et al., Content-based image retrieval at the end of the early years, *IEEE Trans. on PAMI*, Vol. 22, 2000.

[2]   A. Yoshitaka and T. Ichikawa, A survey on content-based retrieval for multimedia databases, *IEEE TKDE*, Vol. 11, 81-93, 1999.

[3]   A. J. M. Traina, et al., The Metric Histogram: A New and Efficient Approach for Content-based Image Retrieval, in Sixth IFIP Working Conference on Visual Database Systems, Brisbane, Australia, 2002

[4]   Y. A. Aslandogan and C. T. Yu, Techniques and systems for image and video retrieval, *IEEE TKDE*, Vol. 11, 56-63, 1999.

[5]   H. Yamamoto, et al., Content-based Similarity Retrieval of Images Based on Spatial Color Distributions, in 10th Intl. Conference on Image Analysis and Processing, 1999, 951-956.

[6]   E. G. M. Petrakis and E. Milios, Efficient Retrieval by Shape Content, in IEEE Intl. Conference on Multimedia Computing and Systems, 1999, 616-621.

[7]   M. Adoram and M. Lew, Irus: Image Retrieval Using Shape, in IEEE Intl. Conference on Multimedia Computing and Systems, IEEE Computer Society, 1999, 597-602.

[8]   G. L. Gimel'farb and A. K. Jain, On retrieving textured images from an image database, *Pattern Recognition*, Vol. 29, 1,461-1,483, 1996.

[9]   W. Ma and B. Manjunath, Image Indexing Using a Texture Dictionary, in SPIE, Conference on Image Storage and Archiving System, 1995, 2606, 288-298.

[10]   H. D. Wactlar et al., Lessons learned from building a terabyte digital video library, *IEEE Computer*, Vol. 32, 66-73, 1999.

[11]   S. Tsekeridou and I. Pitas, Content-based video parsing and indexing based on audio-visual interaction, *IEEE Transactions on Circuits and Systems for Video Technology*, Vol. 11, 522-535, 2001.

[12]   Y. Rubner and C. Tomasi, *Perceptual Metrics for Image Database Navigation*, Kluwer Academic Publishers, Dordrect, 2001.

[13]   E. G. M. Petrakis and C. Faloutsos, Similarity searching in medical image databases, *IEEE TKDE*, Vol. 9, 435-447, 1997.

[14]   E. Albuz, E. Kocalar, and A. A. Khokhar, Scalable color image indexing and retrieval using vector wavelets, *IEEE TKDE*, Vol. 13, 851-861, 2001.

[15]   L. G. Brown, A survey of image registration techniques, *Computing Surveys*, Vol. 24, 325-376, 1992.

[16]   R. Mohr, T. Pavlidis, and A. Sanfeliu, *Structural Pattern Analysis*, World Scientific, Teaneck, NJ, 1990.

[17]   H. Bunke and A. Sanfeliu, *Syntactic and Structural Pattern Recognition Theory and Applications: Theory and Applications*, World Scientific, Teaneck, NJ, 1990.

[18]   C. Faloutsos, *Searching Multimedia Databases by Content*, Kluwer Academic Publishers, Boston, MA, 1996.

[19]   S. Santini and A. Gupta, A Wavelet Data Model for Image Databases, in IEEE Intl. Conf. on Multimedia and Expo, IEEE Computer Society, Tokyo, Japan, 2001

[20]   W. A. Burkhard and R. M. Keller, Some approaches to best-match file searching, *CACM*, Vol. 16, 230-236, 1973.

[21]   J. K. Uhlmann, Satisfying general proximity/similarity queries with metric trees, *Information Processing Letter*, Vol. 40, 175-179, 1991.

[22]   P. N. Yianilos, Data Structures and Algorithms for Nearest Neighbor Search in General Metric Spaces, in Fourth Annual ACM/SIGACT-SIAM Symposium on Discrete Algorithms - SODA, Austin, TX, 1993, 311-321.

[23]   T.-C. Chiueh, *Content-based image indexing*, in VLDB, J. B. Bocca, M. Jarke and C. Zaniolo, Eds., Morgan Kaufmann, Santiago de Chile, Chile, 1994, 582-593.

[24]   R. A. Baeza-Yates, et al., Proximity Matching Using Fixed-queries Trees, in 5th Annual Symp. on Combinatorial Pattern Matching (CPM), Springer Verlag, Asilomar, CA, 1994, 198-212.

[25]   T. Bozkaya and Z. M. Özsoyoglu, Indexing large metric spaces for similarity search queries, *ACM Transactions on Database Systems (TODS)*, Vol. 24, 361-404, 1999.

[26]   S. Brin, *Near neighbor search in large metric spaces*, in VLDB, U. Dayal, P. M. D. Gray and S. Nishio, Eds., Morgan Kaufmann, Zurich, Switzerland, 1995, 574-584.

[27]   D. Shasha and T. L. Wang, New techniques for best-match retrieval, *ACM TOIS*, Vol. 8, 140-158, 1990.

[28]   P. Ciaccia, M. Patella, and P. Zezula, *M-tree: An efficient access method for similarity search in metric spaces*, in VLDB, M. Jarke, Ed., Athens, Greece, 1997, 426-435.

[29]   A. Guttman, *R-tree: A dynamic index structure for spatial searching*, in *ACM SIGMOD*, ACM PRess, Boston, MA, 1984, 47-57.

[30]   E. Chávez et al., Searching in metric spaces, *Computing Surveys*, Vol. 33, 273-321, 2001.

[31]   C. Traina, Jr. et al., Slim-trees: High Performance Metric Trees Minimizing Overlap Between Nodes, in Intl. Conf. on Extending Database Technology, C. Zaniolo, et al., Eds., Springer, Konstanz, Germany, 2000, 51-65.

[32]   C. Traina, Jr. et al., Fast indexing and visualization of metric datasets using slim-trees, *IEEE TKDE*, Vol. 14, 244-260, 2002.

[33]   R. F. Santos, Filho, et al., *Similarity search without tears: The omni family of all-purpose access methods*, in *IEEE ICDE*, IEEE Computer Society, Heidelberg, Germany, 2001, 623-630.

[34]   M. Flickner and et al., Query by image and video content: The QBIC system, *IEEE Computer*, Vol. 28, 23-32, 1995.

[35]   J. R. Smith and S.-F. Chang, Visualseek: A Fully Automated Content-based Image Query System, in ACM Multimedia '96, ACM Press, Boston, MA, 1996, 87-98.

[36]   A. Gupta, A Virage Perspective, in Tech. Report Virage Corp., 1995.

[37]   J. R. Smith et al., Integrating Features, Models, and Semantics for Trec Video Retrieval, in The Tenth Text REtrieval Conference, TREC 2001, NIST, Gaithersburg, MD, 2001, 240-249.

[38]   V. E. Ogle and M. Stonebraker, Chabot: Retrieval from a relational databases of images, *IEEE Computer*, vol. 28, 40-48, 1995.

[39]   X. Cao and H. K. Huang, Current status and future advances of digital radiography and pacs, *IEEE Engineering in Medicine and Biology Magazine*, Vol. 9, 80-88, 2000.

[40]   E. L. Siegel and R. M. Kolodner, *Filmless Radiology*, Springer Verlag, New York City, NY, 1999.

[41]   I. I. 13249-1:2000, *Information technology - database languages - sql multimedia and application packages - part 1: Framework*, International Organization for Standardization - ISO, 2001.

[42]   I. I. 13249-5:2001, *Information technology - database languages - sql multimedia and application packages - part 5: Still image*, International Organization for Standardization - ISO, 2001.

30

SMALL SAMPLE LEARNING ISSUES FOR INTERACTIVE VIDEO RETRIEVAL

Xiang Sean Zhou
Siemens Corporate Research
Princeton, New Jersey, USA
`xzhou@scr.siemens.com`

Thomas S. Huang
Beckman Institute for Advanced Science and Technology
University of Illinois at Urbana-Champaign
Urbana, Illinois, USA
`huang@ifp.uiuc.edu`

1. INTRODUCTION

With the ever-increasing amount of digital image and video data along with faster and easier means for information access and exchange, we are facing a pressing demand for machine-aided content-based image and video analysis, indexing, and retrieval systems [1].

1.1 CONTENT-BASED VIDEO RETRIEVAL

To design a content-based video retrieval system, one should first define a set of features to represent the contents and then select a similarity measure(s) (Figure 30.1).

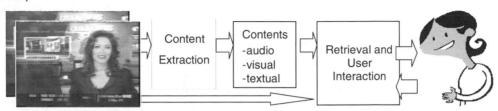

Figure 30.1 Content-based multimedia retrieval: system diagram.

Video contents are embedded in multiple modalities: images, audio, and text. From the video frames a machine can extract features such as color [2], [3], texture [4], shape [5], and structure [6], [7]. For audio, the machine may use features such as Mel Frequency Cepstral Coefficients (MFCC), pitch, energy, etc., to distinguish or recognize various natural sound, music, speech, and noise [8]. Furthermore, text can be extracted from closed caption, or through automatic speech recognition (ASR), or optical character recognition (OCR) from the video frames [9] (Figure 30.2).

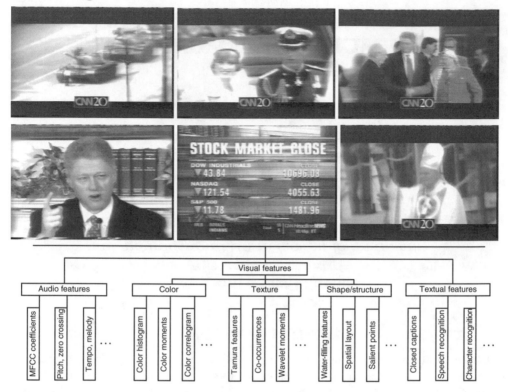

Figure 30.2 Video contents can be represented by a combination of audio, visual, and textual features.

1.2   FEATURE SPACE AND SIMILARITY MEASURE

If the features are recorded into distributions/histograms, histogram intersection [1] or Kullback-Leibler distance [10] is usually employed as the similarity measure.

A more common practice is to formulate a combined feature vector for each *video unit.* (A *video unit* can be one shot, or a "scene" across multiple shots. See Section III on video segmentation for details.) In this case, each video unit can be regarded as a point in a high-dimensional feature space, and a reasonable choice of similarity (or "dissimilarity") measure is Euclidean distance.

Feature components generally have different dynamic ranges. A straightforward and convenient preprocessing step is to normalize each component across the database to a normal distribution with zero mean and unit variance, although other candidates, such as a uniform distribution, can be viable choices as well.

Note that by using Euclidean distance, one assumes that different features have the same contribution in the overall similarity measure. In other words, in the feature space, points of interest are assumed to have a hyper-spherical Gaussian distribution. This, of course, is often not the case. For example, when searching for *gunfight* segments, audio features can be most discriminative, while color may vary significantly. However, if the interest is on *sunset* scenes, color feature should be more distinctive than audio. Visualizing in a normalized feature space of color and audio features, we will not see circularly symmetric distributions for most classes. In addition, when an elongated distribution has its axes not aligned with the feature axes, we cannot treat different features independently any more. This issue was discussed in [11] and [12]. Finally, the distribution may be far from Gaussion assumption; thus the use of non-linear techniques become essential [13], [14]. We can see that an important question is how to dynamically select the best combination of features and distance measures for different queries under various distribution assumptions.

Figure 30.3 Real queries are often specific and personal. A building detector [19] may declare these two images "similar"; however, a user often looks for more specific features such as shape, curvature, materials, and color, etc.

1.3 OFF-LINE TRAINING TO FACILITATE RETRIEVAL

To answer this question, some researchers focus on specific queries one at a time, assuming sufficient amount of labeled data is available beforehand for training. The most important and interesting query is probably a *human face*, which has recently attracted significant research efforts (e.g., [15], [16]). Others try to construct generative or discriminative models to detect *people* [17], *cars*, or *explosion* [18], or to separate *city/buildings* from *scenery* images [19], etc.

With a sufficiently large training set and a good model, a *car* detector or a *building* detector may yield high performance on retrieving *cars* and *buildings*, comparing to schemes not using a training set. However, these detectors are often too general in nature, while the unfortunate real life story [20] is that most database queries are rather specific and personal (Figure 30.3), and their sheer number of possibilities renders off-line preprocessing or tuning impossible, not to mention the sophistication and subtleness of personal interpretations. (Of course, general-purpose detectors are sometimes useful in narrowing down the search scope, and can be good for some specific applications where the number of classes is small and the classes are well defined.) It is therefore important to explore techniques for learning from user interactions in real time. This will be the main topic of this chapter. (This is also discussed by some of the previous chapters, such as the one by Ortega and Mehrotra, with a different approach.)

1.4 ON-LINE LEARNING WITH USER IN THE LOOP

For the purpose of quantitative analysis, we have to impose the *observability assumption* that the features we use possess discriminating power to support the ground-truth classification of the data in the user's mind. In other words, for the machine to be able to help there must be certain degree of correlation or mutual information between the features and the query concept in the user's mind. For example, the machine can help the user to find *buildings* with certain *curvature* (Figure 30.3) only if *curvature* can be and has been reliably measured from the image. On the other hand, a machine will not associate the word "romantic!" with a picture of Paris and the Eiffel Tower—humans might do so because we use additional information outside the picture. (This fact about human intelligence indeed prompted effort to build a vast knowledge base to encode "human knowledge: facts, rules of thumb, and heuristics for reasoning about the objects and events of everyday life." [21])

Note that in reality the *observability assumption* is strong since it is very difficult to find a set of adequate features to represent high-level concepts, and this is likely to be an everlasting research frontier. (Imagine the query for video segments that "convey excitement," or the query for faces that "look British" [20]—it is hardly imaginable that robust numerical descriptors even exist for such high-level, subtle, and sometimes personal concepts in the human minds. I am sure the frames shown in Figure 30.2 can stir up much more "contents" in your memory than a machine can extract and represent.)

Even with good discriminative features, perceptual similarity is not fixed but depends on the application, the person, and the context of usage. A piece of music can invoke different feelings from different people, and "a picture is worth a thousand words" (Figure 30.2). Every query should be handled differently, emphasizing a different *discriminative subspace* of the original feature space.

Early CBIR systems invited the user into the loop by asking the user to provide a feature-weighting scheme for each retrieval task. This proved to be too technical and thus a formidable burden on the user's side. A more natural and friendly way of getting the user in the loop is to ask the user to give feedback regarding the relevance of the current outputs of the system. This technique is referred to as "relevance feedback" ([22]). Though this is an idea initiated in the text retrieval field [23], it seems to work even better in the multimedia domain: it is easier to tell the relevance of an image or video than that of a document—it takes time to read through a document while an image reveals its content instantly.

In content-based image retrieval systems, on-line learning techniques have been shown to provide dramatic performance boost [11], [24], [25], [26], [27].

2. RELEVANCE FEEDBACK IN A FEATURE SPACE

Since different types of multimedia information can be represented in the same form of feature vectors, media type becomes transparent to the machine. Therefore, the algorithms discussed here are applicable for retrieval of different media types. We assume that each meaningful "unit" of multimedia information is represented by a feature vector. For images, the "unit" can be the whole image, image blocks, or segmented regions; and for videos, the "unit" can be shots or keyframes, depending upon the application scenarios.

2.1 BLENDING KEYWORDS AND LOW-LEVEL AUDIO/VISUAL FEATURES

To combine the use of low-level features with keywords, we convert keyword annotations for each image into a vector, with components v_{ij} indicating the appearance or "probability" of keyword j in image i. With $v_{ij} \in [0,1]$, we call it a *soft vector representation* [28].

Another way of modeling relationships among keywords is to apply linear multidimensional scaling (MDS) [29] on the word similarity matrix to arrive at a low-dimensional space; or to use nonlinear techniques such as locally linear embedding to construct such a space [30] in which each word is represented by a point, and their mutual distances are preserved as much as possible. These schemes have the advantage of compact feature representation in a low dimensional feature space, but with the drawback of losing the semantic meaning of the axes. Also, they have poor scalability; i.e., with new keywords the MDS procedure has to be repeated, and the new lower-dimensional embedding can be very different from the previous one, even with a relatively small amount of new data. While with the simple *soft vector representation*, the insertion of new words is a linear incremental process.

2.2 SYSTEM REQUIREMENTS

In the abstraction of the feature space, each "unit" of multimedia data becomes a point. Relevance feedback becomes a supervised classification problem, or an on-line learning problem in a batch mode, *but with unique characteristics*. The uniqueness lies in at least three aspects:

First, the machine needs to learn and respond in real time. The feature subset and the similarity measure should be dynamically determined by the current user for the current task. Therefore real-time response is important.

Second, the number of training samples is very small relative to the dimension of the feature space, and relative to the requirements by popular learning machines such as support vector machines (SVM) [31].

Third, the desired output is not necessarily a binary decision on each point, but rather a rank-ordered top-k return, as a typical search engine will do. This is a less demanding task since we actually do not care about the rank or configuration of the negative points as long as they are far beyond the top-k returns. In fact, algorithms aiming at binary classification are ill fitted to this problem and may perform poorly.

2.3 RELEVANCE FEEDBACK

We use the phrase "relevance feedback" to denote the on-line learning process during multimedia information retrieval, based on the relevance judgments fed back by the user. The scenario is like this:

- *Machine provides initial retrieval results through query-by-keyword, sketch, or example, etc.*

Then, iteratively:

- *User provides judgment on the current results as to whether, and to what degree, they are relevant to her/his request;*

- *Machine learns and tries again.*

The task is to design an algorithm that learns a discriminating subspace from the limited number of examples provided by the user in an interactive fashion. Two cases need to be addressed, ideally *in a unified framework*: first, when only positive examples are given, a transformation, *linear or nonlinear*, shall bring them together in the new metric space. (Feature weighting [24] and whitening transform [11] are the linear solutions, with and without the independence assumption on feature components, respectively.) Second, when negative examples are also given, the transformation shall separate the positive and negative examples in the new space.

3. STATE OF THE ART

Among different media types, image retrieval is the most active in recent years [32]. We give a brief review of the state of the art in relevance feedback techniques in the context of image retrieval. Again, many of these techniques are directly applicable for the retrieval of video.

In its short history, relevance feedback developed along the path from heuristic-based techniques to optimal learning algorithms, with early work inspired by term weighting and relevance feedback techniques in document retrieval [23]. These methods proposed heuristic formulation with empirical parameter adjustment, mainly along the line of independent axis weighting in the feature space [24], [33]. The intuition is to emphasize more the feature(s) that best cluster the positive examples and separate the positive and the negative.

Early work [24], [33] bears the mark of its origins in document retrieval. For example, in [24], learning based on "term frequency" and "inverse document frequency" in the text domain is transformed into learning based on the ranks of the positive and negative images along each feature axis in the continuous feature space. The scheme in [33] quantizes the features and then groups the images or regions into hierarchical trees whose nodes are constructed through single-link clustering. Then weighting on groupings is based on "set operations."

Aside from the lack of optimality claim, the assumption of feature independence imposed in these methods is also artificial, unless independent components can be extracted beforehand (by using, for example, independent component analysis (ICA). However, ICA is not guaranteed to give satisfactory results [34]).

Later on, researchers began to look at this problem from a more systematic point of view by formulating it into an optimization, learning, or classification problem. In [11] and [12], based on the minimization of total distances of positive examples from the new query, the optimal solutions turn out to be the weighted average of the positive examples as the optimal query and a whitening transform (or Mahalanobis distance) as the optimal distance metric. Additionally, [12] adopts a two-level weighting scheme to better cope with singularity due to the small number of training samples. To take into account the negative examples, [35] updates the feature weights along each feature axis by comparing the variance of positive examples to the variance of the union of positive and negative examples. (It will become clear later on that this scheme is actually the reduced diagonal solution of *minimizing the ratio of positive scatter over the overall scatter among both positive and negative examples, with respect to a transformation in the original space.* This is obviously not the best intuition for this problem, *cf.* Section 5.)

Assuming that the user is searching for a particular target, and the feedback is in the form of "relative judgment," [36] proposes the stochastic comparison search as its relevance feedback algorithm.

MacArthur et al. [37] cast relevance feedback as a two-class learning problem and use a decision tree algorithm to sequentially "cut" the feature space until all points within a partition are of the same class. The database is classified by the resulting decision tree: images that fall into a relevant leaf are collected and the nearest neighbors of the query are returned.

While most CBIR systems use well-established image features such as color histogram/moments, texture, shape, and structure features, there are alternatives. Tieu and Viola [25] use more than 45000 "highly selective features" and a boosting technique to learn a classification function in this feature space. The features were demonstrated to be sparse with high kurtosis and were argued to be expressive for high-level semantic concepts. Weak two-class classifiers were formulated based on Gaussian assumption for both the positive and negative (randomly chosen) examples along each feature component, independently. The strong classifier is a weighted sum of the weak classifiers as in AdaBoost [38].

In [26], the Gaussian mixture model on DCT coefficients is used as image representation. Then Bayesian inference is applied for image regional matching and learning over time. Note that in this work each image is represented by a distribution (mixture model) instead of a single feature vector. Richer information captured by the mixture model makes Bayesian inference and image regional matching possible.

Recently, there also have been attempts to incorporate the support vector machine (SVM) into the relevance feedback process [14], [39], [40]. However, SVM as a two-class classifier is not directly suitable for relevance feedback because the training examples are far too few to be representative of the true distributions [39]. However, a kernel-based one-class SVM as density estimator for positive examples has been shown to outperform the whitening transform method in [39].

Without assuming one Gaussion mode for positive examples, Parzen window density estimation can be applied to capture nonlinearity in distribution [41], or an "aggregate dissimilarity" function is used to combine for a candidate image the pair-wise distances to every positive example. The major weakness of these schemes is in their equal treatment of different feature components for different tasks.

Formulated in the transductive learning framework, the D-EM algorithm [42] uses examples from the user feedback (labeled data) as well as other data points (unlabeled data).

4. TRADITIONAL DISCRIMINANT ANALYSIS

A unified way of looking at the above algorithms is to analyze them from the feature space transformation point of view: indeed, the feature-weighting scheme (e.g., [35]) is the simplified diagonal form of a linear transformation in the original feature space, assuming feature independence. Mahalanobis distance or generalized (or weighted) Euclidean distance using the inverse of the covariance matrix of the positive examples ([11], [12]) is a whitening transformation based on the configuration of the positive examples, assuming Gaussian distribution.

From the pattern classification point of view, when only positive examples are to be considered and with Gaussian assumption, the whitening transformation is the optimal choice [43]. When both positive and negative examples are to be considered—instead of the aforementioned various, seemingly plausible heuristics for feature-weighting—two optimal linear transformations based on the traditional discriminant analysis are worth investigating. Of course, the "optimality" depends on the choice of objective function; in this sense, it becomes a problem of formulating the best objective function.

4.1 TWO-CLASS ASSUMPTION

One approach is the two-class fisher discriminant analysis (FDA). The objective is to find a lower-dimensional space in which the ratio of between-class scatter over within-class scatter is maximized.

$$W = \arg \max_{W} \frac{\left| W^T S_b W \right|}{\left| W^T S_w W \right|} \tag{30.1}$$

where

$$S_b = (m_x - m)(m_x - m)^T + (m_y - m)(m_y - m)^T \tag{30.2}$$

$$S_w = \sum_{i=1}^{N_x} (x_i - m_x)(x_i - m_x)^T + \sum_{i=1}^{N_y} (y_i - m_y)(y_i - m_y)^T \tag{30.3}$$

We use $\{x_i, i = 1, ..., N_x\}$ to denote the positive examples, and $\{y_i, i = 1, ..., N_y\}$ to denote the negative examples. m_x, m_y, and m are the mean vectors of the sets $\{x_i\}$, $\{y_i\}$, and $\{x_i\} \cup \{y_i\}$, respectively. (See [43] for details.)

For this two-class discriminant analysis, it is part of the objective that negative examples shall cluster in the discriminating subspace. This is an unnecessary and potentially damaging requirement since the relatively small training sample is a not good representative for the overall population, *especially the negative examples*. Plus, very likely the negative examples will belong to multiple classes. Therefore the effort of rounding up all the negative examples can mislead the resulting discriminating subspace into the wrong direction.

4.2 MULTICLASS ASSUMPTION

Another choice is the multiple discriminant analysis (MDA) [43], where each negative example is treated as from a different class. It becomes an $(N_y + 1)$-class discriminant analysis problem. The reason for the crude assumption on the number of negative classes is that the class labels within the negative examples are not available. One may suggest that the user shall provide this information. However, from a user interface design point of view, it is reasonable for the user to click to indicate items as relevant versus irrelevant (say, "horses" and "nonhorses"), but troublesome and unnatural for the user to further identify for the machine what the negative items really are ("these are tigers, those are zebras, and that is a table...").

For MDA the objective function has the same format as in Equation (30.1). The difference is in the definitions of the scatter matrices:

$$S_b = (m_x - m)(m_x - m)^T + \sum_{i=1}^{N_y} (y_i - m)(y_i - m)^T \qquad (30.4)$$

$$S_w = \sum_{i=1}^{N_x} (x_i - m_x)(x_i - m_x)^T \qquad (30.5)$$

In this setting, it is part of the objective that all negative examples shall be apart from one another in the discriminating subspace. This is again an unnecessary and potentially damaging requirement since there are cases in which several negative examples come from the same class. The effort of splitting them up can lead the resulting discriminating subspace into the wrong direction.

4.3 DIMENSIONALITY REDUCTION MATRIX

For both FDA and MDA, the columns of the optimal W are the generalized eigenvector(s) w_i associated with the largest eigenvalue(s) λ, i.e.,

$$S_b w = \lambda_i S_w w_i \qquad (30.6)$$

The traditional *discriminating dimensionality reduction matrix*, formed by the k eigenvectors associated with the top k eigenvalues, is defined as

$$A = [w_1, ..., w_k] \qquad (30.7)$$

In the new space $x_{new} = A^T x_{old}$, the following "actions" are employed to ensure the optimal ratio in Equation (30.1): for FDA, the positive centroid is "pushed" apart from the negative centroid, while examples of the same label are "pulled" closer to one another; for MDA, the positive centroid and every negative example are "pushed" apart from one another, while positive examples are "pulled" closer to one another.

Note that the effective dimension of the new space is independent of the original dimensionality. Specifically, for FDA, since the rank of S_b is only one, the discriminating subspace has one dimension. For MDA, the effective new dimension is at most $\min\{N_x, N_y\}$. However, after regularization (see Section 5.3) these can be artificially higher.

5. BIASED DISCRIMINANT ANALYSIS

Instead of confining ourselves to the traditional settings of the discriminant analysis, a better way is to use a new form of the discriminant analysis, namely, our proposed biased discriminant analysis (BDA).

5.1 (1+x)-CLASS ASSUMPTION

We first define the (1+ *x*)-*class classification problem* or *biased classification problem* as the learning problem in which there are an unknown number of classes but the user is only interested in one class, i.e., the user is biased toward one class. And the training samples are labeled by the user as only "positive" or "negative" as to whether they belong to the target class or not. Thus, the negative examples can come from an uncertain number of classes.

Much research has addressed this problem simply as a two-class classification problem with symmetric treatment on positive and negative examples, such as FDA. However the intuition is like "all good marriages are good in a similar way, while all bad marriages are bad in their own ways"; or we say "all positive examples are good in the same way, while negative examples are bad in their own ways." Therefore, it is desirable to distinguish a real two-class problem from a (1+x)-class problem. When the negative examples are far from representing their true distributions, which is certainly true in our case, this distinction becomes critical. (Tieu and Viola [25] used a random sampling strategy to increase the number of negative examples and thus their representative power. This is somewhat dangerous since unlabeled positive examples can be included in these "negative" samples.)

5.2 BDA

For a biased classification problem, we ask the following question instead: What is the optimal discriminating subspace in which the positive examples are "pulled" closer to one another while the negative examples are "pushed" away from the positive ones?

Or mathematically: What is the optimal transformation such that the ratio of "the negative scatter with respect to positive centroid" over "the positive within class scatter" is maximized? We call this *biased discriminant analysis* (BDA) due to the biased treatment of the positive examples. We define the biased criterion function

$$W = \arg\max_{w} \frac{\left| W^T S_y W \right|}{\left| W^T S_x W \right|} \qquad (30.8)$$

where

$$S_y = \sum_{i=1}^{N_y} (y_i - m_x)(y_i - m_x)^T \qquad (30.9)$$

$$S_x = \sum_{i=1}^{N_x} (x_i - m_x)(x_i - m_x)^T \qquad (30.10)$$

The optimal solution and transformations are of the same formats as those of FDA or MDA, subject to the differences defined by Equations (30.9) and (30.10).

Note that the discriminating subspace of BDA has an effective dimension of min{N_x, N_y}, the same as MDA and higher than that of FDA.

5.3   REGULARIZATION AND DISCOUNTING FACTORS

It is well known that the sample-based plug-in estimates of the scatter matrices based on Equations (30.2)-(30.5), (30.9), and (30.10) will be severely biased for a small number of training examples, in which case regularization is necessary to avoid singularity in the matrices. This is done by adding small quantities to the diagonal of the scatter matrices. For detailed analysis see [44]. The regularized version of S_x, with n being the dimension of the original space and I being the identity matrix, is

$$S_x^r = (1-\mu)S_x + \frac{\mu}{n}tr[S_x]I \qquad (30.11)$$

The parameter μ controls shrinkage toward a multiple of the identity matrix.

The influence of the negative examples can be tuned down by a discounting factor γ, and the discounted version of S_y is

$$S_y^d = (1-\gamma)S_y + \frac{\gamma}{n}tr[S_y]I \qquad (30.12)$$

With different combinations of the (μ, γ) values, regularized BDA provides a fairly rich set of alternatives. The combination ($\mu = 0$, $\gamma = 1$) gives a subspace that is mainly defined by minimizing the scatters among the positive examples, resembling the effect of a whitening transform. The combination ($\mu = 1$, $\gamma = 0$) gives a subspace that mainly separates the negative from the positive centroid, with minimal effort on clustering the positive examples. The combination ($\mu = 0$, $\gamma = 0$) is the full BDA and ($\mu = 1$, $\gamma = 1$) represents the extreme of discounting all configurations of the training examples and keeping the original feature space unchanged.

BDA captures the essential nature of the problem with minimal assumption. In fact, even the Gaussian assumption on the positive examples can be further relaxed by incorporating kernels.

5.4   DISCRIMINATING TRANSFORM

Similar to FDA and MDA, we first solve for the generalized eigenanalysis problem with generalized square eigenvector matrix V associated with the eigenvalue matrix Λ, satisfying

$$S_y V = S_x V \Lambda \qquad (30.13)$$

However, instead of using the traditional *discriminating dimensionality reduction matrix* in the form of Equation (30.7), we propose the *discriminating transform matrix* as

$$A = V\Lambda^{1/2} \qquad (30.14)$$

As for the transformation A, the weighting of different eigenvectors by the square roots of their corresponding eigenvalues in fact has no effect on the value of the objective function in Equation (30.8). But it will make a difference when a k-nearest neighbor classifier is to be applied in the transformed space.

5.5 GENERALIZING BDA

One generalization of BDA is to take in multiple positive clusters instead of one [45]. For example, the training set may be labeled as *red horse, white horse, black horse,* and *non horse,* then we shall formulate BDA such that *red horses, white horses,* and *black horses* all cluster within their own color, plus all *non-horses* are pushed away from these three. The definition of this generalized BDA will be similar to above, but we need to change equation (30.9) and (30.10) into equation (30.15) and (30.16) below.

$$S_N = \sum_{c=1}^{C} \sum_{i=1}^{N_N} (y_i - m_x^c)(y_i - m_x^c)^T \tag{30.15}$$

$$S_P = \sum_{c=1}^{C} \sum_{i=1}^{N_P^c} (x_i^c - m_x^c)(x_i^c - m_x^c)^T \tag{30.16}$$

where C is the total number of clusters in positive examples, $\{x_i^c\}$, $i = 1,2,..., N_P^c$ are the positive examples in cluster c, and m_x^c is the mean vector of positive cluster c.

If the separation of different colored horses is important to the user, we can add the corresponding scatters into equation (30.15). Note this is not MDA since again it is "biased" toward *horses*: the scatter of *non horses* are not minimized.

This generalization is applicable if the user is willing to give clustering information on the positive examples, which might not be realistic in many applications.

6. NONLINEAR EXTENSION USING KERNELS

The Gaussian assumption implied by BDA does not hold in general for real world problems. In this section, we discuss a non linear extension of BDA using a kernel approach.

6.1 THE KERNEL TRICK

For a comprehensive introduction to kernel machines and SVM, refer to [46]. Here we briefly introduce the key idea.

The original linear BDA algorithm is applied in a *"feature space,"*[1] **F,** which is related to the original space by a nonlinear mapping

$$\phi : C \to F$$
$$x \to \phi(x) \tag{30.17}$$

where C is a compact subset of R^N. However, this mapping is expensive and will not be carried out explicitly, but through the evaluation of a kernel matrix K,

---

[1] A term used in kernel machine literature to denote the new space after the nonlinear transform—this is not to be confused with the *feature space* concept previously used to denote the space for features/descriptors extracted from the media data.

with components $k(x_i, x_j) = \phi^T(x_i)\phi(x_j)$. This is the same idea as that adopted by support vector machine [31], kernel PCA, and kernel discriminant analysis [48], [49].

6.2 BDA IN KERNEL FORM

The trick here is to rewrite the BDA formulae using only dot-products of the form $\phi_i^T\phi_j$.

Using superscript ϕ to denote quantities in the new space, we have the objective function in the following form:

$$w^* = \arg\max_w \frac{w^T S_y^\phi w}{w^T S_y^\phi w} \tag{30.18}$$

where

$$S_y^\phi = \sum_{i=1}^{N_y}(\phi(y_i) - m_x^\phi)(\phi(y_i) - m_x^\phi)^T \tag{30.19}$$

$$S_x^\phi = \sum_{i=1}^{N_x}(\phi(x_i) - m_x^\phi)(\phi(x_i) - m_x^\phi)^T \tag{30.20}$$

Here m_x^ϕ is the positive centroid in the feature space F. We will express it in matrix form for the convenience of later derivations:

$$m_x^\phi = [\phi(x_1),...,\phi(x_{N_x})]\left.\begin{bmatrix}\frac{1}{N_x}\\ ...\\ \frac{1}{N_x}\end{bmatrix}\right\}N_x \tag{30.21}$$

Since the solution for w is the eigenvector(s) corresponding to the nonzero eigenvalues of the scatter matrices formed by the input vectors $\phi(x_i)$ and $\phi(y_j)$ in the feature space F, the optimal w is in the subspace spanned by the input vectors in the feature space. Thus, we can express w as a linear combination of $\phi(x_i)$ and $\phi(y_j)$, and change the problem from finding the optimal w to finding the optimal α:

$$w = \sum_{i=1}^{N_x}\alpha_i\phi(x_i) + \sum_{j=1}^{N_y}\alpha_{j+N_x}\phi(y_j) = \Phi\alpha \tag{30.22}$$

where

$$\Phi = [\phi(x_1),...,\phi(x_{N_x}),\phi(y_1),...,\phi(y_{N_y})] \tag{30.23}$$

The numerator of (30.18), after being expressed as the negative scatter with respect to positive centroid *in the feature space*, can be rewritten as

$$w^T S_y^\phi w = \alpha^T \Phi^T \sum_{j=1}^{N_y} (\phi(y_j) - m_x^\phi)(\phi(y_j) - m_x^\phi)^T \Phi \alpha$$

$$= \alpha^T \sum_{j=1}^{N_y} (K_{y_j} - K_{mx})(K_{y_j} - K_{mx})^T \alpha$$

(30.24)

where

$$K_{y_j} = \Phi^T \phi(y_j) = \begin{bmatrix} \phi^T(x_1)\phi(y_j) \\ \dots \\ \phi^T(x_{N_x})\phi(y_j) \\ \phi^T(y_1)\phi(y_j) \\ \dots \\ \phi^T(y_{N_y})\phi(y_j) \end{bmatrix}$$

$$= \begin{bmatrix} k(x_1, y_j) \\ \dots \\ k(x_{N_x}, y_j) \\ k(y_1, y_j) \\ \dots \\ k(y_{N_y}, y_j) \end{bmatrix}$$

(30.25)

$$K_{mx} = \Phi^T m_x^\phi = \begin{bmatrix} \phi^T(x_1) \\ \dots \\ \phi^T(x_{N_x}) \\ \phi^T(y_1) \\ \dots \\ \phi^T(y_{N_y}) \end{bmatrix} [\phi(x_1), \dots, \phi(x_{N_x})] \left. \begin{bmatrix} \dfrac{1}{N_x} \\ \dots \\ \dfrac{1}{N_x} \end{bmatrix} \right\} N_x \times 1$$

$$= K_x \left. \begin{bmatrix} \dfrac{1}{N_x} \\ \dots \\ \dfrac{1}{N_x} \end{bmatrix} \right\} N_x \times 1$$

(30.26)

Both of these are vectors of dimension $(N_x + N_y) \times 1$, and K_x is of size $(N_x + N_y) \times N_x$, and is defined in Equation (30.30). We can further rewrite the summation term in the middle of Equation (30.24) into a matrix operation; this is done by realizing that for two sets of column vectors v_i and w_i in $V = [\dots, v_i, \dots]$ and $W = [\dots, w_i, \dots]$, we have

$$\sum_i v_i w_i^{\ T} = VW^T \tag{30.27}$$

Therefore, the middle portion of Equation (30.24) is further written as

$$\sum_{j=1}^{N_y}(K_{y_j} - K_{mx})(K_{y_j} - K_{mx})^T \tag{30.28}$$

$$= (K_y - K_x I_{N_x}^y)(K_y - K_x I_{N_x}^y)^T$$

where

$$K_y = \begin{bmatrix} K_{y_1},...,K_{y_{N_y}} \end{bmatrix} = \begin{bmatrix} \phi^T(x_1) \\ ... \\ \phi^T(x_{N_x}) \\ \phi^T(y_1) \\ ... \\ \phi^T(y_{N_y}) \end{bmatrix} [\phi(y_1),...,\phi(y_{N_y})]$$

$$= \begin{bmatrix} k(x_1,y_1) & ... & k(x_1,y_{N_y}) \\ ... & ... & ... \\ k(x_{N_x},y_1) & ... & k(x_{N_x},y_{N_y}) \\ k(y_1,y_1) & ... & k(y_1,y_1) \\ ... & ... & ... \\ k(y_{N_y},y_1) & ... & k(y_{N_y},y_{N_y}) \end{bmatrix} \tag{30.29}$$

$$K_x = \begin{bmatrix} K_{x_1},...,K_{x_{N_x}} \end{bmatrix} = \begin{bmatrix} \phi^T(x_1) \\ ... \\ \phi^T(x_{N_x}) \\ \phi^T(y_1) \\ ... \\ \phi^T(y_{N_y}) \end{bmatrix} [\phi(x_1),...,\phi(x_{N_x})]$$

$$= \begin{bmatrix} k(x_1,x_1) & ... & k(x_1,x_{N_x}) \\ ... & ... & ... \\ k(x_{N_x},x_1) & ... & k(x_{N_x},x_{N_x}) \\ k(y_1,x_1) & ... & k(y_1,x_1) \\ ... & ... & ... \\ k(y_{N_y},x_1) & ... & k(y_{N_y},x_{N_x}) \end{bmatrix} \tag{30.30}$$

and $I_{N_x}^y$ is an $N_x{\times}N_y$ matrix of all elements being $\dfrac{1}{N_x}$.

Similarly, we can rewrite the denominator of (30.18):

$$
\begin{aligned}
w^T S_x^\phi w &= \alpha^T (K_x - K_x I_{N_x}^x)(K_x - K_x I_{N_x}^x)^T \alpha \\
&= \alpha^T K_x (I - I_{N_x}^x)(I - I_{N_x}^x)^T K_x^T \alpha
\end{aligned}
\tag{30.31}
$$

Here I is the identity matrix and $I_{N_x}^x$ is an $N_x{\times}N_x$ matrix of all elements being $\dfrac{1}{N_x}$.

Since $I_{N_x}^x = (I_{N_x}^x)^T$ and $I_{N_x}^x (I_{N_x}^x)^T = I_{N_x}^x$, the middle part of Equation (30.31) can be further expanded and simplified as

$$
\begin{aligned}
(I - I_{N_x}^x)(I - I_{N_x}^x)^T \\
= I - I_{N_x}^x - I_{N_x}^x + I_{N_x}^x \\
= I - I_{N_x}^x
\end{aligned}
\tag{30.32}
$$

6.3 THE NONLINEAR SOLUTION

To this point, by substituting Equations (30.24), (30.28), (30.31), and (30.32) into (30.18)-(30.20), we arrive at a new generalized Rayleigh quotient, and it is again a generalized eigen-analysis problem, where the optimal a's are the generalized eigenvectors associated with the largest eigenvalues λ's, i.e.,

$$
(K_y - K_x I_{N_x}^y)(K_y - K_x I_{N_x}^y)^T \alpha = \lambda K_x (I - I_{N_x}^x) K_x^T \alpha
\tag{30.33}
$$

With optimal a, the projection of a new pattern z onto w, ignoring weighting by square-rooted eigenvalue, is directly given by

$$
\begin{aligned}
w^T \phi(z) &= (\Phi \alpha)^T \phi(z) = \alpha^T (\Phi^T \phi(z)) \\
&= \sum_{i=1}^{N_x} \alpha_i k(x_i, z) + \sum_{j=1}^{N_y} \alpha_{j+N_x} k(y_j, z)
\end{aligned}
\tag{30.34}
$$

6.4 THE ALGORITHM

When applied in a retrieval system as the relevance feedback algorithm, it can be briefly described as follows:

With only one query for the first round, it just retrieves the nearest Euclidean neighbors as the return. During following rounds of relevance feedback with a set of positive and negative examples, and a chosen kernel k, it will

1. Compute K_x and K_y as defined in Equation (30.29) and (30.30);

2. Solve for a's and λ's as in Equation (30.33);

3. With a's ordered in descending order of their eigenvalues, select the subset $\{a_i | \lambda_i > \tau \lambda_1\}$, where τ is a small positive number, say, 0.01;

4. $u_i = u_i \sqrt{\lambda_i}$, for selected u_i in the previous step;

5. Compute the projection of each point z onto the new space as in Equation (30.34).

6. In the new space, return the points corresponding to the Euclidean nearest neighbors from the positive centroid.

7. EXPERIMENTS AND ANALYSIS

Using image retrieval as examples, we compare the three proposed discriminating transforms to the optimal two-level whitening transforms [12], and compare the kernel versions with SVM, on both synthetic data and real world image databases. The scenario is "query by example" followed by several rounds of relevance feedback by the user. The machine first learns an optimal transform, linear or nonlinear, and then all training and testing points are transformed into the new space, where the new query is the mean of the transformed positive examples, and its 20 nearest neighbors are returned for further judgment from the user.

7.1 EXPERIMENTS ON SYNTHETIC DATA

For the non kernel versions of FDA, MDA, and BDA, all the transform matrices are linear, and the decision boundaries are either linear or quadratic.

To test the performance of BDA versus FDA or MDA, we used some toy problems as depicted in Figure 30.4. Original data are in 2-D feature space, and positive examples are "o"s and negative examples are "x"s in the figure. In all cases, it is assumed that the number of modes for negative examples is unknown. FDA, MDA, and BDA are applied to find the best projection direction by their own criterion functions for each case, and the resulting (generalized) eigenvector corresponding to the maximum eigenvalue is drawn in solid, dash-dotted, and dashed lines, respectively. The would-be projections are also drawn as bell-shaped curves to the side of the corresponding eigenvectors, assuming Gaussian distribution for each mode.

Here, FDA treats positive and negative examples equally; i.e., it tries to decrease the scatter among negative examples as part of the effort. This makes it a bad choice in the cases of (b) or (d). Without any prior knowledge about the number of classes to which the negative examples belong, MDA can only treat each example as a separate class/mode. Since MDA has in its criterion function the tendency of increasing the scatter among all classes/modes, which includes the scatter among negative examples, this makes it a bad choice for cases (a) and (c). Notice the two modes of the negative examples moved apart from each other and toward the positive examples, and BDA is able to adapt to the change and gives better class separation in both cases. MDA fails in (c), and FDA fails in (d).

In all cases, BDA yields good separation of negative examples from positive ones, as well as clustering of positive examples (it finds a balance between these two goals). Note from (c) to (d), the two negative modes move apart from each other and toward the positive ones. FDA and MDA yield unchanged results, for (c) FDA gives better separation and for (d) MDA gives better separation. BDA is able to adapt to the change and gives better separation in both cases.

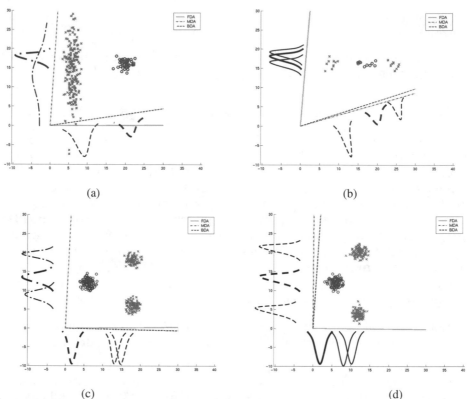

(a)   (b)

(c)   (d)

Figure 30.4 Comparison of FDA, MDA, and BDA for dimensionality reduction from 2-D to 1-D.

FDA and MDA are inadequate in biased classification or biased dimensionality reduction problems because of their forceful assumption on the number of modes. BDA avoids making this assumption by directly modeling the asymmetry and hence gives better results.

To test the ability of the KBDA in dealing with nonlinearly distributed positive examples, six sets of synthetic data in two-dimensional space are used (see Figure 30.5). The circles are positive examples and the crosses negative. A simulated query process is used for training sample selection; i.e., the 20 nearest neighbors of a randomly selected positive point are used as training samples. The bar diagram shows the averaged hit rate in the top 20 returns. A clear boost in hit rates is observed when using KBDA.

7.2 IMAGE DATABASE TESTING

A fully labeled set of 500 images from the Corel image set is used for testing. It consists of five classes, each with 100 images. Each image is represented by a 37-dimensional vector, which consists of 9 color features, 10 texture features, and 18 edge-based structure features [6], [13].

Each round, 10 positive and 10 negative images are randomly drawn as training samples. The hit rate in the top 100 returns is recorded. Five hundred rounds of testing are performed on all five classes, and the averaged hit rates are shown in

Table 30.1. Here, QM, or query movement, a popular technique in information retrieval, is implemented by simply taking the positive centroid as the new query point and returning the nearest neighbors in the order of increasing distance. It is worth pointing out that based on the performance shown in Table 30.1, KBDA outperforms a one-class SVM-based kernel density estimator (*cf.* [39]).

More test results on image databases are available in [50].

7.3 EXPERIMENT ON VIDEOS

We are currently building a large video database with both numerical features and textual annotations. Thus far we could only offer some preliminary testing results, hoping that these may shed some lights on promising future research directions.

We experimented with a small set of 50 news video shots, with keyword annotations manually extracted/assigned based on the transcript. The vocabulary is fixed with 20 keywords ("Clinton," "Middle-East," "Arafat," "Israel," "advertisement," "weather," etc.), resulting in a 20-dimension sub-vector. We also manually assigned a keyword association matrix [28] (e.g., with cross-triggering among "Middle-East," "Arafat," "Israel," and "Clinton"). In a real world scenario, this matrix can be formed through either manual encoding [21], or learning from a large corpus [23], [30], or learning from long-term user interactions [28]. We used the same visual features as the ones used in our image database with 37 dimensions, extracted from the first frame of each shot. Initial testing indicates that relevance feedback algorithms such as BDA can correctly identify the proper subspace within the joint feature space of visual and textual components and thus can facilitate queries across modalities.

One run of the linear BDA algorithm would proceed as follows: the query shot was the anchorperson (similar to the one shown in Figure 30.1) reporting on President Clinton's remarks on Middle East crisis (with keywords "Clinton" and "Middle East"). With one input, the system applies Euclidean distance; and visual features apparently dominated the results: the top ten returns are all anchorperson shots, but fortunately with one of them regarding Israel Defense Force (IDF). Assumed to be interested in Middle East crisis, the user then selected the IDF shot as the second query, and labeled four of the other anchorperson shots as negative examples. By applying the BDA algorithm, the system largely ignored visual features (because of the negative examples) and weighted more on the textual components for "Middle East," "Israel," and "Clinton". The subsequent top ten returns included all the four segments on Middle East crisis, including a shot of Arafat, and a shot on a meeting between Clinton and Arafat.

Table 30.1 Averaged hit rate in top 100 for 500 rounds of testing.

QM	WT ($\mu=0.1$)	BDA ($\mu=0.1$, $\gamma=0$)	KBDA (RBF: $\sigma=0.7$)
62.7%	70.4%	74.2%	79.1%

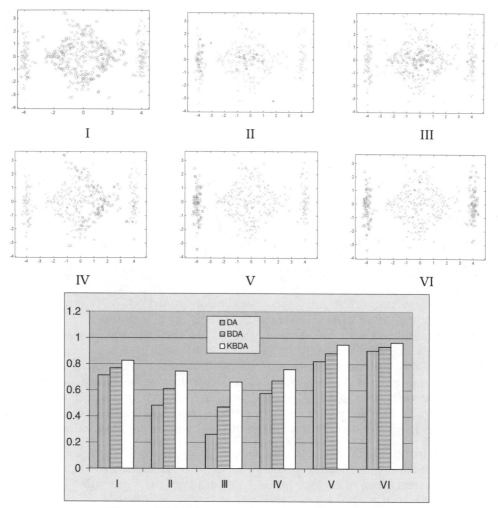

Figure 30.5 Test results on synthetic data.

8. CONCLUSION

In this chapter, we have briefly reviewed the existing relevance feedback techniques. Emphasis is put on the analysis of the unique characteristics of multimedia information retrieval problems and the corresponding on-line learning algorithms. We presented a variant of discriminant analysis that is suited for the asymmetric nature of this small sample learning problem. A kernel-based approach is used to derive its non linear counterpart. We also presented experimental results on synthetic and real world image and video datasets. The proposed algorithm demonstrated promising potential in facilitating complex retrieval tasks using both numerical and textual inputs.

REFERENCES

[1] S. W. Smoliar and H. Zhang, Content-based video indexing and retrieval, *IEEE Multimedia*, vol. 1, no. 2, pp. 62-75, 1994.

[2] M. J. Swain and D. H. Ballard, Color indexing, *Int'l Journal of Computer Vision*, vol. 7, pp. 11-32, 1991.

[3] J. Huang, S. R. Kumar, M. Mitra, W.-J., Zhu, and R. Zabih, Image Indexing Using Color Correlograms, in Proc. IEEE Conf. on Computer Vision and Pattern Recognition, San Juan, Puerto Rico, 1997, pp. 762-768.

[4] R. M. Haralick, K. Shanmugam, and I. Dinstein, Texture feature for image classification, *IEEE Trans. on System, Man, and Cybernetics*, vol. 3, no. 1, pp. 610-621, Nov. 1973.

[5] M. K. Hu, Visual pattern recognition by moment invariants, *IRE Trans. on Information Theory*, vol. 8, pp. 179-187, 1962.

[6] X. S. Zhou and T. S. Huang, Edge-based structural feature for content-based image retrieval, *Pattern Recognition Letters*, vol. 22, no. 5, pp 457-468, Apr. 2001.

[7] Q. Tian, N. Sebe, M. S. Lew, E. Loupias, and T. S. Huang, Image retrieval using wavelet-based salient points, *Journal of Electronic Imaging, Special Issue on Storage and Retrieval of Digital Media*, vol. 10, no. 4, pp 835-849, Oct. 2001.

[8] E. Wold, T. Blum, D. Keislar, and J. Wheaton, Content-based classification, search, and retrieval of audio, *IEEE Multimedia Magazine*, vol. 3, no. 3, pp. 27-36, July-Sept. 1996.

[9] T. Sato, T. Kanade, E. Hughes, and M. Smith, Video OCR for Digital News Archive, in Proc. Workshop on Content-Based Access of Image and Video Libraries, Los Alamitos, CA, Jan. 1998, pp 52-60.

[10] K. Fukunaga, *Introduction to Statistical Patter Recognition*, New York: Academic Press, 1971.

[11] Y. Ishikawa, R. Subramanya, and C. Faloutsos, MindReader: Query Databases Through Multiple Examples, in Proc. The 24th Int'l Conf. on Very Large Data Bases, New York, 1998, pp. 433-438.

[12] Y. Rui and T. S. Huang, Optimizing Learning in Image Retrieval, in Proc. IEEE Conf. on Computer Vision and Pattern Recognition, Hilton Head Island, SC, June 2000, pp. 236-243.

[13] X. S. Zhou and T. S. Huang, Comparing Discriminating Transformations and SVM for Multimedia Retrieval, in Proc. 9th ACM Int'l Conf. on Multimedia, Ottawa, Canada. Sept. 2001, pp. 137-146.

[14] S. Tong and E. Chang, Support Vector Machine Active Learning for Image Retrieval, in Proc. ACM Multimedia, Ottawa, Canada, Sept. 2001

[15] FaceIt SDK, Visionics Corporate Website, http://www.visionics.com.

[16] S. Satoh and T. Kanade, NAME-IT: Association of Face and Name in Video, in Proc. IEEE Conf. on Computer Vision and Pattern Recognition, San Juan, Puerto Rico, 1997.

[17] D. A. Forsyth and M. M. Fleck, Finding People and Animals by Guided Assembly, in Proc. IEEE Int'l Conf. on Image Processing, Santa Barbara, CA, Oct. 1997.

[18] M. R. Naphade, T. Kristjansson, B. Frey, and T. S. Huang, Probabilistic Multimedia Objects Multijects: A Novel Approach to Indexing and Retrieval in Multimedia Systems, in Proc. IEEE Int'l Conf. on Image Processing, vol. 3, Chicago, Oct. 1998, pp. 536-540.

[19] Q. Iqbal and J. K. Aggarwal, Applying Perceptual Grouping to Content-based Image Retrieval: Building Images, in Proc. IEEE Conf. Computer Vision and Pattern Recognition, 1999, pp. 42-48.

[20] P. Enser and C. Sandom, Retrieval of Archival Moving Imagery: CBIR Out of the Frame?, in Proc. Int'l Conf. on Image and Video Retrieval, London, UK, July 2002.

[21] The CYC® Knowledge Base, Cycorp, Inc. website: http://www.cyc.com.

[22] X. S. Zhou and T. S. Huang, Relevance feedback for image retrieval: A comprehensive review, *ACM Multimedia Systems Journal, special issue on CBIR*, in press, 2002.

[23] G. Salton, *Automatic Text Processing*, Addison-Wesley, Reading, MA, 1989.

[24] Y. Rui, T. S. Huang, M. Ortega, and S. Mehrotra, Relevance feedback: A power tool in interactive content-based image retrieval, *IEEE Trans. on Circuits and Systems for Video Technology*, vol. 8, no. 5, pp. 644-655, Sept. 1998.

[25] K. Tieu and P. Viola, Boosting Image Retrieval, in Proc. IEEE Conf. on Computer Vision and Pattern Recognition, Hilton Head Island, SC, June 2000, pp. 228-235.

[26] N. Vasconcelos and A. Lippman, Bayesian Relevance Feedback for Content-based Image Retrieval, in Proc. IEEE Workshop on Content-based Access of Image and Video Libraries, Hilton Head Island, SC, June 2000, pp. 63-67.

[27] X. S. Zhou and T. S. Huang, Small Sample Learning During Multimedia Retrieval Using BiasMap, in Proc. IEEE Conf. on Computer Vision and Pattern Recognition, Hawaii, Dec. 2001.

[28] X. S. Zhou and T. S. Huang, Unifying keywords and contents in image retrieval, *IEEE Multimedia*, April-June Issue, 2002.

[29] T. Cox and M. Cox, *Multidimensional Scaling.* London: Chapman & Hall, 1994.

[30] S. Roweis and L. Saul, Nonlinear dimensionality reduction by locally linear embedding, *Science*, vol. 290, no.5500, pp. 2323-2326, Dec. 2000.

[31] V. Vapnik, *The Nature of Statistical Learning Theory.* New York: Springer, 1995.

[32] A. W. M. Smeulders, M. Worring, S. Santini, A. Gupta, and R. C. Jain, Content-based image retrieval at the end of the early years, *IEEE Trans. on Pattern Analysis and Machine Intelligence*, vol. 22, no. 12, pp. 1349-1380, Dec. 2000.

[33] R. W. Picard, T. P. Minka, and M. Szummer, Modeling User Subjectivity in Image Libraries, in Proc. IEEE Int'l Conf. on Image Processing, Lausanne, Sept. 1996, pp. 777-780.

[34] P. Comon, Independent component analysis – a new concept?, *Signal Processing*, vol. 36, pp. 287-314, 1994.

[35] R. Schettini, G. Ciocca, and I. Gagliardi, Content-based Color Image Retrieval with Relevance Feedback, in Proc. IEEE Int'l Conf. on Image Processing, Kobe, 1999.

[36] I. J. Cox, M. Miller, T. Minka, and P. Yianilos, An Optimized Interaction Strategy for Bayesian Relevance Feedback, in Proc. IEEE Conf. on Computer Vision and Pattern Recognition, Santa Barbara, CA. June 1998, pp. 553-558.

[37] S. D. MacArthur, C. E. Brodley, and C. Shyu, Relevance Feedback Decision Trees in Content-based Image Retrieval, in Proc. of the IEEE Workshop on Content-Based Access to Image and Video Libraries, June 2000, pp. 68-72.

[38] Y. Freund and R. E. Schapire, A short introduction to boosting, *Journal of Japanese Society for Artificial Intelligence*, vol. 14, no. 5, pp. 771-780, Sept. 1999.

[39] Y. Chen, X. S. Zhou, and T. S. Huang, One-class SVM for Learning in Image Retrieval, in Proc. IEEE Int'l Conf. on Image Processing, Thessaloniki, Greece, Oct. 2001.

[40] P. Hong, Q. Tian, and T. S. Huang, Incorporate Support Vector Machines to Content-based Image Retrieval with Relevance Feedback, in Proc. IEEE Int'l Conf. on Image Processing, Vancouver, Canada, Sept. 2000.

[41] C. Meilhac and C. Nastar, Relevance Feedback and Category Search in Image Databases, in Proc. IEEE Int'l Conf. on Multimedia Computing and Systems, Florence, Italy, June 1999, pp. 512-517.

[42] Y. Wu, Q. Tian, and T. S. Huang, Discriminant EM Algorithm with Application to Image Retrieval, in Proc. IEEE Conf. on Computer Vision and Pattern Recognition, Hilton Head Island, SC, June, 2000, pp. 222-227.

[43] R. O. Duda and P. E. Hart, *Pattern Classification and Scene Analysis*. New York: John Wiley & Sons, Inc., 1973.

[44] J. Friedman, Regularized discriminant analysis, *Journal of American Statistical Association*, vol. 84, no. 405, pp. 165-175, 1989.

[45] X. S. Zhou and T. S. Huang, A Generalized Relevance Feedback Scheme for Image Retrieval, in Proc. SPIE Photonics East Vol. 4210: Internet Multimedia Management Systems, Boston, MA, Nov. 2000.

[46] B. Scholkopf, C. Burges, and A. Smola, *Advances in Kernel Methods: Support Vector Learning*, Cambridge, MA: The MIT Press, 1999.

[47] B. Scholkopf, A. Smola, and K.-R. Muller, Nonlinear component analysis as a kernel eigenvalue problem, *Neural Computation*, vol. 10, pp. 1299-1319, 1998.

[48] G. Baudat and F. Anouar. Generalized discriminant analysis using a kernel approach, *Neural Computation*, vol. 12, no. 10, pp. 2385-2404, 2000.

[49] S. Mika, G. Ratsch, and K.-R. Muller, A mathematical programming approach to the kernel Fisher algorithm, in *Advances in Neural Information Processing Systems (NIPS)*, vol. 13, T. Leen, T. Dietterich, and V. Tresp, Eds., Cambridge, MA: The MIT Press, 2001, pp. 591-597.

[50] X. S. Zhou, Content-based access of image and video data, Ph.D. dissertation, University of Illinois, Urbana, IL, USA, 2002.

31

COST EFFECTIVE AND SCALABLE VIDEO STREAMING TECHNIQUES

Kien A. Hua and Mounir Tantaoui

School of Electrical Engineering and Computer Science
University of Central Florida
Orlando, Florida, USA
`kienhua@cs.ucf.edu,tantaoui@cs.ucf.edu`

1. INTRODUCTION

Video on demand (VOD) is a key technology for many important applications such as home entertainment, digital libraries, electronic commerce, and distance learning. A VOD system allows geographically distributed users to play back any video from a large collection stored on one or more servers. Such a system may also support VCR-like interactions such as fast forward, fast rewind, jump forward, jump backward, and pause. To accept a client request, the VOD server must allocate enough resources to guarantee a jitter-free playback of the video. Such resources include storage and network I/O bandwidth. Sufficient storage bandwidth must be available for continuous transfer of data from storage to the network interface card (NIC), which in turn needs enough bandwidth to forward the stream to remote clients. Due to the high bandwidth requirement of video streams (e.g., 4 megabits/second for MPEG-2 videos), server bandwidth determines the number of clients the server is able to support simultaneously [18]. The simplest VOD system dedicates one video stream for each user (*Unicast*). Obviously, this approach is very expensive and not scalable.

To support a large number of users, requests made to the same video can be batched together and serviced with a single stream using *multicast*. This is referred to as *Batching*. This solution is quite effective since applications typically follow the 80-20 rule. That is, 20% of the data are requested 80% of the time. Since majority of the clients request popular videos, these clients can share the video streams and significantly reduce the demand on server bandwidth. A potential drawback of this approach is the long service delay due to the batching period. A long batching period makes the multicast more efficient, but would result in a long wait for many clients. Some may decide to renege on their service request. On the other hand, a batching period too short would defeat the purpose of using multicast. This scheme also has the following limitation. A single video stream is not adaptable to clients with different receiving capability. Supporting VCR-like interactivity is also difficult in this environment.

In this chapter we present several cost-effective techniques to achieve scalable video streaming. We describe a typical architecture for video-on-demand streaming in Section 2. In Sections 3 and 4, we discuss periodic broadcast techniques and multicast techniques, respectively. A new communication paradigm called *Range Multicast* is introduced in Section 5. Techniques to handle VCR interactions are presented in Section 6. Section 7 describes other techniques that deal with user heterogeneity. Finally, we summarize this chapter in Section 8.

2. VOD SYSTEM ARCHITECTURE

A server channel is defined as a unit of server capacity (i.e., server bandwidth and computing resources) required to support a continuous delivery of video data. The number of channels a server can have typically depends on its bandwidth. These channels are shared by all clients. Their requests are queued at the server, and served according to some scheduling policy when a free channel becomes available. When a service is complete, the corresponding channel is returned to the pool to serve future requests. When multicast is used for video delivery, a channel is allocated to load the video from storage and deliver it to a group of clients simultaneously as shown in Figure 31.1. The communication tree illustrates the one-to-many delivery mechanism. At the receiving end, video data are either sent to the video player to be displayed or temporarily stored in a disk buffer for future display.

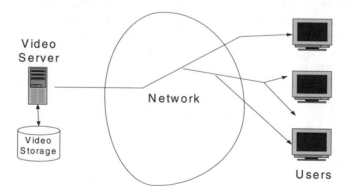

Figure 31.1 A multicast delivery.

2.1 SERVER ARCHITECTURE

A typical architecture of a video server is illustrated in Figure 31.2. The *Coordinator* is responsible for accepting requests from users. To deliver a video, the Coordinator dispatches a *Data Retrieval Handler* to load the data blocks[1] from disk, and a *Video Delivery Handler* to transmit these blocks to the clients. Data retrieved from storage are first staged in a streaming buffer. The *Video Delivery Handler* feeds on this buffer, and arranges each data block into a packet. The header of the packet contains the location of the block in the video file. This information serves as both the timestamp and the sequence number

---

[1] A block is the smallest unit for disk access.

for the client to order the data blocks for correct display. The *Directory Manager* maintains a video directory that keeps information about the videos currently in delivery, such as video title, the multicast address of the channel currently in use, and other important characteristics of the video. The system administrator, through a *graphical user interface* (GUI), can perform various administrative works such as adding or removing a video from the database.

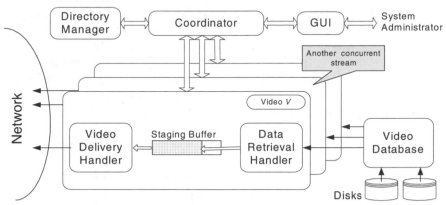

Figure 31.2 Server architecture.

2.2 CLIENT ARCHITECTURE

The components of a typical client software are illustrated in Figure 31.3. The *Coordinator* is responsible for coordinating various server activities. The main role of the *Directory Explorer* is to maintain an up-to-date directory of the videos. To request service, the user can select a video from this catalogue. This action sends a command to the *Coordinator*, which in turn activates the *Loader* to receive data, and the *Video Player* to render the video onto the screen. The Loader and the Video Player communicate through a staging buffer. In some system, the incoming stream can also be saved to disk for future use, or to support VCR-like interactions.

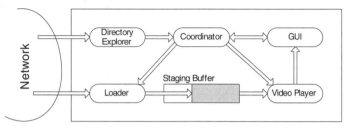

Figure 31.3 Client architecture.

3. PERIODIC BROADCAST TECHNIQUES

Periodic Broadcast is a highly scalable solution for streaming popular videos. In this environment, a video is segmented into several segments, each repeatedly broadcast on a dedicated channel. A client receives a video by tuning to one or more appropriate channels at a time to download the data. The communication protocol ensures that the broadcast of the next video segment is available to the client before the playback of the current segment runs out. Obviously, this scheme is highly scalable. The system can serve a very large community of users with minimal server bandwidth. In fact, the bandwidth requirement is independent of the number of users the system is designed to support. A limitation of this approach is its non-zero service delay. Since each client cannot begin the playback until the next occurrence of the first segment, its broadcast period determines the worst service latency. Many periodic broadcast techniques have been designed to keep this delay, or the size of the first segment, as small as possible to provide near-on-demand services. These techniques can be classified into two major categories. The first group, called *Server-Oriented Approach*, includes techniques that reduce service latency by increasing server bandwidth. Methods in the second category, called *Client-Oriented Approach*, improve latency by requiring more client bandwidth. We discuss these two approaches in this section.

To facilitate the following discussion, we assume a video v of L seconds long. A portion of the server bandwidth, B Mbits/sec, is allocated to v. This dedicated bandwidth is organized into K logical channels by time multiplexing. In other words, we repeatedly broadcast the data segments of v on K channels. The playback rate of v is b Mbits/sec.

3.1 SERVER-ORIENTED APPROACH: INCREASING SERVER BANDWIDTH

We first focus on techniques that allow the users to reduce service latency by increasing only server bandwidth. The rationale for this approach is that server bandwidth, shared by a large community of users, contributes little to the overall cost of the VOD environment. Researchers in this camp argue that this solution is much less expensive than the Client-Oriented Approach which demands each user to equip with substantial client bandwidth, e.g., using a T1 line instead of DSL or a cable modem.

3.1.1 Staggered Broadcasting

Staggered Broadcasting [8][9] is the earliest and simplest video broadcast technique. This scheme staggers the starting times for the video v evenly across K channels. In other words, if the first channel starts broadcasting video v at the playback rate b at time t_0, the second channel starts broadcasting the same video at time $t_0 + L/K$, the third channel at time $t_0 + 2*L/K$, and so on. The difference in the starting times, L/K, is called the *phase offset*. Since a new stream of video v is started every phase offset, it is the longest time any client needs to wait for this video.

Another way to implement the Staggered Broadcasting scheme is as follows. Video v can be fragmented into K segments (S_1, S_2, S_3, .., S_K) of equal size, each of length L/K. Each channel C_i, $1 \leq i \leq K$, repeatedly broadcasts segment S_i at the playback rate b. A client requesting video v tunes to channel C_1 and waits for the beginning of segment S_1. After downloading segment S_1, the client switches to channel C_2 to download S_2 immediately. This process is repeated for the subsequent data segments until segment S_K is downloaded from channel C_k.

The advantage of Staggered Broadcasting is that clients download data at the playback rate. They do not need extra storage space to cache the incoming data.

This simple scheme, however, scales only linearly with increases to the server bandwidth. Indeed, if one wants to cut the client waiting time by half, one has to double the number of channels allocated to the video. This solution is very demanding on server bandwidth. In the following, we present more efficient techniques that can reduce service latency exponentially with increases in server bandwidth.

3.1.2 Skyscraper Broadcasting

In *Skyscraper Broadcasting* [18], the server bandwidth of B Mbit/sec is divided into $\lfloor B/b \rfloor$ logical channels of bandwidth b. Each video is fragmented into K data segments. The size of segment S_n is determined using the following recursive function:

$$f(n) = \begin{cases} 1 & n = 1, \\ 2 & n = 2 \text{ or } 3, \\ 2 \cdot f(n-1) + 1 & n \bmod 4 = 0, \\ f(n-1) & n \bmod 4 = 1, \\ 2 \cdot f(n-1) + 2 & n \bmod 4 = 2, \\ f(n-1) & n \bmod 4 = 3. \end{cases} \tag{31.1}$$

Formula (31.1) expresses the size of each data segment in term of the size of the first segment. Expanding this formula gives us the following series referred to as the *broadcast series*:

$$[1, \ 2, \ 2, \ 5, \ 5, \ 12, \ 12, \ 25, \ 25, \ \ldots \ ]$$

That is, if the size of the first data segment is D_1, the size of the second and third segments are $2 \cdot D_1$, the fourth and fifth are $5 \cdot D_1$, sixth and seventh are $12 \cdot D_1$, and so forth. This scheme limits the size of the biggest segments to W units or $W \cdot D_1$. These segments stack up to a skyscraper, thus the name *Skyscraper Broadcasting*. W is called the *width* of the skyscraper.

Skyscraper Broadcasting repeatedly broadcasts each segment on its dedicated channel at the playback rate b. Adjacent segments having the same size form an *odd* or *even group* depending on whether their sizes are odd or even, respectively. Thus, the first segment is an odd group by itself; the second and third segments form an even group; the fourth and fifth form an odd group; the sixth and seventh form an even group; and so on. To download the video, each client employs two concurrent threads - an *Odd Loader* and *an Even Loader*. They download the odd groups and the even groups, respectively. When a loader reaches the first W-segment, the client uses only this loader to download the remaining W-segments sequentially to minimize the requirement on client buffer space.

Figure 31.4 gives an example of the Skyscraper Broadcasting scheme, where three clients x, y, and z requested the video v just before time slots 5, 10, and 11, respectively. The segments downloaded by each of the three clients are filled with a distinct texture. Let us focus on Client x whose segments are black. Its Odd Loader and Even Loader start downloading the first and second segments, respectively, at the beginning of the fifth time slot. When the second segment is exhausted, the Even Loader switches to download the third segment on Channel 3. Similarly, when the first segment is exhausted, the Odd Loader turns to download first the fourth and then the fifth segments. After the Even Loader has finished downloading the third segment on Channel 3, this loader tunes into Channel 6 to wait for the next occurrence of segment 6 at the beginning of time slot 13. If W is set to 12 in this example, this client will continue to use the Even Loader to download the remaining W-segments. The playback timing of the downloaded segments is illustrated at the bottom of Figure 31.4. We note that each segment is available for download before it is required for the playback.

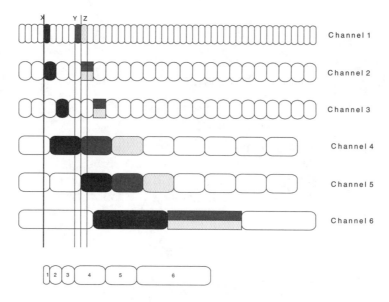

Figure 31.4 Skyscraper downloading scheme.

The worst waiting time experienced by any client, which is also the size of the first segment D_1, is given by the following formula:

$$D_1 = \frac{L}{\sum_{i=1}^{K} \min(\ f(i), W\ )}$$

The major advantage of this approach is the fixed requirement on client bandwidth regardless of the desired service latency. To achieve better service latency, one needs only add server bandwidth. This additional cost is usually negligible because the access latency can be reduced at an exponential rate; and server resources are shared by a very large user community. In practice, it is difficult to provide near-on-demand services using Staggered Broadcasting. Skyscraper technique addresses this problem efficiently. As an example, using 10 channels for a 120-minute video, Staggered Broadcasting has a maximum waiting time of 12 minutes while it is less than one minute for Skyscraper Broadcasting. Adding only a few more channels can further reduce this waiting time.

3.1.3 Client-Centric Approach

The *Client-Centric Approach* (CCA) [16] is another periodic broadcast technique that allows one to improve service delay by adding only server resources once the client capability has been determined. As in the *Skyscraper* technique CCA divides server bandwidth into K logical channels; each repeatedly broadcasts a distinct video segment. The fragmentation function is given in Formula 31.2, where the parameter c denotes the maximum number of channels each client can tune into at one time to download simultaneously c segments:

$$f(n) = \begin{cases} 1 & n=1, \\ 2 \cdot f(n-1) & n \bmod(c) \neq 1, \\ f \cdot (n-1) & n \bmod(c) = 1. \end{cases} \tag{31.2}$$

CCA can be viewed as a generalization of the Skyscraper technique in that each transmission group can have more than two segments, and the number of data loaders is not limited to two as in Skyscraper Broadcasting. CCA enables applications to exploit the available client bandwidth. This approach, however, is not the same as the Client-Oriented Approach presented in Section 3.2, which, given a fixed client bandwidth, cannot improve the service delay by adding only server resources.

Similar to Skyscraper Broadcasting, CCA also limits the sizes of larger data fragments to W. At the client end, the reception of the video is done in terms of *transmission groups*. The video segments are grouped into g transmission groups where $g = \lceil K/c \rceil$. Therefore, each group has c segments except for the last group. To receive the video, a client uses c loaders; each can download data at the playback rate b. When a loader L_i finishes downloading segment S_i in $group_j$, this loader switches to download segment S_{i+c} in $group_{j+1}$. Since segments in the same group (of different sizes) are downloaded at the same time, continuity within the same group is guaranteed. Furthermore, since the size of the last segment in $group_j$ is always the same size as the first segment of the next $group_{j+1}$, continuity across group boundaries is also guaranteed. Comparing to the Skyscraper scheme, CCA uses extra client bandwidth to further reduce the access latency. For L=120 minutes (video length), K=10 channels, and c=3 loaders, CCA cut the access latency by half.

3.1.4 Striping Broadcasting

Striping Broadcasting [26] employs a 2D data fragmentation scheme as follows. Let N be the index of the first W-segment. The size of each segment i, denoted by L_i, is determined using the following equation:

$$L_i = \begin{cases} 2^{i-1} \cdot L_1 & i \in [2, N-1], \\ 2^{N-1} \cdot L_1 & i \in [N, K]. \end{cases}$$

That is, the size of the first *N-1* segments increases geometrically, while the sizes of the other segments (i.e., W-segments) are kept the same. Each of the W-segments is further divided into two equally sized fragments called *stripes*. Compared to Skyscraper technique, Striping Broadcasting employs a faster segment-size progression. This results in a smaller first segment, and therefore better service latency. Since the size of all segments must be equal to the size of the entire video, we can compute the worst access delay as follows:

$$L_1 = \frac{L}{(K-N+2)\cdot 2^{N-1}}$$

To deliver a video, the server periodically broadcasts each of the first N-1 segments, at the playback rate, on its own channel. Each stripe of the W-segments is also periodically broadcast at half the playback rate. To allow the client to download the W-segments as late as possible to save buffer space, phase offsets are employed in the broadcast. Let D_i denote the phase offset for segment i where $1 \le i < N$; and D_{i1} and D_{i2} represent the phase offsets for the first and second stripes, respectively, of segment i where $N \le i \le K$. The phase offsets are calculated as follows:

$$\begin{pmatrix} D_1 \\ D_i \\ D_{i1} \\ D_{i2} \end{pmatrix} = \begin{cases} 0 & \\ 2^{i-2}\cdot L_1 & i\in [2, N-1] \\ 2^{N-2}\cdot L_1 & i\in [N, K] \\ 2^{N-1}\cdot L_1 & i\in [N, K] \end{cases}$$

A broadcast example using the non-uniform phase delay is shown in Figure 31.5. The server broadcasts an index at the beginning of each occurrence of the first segment to inform the client when to download the remaining segments.

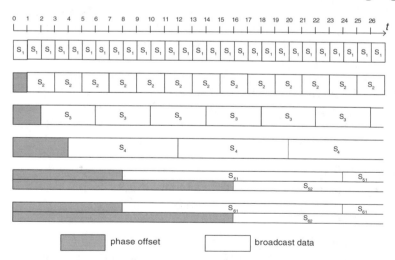

Figure 31.5 Broadcast schedule when N=5 and K=6.

The client uses three concurrent threads to download the video from up to three channels simultaneously. The first loader first downloads the first segment as soon as possible. All three loaders then download the remaining segments in the order specified in the broadcast index.

Striping Broadcasting betters Skyscraper Broadcasting in terms of service latency. The former also requires less client buffer space. As an example, to limit the service delay under 24 seconds for a one-hour video, Striping Broadcasting requires the client to cache no more than 15% the size of the video while Skyscraper Broadcasting requires 33%. Skyscraper Broadcasting, however, is less demanding on client bandwidth since it uses only two concurrent download threads.

A near-video-on-demand system based on Striping Broadcasting was implemented at the University of Central Florida [26]. The prototype runs on Microsoft Windows operating system. The server software lets the content publisher select and specify suitable parameters for videos to be broadcast (see Figure 31.6). Once the selection is done, the publisher can multicast the video directory to a selected multicast group.

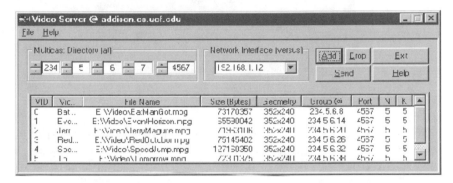

Figure 31.6 Server software: main window.

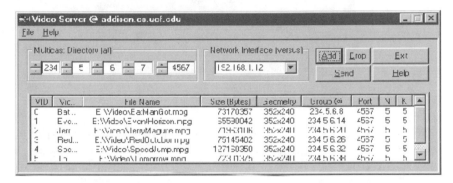

Figure 31.7 Client software: JukeBox.

A user wishing to use the service selects the desired video through a client software called *JukeBox* (see Figure 31.7). In response, a video player and a control panel pop up (see Figure 31.8); and the selected video is subsequently played out. The client software is built on Microsoft DirectShow and currently supports MPEG-1 system files containing both video and audio tracks. Other file formats supported by DirectShow can also be supported without much modification to the software.

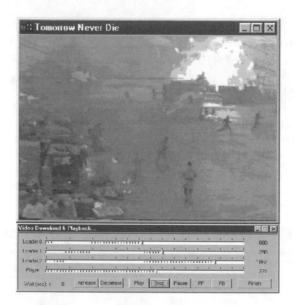

Figure 31.8 Playback of a video.

Experiments with the prototype on a local area network were conducted. Four video clips, each over 5 minutes long, were repeatedly broadcast in these experiments. The number of jitters observed was very small (less than three per playback). This problem was due to packet loss, not a result of the broadcast technique. Each jitter was very brief. The playback quality was comparable to that of commercial streaming systems. The service delays were measured to be less than 13 seconds.

3.2 CLIENT-ORIENTED APPROACH: INCREASING CLIENT BANDWIDTH

All the broadcast techniques, discussed so far, aim at enabling the users to improve service latency by adding only server bandwidth. In this subsection, we discuss techniques that require increases to both server and client bandwidth in order to improve system performance.

3.2.1 Cautious Harmonic Broadcasting

Cautious Harmonic Broadcasting [22] partitions each video into K equally sized segments. The first channel repeatedly broadcasts the first segment S_1 at the playback rate. The second channel alternatively broadcasts S_2 and S_3 at half the playback rate. Each of the remaining segments S_i is repeatedly broadcast on its dedicated channel at $1/(i-1)$ the playback rate. Although this scheme uses many channels to deliver a video, the total bandwidth grow slowly following the harmonic series, typically adding up to only $5b$ or $6b$.

A client plays back a video by downloading all the segments simultaneously. This strategy has the following drawbacks:

- The client must match the server bandwidth allocated to the longest video. The requirement on client bandwidth is therefore very high making the overall system very expensive.

- Improving access delay requires adding bandwidth to both server and client bandwidth. This makes system enhancement very costly.

- Since the client must receive data from many channels simultaneously (e.g., 240 channels are required for a 2-hour video if the latency is kept under 30 seconds), a storage subsystem with the capability to move their read heads fast enough to multiplex among so many concurrent streams would be very expensive.

3.2.2 Pagoda Broadcasting

As in the Harmonic scheme, *Pagoda Broadcasting* [23] also divides each video into equally sized segments. However, it addresses the problems of having too many channels by allowing segments to share channels. The number of segments allocated to each channel is determined according to the following series:

$$\{1, 3, 5, 15, 25, 75, 125, ... \}$$

Since the numbers grow very fast in the above series, this scheme requires much less channels than in the Harmonic Broadcasting. The segments assigned to a channel do not have to be consecutive. An example is given in Figure 31.9. It shows that 19 segments are repeatedly broadcast on four channels. The idea is to broadcast each segment at least once every period in term of the Harmonic Broadcast scheme.

Each channel broadcasts data at the playback rate. A client requesting a video downloads data from all the channels simultaneously. The benefit of Pagoda Broadcasting is to achieve a low server bandwidth requirement as in Harmonic Broadcast without the drawback of using many channels. Pagoda Broadcasting, however, has not addressed the high demand on client bandwidth. For instance, keeping service delay less than 138 seconds for a 2-hour video requires each client to have a bandwidth five times the playback rate. Furthermore, performance enhancement or adding a longer video to the database may require the clients to acquire additional bandwidth. In comparison with the Server-Oriented Approach, presented in Section 3.1, the savings in server bandwidth under Pagoda Broadcasting is not worth the significantly more expensive client hardware. Nevertheless, this scheme can be used for local-scale applications, such as corporate training and campus information systems, which rely on an intranet for data transmission.

1st Stream	S_1	S_1	S_1	S_1	S_1	S_1	S_1	S_1	S_1	S_1
2nd Stream	S_2	S_4	S_2	S_5	S_2	S_4	S_2	S_5	S_2	S_4
3rd Stream	S_3	S_6	S_8	S_3	S_7	S_9	S_3	S_6	S_8	S_3
4th Stream	S_{10}	S_{11}	S_{12}	S_{13}	S_{14}	S_{15}	S_{16}	S_{17}	S_{18}	S_{19}

Figure 31.9 A broadcast example in Pagoda Broadcast.

4. MULTICAST TECHNIQUES

In a Multicast environment, videos are not broadcast repeatedly, but multicast on demand. In this section, we first discuss the *Batching* approach, and then present a more efficient technique called *Patching*.

4.1 BATCHING

In this environment, users requesting the same video, within a short period of time, are served together using the multicast facility. Since there could be several such batches of pending requests, a scheduler selects one to receive service according to some queuing policy. Some scheduling techniques for the Batching approach are as follows:

- **First-Come-First-Serve** [2][10]: As soon as some server bandwidth becomes free, the batch containing the oldest request with the longest waiting time is served next. The advantage of this policy is its fairness. Each client is treated equally regardless of the popularity of the requested video. This technique, however, results in a lower system throughput because it may serve a batch with few requests, while another batch with many requests is pending.

- **Maximum Queue Length First** [2]: This scheme maintains a separate waiting queue for each video. When server bandwidth becomes available, this policy selects the video with the most number of pending requests (i.e., longest queue) to serve first. This strategy maximizes server throughput. However, it is unfair to users of less popular videos.

- **Maximum Factored Queued Length First** [2]: This scheme also maintains a waiting queue for each video. When server resource becomes available, the video v_i selected to receive service is the one with the longest queue weighted by a factor $1/\sqrt{f_i}$, where f_i denotes the access frequency or the popularity of v_i. This factor prevents the system from always favoring more popular videos. This scheme presents a reasonably fair policy without compromising system throughput.

The benefit of periodic broadcast is limited to popular videos. In this sense, Batching is more general. It, however, is much less efficient than periodic broadcast in serving popular videos. A hybrid of these two techniques, called *Adaptive Hybrid Approach* (AHA), was presented in [17] offering the best performance. This scheme periodically assesses the popularity of each video based on the distribution of recent service requests. Popular videos are repeatedly broadcast using Skyscraper Broadcasting while less demanded ones are served using Batching. The number of channels used for periodic broadcast depends on the current mix of popular videos. The remaining channels are allocated to batching. The AHA design allows the number of broadcast channels allocated to each video to change in time without disrupting the on-going playbacks.

4.2 PATCHING

All the techniques discussed so far can only provide near-on-demand services. A multicast technique that can deliver videos truly on demand is desirable. At first sight, making multicast more efficient and achieving zero service delay seem to be two conflicting goals. We discuss in this subsection one such solution called *Patching*.

The *patching* technique [7][15][24] allows a new client to join an on-going multicast and still receive the entire video stream. This is achieved by receiving the missed portion in a separate *patching stream*. As this client displays data arriving in the patching stream, it caches the multicast stream in a buffer. When the patching stream terminates, the client switches to playback the prefetched data in the local buffer while the multicast stream continues to arrive. This strategy is illustrated in Figure 31.10. The diagram on the left shows a Client *B* joining a multicast *t* time units late, and must receive the first portion of the video through a patching stream. The right diagram shows *t* time units later. Client *B* has now just finished the patching stream, and is switching to play back the multicast data previously saved in the local buffer.

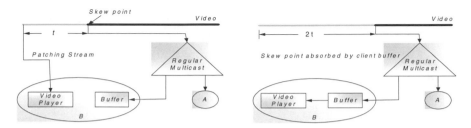

Figure 31.10 Patching.

In a simple Patching environment, a new client can be allowed to join the current multicast if the client has enough buffer space to absorb the time skew; otherwise a new multicast is initiated. This strategy is too greedy in sharing the multicasts, and may result in many long patching streams. A better patching technique should allow only clients arriving within a *patching period* to join the current multicast. The appropriate choice of this period is essential to the performance of Patching. If the patching period is too big, there are many long patching streams. On the other hand, a small patching period would result in many inefficient multicasts. In either case, the benefit of multicast diminishes. A technique for determining the optimal patching period was introduced in [6]. A patching period is optimal if it results in minimal requirement on server bandwidth. In [6], clients arriving within a patching period are said to form a multicast group. This scheme computes D, the mean amount of data transmitted for each multicast group, and τ, the average time duration of a multicast group. The server bandwidth requirement is then given by D/τ which is a function of the patching period. The optimization can then be done by finding the patching period that minimizes this function. It was shown in [6] that under different request inter-arrival times, the optimal patching period is between 5 and 15 minutes for a 90-minute video.

5. BEYOND CONVENTIONAL MULTICAST AND BROADCAST

It has been recognized that standard multicast is inadequate for VOD applications. Patching addresses this drawback by supplementing each multicast stream with patching streams. In this subsection, we consider a new communication paradigm called *Range Multicast* (RM) [19].

The Range Multicast technique employs software routers placed at strategic locations on the *wide-area network* (WAN), and interconnected using unicast paths to implement an overlay structure to support the range multicast paradigm. As a video stream passes through a sequent of such software router nodes on the delivery path, each caches the video data into a fixed-size FIFO buffer. Before it is full (i.e., the first frame is still resident), such a buffer can be used to provide the entire video stream to subsequent clients requesting the same video.

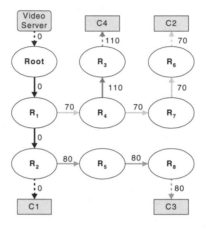

Figure 31.11 A range multicast example.

A range multicast example is given in Figure 31.11. The root node is the front-end note for the server to communicate with the rest of the overlay network. We assume that each node has enough buffer space to cache up to 100 video blocks for the video stream passing through. The label on each link indicates the time stamp of a particular service requested by some client. For instance, label "0" indicates that Client C_1 requests the video at time 0. For simplicity, we assume that there is no transmission delay. Thus, C_1 can make a request at time 0 and receive the first block of the data stream instantaneously. Figure 31.11 illustrates the following scenario. At time 0, a node C_1 requests a video v. Since no node currently caches the data, the root has to allocate a new stream to serve C_1. As the data go toward C_1, all the non-root nodes along the way, R_1 and R_2, cache the data in their local buffer. At time 70, client C_2 requests the same video v. At this time R_1 has not dropped the first video block from its buffer, and can serve C_2. All the nodes along the path from the serving node R_1 to C_2 (i.e., R_4, R_7, and R_6) are asked to cache the video. Similarly, client C_3 requesting video v at time 80 can receive the service from node R_2 which still holds the first video block. At time 101, R_1 and R_2 cast out the first video block in their cache. Nevertheless, client C_4 can still join the multicast group at time 110 by receiving the full service from node R_4 which still has the first block of the video. In this example, four clients join a range multicast at different times, but still receive the entire video stream. This is achieved using only one server stream. This characteristic is not possible with traditional multicast.

Range Multicast is a shift from conventional thinking about multicast where every receiver must obtain the same data packet at all times. In contrast, the data available from a range multicast at any time is not a "data point," but a contiguous segment of the video. In other words, a sliding window over the video is multicast to a range of receivers. This unique characteristic is important to video-on-demand applications in two ways:

- Better service latency: Since clients can join a multicast at their specified time instead of the multicast time, the service delay is zero.

- Less demanding on server bandwidth: Since clients can join an existing range multicast, the server does not need to multicast as frequently to save server resources.

Server bandwidth often dictates the performance of a VOD system. Range multicast enables such a system to scale beyond the physical limitation of the video server.

6. SUPPORTING VCR-LIKE INTERACTIONS

An important functionality of a VOD system is to offer VCR-like interactivity. This section first introduces various interaction functions that a VOD system can support, then it discusses some techniques that handle such functions in both the multicast and the broadcast environments.

5.1 FORMS OF INTERACTIVITY

Videos are sequences of continuous frames. The frame being displayed to the monitor by a client is referred to as the current frame or the *play point*. The client watching the video can change the position of the play point to render another frame that is called the *destination point*. There exist two types of interactive functions, *continuous interactive functions* and *discontinuous functions*. In the first type of interactivity, a client continuously advances the play point one frame at a time at a speed that is different from the playback rate. The frames between the original play point and the destination point are all rendered to the monitor. Such interactions include the fast-forward and fast-reverse actions. In the second type of interactions, the user instantaneously changes the play point to another frame; such interactions are the jump forward and the jump backward interactions.

Furthermore, the interactions are divided into *forward* and *backward interactions*. Let's denote by b the playback rate, Δt the time taken by the interaction, and Δl the video length of the interaction in time unit. The Δl, in other words, is the difference between the current play point and the destination point. The parameter x [11] that represents all type of interactions is then defined as:

$$x = \frac{\Delta l / \Delta t}{b}$$

Table 31.1 gives the potential forward and backward interactions with the possible values of the parameter x.

A play action is therefore an action where its length over its duration equals to the playback rate. A fast forward action is when the length over the duration of the action is greater than the playback rate. In the case of a jump action, since the duration is zero, then the parameter x is infinite. A pause action is of length zero; therefore the parameter x is zero. Fast reverse, jump backward, and play backward are the exact opposite of fast forward, jump forward, and play respectively.

Table 31.1 Backward and forward interactions.

Action	Backward Interactions				Forward Interactions		
	Jump Backward	Fast Reverse	Play Backward	Pause	Play	Fast Forward	Jump Forward
X	-∞	[-x1,-1)	-1	0	1	(1,x1)	+∞

Figure 31.12 shows the position of the play point in the client buffer system after some interactions. The incoming frames fill up the buffer from one side while frames from the other side are discarded. After a play action of 2 frames the play point does not change its position. After a pause action, the play point follows the frame the user had paused in. After a fast forward action, the play point gradually moves toward the newest frame. After a jump backward, the play point instantly jumps towards an older frame in the buffer. From the figure, because a pause action is toward oldest frames, it is considered as a backward action. To complete the canonical world of interactions, there are two more actions, the *slow forward* with a parameter $x \in (0,1)$ and the *slow backward* with $x \in (-1,0)$. Both interactions are backward interaction for the same reason as the pause interaction.

Another parameter, *duration ratio* representing the degree of interactivity, is defined as the portion of time a user spends performing interactions over the time spent on normal play. *Blocked interactions* are interactions that cannot complete due to some limitation of resources.

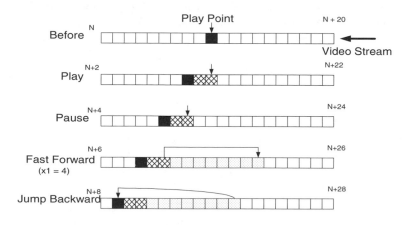

Figure 31.12 Play point position in the client buffer.

6.2 VCR INTERACTION TECHNIQUES

This section presents some techniques that handle VCR-interactions in both the multicast and broadcast environment. These techniques try to take full advantage of the data sharing of multicast delivery while providing VCR functionality to each client individually. When a user requests an interaction, the VOD system not only should provide the interaction but it should also guarantee a jitter-free playback when the user resumes from the interaction. A good interaction technique is the one that provides a small degree of blocked interactions with a high duration ratio.

In the multicast environment, the *interactive multicast* technique [3][4] schedules the delivery of the video in some specified time slots; such time slots could range from 6 seconds to 30 minutes. The video is delivered only if some requests are pending in the queuing server. Requests during the same time slot form one multicast group. Because the period of a slot is known, users that jump forward or backward change their multicast group. A user jumping forward changes its multicast group to a group that has been scheduled earlier, while a user jumping backward changes to a multicast group scheduled after its current group. If such multicast group does not exist, an emergency channel is issued to provide the service. The continuous actions are handled in the client buffer, which contains a multicast threshold. When incoming frames exceed the threshold, a multicast change occurs. Therefore, the continuous actions are performed within the buffer's limitation.

This technique provides only limited discontinuous actions; for example if the period of the slots are 5 minutes, users cannot jump 7 minutes. Furthermore, using emergency channels to accommodate users degrades the performance of the multicast paradigm.

The *split and merge* [20] protocol offers an efficient way to provide VCR interactivity. This protocol uses two types of stream, an S stream for normal playback of the video, and an I stream to provide interactive actions. When a user initiates an interactive operation, the user splits from its multicast group, and uses an I stream. After the completion of the interaction, the user merges from the I stream to a new multicast group. The split and merge scheme uses synchronized buffers located at the access node to merge the clients smoothly. If the merging fails, a new S stream is allocated to the user. The drawback of the split and merge protocol is that it requires an excessive number of I channels. Because all interactions are served using I streams, a blocked interaction is queued to be served as a new S stream, thus causing a high blocking rate of VCR interactions and therefore degrading multicast scalability.

The technique proposed in [1] improves the split and merge protocol by disregarding I channels whenever they are not needed. Some interactions do not need an I stream to be performed and can join directly a new multicast. In this technique the allocation of regular streams is independent from user's interaction behavior, and an I channel is only allocated if it is available and the user's buffer is exhausted by the VCR operation.

A typical merging operation in this scheme is composed of three steps. First, some frames in the client buffer are disregarded to free some buffer space for future frames. Second, the client buffer prefetches frames from the targeted S channel while displaying frames from the I channel. After an offset time, the client merges to the targeted channel and releases the I channel.

In the broadcast situation, because the system can support unlimited number of users, using guard or emergency channels to support VCR interactivity violates the intention of using periodic broadcast. In this environment, it was observed in [11][12] that the play point can be maintained at the middle of the video segment currently in the prefetch buffer in order to accommodate interactive actions in either forward or reverse direction equally well. Because, the position of the play point profoundly influences future interactions (see figure 12), keeping it in the middle enhances the chance of their completion. This is accomplished in [11][12] by selectively prefetching the segments to be loaded depending on the current position of the play point. This scheme is called *Active Buffer Management*; in [12] it extends the staggered broadcasting scheme and in [11] it extends the client centric approach periodic broadcast. In general, Active Buffer Management can be set to take advantage of the user behavior. If the user shows more forward actions than backward actions, the play point can be kept near the beginning of the video segment in the buffer, and vice versa.

The technique proposed in [27] improves on the duration ratio of the active buffer management technique by extending the client centric approach periodic broadcast. The *broadcast-based interaction technique* improves on the overall duration ratio by periodically broadcasting a compressed version of the video. Clients watching the compressed segments at the playback rate will have the impression of fast playing the normal video. The broadcast-based interaction technique retains the most desirable property of the broadcast approach, namely unlimited scalability. The technique divides the client buffer space into two parts, one part holds the normal version of the video, and the second part holds the compressed version. The two play points of the two separate buffers are held on the same frame throughout the download of the entire video. When the user initiates a discontinuous action, the play point of the normal buffer fetches the destination frame; if such frame does not exist, the user downloads the appropriate segment of the normal video. However, when a continuous action is performed, the play point renders the next frame in the interactive buffer. Once the continuous interaction resumes the loader fetches the appropriate normal segments to match the content of the destination point.

7. HANDLING RECEIVER HETEROGENEITY

Stream sharing techniques such as broadcast and multicast enable large-scale deployment of VOD applications. However, clients of such applications might use different forms of end devices ranging from simple palmtop personal digital assistants (PDA), to powerful PCs and high-definition television (HDTV) as receptors. Delivering a uniform representation of a video does not take advantage of the high-bandwidth capabilities nor does it adapt to the low-bandwidth limitation of the receivers. A heterogeneous technique that can adapt to a range of receivers is critical to the overall quality of service (QoS) of the VOD system.

One solution to the heterogeneity problem is the use of *layered media formats*. The basic mechanism is to encode the video data as a series of layers; the lowest layer is called the *base layer* and higher layers are referred to as *enhancement layers* [5][21]. By delivering various layers in different multicast groups, a user can individually mould its service to fit its capacity, independently of other users. In the *Receiver-Driven Layered Multicast* technique [21], a user keeps adding layers until it is congested, then drops the higher layer. Hence, clients look for the optimal number of layers by trying to join and leave different multicast groups.

Another approach is the use of *bandwidth adaptors* [13] between the server and receivers. This technique provides means to adapt to the receiving capabilities within the periodic broadcast framework. The main role of the adaptor is the conversion of incoming video segments into segments suitable for broadcast in the downstream at a lower data rate. Since the video arrives repeatedly at the adaptor at a higher speed than the data rate of the broadcast initiated at the adaptor, an *As Late As Possible* caching policy is used to ensure that an incoming segment is cached only if it will not occur again before it is needed for the broadcast in the downstream. Furthermore, an *As Soon As Possible* policy is used to cast out any segment after its broadcast if this segment will occur again before it will be needed for a future broadcast in the downstream. The adaptor, thus, stores only what it needs from a video, but never the video in its entirety. In this environment, clients can be routed to the right adaptor according to their capabilities. The adaptor sending the video data to the clients becomes their server in a transparent fashion. Compared to techniques relying on multi-resolution encoding, a major advantage of the Adaptor approach is that clients with lesser capability can still enjoy the same video with no loss of quality.

A different technique, called *Heterogeneous Receiver-Oriented* (HeRO) Broadcasting [14], proposes a trade-off between waiting time and service delay in a periodic broadcast environment. HeRO is derived from the observation that even though a periodic broadcast might be designed for a particular client bandwidth, a client, depending on its arrival time, might actually need less bandwidth than what the broadcast scheme was intended for. In fact, clients can do this most of the time in a carefully designed broadcast environment. In HeRO, data fragmentation is based on the geometric series $[1, 2, 2^2, ... 2^{K-1}]$, where K is the number of channels allocated to the video. Each channel i periodically broadcasts the segment of size 2^{i-1}. Users with different capacities are constrained to start their download at some specific times. That is, users with high bandwidth can start the download at the next occurrence of the first segment; but others with less bandwidth must wait for some specific time to start the download. This technique is illustrated in Figure 31.13, where the server repeatedly broadcasts 4 segments on the first set of channels. The numbers at the top indicate the minimum number of loaders a client needs to have in order to start the download at the beginning of that specific time slot. Clients with less than this number of loaders will have to wait for a future slot. For instance, a client with two loaders can only start its download at the beginning of time slots 2, 3, 4, 6, or 7, etc. in order to see a continuous playback. This pattern repeats every broadcast period of the largest segment. Two such periods are shown in Figure 31.13. To reduce service latency for less capable clients, HeRO provides the option to broadcast the longer segments on a second channel with a phase offset equal to half their size. The example in Figure 31.13 uses two such channels. We note that clients with only enough bandwidth for one or two loaders now have more possible slots to join the broadcast. For instance, a client with one loader can now join the HeRO broadcast at the beginning of slot 6. We observe that the HeRO approach, unlike the multi-resolution encoding techniques, does not reduce the playback quality of low-bandwidth clients.

8. SUMMARY

Video on demand is certainly a promising technology for many multimedia applications still to come. Unlike traditional data, delivery of a video clip can take up substantial bandwidth for a long period of time. This chapter describes some cost effective and scalable solutions for large-scale deployment of VOD systems.

A key aspect of designing a scalable VOD system is to leverage multicast and broadcast to facilitate bandwidth sharing. For popular videos or if the video database contains only a few videos, a periodic broadcast technique achieves the best cost/performance. This approach, however, cannot deliver videos without some delay. Two alternatives have been considered for addressing this limitation: the *Server-Oriented Approach* (e.g., Skyscraper Broadcast, Striping Broadcast) reduces service delay by increasing server bandwidth, whereas the *Client-Oriented Approach* (e.g., *Cautious Harmonic Broadcast, Pagoda Broadcast*) requires client to equip with significantly more download bandwidth. For wide-area deployment, the *Server-Oriented Approach* is preferred because the cost of server bandwidth can be shared by a very large community of users, and therefore contributes little to the cost of the overall system. *Client-Oriented Approach* is limited to local deployment on an intranet where client bandwidth is abundant (e.g., receiving the video at five times the playback rate). Applications such as corporate training and campus information systems can benefit from these techniques.

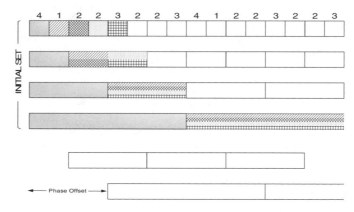

Figure 31.13 A Heterogeneous Receiver-Oriented (HeRO) broadcast example.

For less popular videos, multicast on demand is a better solution than repeatedly broadcasting the videos. Standard multicast, however, makes users wait for the batching period. *Patching* resolves this problem by allowing the clients to join an ongoing multicast while playing back the missing part of the video arriving in a patching stream. For the best performance, a system should use both periodic broadcast and multicast. Techniques, such as AHA, can be used to monitor the popularity of the videos, and apply the best mechanism to deliver them. Range Multicast is a new concept in multicast. It enables clients to join a range multicast at their specified time, and still receive the entire video stream. Since many users actually receive their videos from the network, this new communication paradigm enables the VOD system to scale far beyond the physical limitation of the video server.

VCR-like interaction is a desirable feature for many VOD applications. It provides a convenient environment to browse and search for video content. Techniques such as *Split and Merge* can be used to provide such operations under multicast. For periodic broadcast, the *broadcast-based interaction* technique offers a highly scalable solution. In fact, the amount of server bandwidth required to support interactivity is independent of the number of users currently using this service.

Another important consideration in designing VOD systems is the capability to handle receiver heterogeneity. Multi-resolution encoding techniques provide a good solution for many applications. For those that demand the same high QoS for clients of various capabilities, techniques such as *Bandwidth Adaptor* and HeRO can be used in the periodic broadcast framework. A bandwidth adaptor, as the name implies, receives broadcast data from upstream at a high speed, and broadcasts them to the downstream at a lower rate. Such a device can be placed on the server side to service a wide area of users, or on the client side to support a small group of users. HeRO is a different approach that handles differences in receiving bandwidths using only a single broadcast scheme. This is achieved by indicating in the broadcast when a client with a particular receiving capability can start its download. This solution is simple and effective.

REFERENCES

[1] E. L A. Profeta and K.G. Shin, Providing Unrestricted VCR Functions in Multicast Video-on-Demand Servers, in IEEE Int'l. Conf. on Multimedia Computing Systems (ICMCS'98), Austin, Texas, 1998.

[2] C.C Agarwal, J. L. Wolf, and P.S. Yu, On Optimal Batching Policies for Video on Demand Storage Servers, in Proc. of IEEE ICMCS'96, pp.253-258, Jun. 1996.

[3] K. C. Almeroth and M.H Ammar, The use of multicast delivery to provide a scalable and interactive video-on-demand service, *IEEE Journal of Selected Areas in Communications*, vol. 14, Aug. 1996.

[4] K.C. Almeroth and M. H. Ammar, A Scalable Interactive Video-on-Demand Service Using Multicast Communication, In Proc. of Int'l Conf. on Computer Communication Networks, pp. 292-301,1994.

[5] H. M. Briceo, S. Gortler, and L. McMillan, Naïve-network Aware Internet Video Encoding, in Proc. of the 7th ACM Int'l Multimedia Conf., pp. 251-260, Oct. 1999

[6] Y. Cai and K. A. Hua, Optimizing Patching Performance, in Proc. of SPIE's Conf. on Multimedia Computing and Networking (MMCN'99), pp. 204-246, Jan. 1999.

[7] S. W. Carter and D.D.E. Long, Improving bandwidth efficiency of video-on-demand servers, *Computer Networks and ISDN Systems*, 31(1): 99-111, Mar. 1999.

[8] A. Dan, D. Sitaram, and P. Shahabuddin, Scheduling Policies for an On-demand Video Server with Batching, in Proc. of ACM Multimedia Conference, Oct. 1994.

[9] A. Dan, P. Shahabuddin, D. Sitaram, and D. Towsley, Channel Allocation under batching and VCR control in video-on-demand systems, *Journal of Parallel and Distributed Computing*, 30(2): 168-79, Nov. 1995.

[10]   A. Dan, D. Sitaram, and P. Shahabuddin, Scheduling policies for on demand video server, *Multimedia Systems*, 4(3): 112-121, Jun. 1996.

[11]   Z. Fei, I. Kamel, S. Mukherjee, and M. Ammar, Providing Interactive Functions Through Active Client Buffer Management in Partitioned Video Broadcast, in Proc. of 1st Int'l Workshop on Networked Group Communication, (NGC'99), Nov. 1999.

[12]   Z. Fei, I. Kamal, S. Mukherje, M. Ammar, Providing Interactive Functions for Staggered Multicast Near Video-on-Demand Systems, Proc. of the IEEE Int'l Conf on Multimedia Computing Systems, Jun. 1999.

[13]   K. A. Hua, Olivier Bagouet, and David Oger, Bandwidth adaptors for heterogeneous broadcast-based video-on-demand systems, Technical report.

[14]   K. A. Hua, O. Bagouet, and D. Oger, A Periodic Broadcast Protocol for Heterogeneous Receivers, in SPIE Conf. On Multimedia Computing and Networking. Jan. 2003.

[15]   K. A. Hua, Y. Cai, and S. Sheu, Patching, A Multicast Technique for True Video on Demand Services, in Proc. of ACM Multimedia Conf., Sept. 1998.

[16]   K. A. Hua, Y. Cai, and S. Sheu, Exploiting Client Bandwidth for More Efficient Video Broadcast, in Proc. of the Int'l Conf. on Computer Communication Networks, 1998.

[17]   K. A. Hua, J-H Oh, and K. Vu, An Adaptive Video Multicast Scheme for Varying Workloads, *ACM-Springer Multimedia Systems Journal*, Vol. 8, Issue 4, August 2002, pp. 258-269.

[18]   K. A. Hua and S. Sheu, Skyscraper Broadcasting: A New Broadcasting Scheme for Metropolitan Video-on-Demand Systems, in Proc. of SIGCOMM 97, pp. 89-100, Sept. 1997.

[19]   K. A. Hua, D. A. Tran, and R. Villafane, Caching Multicast Protocol for On-Demand Video Delivery, in the Proc. of ACM/SPIE Conf. on Multimedia Computing and Networking (MMCN 2000), pp. 2-13, Jan. 2000.

[20]   W. Liao and V. O. Li, The split and merge (SAM) protocol for interactive video on demand, *IEEE Multimedia*, vol. 4, pp. 51-62, Oct.-Dec. 1997.

[21]   S. McCanne, V. Jacobson, and M. Vetterli, Receiver-driven Layered Multicast, in ACM SiGCOMM'96, Aug. 1996.

[22]   J.-F. Paris, S. W. Carter, and D. D. E. Long, Efficient Broadcasting Protocols for Video-on-Demand, in Proc. of the int'l Symposium on Modeling, Analysis, and Simulation of Computing and Telecom Systems, pp. 127-32, Jul. 1998.

[23]   J.-F. Paris, A Simple Low-bandwidth Broadcasting Protocol for Video on Demand, in Proc. of the 8th Int'l Conf. on Computer Communications and Networks, 1998.

[24]   S. Sen, L. Gao, J. Rexford, and D. Towsley, Optimal Patching Schemes for Efficient Multimedia Streaming, in Proc. IEEE NOSSDAV'99, June 1999.

[25]   L.-S. Juhn and L.-M. Tseng, Harmonic Broadcasting for video-on-demand service, *IEEE Trans. on Broadcasting*, 43(3): 268-71, Sept. 1997

[26]   S. Sheu and K Hua, Scalable Technologies for distributed multimedia systems, Ph.D. Dissert., SEECS, Univ. of Central Florida, 1999.

[27]   M. Tantaoui, K. A. Hua, and S. Sheu, Interaction in video broadcast, *ACM Multimedia*, Dec. 2002.

32

DESIGN AND DEVELOPMENT OF A SCALABLE END-TO-END STREAMING ARCHITECTURE[1]

Cyrus Shahabi and Roger Zimmermann
Integrated Media Systems Centre
Department of Computer Science
University of Southern California
Los Angeles, California 90089–0781
`[Shahabi,rzimmerm]@usc.edu`

1. INTRODUCTION

In this chapter, we report on the design, implementation and evaluation of a scalable real-time streaming architecture that would enable applications such as news-on-demand, distance learning, e-commerce, corporate training and scientific visualization on a large scale. A growing number of applications store, maintain and retrieve large volumes of real-time data, where the data are required to be available online. We denote these data types collectively as *"continuous media,"* or CM for short. Continuous media is distinguished from traditional textual and record-based media in two ways. First, the retrieval and display of continuous media are subject to real-time constraints. If the real-time constraints are not satisfied, the display may suffer from disruptions and delays termed *hiccups*. Second, continuous media objects are large in size. A two-hour MPEG-2 video with a 4 Megabit per second (Mb/s) bandwidth requirement is 3.6 Gigabytes (GB) in size. Popular examples of CM are video and audio objects, while less familiar examples are haptic, avatar and application coordination data [26].

The first research papers on the design of continuous media servers appeared about a decade ago (e.g. [1, 23]), followed by many papers on this topic during the past decade (here is an example for every year [7, 24, 6, 5, 4, 22, 33, 12, 16, 18, 40]). Our research during this past decade started in early 90's [9] and continued through 2002 [11]. Some of the projects described in these papers resulted in prototype servers, such as Streaming-RAID [35], Oracle Media Server [17], UMN system [15], Tiger [2], Fellini [19], Mitra [13] and RIO [20]. These first generation continuous media servers were primarily focused on the design of different data placement paradigms, buffer management mechanisms and

---

[1] This research has been funded in part by NSF grants EEC-9529152 (IMSC ERC) and IIS-0082826, and unrestricted cash gifts from NCR, Microsoft and Okawa Foundation.

retrieval scheduling techniques to optimize for throughput and/or startup latency time.

There are two major shortcomings with these research prototypes. First, since they were implemented concurrently during the same time frame, each one of them could not take advantage of the findings and proposed techniques of the other projects. For example, UMN, Fellini and Mitra independently proposed a seemingly identical design for their data placement technique based on round-robin assignments of blocks to disk clusters. While this approach was already superior in throughput to RAID striping (used by Streaming-RAID), it resulted in a higher worst case startup latency time. In contrast, RIO's data placement, which is based on random block assignment, is more flexible and results in the same throughput as round-robin with a shorter expected startup latency time. On the other hand, UMN's simple scheduling policy resulted in a good performance without complicating the code as opposed to the constrained scheduling policies of Mitra (round-robin), Fellini (cycle-based) and RIO (round-based). We extended UMN's scheduling (to deadline-driven) and adapted the disk cluster (in Mitra's and Fellini's terms) or *logical volume striping* (in UMN's vocabulary) storage design. We extended RIO's random data placement (to pseudo-random placement for easier bookkeeping and storage scale-up) and instead of the expensive shared-memory architecture of UMN (based on SGI's Onyx), we employed a shared-nothing approach on commodity personal computer hardware.

The second shortcoming is that almost all of these research prototypes were completed before the industry's standardizations for streaming continuous media over the IP networks. Hence, each prototype has its own proprietary media content format, client (and codec) implementation and communication/network protocol. As a matter of fact, some of these prototypes did not even focus on these aspects and never reported on their assumed network and client configurations. They mainly assumed a very fast network with constant-bit-rate media types in their corresponding research publications. Practically, these environment assumptions and the specific content type are not realistic.

Several commercial implementations of continuous media servers are now available in the marketplace. We group them into two broad categories: 1) single-node, consumer oriented systems (e.g., low-cost systems serving a limited number of users) and 2) multi-node, professional oriented systems (e.g., high-end broadcasting and dedicated video-on-demand systems). Table 32.1 lists some examples for both types. RealNetworks, Apple Computer and Microsoft product offerings fit into the first category, while SeaChange and nCUBE offer solutions that are oriented towards the second category. Table 32.1 is a non-exhaustive list summarizing some of the more popular and recognizable industry products.

Commercial systems often use proprietary technology and algorithms; therefore, their design choices and development details are not publicly available and objective comparisons are difficult to achieve.[2] Because these details are not known, it is also unclear as to how much research resulting from academic work have been incorporated into these systems. For example, although a good body

---

[2] One notable exception is Apple Computer, Inc., which has published the source code of its Darwin Streaming Server.

of work has been developed on how to configure a video server (e.g., block size, number of disks) given an application's requirements (e.g., tolerable latency, required throughput), as reviewed in [10], there is no indication that any of the commercial products utilize these results. However, those details that we are aware of are referred to throughout this chapter when comparing them with our design choices.

In this chapter we report on our streaming media architecture called Yima.[3] The focus of our implementation has been on providing a high-performance, scalable system that incorporates the latest research results and is fully compatible with open industry standards.

Table 32.1 A comparison of continuous-media servers. Note that Yima is a prototype system and does not achieve the refinement of the commercial solutions. However, we use it to demonstrate several advanced concepts.

Video Server	Yima	RealNetworks, Microsoft, Apple QuickTime	SeaChange	nCUBE
	Prototype	Consumer Level	Professional Level	
Authoring Suite		Yes	Yes	Yes
MPEG-1 & 2	Yes	Yes[a]	Yes	Yes
MPEG-4	Yes	Yes[b]	Yes	?
HDTV MPEG-2	Yes		?	?
Multi-node Clusters	Yes		Yes	Yes
Multi-node Fault-tolerance[c]	Yes		Yes	Yes
Synchronized streams[d]	Yes			
Selective Retransmissions	Yes	Yes	?	?
Hardware	Commodity PC	Commodity PC	Commodity PC	Proprietary Hypercube
Software	Linux	Windows NT / Mac[e]	Windows NT	Proprietary

[a]These systems support many other codecs that are commonly used in PC/Mac environments.
[b]Supported in Windows Media Player.
[c]Multi-node Fault-Tolerance is defined as the capability of one node taking over from another node without replicating the data.
[d]By synchronized streams we mean the simultaneous playback of multiple independent audio and video streams for, say, panoramic systems.
[e]RealSystem Producer products are also available for several Unix variants.

Specifically, some of the features of the Yima architecture are as follows. It is designed as a completely distributed system (with no single point of failure or bottleneck) on a multi-node, multi-disk platform. It separates physical disks (used to store the data) from the concept of logical disks (used for retrieval scheduling) to support fault tolerance [39] and heterogeneous disk subsystems[4]

---

[3] Yima in ancient Iranian religion is the first man, the progenitor of the human race and son of the sun.

[4] Note that our concept of *logical* disks is very different from the concept of logical volume striping used in the UMN system.

[38]. Data blocks are pseudo-randomly placed on all nodes and the non-deterministic scheduling is performed locally on each node. Yima includes a method to reorganize data blocks in real-time for online adding/removing disk drives [11] (Sec. 4). Also included are a flexible rate-control mechanism between clients and the server to support both variable and constant bit rate media types (see Sec. 3) as well as optimization techniques such as Super-streaming [25]. There are also techniques to resolve stream contentions at the server for ensuring inter-stream synchronization as proposed in [28].

Yima follows open industry standards as proposed by the Internet Streaming Media Alliance (ISMA; www.ism-alliance.org). Content-wise, Yima supports the MPEG-4 file format, MP4, and can stream MPEG-1, MPEG-2, MPEG-4 and Quicktime video formats. It supports the RTP and RTSP communication standards for IP based networks. The clients can be off-the-shelf QuickTime players as well as our own proprietary clients [34] that handle advanced multi-channel audio and video playback and HDTV displays. Although our proprietary clients can support specialized playback of multiple media types, they still follow communication and format standards.

A general description of Yima server and its various clients are provided in [34]. In that paper, we also compared our initial architectural design with the new fully distributed design. In [36] we reported on our experiments in streaming MPEG-4 content from a Yima server located at the USC's campus to a residential location connected via ADSL [36]. In [11], we proposed alternative schemes to redistribute media blocks when disks are added to or removed from the Yima server. These schemes were then compared via a simulation study.

In this chapter, however, we describe the details of our fully distributed architecture and report on several experiments we conducted in real-world settings. We show when different components of a node (i.e., CPU, network and disk) become a bottleneck. We have also implemented the superior redistribution scheme of [11], Randomized SCADDAR, in Yima and evaluated it in these real-world settings. Finally, we describe our novel client-controlled transmission rate smoothing protocol and explain its implementation in Yima. Several experiments have also been conducted to evaluate our smoothing protocol.

The remainder of this chapter is organized as follows. After describing the overall system architecture in Sec. 2 we will elaborate in greater detail on three innovative aspects of Yima: 1) a client-controlled transmission rate smoothing protocol to support variable bit rate media (Sec. 3), 2) efficient online scalability of storage where disks can be added or removed without stream interruption (Sec. 4) and 3) the scalable multi-node architecture with independent, local scheduling and data transmission (Sec. 5). Finally, in Sec. 6, we summarize this chapter and lay out our future plans in this area.

2. SYSTEM ARCHITECTURE

There are two basic techniques to assign the data blocks of a media object, in a load balanced manner, to the magnetic disk drives that form the storage system: in a *round-robin* sequence [3], or in a *random* manner [31]. Traditionally, the round-robin placement utilizes a cycle-based approach for scheduling of resources to guarantee a continuous display, while the random placement utilizes a deadline-driven approach. In general, the round-robin approach provides high throughput with little wasted bandwidth for video objects that are retrieved sequentially. This approach can employ optimized disk scheduling algorithms (such as *elevator* [27]) and object replication and request migration [8] techniques to reduce the inherently high startup latency. The random approach allows for fewer optimizations to be applied, potentially resulting in less throughput. However, there are several benefits that outweigh this drawback, as described in [32], such as 1) support for multiple delivery rates with a single server block size, 2) support for interactive applications and 3) support for data reorganization during disk scaling.

One potential disadvantage of random data placement is the need for a large amount of meta-data: the location of each block must be stored and managed in a centralized repository (e.g., tuples of the form <$node_x$, $disk_y$>). Yima avoids this overhead by utilizing a *pseudo-random* block placement. With pseudo-random number generators, a seed value initiates a sequence of random numbers which can be reproduced by using the same seed. File objects are split into fixed-size blocks and each block is assigned to a random disk. Block retrieval is similar. Hence, Yima needs to store only the seed for each file object, instead of locations for every block, to compute the random number sequence.

The design of Yima is based on a bipartite model. From a client's viewpoint, the scheduler, the RTSP and the RTP server modules are all centralized on a single master node. Yima expands on decentralization by keeping only the RTSP module centralized (again from the client's viewpoint) and parallelizing the scheduling and RTP functions as shown in Figure 32.1. Hence, every node retrieves, schedules and sends data blocks that are stored locally directly to the requesting client, thereby eliminating a potential bottleneck caused by routing all data through a single node. The elimination of this bottleneck and the distribution of the scheduler reduces the inter-node traffic to only control related messages, which is orders of magnitude less than the streaming data traffic. The term "bipartite" relates to the two groups, a server group and a client group (in the general case of multiple clients), such that data flows only between the groups and not between members of a group. Although the advantages of the bipartite design are clear, its realization introduces several new challenges. First, since clients are receiving data from multiple servers, a global order of all packets per session needs to be imposed and communication between the client and servers needs to be carefully designed. Second, an RTSP server node needs to be maintained for client requests along with a distributed scheduler and RTP server for each node. Lastly, a flow control mechanism is needed to prevent client buffer overflow or starvation.

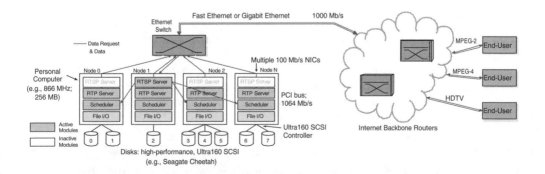

Figure 32.1 The Yima multi-node hardware architecture. Each node is based on a standard PC and connects to one or more disk drives and the network.

Each client maintains contact with one RTSP module for the duration of a session to relay control related information (such as PAUSE and RESUME commands). A session is defined as a complete RTSP transaction for a continuous media stream, starting with the DESCRIBE and PLAY commands and ending with a TEARDOWN command. When a client requests a data stream using RTSP, it is directed to a server node running an RTSP module. For load-balancing purposes each server node may run an RTSP module. For each client, the decision of which RTSP server to contact can be based on either a round-robin DNS or a load-balancing switch. Moreover, if an RTSP server fails, sessions are not lost — instead they are reassigned to another RTSP server and the delivery of data is not interrupted.

In order to avoid bursty traffic and to accommodate variable bitrate media, the client sends slowdown or speedup signals to adjust the data transmission rate from Yima. By periodically sending these signals to the Yima server, the client can receive a smooth flow of data by monitoring the amount of data in its buffer. If the amount of buffer data decreases (increases), the client will issue speedup (slowdown) requests. Thus, the amount of buffer data can remain close to constant to support consumption of variable bitrate media. This mechanism will complicate the server scheduler logic, but bursty traffic is greatly reduced as shown in Sec. 3.

3. VARIABLE BITRATE SMOOTHING

Many popular compression algorithms use variable bitrate (VBR) media stream encoding. VBR algorithms allocate more bits per time to complex parts of a stream and less bits to simple parts to keep the visual and aural quality at near constant levels. For example, an action sequence in a movie may require more bits per second than the credits that are displayed at the end. As a result, different transmission rates may be required over the length of a media stream to avoid starvation or overflow of the client buffer. As a contradictory requirement we would like to minimize the variability of the data transmitted through a network. High variability produces uneven resource utilization and may lead to congestion and exacerbate display disruptions.

To achieve scalability and high resource utilization, both at the server and client sides, it was our goal to reduce the variability of the transmitted data. We designed and implemented a novel technique that adjusts the multimedia traffic based on an end-to-end rate control mechanism in conjunction with an intelligent buffer management scheme. Unlike previous studies [21, 14], we consider multiple signaling thresholds and adaptively predict the future bandwidth requirements. With this *Multi-Threshold Flow Control* (MTFC) scheme, VBR streams are accommodated without a priori knowledge of the stream bitrate. Furthermore, because the MTFC algorithm encompasses server, network and clients, it adapts itself to changing network conditions. Display disruptions are minimized even with few client resources (e.g., a small buffer size).

3.1 APPROACH

MTFC is an adaptive technique that works in a real-world, dynamic environment with minimal prior knowledge of the multimedia streams needing to be served. It is designed to satisfy the following desirable characteristics:

- **Online operation**: This is required for live streaming and it is also desirable for stored streams.

- **Content independence**: An algorithm that is not tied to any particular encoding technique will continue to work when new compression algorithms are introduced.

- **Minimizing feedback control signaling**: The overhead of online signaling should be negligible to compete with offline methods that do not need any signaling.

- **Rate smoothing**: The peak data rate as well as the number of rate changes should be lowered compared with the original, unsmoothed stream. This will improve network transmissions and resource utilization.

The MTFC algorithm incorporates the following: 1) a multi-threshold client buffer model, 2) a consumption prediction model and 3) a rate change Δr computation algorithm. The three components work together as follows. The client playout buffer incorporates multiple thresholds. When the data level sufficiently deviates from the buffer midpoint, a correction message is sent to the server to adjust the sending rate. The sending rate change Δr is computed based on the value of the threshold that was crossed and the estimated data consumption in the near future. MTFC is parameterized in many different ways. For example, thresholds can either be equi-spaced or not (e.g., exponential) and we have tested several consumption prediction algorithms. The full details of the MTFC can be found in [37]

3.2 PERFORMANCE EVALUATION

MTFC is fully implemented in our architecture and we have tested its effectiveness in both LAN and WAN environments. The following is a subset of the test results. Figure 32.2(a) visually illustrates the effectiveness of our MTFC algorithm. Shown are the unsmoothed and the smoothed transmission rate of the movie "Twister" with a playout buffer size $B = 32$ MB, $m = 17$ thresholds and a consumption prediction window size $w_{pred} = 180$ seconds. The variability is clearly reduced as well as the peak rate.

To objectively quantify the effectiveness of our technique with many different parameter sets, we measured the standard deviation of the transmission schedule. Figure 32.2(b) presents the reduction in standard deviation achieved by MTFC, across client buffer sizes ranging from 8 MB to 32 MB and with the number of thresholds ranging from 3 to 17. In all cases, the standard deviation is reduced substantially, by 28-42% for an 8 MB client buffer, 30-47% for a 16 MB client buffer, and 38-59% for a 32 MB client buffer. This figure illustrates what is intuitively clear: an increase in the client buffer size yields smoother traffic. More importantly, it also shows that a higher number of thresholds results in smoother traffic.

4. DATA REORGANIZATION

An important goal for any computer cluster is to achieve a balanced load distribution across all the nodes. Both round-robin and random data placement techniques, described in Sec. 2, distribute data retrievals evenly across all disk drives over time. A new challenge arises when additional nodes are added to a server in order to a) increase the overall storage capacity of the system, or b) increase the aggregate disk bandwidth to support more concurrent streams. If the existing data are not redistributed then the system evolves into a *partitioned* server where each retrieval request will impose a load on only part of the cluster. One might try to even out the load by storing new media files on only the added nodes (or at least skewed towards these nodes), but because future client retrieval patterns are usually not precisely known this solution becomes ineffective.

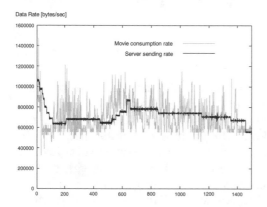

Figure 32.2(a). Real consumption rate versus smoothed server sending rate for a 25 minute segment of a typical movie. The smoothing parameters used are as follows: 32 MB playout buffer size, 17 thresholds and 180 sec. prediction window size.

Figure 32.2(b). Reduction in rate variability of the movie "Twister" with different client buffer sizes and number of thresholds. The transmission schedule becomes smoother as the number of thresholds increases and also with increased buffer size.

Figure 32.2 Smoothing effectiveness of the MTFC algorithm with the movie "Twister."

Redistributing the data across all the disks (new and old) will once again yield a balanced load on all nodes. Reorganizing blocks placed in a random manner requires much less overhead when compared to redistributing blocks placed using a constrained placement technique. For example, with round-robin striping, when adding or removing a disk, almost all the data blocks need to be relocated. More precisely, z_j blocks will move to the old and new disks and $B–z_j$ blocks will stay put as defined in Eq. 1:

$$z_j = B - \frac{B-1}{LCM} \times N_{j-1} - \begin{cases} (B-1 \bmod N_{j-1})+1 & \text{if } (B-1 \bmod LCM) < N_{j-1} \\ N_{j-1} & \text{otherwise} \end{cases} \tag{1}$$

where B is the total number of object blocks, N_j is the number of disks upon the jth scaling operation and LCM is the least common multiple of N_{j-1} and N_j. Instead, with a randomized placement, only a fraction of all data blocks need to be moved. To illustrate, increasing a storage system from four to five disks requires that only 20% of the data from each disk be moved to the new disk. Assuming a well-performing pseudo-random number generator, z_j blocks move to the newly added disk(s) and $B-z_j$ blocks stay on their disks where z_j is approximated as:

$$z_j \approx \frac{|N_j - N_{j-1}|}{\max(N_j, N_{j-1})} \times B \qquad (2)$$

z_j is approximated since it is the theoretical percentage of block moves. z_j should be observed when using a 32-bit pseudo-random number generator and the number of disks and blocks increases. The number of block moves is simulated in Figure 32.3a as disks are scaled one-by-one from 1 to 200. Similarly for disk removal, only the data blocks on the disk that are scheduled for removal must be relocated. Redistribution with random placement must ensure that data blocks are still randomly placed after disk scaling (adding or removing disks) to preserve the load balance.

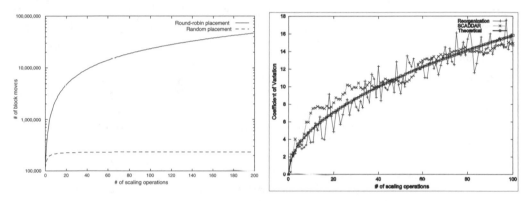

Figure 32.3(a). Block movement (72 GB disks, 256 KB blocks).	Figure 32.3(b). Coefficient of variation.

Figure 32.3 Block movement and coefficient of variation.

As described in Sec. 2, we use a pseudo-random number generator to place data blocks across all the disks of Yima. Each data block is assigned a random number, X, where X mod N is the disk location and N is the number of disks. If the number of disks does not change, then we are able to reproduce the random location of each block. However, upon disk scaling, the number of disks changes and some blocks need to be moved to different disks in order to maintain a balanced load. Blocks that are now on different disks cannot be located using the previous random number sequence; instead a new random sequence must be derived from the previous one.

In Yima we use a composition of random functions to determine this new sequence. Our approach, termed SCAling Disks for Data Arranged Randomly (SCADDAR), preserves the pseudo-random properties of the sequence and results in minimal block movements after disk scaling while it computes the new locations for data blocks with little overhead [11]. This algorithm can support scaling of disks while Yima is online through either an eager or lazy method; the eager method uses a separate process to move blocks while the lazy method moves blocks as they are accessed.

We call the new random numbers, which accurately reflect the new block locations after disk scaling, X_j, where j is the scaling operation (j is 0 for the initial random numbers generated by the pseudo-random number generator). We wish to map X_0 to X_j for every block at every scaling operation, j, such that X_j mod N_j results in the new block locations. We have formulated Eqs. 3 and 4 to compute the new X_j's for an addition of disks and a removal of disks, respectively:

$$X_j = \begin{cases} p\_r(X_{j-1}) \times N_j + r_{j-1} & \text{if } (p\_r(X_{j-1}) \bmod N_j) < N_{j-1} \text{ (a)} \\ p\_r(X_{j-1}) \times N_j + (p\_r(X_{j-1}) \bmod N_j) & \text{otherwise} \quad \text{(b)} \end{cases} \quad (3)$$

$$X_j = \begin{cases} p\_r(X_{j-1}) \times N_j + \text{new}(r_{j-1}) & \text{if } r_{j-1} \text{ is not removed} \quad \text{(a)} \\ p\_r(X_{j-1}) & \text{otherwise} \quad \text{(b)} \end{cases} \quad (4)$$

In order to find X_j (and ultimately the block location at X_j mod N_j) we must first compute X_0, X_1, X_2, ..., X_{j-1}. Eqs. 3 and 4 compute X_j by using X_{j-1} as the seed to the pseudo-random number generator so we can ensure that X_j and X_{j-1} are independent. We base this on the assumption that the pseudo-random function performs well.

We can achieve our goal of a load balanced storage system where similar loads are imposed on all the disks on two conditions: *uniform* distribution of blocks and *random* placement of these blocks. A uniform distribution of blocks on disks means that all disks contain the same (or similar) number of blocks. So, we need a metric to show that SCADDAR achieves a balanced load. We would like to use the uniformity of the distribution as a metric; hence, the standard deviation of the number of blocks across disks is a suitable choice. However, because the averages of blocks per disk will differ when scaling disks, we normalize the standard deviation and use the coefficient of variation (standard deviation divided by average) instead. One may argue that a uniform distribution may not necessarily result in a *random* distribution since, given 100 blocks, the first 10 blocks may reside on the first disk, the second 10 blocks may reside on the second disk, and so on. However, given a *perfect* pseudo-random number generator (one that outputs independent X_j's that are all uniformly distributed between *0* and *R*), SCADDAR is statistically indistinguishable from complete reorganization.[5] This fact can be easily proven by demonstrating a coupling between the two schemes; due to lack of space, we omit the proofs from this chapter version. However, since real life pseudo-random number generators are unlikely to be perfect, we cannot formally analyze the properties of SCADDAR so we use simulation to compare it with complete reorganization instead. Thus, SCADDAR does result in a satisfactory random distribution. For the rest of this section, we use the uniformity of block distribution as an indication of load balancing.

We performed 100 disk scaling operations with 4000 data blocks. Figure 32.3b shows the coefficient of variations of three methods when scaling from 1 to 100 disks (one-by-one). The first curve shows complete block reorganization after adding each disk. Although complete reorganization requires a great amount of block movement, the uniform distribution of blocks is ideal. The second curve

---

[5] This does not mean that SCADDAR and complete reorganization will always give identical results; the two processes are identical in *distribution*.

shows SCADDAR. This follows the trend of the first curve suggesting that the uniform distribution is maintained at a near-ideal level while minimizing block movement. The third curve shows the theoretical coefficient of variation as defined in Def. 32.1. The theoretical coefficient of variation is derived from the theoretical standard deviation of Bernoulli trials. The SCADDAR curve also follows a similar trend as the theoretical curve.

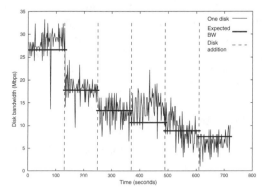
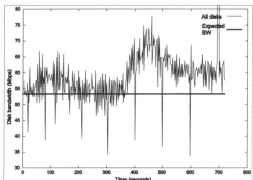

(a). Bandwidth of one disk.   (b). Total disk bandwidth.

Figure 32.4 Disk bandwidth when scaling from 2 to 7 disks. One disk is added every 120 seconds (10 clients each receiving 5.33 Mbps).

Table 32.2 List of terms used repeatedly in this section and their respective definitions.

Term	Definition
N	The number of concurrent clients supported by Yima server
N_{max}	The maximum number of sustainable, concurrent clients
U_{idle}	Idle CPU in percentage
U_{system}	System (or kernel) CPU load in percentage
U_{user}	User CPU load in percentage
B_{aveNet}	Average network bandwidth per client (Mb/s)
B_{net}	Network bandwidth (Mb/s)
B_{disk}	The amount of movie data accessed from disk per second (termed disk bandwidth) (MB/s)
B_{cache}	The amount of movie data accessed from server cache per second (termed cache bandwidth) (MB/s)
$B_{aveNet}[i]$	The B_{aveNet} measured for i-th server node in a multinode experiment
$B_{net}[i]$	The B_{net} measured for i-th server node in a multinode experiment
$B_{disk}[i]$	The B_{disk} measured for i-th server node in a multinode experiment
$B_{cache}[i]$	The B_{cache} measured for i-th server node in a multinode experiment
$R_{\Delta r}$	The number of rate changes per second

Definition 32.1: Theoretical coefficient of variation is a percentage and defined as $\sqrt{\dfrac{D-1}{B}} \times 100$.

When adding disks to a disk array, the average disk bandwidth usage across the array will decrease after data are redistributed using SCADDAR. The resulting load balanced placement leads to more available bandwidth which can be used to support additional streams. We measure the bandwidth usage of a disk within a disk array as the array is scaled. In Figure 32.4a, we show the disk bandwidth of one disk among the array when the array size is scaled from 2 to 7 disks (1 disk added every 120 seconds beginning at 130 seconds). 10 clients are each receiving 5.33 Mbps streams across the duration of the scaling operations. The bandwidth continues to decrease and follows the expected bandwidth (shown as the solid horizontal lines) as the disks are scaled. While the bandwidth usage of each disk decreases, we observe in Figure 32.4b that the total bandwidth of all disks remains fairly level and follows the expected total bandwidth (53.3 Mbps); fluctuations are due to the variable bitrate of the media.

5. SCALABILITY EXPERIMENTS

The goal of every cluster architecture is to achieve close to a linear performance scale-up when system resources are increased. However, achieving this goal in a real-world implementation is very challenging. In this section we present the results of two sets of experiments. First, we compared a single node server with two different network interface transmission speeds: 100 Mbps versus 1 Gbps. In the second set of experiments we scaled a server cluster from 1 to 2 to 4 nodes. We start by describing our measurement methodology. Note that Table 32.2 lists all the terms used in this section.

5.1 METHODOLOGY OF MEASUREMENT

One of the challenges when stress-testing a high-performance streaming media server is the potential support of a large number of clients. For a realistic test environment, these clients should not only be simulated, but rather should be real viewer programs that run on various machines across a network. To keep the number of client machines manageable we actually ran several client programs on a single machine. Since decompressing multiple MPEG-2 compressed DVD-quality streams requires a high CPU load, we changed our client software to not actually decompress the media streams. Such a client is identical to a real client in every respect, except that it does not render any video or audio. Instead, this *dummy* client consumes data according to a movie trace data file, which was the pre-recorded consumption behaviour of a real client with respect to a particular movie. Thus, by changing the movie trace file, each dummy client can emulate a DVD stream (5 Mbps, VBR), a HDTV stream (20 Mbps, CBR) or an MPEG-4 stream (800 Kbps, VBR). For all the experiments in this section, we chose trace data from the DVD movie "Twister" as the consumption load. The dummy client used the following smoothing parameters: 16 MB playout buffer, 9 thresholds and 90-second prediction window. For each experiment, we started clients in a staggered manner (one after the other every 2 to 3 minutes).

On the server side, we recorded the following statistics every 2 seconds: CPU load (u_{idle}, u_{system} and u_{user}) disk bandwidth (B_{disk}), cache bandwidth (B_{cache}), $R_{\Delta r}$, the total network bandwidth (B_{net}) for all clients, the number of clients served and B_{aveNet}. On the client side, the following statistics were collected every second: the stream consumption rate (this could be used as the input movie trace data for dummy clients), the stream receiving rate, the amount data in the buffer, the time and value of each rate change during that period and the sequence number for each received data packet.

Our servers were run without any admission control policies because we wanted to find the maximum sustainable throughput using multiple client sessions. Therefore, client starvation would occur when the number of sessions increased beyond a certain point. We defined that threshold as the maximum number of sustainable, concurrent sessions. Specifically, this threshold marks the point where certain server system resources reach full utilization and becomes a bottleneck, such as the available network bandwidth, the disk bandwidth and the CPU load. Beyond this threshold, the server is unable to provide enough data for all clients.

We first describe the Yima server performance with two different network connections, and we then evaluate Yima in a cluster scale-up experiment.

5.2 NETWORK SCALE-UP EXPERIMENTS

Experimental setup: We tested a single node server with two different network connections: 100 Mb/s and 1 Gb/s Ethernet. Figure 32.1 illustrates our experimental setup. In both cases, the server consisted of a single Pentium III 933 MHz PC with 256 MB of memory. The PC is connected to an Ethernet switch (model Cabletron 6000) via a 100 Mb/s network interface for the first experiment and a 1 Gb/s network interface for the second experiment. Movies are stored on a 73 GB Scagate Cheetah disk drive (model ST373405LC). The disk is attached through an Ultra2 low-voltage differential (LVD) SCSI connection that can provide 80 MB/s throughput. Red-Hat Linux 7.2 is used as the operating system. The clients are based on several Pentium III 933 MHz PCs, which are connected to the same Ethernet switch via 100 Mb/s network interfaces. Each PC could support 10 concurrent MPEG-2 DVD dummy clients (5.3 Mbps for each client). The total number of client PCs involved was determined by the number of clients needed. For both experiments, dummy clients were started every 2 to 3 minutes and use the trace data file pre-recorded from a real DVD client playing the movie "Twister" (see Figure 32.2a).

Experimental results: Figure 32.5 shows the server measurement results for both sets of experiments (100 Mbps, 1 Gbps) in two columns. Figures 32.5(c) and (d) present B_{aveNet} with respect to the number of clients, N. Figure 32.5(c) shows that, for a 100 Mb/s network connection, B_{aveNet} remains steady (between 5.3 and 6 Mbps) when N is less than 13; after 13 clients, B_{aveNet} decreases steadily and falls below 5.3 Mbps (depicted as a dashed horizontal line), which is the average network bandwidth of our test movie. Note that the horizontal dashed line intersects with the B_{aveNet} curve at approximately 12.8 clients. Thus, we consider 12 as the maximum number of clients, N_{max}, supportable by a 100 Mb/s networking interface. An analoguous result can be observed in Figure 32.5(d), indicating that $N_{max} = 35$ with a 1 Gb/s network connection.

Figures 32.5(a) and (b) show the CPU utilization as a function of N for 100 Mb/s and 1 Gb/s network connections. Both figures contain two curves: u_{system} and $u_{system} + u_{user}$. As expected, the CPU load (both u_{system} and u_{user}) increases steadily as N increases. With the 100 Mb/s network connection, the CPU load reaches its maximum at 40% with 12 clients, which is exactly N_{max} suggested by Figure 32.5(c) (vertical dashed line). Similarly, for 1Gb/s, the CPU load levels off at 80% where $N_{max} = 35$ clients. Note that in both experiments, u_{system} accounts for more than 2/3 of the maximum CPU load.

Yima implements a simple yet flexible caching mechanism in the file I/O module (Figure 32.1). Movie data are loaded from disks as blocks (e.g., 1 MB). These blocks are organized into a shared block list maintained by the file I/O module in memory. For each client session, there are at least two corresponding blocks on this list. One is the block currently used for streaming, and the other is the prefetched, next block. Some blocks may be shared because the same block is used by more than one client session simultaneously. Therefore, a session counter is implemented for each block. When a client session requests a block, the file I/O module checks the shared block list first. If the block is found, then the corresponding block counter will be incremented and the block made available; otherwise, the requested block will be fetched from disk and added to the shared block list (with its counter set to one). We define the cache bandwidth, B_{cache}, as the amount of data accessed from the shared block list (server cache) per second.

Figures 32.5(e) and (f) show B_{disk} and B_{cache} as a function of N for 100 Mb/s and 1 Gb/s network connections. In both experiments, the $B_{disk} + B_{cache}$ curves increase linearly until N reaches its respective N_{max} (12 for 100 Mb/s and 35 for 1 Gb/s), and they level off beyond those points. For the 100 Mb/s network connection, $B_{disk} + B_{cache}$ level off at around 8.5 MBps, which equals the 68 Mb/s peak rate, B_{net}, in Figure 32.5(i) with $N = N_{max}$. Similarly, for the 1 Gb/s network connection, B_{disk} and B_{cache} level off at 25 MBps, which corresponds to the 200 Mb/s maximum, B_{net}, in Figure 32.5(j) with $N = N_{max} = 35$. In both cases, B_{cache} contributes little to $B_{disk} + B_{cache}$ when N is less than 15. For $N > 15$, caching becomes increasingly effective. For example, with 1 Gb/s network connection, B_{cache} accounts for 20% of 30% to $B_{disk} + B_{cache}$ with N between 35 and 40. This is because for higher N, the probability that the same cached block is accessed by more than one client increases. Intuitively, caching is more effective with large N.

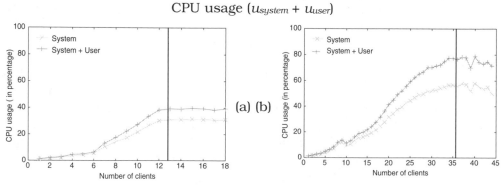

100 Mbps Network Interface   **1 Gbps Network Interface**

CPU usage ($u_{system} + u_{user}$)

(a) (b)

Avg. network bandwidth (B_{aveNet})

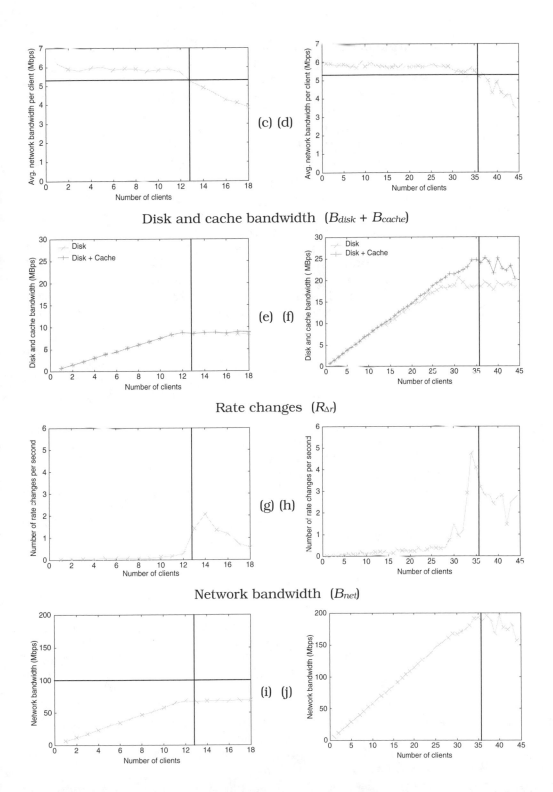

Figure 32.5 Yima single node server performance with 100 Mbps versus 1 Gbps network connection.

Figures 32.5(i) and (j) show the relationship of B_{net} and N for both network connections. Both figures nicely complement Figures 32.5(e) and (f). With the 100 Mbps connection, B_{net} increases steadily with respect to N until it levels off at 68 Mbps with N_{max} (12 clients). For the 1 Gb/s connection, the result is similar except that B_{net} levels off at 200 Mbps with $N = 35$ (N_{max} for 1 Gb/s setup). Notice that the horizontal dashed line in Figure 32.5(i) represents the theoretical bandwidth limit for the 100 Mb/s setup.

Figures 32.5(g) and (h) show $R_{\Delta r}$ with respect to N for 100 Mb/s and 1 Gb/s network connections. Both figures suggest a similar trend: there exists a threshold T where, if N is less than T, $R_{\Delta r}$ is quite small (approximately 1 per second); otherwise, $R_{\Delta r}$ increases significantly to 2 for 100 Mb/s connection and 5 for the 1 Gb/s connection. With the 100 Mb/s setup, T is reached at approximately 12 clients. For the 1 Gb/s case, the limit is 33 clients. In general, T roughly matches N_{max} for both experiments. Note that in both cases, for $N > T$, at some point $R_{\Delta r}$ begins to decrease. This is due to client starvation. Under these circumstances such clients send a maximum stream transmission rate change request. Because this maximum cannot be increased, no further rate changes are sent.

Note that in both the 100 Mb/s and 1 Gb/s experiments, N_{max} is reached when some system resources become a bottleneck. For the 100 Mb/s setup, Figure 32.5(a) and Figure 32.5(e) suggest that the CPU and disk bandwidth are not the bottleneck, because neither of them reaches more than 50% utilization. On the other hand, Figure 32.5(i) indicates that the network bandwidth, B_{net}, reaches approximately 70% utilization for $N = 12$ (N_{max} for 100 Mb/s setup), and hence is most likely the bottleneck of the system. For the 1 Gb/s experiment, Figure 32.5(f) and Figure 32.5(j) show that the disk and network bandwidth are not the bottleneck. Conversely, Figure 32.5(b) shows that the CPU is the bottleneck of the system because it is heavily utilized ($u_{system} + u_{user}$ is around 80%) for $N = 35$ (N_{max} for the 1 Gb/s setup).

5.3 SERVER SCALE UP EXPERIMENTS

Experimental setup: To evaluate the cluster scalability of Yima server, we conducted 3 sets of experiments. Figure 32.1 illustrates our experimental setup. The server cluster consists of the same type of rack-mountable Pentium III 866 MHz PCs with 256 MB of memory. We increased the number of server PCs from 1 to 2 to 4, respectively. The server PCs are connected to an Ethernet switch (model Cabletron 6000) via 100 Mb/s network interfaces. The movies are striped over several 18 GB Seagate Cheetah disk drives (model ST118202LC, one per server node). The disks are attached through an Ultra2 low-voltage differential (LVD) SCSI connection that can provide 80 MB/s throughput. RedHat Linux 7.0 is used as the operating system and the client setup is the same as in Sec. 5.2.

Experimental results: The results for a single node server have already been reported in Section 5.2. So, we will not repeat them here, but refer to them where appropriate. Figure 32.6 shows the server measurement results for the 2 node and 4 node experiments in two columns.

Figures 32.6(c) and (d) present the measured bandwidth B_{aveNet} as a function of N. Figure 32.6(c) shows two curves representing the two nodes: $B_{aveNet}[1]$ and $B_{aveNet}[1] + B_{aveNet}[2]$. Similarly, Figure 32.6(d) shows four curves: $B_{aveNet}[1]$, $B_{aveNet}[1] + B_{aveNet}[2]$, $B_{aveNet}[1] + B_{aveNet}[2] + B_{aveNet}[3]$ and $B_{aveNet}[1] + B_{aveNet}[2] +$

$B_{aveNet}[3]$ + $B_{aveNet}[4]$. Note that each server node contributes roughly the same portion to the total bandwidth B_{aveNet}, i.e., 50% in case of the 2 node system and 25% for the 4 node cluster. This illustrates how well the nodes are load balanced within our architecture. Recall that the same software modules are running on every server node, and the movie data blocks are evenly distributed by the random data placement technique. Similarly to Figures 32.5(c) and (d), the maximum number of supported clients can be derived as N_{max} = 25 for 2 nodes and N_{max} = 48 for 4 nodes. Including the previous results from 1 node (see 100 Mb/s experimental results in Figure 32.5) with 2 and 4 nodes, the maximum number of client streams N_{max} are 12, 25 and 48 respectively, which represents an almost ideal linear scale-up.

Figures 32.6(a) and (b) show the average CPU utilization on 2 and 4 server nodes as a function of N. In both figures, u_{system} and u_{system} + u_{user} are depicted as two curves with similar trends. For 2 nodes the CPU load (u_{system} + u_{user}) increases gradually from 3% with N = 1 to approximately 38% with N = 25 (N_{max} for this setup), and then levels off. With 4 nodes, the CPU load increases from 2% with N = 1 to 40% with N = 48 (N_{max} for this setup).

Figures 32.6(e) and (f) show B_{disk} + B_{cache}. The 4 curves presented in Figure 32.6(c) cumulatively show the disk and cache bandwidth for 2 nodes: $B_{disk}[1]$, $B_{disk}[1]$ + $B_{cache}[1]$, $B_{disk}[2]$ + $B_{disk}[1]$ + $B_{cache}[1]$ and $B_{disk}[2]$ + $B_{cache}[2]$ + $B_{disk}[1]$ + $B_{cache}[1]$. The curves exhibit the same trend as shown in Figures 32.5(e) and (f) for a single node. B_{disk} + B_{cache} reach a peak value of 17 MB/s with N = N_{max} for the 2 node setup and 32 MB/s for the 4 node experiment. Note that B_{disk} + B_{cache} for 4 nodes is nearly doubled compared with 2 nodes, which is double that of the 1 node setup. In both cases, each server contributes approximately the same portion to the total of B_{disk} + B_{cache}, illustrating the balanced load in the Yima cluster. Furthermore, similar to Figures 32.5(e) and (f), caching effects are more pronounced with large N in both the 2 and 4 node experiments.

Figures 32.6(i) and (j) show the achieved network throughput B_{net}. Again, Figures 32.6(i) and (j) nicely complement Figures 32.6(e) and (f). For example, the peak rate, B_{net}, of 136 Mb/s for the 2 node setup is equal to the 17 MB/s peak rate of B_{disk} + B_{cache}. Each node contributes equally to the total served network throughput.

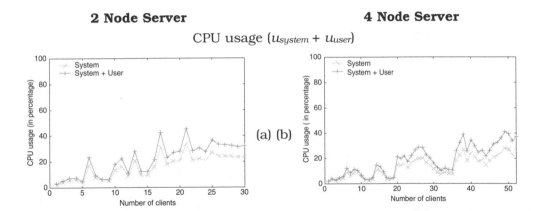

(a) (b)

2 Node Server   **4 Node Server**

CPU usage (u_{system} + u_{user})

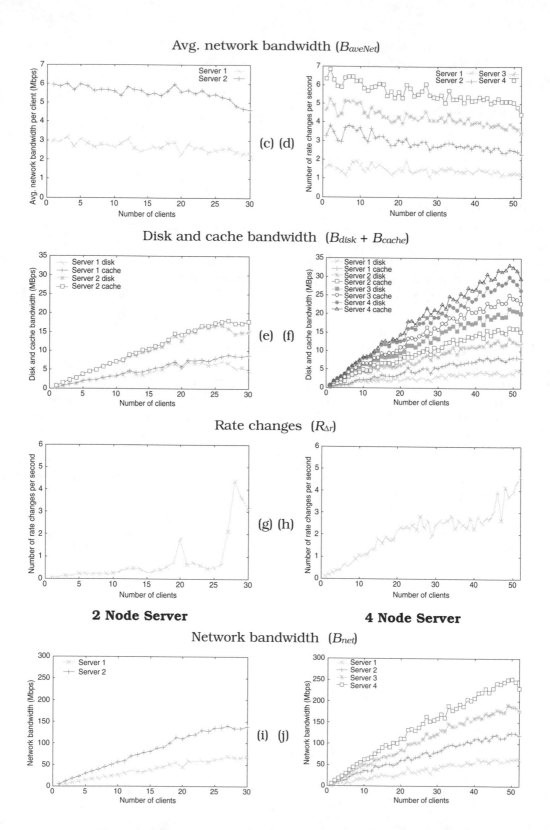

Figure 32.6 Yima server 2 node versus 4 node performance.

Finally, Figures 32.6(g) and (h) show the number of rate changes, $R_{\Delta r}$, that are sent to the server cluster by all clients. Similarly to the 1 node experiment, for the 2 node server $R_{\Delta r}$ is very small (approximately 1 per second) when N is less than 26, and increases significantly above this threshold. For the 4 node experiment, a steady increase is recorded when N is less than 26; after that it remains constant at 2.5 for N between 27 and 45, and finally $R_{\Delta r}$ increases for N beyond 45. Note that for all experiments, with $N < N_{max}$, the rate change messages $R_{\Delta r}$ generate negligible network traffic and server processing load. Therefore, our MTFC smoothing technique (see Sec. 3) is well suited for a scalable cluster architecture.

Overall, the experimental results presented here demonstrate that our current architecture scales linearly to four nodes while at the same time achieves an impressive performance on each individual node. Furthermore, the load is nicely balanced and remains such, even if additional nodes or disks are added to the system (with SCADDAR). We expect that high-performance Yima systems can be built with 8 and more nodes. When higher performing CPUs are used (beyond our 933 MHz Pentium IIIs) each node should be able to eventually reach 300 to 400 Mb/s. With such a configuration almost any currently available network could be saturated (e.g., 8 x 300 Mb/s = 2.4 Gb/s effective bandwidth).

6. CONCLUSIONS AND FUTURE WORK

Yima is a second generation scalable real-time streaming architecture, which incorporates results from first generation research prototypes, and is compatible with industry standards. By utilizing a pseudo-random block assignment technique, we are able to reorganize data blocks efficiently when new disks are added or removed while in a normal mode of operation. Yima has a rate-control *mechanism* that can be utilized by a control *policy* such as Super-streaming [25], to speed-up and slowdown streams. We showed the usefulness of this mechanism in supporting variable-bitrate streams. Finally, the fully distributed design of Yima yields a linear scale up in throughput. We conducted several experiments to verify and evaluate the above design choices in realistic setups.

Our experiments demonstrate graceful scale up of Yima with SCADDAR as disks are added to the storage subsystem. We showed that average load across all disks has uniformly been decreasing as we add disks. The experiments also showed the effectiveness of the MTFC smoothing technique in providing a hiccup-free display of variable-bitrate streams. We showed that the technique is flexible and lightweight enough so that by increasing the number of thresholds a smoother traffic is achieved. Finally, we demonstrated that with Yima single node setup, we can support up to 12 streams each with a 5.3 Mbps bandwidth requirement before the 100 Mbps network card becomes the bottleneck. We installed Quicktime™ on the same hardware and tweaked the software to support the same media format; we could only support 9 streams with Quicktime. The superiority of Yima is attributed to our optimized lightweight scheduler and RTP/RTSP servers.

To demonstrate that the bottleneck was the network card with the single node Yima server, we upgraded the card to a 1 Gbps card and achieved up to 35 streams on a single node before the CPU became the bottleneck. We observed a linear scale-up in throughput with 2 node and 4 node configurations of Yima. A total of 48 streams has been supported on the 4 node configuration when each server carries a 100 Mbps network card.

We intend to extend Yima in three ways. First, we have co-located a 4-node Yima server at one of our industry partner's data centers during 2001. Currently we have a server located across the continental United States that is connected to our campus via Internet 2. We plan to stream high-resolution audio and video content to our campus from distributed sources as part of a demonstration event. This environment provides us with a good opportunity to not only perform several intensive tests on Yima but also collect detailed measurements on hiccup and packet-loss rates as well as synchronization differences. Second, we plan to extend our experiments to support distributed clients from more than one Yima server. We already have four different hardware setups hosting Yima. By co-locating two Yima server clusters at off-campus locations and the other two servers in different buildings within our campus, we have an initial setup to start our distributed experiments. We have some preliminary approaches to manage distributed continuous media servers [29] that we would like to incorporate, experiment with and extend. Finally, we performed some studies on supporting other media types, in particular the haptic [30] data type. Our next step would be to store and stream haptic data.

REFERENCES

[1]   D. Anderson and G. Homsy, A continuous media I/O server and its synchronization, *IEEE Computer*, October 1991.

[2]   W. J. Bolosky, J. S. Barrera, R. P. Draves, R. P. Fitzgerald, G. A. Gibson, M. B. Jones, S. P. Levi, N. P. Myhrvold, and R. F. Rashid. The Tiger Video Fileserver, in 6th Workshop on Network and Operating System Support for Digital Audio and Video, Zushi, Japan, April 1996.

[3]   S. Berson, S. Ghandeharizadeh, R. Muntz, and X. Ju, Staggered Striping in Multimedia Information Systems, in Proceedings of the ACM SIGMOD International Conference on Management of Data, 1994.

[4]   E. Chang and H. Garcia-Molina, Reducing Initial Latency in a Multimedia Storage System, in Proceedings of IEEE International Workshop on Multimedia Database Management Systems, 1996.

[5]   C. S. Freedman and D. J. DeWitt, The SPIFFI Scalable Video-on-Demand System, in Proceedings of the ACM SIGMOD International Conference on Management of Data, pp. 352–363, 1995.

[6]   C. Fedrighi and L. A. Rowe, A Distributed Hierarchical Storage Manager for a Video-on-Demand System, in Storage and Retrieval for Image and Video Databases II, IS&T/SPIE Symp. on Elec. Imaging Science and Tech., pp. 185–197, 1994.

[7]   D. J. Gemmell and S. Christodoulakis, Principles of Delay Sensitive Multimedia Data Storage and Rtrieval, *ACM Trans. Information Systems*, 10(1):51–90, Jan. 1992.

[8]   S. Ghandeharizadeh, S. H. Kim, W. Shi, and R. Zimmermann, On Minimizing Startup Latency in Scalable Continuous Media Servers, in Proceedings of Multimedia Computing and Networking 1997, pp. 144–155, February 1997.

[9]   S. Ghandeharizadeh and C. Shahabi, Management of Physical Replicas in Parallel Multimedia Information Systems, in Proceedings of the

Foundations of Data Organization and Algorithms (FODO) Conference, October 1993.

[10] S. Ghandeharizadeh and C. Shahabi, Distributed Multimedia Systems, in John G. Webster, Editor, *Wiley Encyclopedia of Electrical and Electronics Engineering*, John Wiley and Sons Ltd., 1999.

[11] A. Goel, C. Shahabi, S.-Y. D. Yao, and R. Zimmermann, SCADDAR: An Efficient Randomized Technique to Reorganize Continuous Media Blocks, in Proceedings of the 18th International Conference on Data Engineering, February 2002.

[12] L. Gao and D. F. Towsley, Supplying Instantaneous Video-on-Demand Services Using Controlled Multicast, in ICMCS, Vol. 2, pp. 117–121, 1999.

[13] S. Ghandeharizadeh, R. Zimmermann, W. Shi, R. Rejaie, D. Ierardi, and T.W. Li, Mitra: A Scalable Continuous Media Server, *Kluwer Multimedia Tools and Applications*, 5(1):79–108, July 1997.

[14] J.Y. Hui, E. Karasan, J. Li, and J. Zhang, Client-Server Synchronization and Buffering for Variable Rate Multimedia Retrievals, *IEEE Journal of Selected Areas in Communications*, *14(1)*, pp. 226–237, January 1996.

[15] J. Hsieh, J. Liu, D. Du, T Ruwart, and M. Lin, Experimental Performance of a Mass Storage System for Video-on-Demand, *Special Issue of Multimedia Systems and Technology of Journal of Parallel and Distributed Computing (JPDC)*, 30(2):147–167, November 1995.

[16] P. W.K. Lie, J. C.S. Lui, and L. Golubchik, Threshold-Based Dynamic Replication in Large-Scale Video-on-Demand Systems, *Multimedia Tools and Applications*, 11(1):35–62, 2000.

[17] A. Laursen, J. Olkin, and M. Porter, Oracle Media Server: Providing Consumer Based Interactive Access to Multimedia Data, in Proceedings of the ACM SIGMOD International Conference on Management of Data, pp. 470–477, 1994.

[18] S.H. Lee, K.Y. Whang, Y.S. Moon, and I.Y. Song, Dynamic Buffer Allocation in Video-on-Demand Systems, in Proceedings of the ACM SIGMOD International Conference on Management of Data, 2001.

[19] C. Martin, P. S. Narayan, B. Ozden, R. Rastogi, and A. Silberschatz, The Fellini Multimedia Storage Server, in *Multimedia Information Storage and Management*, Soon M. Chung, Editor, Chapter 5, Kluwer Academic Publishers, Boston, August 1996.

[20] R. Muntz, J. Santos, and S. Berson, RIO: A Real-time Multimedia Object Server, *ACMSigmetrics Performance Evaluation Review*, 25(2):29–35, September 1997.

[21] M. Mielke and A. Zhang, A Multi-Level Buffering and Feedback Scheme for Distributed Multimedia Presentation Systems, in Proceedings of Seventh International Conference on Computer Communications and Networks (IC3N'98), Lafayette, Louisiana, October 1998.

[22] G. Nerjes, P. Muth, and G. Weikum, Stochastic Service Guarantees for Continuous Data on Multi-Zone Disks, in Proceedings of the Principles of Database Systems Conference, pp. 154–160, 1997.

[23] V.G. Polimenis, The Design of a File System that Supports Multimedia, Technical Report TR-91-020, ICSI, 1991.

[24] S. Ramanathan and P. V. Rangan, Feedback Techniques for Intra-Media Continuity and Inter-Media Synchronization in Distributed Multimedia Systems, *The Computer Journal*, 36(1):19–31, 1993.

[25]   C. Shahabi and M. Alshayeji, Super-streaming: A New Object Delivery Paradigm for Continuous Media Servers, *Journal of Multimedia Tools and Applications*, 11(1), May 2000.

[26]   C. Shahabi, G. Barish, B. Ellenberger, N. Jiang, M. Kolahdouzan, A. Nam, and R. Zimmermann, Immersidata Management: Challenges in Management of Data Generated within an Immersive Environment, in Proceedings of the International Workshop on Multimedia Information Systems, October 1999.

[27]   M. Seltzer, P. Chen, and J. Ousterhout, Disk Scheduling Revisited, in Proceedings of the 1990 Winter USENIX Conference, pp. 313–324, Washington, DC, Usenix Association, 1990.

[28]   C. Shahabi, S. Ghandeharizadeh, and S. Chaudhuri, On Scheduling Atomic and Composite Continuous Media Objects, *IEEE Transactions on Knowledge and Data Engineering*, 14(2):447–455, 2002.

[29]   C. Shahabi and F. B. Kashani, Decentralized Resource Management for a Distributed Continuous Media Server, *IEEE Transactions on Parallel and Distributed Systems (TPDS)*, 13(6), June 2002.

[30]   C. Shahabi, M. R. Kolahdouzan, G. Barish, R. Zimmermann, D. Yao, K. Fu, and L. Zhang, Alternative Techniques for the Efficient Acquisition of Haptic Data., in ACM SIGMETRICS/Performance, 2001.

[31]   J. R. Santos and R. R. Muntz, Performance Analysis of the RIO Multimedia Storage System with Heterogeneous Disk Configurations, in ACM Multimedia Conference, Bristol, UK, 1998.

[32]   J. R. Santos, R. Muntz, and B. Ribeiro-Neto, Comparing Random Data Allocation and Data Striping in Multimedia Servers, in Proceedings of ACM SIGMETRICS 2000, pp. 44–55, June 2000.

[33]   H. Shachnai and P. S. Yu, Exploring Wait Tolerance in Effective Batching for VoD Scheduling, *Multimedia Systems*, 6(6):382–394, 1998.

[34]   C. Shahabi, R. Zimmermann, K. Fu, and S.-Y. D. Yao, Yima: A Second Generation of Continuous Media Servers, *IEEE Computer Magazine*, 2002.

[35]   F.A. Tobagi, J. Pang, R. Baird, and M. Gang, Streaming RAID-A Disk Array Management System for Video Files, in First ACM Conference on Multimedia, August 1993.

[36]   R. Zimmermann, K. Fu, C. Shahabi, S.-Y. D. Yao, and H. Zhu, Yima: Design and Evaluation of a Streaming Media System for Residential Broadband Services, in VLDB 2001 Workshop on Databases in Telecommunications (DBTel 2001), Rome, Italy, September 2001.

[37]   R. Zimmermann, K. Fu, and C. Shahabi, A Multi-Threshold Online Smoothing Technique for Variable Rate Multimedia Streams, Technical report, University of Southern California, 2002.
       URL http://www.cs.usc.edu/techreports/technical reports.html.

[38]   R. Zimmermann and S. Ghandeharizadeh, Continuous Display Using Heterogeneous Disk-Subsystems, in Proceedings of the Fifth ACM Multimedia Conference, pp. 227–236, Seattle, November 9-13, 1997.

[39]   R. Zimmermann and S. Ghandeharizadeh, HERA: Heterogeneous Extension of RAID, in Proceedings of the 2000 International Conference on Parallel and Distributed Processing Techniques and Applications (PDPTA 2000), Las Vegas, Nevada, June 2000.

[40]   A. Zhang, Y. Song, and M. Mielke, Netmedia: Streaming Multimedia Presentations in Distributed Environments, *IEEE Multimedia*, 9(1), Jan-Mar 2002.

33

STREAMING MULTIMEDIA PRESENTATIONS IN DISTRIBUTED DATABASE ENVIRONMENTS

Aidong Zhang, Ramazan Savaş Aygün, and Yuqing Song
Department of Computer Science and Engineering
State University of New York at Buffalo
Buffalo, NY, USA
`{azhang,aygun,ys2}@cse.buffalo.edu`

1. INTRODUCTION

There has been increasing interest in flexibly constructing and manipulating heterogeneous presentations from multimedia data resources [34] to support sophisticated applications [26]. In these applications, information from multimedia data resources at one location must be made available at other remote locations for purposes of collaborative engineering, educational learning and tutoring, interactive computer-based training, electronic technical manuals and distributed publishing. Such applications require that basic multimedia objects be stored in multimedia databases/files. The multimedia objects, such as audio samples, video clips, images and animation, are then selectively retrieved, transmitted and composed for customized presentations. In order to support these advanced applications, novel approaches to network-based computing are needed to position adaptive quality-of-service (QoS) management as its central architectural principle. One of the key problems is to design the end system software to be coupled to the network interface. Also, in the best-effort network environment such as Internet, the transmission of media data may experience large variations in available bandwidth, latency and latency variance. Consequently, network delays may occur in the delivery of media data. To meet the demands of the advanced applications over the Internet in the real-world, end system software must have the ability to efficiently store, manage, and retrieve multimedia data. Fundamental principles and advanced techniques need be developed with which one can design an end system software to be integrated with the network to support sophisticated and customized multimedia applications.

Various research has been conducted to develop techniques for distributed multimedia systems. Collaboration between clients and servers has been addressed to support a globally integrated multimedia system [49,30]. In particular, buffering and feedback control for multimedia data transmission in the distributed environment have been studied. In [41,42], server-centric feedback strategies are proposed to maintain synchronized transmission and presentation of media streams. In [38], feedback was used to design robust resource management over the network. In [30], a backward feedback control approach is designed to maintain loosely-coupled synchronization between a server and a client. In [22], an approach is given to use the buffer, a priori information and the current network bandwidth availability to decide whether one should increase the quality of the video stream when more network bandwidth becomes available. An approach on the integration of buffering and feedback control can be found in [37].

Substantial research has been conducted to reduce the burstiness of variable-bit-rate streams by prefetching data [17,43,23]. Especially many bandwidth smoothing algorithms have been proposed for this task. These algorithms *smoothen* the bandwidth requirements for retrieval of the video stream from the video server and transmission across the network. Given a fixed client buffer, it is possible to minimize the peak bandwidth requirement for the continuous delivery of the video stream while minimizing the number of bandwidth rate increases [24]. Another scheme aims at minimizing the total number of bandwidth changes [21]. It is also possible to minimize the variability of the bandwidth [44], the buffer utilization [18] or adhering to a buffer residency constraint [19].

Flow and congestion control for multimedia data transmission in best effort networks has also been investigated [46,8,10,25]. These approaches assume no direct support for resource reservation in the network and attempt to adapt the media streams to current network conditions. Best effort transmission schemes adaptively scale (reduce or increase) the bandwidth requirements of audio/video streams to approximate a connection that is currently sustainable in the network. Best effort schemes split into TCP-friendly and non-TCP-friendly mechanisms. TCP-friendly algorithms avoid starvation of TCP connections by reducing their bandwidth requirements in case of packet loss as an indicator of congestion [46,25,47,13,11]. Non-TCP-friendly schemes reduce rates only to guarantee a higher QoS to its own stream, thereby forcing TCP connections to back off when competing for insufficient network bandwidth [16,31,7,12].

Recent research also addresses the need for a *middleware* framework in transmitting multimedia data. The middleware is located between the system level (operating system and network protocol) and the applications [6]. A good conceptual model of the middleware is described in [40]. It proposes a real-time middleware for asynchronous transfer mode (ATM) systems. It provides virtual connection setup, bandwidth reservation and session synchronization. A middleware control structure is described in [35], which requires applications to implement observation methods for respective system resources and proper adaptation policies within the application. However, there is a lack of a middleware structure, which can effectively put all component techniques together as multimedia middleware services and provide a mapping from the high-level requests of the application to the low-level implementation of the system to support the above mentioned customized multimedia applications.

Multimedia presentation and organization have also been studied at different levels with various models. The proposed synchronization models usually have their own specification, limitations on the user interactions and flexibility on the presentation. Time-based models do not provide the necessary flexibility that is needed for network presentations. Timed Petri Nets are first introduced for multimedia presentations in OCPN [36]. Gibbs [27] proposed a way of composing objects through BLOB stored in the databases and created objects by applying interpretation, derivation and temporal composition to a BLOB. The temporal composition is based on the start times of media objects on a timeline. NSync [5] is a toolkit that manages synchronous and asynchronous interactions, and fine-grained synchronization. The synchronization requirements are specified by synchronization expressions having syntax *When {expression} {action}*. The synchronization expression semantically corresponds to "whenever the *expression* becomes true, invoke the corresponding *action*." Time-based models usually keep the start time and duration of each stream and these models modify the duration and start time after each interaction of the system or the user. In an event-based model, the start of a stream depends on an event signal. SSTS [39] is a combination of Firefly's [9] graph notation and transition rules of OCPN [36]. FLIPS [45] is an event-based model that has barriers and enablers to satisfy the synchronization requirements at the beginning and the end of the streams. PREMO [29] presents an event-based model that also manages time. They have synchronization points, which may also be AND synchronization points to relate several events. Time for media is managed with clock objects and time synchronizable objects, which contain a timer. Multimedia synchronization using ECA rules is covered in [2,3,4]. These projects show how synchronization requirements can be modeled using ECA rules. A hierarchical synchronization model that has events and constraints is given in [14].

This chapter presents the design strategies of a middleware for client-server distributed multimedia applications. The middleware, termed *NetMedia*, provides services to support synchronized presentations of multimedia data to higher level applications. In the *NetMedia* environment, an individual client may access multiple servers to retrieve customized multimedia presentations, and each server simultaneously supports multiple clients. *NetMedia* is capable of flexibly supporting synchronized streaming of continuous media data across best-effort networks. To achieve this, we design a multi-level buffering scheme to control the collaborations between the client and server for flexible data delivery in distributed environments, an end-to-end network delay adaptation protocol, termed DSD, to adjust the sending rate for each stream according to the optimum network delay [50] and a flow and congestion control protocol, termed PLUS, which utilizes probing of the network status to avoid congestion rather than react to it. The DSD and PLUS schemes integrate transmission support strategies and robust software systems at both the server and client ends to dynamically handle adaptive quality-of-service (QoS) management for customized multimedia presentations. The organization of streams and inter-stream synchronization are maintained by using *synchronization rules*, which are based on event-condition-action rules. The synchronization rules are handled by receivers, controllers and actors to provide consistent presentations over networks. This chapter demonstrates how these schemes are used to enforce the interactive control (such as fast forward/backward playback, pause, seek and slow motion) with immediate response for all multimedia streams. This chapter provides an integrated solution to buffer management, congestion control, end-to-end delay variations, interactivity support and synchronization of streams.

The chapter is organized as follows. Section 2 presents the overall architecture of the *NetMedia* and its module design. Section 3 proposes the buffer management at both server and client sites. Section 4 introduces two end-to-end flow control schemes to address network delay and data loss. Section 5 discusses the inter-stream synchronization. Section 6 gives conclusions.

2. *NetMedia* ARCHITECTURE

The *NetMedia* system has three main components: *client* for presentation scheduling, *server* for resource scheduling and *database (file) systems* for data management and storage. Each client supports a GUI (Graphical User Interface) and is responsible for synchronizing images, audio and video packets, and delivering them to an output device such as a PC or a workstation. Meanwhile, each server is superimposed upon a database (file) system, and supports the multi-user aspect of media data caching and scheduling. It maintains timely retrieval of media data from the media database (file) and transfers the data to the client sites through a network. Finally, each database (file) system manages the insertion, deletion and update of the media data stored in the local database (files). As certain media streams may be represented and stored in different formats, the underlying database systems can be heterogeneous. The aim in the proposed middleware is to give the user as much flexibility and individual control as possible. Each media type must be supported individually to allow independent and interactive functionalities as well as QoS requirements.

Both server and client are divided into a pair of the *front end* and the *back end*, which work independently as part of our modular system design and to guarantee an optimum in pipelining of the data.

2.1 SERVER DESIGN

The design of the server permits the individual access of the streams with the support of sharing resources to enhance scalability. The aim of the server functionality can be summarized in the following points:

- *resource sharing* (use of common server disk reading for multiple clients)
- *scalability* (support multiple front end modules)
- *individual QoS and congestion control* (managing individual streams)
- *interactivity*

The server *back end* includes the disk service module which fills in the server buffer according to the request. The server buffer contains a subbuffer for each stream. The *back end* module is shared among all streams and front end modules.

The server *front end* includes a communication module and a packetization module which support reading out the data from the server buffer and delivers it to a network according to the QoS requirements for each stream. It also deals with the admission control of the clients.

Figure 33.1 shows the implementation design of the server component for processing audio and video streams. The disk service module is realized by the DiskReadThread. The Packetization module is realized by the SendThread which reads media units from the server buffer and sends out packets to the network. The Communication module is realized by the ServerTimeThread, AdmissionThread and AdmittedClientSet. The ServerTimeThread reports server's current time and estimates the network delay and time difference between server and client. The AdmittedClientSet keeps track of all admitted clients.   A Server_Probe_Thread is used to get feedback messages and control messages from the probe thread at the client site and initiates the control of probing the network.

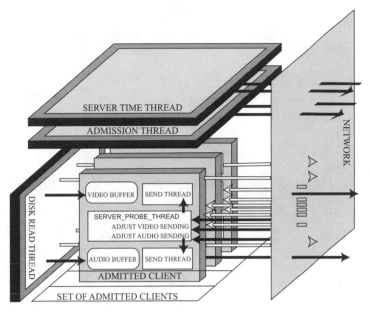

Figure 33.1 Server design for audio and video streams.

2.2 CLIENT DESIGN

The main functionality of a client is to display multiple media streams to the user in the specified format. The client is responsible for synchronizing the audio and video packets and delivering them to the output devices. In addition, the client sends feedbacks to the server which is used for admission control and scheduling in the server. Figure 33. 2 shows the design of the client component.

The client *back end* receives the stream packets from the network and fills in the client cache, which is divided into two components:

- *Communication module* - The communication module provides the interface to the network layer. It primarily consists of two sub-systems to handle UDP and TCP protocols. This enables the client to use faster protocols like UDP for media streams that can tolerate some data loss and reliable protocols like TCP for feedback messages.

- *Buffer Manager* - The buffer manager is responsible for maintaining the client cache in a consistent state.

The client *front end* reads out the data from the client cache and ensures the synchronized presentation of the streams, which is divided into two components:

- *Synchronization Manager* - The synchronization manager schedules the delivery of multimedia data to the output devices. This module controls the delivery of media streams based on QoS parameters defined by the user and the state of the presentation.

- *Presentation Manager* - The presentation manager interacts with the GUI and translates user interactions such as START and STOP into meaningful commands that the other sub-systems can understand.

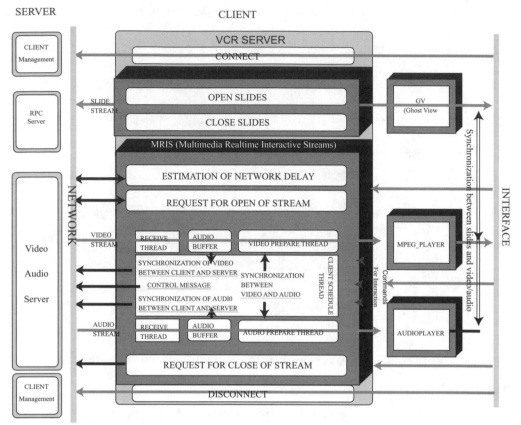

Figure 33.2 Multiple streams synchronized at client.

Note that a stream consisting of transparencies (or images) can be retrieved using conventional transmission methods such as RPC (remote procedure calls) and be synchronized with the audio and video streams at the interface level. The advantage of RPC is that well known file functions such as opening and reading from a certain position can be implemented by standard function calls (which then are transparently executed at the server machine). A RPC request may contain commands to find the file length of a stream, providing directory content or starting the delivery of a slide session. The RPC server retrieves all requested data and sends it back to the client side. Since the scheduling for such data is not as time critical as audio or video data, some delay can be tolerated. In our implementation, the slides are synchronized first according to the audio information. If no audio is available, slides are scheduled with video; otherwise it behaves like a regular postscript document.

To make the access to media streams transparent to the higher levels of the application, we provide an interface, termed MultiMedia RealTime Interactive Stream (MRIS), for requesting the delivery of media streams such as audio, video or any other media forms in a synchronized form from the server. MRIS separates the application from the software drivers needed to access multimedia data from remote sites, and controls specific audio and video output devices. The MRIS Interface has the advantage that the transmission support is independent of the application. Any application that utilizes synchronized multimedia data can rely on this module. We use the MRIS class to implement our virtual-classroom application. The synchronization with slides data can be done at the MRIS interface level.

A crucial design element of a multimedia system is control over the timing requirements of the participating streams so that the streams can be presented in a satisfiable fashion. In *NetMedia*, a robust timing control module, named SleepClock, is designed to handle all timing requirements in the system. The novelty of SleepClock is that it can flexibly control and adjust the invocation of periodic events. We use SleepClock to enforce:

- end-to-end synchronization between server and client,

- smoothed transmission of streams,

- periodic checks in modules and

- timing requirements of media streams.

The detail of the design for SleepClock can be found in [50].

3. BUFFERING STRATEGIES

Buffer management is a crucial part in the *NetMedia* system. There are two main components: server and client buffers.

3.1 SERVER FLOW CONTROL BUFFER

A straightforward server buffer is designed for each stream to perform fast delivery of the stream from the disk to the network. Figure 33.3 shows the structure of a server buffer for a stream. A server stream buffer is a circular list of items. Each item may contain one packet. Each server stream buffer has the following characteristics:

- One producer and one consumer: the producer writes packets to the buffer and the consumer reads from the buffer.

- First in first out: packets which are written in first will be read out first.

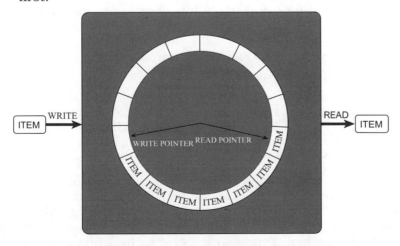

Figure 33.3 Server flow control buffer design.

Specifically, each server stream buffer has two pointers: read pointer (readp) and write pointer(writep). Items from readp to writep are readable; spaces from writep to readp are writable.

Since a server buffer is designed for flow control of the data from the disk to the network, the size of a server buffer can be relatively small.   In the implementation of *NetMedia*, the size of the video buffer is 100 video packets, where each video packet is 1024 bytes, and the size of the audio buffer is 20 audio packets, where each audio packet is 631 bytes.

3.2 CLIENT PACKET BUFFER

Client buffer management is a central part of the *NetMedia* framework. In contrast to the server buffer that must support fast delivery of the stream from the disk to the network, the purpose of a client buffer is to take care of network delays and out of order arriving packets, and to efficiently support user interactions.

The novelty of the client buffer management is the support of interactivity requests like *backward-playback* or *seek*. The difficulties experienced with interactive support in video streams is the video encoding used by most advanced compression techniques [28]. Several works propose methods for implementing forward but do not adequately deal with backward playback. To solve the decoding problem, an approach [48] is proposed to use a separate file for each video stream, preprocessed for backward playback. The advantage is that it does not need special client buffer management. The disadvantage is that a separate file must be created for all video streams, increasing the storage used in the system, and hence the cost.

Our approach is to let the client support interactive requests rather than the server. As frames of video are processed at the client, rather than discard the displayed frames, the client saves the most recent frames to a temporary buffer. The motivation is to reduce the bandwidth requirements of a network in exchange of memory at the client site, because we expect an increase in network load in the next years in comparison to falling memory prices.

The client buffer design for interactivity functions is the use of a circular buffer. The model proposed here allows all interactive functions to occur at any time with an immediate response time. If the buffer is large enough the interactive request for backward playback can be fed directly out of the buffer instead of requesting a retransmission from the server.

The interactive buffer for a stream has two levels: packet and index buffers. The first level is the packet buffer. It is a set of fixed-size packets. Each packet has one pointer, which can be used to point to another packet, so we can easily link any set of packets by pointer manipulation (For example, all packets for a particular frame.) A list of free packet slots is maintained.

The second level for audio buffer is the index buffer of a circular list in which each entry records the playback order of the packet. Each packet in an audio stream has a unique number-PacketNo, which indicates the order of the packets in the whole stream. For the video buffer, each index in *index buffer* points to one frame (which is a list of ordered packets within the frame). We define the point with the minimal packet/frame number as *MinIndex* and the point where we read data for playback as *CurrentIndex*. Both indices traverse the circumference in a clock-wise manner, as shown in Figure 33.4 (Note that in this figure, since the buffer is for a video stream, the two pointers are named as MinFrameIndex and CurrentFrameIndex.)

The distance between *MinIndex* and *CurrentIndex* is defined as *history area*. A packet is not removed immediately after it's displayed. It's removed only when some new packet needs space. So valid indices from MinIndex (inclusive) to CurrentIndex (not inclusive) point to the packets/frames which have been displayed and haven't been released yet. (The data may be used to support backward later.) To keep track of the forward playback capabilities, we introduce the packet/frame with the maximum number as *MaxIndex*. Note that *MaxIndex* is always between *CurrentIndex* and *MinIndex*. The distance between *CurrentIndex* and *MaxIndex* is defined as *future area*. Using the *history and future* areas the user has full interaction capabilities in the two different areas.

When a new packet arrives at the client site, the system checks its packet/frame number (denoted *seqno*) and decides if the packet can be stored. The valid range of the packet/frame numbers is defined as follows:

$$MinIndex \leq seqno \leq CurrentIndex + BuffSize - 1 \qquad (33.1)$$

If the packet has a packet/frame number outside the above range, then it is considered as packet loss and discarded. Otherwise, the packet is stored in a packet slot either taken from the free packet slot list or obtained from releasing some history data. Note that once some history data is released, the *MinIndex* will be updated.

To increase the decoding speed of compressed motion predicted frames we introduce the use of a *decode buffer*. The MPEG compression scheme [28] introduces interframe dependencies among I, P and B type frames. To decode a certain frame, previous or future frames might be needed. For example, given the following frame sequence:

frame-type:	I	B	B	P	B	B	P	B	B	I
frame#:	1	2	3	4	5	6	7	8	9	10

Frames 1,4 and 7 must be decoded before frame 9 can be decoded. To speed up the decoding process for the backward playback, we cache the decoded frames within an I to I frame sequence. In case of dependencies we might be able to feed the needed frames out of the decode buffer instead of calling the decoding routine multiple times.

To implement the above buffer design and allow interactive requests to be fed from their local buffer we need to set a reasonable buffer size for the index and packet buffers. We introduce two consideration factors: MAX_INTERACTIVE_SUPPORT and REASONABLE_DELAY. MAX_INTERACTIVE_SUPPORT defines the longest playback time we expect to be buffered in the client for interactive support, thus defining the size of the index buffer. REASONABLE_DELAY defines the delay within which most packets will pass over the network, thus defining the size of the packet buffer. In the implementation of *NetMedia*, REASONABLE_DELAY is defined to be 2 seconds. The size of MAX_INTERACTIVE_SUPPORT depends on the amount of forward/backward play support (can vary from 10 sec to 5 minutes or more, depending on the memory at the client).

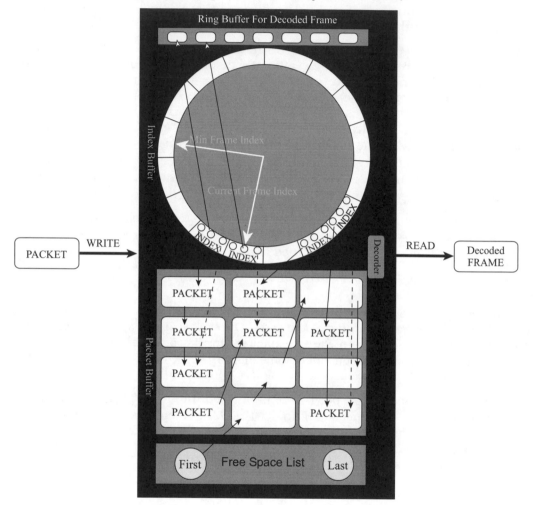

Figure 33.4 Client video buffer.

At the server site, video frames are normally represented in encoding order to make it easier for the decoder to decode in the forward playback. This complicates the decoding of frames for backward playback. Our streams are transmitted in display order to achieve the minimal distance among the dependent frames for both forward and backward playback decoding. The advantage is that our scheme at the client site has uniform design and the same performance for both forward and backward playback. With the use of the decoding buffer and our network control algorithms, we can control the amount of data in the future or backward area in our buffer. This allows the user to specify the amount of interactivity support that he wants for backward or forward playback without immediate request to the server.

Using the above buffer design we can utilize the history and future areas to feed interactive requests directly from the local buffer. If the request does not exceed the cache, the server will not be notified. This will highly reduce the scalability problems faced by the server in case of multiple interactivity requests. Access outside the history or future area must be handled by redirect streaming from the server, but the probability of this case can be drastically reduced by the amount of the buffer size given at the client.

4. END-TO-END FLOW CONTROL

Various delays and congestion in the network may cause the client buffer *underflow/overflow*, which may result in discontinuities in stream playback. We propose the DSD and PLUS schemes to dynamically monitor the network situation and adapt our system to the current situation of the network. The DSD scheme provides full flow control on stream transmission, and the PLUS scheme utilizes probing of the network status and an effective adjustment mechanism to data loss to prevent network congestion. Both schemes work hand in hand to find the optimum adjustment considering network delay and data loss rate, and ensures no overflow/underflow in the client buffer.

4.1 DSD SCHEME

The basic dynamic end-to-end control scheme used in *NetMedia* is the DSD algorithm presented in [50]. The DSD scheme is based on *the Distance in time between Sending and Display* (DSD) media units in a stream. The distance between sending and display at time t is defined to be the difference between the nominal display time of the media unit that is being sent and the media unit that is being displayed. Thus, at a given time, let the media unit being displayed have nominal display time T and the media unit being sent have nominal display time t. Then, $DSD=t-T$. Figure 33.5 illustrates the meaning of the DSD in a stream.

Figure 33.5 Definition of DSD in a stream.

The connection between the DSD and the network delay is as follows. Comparing with the network delay, if DSD is too large, then the overflow data loss at the client buffer could be high; if DSD is too small, then the underflow data loss at the client buffer could also be high. Thus, DSD should be large enough to cover the network delay experienced by most media units; it also should be small enough to make sure the client buffer won't overflow.

By adjusting the current DSD in the system to respond to the current network delays, we can monitor the best DSD dynamically. We can then adjust the sending rate to avoid underflow/overflow in the client buffers and thus alleviate data loss.

Initially, a DSD is selected based on the up-to-date network load, and the sending and display rates are fixed as nominal display rates. The calculation of the current DSD, the best DSD and the adjustment of the DSD is given below.

Calculation of current DSD

At time t, let t_{sent} be the nominal display time of the media unit being sent at the server site and $t_{displayed}$ be the nominal display time of the media unit being displayed at the client site. By definition, the DSD is:

$$DSD_t = t_{sent} - t_{displayed} \qquad (33.2)$$

Suppose at time t, the media unit being displayed has unit number $u^t_{displayed}$, the media unit being sent has unit number u^t_{sent} and nominal duration of one unit is *UnitDuration*. We can then calculate DSD as follows:

$$DSD_t = (u^t_{sent} - u^t_{displayed}) \times UnitDuration \qquad (33.3)$$

Client can always know $u^t_{displayed}$ directly, but it can only estimate u^t_{sent} from the information carried by currently arriving media units. Suppose that the current arriving media unit u was sent at time $SendingTime_u$ and its unit number is n_u. Assume that the server is sending with nominal display rate. At time t, we have

$$u^t_{sent} = n_u + \frac{t - SendingTime_u}{UnitDuration}. \qquad (33.4)$$

We then obtain:

$$DSD_t = (n_u - u^t_{displayed}) \times UnitDuration + (t - SendingTime_u). \qquad (33.5)$$

Calculation of the best DSD

We define an allowable range for *DSD*:

$$MINDSD \leq DSD \leq MAXDSD,$$

where MINDSD = 0 and MAXDSD = MAX_DELAY. We then evenly divide interval [MINDSD,MAXDSD] into k-1 subintervals, with interval points, d_1=MINDSD,d_2,d_3,...d_k=MAXDSD. Here k can be 10, 20 or even 100. The tradeoff is that if k is too small, the best DSD to be found might not be good enough; if k is too big, the overhead in calculating best DSD is too large. For each d_i, we keep track of data loss with respect to it. Following is the definition of relative data loss with respect to d_i.

Let d be the current DSD. Suppose we have a virtual client, which is the same as the actual client, with the same buffer size and the same displaying time for any media unit. The difference is that, suppose for the real client, a real media unit is sent at time t. It then should be displayed at time $t+d$. For the virtual client, this unit is sent at time t and is displayed at time $t+d_i$. So the virtual *DSD* for this unit is d_i. Upon this assumption, the loss rate of the virtual client is defined as the *relative* loss rate with respect to d_i. When a media unit arrives, virtual clients first check if the unit is late or not. Suppose current time is *currentT*, the DSD of the virtual client is d and the sending time of this media unit is *sendT*. If *sendT+d* < *currentT*, then the unit is late. If the unit is late, it's counted as a *relative* loss for this virtual client.

The best DSD finder keeps track of all these virtual clients. The calculation of the best DSD is launched once in a fixed time period (200 milliseconds in the implementation of *NetMedia*). When a new (best) DSD is needed, the best DSD finder browses over all these virtual clients, and the DSD in the virtual client with minimum loss rate is reported as the best DSD. Note that DSD only captures *relative* packet loss, which means that the packet arrived at the client site but was too late for display. Data loss due to network congestion is captured by the PLUS protocol, which will be introduced below.

Adjustment to network delay

At the beginning of an application the DSD is chosen based on the current network load. If the best DSD is different from the current DSD, there are two ways of adjustment. One is to change sending rate and another is to change display rate. The display rate is adjusted by the timing routine at the client site. Only video streams are adjusted.

At any calculation point, if the client calls for adjustment, then a feedback message is sent to the server side to adjust sending rate, so that the system DSD can be changed to the best DSD. The feedback messages will carry the Δd - the difference of the best DSD and the current DSD. A negative Δd means the sending rate should slow down and a positive Δd means the sending rate should speed up. The feedback information also contains information about the loss rate to initiate frame dropping at the server site if necessary.

To slow down, the server stops sending for $|\Delta d|$ time. To speed up, the server always assumes that the current display rate is nominal display rate r. Let the maximum available sending rate over the network be R. The server will speed up using the maximum data rate for a time period. After the reaction time is finished, the server will use the nominal display rate r as the sending rate.

4.2 PLUS SCHEME

The DSD scheme introduced above does not address the congestion problem in the network. In addition, it does not consider the effect of data compression which may introduce burstiness into the data streams. The burstiness results in different bandwidth requirements during transmission of the stream. This makes it difficult to come up with a resource and adaptation scheme because the bandwidth requirements always change.

To address these issues we developed a new flow and congestion control scheme, termed PLUS (Probe-Loss Utilization Streaming protocol), for distributed multimedia presentation systems. This scheme utilizes probing of the network situation and an effective adjustment mechanism to data loss to prevent network congestion. The proposed scheme is also designed to scale with increasing number of PLUS-based streaming traffic and to live in harmony with TCP-based traffic. With the PLUS protocol we address the need to *avoid* congestion rather than *react* to it with the help of a novel probing scheme.

The PLUS protocol eases bandwidth fluctuations by grouping together some number of media units (frames for video) in a window interval ΔW (Figure 33. 6). Easing bandwidth in a given window is defined as *smoothing* [20]. One way of smoothing is done by sending out the data at the average bandwidth requirement for the window (Equation 6):

$$\Delta a = \frac{\Delta W}{number\_of\_packets\_in\_current\_interval} \tag{33.6}$$

By using this method, the client can guarantee that the bandwidth needed is minimal and constant through the interval. The disadvantage of this method is that the feedback received only applies to the current or past bandwidth requirements. It does not take into account that there might be a higher bandwidth request in the future, which the network may not be able to process.

For each interval ΔW, we identify a *critical interval*. The critical interval is an interval in the future playback time that contains the maximum number of packets. The aim of the PLUS probing scheme is to test if the network can support the critical interval. We set ΔW to be a 5 second smoothing window in the implementation of *NetMedia*. This is a reasonable time to reduce the bandwidth through smoothing. It is also granular enough to detect sequences with fast movements (which result in a higher numbers of packets per frame and therefore provide the bottleneck bandwidth of a stream).

Once the critical interval for each sequence is determined we apply our smoothing and probing scheme. The *critical bandwidth* in the future, at a given interval, is provided by its critical interval. To find the minimal bandwidth requirement for the critical interval we apply the averaging algorithm, which spreads the constant sized packets evenly. This leads to a sending difference between consecutive packets, which is defined by:

$$\Delta r = \frac{\Delta W}{pkts\_in\_critical\_interval} \tag{33.7}$$

According to Keshev [33], the bottleneck bandwidth of a connection can be estimated by the packet pair approach at the receiver site. The essential idea is that the inter-packet spacing of two packets will be proportional to the time required for the bottleneck router to process the second packet. The *bottleneck bandwidth* is calculated as:

$$b = \frac{packet\_size}{\Delta r} \tag{33.8}$$

In MPEG encoded streams the loss of I or P frames results in a further loss of dependent frames. Only the loss of B frames does not result in further loss. The loss of a B frame in presentation results only in a short degradation of the quality of service in playback. We use B frames to probe the network if the critical interval can be supported and thereby protect loss of the critical frames like I or P frames (as seen in the lower part of Figure 33.6).

Instead of spreading each packet evenly within our 5 seconds smoothing window we use the bottleneck bandwidth for sending out all B frames. *Critical packets* (belonging to I or P) will still be sent out with the average bandwidth calculated in Equation 33.6. Our scheme will thereby punctually probe the network while acknowledgments will give direct feedback ahead of time if the critical bandwidth can be supported. In case of congestion we will initiate a multiplicative back off of the sending rate. B frames have in our scheme a less chance of survival in case of congestion. This is not a disadvantage because we can proactively provide a bandwidth reduction at the server site by dropping non-critical B frames in time to increase the survival rate of critical packets (and thereby astonishingly increase the survival number of subsequent B frames).

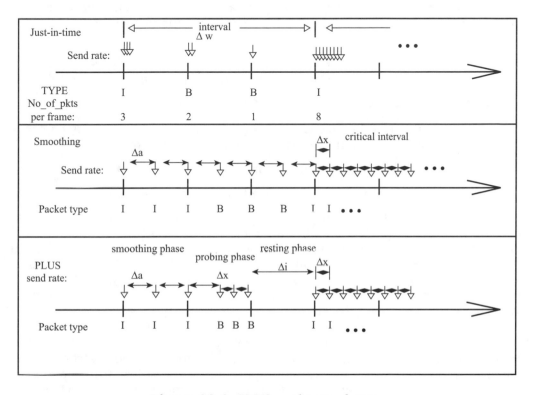

Figure 33.6 PLUS probing scheme.

Concurrent probing of different PLUS streams reports a conservative estimation of the current network situation. The estimation is based on the bandwidth need of the critical intervals of the streams and not necessary of the current average interval. This behavior allows PLUS streams to be very aware of its surrounding and responsive to protect critical packets when the maximal capacity of a connection is reached. If the concurrent probing causes packet loss, PLUS streams back off, to allow living in harmony with TCP and other PLUS streams. Sequential probing of different PLUS streams could report an estimation which leads to data loss in case that multiple PLUS streams send their critical interval at exactly the same time, each assuming the network can support it. The probability of such a situation is neglectably low and can be further reduced with the number of B frames in a stream.

5. INTER-STREAM SYNCHRONIZATION

The multimedia presentations consist of parallel and sequential presentations of streams. Inter-stream synchronization is usually as classified as fine-grained synchronization and coarse-grained synchronization. Fine-grained synchronization is related with the synchronization of parallel streams at fine granules. On the other hand, the coarse-grained synchronization is related to how streams are connected to each other, when they start and end. Coarse-grained synchronization is important to support consistent presentations. Due to the delay and loss of data over networks, synchronization models that support flexible presentations are necessary. For flexible synchronization models, the relationships among streams are declared rather than time-based start and end times of streams. In this section, we explain a flexible synchronization model based on ECA (Event-Condition-Action) rules [15] to manage the coarse-grained synchronization in *NetMedia*.

The synchronization specification lays out media objects in a presentation. The synchronization model should have the following properties:

- It should support composition of media objects in various ways.

- It should support user interactions of VCR type like skip or backward which are helpful in browsing of the multimedia presentations.

- The user interactions should not complicate specifications.

- It should support both time-based and event-based operations.

It has been shown that event-based models have been more robust and flexible for multimedia presentations. The bottleneck of the event-based models is the inapplicability of the model in case there is a change in the course of the presentation (like backwarding and skipping). Most of the previous models are based on event-action relationships. The condition of the presentation and participating streams also influence the actions to be executed. Thus ECA rules, which have been successfully employed in active database systems, are applied to multimedia presentations. Since these rules are used for synchronization, they are termed as *synchronization rules*. Synchronization rules are used in both the synchronization and the organization of streams. Since the structure of a synchronization rule is simple, the manipulation of the rules can be performed easily in existence of user interactions. The synchronization model uses Receiver-Controller-Actor (RCA) scheme to execute the rules. In RCA scheme, receivers, controllers, and actors are objects to receive events, to check conditions and to execute actions, respectively. The synchronization rules can easily be regenerated from SMIL expressions [3].

Synchronization rules form the basis of the management of relationships among the synchronization rules. A *synchronization rule* is composed of an event expression, condition expression and action expression which can be formulated as:

on *event expression* **if** *condition expression* **do** *action expression*

A synchronization rule can be read as: When the *event expression* is satisfied if the *condition expression* is valid, then the actions in the *action expression* are executed. The event expression enables the composition of events that can be created by Boolean operators && and ||. The condition expression enables the composition of conditions using && and ||. The action expression is the list of the actions to be executed when condition expression is satisfied.

5.1 EVENTS, CONDITIONS AND ACTIONS FOR A PRESENTATION

In a multimedia system, the events may be triggered by a media stream, user or the system. Each media stream is associated with events along with its data and it knows when to signal events. When events are received, the corresponding conditions are checked. If a condition is satisfied, the corresponding actions are executed.

The goal in inter-stream synchronization is to determine when to start and end streams. The start and end of streams depend on multimedia events. The hierarchy of multimedia events is given in [3]. The user has to specify information related to the stream events. Allen [1] specifies 13 temporal relationships. Relationships *meets, starts* and *equals* require the *InitPoint* event for a stream. Relationships *finishes* and *equals* require the *EndPoint* event for a stream. Relationships *overlaps* and *during* require *realization* event to start (end) another stream in the mid of a stream. The relationships before and after require temporal events since the gap between two streams can only be determined by time. Temporal events may be *absolute* with respect to a specific point in a presentation (e.g., the beginning of a presentation). Temporal events may also be *relative* with respect to another event.

Definition 1. *An event is represented with source(event_type[,event_data]) where source points the source of the event, event_type represents the type of the event and event_data contains information about the event.*

Event source can be the user or a stream. Optional event data contains information like a realization point. Event type indicates whether the event is *InitPoint, EndPoint* or *realization* if it is a stream event. Each stream has a series of events. Users can also cause events such as start, pause, resume, forward, backward and skip. These events have two kinds of effects on the presentation. Skip and backward change the course of the presentation. Others only affect the duration of the presentation.

Definition 2. A *condition* in a synchronization rule is a 3 tuple

$$C=condition(t_1,\theta,t_2)$$

where θ is a relation from the set $\{=,\neq,<,\leq,>,\geq\}$ and t_i is either a state variable that determines the state of a stream or presentation or a constant.

A condition indicates the status of the presentation and its media objects. The most important condition is whether the direction of the presentation is forward. The receipt of the events matter when the direction is forward or backward. Other types of conditions include the states of the media objects.

Definition 3. An action is represented with

$$action\_type(stream[,action\_data],sleeping\_time)$$

where action_type needs to be executed for stream using action_data as parameters after waiting for sleeping_time. Action_data can be the parameter for speeding, skipping, etc.

An action indicates what to execute when conditions are satisfied. *Starting* and *ending* a stream, and *displaying* or *hiding* images, slides and text are sample actions. For backward presentation, *backwarding* is used to backward and *backend* is used to end in the backward direction. There are two kinds of actions: *Immediate* Action and *Deferred* Action. *Immediate* action is an action that should be applied as soon as the conditions are satisfied. *Deferred* action is associated with some specific time. The deferred action can only start after this *sleeping_time* has been elapsed. If an action has started and has not finished yet, that action is considered as an alive action.

For example, the following synchronization rule,

on   *V1(EndPoint)*   **if** *direction=FORWARD*   **do** *start(V2,4s),*

means that when stream *V1* ends if the direction is forward, start stream *V2* 4 seconds later.

5.2 RECEIVERS, CONTROLLERS AND ACTORS

The synchronization model is composed of three layers, the receiver layer, the controller layer and the actor layer. Receivers are objects to receive events. Controllers check composite events and conditions about the presentation such as the direction. Actors execute the actions once their conditions are satisfied.

Definition 4. *A receiver is a pair R=(e, C), where e is the event that will be received and C is a set of controller objects.*

Receiver *R* can question the event source through its event *e*. When *e* is signaled, receiver *R* will receive *e*. When receiver *R* receives event *e*, it sends information of the receipt of *e* to all its controllers in *C*. A receiver object is depicted in Figure 33. 7. There is a receiver for each single event. The receivers can be set and reset by the system anytime.

Definition 5. *A controller is a 4-tuple C = (R, ee, ce, A) where R is a set of receivers; ee is an event expression; ce is a condition expression; and A is a set of actors.*

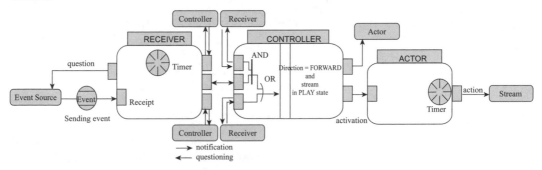

Figure 33.7 The relationships among receiver, controller and actor objects.

Controller *C* has two components to verify, composite events *ee* and conditions *ce* about the presentation. When the controller *C* is notified, it first checks whether the event composition condition, *ee,* is satisfied by questioning the receiver of the event. Once the event composition condition *ee* is satisfied, it verifies the conditions *ce* about the states of media objects or the presentation. After the conditions *ce* are satisfied, the controller notifies its actors in *A*. A controller object is depicted in Figure 33. 7. Controllers can be set or reset by the system anytime.

Definition 6. *An actor is a pair A = (a, t) where a is an action that will be executed after time t passed.*

Once actor *A* is informed, it checks whether it has some sleeping time *t* to wait for. If *t* is greater than 0, actor *A* sleeps for *t* and then starts action *a*. If *t* is 0, action *a* is an immediate action. If *t*>0, action *a* is a deferred action. An actor object is depicted in Figure 33. 7.

5.3 TIMELINE

If multimedia presentations are declared in terms of constraints, synchronization expressions or rules, the relationships among streams are not explicit. Those expressions only keep the relationships that are temporally adjacent or overlapping. The status of the presentation must be known at any instant. The timeline object keeps track of all temporal relationships among streams in the presentation.

Definition 7. *A timeline object is a 4-tuple T =(receiverT, controllerT, actorT, actionT) where receiverT, controllerT, actorT and actionT are time-trackers for receivers, controllers, actors and actions, respectively.*

The time-trackers *receiverT*, *controllerT*, *actorT* and *actionT* keep the expected times of the receipt of events by receivers, the expected times of the satisfaction of the controllers, the expected times of the activation of the actors and the expected times of the start of the actions, respectively. Since skip and backward operations are allowed, alive actions, received or not-received events, sleeping actors and satisfied controllers must be known for any point in the presentation. The time of actions can be retrieved from the timeline object.

The information that is needed to create the timeline is the duration of streams and the relationships among the streams. The expected time for the receipt of *realization*, *InitPoint* and *EndPoint* stream events only depends on the duration of the stream and the start time of the action that starts the stream. Since the nominal duration of a stream is already known, the problem is the determination of the start time of the action. The start of the action depends on the activation of its actor. The activation of the actor depends on the satisfaction of the controller. The expected time when the controller will be satisfied depends on the expected time when the event composition condition of the controller is satisfied.

The expected time for the satisfaction of an event composition condition is handled in the following way: In our model, events can be composed using && and || operators. Assume that ev_1 and ev_2 are two event expressions where $time(ev_1)$ and $time(ev_2)$ give the expected times of satisfaction of ev_1 and ev_2, respectively. Then, the expected time for composite events is found according to the predictive logic for WBT (will become true) in [5]:

$time(ev_1 \&\& ev_2)=maximum(time(ev_1), time(ev_2))$

$time(ev_1 || ev_2)=minimum(time(ev_1), time(ev_2))$

where *maximum* and *minimum* functions return the maximum and minimum of the two values, respectively. The time of the first controller and the receiver is 0, which only depends on the user event to start the presentation.

Another important issue is the enforcement of the smooth presentation of multimedia streams at client sites. This is critical for building an adaptive presentation system that can handle high rate variance of stream presentations and delay variance of networks. In [32], we have observed that the dynamic playout scheduling can greatly enhance the smoothness of the presentations of the media streams. We thus have formulated a framework for various presentation scheduling algorithms to support various Quality of Service (QoS) requirements specified in the presentations [32]. We design a scheduler within the NetMedia-client to support smooth presentations of multimedia streams at the client site in the distributed environment. Our primary goal in the design of playout scheduling algorithms is to create smooth and relatively hiccup-free presentations in the delay-prone environments.

Our algorithms can dynamically adjust any rate variations that are not maintained in the transmission of the data. We define *rendition rate* to be the instantaneous rate of presentation of a media stream. Each stream must report to the scheduler (which implements the synchronization algorithm) its own progress periodically. The scheduler in turn reports to each stream the rendition rate required to maintain the desired presentation. The individual streams must try to follow this rendition rate. The client has several threads running concurrently to achieve the presentation. The detail of these algorithms can be found in [32].

6. CONCLUSION

We have designed and implemented a middleware system infrastructure (termed *NetMedia*) to support the retrieval and transmission of multimedia data in distributed multimedia presentation environments. *NetMedia* gives the application developers a set of services that transparently provide connection management, media data transmission, QoS management and synchronization of multimedia streams. The main features of *NetMedia* include the following:

- A well defined API (Application Programming Interface), which lets applications access remote media streams transparently like a local file access.

- QoS management with the help of the DSD and PLUS protocols.

- A multi-level buffering scheme to control the collaborations between the clients and servers to achieve flexible data-delivery in distributed environments.

- Inter-stream synchronization and organization of streams using synchronization rules.

- Interactive control (like fast forward/backward playback) with immediate response for all multimedia streams.

The middleware can be readily used to facilitate various applications. An interactive learning environment, *NetMedia-VCR*, has been implemented for student tutoring, medical and military training, human resource development, asynchronous distance learning and distributed publishing.

REFERENCES

[1] J. Allen, Maintaining Knowledge about Temporal Intervals, *Communications of ACM*, 26(11):823-843, November, 1983.

[2] R. S. Aygun and A. Zhang, Interactive Multimedia Presentation Management in Distributed Multimedia Systems, in Proc. of Int.Conf. on Information Technology: Coding and Computing, pages 275-279, Las Vegas, Nevada, April 2001.

[3] R. S. Aygun and A. Zhang, Middle-tier for Multimedia Synchronization, in 2001 ACM Multimedia Conference, pages 471-474, Ottawa, Canada, October 2001.

[4] R. S. Aygun and A. Zhang, Management of Backward-skip Interactions Using Synchronization Rules, in The 6th World Conference on Integrated Design & Process Technology, Pasadena, California, June 2002.

[5] B. Bailey, J. Konstan, R. Cooley, and M. Dejong, Nsync – A Toolkit for Building Interactive Multimedia Presentations, in Proceedings of ACM Multimedia, pages 257-266, September 1998.

[6] P.A. Bernstein, Middleware: A model for distributed system services, *Communications of the ACM*, 39(2), pages 86-98, February 1996.

[7] J.-C. Bolot and T. Turletti, A Rate Control Mechanism for Packet Video in the Internet, in IEEE Infocomm, pages 1216-1223, Toronto, Canada, June 1994.

[8] J.-C. Bolot, T. Turletti, and I. Wakeman, Scalable Feedback Control for Multicast Video Distribution in the Internet, in SIGCOM Symposium on Communications Architectures and Protocols, pages 58-67, UK London, Aug. 1994.

[9] M. C. Buchanan and P. T. Zellweger, Scheduling Multimedia Documents Using Temporal Constraints, in Proceedings Third International Workshop on Network and Operating Systems Support for Digital audio and Video, pages 237-249, IEEE Computer Society, November 1992.

[10] I. Busse, B. Deffner, and H. Schulzrinne, Dynamic QoS Control of Multimedia Applications based on RTP, *Computer Communications*, 19: 49-58, January 1996.

[11] S. Cen, C. Pu, and J. Walpole, Flow and Congestion Control for Internet Streaming Applications, in Proceedings of Multimedia Computing and Networking (MMCN98), 1998.

[12] S. Chakrabarti and R. Wang, Adaptive Control for Packet Video, in Proc. IEEE Multimedia and Computer Systems, pages 56-62, 1994.

[13] D.D. Clark, S. Shenker, and L. Zhang, Supporting Real-time Applications in an Integrated Services Packet Network: Architecture and Mechanism, in Proc. SIGCOMM, pages 14-26, 1992.

[14] J. P. Courtiat and R. C. D. Oliveira, Proving Temporal Consistency in a New Multimedia Synchronization Model, in Proceedings of ACM Multimedia, pages 141-152, November 1996.

[15] U. Dayal, Active Database Management Systems, in Proceedings of the 3rd International. Conference on Data and Knowledge Bases, 1988.

[16] L. Delgrossi, C. Halstrick, D. Hehmann, D.G.Herrtwich, O. Krone, J. Sandvoss, and C. Vogt, Media Scaling for Audiovisual Communication with the Heidelberg Transport System, in Proceedings of ACM Multimedia, pages 99-104, Anaheim, CA, August 1993.

[17] W. Feng. *Buffering Techniques for Delivery of Compressed Video in Video-on-Demand Systems*. Kluwer Academic Publishers, 1997.

[18] W. Feng, Rate-constrained Bandwidth Smoothing for the Delivery of Stored Video, in SPIE Multimedia Computing and Networking, Feb. 1997.

[19] W. Feng, Time Constrained Bandwidth Smoothing for Interactive Video-on-Demand Systems, in International Conference on Computer Communications, pages 291-302, Cannes, France, 1997.

[20] W. Feng, F. Jahanian, and S. Sechrest, Providing VCR Functionality in a Constant Quality Video On-Demand Transportation Service, in Proc. IEEE Multimedia 1996, pages 127-135, Hiroshima, Japan, June 1996.

[21] W. Feng, F. Jahanian, and S. Sechrest, An Optimal Bandwidth Allocation Strategy for the Delivery of Compressed Prerecorded Video, *ACM/Springer-Verlag Multimedia Systems Journal*, 5(5): 297-309, 1997.

[22] W. Feng, B. Krishnaswami, and A. Prabhudev, Proactive Buffer Management for the Streamed Delivery of Stored Video, in Proc. ACM Multimedia, pages 285-290, Bristol, UK, 1998.

[23] W. Feng and J. Rexford, A Comparison of Bandwidth Smoothing Techniques for the Transmission of Prerecorded Compressed Video, in IEEE INFOCOM, pages 58-66, Kobe, Japan, April 1997.

[24] W. Feng and S. Sechrest, Smoothing and Buffering for Delivery of Prerecorded Compressed Video, in SPIE Multimedia Computing and Networking, pages 234-242, San Jose, CA, Feb 1995.

[25] S. Floyd and F. Kevin, Router mechanism to support end-to-end congestion control. Technical report, February 1997.

[26] E. Fox and L. Kieffer, Multimedia Curricula, Courses, and Knowledge Modules, *ACM Computing Surveys*, 27(4): 549-551, December 1995.

[27] S. Gibbs, C. Breiteneder, and D. Tsichritzis, Data Modeling of Time-Based Media, in Proceedings of ACM-SIGMOD International Conference on Management of Data, pages 91-102, Minneapolis, Minnesota, May 1994.

[28] B. Haskell, A. Puri, and A. Netravaldi. *Digital Video: An Introduction to MPEG2*. Chapman and Hall, New York, 1996.

[29] I. Herman, N. Correira, D. A. Duce, D. J. Duke, G. J. Reynolds, and J. V. Loo, A standard model for multimedia synchronization: Premo synchronization objects, *Multimedia Systems*, 6(2): 88-101, 1998.

[30] J. Hui, E. Karasan, J. Li, and J. Zhang, Client-Server Synchronization and Buffering for Variable Rate Multimedia Retrievals, *IEEE Journal on Selected Arears in Communications*, 14(1), January 1996.

[31] K. Jeffay, D.L. Stone, and F. Smith, Transport and display mechanisms for multimedia conferencing across packet switched networks, *Networks and ISDN Systems*, 26(10): 1281-1304, July 1994.

[32] T.V. Johnson and A. Zhang, Dynamic Playout Scheduling Algorithms for Continuous Multimedia Streams, *ACM Multimedia Systems*, 7(4): 312-325, July 1999.

[33] S. Keshav, A Control-theoretic Approach to Flow Control, in Proceedings of the Conference on Communications Architecture and Protocols, pages 3-15, Zuerich, Switzerland, September 1991.

[34] T. Lee, L. Sheng, T. Bozkaya, N.H. Balkir, Z.M. Ozsoyoglu, and G. Ozsoyoglu, Querying Multimedia Presentations Based on Content, *IEEE Transactions on Knowledge and Data Engineering*, 11(3): 361-385, May/June 1999.

[35] B. Li and K. Nahrstedt, A Control-based Middleware Framework for Quality of Service Adaptations, *IEEE Journal of Selected Areas in Communications, Special Issue on Service Enabling Platforms*, 1999.

[36] T. Little and A. Ghafoor, Synchronization and Storage Models for Multimedia Objects, *IEEE Journal on Selected Areas in Communications*, 8(3): 413-427, April 1990.

[37] M. Mielke and A. Zhang, A Multi-level Buffering and Feedback Scheme for Distributed Multimedia Presentation Systems, in Proceedings of the Seventh International Conference on Computer Communications and Networks (IC3N'98), Lafayette, Louisiana, October 1998.

[38] K. Nahrstedt and J. Smith, New Algorithms for Admission Control and Scheduling to Support Multimedia Feedback Remote Control Applications, in the Third IEEE International Conference on Multimedia Computing and Systems (ICMCS'96), pages 532-539, Hiroshima, Japan, June 1996.

[39] J. Nang and S. Kang, A New Multimedia Synchronization Specification Method for Temporal and Spatial Events, in Proceedings of International Conference on Multimedia Computing and Systems, pages 236-243. IEEE Computer Society, June 1997.

[40] G.Neufeld, D.Makaroff, and N.Hutchinson, The Design of a Variable Bit Rate Continuous Media Server, Technical Report TR-95-06, Dept. of Computer Science, University of British Columbia, Vancouver, Canada, March 1995.

[41] S. Ramanathan and P.V. Rangan, Adaptive Feedback Techniques for Synchronized Multimedia Retrieval over Integrated Networks, *IEEE/ACM Transactions on Networking*, 1(2): 246-260, April 1993.

[42] S. Ramanathan and P.V. Rangan, Feedback Techniques for Intra-Media Continuity and Inter-Media Synchronization in Distributed Multimedia Systems, *The Computer Journal*, 36(1): 19-31, 1993.

[43] J. Rexford and D. Towsley, Smoothing Variable Bit Rate Video In an Internetwork, in Proceedings of the SPIE Conference, University of Massachusetts at Amherst, Technical Report 97-33, 1997.

[44] J. Salehi, Z. Zhang, J. Kurose, and D. Towsley, Optimal Buffering for the Delivery of Compressed Prerecorded Video, in ACM SIGMETRICS, pages 222-231, May 1996.

[45] J. Schnepf, J. Konstan, and D. Du, FLIPS: Flexible Interactive Presentation Synchronization, *IEEE Selected Areas of Communication*, 14(1): 114-125, 1996.

[46] D. Sisalem and H. Schulzrinne, The Loss-delay Based Adjustment Algorithm: A Tcp-friendly Adaptation Scheme, in Proc. NOSSDAV 98, Cambridge, UK, July 1998.

[47] T.Ott, J. Kempermann, and M. Mathis, Window Size Behaviour in TCP/IP with Congestion Avoidance Algorithm, in IEEE Workshop on the Architecture and Implementation of High Performance Communication Systems, June 1997.

[48] M. Vernick, The Design, Implementation, and Analysis of the Stony Brook Video Server, Ph.D. thesis, SUNY Stony Brook, 1996.

[49] H. M. Vin, M. S. Chen, and T. Barzilai, Collaboration Management in DiCE. *The Computer Journal*, 36(1): 87-96, 1993.

[50] A. Zhang, Y. Song and M. Mielke, NetMedia: Streaming Multimedia Presentations in Distributed Environments, *IEEE Multimedia*, Vol. 9, No. 1, January-March, 2002, pp. 56-73.

34

VIDEO STREAMING: CONCEPTS, ALGORITHMS, AND SYSTEMS

John G. Apostolopoulos, Wai-tian Tan, Susie J. Wee
Streaming Media Systems Group
Hewlett-Packard Laboratories
Palo Alto, CA, USA
{japos,dtan,swee}@hpl.hp.com

1.   INTRODUCTION

Video has been an important media for communications and entertainment for many decades. Initially video was captured and transmitted in analog form. The advent of digital integrated circuits and computers led to the digitization of video, and digital video enabled a revolution in the compression and communication of video. Video compression became an important area of research in the late 1980's and 1990's and enabled a variety of applications including video storage on DVD's and Video-CD's, video broadcast over digital cable, satellite and terrestrial (over-the-air) digital television (DTV), and video conferencing and videophone over circuit-switched networks. The growth and popularity of the Internet in the mid-1990's motivated video communication over best-effort packet networks. Video over best-effort packet networks is complicated by a number of factors including unknown and time-varying bandwidth, delay, and losses, as well as many additional issues such as how to fairly share the network resources amongst many flows and how to efficiently perform one-to-many communication for popular content. This article examines the challenges that make simultaneous delivery and playback, or streaming, of video difficult, and explores *algorithms and systems* that enable streaming of pre-encoded or live video over packet networks such as the Internet.

We continue by providing a brief overview of the diverse range of video streaming and communication applications. Understanding the different classes of video applications is important, as they provide different sets of constraints and degrees of freedom in system design. Section 3 reviews video compression and video compression standards. Section 4 identifies the three fundamental challenges in video streaming: unknown and time-varying bandwidth, delay jitter, and loss. These fundamental problems and approaches for overcoming them are examined in depth in Sections 5, 6, and 7. Standardized media streaming protocols are described in Section 8, and additional issues in video streaming are highlighted in Section 9. We conclude by describing the design of emerging streaming media content delivery networks in Section 10. Further overview articles include [1,2,3,4,5].

2. OVERVIEW OF VIDEO STREAMING AND COMMUNICATION APPLICATIONS

There exist a very diverse range of different video communication and streaming applications, which have very different operating conditions or properties. For example, video communication application may be for point-to-point communication or for multicast or broadcast communication, and video may be pre-encoded (stored) or may be encoded in real-time (e.g. interactive videophone or video conferencing). The video channels for communication may also be static or dynamic, packet-switched or circuit-switched, may support a constant or variable bit rate transmission, and may support some form of Quality of Service (QoS) or may only provide best effort support. The specific properties of a video communication application strongly influence the design of the system. Therefore, we continue by briefly discussing some of these properties and their effects on video communication system design.

Point-to-point, multicast, and broadcast communications
Probably the most popular form of video communication is one-to-many (basically one-to-all) communication or broadcast communication, where the most well known example is broadcast television. Broadcast is a very efficient form of communication for popular content, as it can often efficiently deliver popular content to all receivers at the same time. An important aspect of broadcast communications is that the system must be designed to provide every intended recipient with the required signal. This is an important issue, since different recipients may experience different channel characteristics, and as a result the system is often designed for the worst-case channel. An example of this is digital television broadcast where the source coding and channel coding were designed to provide adequate reception to receivers at the fringe of the required reception area, thereby sacrificing some quality to those receivers in areas with higher quality reception (e.g. in the center of the city). An important characteristic of broadcast communication is that, due to the large number of receivers involved, feedback from receiver to sender is generally infeasible – limiting the system's ability to adapt.

Another common form of communication is point-to-point or one-to-one communication, e.g. videophone and unicast video streaming over the Internet. In point-to-point communications, an important property is whether or not there is a back-channel between the receiver and sender. If a back-channel exists, the receiver can provide feedback to the sender which the sender can then use to adapt its processing. On the other hand, without a back-channel the sender has limited knowledge about the channel.

Another form of communication with properties that lie between point-to-point and broadcast is multicast. Multicast is a one-to-many communication, but it is not one-to-all as in broadcast. An example of multicast is IP-Multicast over the Internet. However, as discussed later, IP Multicast is currently not widely available in the Internet, and other approaches are being developed to provide multicast capability, e.g. application-layer multicast via overlay networks. To communicate to multiple receivers, multicast is more efficient than multiple unicast connections (i.e. one dedicated unicast connection to each client), and overall multicast provides many of the same advantages and disadvantages as broadcast.

Real-time encoding versus pre-encoded (stored) video

Video may be captured and encoded for real-time communication, or it may be pre-encoded and stored for later viewing. Interactive applications are one example of applications which require real-time encoding, e.g. videophone, video conferencing, or interactive games. However real-time encoding may also be required in applications that are not interactive, e.g. the live broadcast of a sporting event.

In many applications video content is pre-encoded and stored for later viewing. The video may be stored locally or remotely. Examples of local storage include DVD and Video CD, and examples of remote storage include video-on-demand (VOD), and video streaming over the Internet (e.g. as provided by RealNetworks and Microsoft). Pre-encoded video has the advantage of not requiring a real-time encoding constraint. This can enable more efficient encoding such as the multi-pass encoding that is typically performed for DVD content. On the other hand, it provides limited flexibility as, for example, the pre-encoded video cannot be significantly adapted to channels that support different bit rates or to clients that support different display capabilities from those used in the original encoding.

Interactive versus Non-interactive Applications

Interactive applications such as videophone or interactive games have a real-time constraint. Specifically the information has a time-bounded usefulness, and if the information arrives, but is late, it is useless. This is equivalent to a maximum acceptable end-to-end latency on the transmitted information, where by end-to-end we mean: capture, encode, transmission, receive, decode, display. The maximum acceptable latency depends on the application, but often is on the order of 150 ms. Non-interactive applications have looser latency constraints, for example many seconds or potentially even minutes. Examples of non interactive applications include multicast of popular events or multicast of a lecture; these applications require timely delivery, but have a much looser latency constraint. Note that interactive applications require real-time encoding, and non-interactive applications may also require real-time encoding, however the end-to-end latency for non-interactive applications is much looser, and this has a dramatic effect on the design of video communication systems.

Static versus Dynamic Channels

Video communication system design varies significantly if the characteristics of the communication channel, such as bandwidth, delay, and loss, are static or dynamic (time-varying). Examples of static channels include ISDN (which provides a fixed bit rate and delay, and a very low loss rate) and video storage on a DVD. Examples of dynamic channels include communication over wireless channels or over the Internet. Video communication over a dynamic channel is much more difficult than over a static channel. Furthermore, many of the challenges of video streaming, as are discussed later in this article, relate to the dynamic attributes of the channels.

Constant-bit-rate (CBR) or Variable-bit-rate (VBR) Channel

Some channels support CBR, for example ISDN or DTV, and some channels support VBR, for example DVD storage and communication over shared packet networks. On the other hand, a video sequence typically has time-varying complexity. Therefore coding a video to achieve a constant visual quality requires a variable bit rate, and coding for a constant bit rate would produce time-varying quality. Clearly, it is very important to match the video bit rate to what the channel can support. To achieve this a buffer is typically used to couple the video encoder to the channel, and a buffer control mechanism provides feedback based on the buffer fullness to regulate the coarseness/fineness of the quantization and thereby the video bit rate.

Packet-Switched or Circuit-Switched Network

A key network attribute that affects the design of media streaming systems is whether they are packet-switched or circuit-switched. Packet-switched networks, such as Ethernet LANs and the Internet, are shared networks where the individual packets of data may exhibit variable delay, may arrive out of order, or may be completely lost. Alternatively, circuit-switched networks, such as the public switched telephone network (PSTN) or ISDN, reserve resources and the data have a fixed delay, arrives in order, however the data may still be corrupted by bit errors or burst errors.

Quality of Service (QoS) Support

An important area of network research over the past two decades has been QoS support. QoS is a vague, and all-encompassing term, which is used to convey that the network provides some type of preferential delivery service or performance guarantees, e.g. guarantees on throughput, maximum loss rates or delay. Network QoS support can greatly facilitate video communication, as it can enable a number of capabilities including provisioning for video data, prioritizing delay-sensitive video data relative to other forms of data traffic, and also prioritize among the different forms of video data that must be communicated. Unfortunately, QoS is currently not widely supported in packet-switched networks such as the Internet. However, circuit-switched networks such as the PSTN or ISDN do provide various guarantees on delay, bandwidth, and loss rate. The current Internet does not provide any QoS support, and it is often referred to as Best Effort (BE), since the basic function is to provide simple network connectivity by best effort (without any guarantees) packet delivery . Different forms of network QoS that are under consideration for the Internet include Differentiated Services (DiffServ) and Integrated Services (IntServ), and these will be discussed further later in this writeup.

3.   REVIEW OF VIDEO COMPRESSION

This section provides a very brief overview of video compression and video compression standards. The limited space precludes a detailed discussion, however we highlight some of the important principles and practices of current and emerging video compression algorithms and standards that are especially relevant for video communication and video streaming. An important motivation for this discussion is that both the standards (H.261/3/4, MPEG-1/2/4) and the most popular proprietary solutions (e.g. RealNetworks [6] and Microsoft Windows Media [7]) are based on the same basic principles and practices, and therefore by understanding them one can gain a basic understanding for both standard and proprietary video streaming systems. Another goal of this section is to describe what are the different video compression standards, what do these standards actual specify, and which standards are most relevant for video streaming.

3.1 BRIEF OVERVIEW OF VIDEO COMPRESSION

Video compression is achieved by exploiting the similarities or redundancies that exists in a typical video signal. For example, consecutive frames in a video sequence exhibit temporal redundancy since they typically contain the same objects, perhaps undergoing some movement between frames. Within a single frame there is spatial redundancy as the amplitudes of nearby pixels are often correlated. Similarly, the Red, Green, and Blue color components of a given pixel are often correlated. Another goal of video compression is to reduce the irrelevancy in the video signal, that is to only code video features that are perceptually important and not to waste valuable bits on information that is not perceptually important or irrelevant. Identifying and reducing the redundancy in a video signal is relatively straightforward, however identifying what is perceptually relevant and what is not is very difficult and therefore irrelevancy is difficult to exploit.

To begin, we consider image compression, such as the JPEG standard, which is designed to exploit the spatial and color redundancy that exists in a single still image. Neighboring pixels in an image are often highly similar, and natural images often have most of their energies concentrated in the low frequencies. JPEG exploits these features by partitioning an image into 8x8 pixel blocks and computing the 2-D Discrete Cosine Transform (DCT) for each block. The motivation for splitting an image into small blocks is that the pixels within a small block are generally more similar to each other than the pixels within a larger block. The DCT compacts most of the signal energy in the block into only a small fraction of the DCT coefficients, where this small fraction of the coefficients is sufficient to reconstruct an accurate version of the image. Each 8x8 block of DCT coefficients is then quantized and processed using a number of techniques known as zigzag scanning, run-length coding, and Huffman coding to produce a compressed bitstream [8]. In the case of a color image, a color space conversion is first applied to convert the RGB image into a luminance/chrominance color space where the different human visual perception for the luminance (intensity) and chrominance characteristics of the image can be better exploited.

A video sequence consists of a sequence of video frames or images. Each frame may be coded as a separate image, for example by independently applying JPEG-like coding to each frame. However, since neighboring video frames are typically very similar much higher compression can be achieved by exploiting the similarity between frames. Currently, the most effective approach to exploit the similarity between frames is by coding a given frame by (1) first predicting it based on a previously coded frame, and then (2) coding the error in this prediction. Consecutive video frames typically contain the same imagery, however possibly at different spatial locations because of motion. Therefore, to improve the predictability it is important to estimate the motion between the frames and then to form an appropriate prediction that compensates for the motion. The process of estimating the motion between frames is known as *motion estimation (ME)*, and the process of forming a prediction while compensating for the relative motion between two frames is referred to as

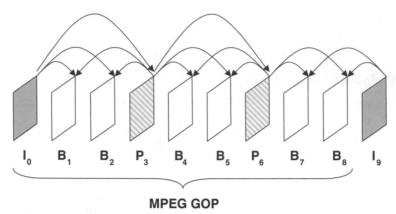

MPEG GOP

Figure 34.1 Example of the prediction dependencies between frames.

motion-compensated prediction (MC-P).   Block-based ME and MC-prediction is currently the most popular form of ME and MC-prediction: the current frame to be coded is partitioned into 16x16-pixel blocks, and for each block a prediction is formed by finding the best-matching block in the previously coded reference frame.   The relative motion for the best-matching block is referred to as the *motion vector.*

There are three basic common types of coded frames: (1) intra-coded frames, or I-frames, where the frames are coded independently of all other frames,   (2) predictively coded, or P-frames, where the frame is coded based on a previously coded frame, and (3) bi-directionally predicted frames, or B-frames, where the frame is coded using both previous and future coded frames.   Figure 34.1 illustrates the different coded frames and prediction dependencies for an example MPEG Group of Pictures (GOP).   The selection of prediction dependencies between frames can have a significant effect on video streaming performance, e.g. in terms of compression efficiency and error resilience.

Current video compression standards achieve compression by applying the same basic principles [9, 10]. The temporal redundancy is exploited by applying MC-prediction, the spatial redundancy is exploited by applying the DCT, and the color space redundancy is exploited by a color space conversion. The resulting DCT coefficients are quantized, and the nonzero quantized DCT coefficients are runlength and Huffman coded to produce the compressed bitstream.

3.2 VIDEO COMPRESSION STANDARDS

Video compression standards provide a number of benefits, foremost of which is ensuring interoperability, or communication between encoders and decoders made by different people or different companies. In this way standards lower the risk for both consumer and manufacturer, and this can lead to quicker acceptance and widespread use. In addition, these standards are designed for a large variety of applications, and the resulting economies of scale lead to reduced cost and further widespread use.

Currently there are two families of video compression standards, performed under the auspices of the International Telecommunications Union-Telecommunications (ITU-T, formerly the International Telegraph and Telephone Consultative Committee, CCITT) and the International Organization for

Standardization (ISO). The first video compression standard to gain widespread acceptance was the ITU H.261 [11], which was designed for videoconferencing over the integrated services digital network (ISDN). H.261 was adopted as a standard in 1990. It was designed to operate at p = 1,2, ..., 30 multiples of the baseline ISDN data rate, or p x 64 kb/s. In 1993, the ITU-T initiated a standardization effort with the primary goal of videotelephony over the public switched telephone network (PSTN) (conventional analog telephone lines), where the total available data rate is only about 33.6 kb/s. The video compression portion of the standard is H.263 and its first phase was adopted in 1996 [12]. An enhanced H.263, H.263 Version 2 (V2), was finalized in 1997, and a completely new algorithm, originally referred to as H.26L, is currently being finalized as H.264/AVC.

The Moving Pictures Expert Group (MPEG) was established by the ISO in 1988 to develop a standard for compressing moving pictures (video) and associated audio on digital storage media (CD-ROM). The resulting standard, commonly known as MPEG-1, was finalized in 1991 and achieves approximately VHS quality video and audio at about 1.5 Mb/s [13]. A second phase of their work, commonly known as MPEG-2, was an extension of MPEG-1 developed for application toward digital television and for higher bit rates [14]. A third standard, to be called MPEG-3, was originally envisioned for higher bit rate applications such as HDTV, but it was recognized that those applications could also be addressed within the context of MPEG-2; hence those goals were wrapped into MPEG-2 (consequently, there is no MPEG-3 standard). Currently, the video portion of digital television (DTV) and high definition television (HDTV) standards for large portions of North America, Europe, and Asia is based on MPEG-2. A third phase of work, known as MPEG-4, was designed to provide improved compression efficiency and error resilience features, as well as increased functionality, including object-based processing, integration of both natural and synthetic (computer generated) content, content-based interactivity [15].

Table 34.1 Current and emerging video compression standards.

Video Coding Standard	Primary Intended Applications	Bit Rate
H.261	Video telephony and teleconferencing over ISDN	p x 64 kb/s
MPEG-1	Video on digital storage media (CD-ROM)	1.5 Mb/s
MPEG-2	Digital Television	2-20 Mb/s
H.263	Video telephony over PSTN	33.6 kb/s and up
MPEG-4	Object-based coding, synthetic content, interactivity, video streaming	Variable
H.264/MPEG-4 Part 10 (AVC)	Improved video compression	10's to 100's of kb/s

The H.26L standard is being finalized by the Joint Video Team, from both ITU and ISO MPEG. It achieves a significant improvement in compression over all prior video coding standards, and it will be adopted by both ITU and ISO and called H.264 and MPEG-4 Part 10, Advanced Video Coding (AVC).

Currently, the video compression standards that are primarily used for video communication and video streaming are H.263 V2, MPEG-4, and the emerging H.264/MPEG-4 Part 10 AVC will probably gain wide acceptance.

What Do The Standards Specify?

An important question is what is the scope of the video compression standards, or what do the standards actually specify. A video compression system is composed of an encoder and a decoder with a common interpretation for compressed bit-streams. The encoder takes original video and compresses it to a bitstream, which is passed to the decoder to produce the reconstructed video. One possibility is that the standard would specify both the encoder and decoder. However this approach turns out to be overly restrictive. Instead, the standards have a limited scope to ensure interoperability while enabling as much differentiation as possible.

The standards do not specify the encoder nor the decoder. Instead they specify the bitstream syntax and the decoding process. The bitstream syntax is the format for representing the compressed data. The decoding process is the set of rules for interpreting the bitstream. Note that specifying the decoding process is different from specifying a specific decoder implementation. For example, the standard may specify that the decoder use an IDCT, but not how to implement the IDCT. The IDCT may be implemented in a direct form, or using a fast algorithm similar to the FFT, or using MMX instructions. The specific implementation is not standardized and this allows different designers and manufacturers to provide standard-compatible enhancements and thereby differentiate their work.

The encoder process is deliberately not standardized. For example, more sophisticated encoders can be designed that provide improved performance over baseline low-complexity encoders. In addition, improvements can be incorporated even after a standard is finalized, e.g. improved algorithms for motion estimation or bit allocation may be incorporated in a standard-compatible manner. The only constraint is that the encoder produces a syntactically correct bitstream that can be properly decoded by a standard-compatible decoder.

Limiting the scope of standardization to the bitstream syntax and decoding process enables improved encoding and decoding strategies to be employed in a standard-compatible manner - thereby ensuring interoperability while enabling manufacturers to differentiate themselves. As a result, it is important to remember that "not all encoders are created equal", even if they correspond to the same standard.

4. CHALLENGES IN VIDEO STREAMING

This section discusses some of the basic approaches and key challenges in video streaming. The three fundamental problems in video streaming are briefly highlighted and are examined in depth in the following three sections.

Video Delivery via File Download

Probably the most straightforward approach for video delivery of the Internet is by something similar to a file download, but we refer to it as video download to keep in mind that it is a video and not a generic file. Specifically, video download is similar to a file download, but it is a LARGE file. This approach allows the use of established delivery mechanisms, for example TCP as the transport layer or FTP or HTTP at the higher layers. However, it has a number of disadvantages. Since videos generally correspond to very large files, the download approach usually requires long download times and large storage spaces. These are important practical constraints. In addition, the entire video must be downloaded before viewing can begin. This requires patience on the viewers part and also reduces flexibility in certain circumstances, e.g. if the viewer is unsure of whether he/she wants to view the video, he/she must still download the entire video before viewing it and making a decision.

Video Delivery via Streaming

Video delivery by video streaming attempts to overcome the problems associated with file download, and also provides a significant amount of additional capabilities. The basic idea of video streaming is to split the video into parts, transmit these parts in succession, and enable the receiver to decode and playback the video as these parts are received, without having to wait for the entire video to be delivered. Video streaming can conceptually be thought to consist of the following steps:
1) Partition the compressed video into packets
2) Start delivery of these packets
3) Begin decoding and playback at the receiver while the video is still being delivered

Video streaming enables *simultaneous delivery and playback* of the video. This is in contrast to file download where the entire video must be delivered before playback can begin. In video streaming there usually is a short delay (usually on the order of 5-15 seconds) between the start of delivery and the beginning of playback at the client. This delay, referred to as the pre-roll delay, provides a number of benefits which are discussed in Section 6.

Video streaming provides a number of benefits including low delay before viewing starts and low storage requirements since only a small portion of the video is stored at the client at any point in time. The length of the delay is given by the time duration of the pre-roll buffer, and the required storage is approximately given by the amount of data in the pre-roll buffer.

Expressing Video Streaming as a Sequence of Constraints

A significant amount of insight can be obtained by expressing the problem of video streaming as a *sequence of constraints*. Consider the time interval between displayed frames to be denoted by Δ, e.g. Δ is 33 ms for 30 frames/s video and 100 ms for 10 frames/s video. Each frame must be delivered and decoded by its playback time, therefore the sequence of frames has an associated sequence of deliver/decode/display deadlines:
- o   Frame N must be delivered and decoded by time T_N
- o   Frame N+1 must be delivered and decoded by time $T_N + \Delta$
- o   Frame N+2 must be delivered and decoded by time $T_N + 2\Delta$
- o   Etc.

Any data that is lost in transmission cannot be used at the receiver. Furthermore, any data that arrives late is also useless. Specifically, any data that arrives after its decoding and display deadline is too late to be displayed. (Note that certain data may still be useful even if it arrives after its display time, for example if subsequent data depends on this "late" data.) Therefore, an important goal of video streaming is to perform the streaming in a manner so that this sequence of constraints is met.

4.1 BASIC PROBLEMS IN VIDEO STREAMING

There are a number of basic problems that afflict video streaming. In the following discussion, we focus on the case of video streaming over the Internet since it is an important, concrete example that helps to illustrate these problems. Video streaming over the Internet is difficult because the Internet only offers best effort service. That is, it provides no guarantees on bandwidth, delay jitter, or loss rate. Specifically, these characteristics are unknown and dynamic. Therefore, a key goal of video streaming is to design a system to reliably deliver high-quality video over the Internet when dealing with unknown and dynamic:

- o Bandwidth
- o Delay jitter
- o Loss rate

The bandwidth available between two points in the Internet is generally unknown and time-varying. If the sender transmits faster than the available bandwidth then congestion occurs, packets are lost, and there is a severe drop in video quality. If the sender transmits slower than the available bandwidth then the receiver produces sub-optimal video quality. The goal to overcome the bandwidth problem is to estimate the available bandwidth and then match the transmitted video bit rate to the available bandwidth. Additional considerations that make the bandwidth problem very challenging include accurately estimating the available bandwidth, matching the pre-encoded video to the estimated channel bandwidth, transmitting at a rate that is fair to other concurrent flows in the Internet, and solving this problem in a multicast situation where a single sender streams data to multiple receivers where each may have a different available bandwidth.

The end-to-end delay that a packet experiences may fluctuate from packet to packet. This variation in end-to-end delay is referred to as the delay jitter. Delay jitter is a problem because the receiver must receive/decode/display frames at a constant rate, and any late frames resulting from the delay jitter can produce problems in the reconstructed video, e.g. jerks in the video. This problem is typically addressed by including a playout buffer at the receiver. While the playout buffer can compensate for the delay jitter, it also introduces additional delay.

The third fundamental problem is losses. A number of different types of losses may occur, depending on the particular network under consideration. For example, wired packet networks such as the Internet are afflicted by packet loss, where an entire packet is erased (lost). On the other hand, wireless channels are typically afflicted by bit errors or burst errors. Losses can have a very destructive effect on the reconstructed video quality. To combat the effect of losses, a video streaming system is designed with error control. Approaches for error control can be roughly grouped into four classes: (1) forward error correction (FEC), (2) retransmissions, (3) error concealment, and (4) error-resilient video coding.

The three fundamental problems of unknown and dynamic *bandwidth, delay jitter, and loss,* are considered in more depth in the following three sections. Each section focuses on one of these problems and discusses various approaches for overcoming it.

5. TRANSPORT AND RATE CONTROL FOR OVERCOMING TIME-VARYING BANDWIDTHS

This section begins by discussing the need for streaming media systems to adaptively control its transmission rate according to prevalent network condition. We then discuss some ways in which appropriate transmission rates can be estimated dynamically at the time of streaming, and survey how media coding has evolved to support such dynamic changes in transmission rates.

5.1 THE NEED FOR RATE CONTROL

Congestion is a common phenomenon in communication networks that occurs when the offered load exceeds the designed limit, causing degradation in network performance such as throughput. Useful throughput can decrease for a number of reasons. For example, it can be caused by collisions in multiple access networks, or by increased number of retransmissions in systems employing such technology. Besides a decrease in useful throughput, other symptoms of congestion in packet networks may include packet losses, higher delay and delay jitter. As we have discussed in Section 4, such symptoms represent significant challenges to streaming media systems. In particular, packet losses are notoriously difficult to handle, and is the subject of Section 7.

To avoid the undesirable symptoms of congestion, control procedures are often employed to limit the amount of network load. Such control procedures are called rate control, sometimes also known as congestion control. It should be noted that different network technologies may implement rate control in different levels, such as hop-to-hop level or network level [16]. Nevertheless, for inter-networks involving multiple networking technologies, it is common to rely on rate control performed by the end-hosts. The rest of this section examines rate control mechanisms performed by the sources or sinks of streaming media sessions.

5.2 RATE CONTROL FOR STREAMING MEDIA

For environments like the Internet where little can be assumed about the network topology and load, determining an appropriate transmission rate can be difficult. Nevertheless, the rate control mechanism implemented in the Transmission Control Protocol (TCP) has been empirically proven to be sufficient in most cases. Being the dominant traffic type in the Internet, TCP is the workhorse in the delivery of web-pages, emails, and some streaming media. Rate control in TCP is based on a simple "Additive Increase Multiplicative Decrease" (AIMD) rule [17]. Specifically, end-to-end observations are used to infer packet losses or congestion. When no congestion is inferred, packet transmission is increased at a constant rate (additive increase). Conversely, when congestion is inferred, packet transmission rate is halved (multiplicative decrease).

Streaming Media over TCP

Given the success and ubiquity of TCP, it may seem natural to employ TCP for streaming media. There are indeed a number of important advantages of using TCP. First, TCP rate control has empirically proven stability and scalability. Second, TCP provides guaranteed delivery, effectively eliminating the much dreaded packet losses. Therefore, it may come as a surprise to realize that streaming media today are often carried using TCP only as a last resort, e.g., to get around firewalls. Practical difficulties with using TCP for streaming media include the following. First, delivery guarantee of TCP is accomplished through persistent retransmission with potentially increasing wait time between consecutive retransmissions, giving rise to potentially very long delivery time. Second, the "Additive Increase Multiplicative Decrease" rule gives rise to a widely varying instantaneous throughput profile in the form of a saw-tooth pattern not suitable for streaming media transport.

Streaming Media over Rate-controlled UDP

We have seen that both the retransmission and the rate control mechanisms of TCP possess characteristics that are not suitable for streaming media. Current streaming systems for the Internet rely instead on the best-effort delivery service in the form of User Datagram Protocol (UDP). This allows more flexibility both in terms of error control and rate control. For instance, instead of relying on retransmissions alone, other error control techniques can be incorporated or substituted. For rate control, the departure from the AIMD algorithm of TCP is a mixed blessing: it promises the end of wildly varying instantaneous throughput, but also the proven TCP stability and scalability.

Recently, it has been observed that the *average* throughput of TCP can be inferred from end-to-end measurements of observed quantities such as round-trip-time and packet losses [18,19]. Such observation gives rise to TCP-friendly rate control that attempts to mimic TCP throughput on a macroscopic scale and without the instantaneous fluctuations of TCP's AIMD algorithm [20,21]. One often cited importance of TCP-friendly rate control is its ability to coexist with other TCP-based applications. Another benefit though, is more predictable stability and scalability properties compared to an arbitrary rate control algorithm. Nevertheless, by attempting to mimic average TCP throughput under the same network conditions, TCP friendly rate control also inherits characteristics that may not be natural for streaming media. One example is the dependence of transmission rate on packet round-trip time.

Other Special Cases

Some media streaming systems do not perform rate control. Instead, media content is transmitted without regard to the prevalent network condition. This can happen in scenarios where an appropriate transmission rate is itself difficult to define, e.g., one-to-many communication where an identical stream is transmitted to all recipients via channels of different levels of congestion. Another possible reason is the lack of a feedback-channel.

Until now we have only considered rate control mechanisms that are implemented at the sender, now we consider an example where rate control is performed at the receiver. In the last decade, a scheme known as *layered multicast* has been proposed as a possible way to achieve rate control in Internet multicast of streaming media. Specifically, a scalable or layered compression scheme is assumed that produces multiple layers of compressed media, with a base layer that offers low but usable quality, and each additional layer provides further refinement to the quality. Each *receiver* can then individually decide how many layers to receive [22]. In other words, rate control is performed at the receiving end instead of the transmitting end. Multicast rate control is still an area of active research.

5.3 MEETING TRANSMISSION BANDWIDTH CONSTRAINTS

The incorporation of rate control introduces additional complexity in a streaming media system. Since transmission rate is dictated by channel conditions, problems may arise if the determined transmission rate is lower than the media bit rate. Client buffering helps to a certain degree to overcome occasional short-term drops in transmission rate. Nevertheless, it is not possible to stream a long 200 *kbps* stream through a 100 *kbps* channel, and the media bit rate needs to be modified to conform with the transmission constraints.

Transcoding

A direct method to modify the media bit rate is recompression, whereby the media is decoded and then re-encoded to the desired bit rate. There are two drawbacks with this approach. First, the media resulting from recompression is generally of lower quality than if the media was coded directly from the original source to the same bit rate. Second, media encoding generally requires extensive computation, making the approach prohibitively expensive. The complexity problem is solved by a technique known as compressed-domain transcoding. The basic idea is to selectively re-use compression decisions already made in the compressed media to reduce computation. Important transcoding operations include bit rate reduction, spatial downsampling, frame rate reduction, and changing compression formats [23].

Multiple File Switching

Another commonly used technique is multi-rate switching whereby multiple copies of the same content at different bit-rates are made available. Early implementations of streaming media systems coded the same content at a few strategic media rates targeted for common connection speeds (e.g. one for dialup modem and one for DSL/cable) and allowed the client to choose the appropriate media rate at the beginning of the session. However, these early systems only allowed the media rate to be chosen once at the beginning of each session. In contrast, multi-rate switching enables dynamic switching between different media rates within a single streaming media session. This mid-session switching between different media rates enables better adaptation to longer-term fluctuations in available bandwidth than can be achieved by the use of the client buffer alone. Examples include Intelligent Streaming from Microsoft and SureStream from Real Networks.

This approach overcomes both limitations of transcoding, as very little computation is needed for switching between the different copies of the stream, and no recompression penalty is incurred. However, there are a number of disadvantages. First, the need to store multiple copies of the same media incurs higher storage cost. Second, for practical implementation, only a small number of copies are used, limiting its ability to adapt to varying transmission rates.

Scalable Compression

A more elegant approach to adapt to longer-term bandwidth fluctuations is to use layered or scalable compression. This is similar in spirit to multi-rate switching, but instead of producing multiple copies of the same content at different bit rates, layered compression produces a set of (ordered) bitstreams (sometimes referred to as layers) and different subsets of these bitstreams can be selected to represent the media at different target bit rates [20]. Many commonly used compression standards, such as MPEG-2, MPEG-4 and H.263 have extensions for layered coding. Nevertheless, layered or scalable approaches are not widely used because they incur a significant compression penalty as compared to non-layered/non-scalable approaches.

5.4 EVOLVING APPROACHES

Rate control at end-hosts avoids congestion by dynamically adapting the transmission rate. Alternatively, congestion can also be avoided by providing unchanging amount of resources to each flow, but instead limiting the addition of new flows. This is similar to the telephone system that provides performance guarantees although with a possibility for call blocking.

With all the difficulties facing streaming media systems in the Internet, there has been work towards providing some Quality of Service (QoS) support in the Internet. The Integrated Services (IntServ) model of the Internet [24], for instance, is an attempt to provide end-to-end QoS guarantees in terms of bandwidth, packet loss rate, and delay, on a per-flow basis. QoS guarantees are established using explicit resource allocation based on the Resource Reservation Protocol (RSVP). The guarantees in terms of bandwidth and packet loss rate would have greatly simplified streaming media systems. Nevertheless, this is only at the expense of additional complexity in the network. The high complexity and cost of deployment of the RSVP-based service architecture eventually led the IETF to consider other QoS mechanisms. The Differentiated Services (DiffServ) model, in particular, is specifically designed to achieve low complexity and easy deployment at the cost of less stringent QoS guarantees than IntServ. Under DiffServ, service differentiation is no longer provided on a per-flow basis. Instead, it is based on the *code-point* or tag in each packet. Thus, packets having the same tags are given the same treatment under DiffServ regardless of where they originate. The cost of easy deployment for DiffServ compared to IntServ is the reduced level of QoS support. Specific ways in which streaming media systems can take advantage of a DiffServ Internet is currently an area of active research.

6.   PLAYOUT BUFFER FOR OVERCOMING DELAY JITTER

It is common for streaming media clients to have a 5 to 15 second buffering before playback starts. As we have seen in Section 4, streaming can be viewed as a sequence of constraints for individual media samples. The use of buffering essentially relaxes all the constraints by an identical amount. Critical to the performance of streaming systems over best-effort networks such as the Internet, buffering provides a number of important advantages:

1. Jitter reduction: Variations in network conditions cause the time it takes for packets to travel between identical end-hosts to vary. Such variations can be due to a number of possible causes, including queuing delays and link-level retransmissions. Jitter can cause jerkiness in playback due to the failure of same samples to meet their presentation deadlines, and have to be therefore skipped or delayed. The use of buffering effectively extends the presentation deadlines for all media samples, and in most cases, practically eliminates playback jerkiness due to delay jitter. The benefits of a playback buffer are illustrated in Figure 34.2, where packets are transmitted and played at a constant rate, and the playback buffer reduces the number of packets that arrive after their playback deadline.
2. Error recovery through retransmissions: The extended presentation deadlines for the media samples allow retransmission to take place when packets are lost, e.g., when UDP is used in place of TCP for transport. Since compressed media streams are often sensitive to errors, the ability to recover losses greatly improves streaming media quality.
3. Error resilience through Interleaving: Losses in some media streams, especially audio, can often be better *concealed* if the losses are isolated instead of concentrated. The extended presentation deadlines with the use of buffering allow interleaving to transform possible burst loss in the channel into isolated losses, thereby enhancing the concealment of the subsequent losses. As we shall discuss in the next section, the extended deadlines also allow other forms of error control schemes such as the use of error control codes, which are particularly effective when used with interleaving.
4. Smoothing throughput fluctuation: Since a time-varying channel gives rise to time varying throughput, the buffer can provide needed data to sustain streaming when throughput is low. This is especially important when streaming is performed using TCP (or HTTP), since the server typically does not react to a drop in channel throughput by reducing media rate.

The benefits of buffering do come at a price though. Besides additional storage requirements at the streaming client, buffering also introduces additional delay before playback can begin or resume (after a pause due to buffer depletion). Adaptive Media Playout (AMP) is a new technique that enables a valuable tradeoff between delay and reliability [25,4].

7.   ERROR CONTROL FOR OVERCOMING CHANNEL LOSSES

The third fundamental problem that afflicts video communication is losses. Losses can have a very destructive effect on the reconstructed video quality, and if the system is not designed to handle losses, even a single bit error can have a catastrophic effect. A number of different types of losses may occur, depending on the particular network under consideration. For example, wired packet networks such as the Internet are afflicted by packet loss, where congestion may cause an entire packet to be discarded (lost). In this case the receiver will either completely receive a packet in its entirety or completely lose a packet. On the other hand, wireless channels are typically afflicted by bit errors or burst errors at the physical layer. These errors may be passed up from the physical layer to the application as bit or burst errors, or alternatively, entire packets may be

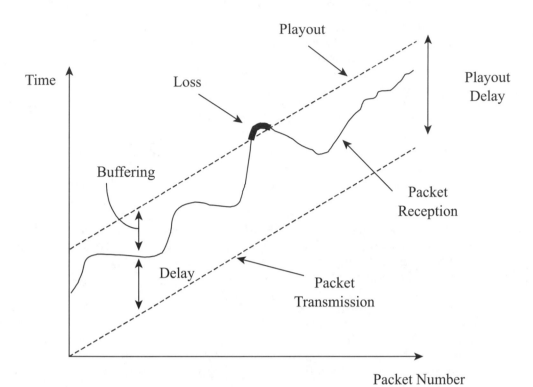

Figure 34.2 Effect of playout buffer on reducing the number of late packets.

discarded when any errors are detected in these packets. Therefore, depending on the interlayer communication, a video decoder may expect to always receive "clean" packets (without any errors) or it may receive "dirty" packets (with errors). The loss rate can vary widely depending on the particular network, and also for a given network depending on the amount of cross traffic. For example, for video streaming over the Internet one may see a packet loss rate of less than 1 %, or sometimes greater than 5-10 %.

A video streaming system is designed with error control to combat the effect of losses. There are four rough classes of approaches for error control: (1) retransmissions, (2) forward error correction (FEC), (3) error concealment, and (4) error-resilient video coding. The first two classes of approaches can be thought of as channel coding approaches for error control, while the last two are source coding approaches for error control. These four classes of approaches are discussed in the following four subsections. A video streaming system is typically designed using a number of these different approaches. In addition, joint design of the source coding and channel coding is very important and this is discussed in Section 7.5. Additional information, and specifics for H.263 and MPEG-4, are available in [26,27,28].

7.1 RETRANSMISSIONS

In retransmission-based approaches the receiver uses a back-channel to notify the sender which packets were correctly received and which were not, and this enables the sender to resend the lost packets. This approach efficiently uses the available bandwidth, in the sense that only lost packets are resent, and it also easily adapts to changing channel conditions. However, it also has some disadvantages. Retransmission leads to additional delay corresponding roughly to the round-trip-time (RTT) between receiver-sender-receiver. In addition, retransmission requires a back-channel, and this may not be possible or practical in various applications such as broadcast, multicast, or point-to-point without a back-channel.

In many applications the additional delay incurred from using retransmission is acceptable, e.g. Web browsing, FTP, telnet. In these cases, when guaranteed delivery is required (and a back-channel is available) then feedback-based retransmits provide a powerful solution to channel losses. On the other hand, when a back-channel is not available or the additional delay is not acceptable, then retransmission is not an appropriate solution.

There exist a number of important variations on retransmission-based schemes. For example, for video streaming of time-sensitive data one may use delay-constrained retransmission where packets are only retransmitted if they can arrive by their time deadline, or priority-based retransmission, where more important packets are retransmitted before less important packets. These ideas lead to interesting scheduling problems, such as which packet should be transmitted next (e.g. [29,4]).

7.2 FORWARD ERROR CORRECTION

The goal of FEC is to add specialized redundancy that can be used to recover from errors. For example, to overcome packet losses in a packet network one typically uses block codes (e.g. Reed Solomon or Tornado codes) that take K data packets and output N packets, where N-K of the packets are redundant packets. For certain codes, as long as *any* K of the N packets are correctly received the original data can be recovered. On the other hand, the added redundancy increases the required bandwidth by a factor of N/K. FEC provides a number of advantages and disadvantages. Compared to retransmissions, FEC does not require a back-channel and may provide lower delay since it does not depend on the round-trip-time of retransmits. Disadvantages of FEC include the overhead for FEC even when there are no losses, and possible latency associated with reconstruction of lost packets. Most importantly, FEC-based approaches are designed to overcome a predetermined amount of loss and they are quite effective *if* they are appropriately matched to the channel. If the losses are less than a threshold, then the transmitted data can be perfectly recovered from the received, lossy data. However, if the losses are greater than the threshold, then only a portion of the data can be recovered, and depending on the type of FEC used, the data may be completely lost. Unfortunately the loss characteristics for packet networks are often unknown and time varying. Therefore the FEC may be poorly matched to the channel -- making it ineffective (too little FEC) or inefficient (too much FEC).

7.3 ERROR CONCEALMENT

Transmission errors may result in lost information. The basic goal of error concealment is to estimate the lost information or missing pixels in order to conceal the fact that an error has occurred. The key observation is that video exhibits a significant amount of correlation along the spatial and temporal dimensions. This correlation was used to achieve video compression, and unexploited correlaton can also be used to estimate the lost information.

Therefore, the basic approach in error concealment is to exploit the correlation by performing some form of spatial and/or temporal interpolation (or extrapolation) to estimate the lost information from the correctly received data.

To illustrate the basic idea of error concealment, consider the case where a single 16x16 block of pixels (a macroblock in MPEG terminology) is lost. This example is not representative of the information typically lost in video streaming, however it is a very useful example for conveying the basic concepts. The missing block of pixels may be assumed to have zero amplitude, however this produces a black/green square in the middle of the video frame which would be highly distracting. Three general approaches for error concealment are: (1) spatial interpolation, (2) temporal interpolation (freeze frame), and (3) motion-compensated temporal interpolation. The goal of spatial interpolation is to estimate the missing pixels by smoothly extrapolating the surrounding correctly received pixels. Correctly recovering the missing pixels is extremely difficult, however even correctly estimating the DC (average) value is very helpful, and provides significantly better concealment than assuming the missing pixels have an amplitude of zero. The goal of temporal extrapolation is to estimate the missing pixels by copying the pixels at the same spatial location in the previous correctly decoded frame (freeze frame). This approach is very effective when there is little motion, but problems arise when there is significant motion. The goal of motion-compensated temporal extrapolation is to estimate the missing block of pixels as a motion compensated block from the previous correctly decoded frame, and thereby hopefully overcome the problems that arise from motion. The key problem in this approach is how to accurately estimate the motion for the missing pixels. Possible approaches include using the coded motion vector for that block (if it is available), use a neighboring motion vector as an estimate of the missing motion vector, or compute a new motion vector by leveraging the correctly received pixels surrounding the missing pixels.

Various error concealment algorithms have been developed that apply different combinations of spatial and/or temporal interpolation. Generally, a motion-compensated algorithm usually provides the best concealment (assuming an accurate motion vector estimate). This problem can also be formulated as a signal recovery or inverse problem, leading to the design of sophisticated algorithms (typically iterative algorithms) that provide improved error concealment in many cases.

The above example where a 16x16 block of pixels is lost illustrates many of the basic ideas of error concealment. However, it is important to note that errors typically lead to the loss of much more than a single 16x16 block. For example, a packet loss may lead to the loss of a significant fraction of an entire frame, or for low-resolution video (e.g. 176x144 pixels/frame) an entire coded frame may fit into a single packet, in which case the loss of the packet leads to the loss of the entire frame. When an entire frame is lost, it is not possible to perform any form of spatial interpolation as there is no spatial information available (all of the pixels in the frame are lost), and therefore only temporal information can be used for estimating the lost frame. Generally, the lost frame is estimated as the last correctly received frame (freeze frame) since this approach typically leads to the fewest artifacts.

A key point about error concealment is that it is performed at the decoder. As a result, error concealment is outside the scope of video compression standards. Specifically, as improved error concealment algorithms are developed they can be incorporated as standard-compatible enhancements to conventional decoders.

7.4 ERROR RESILIENT VIDEO CODING

Compressed video is highly vulnerable to errors. The goal of error-resilient video coding is to design the video compression algorithm and the compressed bitstream so that it is resilient to specific types of errors. This section provides an overview of error-resilient video compression. It begins by identifying the basic problems introduced by errors and then discusses the approaches developed to overcome these problems. In addition, we focus on which problems are most relevant for video streaming and also which approaches to overcome these problems are most successful and when. In addition, scalable video coding and multiple description video coding are examined as possible approaches for providing error resilient video coding.

7.4.1 Basic problems introduced by errors

Most video compression systems possess a similar architecture based on motion-compensated (MC) prediction between frames, Block-DCT (or other spatial transform) of the prediction error, followed by entropy coding (e.g. runlength and Huffman coding) of the parameters. The two basic error-induced problems that afflict a system based on this architecture are:
1) Loss of bitstream synchronization
2) Incorrect state and error propagation

The first class of problems, loss of bitstream synchronization, refers to the case when an error can cause the decoder to become confused and lose synchronization with the bitstream, i.e. the decoder may loss track of what bits correspond to what parameters. The second class of problems, incorrect state and error propagation, refers to what happens when a loss afflicts a system that uses predictive coding.

7.4.2 Overcoming Loss of Bitstream Synchronization

Loss of bitstream synchonization corresponds to the case when an error causes the decoder to loss track of what bits correspond to what parameters. For example, consider what happens when a bit error afflicts a Huffman codeword or other variable length codeword (VLC). Not only would the codeword be incorrectly decoded by the decoder, but because of the variable length nature of the codewords it is highly probably that the codeword would be incorrectly decoded to a codeword of a different length, and thereby all the subsequent bits in the bitstream (until the next resync) will be misinterpreted. Even a *single* bit error can lead to significant subsequent loss of information.

It is interesting to note that fixed length codes (FLC) do not have this problem, since the beginning and ending locations of each codeword are known, and therefore losses are limited to a single codeword. However, FLC's do not provide good compression. VLC's provide significantly better compression and therefore are widely used.

The key to overcoming the problem of loss of bitstream synchronization is to provide mechanisms that enable the decoder to quickly isolate the problem and resynchronize to the bitstream after an error has occurred. We now consider a number of mechanisms that enable bitstream resynchronization.

Resync Markers

Possibly the simplest approach to enable bitstream resynchronization is by the use of resynchronization markers, commonly referred to as resync markers. The basic idea is to place unique and easy to find entry points in the bitstream, so that if the decoder losses sync, it can look for the next entry point and then begin decoding again after the entry point. Resync markers correspond to a bitstream pattern that the decoder can unmistakably find. These markers are designed to be distinct from all codewords, concatenations of codewords, and minor perturbations of concatenated codewords. An example of a resync marker is the three-byte sequence consisting of 23 zeros followed by a one. Sufficient information is typically included after each resync marker to enable the restart of bitstream decoding.

An important question is where to place the resync markers. One approach is to place the resync markers at strategic locations in the compressed video hierarchy, e.g. picture or slice headers. This approach is used in MPEG-1/2 and H.261/3. This approach results in resyncs being placed every fixed number of blocks, which corresponds to a variable number of bits. An undesired consequence of this is that active areas, which require more bits, would in many cases have a higher probability of being corrupted. To overcome this problem, MPEG-4 provides the capability to place the resync markers periodically after every fixed number of bits (and variable number of coding blocks). This approach provides a number of benefits including reduces the probability that active areas be corrupted, simplifies the search for resync markers, and supports application-aware packetization.

Reversible Variable Length Codes (RVLCs)

Conventional VLC's, such as Huffman codes, are uniquely decodable in the forward direction. RVLCs in addition have the property that they are also uniquely decodable in the backward direction. This property can be quite beneficial in recovering data that would otherwise be lost. For example, if an error is detected in the bitstream, the decoder would typically jump to the next resync marker. Now, if RVLCs are used, instead of discarding all the data between the error and the resync, the decoder can start decoding backwards from the resync until it identifies another error, and thereby enables partial recovery of data (which would otherwise be discarded). Nevertheless, RVLC are typically less efficient than VLC.

Data Partitioning

An important observation is that bits which closely follow a resync marker are more likely to be accurately decoded than those further away. This motivates the idea of placing the most important information immediately after each resync (e.g. motion vectors, DC DCT coefficients, and shape information for MPEG-4) and placing the less important information later (AC DCT coefficients). This approach is referred to as data partitioning in MPEG-4. Note that this approach is in contrast with the conventional approach used in MPEG-1/2 and H.261/3, where the data is ordered in the bitstream in a consecutive macroblock by macroblock manner, and without accounting for the importance of the different types of data.

To summarize, the basic idea to overcome the problem of loss of bitstream synchronization is to first isolate (localize) the corrupted information, and second enable a fast resynchronization.

Application-Aware Packetization: Application Level Framing (ALF)

Many applications involve communication over a packet network, such as the Internet, and in these cases the losses have an important structure that can be exploited. Specifically, either a packet is accurately received in its entirety or it is completely lost. This means that the boundaries for lost information are exactly determined by the packet boundaries. This motivates the idea that to combat packet loss, one should *design (frame) the packet payload to minimize the effect of loss*. This idea was crystallized in the Application Level Framing (ALF) principle presented in [30], who basically said that the "application knows best" how to handle packet loss, out-of-order delivery, and delay, and therefore the application should design the packet payloads and related processing. For example, if the video encoder knows the packet size for the network, it can design the packet payloads so that each packet is independently decodable, i.e. bitstream resynchronization is supported at the packet level so that each correctly received packet can be straightforwardly parsed and decoded. MPEG-4, H.263V2, and H.264/MPEG-4 AVC support the creation of different forms of independently decodable *video packets*. As a result, the careful usage of the application level framing principle can often overcome the bitstream synchronization problem. Therefore, the major obstacle for reliable video streaming over lossy packet networks such as the Internet, is the error propagation problem, which is discussed next.

7.4.3 Overcoming Incorrect State and Error Propagation

If a loss has occurred, and even if the bitstream has been resynchronized, another crucial problem is that the state of the representation at the decoder may be different from the state at the encoder. In particular, when using MC-prediction an error causes the reconstructed frame (state) at the decoder to be incorrect. The decoder's state is then different from the encoder's, leading to incorrect (mismatched) predictions and often significant error propagation that can afflict many subsequent frames, as illustrated in Figure 34. 3. We refer to this problem as having incorrect (or mismatched) state at the decoder, because the state of the representation at the decoder (the previous coded frame) is not the same as the state at the encoder. This problem also arises in other contexts (e.g. random access for DVD's or channel acquisition for Digital TV) where a decoder attempts to decode beginning at an arbitrary position in the bitstream.

A number of approaches have been developed over the years to overcome this problem, where these approaches have the common goal of trying to limit the effect of error propagation. The simplest approach to overcome this problem is by using I-frame only. Clearly by not using any temporal prediction, this approach avoids the error propagation problem, however it also provides very poor compression and therefore it is generally not an appropriate streaming solution. Another approach is to use periodic I-frames, e.g. the MPEG GOP. For example, with a 15-frame GOP there is an I-frame every 15 frames and this periodic reinitialization of the prediction loop limits error propagation to a

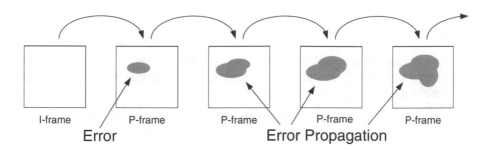

Figure 34.3 Example of error propagation that can result from a single error.

maximum of one GOP (15 frames in this example). This approach is used in DVD's to provide random access and Digital TV to provide rapid channel acquisition. However, the use of periodic I-frames limits the compression, and therefore this approach is often inappropriate for very low bit rate video, e.g. video over wireless channels or over the Internet.

More sophisticated methods of intra coding often apply partial intra-coding of each frame, where individual macroblocks (MBs) are intra-coded as opposed to entire frames. The simplest approach of this form is periodic intra-coding of all MBs: $1/N$ of the MB's in each frame are intra-coded in a predefined order, and after N frames all the MBs have been intra-coded. A more effective method is pre-emptive intra-coding, where one optimizes the intra-inter mode decision for each macroblock based on the macroblocks's content, channel loss model, and the macroblock's estimated vulnerability to losses.

The use of intra-coding to reduce the error propagation problem has a number of advantages and disadvantages. The advantages include: (1) intra coding does successfully limit error propagation by reinitializing the prediction loop, (2) the sophistication is at the encoder, while the decoder is quite simple, (3) the intra-inter mode decisions are outside the scope of the standards, and more sophisticated algorithms may be incorporated in a standard-compatible manner. However, intra-coding also has disadvantages including: (1) it requires a significantly higher bit rate than inter coding, leading to a sizable compression penalty, (2) optimal intra usage depends on accurate knowledge of channel characteristics. While intra coding limits error propagation, the high bit rate it requires limits its use in many applications.

Point-to-Point Communication with Back-Channel
The special case of point-to-point transmission with a back-channel and with real-time encoding facilitates additional approaches for overcoming the error propagation problem [31]. For example, when a loss occurs the decoder can notify the encoder of the loss and tell the encoder to reinitialize the prediction loop by coding the next frame as an I-frame. While this approach uses I-frames to overcome error propagation (similar to the previous approaches described above), the key is that I-frames are *only used* when necessary. Furthermore, this approach can be extended to provide improved compression efficiency by using P-frames as opposed to I-frames to overcome the error propagation. The basic idea is that both the encoder and decoder store multiple previously coded frames. When a loss occurs the decoder notifies the encoder which frames were correctly/erroneously received and therefore which frame should be used as the

reference for the next prediction. These capabilities are provided by the Reference Picture Selection (RPS) in H.263 V2 and NewPred in MPEG-4 V2. To summarize, for point-to-point communications with real-time encoding and a reliable back-channel with a sufficiently short round-trip-time (RTT), feedback-based approaches provide a very powerful approach for overcoming channel losses. However, the effectiveness of this approach decreases as the RTT increases (measured in terms of frame intervals), and the visual degradation can be quite significant for large RTTs [32].

Partial Summary: Need for Other Error-Resilient Coding Approaches
This section discussed the two major classes of problems that afflict compressed video communication in error-prone environments: (1) bitstream synchronization and (2) incorrect state and error propagation. The bitstream synchronization problem can often be overcome through appropriate algorithm and system design based on the application level framing principle. However, the error propagation problem remains a major obstacle for reliable video communication over lossy packet networks such as the Internet. While this problem can be overcome in certain special cases (e.g. point-to-point communication with a back-channel and with sufficiently short and reliable RTT), many important applications do not have a back-channel, or the back-channel may have a long RTT, thereby severely limiting effectiveness. Therefore, it is important to be able to overcome the error propagation problem in the *feedback-free* case, when there does not exist a back-channel between the decoder and encoder, e.g. broadcast, multicast, or point-to-point with unreliable or long-RTT back-channel.

7.4.4 Scalable Video Coding for Lossy Networks

In scalable or layered video the video is coded into a base layer and one or more enhancement layers. There are a number of forms of scalability, including temporal, spatial, and SNR (quality) scalability. Scalable coding essentially prioritizes the video data, and this prioritization effectively supports intelligent discarding of the data. For example, the enhancement data can be lost or discarded while still maintaining usable video quality. The different priorities of video data can be exploited to enable reliable video delivery by the use of unequal error protection (UEP), prioritized transmission, etc. As a result, scalable coding is a nice match for networks which support different qualities of service, e.g. DiffServ.

While scalable coding prioritizes the video data and is nicely matched to networks that can exploit different priorities, many important networks do not provide this capability. For example, the current Internet is a best-effort network. Specifically, it does not support any form of QoS, and all packets are equally likely to be lost. Furthermore, the base layer for scalable video is critically important – if the base layer is corrupted then the video can be completely lost. Therefore, there is a fundamental mismatch between scalable coding and the best-effort Internet: Scalable coding produces multiple bitstreams of differing importance, but the best-effort Internet does not treat these bitstreams differently – every packet is treated equally. This problem motivates the development of Multiple Description Coding, where the signal is coded into multiple bitstreams, each of roughly equal importance.

7.4.5   Multiple Description Video Coding

In Multiple Description Coding (MDC) a signal is coded into two (or more) separate bitstreams, where the multiple bitstreams are referred to as multiple descriptions (MD). MD coding provides two important properties: (1) each description can be independently decoded to give a usable reproduction of the original signal, and (2) the multiple descriptions contain complementary information so that the quality of the decoded signal improves with the number of descriptions that are correctly received. Note that this first property is in contrast to conventional scalable or layered schemes, which have a base layer that is critically important and if lost renders the other bitstream(s) useless. MD coding enables a useful reproduction of the signal when *any* description is received, and provides increasing quality as more descriptions are received.

A number of MD video coding algorithms have recently been proposed, which provide different tradeoffs in terms of compression performance and error resilience [33,34,35,36]. In particular, the MD video coding system of [36,38] has the importance property that it enables repair of corrupted frames in a description using uncorrupted frames in the other description so that usable quality can be maintained even when *both* descriptions are afflicted by losses, as long as both descriptions are not *simultaneously* lost. Additional benefits of this form of an MD system include high compression efficiency (achieving MDC properties with only slightly higher total bit rate than conventional single description (SD) compression schemes), ability to successfully operate over paths that support different or unbalanced bit rates (discussed next) [37], and this MD video coder also corresponds to a *standard-compatible enhancement* to MPEG-4 V2 (with NEWPRED), H.263 V2 (with RPS), and H.264/MPEG-4 Part 10 (AVC).

Multiple Description Video Coding and Path Diversity
MD coding enables a useful reproduction of the signal when any description is received -- and specifically this form of MD coding enables a useful reproduction when at least one description is received *at any point in time.* Therefore it is beneficial to increase the probability that at least one description is received correctly at any point in time.   This can be achieved by combining MD video coding with a path diversity transmission system [38], as shown in Figure 34. 4, where different descriptions are explicitly transmitted over different network paths (as opposed to the default scenarios where they would proceed along a single path).   Path diversity enables the end-to-end video application to effectively see a virtual channel with improved loss characteristics [38].   For example, the application effectively sees an average path behavior, which generally provides better performance than seeing the behavior of any individual random path.   Furthermore, while any network path may suffer from packet loss, there is a much smaller probability that all of the multiple paths *simultaneously* suffer from losses. In other words, losses on different paths are likely to be uncorrelated. Furthermore, the path diversity transmission system and the MD coding of [36,38] complement each other to improve the effectiveness of MD coding: the path diversity transmission system reduces the probability that both descriptions are simultaneously lost, and the MD decoder enables recovery from losses as long as both descriptions are not simultaneously lost. A path diversity transmission system may be created in a number of ways, including by source-based routing or by using a relay infrastructure.   For example, path diversity may be achieved by a relay infrastructure, where each

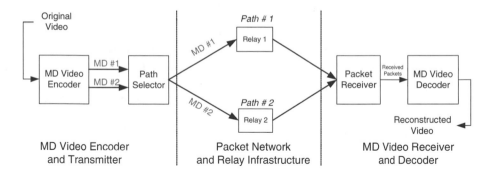

Figure 34.4 Multiple description video coding and path diversity for reliable
communication over lossy packet networks.

stream is sent to a different relay placed at a strategic node in the network, and
each relay performs a simple forwarding operation. This approach corresponds
to an application-specific overlay network on top of the conventional Internet,
providing a service of improved reliability while leveraging the infrastructure of
the Internet [38].

Multiple Description versus Scalable versus Single Description
The area of MD video coding is relatively new, and therefore there exist many
open questions as to when MD coding, or scalable coding, or single description
coding is preferable. In general, the answer depends crucially on the specific
context, e.g. specific coder, playback delay, possible retransmits, etc. A few
works shed light on different directions. [39,40] proposed MD image coding sent
over multiple paths in an ad-hoc wireless network, and [41] examined MD versus
scalable coding for an EGPRS cellular network. Analytical models for accurately
predicting SD and MD video quality as a function of path diversity and loss
characteristics are proposed in [42]. Furthermore, as is discussed in Section 10,
path diversity may also be achieved by exploiting the infrastructure of a content
delivery network (CDN), to create a Multiple Description Streaming Media CDN
(MD-CDN) [43]. In addition, the video streaming may be performed in a channel-
adaptive manner as a function of the path diversity characteristics [44,4].

7.5 JOINT SOURCE/CHANNEL CODING
Data and video communication are fundamentally different. In data
communication all data bits are equally important and *must* be reliably
delivered, though timeliness of delivery may be of lesser importance. In
contrast, for video communication some bits are more important than other bits,
and often it is *not* necessary for all bits to be reliably delivered. On the other
hand, timeliness of delivery is often critical for video communication. Examples
of coded video data with different importance include the different frames types
in MPEG video (i.e. I-frames are most important, P-frames have medium
importance, and B-frames have the least importance) and the different layers in
a scalable coding (i.e. base layer is critically important and each of the
enhancement layers is of successively lower importance). A basic goal is then to
exploit the differing importance of video data, and one of the motivations of joint
source/channel coding is to jointly design the source coding and the channel
coding to exploit this difference in importance. This has been an important area
of research for many years, and the limited space here prohibits a detailed

discussion, therefore we only present two illustrative examples of how error-control can be adapted based on the importance of the video data. For example, for data communication all bits are of equal importance and FEC is designed to provide equal error protection for every bit. However, for video date of unequal importance it is desirable to have unequal error protection (UEP) as shown in Table 34.2. Similarly, instead of a common retransmit strategy for all data bits, it is desirable to have unequal (or priortized) retransmit strategies for video data.

Table 34.2 Adapting error control based on differing importance of video data: unequal error protection and unequal (prioritized) retransmission based on coded frame type.

	I-frame	P-frame	B-frame
FEC	Maximum	Medium	Minimum (or none)
Retransmit	Maximum	Medium	Can discard

8.   MEDIA STREAMING PROTOCOLS AND STANDARDS

This section briefly describes the network protocols for media streaming over the Internet. In addition, we highlight some of the current popular specifications and standards for video streaming, including 3GPP and ISMA.

8.1 PROTOCOLS FOR VIDEO STREAMING OVER THE INTERNET

This section briefly highlights the network protocols for video streaming over the Internet. First, we review the important Internet protocols of IP, TCP, and UDP. This is followed by the media delivery and control protocols.

Internet Protocols: TCP, UDP, IP

The Internet was developed to connect a heterogeneous mix of networks that employ different packet switching technologies. The Internet Protocol (IP) provides baseline best-effort network delivery for all hosts in the network: providing addressing, best-effort routing, and a global format that can be interpreted by everyone. On top of IP are the end-to-end transport protocols, where Transmission Control Protocol (TCP) and User Datagram Protocol (UDP) are the most important. TCP provides reliable byte-stream services. It guarantees delivery via retransmissions and acknowledgements. On the other hand, UDP is simply a user interface to IP, and is therefore unreliable and connectionless. Additional services provided by UDP include checksum and port-numbering for demultiplexing traffic sent to the same destination. Some of the differences between TCP and UDP that affects streaming applications are:

o   TCP operates on a byte stream while UDP is packet oriented.
o   TCP guarantees delivery via retransmissions, but because of the retransmissions its delay is unbounded. UDP does not guarantee delivery, but for those packets delivered their delay is more predictable (i.e. one-way delay) and smaller.
o   TCP provides flow control and congestion control. UDP provides neither. This provides more flexibility for the application to determine the appropriate flow control and congestion control procedures.
o   TCP requires a back-channel for the acknowledgements. UDP does not require a back-channel.

Web and data traffic are delivered with TCP/IP because guaranteed delivery is far more important than delay or delay jitter. For media streaming the uncontrollable delay of TCP is unacceptable and compressed media data is usually transmitted via UDP/IP despite control information, which is usually transmitted via TCP/IP.

Media Delivery and Control Protocols

The IETF has specified a number of protocols for media delivery, control, and description over the Internet.

Media Delivery

The Real-time Transport Protocol (RTP) and Real-time Control Protocol (RTCP) are IETF protocols designed to *support streaming media*. RTP is designed for data transfer and RTCP for control messages. Note that these protocols do *not* enable real-time services, only the underlying network can do this, however they provide functionalities that support real-time services. RTP does not guarantee QoS or reliable delivery, but provides support for applications with time constraints by providing a standardized framework for common functionalities such as time stamps, sequence numbering, and payload specification. RTP enables detection of lost packets. RTCP provides feedback on quality of data delivery. It provides QoS feedback in terms of number of lost packets, inter-arrival jitter, delay, etc. RTCP specifies periodic feedback packets, where the feedback uses no more than 5 % of the total session bandwidth and where there is at least one feedback message every 5 seconds. The sender can use the feedback to adjust its operation, e.g. adapt its bit rate. The conventional approach for media streaming is to use RTP/UDP for the media data and RTCP/TCP or RTCP/UDP for the control. Often, RTCP is supplemented by another feedback mechanism that is explicitly designed to provide the desired feedback information for the specific media streaming application. Other useful functionalities facilitated by RTCP include inter-stream synchronization and round-trip time measurement.

Media Control

Media control is provided by either of two session control protocols: Real-Time Streaming Protocol (RTSP) or Session Initiation Protocol (SIP). RTSP is commonly used in video streaming to establish a session. It also supports basic VCR functionalities such as play, pause, seek and record. SIP is commonly used in voice over IP (VoIP), and it is similar to RTSP, but in addition it can support user mobility and a number of additional functionalities.

Media Description and Announcement

The Session Description Protocol (SDP) provides information describing a session, for example whether it is video or audio, the specific codec, bit rate, duration, etc. SDP is a common exchange format used by RTSP for content description purposes, e.g., in 3G wireless systems. It has also been used with the Session Announcement Protocol (SAP) to announce the availability of multicast programs.

8.2 VIDEO STREAMING STANDARDS AND SPECIFICATIONS

Standard-based media streaming systems, as specified by the 3rd Generation Partnership Project (3GPP) for media over 3G cellular [45] and the Internet

Streaming Media Alliance (ISMA) for streaming over the Internet [46], employ the following protocols:
- Media encoding
    - MPEG-4 video and audio (AMR for 3GPP), H.263
- Media transport
    - RTP for data, usually over UDP/IP
    - RTCP for control messages, usually over UDP/IP
- Media session control
    - RTSP
- Media description and announcement
    - SDP

The streaming standards do not specify the storage format for the compressed media, but the MP4 file format has been widely used. One advantage of MP4 file format is the ability to include "hint tracks" that simplify various aspects of streaming by providing hints such as packetization boundaries, RTP headers and transmission times.

9. ADDITIONAL VIDEO STREAMING TOPICS

MULTICAST

Multicast or one-to-many communication has received much attention in the last few years due to the significant bandwidth savings it promises, and the challenges it presents. Consider the multicast extension of the Internet, or IP multicast, as an example. When multiple clients are requesting the same media stream, IP multicast reduce network resource usage by transmitting only one copy of the stream down shared links, instead of one per session sharing the link. Nevertheless, besides the many practical difficulties in supporting IP multicast for the wide-area Internet, the basic properties of multicast communication present a number of challenges to streaming media systems. First and foremost is the problem of heterogeneity: different receivers experience different channel conditions and may have conflicting requirements, e.g. in terms of maximum bit-rate that can be supported, and the amount of error protection needed. Heterogeneity is typically solved by using multiple multicast to provide choices for the receivers. For instance, it is possible to establish different multicasts for different ranges of intended bit-rates [47]. Alternatively, the different multicasts can contain incremental information [22, 48]. A second challenge that multicast presents is the more restricted choice for error control. While retransmission has been the error control mechanism of choice for many streaming applications, its applicability in multicast has been limited by a number of challenges. Using IP multicast for retransmission, for instance, requires that both the retransmission request and the actual retransmission be transmitted to all the receivers in the multicast, an obviously inefficient solution. Even when retransmissions are handled by unicast communication, scalability concerns still remain, since a single sender will have to handle the requests of potentially many receivers.

END-TO-END SECURITY AND TRANSCODING

Encryption of media is an effective tool to protect content from eavesdroppers. Transcoding at intermediate nodes within a network is also an important technique to adapt compressed media streams for particular client capabilities or network conditions. Nevertheless, network transcoding poses a serious threat to the end-to-end security because transcoding encrypted streams generally requires decrypting the stream, transcoding the decrypted stream, and then re-encrypting the result. Each transcoding node presents a possible breach to the security of the system. This problem can be overcome by Secure Scalable Streaming (SSS) which enables downstream transcoding without decryption [49,50]. SSS uses jointly designed scalable coding and progressive encryption techniques to encode and encrypt video into secure scalable packets that are transmitted across the network. These packets can be transcoded at intermediate, possibly untrusted, network nodes by simply truncating or discarding packets and without compromising the end-to-end security of the system. The secure scalable packets have unencrypted headers that provide hints, such as optimal truncation points, which the downstream transcoders use to achieve rate-distortion (R-D) optimal fine-grain transcoding *across* the encrypted packets.

STREAMING OVER WIRED AND WIRELESS LINKS

When the streaming path involves both wired and wireless links, some additional challenges evolve. The first challenge involves the much longer packet delivery time with the addition of a wireless link. Possible causes for the long delay include the employment of FEC with interleaving. For instance, round-trip propagation delay in the 3G wireless system is in the order of 100 ms even before link-level retransmission. With link-level retransmission, the delay for the wireless link alone can be significant. The long round-trip delay reduces the efficiency of a number of end-to-end error control mechanisms: the practical number of end-to-end retransmissions is reduced, and the effectiveness of schemes employing RPS and NewPred is also reduced. The second challenge is the difficulty in inferring network conditions from end-to-end measurements. In high-speed wired networks, packet corruption is so rare that packet loss is a good indication of network congestion, the proper reaction of which is congestion control. In wireless networks however, packet losses may be due to corruption in the packet, which calls for stronger channel coding. Since any end-to-end measurements contain aggregate statistics across both the wired and wireless links, it is difficult to identify the proper cause and therefore perform the proper reaction.

10.   STREAMING MEDIA CONTENT DELIVERY NETWORKS

The Internet has rapidly emerged as a mechanism for users to find and retrieve content, originally for webpages and recently for streaming media. Content delivery networks (CDNs) were originally developed to overcome performance problems for delivery of static web content (webpages). These problems include network congestion and server overload, which arise when many users access popular content. CDNs improve end-user performance by caching popular content on edge servers located closer to users. This provides a number of advantages. First, it helps prevent server overload, since the replicated content can be delivered to users from edge servers. Furthermore, since content is delivered from the closest edge server and not from the origin server, the content is sent over a shorter network path, thus reducing the request response time, the probability of packet loss, and the total network resource usage. While CDNs were originally intended for static web content, recently, they are being designed for delivery of streaming media as well.

10.1 STREAMING MEDIA CDN DESIGN

A streaming media CDN is a CDN that is explicitly designed to deliver streaming media, as opposed to static webpages. Streaming media CDN design and operation is similar in many ways to conventional (webpage) CDN design and operation. For example, there are three key problems that arise in general CDN design and operation. The first is the server placement problem: Given N servers, where should these servers be placed on the network? The second problem is the content distribution problem: On which servers should each piece of content be replicated? The third problem is the server selection problem: For each request, which is the optimal server to direct the client to for delivery of the content?

Many aspects of a streaming media CDN are also quite different from a conventional CDN. For example, client-server interaction for a conventional (webpage) CDN involves a short-lived (fraction of a second) HTTP/TCP session(s). However, a streaming session generally has a long duration (measured in minutes) and uses RTSP and RTP/UDP. While congestion and packet loss may lead to a few second delay in delivering a webpage and is often acceptable, the corresponding effect on a streaming media session would be an interruption (stall or visual artifacts) that can be highly distracting. Clearly, delay, packet loss, and any form of interruption can have a much more detrimental effect on video streaming then on static webpage delivery. In addition, in a conventional CDN each piece of content (webpage) is relatively small, on the order of 10's of kilobytes, and therefore it can be replicated in its entirety on each chosen server. However, streaming media, such as movies, have a long duration and require a significant amount of storage, on the order of megabytes or gigabytes, and therefore it is often not practical or desirable to replicate an entire video stream on each chosen server. Instead, the video can be partitioned into parts, and only a portion of each video is cached on each server. There are many interesting problems related to caching of video, e.g. see [51] and references therein.

Two other capabilities that are important for streaming media CDN, and are of lesser importance for a conventional CDN for webpage distribution, are multicast and server hand-off. Multicast is clearly a highly desirable capability for streaming of popular media. While wide-area IP Multicast is currently not available in the Internet, a streaming media CDN can be explicitly designed to provide this capability via application-layer multicast: the infrastructure in the streaming media CDN provide an overlay on the Internet and are used as the nodes for the multicast tree. Communication between nodes employs only simple ubiquitous IP service, thereby avoiding the dependence of IP multicast. Another important capability is hand-off between streaming servers. Because streaming media sessions are long lived, it is sometimes required to perform a midstream hand-off from one streaming server to another. This functionality is not required for webpage delivery where the sessions are very short in duration. Furthermore, when the streaming session involves transcoding, mid-stream hand-off of the transcoding session is also required between servers [52].

10.2 MULTIPLE DESCRIPTION STREAMING MEDIA CDN (MD-CDN)

CDNs have been widely used to provide low latency, scalability, fault tolerance, and load balancing for the delivery of web content and more recently streaming media. Another important advantage offered by streaming media CDNs is their

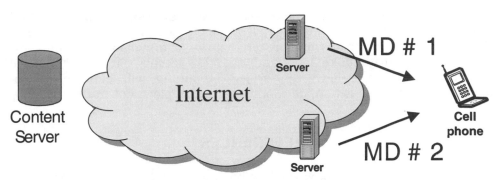

Figure 34.5 An MD-CDN uses MD coding and path diversity to provide improved reliability for packet losses, link outages, and server failures.

distributed infrastructure. The distributed infrastructure of a CDN can be used to explicitly achieve *path diversity* between each client and multiple nearby edge servers. Furthermore, appropriately coupling MD coding with this path diversity can provide improved reliability to packet losses, link outages, and server failures. This system is referred to as a *Multiple Description Streaming Media Content Delivery Network* [43] or an MD-CDN for short.

An MD-CDN operates in the following manner: (1) MD coding is used to code a media stream into multiple complementary descriptions, (2) the different descriptions are distributed across different edge servers in the CDN, (3) when a client requests a media stream, it is directed to multiple nearby servers that host complementary descriptions, and (4) the client simultaneously receives the different complementary descriptions through different network paths from different servers. That is, the existing CDN infrastructure is exploited to achieve path diversity between multiple servers and each client. In this way, disruption in streaming media occurs only in the less likely case when simultaneous losses afflict both paths. This architecture also reaps the benefits associated with CDNs, such as reduced response time to clients, load balancing across servers, robustness to network and server failures, and scalability to number of clients.

Further information about MD-CDN design and operation is available in [43]. Other related works include distributing MD coded data in peer-to-peer networks [53], streaming a conventional single description stream to a single client from multiple servers [54], and the use of Tornado codes and multiple servers to reduce download time for bulk data (not video) transfer [55].

11. SUMMARY

Video communication over packet networks has witnessed much progress in the past few years, from download-and-play to various adaptive techniques, and from direct use of networking infrastructure to the design and use of overlay architectures. Developments in algorithms and in computer, communication and network infrastructure technologies have continued to change the landscape of streaming media, each time simplifying some of the current challenges and spawning new applications and challenges. For example, the emergence of streaming media CDNs presents a variety of conceptually exciting and practically important opportunities to not only mitigate existing problems, but create new applications as well. The advent of high bandwidth wireless networking technologies calls for streaming media solutions that support not only wireless environments, but user mobility as well. Possible QoS support in the Internet, on the other hand, promises a more predictable channel for streaming media applications that may make low-bandwidth low-latency streaming over IP a reality. Therefore, we believe that video streaming will continue to be a compelling area for exploration, development, and deployment in the future.

REFERENCES

[1]   M.-T. Sun and A. Reibman, eds, *Compressed Video over Networks*, Marcel Dekker, New York, 2001.

[2]   G. Conklin, G. Greenbaum, K. Lillevold, A. Lippman, and Y. Reznik, Video Coding for Streaming Media Delivery on the Internet, *IEEE Trans. Circuits and Systems for Video Technology*, March 2001.

[3]   D. Wu, Y. Hou, W. Zhu, Y.-Q. Zhang, and J. Peha, Streaming Video over the Internet: Approaches and Directions, *IEEE Transactions on Circuits and Systems for Video Technology*, March 2001.

[4]   B. Girod, J. Chakareski, M. Kalman, Y. Liang, E. Setton, and R. Zhang, Advances in Network-Adaptive Video Streaming, *2002 Tyrrhenian Inter. Workshop on Digital Communications*, Sept 2002.

[5]   Y. Wang, J. Ostermann, and Y.-Q. Zhang, *Video Processing and Communications*, New Jersey, Prentice-Hall, 2002.

[6]   www.realnetworks.com

[7]   www.microsoft.com/windows/windowsmedia

[8]   G. K. Wallace, The JPEG Still Picture Compression Standard, *Communications of the ACM*, April, 1991.

[9]   V. Bhaskaran and K. Konstantinides, *Image and Video Compression Standards: Algorithms and Architectures*, Boston, Massachusetts: Kluwer Academic Publishers, 1997.

[10]   J. Apostolopoulos and S. Wee, Video Compression Standards, *Wiley Encyclopedia of Electrical and Electronics Engineering*, John Wiley & Sons, Inc., New York, 1999.

[11]   Video codec for audiovisual services at px64 kbits/s, ITU-T Recommendation H.261, Inter. Telecommunication Union, 1993.

[12]   Video coding for low bit rate communication, ITU-T Rec. H.263, Inter. Telecommunication Union, version 1, 1996; version 2, 1997.

[13]   ISO/IEC 11172, Coding of moving pictures and associated audio for digital storage media at up to about 1.5 Mbits/s. International Organization for Standardization (ISO), 1993.

[14]   ISO/IEC 13818. Generic coding of moving pictures and associated audio information. International Organization for Standardization (ISO), 1996.

[15]   ISO/IEC 14496. Coding of audio-visual objects. International Organization for Standardization (ISO), 1999.

[16]   M. Gerla and L. Kleinrock, Flow Control: A Comparative Survey, *IEEE Trans. Communications*, Vol. 28 No. 4, April 1980, pp 553-574.

[17]   V. Jacobson, Congestion Avoidance and Control, *ACM SIGCOMM*, August 1988.

[18]   M. Mathis et al., The Macroscopic Behavior of the TCP Congestion Avoidance Algorithm, *ACM Computer Communications Review*, July 1997.

[19] J. Padhye et al., Modeling TCP Reno Performance: A Simple Model and its Empirical Validation, *IEEE/ACM Trans. Networking*, April 2000.

[20] W. Tan and A. Zakhor, Real-time Internet Video using Error-Resilient Scalable Compression and TCP-friendly Transport Protocol, *IEEE Trans. on Multimedia*, June 1999.

[21] S. Floyd et al., Equation-based Congestion Control for Unicast Applications, *ACM SIGCOMM*, August 2000.

[22] S. McCanne, V. Jacobsen, and M. Vetterli, Receiver-driven layered multicast, *ACM SIGCOMM*, Aug. 1996.

[23] S. Wee, J. Apostolopoulos and N. Feamster, Field-to-Frame Transcoding with Temporal and Spatial Downsampling, *IEEE International Conference on Image Processing*, October 1999.

[24] P. White, RSVP and Integrated Services in the Internet: A Tutorial, *IEEE Communications Magazine*, May 1997.

[25] M. Kalman, E. Steinbach, and B. Girod, Adaptive Media Playout for Low Delay Video Streaming over Error-Prone Channels, preprint, to appear *IEEE Trans. Circuits and Systems for Video Technology*.

[26] Y. Wang and Q. Zhu, Error control and concealment for video communications: A review, *Proceedings of the IEEE*, May 1998.

[27] N. Färber, B. Girod, and J. Villasenor, Extension of ITU-T Recommendation H.324 for error-resilient video transmission, *IEEE Communications Magazine*, June 1998.

[28] R. Talluri, Error-resilient video coding in the ISO MPEG-4 standard, *IEEE Communications Magazine*, June 1998.

[29] P. Chou and Z. Miao, Rate-distortion optimized streaming of packetized media, *IEEE Trans. on Multimedia*, submitted Feb. 2001.

[30] D. Clark and D. Tennenhouse, Architectural Considerations for a New Generation of Protocols, *ACM SIGCOMM*, September 1990.

[31] B. Girod and N. Färber, Feedback-based error control for mobile video transmission, *Proceedings of the IEEE*, October 1999.

[32] S. Fukunaga, T. Nakai, and H. Inoue, Error resilient video coding by dynamic replacing of reference pictures, *GLOBECOM*, Nov. 1996.

[33] S. Wenger, Video Redundancy Coding in H.263+, *Workshop on Audio-Visual Services for Packet Networks*, September 1997.

[34] V. Vaishampayan and S. John, Interframe balanced-multiple-description video compression, *IEEE Inter Conf. on Image Processing*, Oct.1999.

[35] A. Reibman, H. Jafarkhani, Y. Wang, M. Orchard, and R. Puri, Multiple description coding for video using motion compensated prediction, *IEEE Inter. Conf. Image Processing* , October 1999.

[36] J. Apostolopoulos, Error-resilient video compression via multiple state streams, *Proc. International Workshop on Very Low Bitrate Video Coding (VLBV'99)*, October 1999.

[37] J. Apostolopoulos and S. Wee, Unbalanced Multiple Description Video Communication Using Path Diversity, *IEEE International Conference on Image Processing*, October 2001.

[38] J. Apostolopoulos, Reliable Video Communication over Lossy Packet Networks using Multiple State Encoding and Path Diversity, *Visual Communications and Image Processing*, January 2001.

[39]   N. Gogate and S. Panwar, Supporting video/image applications in a mobile multihop radio environment using route diversity, *Proceedings Inter. Conference on Communications*, June 1999.

[40]   N. Gogate, D. Chung, S.S. Panwar, and Y. Wang, Supporting image/video applications in a mobile multihop radio environment using route diversity and multiple description coding, Preprint.

[41]   A. Reibman, Y. Wang, X. Qiu, Z. Jiang, and K. Chawla, Transmission of Multiple Description and Layered Video over an (EGPRS) Wireless Network, *IEEE Inter. Conf. Image Processing*, September 2000.

[42]   J. Apostolopoulos, W. Tan, S. Wee, and G. Wornell, Modeling Path Diversity for Multiple Description Video Communication, *IEEE Inter. Conference on Acoustics, Speech, and Signal Processing*, May 2002.

[43]   J. Apostolopoulos, T. Wong, W. Tan, and S. Wee, On Multiple Description Streaming with Content Delivery Networks, *IEEE INFOCOM*, July 2002.

[44]   Y. Liang, E. Setton and B. Girod, Channel-Adaptive Video Streaming Using Packet Path Diversity and Rate-Distortion Optimized Reference Picture Selection, to appear *IEEE Fifth Workshop on Multimedia Signal Processing*, Dec. 2002.

[45]   www.3gpp.org

[46]   www.isma.tv

[47]   S. Cheung, M. Ammar and X. Li, On the use of Destination Set Grouping to Improve Fairness in Multicast Video Distribution, *IEEE INFOCOM*, March 1996.

[48]   W. Tan and A. Zakhor, Video Multicast using Layered FEC and Scalable Compression, *IEEE Trans. Circuits and Systems for Video Technology*, March 2001.

[49]   S. Wee, J. Apostolopoulos, Secure Scalable Video Streaming for Wireless Networks, *IEEE International Conference on Acoustics, Speech, and Signal Processing*, May 2001.

[50]   S. Wee, J. Apostolopoulos, Secure Scalable Streaming Enabling Transcoding without Decryption, *IEEE International Conference on Image Processing*, October 2001.

[51]   Z. Miao and A. Ortega, Scalable Proxy Caching of Video under Storage Constraints, *IEEE Journal on Selected Areas in Communications*, to appear 2002.

[52]   S. Roy, B. Shen, V. Sundaram, and R. Kumar, Application Level Hand-off Support for Mobile Media Transcoding Sessions, *ACM NOSSDAV*, May, 2002.

[53]   V. N. Padmanabhan, H. J. Wang, P. A. Chou, and K. Sripanidkulchai, Distributing streaming media content using cooperative networking, *ACM NOSSDAV*, May 2002.

[54]   T. Nguyen and A. Zakhor, Distributed video streaming over internet SPIE Multimedia Computing and Networking 2002, January 2002.

[55]   J. Byers, M. Luby, and M. Mitzenmacher, Accessing multiple mirror sites in parallel: Using tornado codes to speed up downloads, *IEEE INFOCOM*, 1999.

35

CONTINUOUS DISPLAY OF VIDEO OBJECTS USING HETEROGENEOUS DISK SUBSYSTEMS

Shahram Ghandeharizadeh
Department of Computer Science
University of Southern California
Los Angeles, California, USA
shahram@cs.usc.edu

Seon Ho Kim
Department of Computer Science
University of Denver
Denver, Colorado, USA
seonkim@cs.du.edu

1. INTRODUCTION

Continuous media objects, audio and video clips, are large in size and must be retrieved at a pre-specified rate in order to ensure their hiccup-free display [9, 29]. Even with 100 gigabyte disk drives, a video library consisting of 1000 MPEG-2 titles (with an average display time of 90 minutes) requires thirty such disks for data storage.[1] Over time such a storage system will evolve to consist of a heterogeneous collection of disk drives. There are several reasons why a system administrator might be forced to buy new disk drives over time. First, the application might require either a larger storage capacity due to introduction of new titles or a higher bandwidth due to a larger number of users accessing the library. Second, existing disks might fail and need to be replaced.[2] The system administrator may not be able to purchase the original disk models due to the technological trends in the area of magnetic disks: Approximately every 12 to 18 months, the cost per megabyte of disk storage drops by 50%, its storage space doubles in size, and its average transfer rate increases by 40% [17, 24]. Older disk models are discontinued because they cannot compete in the market place. For example, a single manufacturer introduced three disk models in the span of six years, a new model every two years; see Table 35.1. The oldest model

---

[1] Assuming an average bandwidth requirement of 4 Mbps for each clip, the system designer might utilize additional disk drives to satisfy the bandwidth requirement of this library, i.e., number of simultaneous users accessing the library.

[2] Disks are so cheap and common place that replacing failed disks is cheaper than fixing them.

(introduced in 1994) costs more than the other two while providing both a lower storage capacity and a lower bandwidth.

With a heterogeneous disk subsystem, a continuous media server must continue to deliver the data to a client at the bandwidth pre-specified by the clip. For example, if a user references a movie that requires 4 megabits per second (Mbps) for its continuous display, then, once the system initiates its display, it must be rendered at 4 Mbps for the duration of its presentation.[3] Otherwise, a display may suffer from frequent disruptions and delays, termed hiccups. In this paper, we investigate techniques that ensure continuous display of audio and video clips with heterogeneous disk drives. These are categorized into two groups: *partitioning* and *non-partitioning* techniques. With the partitioning techniques, disks are grouped based on their model. To illustrate, assume a system that has evolved to consist of three types of disks: Seagate Barracuda 4LP, Seagate Cheetah 4LP, Seagate Barracuda 18; see Table 35.1. With this approach, the system constructs three disk groups. Each group is managed independently. A frequently accessed (hot) clip might be replicated on different groups in order to avoid formation of hot spots and bottlenecks [7, 16, 28]. With the non-partitioning techniques, the system constructs a logical representation of the physical disks. This logical abstraction provides the illusion of a homogeneous disk subsystem to those software layers that ensure a continuous display.

In general, the non-partitioning schemes are superior because the resources (i.e., bandwidth and storage space) are combined into a unified pool, eliminating the need for techniques to detect bottlenecks and replicate data in order to eliminate detected bottlenecks. Hence, non-partitioning techniques are sensitive to neither the frequency of access to objects nor the distribution of requests as a function of time. Moreover, scheduling of resources is simple with non-partitioning schemes. With the partitioning techniques, the system must monitor the load on each disk partition when activating a request in order to balance the load across partitions evenly. This becomes a difficult task when all partitions are almost completely utilized. The disadvantage of non-partitioning techniques is as follows. First, the design and implementation of availability techniques that guarantee a continuous display in the presence of disk failures becomes somewhat complicated. Second, deciding on the configuration parameters of a system with a non-partitioning technique is not a trivial task.

Figure 35.1 shows a taxonomy of the possible non-partitioning approaches. Among them, many studies [20, 23, 2, 27, 3, 13, 14] have described techniques in support of a hiccup-free display assuming a fixed transfer rate for each disk model. One may apply these studies to multi-zone disk drives by assuming the average transfer rate (weighted by space contributed by each zone) for each disk model. The advantage of this approach is its simplicity and straightforward implementation. However, its performance may degrade when requests arrive in a manner that they reference the data residing in the slowest disk zones.

Disk Grouping and Disk Merging techniques [30] support heterogeneous collection of single-zone disks with deterministic performance. Another approach

---

[3] This study assumes constant bit rate media types. Extensions of this work in support of variable bit rate can be accomplished by extending our proposed techniques with those surveyed in [1].

is to utilize multi-zone disks[4] as they are. For example, the Seagate Barracuda 18 provides 18 gigabyte of storage and consists of 9 zones with each zone providing a different data transfer rate (see Table 35.1). To the best of our knowledge, there are only four techniques in support of hiccup-free display with multi-zone disk drives: IBM's Logical Track [25], Hewlett Packard's Track Pairing [4], and USC's FIXB [11] and deadline driven [12] techniques. Logical Track, Track Pairing, and FIXB provide deterministic performance guarantee while deadline driven approach only stochastically guarantees a hiccup-free display. Studies that investigate stochastic analytical models in support of admission control with multi-zone disks, e.g., [22], are orthogonal because they investigate only admission control (while the above four techniques describe how the disk bandwidth should be scheduled, the block size for each object, and admission control). Moreover, we are not aware of a single study that investigates hiccup-free display using a heterogeneous collection of multi-zone disk drives.

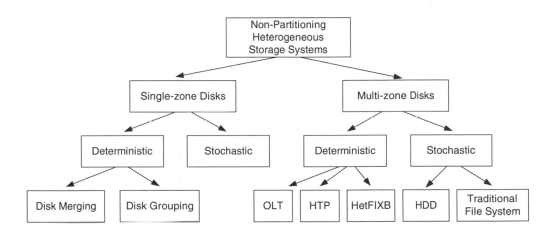

Figure 35.1 Taxonomy of techniques.

This study extends the four techniques to a heterogeneous disk subsystem. While these extensions are novel and a contribution in their own right, we believe that the primary contribution of this study is the performance comparison of these techniques and quantification of their tradeoffs. This is because three of the described techniques assume certain characteristics about the target platform. Our performance results enable a system designer to evaluate the appropriateness of a technique in order to decide whether it is worthwhile to refine its design by eliminating its assumptions.

The rest of this chapter is organized as follows. Section 2 introduces hiccup-free display techniques for a heterogeneous disk subsystem. Section 3 quantifies the performance tradeoffs associated with these techniques. Our results demonstrate tradeoffs between cost per simultaneous stream supported by a technique, its startup latency, throughput, and the amount of disk space that it wastes.

---

[4] More detailed discussion about multi-zone disks can be found in [18, 10].

Table 35.1 Three different Seagate disk models and their zone characteristics.

Seagate Barracuda 4LP				
Introduced in 1994, 2 Gbytes capacity, with a $1,200 price tag				
Zone Id	Size (MB)	Track Size (MB)	No. of Tracks	Rate (MB/s)
0	506.7	0.0908	5579	10.90
1	518.3	0.0903	5737	10.84
2	164.1	0.0864	1898	10.37
3	134.5	0.0830	1620	9.96
4	116.4	0.0796	1461	9.55
5	121.1	0.0767	1579	9.20
6	119.8	0.0723	1657	8.67
7	103.2	0.0688	1498	8.26
8	101.3	0.0659	1536	7.91
9	92.0	0.0615	1495	7.38
10	84.6	0.0581	1455	6.97

Seagate Cheetah 4LP				
Introduced in 1996, 4 Gbytes capacity, with a $1,100 price tag				
Zone Id	Size (MB)	Track Size (MB)	No. of Tracks	Rate (MB/s)
0	1017.8	0.0876	11617	14.65
1	801.6	0.0840	9540	14.05
2	745.9	0.0791	9429	13.23
3	552.6	0.0745	7410	12.47
4	490.5	0.0697	7040	11.65
5	411.4	0.0651	6317	10.89
6	319.6	0.0589	5431	9.84

Seagate Barracuda 18				
Introduced in 1998, 18 Gbytes capacity, with a $900 price tag				
Zone Id	Size (MB)	Track Size (MB)	No. of Tracks	Rate (MB/s)
0	5762	0.1268	45429	15.22
1	1743	0.1214	14355	14.57
2	1658	0.1157	14334	13.88
3	1598	0.1108	14418	13.30
4	1489	0.1042	14294	12.50
5	1421	0.0990	14353	11.88
6	1300	0.0923	14092	11.07
7	1268	0.0867	14630	10.40
8	1126	0.0807	13958	9.68

For example, while USC's FIXB results in the best cost/performance ratio, the potential maximum latency incurred by each user is significantly larger than the other techniques. The choice of a technique is application dependent: One must analyze the requirements of an application and choose a technique accordingly. For example, with nonlinear editing systems, the deadline driven technique is more desirable than the other alternatives because it minimizes the latency incurred by each user [19]. Our conclusions and future research directions are contained in Section 4.

Table 35.2 List of terms used repeatedly in this chapter.

Terms	Definition
K	Number of disk models
D_i	Disk model i, $0 \leq i < K$
q_i	Number of disks for disk model D_i
d_j^i	jth disk drive of disk model D_i, $0 \leq j < q_i$
m_i	Number of zones for each disk of disk model D_i
$Z_i(d_k^l)$	Zone i of disk d_k^l, $0 \leq i < m_i$
$\#TR_i$	Number of tracks in disk model i
$NT(Z_i())$	Number of tracks in zone i
$PT_i(Z_j())$	Track i of zone j, $0 \leq i < NT(Z_j())$
LT_i	Logical track i
$AvgR_i$	Average transfer rate of disk model i
B_i	Block size for disk model i
T_{W_Seek}	Worst seek time of a zone (including the maximum rotational latency)
T_{cseek}	Seek time required to make a complete span
$R(Z_i)$	Transfer rate of Z_i
$S(Z_i)$	Storage capacity of Z_i
T_{scan}	Time to perform one sweep of m zones
$T_{MUX}(Z_i)$	Time to read N blocks from zone Z_i
R_C	Display bandwidth requirement, *consumption rate*
N	Max. number of simultaneous displays, *throughput*
l	Max. startup latency

2. TECHNIQUES

In order to describe the alternative techniques, we assume a configuration consisting of K disk models: $D_0, D_1,..., D_{K-1}$. There are q_i disks belonging to disk model i : $d_0^i, d_1^i,..., d_{q_i-1}^i$. A disk drive of model D_i consists of m_i zones. To illustrate, Figure 35.2 shows a configuration consisting of two disk models D_0 and D_1 ($K=2$). There are two disks of each model ($q_0=q_1=2$), numbered $d_0^0, d_1^0, d_0^1, d_1^1$. Disks of model 0 consist of 2 zones ($m_0=2$) while those of model 1 consist of 3 zones ($m_1=3$). Zone j of a disk (say d_0^0) is denoted as $Z_j(d_0^0)$. Figure 35.2 shows a total of 10 zones for the 4 disk drives and their unique indexes. The kth physical track of a specific zone is indexed as $PT_k(Z_j(d_0^i))$.

We use the set notation, { :   }, to refer to a collection of tracks from different zones of several disk drives. This notation specifies a variable before the colon and the properties that each instance of the variable must satisfy after the colon. For example, to refer to the first track from all zones of the disk drives that belong to disk model 0, we write:

$$\{PT_0(Z_j(d_i^0)): \quad i, \forall j \text{ where } 0 \quad \leq \quad m_0 \text{ and } 0 \quad \leq < q_0\}.$$

With the configuration of Figure 35.2, this would expand to:

$$\{PT_0(Z_0(d_0^0)), PT_0(Z_0(d_1^0)), PT_0(Z_1(d_0^0)), PT_0(Z_1(d_1^0))\}.$$

2.1 IBM'S LOGICAL TRACK [25]

This section starts with a description of this technique for a single multi-zone disk drive. Subsequently, we introduce two variations of this technique, OLT1 and OLT2, for a heterogeneous collection of disk drives. While OLT1 constructs several logical disks from the physical disks, OLT2 provides the abstraction of only one logical disk. We describe each in turn.

With a single multi-zone disk drive, this technique constructs a logical track from each distinct zone provided by the disk drive. Conceptually, this approach provides equi-sized logical tracks with a single data transfer rate such that one can apply traditional continuous display techniques [2, 27, 3, 13, 21, 23]. With K different disk models, a naive approach would construct a logical track LT_k by utilizing one track from each zone: $LT_k \quad \{PT_k(Z_j(d_p^i)): \forall i, j, p\}$ where $0 \leq j < m_i$ and $0 \leq i < K$ and $0 \leq p < q_i$.

With this technique, the value of k is bounded by the zone with the fewest physical tracks, i.e., $0 \leq k < Min[NT(Z_j(d_{q_i}^i))]$, where $NT(Z_j(d_{q_i}^i))$ is the number of physical tracks in zone j of disk model D_i. Large logical tracks result in a significant amount of memory per simultaneous display, rendering a continuous media server economically unavailable. In the next section, we describe two optimized versions of this technique that render its memory requirements reasonable.

2.2 OPTIMIZED LOGICAL TRACK 1 (OLT1)

Assuming that a configuration consists of the same number of disks for each model,[5] OLT1 constructs logical disks by grouping one disk from each disk model (q logical disks). For each logical disk, it constructs a logical track consisting of one track from each physical zone of a disk drive. To illustrate, in Figure 35.2, we pair one disk from each model to form a logical disk drive. The two disks that constitute the first logical disk in Figure 35.2, i.e., disks d_0^0 and d_0^1, consist of a different number of zones (d_0^0 has 2 zones while d_0^1 has 3 zones). Thus, a logical track consists of 5 physical tracks, one from each zone.

---

[5] This technique is applicable as long as the number of disks for each model is a multiple of the model with the fewest disk drives: if $min(q_i)$, $0 \leq i < K$, denotes the model with fewest disks, then q_j is a multiple of $min(q_i)$.

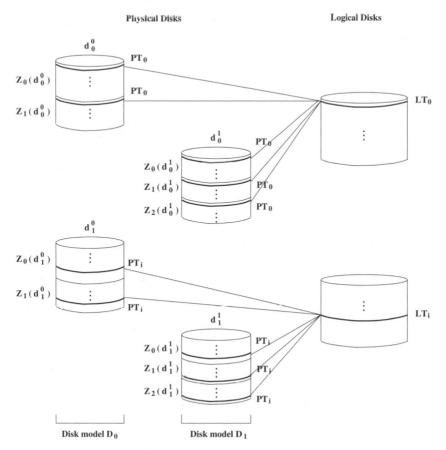

Figure 35.2 OLT1.

Logical disks appear as a homogeneous collection of disk drives with the same bandwidth. There are a number of well known techniques that can guarantee hiccup-free display given such an abstraction, see [2, 27, 3, 13, 21, 23, 26]. Briefly, given a video clip X, these techniques partition X into equi-sized blocks that are striped across the available logical disks [3, 26, 27]: one block per logical disk in a round-robin manner. A block consists of either one or several logical tracks.

Let T_i denote the time to retrieve m_i tracks from a single disk of model D_i consisting of m_i zones: $T_i = m_i \times$ *(a revolution time + seek time)*. Then, the transfer rate of a logical track (R_{LT}) is: $R_{LT} =$ *(size of a logical track)*$/Max[T_i]$ for all i, $0 \le i < K$.

In Figure 35.2, to retrieve a LT from the first logical disk, d_0^0 incurs 2 revolution times and 2 seeks to retrieve two physical tracks, while disk d_0^1 incurs 3 revolutions and 3 seeks to retrieve three physical tracks. Assuming a revolution time of 8.33 milliseconds (7200 rpm) and the average seek time of 10 milliseconds for both disk models, d_0^0 requires 36.66 milliseconds ($T_0 = 36.66$) while d_0^1 requires 54.99 ($T_1 = 54.99$) milliseconds to retrieve a LT. Thus, the transfer rate of the LT is determined by disk model D_1. Assuming that a LT is 1 megabyte in size, its transfer rate is *(size of a logical track)/Max[T_0,T_1]* = 1 megabyte/54.99 milliseconds = 18.19 megabytes per second.

This example demonstrates that OLT1 wastes disk bandwidth by requiring one disk to wait for another to complete its physical track retrievals. In our example, this technique wastes 33.3 % of D_0's bandwidth. In addition, this technique wastes disk space because the zone with the fewest physical tracks determines the total number of logical tracks. In particular, this technique eliminates the physical tracks of those zones that have more than NT_{min} tracks, $NT_{min}=Min[$ $NT(Z_j(d^i_{q_i}))$]], i.e., it eliminates $PT_k(Z_j(d^i_{q_i}))$ that have $NT_{min} \le k < NT(Z_j(d^i_{q_i}))$, for all i and j, $0 \le i < K$ and $0 \le j < m_i$.

2.3 OPTIMIZED LOGICAL TRACK 2 (OLT2)

OLT2 extends OLT1 with the following additional assumption: each disk model has the same number of zones, i.e., m_i is identical for all disk models, $0 \le i < K$. Using this assumption, it constructs logical tracks by pairing physical tracks from zones that belong to different disk drives. This is advantageous for two reasons. First, it eliminates the seeks required per disk drive to retrieve the physical tracks. Second, assuming an identical revolution rate of all heterogeneous disks, it prevents one disk drive to wait for another.

The details of OLT2 are as follows. First, it reduces the number of zones of each disk to that of the disk with fewest zones: $m_{min} = Min[m_i]$ for all i, $0 \le i < K$. Hence, we are considering only zones, $Z_j(d^i_k)$ for all i, j, and k ($0 \le i < K$, $0 \le j < m_{min}$, and $0 \le k < q$). For example, in Figure 35.3, the slowest zone of disks of d^1_0 and d^1_1 (Z_2) are eliminated such that all disks utilize only two zones. This technique requires m_{min} disks of each disk model (totally $m_{min} \times K$ disks). Next, it constructs LTs such that no two physical tracks (from two different zones) in a LT belong to one physical disk drive. A logical track LT_k consists of a set of physical tracks:

$$LT_k = \{PT_{k \bmod NT_{min}}(Z_{l \bmod m_{min}}(d^i_{j \bmod m_{min}})):$$
$$i, \forall j \quad where \ 0 \ i \le K \quad and \quad 0 \ j < m_{min}\} \quad (35.1)$$

where $l = \lfloor k/NT_{min} \rfloor + j$. The total number of LTs is $m_{min} \times NT_{min}$, thus $0 \le k < m_{min} \times NT_{min}$.

OLT2 may use several possible techniques to force all disks to have the same number of zones. For each disk with δ_z zones more than m_{min}, it can either (a) merge two of its physically adjacent zones into one, repeatedly, until its number of logical zones is reduced to m_{min}, (b) eliminate its innermost δ_z zones, or (c) a combination of (a) and (b). With (a), the number of simultaneous displays is reduced because the bandwidth of two merged zones is reduced to the bandwidth of the slower participating zone. With (b), OLT2 wastes disk space while increasing the average transfer rate of the disk drive, i.e., number of simultaneous displays. In [11], we describe a configuration planner that empowers a system administrator to strike a compromise between these two factors for one of the techniques described in this study (HetFIXB). The extensions of this planner in support of OLT2 is a part of our future research direction.

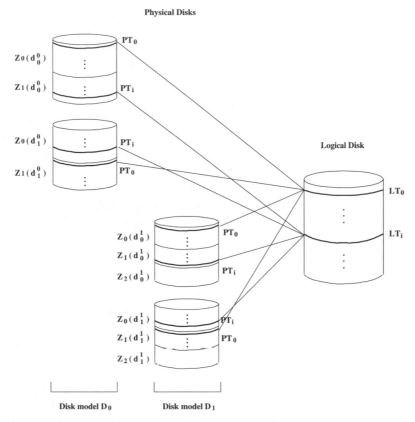

Figure 35.3 OLT2.

2.4 HETEROGENEOUS TRACK PAIRING (HTP)

We describe this technique in two steps. First, we describe how it works for a single multi-zone disk drive. Next, we extend the discussion to a heterogeneous collection of disk drive. Finally, we discuss the tradeoff associated with this technique.

Assuming a single disk drive (say d_0^0) with #TR_0 tracks, Track Pairing [4] pairs the innermost track ($TR_{\#TR_0 - 1}(d_0^0)$) with the outermost track ($TR_0(d_0^0)$), working itself towards the center of the disk drive. The result is a logical disk drive that consists of #$TR_0/2$ logical tracks that have approximately the same storage capacity and transfer rate. This is based on the (realistic) assumption that the storage capacity of tracks increases linearly as one moves from the innermost track to the outermost track. Using this logical disk drive, the system may utilize one of the traditional continuous display techniques in support of hiccup-free display.

Assuming a heterogeneous configuration consisting of K disk models, HTP utilizes Track Pairing to construct track pairs for each disk. If the number of disks for each disk model is identical ($q_0 = q_1 = ... = q_{K-1}$), HTP constructs q_i groups of disk drives consisting of one disk from each of the K disk models. Next, it realize a logical track that consists of K track pairs, one track pair from each disk drive in the group. These logical tracks constitute a logical disk. Obviously, the disk with the fewest number of tracks determines the total number of logical tracks for each logical disk. With such a collection of homogeneous logical disks, one can use one of the popular hiccup-free display techniques. For example, similar to both OLT1 and OLT2, one can stripe a video clip into blocks and assign the blocks to the logical disks in a round-robin manner.

HTP wastes disk space in two ways. First, the number of tracks in a logical disk is determined by the physical disk drive with fewest track pairs. For example, if a configuration consists of two heterogeneous disks, one with 20,000 track pairs and the other with 15,000 track pairs, then the resulting logical disk will consist of 15,000 track pairs. In essence, this technique eliminates 5,000 track pairs from the first disk drive. Second, while it is realistic to assume that the storage capacity of each track increases linearly from the innermost track to the outermost one, it is not 100% accurate [4]. Once the logical tracks are realized, the storage capacity of each logical track is determined by the track with the lowest storage capacity.

2.5 HETEROGENEOUS FIXB

In order to describe this technique, we first describe how the system guarantees continuous display with a single multi-zone disk drive. Next, we describe the extensions of this technique to a heterogeneous disk drive.

2.5.1 FIXB with One Multi-zone Disk [11]

With this technique, the blocks of an object X are rendered equi-sized. Let B denote the size of a block. The system assigns the blocks of X to the zones in a round-robin manner starting with an arbitrary zone. FIXB configures the system to support a fixed number of simultaneous displays, N. This is achieved by requiring the system to scan the disk in one direction, say starting with the outermost zone moving inward, visiting one zone at a time and multiplexing the bandwidth of that zone among N block reads. Once the disk arm reads N blocks from the innermost zone, it is repositioned to the outermost zone to start another sweep of the zones. The time to perform one such sweep is denoted as T_{scan}. The system is configured to produce and display an identical amount of data per T_{scan} period. The time required to read N blocks from zone i, denoted $T_{MUX}(Z_i)$, is dependent on the transfer rate of zone i. This is because the time to read a block ($T_{disk}(Z_i)$) during one $T_{MUX}(Z_i)$ is a function of the transfer rate of a zone.

During each T_{MUX} period, N active displays might reference different objects. This would force the disk to incur a seek when switching from the reading of one block to another, termed T_{W_Seek}. T_{W_Seek} also includes the rotational latency. At the end of a T_{scan} period, the system observes a long seek time (T_{cseek}) attributed to the disk repositioning its arm to the outermost zone. The disk produces m blocks of X during one T_{scan} period ($m \times B$ bytes). The number of bytes required to guarantee a hiccup-free display of X during T_{scan} should either be lower than or equal to the number of bytes produced by the disk. This constraint is formally stated as:

$$R_C \quad \times T_{cseek} \quad + \sum_{i=0}^{m-1} T_{MUX}(Z_i)) \quad \leq n \times B \qquad (35.2)$$

The amount of memory required to support a display is minimized when the left hand side of Eq. 35.2 equals its right hand side.

During a T_{MUX}, N blocks are retrieved from a single zone, Z_{Active}. In the next T_{MUX} period, the system references the next zone $Z_{(Active+1) \bmod m}$. When a display references object X, the system computes the zone containing X_0, say Z_i. The transfer of data on behalf of X does not start until the active zone reaches Z_i. One block of X is transferred into memory per T_{MUX}. Thus, the retrieval of X requires f such periods. (The display of X may exceed $\sum_{j=0}^{f-1} T_{MUX}(Z_{(i+j) \bmod m})$ seconds as described below.) The memory requirement for displaying object X varies due to the variable transfer rate. This is best illustrated using an example. Assume that the blocks of X are assigned to the zones starting with the outermost zone, Z_0. If Z_{Active} is Z_0 then this request employs one of the idle $T_{disk}(Z_0)$ slots to read X_0. Moreover, its display can start immediately because the outermost zone has the highest transfer rate. The block size and N are chosen such that the data accumulates in memory when accessing outermost zones and decreases when reading data from innermost zones on behalf of a display. In essence, the system uses buffers to compensate for the low transfer rates of innermost zones using the high transfer rates of outermost zones, harnessing the average transfer rate of the disk. Note that the amount of required memory reduces to zero at the end of one T_{scan} in preparation for another sweep of the zones.

The display of an object may not start upon the retrieval of its block from the disk drive. This is because the assignment of the first block of an object may start with an arbitrary zone while the transfer and display of data is synchronized relative to the outermost zone, Z_0. In particular, if the assignment of X_0 starts with a zone other than the outermost zone (say Z_i, $i \neq 0$) then its display might be delayed to avoid hiccups. The duration of this delay depends on: 1) the time elapsed from retrieval of X_0 to the time that block X_{m-1} is retrieved from zone Z_0, termed $T_{accessZ0}$, and 2) the amount of data retrieved during $T_{accessZ0}$. If the display time of data corresponding to item 2 ($T_{display(m-i)}$) is lower than $T_{accessZ0}$, then the display must be delayed by $T_{accessZ0} - T_{display(m-i)}$. To illustrate, assume that X_0 is assigned to the innermost zone Z_{m-1} (i.e., $i=m-1$) and the display time of each of its block is 4.5 seconds, i.e., $T_{display(1)}=4.5$ seconds. If 10 seconds elapse from the time X_0 is read until X_1 is read from Z_0 then the display of X must be delayed by 5.5 seconds relative to its retrieval from Z_{m-1}. If its display is initiated upon retrieval, it may suffer from a 5.5 second hiccup. This delay to avoid a hiccup is shorter than the duration of a T_{scan}. Indeed, the maximum latency observed by a request is T_{scan} when the number of active displays is less than N.

$$l = T_{scan} = T_{cseek} + \sum_{i=0}^{m-1} T_{MUX}(Z_i) \qquad (35.3)$$

This is because at most N displays might be active when a new request arrives referencing object X. In the worst case scenario, these requests might be retrieving data from the zone that contains X_0 (say Z_i) and the new request arrives too late to employ the available idle slot. (Note that the display may not employ the idle slot in the next T_{MUX} because Z_{i+1} is now active and it contains X_1 instead of X_0.) Thus, the display of X must wait one T_{scan} period until Z_i becomes active again.

$T_{MUX}(Z_i)$ can be defined as:

$$T_{MUX}(Z_i) = N \times (\frac{B}{R(Z_i)} + T_{W\_Seek}) \qquad (35.4)$$

Substituting this into Eq. 35.2, the block size is defined as:

$$B = \frac{R_C \ (\mathcal{R}_{cseek} \quad m+ \ N \times T_{W\_Seek})}{m \ -R_C \times \sum_{i=0}^{m-1} \frac{N}{R(Z_i)}} \qquad (35.5)$$

Observe that FIXB wastes disk space when the storage capacity of the zones is different. This is because once the storage capacity of the smallest zone is exhausted then no additional objects can be stored as they would violate a round-robin assignment.

2.5.2 Extensions of FIXB (HetFIXB)

With a heterogeneous collection of disks, we continue to maintain a T_{scan} per disk drive. While the duration of a T_{scan} is identical for all disk drives, the amount of data produced by each T_{scan} is different. We compute the block size for each disk model (recall that blocks are equi-sized for all zones of a disk) such that the faster disks compensate for the slower disks by producing more data during their T_{scan} period. HetFIXB aligns the T_{scan} of each individual disk drive with one another such that they all start and end in a T_{scan}.

To support N simultaneous displays, HetFIXB must satisfy the following equations.

$$M = \sum_{i=0}^{K-1} M_i, \quad where \quad M_i = n_i \times B_i \qquad (35.6)$$

$$AvgR_i : AvgR_j = M_i : M_j, \quad 0 \leq i, j < K \qquad (35.7)$$

$$T_{scan} = T_p / K, \quad where \quad T_p = \frac{M}{R_C} \qquad (35.8)$$

$$T_{scan_i} = T_{cseek} + \sum_{j=0}^{m_i-1} N(\frac{B_i}{R(Z_j(D_i))} + seek_i) \leq T_{scan} \qquad (35.9)$$

where $0 \leq i < K$.

To illustrate, assume a configuration consisting of 3 disks; see Figure 35.4. Assume the average transfer rates of disks, $AvgR_0 = 80$ Mbps, $AvgR_1 = 70$ Mbps, and $AvgR_2 = 60$ Mbps, respectively. When $R_C = 4$ Mbps, 1.5 Mbytes of data ($M =$

1.5 MB) is required every 3 seconds (T_p = 3 sec) to support a hiccup-free display. Based on the ratio among the average transfer rates of disk models, M_0 = 0.5715 MB, M_1 = 0.5 MB, and M_2 = 0.4285 MB. Thus, $B_0 = M_0 / m_0$ = 0.19 MB, $B_1 = M_1 / m_1$ = 0.25 MB, $B_2 = M_2 / m_2$ = 0.14 MB. An object X is partitioned into blocks and blocks are assigned into zones in a round-robin manner. When a request for X arrives, the system retrieves X_0, X_1, and X_2 ($M_0 = 3 \times B_0$ amount of data) from D_0 during the first T_{scan}. A third of M (0.5 MB) is consumed during the same T_{scan}. Hence, some amount of data, 0.0715 MB, remains un-consumed in the buffer. In the next T_{scan}, the system retrieves X_3 and X_4 ($M_1 = 2 \times B_1$ amount of data) from D_1. While the same amount of data (0.5 MB) is retrieved and consumed during this T_{scan}, the accumulated data (0.0715 MB) still remains in the buffer. Finally, during the last T_{scan}, the system retrieves X_5, X_6, and X_7 ($M_2 = 3 \times B_2$ amount of data) from D_2. Even though the amount of data retrieved in this T_{scan} (0.4285 MB) is smaller than the amount of data displayed during a T_{scan} (0.5 MB), there is no starvation because 0.0715 megabytes of data is available in the buffer. This process is repeated until the end of display.

2.6 HETEROGENEOUS DEADLINE DRIVEN (HDD)

With this technique, blocks are assigned across disk drives in a random manner. A client issues block requests, each tagged with a deadline. Each disk services block requests with the Earliest Deadline First policy. In [12], we showed that the assignment of blocks to the zones in a disk should be independent of the frequency of access to the blocks. Thus, blocks are assigned to the zones in a random manner. The size of the blocks assigned to each disk model is different. They are determined based on the average weighted transfer rate of each disk model. Let WR_i denote the weighted average transfer rate of disk model i:

$$WR_i = \sum_{j=0}^{m_i-1}[S(Z_j(D_i)) \times R(Z_j(D_i)) / \sum_{k=0}^{m_i-1}S(Z_k(D_i))] \qquad (35.10)$$

$$WR_i : WR_j = B_i : B_j, \quad 0 \leq i,j < K \qquad (35.11)$$

Assuming $B_i \geq B_j$ where $i < j$ and $0 \leq i,j < K$, an object X is divided into blocks such that the size of each block X_i is $B_{i \bmod K}$. Blocks with the size of B_i are randomly assigned to disks belonging to model i. A random placement may incur hiccups that are attributed to the statistical variation of the number of block requests per disk drive, resulting in varying block retrieval time. Traditionally, double buffering has been widely used to absorb the variance of block retrieval time: while a block in a buffer is being consumed, the system fills up another buffer with data. However, we generalize double buffering to **N** buffering and prefetching **N-1** buffers before initiating a display. This minimizes the hiccup probability by absorbing a wider variance of block retrieval time, because data retrieval is N-1 blocks ahead of data consumption.

We assume that, upon a request for a video clip X, a client: (1) concurrently issues requests for the first N-1 blocks of X (to prefetch data), (2) tags the first N-1 block requests, X_i ($0 \leq i < N$), with a deadline, $\sum_{j=0}^{i} sizeof(X_j)/R_C$, (3) starts display as soon as the first block (X_0) arrives. For example, when N=4, first three blocks are requested at the beginning. Then, the next block request is issued immediately after the display is initiated. Obviously, there are other ways of deciding both the deadline of the prefetched blocks and when to initiate display blocks. In [12], we analyzed the impact of these alternative decisions and demonstrated that the combination of the above two choices enhances system performance.

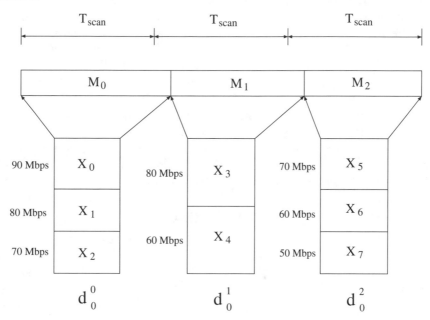

Figure 35.4 HetFIXB.

2.7 DISK GROUPING (DG)

Assuming a fixed data transfer rate of a disk (single-zone disk model), this technique groups physical disks into logical ones and assumes a uniform characteristic for all logical disks. To illustrate, the six physical disks of Figure 35.5 are grouped to construct two homogeneous logical disks. In this figure, a larger physical disk denotes a newer disk model that provides both a higher storage capacity and a higher performance. The blocks of a movie X are assigned to the logical disks in a round-robin manner to distribute the load of a display evenly across the available resources [2, 23, 8]. A logical block is declustered [15] across the participating physical disks. Each piece is termed a fragment (e.g., $X_{0.0}$ in Figure 35.5). The size of each fragment is determined such that the service time ($T_{service}$) of all physical disks is identical.

When the system retrieves a logical block into memory on behalf of a client, all physical disks are activated simultaneously to stage their fragments into memory. This block is then transmitted to the client. To guarantee a hiccup-free display, the display time of a logical block (B^l) must be equivalent to the duration of a time period: $T_p = B^l / R_C$. Moreover, if a logical disk services N simultaneous displays then the duration of each time period must be greater or equal to the service time to read N fragments from every physical disk drive i that is a part of the logical disk. Thus, the following constraint must be satisfied:

$$T_p \not\geq T_{service_i} = N \times (\frac{B_i}{R_{D_i}} + T_{seek_i}) \qquad (35.12)$$

Because the time period must be equal for all physical drives (*K*) that constitutes a logical disk, we obtain the following equations:

$$B^l = \sum_{i=0}^{K-1} B_i \qquad (35.13)$$

$$\frac{B^l}{R_C \times N} = \frac{B_0}{R_{D_0}} + T_{seek_0} = \cdots = \frac{B_{K-1}}{R_{D K-1}} + T_{seek_{K-1}} \qquad (35.14)$$

To support *N* displays with one logical disk, the system requires 2×*N* memory buffers. Thus the total number of displays supported by the system (*N_tot*) is *N* × *the number of logical disks*, and the required memory space is $2 \times N_{tot} \times B^l$.

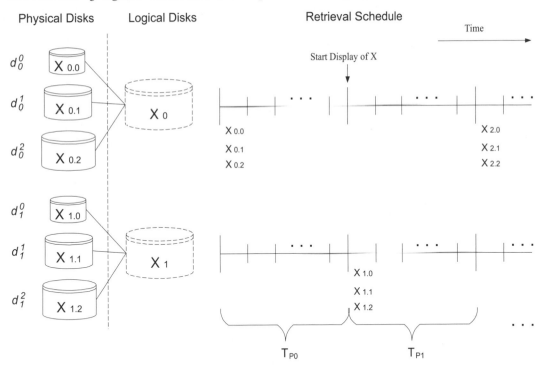

Figure 35.5 Disk grouping.

2.8 DISK MERGING (DM)

Using a fixed data transfer rate of a disk as in DG, this technique separates the concept of logical disk from the physical disks all together. Moreover, it forms logical disks from fractions of the available bandwidth and storage capacity of the physical disks; see Figure 35.6. These two concepts are powerful abstractions that enable this technique to utilize an arbitrary mix of physical disk models. The design of this technique is as follows. First, it chooses how many logical disks should be mapped to each of the *slowest physical disks* and denotes this factor with *p0*. Note that we can choose any number of logical disks from the slowest physical disk by taking a fraction of disk bandwidth and storage capacity of the disk. For example, in Figure 35.6, the two slowest disks d_2 and d_3 each represent 1.5 logical disks, i.e., p_0=1.5. Then, the time period T_p and the block size necessary to support p_0 logical disks on a physical disk can be established as follows:

$$T_p \geq N \times (\frac{B^l \times p_0}{R_{D_0}} + \lceil p_0 \rceil \times T_{seek_0})$$ (35.15)

$$\frac{B^l}{N \times R_C} = \frac{B^l \times p_i}{R_{D_i}} + \lceil p_i \rceil \times T_{seek_i}$$ (35.16)

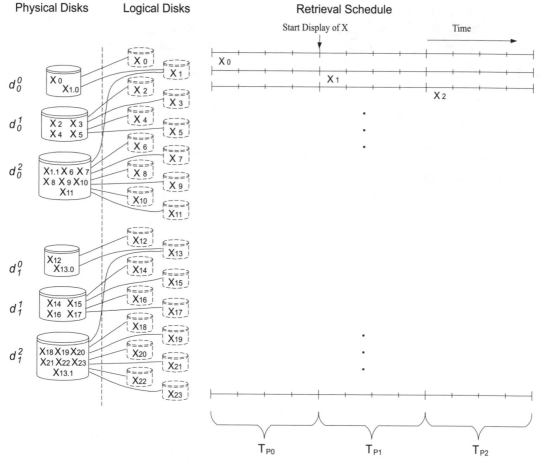

Figure 35.6 Disk merging.

Because of the ceiling-function there is no closed formula solution for p_i from the above equation. However, numerical solutions can be easily found. An initial estimate for p_i may be obtained from the bandwidth ratio of the two different disk types involved. When iteratively refined, this value converges rapidly towards the correct ratio. In the example configuration of Figure 35.6, the physical disks d_1 and d_4 realize 2.5 logical disks each (p_1=2.5) and the disks d_0

and d_3 support 3.5 logical disks (p_2=3.5). Note that fractions of logical disks (e.g., 0.5 of d_0 and 0.5 of d_3) are combined to form additional logical disks (e.g., ld_7, which contains block X_7 in Figure 35.6). The total number of logical disks (K^l_{Tot}) is defined as:

$$K^l_{Tot} = \left\lfloor \sum_{i=0}^{K-1} p_i \times q_i \right\rfloor \qquad (35.17)$$

Once a homogeneous collection of logical disks is formed, the block size B^l of a logical disk determines the total amount of memory with the total number of simultaneous displays ($N_{Tot} = N \times K^l_{Tot}$); see [30].

3. EVALUATION

First, we quantify the performance tradeoffs associated with alternative techniques that support a heterogeneous collection of multi-zone disks. While OLT1, OLT2, HTP, and HetFIXB were quantified using analytic models, HDD was quantified using a simulation study. We conducted numerous experiments analyzing different configurations with different disk models from Quantum and Seagate. Here, we report on a subset of our results in order to highlight the tradeoffs associated with different techniques. In all results presented here, we used the three disk models shown in Table 35.1. Both Barracuda 4LP and 18 provide a 7200 rpm while the Cheetah provides a 10,000 rpm. Moreover, we assumed that all objects in the database require a 4 Mbps bandwidth for their continuous display.

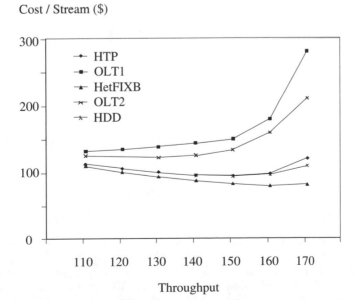

Figure 35.7 One Seagate disk model (homogeneous).

Cost / Stream ($)

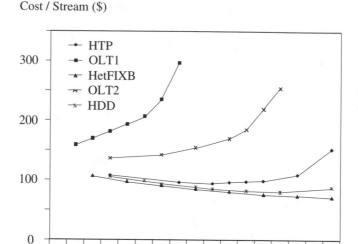

Figure 35.8 Two Seagate disk models (heterogeneous).

Cost / Stream ($)

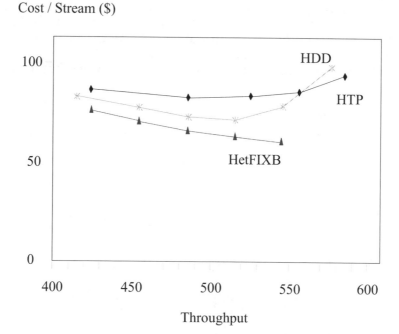

Figure 35.9 Three Seagate disk models (heterogeneous).

Figures 35.7, 35.8, and 35.9 show the cost per stream as a function of the number of simultaneous displays supported by the system (throughput) for three different configurations. Figure 35.7 shows a system that is installed in 1994 and consists of 10 Barracuda 4LP disks. Figure 35.8 shows the same system two years later when it is extended with 10 Cheetah disks. Finally, Figure 35.9 shows this system in 1998 when it is extended with 10 Barracuda 18 disks. To estimate system cost, we assumed: a) the cost of each disk at the time when they were purchased with no depreciation cost, and, b) the system is configured with sufficient memory to support the number of simultaneous displays shown on the x-axis. We assumed that the cost of memory is $7/MB, $5/MB, and $3/MB in 1994, 1996, and 1998, respectively. Additional memory is purchased at the time of disk purchases in order to support additional users. (Once again, we assume no depreciation of memory.) While one might disagree with our assumptions for computing the cost of the system, note that the focus of this study is to compare the different techniques. As long as the assumptions are kept constant, we can make observations about the proposed techniques and their performance tradeoffs.

In these experiments, OLT2 constructed logical zones in order to force all disk models to consist of the same number of zones. This meant that OLT2 eliminated the innermost zone (zone 10) of Barracuda 4LP, splitting the fastest three zones of Cheetah into six zones, and splitting the outermost zone of Barracuda 18 into two. Figure 35.9 does not show OLT1 and OLT2 because: a) their cost per stream is almost identical to that shown in Figure 35.8, and b) we wanted to show the difference between HetFIXB, HDD, and HTP.

Figures 35.7, 35.8, and 35.9 show that HetFIXB is the most cost effective technique; however, it supports fewer simultaneous displays as a function of heterogeneity. For example, with one disk model, it provides a throughput similar to the other techniques. However, with 3 disk models, its maximum throughput is lower than those provided by HDD and HTP. This is dependent on the physical characteristics of the zones because HetFIXB requires the duration of T_{scan} to be identical for all disk models. This requirement results in fragmentation of the disk bandwidth which in turn limits the maximum throughput of the system. Generally speaking, the greater the heterogeneity, the greater the degree of fragmentation. However, the zone characteristics ultimately decide the degree of fragmentation. One may construct logical zones in order to minimize this fragmentation; see [11]. This optimization is not reported because of strict space limitations imposed by the call for paper. It raises many interesting issues that are not presented here. Regardless, the comparison shown here is fair because our optimizations are applicable to all techniques.

OLT1 provides inferior performance as compared to the other techniques because it wastes a significant percentage of the available disk bandwidth. To illustrate, Figure 35.10 shows the percentage of wasted disk bandwidth for each disk model with each technique when the system is fully utilized (the trend holds true for less than 100% utilization). OLT1 wastes 60% of the bandwidth provided by Cheetah and approximately 30% of Barracuda 18. Most of the wasted bandwidth is attributed to these disks sitting idle. Cheetahs sit idle because they provide a 10,000 rpm as compared to 7200 rpm provided by the Barracudas. Barracuda 4LP and 18 disks sit idle because of their zone characteristics. In passing, while different techniques provide approximately similar cost per performance ratios, each wastes bandwidth in a different manner. For example, both HTP and HetFIXB provide approximately similar cost per performance ratios, HTP wastes 40% of Cheetah's bandwidth while HetFIXB wastes only 20% of the bandwidth provided by this disk model. HTP makes up for this limitation by harnessing a greater percentage of the bandwidth provided by Barracuda 4LP and 18.

Figure 35.11 shows the maximum latency incurred by each technique as a function of the load imposed on the system. In this figure, we have eliminated

OLT1 because of its prohibitively high latency. (One conclusion of this study is that OLT1 is not a competitive strategy.) The results show that HetFIXB provides the worst latency while HDD's maximum latency is below 1 second. This is because HetFIXB forces a rigid schedule with a disk zone being activated in an orderly manner (across all disk drives). If a request arrives and the zone containing its referenced block is not active then it must wait until the disk head visits that zone (even if idle bandwidth is available). With HDD, there is no such rigid schedule in place. A request is serviced as soon as there is available bandwidth. Of course, this is at the risk of some requests missing their deadlines. This happens when many requests collide on a single disk drive due to random nature of requests to the disks. In these experiments, we ensured that such occurrences impacted one in a million requests, i.e., a hiccup probability is less than one in a million block requests.

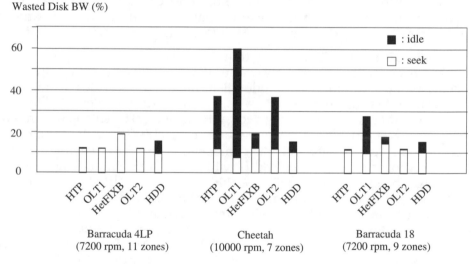

Figure 35.10 Wasted disk bandwidth.

OLT2 and HTP provide a better latency as compared to HetFIXB because they construct fewer logical disks [2, 14]. While OLT2 constructs a single logical disk, HTP constructs 10 logical disks, and HetFIXB constructs 30 logical disks. In the worst case scenario (assumed by Figure 35.11), with both HTP and HetFIXB, all active requests collide on a single logical disk (say $d_{bottleneck}$). A small fraction of them are activated while the rest wait for this group of requests to shift to the next logical disk (in the case of HetFIXB, they wait for one T_{scan}). Subsequently, another small fraction is activated on $d_{bottleneck}$. This process is repeated until all requests are activated. Figure 35.11 shows the incurred latency by the last activated request.

With three disk models (Figure 35.9), OLT1 and OLT2 waste more than 80% of disk space, HTP and HDD waste approximately 70% of disk space, while HetFIXB wastes 44% of the available disk space. However, this does NOT mean that HetFIXB is more space efficient than other techniques. This is because the percentage of wasted disk space is dependent on the physical characteristics of the participating disk drives: number of disk models, number of zones per disk, track size of each zone, storage capacity of individual zones and disk drives. For example, with two disk models (Figure 35.8), HetFIXB wastes more disk space when compared with the other techniques.

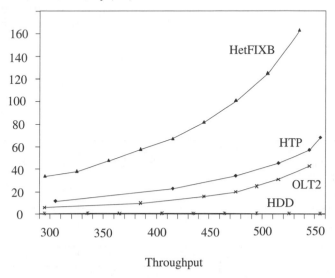

Throughput

Figure 35.11 Maximum startup latency.

So far, we have reported experimental results using three real disk models. In the next experiments, we further evaluate the techniques using three imaginary multi-zone disk models that could represent disks in the near future (see Table 35.3). First, we performed similar experiments to quantify the cost per stream. Figure 35.12 shows a system that was installed in 2000 using ten type 0 disks in Table 35.3. Figure 35.13 shows the same system two years later when it is extended with 10 type 1 disks. Finally, Figure 35.14 shows this system in 2004 when it is extended with 10 type 2 disks. We assumed that the cost of memory is $1/MB, $0.6/MB, and $0.35/MB in 2000, 2002, and 2004, respectively.

Figures 35.12, 35.13, and 35.14 compare all deterministic approaches in non-partitioning techniques. For Disk Grouping (DG) and Merging (DM) techniques, we assumed two different fixed data transfer rates: one with a pessimistic view, the other with an optimistic view. For example, DG(Avg) means disk grouping technique using the average data transfer rate of a multi-zone disk while DG(Min) uses the transfer rate of the slowest zone. Note that there would be no hiccups when the minimum rate is assumed in the above techniques because the technique decides system parameters in a manner that ensures continuous display. However, if the average rate is assumed, hiccups might occur because the active requests might reference the data from the slowest zone, observing a bandwidth lower than the assumed average transfer rate.

Figure 35.12 shows a similar result as in Figure 35.7. HetFIXB and HTP are superior to OLT2. DG(Avg)[6] also provides a very high throughput with the second lowest cost/stream. However, DG(Min) supports only 140 at maximum because of the lowered disk bandwidth to guarantee a continuous display. Figure 35.13

---

[6] For single model, DG(Avg) is equal to DM(Avg) and DG(Min) is equal to DM(Min).

and 14 also show a similar trend. DG(Avg) and DM(Avg) provide the best cost/stream being followed by HetFIXB and HTP.

Figure 35.15 shows the percentage of wasted disk space with each technique. It is important to note that the obtained results are dependent on the characteristics of the assumed disk drives, namely the discrepancy between the bandwidth and storage ratios of different disk models. For example, the bandwidth of type 2 disk is twice that of type 0 disk. However, the storage capacity of type 2 disk is five times that of type 0 disk. Because all techniques construct logical blocks based on the disk bandwidth ratio, they waste a considerable amount of storage provided by type 1 and 2 disks.

Table 35.3 Three imaginary disk models and their zone characteristics.

	Type 0 30 GB, 2000, $300		Type 1 80 GB, 2002, $300		Type 2 150 GB, 2004, $300	
Zone Id	Size (GB)	Rate (Mb/s)	Size (GB)	Rate (Mb/s)	Size (GB)	Rate (Mb/s)
0	10	100	30	150	60	200
1	8	90	12	120	50	150
2	6	80	9	110	40	100
3	4	70	9	100	-	-
4	2	60	8	90	-	-
5	-	-	6	80	-	-
6	-	-	6	70	-	-

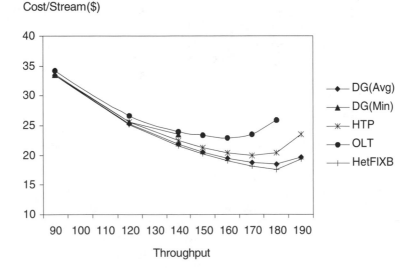

Figure 35.12 One disk model, type 0 (homogeneous).

Cost/Stream($)

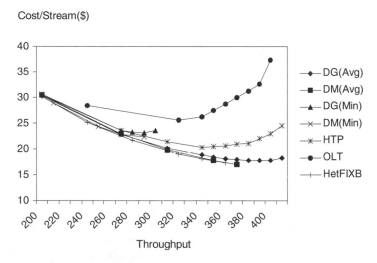

Figure 35.13 Two disk models, type 0 and 1 (heterogeneous).

Cost/Stream($)

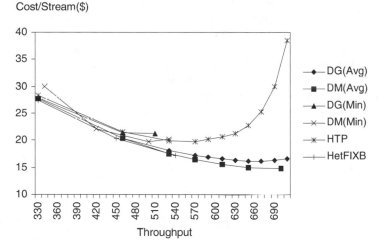

Figure 35.14 Three disk models, type 0, 1, 2 (heterogeneous).

% of Wasted Space

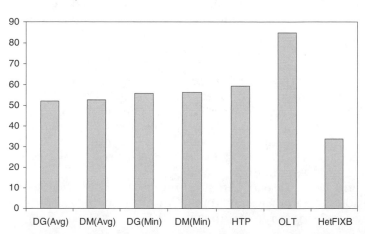

Figure 35.15 Wasted disk space with three disk models.

4. CONCLUSION

In this study, we quantified the tradeoff associated with alternative multi-zone techniques when extended to a configuration consisting of heterogeneous disk drives. Ignoring OLT1, our principle result is that no single strategy dominates on all metrics, i.e., throughput, startup latency, cost per simultaneous display, and wasted disk space. All proposed techniques strive to distribute the load of a single display evenly across the available disk bandwidth in order to prevent formation of bottlenecks. They belong to a class of algorithm that is commonly termed non-partitioning. An alternative approach might have been to partition resources into clusters and treat each cluster as an independent server. For example, with a configuration consisting of 3 disk models, we could have constructed three servers and assigned objects to different servers with the objective to distribute the workload of the system as evenly as possible [7]. The system would replicate popular clips across multiple servers in order to prevent formation of bottlenecks [16, 6]. Using this approach, one could optimize system parameters (such as block size) for each configuration independently in order to maximize the performance of each subserver. This is ideal for static workloads that do not change overtime. However, for dynamic workloads, one must employ detective techniques that monitor the frequency of access to objects and replicate popular objects in order to prevent formation of bottlenecks. [30] utilizes a simulation model to show that this approach is generally inferior to a non-partitioning scheme. This is because detective techniques must wait for formation of bottlenecks prior to eliminating them [5].

As a future research direction, one may develop a configuration planner that consumes the requirements of an application and computes system parameters (along with a choice of a non-partitioning technique) that meets the application requirements.

REFERENCES

[1] J. Al-Marri and S. Ghandeharizadeh, An Evaluation of Alternative Disk Scheduling Techniques in Support of Variable Bit Rate Continuous Media, in Proceedings of the International Conference on Extending Database Technology (EDBT), Valencia, Spain, March 23-27, 1998.

[2] S. Berson, S. Ghandeharizadeh, R. Muntz, and X. Ju, Staggered Striping in Multimedia Information Systems, in Proceedings of the ACM SIGMOD International Conference on Management of Data, pages 79-89, 1994.

[3] S. Berson, L. Golubchik, and R. R. Muntz, A Fault Tolerant Design of a Multimedia Server, in Proceedings of the ACM SIGMOD International Conference on Management of Data, pages 364-375, 1995.

[4] Y. Birk, Track-pairing: A novel Data Layout for VOD Servers with Multi-zone-recording Disks, in IEEE International Conference on Multimedia Computing and System, June 1995.

[5] H. Chou and D. J. DeWitt, An Evaluation of Buffer Management Strategies for Relational Database Systems, in Proceedings of the International Conference on Very Large Databases, 1985.

[6] A. Dan, M. Kienzle, and D. Sitaram, A Dynamic Policy of Segment Replication for Load-balancing in Video-on-Demand Computer Systems, in *ACM Multimedia Systems*, Number 3, pages 93-103, 1995.

[7] A. Dan and D. Sitaram, An Online Video Placement Policy based on Bandwidth to Space Ratio (BSR), in Proceedings of ACM SIGMOD, 1995.

[8] R. Flynn and W. Tetzlaff, Disk Striping and Block Replication Algorithms for Video File Servers, in Proceedings of the International Conference on Multimedia Computing and Systems, June 1996.

[9] C. Freedman and D. DeWitt, The SPIFFI Scalable Video-on-Demand System, in SIGMOD Conference, 1995.

[10] S. Ghandeharizadeh and S. H. Kim, A Comparison of Alternative Continuous Display Techniques with Heterogeneous Multi-Zone Disks, in the Proceedings of the 8th International Conference on Information Knowledge Management (CIKM'99), Nov. 1999.

[11] S. Ghandeharizadeh, S. H. Kim, C. Shahabi, and R. Zimmermann, Placement of Continuous Media in Multi-Zone Disks, in *Multimedia Information Storage and Management*, Soon M. Chung Ed., Chapter 2. Kluwer Academic Publishers, Boston, 1996.

[12] S. Ghandeharizadeh and S.H. Kim, Design of Multi-user Editing Servers for Continuous Media, in *the Journal of Multimedia Tools and Applications*, 11(1), May 2000.

[13] S. Ghandeharizadeh and S.H. Kim, Striping in Multi-disk Video Servers, in High-Density Data Recording and Retrieval Technologies, Proc. SPIE 2604, pages 88-102, October 1995.

[14] S. Ghandeharizadeh, S.H. Kim, W. Shi, and R. Zimmermann, On Minimizing Startup Latency in Scalable Continuous Media Servers, in Proceedings of Multimedia Computing and Networking, Proc. SPIE 3020, pages 144-155, Feb. 1997.

[15] S. Ghandeharizadeh and L. Ramos, Continuous Retrieval of Multimedia Data Using Parallelism, IEEE *Transactions on Knowledge and Data Engineering*, 5(4), Aug. 1993.

[16] S. Ghandeharizadeh and C. Shahabi. Management of Physical Replicas in Parallel Multimedia Information Systems, in Proceedings of the Foundations of Data Organization and Algorithms (FODO) Conference, October 1993.

[17] E. Grochowski, Disk Drive Price Decline, in *IBM Almaden Research Center*, 1997.

[18] S.H. Kim, Design of Continuous Media Servers with Multi-zone Disk Drives, Ph.D. Dissertation, University of Southern California, Aug. 1999.

[19] S.H. Kim and S. Ghandeharizadeh, Design of Multi-user Editing Servers for Continuous Media, in International Workshop on the Research Issues in Data Engineering (RIDE'98), Feb. 1998.

[20] M. Leung, J. C. Lui, and L. Golubchik, Buffer and I/O Resource Pre-allocation for Implementing Batching and Buffering Techniques for Video-On-Demand Systems, in Proceedings of the International Conference on Data Engineering, 1997.

[21] R. Muntz, J. Santos, and S. Berson, RIO: A Real-time Multimedia Object Server, *ACM Sigmetrics Performance Evaluation Review*, 25(2), Sep. 1997.

[22] G. Nerjes, P. Muth, and G. Weikum, Stochastic Service Guarantees for Continuous Data on Multi-zone Disks, in the 16th Symposium on Principles of Database Systems (PODS'97), May 1997.

[23] B. Ozden, R. Rastogi, and A. Silberschatz, Disk Striping in Video Server Environments, in IEEE International Conference on Multimedia Computing and System, June 1996.

[24] D. A. Patterson, Terabytes ?? Teraflops (or Why Work on Processors When I/O is Where the Action is?, in Key note address at the ACM-SIGMETRICS Conference, 1993.

[25] M.F. Mitoma S.R. Heltzer, J.M. Menon, Logical data tracks extending among a plurality of zones of physical tracks of one or more disk devices, U.S. Patent No. 5,202,799, April 1993.

[26] F.A. Tobagi, J. Pang, R. Baird, and M. Gang, Streaming RAID-A Disk Array Management System for Video Files, in Proceedings of the First ACM Conference on Multimedia, August 1993.

[27] H.M. Vin, S.S. Rao, and P. Goyal, Optimizing the Placement of Multimedia Objects on Disk Arrays, in IEEE International Conference on Multimedia Computing and System, May 1995.

[28] J.L Wolf, P.S. Yu, and H. Shachnai, DASD Dancing: A Disk Load Balancing Optimization Scheme for Video-on-Demand Computer Systems, in Proceedings of ACM SIGMETRICS, May 1995.

[29] J. Wong, K. Lyons, D. Evans, R. Velthuys, G. Bochmann, E. Dubois, N. Georganas, G. Neufeld, M. Ozsu, J. Brinskelle, A. Hafid, N. Hutchinson, P. Iglinski, B. Kerherve, L. Lamont, D. Makaroff, and D. Szafron, Enabling Technology for Distributed Multimedia Applications, *IBM System Journal*, 1997.

[30] R. Zimmermann and S. Ghandeharizadeh, Continuous Display Using Heterogeneous Disk-Subsystems, in Proceedings of ACM Multimedia Conference, Nov. 1997.

36

TECHNOLOGIES AND STANDARDS FOR UNIVERSAL MULTIMEDIA ACCESS

Anthony Vetro and Hari Kalva
Mitsubishi Electric Research Labs
Murray Hill, NJ, USA
`avetro@merl.com, hk168@columbia.edu`

1. INTRODUCTION

Three major trends have emerged over the past five years in the way communication, information and entertainment services are provided to consumers: wireless communication, Internet technologies and digital entertainment (audio/video/games). Mobile telephony and Internet technologies have seen tremendous global growth over this period. The scale of the growth and the market needs have resulted in the basic Internet services over mobile telephones. The third major trend, digital entertainment, can be seen in the rapid adoption of digital TV and DVD products. Digital entertainment over the Internet has so far been mainly digital music and, to a lesser extent, streaming video, and online video games. New standards and technologies are making ways for offering applications and services that combine these three trends. A number of technologies have emerged that make possible access to digital multimedia content over wired and wireless networks on a range of devices with varying capabilities such as mobile phones, personal computers and television sets. This universal multimedia access (UMA), enabled by new technologies and standards, poses new challenges and requires new solutions. In this chapter we discuss the technologies, standards, and challenges that define and drive UMA.

The main elements of a system providing UMA services are 1) the digital content, 2) the sending terminal, 3) the communication network and 4) the receiving terminal. These seemingly simple and few elements represent a myriad of choices with multiple content formats, video frame rates, bit rates, network choices, protocol choices and receiving terminal capabilities. The operating environment of the service adds a dynamically varying factor that affects the operation of the end-to-end system. The selection of appropriate content formats, networks and terminals and adapting these elements, if and when necessary, to deliver multimedia content is the primary function of UMA services.

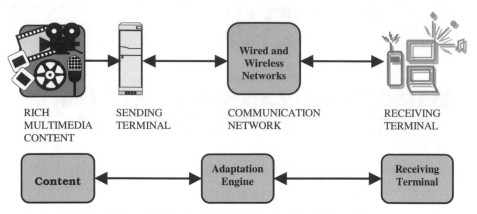

Figure 36.1 Capability mismatch and adaptation for universal access.

Figure 36.1 shows the components of a general purpose multimedia communications system. Rich multimedia content is communicated over communication networks to the receiving terminals. In practice, the multimedia content includes various content encoding formats such as MPEG-2, MPEG-4, Wavelets, JPEG, AAC, AC-3, etc.; all these encoding formats at various frame rates and bit rates. The communication networks also have different characteristics such as bandwidth, bit error rate, latency and packet loss rate depending on the network infrastructure and load. Likewise, the receiving terminals have different content playback and network capabilities as well as different user preferences that affect the type of content that can be played on the terminals. *The communication network characteristics together with the sending terminal and the receiving terminal capabilities and their current state constitute the content playback environment.* The mismatch between the rich multimedia content and the content playback environment is the primary barrier for the fulfilment of the UMA promise. The adaptation engine is the entity that bridges this mismatch by either adapting the content to fit to the content playback environment or adapting the content playback environment to accommodate the content.

The content playback environment can consist of a number of elements such as routers, switches, wired-to-wireless gateways, relay servers, video server proxies, protocol converters, and media translators. Multimedia content on its way to the receiver terminal passes through one or more of these elements and offers possibilities of *adapting* the content within one of these elements to enable playback with acceptable quality of service. The adaptation performed can be 1) content adaptation and 2) content playback environment adaptation. Content adaptation means adapting the content to fit the playback environment. The nature of content determines the operations involved in the actual adaptation. For object-based content such as MPEG-4 content, adaptation at the highest level, the content level, involves adding or dropping objects in the content in response to the varying playback environment. See Section 2 for details on MPEG-4 and object based content representation. Adaptation at a lower level, e.g., the object level, is done by modifying the bit rate, frame rate, resolution or the encoding format of the objects in the presentation. For stored content, object level adaptation generally means reduction in quality compared with the source; for live content, adaptation can also result in the improvement in quality. Adapting the content playback environment involves acquiring additional resources to handle the content. The resources acquired can be session bandwidth, computing resources at the sending and receiving terminals, decoders in the receiver terminals, or improving the network latency and packet loss. The playback environment can change dynamically and content adaptation should match the changing environment to deliver content at the best quality possible.

While there has been significant work published in adapting the Web content, in this chapter we emphasize digital video adaptation, as the primary focus of research is digital video communication.

1.1 UMA INFRASTRUCTURE

The digital audio-visual services (AV) infrastructure has to be augmented to support UMA. The AV infrastructure consists of video servers, proxies, network nodes, wired-to-wireless gateways and additional nodes depending on the delivery network used. This raises the possibility of applying adaptation operations at one of more elements between a sending terminal and the receiving terminal. With UMA applications, new adaptation possibilities as well as additional adaptation locations have to be considered. In the following we consider some of the important aspects of the UMA infrastructure.

1.1.1 Cooperative Playback

The mismatch between the rich multimedia content and small devices such as mobile phones and PDAs can also be addressed using additional receiver side resources without having to adapt the content to a lower quality. The resources available to play back the content can include the resources available in the devices' operating environment. The receiving terminal can cooperatively play the rich content together with the devices in its immediate proximity. For example, when a high quality MPEG-2 video is to be played, the receiving PDA can identify a device in its current environment that can play the MPEG-2 video and coordinate the playback of the content on that device. The emerging technologies such as Bluetooth and UPnP make the device discovery in the receiving terminal's environment easier. In [2] Pham et al. propose a situated computing framework that employs the capabilities of the devices in the receiver's environment to play back multimedia content with small screen devices. Cooperative playback raises another important issue of session migration between clients. Depending on the session requirements and the devices in a receiving terminal's environment, sessions may have to be migrated from one client to another. An architecture for application session handoff in receivers is presented in [5,6]. Their work primarily addresses applications involving image and data retrieval from databases. Application session handoff involving multimedia and video intensive applications poses new challenges. The problem becomes more complex when user interaction is involved and when content adaptation is performed at intermediate nodes.

1.1.2 Real-time Transcoding vs Multiple Content Variants

An alternative to real-time transcoding is creating content in multiple formats and at multiple bitrates and making an appropriate selection at delivery time. This solution is not compute-intensive but highly storage-intensive. Smith et al. propose a Infopyramid approach to enable UMA [7]. The approach is based on creating content with different fidelities and modalities and selecting the appropriate content based on a receiver's capabilities. A tool that supports this functionality within the context of MPEG-7 is described further in Section 2.3.2. With so many encoding formats, bitrates, and resolutions possible, though, storing content in multiple formats at multiple bitrates becomes impractical for certain applications. However, for cases in which the types of receiving terminals are limited, storing multiple variants of the media is sufficient. For the general case, where no limitations exist on the terminal or access network, real-time media transcoding is necessary to support a flexible content adaptation framework.

In the case of video streams, to meet certain network bandwidth constraints, the encoding rate has to be reduced. The desired bitrate can be achieved either by reducing the spatial resolution, by reducing the frame rate, or by reducing the bitrate. The perceived quality of the delivered video depends on the target application as well as the target device capabilities such as the receiving terminal's display capability. The mode of adaptation chosen is also influenced by the resources available to perform the adaptation. Frame dropping has the lowest complexity of the three. In general, the complexity of rate reduction is lower than that of resolution reduction. The problem of determining an optimal choice to maximize the quality of the delivered video is difficult and is an open research issue.

1.1.3 Adaptation in Wireless Networks

The freedom of un-tethered connected devices is making wireless and mobile networks popular for commercial as well as consumer applications. The emergence of 802.11 wireless Ethernet standards is creating new possibilities at the same time posing new challenges for multimedia access. The mobile network topology offers additional locations in the network to perform content adaptation. The wired-to-wireless gateway is one location to adapt the content to suit the wireless network and the mobile device. The mobility of the receiver poses new challenges to perform content adaptation. The mobility of the users causes the users to move to different wireless networks. The receiver handoff becomes difficult when a content adaptation session is in progress. The gateways in question have to communicate the state of the adaptation session in order to seamlessly deliver the content. The problem becomes difficult with content that includes multiple media objects that have to be transcoded. Roy et al. [8] discuss an approach to handing-off media transcoding sessions. The approach of applying content adaptation operations inside a mobile network may prove to be costly because of the cost of session migration and the cost of computing resources for adaptation.

1.1.4 P2P Networks

Peer to peer networking was made popular by Gnutella and Napster, the popular software programs for file sharing on the Internet. The main reason for the popularity of this P2P application was the availability of desirable content – music and movies. P2P networks for content distribution have the potential to alleviate the bandwidth bottlenecks suffered by content distribution networks. Content can be distributed from the closest peers to localise the traffic and minimize bandwidth bottlenecks in the content distribution network. However, the P2P networks are still in their infancy and a number of issues are yet to be addressed in the P2P content distribution networks. Since P2P networks are possible primarily through voluntary user participation, one of the issues to address is the *source migration* when a user that is currently the source for a media distribution session withdraws. In such cases, the session has to be migrated to a new source with little or no disruption to the end user. The CoopNet project discussed in [3] proposes the use of multi description coding in conjunction with a central server to address the issue of source migration. Another important issue in P2P networking is content discovery. The easiest solution is a central server that indexes the content available on the P2P network. With content adaptation and user preferences in place, multiple variants of the same content will be present on the network. Managing the content variations and delivering the appropriate content to the users is a difficult problem. The simplest cases for P2P delivery are simple audio/video streaming applications. With multi-object and interactive applications, the content distribution as well as source migration becomes problematic. The server functionality required to support interactive content makes P2P networks unsuitable for delivering interactive content. Similar arguments hold for content adaptation operations in a P2P network. When multiple variants of the content are available on the P2P network, the problem becomes one of locating an appropriate variant to fit the receiver capabilities. The Digital Item Declaration Language, specified as Part 2 of the MPEG-21 standard, is targeted to overcome such a problem. Essentially, a language has been defined that can be used to structure and describe variants of the multimedia content and allow one to customize the delivery and/or presentation of the content to a particular receiver or network.

2. CONTENT REPRESENTATION

Content representation plays a critical role in determining the content adaptability. Content representation as used in this chapter refers to the media encoding types, presentation format, the storage format and the meta data format used. The media encoding type specifies the encoding format of the elementary streams in the content, e.g., MPEG-2 video, MPEG-1 audio, JPEG images, etc. The encoding format determines the basic media encoding properties: the encoding algorithm, scalability, bitrate, spatial resolution and temporal resolution (frame rate). Not all the media types are characterized by these parameters; e.g., a JPEG image does not have any temporal resolution. The presentation format specifies the inter-relationships among the objects in the presentation and determines the presentation order of the content. The inter-relationships are described using technologies such as MPEG-4 binary representation for scenes (BIFS) [9] and Synchronised Multimedia Integration Language (SMIL) [15]. The storage format specifies the stored representation of the content. MPEG-4 file format specifies a format for efficient storage of object based presentations and lends itself nicely to content adaptation and UMA. The meta data specifies the additional information about the content that can be used to efficiently process the content including content adaptation. MPEG-7 specifies the formats for content meta data descriptions. In this section we give a brief overview of the content representation primarily covering MPEG-4 and MPEG-7 technologies.

2.1 MPEG-4 SYSTEMS

Traditionally the visual component of multimedia presentations has mainly been rectangular video, graphics and text. Advances in image and video encoding and representation techniques [9,10,11] have made possible encoding and representation of audio-visual scenes with semantically meaningful *objects*. The traditionally rectangular video can now be coded and represented as a collection of arbitrarily shaped visual objects. The ability to create object-based scenes and presentations creates many possibilities for a new generation of applications and services. The MPEG-4 series of standards specify tools for such object-based audio-visual presentations. While Audio and Video parts of the standard specify new and efficient algorithms for encoding media, the Systems part of MPEG-4 makes the standard radically different by specifying the tools for object-based representation of presentations. The most significant addition to the MPEG-4 standards, compared with MPEG-1 and MPEG-2, is the scene composition at the user terminal. The individual objects that make up a scene are transmitted as separate elementary streams and composed upon reception according to the composition information delivered along with the media objects. These new representations techniques will make flexible content adaptation possible.

2.1.1 Components of Object Based Presentations

An object-based presentation consists of *objects* that are composed to create a scene; a sequence of scenes forms a presentation. There is no clear-cut definition of what constitutes an *object*. When used in the sense of audio-visual (AV) objects, an object can be defined as *something* that has semantic and structural significance in an AV presentation. An object can thus be broadly defined as a building block of an audio-visual scene. When composing a scene of a city, buildings can be the objects in the scene. In a scene that shows the interior of a building, the furniture and other items in the building are the objects. The granularity of objects in a scene depends on the application and context. The main advantage of breaking up a scene into objects is the coding efficiency gained by applying appropriate compression techniques to different objects in a scene. In addition to coding gains, there are several other benefits of object-based representation: modularity, reuse of content, ease of manipulation, object annotation, as well as the possibility of playing appropriate objects based on the network and the receiver resources available. To appreciate the efficiency of object-based presentations, consider a home shopping channel such as the ones currently available on TV. The information on the screen consists mostly of text, images of products, audio and video (mostly quarter-screen and sometimes full screen). All this information is encoded using MPEG-2 video/audio at 30 fps. However, if this content is created using object-based technology, all static information such as text and graphics is transmitted only at the beginning of a scene and the rest of the transmission consists of only audio, video and text and image updates that take up significantly less bandwidth. In addition to this, the ability to interact with individual objects makes applications such as e-commerce possible.

The key characteristic of the object-based approach to audio-visual presentations is the composition of scenes from individual objects at the receiving terminal, rather than during content creation in the production studio (e.g., MPEG-2 video). This allows prioritising objects and delivering individual objects with the QoS required for that object. Multiplexing tools such as FlexMux [9] allow multiplexing of objects with similar QoS requirements in the same FlexMux stream. Furthermore, static objects such as a scene background are transmitted only once and result in significant bandwidth savings. The ability to dynamically add and remove objects from scenes at the individual user terminal even in broadcast systems makes a new breed of applications and services possible. Frame-based systems do not have this level of sophistication and sometimes use makeshift methods such as image mapping to simulate simple interactive behaviour. This paradigm shift while creating new possibilities for applications and services makes content creation and delivery complex. The end user terminals that process object-based presentations are now more complex but also more capable.

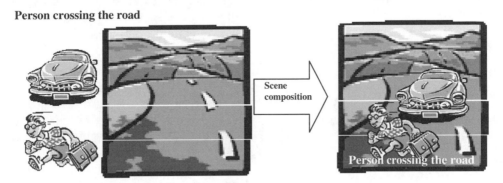

Figure 36.2 Example of an object-based scene.

Figure 36.2 shows a scene with four visual objects, a person, a car, the background and the text. In object-based representation, each of these visual objects is encoded separately in a compression scheme that gives the best quality for that object. The final scene as seen on a user terminal would show a person running across a road and the text at the bottom of the scene, just like in a frame-based system. To compose a scene, object-based systems must also include the composition data for the objects that a terminal uses for spatio-temporal placement of objects in a scene. The scene may also have audio objects associated with (or independent of) visual objects. The compressed objects are delivered to a terminal along with the composition information. Since scenes are composed at the user end of the system, users may be given control on which objects are played. If a scene has two audio tracks (in different languages) associated with it, users can choose the track they want to hear. Whether the system continues to deliver the two audio tracks even though only one track is played is system dependent; broadcast systems may deliver all the available tracks while remote interactive systems with upstream channels may deliver objects as and when required. Since even text is treated and delivered as a separate object, it requires far less bandwidth than transmitting the encoded image of the rendered text. However the delivered text object now has to include font information and the user terminals have to know how to render fonts. User terminals could be designed to download fonts or decoders necessary to render the objects received.

2.1.2 Scene Composition

Scene composition can simply be defined as spatio-temporal placement of objects in the scene. Spatial composition determines the position of objects in a scene while temporal composition determines its position over time. Operations such as object animation, addition, and removal can be accomplished by dynamically updating the composition parameters of objects in a scene. All the composition data that is provided to a terminal can itself be treated as a separate object.

Since the composition is the most critical part of object-based scenes, the composition data stream has very strict timing constraints and is usually not loss tolerant. Any lost or even delayed composition information could distort the content of a presentation. Treating the composition data as a separate data object allows the system to deliver it over a reliable channel. Figure 36.3 shows the parameters for spatial composition of objects in a scene. The gray lines are not part of the scene; x and y are horizontal and vertical displacements from the top left corner. The cube is rotated by an angle of θ radians. The relative depth of

Figure 36.3 Composition parameters.

the ellipse and a cylinder are also shown. The ellipse is closer to the viewer ($z < z'$) and hence is displayed on top of the cylinder in the final rendered scene. Even audio objects may have spatial composition associated with them. An object can be animated by continuously updating the necessary composition parameters.

2.2 OVERVIEW OF MPEG-4 SYSTEMS

The MPEG-4 committee has specified tools to encode individual objects, compose presentations with objects, store these object-based presentations and access these presentations in a distributed manner over networks [9]. The MPEG-4 Systems specification provides the glue that binds the audio-visual objects in a presentation [12,13]. The basis for the MPEG-4 Systems architecture is the separation of the media and data streams from the stream descriptions. The scene description stream, also referred to as BIFS, describes a scene in terms of its composition and evolution over time and includes the scene composition and scene update information. The other data stream that is part of the MPEG-4 systems is the object description or OD stream, which describes the properties of data and media streams in a presentation. The description contains a sequence of object descriptors, which encapsulate the stream properties such as scalability, QoS required to deliver the stream and the decoders and buffers required to process the stream. The object descriptor framework is an extensible framework that allows separation of an object and the object's properties. This separation allows for providing different QoS for different streams; for example, scene description streams which have very low or no loss tolerance and the associated media streams, which are usually loss tolerant. These individual streams are referred to as elementary streams at the system level. The separation of media data and meta data also makes it possible to use different media data (MPEG-1 or H.263 video) without modifying the scene description. The media and media descriptions are communicated to the receivers on separate channels. The receivers can first receive the media descriptions and then request appropriate media for delivery based on its capabilities.

An elementary stream is composed of a sequence of access units (e.g., frames in an MPEG-2 video stream) and is carried across the Systems layer as sync-layer (SL) packetized access units. The sync-layer is configurable and the configuration for a specific elementary stream is specified in its elementary stream (ES) descriptor. The ES descriptor for an elementary stream can be found in the object descriptor for that stream which is carried separately in the OD stream. The sync layer contains the information necessary for inter-media synchronization. The sync-layer configuration indicates the mechanism used to synchronize the objects in a presentation by indicating the use of timestamps or implicit media specific timing. Unlike MPEG-2, MPEG-4 Systems does not specify a single clock for the elementary streams. Each elementary stream in an MPEG-4 presentation can potentially have a different clock speed. This puts additional burden on a terminal, as it now has to support recovery of multiple clocks. In addition to the scene description and object description streams, an MPEG-4 session can contain Intellectual Property Management and Protection (IPMP) streams to protect media streams, or Object Content Information (OCI) streams that describe the contents of the presentation, and a clock reference stream. All the data flows between a client and a server are SL-packetized.

The data communicated to the client from a server includes at least one scene description stream. The scene description stream, as the name indicates, carries the information that specifies the spatio-temporal composition of objects in a scene. The MPEG-4 scene description is based on the VRML specification. VRML was intended for 3D modelling and is a static representation (a new object cannot be dynamically added to the model). MPEG-4 Systems extended the VRML specification with additional 2D nodes, a binary representation, dynamic updates to scenes, and new nodes for server interaction and flex-timing [14]. A scene is represented as a graph with media objects associated with the leaf nodes. The elementary streams carrying media data are bound to these leaf nodes by means of BIFS URLs. The URLs can either point to object descriptors in the object descriptor stream or media data directly at the specified URL. The intermediate nodes in the scene graph correspond to functions such as transformations, grouping, sensors, and interpolators.

The VRML event model adopted by MPEG-4 systems has a mechanism called *ROUTEs* that propagates events in a scene. A ROUTE is a data path between two fields; a change in the value of the source field will effect a change in the destination field. Through intermediate nodes, this mechanism allows user events such as mouse clicks to be translated into actions that transform the content displayed on the terminal. In addition to VRML functionality, MPEG-4 includes features to perform server interaction, polling terminal capability, binary encoding of scenes, animation and dynamic scene updates. MPEG-4 also specifies a Java interface to access a scene graph from an applet. These features make possible content with a range of functionality blurring the line between applications and content.

Figure 36.4 shows the binding of elementary streams in MPEG-4 Systems. The figure shows a scene graph with a group node (G), a transform node (T), an image node (I), an audio node (A) and a video node (V). Elementary streams are shown in the figure with a circle enclosing the components of the stream. The scene description forms a separate elementary stream. The media nodes in a scene description are associated with a media object by means of object IDs (OD ID). The object descriptors have a unique ID in the scene and are carried in an object descriptor stream. An object descriptor is associated with one or more elementary streams. The elementary streams are packetized and carried in separate channels. A receiver processes the scene description stream first and determines the objects it needs to render the scene. The receiver then retrieves the corresponding object descriptors from the object descriptor stream and determines the elementary streams it has to request from the sender. Since the receiver knows the types of the objects before it requests them from the sender, it can avoid requesting any object it cannot process or it can initiate content adaptation negotiation to request the object in a format that the receiver can process.

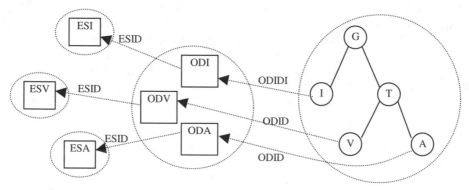

Figure 36.4 Stream Association in MPEG-4 Systems.

The MP4 file format that is part of the MPEG-4 specification offers a versatile container to store content in a form that is suitable for streaming. The MP4 file format can be used to store multiple variations of the same content and select an appropriate version for delivery. The hint track mechanism supported by the MP4 file format allows supporting multiple transport formats, e.g., RTP and MPEG-2 TS, with minimal storage overhead. In summary, MPEG-4 specifies a flexible framework to develop content and applications suitable for UMA services.

2.3 MPEG-7 TOOLS

The description of multimedia content is an extremely important piece of the UMA puzzle as it provides an essential understanding of the source material to be distributed. MPEG-7 provides a standardized set of description tools that allow for the description of rich multimedia content. Among the many tools available, a subset of these tools is particularly targeted towards supporting the UMA concept and framework. These tools are highlighted here and their use in the context of UMA applications is described. In particular, we describe the tools for data abstraction, tools that facilitate the description of multiple versions of the content, tools that provide transcoding hints and tools that describe user preferences. Complete coverage of the MPEG-7 standard is provided in Chapter 29 and as well as the standard itself [16].

2.3.1 Data Abstraction

In general, summaries provide a compact representation, or an abstraction, of the audio-visual content to enable discovery, browsing, navigation, and visualization. The Summary Description Schemes (DS) in MPEG-7 enable two types of navigation modes: hierarchical and sequential. In the hierarchical mode, the information is organized into successive levels, each describing the audio-visual content at a different level of detail. The levels closer to the root of the hierarchy provide more coarse summaries, while levels further from the root provide more detailed summaries. On the other hand, the sequential summary provides a sequence of images or video frames, possibly synchronized with audio, which may compose a slide-show or audio-visual skim. There are many existing methods for key frame extraction and visual summarization; some examples can be found in [18,57,58,59].

The description of summaries can be used for adaptive delivery of content in a variety of cases in which limitations exist on the processing power of a terminal, the bandwidth of a network, or even an end-user's time. For example, in [19], a system for delivering content to multiple clients with varying capabilities was proposed. In this system, video summarization was performed prior to transcoding based on inputs from the user.

2.3.2 Multiple Variations

Variations provide information about different variations of audio-visual programs, such as summaries and abstracts; scaled, compressed and low-resolution versions; and versions with different languages and modalities, e.g., audio, video, image, text and so forth. One of the targeted functionalities of MPEG-7's Variation DS is to allow a server or proxy to select the most suitable variation of the content for delivery according to the capabilities of terminal devices, network conditions, or user preferences. The Variations DS describes the different alternative variations. The variations may refer to newly authored content, or correspond to content derived from another source. A variation fidelity value gives the quality of the variation compared to the original. The variation type attribute indicates the type of variation, such as summary, abstract, extract, modality translation, language translation, colour reduction, spatial reduction, rate reduction, compression, and so forth. Further details on the use of variations for adaptable content delivery can be found in [17].

2.3.3 Transcoding Hints

Transcoding hints provide a means to specify information about the media to improve the quality and reduce the complexity for transcoding applications. The generic use of these hints is illustrated in Figure 36.5.

Among the transcoding hints that have been standardized by MPEG-7 are the Difficulty Hint, Shape Hint, Motion Hint and Coding Hint. The Difficulty Hint describes the bit-rate coding complexity. This hint can be used for improved bit rate control and bit rate conversion, e.g., from constant bit-rate (CBR) to variable bit-rate (VBR). The Shape Hint specifies the amount of change in a shape boundary over time and is proposed to overcome the composition problem when encoding or transcoding multiple video objects with different frame-rates. The Motion Hint describes: i) the motion range, ii) the motion uncompensability and iii) the motion intensity. This metadata can be used for a number of tasks including, anchor frame selection, encoding mode decisions, frame-rate and bit-rate control, as well as bitrate allocation among several video objects for MPEG-4 object-based transcoding. The Coding Hint provides generic information contained in a video bitstream, such as the distance between anchor frames and the average quantization parameter used for coding. These transcoding hints, especially the search range hint, aim to reduce the computational complexity of the transcoding process. Further information about the use of these hints may be found in [20,21,53].

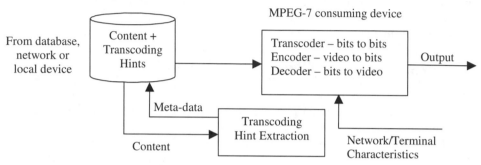

Figure 36.5 Generic illustration of the transcoding hints application.

2.3.4 User Preferences

The UserInteraction DS defined by MPEG-7 describe preferences of users pertaining to the consumption of the content, as well as usage history. The MPEG-7 content descriptions can be matched to the preference descriptions in order to select and personalize content for more efficient and effective access, presentation and consumption. The UserPreference DS describes preferences for different types of content and modes of browsing, including context dependency in terms of time and place. The UserPreference DS describes also the weighting of the relative importance of different preferences, the privacy characteristics of the preferences and whether preferences are subject to update, such as by an agent that automatically learns through interaction with the user. The UsageHistory DS describes the history of actions carried out by a user of a multimedia system. The usage history descriptions can be exchanged between consumers, their agents, content providers, and devices, and may in turn be used to determine the user's preferences with regard to content.

It should be obvious that the above descriptions of a user are very useful to customize and personalize content according to the particular person, which is a key target of the UMA framework. Extensions to these descriptions of the user are being explored within the context of MPEG-21. Further details may be found in Section 4 of this chapter.

3. CONTENT ADAPTATION

Content adaptation can generally be defined as the conversion of a signal from one format to another, where the input and output signals are typically compressed bitstreams. This section focuses mainly on the adaptation, or transcoding, of video signals. In the earliest work, this conversion corresponded to a reduction in bitrate to meet available channel capacity. Additionally, researchers have investigated conversions between constant bitrate (CBR) streams and variable bit-rate (VBR) streams to facilitate more efficient transport of video. To improve the performance, techniques to refine the motion vectors have been proposed.

In this section, we first provide an overview of the techniques used for bitrate reduction and the corresponding architectures that have been proposed. Then, we describe recent advances with regard to spatial and temporal resolution reduction techniques and architectures. This section ends with a summary of the work covered and provides reference to additional adaptation techniques.

3.1 BITRATE REDUCTION

The objective in bitrate reduction is to reduce the bitrate accurately, while maintaining low complexity and achieving the highest quality possible. Ideally, the quality of the reduced rate bitstream should have the quality of a bitstream directly generated with the reduced rate. The most straightforward means to achieve this is to decode the video bitstream and fully re-encode the reconstructed signal at the new rate. This is illustrated in Figure 36.6. The best performance can be achieved by calculating new motion vectors and mode decisions for every macroblock at the new rate. However, significant complexity saving can be achieved, while still maintaining acceptable quality, by reusing information contained in the original incoming bitstreams and also considering simplified architectures [22,23,24,25,26].

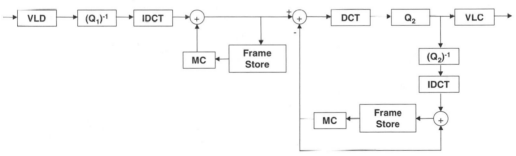

Figure 36.6 Reference transcoding architecture for bit rate reduction.

In the following, we review progress made over the past few years on bit-rate reduction architectures and techniques, where the focus has been centered on two specific aspects, complexity and drift reduction. Drift can be explained as the blurring or smoothing of successively predicted frames. It is caused by the loss of high frequency data, which creates a mismatch between the actual reference frame used for prediction in the encoder, and the degraded reference frame used for prediction in the transcoder and decoder. We will consider two types of systems, a closed-loop and an open-loop system. Each demonstrates the trade-off between complexity and quality.

3.1.1 Transcoding Architectures

In Figure 36.7, the open-loop and closed-loop systems are shown. The architecture in Figure 36.7(a) is open-loop systems, while the architecture in Figure 36.7(b) is a closed-loop system. In the open-loop system, the bitstream is variable-length decoded (VLD) to extract the variable-length codewords corresponding to the quantized DCT coefficients, as well as macroblock data corresponding to the motion vectors and other macroblock-type information. In this scheme, the quantized coefficients are inverse quantized, then simply re-quantized to satisfy the new output bitrate. Finally, the re-quantized coefficients and stored macroblock information are variable length coded (VLC). An alternative open-loop scheme, which is not illustrated here, but is even less complex than the one shown in Figure 36.7(a), is to directly cut high frequency data from each macroblock [29]. To cut the high frequency data without actually doing the VLD, a bit-profile for the AC coefficients is maintained. As macroblocks are processed, codewords corresponding to high-frequency coefficients are eliminated as needed so that the target bitrate is met. In terms of complexity, both open-loop schemes are relatively simple since a frame memory is not required and there is no need for an IDCT. In terms of quality, better coding efficiency can be obtained by the re-quantization approach since the variable-length codes that are used for the re-quantized data will be more efficient. However, both architectures are subject to drift.

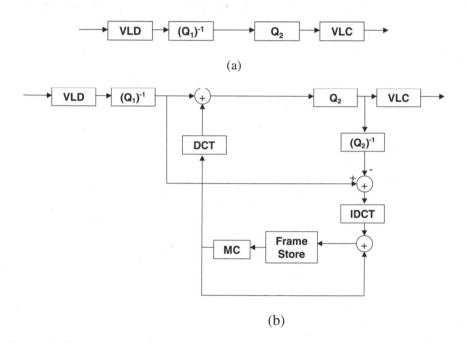

Figure 36.7 Simplified transcoding architectures for bitrate reduction. (a) open-loop, partial decoding to DCT coefficients, then re-quantize, (b) closed-loop, drift compensation for re-quantized data.

In general, the reason for drift is mainly due to the loss of high-frequency information. Beginning with the I-frame, which is a reference for the next P-frame, high-frequency information is discarded by the transcoder to meet the new target bit-rate. Incoming residual blocks are also subject to this loss. When a decoder receives this transcoded bitstream, it will decode the I-frame with reduced quality and store it in memory. When it is time to decode the next P-frame, the degraded I-frame is used as a predictive component and added to a degraded residual component. Since both components are different than what was originally derived by the encoder, a mismatch between the predictive and residual components is created. As time goes on, this mismatch progressively increases.

The architecture shown in Figure 36.7(b) is a closed-loop system and aims to eliminate the mismatch between predictive and residual components by approximating the cascaded decoder-encoder approach [27]. The main difference in structure between the reference architecture and this simplified scheme is that reconstruction in the reference is performed in the spatial domain, thereby requiring two reconstruction loops with one DCT and two IDCT's. On the other hand, in the simplified structure that is shown, only one reconstruction loop is required with one DCT and one IDCT. In this structure, some arithmetic inaccuracy is introduced due to the non-linear nature in which the reconstruction loops are combined. However, it has been found the approximation has little effect on the quality [27]. With the exception of this slight inaccuracy, this third architecture is mathematically equivalent to a cascaded decoder-encoder approach. In [32], additional causes of drift, e.g., due to floating-point inaccuracies, have been further studied. Overall, though, in comparison to the open-loop architectures discussed earlier, drift is eliminated since the mismatch between predictive and residual components is compensated for.

3.1.2 Simulation Results

In Figure 36.8, a frame-based comparison of the quality between the reference, open-loop and closed-loop architectures is shown. The input to the transcoders is the Foreman sequence at CIF resolution coded at 2Mbps with GOP structure N=30 and M=3. The transcoded output is re-encoded with a fixed quantization parameter of 15. To illustrate the effect of drift in this plot, the decoded quality of only the I- and P-frames is shown. It is evident that the open-loop architecture suffers from severe drift, and the quality of the simplified closed-loop architecture is very close to that of the reference architecture.

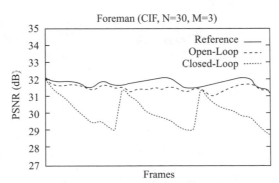

Figure 36.8 Frame-based comparison of PSNR quality for reference, open-loop and closed-loop architectures for bitrate reduction.

3.1.3 Motion-Compensation in the DCT Domain

The closed-loop architecture described above provides an effective transcoding structure in which the macroblock reconstruction is performed in the DCT domain. However, since the memory stores spatial domain pixels, the additional DCT/IDCT is still needed. In an attempt to further simplify the reconstruction process, structures that reconstruct references frames completely in the compressed-domain have been proposed [29,30,27]. All of these architectures are based on the compressed-domain methods for motion compensation proposed by Chang and Messerschmidt [28]. It was found that decoding completely in the compressed-domain could yield equivalent quality to spatial-domain decoding [29]. However, this was achieved with floating-point matrix multiplication and proved to be quite costly. In [27], simplification of this computation was achieved by approximating the floating-point elements by power-of-two fractions, so that shift operations could be used, and in [31], simplifications have been achieved through matrix decomposition techniques.

Regardless of which simplification is applied, once the reconstruction has been accomplished in the compressed domain, one can easily re-quantize the drift-free blocks and VLC the quantized data to yield the desired bitstream. In [27], the bit re-allocation has been accomplished using the Lagrangian multiplier method. In this formulation, sets of quantizer steps are found for a group of macroblocks so that the average distortion caused by transcoding error is minimized.

3.1.4 Motion Vector Refinement

In all of the above-mentioned methods, significant complexity is reduced by assuming that the motion vectors computed at the original bit-rate are simply reused in the reduced rate bitstream. It has been shown that re-using the motion vectors in this way leads to non-optimal transcoding results due to the mismatch between prediction and residual components [25,32]. To overcome this loss of quality without performing a full motion re-estimation, the motion vector refinement schemes have been proposed. Such schemes can be used with any of the above bit-rate reduction architectures above for improved quality, as well as the spatial and temporal resolution reduction architectures described below.

3.1.5 CBR-to-VBR Conversion

While the above architectures have mainly focused on general bit-rate reduction techniques for the purpose of transmitting video over band-limited channels, there has also been study on the conversion between constant bitrate (CBR) streams and variable bitrate (VBR) streams to facilitate more efficient transport of video [33]. In this work, the authors exploit the available channel bandwidth of an ATM network and adapt the CBR encoded source accordingly. This is accomplished by first reducing the bitstream to a VBR stream with a reduced average rate, then segmenting the VBR stream into cells and controlling the cell generation rate by a traffic shaping algorithm.

3. 2 SPATIAL RESOLUTION REDUCTION

Similar to bitrate reduction, the Reference architecture for reduced spatial resolution transcoding refers to the cascaded decoding, spatial-domain down-sampling, followed by a full-re-encoding. Several papers have addressed this problem of resolution conversion, e.g., [36,25,37,34,38]. The primary focus of the work in [36] was on motion vector scaling techniques. In [25], the problems associated with mapping motion vector and macroblock type data were addressed. The performance of motion vector refinement techniques in the context of resolution conversion was also studied in this work. In [37], the authors propose to use DCT-domain down-scaling and motion compensation for transcoding, while also considering motion vector scaling and coding mode decisions. With the proposed two-loop architecture, computational savings of

40% have been reported with a minimal loss in quality. In [34], a comprehensive study on the transcoding to lower spatio-temporal resolutions and to different encoding formats has been provided based on the reuse of motion parameters. In this work, a full decoding and encoding loop were employed, but with the reuse of information, processing time was improved by a factor of 3. While significant savings can be achieved with the reuse of macroblock data, further simplifications of the Reference transcoding architecture were investigated in [35] based on an analysis of drift errors. Several new architectures, including an intra refresh architecture, were proposed.

In the following, the key points from the above works are reviewed, including the performance of various motion vector scaling algorithms, DCT-domain down-conversion and the mapping of macroblock-type information to the lower resolution. Also, the concepts of the intra-refresh architecture will be discussed.

3.2.1 Motion Vector Mapping

When down-sampling four macroblocks to one macroblock, the associated motion vectors have to be mapped. Several methods suitable for frame-based motion vector mapping have been described in past works [36,34,38]. To map from four frame-based motion vectors, i.e., one for each macroblock in a group, to one motion vector for the newly formed macroblock, a weighted average or median filters can be applied. This is referred to as a 4:1 mapping. However, with certain compression standards, such as MPEG-4 and H.263, there is support in the syntax for advanced prediction modes that allow one motion vector per 8x8 block. In this case, each motion vector is mapped from a 16x16 macroblock in the original resolution to an 8x8 block in the reduced resolution macroblock with appropriate scaling by 2. This is referred as a 1:1 mapping. While 1:1 mapping provides a more accurate representation of the motion, it is sometimes inefficient to use since more bits must be used to code four motion vectors. An optimal mapping would adaptively select the best mapping based on an R-D criterion. A good evaluation of the quality that can be achieved using the different motion vector mapping algorithms can be found in [34].

Because MPEG-2 supports interlaced video, we also need to consider field-based MV mapping. In [39], the top-field motion vector was simply used. An alternative scheme that averages the top and bottom field motion vectors under certain conditions was proposed in [35]. However, it should be noted that the appropriate motion vector mapping technique is dependent on the down-conversion scheme used. This is particularly important for interlaced data, where the target output may be a progressive frame. Further study on the relation between motion vector mapping and the texture down-conversion is needed.

3.2.2 DCT-Domain Down-Conversion

The most straightforward way to perform down-conversion in the DCT-domain is to cut the low-frequency coefficients of each block and recompose the new macroblock using the composting techniques proposed [28]. A set of DCT-domain filters can be derived by cascading these two operations. More sophisticated filters that attempt to retain more of the high frequency information, such as the frequency synthesis filters used in [40] and derived in the references therein, may also be considered. The filters used in this work perform the down-conversion operations on the rows and columns of the macroblock using separable 1D filters. These down-conversion filters can be applied in both the horizontal and vertical directions, and to both frame-DCT and field-DCT blocks. Variations of this filtering approach to convert field-DCT blocks to frame-DCT blocks, and vice versa, have also been derived [29].

3.2.3 Conversion of Macroblock Type

In transcoding video bitstreams to a lower spatial resolution, a group of four macroblocks in the original video corresponds to one macroblock in the transcoded video. To ensure that the down-sampling process will not generate an output macroblock in which its sub-blocks have different coding modes, e.g., both inter- and intra-sub-blocks within a single macroblock, the mapping of MB modes to the lower resolution must be considered. Three possible methods to overcome this problem when a so-called mixed-block is encountered are outlined below.

In the first method, ZeroOut, the MB modes of the mixed macroblocks are all modified to inter-mode. The MV's for the intra-macroblocks are reset to zero and so are corresponding DCT coefficients. In this way, the input macroblocks that have been converted are replicated with data from corresponding blocks in the reference frame. The second method, IntraInter, maps all MB's to inter-mode, but the motion vectors for the intra-macroblocks are predicted. The prediction can be based on the data in neighboring blocks, which can include both texture and motion data. As an alternative, we can simply set the motion vector to be zero, depending on which produces less residual. In an encoder, the mean absolute difference of the residual blocks is typically used for mode decision. The same principles can be applied here. Based on the predicted motion vector, a new residual for the modified macroblock must be calculated. In the third method, InterIntra, the MB modes are all modified to intra-mode. In this case, there is no motion information associated with the reduced-resolution macroblock; therefore all associated motion vector data are reset to zero and the intra-DCT coefficients are generated to replace the inter-DCT coefficients.

It should be noted that to implement the IntraInter and InterIntra methods, we need a decoding loop to reconstruct full-resolution picture. The reconstructed data is used as a reference to convert the DCT coefficients from intra-to-inter, or inter-to-intra. For a sequence of frames with a small amount of motion and a low-level of detail, the low complexity strategy of ZeroOut can be used. Otherwise, either IntraInter or InterIntra should be used. The performance of InterIntra is a little better than IntraInter, because InterIntra can stop drift propagation by transforming inter-blocks to intra-blocks.

3.2.4 Intra-Refresh Architecture

In reduced resolution transcoding, drift error is caused by many factors, such as requantization, motion vector truncation and down-sampling. Such errors can only propagate through inter-coded blocks. By converting some percentage of inter-coded blocks to intra-coded blocks, drift propagation can be controlled. In the past, the concept of intra-refresh has successfully been applied to error-resilience coding schemes [41], and it has been found that the same principle is also very useful for reducing the drift in a transcoder [35].

The intra-refresh architecture for spatial resolution reduction is illustrated in Figure 36.9. In this scheme, output macroblocks are subject to a DCT-domain down-conversion, requantization and variable-length coding. Output macro-blocks are either derived directly from the input bitstream, i.e., after variable-length decoding and inverse quantization, or retrieved from the frame store and subject to a DCT operation. Output blocks that originate from the frame store are independent of other data, hence coded as intra-blocks; there is no picture drift associated with these blocks.

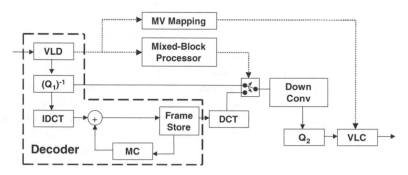

Figure 36.9 Illustration of intra-refresh architecture for reduced spatial resolution transcoding.

The decision to code an intra-block from the frame store depends on the macroblock coding modes and picture statistics. In a first case based on the coding mode, an output macroblock is converted if the possibility of a mixed-block is detected. In a second case based on picture statistics, the motion vector and residual data are used to detect blocks that are likely to contribute to larger drift error. For this case, picture quality can be maintained by employing an intra-coded block in its place. Of course, the increase in the number of intra-blocks must be compensated for by the rate control. Further details on the rate control can be found in [35], and a complexity-quality evaluation of this architecture compared to the reference method can be found in [43].

3.3 TEMPORAL RESOLUTION REDUCTION

Reducing the temporal resolution of a video bitstream is a technique that may be used to reduce the bitrate requirements imposed by a channel. However, this technique may also be used to satisfy processing limitations imposed by a terminal. As discussed earlier, motion vectors from the original bitstream are typically reused in bitrate reduction and spatial resolution reduction transcoders to speed up the re-encoding process. In the case of spatial resolution reduction, the input motion vectors are mapped to the lower spatial resolution. For temporal resolution reduction, we are faced with a similar problem in which it is necessary to estimate the motion vectors from the current frame to the previous non-skipped frame that will serve as a reference frame in the receiver. Solutions to this problem have been proposed in [32,44,45]. Assuming a pixel-domain transcoding architecture, this re-estimation of motion vectors is all that needs to be done since new residuals corresponding to the re-estimated motion vectors will be calculated. However, if a DCT-domain transcoding architecture is used, a method of re-estimating the residuals in the DCT-domain is needed. A solution to this problem has been described in [46]. In [47], the issue of motion vector and residual mapping has been addressed in the context of a combined spatio-temporal reduction in the DCT-domain based on the intra-refresh architecture described earlier. The key points of these techniques will be discussed in the following.

3.3.1 Motion Vector Re-Estimation

As described in [32,44,45], the problem of re-estimating a new motion vector from the current frame to a previous non-skipped frame can be solved by tracing the motion vectors back to the desired reference frame. Since the predicted blocks in the current frame are generally overlapping with multiple blocks, bilinear interpolation of the motion vectors in the previous skipped frame has been proposed, where the weighting of each input motion vector is proportional to the amount of overlap with the predicted block. In the place of this bilinear interpolation, a majority-voting or dominant vector selection scheme as proposed in [32,47] may also be used, where the motion vector associated with the largest overlapping region is chosen.

In order to trace back to the desired reference frame in the case of skipping multiple frames, the above process can be repeated. It is suggested, however, that some refinement of the resulting motion vector be performed for better coding efficiency. In [45], an algorithm to determine an appropriate search range based on the motion vector magnitudes and the number of frames skipped has been proposed. To dynamically determine the length of skipped frames and maintain smooth playback, frame rate control based on characteristics of the video content has also been proposed [45,46].

3.3.2 Residual Re-Estimation

The problem of estimating a new residual for temporal resolution reduction is more challenging for DCT-domain transcoding architectures than for pixel-domain transcoding architectures. With pixel-domain architectures, the residual between the current frame and the new reference frame can be easily computed given the new motion vector estimates. For DCT-domain transcoding architectures, this calculation should be done directly using DCT-domain motion compensation techniques [28]. Alternative methods to compute this new residual have been presented in [46,47].

3.4 SCALABLE CODING

An important area in the context of content adaptation is scalable video coding. Various forms of scalable video coding have been explored over the past decade, i.e., SNR scalability, frequency scalability, spatial scalability and temporal scalability. The advantage of scalable coding schemes in general is that multiple qualities and/or resolution can be encoded simultaneously. The ability to reduce bitrate or spatio-temporal resolutions is built into the coding scheme itself. Typically, the signal is encoded into a base layer and enhancement layers, where the enhancement layers add spatial, temporal and/or SNR quality to the reconstructed base layer.

Recently, a new form of scalability, known as Fine Granular Scalability (FGS), has been developed and adopted by the MPEG-4 standard. In contrast to conventional scalable coding schemes, FGS allows for a much finer scaling of bits in the enhancement layer [48]. In contrast to traditional scalable coding techniques, FGS provides an enhancement layer that is continually scalable. This is accomplished through a bit-plane coding method of DCT coefficients in the enhancement layer, which allows the enhancement layer bitstream to be truncated at any point. In this way, the quality of the reconstructed frames is proportional to the number of enhancement bits received.

The standard itself does not specify how the rate allocation, or equivalently, the truncation of bits on a per frame basis is done. The standard only specifies how a truncated bitstream is decoded. In [50,49], optimal rate allocation strategies that essentially truncate the FGS enhancement layer have been proposed. The truncation is performed in order to achieve constant quality over time under a dynamic rate budget constraint.

Embedded image coders, such as the JPEG-2000 coder [51], also employ bit-plane coding techniques to achieve scalability. Both SNR and spatial scalability are supported. In contrast to most scalable video coding schemes that are based on a Discrete Cosine Transform (DCT), most embedded image coders are based on a Discrete Wavelet Transform (DWT). The potential to use the wavelet transform for scalable video coding is also being explored [52].

3.5 SUMMARY

This section has reviewed work related to the transcoding of video bitstreams, including bitrate, spatial resolution and temporal resolution reduction. Also, some transcoding considerations with regard to scalable coding have been discussed.

It should be noted that there are additional content adaptation schemes that have been developed and proposed, but have not been covered here. Included in these additional adaptation schemes are object-based transcoding [53], transcoding between FGS and single-layer [54], and syntax conversions [55,56]. Key frame extraction and video summarization is also considered a form of content adaptation in the sense that the original video has been abstracted or reduced in time [57,58,59]. Joint transcoding of multiple streams has been presented in [60] and system layer transcoding has been described in [61].

4. ENVIRONMENT DESCRIPTION

The description of a usage environment is a key component for Universal Multimedia Access. This section focuses on standardization activities in this area. First, an overview of the MPEG-21 Multimedia Framework is provided. Then, the specific part of this standard, Part 7: Digital Item Adaptation, which actually targets the description of usage environments, is introduced. This follows with a more detailed overview of the specific usage environment descriptions under development by this part of the standard. We conclude this section with references to related standardization activities and an overview of the issues in capability understanding.

4.1 MPEG-21 MULTIMEDIA FRAMEWORK

Moving Picture Experts Group (MPEG) is a working group of ISO/IEC in charge of the development of standards for coded representation of digital audio and video. Established in 1988, the group has produced four important standards. Among them are MPEG-1, the standard on which such products as Video CD and MP3 are based, MPEG-2, the standard on which such products as Digital Television set top boxes and DVD are based, MPEG-4, the standard for multimedia for the fixed and mobile web and MPEG-7, the standard for description and search of audio and visual content. Work on the new standard, MPEG-21, formally referred to as ISO/IEC 21000 "Multimedia Framework," was started in June 2000. So far a Technical Report has been produced [62] and several normative parts of the standard have been developed. An overview of the parts that have been developed or are under development can be found in [63].

Today, many elements exist to build an infrastructure for the delivery and consumption of multimedia content. There is, however, no 'big picture' to describe how these elements, either in existence or under development, relate to each other. The aim for MPEG-21 is to describe how these various elements fit together. Where gaps exist, MPEG-21 will recommend which new standards are required. MPEG will then develop new standards as appropriate, while other bodies may develop other relevant standards. These specifications will be integrated into the multimedia framework through collaboration between MPEG and these bodies.

The result is an open framework for multimedia delivery and consumption, with both the content creator and content consumer as focal points. This open framework provides content creators and service providers with equal opportunities in the MPEG-21 enabled open market. This will also be to the benefit of the content consumer providing them access to a large variety of contents in an interoperable manner.

The vision for MPEG-21 is to define a multimedia framework *to enable transparent and augmented use of multimedia resources across a wide range of networks and devices* used by different communities. MPEG-21 introduces the concept of *Digital Item*, which is an abstraction for a multimedia content, including the various types of relating data. For example, a musical album may consist of a collection of songs, provided in several encoding formats to suit the capabilities (e.g., bitrate, CPU, etc.) of the device on which they will be played. Furthermore, the album may provide the lyrics, some bibliographical information about the musicians and composers, a picture of the album, information about the rights associated to the songs and pointers to a web site where other related material can be purchased. All this aggregation of content is considered by MPEG-21 as a Digital Item, where *Descriptors* (e.g., the lyrics) are associated to the *Resources*, i.e., the songs themselves.

Conceptually, a Digital Item is defined as a structured digital object with a standard representation, identification and description. This entity is also the fundamental unit of distribution and transaction within this framework.

Seven key elements within the MPEG-21 framework have been defined. One of these elements is Terminals and Networks. Details on the other elements in this framework can be found in [62,63]. The goal of the MPEG-21 Terminals and Networks is to achieve interoperable transparent access to advanced multimedia content by shielding users from network and terminal installation, management and implementation issues. This will enable the provision of network and terminal resources on demand to form user communities where multimedia content can be created and shared. Furthermore, the transactions should occurs with an agreed/contracted quality, while maintaining reliability and flexibility, and also allowing the multimedia applications to connect diverse sets of Users, such that the quality of the user experience will be guaranteed.

4.2 DIGITAL ITEM ADAPTATION

Universal Multimedia Access is concerned with the access to any multimedia content from any type of terminal or network and thus it is closely related to the target mentioned above of "achieving interoperable transparent access to (distributed) advanced multimedia content." Toward this goal, and in the context of MPEG-21, we target the adaptation of Digital Items. This concept is illustrated in Figure 36.10. As shown in this conceptual architecture, a Digital Item is subject to a resource adaptation engine, as well as a descriptor adaptation engine, which together produce the modified Digital Item.

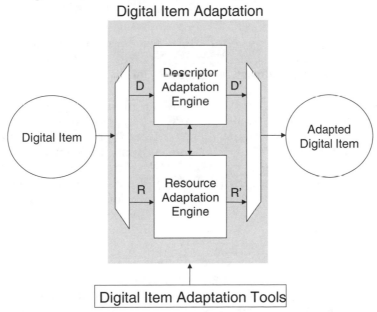

Figure 36.10 Concept of Digital Item Adaptation.

The adaptation engines themselves are non-normative tools of Digital Item Adaptation. However, tools that provide support for Digital Item Adaptation in terms of resource adaptation, descriptor adaptation and/or Quality of Service management are within the scope of the Requirements [64].

With this goal in mind, it is essential to have available not only the description of the content, but also a description of its format and of the usage environment. In this way, content adaptation may be performed to provide the user the best content experience for the content requested with the conditions available. The dimensions of the usage environment are discussed in more detail below, but it essentially includes the description of terminal and networks resources, as well as user preferences and characteristics of the natural environment. While the content description problem has been addressed by MPEG-7, the description of content format and usage environments has not been addressed and it is now the target of Part 7 of the MPEG-21 standard, Digital Item Adaptation.

It should be noted that there are other tools besides the usage environment description that are being specified by this part of the standard; however they are not covered in this section. Please refer to [65] for complete information on the tools specified by Part 7, Digital Item Adaptation.

4.3 USAGE ENVIRONMENT DESCRIPTIONS

Usage environment descriptions include the description of terminal and networks resources, as well as user preferences and characteristics of the natural environment. Each of these dimensions, according to the latest Working Draft (WD) v2.0 released in July 2002 [65], are elaborated on further below. This part of the standard will be finalized in July 2003; so changes to the information provided below are expected.

4.3.1 User Characteristics

The specification of a User's characteristics is currently divided into five subcategories, including User Type, Content Preferences, Presentation Preferences, Accessibility Characteristics and Mobility Characteristics.

The first two subcategories, User Type and Content Preferences, have adopted Description Schemes (DS's) from the MPEG-7 specification as starting points [66]. Specifically, the Agent DS specifies the User type, and the UserPreferences DS and the UsageHistory DS specify the content preferences. The Agent DS describes general characteristics of a User such as name and contact information. A User can be a person, a group of persons or an organization. The UserPreferences descriptor is a container of various descriptors that directly describe the preferences of an End User. Specifically, these include descriptors of preferences related to the:

- creation of Digital Items (e.g., created when, where, by whom)
- classification of Digital Items (e.g., form, genre, languages)
- dissemination of Digital Items (e.g., format, location, disseminator)
- type and content of summaries of Digital Items (e.g., duration of an audio-visual summary)

The UsageHistory descriptor describes the history of actions on Digital Items by an End User (e.g., recording a video program, playing back a music piece); as such, it describes the preferences of an End User indirectly.

The next subcategory, Presentation Preferences, defines a set of user preferences related to the means by which audio-visual information is presented or rendered for the user. For audio, descriptions for preferred audio power, equaliser settings, and audible frequencies and levels are specified. For visual information, display preferences, such as the preferred color temperature, brightness, saturation and contrast, are specified.

Accessibility Characteristics provide descriptions that would enable one to adapt content according to certain auditory or visual impairments of the User. For audio, an audiogram is specified for the left and right ear, which specifies the hearing thresholds for a person at various frequencies in the respective ears. For visual related impairments, colour vision deficiencies are specified, i.e., the type and degree of the deficiency. For example, given that a User has a severe green-deficiency, an image or chart containing green colours or shades may be adapted accordingly so that the User can distinguish certain markings. Other types of visual impairments are being explored, such as the lack of visual acuity.

Mobility Characteristics aim to provide a concise description of the movement of a User over time. In particular, the directivity and location update intervals are specified. Directivity is defined to be the amount of angular change in the direction of the movement of a User compared to the previous measurement. The location update interval defines the time interval between two consecutive location updates of a particular User. Updates to the location are received when the User crosses a boundary of a pre-determined area (e.g., circular, elliptic, etc.) centered at the coordinate of its last location update. These descriptions can be used to classify Users, e.g., as pedestrians, highway vehicles, etc, in order to provide adaptive location-aware services.

4.3.2 Terminal Capabilities

The description of a terminal's capabilities is needed in order to adapt various forms of multimedia for consumption on a particular terminal. There are a wide variety of attributes that specify a terminal's capabilities; however we limit our discussion here to some of the more significant attributes that have been identified. Included in this list are encoding and decoding capabilities, display and audio output capabilities, as well as power, storage and input-output characteristics.

Encoding and decoding capabilities specify the format a particular terminal is capable of encoding or decoding, e.g., MPEG-2 MP@ML video. Given the variety of different content representation formats that are available today, it is necessary to be aware of the formats that a terminal is capable of.

Regarding output capabilities of the terminal, display attributes, such as the size or resolution of a display, its refresh rate, as well as if it is colour capable or not, are specified. For audio, the frequency range, power output, signal-to-noise ratio and number of channels of the speakers are specified.

Power characteristics describe the power consumption, battery size and battery time remaining of a particular device. Storage characteristics are defined by the transfer rate and a Boolean variable that indicates whether the device can be written to or not. Also, the size of the storage is specified. Input-Output characteristics describe bus width and transfer speeds, as well as the minimum and maximum number of devices that can be supported.

4.3.3 Network Characteristics

Two main categories are considered in the description of network characteristics, Network Capabilities and Network Conditions. Network Capabilities define static attributes of a particular network link, while Network Conditions describe more dynamic behaviour. This characterisation enables multimedia adaptation for improved transmission efficiency.

Network Capabilities include the maximum capacity of a network and the minimum guaranteed bandwidth that a particular network can provide. Also specified are attributes that indicate if the network can provide in-sequence packet delivery and how the network deals with erroneous packets, e.g., does it forward them on to the next node or discard them.

Network Conditions specify error, delay and utilisation. The error is specified in terms of packet loss rate and bit error rate. Several types of delay are considered including one-way and two-way packet delay, as well as delay variation. Utilisation includes attributes to describe the instantaneous, maximum and average utilisation of a particular link.

4.3.4 Natural Environment Characteristics

The description of the natural environment is currently defined by Location, Time and Audio-Visual Environment Characteristics.

In MPEG-21, Location refers to the location of usage of a Digital Item, while Time refers to the time of usage of a Digital Item. Both are specified by MPEG-7 DS's, the Place DS for Location and the Time DS for Time [66]. The Place DS describes existing, historical and fictional places and can be used to describe precise geographical location in terms of latitude, longitude and altitude, as well as postal addresses or internal coordinates of the place (e.g., an apartment or room number, the conference room, etc.). The Time DS describes dates, time points relative to a time base and the duration of time.

Audio-Visual Environment Characteristics target the description of audio-visual attributes that can measured from the natural environment and affect the way content is delivered and/or consumed by a User in this environment. For audio, the description of the noise levels and a noise frequency spectrum is specified. With respect to the visual environment, illumination characteristics that may affect the perceived display of visual information are specified.

4.4 RELATED STANDARDIZATION ACTIVITIES

The previous sections have focused on the standardization efforts of MPEG in the area of environment description. However, it was the Internet Engineering Task Force (IETF) that initiated some of the earliest work in this area. The result of their work provided many of the ideas for future work. Besides MPEG, the World Wide Web Consortium (W3C) and Wireless Application Protocol (WAP) Forum have also been active in this area. The past and current activity of each group is described further below.

4.4.1 CONNEG – Content Negotiation

The first work on content negotiation was initiated by the IETF and resulted in the Transparent Content Negotiation (TCN) framework [67]. This first work supplied many of the ideas for future work within the IETF and by other bodies. The CONNEG working group was later established by the IETF to develop a protocol-independent content negotiation framework (referred to as CONNEG), which was a direct successor of the TCN framework. In CONNEG, the entire process of content negotiation is specified, keeping the independence between the protocol performing the negotiation and the characteristics concerning the server and the User. The CONNEG framework addresses three elements:

- Description of the capabilities of the server and contents
- Description of the capabilities of the user terminal
- Negotiation process by which capabilities are exchanged

Within CONNEG, expressing the capabilities (Media Features) of server and user terminal is covered by negotiation metadata. Media Feature names and data types are implementation-specific. The protocol for exchanging descriptions is covered by the abstract negotiation framework and is necessary for a specific application protocol. Protocol independence is addressed by separating the negotiation process and metadata from concrete representations and protocol bindings. The CONNEG framework specifies the following three elements:

- *Abstract negotiation process*: This consists of a series of information exchanges (negotiations) between the sender and the receiver that continue until either party determines a specific version of the content to be transmitted by the server to the user terminal.
- *Abstract negotiation metadata*: Description tools to describe the contents/server and the terminal capabilities/characteristics.
- *Negotiation metadata representation*: A textual representation for media features names and values, for media feature set descriptions and for expressions of the media feature combining algebra, which provide a way to rank feature sets based upon sender and receiver preferences.

A format for a vocabulary of individual media features and procedures for feature registration are presented in [68]. It should be noted that the vocabulary is mainly limited to some basic features of text and images.

4.4.2 CC/PP – Composite Capabilities / Profile Preferences

The W3C has established a CC/PP working group with the mission to develop a framework for the management of device profile information. Essentially, CC/PP is a data structure that lets you send device capabilities and other situational parameters from a client to a server. A CC/PP profile is a description of device capabilities and user preferences that can be used to guide the adaptation of content presented to that device.

CC/PP is based on RDF, the Resource Description Framework, which was designed by the W3C as a general-purpose metadata description language. The foundation of RDF is a directed labelled graph used to represent entities, concepts and relationships between them. There exists a specification that describes how to encode RDF using XML [69], and another that defines an RDF schema description language using RDF [70].

The structure in which terminal characteristics are organized is a tree-like data model. Top-level branches are the components described in the profile. Each major component may have a collection of attributes or preferences. Possible components are the hardware platform, the software platform and the application that retrieves and presents the information to the user, e.g., a browser. A simple graphical representation of the CC/PP tree based on these three components is illustrated in Figure 36.11.

The description of each component is a sub-tree whose branches are the capabilities or preferences associated with that component. RDF allows modelling a large range of data structures, including arbitrary graphs.

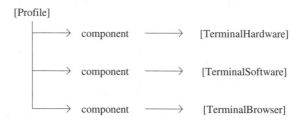

Figure 36.11 Graphical representation of CC/PP profile components.

A CC/PP profile contains a number of attribute names and associated values that are used by a server to determine the most appropriate form of a resource to deliver to a client. It is structured to allow a client and/or proxy to describe their capabilities by reference to a standard profile, accessible to an origin server or other sender of resource data, and a smaller set of features that are in addition to or different than the standard profile. A set of CC/PP attribute names, permissible values and associated meanings constitute a CC/PP vocabulary. More details on the CC/PP structure and vocabulary can be found in [71].

4.4.3 UAProf – User Agent Profile

The WAP Forum has defined an architecture optimized for connecting terminals working with very low bandwidths and small message sizes. The WAP Forum architecture is based on a proxy server, which acts as a gateway to the optimised protocol stack for the mobile environment. The mobile terminal connects to this proxy. On the wireless side of the communication, it uses an optimised protocol to establish service sessions via the Wireless Session Protocol (WSP), and an optimised transmission protocol, Wireless Transaction Protocol (WTP). On the fixed side of the connection, HTTP is used, and the content is marked up using the Wireless Markup Language (WML), which is defined by the WAP Forum.

The User Agent Profile (UAProf) defined by the WAP Forum is intended to supply information that shall be used to format the content according to the terminal. CC/PP is designed to be broadly compatible with the UAProf specification [72]. That is, any valid UAProf profile is intended to be a valid CC/PP profile.

The following areas are addressed by the UAProf specification:

- *HardwarePlatform:* A collection of properties that adequately describe the hardware characteristics of the terminal device. This includes the type of device, model number, display size, input and output methods, etc.
- *SoftwarePlatform:* A collection of attributes associated with the operating environment of the device. Attributes provide information on the operating system software, video and audio encoders supported by the device and user's preference on language.
- *BrowserUA:* A set of attributes to describe the HTML browser.
- *NetworkCharacteristics:* Information about the network-related infrastructure and environment such as bearer information. These attributes can influence the resulting content, due to the variation in capabilities and characteristics of various network infrastructures in terms of bandwidth and device accessibility.
- *WapCharacteristics:* A set of attributes pertaining to WAP capabilities supported on the device. This includes details on the capabilities and characteristics related to the WML Browser, Wireless Telephony Application (WTA), etc.
- *PushCharacteristics:* A set of attributes pertaining to push specific capabilities supported by the device. This includes details on supported MIME-types, the maximum size of a push-message shipped to the device, the number of possibly buffered push-messages on the device, etc.

Further details on the above can be found in [72].

4.5 CAPABILITY UNDERSTANDING

For a given multimedia session, it is safe to assume that the sending and the receiving terminals have fixed media handling capabilities. The senders can adapt content in a fixed number of ways and can send content in a fixed number of formats. Similarly, the receivers can handle only a fixed number of content formats. The resources that vary during a session are available bandwidth, QoS, computing resources (CPU and memory), and battery capacity of the receiver. During the session set up, the sending and receiving terminals have to understand each other's capabilities and agree upon a content format for the session.

Content adaptation has two main components: 1) capability understanding and 2) content format negotiation. Capability understanding involves having knowledge of a client's media handling capabilities and the network environment and content format negotiation involves agreeing on a suitable content format based on the receiver's capabilities and content playback environment. From a content delivery point of view, the receiver capabilities of interest are the ability to handle content playback. The sender and content adaptation engines have little or no use for client device specifications such as CPU speed and available memory.

Capability understanding and content adaptation is performed during session set up and may also continue during the session. The capability understanding is typically happens in two ways: 1) receiver submits its capabilities to the sender and the sender determines the suitable content (suitable content request) 2) sender proposes a content description that the receiver can reject until the server sends an acceptable content description or a session is disconnected (suitable content negotiation).

In suitable Content Request method, receivers submit content requests along with their capability description. A sender has to understand the capabilities and adapt the content to fit the receiver's capabilities. Receiver capabilities vary from device to device and a sender has to understand all of the receivers' capabilities and adapt the content suitably. In Suitable Content Negotiation method, receiver requests specific content and sender responds with a content description that it can deliver. The receiver then has to determine if the content description (including format, bitrate, etc.) proposed by the sender is acceptable. The receiver can then counter with a content description that it can handle to which sender responds and so on.

The disadvantage of the first method is that a sender has to understand receivers' capabilities. The advantage is that the sender has sufficient knowledge to perform adaptation and avoid negotiation with the receiver. The advantage of the second method is that the sender has to know what it can deliver and doesn't have to know receiver capabilities. Receivers describe their capabilities in terms of the content they can handle and let the sender determine if it can deliver in an acceptable format. When a client requests content (e.g., by clicking on a URL), the receiver does not know what media is part of the requested content. The receiver does not have knowledge of the content adaptation the sender employs and may have to communicate all of its capabilities. This imposes a burden on both the sender and receiver processing unnecessary information.

4.5.1 Retrieving Terminal Capabilities

To adapt content for a particular device, a sender only needs to know the media the receiver can handle and the media formats the server can support for that particular piece of content. When a receiver requests content (e.g., by clicking on a URL), the receiver does not know what media is part of the requested content. The sender has complete knowledge of the content and the possible adaptations. The sender can selectively request whether the receiver supports certain media types and purpose the content accordingly. The alternative of a client communicating its *capabilities* is expensive, as the server doesn't need to know of and may not understand all of the client's capabilities.

A solution that addresses these issues and is extensible is using a Terminal Capability Information Base (TCIB). TCIB is a virtual database of receiver capabilities that is readable by a sender. TCIB would contain parameters such as media decoding capabilities, software and hardware available in the terminal. Senders can selectively query the capabilities they are interested in using a protocol such as SNMP.[1] The senders can also set *session traps* to receive messages when an event happens (e.g., battery level falls below a threshold). The advantage of this approach is that the protocol is simple and scalable and can be extended as server needs and terminal capabilities change. The protocol also allows easy to extend private object identifiers to define implementation-specific variables. The strategy implicit in this mechanism is that understanding client's

---

[1] Simple Network Management Protocol is used to manage network elements on the Internet.

capabilities to any significant level of detail is accomplished primarily by polling for appropriate information.

5. CONCLUDING REMARKS

The UMA promise of delivering multimedia content to a range of receivers with a range of capabilities is a challenging task. The diversity of receiver terminals, networks and content formats necessitates standardization. Representing content in all possible modalities and fidelities is impractical and resource intensive. Real time techniques to transcode content to different fidelities and transform content to different modalities is essential. Taking advantage of object-based representation of content is necessary to be able to *scale* content to suit the content playback environment. Efficient media transcoding techniques are needed to transcode media streams to appropriate formats. Content distribution should take into account distributed adaptation operations. Real time adaptation may not always be possible because of the processing overhead required for adaptation. The choice of real time adaptation vs. storing multiple variants of the content is largely application dependent. For large, general-purpose systems, real time adaptation is desirable. Session mobility is an important issue to enable seamless playback across clients as well as servers. Relatively small amount of research has been done in this area. Environment description and capability understanding is another key area. While there has been significant work in capability description, the amount of work in capability understanding/negotiation is limited.

Standardization, cooperative playback, content representation techniques, media transcoding and network resource renegotiation technologies are important factors determining how soon we can deliver the UMA promise.

REFERENCES

1. M. Handley and V. Jacobson, SDP: Session Description Protocol, RFC 2327, IETF, April 1998.
2. T.L. Pham, G. Schneider, and S, Goose, A Situated Computing Framework for Mobile and Ubiquitous Multimedia Access using Small Screen and Composite Devices, Proceedings of the ACM Multimedia 2000, pp. 323-331.
3. V.N. Padmanabhan et. al., Distributing Streaming Media Content Using Cooperative Networking,' Proceedings of the NOSDAV '02, May 2002, pp. 177-186.
4. Z. Lei and N. Georganas, Context-based media adaptation for pervasive computing, Proceedings of the Canadian Conference on Electrical and Computer Engineering, Toronto, May 2001.
5. T. Phan et. al., Handoff of application sessions across time and space, Proceedings of the IEEE International Conference on Communications, June 2001.
6. T Phan et. al., A scalable distributed middleware architecture to support mobile Internet applications, Proceedings of the ACM 1st Workshop on Mobile and Wireless Internet, 2001.
7. [J. Smith, R. Mohan, S.-S. Lee, Scalable Multimedia Delivery for Pervasive Computing, Proceedings of the ACM Multimedia 1999, pp. 131-140.

8. S. Roy, B. Shen, and V. Sundaram, Application Level Hand-off support for Mobile Media Transcoding Sessions, Proceedings of the NOSSDAV, May 2002, pp. 95-104.

9. ISO/IEC/SC29/WG11, Generic Coding of Moving Pictures and Associated Audio (MPEG-4 Systems) - ISO/IEC 14386-1, International Organization for Standardization, April 1999.

10. ISO/IEC/SC29/WG11, Generic Coding of Moving Pictures and Associated Audio (MPEG-4 Video) - ISO/IEC 14386-2, International Organization for Standardization, April 1999.

11. M. Kunt, A. Ikonomopoulos, M. Kocher, Second-generation image coding techniques, IEEE Proceedings, Vol. 73, No. 4, pp. 549–574, April 1985.

12. R. Koenen, MPEG-4: Multimedia for our time, IEEE Spectrum, Vol. 36, No. 2, pp. 26-33, February 1999.

13. A. Puri and A. Eleftheriadis, MPEG-4: A Multimedia Coding Standard Supporting Mobile Applications ACM Mobile Networks and Applications Journal, Special Issue on Mobile Multimedia Communications, Vol. 3, No. 1, June 1998, pp. 5-32 (invited paper).

14. M. Kim, P. Westerink, and W. Belknap, MPEG-4 Advanced Synchronization Model (FlexTime Model), Contribution ISO-IEC JTC1/SC29/WG11 MPEG99/5307, December 1999, (50th MPEG meeting).

15. W3C Recommendation, Synchronized Multimedia Integration Language (SMIL 2.0), http://www.w3.org/TR/2001/REC-smil20-20010807/, August 2001.

16. ISO/IEC IS 15938-5, Information Technology – Multimedia Content Description Interface – Multimedia Description Schemes, 2001.

17. R. Mohan, J.R. Smith and C.S. Li, Adapting multimedia internet content for universal access, IEEE Trans. Multimedia, vol. 1, no. 1, pp. 104-114, March 1999.

18. B. Shahraray and D.C. Gibbon, Automatic generation of pictorial transcripts, Proc. SPIE Conf. Multimedia Computing and Networking, pp. 512-518, 1995.

19. A. Vetro, A. Divakaran, and H. Sun, Providing Multimedia Services to a Diverse Set of Consumer Devices, Proc. IEEE Int'l Conf. Consumer Electronics, pp. 32-33, June 2001.

20. P. Kuhn and T. Suzuki, MPEG-7 Meta-data for Video Transcoding: Motion and Difficulty Hints, Proc. SPIE Conf. Storage and Retrieval for Multimedia Databases, SPIE 4315, San Jose, CA, Jan. 2001.

21. P. Kuhn, T. Suzuki, and A. Vetro, MPEG-7 Transcoding Hints for Reduced Complexity and Improved Quality, Int'l Packet Video Workshop, Kyongju, Korea, April 2001.

22. Y. Nakajima, H. Hori, and T. Kanoh, Rate conversion of MPEG coded video by re-quantization process, Proc. IEEE Int'l Conf. Image Processing, vol. 3, pp. 408-411, Washington, DC, Oct. 1995.

23. H. Sun, W. Kwok, and J. Zdepski, Architectures for MPEG compressed bitstream scaling, IEEE Trans. Circuits Syst. Video Technol., Apr. 1996.

24. G. Kessman, R. Hellinghuizen, F. Hoeksma and G. Heidman, Transcoding of MPEG bitstreams, Signal Processing: Image Communications, vol. 8, pp. 481-500, Sept. 1996.

25. N. Bjork and C. Christopoulos, Transcoder architectures for video coding, IEEE Trans. Consumer Electronics, vol. 44, no. 1, pp.88-98, Feb. 1998.

26. P. Assuncao and M. Ghanbari. Optimal transcoding of compressed video, Proc. IEEE Int'l Conf. Image Processing, pp. 739-742, Santa Barbara, CA, Oct. 1997.

27. P. Assuncao and M. Ghanbari, A frequency-domain video transcoder for dynamic bit-rate reduction of MPEG-2 bitstreams, IEEE Trans. Circuits Syst. Video Technol., Dec. 1998.

28. S.F. Chang and D.G. Messerschmidt, Manipulation and composting of MC-DCT compressed video, IEEE Journal Selected Area Communications, vol. 13, no. 1, Jan 1995.

29. H. Sun, A. Vetro, J. Bao, and T. Poon, A new approach for memory-efficient ATV decoding, IEEE Trans. on Consumer Electronics, vol. 43, no. 3, pp. 517-5525, Aug. 1997.

30. J. Wang and S. Yu, Dynamic rate scaling of coded digital video for IVOD applications, IEEE Trans. on Consumer Electronics, vol. 44, no. 3, pp. 743-749, Aug. 1998.

31. N. Merhav, Multiplication-free approximate algorithms for compressed-domain linear operations on images, IEEE Trans. Image Processing, vol. 8, no. 2, pp. 247-254, Feb. 1999.

32. J. Youn, M. T. Sun, and C. W. Lin. Motion vector refinement for high performance transcoding, IEEE Trans. Multimedia, vol. 1, no. 1, pp. 30-40, Mar. 1999.

33. M. Yong, Q.F. Zhu, and V. Eyuboglu, VBR transport of CBR-encoded video over ATM networks, Proc. Sixth Int'l Workshop on Packet Video, Sept. 1994.

34. T. Shanableh and M. Ghanbari, Heterogeneous video transcoding to lower spatio-temporal resolutions and different encoding formats," IEEE Trans. Multimedia, vol.2, no.2, June 2000.

35. P. Yin. A. Vetro, B. Lui and H. Sun, Drift compensation for reduced spatial resolution transcoding, IEEE Trans. Circuits Syst. Video Technol., to appear.

36. B. Shen, I.K. Sethi, and B. Vasudev, Adaptive motion vector re-sampling for compressed video downscaling," IEEE Trans. on Circuits and Systems for Video Technology, pp. 929-936, April 1996.

37. W. Zhu, K. H. Yang and M. J. Beacken, CIF-to-QCIF video bitstream down-conversion in the DCT domain," Bell Labs Technical Journal, 3(3), July-September 1998.

38. P. Yin, M. Wu, and B. Lui. Video transcoding by reducing spatial resolution, Proc. IEEE Int'l Conf. Image Processing, Vancouver, BC, Oct. 2000.

39. S. J. Wee, J. G. Apostolopoulos and N. Feamster, Field-To-Frame Transcoding with Spatial and Temporal Downsampling," Proc. IEEE Int'l Conf. Image Processing, Kobe, Japan, Oct. 1999.

40. A. Vetro, H. Sun, P. DaGraca, and T. Poon, Minimum drift architectures for three-layer scalable DTV decoding, IEEE Trans. Consumer Electronics, Aug 1998.

41. K. Stuhlmuller, N. Farber, M. Link and B. Girod, ``Analysis of video transmission over lossy channels," J. Select Areas of Commun., June 2000.

42. T. Shanableh and M. Ghanbari, Transcoding architectures for DCT-domain heterogeneous video transcoding, Proc. IEEE Int'l Conf. Image Processing, Thessaloniki, Greece, Sept. 2001.

43. A. Vetro, T. Hata, N. Kuwahara, H. Kalva and S. Sekiguchi, Complexity-quality evaluation of transcoding architecture for reduced spatial resolution, Proc. IEEE Int'l Conf. Consumer Electronics, Los Angeles, CA, June 2002.

44. A. Lan and J.N. Hwang, Context dependent reference frame placement for MPEG video coding, Proc. IEEE Int'l Conf. Acoustics, Speech, Signal Processing, vol. 4 pp. 2997-3000, April 1997.

45. J.N. Hwang, T.D. Wu and C.W. Lin, Dynamic frame-skipping in video transcoding, Proc. IEEE Workshop on Multimedia Signal Processing, pp. 616-621, Redonda Beach, CA, Dec. 1998.

46. K.T. Fung, Y.L. Chan and W.C. Siu, New architecture for dynamic frame-skipping transcoder, IEEE Trans. Image Processing, vol. 11, no. 8, pp. 886-900, Aug. 2002.

47. A. Vetro, P. Yin, B. Liu and H. Sun, Reduced spatio-temporal transcoding using an intra-refresh technique, Proc. IEEE Int'l Symp. Circuits and Systems, Scottsdale, AZ, May 2002.

48. W. Li, Overview of Fine Granularity Scalability in MPEG-4 video standard, IEEE Trans. Circuits and Systems for Video Technology, vol. 11, no. 3, pp. 301–317, 2001.

49. X.M. Zhang, A. Vetro, Y.Q. Shi and H. Sun, Constant quality constrained rate allocation for FGS coded video, Proc. SPIE Conf. on Visual Communications and Image Processing, San Jose, CA, Jan. 2002.

50. Q. Wang, F. Wu, S. Li, Z. Xiong, Y.Q. Zhang, and Y. Zhong, A new rate allocation scheme for progressive fine granular scalable coding, Proc. IEEE Int'l Symp. Circuits and Systems, 2001.

51. A. Skodras, C. Christopoulos, and T. Ebrahimi, The JPEG 2000 Still Image Compression Standard, IEEE Signal Processing Magazine, vol. 18, no. 5, pp. 36-58, Sept. 2001.

52. Video Group, The Status of Interframe Wavelet Coding Exploration in MPEG, ISO/IEC JTC1/SC29/WG11 N4928, Klagenfurt, Austria, July 2002.

53. A. Vetro, H. Sun and Y. Wang, Object-based transcoding for adaptive video content delivery, IEEE Trans. Circuits Syst. Video Technol., March 2001.

54. Y.C. Lin, C.N. Wang, T. Chiang, A. Vetro and H. Sun, Efficient FGS-to-Single layer transcoding, Proc. IEEE Int'l Conf. Consumer Electronics, Los Angeles, CA, June 2002.

55. J.L. Wu, S.J. Huang, Y.M. Huang, C.T. Hsu and J. Shiu, An efficient JPEG to MPEG-1 transcoding algorithm, IEEE Trans. Consumer Electronics, vol. 42, no. 3, pp. 447-457, Aug. 1996.

56. N. Memon and R. Rodilia, Transcoding GIF images to JPEG-LS, IEEE Trans. Consumer Electronics, vol. 43, no. 3, pp. 423-429, Aug. 1997.

57. H.S. Chang, S. Sull and S. U. Lee, Efficient video indexing scheme for content-based retrieval, IEEE Trans. Circuits Syst. Video Technol., Vol. 9, No. 8, pp. 1269-1279, December 1999.

58. A Hanjalic and H. Zhang, An Integrated Scheme for Automated Video Abstraction Based on Unsupervised Cluster-Validity Analysis, IEEE Trans. Circuits Syst. Video Technol., Vol. 9, No. 8, December 1999.

59. A. Divakaran, R. Regunathan, and K. A. Peker, Video summarization using descriptors of motion activity, Journal of Electronic Imaging, vol. 10, no. 4, Oct. 2001.

60. H. Sorial, W.E. Lynch and A. Vincent, Joint transcoding of multiple MPEG video bitstreams, Proc. IEEE Int'l Symp. Circuits and Systems, Orlando, FL, May 1999.

61. S. Gopalakrishnan, D. Reininger, and M. Ott, Realtime MPEG system stream transcoder for heterogeneous networks, Proc. Packet Video Workshop, New York, NY, April 1999.

62. ISO/IEC TR 21000-1 Information Technology – Multimedia Framework – Part 1: Vision, Technologies and Strategies, 2001.

63. Requirements Group, MPEG-21 Overview v4, J. Bormans, Ed., ISO/IEC JTC1/SC29/WG11 N4801, Fairfax, USA, May 2002.

64. Requirements Group, MPEG-21 Requirements v1.2, J. Bormans, Ed., ISO/IEC JTC1/SC29/WG11 N4988, Klagenfurt, Austria, July 2002.

65. Multimedia Description Schemes Group, MPEG-21 Digital Item Adaptation WD v2.0, A. Vetro, A. Perkis, S. Devillers, Eds., ISO/IEC JTC1/SC29/WG11 N4944, Klagenfurt, Austria, July 2002.

66. ISO/IEC IS 15938-5, Information Technology – Multimedia Content Description Interface – Part 5: Multimedia Description Schemes, 2001.

67. K. Holtman, A. Mutz, Transparent Content Negotiation in HTTP, RFC 2295, March 1998.

68. K. Holtman, A.Mutz, T. Hardie, Media Feature Tag Registration Procedure, RFC 2506, March 1999.

69. Resource Description Framework (RDF) Model and Syntax Specification; Ora Lassila, Ralph Swick; World Wide Web Consortium Recommendation: http://www.w3.org/TR/REC-rdf-syntax, February 1999.

70. RDF Vocabulary Description Language 1.0: RDF Schema; Dan Brickley, R. V. Guha; World Wide Web Consortium Working Draft: http://www.w3.org/TR/PR-rdf-schema, April 2002.

71. Composite Capability/Preference Profiles (CC/PP): Structure and Vocabularies; G. Klyne, F. Reynolds, C. Woodrow, H. Ohto; World Wide Web Consortium Working Draft: http://www.w3.org/TR/CCPP-struct-vocab/, March 2001.

72. WAP-248: WAG UAProf: User Agent Profile Specification; Wireless Application Protocol; http://www1.wapforum.org/tech/documents/ WAP-248-UAProf-20011020-a.pdf, October 2001.

37

SERVER-BASED SERVICE AGGREGATION SCHEMES FOR INTERACTIVE VIDEO-ON-DEMAND

Prithwish Basu, Thomas D. C. Little, Wang Ke, and Rajesh Krishnan

Department of Electrical and Computer Engineering
Boston University
8 Saint Mary's St., Boston, MA 02215, USA
{pbasu,tdcl,ke,krash}@bu.edu

1. INTRODUCTION

Interactive video-on-demand (i-VoD or VoD) describes one of the most attractive applications that has been proposed since the rapid advances in high-speed networking and digital video technology. Principal deterrents to the widespread deployment of i-VoD services are the cost of data storage, server and network capacity to the end user. In the worst case, support of independent VCR-like interactions requires a separate channel for each user. Because of the data-intensive and streaming nature of video data, fully provisioned interactive channels are prohibitively expensive to deploy. Service aggregation schemes attempt to alleviate these problems by sharing both server and network resources at the expense of small overheads in terms of delay and/or buffer space. These techniques propose to reduce resource requirements by aggregating users into groups. Many such techniques exist in the VoD literature (some of them only support non-interactive VoD) – batching [6], server caching [22,7], client caching or bridging [1] and chaining [20] (a limited form of distributed caching), adaptive piggybacking or rate adaptive merging [12], content insertion [17,23], content excision [23], periodic broadcasting [14], patching [13,8], and catching [10] to name a few. Table 37. 1 illustrates the relative merits of these schemes.

A major bottleneck in the delivery of video is on the disk-to-server path. Interval-Caching solutions [7] attempt to minimize the bandwidth required on this path by caching intervals of video between successive disk accesses and then by serving subsequent requests from the cache. These schemes do not, however, attempt to minimize the network bandwidth on the subsequent server-to-client path.

Table 37.1 Taxonomy of Service Aggregation Schemes in VoD

Service Aggregation Scheme	Entity that performs aggregation	Initial Start-up Latency	Interactions Allowed?	Client Buffering Required?	Network Bandwidth Gains
Batching	Server	Significant	No (*possible w/ contingency channels*)	Minimal	High
Interval Caching	Server	No	Possible	Minimal	Saves disk b/w only
Chaining	Client	Yes	Yes	Yes (*distributed*)	Moderate to High
Adaptive Piggyback (*RateAdaptive Merging*)	Server (*client only switches multicast gr.*)	No	Yes (*unrestricted*)	Minimal	Moderate
Periodic Broadcast	Server/Client (*only for very popular movies*)	Yes *broadcast series dependent*	No	Significant	Moderate to High *broadcast series dependent*
Patching	Server/Client	Yes	Yes	Significant	Moderate
Selective Catching	Server/Client	No	Yes	Significant	Moderate to High

Clustering schemes, in contrast, seek to minimize this network bandwidth by aggregating channels that are identical except for their potentially disparate start times. That is, they are "aggregatable" if their skew is within some bound, or their skew can be eliminated.[1] Batching is a clustering technique in the near-VoD scenario, in which users must wait until the next launch of a stream occurs. A more attractive technique for the end user is a technique called rate adaptive merging [12,16] in which a user acquires a stream without any waiting time; however, the system attempts to aggregate streams within a skew bound by *minimally* altering the "content progression rates" of a subset of these streams. Experiments have shown that speeding up a video stream by approximately 7% does not result in a perceivable difference in visual quality [17,4], giving credibility to this approach. A trailing stream, or streams, can be accelerated towards a "leading" stream, or streams, until both of them are at the same position in the program (or equivalently, their skews are zero) [16]. At this instant of zero skew, all users attached to the stream (e.g., via multicasting to multiple destinations and users) are served by a single stream from the video source.

Both batching and rate adaptive merging have the advantage of having the burden of implementation fall mainly on the server and partly on the

---

[1] *Skew* is defined as the difference in program positions at current instant of time of two video streams containing the same program.

network (through the ability to support multicast), with no extra requirement on the client except that it be able to display the data received and send back control signals (to make requests and perform VCR-like operations). This means that potentially "true-VoD" service can be extended even to wireless handheld devices that do not have a large storage capacity but which are connected to the network. Client based aggregation schemes that require clients to buffer video from multiple staggered broadcast channels [14,13,8] are thus unsuitable for heterogeneous pervasive computing environments.

It is also desirable to reduce the buffering requirement for interval caching schemes. This requirement is dependent on the skews between streams carrying the same content. Therefore, bridging the skew and/or reducing the number of streams by clustering can also reduce the memory requirement of a caching scheme. A continuum of aggregation schemes that lie between pure caching and pure clustering exists. Tsai and Lee [23] explore this relationship in the near-VoD scenario. Therefore, one can extend solutions for clustering to buffer management and vice versa. In this chapter, we focus on *clustering* schemes.

In hybrid CATV architectures that consist of the standard broadcast delivery mechanism augmented with on-demand channels, stream merging is used in schemes like *Catching* [10] in order to reclaim the on-demand channels. The described clustering algorithms are applicable in such scenarios as well.

The scope for aggregation of video streams is ameliorated by the long lived nature of video traffic and the high density of accesses on popular titles (e.g., hit movies) during certain times of the day. By clustering streams during peak periods, a system can accommodate a much larger number of users than without aggregation, thus increasing potential revenue earned. Therefore, in any VoD system where peak demands are higher than the available bandwidth, aggregation schemes are attractive. This is particularly true of scenarios that commonly arise with Internet traffic and content. For example, when a popular news item is made available by a news provider, the peak demand can far outstrip the available network bandwidth, thus making aggregation increasingly attractive.

Because resource sharing by aggregating users is a promising paradigm for VoD systems supporting large user populations, it is useful to study its inherent complexity. Irrespective of whether we are trying to use rate adaptive merging or content insertion or using "shrinking" to reduce buffer usage for caching, the underlying clustering problem remains the same. In this chapter, we take a systematic approach to formulating these clustering problems and discuss the existence of optimal solutions. These algorithms are general across a large class of implementation architectures as well as aggregation techniques.

From a performance engineering standpoint, one must not lose sight of the fact that simpler solutions while provably sub-optimal can at times be preferable due to simplicity, elegance or speed. Often, they can be necessitated by the fact that optimal solutions are computationally expensive or intractable. In such cases, the engineering approach is to seek good approximate or heuristic alternatives that provide near-optimal solutions. With this in mind, we present optimal solution approaches and also investigate when such sophisticated algorithms are warranted by the application under consideration. In the "static" version of the problem, clusters of users receiving the same video content are formed but it is assumed that the users cannot break away after their streams have been merged with others in a cluster. If we allow the users to break away from their clusters by interacting, the gains due to aggregation will be lost. Hence, dynamic approaches are necessary for handling these interactive situations. We present some such approaches and also explore how far the gains from an optimal solution in the static case are retained in dynamic scenarios. Our performance results can be readily applied to the related capacity planning and design problem; that is, given the mean arrival (and interaction) rate and distribution, what should the design capacity be, in number of streams, of a video-server network which actively uses service aggregation?

We transform the static clustering problem (with the objective of minimizing average bandwidth) to the RSMA-slide problem (terminology used by Rao et al. [19]). An important observation here is that an RSMA-slide on n sinks is isomorphic to an optimal binary tree on n leaf nodes. Aggarwal et al. have described an optimal binary tree formulation in reference [2], but we describe a more general approach that can model additional interactive scenarios. By periodically recomputing the RSMA-slide with a small period, we can reclaim up to 33% of channel bandwidth when there are 1,250 interacting users in the system watching 100 programs, in steady state. We compare the RSMA-slide algorithm with other heuristic approaches and find it to be the best overall policy for varying arrival and interaction rates of users. We also find that under server overload situations, EMCL-RSMA (forming maximal clusters followed by merging by RSMA-slide in the clusters) is the best policy as it releases the maximum number of channels in a given time budget, consuming minimum bandwidth during the process. We also investigate some variants of the RSMA-slide algorithm such as event-triggered RSMA, Stochastic RSMA, etc. and compare their performance with the RSMA-slide.

Related work

The static stream clustering problem has been studied by Aggarwal et al. [2] and Lau et al. [18]. The latter provides a heuristic solution for the problem while the former discusses an optimal solution based on binary trees. Neither discuss how to accommodate user interactions although Aggarwal et al. consider implications of user arrivals. Basu et al. investigate the stream clustering problem from a user arrival/interaction standpoint with early performance results [3].

An important contribution of this work is the inclusion of user arrivals and interactions into the stream clustering model and showing that iterative application of optimal static algorithms in a controlled fashion can preserve most of the gains in dynamic interactive situations too. We use a different problem formulation that allows us to model most interactive situations although in the static case it reduces to the optimal binary tree formulation in [2]. We point out that the problem of freeing maximum number of channels under server-overload is different from the minimum bandwidth clustering problem and that our algorithm Earliest Maximal Clustering, also known as EMCL [17] is optimal in the static case. We show performance simulations over a wide range of parameters and a number of different clustering policies. We also demonstrate that the optimal re-computation period depends not only on the rates of arrival and interaction but also on the specific policy used.

Another contribution of this work is an attempt to formulate the dynamic stream clustering problem as a Stochastic Dynamic Programming problem, although that resulted in a few negative results. To the best of our knowledge, this is the first approach of its kind in this domain, and we believe that our preliminary findings will spur the VoD research community to find better solutions in this space. We also present an approximate analytical technique for predicting the average number of streams in such an aggregation system in steady state, and demonstrate that the analytical results are reasonably close to the simulation results.

The remainder of the chapter is organized as follows. In Section 2, we formulate the clustering problem mathematically, and consider optimal algorithmic solutions. Section 3 introduces a stochastic dynamic programming approach to tackle the *dynamic* stream clustering problem. In Section 4 we present the results of simulations of heuristic and approximate algorithms for clustering. In Section 5 we conclude with our recommendations of clustering algorithms that should be used for aggregation in interactive VoD systems.

2. STREAM CLUSTERING PROBLEMS

Here we characterize the space of on-demand video stream clustering problems by formulating a series of sub-problems. We provide optimal solutions for some of these sub-problems and present heuristic solutions derived from these optimal approaches for the remainder.

2.1 GENERAL PROBLEM DESCRIPTION

A centralized video-server stores a set of k movies $M = \{M_1, M_2, \ldots, M_k\}$, each having lengths $\ell_1, \ell_2, \ldots, \ell_k$, respectively. It accepts requests from a pool of subscribers for a subset of these stored movies and disseminates the content to the customers over a delivery network which supports group communication (e.g., multicast). Assume that the movie popularity distribution is skewed (non-uniform) obeying a probability distribution $P(\cdot)$. Each *popular* movie can be played at two different content progression rates, r (normal speed) and $r + \Delta r$ (accelerated speed). Accelerated versions of these movies are created by discarding some of the less important video frames (e.g., B frames in the MPEG data stream) by the server on the fly. If the frame dropping rate is maintained below a level of $\approx 7\%$, the minor degradation of quality is not readily perceivable by the viewer [4,16].

There is an opportunity to reclaim streaming data bandwidth if skewed streams of identical movies are re-aligned and consolidated into a single stream. This can be achieved by the server attempting to change the rate of content progression (the playout rate) of one or both of the streams.

There are several means of delivering the same content to a group of users using a single channel. Multicast is a major vehicle for achieving this over the IP networks. But because it has not been widely deployed in the general Internet, another scheme called IP Simulcast can be used to achieve the same goal by relaying or repeating the streams along a multicast tree [9]. If the medium has broadcasting capabilities (such as cable TV channels) multicast groups can be realized by means of restricted broadcast.

The term *clustering* refers to the action of creating a schedule for merging a set of skewed identical-content streams into one. The schedule consists of content progression directives for each stream at specific time instants in the future. For large sets of streams, the task of generating this schedule is computationally complex and is exacerbated by the dynamics of user interactions to achieve VCR functions such as stop, rewind, fast-forward and quit. It is clear that efficient algorithms for performing this clustering are needed for large sets of streams. The term *merging* refers to the actual process of following the clustering algorithm to achieve the goal.

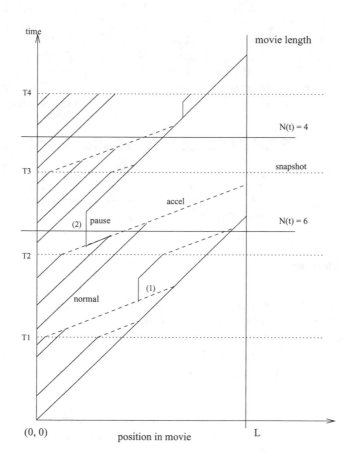

Figure 37.1 Aggregation of streams.

A general scenario for aggregation of video streams with clustering is outlined below. At time instants T_i the server captures a snapshot of the stream positions – the relative skew between streams of identical content – and tries to calculate a clustering schedule using a clustering algorithm **CA** for aggregating the existing streams. Aggregation is achieved by accelerating a trailing stream towards a leading one and then merging whenever possible. Changing the rates of the streams is assumed to be of zero cost and the merging of two streams is assumed to be seamless. Viewers, however, are free to interact via VCR functions and cause breakouts of the existing streams. That is, new streams usually need to be created to track the content streaming required for the VCR function. Sometimes these breakouts are clusters themselves when more than one viewer initiates an identical breakout. The interacting customers are allocated individual streams, if available, during their interaction phase but are attempted to be aggregated at the next snapshot epoch. We restrict our discussion to the VCR actions of fast-forward, rewind, pause, quit.

Figure 37.1 illustrates the merging process. Here the lifespans of several streams receiving the same content are plotted over time. The horizontal axis denotes the position of a stream in the program and the vertical axis is the time line. The length of the movie is assumed to be L as indicated by the vertical line on the right. Streams are shown in one of the following three states: normal playback, accelerated or paused. Accelerated streams have a steeper slope than the normal streams because they move faster than the latter. Paused streams have infinite slope because they do not change their position with time. Between time 0 and T_1 all arriving streams progress at a normal rate.

The first snapshot capturing positions of existing streams occurs at time T_1. At this time a clustering algorithm is used to identify the clusters of streams that should be merged into a single stream (per cluster). The merging schedule is also computed within each cluster. Different algorithms can be applied for performing the clustering and merging (see Section 4). When new users arrive in between snapshots they are advanced at a normal rate until the next snapshot when the merging schedule is computed again and the streams are advanced according to the new clustering schedule. The clusters and the merging schedule are computed under an assumption that no interactions will occur before all the streams merge. But because the system is interactive, users can interact in the middle. If a user interacts, on a stream that is carrying multiple users, a new stream is allocated while the others on the stream continue. Case (1) in the figure demonstrates this phenomenon. However, if the only user on a stream interacts, the server stops sending data on the previous stream because it does not carry any user anymore. This is illustrated by case (2) in the figure. $N(t)$ denotes the number of streams existing in the system at a time instant t.

A state of a non-interacting stream S_i can be represented by a position-rate tuple (p_i, r_i), where $p_i \in \Re^+$ is the program position of the ith stream, measured in seconds from the beginning of the movie and $r_i \in \{r, r + \Delta r\}$ is the content progression rate. The state equation is given by:

$$p_i(t + \Delta t) = p_i(t) + r_i(t)\Delta t \qquad (37.1)$$

and $r_i(t)$ is determined by the clustering algorithm **CA**. Note that in this discussion, time is measured in continuous units for simplicity. In reality, time is discrete as the smallest unit of time is that required for the play-out of one video frame (approximately 33 ms for NTSC-encoded video).

As an example of a merge, consider streams S_i (rate r) and S_j (rate $r + \Delta r$). If they can merge in time t, then the following will hold:

$$p_i + rt = p_j + (r + \Delta r)t \qquad (37.2)$$

$$t = \frac{p_i - p_j}{\Delta r}. \qquad (37.3)$$

The most general problem is to find an algorithm **CA** that minimizes $\frac{1}{T} \int_{\tau}^{\tau+T} N(t)dt$, the average number of streams (aggregated or non-aggregated) in the system in a time interval $[\tau, \tau + T]$ for given arrival and user interaction rates. The space of optimization problems in this aggregation context is explored next.

2.2 CLUSTERING UNDER A TIME CONSTRAINT

There is a streaming bandwidth overload scenario that exists when too many users need unique streams from a video server. This scenario arises when many users interact simultaneously, but distinctly, or when a portion of a video server cluster fails or is taken offline. The objective of a solution for this scenario is to recover the maximum number of channels possible within a finite time budget, by clustering outstanding streams. We assume that interactions are not honored during the recovery period because the system is already under overload.

Maximizing the number of channels recovered is desirable because it enables us to restore service to more customers and possibly to admit new requests. Recovery within a short period of time is necessary since customers will not wait indefinitely on failure. It is possible to mask failures for a short period of time by delivering secondary content such as advertisements [17]. This period can be exploited to reconfigure additional resources. However reconfigurability adds substantially to system costs and additional resources may not be available.

Optimal stream clustering under failure can be solved easily in polynomial time and a linear-time algorithm using either content insertion or rate adaptive merging is described in reference [16]. This algorithm, called EMCL (Earliest Maximal CLuster), starts from the leading stream and clusters together all streams within the time budget, then repeats the process by starting from the next stream ahead until the final, trailing stream is reached. A more general class of one-dimensional clustering problems has been shown to be polynomially solvable [12].

Three points are worth noting here. Firstly, the EMCL algorithm can be applied when both content insertion and rate adaptation are used. The number of channels recovered by the algorithm is non-decreasing with increase in the time budget. Using both techniques has the same effect as increasing the time budget when using one technique alone. Therefore we can apply the same algorithm. We will see later that this property does not generalize to other optimization criteria.

Secondly, this algorithm is not affected if some leading streams will reach the end of the movie before clustering. By accelerating all streams that are within the time budget from the end of the program and then computing the clustering for the remaining streams, we can achieve an optimal clustering.

Thirdly, iterative application of this algorithm does not provide further gains in the absence of interactions, exits or new arrivals. In the dynamic case, the average bandwidth, in number of channels, used during clustering is of consequence. We consider this average bandwidth minimizing constraint and dynamicity in the following sections.

2.3 CLUSTERING TO MINIMIZE BANDWIDTH

Another goal for clustering is to minimize the average bandwidth required in the provisioning of a system. This can be measured by the number of channels per subscriber and is more interesting and applicable from a service aggregation perspective. A special case of this problem occurs when we wish to construct a merging schedule at some instant that uses minimum bandwidth to merge all streams into one to carry a group of users. Ignoring arrivals, exits and break-aways due to interaction, the problem has been approached heuristically by Lau et al. [18] and analytically by Basu et al. [3].

The problem can be readily transformed into a special case of the Rectilinear Steiner Minimal Arborescence (RSMA) problem [3], which is defined by Rao et al. [19] as:

Given a set N of n nodes lying in the first quadrant of E^2, find a minimum length directed tree (called Rectilinear Steiner Minimal Arborescence, or RSMA) rooted at the origin and containing all nodes in N, composed solely of horizontal and vertical arcs oriented from left to right and from bottom to top.

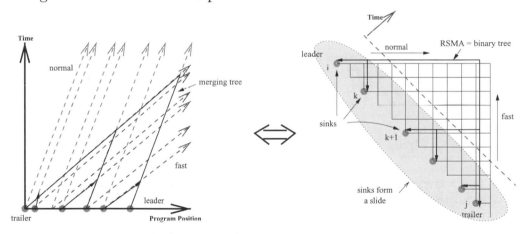

Figure 37.2 Minimum bandwidth clustering.

The transformation is illustrated in Figure 37.2 with a minor difference that an *up arborescence* – a directed tree with arcs directed towards the root – is used instead. In the figure on the left, the shaded dots denote the stream positions at the time instant corresponding to the beginning of the snapshot. Each stream has a choice of either progressing at a normal speed (larger slope in the figure) or at an accelerated speed (smaller slope). In the optimal tree, the absolute trailer must always be accelerated and the absolute leader must always be retarded, and the two lines corresponding to them will meet at the root. In Steiner tree terminology the shaded dots ($\in N$) are called *sinks*. The RSMA is denoted in the figure on the right as a rooted directed tree (the arrow directions have been reversed for a better intuitive feel).

In the RSMA formulation, *normal* streams move horizontally whereas the *accelerated* streams move vertically. The leader always remains horizontal and the trailer vertical. The sinks lie on a straight line of slope $= -1$ and the final merge point which is the root of the RSMA is at the origin. This transformation leads to a special case of the RSMA problem in which the sinks form a *slide*. The slide is a configuration without a directed path from one sink to another (i.e., there are no sinks in the interior of the rectilinear grid (called a *Hanan grid*). Note that this special case of the problem has an *optimal sub-structure* that is lacking in the general case, and finding the min-cost RSMA spanning n sinks is equivalent to finding an optimal cost binary tree on n leaf-nodes or an optimal bracketing sequence. The number of binary trees on n leaf-nodes (or RSMAs with n sinks forming a slide) is given by the $(n-1)^{st}$ Catalan number, $\frac{1}{n}\binom{2n-2}{n-1}$.

Although this constitutes an exponential search space, the special structure of the problem allows us to compute an optimal binary tree using a dynamic programming (DP) algorithm of complexity $O(n^3)$ [19]. One can easily verify that the RSMA for the particular placement of sinks in Figure 37.2 is a minimum length binary tree. We call the DP algorithm **RSMA-slide** in this chapter.

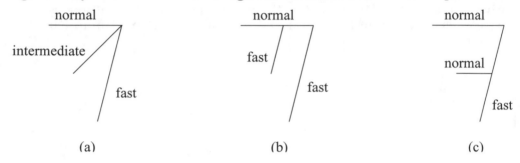

    (a)   (b)   (c)

Figure 37.3 Merging with more than two rates.

One point to be noted here is that from an optimality standpoint, more than two content progression rates are not needed. Figure 37.3 illustrates the situation. A line segment with a higher Cartesian slope has a higher content progression rate. In Figure 37.3(a), three rates are used for content progression and the streams yield a single meeting point. Clearly the merging trees in Figures 3(b-c) have lower cost in terms of bandwidth used than the previous one, because a merge occurs earlier in those two cases. Therefore, two content progression rates are sufficient for rate adaptation.

Now, we describe the RSMA-slide algorithm which is based on optimal binary trees [2]. Let L be the length of the movie that is served to streams $i,...,j$, with $p_i > p_j$. $P(i,j)$ denotes the optimal program position where streams i and j would merge and is given by:

$$P(i,j) = p_i, \qquad i = j \qquad\qquad\qquad (37.4)$$

$$= p_i + r \times \frac{p_i - p_j}{\Delta r}, \qquad i \neq j . \qquad\qquad (37.5)$$

Let $T(i,j)$ denote an optimal binary tree for merging streams $i,...,j$. Let $\text{Cost}(i,j)$ denote the cost of this tree. Because this is a binary tree, there exists a point k such that the right subtree contains the nodes $i,...,k$ and the left subtree contains the nodes $k+1,\ldots,j$. From the principle of optimality, if $T(i,j)$ is optimal (has minimum cost) then both the left and right subtrees must be optimal. That is, the right and left subtrees of $T(i,j)$ must be $T(i,k)$ and $T(k+1,j)$. The cost of this tree is given by

$$\text{Cost}(i,j) = L - p_i, \qquad i = j \qquad\qquad (37.6)$$

$$= \text{Cost}(i,k) + \text{Cost}(k+1,j) - \max(L - P(i,j), 0), \qquad i \neq j \qquad (37.7)$$

and the optimal policy merges $T(i,k^*)$ and $T(k^*+1,j)$ into $T(i,j)$, where k^* is given by

$$k^* = \text{argmin}_{i \leq k < j} \{\text{Cost}(i,k) + \text{Cost}(k+1,j) - \max(L - P(i,j), 0)\}. \quad (37.8)$$

Here $\text{Cost}(i,k)$ and $\text{Cost}(k+1,j)$ are the costs of the right and the left subtrees respectively, calculated until the end of the movie.

We begin by calculating $T(i,i)$ and $\text{Cost}(i,i)$ for all i. Then, we calculate $T(i,i+1)$ and $\text{Cost}(i,i+1)$, then $T(i,i+2)$ and $\text{Cost}(i,i+2)$ and so on, until we find $T(1,n)$ and $\text{Cost}(1,n)$. This gives us our optimal cost. The algorithm can be summarized as follows:

Algorithm RSMA-slide
{
    for $(i = 1$ to $n)$
        Initialize $P(i,i)$, $\text{Cost}(i,i)$ and $T(i,i)$ from eqns. 4 and 6

    for $(p = 1$ to $n-1)$
      for $(q = 1$ to $n-p)$
        Compute $P(q,q+p)$, $\text{Cost}(q,q+p)$ and $T(q,q+p)$ from eqns. 5, 7 and 8
}

There are $O(n)$ iterations of the outer loop and $O(n)$ iterations of the inner loop. Additionally, determination of $\text{Cost}(i,j)$ requires $O(n)$ comparisons in the *argmin* step. Hence, the algorithm **RSMA_Slide** has a complexity of $O(n^3)$. Therefore, we conclude that the static clustering problem has an optimal polynomial time solution, in contrast to the conclusion of Lau et al. [18]. Note that in practice, n is not likely to be very high, thus making the complexity acceptable.

Although optimal binary trees are a direct way to model this particular problem (where there are no interactions), the RSMA formulation offers more generality as we shall explain in later sections. Recently, Shi et al. proved that the min-cost RSMA problem is NP-Complete [21]. However, before RSMA was proved to be NP-Complete, Rao et al. had proposed a 2-approximation algorithm of time complexity $O(n \log n)$ for solving the RSMA [19].

Minimum bandwidth clustering can be used as a sub-scheme to clustering under a time budget, so that channels are released quickly thereby accommodating some interactivity. This technique can speed up computations in a dynamic scenario, in which cluster re-computation is performed on periodic snapshots. The time-constrained clustering algorithm has linear complexity and the dynamic programming algorithm of cubic complexity needs to be run on a smaller number of streams. Simulation results for these are shown in Section 4.

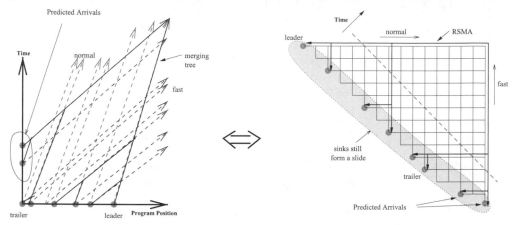

Figure 37.4 Clustering with predicted arrivals.

2.4 CLUSTERING WITH ARRIVALS

Relaxing the preceding problem to include stream arrivals, but not exits nor other interactions, we consider continuing the investigation of the problem space. Suppose we could have a perfect prediction of new stream arrivals. The problem again transforms to the RSMA-slide formulation. Although the sinks do not lie on a straight line, they would still form a slide (e.g., the two points at the bottom right corner of Figure 37.4 form a slide along with the points lying towards their left in the exterior of the grid). Therefore, a variant of the RSMA-slide algorithm of Section 2.3 can be used here. The cost calculation can differ in this case because two points can merge by following paths of unequal length. For example, the two predicted points in Figure 37.4(b) are merged by line segments of different lengths. After the RSMA is calculated, a corresponding merging tree can be found for the original formulation as shown in Figure 37.4(a).

Alternatively, we can batch streams for a period equal to or greater than the maximum window for merging. In other words, the users that have newly arrived can be made to wait for an interval of time for which the previously arrived streams are being merged. Although this can reduce the problem to the preceding one, in practice, this can lead to loss of revenue from customer reneging due to excessive waiting times.

Aggarwal et al. [2] instead use a periodic re-computation approach with an initial greedy merging strategy. They show that if the arrival rate is fairly constant, a good re-computation interval can be determined. If a perfect predictor were available (e.g., via advance reservations), then better performance can be achieved by using the exact RSMA-slide algorithm.

2.5 CLUSTERING WITH INTERACTIONS AND EXITS

Here we show how a highly restricted version of the problem with certain unrealistic assumptions is transformable to the general RSMA problem. We assume a jump-type interaction model (i.e., a user knows where to fast forward or rewind to, and resumes instantaneously). We also assume that the leading stream does not exit and that interactions do not occur in singleton streams (which are alone in their clusters). Furthermore we assume that a perfect prediction oracle exists which predicts the exact jump interactions. Our "perfect predictor" identifies the time and the program position of all user interactions and their points of resumption.

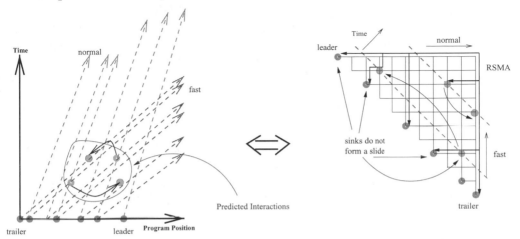

Figure 37.5 Clustering under perfect prediction.

Figure 37.5 shows a transformation to the RSMA problem in the general case. The shaded points in the interior of the grid depict predicted interactions. The problem here is to find a schedule that merges all streams taking into account all the predicted interactions and consumes minimum bandwidth during the process. This problem is isomorphic to the RSMA problem, which has recently been shown to be NP-Complete [21]. Also, the illustration shows that the RSMA is no longer a binary tree in this case.

Stream exits can occur due to the ending of a movie or a user quitting from a channel which has only one associated user. Exits and interactions by a user, alone on a stream, result in truncation and deletion of streams, respectively, and cause the RSMA tree to be disconnected and form a forest of trees. The above model is not powerful enough to capture these scenarios. Perhaps a generalized RSMA forest can be used to model these situations. However, the RSMA forest problem is expected to be at least as hard as the RSMA problem.

An approximation algorithm for RSMA that performs within twice the optimal is available using a "plane sweep" technique. But in a practical situation in which a perfect prediction oracle does not exist, knowing the arrivals, and especially interactions, beforehand is not feasible. Therefore the practical solution is to periodically capture snapshots of the system and base clustering on these snapshots. The source of sub-optimality here is due to the possibility that a stream could keep on following its path as specified by the algorithm at the previous snapshot even though a stream with which it was supposed to merge has interacted and does not exist nearby anymore. Only at the next snapshot does the situation get clarified as the merging schedule is computed again. Also note that causal event-driven cluster recomputation will perform worse than the perfect oracle-based solution. For any stochastic solution to be optimal and computationally tractable, the RSMA must be solvable by a low order polynomial time algorithm. The important point here is that because we do not have a low cost predictive solution in interactive situations, we must recompute the merging schedules periodically to remain close to the optimal.

Note that an algorithm with time complexity $O(n^3)$ (for RSMA-slide) may be expensive for an on-line algorithm and it is worthwhile comparing it with its approximation algorithm that runs faster, producing worst-case results with cost within twice the optimal. In the simulations section (Section 4), we compare both the algorithms under dynamic situations. Rao et al. [19] give a polynomial-time approximation algorithm based on a plane-sweep technique which computes an RSMA with cost within twice the optimal. The algorithm is outlined below.

If the problem is set in the first quadrant of E^2 and the root of the RSA is at the origin, then for points p and q belonging to the set of points on the rectilinear grid of horizontal and vertical lines passing through all the sinks, the parent node of points p and q is defined as $parent(p,q) = (\min(p.x, q.x), \min(p.y, q.y))$, and the rectilinear norm of a point $p = (x, y)$ is defined by $\|(x, y)\| = x + y$.

Algorithm RSMA- heuristic(S :*sorted stream list*)
{
  Place the points in S on the RSA grid
  while (S $\neq \phi$)

    Find 2 nodes $p, q \in S$ such that $\|parent(p,q)\|$ is maximum

    $S \leftarrow S - \{p, q\} \cup parent(p, q)$

    Append branches $parent(p,q) \rightarrow p$ and $parent(p,q) \rightarrow q$ to the RSA tree.
  end
}

The algorithm yields an heuristic tree upon termination. Because it is possible to store the points in a *heap* data structure, we can find the parent of p and q with the maximum norm in $O(\log n)$ time and hence the total time complexity is $O(n \log n)$.

The fact that the static framework breaks down in the dynamic scenario is not surprising. Wegner [25] suggests that interaction is more powerful than algorithms and such scenarios may be difficult to model using an algorithmic framework.

2.6 CLUSTERING WITH HARDER CONSTRAINTS AND HEURISTICS

In practice, stream clustering in VoD would entail additional constraints. We list a few of them here:

- Limits on total duration of rate adaptation per customer
- Critical program sections during which rate adaptation is prohibited
- Limited frequency of rate changes per customer
- Continuous rather than "jump" interactions
- Multiple classes of customers with varying pricing policies

The inherent complexity of computing the optimal clustering under dynamicity, made even harder by these additional constraints, warrants search for efficient heuristic and approximation approaches. We describe a number of heuristic approaches which are worthy of experimental evaluation. We evaluate these approaches experimentally in Section 4.6.

- **Periodic** Periodically take a snapshot of the system and compute the clustering. The period is a fixed system parameter determined by experimentation. Different heuristics can be used for the clustering computation:

  - Recompute **RSMA-slide** (optimal binary tree) using $O(n^3)$ dynamic programming algorithm.

  - Use the $O(n \log n)$ RSMA heuristic algorithm.

  - Use the $O(n)$ EMCL algorithm to release the maximum number of channels possible.

  - Follow EMCL algorithm by **RSMA-slide**, **heuristic RSMA-slide** or all merge with leader.

- **Event-triggered** Take a snapshot of the system whenever the number of new events (arrivals or interactions) exceeds a given threshold. In the extreme case, recomputation can be performed at the occurrence of every event.

- **Predictive** Predict arrivals and interactions based on observed behavior. Compute the RSMA based on predictions.

2.7 DEPENDENCE OF MERGING GAINS ON RECOMPUTATION INTERVAL

In this section, we use analytical techniques to investigate the channel usage due to *RSMA* in the steady state. The modeling approach is more completely described in reference [15].

First, we derive an approximate expression for the number of streams needed to support users watching the same movie in a periodic snapshot RSMA-based aggregation system as a function of time. We consider the non-interactive case here. In a typical recomputation interval R, new streams arrive (one user per stream) at a rate λ_a; the older streams are included in the previous snapshot captured at the beginning of the current recomputation interval, and are merged using RSMA. Now, to make the calculation tractable, we assume that the process of capturing snapshots enforces a "natural clustering" of streams in the sense that the streams being merged in the current recomputation interval are less likely to merge with streams in the previous recomputation interval, at least for large values of R. This assumption is reasonable because the last stream of the current cluster (the trailer) accelerates while the leading stream of the previous cluster (leader) progresses normally, thus resulting in a widening gap between end-points of clusters.

We model one recomputation interval R, and then consider the effect of a sequence of R's by a summation. Again, we make a very simplistic assumption to make the problem tractable. We assume that the streams in a snapshot are equally spaced and the separation is $1/\lambda_a$. In this situation, the RSMA is equivalent to progressively merging streams in a pair wise fashion. If L is the length of the movie and $K = 15$ is a factor arising due to the speed-up, the number of streams as a function of time from the beginning of the snapshot is given by:

$$C(t) = \begin{cases} 0 & : t < 0 \\ \lambda_a t & : 0 \le t < R \\ \dfrac{\lambda_a R}{2^{\left\lceil \log\left(\frac{\lambda_a(t-R)}{K}+1\right)\right\rceil}} \approx \dfrac{KR}{t - R + \dfrac{K}{\lambda_a}} & : R \le t < R + L \qquad (37.9) \\ 0 & : t \ge R + L. \end{cases}$$

At any time instant t, the total number of streams $N(t)$ is built from contributions of a sequence of recomputation intervals. It is given by:

$$N(t) = \sum_{n=0}^{\infty} C(t - nR). \qquad (37.10)$$

Note that only some of the terms in the above summation have a non-zero contribution to the total sum. We are interested in calculating the average number of streams in steady state, in other words, in the value of $\lim_{T \to \infty} \frac{1}{T} \int_{t_0}^{t_0+T} N(t)dt$. But, it can be shown that $N(t)$ is periodic with period R, hence we calculate the value of $\frac{1}{R} \int_{t_0}^{t_0+R} N(t)dt$, where t_0 signifies a time instant after steady state has been reached.

As we indicated before, $N(t)$ is non-zero only for a sequence of recomputation intervals; hence by appropriately taking contributions from relevant intervals, we get an expression for the average number of streams in the system in steady state:

$$\tilde{N}(R) = \frac{1}{R} \int_{t_0}^{t_0+R} N(t)dt \approx \frac{1}{2}\lambda_a R + K \ln\left(\frac{\lambda_a L + \lambda_a R + K}{K}\right). \qquad (37.11)$$

We observe that $\tilde{N}$ has a linear component in R and a logarithmic (sub-linear) component in R, thus making the overall effect linear in R, especially for large value of R. We shall see in Section 4.6 that this agrees in behavior with the simulation results, although there are inaccuracies in the exact values of $N(t)$ due to the approximate nature of the model developed here.

3. STOCHASTIC DYNAMIC PROGRAMMING APPROACH

In this section, we attempt to model the interactive VoD system using stochastic dynamic programming (DP). We model the arrival, departure and interactions in the systems as *events*, and as in a standard DP based approach, we apply an optimization "control" at the occurrence of each event. Let us consider the system at time $t_0 = 0$. Suppose there are N users $u_1, u_2, ..., u_N$ in the system (for simplicity of analysis we assume all users receive the same program) at that particular time instant, and their program positions are given by a sorted position vector $P = (p_1, p_2, ..., p_N)$, respectively, where $p_1 \geq p_2 \geq \cdots \geq p_N$. Because this is an aggregation based system, the number of streams in the system, n, is less than the number of users. Because the user position vector P is sorted, the stream position vector can be easily constructed from P in linear time. Therefore, the state of the system can be denoted in two ways: (1) by the user position vector P as described in the previous paragraph, or, (2) by a stream position vector $S = (s_1, s_2, ..., s_n)$ with ID's of users that each stream is carrying. We choose the latter representation for our analysis in this section.

In the following analysis, we assume that the Nth user arrived at time, t_0. Hence, $p_N = 0$, $s_n = 0$ and we must apply an optimization control at this time instant. An optimization control is nothing but a merging schedule which results in the minimization of the total number of streams in the long run under steady state. A merging schedule is a binary sequence $M = (r_0, r_1, ..., r_n)$ of length n, where r_i denotes the content progression rate of stream s_i, and $r_i \in \{r, r + \Delta r\}$, $\forall i$.

We describe a model for the events in the system. Suppose, the next event occurs at time instant $t_0 + \tau$. The streams in the system progress according to the merging schedule M, which was prescribed at time t_0. The new position vector S' will be a function of the old position vector S, the merging schedule M, the inter-event time τ and also the type of event. The size of the position vector S, increases, decreases or stays the same depending on the type of the event:

- **Arrival:** size increases by 1

- **Departure due to a user initiated quit:** size decreases by 1 (if the quit is initiated in a single-user stream) or stays the same (if the quit is initiated in a multi-user stream)

- **Exit due to the end of the program:** size reduces by 1, and number of users on the exiting stream.

- **Interaction:** size increases by 1 (if initiated in a multi-user stream) or stays the same (if initiated in a single-user stream)

3.1 COST FORMULATION FOR STOCHASTIC DYNAMIC PROGRAMMING

We begin with a brief introduction to the stochastic DP approach [5] that we decided to utilize to model the system. The quantity that we want to minimize in the system is the number of streams, n, in steady state. Using the DP terminology, we refer to this quantity as average *cost* incurred per unit time. The *total* average cost incurred can be represented by a sum of *short-term* cost and *long-term* cost per unit time. The former reflects the immediate gains achieved by the merging schedule M (i.e., before the next event occurs), whereas the latter reflects the long term gains due to M. If the current state (basically the position vector S at time t_0) is represented by i and the long-run cost incurred at state i is denoted by a function $h(i)$, then the total average cost per unit time at the current time instant t_0 is given by the Bellman's DP equation:

$$J^* + h(i) = \min_{u \in U(i)} \left[E[g(i,u)] + \sum_{j=1}^{t} p_{ij}(u)h(j) \right], \tag{37.12}$$

where $U(i)$ is the set of possible controls (merging schedules, in our setting) that can be applied in state i, $g(i,u)$ is the short-term cost incurred due to the application of control u in state i, $E[g]$ denotes the expected value of the short-term cost, $p_{ij}(u)$ denotes the transition probabilities from state i to state j under control u and $h(j)$ is the long-run average cost at state j. t is a state that can be reached from all initial states and under all policies with a positive probability, and J^* is the optimal average cost per unit time starting from state t.

The Curse of Dimensionality

Under this formulation, we quickly run into some problems. First, the state space is very large. If the maximum number of users in the system is N and the program contains L distinct program positions, the number of states is clearly exponential in N, no matter how coarsely the program positions are represented (obviously, this depends on the coarseness of representation). For fine grain state representations, the size of the state space grows exponentially as well. Also, the space of controls is exponential in n, where n denotes the number of streams in the system, because each stream has two choices for the content progression rate. Furthermore, the long-run cost function $h(\cdot)$ is very hard to characterize for all states especially in the absence of a formula based technique. Currently we believe that $h(\cdot)$ can be characterized by running a simulation for long enough that states start to recur and then the long run average cost values can be updated for every state over a period of time. However, this technique is extremely cumbersome and impractical owing to the hugeness of the state space in a VoD system with a large dimension (N). Therefore, due to this *curse of dimensionality*, we believe that the DP method is impractical in its actual form and attempt to use it in a partial manner, which we explain in detail next.

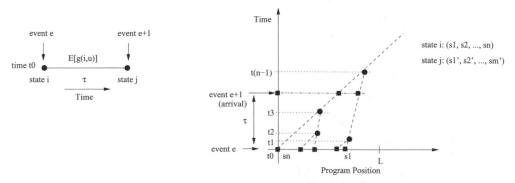

Figure 37.6 Event model and controls.

Expected Cost Calculation

Figure 37.6 demonstrates the basic model of events and controls. The left side depicts the occurrence of events e and $e+1$ for a section of the timeline. The inter-event time is τ and the system moves from state i to state j under control u and incurs a short-term expected cost denoted by $E[g(i, u)]$ which depends on the distribution of the random variable τ. We illustrate a technique for the calculation of this quantity below.

The control space is chosen carefully to prevent an exponential search and to yield an optimal substructure within a control. u is considered to be a feasible control if it represents a binary tree with $s_1, s_2, \ldots, s_n$ as leaves. Hence u is a function of the state i. Although the control space $U(i)$ grows combinatorially with n, the use of the principle of optimality (as demonstrated later) results in a polynomial time algorithm for choosing the optimal control. On the right side of Figure 37.6, a control is shown as a binary tree which is a function of state i. We depict the intermediate merge points in the binary tree by $t_k : 1 \le k \le n - 1$; $t_1 \le t_2 \le \cdots \le t_{n-1}$. $t_n = \frac{L - s_1}{r + \Delta}$ denotes the time that the leading stream will run. Here we assume that the leading stream goes fast after the final merge and that all n streams merge before the movie ends (i.e., $t_{n-1} < \frac{L - s_1}{r + \Delta}$. t_k's can be determined easily from the stream position vector (essentially state i) and the control under consideration, u.

We formulate the short-term cost function for a given state i and a control u as the cost per unit time between events e and $e+1$. For simplicity, we assume that $t_0 = 0$; because this contributes to an additive term in the expected cost formulation and it does not affect the comparison of costs under two different controls.

$$
g(i, u) = \begin{cases}
n & : \; 0 = t_0 < \tau < t_1 \\
\dfrac{1}{\tau}(nt_1 + (n-1)(\tau - t_1)) & : \; t_1 \le \tau < t_2 \\
\vdots & : \; \vdots \\
\dfrac{1}{\tau}(nt_1 + (n-1)(t_2 - t_1) + \cdots + (n-k)(\tau - t_k)) & : \; t_k \le \tau < t_{k+1} \\
\vdots & : \; \vdots \\
\dfrac{1}{\tau}(nt_1 + (n-1)(t_2 - t_1) + \cdots + 2(t_{n-1} - t_{n-2}) + (\tau - t_{n-1})) & : \; t_{n-1} \le \tau < t_n = \dfrac{L - s_1}{r + \Delta}.
\end{cases}
$$

$$(37.13)$$

The general (k^{th}) branch in the above cost formulation simplifies to the following:

$$
\begin{aligned}
g_k(i, u) &= \frac{1}{\tau}(nt_1 + (n-1)(t_2 - t_1) + \cdots + (n-k)(\tau - t_k)) \\
&= \frac{1}{\tau}\sum_{l=0}^{k} t_l + (n-k), \qquad t_k \le \tau < t_{k+1}.
\end{aligned}
$$

$$(37.14)$$

The only random variable in the above analysis is the inter-event time τ which is exponentially distributed because the occurrence of various events can be modeled by the Poisson distribution. In particular, if the arrivals occur with a mean rate λ_a, quits occur with a mean rate λ_q, and interactions occur with a mean rate λ_i, and each is modeled by the Poisson distribution, then the aggregate event process is also Poisson with a rate $\lambda = \lambda_a + \lambda_q + \lambda_i$ (i.e., the mean inter-event times (τ) are exponentially distributed with mean λ). The expected cost under control u can be calculated in the following manner.

$$E[g(i, u)] = \int_0^\infty g(i, u) f(\tau) d\tau = \int_0^\infty \lambda e^{-\lambda\tau} g(i, u) dt = \int_0^\infty \lambda e^{-\lambda\tau} g(i, u) d t$$
$$= \sum_{k=0}^{n-1} \left[\int_{t_k}^{t_{k+1}} \lambda e^{-\lambda\tau} g_k(i, u) d\tau \right] = \sum_{k=0}^{n-1} E[g_k(i, u)], \tag{37.15}$$

where $E[g_k(i, u)]$ is given by

$$E[g_k(i, u)] = \int_{t_k}^{t_{k+1}} \lambda e^{-\lambda\tau} \left[\frac{1}{\tau} \sum_{l=0}^{k} t_l + (n - k) \right] d\tau$$
$$= (n - k) \int_{t_k}^{t_{k+1}} \lambda e^{-\lambda\tau} d\tau + \left(\sum_{l=0}^{k} t_l \right) \int_{t_k}^{t_{k+1}} \frac{1}{\tau} \lambda e^{-\lambda\tau} dt \tag{37.16}$$
$$= (n - k)[e^{-\lambda t_k} - e^{-\lambda t_{k+1}}] + \lambda \left(\sum_{l=0}^{k} t_l \right) [E_i(\lambda t_k) - E_i(\lambda t_{k+1})],$$

where $E_i(t) = \int_t^\infty \frac{e^{-x}}{x} dx$. From Equation 37.15, after algebraic simplification, we yield a compact expression for the expected cost:

$$E[g(i, u)] = n - \sum_{k=1}^{n} e^{-\lambda t_k} + \lambda \sum_{k=1}^{n} t_k E_i(\lambda t_k) \tag{37.17}$$

Stochastic RSMA

Having derived an expression for the expected short-term cost, we attempted to use it for choosing the optimal control. In other words, given that we are in state i, we find the control $u(i)$ that minimizes $E[g(i, u(i))]$. Instead of direct constrained minimization, we use the principle of optimality and the separability of terms in Equation 37.17 to represent the cost recursively as follows.

$$E[g(i, u)] = 1, \qquad n = 1$$
$$= E[g(i_L, u_L)] + E[g(i_R, u_R)] - (e^{-\lambda t_{n-1}} - e^{-\lambda t_n}) \tag{37.18}$$
$$+ \lambda(t_{n-1} E_i(\lambda t_{n-1}) - t_n E_i(\lambda t_n)), \qquad n > 1,$$

where u_L and u_R represent the left and the right subtrees of u, respectively, and i_L and i_R represent the corresponding portions of the state i. Because t_{n-1} is fixed for a given state i, a static DP approach that is very similar to the RSMA-slide approach described in Section 2 can be applied to produce the optimal u. The time complexity of this algorithm, called Stochastic RSMA, is $O(n^3)$. Simulation results for the algorithm are shown in Section 4.6.

4. PERFORMANCE

In this section, we show simulations of the clustering algorithms for rate adaptive merging described earlier, and compare them with some heuristic algorithms that are described later in this section.

4.1 OBJECTIVE

Some of the algorithms described in Section 2 are optimal in the static case (i.e., in the absence of user interactions). In addition, no optimal algorithms are known for the dynamic scenario in which users arrive into the VoD system, interact and depart. However, the behavior of such a system with the application of the aforementioned clustering and merging techniques can be studied by simulations. In these simulations, we investigate the gains offered by an ensemble of techniques under varying conditions. We attempt to answer the following questions:

- Which policy is best in the non-interactive scenario?

- Which policy is best under dynamic conditions in the long run?

- What is the effect of high arrival and interaction rates on these algorithms?

- What algorithm should be used when facing server overload?

4.2 THE SIMULATION SETUP

We treat the simulation as a discrete-time control problem. As outlined in Section 2, there are various ways in which we can control the system including: apply control with a fixed frequency, apply control with a fixed event frequency or adapt control frequency depending on the event frequency. The last technique is the hardest to implement. In this work, we have considered only the first method of periodic application of control.

The simulation set up consists of two logical modules: a discrete event simulation driver and a clustering unit. The simulation driver maintains the state of all streams and generates events for user arrivals, departures and VCR actions (fast-forward, rewind, pause and quit) with inter-event times obeying an exponential probability distribution. Here we have not considered a bursty model of interactions. A merging window (also called the re-computation period) is an interval of time within which a clustering algorithm attempts to release channels. Once every re-computation period, the simulation driver conveys the status of each stream to the clustering unit and then queries it after every simulation tick to obtain each stream's status according to the specified clustering algorithm. It then advances each stream accordingly. For instance, if the clustering unit reports the status of a particular stream to be accelerating at a given instant of time, the simulation driver advances that stream at a slightly faster rate of $^{16}\!/_{15}$ times that of a normal-speed stream.

The clustering unit obtains the status of all running streams from the simulation driver and computes and stores the complete merging tree for every cluster in its internal data structures until it is asked to recompute the clusters. On this basis, it supports querying on the status of any running non-interacting stream by the simulation driver after every simulation time unit.

4.3 ASSUMPTIONS AND VARIABLES

The main parameters in our simulations are listed in Table 37. 2. We simulated three types of interactions: *fast forward, rewind* and *pause*. λ_{int} is an individual parameter that denotes the rate of occurrence of each of the above interaction types (e.g., if $\lambda_{int} = 0.02$ sec^{-1}, it means a fast-forward or a rewind, etc. occurs in the system once every 50 seconds). For simplicity we maintain λ_{int} the same for each type of interaction.

Table 37.2 Simulation Parameters

Parameter	Meaning	Value
M	Number of movies[2]	100
L	Length of a movie	30 min
W	Re-computation window	10–1000 sec
λ_{arr}	Mean arrival rate[3]	0.1, 0.3, 0.7, 1.0 sec^{-1}
λ_{int}	Mean interaction Rate[3,4]	0, 0.01, 0.1 sec^{-1}
λ_{dur}	Mean interaction time[3]	5 sec

λ_{arr} and λ_{int} determine the number of users in the system at any given time. The user pool is assumed to be large. Also, we do not consider the probability of *blocking* of a requesting user here as we assume that the capacity of the system is greater than the number of occupied channels.

The content progression rate of a normal speed stream is set at 30 fps and that of an accelerated stream is 32 fps. The accelerated rate can be achieved in practice by dropping a B picture from every MPEG group of pictures while maintaining a constant frame play-out rate. Fast-forward and fast-rewind are fixed at 5X normal speed. Finally, the simulation granularity is 1 sec and the simulation duration is 8,000 sec.

We investigated the channel and bandwidth reclamation rates for each clustering algorithm and compared the results (Section 4.5).

4.4 CLUSTERING ALGORITHMS AND HEURISTICS

We simulated six different aggregation policies and compared them:

EMCL-RSMA Clusters are formed by the EMCL algorithm, and then merging is performed in each cluster by the RSMA-slide algorithm (Section 2).

EMCL-AFL Clusters are formed by the EMCL algorithm, and then merging is performed by forcing all trailers in a cluster to chase their cluster leader.

EMCL-RSMA-heuristic Clusters are formed by the EMCL algorithm, and then merging is performed by a RSMA-heuristic which is computationally less expensive than the RSMA-slide algorithm.

---

[2] Movie popularities obey Zipf's law.
[3] Exponential distribution assumed.
[4] for each type of interaction.

RSMA The RSMA-slide algorithm is periodically applied to the entire set of streams over the full length of the movie, but without performing EMCL for clustering.

RSMA-heuristic The RSMA-heuristic algorithm is periodically applied to the entire set of streams over the full length of the movie, but without performing EMCL for clustering.

Greedy-Merge-heuristic Merging is done in pairs starting from the top (leading stream) in the sorted list. If a trailer cannot catch up to its immediate leader within the specified window, the next stream is considered. At every merge, interaction, or arrival event, a check is performed to determine if the immediate leader is a normal speed stream. If so, and merging is possible within a specified window, the trailing stream is accelerated towards the immediate leader. This process is continued throughout the window. GM is similar to Golubchik's heuristic [11] apart from the fact that it re-evaluates the situation at every interaction event as well.

Table 37.3 Arrival and Interaction Patterns

Arrivals/ Interactions	NO	LOW	HIGH
LOW	λ_{arr} =0.1, λ_{int} =0	λ_{arr} =0.1, λ_{int} =0.01	λ_{arr} =0.1, λ_{int} =0.1
LOW-MEDIUM	λ_{arr} =0.3, λ_{int} =0	λ_{arr} =0.3, λ_{int} =0.01	λ_{arr} =0.3, λ_{int} =0.1
HIGH-MEDIUM	λ_{arr} =0.7, λ_{int} =0	λ_{arr} =0.7, λ_{int} =0.01	λ_{arr} =0.7, λ_{int} =0.1
HIGH	λ_{arr} =1.0, λ_{int} =0	λ_{arr} =1.0, λ_{int} =0.01	λ_{arr} =1.0, λ_{int} =0.1

The first three policies release the same number of channels at the end of an interval because they use the EMCL algorithm for creating mergeable clusters. But, the amount of bandwidth saved in the merging process is different in each case. EMCL-RSMA proves to be optimal over a given merging window, so it conserves the maximum bandwidth. EMCL-AFL is a heuristic and can perform poorly in many situations. EMCL-RSMA-heuristic, in contrast, uses an approximation algorithm for the RSMA-slide that guarantees a solution within twice the optimal [19]. Although our problem falls within the *slide* category and has an $O(n^3)$ exact solution, we experiment with the approximation algorithm that runs in $O(n \log n)$ time.

4.5 RESULTS AND ANALYSIS

Here we discuss clustering gains (channel reclamation) and merging gains (bandwidth reclamation) due to the aforementioned algorithms. We classify the simulations into two categories:

4.5.1 Static Snapshot Case

The static snapshot case is applicable in overload situations. That is, when the server detects an overload, it wants to release as many channels as possible in a static merging window (one in which there are no user interactions). We vary the window size for a fixed number of users. Figure 37.7(a) shows that the clustering gain increases as *W* increases. Also visible is that EMCL outperforms GreedyMerge over any static window of size *W*. Figure 37.7(b) compares the bandwidth gains due to the three merging algorithms. The merging gain increases with increase in *W*. Also, over any static window in the graph, the gains due to EMCL-RSMA and EMCL-RSMA-heur are almost identical! Although EMCL-RSMA is provably optimal over a static window, the RSMA heuristic does as well in most cases (the reason for this may be attributed to the parameters of the particular inter-arrival distribution). Although cases can be constructed in which the heuristic produces a result that is twice the optimal value, in most practical situations it performs as well as the exact algorithm at a lower computational cost.

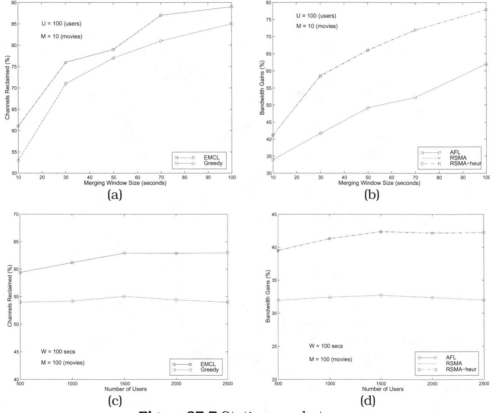

(a) (b)

(c) (d)

Figure 37.7 Static snapshot case.

Figure 37.7(c) shows the number of users varied for a fixed window size of 100 *s* comparing GM and EMCL. Clearly, EMCL performs better than GM and their clustering gains remain almost constant throughout the curves although the gap between the two appears to increase as *U* increases. Figure 37.7(d) shows that EMCL-RSMA and its heuristic version perform equally well and clearly reclaim more bandwidth than EMCL-AFL. Even in this case, the percentage gains appear to remain constant as *U* increases.

4.5.2 Dynamic Case

The dynamic case is the general scenario in which users come into the system, interact and then leave. Here, we simulate over a time equivalent of approximately $4\,1/2$ lengths of the movies.

We study the behavior of cumulative channel reclamation gains due to the above algorithms as a function of time for cases with varying rates of user arrival and interactions. We considered 12 different arrival-interaction patterns as shown in Table 37. 3.

For each of the above cases and for each of the six different aggregation policies, we ran the simulations for eight different re-computation window sizes(W): 10, 25, 50, 100, 200, 500, 750 and 1,000 seconds. Figure 37.8 shows the steady-state behavior of the system for λ_{arr} = 0.7 and λ_{int} = 0.1. We can clearly see that RSMA outperforms all other policies by a large margin for small and medium values of W. For W = 100 sec, RSMA reclaims about 400 channels from 1,250 channels after the steady state is reached thus yielding gains of about 33%. This essentially means that for the particular type of traffic (λ_{arr} = 0.7 and λ_{int} = 0.1), we do not need more than 850 channels to serve 1,250 streams.

For smaller W, other policies provide diminished channel gains. But as the re-computation interval is increased, RSMA begins to suffer and EMCL based policies and GM begins to perform well. This is because RSMA takes a snapshot of the whole system at the beginning of a recomputation interval and advances streams according to the *RSMA tree* computed from that snapshot. In the situation in which the user arrival rate is moderately high and the interaction rate is high as well, for high values of W, many streams come in and interact between two consecutive snapshots and thus the gains due to RSMA drop. In contrast, GM tries to re-evaluate the situation at every arrival and interaction event, so it performs very well.

Our speculation and conclusion is that different policies will perform at their best for different values of W and, indeed, we found this to be true. Figure 37.9 shows how the channel gains vary with the value of the recomputation window. For consistency, we consider the case with λ_{arr} = 0.7 and λ_{int} = 0.1. The graphs for the other cases are similar in shape and have not been shown due to space limitations. In the steady state, the system has 1,250 users on average. For small values of W ($\leq$ 500 sec.), EMCL-based schemes and GM do not perform well, whereas, RSMA and RSMA-heur perform well. However, for larger values of W (> 500 sec.), EMCL-based schemes and GM perform well and RSMA-heur is highly sub-optimal. Although RSMA shows reduced gains for higher values of W, it is still superior to EMCL-based algorithms. GM outperforms all algorithms, including RSMA for $W \geq 750$ due to the reason mentioned in the previous paragraph. But GM is more CPU intensive as it reacts to every arrival, interaction, and merge event. In that sense, it is not a true snapshot algorithm like the others, and is not a scalable solution under heavy rates of arrival and interaction although it may perform better than the other algorithms in such situations.

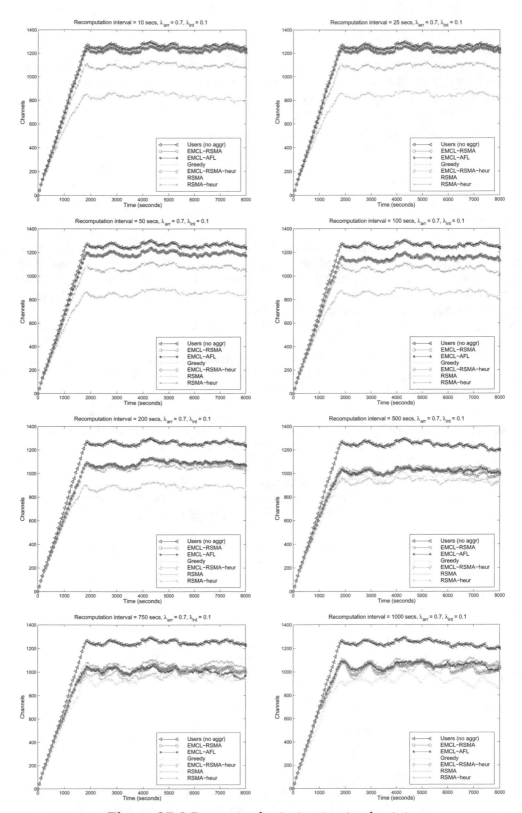

Figure 37.8 Dynamic clustering in steady state.

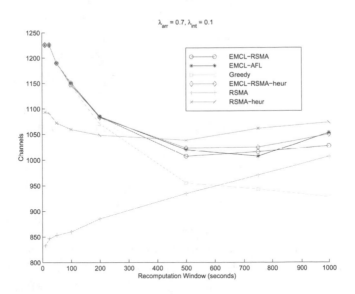

Figure 37.9 Dependence of channel gains on *W*.

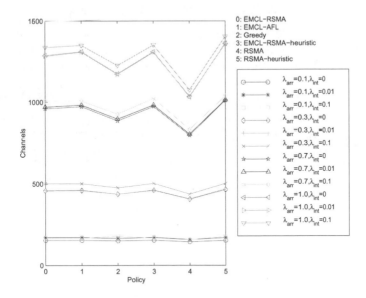

Figure 37.10 Comparison: With best *W* for each policy.

Each policy attains its best results for different values of *W*. RSMA performs best for $W = 10$ sec; GM performs best for $W = 1,000$ sec. and the EMCL based algorithms perform best for $500 \leq W \leq 750$. Figure 37.10 plots the best channel gain achieved by a policy for every class of traffic (see Table 37. 3). In the figure, there are four sets of curves with three curves each. Each set corresponds to an arrival rate because λ_{arr} is the most dominant factor affecting the number of users in the system. The mean numbers of users in the system for λ_{arr} = 0.1, 0.3, 0.7 and 1.0 are 180, 540, 1,250 and 1,800, respectively. The mean clustering gains increase significantly with increase in the arrival rate.

For each of the four sets, the channel gains are almost identical for λ_{int} = 0 and λ_{int} = 0.01 but are less for λ_{int} = 0.1. That is not far from expectations as a high degree of interaction causes each algorithm to perform worse. More importantly, the RSMA algorithm emerges as the best overall policy. As λ_{arr} increases, the gap between RSMA and the other algorithms widens.

Also, there is little difference in the gains due to EMCL-RSMA and its heuristic counterpart, although there is a vast difference between the gains due to RSMA and those due to the RSMA heuristic. This is because EMCL breaks a set of streams into smaller groups and the EMCL-RSMA heuristic performs almost identically well as EMCL-RSMA for smaller groups. But, in case of RSMA, the number of streams is large (EMCL has not been applied to the stream cluster), so the sub-optimality of the heuristic appears. Note that the RSMA-slide algorithm of complexity $O(n^3)$ is affordable in practical situations because it is invoked only once in every recomputation period. As a reference point, the running time of the algorithm on an Intel Pentium Pro 200 MHz PC running the Linux operating system is well below the time required to play out a video frame; therefore no frame jitter is expected on modern playback hardware.

From the above simulations, we conclude the following:

- EMCL-RSMA is the optimal policy in the static snapshot case as it reclaims the maximum number of channels with minimum bandwidth usage during merging. Thus it is the ideal policy for handling server overloads.

- In a VoD system showing 100 movies, for dynamic scenarios with moderate rates of arrival and interaction, periodically invoking RSMA with low re-computation periods yields the best results.

- In realistic VoD settings, the arrival rate plays a greater role than the interaction rate (except for interactive game settings where aggregation is not a feasible idea). High arrival rates result in more streams in the system thus creating better scope for aggregation.

- Higher degrees of interaction result in lesser clustering gains for every algorithm.

- With increasing rates of arrival, RSMA's edge over other algorithms increases.

- The EMCL-RSMA heuristic performs almost as well as EMCL-RSMA although the RSMA-heuristic performs very badly as compared to RSMA.

4.6 COMPARISONS BETWEEN RSMA VARIANTS

Here we compare the performance of the various forms of the RSMA-slide algorithm, namely, periodic snapshot RSMA (**Slide**), event-triggered RSMA (Section 2.6), stochastic RSMA (Section 3) and predictive RSMA (Section 2.6). We consider a single movie of length $L = 2$ hours.

First, we evaluate the performance of **Slide** in the "arrival-only" situation because it lends itself to easy comparison with the analytical framework described in Section 2.7. The behavior of the channel gains is depicted in Figure 37.11(a) as a function of the recomputation interval R. The simulation results show that as R is increased, the channel gain degrades as a linear function of R. The analytically estimated channel gains as described in Section 2.7 are plotted on the same graph. From the graph we see that the analytical estimation comes reasonably close to the simulation results. Although the analytical technique underestimates the simulation results by approximately 10 streams, the slopes of the two curves are almost the same. The slope is very close to the analytical estimate of $\frac{1}{2}\lambda_a$ (Equation 37.11). The underestimation can be attributed primarily to the assumption that the recomputation results in formation of natural clusters, and the streams in two different clusters do not merge. This assumption may not be valid due to a possible effect of streams in one recomputation interval on those in the next or previous one.

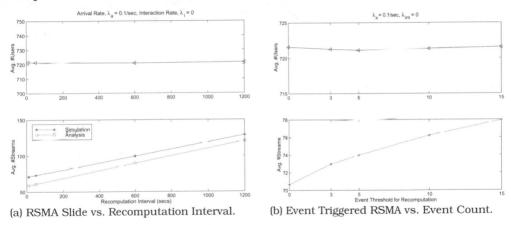

(a) RSMA Slide vs. Recomputation Interval.   (b) Event Triggered RSMA vs. Event Count.

Figure 37.11 Two variants of RSMA.

Next, we consider the threshold based, event-triggered RSMA, in which recomputation is not done periodically but only when the number of events exceeds a threshold. Figure 37.11(b) shows the variation of the gains with the increase in the event count threshold. The case in which the threshold is 0 corresponds to the pure event-triggered RSMA where recomputation happens at the occurrence of each event. As expected, the gains decrease as the threshold is increased.

In summary, Table 37.4 shows the number of users and streams for each variant of RSMA in steady state. The **Slide** refers to the **RSMA-slide** with $R = 10$ sec. The rationale for using this value is that the arrival rate considered is 0.1 sec, and hence on average there will be an arrival every 10 seconds. By "perfect prediction," we intend the following: to evaluate whether gains are achieved with this technique that warrant obtaining the difficult prediction of future events. Note that the RSMA can be computed optimally under perfect prediction if the events comprise arrivals only, as shown in Section 2.4. When the actual stream arrives later, it just follows the path computed by the predictive RSMA algorithm. Stochastic RSMA reacts to every arrival event and the RSMA tree is determined using the technique described in Section 3. The event-triggered

RSMA algorithm corresponds to the case in which the event count threshold is 0.

Table 37.4 Comparison between Variants of RSMA ($\lambda_{arr} = 0.1$, $\lambda_{int} = 0$, $\lambda_q = 0$)

RSMA Variant	Avg. #Users	Avg. #Streams	Avg. #Users per Stream
RSMA Slide ($R = 10\,\mathrm{sec}$)	722	71	10.17
Event Triggered RSMA	722	71	10.17
Predictive RSMA	721	71	10.15
Stochastic RSMA	732	137	5.34
GreedyMerge	720	373	1.93

We see from Table 37. 4 that with the exception of the stochastic RSMA, the other variants perform almost identically under the current set of conditions. Because the merging algorithm alters the effective service time of a user by manipulation of the content progression rates, it affects the number of users in the system directly. The random number generator in the simulator produced user arrivals with mean $\lambda_{effective} = 0.1033/\mathrm{sec}$, which is slightly higher than the intended mean of $\lambda_{arr} = 0.1/\mathrm{sec}$. Hence the average number of users in steady state in a non-aggregated system should be $\lambda_{effective} \times L = 0.1033 \times 7200 = 744$, but due to aggregation, the effective service time for many streams is reduced when they progress in an accelerated fashion; hence the number of users in the system is reduced. Stochastic RSMA performs worse than the other RSMA variants because only the short-term cost is considered in the formulation. For comparison, we show the gains due to the deterministic "greedy" algorithm. It performs much worse than its stochastic greedy counterpart (i.e., stochastic RSMA) as well as the other RSMA variants.

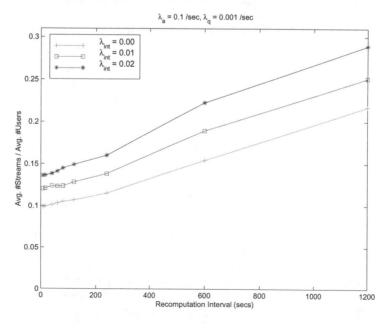

Figure 37.12 Effect of varying degrees of interactivity.

Among all RSMA variants, the **Slide** is the simplest to implement and is computationally less expensive than its event-triggered counterpart because it does not react to every event. Predictive RSMA, in contrast, does not provide additional gains even with perfect prediction; hence the computational overhead involved in prediction is not justified. We conclude that the **Slide** algorithm yields the best results with the least computational overhead than its variants.

Figure 37.12 shows the effect of varying degrees of interactivity on the RSMA slide algorithm as a function of R in steady state. We considered three different values of interaction rates: 0, 0.01 and 0.02 per second for each type of interaction, respectively. The quit rate is held at a low constant value (0.001 per second) in order to prevent the number of users in the system from substantially varying. The linearity dependence on R is lost as the interaction rate increases – for $\lambda_{int} = 0$, the curve is practically linear, whereas for $\lambda_{int} = 0.02$, the curve is not linear. This can be attributed to random interactions such as fast-forward and rewind result in greater fluctuations in stream positions at snapshots, thus making the merging trees of adjacent snapshots less correlated. Non-linearity arises in all three curves for lower values of R because there is no natural clustering due to taking snapshots with low R, and hence the streams in adjacent temporal clusters can merge later. Our analytical estimation framework does not take such behavior into account. As R increases, the gap between the interactive case and the non-interactive case widens significantly. This phenomenon is not unexpected as a larger R means more interaction events in that interval, which in turn translates to the increase in the loss of gains that were achieved due to merging. Hence, as observed before, high degrees of interactivity can be handled by recomputing the RSMA frequently.

Table 37.5 Comparison between Variants of RSMA
($\lambda_{arr} = 0.1$, $\lambda_{int} = 0.01$, $\lambda_q = 0.001$)

RSMA Variant	Avg. #Users per Stream
RSMA Slide ($R = 6\,\mathrm{sec}$)	8.33
RSMA Slide ($R = 16\,\mathrm{sec}$)	8.23
Event Triggered RSMA	8.62
Stochastic RSMA	4.23

Table 37.5 shows the performance comparison of the different RSMA variants in the dynamic, interactive cases. In comparison with Table 37.4, the average number of users per stream is lower due to user interactions. The aggregate event rate is $\lambda_{arr} + \lambda_{int} + \lambda_q = 0.161$, and hence the mean inter-event time is $1/0.161 \approx 6\,\mathrm{sec}$. Hence we use $R = 6$ in the RSMA-slide algorithm. The event-triggered RSMA performs slightly better than its slide counterpart, but not by a substantial margin. Expectedly, RSMA-slide with a slightly higher R (16 sec) yields slightly worse results than RSMA-slide with $R = 6$. However, Stochastic RSMA performs much worse as compared to the other variants because it does not take the long-term cost into account. The event triggered version of RSMA starts performing better than the periodic version as interactivity increases, but it has higher computational overhead. Therefore, a system designer should evaluate the system parameters and evaluate the trade-offs between the desired

level of performance and computational overhead, and choose a clustering algorithm accordingly.

5. SUMMARY AND CONCLUSION

Server based aggregation schemes are important for scalable delivery of interactive VoD content to heterogeneous end-clients with varying memory and storage capacities. In this chapter, we described the modeling of *service aggregation by server-side stream clustering* in i-VoD systems. We investigated various optimality criteria arising in stream clustering in video-on-demand, and then identified an optimal algorithm (RSMA-slide) with time complexity $O(n^3)$ for the "static stream clustering" problem. We also modeled a restricted version of the dynamic clustering problem (with user arrivals and interactions) by a general RSMA which has been recently proven to be NP-complete. Because the general RSMA formulation in the interactive case requires unreasonable assumptions and does not yield a polynomial time solution, we have modeled the dynamic problem using iterative invocation of the optimal algorithms applicable in the static case.

We demonstrated full-scale simulations of the dynamic scenario with users entering, interacting and leaving the system. In the simulations, we performed six different aggregation policies and compared their clustering gains in the steady-state. We investigated each policy for several re-computation intervals (W), arrival and interaction rates (λ_{arr} and λ_{int} respectively), and observed that a particular policy performs at its best at different values of W, λ_{arr} and λ_{int}. Although EMCL_RSMA and its heuristic counterpart perform equally well in most situations, RSMA-slide outperforms its heuristic counterpart by a large margin because of the larger number of streams in a snapshot.

Also shown was an attempt to use stochastic dynamic programming to model the dynamic, interactive scenario, but ran into difficulties due to computational complexity. Although we were able to craft a closed-form representation of the short-term cost due to merging between two successive events, we were unable to characterize the long-term cost in this study. Unfortunately, optimization with short-term cost as the objective does not yield satisfactory results. However, we remain hopeful that further investigation in this direction can result in the discovery of a good long-term cost formulation.

We observed that periodic invocation of the RSMA-slide clustering algorithm with a small recomputation interval produces best results under moderately high rates of arrival and interactions. Its $O(n^3)$ time-complexity is also justified because it is invoked only once in a recomputation period that is of the order of tens of seconds to a few minutes, and also because it yields much better gains than does its heuristic version. We analytically approximated channel usage due to RSMA-slide in non-interactive situations as a function of R, and then showed by simulation that its performance degrades linearly with increase in R; the analysis and simulation results were in reasonable agreement with each other.

Under high rates of arrival and interaction, we showed that RSMA with large re-computation periods yields highly suboptimal results, and heuristics that react to every event have superior performance. A greedy heuristic like GM may result in a much larger number of undesirable rate changes in the system. We also evaluated the performance of other variants of the RSMA-slide algorithm and found that the event-triggered RSMA and RSMA-Slide with $R = 1/\lambda_a$ perform almost equally well. The event-triggered version performs well at the cost of a higher computational overhead. Hence we choose RSMA-slide with low to medium re-computation intervals to be the best overall clustering policy in moderately interactive situations. R can be tuned by the system designer to balance the trade-off between optimality and computational overhead.

There are a variety of unsolved problems related to the work described herein. Future work will consider the feasibility and complexity of distributed clustering schemes and the clustering of streams served from geographically displaced servers. If logical channels are not independent and share a common medium like IP multicast, new stream creation can impact existing streams. Specifically we have assumed QoS in-elasticity; interaction of these algorithms in the presence of heterogeneous and layered streams is interesting. Another important issue that remains to be resolved before clustering schemes can be deployed in practice is seamless splicing of video frames compounded by network delays and delays introduced by inter-frame coding as in MPEG. Moreover, blocking of users due to depletion of channels is not considered here; however, blocking and service denial is inevitable in a system with finite resources and its probability has to be minimized.

REFERENCES

[1]   Almeroth, K. C. and Ammar, M. H., On the use of multicast delivery to provide a scalable and interactive video-on-demand service, IEEE JSAC, 14, 1110, 1996.

[2]   Aggarwal, C. C., Wolf, J. L., and Yu, P. S., On optimal piggyback merging policies for video-on-demand systems, in Proc. ACM SIGMETRICS '96, Philadelphia, PA, USA, 1996, 200-209.

[3]   Basu, P., Krishnan, R., and Little, T. D. C., Optimal stream clustering problems in video-on-demand, in Proc. PDCS '98 - Special Session on Distributed Multimedia Computing, Las Vegas, NV, USA, 1998, 220-225.

[4]   Basu, P., et al., An implementation of dynamic service aggregation for interactive video delivery, in Proc. SPIE MMCN '98, San Jose, CA, USA, 1998, 110-122.

[5]   Bertsekas, D. P., Dynamic Programming and Optimal Control, Vols I and II, Athena Scientific, Massachusetts, 1995.

[6]   Dan, A., et al., Channel allocation under batching and vcr control in video-on-demand systems, J. Parallel and Distributed Computing (Special Issue on Multimedia Processing and Technology), 30, 168, 1995.

[7]   Dan, A. and Sitaram, D., Multimedia caching strategies for heterogeneous application and server environments, in Multimedia Tools and Applications, Vol. 4, Kluwer Academic Publishers, 1997.

[8]   Eager, D., Vernon, M., and Zahorjan, J., Optimal and efficient merging schedules for video-on-demand servers, in Proc. ACM Multimedia '99, Orlando, FL, USA, 1999, 199-202.

[9]   Furht, B., Westwater, R., and Ice, J., IP simulcast: A new technique for multimedia broadcasting over the internet, Journal of Computing and Information Technology, 6, 245, 1998.

[10]   Gao, L., Zhang, Z. L., and Towsley, D., Catching and Selective Catching: efficient latency reduction techniques for delivering continuous multimedia streams, in Proc. ACM Multimedia '99, Orlando, FL, USA, 1999, 203-206.

[11]   Golubchik, L., Lui, J. C. S., and Muntz, R. R., Adaptive piggybacking: a novel technique for data sharing in video-on-demand storage servers, in Multimedia Systems, Vol. 4, ACM/Springer-Verlag, 1996, 140-155.

[12]   Gonzalez, T. F., On the computational complexity of clustering and related problems, in Proc. IFIP Conference on System Modeling and Optimization, New York, NY, USA, 1981, 174-182.

[13]   Hua, K. A., Cai, Y., and Sheu, S., Patching: A multicast technique for true video-on-demand services, in Proc. ACM Multimedia '98, Bristol, U.K, 1998, 191-200.

[14]   Hua, K. A. and Sheu, S., Skyscraper broadcasting: A new broadcasting scheme for metropolitan video-on-demand systems, in Proc. ACM SIGCOMM '97, 1997, 89-100.

[15]   Ke, W., Basu, P., and Little, T. D. C., Time domain modeling of batching under user interaction and dynamic piggybacking schemes, in Proc. SPIE MMCN '02, San Jose, CA, USA, 2002.

[16]   Krishnan, R. and Little, T. D. C., Service aggregation through a novel rate adaptation technique using a single storage format, in Proc. NOSSDAV '97, St. Louis, MO, USA, 1997.

[17]   Krishnan, R., Ventakesh, D., and Little, T. D. C., A failure and overload tolerance mechanism for continuous media servers, in Proc. ACM Multimedia '97, Seattle, WA, USA, 1997, 131-142.

[18]   Lau, S.-W., Lui, J. C. S., and Golubchik, L., Merging video streams in a multimedia storage server: complexity and heuristics, Multimedia Systems, 6, 29, 1998.

[19]   Rao, S. K., et al., The rectilinear steiner arborescence problem, Algorithmica, 7, 277, 1992.

[20]   Sheu, S. and Hua, K. A., Virtual batching: A new scheduling technique for video-on-demand servers, in Fifth International Conference on Database Systems for Advanced Applications, Melbourne, Australia, 1997.

[21]   Shi, W. and Su, C., The rectilinear steiner arborescence problem is NP-complete, in Proc. ACM/SIAM SODA '00, San Francisco, CA, USA, 2000.

[22]   Sincoskie, W. D., System architecture for a large scale video on demand service, Computer Networks and ISDN systems, 22, 155, 1991.

[23]   Tsai, W.-J. and Lee, S.-Y., Dynamic buffer management for near video-on-demand systems, in Multimedia Tools and Applications, Vol. 6 Kluwer Academic Publishers, 1998, 61-83.

[24]   Venkatesh, D. and Little, T. D. C., Dynamic Service Aggregation for Efficient Use of Resources in Interactive Video Delivery, in Proc. NOSSDAV '95, Lecture Notes in Computer Science, Vol. 1018, Little, T. D. C. and Gusella, R., Eds., Springer-Verlag, 1995, 113-116.

[25]   Wegner, P., Why interaction is more powerful than algorithms, CACM, 40, 80, 1997.

38

CHALLENGES IN DISTRIBUTED VIDEO MANAGEMENT AND DELIVERY

Rainer Lienhart[†], **Igor Kozintsev**[†], **Yen-Kuang Chen**[†],
Matthew Holliman[†], **Minerva Yeung**[†],
Andre Zaccarin[†1], **and Rohit Puri**[2]

[†] *Intel Corporation, Santa Clara, California, USA*
[†] *EECS Dept, UC Berkeley, USA.*

Emails: *{Rainer.Lienhart, Igor.v.Kozintsev, Matthew.Holliman, Yen-Kuang.Chen, Minerva.Yeung}@intel.com, Andre.Zaccarin@gel.ulaval.ca, rpuri@eecs.berkeley.edu.*

1. INTRODUCTION

In recent years, we have witnessed the evolution of computing paradigms with the advent of the Internet as a dominant application, content and service delivery platform as well as with the gradual migration of some applications and services from client-server, to edge services, and then to peer-to-peer (P2P) and distributed computing systems. P2P computing systems are capable of accommodating a vast number of users – as publishers as well as consumers. Distributed video storage, distributed video search and retrieval, distributed video processing, and distributed video distribution, among others, bear many opportunities but along also many technical challenges.

The computing platforms and devices connected to a P2P network, however, can differ significantly in their computing capabilities as well as in the speeds and types of network connectivity. This poses serious practical challenges especially for video [16][24]. In the future, the diversity of these capabilities will only increase as the infrastructure must accommodate wireless connections and devices like personal digital assistants (PDAs), tablets, compact media players, and even cell phones, as illustrated in Figure 38.1. Combined digital devices like mobile phones with attached cameras are a reality, aggressively pushed by mobile services providers such as NTT DoCoMo and others. Together with the expected emergence of low-power surveillance and multimedia sensor networks, the days

---

[1] Present affiliation: ECE Dept, Université Laval, Quebec City, Quebec, Canada.

[2] Dr. R. Puri's research was supported in part by the grant from Intel Corporation during years 2000 to 2002. The authors would also like to acknowledge Professor Kannan Ramchandran and Abhik Majumdar of the University of California at Berkeley for their insightful comments and discussions, as well as for their help in improving the quality of this document.

of media transmission as a "downlink" experience (e.g., TV broadcast) only will soon be over.

Many interesting questions arise in distributed video management, storage and retrieval, and delivery. Basically, the key question is centred on: **How should video data be stored, managed, and delivered to accommodate for the connectivity and bandwidth dynamics as well as variances of the networks, computing capabilities of the clients, and battery life of multiple, and mostly mobile, devices?**

There are many technical challenges, specifically:
- Given that popular content is replicated in multiple locations in a P2P network, can we design efficient coding and transmission algorithms that operate in a distributed manner enabling efficient downloads?
- How should the P2P system deal with the limited storage capacity since more and more video is added every day?
- How can the videos be indexed to enable efficient search in distributed video databases?
- How can we accommodate the varying bandwidths and provide reliable delivery to distributed devices?
- Where and how should we incorporate "transcoding"? Where in the multi-device system should computations such as video transcoding or channel coding for robust wireless transmission be performed? How can video transcoding be done efficiently?
- How can robust delivery be achieved over wireless LANs? How can real-time streaming media resources be handled in unicast and multicast environments?
- How can we monitor the quality of service (QoS) of the video streams delivered? Can we quantify the perceived visual quality or client experiences especially in the event of a paid service?

We will address most of the above questions in the subsequent sections, present some current solutions, and highlight future directions. We first present a general architecture for multi-device distributed computing systems, and propose a P2P computing and networking framework, the design principles and system architecture of a video-adaptive and aware peer service platform, in Section 2. Section 3 is on distributed video storage and data management. We focus on three important aspects: multi-rate differential encoding (Section 3.1), distributed video coding (Section 3.2), and adaptive storage and caching management (Section 3.3). Automatic indexing and support for distributed search are addressed in Section 4. We show in Section 5 two important video delivery mechanisms: optimal delivery with dynamic transcoding (Section 5.1) to adapt content to network connectivity and client resources, and robust video unicast and multicast over wireless LANs (Section 5.2). We will touch upon the quality of service monitoring in Section 6. The sequence of the presentation is intended to give readers an overview of the layers and aspects of technical challenges lying ahead, as well as some promising directions of research to provide some answers to the aforementioned questions.

2. SYSTEM ARCHITECTURE

Figure 38.1 shows a potential scenario of a distributed video storage, management and delivery system. We shall use this to illustrate some interesting aspects of distributed video production, distributed video coding, distributed video processing, and distributed video distribution. In this base model, every user/client can be a video producer/provider, intermediate service provider (e.g., such as of video transcoding), and video consumer.

To further illustrate the distributed video database scenario depicted in Figure 38.1, we use the following example: When a user with a handheld device requests a media file (or a life stream) from the distributed video database system by specifying desired attribute, the underlying P2P infrastructure (MAPS - Media Accelerated Peer Services) issues a distributed query, and returns a list of possible downloads together with their associated transmission and local decoding costs as well as their compression format description such as standard MPEG-1, high-quality MPEG-2, and low bit-rate MPEG-4. The local P2P client aggregates the return list with respect to video quality. It also extends the list by possible lower video formats, which could be achieved by transcoding, and which may be more adequate for the client device. From this list the user selects the desired (and feasible) quality of the video. If the version of the video the user selects already exists in the network, the file is streamed to its destination; if not, the P2P infrastructure transcodes the content. For instance, suppose the user requests an MPEG-4 version of a video clip that exists only in MPEG-2 format, possibly at a different rate and resolution. After performing the search, the P2P system discovers the MPEG-2 version of the requested clip and returns the cost associated with streaming it to the client and decoding it there. In this scenario, it is, firstly, costly to transfer the MPEG-2 sequence over the wireless link and, secondly, the handheld device does not have the computational power to decode and render the sequence in real time. Moreover the device has limited battery life and should not perform transcoding itself even if it is possible. Hence, the P2P infrastructure automatically assigns a peer to perform the transcoding based on a cost evaluation, so that only a low bitrate MPEG-4 version of the video is transferred over the wireless link to the user. This example demonstrates that the system not only takes into account storage capacity and network bandwidth, but also CPU power and battery capacity when needed.

Our proposed system architecture to enable scenarios as described above is depicted in Figure 38.2. Existing P2P infrastructures are extended with several modules. These modules are collated into the **Media Accelerating Peer Services (MAPS)** [14]. In detail, the software layers in the system can be classified as follows, from bottom to top:

1. The traditional underlying operating system and related services, viz. sockets, RPC, DCOM, etc.
2. The P2P service layer providing basic functionality such as peer discovery, user and group management, name resolution, and delivery primitives. Examples of systems providing some or all of these services are Gnutella, FreeNet [4], and JXTA [9]. MAPS extends this layer with a plug-in architecture for customization of the basic operations of search and delivery. MAPS currently provides two extensions: a module enabling

real-time media streaming (via RTP over UDP) and a module for enhanced search capabilities (via distributed SQL).

3. The MAPS video support modules, which use the underlying service layer for network and resource access and which provide transformation and computation services on data obtained from/sent to the layer below. Examples include a support module for MPEG video that provides transcoding support, and a module that automatically analyzes images, videos, and music files on each peer to provide better search capabilities.

4. The application layer.

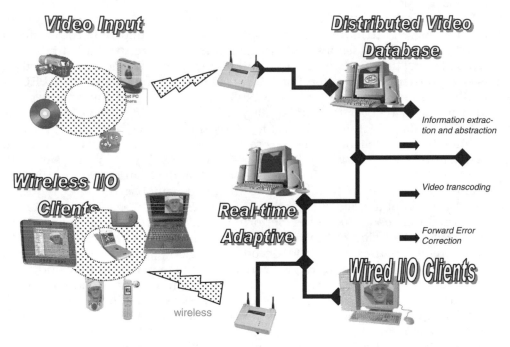

Figure 38.1   A distributed video management and delivery system. Every user can be a video producer/provider, intermediate service provider (e.g., such as of transcoding), and video consumer.

Similar to other P2P environments, a key characteristic of the architecture is the transparency of the physical location of a resource. That is, the location of data, services, and other resources need not be determined explicitly by applications using the provided services. Instead, data and services can exist anywhere in the network, and the system handles their discovery and delivery to the requesting application.

3. DISTRIBUTED STORAGE

The computing devices in a P2P network differ significantly in their computing power and link bandwidth capabilities. This heterogeneity must be adequately considered by advanced media archiving techniques, so that the content can be seamlessly served to the end-clients. Section 3.1 explains how multiple versions of the same video at different bitrates can be created and stored efficiently. It is followed by the VISDOM framework [18] – a fully distributed video coding and transmission scheme. VISDOM demonstrates that content can be replicated in disparate locations in order to provide load balancing of servers and fault tolerance. By nature, its transmission scheme falls semantically under Section 5; however, it is presented here together with the coding scheme for clarity. Finally, Section 3.3 addresses how individual peers can deal with their limited storage capacity in a distributed fashion.

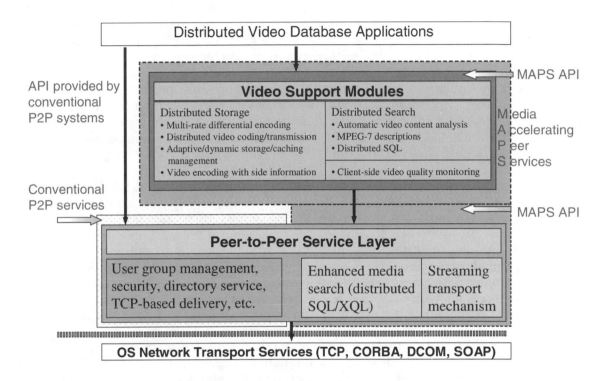

Figure 38.2 System architecture.

3.1   MULTI-RATE DIFFERENTIAL ENCODING

In the P2P scenario of a distributed video database, video data has to be encoded and stored to accommodate for the different connectivity and bandwidth of the network, and computing capabilities of the clients. As described in the previous section, transcoding is the principal means through which video content is adapted to the client specifications. By default, our P2P system keeps the transcoded content for future retrieval since transcoding is a very expensive operation; however, storage is also a precious resource. Therefore, Section 3.1.2 describes our approach of saving storage space by differentially encoding the transcoded bit stream of the video clip with respect to the compressed bit stream of the same clip, but at a different bit rate.

This computational trade-off can also be done at encoding time. Instead of encoding a clip at a single rate and transcoding it at other rates at later times, it is possible to simultaneously encode it at different rates that are likely to be targeted rates by the users. Compared to transcoding, this encoding is better from a rate-distortion point of view since it avoids the successive re-quantization done in transcoding. Although the amount of storage gets increased, it is possible to combine the multi-rate encoder with the differential encoder mentioned above to compensate for this increase.

3.1.1   Multi-Rate Encoding of Video

Figure 38.3 shows the block diagrams of the multi-rate encoder we have developed [27]. The encoder is used to generate N streams at different rates from the same video sequence. For each frame f_n, the encoder computes the compressed data for all streams. The first stream is the reference stream, and the others are dependent streams. There are three main characteristics to this encoder. First, most of the processing is done in the DCT domain. This avoids going back and forth from the image space to the DCT space and therefore can significantly reduce memory operations. Second, motion estimation is done once. As this is the most computationally heavy block of the encoder, this significantly reduces the computational load. Finally, with this architecture, motion compensation has to be performed in the DCT domain.

Fast algorithms for motion compensation in the DCT domain (MC-DCT) can be used within our encoder. Also, since we are sequentially encoding a video sequence at different rates, we have the possibility to re-use data that was computed for the reference stream to reduce computation, and also determine how much loss we incur if no motion compensation or only part of it is performed on the dependent stream. The adaptation of the motion compensation is done on a macro block level and three modes can be used. The first of these modes is the usual mode for which the previously reconstructed frame of the same stream is motion compensated to generate the current predicted frame. In the second mode, the motion compensation is computed on the difference between the $(n\text{-}1)$th reconstructed frames of the current dependent stream and the reference stream. Typically, this difference will be small which, in the DCT domain, translates by a higher number of small or zero coefficients. An appropriate implementation of the MC-DCT can then take advantage of this by not computing the motion compensated DCT coefficients where we expect them to be zero or small. This can be determined by using the quantized residuals of the reference stream $(\tilde{E}_n^1)$. Of course, if a DCT coefficient of the motion compensated frame is not correctly computed at the encoder, it will contribute to a drift (mismatch) error at the decoding time. Finally, in the third mode, we simply use the data from the motion compensated frame of the reference stream. This is more appropriate for B frames since the mismatch errors that are introduced do not propagate to other frames.

3.1.2   Differential Encoding of DCT Coefficients

Figure 38.4 shows a high level diagram of the differential encoder. We assume that a video sequence is simultaneously encoded at multiple rates or quality levels to generate n independent standard compliant (e.g., MPEG-2) video streams. This can be done using the encoder described in Section 3.1.1. One of the video streams is chosen to be the reference stream, in this example, the ith stream, and the other single layer streams are encoded **without loss** with respect to that reference stream to form differentially encoded streams. The output of the differential video encoder is therefore a reference stream, which is a standard compliant video stream, and a number of differentially encoded streams.

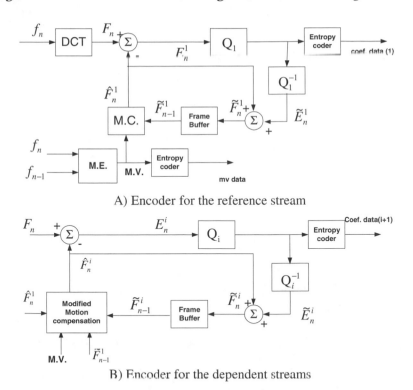

A) Encoder for the reference stream

B) Encoder for the dependent streams

Figure 38.3 Block diagram of multi-rate encoder.

As illustrated in the bottom of Figure 38.4, those differential bit streams can then be used to reconstruct without loss the single layer video streams when the reference stream is available. We emphasize that the video stream generated by the differential decoder is one of the streams generated by the multi-rate encoder, and therefore, it has the rate-distortion characteristics of the single layer encoder used to generate it. This attribute differentiates this approach from scalable codecs and transcoders.

The differential lossless encoder we have developed works at the level of the quantized DCT coefficients. Basically, for a given block of 8x8 pixels, the values of the DCT coefficients of the reference stream are used to predict the values of the DCT coefficients of the target stream. The difference between the predicted and actual values of the target stream DCT coefficients is **losslessly** encoded using an entropy encoder. Coefficients from inter- and intra-coded blocks are treated differently as we briefly explain next. For intra-coded blocks, the predicted coefficient values of the target stream are simply the quantized coefficient values of the reference stream which are re-quantized with the quantization parameters used for the target stream. The same simple prediction mode could be used for inter-coded blocks. For inter-coded blocks, however, the DCT coefficients encode the difference between the video frame and the motion compensated prediction. As the motion compensated data is not the same at different bit rates, the motion compensated difference will vary with different bit streams. The inter-coefficient predictor we use takes into account the mismatch between the motion compensated frame difference of both streams. It adds to the quantized coefficient values of the reference stream the difference between the motion compensated coefficients of the reference and target streams.

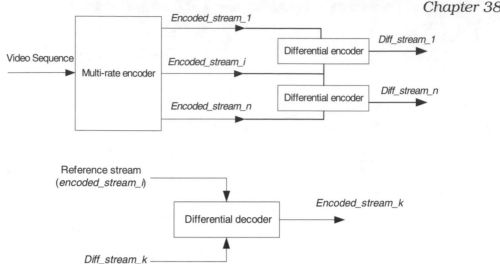

Figure 38.4   High-level representation of the differential video encoder (top) and decoder (bottom).

Using the multi-rate encoder described previously, we encoded video sequences without rate control at multiple bit rates. Table 38. 1 shows the results for the sequence Football (CIF, 352x240, 30fps) encoded at 5 different rates from 750Kbits/sec to 2Mbits/sec using a quantization parameter value ranging from 7 to 17. The results show that the differential encoding of the coefficients allows saving 30% of the storage space when compared to independently storing the five bit streams at a cost of slightly increasing the CPU load (required to perform differential decoding). The results also show that the performance is independent of the choice of the reference stream. In this specific case, the reference stream was alternatively chosen to be the stream encoded at 1.5 Mbits/sec, 1.2 Mbits/sec, and 1 Mbits/sec with very small difference in performance.

Table 38.1   Bitrate of differentially encoded streams and storage savings with respect to independent coding of the streams for the CIF sequence Football.

Quant. scale value	Bit rate of differential and reference streams (bits/sec)			Bit rate of inde-pendently coded streams
	Ref Q=9	**Ref Q=11**	**Ref Q=13**	
7	1,035,357	1,200,314	1,336,536	1,957,853
9	1,492,394	785,261	926,537	1,492,394
11	670,705	1,119,814	625,793	1,193,814
13	626,852	536,158	992,703	992,703
17	435,082	586,105	481,483	741,834
Total bit rate	**4,260,390**	**4,301,652**	**4,363,052**	**6,378,599**
Savings	**33.2%**	**32.6%**	**31.6%**	

3.2   DISTRIBUTED VIDEO CODING AND TRANSMISSION

Storage of popular content at a single node in a distributed video database can lead to fragility in the system due to the "single point of failure" phenomenon. Situations when a large number of clients attempt to access data from this single popular storage can lead to excessive loading and consequently content node failure.

A simple and reasonable, but not necessary efficient and always sufficient approach to combat the above problem and to ensure reliability is to replicate or mirror the data at multiple locations, as it usually happens in P2P networks. Therefore, we have developed the VISDOM framework [18], which uses algorithms for efficient coding and transmission of video content that are distributed in nature. VISDOM enables download of real-time content in parallel from multiple sources, with no need for any communication between them, by leveraging the latest advances in the fields of video compression, distributed source coding theory, and networking. Although VISDOM is distributed, there is no loss in performance with respect to an equivalent centralized system. It also leads to load balancing of servers and fault tolerance. That is, if there is a link/server outage, there will be graceful quality degradation at the client end (the decoded quality decreases smoothly with the reduction in the number of received packets).

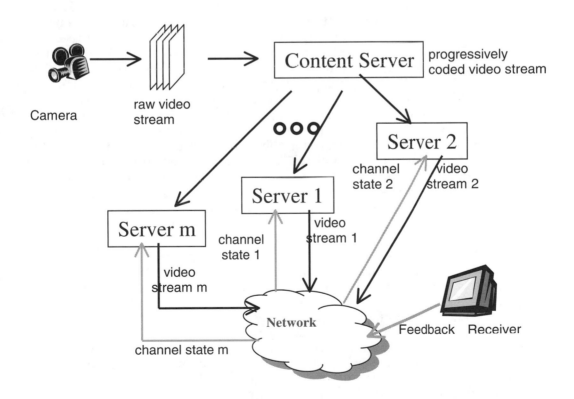

Figure 38.4 VISDOM system block diagram.

The VISDOM approach seeks to deliver a nearly *constant perceptual quality* to the client. Given that the receiver is connected to multiple sources, VISDOM accomplishes this task while *minimizing the net aggregate bandwidth* used from all the sources put together. The striking aspect of VISDOM approach is that the net aggregate bandwidth used in VISDOM is the same as that would be used if content were to be streamed from a single source. In this sense, there is no loss in performance with respect to a centralized system even though the VISDOM system is distributed in nature. In fact, the natural diversity effect that arises because of streaming from multiple sources provides load balancing and fault tolerance automatically. The three main aspects of VISDOM, namely video compression, distributed coding, and networking, are operating in synergism leading to a satisfying end user experience. A block diagram of the system is shown in Figure 38.4.

Video Compression: We use a multi-resolution encoded video bit-stream, such as H.263+ or MPEG4-FGS [6]. The advantage of using such a bit-stream is that a single stream can be served over a dynamic network with varying bandwidths and/or cater to diverse receivers.

Robust (Distributed) Coding: A multi-resolution bit stream, as described above, is sensitive to the position of packet loss, i.e., a loss of more "important" bits early on in the bit stream can render the remaining bits useless from the point of view of decoding. Such a prioritized bit stream is inherently mismatched to the Internet, which offers equal treatment to all packets. So, a transcoding mechanism to transform a prioritized bit stream into an un-prioritized one, in which the decoded quality is only a function of the number of packets received, is required. The MDFEC (Multiple Descriptions through Forward Error Correction codes) algorithm offers a computationally efficient solution to this problem by trading off bandwidth for quality [21]. The MDFEC algorithm is discussed in more details in Section 5.2. Essentially, the algorithm uses Forward Error Correction (FEC)[3] codes of varying strengths to differentially protect (the more important parts of the bit stream are protected more than the less important parts) the video bit stream. The parameters for encoding are decided based on both the state of the transmission channel as well as the rate-distortion characteristics of the source content. Given these two factors, these parameters are optimized so as to give the best end-to-end performance using a computationally efficient, Lagrangian approach based algorithm. The computational ease of the approach enables dynamic adaptation to changes in network/content characteristics.

Figure 38.5 illustrates the operation of the MDFEC framework. The source content that has been encoded in the multi-resolution format in N layers is unequally protected using channel codes of varying strength. The first layer is protected using an n=N, k = 1 channel code, the second layer is protected using n=N, k =2 code and so on. Thus, with the reception of any one packet, quality commensurate with the first multi-resolution layer is decoded and with the reception of any two packets quality commensurate with the first two layers is obtained and so on. Unlike the multi-resolution bit stream where the decoded quality is sensitive to the position of packet loss, in the MDFEC stream the decoded quality is dependent only on the number of losses.

---

[3] FECs introduce redundancy at the encoder by converting **k** message symbols into **n>k** code symbols. This enables the decoder to recover the input **k** message symbols correctly even if there are some losses in transmission and only some **m** $\geq$ **k** code symbols are received.

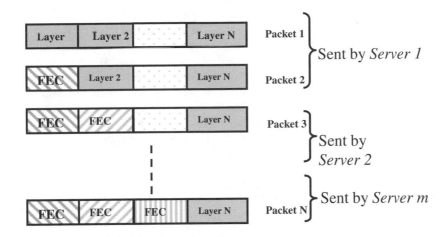

Figure 38.5 MDFEC: The arrangement of FECs and video layers enables successful decoding of layer 1 with reception of any one packet and layers 1 and 2 with the reception of any two packets and so on. The amount of data and protection for each layer can be matched to the transmission channel at hand.

While MDFEC was originally designed for a single channel between the sender and the receiver, an interesting property of this algorithm is that it is inherently distributed. There is a notion of an "aggregate" channel between the client and all the servers, and in this case the MDFEC algorithm is applied to this aggregate channel. How this is achieved in a manner that the servers do not need to communicate with each other is explained in the next paragraph.

Networking: Streaming real-time content from multiple senders to the client in a scalable way without any communication between the various remote servers is accomplished through the client, which acts as the synchronization "master" coordinating all the servers. Thus the client is the controlling agent in this framework ensuring that all the servers operate in harmony.

Further, the same idea is used in ensuring the distributed MDFEC encoding at the servers. The client having access to the aggregate channel state conveys that to all the servers. The individual servers use this information along with the content rate-distortion characteristics to run the MDFEC algorithm independently and hence come up with identical encoding strategies. Then based on a pre-decided allocation, which is also conveyed by the client to all the servers, each of the servers send the pre-assigned portion of the MDFEC bit-stream. See Figure 38.5 for a sample allocation of packets. The dynamic, adaptive nature of MDFEC allows us to run the VISDOM algorithm over coarse time scales to adapt to the varying traffic conditions in the network. The system is fault-tolerant since if one of the servers goes down, the graceful quality degradation of MDFEC with the number of packets as well as the dynamic nature of the adaptation absorbs these variations leading to a nearly imperceptible end-user experience.

Simulation Results: The validity of the VISDOM framework was confirmed in an Internet setting where a content client was placed at University of California, Berkeley and three content servers at University of Illinois, Urbana-Champaign, Rice University, Houston, Texas and EPFL (Ecole Polytechnique Federale de

Lausanne) Lausanne, Switzerland respectively. The Football video sequence was streamed to the client in three settings: single server, two servers, and three servers. Figure 38.6 (a),(c),(e) show the delivered picture quality measured in PSNR (dB) as a function of time for the single, two, and three server case respectively. Figure 38.6 (b), (d), (f) show the total transmission rate measured in Kbytes/sec as a function of time for the single, two and three server case. We notice from these plots that for nearly the same transmission rate, the delivered quality becomes more and more smooth as the number of servers is increased. This highlights the diversity effect of using multiple servers.

3.3   ADAPTIVE/DYNAMIC STORAGE AND CACHING MANAGEMENT

The storage or caching operations in current database systems—centralized or distributed—are performed on a binary level; either some data is stored/cached or it is deleted. There is no concept of trading in gradual degradation for storage efficiency. For multimedia data such as video "gradual" states of data storage are much more suitable. Media data is bulky but possesses the unique property that it is *malleable*.

The MAPS Enhanced Media Storage module offers a novel cache management strategy for media data by going beyond binary caching decisions. It exploits transcoding of media data to achieve a higher utilization of cache space. Less frequently accessed media data is cached at a lower quality level in order to consume less cache space. Consequently, more data will fit into the cache, which in turn will lead to a higher cache hit rate.

For every cache entry, a recall value is maintained indicating its popularity. An entry's recall value is calculated based on its access pattern, using heuristics with the following properties:
- The more often a media cache entry is accessed, the larger its recall value.
- The recall value of a cache entry that is never accessed approaches zero over time, and will thus be deleted if the cache runs out of space.
- Often-recalled data will have a higher recall value than newly accessed data.

The cache management system uses entries' recall values to rank the media data in descending order (see Figure 38.7) and to assign class numbers to them according to a local mapping table (see Table 38. 2). The class number of a data entry determines its maximum allowed bitrate. If the compression ratio of a cached entry is higher than its allowed bit rate, the entry is added to a background transcoding queue.

The amount of transcoding possible at each cache node in a distributed storage system depends on the computational capabilities of the node as well as its current load. Thus, different transcoding tables and cache-partitioning tables should be available at each node for different workloads as well as for nodes with different performance. For instance, a very slow node should not use any transcoding, but instead support only a binary caching decision. Consequently, there would be only one data class assigned to the whole cache. A powerful node, on the other hand, can use a very fine granularity scale transcoding scheme.

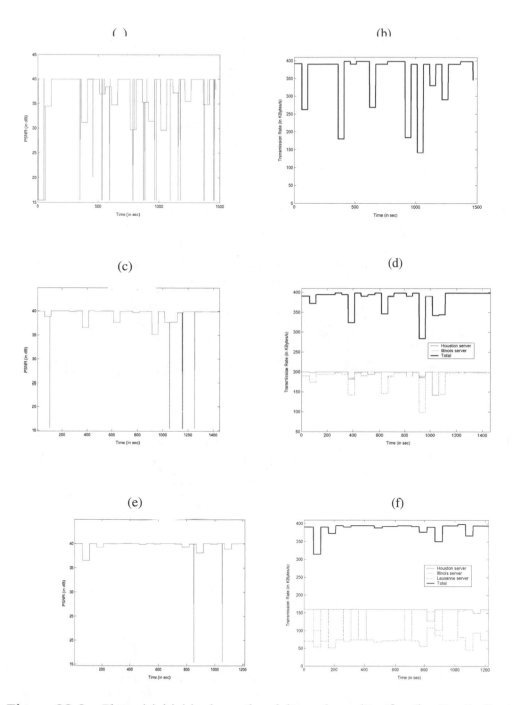

Figure 38.6 Plots (a),(c),(e) show the delivered quality for the Football video sequence (measured in PSNR (dB)) as a function of time for one, two and three servers respectively. Plots (b),(d),(f) show the total transmission rate as a function of time for one, two and three servers respectively. Even though the transmission rate is nearly the same for the three cases, the delivered quality becomes more and more smooth as the number of content servers is increased.

Table 38.2   Mapping data to data classes based on its position in the cache. The position is determined by the data's recall value.

Class	From /To
0	$[0, N_0]$
1	$]N_0, N_1]$
2	$]N_1, N_2]$
...	...
N	$]N_{N-1}, N_N]$
N+1	$] N_N, N_{N+1}]$

Table 38. 3 gives an example of a possible mapping from class numbers to target compression qualities.

Table 38.3   Mapping data class numbers to best allowed compression quality.

Class	Video Target Quality	Audio Target Quality
0	Unchanged original quality	Unchanged original quality
1	MPEG4 320x240 at 1.5 Mbits/s	MP3 128Kbits/s
2	MPEG4 320x240 at 700 Kbits /s	MP3 96Kbits/s
...	...	...
N	MPEG4 160x120 at 100 Kbits/s	MP3 mono 11.2 KHz at 16 Kbits/s

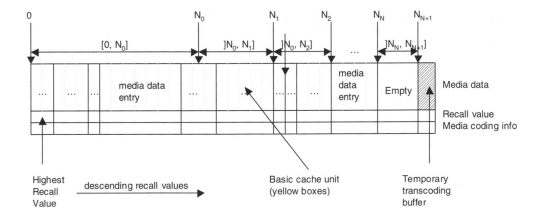

Figure 38.7   Media data ordering in the adaptive/dynamic storage and caching management system based on recall values. The total cache size is N_{N+1}.

4. DISTRIBUTED SEARCH

Video — a composition of moving pictures and audio clips — possess many more dimensions of searchable attributes than text or web documents. Consequently, a video-centric P2P system must allow users to search for video based on a large and diverse set of syntactic and semantic attributes. While costly manual annotations may exist in special — but rare — cases, a video-catering peer service must be able to extract information automatically from media content to assist search, browsing, and indexing.

MAPS Distributed Search subsystem therefore provides
- A toolbox to extract syntactic and semantic attributes automatically from media (including MPEG-7 audio and video descriptors), and
- A powerful search mechanism beyond hash keys, file names and keywords.

Enhanced search specification: For search, the MAPS Media Search subsystem allows each node to have different search capabilities. Besides standard search methods based on keyword matches, SQL queries are used. If a node does not support SQL queries, it either ignores the request or maps it automatically to a keyword-based search.

Automatic extraction of Meta-descriptions: The MAPS Distributed Search subsystem exploits two sources for acquiring Meta information:
- Meta information encoded in the media files or the media files accompanying description files,
- Information derived by automatically analyzing the media content.

The most prominent examples of standardized meta information extracted by MAPS from media files are ID3 tags [11] from audio clips, e.g., describing album, title, and lyric information, Exif (Exchangeable Image File) tags from images captured with digital cameras, e.g., describing camera make and model [8], and general information such as frame size, bitrate, duration, and, if available, the date and time of recording [2]. In the future, MPEG-7 media descriptions will be parsed, too.

Figure 38.8   Example of text extraction in videos.

Furthermore, by extracting semantic information automatically from media, our MAPS Video Content Analysis module supports search beyond standard Meta information. Currently, the MAPS Video Content Analysis module incorporates several types of automatic information extraction, for example, visible text in images and video (see Figure 38.8) [13], faces and their positions, key frames for efficient video summarization [12][26], and colour signatures for similarity search (Figure 38.9). The list of media content analysis tools is not exhaustive as automatic content analysis algorithms are ongoing research topics [23]. Many of the state-of-the-art tools can be plugged into the MAPS Video Content Analysis module.

5. DISTRIBUTED DELIVERY MECHANISMS

Once the requested video has been located, the P2P system optimises its delivery. In conventional file sharing systems media data is typically delivered asynchronously using standard network routing methods, and reliable network protocols such as TCP. This, however, may not be the best solution in situations when **real-time** streaming is desired and/or video transcoding is required. In the following sections we describe mechanisms that facilitate the delivery of video data over heterogeneous networks to heterogeneous peers. Section 5.1 describes our novel scheme of dynamic transcoding along a network path, while Section 5.2 focuses on robust video unicast and multicast in wireless LANs.

5.1   OPTIMAL DELIVERY WITH DYNAMIC TRANSCODING

In a network in which data is shared and proliferated, typically multiple copies of a given file will exist throughout the system. When the data of interest is multimedia, in general, and video in particular, a larger array of possibilities arises. For example, different versions of the same video are likely to exist throughout the network at multiple bit rates and resolutions. Furthermore, content can be transcoded or transformed, even on the fly, e.g., to generate versions at new bit rates and resolutions or to incorporate additional redundancy for error resiliency over noisy channels—in other words, to make the content adaptive to client device capabilities, storage capacities, and bandwidth constraints.

In such a scenario, obtaining an exact version of data from a far-away source may be a poorer choice than obtaining a different version of the data from a nearby source and transcoding it into the required version. The "best" means of getting a version of an object (e.g., a video or audio clip) on the network can be estimated by a metric that attempts to rank and prioritize available paths to an object. In this context, a path consists of both a conventional network route and an optional sequence of transformations throughout the network that are necessary to deliver the object to the destination node so as to satisfy the requester's constraints. We will define a cost function to measure the aggregate computational complexity and transmission overhead required to deliver a given object. MAPS attempts to minimize the cost function over the set of possible paths to find the most efficient sequence of operations to employ.

To support cost-function-based optimization, nodes exchange periodic link and node state information through intermittent broadcast updates and as additional information appended to query response packets. Node state information includes an indication of the node's level of activity (based on current reservations and any queued best-effort requests), its transcoding capabilities (reflective of computational power), and an advertised cost associated with each transformation type. Each node assigns its own values to transformations' costs. The cost function may therefore vary from node to node, e.g., a node may penalize computation more than storage depending on its capabilities, or vice versa. Since transcoding is potentially a lengthy operation, MAPS uses a resource-reservation type approach, so that nodes do not become overloaded with concurrent requests.

Figure 38.9   Sample results of color-based similarity search. Similar videos are showed in small icons surrounding the displaying video in the center.

The cost of the most efficient sequence of operations to obtain a version of an object O satisfying some set of required parameter constraints P at a given node N_D can be expressed (recursively) as follows:

$$c(N_D, O, P) = \min_j \min_k \{ c(N_k, O, P_j) + c_t(N_k, N_D, O, P_j) + c_x(N_D, O, P, P_j) \},$$

where $\{ N_k \}$ is the set of "neighbouring" nodes (viz. peer nodes from which the object may be delivered to N_D, including itself), $\{ P_j \}$ is the set of possible constraints for which a version of the object can be delivered to the local node N_D from its neighbouring nodes, $c_t(N_k, N_D, O, P_j)$ is the cost of transmission of the intermediate object from the neighbouring node N_k to the local node N_D, and $c_x(N_D, O, P, P_j)$ is the cost of locally converting a version of O possessing initial parameters P_j to one that satisfies the node's target constraints P.

For example, if a node N_D has a version of the object that satisfies its target constraints P, then the cost of retrieving it, $c(N_D, R, P)$, is zero; $c_t(N_D, N_D, R, P)$ is zero since no transmission is required, and $c_x(N_D, R, P, P)$ is zero since $P=P_j$ making transcoding unnecessary. On the other hand, if N_D does not have a version of the resource that satisfies the constraints P, then the cost $c(N_D, R, P)$ can include a transcoding cost at node N_D, the cost of transmission from a peer, as well as the cost of any similar operations (i.e. other transcoding and transmissions) required on its neighbours to obtain the needed object. The best way to obtain a media object can be determined by minimizing the preceding cost function.

Figure 38.10 shows how path selection is determined for an example where a node Peer A requests an MPEG-4 version of a stream ("example") from the network. The requesting node constructs a directed graph representing the relationships between participating nodes (viz. nodes containing the requested stream, and nodes capable of transforming the content to the needed form), and their advertised costs for carrying out any needed transformations. Transmission and transcoding costs are expressed as edge weights, while peer nodes and content transformations are mapped to vertices. While Peer B transcodes "example" from MPEG-2 to MPEG-4, a virtual vertex (Peer B) is created.

In reality, minimizing the proceeding cost function is computationally infeasible. Each peer node in the network can potentially be used to process different versions of the object. The graph of the problem will become more complicated if continuous transcoding of different bitrates is allowed. In practice, we simplify the formulation by some assumptions. We trade the optimality of the cost with the computational complexity of evaluating the formulation. To find a computationally feasible solution to the problem of transcoding and routing of multimedia data we assume that the transcoding, if needed, is performed firstly in a single peer on the network, and, secondly, in only one peer. Thus, for a network consisting of n peers, there are only n possible transcoding positions. For each transcoding position N_k, we calculate the total cost of taking the route as

$$c^k(N_D,O,P) = \min_j \{c(N_k,O,P_j) + c_x(N_k,O,P,P_j)\} + c_t(N_k,N_D,O,P),$$

where the delivery cost from Node N_k to Node N_D $C_t(...)$ can be found by using well-known network routing optimisation via dynamic programming. In other words, we simplify the optimization problem to two network routing problems – one from the source to the transcoder and the other one from the transcoder to the destination.

After we have the cost of transcoding at each note, we minimize the overall cost by:

$$c(N_D,O,P) = \min_k \{c^k(N_D,O,P)\},$$

i.e., after solving at most n dynamic programming problems, the optimal path can be determined. This approach is possible in the situations where N is not restrictively large. Note that most of the times only a limited number of nodes in the peer network are able to actually handle transcoding efficiently, and, therefore, the actual number of transcoding nodes is significantly smaller than N.

The path MAPS uses to deliver a video is determined by minimizing the preceding cost function. One point to note here is that although "cost" is advertised based on a local perspective by peers, the cumulative cost of an operation is inherently global in nature as it spans multiple nodes. The assumption is that peers behave "fairly," by trading individual for global benefits.

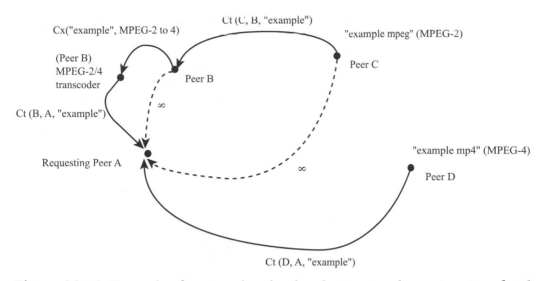

Figure 38.10 Example of transcoding/path selection graph construction for delivery of a video stream "example" to a client Peer A possessing only an MPEG-4 decoder. The content is available in MPEG-2 format from Peer C, or directly as an MPEG-4 sequence from Peer D. The requesting node compares the cost of obtaining the clip along several possible paths: directly from D, which involves a cost $c_t(D, A)$ of transmission and from C to B to A, which involves the cost of transmission, $c_t(C, B) + c_t(B, A)$, plus the cost of transcoding at B, $c_x("example", MPEG\text{-}2\ to\ MPEG\text{-}4)$. Because Peer A can only decode MPEG-4, it is useless to get an MPEG-2 version of the object to Peer A. In this case, which cannot satisfy the required constraints, some edges are marked with infinite cost.

5.2 ROBUST VIDEO UNICAST AND MULTICAST IN WIRELESS LANS

With wireless networks gaining prominence and acceptance, especially the LANs based on the IEEE 802.11 standards, it is foreseeable that wireless streaming of audio/video will be a critical part of the P2P infrastructure. Therefore, we investigate in detail the specific features of real-time video streaming in wireless LANs.

There are two major challenges for video streaming over wireless LANs: (1) fluctuations in channel quality, and (2) higher bit/packet error rates compared with wired links. In addition to wireless channel-related challenges, when there are multiple clients, we have the problem of heterogeneity among receivers, since each user will have different channel conditions, power limitations, processing capabilities, etc., and only limited feedback channel capabilities. (This is the case when, for example, multiple peers request a resource that is a real-time streaming.) In the following we analyze both the single-user and multi-user cases in detail, and propose practical solutions to address the aforementioned challenges.

There has been a substantial amount of prior work in the area of video streaming. The single user scenario has been addressed by a number of researchers (see, e.g., [1] [20] [3] and references therein). In the following we introduce a new method that combines the reliability and fixed delay advantages of Forward Error Control (FEC) coding with the bandwidth-conserving channel-adaptive properties of Automatic Repeat ReQuest (ARQ) protocol. In a multicast scenario, to tackle the problem of heterogeneity and to ensure graceful quality degradation the use of multi-resolution based scalable bit streams has been previously suggested in [19][25]. However, such a bit stream is sensitive to the position of packet loss, i.e., the received quality is a function of which packets are erased. To overcome this problem, the Priority Encoding Transmission scheme was proposed in [1], which allows for different resolution layers to be protected by different channel codes based on their importance. Substantial work has been done towards finding an algorithm for optimizing the amount of parity bits used to protect each resolution layer [20][21]. A near-optimal, O(N) (where N is the number of packets) complexity, algorithm was proposed in [21] to solve this problem.

5.2.1 802.11 Wireless LAN as Packet Erasure Channel

To address the problem of wireless video streaming we need to model the communication channel. The IEEE 802.11 Media Access Control (MAC)/Logical Link Control (LLC) and physical (PHY) layers represent two lower layers in the Open System Interconnect (OSI) reference model, i.e., the Data link and Physical layers. In real life, we do not have direct access to the Physical (or even MAC) layer. Further, most of the successful wireless networks adopt the Internet Protocol (IP) as a network layer simplifying the integration of wireless networks into the Internet networks. In this scenario, user applications see the wireless channel as an IP packet channel with erasures — much like the wired Ethernet. Therefore, in designing our algorithms (which run at the application layer) we model the wireless network channel as a packet erasure channel at the network layer level. There is a very close connection between the problem discussed in Section 3.2 and the problem we consider here. In fact having multiple unreliable servers streaming data to a client is essentially analogous to having a single server sending information over a packet erasure channel. It is not a surprise, therefore, that we use a similar MDFEC-based solution for streaming over wireless LANs.

In order to reliably communicate over packet erasure channels, it is necessary to exert some form of error control. Asynchronous communication protocols, such as ARQ, are reliable but have unbounded delay. In synchronous protocols, the data are transmitted with a bounded delay but generally not in a channel adaptive manner. To provide for some measure of reliability, Forward Error Control coding is employed (see also S=section 3.2). If the number of erased packets is less than the decoding threshold for the FEC code, the original data can be recovered perfectly. However, FEC techniques cannot guarantee that the receiver receives all the packets without error. Popular Reed-Solomon (RS) codes are described by two numbers **(n,k)**, where **n** is the length of the codeword and **k** is the number of data symbols in the codeword. Each symbol is drawn from a finite field of 2^s elements, where **s** (we use **8**) is the number of bits to be represented in each symbol. The total number of words in the code equals $2^s - 1$. RS codes can be used to correct errors, erasures, or both. Particularly efficient decoding algorithms based on Vandermonde matrices [22] exist if only erasures are to be corrected. In this case, each parity symbol can correct <u>any</u> one missing data symbol. This means that we can recover the original codeword, and hence the original data, if at least **k** of the original **n** symbols is received.

5.2.2 Robust Video Unicast over Wireless LANs

In a unicast scenario there is only one recipient of the video data. The cost function to be optimized is a reconstruction distortion subject to rate and delay constraints. The first approach we describe is the purely FEC-based MDFEC algorithm [21], which assumes as input a scalable video bit stream. On the other hand, the Hybrid ARQ protocol [17] is a combination of FEC and ARQ techniques and does not assume a scalable video bit stream as input, and therefore is readily applicable to existing video content stored on DVDs and VCDs. The problem we address in this section is that of finding the parameters for source and channel coding schemes for a single server and a single client, to maximize the overall data quality (or, equivalently, minimize the distortion) subject to a communication delay constraint.

MDFEC

MDFEC is a transcoding mechanism to convert a prioritized multiresolution bit stream into a non-prioritized multiple description bit stream (see Figure 38.11) using efficient FEC codes.

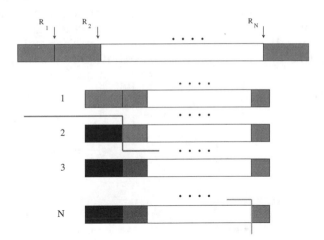

Figure 38.11 Mapping progressive stream to packets using MDFEC codes. Progressive stream (top) is partitioned by R_i and encoded with (N,i) codes. R_i corresponds to layer i in Figure 38.5.

Let **d** be an N-dimensional distortion vector (also called the distortion profile) where d_k reflects the distortion attained when k out of N packets are received. The progressive bit stream is marked at N different positions (that form N resolution layers), which correspond to achieving the distortion levels d_k as shown in Figure 38.11. The ith resolution layer is split into i equal parts and an (N,i) RS code is applied to it to form the N packets as shown in Figure 38.11. Since every packet contains information from all the N resolution layers, they are of equal priority. The RS code ensures that the ith resolution layer can be decoded on the reception of at least i packets. Since the distortion-rate function $D(r)$ for a source is a one-to-one function of the rate r, finding the N dimensional distortion vector **d** corresponds to finding the rate partition $\mathbf{R} = (R_1,...,R_N)$ of the multiresolution bit stream (see Figure 38.11). A fast, near-optimal algorithm, of complexity O(N) (where N is the number of packets), based on Lagrangian principles, to solve this problem is described in [21].

Hybrid ARQ

MDFEC method is an attractive solution but it requires progressive video input. Here we propose a way to combine the ARQ and FEC error control methods to improve the performance of unicast communications of single resolution video over packet erasure channels. Hybrid ARQ schemes have been extensively studied in the literature for various communication channels. We do not attempt to survey all of them (the reader is referred to [15] for a textbook treatment) but rather we propose a scheme that specifically addresses the problem of video streaming over 802.11 networks. The idea is illustrated in Figure 38.12. We start by splitting our multimedia data into "packet groups," consisting of k packets each, and then, for each packet group, append n-k RS parity packets to the group as in the FEC coding scheme described above. However, unlike in the pure FEC scheme, we initially send only the first k data packets to the receiver. Then transmitter starts sending parity packets until one of the following two events occurs: either an acknowledgment from the receiver arrives, or the deadline for the transmission is reached. Once at least k packets are received intact, the receiver sends an acknowledgment. Once the acknowledgment is received, the transmitter continues with the next k data packets. One significant advantage of this algorithm is that it does not break down even when acknowledgments are lost. Instead, the transmitter simply assumes that more parity is needed.

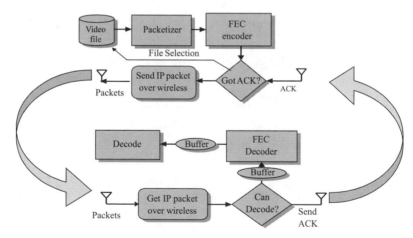

Figure 38.12 Hybrid ARQ algorithm data flow.

The Hybrid ARQ scheme is a general algorithm and can be adjusted to fit specific cases as is appropriate. For further details on Hybrid ARQ algorithm and its theoretical performance evaluation we refer to [17].

5.2.3 Robust Video Multicast over Wireless LANs

ARQ-based schemes are less appropriate for the multicast case for two reasons: ACK explosions and the requirement to retransmit different packets to all users. For significant packet loss rates, each user will require frequent packet replacement, and different users are most likely to require different packets. To respond to requests by multiple users, we may have to resend a significant fraction of the original data even for small loss rates. However, for small multicast networks, the hybrid ARQ scheme can alleviate the problem of sending different correction packets to each user. Because each parity packet can replace any missing data packet, there is no need for each user to identify which packet is missing. Instead, each user can simply transmit an acknowledgment when it receives enough parity to decode the transmitted data. When acknowledgment packets have been received from all known multicast users, the transmitter can move on to the next packet group.

An alternative attractive solution is based on the MDFEC approach. In order to address the multiuser setup we propose to minimize the following criterion: $\Delta(\mathbf{R}) = \max_i (E[d_i(\mathbf{R})] - E[d_i]_{min})$, where $\mathbf{R}$ is the rate partition as defined above, $E[d_i]_{min}$ is the minimum expected distortion for the i-th client, achieved by using the optimal coding scheme when it is the only client, and $E[d_i(\mathbf{R})]$ is the expected distortion for the particular coding scheme being used. Such an overall quality criterion is fair in the sense that it minimizes the maximum penalty that any client suffers. It then can be shown that the problem of finding the optimal partitions $\mathbf{R}$ is the problem of minimizing the intersection of convex surfaces [17]. For example, in a two-client case our solution is to use the single user version of MDFEC algorithm for each client, and then find the lowest intersection of two distortion-rate surfaces (convex) as the point of equal disappointment. Similarly, for more than two users we can perform the outlined optimization pairwise and select the lowest point among all possible pairs.

To actually find the rate partition that maximizes the overall quality criterion, we use a simplex optimization method although other standard techniques may be used. While the computational complexity of the simplex algorithm may be high, we do not expect that the algorithm will be run for a very large number of users. This is due to the fact that the algorithm runs at each access point in the wireless LAN network and we do not expect many users present in the vicinity of any access point.

Table 38. 4 shows the difference in the MSE penalty ($\Delta(\mathbf{R})$) for users with packet erasure channels having different probabilities p_i for the case when the parameters were optimized for the worst channel case and for the proposed minimax criterion. In the first case the user 1 suffers significant quality degradation (compared to a system optimized for user 1 only), while in the second case users 1 and 3 "share" the largest degradation in MSE.

Table 38.4   Comparison of penalty in distortion for three users for MDFEC designed for the worst-case channel, and for MDFEC designed for the minimax cost.

user	p_i	$E[d_i]_{min}$	Worst case MSE penalty	Minimax MSE penalty
1	0.113	137.5	32.3	7.0
2	0.156	162.9	14.9	0.3
3	0.197	193.5	0	7.0

To summarize this section, we point out that media streaming over wireless networks in a single or multiple user scenario poses specific problems to P2P delivery mechanisms that can be efficiently addressed by channel coding for packet erasures (FEC, ARQ, and their combinations) and progressive source coding.

6. QUALITY-OF-SERVICE MONITORING AT CLIENT

In the previous sections we described how the delivery of media data can be efficiently and robustly delivered over distributed networks. However, to many end-users (clients) the only thing that matters is the quality-of-service (QoS) of the received video. QoS can be in the form of perceived quality of the received content, the response time, the latency/delay, and the playback experiences. For example, received content quality can vary extensively depending on a number of factors, including encoder sophistication, server load, intermediate transformations (re-encoding and transcoding), and channel characteristics (packet loss, jitter, etc).

The optimized delivery and robust wireless streaming mechanisms proposed in Sections 5.1 and 5.2 are aimed to enhance QoS from system resource optimization perspectives, taking into account the practical constraints such as unreliable and error-prone channels. However, in order to address QoS end-to-end, we need to evaluate the QoS at the client side to validate the results. This is especially important in the events of paid delivery services to guarantee the QoS at the receiving end.

6.1   QUALITY METRICS FOR VIDEO DELIVERY

For QoS investigations, many in the networking research communities have relied on metrics such as bit and/or packet erasure characteristics. It is more realistic to investigate errors in the form of packet drops if we focus on video delivery over common TCIP/IP infrastructure. Unlike ASCII data, multimedia data, when encoded (compressed), can pose special challenges when the packets are dropped – the degradation of the quality is not uniform. The image/video quality can vary significantly based on loss packets – some packets would render a significant portion of the image fail to display, while others would only cause slight degradation to some small locally confined pixel areas. The quantitative measurements like bit/packet erasure rates and variances are not representative of perceived visual quality, and, of course, QoS in general. to the end users. While scalable encoding/compression aim to solve the fluctuations of quality caused by packet losses, it does not lessen the needs for client side quality monitoring.

While QoS can embrace many different metrics to describe user perceived level of services, we will focus on image/video visual quality in this section.

Figure 38.13 shows the extent of quality fluctuations versus packet erasure measurements for standard encoding and also for scalar (progressive) encoding, in one example image.   We use PSNR (peak-signal-to-noise-ratio) as the objective quality metrics, although other metrics can be used for perceptual quality measurements. Here we simulate the characteristics of packet erasure channels (Section 5.2). In the event that we have an average of 1 packet erasure (1024 bytes) per image (JEPG encoded), the PSNR of the received image frame can vary from under 10dB to close to 20dB – a 10dB differential! Scalable encoding improves quality despite the packet drops, but the PSNR can still show a 5 to 10dB differential for most erasure levels. (Similar characteristics have been observed for other images in our experiments.) This example demonstrates the needs for finding good quality metrics and tools for assessing and monitoring perceived quality (visual in this case) of the received content.

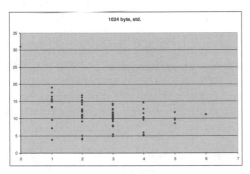
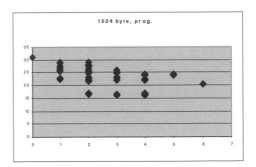

(a) PSNR vs. packet erasure characteristics on test image (Standard compression, packet size = 1024 bytes)

(b) PSNR vs. packet erasure characteristics on test image (Progressive compression, packet size = 1024 bytes)

Figure 38.13 PSNR vs. packet erasure characteristics.

In order to verify that the results of delivery (through a combination of packet loss, transcoding, or re-encoding, among other operations), we need to first focus on how to reliably assess a client's received content quality. For example, distortion assessment tools could be applied to quality-based real-time adaptation of streaming services. A streaming server could increase or decrease the bandwidth/error correction assigned to a stream based on a client's estimated perceptual quality, while given a corrupted frame, a streaming client could decide whether to attempt to render that frame, or instead repeat a previous uncorrupted frame. Similarly, encoders or transcoders could use automated quality monitoring to ensure that certain quality bounds are maintained during content transformations such as those of Section 5.1.

6.2 WATERMARKING FOR AUTOMATIC QUALITY MONITORING

For client side applications, it is often not feasible for the client devices to have access to the original content for quality comparison or assessment purposes. Thus, a general quality assessment scheme should require no use of the original uncorrupted signal. To tackle the challenge, we propose the use of *digital watermarks* as a means of quality assessment and monitoring [10].

Digital watermarking has previously been proposed as a key technology for copyright protection and content authentication. Watermarking considers the problem of embedding an imperceptible signal (the watermark) in a cover signal (most commonly image, video, or audio content) such that the embedded mark can later be recovered and retrieved, generally to make some assertion about the content, e.g., to verify copyright ownership, content authenticity, and authorization of use, among others [7][5]. Typically, a watermark is embedded by perturbing the host content by some key-dependent means, e.g., by additive noise, quantization, or a similar process.

The motivation behind the approach lies in the intuition that if a host signal is distorted or perturbed, a watermark embedded within the host signal can be designed to degrade, and reflect the distortion by the degradation. By embedding known information (the watermark) in the content itself, a detector without access to the original material may be able to deduce the distortions to which the host content has been subjected, based on the distortions the embedded data appears to have encountered.

Figure 38.14 shows an overview of the general approach. If a watermark is designed well, it is possible to make a variety of assertions about the distortion to which the host signal has been subjected. In fact, the watermark can be designed to bear quantitative and qualitative correlations with perceived quality, and other QoS measurements. One approach is to embed perceptual weighting into a host signal (for example, an image) that reflects the human perception of the signals.

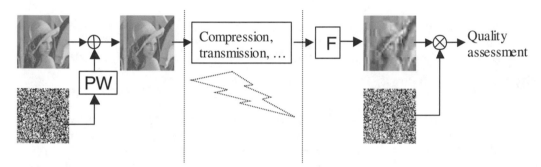

Figure 38.14 The system overview (PW: perceptual weighting; F: filtering)

However, there is a tradeoff. The watermark embedded has to be sufficiently robust, such that small distortions cannot render the watermark to disappear or be undetectable. Yet it has to be sufficiently "fragile" to reflect the quality degradations. To achieve the goals, we propose using the class of robust detectors with bounded distortions, such that the detector response degrades with distortion magnitude, in a predictable manner. The detected results can then correlate with quality (quantitative and/or qualitative) metrics. In addition, the distortions can be localized and quantified with perceptual impacts. We call this "distortion-dependent watermarking."

There are few studies into this area and there are many interesting questions and opportunities for further research. [10] provides some basic problem formulations and investigation results. To illustrate, and as a basic formulation, we define the quality watermark signal w as

$$w_n = g_n(Q(X_n + r_n)),$$

where Q corresponds to a scalar quantizer with step size L, X_n is the nth host signal element, r_n is chosen pseudo-randomly and uniformly over the range [0, L), and $g_n(\circ)$ corresponds to the output of the nth pseudo-random number generator seeded with the specified argument, i.e., quantizer output. The quality watermark signal w and a Gaussian-distributed reference watermark w_{ref} are summed and normalized, and embedded into the content using one of the many watermarking algorithms reported in the literature. This essentially constitutes a pre-processing step that is applied before the content is manipulated and delivered. In [10], a simple spatial-domain white-noise watermark was used as the underlying embedding/detection mechanism. Several related and widely used objective quality metrics can subsequently be estimated from the distorted content, for example, the sum of absolute differences (SAD). From our investigations the predicted distortion closely tracks the actual distortion measured from the experiments. Similar approaches can also be used to construct other perceptually weighted distortion metrics.

To our knowledge the problem of blind signal quality assessment has not been studied extensively, and many more sophisticated solutions are no doubt possible. As digital delivery of content becomes more prevalent in the coming years, we believe that such solutions will collectively become important enabling components in media delivery infrastructure.

7. SUMMARY AND DISCUSSIONS

In this chapter we have presented many important aspects of distributed video management, computing, and delivery systems in a peer-to-peer setting, beyond the existing infrastructure. We, however, believe the issues discussed here, and some of the solutions presented, are just a beginning, rather than a conclusion, of related topics. Many interesting questions are not yet answered or have been extensively studied.

Nonetheless, this is a start. We have introduced novel schemes for multi-rate differential encoding, distributed video coding, and distributed adaptive storage and cache management, for data storage and management. We have also discussion distributed search mechanisms. For content delivery over distributed networks, we have presented new formulations and schemes for optimal dynamic transcoding as well as robust unicast and multicast over wireless LANs. In addition, we have introduced the concept of perceived quality monitoring via embedded watermarks to assess or validate the content delivery results. And more importantly, we have generalized the various components and designed an architecture on top of standard peer-to-peer service layer, and built a prototype system MAPS to illustrate the concepts. The technology components and the integrated system architecture are aimed to enable an array of emerging applications in distributed video management and delivery.

REFERENCES

[1]   A. Albanese, J. Blomer, J. Edmonds, M. Luby, and M. Sudan, Priority encoding transmission, IEEE Trans. Inf. Theory, no. 42, pp. 1737--1744, November 1996.

[2]   AV/C Digital Interface Command Set VCR Subunit Specification. 1394 Trade Association Steering Committee. Version 2.0.1, Jan. 1998.

[3]   P. Chou and Z. Miao, Rate-distortion optimized streaming of packetized media, submitted to IEEE Transactions on Multimedia, 2001.

[4]   I. Clarke, O. Sandberg, B. Wiley, and T.W. Hong, Freenet: A Distributed Anonymous Information Storage and Retrieval System, in Designing Privacy Enhancing Technologies: International Workshop on Design Issues in Anonymity and Unobservability, ed. by H. Federrath. Springer: New York (2001).

[5]   Communications of the ACM, Special Issue on Digital Watermarking, 1998.

[6]   G. Cote, B. Erol, M. Gallant and F. Kossentini, H.263+: Video Coding at Low Bit Rates, IEEE Transactions on Circuits and Systems for Video Technology, vol.8, no.7, pp.849-66, November 1998.

[7]   I. Cox, M. Miller, and J. Bloom, Digital Watermarking, Morgan Kaufmann, 2002.

[8]   EXIF - Exchangeable Image File tag specification. http://www.pima.net/standards/it10/PIMA15740/exif.htm.

[9]   L. Gong, JXTA: A Network Programming environment, IEEE Internet Computing, pp. 88-95, May 2001.

[10]   M. Holliman and M. Yeung, Watermarking for automatic quality monitoring, Proceedings of SPIE: Security and Watermarking of Multimedia Contents IV, Volume 4675, pp. 458-469, 2002.

[11]   ID3 tag specification. http://www.id3.org/.

[12]   R. Lienhart, Reliable Transition Detection in Videos: A Survey and Practitioner's Guide, International Journal of Image and Graphics (IJIG), Vol. 1, No. 3, pp. 469-486, 2001.

[13]   R. Lienhart and A. Wernicke, Localizing and Segmenting Text in Images, Videos and Web Pages, IEEE Transactions on Circuits and Systems for Video Technology, Vol. 12, No.4, pp. 256 -268, April 2002.

[14]   R. Lienhart, M. Holliman, Y.-K. Chen, I. Kozintsev, and M. Yeung, Improving Media Services on P2P Networks, IEEE Internet Computing, pp. 73-77, Jan./Feb. 2002.

[15]   S. Lin and D. Costello, Error Control Coding: Fundamentals and Applications, Prentice-Hall, 1983.

[16]   M. Macedonia, Distributed file sharing: barbarians at the gates? IEEE Computer, Vol. 33(8), pp. 99-101, Aug. 2000.

[17]   A. Majumdar, D. Sachs, I. Kozintsev, K. Ramchandran and M. Yeung, Multicast and unicast real-time video streaming over wireless LANs, IEEE Transactions on Circuits and Systems for Video Technology, Volume: 12 Issue: 6 , Jun 2002, Page(s): 524 -534.

[18]   A. Majumdar, R. Puri and K. Ramchandran, Rate-Distortion Efficient Video Transmission from Multiple Servers, International Conference on Multimedia and Expo (ICME), Lausanne, Switzerland, August 2002.

[19]   S. McCanne, V. Jacobson, and M. Vetterli, Receiver-driven layered multicast, ACM SIGCOMM 96, 1996.

[20]   A. E. Mohr, E. A. Riskin, and R. E. Ladner, Graceful degradation over packet erasure channels through forward error correction, Proc. Data Compression Conference, Snowbird, UT, Mar. 1999.

[21]   R. Puri, K. W. Lee, K. Ramchandran, and V. Bharghavan, An Integrated Source Transcoding and Congestion Control Framework for Video Streaming in the Internet, IEEE Transactions on Multimedia, vol. 3, no. 1, pp. 18--32, March, 2001.

[22]   L. Rizzo, On the feasibility of software FEC, DEIT Technical Report LR-970131. Available at http://www.iet.unipi.it/~luigi/softfec.ps.

[23]   Video Content Analysis Homepage. http://www.videoanalysis.org.

[24]   S. Tilley and M. DeSouza, Spreading Knowledge about Gnutella: A Case Study in Understanding Net-centric Applications, in Proceedings of the 9th International Workshop on Program Comprehension, pp. 189 –198, 2001.

[25]   D. Wu, T. Hou, and Y.-Q. Zhang, Scalable video transport over wireless IP networks, IEEE International Symposium on Personal, Indoor and Mobile Radio Communication, Sept. 2000, pp. 1185-1191.

[26]   B.-L. Yeo and B. Liu, Rapid Scene Analysis on Compressed Video, IEEE Transactions on Circuits and Systems for Video Technology, Vol. 5, No. 6, Dec. 1995.

[27]   A. Zaccarin & B.L. Yeo, Multi-rate encoding of a video sequence in the DCT domain, Proceedings of IEEE Int. Symposium on Circuits and Systems, Phoenix, AZ, USA. May 2002.

39

VIDEO COMPRESSION: STATE OF THE ART AND NEW TRENDS

Luis Torres

Department of Signal Theory and Communications
Technical University of Catalonia
Barcelona, Spain
`luis@gps.tsc.upc.es`

Edward Delp

Purdue University
School of Electrical Engineering
West Lafayette, IN, USA
`ace@ecn.purdue.edu`

1. INTRODUCTION

Image and video coding is one of the most important topics in image processing and digital communications. During the last thirty years we have witnessed a tremendous explosion in research and applications in the visual communications field. The field is now mature as is proven by the large number of applications that make use of this technology. Digital Video Broadcasting, Digital Versatile Disc, and Internet streaming are only a few of the applications that use compression technology. There is no doubt that the beginning of the new century revolves around the "information society." Technologically speaking, the information society will be driven by audio and visual applications that allow instant access to multimedia information. This technological success would not be possible without image and video compression. Image and video standards have played a key role in this deployment. The advent of coding standards, adopted in the past years, has allowed people around world to experience the "digital age." Each standard represents the state of the art in compression at the particular time that it was adopted. Now is the time to ask: are there any new ideas that may advance the current technology? Have we reached a saturation point in image and video compression research? Although the future is very difficult to predict, this chapter will try to provide a brief summary of the state of the art and an overview to where this exciting area is heading. Section 2 presents a brief summary of the technology related to present still image and video compression standards. Further developments in the standards will also be presented.

Section 3 presents ideas as to how compression techniques will evolve and where the state of the art will be in the future. We will also describe new trends in compression research such as joint source/channel coding and scalable compression. Section 4 will introduce preliminary results of face coding in which a knowledge-based approach will be shown as a promising technique for very low bit rate video coding. Section 5 describes media streaming which is a new and exciting area for compression research.

2. STANDARDS AND STATE OF THE ART

2.1 STILL IMAGE CODING

For many years the Discrete Cosine Transform (DCT) has represented the state of the art in still image coding. JPEG is the standard that has incorporated this technology [1,2]. JPEG has been a success and has been deployed in many applications reaching worldwide use. However, for some time it was very clear that a new still image coding standard needed to be introduced to serve the new range of applications which have emerged in the last years. The result is JPEG2000 that has already been standardized [3,4]. The JPEG2000 standard uses the Discrete Wavelet Transform. Tests have indicated that at low data rates JPEG2000 provides about 20% better compression efficiency for the same image quality than JPEG. JPEG2000 also offers a new set of functionalities. These include error resilience, arbitrarily shaped region of interest, random access, lossless, and lossy coding as well as a fully scalable bit stream. These functionalities introduce more complexity for the encoder. MPEG-4 has a "still image" mode known as Visual Texture Coding (VTC) which also uses wavelets but supports less functionalities than JPEG2000 [5,6]. For a comparison between the JPEG2000 standard, JPEG, MPEG-4 VTC, and other lossless JPEG schemes see [7]. For further discussion on the role of image and video standards see [8].

2.2 VIDEO CODING

During the last ten years, the hybrid scheme combining motion compensated prediction and DCT has represented the state of the art in video coding. This approach is used by the ITU H.261 [9] and H.263 [10], [11] standards as well as for the MPEG-1 [12] and MPEG-2 [13] standards. However in 1993, the need to add new content-based functionalities and to provide the user the possibility to manipulate the audio-visual content was recognized and a new standard effort known as MPEG-4 was launched. In addition to these functionalities, MPEG-4 provides also the possibility of combining natural and synthetic content. MPEG-4 phase 1 became an international standard in 1999 [5,6]. MPEG-4 is having difficulties finding widespread use, mainly due to the protection of intellectual property and to the need to develop automatic and efficient segmentation schemes.

The frame-based part of MPEG-4, which incorporates error resilience tools, is finding its way in the mobile communications and Internet streaming. H.263, and several variants of it [11], is also very much used in mobile communication and streaming and it will be interesting to see how these two standards compete in these applications.

The natural video part of MPEG-4 is also based in motion compensation prediction followed by the DCT; the fundamental difference is that of adding the coding of the object shape. Due to its powerful object-based approach, the use of the most efficient coding techniques, and the large variety of data types that it incorporates, MPEG-4 represents today the state of the art in terms of visual data coding technology [6]. How MPEG-4 is deployed and what applications will make use of its many functionalities is still an open question.

2.3 THE NEW STANDARD JVT/H.26L

The Joint Video Team (JVT) standard development project is a joint project of the ITU-T Video Coding Experts Group (VCEG) and the ISO/IEC Moving Picture Experts Group (MPEG) for the development of a new video coding standard [14], [15]. The JVT project was created in December of 2001 to take over the work previously under way in the ITU-T H.26L project of VCEG and create a final design for standardization in both the ITU-T and MPEG [16].

The main goals of the JVT/H.26L standardization effort are the definition of a simple and straightforward video coding design to achieve enhanced compression performance and provision of a "network-friendly" packet-based video representation addressing "conversational" (i.e., video telephony) and "non-conversational" (i.e., storage, broadcast, or streaming) applications. Hence, the JVT/H.26L design covers a Video Coding Layer (VCL), which provides the core high compression representation of the video picture content, and a Network Adaptation Layer (NAL), which packages that representation for delivery over each distinct class of networks. The VCL design has achieved a significant improvement in rate-distortion efficiency - providing nearly a factor of two in bitrate savings against existing standards. The NAL designs are being developed to transport the coded video data over existing and future networks such as circuit-switched wired networks, MPEG-2/H.222.0 transport streams, IP networks with RTP packetization, and 3G wireless systems.

A key achievement expected from the JVT project is a substantial improvement in video coding efficiency for a broad range of application areas. The JVT goal is a capability of 50% or greater bit rate savings from H.263v2 or MPEG-4 Advanced Simple Profile at all bit rates. The new standard will be known as MPEG-4 part 10 AVC and ITU-T Recommendation H.264.

2.3.1 Technical Overview

The following description of the new JVT/H.26L standard has been extracted from the excellent review presented in [17].

The JVT/H.26L design supports the coding of video (in 4:2:0 chrominance format) that contains either progressive or interlaced frames, which may be mixed together in the same sequence. Each frame or field picture of a video sequence is partitioned into fixed size macroblocks that cover a rectangular picture area of 16×16 luminance and 8×8 chrominance samples. The luminance and chrominance samples of a macroblock are generally spatially or temporally predicted, and the resulting prediction error signal is transmitted using transform coding. The macroblocks are organized in slices, which generally represent subsets of a given picture that can be decoded independently. Each macroblock can be transmitted in one of several coding modes depending on the slice-coding type. In all slice-coding types, two classes of intra-coding modes are supported. In addition to the intra-modes, various predictive or motion-compensated modes are provided for P-slice macroblocks. The JVT/H.26L syntax supports quarter- and eighth-pixel accurate motion compensation. JVT/H.26L generally supports multi-picture motion-compensated prediction. That is, more than one prior coded picture can be used as reference for building the prediction signal of predictive coded blocks.

In comparison to prior video coding standards, the concept of B-slices/B-pictures (B for bi-predictive) is generalized in JVT/H.26L. For example, other pictures can reference B-pictures for motion-compensated prediction depending on the memory management control operation of the multi-picture buffering. JVT/H.26L is basically similar to prior coding standards in that it utilizes transform coding of the prediction error signal. However, in JVT/H.26L the transformation is applied to 4×4 blocks unless the Adaptive Block size Transform (ABT) is enabled. For the quantization of transform coefficients, JVT/H.26L uses scalar quantization. In JVT/H.26L, two methods of entropy coding are supported.

2.3.2 JVT/H.26L Tools

The following tools are defined in JVT/H.26L:

Table 39.1 Tools supported in JVT/H.26L

Coding tools		Baseline Profile	Main Profile
Picture formats	Progressive pictures	X	X
	Interlaced pictures	Level 2.1 and above	Level 2.1 and above
Slice/ picture types	I and P coding types	X	X
	B coding types		X
	SI and SP coding types		
Macro block prediction	Tree-structured motion compensation	X	X
	Intra blocks on 8×8 basis		
	Multi-picture motion comp.	X	X
	1/4-pel accurate motion compensation	X	X
	1/8-pel accurate motion compensation		
Transform coding	Adaptive block size transform		X
Entropy coding	VLC-based entropy coding	X	X
	CABAC		X
In-loop filtering	In-loop deblocking filter	X	X

2.3.3 JVT/H.26L Comparisons with Prior Coding Standards

For demonstrating the coding performance of JVT/H.26L, the new standard has been compared to the successful prior coding standards MPEG-2 [13], H.263 [10], and MPEG-4 [5], for a set of popular QCIF (10Hz and 15 Hz) and CIF (15Hz and 30Hz) sequences with different motion and spatial detail information. The QCIF sequences are: Foreman, News, Container Ship, and Tempete. The CIF sequences are: Bus, Flower Garden, Mobile and Calendar, and Tempete. All video encoders have been optimised with regards to their rate-distortion efficiency using Lagrangian techniques [18], [19]. In addition to the performance gains, the use of a unique and efficient coder control for all video encoders allows a fair comparison between them in terms of coding efficiency. For details see [17]. Table 39.2 shows the average bitrate savings of JVT/H.26L with respect to MPEG-4 ASP (Advanced Simple Profile), H.263 HLP (High Latency Profile), and MPEG-2.

Table 39.2 Average bitrate savings of JVT/H.26L

Coder	MPEG-4 ASP	H.263 HLP	MPEG-2
JVT/H.26L	38.62%	48.80%	64.46%
MPEG-4 ASP	-	16.65%	42.95%
H.263 HLP	-	-	30.61%

2.4 WHAT CAN BE DONE TO IMPROVE THE STANDARDS?

Can something be done to "significantly" improve the performance of compression techniques? How will this affect the standards? We believe that no significant improvements are to be expected in the near future. However, compression techniques that require new types of functionalities driven by applications will be developed. For example, Internet applications may require new types of techniques that support scalability modes tied to the network transport (see JVT/H.26L standard for some of this). We may also see proprietary methods developed that use variations on standards, such as the video compression technique used by RealNetworks, for applications where the content provider wishes the user to obtain both the encoder and decoder from them so that the provider can gain economic advantage.

2.4.1 Still Image Coding

JPEG2000 represents the state of the art with respect to still image coding standards. This is mainly due to the 20% improvement in coding efficiency with respect to the DCT as well as the new set of functionalities incorporated. Non-linear wavelet decomposition may bring further improvement [20]. Other improvements will include the investigation of color transformations for color images [20] and perceptual models [22]. Although other techniques, such as fractal coding or vector quantization have also being studied, they have not found their way into the standards. Other alternate approaches such as "second generation techniques" [23] raised a lot of interest for the potential of high compression ratios. However, they have not been able to provide very high quality. Second generation techniques and, in particular, segmentation-based image coding schemes have produced a coding approach more suitable for content access and manipulation than for strictly coding applications. These schemes are the basis of MPEG-4 object-based schemes.

There are many schemes that may increase the coding efficiency of JPEG2000. But all these schemes may only improve by a small amount. We believe that the JPEG2000 framework will be widely used for many applications.

2.4.2 Video Coding

All the video coding standards based on motion prediction and the DCT produce block artifacts at low data rate. There has been a lot of work using post-processing techniques to reduce blocking artifacts [24], [25], [26]. A great deal of work has been done to investigate the use of wavelets in video coding. This work has taken mainly two directions. The first one is to code the prediction error of the hybrid scheme using the DWT [27]. The second one is to use a full 3-D wavelet decomposition [28], [29]. Although these approaches have reported coding efficiency improvements with respect to the hybrid schemes, most of them are intended to provide further functionalities such as scalability and progressive transmission. See the later developments in fully scalable 3D subband coding schemes in [30], [31].

One of the approaches that reports major improvements using the hybrid approach is the one proposed in [32]. Long-term memory prediction extends motion compensation from the previous frame to several past frames with the result of increased coding efficiency. The approach is combined with affine motion compensation. Data rate savings between 20 and 50% are achieved using the test model of H.263+. The corresponding gains in PSNR are between 0.8 and 3 dB.

It can be said that MPEG-4 version 10 also known as JVT/H.26L represents the state of the art in video coding. In addition, MPEG-4 combines frame-based and segmentation-based approaches along with the mixing of natural and synthetic content allowing efficient coding as well as content access and manipulation. There is no doubt that other schemes may improve the coding efficiency established in JVT/H.26L but no significant breakthrough is expected. The basic question remains: what is next? The next section will try to provide some clues.

3. NEW TRENDS IN IMAGE AND VIDEO COMPRESSION

Before going any further, the following question has to be raised: if digital storage is becoming so cheap and so widespread and the available transmission channel bandwidth is increasing due to the deployment of cable, fiber optics, and ADSL modems, why is there a need to provide more powerful compression schemes? The answer is, with no doubt, mobile video transmission channels and Internet streaming. For a discussion on the topic see [33], [34].

3.1 IMAGE AND VIDEO CODING CLASSIFICATION

In order to have a broad perspective, it is important to understand the sequence of image and video coding developments expressed in terms of "generation-based" coding approaches. Table 39.3 shows this classification according to [35]. It can be seen from this classification that the coding community has reached third generation image and video coding techniques. MPEG-4 provides segmentation-based approaches as well as model-based video coding in the facial animation part of the standard.

Table 39.3 Image and video coding classification

Coding generation	Approach	Technique
0th Generation	Direct waveform coding	PCM
1st Generation	Redundancy removal	DPCM,DCT DWT,VQ
2nd Generation	Coding by structure	Image segmentation
3rd Generation	Analysis and Synthesis	Model-based coding
4rd Generation	Recognition and reconstruction	Knowledge-based coding
5th Generation	Intelligent coding	Semantic coding

3.2 CODING THROUGH RECOGNITION AND RECONSTRUCTION

Which techniques fall within the "recognition and reconstruction" fourth generation approaches? The answer is coding through the understanding of the content. In particular, if we know that an image contains a face, a house, and a car we could develop recognition techniques to identify the content as a previous step. Once the content is recognized, content-based coding techniques can be applied to encode each specific object. MPEG-4 provides a partial answer to this approach by using specific techniques to encode faces and to animate them [5]. Some researchers have already addressed this problem. For instance, in [36] a face detection algorithm is presented which helps to locate the face in a videoconference application. Then, bits are assigned in such a way that the face is encoded with more quality than the background.

3.3 CODING THROUGH METADATA

If it is clear that understanding the visual content helps provide advanced image and video coding techniques then the efforts of MPEG-7 may also help in this context. MPEG-7 strives at specifying a standard way of describing various types of audio-visual information. Figure 39.1 gives a very simplified picture of the elements that define the standard. The elements that specify the description of the audio-visual content are known as *metadata.*

Once the audio-visual content is described in terms of the metadata, the image is ready to be coded. Notice that what is coded is not the image itself but the description of the image (the metadata). An example will provide further insight.

Let us assume that automatic tools to detect a face in a video sequence are available. Let us further simplify the visual content by assuming that we are interested in high quality coding of a videoconference session. Prior to coding, the face is detected and represented using metadata. In the case of faces, some core experiments in MPEG-7 show that a face can be well represented by a few coefficients, for instance by using the projection of the face on an eigenspace previously defined. The image face can be well reconstructed, up to a certain quality, by coding only a very few coefficients. In the next section, we will provide some very preliminary results using this approach.

Once the face has been detected and coded, the background remains to be coded. This can be done in many different ways. The simplest case is when the background is roughly coded using conventional schemes (1st generation coding). If the background is not important, then it can not even be transmitted and the decoder adds some previously stored background to the transmitted image face.

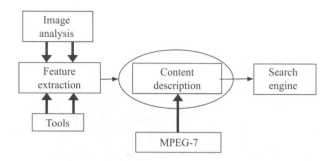

Figure 39.1 MPEG-7 standard.

For more complicated video sequences, we need to recognize and to describe the visual content. If this is available, then coding is "only" a matter of assigning bits to the description of each visual object.

MPEG-7 will provide mechanisms to fully describe a video sequence (in this section, a still image is considered a particular case of video sequence). This means that knowledge of color and texture of objects, shot boundaries, shot dissolves, shot fading and even scene understanding of the video sequence will be known prior to encoding. All this information will be very useful to the encoding process. Hybrid schemes could be made much more efficient, in the motion compensation stage, if all this information is known in advance. This approach to video coding is quite new. For further information see [37], [33].

It is also clear that these advances in video coding will be possible only if sophisticated image analysis tools (not part of the MPEG-7 standard) are developed. The deployment of new and very advanced image analysis tools are one of the new trends in video coding. The final stage will be intelligent coding implemented through semantic coding. Once a complete understanding of the scene is achieved, we will be able to say (and simultaneously encode): this is a scene that contains a car, a man, a road, and children playing in the background. However we have to accept that we are still very far from this 5th generation scheme.

3.4 CODING THROUGH MERGING OF NATURAL AND SYNTHETIC CONTENT

In addition to the use of metadata, future video coding schemes will merge natural and synthetic content. This will allow an explosion of new applications combining these two types of contents. MPEG-4 has provided a first step towards this combination by providing efficient ways of face encoding and animation. However, more complex structures are needed to model, code, and animate any kind of object. The needs arisen in [38] are still valid today. No major step has been made concerning the modeling of any arbitrary-shaped object. For some related work see [39].

Video coding will become multi-modal and cross-modal. Speech and audio will come to the rescue of video (or vice versa) by combining both fields in an intelligent way. To the best of our knowledge, the combination of speech and video for video coding purposes has not yet been reported. Some work has been done with respect to video indexing [40], [41], [42].

3.5 OTHER TRENDS IN VIDEO COMPRESSION: STREAMING AND MOBILE ENVIRONMENTS

The two most important applications in the future will be wireless or mobile multimedia systems and streaming content over the Internet. While both MPEG-4 and H.263+ have been proposed for these applications, more work needs to be done.

In both mobile and Internet streaming one major problem that needs to be addressed is: how does one handle errors due to packet loss and should the compression scheme adapt to these types of errors? H.263+ [43] and MPEG-4 [44] both have excellent error resilience and error concealment functionalities.

The issue of how the compression scheme should adapt is one of both scalability and network transport design. At a panel on the "Future of Video Compression" at the Picture Coding Symposium held in April 1999, it was agreed that rate scalability and temporal scalability were important for media streaming applications. It also appears that one may want to design a compression scheme that is tuned to the channel over which the video will be transmitted. We are now seeing work done in this area with techniques such as multiple description coding [45], [46].

MPEG-4 is proposing a new "fine grain scalability" mode and H.263+ is also examining how multiple description approaches can be integrated into the standards. We are also seeing more work in how the compression techniques should be "matched" to the network transport [47], [48], [49].

3.6 PROTECTION OF INTELLECTUAL PROPERTY RIGHTS

While the protection of intellectual property rights is not a compression problem, it will have impact on the standards. We are seeing content providers demanding that methods exist for both conditional access and copy protection. MPEG-4 is studying watermarking and other techniques. The newly announced MPEG-21 will address this in more detail [50].

4. FACE CODING USING RECOGNITION AND RECONSTRUCTION

This section presents an example of fourth generation video coding techniques, using recognition and reconstruction of visual data. The approach has to be considered as a validation of the fourth generation video coding techniques and to provide a framework worth exploring. It is not intended to provide any complete new video coding scheme. For more details in eigenface coding approaches using adaptive techniques see [53], [54].

The scheme is based on the well-known eigenspace concepts used in face recognition systems, which have been modified to cope with the video compression application. Let us simplify the visual content by assuming that we are interested in the coding of faces in a videoconference session. Let us assume that automatic tools to detect a face in a video sequence are available [56]. Then, some experiments show that a face can be well represented by very few coefficients found through the projection of the face on an eigenspace previously defined. The image face can be well reconstructed (decoded), up to a certain quality, by coding only very few coefficients [51].

Our coding technique is based on a face recognition approach, which has been modified to cope with the coding application [52]. It assumes that a set of training images for *each* person contained in the video sequence is previously known. Once these training images have been found (usually coming from an image database or from a video sequence), a Principal Component Analysis (PCA) is performed for each individual using the corresponding training set of each person. This means that a PCA decomposition for every face image to be coded is obtained. The PCA is done previously to the encoding process.

After the PCA, the face to be coded is projected and reconstructed using each set of different eigenvectors (called eigenfaces) obtained in the PCA stage. If the reconstruction error using a specific set of eigenfaces is below a threshold, then the face is said to match the training image which generated this set of eigenfaces. In this case the recognized face is coded by quantizing only the most important coefficients used in the reconstruction. The size of the coded image has to be previously normalized for PCA purposes and then denormalized at the decoder. It is clear that the corresponding eigenfaces of each person have to be transmitted previously to the decoder. However this can be done using *conventional* still image coding techniques such as JPEG and no significant increment in bit rate is generated.

Figure 39.2 shows five views of the image *Ana* and Figure 39.3 five views of the image *José Mari*. These images come from the test sequences accepted in MPEG-7.

Figure 39.2 Five training views of the image *Ana.*

Figure 39.3 Five training views of the image *José Mari.*

Figure 39.4 shows the original image Ana, the reconstruction of the detected face image Ana using the eigenvectors, and corresponding projected coefficients of the PCA using the training images of Ana and the error done. Figure 39.5 shows the equivalent result for José Mari. Only 5 real numbers have been used to decode the shown images which means a very high compression ratio. The image size is 50x70 and the original image has been encoded using 8 bits/pixel. The compression ratio of the decoded images is 350.

Figure 39.4 Decoded (reconstructed) image *Ana*. Left: original image. Center: reconstructed image. Compression factor 350. Right: Error image.

Figure 39.5 Decoded (reconstructed) image *José Mari*. Left: original image. Center: reconstructed image. Compression factor 350. Right: Error image.

The presented results show that image coding using recognition and reconstruction may be the next step forward in video coding. Good object models will be needed, though, to encode any kind of object following this approach.

5. CONCLUSION

We feel that any advances in compression techniques will be driven by applications such as databases, and wireless and Internet streaming. New semantic-based techniques so far have promised much but have delivered little new results. Much work needs to be in the area of segmentation of video. Fourth generation video coding techniques, using recognition and reconstruction of visual data, may be the next step forward in video coding.

REFERENCES

[1]   ISO/IEC 10918. Information technology – Digital compression and coding of continuous-tone still images. 1994.

[2]   W. B. Pennebaker and J. L. Mitchell, JPEG Still Image Data Compression Standard, Van Nostrand Reihold, New York, 1992.

[3]   ISO/IEC 15444-1:2000. Information technology – JPEG 2000 image coding system – Part 1: Core coding system.

[4]   David S. Taubman, Michael W. Marcellin, JPEG2000: Image Compression Fundamentals, Standards, and Practice. Kluwer International Series in Engineering and Computer Science, November 2001.

[5]   ISO/IEC ISO/IEC 14496-2: 1999: Information technology – Coding of audio visual objects – Part 2: Visual, December 1999.

[6]   F. Pereira, T. Ebrahimi, The MPEG-4 Book. Prentice Hall PTR, July 2002.

[7]   D. Santa Cruz and T. Ebrahimi, A study of JPEG2000 still image coding versus other standards, Proceedings of the European Signal Processing Conference (EUSIPCO), Tampere, Finland, September 5-8, 2000.

[8]   F. Pereira, Visual data representation: recent achievements and future developments, Proceedings of the European Signal Processing Conference (EUSIPCO), Tampere, Finland, September 5-8, 2000.

[9]   ITU-T Recommendation H.261. Video codec for audiovisual services at p x 64 kbit/s. 1993.

[10]   ITU-T Recommendation H.263. Video coding for low bit rate communication. 1998.

[11]   G. Côté, B. Erol, M. Gallant, and F. Kossentini, H.263+: Video coding at low bit rates, IEEE Transactions on Circuit and Systems for Video Technology, vol. 8, no. 7, November 1998.

[12]   ISO/IEC 11172 (MPEG-1). Information Technology – Coding of moving pictures and associated audio for digital storage media up to about 1.5 Mbit/s. 1993.

[13]   ISO/IEC 13818 (MPEG-2). Information Technology – Generic coding of moving pictures and associated audio. ITU-T Recommendation H.262. 1994.

[14]   http://bs.hhi.de/~wiegand/JVT.html

[15]   http://mpeg.telecomitalialab.com/

[16]   ISO/IEC/JTC1/SC29/WG11. MPEG 2001 N4558. Pataya, December 2001.

[17]   H. Schwarz and T. Wiegand, The Emerging JVT/H.26L video coding standard, in Proceedings of IBC 2002, Amsterdam, NL, September 2002.

[18]   T. Wiegand and B. D. Andrews, An Improved H.263 coder using rate-distortion optimization, ITU-T/SG16/Q15-D-13. April 1998, Tampere, Finland.

[19]   G. J. Sullivan and T. Wiegand, Rate-Distortion optimization for video compression, IEEE Signal Processing Magazine, vol. 15, pp. 74-90, November 1998.

[20]   D. Wajcer, D. Stanhill, and Y. Zeevi, Representation and coding of images with nonsepa-rable two-dimensional wavelet, Proceedings of the IEEE International Conference on Image Processing, Chicago, USA, October 1998.

[21]   M. Saenz, P. Salama, K. Shen and E. J. Delp, An evaluation of color embedded wavelet image compression techniques, Proceedings of the SPIE/IS&T Conference on Visual Communications and Image Processing (VCIP), January 23-29, 1999, San Jose, California, pp. 282-293.

[22]   N. S. Jayant, J. D. Johnston and R. J. Safranek, Signal compression based on models of human perception, Proceedings of the IEEE, vol. 81, no. 10, October 1993, pp. 1385-1422.

[23]   L. Torres and M. Kunt, Editors, Video coding: the second generation approach, Kluwer Academic Publishers, Boston, USA, January 1996.

[24]   K. K. Pong and T. K. Kan, Optimum loop filter in hybrid coders, IEEE Transactions on Circuits and Systems in Video Technology, vol. 4, no.2, pp. 158-167, 1997.

[25] T. O'Rourke and R. L. Stevenson, Improved image decompression for reduced transform coding artifacts, IEEE Transactions on Circuits and Systems for Video Technology, vol. 5, no. 6, pp. 490-499, December 1995.

[26] R. Llados-Bernaus, M. A. Robertson and R. L. Stevenson, A stochastic technique for the removal of artifacts in compressed images and video, in Recovery Techniques for Image and Video Compression and Transmission, Kluwer, 1998.

[27] K. Shen and E. J. Delp, Wavelet based rate scalable video compression, IEEE Transac-tions on Circuits and Systems for Video Techno-logy, vol. 9, no. 1, February 1999, pp. 109-122.

[28] C. I. Podilchuk, N. S. Jayant, and N. Farvardin, Three-dimensional subband coding of video, IEEE Transactions on Image Processing, vol. 4, no. 2, pp. 125-139, February 1995.

[29] D. Taubman and A. Zakhor, Multirate 3-D subband coding of video, IEEE Transactions on Image Processing, vol. 3, no. 5, pp. 572-588, September 1994.

[30] Kim, Z. Xiong, W.A. Pearlman, Low Bit-Rate Scalable Video Coding with 3D Set Partitioning in Hierarchical Trees (3D SPIHT), IEEE Trans. on Circuits and Systems for Video Technology, vol. 10, no. 8, pp. 1374-1387, Dec. 2000.

[31] V. Bottreau, M. Bénetière and B. Felts, A fully scalable 3d subband video codec, Proceedings of IEEE International Conference on Image Processing, Thessaloniki, Greece, October 7-10, 2001.

[32] T. Wiegand, E. Steinbach, and B. Girod, Long term memory prediction using affine compensation, Proceedings of the IEEE International Conference on Image Processing, Kobe, Japan, October 1999.

[33] R. Schäfer, G. Heising, and A. Smolic, Improving image compression - is it worth the effort? Proceedings of the European Signal Processing Conference (EUSIPCO), Tampere, Finland, September 5-8, 2000.

[34] M. Reha Civanlar and A. Murat Teklap, Real-time Video over the Internet, Signal Processing: Image Communication, vol. 15, no. 1-2, pp. 1-5, September 1999 (Special issue on streaming).

[35] H. Harashima, K.Aizawa, and T. Saito, Model-based analysis synthesis coding of video-telephone images – conception and basic study of intelligent image coding, Transactions IEICE, vol. E72, no. 5, pp. 452-458, 1989.

[36] J. Karlekar and U. B. Desai, Content-based very low bit-rate video coding using wavelet transform, Proceedings of the IEEE Interna-tional Conference on Image Processing, Kobe, Japan, October 1999.

[37] P. Salembier and O. Avaro, MPEG-7: Multimedia content description interface, Workshop on MPEG-21, Noordwijkerhout, the Netherlands, March 20-21, 2000. http://www.cselt.it/mpeg/events/mpeg21/

[38] D. Pearson, Developments in model-based coding, Proceedings of the IEEE, vol. 86, no. 6, pp. 892-906, June 1995.

[39] V. Vaerman, G. Menegaz, and J. P. Thiran, A parametric hybrid model used for multidimensional object representation, Proceedings of the IEEE International Conference on Image Processing, Kobe, Japan, October 1999.

[40] T. Huang, From video indexing to multimedia understanding, Proceedings of the 1999 International Workshop on Very Low Bitrate Video Coding, Kyoto, Japan, October 1999. (Keynote speech.)

[41]   A. Albiol, L. Torres, E. Delp, Combining audio and video for video sequence indexing applications, IEEE International Conference on Multimedia and Expo, Lausanne, Switzerland, August 2002.

[42]   G. Iyengar, H. Nock, C. Neti, M. Franz, Semantic indexing of multimedia using audio, text and visual cues, IEEE International Conference on Multimedia and Expo, Lausanne, Switzerland, August 2002.

[43]   S. Wenger, G. Knorr, J. Ott, and F. Kossentini, Error resilience support in H.263+, IEEE Transactions on Circuits and Systems for Video Technology, vol. 8, no. 7, pp. 867-877, Novem-ber 1998.

[44]   R. Talluri, Error-resilient video coding in the ISO MPEG-4 Standard, IEEE Communications Magazine, vol. 2, no. 6, pp. 112-119, June 1999.

[45]   S. D. Servetto, K. Ramchandran, V.A. Vaishampayan, and K. Nahrstedt, Multiple description wavelet based image coding, IEEE Transactions on Image Processing, vol. 9, no. 5, pp. 813-826, May 2000.

[46]   S. D. Servetto and K. Nahrstedt, Video streaming over the public Internet: Multiple description codes and adaptive transport protocols, Proceedings of the 1999 Inter-national Conference on Image Processing, Kobe, Japan, October 1999.

[47]   W. Tan and A. Zakhor, Real-time Internet video using error resilient scalable compression and TCP-friendly transport protocol, IEEE Transactions on Multimedia, vol. 1, no. 2, pp. 172-186, June 1999.

[48]   H. Radha, Y. Chen, K. Parthasarathy and R. Cohen, Scalable Internet video using MPEG-4, Signal Processing: Image Communication, vol. 15, no. 1-2, pp. 95-126, September 1999.

[49]   U. Horn, K. Stuhlmüller, M. Link and B. Girod, Robust Internet video transmission based on scalable coding and unequal error protection, Signal Processing: Image Communication, vol. 15, no. 1-2, pp. 77-94, September 1999.

[50]   J. Bormans, K. Hill, MPEG-21 Overview v.4, ISO/IEC JTC1/SC29/WG11/N4801, Fairfax, May 2002.

[51]   B. Moghaddam, A. Pentland, Probabilistic visual learning for object representation, IEEE Transactions on Pattern Analysis and Machine Intelligence, Vol. 19, no. 7, pp. 696-710, July 1997.

[52]   L. Torres, J. Vilà, Automatic face recognition for video indexing applications, Invited paper, Pattern Recognition. Vol 35/3, pp 615-625, December 2001.

[53]   W E. Vieux, K. Schwerdt and J.L. Crowley, Face-tracking and Coding for Video-Compression, First International Conference on Computer Vision Systems, Las Palmas, Spain, January, 1999.

[54]   L. Torres, D. Prado, A proposal for high compression of faces in video sequences using adaptive eigenspaces, IEEE International Conference on Image Processing, Rochester, USA, September 22-25, 2002.

[55]   B. Moghaddam, A. Pentland, Probabilistic visual learning for object representation, IEEE Transactions on Pattern Analysis and Machine Intelligence, vol. 19, no. 7, pp. 696-710, July 1997.

[56]   F. Marqués, V. Vilaplana, and A. Buxes, Human face segmentation and tracking using connected operators and partition projection, Proceedings of the IEEE International Confere-nce on Image Processing, Kobe, Japan, October 1999.

[57]   Special Session on Robust Video, Proceedings of the IEEE International Conference on Image Processing, Kobe, Japan, October 1999.

40

COMPRESSED-DOMAIN VIDEO PROCESSING

Susie Wee, Bo Shen and John Apostolopoulos
Streaming Media Systems Group
Hewlett-Packard Laboratories
Palo Alto, CA, USA
{swee,boshen,japos}@hpl.hp.com

1. INTRODUCTION

Video compression algorithms are being used to compress digital video for a wide variety of applications, including video delivery over the Internet, advanced television broadcasting, video streaming, video conferencing, as well as video storage and editing. The performance of modern compression algorithms such as MPEG-1, MPEG-2, MPEG-4, H.261, H.263, and H.264/MPEG-4 AVC is quite impressive -- raw video data rates often can be reduced by factors of 15-80 or more without considerable loss in reconstructed video quality. This fact, combined with the growing availability of video encoders and decoders and low-cost computers, storage devices, and networking equipment, makes it evident that between video capture and video playback, video will be handled in compressed video form.

End-to-end compressed digital video systems motivate the need to develop algorithms for handling compressed digital video. For example, algorithms are needed to adapt compressed video streams for playback on different devices and for robust delivery over different types of networks. Algorithms are needed for performing video processing and editing operations, including VCR functionalities, on compressed video streams. Many of these algorithms, while simple and straightforward when applied to raw video, are much more complicated and computationally expensive when applied to compressed video streams. This motivates the need for developing efficient algorithms for performing these tasks on compressed video streams.

In this chapter, we describe compute- and memory-efficient, quality-preserving algorithms for handling compressed video streams. These algorithms achieve efficiency by exploiting coding structures used in the original compression process. This class of efficient algorithms for handling compressed video streams is called compressed-domain processing (CDP) algorithms. CDP algorithms that change the video format and compression format of compressed video streams are called compressed-domain transcoding algorithms, and CDP algorithms that perform video processing and editing operations on compressed video streams are called compressed-domain editing algorithms.

These CDP algorithms are useful for a number of applications. For example, a video server transmitting video over the Internet may be restricted by stringent bandwidth requirements. In this scenario, a high-rate compressed bitstream may need to be transcoded to a lower-rate compressed bitstream prior to transmission; this can be achieved by lowering the spatial or temporal resolution of the video or by more coarsely quantizing the MPEG data. Another application may require MPEG video streams to be transcoded into streams that facilitate video editing functionalities such as splicing or fast-forward and reverse play; this may involve removing the temporal dependencies in the coded data stream. Finally, in a video communication system, the transmitted video stream may be subject to harsh channel conditions resulting in data loss; in this instance it may be useful to create a standard-compliant video stream that is more robust to channel errors and network congestion.

This chapter presents a series of compressed-domain image and video processing algorithms that were designed with the goal of achieving high performance with computational efficiency. It focuses on developing transcoding algorithms for bitstreams that are based on video compression algorithms that rely on the block discrete cosine transform (DCT) and motion-compensated prediction. These algorithms are applicable to a number of predominant image and video coding standards including JPEG, MPEG-1, MPEG-2, MPEG-4, H.261, H.263, and H.264/MPEG-4 AVC. Much of this discussion will focus on MPEG; however, many of these concepts readily apply to the other standards as well.

This chapter proceeds as follows. Section 2 defines the compressed-domain processing problem. Section 3 gives an overview of MPEG basics and it describes the CDP problem in the context of MPEG. Section 4 describes the basic methods used in CDP algorithms. Section 5 describes a series of CDP algorithms that use the basic methods of Section 4. Finally, Section 6 describes some advanced topics in CDP.

2. PROBLEM STATEMENT

Compressed-domain processing performs a user-defined operation on a compressed video stream without going through a complete decompress/process/re-compress cycle; the processed result is a new compressed video stream. In other words, the goal of compressed-domain processing (CDP) algorithms is to efficiently process one standard-compliant compressed video stream into another standard-compliant compressed video stream with a different set of properties. Compressed-domain transcoding algorithms are used to change the video format or compression format of compressed streams, while compressed-domain editing algorithms are used to perform processing operations on compressed streams. CDP differs from the encoding and decoding processes in that both the input and output of the transcoder are compressed video streams.

A conventional solution to the problem of processing compressed video streams, shown in the top path of Figure 40.1, involves the following steps: first, the input compressed video stream is completely decompressed into its pixel-domain representation; this pixel-domain video is then processed with the appropriate operation; and finally the processed video is recompressed into a new output compressed video stream. Such solutions are computationally expensive and have large memory requirements. In addition, the quality of the coded video can deteriorate with each re-coding cycle.

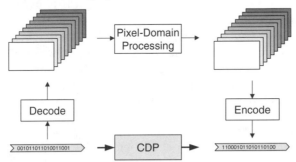

Figure 40.1 Processing compressed video: the conventional pixel-domain solution (top path) and the compressed-domain processing solution (bottom path).

Compressed-domain processing methods can lead to a more efficient solution by only partially decompressing the bitstream and performing processing directly on the compressed-domain data. The resulting CDP algorithms can have significant savings over their conventional pixel-domain processing counterparts. Roughly speaking, the degree of savings will depend on the particular operation, the desired performance, and the amount of decompression required for the particular operation. This is discussed further in Subsection 3.4 within the context of MPEG compression.

3. MPEG CODING AND COMPRESSED-DOMAIN PROCESSING

3.1 MPEG FRAME CODING

Efficient CDP algorithms are designed to exploit various features of the MPEG video compression standards. Detailed descriptions of the MPEG video compression standards can be found in [1][2]. This section briefly reviews some aspects of the MPEG standards that are relevant to CDP.

MPEG codes video in a hierarchy of units called sequences, groups of pictures (GOPs), pictures, slices, macroblocks, and blocks. 16x16 blocks of pixels in the original video frames are coded as a macroblock, which consists of four 8x8 blocks. The macroblocks are scanned in a left-to-right, top-to-bottom fashion, and series of these macroblocks form a slice. All the slices in a frame comprise a picture, contiguous pictures form a GOP, and all the GOPs form the entire sequence. The MPEG syntax allows a GOP to contain any number of frames, but typical sizes range from 9 to 15 frames. Each GOP refreshes the temporal prediction by coding the first frame in intraframe mode, i.e., without prediction. The remaining frames in the GOP can be coded with intraframe or interframe (predictive) coding techniques.

The MPEG algorithm allows each frame to be coded in one of three modes: intraframe (I), forward prediction (P), and bidirectional prediction (B). A typical IPB pattern in *display order* is:

$$B_7 \ B_8 \ P_9 \ B_{10} \ B_{11} \ I_0 \ B_1 \ B_2 \ P_3 \ B_4 \ B_5 \ P_6 \ B_7 \ B_8 \ P_9 \ B_{10} \ B_{11} \ I_0 \ B_1 \ B_2 \ P_3$$

The subscripts represent the index of the frame within a GOP. I frames are coded independently of other frames. P frames depend on a prediction based on the preceding I or P frame. B frames depend on a prediction based on the preceding and following I or P frames. Notice that each B frame depends on data from a future frame, i.e., future frame must be (de)coded before a current B frame can be (de)coded. For this reason, the coding order is distinguished from the display order. The *coding order* for the sequence shown above is:

$$P_9 \ B_7 \ B_8 \ G \ I_0 \ B_{10} \ B_{11} \ P_3 \ B_1 \ B_2 \ P_6 \ B_4 \ B_5 \ P_9 \ B_7 \ B_8 \ G \ I_0 \ B_{10} \ B_{11} \ P_3 \ B_1 \ B_2$$

MPEG requires the coded video data to be placed in the data stream in coding order. G represents a GOP header that is placed in the compressed bitstream.

A GOP always begins with an I frame. Typically, it includes the following (display order) P and B frames that occur before the next I frame, although the syntax also allows a GOP to contain multiple I frames. The GOP header does not specify the number of I, P, or B frames in the GOP, nor does it specify the structure of the GOP -- these are completely determined by the order of the data in the stream. Thus, there are no rules that restrict the size and structure of the GOP, although care should be taken to ensure that the buffer constraints are satisfied.

MPEG uses block motion-compensated prediction to reduce the temporal redundancies inherent to video. In block motion-compensated prediction, the current frame is divided into 16x16 pixel units called macroblocks. Each macroblock is compared to a number of 16x16 blocks in a previously coded frame. A single motion vector (MV) is used to represent this block with the best match. This block is used as a prediction of the current block, and only the error in the prediction, called the residual, is coded into the data stream.

The frames of a video sequence can be coded as an I, P, or B frame. In I frames, every macroblock must be coded in intraframe mode, i.e., without prediction. In P frames, each macroblock can be coded with forward prediction or in intraframe mode. In B frames, each macroblock can be coded with forward, backward, or bidirectional prediction or in intraframe mode. One MV is specified for each forward- and backward-predicted macroblock while two MVs are specified for each bidirectionally predicted macroblock. Thus, each P frame has a forward motion vector field and one anchor frame, while each B frame has a forward and backward motion vector field and two anchor frames. In some of the following sections, we define B_{for} and B_{back} frames as B frames that use only forward or only backwards prediction. Specifically, B_{for} frames can only have intra- and forward-predicted macroblocks while B_{back} frames can only have intra- and backward-predicted macroblocks.

MPEG uses discrete cosine transform (DCT) coding to code the intraframe and residual macroblocks. Specifically, four 8x8 block DCTs are used to encode each macroblock and the resulting DCT coefficients are quantized. Quantization usually results in a sparse representation of the data, i.e., one in which most of the amplitudes of the quantized DCT coefficients are equal to zero. Then, only the amplitudes and locations of the nonzero coefficients are coded into the compressed data stream.

3.2 MPEG FIELD CODING

While many video compression algorithms, including MPEG-1, H.261, and H.263, are designed for progressive video sequences, MPEG-2 was designed to support both progressive and interlaced video sequences, where two fields, containing the even and odd scanlines, are contained in each frame. MPEG-2 provides a number of coding options to support interlaced video. First, each interlaced video frame can be coded as a frame picture in which the two fields are coded as a single unit or as a field picture in which the fields are coded sequentially. Next, MPEG-2 allows macroblocks to be coded in one of five motion compensation modes: frame prediction for frame pictures, field prediction for frame pictures, field prediction for field pictures, 16x8 prediction for field pictures, and dual prime motion compensation. The frame picture and field picture prediction dependencies are as follows. For frame pictures, the top and bottom reference fields are the top and bottom fields of the previous I or P frame. For field pictures, the top and bottom reference fields are the most recent top and bottom fields. For example, if the top field is specified to be first, then MVs from the top field can point to the top or bottom fields in the previous frame, while MVs from the bottom field can point to the top field of the current frame or the bottom field of the previous frame. Our discussion focuses on P-frame prediction because the transcoder described in Subsection 5.1.5 only processes the MPEG I and P frames. We also focus on field picture coding of interlaced video, and do not discuss dual prime motion compensation.

In MPEG field picture coding, each field is divided into 16x16 macroblocks, each of which can be coded with field prediction or 16x8 motion compensation. In field prediction, the 16x16 field macroblock will contain a field selection bit which indicates whether the prediction is based on the top or bottom reference field and a motion vector which points to the 16x16 region in the appropriate field. In 16x8 prediction, the 16x16 field macroblock is divided into its upper and lower halves, each of which contains 16x8 pixels. Each half has a field selection bit which specifies whether the top or bottom reference field is used and a motion vector which points to the 16x8 pixel region in the appropriate field.

3.3 MPEG BITSTREAM SYNTAX

The syntax of the MPEG-1 data stream has the following structure: A **Sequence header** consists of a *sequence start code* followed by *sequence parameters*. Sequences contain a number of GOPs. Each **GOP header** consists of a *GOP start code* followed by *GOP parameters*. GOPs contain a number of pictures. Each **picture header** consists of a *picture start code* followed by *picture parameters*. Pictures contain a number of slices. Each **slice header** consists of a *slice start code* followed by *slice parameters*. The slice header is followed by **slice data**, which contains the coded macroblocks.

The *sequence header* specifies the picture height, picture width, and sample aspect ratio. In addition, it sets the frame rate, bitrate, and buffer size for the sequence. If the default quantizers are not used, then the quantizer matrices are also included in the sequence header. The *GOP header* specifies the time code and indicates whether the GOP is open or closed. A GOP is open or closed depending on whether or not the temporal prediction of its frames require data from other GOPs. The *picture header* specifies the temporal reference parameter, the picture type (I, P, or B), and the buffer fullness (via the vbv_delay parameter). If temporal prediction is used, it also describes the motion vector precision (full or half pixel) and the motion vector range. The *slice header* specifies the macroblock row in which slice starts and the initial quantizer scale factor for the DCT coefficients. The *macroblock header* specifies the relative position of the macroblock in relation to the previously coded macroblock. It contains a flag to indicate whether intra- or interframe coding is used. If interframe coding is used, it contains the coded motion vectors, which may be differentially coded with respect to previous motion vectors. The quantizer scale factor may be adjusted at the macroblock level. One bit is used to specify whether the factor is adjusted. If it is, the new scale factor is specified. The macroblock header also specifies a coded block pattern for the macroblock. This describes which of the luminance and chrominance DCT blocks are coded. Finally, the DCT coefficients of the coded blocks are coded into the bitstream. The DC coefficient is coded first, followed by the runlengths and amplitudes of the remaining nonzero coefficients. If it is an intra-macroblock, then the DC coefficient is coded differentially.

The sequence, GOP, picture, and slice headers begin with start codes, which are four-byte identifiers that begin with 23 zeros and a one followed by a one byte unique identifier. Start codes are useful because they can be found by examining the bitstream; this facilitates efficient random access into the compressed bitstream. For example, one could find the coded data that corresponds to the 2nd slice of the 2nd picture of the 22nd GOP by simply examining the coded data stream, without parsing and decoding the data. Of course, reconstructing the actual pixels of that slice may require parsing and decoding additional portions of the data stream because of the prediction used in conventional video coding algorithms. However, computational benefits could still be achieved by locating the beginning of the 22nd GOP and parsing and decoding the data from that point on, thus exploiting the temporal refresh property inherent to GOPs.

3.4 COMPRESSED-DOMAIN PROCESSING FOR MPEG

The CDP problem statement was described in Section 2. In essence, the goal of CDP is to develop efficient algorithms for performing processing operations on compressed bitstreams. While the conventional approach requires decompressing the bitstream, processing the decoded frames, and re-encoding the result, improved efficiency, with respect to compute and memory requirements, can be achieved by exploiting structures used in the compression algorithms and using this knowledge to avoid the complete decode and re-encode cycle. In the context of MPEG transcoding, improved efficiency can be achieved by exploiting the structures used in MPEG coding. Furthermore, a decode/process/re-encode cycle can lead to significant loss of quality (even if *no processing* is performed besides the decode and re-encode) -- carefully designed CDP algorithms can greatly reduce and in some cases prevent this loss in quality.

MPEG coding uses a number of structures, and different compressed-domain processing operations require processing at different levels of depth. From highest to lowest level, these levels include:

- Sequence-level processing
- GOP-level processing
- Frame-level processing
- Slice-level processing
- Macroblock-level processing
- Block-level processing

Generally speaking, deeper-level operations require more computations. For example, some processing operations in the time domain require less computation if no information below the frame level needs to be adjusted. Operations of this kind include fast forward recoding and cut-and-paste or splicing operations restricted to cut points at GOP boundaries. However, if frame-accurate splicing [3] is required, frame and macroblock level information may need to be adjusted for frames around the splice point, as described in Section 5. In addition, in frame rate reduction transcoding, if the transcoder chooses to only drop non-reference frames such as B frames, a frame-level parsing operation could suffice.

On the other hand, operations related to the modification of content within video frames have to be performed below the frame level. Operations of this kind include spatial resolution reduction transcoding [4], frame-by-frame video reverse play [5], and many video-editing operations such as fading, logo insertion, and video/object overlaying [6,7]. As expected, these operations require significantly more computations; so for these operations efficient compressed-domain methods can lead to significant improvements.

4. COMPRESSED-DOMAIN PROCESSING METHODS

In this section, we examine the basic techniques of compressed-domain processing methods. Since the main techniques used in video compression include spatial to frequency transformation, particularly DCT, and motion-compensated prediction, we focus the investigation on compressed domain methods in these two domains, namely, in the DCT domain and the motion domain.

4.1 DCT-DOMAIN PROCESSING

As described in Section 3, the DCT is the transformation used most often in image and video compression standards. It is therefore important to understand some basic operations that can be performed directly in the DCT domain, i.e., without an inverse DCT/forward DCT cycle.

The earliest work on direct manipulation of compressed image and video data expectedly dealt with point processing, which consists of operations such as contrast manipulation and image subtraction where a pixel value in the output image at position p depends solely on the pixel value at the same position p in the input image. Examples of such work can be found in Chang and Messerschmitt [8], who developed some special functions for video compositing, and in Smith and Rowe [9], who developed a set of algorithms for basic point operations. When viewing compressed domain manipulation as a matrix operation, point processing operations on compressed images and video can be characterized as inner-block algebra (IBA) operations since the information in the output block, i.e. the manipulated block, comes solely from information in the corresponding input block. These operations are listed in Table 40.1.

Table 40.1 Mathematical expression of spatial vs. DCT domain algebraic operations

	Spatial domain signal – x	Transform domain signal – X
Scalar addition	$[f]+\alpha$	$[F]+\begin{bmatrix} 8\alpha/Q_{00} & 0 \\ 0 & 0 \end{bmatrix}$
Scalar Multiplication	$\alpha[f]$	$\alpha[F]$
Pixel Addition	$[f]+[g]$	$[F]+[G]$
Pixel Multiplication	$[f]\bullet[g]$	$[F]\otimes[G]$

In this table, lower case f and g are used to represent spatial domain signals, while upper case F and G represent their corresponding DCT domain signals. Since compression standards typically use block-based schemes, each block can be treated as a matrix. Therefore, the operations can be expressed in forms of matrix operations. In general, the relationship holds as:

$DCT(x) = X$,

where $DCT(\ )$ represents the DCT function.

Because of the large number of zeros in the block in the DCT domain, the data manipulation rate is heavily reduced. The speedup of the first three operations in Table 40. 1 is quite obvious given that the number of non-zero coefficients in F and G is quite small. As an example of these IBA operations, consider the compositing operation where foreground f is combined with background b with a factor of α to generate an output R in DCT representation. In spatial domain, this operation can be expressed as: $R = DCT(\alpha[f]+(1-\alpha)[b])$. Given the DCT representation of f and b in the compressed domain, F and B, the operation can be conveniently performed as: $R = \alpha[F]+(1-\alpha)[B]$. The operation is based on the linearity of the DCT and corresponds to a combination of some of the above-defined image algebra operations; it can be done in DCT domain efficiently with significant speedup. Similar compressed domain algorithms for subtitling and dissolving applications can also be developed based on the above IBA operations with computational speedups of 50 or more over the corresponding processing of the uncompressed data [9].

These methods can also be used for color transformation in the compressed domain. As long as the transformation is linear, it can be derived in the compressed domain using a combination of these IBA operations.

Pixel multiplication can be achieved by a convolution in the DCT domain. Compressed-domain convolution has been derived in [9] by mathematically combining the decompression, manipulation, and re-compression processes to obtain a single equivalent local linear operation where one can easily take advantage of the energy compaction property in quantized DCT blocks. A similar approach was taken by Smith [10] to extend point processing to global processing of operations where the value of a pixel in the output image is an arbitrary linear combination of pixels in the input image. Shen et al. [11] have studied the theory behind DCT domain convolution based on the orthogonal property of DCT. As a result, an optimized DCT domain convolution algorithm is proposed and applied to the application of DCT domain alpha blending. Specifically, given foreground f to be blended with the background b with an alpha channel a to indicate the transparency of each pixel in f, the operation can be expressed as: $R = DCT([a] \cdot [f] + (1 - [a]) \cdot [b])$. The DCT domain operation is performed as: $R = [A] \otimes [F] + (1 - [A]) \otimes [B]$, where A is the DCT representation of a. A masking operation can also be performed in the same fashion with A representing the mask in the DCT domain. This operation enables the overlay of an object in the DCT domain with arbitrary shape. An important application for this is logo-insertion. Another example where processing of arbitrarily shaped objects arise is discussed in Section 6.1.

Many image manipulation operations are local or neighborhood operations where the pixel value at position p in the output image depends on neighboring pixels of p in the input image. We characterize methods to perform such operations in the compressed domain as inner-block rearrangement or resampling (IBR) methods. These methods are based on the fact that DCT is a unitary orthogonal transform and is distributive to matrix multiplication. It is also distributive to matrix addition, which is actually the case of pixel addition in Table 40. 1. We group these two distributive properties of DCT in Table 40.2.

Table 40.2 Mathematical expression of distributiveness of DCT

	Spatial domain signal – x	Transform domain signal – X
Matrix Addition	$[f] + [g]$	$[F] + [G]$
Matrix Multiplication	$[f][g]$	$[F][G]$

Based on above, Chang and Messerschmitt [8] developed a set of algorithms to manipulate images directly in the compressed domain. Some of the interesting algorithms they developed include the translation of images by arbitrary amounts, linear filtering, and scaling. In general, a manipulation requiring uniform and integer scaling, i.e. certain forms of filtering, is easy to implement in the DCT domain using the resampling matrix. Since each block can use the same resampling matrix in space invariant filtering, these kinds of manipulations require little overhead in the DCT domain. In addition, translation of images by arbitrary amounts represents a shifting operation that is often used in video coding. We defer a detailed discussion of this particular method to Section 4.3.

Another set of algorithm has also been introduced to manipulate the orientation of DCT blocks [12]. These methods can be employed to flip-flop a DCT frame as well as rotate a DCT frame at multiples of 90 degrees, simply by switching the location and/or signs of certain DCT coefficients in the DCT blocks. For example, the DCT transform result of a transposed pixel block f is equivalent of the transpose of the corresponding DCT block. This operation is expressed mathematically as:

$$DCT([f]^t) = [F]^t.$$

A horizontal flip of a pixel block ($[f]^h$) can be achieved in the DCT domain by performing an element-by-element multiplication with a matrix composed of only two values: 1 or −1. The operation is therefore just sign reversal on some non-zero coefficients. Mathematically, this operation is expressed as:

$$DCT([f]^h) = [F] \bullet [H],$$

where H is defined as follows assuming an 8x8 block operation,

$$H_{ij} = \begin{cases} -1 & j = 1,3,5,7 \\ 1 & j = 0,2,4,6 \end{cases}.$$

For the full set of operations of this type, please refer to [12]. Note that for all the cases, the DC coefficient remains unchanged because of the fact that each pixel maintains its gray level while its location within the block is changed. These flip-flop and special angle rotation methods are very useful in applications such as image orientation manipulation that is used often in copy machines, printers, and scanners.

4.2 MOTION VECTOR PROCESSING (MV RESAMPLING)

From a video coding perspective, motion vectors are estimated through block matching in a reference frame. This process is often compute intensive. The key of compressed-domain manipulation of motion vectors is to derive new motion vectors out of existing motion vector information contained in the input compressed bitstream.

Consider a motion vector processing scenario that arises in a spatial resolution reduction transcoder. Given the motion vectors for a group of four 16x16 macroblocks of the original video (NxM), how does one estimate the motion vectors for the 16x16 macroblocks in the downscaled video (e.g., N/2xM/2)? Consider forward-predicted macroblocks in a forward-predicted (P) frame, wherein each macroblock is associated with a motion vector and four 8x8 DCT blocks that represent the motion-compensated prediction residual information. The downscale-by-two operation requires four input macroblocks to form a single new output macroblock. In this case, it is necessary to estimate a single motion vector for the new macroblock from the motion vectors associated with the four input macroblocks.

The question asked above can be viewed as a motion vector resampling problem. Specifically, given a set of motion vectors MV in the input compressed bitstream, how does one compute the motion vectors MV* of the output compressed bitstream? Motion vector resampling algorithms can be classified into 5 classes as shown in Figure 40.2 [5]. The most accurate, but least efficient algorithm is Class V, in which one decompresses the original frames into their full pixel representation; and then one performs full search motion estimation on the decompressed frames. Since motion estimation is by far the most compute-intensive part of the transcoding operation, this is a very expensive solution. Simpler motion vector resampling algorithms are given in classes I through IV in order of increasing computational complexity, where increased complexity typically results in more accurate motion vectors. Class I MV resampling algorithms calculate each output motion vector based on its corresponding input motion vector. Class II algorithms calculate each output motion vector based on a neighbourhood of input motion vectors. Class III algorithms also use a neighbourhood of input motion vectors, but also consider other parameters from the input bitstream such as quantization parameters and coding modes when processing them. Class IV algorithms use a neighbourhood of motion vectors and other input bitstream parameters, but also use the decompressed frames. For example, the input motion vectors may be used to narrow the search range used when estimating the output motion vectors. Finally, Class V corresponds to full search motion estimation on the decompressed frames.

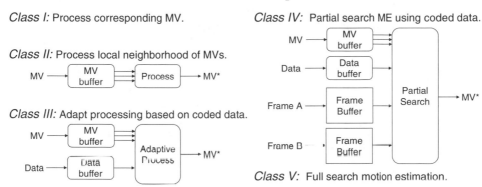

Figure 40.2 Classes of motion vector resampling methods.

The conventional spatial-domain approach of estimating the motion vectors for the downscaled video is to first decompress the video, downscale the video in the spatial domain then use one of the several widely known spatial-domain motion-estimation techniques (e.g., [13]) to recompute the motion vectors. This is computationally intensive. A class II approach might be to simply take the average of the four motion vectors associated with the four macroblocks and divide it by two so that the resulting motion vector can be associated with the 16x16 macroblock of the downscaled-by-two video. While this operation requires little processing, the motion vectors obtained in this manner are not optimal in most cases.

Adaptive motion vector resampling (AMVR) is a class III approach proposed in [4] to estimate the output motion vectors using the original motion information from the MPEG or H.26x bitstream of the original NxN video sequence. This method uses the DCT blocks to derive the block-activity information for the motion-vector estimation. When comparing the compressed-domain AMVR method to the conventional spatial-domain method, the results suggest that AMVR generates, with significantly less computation, motion vectors for the N/2xM/2 downscaled video that are very close to the optimal motion vector field that would be derived from an N/2xM/2 version of the original video sequence.

This weighted average motion vector scheme can also be extended to motion vector downsampling by arbitrary factors. In this operation, the number of participating macroblocks is not an integer. Therefore, the portion of the area of the participating macroblock is used to weight the contributions of the existing motion vectors.

A class IV method for performing motion vector estimation out of existing motion vectors can be found in [14]. In frame rate reduction transcoding, if a P-picture is to be dropped, the motion vectors of macroblocks on the next P-picture should be adjusted since the reference frame is now different. Youn et al. [14] proposed a motion vector composition method to compute a motion vector from the incoming motion vectors. In this method, the derived motion vector can be refined by performing partial search motion estimation within a narrow search range.

The MV resampling problem for the compressed-domain reverse play operation was examined in [5]. In this application, the goal was to compute the "backward" motion vectors between two frames of a sequence when given the "forward" motion vectors between two frames. Perhaps contrary to intuition, the resulting forward and backwards motion vector fields are not simply inverted versions of one another because of the block-based motion-compensated processing used in typical compression algorithms. A variety of MV resampling algorithms are presented in [5], and experimental results are given that illustrate the tradeoffs in complexity and performance.

4.3 MC+DCT PROCESSING

The previous section introduced methodologies for deriving output motion vectors from existing input motion vectors in the compressed domain. However, if the newly derived motion vectors are used in conjunction with the original residual data, the result is imperfect and will result in a drift error. To avoid drift error, it is important to reconstruct the original reference frame and re-compute the residual data. This renders the IDCT process as the next computation bottleneck since the residual data is in the DCT domain. Alternatively, DCT domain motion compensation methods, such as the one introduced in [8], can be employed where the reference frames are converted to a DCT representation so that no IDCT is needed.

Inverse motion compensation is the process of extracting a 16x16 block given a motion vector in the reference frame. It can be characterised by group matrix multiplications. Due to the distributive property of the DCT, this operation can be achieved by matrix multiplications of DCT blocks. Mathematically, consider a block g of size 8x8 in a reference frame pointed by a motion vector (x,y). Block g may lie in an area covered by a 2x2 array of blocks (f_1, f_2, f_3, f_4) in the reference frame. g can then be calculated as:

$$g = \sum_{i=1}^{4} m_{xi} f_i m_{yi} \text{ , where } 0 \leq x, y \leq 7 \text{ .}$$

The *shifting* matrices m_{xi} and m_{yi} are defined as:

$$m_{xi} = \begin{cases} \begin{bmatrix} 0 & I_x \\ 0 & 0 \end{bmatrix} & i = 1,2 \\ \begin{bmatrix} 0 & I_{8-x} \\ 0 & 0 \end{bmatrix} & i = 3,4 \end{cases}, \text{ and } m_{yi} = \begin{cases} \begin{bmatrix} 0 & 0 \\ I_y & 0 \end{bmatrix} & i = 1,3 \\ \begin{bmatrix} 0 & 0 \\ I_{8-y} & 0 \end{bmatrix} & i = 2,4 \end{cases},$$

where I_z are identity matrices of size 8x8. In the DCT domain, this operation can be expressed as:

$$G = \sum_{i=1}^{4} M_{xi} F_i M_{yi} , \qquad (40.1)$$

where M_{xi} and M_{yi} are the DCT representations of m_{xi} and m_{yi} respectively. Since these shifting matrices are constant, they can be pre-computed and stored in the memory. However, the computing of Eq (40.1) may still be CPU-intensive since the shifting matrices may not be sparse enough. To this end, various authors have proposed different methods to combat this problem.

Merhav and Bhaskaran [15] proposed to decompose the DCT domain shifting matrices. Matrix decomposition methods are based on the sparseness of the factorized DCT transform matrices. Factorization of DCT transform matrix is introduced in a fast DCT algorithm [16]. The goal of the decomposition is to replace the matrix multiplication with a product of diagonal matrices, simple permutation matrices and more sparse matrices. The multiplication with a diagonal matrix can be absorbed in the quantization process. The multiplication with a permutation matrix can be performed by coefficient permutation. And finally, the multiplication with a sparse matrix requires fewer multiplications. Effectively, the matrix multiplication is achieved with less computation.

In an alternative approach, the coefficients in the shifting matrix can be approximated so that floating point multiplication can be replaced by integer shift and add operation. Work of this kind is introduced in [17]. Effectively, fewer basic CPU operations are needed since multiplication operations are avoided. A similar method is also used in [18] for DCT domain downsampling of images by employing the approximated downsampling matrices.

To further reduce the computation complexity of the DCT domain motion compensation process, a look-up-table (LUT) based method [19] is proposed by modelling the statistical distribution of DCT coefficients in compressed images and video sequences and precomputing all possible combinations of $M_{xi} F_i M_{yi}$ as in Eq (40.1). As a result, the matrix multiplications are reduced to simple table look-ups. Using around 800KB of memory, the LUT-based method can save more than 50% of computing time.

4.4 RATE CONTROL/BUFFER REQUIREMENTS

Rate control is another important issue in video coding. For compressed domain processing, the output of the process should also be confined to a certain bitrate so that it can be delivered in a constant transmission rate. Eleftheriadis and Anastassiou [20] have considered rate reduction by an optimal truncation or selection of DCT coefficients. Since fewer coefficients are coded, a lower number of bits are spent in coding them. Nakajima et al. [21] achieve the similar rate reduction by re-quantization using a larger quantization step size.

For compressed domain processing, it is important for the rate control module to use compressed domain information existing in the original stream. This is a challenging problem, since the compressed bitstream lacks information that was available to the original encoder. To illustrate this problem, consider TM5 rate control, which is used in many video coding standards. This rate controller begins by estimating the number of bits available to code the picture, and computes a reference value of the quantization parameter based on the buffer fullness and target bitrate. It then adaptly raises or lowers the quantization parameter for each macroblock based on the spatial activity of that macroblock. The spatial activity measure as defined in TM5 as the variance of each block:

$$V = \frac{1}{64}\sum_{i=1}^{64}\left(P_i - P_{mean}\right)^2 \quad , \text{ where } \quad P_{mean} = \frac{1}{64}\sum_{i=1}^{64}P_i \; .$$

However, the pixel domain information P_i may not be available in the compressed domain processing. In this case, the activity measure has to be derived from the DCT coefficients instead of the pixel domain frames. For example, the energy of quantized AC coefficients in the DCT block can be used as a measure of the variance. It has been shown in [4] that this approach achieves satisfactory rate control. In addition, the target bit budget for a particular frame can be derived from the bitrate reduction factor and the number of bits spent for the corresponding original frame, which is directly available from the original video stream.

4.5 FRAME CONVERSIONS

Frame conversions are another basic tool that can be used in compressed-domain video processing operations [22]. They are especially useful in frame-level processing applications such as splicing and reverse play. Frame conversions are used to convert coded frames from one prediction mode to another to change the prediction dependencies in coded video streams. For example, an original video stream coded with I, P, and B frames may be temporarily converted to a stream coded with all I frames, i.e. a stream without temporal prediction, to facilitate pixel-level editing operations. Also, an IPB sequence may be converted to an IB sequence in which P frames are converted to I frames to facilitate random access into the stream. Also, when splicing two video sequences together, frame conversions can be used to remove prediction dependencies from video frames that are not included in the final spliced sequence. Furthermore, one may wish to use frame conversions to add prediction dependencies to a stream, for example to convert from an all I-frame compressed video stream to an I and P frame compressed stream to achieve a higher compression rate.

A number of frame conversion examples are shown in Figure 40.3. The original IPB sequence is shown in the top. Examples of frame conversions that remove temporal dependencies between frames are given: specifically P-to-I frame, B-to-B_{for} conversion, and B-to-B_{back} conversion. These operations are useful for editing operations such as splicing. Finally, an example of I-to-P conversion is shown in which prediction dependencies are added between frames. This is useful in applications that require further compression of a pre-compressed video stream.

Frame conversions require macroblock level and block-level processing because they modify the motion vector and DCT coefficients of the compressed stream. Specifically, frame conversions require examining each macroblock of the compressed frame, and when necessary changing its coding mode to an appropriate dependency. Depending on the conversion, some, but not all, macroblocks may need to be processed. An example in which a macroblock may not need to be processed is in a P-to-I frame conversion. Since P frames contain i- and p- type macroblocks and I frames contain only i-type macroblocks, a P-to-I conversion requires converting all p-type input macroblocks to i-type output macroblocks; however, note that i-type input macroblocks do not need to be converted. The list of frame types and allowed macroblock coding modes are shown in the upper right table in Figure 40.3. The lower right table shows macroblock conversions needed for some frame conversion operations. These conversions will be used in the splicing and reverse play applications described in Section 5. The conversion of a macroblock from p-type to i-type can be performed with the inverse motion compensation process introduced in Subsection 4.3.

One should note that in standards like MPEG, frame conversions performed on one frame may affect the prediction used in other frames because of the prediction rules specified by I, P, and B frames. Specifically, I-to-P and P-to-I frame conversions do not affect other coded frames. However, I-to-B, B-to-I, P-to-B, and B-to-P frame conversions do affect other coded frames. This can be understood by considering the prediction dependency rules of MPEG. Specifically, since P frames are specified to depend on the nearest preceding I or P frame and B frames are specified to depend on the nearest surrounding I or P frames, it is understandable that frame conversions of certain types will affect the prediction dependency tree inferred from the frame coding types.

5. APPLICATIONS

This section shows how the compressed domain processing methods described in Section 4 can be applied to video transcoding and video processing/editing applications. Algorithms and architectures are described for a number of CDP operations.

5.1 COMPRESSED-DOMAIN TRANSCODING APPLICATIONS

With the introduction of the next generation wireless networks, mobile devices will access an increasing amount of media-rich content. However, a mobile device may not have enough display space to render content that was originally created for desktop clients. Moreover, wireless networks typically support lower bandwidths than wired networks, and may not be able to carry media content made for higher-bandwidth wired networks. In these cases, transcoders can be used to transform multimedia content to an appropriate video format and bandwidth for wireless mobile streaming media systems.

A conceptually simple and straightforward method to perform this transcoding is to decode the original video stream, downsample the decoded frames to a smaller size, and re-encode the downsampled frames at a lower bitrate. However, a typical CCIR601 MPEG-2 video requires almost all the cycles of a 300Mhz CPU to perform real-time decoding. Encoding is significantly more complex and usually cannot be accomplished in real time without the help of dedicated hardware or a high-end PC. These factors render the conceptually simple and straightforward transcoding method impractical. Furthermore, this simple approach can lead to significant loss in video quality. In addition, if transcoding is provided as a network service in the path between the content provider and content consumer, it is highly desirable for the transcoding unit to handle as many concurrent sessions as possible. This scalability is critical to enable wireless networks to handle user requests that may be very intense at high load times. Therefore, it is very important to develop fast algorithms to reduce the compute and memory loads for transcoding sessions.

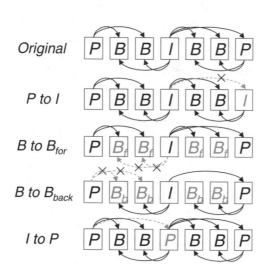

Frame Type	MB Type
I	i
P	i, p
B	i, b_{forw}, b_{back}, b
B_{intra}	i
B_{forw}	i, b_{forw}
B_{back}	i, b_{back}

Frame Conversion	MB Conversion
$P \rightarrow I$	$p \rightarrow i$
$B \rightarrow B_{forw}$	$b \rightarrow b_{forw}$ $b_{back} \rightarrow i$
$B \rightarrow B_{back}$	$b \rightarrow b_{back}$ $b_{forw} \rightarrow i$
$B \rightarrow B_{intra}$	$b \rightarrow i$ $b_{back} \rightarrow i$ $b_{forw} \rightarrow i$

Figure 40.3 Frame conversions and required macroblock conversions.

5.1.1 Compressed-Domain Transcoding Architectures

Video processing applications often involve a combination of spatial and temporal processing. For example, one may wish to downscale the spatial resolution and lower the frame rate of a video sequence. When these video processing applications are performed on compressed video streams, a number of additional requirements may arise. For example, in addition to performing the specified video processing task, the output compressed video stream may need to satisfy additional requirements such as maximum bitrate, buffer size, or particular compression format (e.g. MPEG-4 or H.263). While conventional approaches to applying traditional video processing operations on compressed video streams generally have high compute and memory requirements, the algorithmic optimizations described in Section 4 can be used to design efficient compressed-domain transcoding algorithms with significantly reduced compute and memory requirements. A number of transcoding architectures were discussed in [23][24][25][26].

Figure 10.1 shows a progression of architectures that reduce the compute and memory requirements of such applications. These architectures are discussed in the context of lowering the spatial and temporal resolution of the video from S_0, T_0 to S_1, T_1 and lowering the bitrate of the bitstream from R_0 to R_1. The top diagram shows the conventional approach to processing the compressed video stream. First the input compressed bitstream with bitrate R_0 is decoded into its decompressed video frames, which have a spatial resolution and temporal frame rate of S_0 and T_0. These frames are then processed temporally to a lower frame rate $T_1 < T_0$ by dropping appropriate frames. The spatial resolution is then reduced to $S_1 < S_0$ by spatially downsampling the remaining frames. The resulting frames with resolution S_1, T_1 are then re-encoded into a compressed bitstream with a final bitrate of $R_1 < R_0$. The memory requirements of this approach are high because of the frame stores required to store the decompressed video frames at resolution S_0, T_0. The computational requirements are high because of the operations needed to decode, process, and re-encode the frames; in particular, motion estimation performed during re-encoding can be quite compute intensive.

The bottom diagram shows an improved approach for this transcoding operation. Once again, the temporal frame rate is reduced at the bitstream layer by exploiting the picture start codes and picture headers. Furthermore, deriving the output coding parameters from those given in the input bitstream can significantly reduce the compute requirements of the final encode operation. This is advantageous because some of the computations that need to be performed in the encoder, such as motion estimation, may have already been performed by the original encoder and may be represented by coding parameters, such as motion vectors, given in the input bitstream. Rather than blindly recomputing this information from the decoded, downsampled video frames, the encoder can exploit the information contained in the input bitstream. In other words, much of the information that is derived in the original encoder can be reused in the transcoder. Specifically, the motion vectors, quantization parameters, and prediction modes contained in the input compressed bitstream can be used to calculate the motion vectors, quantization parameters, and prediction modes used in the encoder, thus largely bypassing the expensive operations performed in the conventional encoder.

Also, when transcoding to reduce the spatial resolution, the number of macroblocks in the input and output frames can differ; the bottom architecture can be further improved to consider this difference and achieve a better tradeoff in complexity and quality [23]. Note that the DCT-domain methods discussed in Section 4 can be used for further improvements.

5.1.2 Intraframe Transcoding

Images and video frames coded with intraframe methods are represented by sets of block DCT coefficients. When using intraframe DCT coding, the original video frame is divided into 8x8 blocks, each of which is independently transformed with an 8x8 DCT. This imposes an artificial block structure that complicates a number of spatial processing operations, such as translation, downscaling, and filtering, that were considered straightforward in the pixel domain.

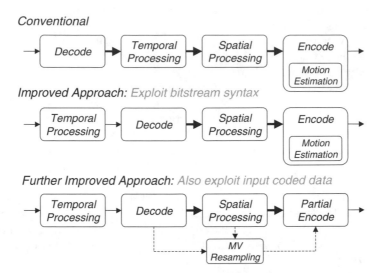

Figure 40.4 Architectural development of CDP algorithms.

For spatial downsampling or resolution reduction on an intra-coded frame, one 8x8 DCT block of the downscaled image is determined from multiple 8x8 DCT blocks of the original image. Efficient downsampling algorithms can be derived in the DCT domain. Based on the distributed property of the DCT discussed in Subsection 4.1, DCT-domain downsampling can be achieved by matrix multiplication. Merhav and Bhaskaran [28] have developed an efficient matrix multiplication for downscale of DCT blocks. Natarajan and Bhaskaran [18] also used approximated DCT matrices to achieve the same goal. The approximated DCT matrices contain only elements of value 0, 1, or a power of ½. Effectively, the matrix multiplication can be achieved by integer shifts and additions, leading to a multiplication free implementation.

Efficient algorithms have also been developed for filtering images in the DCT domain. For example, [29] proposes a method to apply two-dimensional symmetric, separable filters to DCT-coded images.

5.1.3 Interframe Transcoding

Video frames coded with interframe coding techniques are represented with motion vectors and residual DCT coefficients. These frames are coded based on a prediction from one or more previously coded frames; thus, properly decoding one frame requires first decoding one or more other frames. This temporal dependence among frames severely complicates a number of spatial and temporal processing techniques such as translation, downscaling, and splicing.

To facilitate efficient transcoding in the compressed domain, one wants to reuse as much information as possible in the origin video bitstream. The motion vector information of the transcoded video can be derived using the motion vector processing method introduced in Subsection 4.2. The computing of the residual DCT data can follow the guidelines provided in Subsection 4.3. Specifically, an interframe representation can be transcoded to an intraframe representation in the DCT domain. Subsequently, the DCT domain residual data can be obtained based on the derived motion vector information.

5.1.4 Format Conversion: Video Downscaling

Downscaling, or reducing the spatial resolution, of compressed video streams is an operation that benefits from the compressed-domain methods described in Section 4 and the compressed-domain transcoding architectures presented in Subsection 5.1.1. A block diagram of the compressed-domain downscaling algorithm is shown in Figure 40.5. The input bitstream is partially decoded into its motion vector and DCT domain representation. The motion vectors are resampled with the MV resampling methods described in Subsection 4.2. The DCT coefficients are processed with the DCT-domain processing techniques described in Subsections 4.1 and 4.3. A number of coding parameters from the input bitstream are extracted and used in the MV resampling and partial encoding steps of the transcoder. Rate control techniques, like those described in Section 4.4, are used to adapt the bitrate of the output stream. This is discussed in more detail below.

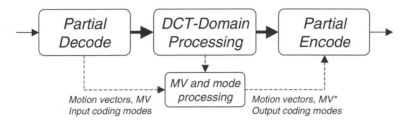

Figure 40.5 Compressed-domain downscaling algorithm.

The compressed-domain downscaling operation is complicated by the prediction dependencies used between frames during compression. Specifically, there are two tracks of dependencies in such a transcoding session. The first dependency is among frames in the original input video stream, while the second is among frames in the output downsampled video stream. The motion vectors for the downsampled version can be estimated based on the motion vectors in the original video. However, even when the motion information in the original video is reused, it is necessary to reconstruct the reference frames to avoid drift error due to imperfect motion vector estimation. As described in Subsection 4.3, the reconstruction may be performed using a DCT domain motion compensation method.

The selection of coding type for macroblock in the interframes is also an important issue. In the downsampling-by-two case, there may be four macroblocks each with a different coding type involved in the creation of each output macroblock; the transcoder may choose the dominant coding type as the coding type for the output macroblock. In addition, rate control must be used to control the bitrate of the transcoding result.

5.1.5 Format Conversion: Field-to-Frame Transcoding

This section focuses on the problem of transcoding a field-coded compressed bitstream to a lower-rate, lower-resolution frame-coded compressed bitstream [26]. For example, conversions between interlaced MPEG-2 sequences to progressive MPEG-1, H.261, H.263, or MPEG-4 simple profile streams lie within this space. To simplify discussion, this section focuses on transcoding a given MPEG-2 bitstream to a lower-rate H.263 or MPEG-4 simple profile bitstream

[26][30][31]. This is a practically important transcoding problem for converting MPEG-2 coded DVD and Digital TV video, which is often interlaced, to H.263 or MPEG-4 video for streaming over the Internet or over wireless links (e.g. 3G cellular) to PCs, PDAs, and cell phones that usually have progressive displays. For brevity, we refer to the output format as H.263; however it can be H.261, H.263, MPEG-1, or MPEG-4.

The conventional approach to the problem is as follows. An MPEG bitstream is first decoded into its decompressed interlaced video frames. These high-resolution interlaced video frames are then downsampled to form a progressive video sequence with a lower spatial resolution and frame rate. This sequence is then re-encoded into a lower-rate H.263 bitstream. This conventional approach to transcoding is inefficient in its use of computational and memory resources. It is desirable to have computation- and memory-efficient algorithms that achieve MPEG-2 to H.263 transcoding with minimal loss in picture quality.

A number of issues arise when designing MPEG-2 to H.263 transcoding algorithms. While both standards are based on block motion compensation and the block DCT, there are many differences that must be addressed. A few of these differences are listed below:

- *Interlaced vs. progressive video format:* MPEG-2 allows interlaced video formats for applications including digital television and DVD. H.263 only supports progressive formats.
- *Number of I frames:* MPEG uses more frequent I frames to enable random access into compressed bitstreams. H.263 uses fewer I frames to achieve better compression.
- *Frame coding types:* MPEG allows pictures to be coded as I, P, or B frames. H.263 has some modes that allow pictures to be coded as I, P, or B frames; but has other modes that only allow pictures to be coded as I, P, or optionally PB frames. Traditional I, P, B frame coding allows any number of B frames to be included between a pair of I or P frames, while H.263 I, P, PB frame coding allows at most one.
- *Prediction modes:* In support of interlaced video formats, MPEG-2 allows field-based prediction, frame-based prediction, and 16x8 field-based prediction. H.263 only supports frame-based prediction but optionally allows an advanced prediction mode in which four motion vectors are allowed per macroblock.
- *Motion vector restrictions:* MPEG motion vectors must point inside the picture, while H.263 has an unrestricted motion vector mode that allows motion vectors to point outside the picture. The benefits of this mode can be significant, especially for lower-resolution sequences where the boundary macroblocks account for a larger percentage of the video.

A block diagram of the MPEG-2 to H.263 transcoder [26][30] is shown in Figure 40.6. The transcoder accepts an MPEG IPB bitstream as input. The bitstream is scanned for picture start codes and the picture headers are examined to determine the frame type. The bits corresponding to B frames are discarded, while the remaining bits are passed on to the MPEG IP decoder. The decoded frames are downsampled to the appropriate spatial resolution and then passed to the modified H.263 IP encoder.

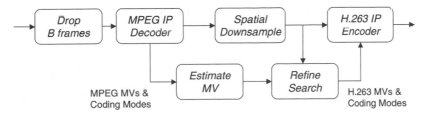

Figure 40.6 MPEG-2 to H.263 transcoder block diagram.

This encoder differs from a conventional H.263 encoder in that it does not perform conventional motion estimation; rather, it uses motion vectors and coding modes computed from the MPEG motion vectors and coding modes and the decoded, downsampled frames. There are a number of ways that this motion vector resampling can be done [4][5]. The class IV partial search method described in Subsection 4.2 was chosen. Specifically, the MPEG motion vectors and coding modes are used to form one or more initial estimates for each H.263 motion vector. A set of candidate motion vectors is generated; this set may include each initial estimate and its neighbouring vectors, where the size of the neighbourhood can vary depending on the available computational resources. The set of candidate motion vectors is tested on the decoded, downsampled frames and the best vector is chosen based on a criterion such as residual energy. A half-pixel refinement may be performed and the final mode decision (inter or intra) is then made.

Design considerations

Many degrees of freedom exist when designing an MPEG-2 to H.263 transcoder. For instance, a designer can make different choices in the mapping of input and output frame types; and the designer can choose how to vary the temporal frame rate and spatial resolution. Each of these decisions has a different impact on the computational and memory requirements and performance of the final algorithm. This section presents a very simple algorithm that makes design choices that naturally match the characteristics of the input and output bitstreams.

The target format of the transcoder can be chosen based on the format of the input source bitstream. A careful choice of source and target formats can greatly reduce the computational and memory requirements of the transcoding operation.

Spatial and temporal resolutions: The chosen correspondence between the input and output coded video frames is shown in Figure 40.7. The horizontal and vertical spatial resolutions are reduced by factors of two because the MPEG-2 interlaced field format provides a natural factor of two reduction in the vertical spatial resolution. Thus, the spatial downsampling is performed by simply extracting the top field of the MPEG-2 interlaced video frame and horizontally downsampling it by a factor of two. This simple spatial downsampling method allows the algorithm to avoid the difficulties associated with interlaced to progressive conversions. The temporal resolution is reduced by a factor of three, because MPEG-2 picture start codes, picture headers, and prediction rules make it possible to efficiently discard B-frame data from the bitstream without impacting the remaining I and P frames. Note that even though only the top fields of the MPEG I and P frames are used in the H.263 encoder, both the top and bottom fields must be decoded because of the prediction dependencies that result from the MPEG-2 interlaced field coding modes.

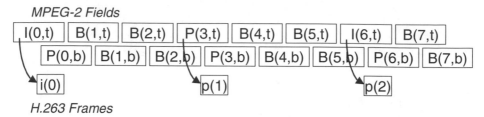

Figure 40.7 Video formats for MPEG-2 to H.263 transcoding.

Frame coding types: MPEG-2 allows I, P, and B frames while H.263 allows I and P frames and optionally PB frames. With sufficient memory and computational capabilities, an algorithm can be designed to transcode from any input MPEG coding pattern to any output H.263 coding pattern as in [31]. Alternatively, one may take the simpler approach of determining the coding pattern of the target H.263 bitstream based on the coding pattern of the source MPEG-2 bitstream. By aligning the coding patterns of the input and output bitstreams and allowing temporal downsampling, a significant improvement in computational efficiency can be achieved.

Specifically, a natural alignment between the two standards can be obtained by dropping the MPEG B frames and converting the remaining MPEG I and P frames to H.263 I and P frames, thus exploiting the similar roles of P frames in the two standards and exploiting the ease in which B frame data can be discarded from an MPEG-2 bitstream without affecting the remaining I and P frames. Since MPEG-2 sequences typically use an IBBPBBPBB structure, dropping the B frames results in a factor of three reduction in frame rate. While H.263 allows an advanced coding mode of PB pictures, it is not used in this algorithm because it does not align well with MPEG's IBBPBBPBB structure.

The problem that remains is to convert the MPEG-coded interlaced I and P frames to the spatially downsampled H.263-coded progressive I and P frames. The problem of frame conversions can be thought of as manipulating prediction dependencies in the compressed data; this topic was addressed in [22] and in Subsection 4.5 for MPEG progressive frame conversions. This MPEG-2 to H.263 transcoding algorithm requires three types of frame conversions: (1) MPEG I field to H.263 I frame, (2) MPEG I field to H.263 P frame, and (3) MPEG P field to H.263 P frame. The first is straightforward. The latter two require the transcoder to efficiently calculate the H.263 motion vectors and coding modes from those given in the MPEG-2 bitstream. When using the partial search method described in Subsection 4.3, the first step is to create one or more initial estimates of each H.263 motion vector from the MPEG-2 motion vectors. In the following two sections, we discuss the methods used to accomplish this for MPEG I field to H.263 P frame conversions and for MPEG P field to H.263 P frame conversions. Further details of the MPEG-2 to H.263 transcoder, including the progressive to interlace frame conversions, are given in [26][30]. These conversions address the differences between the MPEG-2 and H.263 standards described at the beginning of the section, and exploit the information in the input video stream to greatly reduce the computational and memory requirements of the transcoder with little loss in video quality.

5.2 EDITING

This section describes a series of compressed-domain editing applications. It begins with temporal mode conversion, which can be used to transcode an MPEG sequence into a format that facilitates video editing operations. It then describes two frame-level processing operations, frame-accurate splicing and frame-by-frame reverse play. All these operations use the frame conversion methods described in Subsection 4.5 to manipulate the prediction dependencies of compressed frames [22].

5.2.1 Temporal Mode Conversion

The ability to transcode between arbitrary temporal modes adds a great deal of flexibility and power to compressed-domain video processing. In addition, it provides a method of trading off parameters to achieve various rate/robustness profiles. For example, an MPEG sequence consisting of all I frames, while least efficient from a compression viewpoint, is most robust to channel impairments in a video communication system. In addition, the all I-frame MPEG video stream best facilitates many video-editing operations such as splicing, downscaling, and reverse play. Finally, once an I-frame representation is available, the intraframe transcoding algorithms described in Subsection 5.1 can be applied to each frame of the sequence to achieve the same effect on the entire sequence.

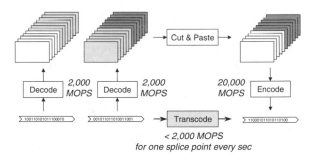

Figure 40.8 Splicing operation.

In general, temporal mode conversions can be performed with the frame conversion method described in Subsection 4.5. For frames that need to be converted to different prediction modes, macroblock and block level processing can be used to convert the appropriate macroblocks between different types.

The following steps describe a DCT-domain approach to transcoding an MPEG video stream containing I, P, and B frames into an MPEG video stream containing only I frames. This processing must be performed for the appropriate macroblocks of the converted frames.

1. *Calculate the DCT coefficients of the motion-compensated prediction.* This can be calculated from the intraframe coefficients of the previously coded frames by using the compressed-domain inverse motion compensation routine described in Subsection 4.3.
2. *Form the intraframe DCT representation of each frame.* This step simply involves adding the predicted DCT coefficients to the residual DCT coefficients.

3. *Requantize the intraframe DCT coefficients.* This step must be performed to ensure that the buffer constraints of the new stream are satisfied. Requantization may be used to control the rate of the new stream.

4. *Reorder the coded data and update the relevant header information.* If B-frames are used, the coding order of the IPB MPEG stream will differ from the coding order of the I-only MPEG stream. Thus, the coded data for each frame must be shuffled appropriately. In addition, the appropriate parameters of the header data must be updated.

5.2.2 Frame-Accurate Splicing

The goal of the splicing operation is to form a video data stream that contains the first N_{head} frames of one video sequence and the last N_{tail} frames of another video sequence. For uncoded video, the solution is obvious: simply discard the unused frames and concatenate the remaining data. Two properties make this solution obvious: (1) the data needed to represent each frame is self-contained, i.e. it is independent of the data from other frames; and (2) the uncoded video data has the desirable property of original ordering, i.e. the order of the video data corresponds to the display order of the video frames. MPEG-coded video data does not necessarily retain these properties of temporal independence or original ordering (although it can be forced to do so at the expense of compression efficiency). This complicates the task of splicing two MPEG-coded data streams.

This section describes a flexible algorithm that splices two streams directly in the compressed domain [3]. The algorithm allows a natural tradeoff between computational complexity and compression efficiency, thus it can be tailored to the requirements of a particular system. This algorithm possesses a number of attributes. A minimal number of frames are decoded and processed, thus leading to low computational requirements while preserving compression efficiency. In addition, the head and tail data streams can be processed separately. Finally, if desired, the processing can be performed so that the final spliced data stream is a simple concatenation of the two streams and so that the order of the coded video data remains intact.

The conventional splicing solution is to completely decompress the video, splice the decoded video frames, and recompress the result. With this method, every frame in the spliced video sequence must be recompressed. This method has a number of disadvantages, including high computational requirements, high memory requirements, and low performance, since each recoding cycle can deteriorate the video data.

An improved compressed-domain splicing algorithm is shown in Figure 40.9. The computational requirements are reduced by only processing the frames affected by the splice, and by only decoding the frames needed for that processing. This is also shown in Figure 40.9. Specifically, the only frames that need to be recoded are within the GOPs affected by the head and tail cut points; at most, there will be one such GOP in the head data stream and one in the tail data stream. Furthermore, the only additional frames that need to be decoded are the I and P frames in the two GOPs affected by the splice.

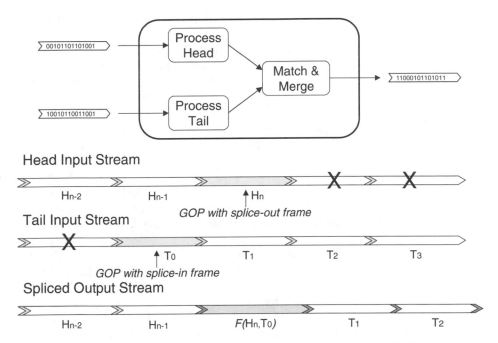

Figure 40.9 Compressed-domain splicing and processed bitstreams.

The algorithm results in an MPEG-compliant data stream with variable-sized GOPs. This exploits the fact that the GOP header does not specify the number of frames in the GOP or its structure; rather these are fully specified by the order of the data in the coded data stream.

Each step of the splicing operation is described below. Further discussion is included in [3].

1. Process the head data stream. This step involves removing any backward prediction dependencies on frames not included in the splice. The simplest case occurs when the cut for the head data occurs immediately after an I or P frame. When this occurs, there are no prediction dependencies on cut frames and all the relevant video data is contained in one contiguous portion of the data stream. The irrelevant portion of the data stream can simply be discarded, and the remaining relevant portion does not need to be processed. When the cut occurs immediately after a B frame, some extra processing is required because one or more B-frame predictions will be based on an anchor frame that is not included in the final spliced video sequence. In this case, the leading portion of the data stream is extracted up to the last I or P frame included in the splice, then the remaining B frames should be converted to B_{for} frames or P frames.

2. Form the tail data stream. This step involves removing any forward prediction dependencies on frames not included in the splice. The simplest case occurs when the cut occurs immediately before an I frame. When this occurs, the video data preceding this frame may be discarded and the remaining portion does not need to be processed. When the cut occurs before a P frame, the P frame must be converted to an I frame and the remaining data remains intact. When the cut occurs before a B frame, extra processing is required because one of the anchor frames is not included in the spliced sequence. In this case, if the first non-B frame is a P frame, it must be converted to an I frame. Then, each of the first consecutive B frames must be converted to B_{back} frames.

3. Match and merge the head and tail data streams. The IPB structure and the buffer parameters of the head and tail data streams determine the complexity of the matching operation. This step requires concatenating the two streams and then processing the frames near the splice point to ensure that the buffer constraints are satisfied. This requires *matching* the buffer parameters of the pictures surrounding the splice point. In the simplest case, a simple requantization will suffice. However, in more difficult cases, a frame conversion will also be required to prevent decoder buffer underflow. Furthermore, since prediction dependencies are inferred from the coding order of the compressed stream, when the merging step is performed the coded frames must be interleaved appropriately. The correct ordering will depend on the particular frame conversions used to remove the dependencies on cut frames.

The first two steps may require converting frames between the I, P, and B prediction modes. Converting P or B frames to I frames is quite straightforward as is B-to-B_{for} conversion and B-to-B_{back} conversion; however, conversion between any other set of prediction modes can require more computations to compute new motion vectors. Exact algorithms involve performing motion estimation on the decoded video -- this process can dominate the computational requirements of the algorithm. Approximate algorithms such as motion vector resampling can significantly reduce the computations required for these conversions.

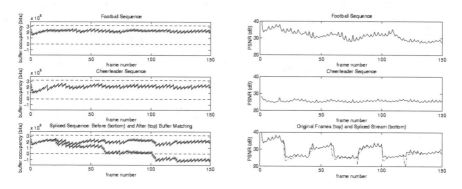

Figure 40.10 Performance of compressed-domain splicing algorithm.

Results of a spliced video sequence are shown in Figure 40.10. The right side of the figure plots the frame quality (in peak signal-to-noise ratio) for original compressed football and cheerleader sequences, and the spliced result when splicing between the two sequences every twenty frames. In the spliced result, the solid line contains the original quality values from the corresponding frames in the original coded football and cheerleader sequences, while the dotted line represents the quality of the sequence resulting from the compressed-domain splicing operation. Note that the spliced sequence has a slight degradation in quality at the splice points. This slight loss in quality is due to the removal of prediction dependencies in the compressed video in conjunction with the rate matching needed to satisfy buffer requirements. However, note that it returns to full quality a few frames after the splice point (within one GOP). The plots on the left show the buffer occupancy for the original input sequences and the output spliced sequence. In the bottom plot, the bottom line shows the buffer usage if the rate matching operation is not performed; this results in an eventual decoder buffer underflow. The top line shows the result of the compressed-domain splicing algorithm with appropriate rate matching. In this case, the buffer occupancy levels stay consistent with the original streams except in small areas surrounding the splice points. However, as we saw in the quality plots, the quality and buffer occupancy levels match those of the input sequences within a few frames.

5.2.3 Frame-by-Frame Reverse Play

The goal of the compressed-domain reverse-play operation is to create a new MPEG data stream that, when decoded, displays the video frames in the reverse order from the original MPEG data stream. For uncoded video the solution is simple: reorder the video frame data in reverse order. The simplicity of this solution relies on two properties: the data for each video frame is self-contained and it is independent of its placement in the data stream. These properties typically do not hold true for MPEG-coded video data.

Compressed-domain reverse-play is difficult because MPEG compression is not invariant to changes in frame order, e.g. reversing the order of the input frames will not simply reverse the order of the output MPEG stream. Furthermore, reversing the order of the input video frames does not result in a "reversed" motion vector field. However, if the processing is performed carefully, much of the motion vector information contained in the original MPEG video stream can be reused to save a significant amount of computations.

This section describes a reverse-play transcoding algorithm that operates directly on the compressed-domain data [32][33]. This algorithm is simple and achieves high performance with low computational and memory requirements. This algorithm only decodes the following data from the original MPEG data stream: I frames must be partially decompressed into their DCT representation and P frames must be partially decompressed to their MV/DCT representation, while for B frames only the forward and backward motion vector fields need to be decoded, i.e. only bitstream processing is needed.

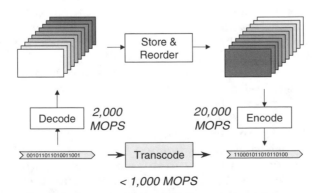

Figure 40.11 Reverse play operation.

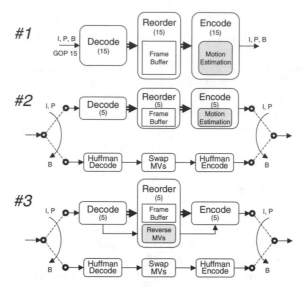

Figure 40.12 Architectures for compressed-domain reverse play.

The development of the compressed-domain reverse play algorithm is shown in Figure 40.12. In the conventional approach shown in the top of the figure, each GOP in the MPEG stream, starting from the end of the sequence, is completely decoded into uncompressed frames and stored in a frame buffer. The uncompressed frames are reordered, and the resulting frames are re-encoded into an output MPEG stream that contains the original frames in reverse order.

The middle figure shows an improved approach to the algorithm. This improvement results from exploiting the symmetry of B frames. Specifically, it uses the fact that the coding of the reverse-ordered sequence can be performed so that the same frames are coded as B frames and thus will have the same surrounding anchor frames. The one difference will be that the forward and backward anchors will be reversed. In this case, major computational savings can be achieved by performing simplified processing on the B frames. Specifically, for B frames only a bitstream-level decoding is used to efficiently decode the motion vectors and coding modes, swap them between forward and backward modes, and repackage the results. This greatly reduces the computational requirements because 2/3 of the frames are B frames and because typically the processing required for B frames is greater than that required for P frames, which in turn is much greater than that required for I frames. Also, note that the frame buffer requirements are reduced by a factor of three because the B frames are not decoded.

The bottom figure shows a further improvement that can be had by using motion vector resampling, as described in Subsection 4.2, on the I and P frames. In this architecture, the motion vectors given in the input bitstream are used to compute the motion vectors for the output bitstream, thereby avoiding the computationally expensive motion estimation process in the re-encoding process. The computational and performance tradeoffs of these architectures are discussed in detail in [5].

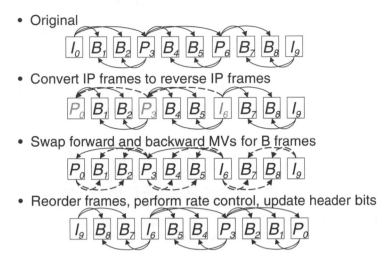

Figure 40.13 MPEG compressed-domain reverse play algorithm.

The resulting compressed-domain reverse-play algorithm shown in Figure 40.13 has the following steps:

1. *Convert the IP frames to reverse IP frames.* While the input motion vectors were originally computed for forward prediction between the I and P frames, the reverse IP frames require output motion vectors to be converted in the reverse order. Motion vector resampling methods described in Subsection 4.2 and in [5] can be used to calculate the new reversed motion vectors. Once the motion vectors are computed, the new output DCT coefficients can be computed directly in the DCT-domain by using the compressed-domain inverse motion compensation algorithm described in Subsection 4.3.

2. *Exchange the forward and backward motion vector fields used in each B frame.* This step exploits the symmetry of the B frame prediction process. In the reversed stream, the B frames will have the same two anchor frames, but in the reverse order. Thus, the forward prediction field can simply be exchanged with the backward prediction field, resulting in significant computational savings. Notice that only the motion vector fields need to be decoded for the B frames.

3. *Requantize the DCT coefficients.* This step must be performed to ensure that the buffer constraints of the new stream are satisfied. Requantization may be used to control the rate of the new stream.

4. *Properly reorder the frame data and update the relevant header information.* If no B frames are used, then the reordering process is quite straightforward. However, when B frames are used, care must be taken to properly reorder the data from the original coding order to the appropriate reverse coding order. In addition, the parameters in the header data must be updated appropriately.

6. ADVANCED TOPICS

6.1 OBJECT-BASED TO BLOCK-BASED TRANSCODING

This chapter focused on compressed-domain processing and transcoding algorithms for block-based compression schemes such as MPEG-1, MPEG-2, MPEG-4 simple profile, H.261, H.263, and H.264/MPEG-4 AVC. These compression standards represent each video frame as a rectangular array of pixels, and perform compression based on block-based processing, e.g. the block DCT and block-based motion estimation and motion compensated prediction. These compression algorithms are referred to as block- or frame-based schemes. Recently, object-based representations and compression algorithms have been developed -- the object-based coding part of MPEG-4 is the most well known example. These object-based representations decompose the image or video into arbitrarily shaped (non-rectangular) objects, unlike the block-based representations discussed above.

Object-based representations provide a more natural representation than square blocks, and can facilitate a number of new functionalities such as interactivity with objects in the video and greater content-creation flexibility. The object-based profiles of MPEG-4 are especially appealing for content creation and editing. For example, it may be useful to separately represent and encode different objects, such as different people or foreground or background objects, in a video scene in order to simplify manipulation of the scene. Therefore, object-based coding, such as MPEG-4, may become a natural approach to create, manipulate, and distribute new content. On the other hand, most clients may have block-based decoders, especially thin clients such as PDAs or cell phones. Therefore, it may become important to be able to efficiently transcode from object-based coding to block-based coding, e.g. from object-based MPEG-4 to block-based MPEG-2 or MPEG-4 simple profile. Efficient object-based to block-based transcoding algorithms were developed for intraframe (image) and interframe (video) compression in [7]. These efficient transcoding algorithms use many of the compressed-domain methods described in Section 4.

At each time instance (or frame), a video object has a shape, an amplitude (texture) within the shape, and a motion from frame to frame. In object-based coding, the shape (or support region) of the arbitrarily shaped object is often represented by a binary mask, and the texture of the object is represented by DCT transform coefficients. The object-based coding tools are often designed based on block-based coding tools. Typically in object-based image coding, such as in MPEG-4, a bounding box is placed around the object and the box is divided into blocks. The resulting blocks are classified as interior, boundary, or exterior blocks based on whether the block is completely within, partially within, or completely outside the object's support. For intraframe coding, a conventional block-DCT is applied to interior blocks and a modified block transform is applied to boundary blocks. For interframe coding, a macroblock and transform block structure similar to block-based video coding is used, where motion vectors are computed for macroblocks and conventional or modified block transforms are applied to interior and boundary blocks.

Many of the issues that arise in intraframe object-based to block-based transcoding algorithms can be understood by considering the simplified problem of overlaying an arbitrarily shaped object onto a fixed rectangular image, and producing the output compressed image that contains the rectangular image with the arbitrarily shaped overlay.

The simplest case occurs when the block boundaries of the fixed rectangular image and of the overlaid object are aligned. In this case, the output blocks can be computed in one of three cases. First, output image blocks that do not contain any portion of the overlay object may be simply copied from the corresponding block in the fixed rectangular image. Second, output image blocks that are completely covered by the overlaid object are replaced with the object's corresponding interior block. Finally, output image blocks that partially contain pixels from the rectangular image and the overlaid object are computed from the corresponding block from the fixed rectangular image and the corresponding boundary block from the overlaid object. Specifically, the new output coded block can be computed by properly masking the two blocks according to the object's segmentation mask. This can be computed in the spatial domain by inverse transforming the corresponding blocks in the background image and object, appropriately combining the two blocks with a spatial-domain masking operation, and transforming the result. Alternatively, it can be computed with compressed-domain masking operations, as described in Subsections 4.1, to reduce the computational requirements of the operation.

If the block boundaries of the object are not aligned with the block boundaries of the fixed rectangular image, then the affected blocks need additional processing. In this scenario, a shifting operation and a combined shifting/masking operation are needed for the unaligned block boundaries. Once again, output blocks that do not contain any portion of the overlaid object are copied from the corresponding input block in the rectangular image. Each remaining output block in the original image will overlap with 2 to 4 of the overlaid object's coded blocks (depending on whether one or both of the horizontal and vertical axes are misaligned). For image blocks with full coverage of the object and for which all the overlapping object's blocks are interior blocks, a shifting operation can be used to compute the new output "shifted" block. For the remaining blocks, a combined shifting/masking operation can be used to compute the new output block. As in the previous example, these computations can be performed in the spatial domain, or possibly more efficiently in the transform domain using the operations described in Subsections 4.1 and 4.3.

The object-to-block based interframe (video) transcoding algorithm shares the issues that arise in the intraframe (image) transcoding algorithm with regard to the alignment of macroblock boundaries between the rectangular video and overlaid video object, or between multiple arbitrarily shaped video objects. Furthermore, a number of important problems arise because of the different prediction dependencies that exist for the multiple objects in the object-coded video and the desired single dependency tree for the block-based coded video. This requires significant manipulation of the temporal dependencies in the coded video. Briefly speaking, given multiple arbitrarily shaped objects described by shape parameters and motion and DCT coefficients, the transcoding algorithm requires the computation of output block-based motion vectors and DCT coefficients. The solution presented computes output motion vectors with motion vector resampling techniques and computes output DCT coefficients with efficient transform-domain processing algorithms for combinations of the shifting, masking, and inverse motion compensation operations. Furthermore, the algorithm uses macroblock mode conversions, similar to those described in Subsection 4.5 and [22], to appropriately compensate for prediction dependencies that originally may have relied upon areas now covered by the overlaid object. The reader is referred to [7] for a detailed description of the transcoding algorithm.

6.2 SECURE SCALABLE STREAMING

It should now be obvious that transcoding is a useful capability in streaming media and media communication applications, because it allows intermediate network nodes to adapt compressed media streams for downstream client capabilities and time-varying network conditions. An additional issue that arises in some streaming media and media communication applications is security, in that an application may require the transported media stream to remain encrypted at all times. In applications where this type of security is required, the transcoding algorithms described earlier in this chapter can only be applied by decrypting the stream, transcoding the decrypted stream, and encrypting the result. By requiring decryption at transcoding nodes, this solution breaks the end-to-end security of the system.

Secure Scalable Streaming (SSS) is a solution that achieves the challenge of simultaneously enabling security and transcoding; specifically it enables transcoding without decryption [34][35]. SSS uses jointly designed scalable coding and progressive encryption techniques to encode and encrypt video into secure scalable packets that are transmitted across the network. The joint encoding and encryption is performed such that these resulting secure scalable packets can be transcoded at intermediate, possibly untrusted, network nodes by simply truncating or discarding packets and without compromising the end-to-end security of the system. The secure scalable packets may have unencrypted headers that provide hints, such as optimal truncation points, which the downstream transcoders use to achieve rate-distortion (R-D) optimal fine-grain transcoding across the encrypted packets.

The transcoding methods presented in this chapter are very powerful in that they can operate on most standard-compliant streams. However, in applications that require end-to-end security (where the transcoder is not allowed to see the bits), SSS can be used with certain types of scalable image and video compression algorithms to simultaneously provide security and scalability by enabling transcoding without decryption.

6.3 APPLICATIONS TO MOBILE STREAMING MEDIA SYSTEMS

The increased bandwidth of next-generation wireless systems will make streaming media a critical component of future wireless services. The network infrastructure will need to be able to handle the demands of mobility and streaming media, in a manner that scales to large numbers of users. Mobile streaming media (MSM) systems can be used to enable media delivery over next-generation mobile networks. For example, a mobile streaming media content delivery network (MSM-CDN) can be used to efficiently distribute and deliver media content to large numbers of mobile users [36]. These MSM systems need to handle large numbers of compressed media streams; the CDP methods presented in this chapter can be used to do so in an efficient and scalable manner. For example, compressed-domain transcoding can be used to adapt media streams originally made for high-resolution display devices such as DVDs into media streams made for lower-resolution portable devices [26], and to adapt streams for different types of portable devices. Furthermore, transcoding can be used to adaptively stream content over error-prone, time-varying wireless links by adapting the error-resilience based on channel conditions [37]. When using transcoding sessions in mobile environments, a number of system-level technical challenges arise. For example, user mobility may cause a server handoff in an MSM-CDN, which in turn may require the midstream handoff of a transcoding session [38]. CDP is likely to play a critical and enabling role in next-generation MSM systems that require scalability and performance, and in many cases CDP will enable next-generation wireless, media-rich services.

ACKNOWLEDGMENTS

The authors gratefully acknowledge Dr. Fred Kitson and Dr. Mark Smith of HP Labs for their support of this technical work throughout the years.

REFERENCES

[1]   J. Mitchell, W. Pennebaker, C. Fogg, and D. LeGall, MPEG Video Compression Standard, Digital Multimedia Standards Series, Chapman and Hall, 1997.

[2]   V. Bhaskaran and K. Konstantinides, Image and Video Compression Standards: Algorithms and Architectures, Kluwer Academic Publishers, Second Edition, June 1997.

[3]   S. Wee and V. Bhaskaran, Splicing MPEG video streams in the compressed-domain, IEEE Workshop on Multimedia Signal Processing, June 1997.

[4]   B. Shen, I.K. Sethi, and V. Bhaskaran, Adaptive Motion Vector Resampling for Compressed Video Down-scaling, IEEE Transactions on Circuits and Systems for Video Technology, vol.9, no.6, pp.929-936, Sept. 1999.

[5]   S. Wee, Reversing motion vector fields, IEEE International Conference on Image Processing, Oct. 1998.

[6]   B. Shen, I.K.Sethi and V. Bhaskaran, Closed-Loop MPEG Video Rendering, International Conference on Multimedia Computing and Systems, pp. 286-293, Ottawa, Canada, June1997.

[7]   J.G. Apostolopoulos, Transcoding between object-based and block-based image/video representations, e.g. MPEG-4 to MPEG-2 transcoding, HP Labs technical report, May 1998.

[8]   S.-F. Chang and D. Messerschmitt, Manipulation and compositing of MC-DCT compressed video, IEEE Journal on Selected Areas in Communications, vol. 13, Jan. 1995.

[9]   B.C. Smith and L. Rowe Algorithms for Manipulating Compressed Images, IEEE Computer Graphics and Applications, Sept. 1993.

[10] B.C. Smith, Fast Software Processing of Motion JPEG Video, in Proc. of the Second ACM International Conference on Multimedia, ACM Press, pp. 77-88, San Francisco, Oct. 1994.

[11] B. Shen, I.K.Sethi and V.Bhaskaran, DCT convolution and its applications in compressed video editing, IEEE Trans. Circuits and Systems for Video Technology, vol.8, no.8, pp.947-952, Dec. 1998.

[12] B. Shen, Block Based Manipulations on Transform Compressed Images and Video, Multimedia Systems Journal, Vol. 6, No. 1, March 1998.

[13] Jaswant R. Jain and Anil K. Jain, Displacement Measurement and Its Application in Interframe Image Coding, IEEE Trans. Communications, vol. com-29, no. 12, pp. 1799-1808, Dec. 1981.

[14] Jeongnam Youn, Ming-Ting Sun, Chia-Wen Lin, Motion vector refinement for high-performance transcoding, IEEE Transactions on Multimedia, vol.1, no.1, pp. 30-40, March 1999.

[15] N. Merhav and V. Bhaskaran, Fast algorithms for DCT-domain image downsampling and for inverse motion compensation, IEEE Transactions on Circuits and Systems for Video Technology, vol. 7, June 1997.

[16] Y. Arai, T. Agui and M. Nakajima, A Fast DCT-SQ Scheme for Images, Trans. of The IEICE, vol. E71, no. 11, Nov. 1988.

[17] P. A. Assuncao and M. Ghanbari, A frequency-domain video transcoder for dynamic bit-rate reduction of MPEG-2 bit streams, IEEE Trans. On Circuits and Systems for Video Technology, vol. 8, no. 8, pp. 953-967, Dec. 1998.

[18] B. Natarajan and V. Bhaskaran, A fast approximate algorithm for scaling down digital images in the DCT domain, IEEE International. Conference On Image Processing, Washington DC. Oct. 1995.

[19] S. Liu, A.C. Bovik, Local Bandwidth Constrained Fast Inverse Motion Compensation for DCT-Domain Video Transcoding, IEEE Trans. On Circuits and Systems for Video Technology, vol. 12, no. 5, May 2002.

[20] A. Eleftheriadis and D. Anastassiou, Constrained and general dynamic rate shaping of compressed digital video, IEEE International Conference on Image Processing, Washington, D.C., 1995.

[21] Y. Nakajima, H, Hori and T. Kanoh, Rate conversion of MPEG coded video by re-quantization process, IEEE International Conference on Image Processing, Washington, D.C., 1995.

[22] S.J. Wee, Manipulating temporal dependencies in compressed video data with applications to compressed-domain processing of MPEG video, IEEE International Conference on Acoustics, Speech, and Signal Processing, Phoenix, Arizona, March 1999.

[23] H. Sun, W. Kwok, J. Zdepski, Architectures for MPEG compressed bitstream scaling, IEEE Transactions on Circuits Systems and Video Technology, April 1996.

[24] G. Keesman, R. Hellinghuizen, F. Hoeksema, and G. Heideman, Transcoding MPEG bitstreams, Signal Processing: Image Communication, September 1996.

[25] N. Bjork and C. Christopoulos, Transcoder architectures for video coding, IEEE International Conference on Image Processing, May 1998.

[26] S.J. Wee, J.G. Apostolopoulos, and N. Feamster, Field-to-Frame Transcoding with Temporal and Spatial Downsampling, IEEE International Conference on Image Processing, Kobe, Japan, October 1999.

[27] B. Shen and S. Roy, A Very Fast Video Spatial Resolution Reduction Transcoder, International Conference On Acoustics, Speech, and Signal Processing, May2002.

[28] N. Merhav and V. Bhaskaran, A fast algorithm of DCT-domain image downscaling, International Conference On Acoustics, Speech, and Signal Processing, Atlanta GA, May 1996.

[29] N. Merhav and R. Kresch, Approximate convolution using DCT coefficient multipliers, IEEE Transactions on Circuits and Systems for Video Technology, vol. 8, Aug. 1998.

[30] N. Feamster and S. Wee, An MPEG-2 to H.263 transcoder, SPIE Voice, Video, and Data Communications Conference, September 1999.

[31] T. Shanableh and M. Ghanbari, Heterogeneous video transcoding MPEG:1,2 to H.263, Packet Video Workshop, April 1999.

[32] S.J. Wee and B. Vasudev, Compressed-Domain Reverse Play of MPEG Video Streams, SPIE Voice, Video, and Data Communications Conference, Boston, MA, November 1998.

[33] M.-S. Chen and D. Kandlur, Downloading and stream conversion: supporting interactive playout of videos in a client station, International Conference on Multimedia Computing, May 1995.

[34] S.J. Wee and J.G. Apostolopoulos, Secure Scalable Video Streaming for Wireless Networks, IEEE International Conference on Acoustics, Speech, and Signal Processing, Salt Lake City, Utah, May 2001.

[35] S.J. Wee and J.G. Apostolopoulos, Secure Scalable Streaming Enabling Transcoding without Decryption, IEEE International Conference on Image Processing, Thessaloniki, Greece, October 2001.

[36] T. Yoshimura, Y. Yonemoto, T. Ohya, M. Etoh, S. Wee, Mobile Streaming Media CDN enabled by Dynamic SMIL, Eleventh International World Wide Web Conference, May 2002.

[37] G. De Los Reyes, A.R. Reibman, S.-F. Chang, J.C.-I Chuang, Error-resilient transcoding for video over wireless channels, IEEE Journal on Selected Areas in Communications, June 2000.

[38] S. Roy, B. Shen, V. Sundaram, R. Kumar, Application Level Hand-Off Support for Mobile Media Transcoding Sessions, Workshop on Network and Operating System Support for Digital Audio and Video, Miami Beach, Florida, May 2002.

41

OBJECTIVE VIDEO QUALITY ASSESSMENT

Zhou Wang, Hamid R. Sheikh, and Alan C. Bovik

Department of Electrical and Computer Engineering
The University of Texas at Austin
Austin, Texas, USA
`zhouwang@ieee.org, hamid.sheikh@ieee.org,`
`bovik@ece.utexas.edu`

1. INTRODUCTION

Digital video data, stored in video databases and distributed through communication networks, is subject to various kinds of distortions during acquisition, compression, processing, transmission and reproduction. For example, lossy video compression techniques, which are almost always used to reduce the bandwidth needed to store or transmit video data, may degrade the quality during the quantization process. For another instance, the digital video bitstreams delivered over error-prone channels, such as wireless channels, may be received imperfectly due to the impairment occurred during transmission. Package-switched communication networks, such as the Internet, can cause loss or severe delay of received data packages, depending on the network conditions and the quality of services. All these transmission errors may result in distortions in the received video data. It is therefore imperative for a video service system to be able to realize and quantify the video quality degradations that occur in the system, so that it can maintain, control and possibly enhance the quality of the video data. An effective image and video quality metric is crucial for this purpose.

The most reliable way of assessing the quality of an image or video is subjective evaluation, because human beings are the ultimate receivers in most applications. The mean opinion score (MOS), which is a subjective quality measurement obtained from a number of human observers, has been regarded for many years as the most reliable form of quality measurement. However, the MOS method is too inconvenient, slow and expensive for most applications.

The goal of objective image and video quality assessment research is to design quality metrics that can predict perceived image and video quality automatically.

Generally speaking, an objective image and video quality metric can be employed in three ways:

1. It can be used to *monitor* image quality for quality control systems. For example, an image and video acquisition system can use the quality metric to monitor and automatically adjust itself to obtain the best quality image and video data. A network video server can examine the quality of the digital video transmitted on the network and control video streaming.

2. It can be employed to *benchmark* image and video processing systems and algorithms. If multiple video processing systems are available for a specific task, then a quality metric can help in determining which one of them provides the best quality results.

3. It can be embedded into an image and video processing system to *optimize* the algorithms and the parameter settings. For instance, in a visual communication system, a quality metric can help optimal design of the prefiltering and bit assignment algorithms at the encoder and the optimal reconstruction, error concealment and postfiltering algorithms at the decoder.

Objective image and video quality metrics can be classified according to the availability of the original image and video signal, which is considered to be distortion-free or perfect quality, and may be used as a reference to compare a distorted image or video signal against. Most of the proposed objective quality metrics in the literature assume that the undistorted reference signal is fully available. Although "image and video quality" is frequently used for historical reasons, the more precise term for this type of metric would be image and video *similarity* or *fidelity* measurement, or full-reference (FR) image and video quality assessment. It is worth noting that in many practical video service applications, the reference images or video sequences are often not accessible. Therefore, it is highly desirable to develop measurement approaches that can evaluate image and video quality blindly. Blind or no-reference (NR) image and video quality assessment turns out to be a very difficult task, although human observers usually can effectively and reliably assess the quality of distorted image or video without using any reference. There exists a third type of image quality assessment method, in which the original image or video signal is not fully available. Instead, certain features are extracted from the original signal and transmitted to the quality assessment system as side information to help evaluate the quality of the distorted image or video. This is referred to as reduced-reference (RR) image and video quality assessment.

Currently, the most widely used FR objective image and video distortion/quality metrics are mean squared error (MSE) and peak signal-to-noise ratio (PSNR), which are defined as:

$$\text{MSE} = \frac{1}{N} \sum_{i=1}^{N} (x_i - y_i)^2 \tag{41.1}$$

$$\text{PSNR} = 10 \log_{10} \frac{L^2}{\text{MSE}} \tag{41.2}$$

where N is the number of pixels in the image or video signal, and x_i and y_i are the i-th pixels in the original and the distorted signals, respectively. L is the dynamic range of the pixel values. For an 8bits/pixel monotonic signal, L is equal to 255. MSE and PSNR are widely used because they are simple to calculate, have clear physical meanings, and are mathematically easy to deal with for optimization purposes (MSE is differentiable, for example). However, they have been widely criticized as well for not correlating well with perceived quality measurement [1-8]. In the last three to four decades, a great deal of effort has been made to develop objective image and video quality assessment methods (mostly for FR quality assessment), which incorporate perceptual quality measures by considering human visual system (HVS) characteristics. Some of the developed models are commercially available. However, image and video quality assessment is still far from being a mature research topic. In fact, only limited success has been reported from evaluations of sophisticated HVS-based FR quality assessment models under strict testing conditions and a broad range of distortion and image types [3,9-11].

This chapter will mainly focus on the basic concepts, ideas and approaches for FR image and video quality assessment. It is worth noting that a dominant percentage of proposed FR quality assessment models share a common error sensitivity based philosophy, which is motivated from psychophysical vision science research. Section 2 reviews the background and various implementations of this philosophy and also attempts to point out the limitations of this approach. In Section 3, we introduce a new way to think about the problem of image and video quality assessment and provide some preliminary results of a novel structural distortion based FR quality assessment method. Section 4 introduces the current status of NR/RR quality assessment research. In Section 5, we discuss the issues that are related to the validation of image and video quality metrics, including the recent effort by the video quality experts group (VQEG) in developing, validating and standardizing FR/RR/NR video quality metrics for television and multimedia applications. Finally, Section 6 makes some concluding remarks and provides a vision for future directions of image and video quality assessment.

2. FULL-REFERENCE QUALITY ASSESSMENT USING ERROR SENSITIVITY MEASURES

An image or video signal whose quality is being evaluated can be thought of as a sum of a perfect reference signal and an error signal. We may assume that the loss of quality is directly related to the strength of the error signal. Therefore, a natural way to assess the quality of an image is to quantify the error between the distorted signal and the reference signal, which is fully available in FR quality assessment. The simplest implementation of the concept is the MSE as given in (1). However, there are a number of reasons why MSE may not correlate well with the human perception of quality:

1. Digital pixel values on which the MSE is typically computed, may not exactly represent the light stimulus entering the eye.

2. The sensitivity of the HVS to the errors may be different for different types of errors, and may also vary with visual context. This difference may not be captured adequately by the MSE.

3. Two distorted image signals with the same amount of error energy may have very different types of errors.

4. Simple error summation, like the one implemented in the MSE formulation, may be markedly different from the way the HVS and the brain arrives at an assessment of the perceived distortion.

In the last three decades, most of the proposed image and video quality metrics have tried to improve upon the MSE by addressing the above issues. They have followed an error sensitivity based paradigm, which attempts to analyze and quantify the error signal in a way that simulates the characteristics of human visual error perception. Pioneering work in this area was done by Mannos and Sakrison [12], and has been extended by other researchers over the years. We shall briefly describe several of these approaches in this section. But first, a brief introduction to the relevant physiological and psychophysical components of the HVS will aid in the understanding of the algorithms better.

2.1 THE HUMAN VISUAL SYSTEM

Figure 41.1 schematically shows the early stages of the HVS. It is not clearly understood how the human brain extracts higher-level cognitive information from the visual stimulus in the later stages of vision, but the components of the HVS depicted in Figure 41.1 are fairly well understood and accepted by the vision science community. A more detailed description of the HVS may be found in [13-15].

2.1.1 Anatomy of the HVS

The visual stimulus in the form of light coming from objects in the environment is focussed by the optical components of the eye onto the retina, a membrane at the back of the eyes that contains several layers of neurons, including photoreceptor cells. The optics consists of the cornea, the pupil (the aperture that controls the amount of light entering the eye), the lens and the fluids that fill the eye. The optical system focuses the visual stimulus onto the retina, but in doing so blurs the image due to the inherent limitations and imperfections. The blur is low-pass, typically modelled as a linear space-invariant system characterized by a point spread function (PSF). Photoreceptor cells in the retina sample the image that is projected onto it.

There are two types of photoreceptor cells in the retina: the cone cells and the rod cells. The cones are responsible for vision in normal light conditions, while the rods are responsible for vision in very low light conditions, and hence are generally ignored in the modelling. There are three different types of cones, corresponding to three different light wavelengths to which they are most sensitive. The L-cones, M-cones and S-cones (corresponding to the Long, Medium and Short wavelengths at which their respective sensitivities peak) split the image projected onto the retina into three visual streams. These visual streams can be thought of as the Red, Green and Blue color components of the visual stimulus, though the approximation is crude. The signals from the photoreceptors pass through several layers of interconnecting neurons in the retina before being carried off to the brain by the optic nerve.

The photoreceptor cells are non-uniformly distributed over the surface of the retina. The point on the retina that lies on the visual axis is called the fovea (Figure 41.1), and it has the highest density of cone cells. This density falls off rapidly with distance from the fovea. The distribution of the ganglion cells, the neurons that carry the electrical signal from the eye to the brain through the optic nerve, is also highly non-uniform, and drops off even faster than the density of the cone receptors. The net effect is that the HVS cannot perceive the entire visual stimulus at uniform resolution.

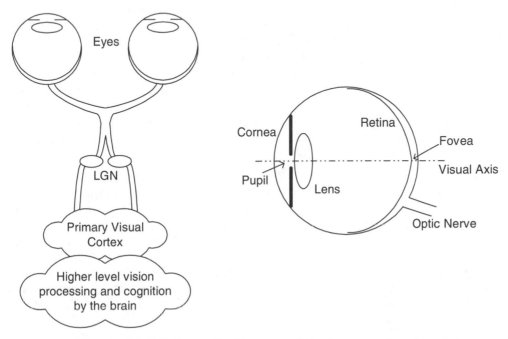

Figure 41.1 Schematic diagram of the human visual system.

The visual streams originating from the eye are reorganized in the optical chiasm and the lateral geniculate nucleus (LGN) in the brain, before being relayed to the primary visual cortex. The neurons in the visual cortex are known to be tuned to various aspects of the incoming streams, such as spatial and temporal frequencies, orientations, and directions of motion. Typically, only the spatial frequency and orientation selectivity is modelled by quality assessment metrics. The neurons in the cortex have receptive fields that are well approximated by two-dimensional Gabor functions. The ensemble of these neurons is effectively modelled as an octave-band Gabor filter bank [14,15], where the spatial frequency spectrum (in polar representation) is sampled at octave intervals in the radial frequency dimension and uniform intervals in the orientation dimension [16]. Another aspect of the neurons in the visual cortex is their saturating response to stimulus contrast, where the output of a neuron saturates as the input contrast increases.

Many aspects of the neurons in the primary visual cortex are not modelled for quality assessment applications. The visual streams generated in the cortex are carried off into other parts of the brain for further processing, such as motion sensing and cognition. The functionality of the higher layers of the HVS is currently an active research topic in vision science.

2.1.2 Psychophysical HVS Features

Foveal and Peripheral Vision

As stated above, the densities of the cone cells and the ganglion cells in the retina are not uniform, peaking at the fovea and decreasing rapidly with distance from the fovea. A natural result is that whenever a human observer fixates at a point in his environment, the region around the fixation point is resolved with the highest spatial resolution, while the resolution decreases with distance from fixation point. The high-resolution vision due to fixation by the observer onto a region is called *foveal* vision, while the progressively lower resolution vision is called *peripheral* vision. Most image quality assessment models work with foveal vision; a few incorporate peripheral vision as well [17-20]. Models may also resample the image with the sampling density of the receptors in the fovea in order to provide a better approximation of the HVS as well as providing more robust calibration of the model [17,18].

Light Adaptation

The HVS operates over a wide range of light intensity values, spanning several orders of magnitude from a moonlit night to a bright sunny day. It copes with such a large range by a phenomenon known as *light adaptation*, which operates by controlling the amount of light entering the eye through the pupil, as well as adaptation mechanisms in the retinal cells that adjust the gain of post-receptor neurons in the retina. The result is that the retina encodes the contrast of the visual stimulus instead of coding absolute light intensities. The phenomenon that maintains the contrast sensitivity of the HVS over a wide range of background light intensity is known as *Weber's Law*.

Contrast Sensitivity Functions

The contrast sensitivity function (CSF) models the variation in the sensitivity of the HVS to different spatial and temporal frequencies that are present in the visual stimulus. This variation may be explained by the characteristics of the receptive fields of the ganglion cells and the cells in the LGN, or as internal noise characteristics of the HVS neurons. Consequently, some models of the HVS choose to implement CSF as a filtering operation, while others implement CSF through weighting factors for subbands after a frequency decomposition. The CSF varies with distance from the fovea as well, but for foveal vision, the spatial CSF is typically modelled as a space-invariant band-pass function (Figure 41.2). While the CSF is slightly band-pass in nature, most quality assessment algorithms implement a low-pass version. This makes the quality assessment metrics more robust to changes in the viewing distance. The contrast sensitivity is also a function of temporal frequency, which is irrelevant for image quality assessment but has been modelled for video quality assessment as simple temporal filters [21-24].

Masking and Facilitation

Masking and facilitation are important aspects of the HVS in modelling the interactions between different image components present at the same spatial location. Masking/facilitation refers to the fact that the presence of one image component (called the *mask*) will decrease/increase the visibility of another image component (called the *test* signal). The mask generally reduces the visibility of the test signal in comparison with the case that the mask is absent. However, the mask may sometimes facilitate detection as well. Usually, the masking effect is the strongest when the mask and the test signal have similar frequency content and orientations. Most quality assessment methods incorporate one model of masking or the other, while some incorporate facilitation as well [1,18,25].

Pooling

Pooling refers to the task of arriving at a single measurement of quality, or a decision regarding the visibility of the artifacts, from the outputs of the visual streams. It is not quite understood as to how the HVS performs pooling. It is quite obvious that pooling involves cognition, where a perceptible distortion may be more annoying in some areas of the scene (such as human faces) than at others. However, most quality assessment metrics use Minkowski pooling to pool the error signal from the different frequency and orientation selective streams, as well as across spatial coordinates, to arrive at a fidelity measurement.

2.1.3 Summary

Summarizing the above discussion, an elaborate quality assessment algorithm may implement the following HVS features:

1. Eye optics modelled by a low-pass PSF.
2. Color processing.
3. Non-uniform retinal sampling.
4. Light adaptation (luminance masking).
5. Contrast sensitivity functions.
6. Spatial frequency, temporal frequency and orientation selective signal analysis.
7. Masking and facilitation.
8. Contrast response saturation.
9. Pooling.

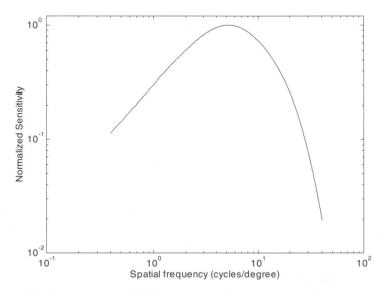

Figure 41.2 Normalized contrast sensitivity function.

2.2 GENERAL FRAMEWORK OF ERROR SENSITIVITY BASED METRICS

Most HVS based quality assessment metrics share an error-sensitivity based paradigm, which aims to quantify the strength of the errors between the reference and the distorted signals in a perceptually meaningful way. Figure 41.3 shows a generic error-sensitivity based quality assessment framework that is based on HVS modelling. Most quality assessment algorithms that model the HVS can be explained with this framework, although they may differ in the specifics.

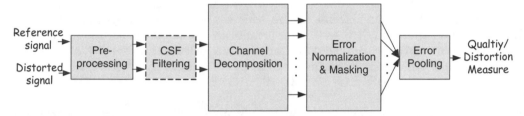

Figure 41.3 Framework of error sensitivity based quality assessment system. Note: the CSF feature can be implemented either as "CSF Filtering" or within "Error Normalization."

Pre-processing

The pre-processing stage may perform the following operations: alignment, transformations of color spaces, calibration for display devices, PSF filtering and light adaptation. First, the distorted and the reference signals need to be properly aligned. The distorted signal may be misaligned with respect to the reference, globally or locally, for various reasons during compression, processing, and transmission. Point-to-point correspondence between the reference and the distorted signals needs to be established. Second, it is sometimes preferable to transform the signal into a color space that conforms better to the HVS. Third, quality assessment metrics may need to convert the digital pixel values stored in the computer memory into luminance values of pixels on the display device through point-wise non-linear transformations. Fourth, a low-pass filter simulating the PSF of the eye optics may be applied. Finally, the reference and the distorted videos need to be converted into corresponding contrast stimuli to simulate light adaptation. There is no universally accepted definition of contrast for natural scenes. Many models work with band-limited contrast for complex natural scenes [26], which is tied with the channel decomposition. In this case, the contrast calculation is implemented later in the system, during or after the channel decomposition process.

CSF Filtering

CSF may be implemented before the channel decomposition using linear filters that approximate the frequency responses of the CSF. However, some metrics choose to implement CSF as weighting factors for channels after the channel decomposition.

Channel Decomposition

Quality metrics commonly model the frequency selective channels in the HVS within the constraints of application and computation. The channels serve to separate the visual stimulus into different spatial and temporal *subbands*. While some quality assessment algorithms implement sophisticated channel decompositions, simpler transforms such as the wavelet transform, or even the Discrete Cosine Transform (DCT) have been reported in the literature primarily due to their suitability for certain applications, rather than their accuracy in modelling the cortical neurons.

While the cortical receptive fields are well represented by 2D Gabor functions, the Gabor decomposition is difficult to compute and lacks some of the mathematical conveniences that are desired for good implementation, such as invertibility, reconstruction by addition, etc. Watson constructed the cortex transform [27] to model the frequency and orientation selective channels, which have similar profiles as 2D Gabor functions but are more convenient to implement. Channel decomposition models used by Watson, Daly [28,29], Lubin [17,18] and Teo and Heeger [1,25,30] attempt to model the HVS as closely as possible without incurring prohibitive implementation difficulties. The subband configurations for some of the models described in this chapter are given in Figure 41.4. Channels tuned to various temporal frequencies have also been reported in the literature [5,22,31,32].

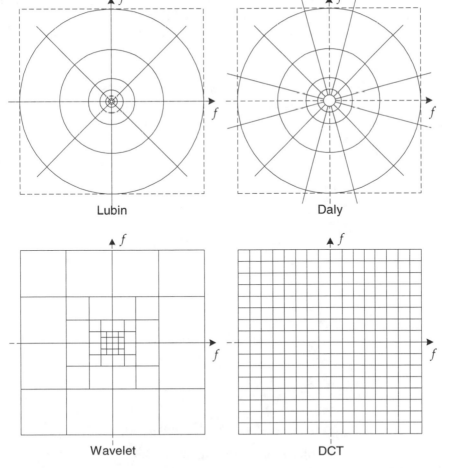

Figure 41.4 Frequency decompositions of various models.

Error Normalization and Masking

Error normalization and masking is typically implemented within each channel. Most models implement masking in the form of a gain-control mechanism that weights the error signal in a channel by a space-varying *visibility threshold* for that channel [33]. The visibility threshold adjustment at a point is calculated based on the energy of the reference signal (or both the reference and the distorted signals) in the neighbourhood of that point, as well as the HVS sensitivity for that channel in the absence of masking effects (also known as the *base-sensitivity*). Figure 41.5(a) shows how masking is typically implemented in a channel. For every channel the base error threshold (the minimum visible contrast of the error) is elevated to account for the presence of the masking signal. The threshold elevation is related to the contrast of the reference (or the distorted) signal in that channel through a relationship that is depicted in Figure 41.5(b). The elevated visibility threshold is then used to normalize the error signal. This normalization typically converts the error into units of Just Noticeable Difference (JND), where a JND of 1.0 denotes that the distortion at that point in that channel is just at the threshold of visibility. Some methods implement masking and facilitation as a manifestation of contrast response saturation. Figure 41.6 shows a set of curves each of which may represent the saturation characteristics of neurons in the HVS. Metrics may model masking with one or more of these curves.

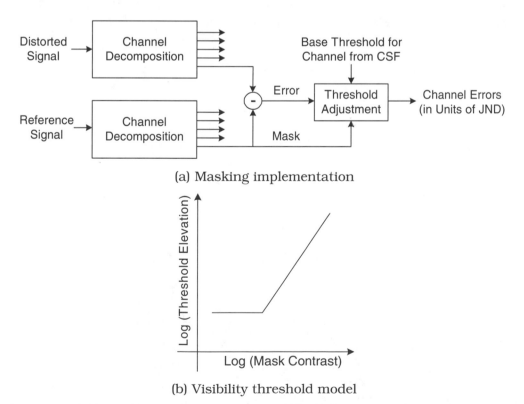

(a) Masking implementation

(b) Visibility threshold model

Figure 41.5 (a) Implementation of masking effect for channel based HVS models. (b) Visibility threshold model (simplified): threshold elevation versus mask contrast.

Error Pooling

Error pooling is the process of combining the error signals in different channels into a single distortion/quality interpretation. For most quality assessment methods, pooling takes the form:

$$E = \left(\sum_l \sum_k |e_{l,k}|^\beta \right)^{1/\beta}$$ (41.3)

where $e_{l,k}$ is the normalized and masked error of the k-th coefficient in the l-th channel, and β is a constant typically with a value between 1 and 4. This form of error pooling is commonly called Minkowski error pooling. Minkowski pooling may be performed over space (index k) and then over frequency (index l), or vice versa, with some non-linearity between them, or possibly with different exponents β. A spatial map indicating the relative importance of different regions may also be used to provide spatially variant weighting to different $e_{l,k}$ [31,34,35].

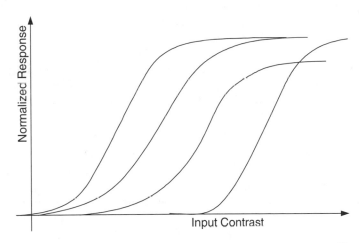

Figure 41.6 Non-linear contrast response saturation effects.

2.3 IMAGE QUALITY ASSESSMENT ALGORITHMS

Most of the efforts in the research community have been focussed on the problem of image quality assessment, and only recently has video quality assessment received more attention. Current video quality assessment metrics use HVS models similar to those used in many image quality assessment metrics, with appropriate extensions to incorporate the temporal aspects of the HVS. In this section, we present some image quality assessment metrics that are based on the HVS error sensitivity paradigm. Later we will present some video quality assessment metrics. A more detailed review of image quality assessment metrics may be found in [2,36].

The visible differences predictor (VDP) by Daly [28,29] aims to compute a probability-of-detection map between the reference and the distorted signal. The value at each point in the map is the probability that a human observer will perceive a difference between the reference and the distorted images at that point. The reference and the distorted images (expressed in luminance values instead of pixels) are passed through a series of processes: point non-linearlity, CSF filtering, channel decomposition, contrast calculation, masking effect modelling, and probability-of-detection calculation. A modified cortex transform [27] is used for channel decomposition, which transforms the image signal into five spatial levels followed by six orientation levels, leading to a total of 31 independent channels (including the baseband). For each channel, a threshold elevation map is computed from the contrast in that channel. A psychometric function is used to convert error strengths (weighted by the threshold elevations) into a probability-of-detection map for each channel. Pooling is then carried out across the channels to obtain an overall detection map.

Lubin's algorithm [17,18] also attempts to estimate a detection probability of the differences between the original and the distorted versions. A blur is applied to model the PSF of the eye optics. The signals are then re-sampled to reflect the photoreceptor sampling in the retina. A Laplacian pyramid [37] is used to decompose the images into seven resolutions (each resolution is one-half of the immediately higher one), followed by band-limited contrast calculations [26]. A set of orientation filters implemented through steerable filters of Freeman and Adelson [38] is then applied for orientation selectivity in four orientations. The CSF is modelled by normalizing the output of each frequency-selective channel by the base-sensitivity for that channel. Masking is implemented through a sigmoid non-linearity, after which the errors are convolved with disk-shaped kernels at each level before being pooled into a distortion map using Minkowski pooling across frequency. An additional pooling stage may be applied to obtain a single number for the entire image.

Teo and Heeger's metric [1,25,30] uses PSF modelling, luminance masking, channel decomposition, and contrast normalization. The channel decomposition process uses quadrature steerable filters [39] with six orientation levels and four spatial resolutions. A detection mechanism is implemented based on squared error. Masking is modelled through contrast normalization and response saturation. The contrast normalization is different from Daly's or Lubin's method in that they take the outputs of channels at all orientations at a particular resolution to perform the normalization. Thus, this model does not assume that the channels at the same resolution are independent. Only channels at different resolutions are considered to be independent. The output of the channel decomposition after contrast normalization is decomposed four-fold by passing through four non-linearities of shapes as illustrated in Figure 41.6, the parameters for which are optimized to fit the data from psychovisual experiments.

Watson's DCT metric [40] is based on an 8×8 DCT transform commonly used in image and video compression. Unlike the models above, this method partitions the spectrum into 64 uniform subbands (8 in each Cartesian dimension). After the block-based DCT and the associated subband contrasts are computed, a visibility threshold is calculated for each subband coefficient within each block using the base-sensitivity for that subband. The base sensitivities are derived empirically. The thresholds are corrected for luminance and texture masking. The error in each subband is weighted by the corresponding visibility threshold and pooled using Minkowski pooling spatially. Pooling across subbands is then performed using the Minkowski formulation with a different exponent.

Safranek-Johnston's perceptual image coder [41] incorporates a quality metric using a similar strategy as in Watson's DCT metric. The channel decomposition uses a generalized quadrature mirror filter (GQMF) [42] for analysis and synthesis. This transform splits the spectrum into 16 uniform subbands (four in each Cartesian dimension). Masking and pooling methods are similar to those in Watson's DCT metric.

Bradley [43] reports a wavelet visible difference predictor (WVDP), which is a simplification of Daly's VDP described above. He uses Watson's derivation of 9/7 Wavelet quantization-noise detection thresholds [44] for a 9/7 biorthogonal wavelet [45] and combines it with a threshold elevation and psychometric detection probability scheme similar to Daly's. Another wavelet based metric has been proposed by Lai and Kuo [6]. Their metric is based on the Haar Wavelet and their masking model can account for channel interactions as well as suprathreshold effects.

The quality metrics proposed above are scalar valued metrics. Damera-Venkata *et al.* proposed a metric for quantifying performance of image restoration systems, in which the degradation is modelled as a linear frequency distortion and additive noise injection [46]. Two complementary metrics were developed to separately quantify these distortions. They observed that if the additive noise is uncorrelated with the reference image, then an error measure from an HVS based metric will correlate well with the subjective judgement. Using a spatially adaptive restoration algorithm [47] (which was originally designed for inverse-halftoning), they isolate the effects of noise and linear frequency distortion. The noise is quantified using a multichannel HVS based metric. A distortion measure quantifies the spectral distortion between the reference and the model restored image.

Some researchers have attempted to measure image quality using single-channel models with the masking-effect models specifically targeting certain types of distortions, such as the blocking artifact. Blocking is recognized as one of the most annoying artifacts in block-DCT based image/video compression such as JPEG, especially at high compression ratios. In [48] and [49], Karunasekera and Kingsbury proposed a quality metric for blocking artifacts. Edge detection is performed first on the error image. An activity map is calculated from the reference image in the neighbourhood of the edges, and an activity-masked edge image is computed such that edges that occur in high activity areas are de-emphasized. The activity-masked edge image is adjusted for luminance masking. A non-linear transformation is applied before pooling. The parameters for the model are obtained from experiments that measure the sensitivity of human observers to edge artifacts embedded in narrow-band test patterns.

In [50], Chou and Li defined a peak signal to perceptible noise ratio (PSPNR), which is a single-channel metric. They model luminance masking and activity masking to obtain a JND profile. The PSPNR has the same definition as given in (2) except that the MSE expression is adjusted for the JND profile.

Another single-channel metric is the objective picture quality scale (PQS) by Miyahara [51]; a number of features that can capture various distortions are combined into one score. The method has also been extended for video quality assessment [52].

2.4 VIDEO QUALITY ASSESSMENT ALGORITHMS

One obvious way to implement video quality metrics is to apply a still image quality assessment metric on a frame-by-frame basis. However, a more sophisticated approach would model the temporal aspects of the HVS in the design of the metric. A number of algorithms have been proposed to extend the HVS features into the dimensions of time and motion [5,22,24,32,52-54]. A survey of video coding distortions can be found in [55]. A review of HVS modelling for video quality metrics is presented in [56].

In [53], Tan *et al.* implemented a Video Distortion Meter by using an image quality assessment metric followed by a "cognitive emulator" that models temporal effects such as smoothing and temporal masking of the frame quality measure, saturation and asymmetric tracking. Asymmetric tracking models the phenomenon that humans tend to notice a quality transition from good to poor more readily than a quality transition from poor to good.

Van den Branden Lambrecht *et al.* has extended the HVS modelling into the time dimension by modelling the temporal dimension of the CSF, and by generating two visual streams tuned to different temporal aspects of the stimulus from the output of each spatial channel [21,22,31,32]. The two streams model the transient and the sustained temporal mechanisms in the HVS. His proposed moving picture quality metric (MPQM) consists of a channel decomposition into four scales, four orientations and two temporal streams. The resulting channel outputs are subtracted to create the error signal. Masking is implemented by normalization of the channel errors by the stimulus dependent visibility thresholds (similar to those used in still image quality assessment metrics). Motion rendering quality assessment has also been proposed by extending the MPQM by extraction of motion information [32].

In [5], Winkler presented a quality assessment metric for color video. The algorithm uses a color space transformation and applies the quality assessment metric on each transformed color channel. Two temporal streams are generated using IIR filters, with spatial decomposition into five subband levels and four orientations. Channels are weighted by the corresponding CSF, and masking is implemented based on the excitatory-inhibitory masking model proposed by Watson and Solomon [33].

Watson's digital video quality (DVQ) metric operates in the DCT domain and is therefore more attractive from an implementation point of view [24,57] since the DCT is efficient to implement and most video coding standards are based on the DCT. A three-dimensional visibility threshold model for spatiotemporal DCT channels was proposed. The DVQ algorithm first takes the DCT of the reference and the distorted signals, respectively. It then computes local contrast, applies temporal CSF filtering, and converts the results into JND units by normalizing them with the visibility thresholds, following which the error signal is computed. Finally, masking and pooling are applied to the error signal. DVQ implements color transformation before applying the metric to each of the chrominance dimensions.

Another metric that models the temporal aspects of HVS is presented by Tan and Ghanbari [54], which aims to evaluate the quality of MPEG video and combines a typical error-sensitivity based perceptual model with a blockiness measurement model. The perceptual model consists of display gamma correction, point non-linearity, contrast calculation, spatial CSF filtering, temporal filtering, frequency decomposition into two channels (diagonal and horizontal/vertical), contrast response non-linearity, error averaging, masking, pooling, temporal averaging and motion masking. The blockiness detector is based on harmonic analysis of the block-edge signal, combined with a visual masking model. The final quality score is either the perceptual model score or the blockiness detector score, based on the amount of blockiness artifact detected.

In [58], Yu *et al.* propose a video quality metric based on the extension of the perceptual distortion metric by Winkler [5] to a perceptual blocking distortion metric. The parameters for the models are obtained by minimizing the error in quality predictions for video sequences obtained from VQEG subjective testing database. This is in contrast to most methods that obtain parameters to fit threshold psychovisual experiments with simple patterns. They specifically address the blocking artifact by pooling spatially over those areas where blocking effects are dominant.

There are several implementation issues that need to be considered before developing a practical video quality assessment system. One important factor affecting the feasibility of a quality metric for video is its computational complexity. While complex quality assessment methods may model the HVS more accurately, their computational complexity may be prohibitively large for many platforms, especially for real-time quality assessment of high-resolution video. Memory requirements are another important issue. For example, in order to implement temporal filtering, a large memory space may be needed to store a number of video frames, which is expensive on many platforms. Another problem of HVS based metrics might be their dependence on viewing configurations, which include the resolution of the display devices, the non-linear relationships between the digital pixel values and the output luminance values, and the viewing distance of the observers. Most models either require that viewing configurations be known or simply assume a fixed set of configurations. How these metrics would perform when the configurations are unknown or the assumptions about the configurations do not hold is another issue that needs to be studied.

2.5 LIMITATIONS

The underlying principle of visual error sensitivity based algorithms is to predict perceptual quality by quantifying perceptible errors. This is accomplished by simulating the perceptual quality related functional components of the HVS. However, the HVS is an extremely complicated, highly non-linear system, and the current understanding of the HVS is limited. How far the error sensitivity based framework can reach is a question that may need many years to answer.

It is worth noting that most error sensitivity based approaches, explicitly or implicitly, make a number of assumptions. The following is an incomplete list (Note: a specific model may use a subset of these assumptions):

1. The reference signal is of perfect quality.

2. Light adaptation follows the Weber's law.

3. After light adaptation, the optics of the eye can be modelled as a linear time-invariant system characterized by a PSF.

4. There exist frequency, orientation and temporal selectivity visual channels in the HVS, and the channel responses can be modelled by a discrete set of linear decompositions.

5. Although the contrast definitions of simple patterns used in psychovisual experiments and the contrast definitions of complex natural images may be different, they are consistent with each other.

6. The relative visual error sensitivity between different spatial and/or temporal frequency channels can be normalized using a bandpass or lowpass CSF.

7. The channel decomposition is lossless or nearly lossless in terms of visual importance, in the sense that the transformed signals maintain most of the information needed to assess the image quality.

8. The channel decomposition effectively decorrelates the image structure, such that the inter- and intra-channel interactions between transformed coefficients can be modelled using a masking model, in which the strength of the mask is determined by the magnitudes (not structures) of the coefficients. After masking, the perceived error of each coefficient can be evaluated individually.

9. For a single coefficient in each channel, after error normalization and masking, the relationship between the magnitude of the error, $e_{l,k}$, and the distortion perceived by the HVS, $d_{l,k}$, can be modelled as a non-linear function: $d_{l,k} = \left| e_{l,k} \right|^{\beta}$.

10. The overall perceived distortion monotonically increases with the summation of the perceived errors of all coefficients in all channels.

11. The overall framework covers a complete set of dominant factors (light adaptation, PSF of the eye optics, CSF of the frequency responses, masking effects, etc.) that affect the perceptual quality of the observed image.

12. Higher level processes happening in the human brain, such as pattern matching with memory and cognitive understanding, are less important for predicting perceptual image quality.

13. Active visual processes, such as the change of fixation points and the adaptive adjustment of spatial resolution because of attention, are less important for predicting perceptual image quality.

Depending on the application environment, some of the above assumptions are valid or practically reasonable. For example, in image and video compression and communication applications, assuming a perfect original image or video signal (Assumption 1) is acceptable. However, from a more general point of view, many of the assumptions are arguable and need to be validated. We believe that there are several problems that are critical for justifying the usefulness of the general error-sensitivity based framework.

The Suprathreshold Problem
Most psychophysical subjective experiments are conducted near the threshold of visibility, typically using a 2-Alternative Forced-Choice (2AFC) method [14,59].

The 2AFC method is used to determine the values of stimuli strength (also called the *threshold strength*) at which the stimuli are *just visible*. These measured threshold values are then used to define visual error sensitivity models, such as the CSF and the various masking effect models. However, there is not sufficient evidence available from vision research to support the presumption that these measurement results can be generalized to quantify distortions much larger than *just visible*, which is the case for a majority of image processing applications. This may lead to several problems with respect to the framework. One problem is that when the error in a visual channel is larger than the threshold of visibility, it is hard to design experiments to validate Assumption 9. Another problem is regarding Assumption 6, which uses the just noticeable visual error threshold to normalize the errors between different frequency channels. The question is: when the errors are much larger than the thresholds, can the relative errors between different channels be normalized using the visibility thresholds?

The Natural Image Complexity Problem

Most psychovisual experimental results published in the literature are conducted using relatively simple patterns, such as sinusoidal gratings, Gabor patches, simple geometrical shapes, transform basis functions, or random noise patterns. The CSF is obtained from threshold experiments using single frequency patterns. The masking experiments usually involve two (or a few) different patterns. However, all such patterns are much simpler than real world images, which can usually be thought of as a superposition of a large number of different simple patterns. Are these simple-pattern experiments sufficient for us to build a model that can predict the quality of complex natural images? Can we generalize the model for the interactions between a few simple patterns to model the interactions between tens or hundreds of patterns?

The Minkowski Error Pooling Problem

The widely used Minkowski error summation formula (3) is based on signal differencing between two signals, which may not capture the structural changes between the two signals. An example is given in Figure 41.7, where two test signals, test signals 1 (up-left) and 2 (up-right), are generated from the original signal (up-center). Test signal 1 is obtained by adding a constant number to each sample point, while the signs of the constant number added to test signal 2 are randomly chosen to be positive or negative. The structural information of the original signal is almost completely lost in test signal 2, but preserved very well in test signal 1. In order to calculate the Minkowski error metric, we first subtract the original signal from the test signals, leading to the error signals 1 and 2, which have very different structures. However, applying the absolute operator on the error signals results in exactly the same absolute error signals. The final Minkowski error measures of the two test signals are equal, no matter how the β value is selected. This example not only demonstrates that "structure-preservation" ability is an important factor in measuring the similarity between signals, but also shows that Minkowski error pooling is inefficient in capturing the structures of errors and is a "structural information lossy" metric. Obviously, in this specific example, the problem may be solved by applying a spatial frequency channel decomposition on the error signals and weighting the errors differently in different channels with a CSF. However, the decomposed signals may still exhibit different structures in different channels (for example, assume that the test signals in Figure 41.7 are from certain channels instead of the spatial domain), then the "structural information lossy" weakness of the Minkowski metric may still play a role, unless the decomposition process strongly decorrelates the image structure, as described by Assumption 8, such that the correlation between adjacent samples of the decomposed signal is very small (in that case, the decomposed signal in a channel would look like random noise). However, this is apparently not the case for a linear channel decomposition method such as the wavelet transform. It has been shown that a strong correlation or dependency exists between intra- and inter-channel wavelet coefficients of natural images [60,61]. In fact, without exploiting this strong dependency, state-of-the-art wavelet image compression techniques, such as embedded zerotree wavelet (EZW) coding [62], set partitioning in hierarchical trees (SPIHT) algorithm [63], and JPEG2000 [64] would not be successful.

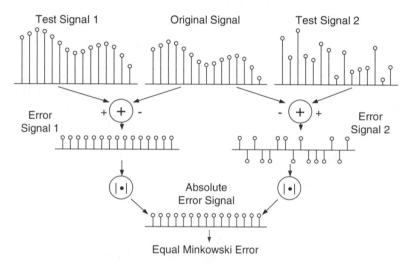

Figure 41.7 Illustration of Minkowski error pooling.

The Cognitive Interaction Problem

It is clear that cognitive understanding and active visual process (e.g., change of fixations) play roles in evaluating the quality of images. For example, a human observer will give different quality scores to the same image if s/he is instructed with different visual tasks [2,65]. Prior information regarding the image content, or attention and fixation, may also affect the evaluation of the image quality [2,66]. For example, it is shown in [67] that in a video conferencing environment, "the difference between sensitivity to foreground and background degradation is increased by the presence of audio corresponding to speech of the foreground person" [67]. Currently, most image and video quality metrics do not consider these effects. How these effects change the perceived image quality, how strong these effects compare with other HVS features employed in the current quality assessment models, and how to incorporate these effects into a quality assessment model have not yet been deeply investigated.

3. FULL-REFERENCE QUALITY ASSESSMENT USING STRUCTURAL DISTORTION MEASURES

The paradigm of error sensitivity based image and video quality assessment considers any kind of image distortions as being certain types of *errors*. Since different error structures will have different effects on perceived image quality, the effectiveness of this approach depends on how the structures of the errors are understood and represented. Linear channel decomposition is the most commonly used way to decompose the error signals into a set of elementary components, and the visual error sensitivity models for these elementary components are relatively easily obtained from psychovisual experiments. As described in Section 2.5, because linear channel decomposition methods cannot fully decorrelate the structures of the signal, the decomposed coefficients still exhibit strong correlations with each other. It has been argued in Section 2.5 that the Minkowski error metric cannot capture these structural correlations. Therefore, the error sensitivity based paradigm relies on a very powerful masking model, which must cover various kinds of intra- and inter-channel interactions between the decomposed coefficients. Current knowledge about visual masking effects is still limited. At this moment, it is not clear whether building a comprehensive masking model is possible or not, but it is likely that even if it were possible, the model would be very complicated.

In this section, we propose an alternative way to think about image quality assessment: it is not necessary to consider the difference between an original image and a distorted image as a certain type of error. What we will now describe as structural distortion measurement may lead to more efficient and more effective image quality assessment methods.

3.1 NEW PHILOSOPHY

In [8] and [68], a new philosophy in designing image and video quality metrics has been proposed:

The main function of the human visual system is to extract structural information from the viewing field, and the human visual system is highly adapted for this purpose. Therefore, a measurement of structural distortion should be a good approximation of perceived image distortion.

The new philosophy can be better understood by comparison with the error sensitivity based philosophy:

First, a major difference of the new philosophy from the error sensitivity based philosophy is the switch from *error measurement* to *structural distortion measurement*. Although error and structural distortion sometimes agree with each other, in many circumstances the same amount of error may lead to significantly different structural distortion. A good example is given in Figures 8 and 9, where the original "Lena" image is altered with a wide variety of distortions: impulsive salt-pepper noise, additive Gaussian noise, multiplicative speckle noise, mean shift, contrast stretching, blurring, and heavy JPEG compression. We tuned all the distorted images to yield the same MSE relative to the original one, except for the JPEG compressed image, which has a slightly smaller MSE. It is interesting to see that images with nearly identical MSE have drastically different perceptual quality. Our subjective evaluation results show that the contrast stretched and the mean shifted images provide very high perceptual quality, while the blurred and the JPEG compressed images have the lowest subjective scores [7,68]. This is no surprise with a good understanding of the new philosophy since the structural change from the original to the contrast stretched and mean shifted images is trivial, but to the blurred and JPEG compressed images the structural modification is very significant.

Figure 41.8 Evaluation of "Lena" images with different types of noise. Top-left: Original "Lena" image, 512×512, 8bits/pixel; Top-right: Impulsive salt-pepper noise contaminated image, MSE=225, Q=0.6494; Bottom-left: Additive Gaussian noise contaminated image, MSE=225, Q=0.3891; Bottom-right: Multiplicative speckle noise contaminated image, MSE=225, Q=0.4408.

Second, another important difference of the new philosophy is that it considers image degradation as *perceived structural information loss*. For example, in Figure 41.9, the contrast stretched image has a better quality than the JPEG compressed image simply because almost all the structural information of the original image is preserved, in the sense that the original image can be recovered via a simple pointwise inverse linear luminance transform. Apparently, a lot of information in the original image is permanently lost in the JPEG compressed image. The reason that a structural information loss measurement can be considered as a prediction of visual perception is based on the assumption that the HVS functions similarly — it has adapted to extract structural information and to detect changes in structural information. By contrast, an error sensitivity based approach estimates *perceived errors* to represent image degradation. If it works properly, then a significant perceptual error should be reported for the contrast stretched image because its difference (in terms of error) from the original image is easily discerned.

Figure 41.9 Evaluation of "Lena" images with different types of distortions. Top-left: Mean shifted image, MSE=225, Q=0.9894; Top-right: Contrast stretched image, MSE=225, Q=0.9372; Bottom-left: Blurred image, MSE=225, Q=0.3461; Bottom-right: JPEG compressed image, MSE=215, Q=0.2876.

Third, the new philosophy uses a *top-down* approach, which starts from the very top level — simulating the hypothesized functionality of the overall HVS. By comparison, the error sensitivity based philosophy uses a *bottom-up* approach, which attempts to simulate the function of each relevant component in the HVS and combine them together, in the hope that the combined system will perform similarly to the overall HVS.

How to apply the new philosophy to create a concrete image and video quality assessment method is an open issue. There may be very different implementations, depending on how the concepts of "structural information" and "structural distortion" are interpreted and quantified. Generally speaking, there may be two ways of implementing a quality assessment algorithm using the new philosophy. The first is to develop a feature description framework of natural images, which covers most of the useful structural information of an image signal. Under such a description framework, structural information changes between the original and the distorted signals can be quantified. The second is to design a structure comparison method that can compare structural similarity or structural difference between the original and the distorted signals directly. As a first attempt to implement this new philosophy, a simple image quality indexing approach was proposed in [7,68], which conforms to the second approach.

3.2 AN IMAGE QUALITY INDEXING APPROACH

Let $x = \{x_i \mid i = 1, 2, \cdots N\}$ and $y = \{x_i \mid i = 1, 2, \cdots N\}$ be the original and the test image signals, respectively. The proposed quality index is defined as:

$$Q = \frac{4\sigma_{xy}\,\bar{x}\,\bar{y}}{(\sigma_x^2 + \sigma_y^2)[(\bar{x})^2 + (\bar{y})^2]}$$

(41.4)

where

$$\bar{x} = \frac{1}{N}\sum_{i=1}^{N} x_i, \quad \bar{y} = \frac{1}{N}\sum_{i=1}^{N} y_i,$$

$$\sigma_x^2 = \frac{1}{N-1}\sum_{i=1}^{N}(x_i - \bar{x})^2, \quad \sigma_y^2 = \frac{1}{N-1}\sum_{i=1}^{N}(y_i - \bar{y})^2,$$

$$\sigma_{xy} = \frac{1}{N-1}\sum_{i=1}^{N}(x_i - \bar{x})(y_i - \bar{y}).$$

The dynamic range of Q is $[-1, 1]$. The best value 1 is achieved if and only if $y_i = x_i$ for all $i = 1, 2, \cdots N$. The lowest value of -1 occurs when $y_i = 2\bar{x} - x_i$, for all $i = 1, 2, \cdots N$.

This quality index models any distortion as a combination of three factors: loss of correlation, mean distortion and contrast distortion. In order to understand this, we rewrite the definition of Q as the product of three components:

$$Q = \frac{\sigma_{xy}}{\sigma_x \sigma_y} \cdot \frac{2\bar{x}\,\bar{y}}{(\bar{x})^2 + (\bar{y})^2} \cdot \frac{2\sigma_x \sigma_y}{\sigma_x^2 + \sigma_y^2}$$

(41.5)

The first component is the correlation coefficient between x and y, which measures the degree of linear correlation between x and y, and its dynamic range is $[-1, 1]$. The best value 1 is obtained when $y_i = ax_i + b$ for all $i = 1, 2, \cdots N$, where a and b are constants and $a > 0$. We consider the linear correlation coefficient as a very important factor in comparing the structures of two signals. Notice that a pointwise linearly changed signal can be recovered exactly with a simple pointwise inverse linear transform. In this sense, the "structural information" is preserved. Furthermore, a decrease in the linear correlation coefficient gives a quantitative measure of how much the signal is changed non-linearly. Obviously, even if x and y are linearly correlated, there still may be relative distortions between them, which are evaluated in the second and third components. The second component, with a range of [0, 1], measures how similar the mean values of x and y are. It equals 1 if and only if $\bar{x} = \bar{y}$. σ_x and σ_y can be viewed as rough estimate of the contrast of x and y, so the third component measures how similar the contrasts of the images are. Its range of values is also [0, 1], where the best value 1 is achieved if and only if $\sigma_x = \sigma_y$.

Image signals are generally non-stationary and image quality is often spatially variant. In practice it is usually desired to evaluate an entire image using a single overall quality value. Therefore, it is reasonable to measure statistical features locally and then combine them together. We apply our quality measurement method to local regions using a sliding window approach. Starting from the top-left corner of the image, a sliding window of size B×B moves pixel by pixel horizontally and vertically through all the rows and columns of the image until the bottom-right corner is reached. At the j-th step, the local quality index Q_j is computed within the sliding window. If there are a total of M steps, then the overall quality index is given by

$$Q = \frac{1}{M} \sum_{j=1}^{M} Q_j \ .$$ (41.6)

It has been shown that many image quality assessment algorithms work consistently well if the distorted images being compared are created from the same original image and the same type of distortions (e.g., JPEG compression). In fact, for such comparisons, the MSE or PSNR is usually sufficient to produce useful quality evaluations. However, the effectiveness of image quality assessment models degrades significantly when the models are employed to compare the quality of distorted images originating from different types of original images with different types of distortions. Therefore, cross-image and cross-distortion tests are very useful in evaluating the effectiveness of an image quality metric.

The images in Figures 41.8 and 41.9 are good examples for testing the cross-distortion capability of the quality assessment algorithm. Obviously, the MSE performs very poorly in this case. The quality indices of the images are calculated and given in Figures 41.8 and 41.9, where the sliding window size is fixed at B=8. The results exhibit surprising consistency with the subjective measures. In fact, the ranks given by the quality index are the same as the mean subjective ranks of our subjective evaluations [7,68]. We noticed that many subjects regard the contrast stretched image to have better quality than the mean shifted image and even the original image. This is no surprise because contrast stretching is often an image enhancement process, which often increases the visual quality of the original image. However, if we assume that the original image is the perfect one (as our quality measurement method does), then it is fair to give the mean shifted image a higher quality score.

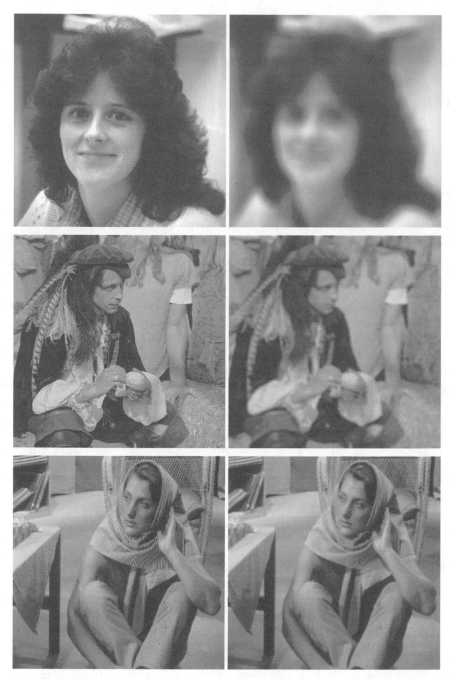

Figure 41.10 Evaluation of blurred image quality. Top-left: Original "Woman" image; Top-right: Blurred "Woman" image, MSE=200, Q=0.3483; Middle-left: Original "Man" image; Middle-right: Blurred "Man" image, MSE=200, Q=0.4123; Bottom-left: Original "Barbara" image; Bottom-right: Blurred "Barbara" image, MSE=200, Q=0.6594.

Figure 41.11 Evaluation of JPEG compressed image quality. Top-left: Original "Tiffany" image; Top-right: compressed "Tiffany" image, MSE=165, Q=0.3709; Middle-left: Original "Lake" image; Middle-right: compressed "Lake" image, MSE=167, Q=0.4606; Bottom-left: Original "Mandrill" image; Bottom-right: compressed "Mandrill" image, MSE=163, Q=0.7959.

In Figures 41.10 and 41.11, different images with the same distortion types are employed to test the cross-image capability of the quality index. In Figure 41.10, three different images are blurred, such that they have almost the same MSE with respect to their original ones. In Figure 41.11, three other images are compressed using JPEG, and the JPEG compression quantization steps are selected so that the three compressed images have similar MSE in comparison with their original images. Again, the MSE has very poor correlation with perceived image quality in these tests, and the proposed quality indexing algorithm delivers much better consistency with visual evaluations.

Interested users may refer to [69] for more demonstrative images and an efficient MATLAB implementation of the proposed quality indexing algorithm.

The proposed quality indexing method is only a rudimentary implementation of the new paradigm. Although it gives promising results under the current limited testings, more extended experiments are needed to validate and optimize the algorithm. More theoretical and experimental connections with respect to human visual perception need to be established. Another important issue that needs to be explored is how to apply it for video quality assessment. In [70], the quality index was calculated frame by frame for a video sequence and combined with other image distortion features such as blocking to produce a video quality measure.

4. NO-REFERENCE AND REDUCED-REFERENCE QUALITY ASSESSMENT

The quality metrics presented so far assume the availability of the reference video to compare the distorted video against. This requirement is a serious impediment to the feasibility of video quality assessment metrics. Reference videos require tremendous amounts of storage space, and, in many cases, are impossible to provide for most applications.

Reduced-reference (RR) quality assessment does not assume the complete availability of the reference signal, only that of partial reference information that is available through an ancillary data channel. Figure 41.12 shows how a RR quality assessment metric may be deployed. The server transmits side-information with the video to serve as a reference for an RR quality assessment metric down the network. The bandwidth available to the RR channel depends upon the application constraints. The design of RR quality assessment metrics needs to look into what information is to be transmitted through the RR channel so as to provide minimum prediction errors. Needless to say, the feature extraction from the reference video at the server would need to correspond to the intended RR quality assessment algorithm.

Perhaps the earliest RR quality assessment metric was proposed by Webster *et al.* [71] and is based on extracting localized spatial and temporal activity features. A spatial information (SI) feature measures the standard deviation of edge-enhanced frames, assuming that compression will modify the edge statistics in the frames. A temporal information (TI) feature is also extracted, which is the standard deviation of difference frames. Three comparison metrics are derived from the SI and TI features of the reference and the distorted videos. The features for the reference video are transmitted over the RR channel. The metrics are trained on data obtained from human subjects. The size of the RR data depends upon the size of the window over which SI & TI features are calculated. The work was extended in [72], where different edge enhancement filters are used, and two activity features are extracted from 3D windows. One feature measures the strength of the edges in the horizontal/vertical directions, while the second feature measures the strength of the edges over other directions. Impairment metric is defined using these features. Extensive subjective testing is also reported.

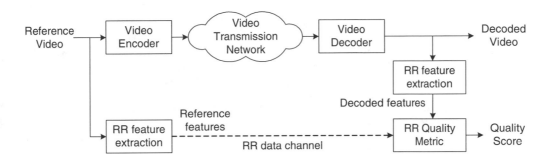

Figure 41.12 Deployment of a reduced-reference video quality assessment metric. Features extracted from the reference video are sent to the receiver to aid in quality measurements. The video transmission network may be lossy but the RR channel is assumed to be lossless.

Another approach that uses side-information for quality assessment is described in [73], in which marker bits composed of random bit sequences are hidden inside frames. The markers are also transmitted over the ancillary data channel. The error rate in the detection of the marker bits is taken as an indicator of the loss of quality. In [74], a watermarking based approach is proposed, where a watermark image is embedded in the video, and it is suggested that the degradation in the quality of the watermark can be used to predict the degradation in the quality of the video. Strictly speaking, both these methods are not RR quality metrics since they do not extract any features from the reference video. Instead, these methods gauge the distortion processes that occur during compression and the communication channel to estimate the perceptual degradation incurred during transmission in the channel.

Given the limited success that FR quality assessment has achieved, it should come as no surprise that designing objective no-reference (NR) quality measurement algorithms is very difficult indeed. This is mainly due to the limited understanding of the HVS and the corresponding cognitive aspects of the brain. Only a few methods have been proposed in the literature [75-80] for objective NR quality assessment, yet this topic has attracted a great deal of attention recently. For example, the video quality experts group (VQEG) [81] considers the standardization of NR and RR video quality assessment methods as one of its future working directions, where the major source of distortion under consideration is block DCT-based video compression.

The problem of NR quality assessment (sometimes called *blind* quality assessment) is made even more complex due to the fact that many unquantifiable factors play a role in human assessment of quality, such as aesthetics, cognitive relevance, learning, visual context, etc., when the reference signal is not available for MOS evaluation. These factors introduce variability among human observers based on each individual's subjective impressions. However, we can work with the following philosophy for NR quality assessment: *all images and videos are perfect unless distorted during acquisition, processing or reproduction.* Hence, the task of blind quality measurement simplifies into blindly measuring the distortion that has possibly been introduced during the stages of acquisition, processing or reproduction. The reference for measuring this distortion would be the statistics of "perfect" natural images and videos, measured with respect to a model that best suits a given distortion type or application. This philosophy effectively decouples the unquantifiable aspects of image quality mentioned above from the task of objective quality assessment. All "perfect images" are treated equally, disregarding the amount of cognitive information in the image, or its aesthetic value [82,83].

The NR metrics cited above implicitly adhere to this philosophy of quantifying quality through blind distortion measurement. Assumptions regarding statistics of "perfect natural images" are made such that the distortion is well separated from the "expected" signals. For example, natural images do not contain blocking artifacts, and any presence of periodic edge discontinuity in the horizontal and vertical directions with a period of 8 pixels, is probably a distortion introduced by block-DCT based compression techniques. Some aspects of the HVS, such as texture and luminance masking, are also modelled to improve prediction. Thus NR quality assessment metrics need to model not only the HVS but also natural scene statistics.

Certain types of distortions are quite amenable to blind measurement, such as blocking artifacts. In wavelet-based image coders, such as the JPEG2000 standard, the wavelet transform is often applied to the entire image (instead of image blocks), and the decoded images do not suffer from blocking artifact. Therefore, NR metrics based on blocking artifacts would obviously fail to give meaningful predictions. The upcoming H.26L standard incorporates a powerful de-blocking filter. Similarly post-processing may reduce blocking artifacts at the cost of introducing blur. In [82], a statistical model for natural images in the wavelet domain is used for NR quality assessment of JPEG2000 images. Any NR metric would therefore need to be specifically designed for the target distortion system. More sophisticated models of natural images may improve the performance of NR metrics and make them more robust to various distortion types.

5. VALIDATION OF QUALITY ASSESSMENT METRICS

Validation is an important step towards successful development of practical image and video quality measurement systems. Since the goal of these systems is to predict perceived image and video quality, it is essential to build an image and video database with subjective evaluation scores associated with each of the images and video sequences in the database. Such a database can then be used to assess the prediction performance of the objective quality measurement algorithms.

In this section, we first briefly introduce two techniques that have been widely adopted in both the industry and the research community for subjective evaluations for video. We then review the quality metric comparison results published in the literature. Finally, we introduce the recent effort by the video quality experts group (VQEG) [81], which aims to provide industrial standards for video quality assessment.

5.1 SUBJECTIVE EVALUATION OF VIDEO QUALITY

Subjective evaluation experiments are complicated by many aspects of human psychology and viewing conditions, such as observer vision ability, translation of quality perception into ranking score, preference for content, adaptation, display devices, ambient light levels etc. The two methods that we will present briefly are single stimulus continuous quality evaluation (SSCQE) and double stimulus continuous quality scale (DSCQS), which have been demonstrated to have repeatable and stable results, provided consistent viewing configurations and subjective tasks, and have consequently been adopted as parts of an international standard by the international telecommunications union (ITU) [84]. If the SSCQE and DSCQS tests are conducted on multiple subjects, the scores can be averaged to yield the mean opinion score (MOS). The standard deviation between the scores may also be useful to measure the consistency between subjects.

Single Stimulus Continuous Quality Evaluation

In the SSCQE method, subjects continuously indicate their impression of the video quality on a linear scale that is divided into five segments, as shown in Figure 41.13. The five intervals are marked with adjectives to serve as guides. The subjects are instructed to move a slider to any point on the scale that best reflects their impression of quality at that instant of time, and to track the changes in the quality of the video using the slider.

EXCELLENT

GOOD

FAIR

POOR

BAD

Figure 41.13 SSCQE sample quality scale.

Double Stimulus Continuous Quality Scale

The DSCQS method is a form of discrimination based method and has the extra advantage that the subjective scores are less affected by adaptation and contextual effects. In the DSCQS method, the reference and the distorted videos are presented one after the other in the same session, in small segments of a few seconds each, and subjects evaluate both sequences using sliders similar to those for SSCQE. The difference between the scores of the reference and the distorted sequences gives the subjective impairment judgement. Figure 41.14 demonstrates the basic test procedure.

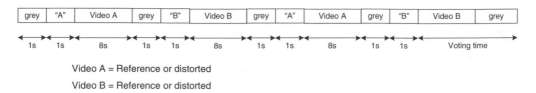

Video A = Reference or distorted

Video B = Reference or distorted

Figure 41.14 DSCQS testing procedure recommended by VQEG FR-TV Phase-II test.

5.2 COMPARISON OF QUALITY ASSESSMENT METRICS

With so many quality assessment algorithms proposed, the question of their relative merits and demerits naturally arises. Unfortunately, not much has been published in comparing these models with one another, especially under strict experimental conditions over a wide range of distortion types, distortion strengths, stimulus content and subjective evaluation criterion. This is compounded by the fact that validating quality assessment metrics comprehensively is time-consuming and expensive, not to mention that many algorithms are not described explicitly enough in the literature to allow reproduction of their reported performance. Most comparisons of quality assessment metrics are not broad enough to be able to draw solid conclusions, and their results should only be considered in the context of their evaluation criterion.

In [3] and [65], different mathematical measures of quality that operate without channel decompositions and masking effect modelling are compared against subjective experiments, and their performance is tabulated for various test conditions. Li *et al.* compare Daly's and Lubin's models for their ability to detect differences [85] and conclude that Lubin's model is more robust than Daly's given their experimental procedures. In [86] three metrics are compared for JPEG compressed images: Watson's DCT based metric [87], Chou and Li's method [50] and Karanusekera and Kingsbury's method [49]. They conclude that Watson's method performed best among the three.

Martens and Meesters have compared Lubin's metric (also called the Sarnoff model) with the root mean squared error (RMSE) [9] metric on transformed luminance images. The metrics are compared using subjective experiments based on images corrupted with noise and blur, as well as images corrupted with JPEG distortion. The subjective experiments are based on *dissimilarity measurements*, where subjects are asked to assess the dissimilarity between pairs of images from a set that contains the reference image and several of its distorted versions. Multidimensional scaling (MDS) technique is used to compare the metrics with the subjective experiments. MDS technique constructs alternate spaces from the dissimilarity data, in which the positions of the images are related to their dissimilarity (subjective or objective) with the rest of the images in that set. Martens and Meesters then compare RMSE and Lubin's method with subjective experiments, with and without MDS, and report that "in none of the examined cases could a clear advantage of complicated distance metrics (such as the Sarnoff model) be demonstrated over simple measures such as RMSE" [9].

5.3 VIDEO QUALITY EXPERTS GROUP

The video quality experts group [81] was formed in 1997 to develop, validate and standardize new objective measurement methods for video quality. The group is composed of experts from various backgrounds and organizations around the world. They are interested in FR/RR/NR quality assessment for various bandwidth videos for television and multimedia applications.

VQEG has completed its Phase I test for FR video quality assessment for television in 2000 [10,11]. In Phase I test, 10 proponent video quality models (including several well-known models and PSNR) were compared with the subjective evaluation results on a video database, which contains video sequences with a wide variety of distortion types and stimulus content. A systematic way of evaluating the prediction performance of the objective models was established, which is composed of three components:

1. Prediction accuracy — the ability to predict the subjective quality ratings with low error. (Two metrics, namely the variance-weighted regression correlation [10] and the non-linear regression correlation [10], were used.)

2. Prediction monotonicity — the degree to which the model's predictions agree with the relative magnitudes of subjective quality ratings. (The Spearman rank order correlation [10] was employed.)

3. Prediction consistency — the degree to which the model maintains prediction accuracy over the range of video test sequences. (The outlier ratio [10] was used.)

The result was, in some sense, surprising, since except for 1 or 2 proponents that did not perform properly in the test, the other proponents performed statistically equivalent, including PSNR [10,11]. Consequently, VQEG did not recommend any method for an ITU standard [10]. VQEG is continuing its work on Phase II test for FR quality assessment for television, and RR/NR quality assessment for television and multimedia.

Although it is hard to predict whether VQEG will be able to supply one or a few successful video quality assessment standards in the near future, the work of VQEG is important and unique from a research point of view. First, VQEG establishes large video databases with reliable subjective evaluation scores (the database used in the FR Phase I test is already available to the public [81]), which will prove to be invaluable for future research on video quality assessment. Second, systematic approaches for comparing subjective and objective scores are being formalized. These approaches alone could become widely accepted standards in the research community. Third, by comparing state-of-the-art quality assessment models in different aspects, deeper understanding of the relative merits of different methods will be achieved, which will have a major impact on future improvement of the models. In addition, VQEG provides an ideal communication platform for the researchers who are working in the field.

6. CONCLUSIONS AND FUTURE DIRECTIONS

There has been increasing interest in the development of objective quality measurement techniques that can automatically predict the perceptual quality of images and video streams. Such methods are useful tools for video database systems and are also desired for a broad variety of applications, such as video acquisition, compression, communication, displaying, analysis, watermarking, restoration and enhancement. In this chapter, we reviewed the basic concepts and methods in objective image and video quality assessment research and discussed various quality prediction approaches. We also laid out a new paradigm for quality assessment based on structural distortion measures, which has some promising advantages over traditional perceptual error sensitivity based approaches.

After decades of research work, image and video quality assessment is still an active and evolving research topic. An important goal of this chapter is to discuss the challenges and difficulties encountered in developing objective quality assessment approaches, and provide a vision for future directions.

For error sensitivity based approaches, the four major problems discussed in Section 2.5 also laid out the possible directions that may be explored to provide improvements of the current methods. One of the most important aspects that requires the greatest effort is to investigate various masking/facilitation effects, especially the masking and facilitation phenomena in the suprathreshold range and in cases where the background is composed of broadband natural image (instead of simple patterns). For example, the contrast matching experimental study on center-surround interactions [88] may provide a better way to quantitatively measure the image distortions at the suprathreshold [89] level. The Modelfest phase one dataset [90] collected simple patterns such as Gabor patches as well as some more complicated broadband patterns including one natural image. Comparing and analysing the visual error prediction capability of different error sensitivity based methods with these patterns may help researchers to better understand whether HVS features measured with simple patterns can be extended to predict the perceptual quality of complex natural images.

The structural distortion based framework is at a very preliminary stage. The newly proposed image quality indexing approach is attractive not only because of its promising results, but also its simplicity. However, it is perhaps too simple and the combination of the three factors is *ad hoc*. More theoretical analysis and subjective experimental work is needed to provide direct evidence on how it is connected with visual perception and natural image statistics. Many other issues may also be considered, such as multiscale analysis, adaptive windowing and space-variant pooling using a statistical fixation model. Furthermore, under the new paradigm of structural distortion measurement, other approaches may emerge that could be very different from the proposed quality indexing algorithm. The understanding of "structural information" would play a key role in these innovations.

Another interesting point is the possibility of combining the advantages of the two paradigms. This is a difficult task without a deeper understanding of both. One possible connection may be as follows: use the structural distortion based method to measure the amount of structural information loss, and the error sensitivity based approach to help determine whether such information loss can be perceived by the HVS.

The fields of NR and RR quality assessment are very young, and there are many possibilities for the development of innovative metrics. The philosophy of doing

NR or RR quality assessment will continue to be that of blind distortion measurement with respect to features that best separate the undistorted signals from the distortion. The success of statistical models for natural scenes that are more suited to certain distortion types and applications will drive the success of NR and RR metrics in the future. A combination of natural scene models with HVS models may also prove beneficial for NR and RR quality assessment.

REFERENCES

[1]   P. C. Teo and D. J. Heeger, Perceptual image distortion, in Proc. IEEE Int. Conf. Image Processing, pp. 982-986, 1994.

[2]   M. P. Eckert and A. P. Bradley, Perceptual quality metrics applied to still image compression, Signal Processing, vol. 70, pp. 177-200, Nov. 1998.

[3]   A. M. Eskicioglu and P. S. Fisher, Image quality measures and their performance, IEEE Trans. Communications, vol. 43, pp. 2959-2965, Dec. 1995.

[4]   B. Girod, What's wrong with mean-squared error, in Digital Images and Human Vision, A. B. Watson, ed., pp. 207-220, MIT Press, 1993.

[5]   S. Winkler, A perceputal distortion metric for digital color video, Proc. SPIE, vol. 3644, pp. 175-184, 1999.

[6]   Y. K. Lai and C.-C. J. Kuo, A Haar wavelet approach to compressed image quality measurement, Journal of Visual Communication and Image Understanding, vol. 11, pp. 17-40, Mar. 2000.

[7]   Z. Wang and A. C. Bovik, A universal image quality index, IEEE Signal Processing Letters, vol. 9, no. 3, pp. 81-84, March 2002.

[8]   Z. Wang, A. C. Bovik and L. Lu, Why is image quality assessment so difficult? Proc. IEEE Int. Conf. Acoustics, Speech, and Signal Proc., vol. 4, pp. 3313-3316, May 2002.

[9]   J.-B. Martens and L. Meesters, Image dissimilarity, Signal Processing, vol. 70, pp. 1164-1175, Aug. 1997.

[10]   VQEG, Final report from the video quality experts group on the validation of objective models of video quality assessment, http://www.vqeg.org/, Mar. 2000.

[11]   P. Corriveau, et al., Video quality experts group: Current results and future directions, Proc. SPIE Visual Comm. and Image Processing, vol. 4067, June 2000.

[12]   J. L. Mannos and D. J. Sakrison, The effects of a visual fidelity criterion on the encoding of images, IEEE Trans. Information Theory, vol. 4, pp. 525-536, 1974.

[13]   W. S. Geisler and M. S. Banks, Visual performance, in Handbook of Optics (M. Bass, ed.), McGraw-Hill, 1995.

[14]   B. A. Wandell, Foundations of Vision, Sinauer Associates, Inc., 1995.

[15]   L. K. Cormack, Computational models, of early human vision, in Handbook of Image and Video Processing (A. Bovik, ed.), Academic Press, May 2000.

[16]   A. C. Bovik, M. Clark, and W. S. Geisler, Multichannel texture analysis using localized spatial filters, IEEE Trans. Pattern Analysis and Machine Intelligence, vol. 12, pp. 55-73, Jan. 1990.

[17]   J. Lubin, The use of psychophysical data and models in the analysis of display system performance, in Digital Images and Human Vision (A. B.

Watson, ed.), pp. 163-178, Cambridge, Massachusetts: The MIT Press, 1993.

[18]   J. Lubin, A visual discrimination model for image system design and evaluation, in Visual Models for Target Detection and Recognition, E. Peli, ed., pp. 207-220, Singapore: World Scientific Publisher, 1995.

[19]   S. Lee, M. S. Pattichis, and A. C. Bovik, Foveated video quality assessment, IEEE Trans. Multimedia, vol. 4, pp. 129-132, Mar. 2002.

[20]   Z. Wang and A. C. Bovik, Embedded foveation image coding, IEEE Trans. Image Processing, vol. 10, pp. 1397-1410, Oct. 2001.

[21]   C. J. van den Branden Lambrecht, Perceptual models and architectures for video coding applications, PhD thesis, Swiss Federal Institute of Technology, Aug. 1996.

[22]   C. J. van den Branden Lambrecht, A working spatio-temporal model of the human visual system for image restoration and quality assessment applications, in Proc. IEEE Int. Conf. Acoust., Speech, and Signal Processing, pp. 2291-2294, 1996.

[23]   C. J. van den Branden Lambrecht and O. Verscheure, Perceptual quality measure using a spatio-temporal model of the human visual system, in Proc. SPIE, vol. 2668, (San Jose, LA), pp. 450461, 1996.

[24]   A. B. Watson, J. Hu, and J. F. III. McGowan, DVQ: A digital video quality metric based on human vision, Journal of Electronic Imaging, vol. 10, no. 1, pp. 20-29, 2001.

[25]   P. C. Teo and D. J. Heeger, Perceptual image distortion, in Proc. SPIE, vol. 2179, pp. 127-141, 1994.

[26]   E. Peli, Contrast in complex images, Journal of Optical Society of America, vol. 7, pp. 2032-2040, Oct. 1990.

[27]   A. B. Watson, The cortex transform: rapid computation of simulated neural images, Computer Vision, Graphics, and Image Processing, vol. 39, pp. 311-327, 1987.

[28]   S. Daly, The visible difference predictor: An algorithm for the assessment of image fidelity, in Proc. SPIE, vol. 1616, pp. 2-15, 1992.

[29]   S. Daly, The visible difference predictor: An algorithm for the assessment of image fidelity, in Digital Images and Human Vision (A. B. Watson, ed.) pp. 179-206, Cambridage, Massachusetts: The MIT Press, 1993.

[30]   D. J. Heeger and T. C. Teo, A model of perceptual image fidelity, in Proc. IEEE Int. Conf. Image Proc., pp. 343-345, 1995.

[31]   C. J. van den Branden Lambrecht and O. Verscheure, Perceptual quality measure using a spatio-temporal model of the human visual system, in Proc. SPIE, vol. 2668, (San Jose, LA), pp. 450461, 1996.

[32]   C. J. van den Branden Lambrecht, D. M. Costantini, G. L. Sicuranza, and M. Kunt, Quality assessment of motion rendition in video coding, IEEE Trans. Circuits and Systems for Video Tech., vol. 9, pp. 766-782. Aug. 1999.

[33]   A. B. Watson and J. A. Solomon, Model of visual contrast gain control and pattern masking, Journal of Optical Society of America, vol. 14, no. 9, pp. 2379-2391, 1997.

[34]   W. Xu and G. Hauske, Picture quality evaluation based on error segmentation, Proc. SPIE, vol. 2308, pp. 1454-1465, 1994.

[35]   W. Osberger, N. Bergmann, and A. Maeder, An automatic image quality assessment technique incorporating high level perceptual factors, in Proc. IEEE Int. Conf. Image Proc., pp. 414-418, 1998.

[36]   T. N. Pappas and R. J. Safranek, Perceptual criteria for image quality evaluation, in Handbook of Image and Video Processing (A. Bovik, ed.), Academic Press, May 2000.

[37]   P. J. Burt and E. H. Adelson, The Laplacian pyramid as a compact image code, IEEE Trans. Communications, vol. 31, pp. 532-540, Apr. 1983.

[38]   W. T. Freeman and E. H. Adelson, The design and use of steerable filters, IEEE Trans. Pattern Analysis and Machine Intelligence, vol. 13, pp. 891-906. 1991.

[39]   E. P. Simoncelli, W. T. Freeman, E. H. Adelson, and D. J. Heeger, Shiftable multi-scale transforms, IEEE Trans. Information Theory, vol. 38, pp. 587-607, 1992.

[40]   A. B. Watson, DCTune: A technique for visual optimization of DCT quantization matrices for individual images, in Society for Information Display Digest of Technical Papers, vol. XXIV, pp. 946-949, 1993.

[41]   R. J. Safranek and J. D. Johnston, A perceptually tuned sub-band image coder with image dependent quantization and post-quantization data compression, in Proc. IEEE Int. Conf. Acoust. Speech, and Signal Processing, pp. 1945-1948, May 1989.

[42]   J. W. Woods and S. D. O'Neil, Subband coding of images, IEEE Trans. Acoustics, Speech and Signal Processing, vol. 34, pp. 1278-1288, Oct. 1986.

[43]   A. P. Bradley, A wavelet difference predictor, IEEE Trans. Image Processing, vol. 5, pp. 717-730, May 1999.

[44]   A. B. Watson, G. Y. Yang, J. A. Solomon, and J. Villasenor, Visibility of wavelet quantization noise, IEEE Trans. Image Processing, vol. 6, pp. 1164-1175, Aug. 1997.

[45]   M. Antonini, M. Barlaud, P. Mathieu, and I. Daubechies, Image coding using the wavelet transform, IEEE Trans. Image Processing, vol. 1, pp. 205-220, Apr. 1992.

[46]   N. Damera-Venkata, T. D. Kite, W. S. Geisler, B. L. Evans, and A. C. Bovik, Image quality assessment based on a degradation model, IEEE Trans. Image Processing, vol. 4, pp. 636-650, Apr. 2000.

[47]   T. D. Kite, N. Damera-Venkata, B. L. Evans, and A. C. Bovik, A high quality, fast inverse halftoning algorithm for error diffused halftones, in Proc. IEEE Int. Conf. Image Proc., vol. 2, pp. 59-63, Oct. 1998.

[48]   S. A. Karunasekera and N. G. Kingsbury, A distortion measure for blocking artifacts in images based on human visual sensitivity, IEEE Trans. Image Processing, vol. 4, pp. 713-724, June 1995.

[49]   S. A. Karunasekera and N. G. Kingsbury, A distortion measure for image artifacts based on human visual sensitivity, in Proc. IEEE Int. Conf. Acoust., Speech, and Signal Processing, vol. 5, pp. 117-120, 1994.

[50]   C. H. Chou and Y. C. Li, A perceptually tuned subband image coder based on the measure of just-noticeable-distortion profile, IEEE Trans. Circuits and Systems for Video Tech., vol. 5, pp. 467-476, Dec. 1995.

[51]   M. Miyahara, K. Kotani, and V. R. Algazi, Objective picture quality scale (PQS) for image coding, IEEE Trans. Communications, vol. 46, pp. 1215-1225, Sept. 1998.

[52]   T. Yamashita, M. Kameda and M. Miyahara, An objective picture quality scale for video images (PQS$_{video}$) - definition of distortion factors, Proc. SPIE, vol. 4067, pp. 801-809, 2000.

[53]   K. T. Tan, M. Ghanbari, and D. E. Pearson, An objective measurement tool for MPEG video quality, Signal Processing, vol. 70, pp. 279-294, Nov. 1998.

[54]   K. T. Tan and M. Ghanbari, A multi-metric objective picture-quality measurement model for MPEG video, IEEE Trans. Circuits and Systems for Video Tech., vol. 10, pp. 1208-1213, Oct. 2000.

[55]   M. Yuen and H. R. Wu, A survey of hybrid MC/DPCM/DCT video coding distortions, Signal Processing, vol. 70, pp. 247-278, Nov. 1998.

[56]   S. Winker, Issues in vision modeling for perceptual video quality assessment, Signal Processing, vol. 78, pp. 231-252, 1999.

[57]   A. B. Watson, Toward a perceptual video quality metric, in Proc. SPIE Human Vision and Electronic Imaging III, vol. 3299, pp. 139-147, Jan. 1998.

[58]   Z. Yu, H. R. Wu, S. Winkler, and T. Chen, Vision-model-based impairment metric to evaluate blocking artifact in digital video, Proceedings of the IEEE, vol. 90, pp. 154-169, Jan. 2002.

[59]   N. Graham, J. G. Robson, and J. Nachmias, Grating summation in fovea and periphery, Vision Research, vol. 18, pp. 815-825, 1978.

[60]   E. P. Simoncelli, Modeling the joint statistics of images in the wavelet domain, in Proc. SPIE, vol. 3813, pp. 188-195, July 1999.

[61]   J. Liu and P. Moulin, Information-theoretic analysis of interscale and intrascale dependencies between image wavelet coefficients, IEEE Trans. Image Processing, vol. 10, pp. 1647-1658, Nov. 2001.

[62]   J. M. Shapiro, Embedded image coding using zerotrees of wavelets coefficients, IEEE Trans. Signal Processing, vol. 41, pp. 2445-3462, Dec. 1993.

[63]   A. Said and W. A. Pearlman, A new, fast, and efficient image codec based on set partitioning in hierarchical trees, IEEE Trans. Circuits and Systems for Video Tech., vol. 6, pp. 243-250, June 1996.

[64]   D. S. Taubman and M. W. Marcellin, JPEG2000: Image Compression Fundamentals, Standards, and Practice, Kluwer Academic Publishers, 2001.

[65]   D. R. Fuhrmann, J. A. Baro, and J. R. Cox Jr., Experimental evaluation of psychophysical distortion metrics for JPEG-encoded images, Journal of Electronic Imaging, vol. 4, pp. 297-406, Oct. 1995.

[66]   W. F. Good, G. S. Maitz, and D. Gur, Joint photographic experts group (JPEG) compatible data compression of mammograms, Journal of Digital Imaging, vol. 7, no. 3, pp. 123-132, 1994.

[67]   M. R. Frater, J. F. Arnold, and A. Vahedian, Impact of audio on subjective assessment of video quality in videoconferencing applications, IEEE Journal of Selected Areas in Comm., vol. 11, pp. 1059-1062, Sept. 2001.

[68]   Z. Wang, Rate Scalable Foveated Image and Video Communications, PhD thesis, Dept. of ECE, The University of Texas at Austin, Dec. 2001.

[69]   Z. Wang, Demo imaegs and free software for 'A Universal Image Quality Index', in http://anchovy.ece.utexas.edu/zwang/research/quality_index/demo.html, 2001.

[70]   Z. Wang, L. Lu and A. C. Bovik, Video quality assessment using structural distortion measurement, Proc. IEEE Int. Conf. Image Proc., Sept. 2002.

[71] A. A. Webster, C. T. Jones, M. H. Pinson, S. D. Voran, and S. Wolf, An objective video quality assessment system based on human perception, Proc. SPIE, vol. 1913, pp. 15-26, 1993.

[72] S. Wolf and M. H. Pinson, Spatio-temporal distortion metrics for in-service quality monitoring of any digital video system, Proc. SPIE, vol. 3845, pp. 266-277, 1999.

[73] O. Sugimoto, R. Kawada, M. Wada, and S. Matsumoto, Objective measurement scheme for perceived picture quality degradation caused by MPEG encoding without any reference pictures, Proc. SPIE, vol. 4310, pp. 932-939, 2001.

[74] M. C. Q. Farias, S. K. Mitra, M. Carli, and A. Neri, A comparison between an objective quality measure and the mean annoyance values of watermarked videos, in Proc. IEEE Int. Conf. Image Proc., Sept. 2002.

[75] Z. Wang, A. C. Bovik and B. L. Evans, Blind measurement of blocking artifacts in images, Proc. IEEE Int. Conf. Image Proc., vol. 3, pp. 981-984, Sept. 2000.

[76] A. C. Bovik and S. Liu, DCT-domain blind measurement of blocking artifacts in DCT-coded images, Proc. IEEE Int. Conf. Acoust., Speech, and Signal Proc., vol. 3, pp. 1725-1728, May 2001.

[77] P. Gastaldo, S. Rovetta and R. Zunino, Objective assessment of MPEG-video quality: a neural-network approach, in Proc. IJCNN, vol. 2, pp. 1432-1437, 2001.

[78] M. Knee, A robust, efficient and accurate single-ended picture quality measure for MPEG-2, available at http://www-ext.crc.ca/vqeg/frames.html, 2001.

[79] H. R. Wu and M. Yuen, A generalized block-edge impairment metric for video coding, IEEE Signal Processing Letters, vol. 4, pp. 317-320, Nov. 1997.

[80] J. E. Caviedes, A. Drouot, A. Gesnot, and L. Rouvellou, Impairment metrics for digital video and their role in objective quality assessment, Proc. SPIE, vol. 4067, pp. 791-800, 2000.

[81] VQEG: The Video Quality Experts Group, http://www.vqeg.org.

[82] H. R. Sheikh, Z. Wang, L. Cormack and A. C. Bovik, Blind quality assessment for JPEG2000 compressed images, Proc. IEEE Asilomar Conference on Signals, Systems and Computers, Nov. 2002.

[83] Z. Wang, H. R. Sheikh and A. C. Bovik, No-reference perceptual quality assessment of JPEG compressed images, Proc. IEEE Int. Conf. Image Proc., Sept. 2002..

[84] ITU-R Rec. BT. 500-10, Methodology for the Subjective Assessment of Quality for Television Pictures.

[85] B. Li, G. W. Meyer, and R. V. Klassen, A comparison of two image quality models, in Proc. SPIE, vol. 3299, pp. 98-109, 1998.

[86] A. Mayache, T. Eude, and H. Cherifi, A comprison of image quality models and metrics based on human visual sensitivity, in Proc. IEEE Int. Conf. Image Proc., pp. 409-413, 1998.

[87] A. B. Watson, DCT quantization matrices visually optimized for individual images, in Proc. SPIE, vol. 1913, pp. 202-216, 1993.

[88] J. Xing and D. J. Heeger, Measurement and modeling of center-surround suppression and enhancement, Vision Research, vol. 41, pp. 571-583, 2001.

[89]   J. Xing, An image processing model of contrast perception and discrimination of the human visual system, in SID Conference, (Boston), May 2002.

[90]   A. B. Watson, Visual detection of spatial contrast patterns: Evaluation of five simple models, Optics Express, vol. 6, pp. 12-33, Jan. 2000.

42

VIDEO WATERMARKING
OVERVIEW AND CHALLENGES

Gwenaël Doërr and Jean-Luc Dugelay

Multimedia Communications, Image Group
Eurécom Institute
Sophia-Antipolis, France
`jean-luc.dugelay@eurecom.fr`

1. INTRODUCTION

If you hold a common banknote up to the light, a watermarked drawing appears. This watermark is invisible during normal use and carries some information about the object in which it is embedded. The watermarks of two different kind of banknotes are indeed different. This watermark is directly inserted into the paper during the papermaking process. This very old technique is known to prevent common methods of counterfeiting. In the past few years, the use and distribution of digital multimedia data has exploded. Because it appeared that traditional protection mechanisms were not anymore sufficient, content owners requested new means for copyright protection. The previous paper watermark philosophy has been transposed to digital data. Digital watermarking, the art of hiding information in a robust and invisible manner, was born. The recent interest regarding digital watermarking is demonstrated in Table 42.1, which reports the increasing number of scientific papers dealing with this subject. Today, entire scientific conferences are dedicated to digital watermarking e.g. "SPIE: Security and Watermarking of Multimedia Content". Moreover, even if it is a relatively new technology, some industries have already commercialised watermarking products e.g. the widespread Digimarc.

Table 42.1 Number of publications having the keyword *watermarking* as their main subject according to INSPEC database, July 2002.

Year	1995	1996	1997	1998	1999	2000	2001
Publications	2	21	54	127	213	334	376

Digital watermarking has first been extensively studied for still images. Today ,however, many new watermarking schemes are proposed for other types of digital multimedia data, so called as *new objects*: audio, video, text, 3D meshes.... This chapter is completely devoted to digital video watermarking. Since the main subject of this book is video databases, the reader is assumed not to be familiar with the concept of digital watermarking. Consequently, the fundamentals of the theory are presented in Section 2. Many applications of digital watermarking in the context of the video are presented in Section 3 in order to give an overview of the possible benefits that technology can bring. Section 4 lists the main challenges that have to be taken up when designing a new video watermarking system. Finally, the major trends in the domain, to the best knowledge of the authors, are reported in Section 5.

2. WHAT IS DIGITAL WATERMARKING?

The end of the previous millennium has seen the transition from the analog to the digital world. Nowadays, audio CDs, Internet and DVDs are more and more widespread. However film and music content owners are still reluctant to release digital content. This is mainly due to the fact that if digital content is left unprotected, it can be copied rapidly, perfectly, at large scale, without any limitation on the number of copies and distributed easily e.g. via Internet. Protection of digital content has relied for a long time on encryption but it appeared that encryption alone is not sufficient enough to protect digital data all along its lifetime. Sooner or later, digital content has to be decrypted in order to be eventually presented to the human consumer. At this very moment, the protection offered by encryption no longer exists and an indelicate user may duplicate or manipulate it.

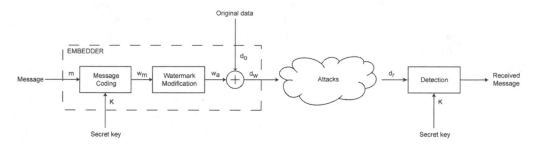

Figure 42.1 Simple watermarking scheme

Digital watermarking has consequently been introduced as a complementary protection technology. The basic idea consists in hiding information imperceptibly into digital content. This watermarked signal should survive most common signal processings and even malicious ones if possible. The hidden information is inherently tied to digital content and protects it when encryption has disappeared. It is important to understand that digital watermarking does not replace encryption. Those are two complementary techniques. On one hand, encryption prevents an unauthorised user from accessing digital content in the clear during its transport. On the other hand, digital watermarking leaves an underlying invisible piece of evidence in digital data if an indelicate authorised user, who had access to the data in the clear after decryption, starts using digital data illegally (reproduction, alteration).

Depending on what information is available during the extraction process, two separate classes of watermark detectors have been defined. If the detector has access to the original data additionally to the watermarked data, the watermark detector is called non-blind. However this kind of algorithm is less and less represented nowadays. Keeping an original version of each released digital data is indeed a very strong constraint for digital content owners in terms of storage space. As a result, most of the watermark detectors are actually considered as blind: the detector has only access to the watermarked data in order to extract the hidden message.

The Figure 42.1 depicts a simple watermarking scheme with blind detection. The goal is to embed the message m into some original data d_o. The first step consists in encoding the message to be hidden with a secret key K. Typically the message is over sampled in order to match the dimension of the original data and is XORed with a pseudo-random noise generated thanks to a pseudo-random number generator which takes the secret key K as an input seed. Next, the generated watermark signal w_m is modified e.g. it is scaled by a given watermarking strength. The final step simply adds the obtained watermark w_a to the original data in order to obtain the watermarked data d_w. This watermark embedding could be performed in whatever desired domain (spatial, Fast Fourier Transform (FFT), Discrete Cosine Transform (DCT), Fourier-Mellin). Watermarked data is then transmitted and is likely to be submitted to various processings (lossy compression, noise addition, filtering) which can be seen as attacks altering the watermark signal. If at some moment, someone wants to check if a watermark has been embedded with the secret key K in some received digital data d_r, the data is simply sent through a detector. The majority of the existing detection algorithms can be seen as the computation of a correlation score between received data d_r and the generated watermark w_m. This correlation score is then compared to a threshold in order to assert the presence of the watermark or not.

There exists a complex trade-off in digital watermarking between three parameters: data payload, fidelity and robustness. It is illustrated in Figure 42.2 and further presented below.

Payload
Data payload can be defined by the number of bits that can be hidden in digital data, which is inherently tied to the number of alternative messages that can be embedded thanks to the watermarking algorithm. It should be noted that, most of the time, data payload depends on the size of the host data. The more host samples are available, the more bits can be hidden. The capacity is consequently often given in terms of bits per sample.

Fidelity
Watermarking digital content can be seen as an insertion of some watermark signal in the original content and this signal is bound to introduce some distortion. As in lossy compression, one of the requirements in digital watermarking is that this distortion should remain imperceptible. In other terms, a human observer should not be able to detect if some digital data has been watermarked or not. The watermarking process should not introduce suspicious perceptible artefacts. The fidelity can also be seen as the perceptual similarity between watermarked and unwatermarked data.

Robustness
The robustness of a watermarking scheme can be defined as the ability of the detector to extract the hidden watermark from some altered watermarked data. The alteration can be malicious or not i.e. the alteration can result from a common processing (filtering, lossy compression, noise addition) or from an attack attempting to remove the watermark (Stirmark [40], dewatermarking

attack [44]). As a result, the robustness is evaluated via the survival of the watermark after attacks.

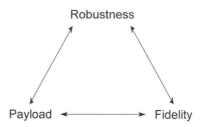

Figure 42.2 Trade-off in digital watermarking

It is quite easy to see that those three parameters are conflicting. One may want to increase the watermarking strength in order to increase the robustness but this results in a more perceptible watermark on the other hand. Similarly, one can increase the data payload by decreasing the number of samples allocated to each hidden bit but this is counterbalanced by a loss of robustness.

As a result, a trade-off has to be found and it is often tied to the targeted application. It is useless to design a high capacity algorithm if there are only a few different messages to be hidden in practice. This is typically the case in a copy control application where two bits are enough to encode the three messages *copy-always*, *copy-once* and *copy-never*. Most of the time, the watermark signal should have a low energy so that the induced distortion remains imperceptible. However in a high degrading environment (e.g. television broadcasting), it is sometimes necessary to embed a strong watermark in order to survive the transmission. Finally some applications do not require the watermark to be robust. In fact the weakness of a fragile watermark can even be exploited in order to ensure the integrity of digital data [43]. If no watermark is found, digital data is not considered legitimate and is discarded. There is not consequently *one* optimal watermarking algorithm. Each watermarking scheme is based on a different trade-off and one has to be cautious when benchmarking various algorithms. It should be ensured that the methods under investigation are evaluated under quite similar conditions [29]. In other terms, in order to perform a fair performance comparison in terms of robustness, the evaluated watermarking algorithm should have roughly the same capacity and introduce approximately the same visual distortion.

The last few years have seen the emergence of a new trend in the watermarking community. The watermarking process is now seen as the transmission of a signal through a noisy channel. Original data is then seen as interfering noise which reduces significantly the amount of reliably communicable watermark information. In this new perspective, Chen and Wornell noticed a precious paper written by Costa [8]. He showed that, if a message is sent through a channel corrupted by two successive additive white Gaussian noise sources and if the transmitter knows the first noise source, interference from the first noise source can be entirely cancelled. From a watermarking point of view, the message can be seen as the watermark, the first known noise source as the original data and the second unknown noise source as the attacks. Even if Costa's model is substantially different from a real watermarking system, it means that side information at the embedder enables to reduce interference from the original data. This implication has received further support from subsequent theoretical work.

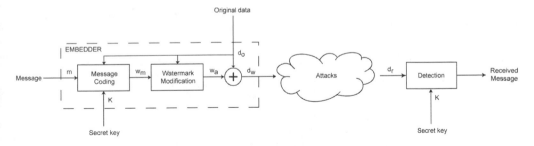

Figure 42.3 Informed watermarking scheme

In Figure 42.1, the embedder can be seen as blind. Information contained in the original data is not exploited during the *message coding* and *watermark modification* steps. Costa's work encourages designing new algorithm based on Figure 42.3 where side information is taken into account during those two steps. Informed watermarking can be done during message coding (informed coding) and/or watermark modification (informed embedding). With informed coding, for a given message, a pool of different alternative watermarks is available and the embedder chooses the one for which the interference introduced by the original data will be minimised. With informed embedding, the goal is to optimally modify the watermark so that the detector extracts the expected message. A typical example is to perceptually shape the watermark accordingly to the original data so that fidelity is increased while robustness is maintained.

Since presenting the whole theory behind digital watermarking is far beyond the scope of this chapter, the interested reader is invited to read the various books devoted to the subject. An introducing overview of digital watermarking can be found in [26]. Further details are developed in [9] where the authors even provide samples of source code. Finally an in depth discussion on informed watermarking is conducted in [19].

3. APPLICATIONS OF WATERMARKING VIDEO CONTENT

If the increasing interest concerning digital watermarking during the last decade is most likely due to the increase in concern over copyright protection of digital content, it is also emphasised by its incredible commercial potential. The following section is consequently completely dedicated to the presentation of various applications in which digital watermarking can bring a valuable support in the context of video. Digital video watermarking may indeed be used in many various applications and some of them are far from the original copyright enforcement context. The applications presented in this section have been gathered in Table 42.2. This is not an exhaustive list and many applications are still to be imagined.

3.1 COPY CONTROL

The Digital Versatile Disk (DVD) and DVD players appeared on the consumer market in late 1996. This new technology was enthusiastically welcomed since DVD players provide a very high-quality video signal. However, the advantages of digital video are counterbalanced by an increased risk of illegal copying. In contrast to traditional VHS tape copying, each copy of digital video data is a perfect reproduction. This raised the concern of copyright owners and Hollywood studios requested that several levels of copy protection should be investigated before any device with digital video recording capabilities could be introduced.

Table 42.2 Video watermarking: applications and associated purpose

Applications	Purpose of the embedded watermark
Copy control	Prevent unauthorised copying.
Broadcast monitoring	Identify the video item being broadcasted.
Fingerprinting	Trace back a malicious user.
Video authentication	Insure that the original content has not been altered.
Copyright protection	Prove ownership.
Enhance video coding	Bring additional information e.g. for error correction.

The Copy Protection Technical Working Group (CPTWG) has consequently been created in order to work on copy protection issues in DVD. A standard has not been defined yet. However a system, which could become the future specification for DVD copy protection, has been defined [5]. The three first components are already built in consumer devices and the other three are still under development.

- **The Content Scrambling System (CSS)**. This method developed by Matsushita scrambles MPEG-2 video. A pair of keys is required for descrambling: one is unique to the disk and the other is specific to the MPEG file being descrambled. Scrambled content is not viewable.

- **The Analog Protection System (APS)**. This system developed by Macrovision modifies NTSC/PAL. The resulting video signal can be displayed on televisions but cannot be recorded on VCR's. However, the data on a disk are not NTSC/PAL encoded and APS has to be applied after encoding in the DVD player. Some bits are consequently stored in the MPEG stream header and give the information of whether and how APS should be applied.

- **The Copy Generation Management System (CGMS)**. This is a pair of bits stored in the header of an MPEG stream encoding one of three possible rules for copying: *copy-always*, *copy-never* and *copy-once*. The copy-once case is included so that time-shifting is allowed i.e. a copy of broadcast media is made for later viewing.

- **5C**. A coalition of five companies designs this mechanism. It allows several compliant devices, connected to the same digital video bus e.g. IEEE1394 (firewire), to exchange keys in an authenticated manner so that encrypted data can be sent over the bus. Noncompliant devices do not have access to the keys and cannot decrypt the data.

- **Watermarking**. The main purpose of watermarking is to provide a more secure solution than storing bits in the MPEG stream header. In DVD, digital watermarking is primarily intended for the CGMS bits and secondary for the APS bits.

- **Physical identifiers**. The idea is to design secure physical media identifiers in order to be able to distinguish between original media and copies.

Figure 42.4 shows how those mechanisms have been put together in the DVD so that copy protection is enforced. The additional performance brought by watermarking is emphasized by the dark walls.

Everything starts when Hollywood studios release a new copyrighted DVD with CGMS bits encoding the message *copy-never*. Both CSS keys are stored on the lead-in area of the DVD. This area is only read by compliant players. This prevents factory-pressed legal disks from being displayed by noncompliant players. Moreover bit-for-bit illegal copies will contain CSS scrambled content, but not the keys. As a result, such illegal copies cannot be displayed by any player, compliant or not. If the output signal given by compliant players is digital, CGMS bits prevent copying in the compliant world while 5C will avoid any communication with any noncompliant devices. However, to date, analog monitors are still widespread and even compliant players output an analog signal for compatibility. Since CGMS bits do not survive digital to analog conversion, watermarking is introduced in order to avoid copying in the compliant world. Unfortunately, in the noncompliant world, APS only disables copying of analog NTSC/PAL signals on VHS tapes. Disks without CSS or CGMS can then be easily generated e.g. thanks to a simple PC with a video capture card.

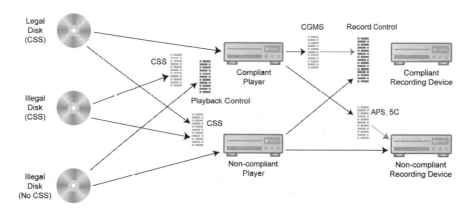

Figure 42.4 DVD copy-protection system

Now illegal disks containing unscrambled content without CSS or CGMS are available. They may have been generated as seen previously. But they can also be generated directly from an original legal disk since CSS was cracked in 1999 [39]. The remaining CGMS bits can then be trivially stripped from the MPEG stream. Such illegal copies can of course be displayed by noncompliant players but watermarking has to be introduced in order to prevent those copies to enter the compliant world. Compliant players will detect the *copy-never* watermark embedded in *unscrambled* DVD-RAM and will refuse playback. The video signal given by a noncompliant player can be recorded by noncompliant recording devices. However watermarking prevents copying with compliant devices. The whole protection system results in two hermetically separated worlds. A consumer should have both types of players in order to display legal and illegal disks. The expense of such a strategy will help to "keep honest people honest".

It is important for DVD recorders to support the *copy-once* case in order to allow time shifting. When the recorder detects the *copy-once* message, it should modify the stream so that the hidden message becomes *copy-never*. This can be easily done in the case of stored bits in the MPEG header but it is less straightforward when using watermarking. Two proposals are investigated. The first one consists in superimposing a second watermark when a *copy-once* watermark is detected. The two watermarks together will then encode the message *copy-never*. The second proposal avoids remarking and exploits the ticket concept [34]. The idea is to use two hidden signals: an embedded watermark W and a physical ticket T. There exists a relationship between the two signals which can be written $F^n(T)=W$, where $F(.)$ is a one way hash function and n is the number of allowed passages through compliant devices. The ticket is decremented each time the data go through a compliant player or recorder. In other terms, the ticket is modified according to the relation $T'=F(T)$. During playback, the ticket in transit can be embedded in MPEG *user_data* bits or in the blanking intervals of the NTSC/PAL standard. During recording, the ticket can be physically marked in the *wobble*[1] in the lead-in of optical disks.

3.2 BROADCAST MONITORING

Many valuable products are distributed over the television network. News items, such as those sold by companies like Reuters or Associated Press, can be worth over 100,000 USD. In France, during the final of the 2002 FIFA World Cup Korea Japan™, advertisers had to pay 100,000 Euros in order to broadcast a thirty-second commercial break shot on television. The same commercial would even have been billed 220,000 Euros if the French national team had played during the final. Owners of copyrighted videos want to get their royalties each time their property is broadcasted. The whole television market is worth many billions of dollars and Intellectual Property Rights violations are likely to occur. As a result, a broadcast surveillance system has to be built in order to check all broadcasted channels. This will help verifying that content owners get paid correctly and that advertisers get what they have paid for. Such a mechanism will prevent confidence tricks such as the one discovered in Japan in 1997 when two TV stations were convicted of overbooking air time [27].

The most naive approach of broadcast monitoring consists of a pool of human observers watching the broadcasts and recording whatever they see. However, this low-cost method is far from being optimal. Human employees are expensive and are not foolproof. As a result, research has been conducted in order to find a way of automating broadcast monitoring. The first approach, referred to as *passive monitoring*, basically makes a computer simulate a human observer: it monitors the broadcasts and compares the received signals with a database of known videos. This approach is nonintrusive and does not require cooperation from advertisers or broadcasters. However such a system has two major drawbacks. First, it relies on the comparison between received signals against a large database, which is nontrivial in practice. Pertinent signatures, clearly identifying each video, have to be defined and an efficient search for nearest neighbours in a large database has to be designed. This results in a system that

---

[1] The wobble is a radial deviation of the position of pits and lands relative to the ideal spiral. Noncompliant recorders will not insert a ticket and the illegal disk will not enter the compliant world.

is not fully reliable. This may be accurate for acquiring competitive market research data i.e. when a company wants to know how much its competitors spend in advertising. On the contrary, a small error rate (5%) is dramatic for verification services because of the large amount of money at stake. The second con is that the reference database is likely to be large and the storage and management costs might become rapidly prohibitive.

In order to reach the accuracy required for verification services, a new kind of systems, referred as *active monitoring*, has been designed. The underlying idea is to transmit computer-recognizable identification information along with the data. Such identification information is straightforward to decode reliably and to interpret correctly. This approach is known to be simpler to implement than passive monitoring. First implementations of active monitoring placed the identification information in a separate area of the broadcast signal e.g. the Vertical Blanking Interval (VBI) of an analog NTSC/PAL video signal. However dissimulating identification data into other data is exactly the purpose of digital watermarking. Even if watermark embedding is more complicated than storing information in some unused part of a video stream, digital watermarking can be considered as a robust way to implement active monitoring. The European project VIVA (Visual Identity Verification Auditor) proved the feasibility of such a system [14]. The participants used a real-time watermarking scheme which provides active monitoring services over a satellite link. The complexity of the detection algorithm is moderate enough to allow simultaneous monitoring of many channels.

3.3 FINGERPRINTING

The explosion of the Internet has created a new way of acquiring copyrighted content. When a user wants to obtain a new video clip or a new movie, the simplest strategy is to log on the Internet and to use one of the popular peer-to-peer systems e.g. Napster, Gnutella, KaZaA, Morpheus. Multimedia digital contents, stored throughout the world on thousands of computers logged on at the same moment, will instantly get accessible. As a result, European engineering students often download and watch the most recent Hollywood films a long time before they are released in their own country. The situation is even worse in audio with the exchange of MP3[1] files. As a result, copyright owners lose a large amount of royalties [32]. Legal action has been taken to ban such distributed systems but, when Napster was sentenced guilty, two other systems appeared. The basic problem does not come from peer-to-peer systems. It would be a great tool if only legal data was transiting on such distributed networks. The problem is that a *traitor* has made available copyrighted material without any kind of permission. The basic idea would consequently be to be able to identify the traitor when an illegal copy is found in order to sue him in court. This can be done by embedding an indelible and invisible watermark identifying the customer.

In the near future, the way people are looking at TV will be significantly modified. Video streaming is indeed likely to become more and more widespread. It is consequently necessary to find a way of protecting digital video content and

---

[1] The MPEG-1 audio layer 3 (MP3) is a popular audio format for transmitting audio files across the Internet.

digital watermarking seems to be a potential candidate [33]. Two major applications are liable to take the homes by storm: Pay-Per-View (PPV) and Video-On-Demand (VOD). In both applications, digital watermarking can be used in order to enforce a fingerprinting policy. The customer ID is embedded into the delivered video data in order to trace back any user breaking his/her licence agreement. The main difference resides in the watermarking strategy as depicted in Figure 42.5. Embedding the watermark on the customer side has been suggested [20] but it should be avoided if possible in order to prevent reverse engineering. In a PPV environment, a video server multicasts some videos and customers have only to connect to the server in order to obtain the video. The video server is passive. At a given moment, it delivers the same video stream to multiple users. In order to enforce fingerprinting, a proposed method [7] is to have each network element (router, node or whatever) embed a piece of watermark as the video stream is relayed. The resulting watermark will contain a trace of the route followed by the video stream. Such a strategy requires support from network providers, who might not be forthcoming about it. In a VOD framework, the video server is active. It receives a request from a customer and sends the requested video. It is a multi-unicast strategy. This time, the video server can insert a watermark identifying the customer since each connection is dedicated to only one customer.

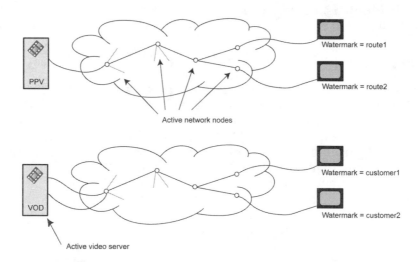

Figure 42.5 Alternative watermarking strategies for video streaming.

Another fingerprinting application has been considered with the apparition of a new kind of piracy. Nowadays illegal copying of brand new movies projected onto cinema screen by means of a handheld video camera has become a common practice. The most memorable example is surely when, one week after its US release, the very anticipated "Starwars Episode I: The Phantom Menace" was available on the Internet in a low quality version, with visible head shadows of audience members. Although the quality of such copies is usually very low, their economical impact can be enormous. Moreover, the upcoming digital cinema format to be introduced in theatres raises some concern. With higher visual quality, the threat becomes larger and Hollywood studios want to oblige cinema owners to prevent the presence of video cameras in their premises. Once again, digital watermarking could provide a solution [21]. A watermark can be embedded during show time identifying the cinema, the presentation date and time. If an illegal copy created with a video camera is found, the watermark is extracted and the cinema to blame is identified. After many blames, the cinema is sanctioned with a ban on the availability of content.

3.4 VIDEO AUTHENTICATION

Large amounts of video data are distributed throughout the Internet every day. More and more video cameras are installed in public facilities for surveillance purposes. However, popular video editing softwares permit today to easily tamper with video content, as shown in Figure 42.6, and video content is no more reliable. For example, in some countries, a video shot from a surveillance camera cannot be used as a piece of evidence in a courtroom because it is not considered trustworthy enough. When someone is emailed a somewhat unusual video, it is quite impossible to determine if it is an original or a hoax. Authentication techniques are consequently needed in order to ensure authenticity of video content. Methods have to be designed for verifying the originality of video content and preventing forgery. When a customer purchases video content via electronic commerce, he wants to be sure that it comes from the alleged producer and that no one has tampered with the content. The very first research efforts for data authentication used cryptography. The major drawback of such an approach is that it provides a *complete verification*. In other terms, the data is considered as untouchable and the data for authentication has to be exactly the same one as the original one. But this strong constraint might be too restricting. One might prefer to allow some distortions on the digital data if the original content has not been significantly modified. This is typically the case in wireless environment where some noise is added to the data. This approach is referred to as *content verification*.

Figure 42.6 Original and tampered video scenes

Researchers have investigated the use of digital watermarking in order to verify the integrity of digital video content. A basic approach consists in regularly embedding an incremental timestamp in the frames of the video [37]. As a result, frame cuts, foreign frame insertion, frame swapping and frame rate alteration can be easily detected. This approach is very efficient for detecting temporal alteration of the video stream. However, it might fail in detecting alterations of the content itself e.g. a character is completely removed from a movie.

Investigations have consequently been conducted in order to prevent modifications of the content of the video itself. One proposal [17] embeds the edge map of each frame in the video stream. During the verification process, if the video content has been modified, there will be a mismatch between the extracted edge map from the verified video and the watermarked edge map. The detector will consequently report content tampering. Another proposal exploits the idea that a movie is made up of one audio and one video stream and that both need to be protected against unauthorised tampering. The fundamental idea is then to combine video and audio watermarking [18] in order to obtain an efficient authenticating system. Features of both streams are embedded one into another. Modification from either the sound track, or the video track, is immediately spotted by the detector, since the extracted and watermarked features will differ.

3.5 COPYRIGHT PROTECTION

Copyright protection is historically the very first targeted applications for digital watermarking. The underlying strategy consists in embedding a watermark, identifying the copyright owner, in digital multimedia data. If an illegal copy is found, the copyright owner can prove his/her paternity thanks to the embedded watermark and can sue the indelicate user in court. This perfect scenario is however likely to be disturbed by malicious users in the real world [10]. If an attacker adds a second watermark into a video clip, both the original owner and the attacker can claim ownership and therefore defeat the purpose of using watermarking. Using the original video clip during the verification procedure happens to prevent the multiple ownership problems in some cases. However, this problem still holds if the watermarking algorithm is invertible because it allows the attacker to produce his/her own counterfeited original video clip. In this case, both the original owner and the attacker have an original video clip which contains the watermark of the other one. As a result, nobody can claim ownership! This situation is referred to as the *deadlock* problem in the watermarking community. Watermark algorithms are consequently required to be noninvertible in order to provide copyright protection services and they are often backed up by an elaborated protocol with a trusted third party. Copyright protection has been investigated for video watermarking [42] even if this is not the most targeted application.

Instead of protecting the whole video stream, copyright owners might rather want to protect only a part of the video content. The commercial value in a video is indeed often concentrated in a small number of video objects e.g. the face of an actor. Moreover, future video formats will distinguish the different objects in a video. This will be the case with the upcoming MPEG-4 format. Recent research has consequently investigated digital watermarking of video objects [41]. Watermarking video objects prevents unauthorised reuse in other video clips. However video objects are likely to be submitted to various video editing such as scaling, rotation, shifting and flipping. As a result, special care must be taken regarding the resilience of the watermark against such processings. This can be quite easily obtained thanks to a geometrical normalisation [4], according to the moments and axes of the video object, prior to embedding and extraction.

3.6 ENHANCED VIDEO CODING

The attentive reader may have noticed that video watermarking and video coding are two conflicting technologies. A perfect video codec should remove any extra redundant information. In other terms, two visually similar videos should have the same compressed representation. If one day, such an optimal video codec is designed, then video watermarking will disappear since unwatermarked and watermarked data would have the same compressed representation. Digital watermarking can be consequently seen as the exploitation of the features of the compression algorithms in order to hide information. However recent research has shown that digital watermarking can benefit the coding community. The video coding process can be sequenced in two steps. During *source coding*, any redundant information is removed in order to obtain the most possible compressed representation of the data while keeping its original visual quality. This compressed representation is then submitted to *channel coding*, where extra redundant information is added for error correction. Channel coding is mandatory since errors are likely to occur during the transmission, e.g. in a wireless environment. Digital watermarking can be introduced as an alternative solution for introducing error correcting information after source coding, without inducing any overhead [3]. Experiments have demonstrated the feasibility of such an approach and results are even reported showing that digital watermarking can have better performances than traditional error correction mechanisms [45].

Embedding useful data directly into the video stream can spare much storage space. A typical video stream is made up of two different parallel streams: the audio and video streams. Those two streams need to be synchronised during playback for pleasant viewing, which is difficult to maintain during cropping operations. Hiding the audio stream into the video one [38] will implicitly provide efficient and robust synchronisation, while significantly reducing the required storage need or available bandwidth. In the same fashion, the actual Picture-in-Picture system can be improved by hiding a video stream into another one [48]. This technology, present in many television sets, uses separate data streams in order to superimpose a small video window over the full-size video displayed on the television set. Digital watermarking allows embedding the secondary video stream into the carrier one. During playback, the watermark is extracted and the embedded video is displayed in a window within the host video. With such an approach, only one stream needs to be transmitted. This approach can be extended so that a user can switch to the *PG* version of an *R* rated movie, with alternative dialogs and scenes replacing inappropriate content.

4. CHALLENGES IN VIDEO WATERMARKING

Digital watermarking has focused on still images for a long time but nowadays this trend seems to vanish. More and more watermarking algorithms are proposed for other multimedia data and in particular for video content. However, even if watermarking still images and video is a similar problem, it is not identical. New problems, new challenges show up and have to be addressed. This section points out three major challenges for digital video watermarking. First, there are many nonhostile video processings, which are likely to alter the watermark signal. Second, resilience to collusion is much more critical in the context of video. Third, real-time is often a requirement for digital video watermarking.

4.1 VARIOUS NONHOSTILE VIDEO PROCESSINGS

Robustness of digital watermarking has always been evaluated via the survival of the embedded watermark after attacks. Benchmarking tools have even been developed in order to automate this process [1]. In the context of video, the possibilities of attacking the video are multiplied. Many different nonhostile video processings are indeed available. Nonhostile refers to the fact that even content provider are likely to process a bit their digital data in order to manage efficiently their resources.

Photometric attacks

This category gathers all the attacks which modify the pixel values in the frames. Those modifications can be due to a wide range of video processings. Data transmission is likely to introduce some noise for example. Similarly, digital to analog and analog to digital conversions introduce some distortions in the video signal. Another common processing is to perform a gamma correction in order to increase the contrast. In order to reduce the storage needs, content owners often transcode, i.e. re-encode with a different compression ratio, their digital data. The induced loss of information is then susceptible to alter the performances of the watermarking algorithm. In the same fashion, customers are likely to convert their videos from a standard video format such as MPEG-1, MPEG-2 or MPEG-4 to a *popular* format e.g. DivX. Here again, the watermark signal is bound to undergo some kind of interferences. Spatial filtering inside each frame is often used to restore a low-quality video. Inter-frames filtering, i.e. filtering between adjacent frames of the video, has to be considered too. Finally, chrominance resampling (4:4:4, 4:2:2, 4:2:0) is commonly used processing to reduce storage needs.

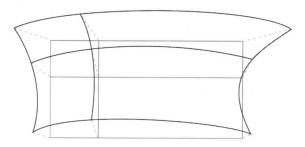

Figure 42.7 Example of distortion created by a handheld camera (exaggerated)

Spatial desynchronisation

Many watermarking algorithms rely on an implicit spatial synchronisation between the embedder and the detector. A pixel at a given location in the frame is assumed to be associated with a given bit of the watermark. However, many nonhostile video processings introduce spatial desynchronisation which may result in a drastic loss of performance of a watermarking scheme. The most common examples are changes across display formats (4/3, 16/9 and 2.11/1) and changes of spatial resolution (NTSC, PAL, SECAM and usual movies standards). Alternatively the pixel position is susceptible to jitter. In particular, positional jitter occurs for video over poor analog links e.g. broadcasting in a wireless environment. In the digital cinema context, distortions brought by the handheld camera can be considered as nonhostile since the purpose of the camera is not explicitly to remove the embedded watermark. It has been shown that the handheld camera attack can be separated into two geometrical distortions [13]: a bilinear transform, due to the misalignment between the camera and the cinema screen, and a curved transform, because of the lens deformations. This results in a curved–bilinear transform depicted in Figure 42.7 which can be modelled with twelve parameters.

Temporal desynchronisation

Similarly temporal desynchronisation may affect the watermark signal. For example, if the secret key for embedding is different for each frame, simple frame rate modification would make the detection algorithm fail. Since changing frame rate is a quite common process, watermarks should be designed so that they survive such an operation.

Video editing

The very last kind of nonhostile attacks gathers all the operation that a video editor may perform. Cut-and-splice and cut-insert-splice are two very common processings used during video editing. Cut-insert-splice is basically what happens when a commercial is inserted in the middle of a movie. Moreover, transition effects, like fade-and-dissolve or wipe-and-matte, can be used in order to smooth the transition between to scenes of the video. Such kind of editing can be seen as temporal editing in contrast to spatial editing. Spatial editing refers to the addition of a visual content in each frame of the video stream. This includes for example graphic overlay, e.g. logos or subtitles insertion, and video stream superimposition, like in the Picture-in-Picture technology. The detector sees such operation as a cropping of some part of the watermark. Such a severe attack is susceptible to induce a high degradation of the detection performances.

Table 42.3 Examples of nonhostile video processings

Photometric	- Noise addition, DA/AD conversion - Gamma correction - Transcoding and video format conversion - Intra- and inter-frames filtering - Chrominance sampling (4:4:4, 4:2:2, 4:2:0)
Spatial desynchronisation	- Changes across display formats (4/3, 16/9, 2.11/1) - Changes of spatial resolution (NTSC, PAL, SECAM) - Positional jitter - Handheld camera attack
Temporal desynchronisation	- Changes of frame rate
Video editing	- Cut-and-splice and cut-insert-splice - Fade-and-dissolve and wipe-and-matte - Graphic overlay (subtitles, logo) - Picture-in-Picture

There are many various attacks to be considered as shown in Table 42.3 and it may be useful to insert countermeasures [12] in the video stream in order to cope with the distortions introduced by such video processings. Moreover, the reader should be aware that many other hostile attacks are likely to occur in the real world. Indeed, it is relatively easy today to process a whole movie thanks to the powerful available personal computers. It is virtually possible to do whatever transformation on a video stream. For example, for still images, Stirmark introduces random local geometric distortions which succeed in trapping the synchronisation of the detector. This software has been optimised for still images and, when used on each frame of the video stream, visible artefacts can be spotted when moving objects go through the fixed geometric distortion. However future versions of Stirmark will surely address this visibility issue.

4.2 RESILIENCE AGAINST COLLUSION

Collusion is a problem that has already been pointed out for still images some time ago. It refers to a set of malicious users who merge their knowledge, i.e. different watermarked data, in order to produce illegal content, i.e. unwatermarked data. Such collusion is successful in two different distinct cases.

- **Collusion type I:** The *same watermark* is embedded into different copies of *different data*. The collusion can estimate[1] the watermark from each watermarked datum and obtain a refined estimate of the watermark by linear combination, e.g. the average, of the individual estimations. Having a good estimate of the watermark permits to obtain unwatermarked data with a simple subtraction with the watermarked one.

- **Collusion type II:** *Different watermarks* are embedded into different copies of the *same data*. The collusion only has to make a linear combination of the different watermarked data, e.g. the average, to produce unwatermarked data. Indeed, generally, averaging different watermarks converges toward zero.

Collusion is a very important issue in the context of digital video since there are twice more opportunities to design a collusion than with still images. When video is considered, the origin of the collusion can be twofold.

- **Inter-videos collusion:** This is the initial origin considered for still images. A set of users have a watermarked version of a video which they gather in order to produce unwatermarked video content. In the context of copyright protection, the same watermark is embedded in different videos and collusion type I is possible. Alternatively, in a fingerprinting application, the watermark will be different for each user and collusion type II can be considered. Inter-videos collusion requires different watermarked videos to produce unwatermarked video content.

- **Intra-video collusion:** This is a video-specific origin. As will be detailed later, many watermarking algorithms consider a video as a succession of still images. Watermarking video comes then down to watermarking series of still images. Unfortunately this opens new opportunities for collusion. If the same watermark is inserted in each frame, collusion type I can be enforced since different images can be obtained from moving scenes. On the other hand, if alternative watermarks are embedded in each frame, collusion type II becomes a danger in static scenes since they produce similar images. As a result, the watermarked video alone permits removing the watermark from the video stream.

---

[1] The watermark is often considered as noise addition. A simple estimation consequently consists in computing the difference between the watermarked data and the low-pass filtered version of it.

Even if collusion is not really of interest depending on the targeted application e.g. broadcast monitoring, it often raises much concern in digital video watermarking. It gives indeed opportunities for forgery if the watermarking algorithm is weak against intra-video collusion.

The reader will have understood that the main danger is intra-frame collusion i.e. when a watermarked video alone is enough to remove the watermark from the video. It has been shown that both strategies *always insert the same watermark in each frame* and *always insert a different watermark in each frame* making collusion attacks conceivable. As a result, an alternative strategy has to be found. A basic rule has been enounced so that intra-video collusion is prevented [47]. The watermarks inserted into two different frames of a video should be as similar, in terms of correlation, as the two frames are similar. In other terms, if two frames look like quite the same, the embedded watermarks should be highly correlated. On the contrary, if two frames are really different, the watermark inserted into those frames should be unalike. This rule is quite straightforward when regarding attentively the definition of the two types of collusion. This can be seen as a form of informed watermarking since this rule implies a dependency between the watermark and the host frame content. A relatively simple implementation of this approach can be done by embedding a spatially localised watermark according to the content of each frame of the video [46]. A small watermark pattern can be embedded in some key locations of each frame, e.g. salient points. During the extraction process, the detector can easily detect the position of the salient points and look for the presence or the absence of a watermark.

The problem of inter-video collusion still holds. Concerning collusion type I, this issue can be prevented by inserting a Trusted Third Party (TTP) which gives the message to be embedded. This message is often a function of the encrypted message that the copyright owner wants to hide and a hash of the host data. Different videos give different messages to be hidden and consequently different embedded watermarks. The TTP also acts as a repository. When an illegal copy is found, the copyright owner extracts the embedded message and transmits it to the TTP, which in turn gives the associated original encrypted message. If the copyright owner can successfully decrypt it, he can claim ownership. Regarding collusion type II, results obtained for still images can easily be extended to digital video. The problem arises when a coalition of malicious users, having each one a copy of the same data but with a different embedded watermark, colludes in order to produce illegal unwatermarked data. They compare their watermarked data, spot the locations where the different versions differ and modify the data in those locations. A traditional countermeasure [6] consists in designing the set of distributed watermarks so that a coalition, gathering at most *c* users, will not succeed in removing the whole watermark signal. It should be noted that *c* is generally very small in comparison with the total number *n* of users. Moreover the set of watermarks is built in such a way that no coalition of users can produce a document which will make an innocent user, i.e. not in the illegal coalition, be framed. In other terms, colluding creates still watermarked video content and the remaining watermark clearly identifies the malicious colluding users, without ever accusing any innocent customer. Implementations of such set of watermark have already been proposed for still images which are based on the projective geometry [16] or the theory of combinatorial designs [51].

4.3 REAL-TIME WATERMARKING

Real-time can be an additional specification for video watermarking. It was not a real concern with still images. When a person wants to embed a watermark or to check the presence of a watermark in an image, a few seconds is an acceptable delay. However, such a delay is unrealistic in the context of the video. Frames are indeed sent at a fairly high rate, typically 25 frames per second, to obtain a smooth video stream. At least the embedder or the detector, and even sometimes both of them, should be able to handle such a rate. In the context of broadcast monitoring, the detector should be able to detect an embedded watermark in real-time. In a VOD environment, the video server should be able to insert the watermark identifying the customer at the same rate that the video is streamed. In order to meet the real-time requirement, the complexity of the watermarking algorithm should obviously be as low as possible. Moreover, if the watermark can be inserted directly into the compressed stream, this will prevent full decompression and recompression and, consequently, it will reduce computational needs. This philosophy has led to the design of very simple watermarking schemes. Exploiting the very specific part of a video compression standard can lead to very efficient algorithms. An MPEG encoded video stream basically consists of a succession of Variable Length Code (VLC). A watermark can consequently be embedded in the stream by modifying those VLC code words [31]. The MPEG standard uses indeed similar VLC code words i.e. with the same run length, the same VLC size and a quantized level difference of one. Such VLC code words can be used alternatively in order to hide a bit.

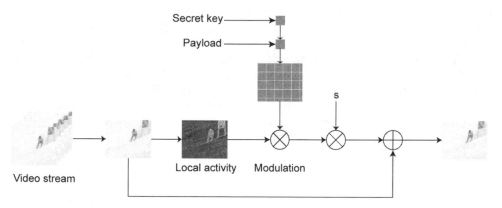

Figure 42.8 Description of JAWS embedding

Just Another Watermarking System (JAWS)

When considering real-time, the watermarking algorithm designed by Philips Research is often considered as a reference. The JAWS algorithm was originally designed for broadcast monitoring and is actually one of the leading candidates for watermarking in DVD. The real-time requirement is met by using simple operations at video rate and only a few complex ones at a much lower rate [25].

The embedding process is depicted in Figure 42.8. First of all, an MxM normally distributed reference pattern p_r is generated with a secret key. In a second step, a reference watermark w_r is created according to the following equation:

$$w_r = p_r - shift(p_r, message) \tag{42.1}$$

where the *shift(.)* function returns a cyclically shifted version of the reference pattern p_r. In JAWS, the message is completely encoded by the shift between the two reference patterns. This reference watermark is then tiled, possibly with truncation, to obtain the full-size watermark w. For each frame, this watermark is then perceptually shaped so that the watermark insertion remains imperceptible. Each element i of the watermark is scaled by the local activity $\lambda(i)$ of the frame, given by Laplacian filtering. The flatter the region is, the lower the local activity is. This is coherent with the fact that the human eye is more sensitive to noise addition in flat regions of an image. Finally, the watermark is scaled by a global factor s and added to the frame F in order to obtain the watermarked frame F_w. As a result, the overall embedding process can be expressed as:

$$F_w(i) = F(i) + s.\lambda(i).w(i) \tag{42.2}$$

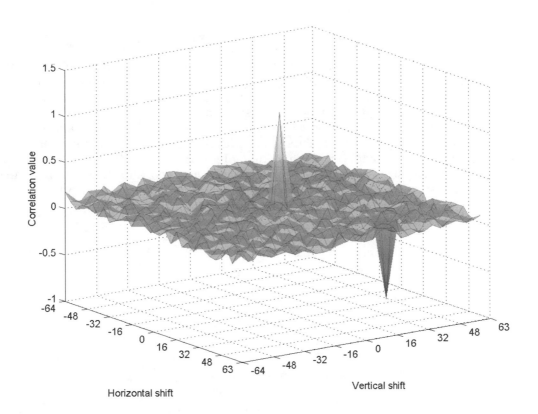

Figure 42.9 Example of SPOMF detection

On the detector side, the incoming frames are folded, summed and stored in an MxM buffer B. The detectors look then for all the occurrences of the reference pattern p_r in the buffer with a two dimensional cyclic convolution. Since such an operation is most efficiently computed in the frequency domain, this leads to Symmetrical Phase Only Matched Filtering (SPOMF) detection which is given by the following equation:

$$SPOMF(B, p_r) \triangleq IFFT\left[\varphi(FFT(B)).\varphi\left(FFT(p_r)^*\right)\right] \quad with \; \varphi(x) = \begin{cases} x/|x| & if \; x \neq 0 \\ 1 & if \; x = 0 \end{cases} \quad (42.3)$$

where *FFT(.)* and *IFFT(.)* are respectively the forward and inverse Fourier transforms and x^* denotes the complex conjugation. Figure 42.9 shows the result of such a detection. Two peaks can be isolated which correspond to the two occurrences of p_r in w_r. The peaks are oriented accordingly to the sign before their associated occurrence of p_r in Equation (1). Because of possible positional jitter, all the relative positions between the peaks cannot be used and relative positions are forced to be multiple of a grid size G. Once the detector has extracted the peaks, the hidden payload can be easily retrieved. The attentive reader would have noticed that this scheme is inherently shift invariant since a shifting operation does not modify the relative position of the peaks. Significant improvements have been added to this scheme afterwards. For example, shift invariance has been further exploited in order to increase the payload [35] and simple modifications permitted to obtain scale invariance [50].

5. THE MAJOR TRENDS IN VIDEO WATERMARKING

Digital watermarking for video is a fairly new area of research which basically benefits from the results for still images. Many algorithms have been proposed in the scientific literature and three major trends can be isolated. The most simple and straightforward approach is to consider a video as a succession of still images and to reuse an existing watermarking scheme for still images. Another point of view considers and exploits the additional temporal dimension in order to design new robust video watermarking algorithms. The last trend basically considers a video stream as some data compressed according to a specific video compression standard and the characteristics of such a standard can be used to obtain an efficient watermarking scheme. Each of those approaches has its pros and cons as detailed in Table 42.4.

Table 42.4 Pros and cons of the different approaches for video watermarking.

	Pros	Cons
Adaptation image → video	Inherit from all the results for still images	Computationally intensive
Temporal dimension	Video-driven algorithms which often permit higher robustness	Can be computationally intensive
Compression standard	Simple algorithms which make real-time achievable	Watermark may be inherently tied to the video format

5.1 FROM STILL IMAGE TO VIDEO WATERMARKING

In its very first years, digital watermarking has been extensively investigated for still images. Many interesting results and algorithms were found and when new areas, such as video, were researched, the basic concern was to try to reuse the previously found results. As a result, the watermarking community first considered the video as a succession of still images and adapted existing watermarking schemes for still images to the video. Exactly the same phenomenon occurred when the coding community switched from image coding to video coding. The first proposed algorithm for video coding was indeed Moving JPEG (M-JPEG), which simply compresses each frame of the video with the image compression standard JPEG. The simplest way of extending a watermarking scheme for still images is to embed the same watermark in the frames of the video at a regular rate. On the detector side, the presence of the watermark is checked in every frame. If the video has been watermarked, a regular pulse should be observed in the response of the detector [2]. However, such a scheme has no payload. The detector only tells if a given watermark is present or not but it does not extract any hidden message. On the other hand, the host data is much larger in size than a single still image. Since one should be able to hide more bits in a larger host signal, high payload watermarks for video could be expected. This can be easily done by embedding an independent multi-bits watermark in each frame of the video [15]. However one should be aware that this gain in payload is counterbalanced by a loss of robustness.

Differential Energy Watermarks (DEW)

The DEW method was initially designed for still images and has been extended to video by watermarking the I-frames of an MPEG stream [31]. It is based on selectively discarding high frequency DCT coefficients in the compressed data stream. The embedding process is depicted in Figure 42.10. The 8x8 pixel blocks of the video frame are first pseudo randomly shuffled. This operation forms the secret key of the algorithm and it spatially randomizes the statistics of pixel blocks i.e. it breaks the correlation between neighbouring blocks. The obtained shuffled frame is then split into n 8x8 blocks. In Figure 42.10, n is equal to 16. One bit is embedded into each one of those blocks by introducing an energy difference between the high frequency DCT-coefficients of the top half of the block (region A) and the bottom half (region B). This is the reason why this technique is called a differential energy watermark.

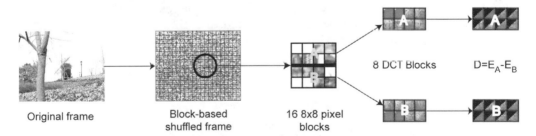

Figure 42.10 Description of DEW embedding

In order to introduce an energy difference, the block DCT is computed for each n 8x8 block and the DCT-coefficients are prequantized with quality factor Q_{jpeg} using the standard JPEG quantization procedure. The obtained coefficients are then separated in two halves and the high frequency energy for each region is computed according to the following equation:

$$E(c, n, Q_{jpeg}) = \sum_{b=0}^{n/2-1} \sum_{i \in S(c)} \left([\theta_{i,b}]_{Q_{jpeg}} \right)^2 \quad with \ S(c) = \{i = \{0,63\} | (i > c)\} \tag{42.4}$$

where $\theta_{i,b}$ is the DCT coefficient with index i in the zig-zag order in the bth DCT block, [.] indicates the prequantization with quality factor Q_{jpeg} and c is a given cut-off index which was fixed to 27 in Figure 42.10. The value of the embedded bit is encoded as the sign of the energy difference $D=E_A-E_B$ between the two regions A and B. All the energy after the cut-off index c in either region A or region B is eliminated by setting the corresponding DCT coefficients to zero to obtain the appropriate sign for the difference D. It should be noted that this can be easily done directly in the compressed domain by shifting the End Of Block (EOB) marker of the corresponding 8x8 DCT blocks toward the DC-coefficient up to the cut-off index. Finally, the inverse block DCT is computed and the shuffling is inversed in order to obtain the watermarked frame. On the detector side, the energy difference is computed and the embedded bit is determined according to the sign of the difference D. This algorithm has been further improved to adapt the cut-off index c to the frequency content of the considered n 8x8 block and so that the energy difference D is greater than a given threshold D_{target} [30].

5.2 INTEGRATION OF THE TEMPORAL DIMENSION

The main drawback of considering a video as a succession of independent still images is that it does not satisfactorily take into account the new temporal dimension. The coding community has made a big step forward when they decided to incorporate the temporal dimension in their coding schemes and it is quite sure that it is the advantage of the watermarking community to investigate such a path. Many researchers have investigated how to reduce the visual impact of the watermark for still image by considering the properties of the Human Visual System (HVS) such as frequency masking, luminance masking and contrast masking. Such studies can be easily exported to video with a straightforward frame-per-frame adaptation. However, the obtained watermark is not optimal in terms of visibility since it does not consider the temporal sensitivity of the human eye. Motion is indeed a very specific feature of the video and new video-driven perceptual measures need to be designed in order to be exploited in digital watermarking [28]. This simple example shows that the temporal dimension is a crucial point in video and that it should be taken into account to design efficient algorithms.

Figure 42.11 Line scan of a video stream

Spread-Spectrum (SS)

One of the pioneer works in video watermarking considers the video signal as a one dimensional signal [22]. Such a signal is acquired by a simple line-scanning as shown in Figure 42.11. Let the sequence $a(j)\in\{-1,1\}$ represent the watermark bits to be embedded. This sequence is spread by a chip-rate cr according to the following equation:

$$b(i) \quad a(j)= \quad j.cr \quad i \leqslant (j+1).cr, \quad i\in N \qquad (42.5)$$

The spreading operation permits to add redundancy by embedding one bit of information into cr samples of the video signal. The obtained sequence $b(i)$ is then amplified locally by an adjustable factor $\lambda(i)\geq0$ and modulated by a pseudo-random binary sequence $p(i)\in\{-1,1\}$. Finally, the spread spectrum watermark $w(i)$ is added to the line-scanned video signal $v(i)$, which gives the watermarked video signal $v_w(i)$. The overall embedding process is consequently described by the following equation:

$$v_w(i) \quad v(i)=w(i)+ \quad v(i) \quad \lambda(i).b(i).p(i), \quad i\in N \qquad (42.6)$$

The adjustable factor $\lambda(i)$ may be tuned according to local properties of the video signal, e.g. spatial and temporal masking of the HVS, or kept constant depending on the targeted application.

On the detector side, recovery is easily accomplished with a simple correlation. However, in order to reduce cross-talk between watermark and video signals, the watermarked video sequence is high-pass filtered, yielding a filtered watermarked video signal $\underline{v_w(i)}$, so that major components of the video signal itself are isolated and removed. The second step is demodulation. The filtered watermarked video signal is multiplied by the pseudo-random noise $p(i)$ used for embedding and summed over the window for each embedded bit. The correlation sum $s(j)$ for the jth bit is given by the following equation:

$$s(j) \quad \sum_{i=j.cr}^{(j+1).cr-1} p(i).\underline{v_w(i)} \quad \sum_{i=j.cr}^{(j+1).cr-1} p(i).\underline{v(i)} \quad \sum_{i=j.cr}^{(j+1).cr-1} p(i).\lambda(i).b(i).p(i) \quad {}_1\Sigma\Sigma_2 \qquad (42.7)$$

The correlation consists of two terms Σ_1 and Σ_2. The main purpose of filtering was to leave Σ_2 untouched while reducing Σ_1 down to 0. As a result, the correlation sum becomes:

$$s(j) \quad \underset{\Sigma_2}{} \approx \sum_{i=j.cr}^{(j+1).cr-1} p(i)^2.\lambda(i).b(i) = a(j).cr.mean(\lambda(i)) \qquad (42.8)$$

The hidden bit is then directly given by the sign of $s(j)$. This pioneer method offers a very flexible framework, which can be used as a basic root of a more elaborate video watermarking scheme.

Other approaches have been investigated to integrate the temporal dimension. Temporal wavelet decomposition can be used for example in order to separate static and dynamic components of the video [49]. A watermark is then embedded in each component to protect them separately. The video signal can also be seen as a three dimensional signal. This point of view has already been considered in the coding community and can be extended to video watermarking. 3D DFT can be used as an alternative representation of the video signal [11]. The HVS is considered on one hand to define an embedding area which will not result in a visible watermark. On the other hand, the obtained embedding area is modified so that it becomes immune to MPEG compression. Considering video as a three dimensional signal may be inaccurate. The three considered dimensions are indeed not homogeneous: there are two *spatial* dimensions and one *temporal* one. This consideration and the computational cost may have hampered further work in this direction. However this approach remains pertinent in some specific cases. In medical imaging for example, different slices of a scanner can be seen as different frames of a video. In this case, the three dimensions are homogeneous and a 3D-transform can be used.

5.3 EXPLOITING THE VIDEO COMPRESSION FORMATS

The last trend considers the video data as some data compressed with a video specific compression standard. Indeed, most of the time, a video is stored in a compressed version in order to spare some storage space. As a result, watermarking methods have been designed, which embed the watermark directly into the compressed video stream. The first algorithm presented in Section 4.3 is a very good example. It exploits a very specific part of the video compression standard (run length coding) in order to hide some information.

Watermarking in the compressed stream can be seen as a form of video editing in the compressed domain [36]. Such editing is not trivial in practice and new issues are raised. The previously seen SS algorithm has been adapted so that the watermark can be directly inserted in the nonzero DCT coefficients of an MPEG video stream [22]. The first concern was to ensure that the watermarking embedding process would not increase the output bit-rate. Nothing ensures indeed that a watermarked DCT-coefficient will be VLC-encoded with the same number of bits than when it was unwatermarked. A straightforward strategy consists then to watermark only the DCT coefficients which do not require more bits to be VLC encoded. The second issue was to prevent the introduced distortion with the watermark to propagate from one frame to another one. The MPEG standard relies indeed on motion prediction and any distortion is likely to be propagated to neighbour frames. Since the accumulation of such propagating signals may result in a poor quality video, a drift compensation signal can be added if necessary. In this case, motion compensation can be seen as a constraint. However it could also be exploited so that the motion vectors of the MPEG stream carry the hidden watermark [24]. The components of the motion vector can be quantised according to a rule which depends on the bit to be hidden. For example, the horizontal component of a motion vector can be quantized to an even value if the bit to be hidden is equal to 0 and to an odd value otherwise.

All the frames of an MPEG coded video are not encoded in the same way. The intra-coded (I) frames are basically compressed with the JPEG image compression standard while the inter-coded (B and P) frames are predicted from other frames of the video. As a result, alternative watermarking strategies can be used depending on the type of the frame to be watermarked [23]. Embedding the watermark directly in the compressed video stream often allows real-time processing of the video. However the counterpart is that the watermark is inherently tied to a video compression standard and may not survive video format conversion.

6. CONCLUSION

Digital watermarking has recently been extended from still images to video content. Further research in this area is strongly motivated by an increasing need from the copyright owners to reliably protect their rights. Because of the large economic stakes, digital watermarking promises to have a great future. New applications are likely to emerge and may combine existing approaches. For example, a watermark can be separated into two parts: one for copyright protection and the other for customer fingerprinting. However many challenges have to be taken up. Robustness has to be considered attentively. There are indeed many nonhostile video processings which might alter the watermark signal. It might not even be possible to be immune against all those attacks and detailed constraints have to be defined according to the targeted application. Since collusion is far more critical in the context of video, it must be seriously considered. Finally the real-time constraint has to be met in many applications. In spite of all those challenges, many algorithms have already been proposed in the literature. It goes from the simple adaptation of a watermarking algorithm for still images to the really video specific watermarking scheme.

Open paths still remain in video watermarking. This technology is indeed in its infancy and is far from being as mature as for still images. Quite all possible image processings have been investigated for still images watermarking. On their side, the proposed algorithms for video have remained relatively simple. Many video processings have not been tried and the line is consequently not exhausted. Moreover, introduction of perceptual measures has significantly improved the performances of algorithms for still images. This approach has not been fully extended to video yet. Perceptual measures for video exist but the major challenge consists in being able to exploit them in real-time. Finally, the second generation of watermarking algorithms has only given its first results. Future discoveries in this domain are likely to be of great help for digital video watermarking.

REFERENCES

[1]   http://www.cl.cam.ac.uk/~fapp2/watermarking/stirmark/
[2]   M. Barni, et al., A Robust Watermarking Approach for Raw Video, in Proceedings of the Tenth International Packet Video Workshop, 2000.
[3]   F. Bartolini, et al., A Data Hiding Approach for Correcting Errors in H.263 Video Transmitted Over a Noisy Channel, in Proceedings of the IEEE Fourth Workshop on Multimedia Signal Processing, pp. 65-70, 2001.
[4]   P. Bas and B. Macq, A New Video-Object Watermarking Scheme Robust to Object Manipulation, in Proceedings of the IEEE International Conference on Image Processing, 2:526-529, 2001.
[5]   J. Bloom, et al., Copy Protection for DVD Video, in Proceedings of the IEEE, 87(7):1267-1276, 1999.
[6]   D. Boneh and J. Shaw, Collusion-Secure Fingerprinting for Digital Data, in IEEE Transactions on Information Theory, 44(5):1897-1905, 1998.
[7]   I. Brown, C. Perkins, and J. Crowcroft, Watercasting: Distributed Watermarking of Multicast Media, in Proceedings of the First International Workshop on Networked Group Communication, vol. 1736 of Lecture Notes in Computer Science, Springer-Verlag, pp. 286-300, 1999.
[8]   M. Costa, Writing on Dirty Paper, in IEEE Transactions on Information Theory, 29(3):439-441, 1983.

[9] I. Cox, M. Miller and J. Bloom, Digital Watermarking, Morgan Kaufmann Publishers, ISBN 1-55860-714-5, 2001.

[10] S. Craver, et al., Can Invisible Watermarks Resolve Rightful Ownerships?, Technical Report RC 20509, IBM Research Division, 1996.

[11] F. Deguillaume, et al., Robust #D DFT Video Watermarking, in Proceddings of SPIE 3657, Security and Watermarking of Multimedia Content, pp. 113-124,1999.

[12] F. Deguillaume, G. Csurka, and T. Pun, Countermeasures for Unintentionnal and Intentionnal Video Watermarking Attacks, in Proceedings of SPIE 3971, Security and Watermarking of Multimedia Content II, pp. 346-357, 2000.

[13] D. Delannay, et al., Compensation of Geometrical Deformations for Watermark Extraction in the Digital Cinema Application, in Proceedings of SPIE 4314, Security and Watermarking of Multimedia Content III, pp. 149-157, 2001.

[14] G. Depovere, et al., The VIVA Project: Digital Watermarking for Broadcast Monitoring, in Proceedings of the IEEE International Conference on Image Processing, 2:202-205, 1999.

[15] J. Dittmann, M. Stabenau, and R. Steinmetz, Robust MPEG Video Watermarking Technologies, in Proceedings of ACM Multimedia, pp. 71-80, 1998.

[16] J. Dittmann, et al., Combining Digital Watermarks and Collusion Secure Fingerprints for Digital Images, Proceedings of SPIE 3657, Security and Watermarking of Multimedia Content, pp. 171-182, 1999.

[17] J. Dittmann, A. Steinmetz, and R. Steinmetz, Content-Based Digital Signature for Motion Pictures Authentication and Content Fragile Watermarking, in Proceedings of the IEEE International Conference on Multimedia Computing and Systems, 2:209-213, 1999.

[18] J. Dittmann, et al., Combined Audio and Video Watermarking: Embedding Content Information in Multimedia Data, in Proceedings of SPIE 3971, Security and Watermarking of Multimedia Content II, pp. 176-185, 2000.

[19] J. Eggers and B. Girod, Informed Watermarking, The Kluwer International Series in Engineering and Computer Science, ISBN 1-4020-7071-3, 2002.

[20] C. Griwodz, et al., Protecting VoD the Easier Way, in Proceedings of ACM Multimedia, pp.21-28, 1998.

[21] J. Haitsma and T. Kalker, A Watermarking Scheme for Digital Cinema, in Proceedings of the IEEE International Conference on Image Processing, 2001.

[22] F. Hartung and B. Girod, Watermarking of Uncompressed and Compressed Video, in Signal Processing, 66(3):283-301, 1998.

[23] C.T. Hsu and J.-L. Wu, DCT-based Watermarking for Video, in IEEE Transactions on Consumer Electronics, 44(1):206-216, 1998.

[24] F. Jordan, M. Kutter, and T. Ebrahimi, Proposal of Watermarking Technique for Hiding/Retrieving Data in Compressed and Decompressed Video, in ISO/IEC JTC1/SC29/WG11, 1997.

[25] T. Kalker, et al., A Video Watermarking System for Broadcast Monitoring, in Proceedings of SPIE 3657, Security and Watermarking of Multimedia Content, pp. 103-112, 1999.

[26] S. Katzenbeisser and F. Petitcolas, Information Hiding: Techniques for Steganography and Digital Watermarking, Artech House, ISBN 1-58053-035-4, 1999.

[27] D. Kilburn, Dirty Linen, Dark Secrets, Adweek, 38(40):35-40, 1997.

[28] S.-W. Kim, et al., Perceptually Tuned Robust Watermarking Scheme for Digital Video Using Motion entropy Masking, in Proceedings of the IEEE International Conference on Consumer Electronics, pp. 104-105, 1999.

[29] M. Kutter and F. Petitcolas, Fair Benchmarking for Image Watermarking Systems, in Proceedings of SPIE 3657, Security and Watermarking of Multimedia Content, pp. 226-239, 1999.

[30] G. Langelaar and R. Lagendijk, Optimal Differential Energy Watermarking of DCT Encoded Images and Video, in IEEE Transactions on Image Processing, 10(1):148-158, 2001.

[31] G. Langelaar, R. Lagendijk, and J. Biemond, Real-Time Labelling of MPEG-2 Compressed Video, in Journal of Visual Communication and Image Representation, 9(4):256-270, 1998.

[32] J. Lewis, Power to the Peer, LAWeekly, 2002.

[33] E. Lin, et al., Streaming Video and Rate Scalable Compression: What Are the Challenges for Watermarking?, in Proceedings of SPIE 4314, Security and Watermarking of Multimedia Content III, pp. 116-127, 2001.

[34] J.-P. Linnartz, The Ticket Concept for Copy Control Based on Embedded Signalling, in Proceedings of the Fifth European Symposium on Research in Computer Security, vol. 1485 of Lecture Notes in Computer Science, Springer, pp. 257-274, 1998.

[35] M. Maes, et al., Exploiting Shift Invariance to Obtain High Payload Digital Watermarking, in Proceedings of the International Conference on Multimedia Computing and Systems, 1:7-12, 1999.

[36] J. Meng and S. Chang, Tools for Compressed-Domain Video Indexing and Editing, in Proceedings of SPIE 2670, Storage and Retrieval for Image and Video Database, pp. 180-191, 1996.

[37] B. Mobasseri, M. Sieffert, and R. Simard, Content Authentication and Tamper Detection in Digital Video, in Proceedings of the IEEE International Conference on Image Processing, 1:458-461, 2000.

[38] D. Mukherjee, J. Chae, and S. Mitra, A Source and Channel Coding Approach to Data Hiding with Applications to Hiding Speech in Video, in Proceedings of the IEEE International Conference on Image Processing, 1:348-352, 1998.

[39] A. Patrizio, Why the DVD Hack was a Cinch, Wired, 1999.

[40] F. Petitcolas, R. Anderson, and M. Kuhn, Attacks on Copyright Marking Systems, in Proceedings of the Second International Workshop on Information Hiding, vol. 1525 of Lecture Notes in Computer Science, Springer, pp. 218-238, 1999.

[41] A. Piva, R. Caldelli, and A. De Rosa, A DWT-Based Object Watermarking System for MPEG-4 Video Streams, in Proceedings of the IEEE International Conference on Image Processing, 3:5-8, 2000.

[42] L. Qiao and K. Nahrstedt, Watermarking Methods for MPEG Encoded Video: Toward Resolving Rightful Ownership, in Proceedings of the IEEE International Conference on Multimedia Computing and Systems, pp. 276-285, 1998.

[43] C. Rey and J.-L. Dugelay, A survey of Watermarking Algorithms for Image Authentication, in EURASIP Journal on Applied Signal Processing, 6:613-621, 2002.

[44] C. Rey, et al., Toward Generic Image Dewatermarking?, in Proceedings of the IEEE International Conference on Image Processing, 2002.

[45] D. Robie and R. Mersereau, Video Error Correction using Data Hiding Techniques, in Proceedings of the IEEE Fourth Workshop on Multimedia Signal Processing, pp. 59-64, 2001.

[46] K. Su, D. Kundur, and D. Hatzinakos, A Content-Dependent Spatially Localized Video Watermark for Resistance to Collusion and Interpolation Attacks, in Proceedings of the IEEE International Conference on Image Processing, 1:818-821, 2001.

[47] K. Su, D. Kundur, and D. Hatzinakos, A Novel Approach to Collusion-Resistant Video Watermarking, in Proceedings of SPIE 4675, Security and Watermarking of Multimedia Content IV, pp. 491-502, 2002.

[48] M. Swanson, B. Zhu, and A. Tewfik, Data Hiding for Video-in-Video, in Proceedings of the IEEE International Conference on Image Processing, 2:676-679, 1997.

[49] M. Swanson, B. Zhu, and A. Tewfik, Multiresolution Scene-Based Video Watermarking Using Perceptual Models, in IEEE Journal on Selected Areas in Communications, 16(4):540-550, 1998.

[50] P. Termont, et al., How to Achieve Robustness Against Scaling in a Real-Time Digital Watermarking System for Broadcast Monitoring, in Proceedings of the IEEE International Conference on Image Processing, 1:407-410, 2000.

[51] W. Trappe, M. Wu, and K. Ray Liu, Collusion-Resistant Fingerprinting for Multimedia, in Proceedings of the IEEE International Conference on Acoustics, Speech, and Signal Processing, 4:3309-3312, 2002.

43

CREATING PERSONALIZED VIDEO PRESENTATIONS USING MULTIMODAL PROCESSING

David Gibbon, Lee Begeja, Zhu Liu,
Bernard Renger, and Behzad Shahraray
Multimedia Processing Research
AT&T Labs – Research
Middletown, New Jersey, USA
`{dcg,lee,zliu,renger,behzad}@research.att.com`

1. INTRODUCTION

At a minimum, a multimedia database must contain the media itself, plus some level of metadata which describes the media. In early systems, this metadata was limited to simple fields containing basic attributes such as the media title, author, etc. and the user would select media using relational queries on the fields. Video on demand systems are one example of this. More recently, systems have been developed to analyze the content of the media to create rich streams of features for indexing and retrieval. For example, by including a transcription of the dialog (along with the temporal information) in the database, the advances that have been made in the field of text information retrieval can be applied to the multimedia information retrieval domain. Typically these systems support random access to the linear media using keyword searches [3] and in some cases, they support queries based on media content [5].

While media content analysis algorithms can provide a wealth of data for searching and browsing, often this data has a significant amount of error. For example, speech recognition word error rates for broadcast news tasks are approximately 30% [26] and shot boundary detection algorithms may produce false positives or miss shots [8]. Multimodal processing of multimedia content can be used to alleviate the problems introduced by such imperfect media analysis algorithms. Also, multimodal processing can yield higher level semantics than processing individual media in isolation. Higher level semantics will, in turn, make information retrieval from multimedia databases more accurate.

Multimodal processing can also be used to go beyond random access to media. For some classes of content, it is possible to recover the boundaries of topics and extract segments of interest to the user. With this, there is less burden on the user. Rather than being presented with a list of points of interest in a particular media stream, a logically cohesive content unit can be extracted and presented to the user with little or no interaction.

This chapter will focus on using multimodal processing in the domain of broadcast television content with the goal of automatically producing customized video content for individual users. Multimodal topic segmentation is used to extract video clips from a wide range of content sources which are of interest to users as indicated by the user's profile. We will primarily use the Closed Caption as the text source although the textual information may come from a variety of sources including post-production scripts, very large vocabulary automatic speech recognition, and transcripts which are aligned with audio using speech processing methods [7].

2. MULTI-MODAL PROCESSING FOR CLIP SEGMENTATION

2.1   PROBLEM DEFINITION

The goal of the clip segmentation algorithm is to break long, multi-topic video programs into smaller segments where each segment corresponds to a single topic. A problem arises, however, because the notion of topic is ill-defined. Consider for example retrieving a news program that aired on the day that John Glenn returned to space aboard the space shuttle as the oldest astronaut in history. Suppose that the news program covered the launch and several background stories including one about John Glenn's first trip to space, one about his career as a U.S. Senator, and so on. If a user requests all video related to John Glenn, then it would be correct to include all three segments, whereas if they specified that they were interested in John Glenn's politics, then the correct system response would be to return the much shorter segment confined to that topic. Further, it is possible that the desired topic is "John Glenn's entrance into politics" in which case the correct segment is shorter still.

2.2   A TWO PHASED APPROACH

Given that it cannot be determined a priori for which topics users will query the system, we cannot define a single, static, topic segmentation for each video stored in the database. Instead, we use a two phase scheme which combines both pre-processing (per content) and dynamic (per query) processing. The pre-processing segmentation is done once as a particular video program is inserted into the database. In contrast, the dynamic phase occurs at query time and uses the query terms as inputs. The two phases are independent, so that in different applications, or for different content types within an application, different optimized algorithms for the first phase may be employed. By storing the results of the phase one processing in a persistent data store in a standard extensible markup language (XML) format, this approach supports applications where metadata indicating the topic segmentation accompanies the content. For example, in a fully digital production process the edit decision list as well as higher level metadata may be recorded and maintained with the content data. Alternatively, the topic segmentation may be entered manually in a post-production operation. The second phase of the processing may be viewed as dynamically (at query time) splitting or merging segments from the first phase. As explained further below, the second phase also has the property that reasonable system performance will be achieved even in the absence of story segmentation from phase one. This may be necessary because typically multimodal clip segmentation algorithms are domain specific and don't generalize well to genres other than those for which they are designed.

2.3   CONTENT PREPROCESSING

We will first describe general purpose (largely genre-independent) media preprocessing necessary for subsequent processing stages, or for content representation for multiple applications.

2.3.1   Video Segmentation

In comparison to unimodal information sources that are limited to textual, or auditory information, multimodal information sources, such as video programs, provide a much richer environment for the identification of topic/story boundaries. The information contained in the video frames can be combined with information from the text and audio to detect story transition points more accurately.   Such transition points in the video can be detected using video segmentation techniques.

Video segmentation is the process of partitioning the video program into segments (scenes) with similar visual content.   Once individual scenes have been identified, one or more images are selected from the large number of frames comprising each scene to represent the visual information that is contained in the scene. The set of representative images is often used to provide a visual index for navigating the video program [30], or is combined with other information sources, such as text, to generate compact representations of the video content [24]. In the context of multimodal story segmentation, the transition points are used to detect and adjust the story boundaries, while the representative images are used to generate visual representations, or storyboards for individual stories.

Ideally the "similarity" measures that are used in the segmentation process should be based on high-level semantic information extracted from the video frames. Such high-level information about the visual contents of the scene would go a long way towards extracting video clips that are relevant to a particular topic. However, most of the algorithms that possess sufficient computational efficiency to be applicable to large volumes of video information are only concerned with the extraction of information pertaining to the structure of the video program. At the lowest level, such algorithms detect well-pronounced, sudden changes in the visual contents of the video frames that are indicative of editing cuts in the video. For many informational video programs (e.g., television news broadcasts) the beginning of a new story or topic is often coincident with a transition in the video content. Therefore, video segmentation can serve as a powerful mechanism for providing candidate points for topic/story boundaries.   Video programs also include other transitions that involve gradual changes in the visual contents of the scene. These can be put into two categories. The first category consists of the gradual transitions that are inserted during the video editing process and include fade, dissolve, and a large number of digital transitions. The second category consists of the changes in the image contents due to camera operations such as pan, tilt, and zoom.

Gradual transitions, such as fade, are often systematically used to transition between main programming and other segments (e.g., commercials) that are inserted within the program. Therefore, detection and classification of these transitions would help isolate such segments. The detection of visual changes induced by camera operations further divides the program into segments from which representative images are taken. The transition points that are generated by this process are usually not coincident with story boundaries, and therefore do not contribute to the story segmentation process. However, the representative images retained by this process generate a more complete representation of the visual contents of the program from which a better visual presentation can be generated.

Many algorithms for shot boundary detection have been reported in the literature [1][8][21][23]. We use a method similar to the one reported in [23] that combines the detection of abrupt and gradual transitions between video shots with the detection of intra-shot changes in the visual contents resulting from camera motions to detect and classify the shot transitions. The method also performs a content-based sampling of the video frames. The subset of frames selected by this process is used to generate the pictorial representation of the stories. The sampling process also serves as a data-reduction process to provide a small subset of video frames on which more computationally expensive processing algorithms (e.g., face detection) operate.

2.3.2   Video OCR

If the system can identify video frames containing text regions, it could provide additional metadata for the query mechanism to take into account. Using this type of text as part of the query search mechanism is not well studied but it can be used to improve the clip selection process. The general process would be as follows [10]:

- Identify video frames containing probable text regions, in part through horizontal differential filters with binary thresholding,
- Filter the probable text region across multiple video frames where that region is identified as containing approximately the same data. This should improve the quality of the image used as input for OCR processing,
- Use OCR software to process the filtered image into text.

2.3.3   Closed Caption Alignment

Many television programs are closed captioned in "real-time." As they are being broadcast, a highly skilled stenographer transcribes the dialog and the output of the stenographic keypad is connected to a video caption insertion system. As a result of this arrangement, there is a variable delay of from roughly three to ten seconds from the time that a word is uttered, and the time that the corresponding closed caption text appears on the video. Further, it is often the case that the dialog is not transcribed verbatim. The timing delay presents a problem for clip extraction, since the clip start and end times are derived from the timing information contained in the closed caption text stream.

We correct the inherent closed caption timing errors using speech processing techniques. The method operates on small segments of the closed caption text stream corresponding to sentences as defined by punctuation and the goal is to determine sentence (not word) start and end times. An audio segment is extracted which corresponds to the closed caption text sentence start and end time. The audio clip start time is extended by 10 seconds back in time to allow for the closed caption delay. The phonetic transcription of the text is obtained using text-to-speech (TTS) techniques and the speech processing engine attempts to find the best match of this phoneme string to the given audio segment. The method works well when the closed caption transcription is accurate (it is even tolerant to acoustic distortions such as background music). When the transcription is poor, the method fails to find the proper alignment. We detect failures with a second pass consistency check, which insures that all sentence start and end times are sequential. If overlaps are detected, we use the mean closed caption delay as an estimate.

If a manual transcription is available, it can be used to yield a higher quality presentation than the closed caption text by using parallel text alignment with the best transcription of a very large vocabulary speech recognizer [7].

2.3.4   Closed Caption Case Restoration

Closed captioning is typically all upper-case to make the characters more readable when displayed on low-resolution television displays and viewed from a distance. When the text is to be used in other applications, however, we must generate the proper capitalization to improve readability and appearance. We employ three linguistic sources for this as follows: 1) a set of deterministic syntactic rules, 2) a dictionary of words that are always capitalized, and 3) a source which gives likelihood of capitalization for ambiguous cases [24]. The generation of the dictionary and the statistical analysis required to generate the likelihood scores are performed using a training corpus similar to the content to be processed. Processed newswire feeds are suitable for this purpose.

2.3.5   Speaker Segmentation and Clustering

In several genres including broadcast news, speaker boundaries provide landmarks for detecting the content boundaries so it is important to identify speaker segments during automatic content-based indexing [18]. *Speaker segmentation* is finding the speaker boundaries within an audio stream, and the *speaker clustering* is grouping speaker segments into clusters that correspond to different speakers.

Mel-Frequency Cepstral Coefficients (MFCC) are widely used audio features in speech domain and provide a smoothed version of spectrum that considers the non-linear properties of human hearing. The degree of smoothness depends on the order of MFCC being employed. In our study, we employed 13th order MFCC features and their temporal derivatives.

The speaker segmentation algorithm consists of two steps: splitting and merging. During splitting, we identify possible speaker boundaries. During merging, neighboring segments are merged if their acoustic characteristics are similar. In the first step, low energy frames, which are local minimum points on the volume contour, are located as boundary candidates. For each boundary candidate, the difference between it and its neighbors is computed. The neighbors of a frame are the two adjacent windows that are before and after the frame respectively, each with duration L seconds (typically L=3 seconds). If the difference is higher than certain threshold and it is the maximum in the surrounding range, we declare that the corresponding frame is a possible speaker boundary.

We adopt divergence to measure the difference [25]. Divergence is computed by numerical integration, making it computationally expensive. However, in some special cases, it can be simplified. For example, when F and G are one dimensional Gaussians, the divergence between F and G can be simplified as a computation directly from the Gaussian parameters. If the means and standard deviations of F and G are (μ_F, σ_F) and (μ_G, σ_G), respectively, the divergence is computed as,

$$D(F,G) = \frac{\sigma_F^2}{\sigma_G^2} + \frac{\sigma_G^2}{\sigma_F^2} - 2 + \left(\frac{\sigma_F^2 + \sigma_G^2}{\sigma_F^2 \sigma_G^2} \right) (\mu_F - \mu_G)^2$$

To simplify the computation, we assume different audio features are independent, and they follow Gaussian distribution. Then, the overall difference between two audio windows is simply the summation of the divergence of each feature.

The above described splitting process yields, in general, over-segmentation. A merging step is necessary to group similar neighboring segments together to reduce the amount of false segmentation. This is achieved by comparing the statistical properties of adjacent segments. The divergence is computed based on the mean and standard deviation vectors of adjacent segments. If it is lower than a preset threshold, the two segments are merged.

Speaker clustering is realized by agglomerative hierarchical clustering [13]. Each segment is initially treated as a cluster on its own. In each iteration, two clusters with the minimum dissimilarity value are merged. This procedure continues until the minimum cluster dissimilarity exceeds a preset threshold. In general, the clustering can also be performed across different programs. This is especially useful in some scenarios in which the speech segments of a certain speaker (e.g., President Bush) on different broadcasts can be clustered together so that users can easily retrieve such content-based clusters.

2.3.6 Anchorperson Detection

The appearance of anchorpersons in broadcast news and other informational programs often indicates a semantically meaningful boundary for reported news stories. Therefore, detecting anchorperson speech segments in news is desirable for indexing news content. Previous efforts are mostly focused on audio information (e.g., acoustic speaker models) or visual information (e.g., visual anchor models such as face) alone for anchor detection using either model-based methods via off-line trained models or unsupervised clustering methods. The inflexibility of the off-line model-based approach (allows only fixed target) and the increasing difficulty in achieving detection reliability using clustering approach lead to an adaptive approach adopted in our system [17].

To adaptively detect an unspecified anchor, we use a scheme depicted in Figure 43.1. There are two main parts in this scheme. One is visual-based detection (top part) and the other is integrated audio/visual-based detection. The former serves as a mechanism for initial on-line training data collection where possible anchor video frames are identified by assuming that the personal appearance (excluding the background) of the anchor remains constant within the same program.

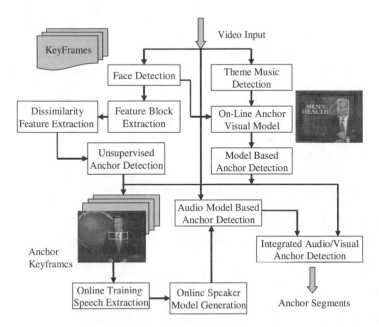

Figure 43.1 Diagram of integrated algorithm for adaptive anchorperson detection.

Two different methods of visual-based detection are described in this diagram. One is along the right column where audio cues are first exploited that identify the theme music segment of the given news program. From that, an anchor frame can be reliably located, from which a feature block is extracted to build an on-line visual model for the anchor. A feature block is a rectangular block, covering the neck-down clothing part of a person, which captures both the style and the color of the clothes. By properly scaling the features extracted from such blocks, the on-line anchor visual model built from such features is invariant to location, scale, and background. Once the model has been generated, all other anchor frames can be identified by matching against it.

The other method for visual-based anchor detection is used when there is no acoustic cue such as theme music present so that no first anchor frame can be reliably identified to build an on-line visual model. In this scenario, we utilize the common property of human facial color across different anchors. Face detection [2] is applied and then feature blocks are identified in a similar fashion for every detected human face. Once invariant features are extracted from all the feature blocks, similarity measures are computed among all possible pairs of detected persons. Agglomerative hierarchical clustering is applied to group faces into clusters that possess similar features (same cloth with similar colors). Given the nature of the anchor's function, it is clear that the largest cluster with the most scattered appearance time corresponds to the anchor class. Either of the above described methods can be used to detect the anchorperson. When the theme music is present, the combination of both methods can be used to detect the anchorperson with higher confidence.

Visual-based anchor detection by itself is not adequate because there are situations where the anchor speech is present but where the anchor does not appear. To precisely identify all anchor segments, we need to recover these segments as well. This is achieved by adding audio-based anchor detection. The visually detected anchor keyframes from the video stream identify the locations of the anchor speech in audio stream. Acoustic data at these locations is gathered to serve as the training data to build an on-line speaker model for the anchor, which can then be applied, together with the visual detection results, to extract all the segments from the given video where the anchor is present.

In our system, we use a Gaussian Mixture Model (GMM) with 13 MFCC features and their first and second order derivatives (for a total of 39 features as opposed to 26 used for speaker segmentation) to model the acoustic property of anchorperson and non-anchorperson audio. Maximum likelihood method is applied to detect the anchorperson segments.

2.3.7 Commercial Detection

Detection of commercials is an important task in news broadcast analysis, and many techniques have been developed for this purpose [9]. For example, to obtain transcription from automatic speech recognition (ASR) or to recognize a speaker from speech signals, it is more efficient to filter out the non-speech or noisy speech signals prior to the recognition. Since most commercials are accompanied with prominent music background, it is necessary to detect commercials and remove them before ASR and speaker identification. Conventionally, this task is performed mainly by relying on only auditory properties. In our system, we explore the solution that utilizes information from different media in the classification [16].

Commercials in national broadcast news can usually be characterized as being louder, faster, and more condensed in time compared with the rest of the broadcast. In our current solution for distinguishing between commercials and news reporting, eighteen clip-based audio and visual features are extracted and utilized in the classification. Each clip is about 3 seconds long, and fourteen clip features are from the audio domain and the rest are from the visual domain. Audio features capture the acoustic characteristics including volume, zero crossing rate, pitch, and spectrum. Visual features capture color and motion related visual properties. Such a combination of audio/visual features is designed to capture the discriminating characteristics between commercials and news reporting.

A Gaussian Mixture Model classifier is applied to detect commercial segments. Based on a collected training dataset, we train GMM models for both news reporting and commercials by the expectation maximization (EM) method. In the detection state, we compute the likelihood values of each video clip with regard to these two GMM models, and then we assign the video clip to the category with larger likelihood value. It is possible that news reporting may also contain music or noisy backgrounds and fast motion, especially in live reporting, which make the decision wrong since it is aurally or visually similar with commercials. This type of error can be easily corrected by smoothing the clip-wise classification decision using contextual information from neighboring clips.

2.4   MULTI-MODAL STORY SEGMENTATION

We now present a phase one algorithm suitable for broadcast news applications. In a typical national news program, news is composed of several headline stories, each of which is usually introduced and summarized by the anchorperson prior to and following the detailed reporting conducted by correspondents and others. Since news stories normally start with an anchorperson introduction, anchorperson segments provide a set of hypothesized story boundaries [11]. Then the text information, closed caption or speech transcription by automatic speech recognition, is used to verify whether adjacent hypothesized stories cover the same topic or not.

Formally, our input data for text analysis is $T = \{T_1, T_2, ..., T_M\}$, where each T_i, $1 \le i \le M$, is a text block that begins with the anchor person's speech and ends before the next anchor's speech, and M is the number of hypothesized stories. To verify story boundaries, we evaluate a similarity measure $sim()$ between every pair (T_a, T_b) of adjacent blocks:

$$sim(T_a, T_b) = \frac{\sum_w f_{w,a} \times f_{w,b}}{\sqrt{\sum_w f_{w,a}^2 \times f_{w,b}^2}}$$

Here, w enumerates all the token words in each text block; $f_{w,*}$ is the weighted frequency of word w in the corresponding block. Stop words, for example, "a, the", etc, are excluded from token word list. We also applied stemming to reduce the number of token words [22]. Each word frequency is weighted by the standard frequency of the same word computed from a database of broadcast news data collected from NBC Nightly News from 1997 to 2001. The higher the frequencies of the common words in the two involved blocks are, the more similar the content of the blocks is. We experimentally determined a threshold to verify the story boundaries.

For each news story, we extract a list of keywords and key images as textual and visual representation of the story. Keywords are the ten token words with the highest weighted frequency within the story. Choosing key images is more sophisticated. We need to choose the most representative keyframes extracted from video shot detection. Anchorpersons' keyframes should be excluded, since they don't give story dependent information. The method we used is to select top five non-anchorperson keyframes that cover maximum number of chosen keywords in corresponding text blocks.

After story segmentation, a higher level news abstract can be extracted, for example, a news summary of the program. We can group all the anchorperson introduction segments, which are the leading anchorperson segments within corresponding stories, as the news summary since they are normally the most concise snapshot of the entire program.

2.5   QUERY-BASED SEGMENTATION

As mentioned previously, the detected story boundaries may be inappropriate for some user queries. We may need to split long stories into sub-segments, or on the other hand, join stories together to form suitable video presentations.

The second phase of processing to determine the clip start and end times for a given program takes place at query time. This processing is independent of the multimodal story segmentation described above. That method will have limited success for genres that are very different from broadcast news, in which case this second phase may be used by itself.

The sentences from the text for a given program are compared against a *query string* that is formed from a user's profile (see Section 3). The query string contains single word terms as well as phrase terms and typically contains from three to five such items. If phrase terms are used, the entire phrase must match in order for it to be counted as a *hit*. A *hit* is defined as a location in the text that matches the query string. The match may not be limited to a lexical match, but may include synonyms or alternate forms obtained via stemming. Typically the query string contains logical 'or' operators, so hits correspond to occurrences of single terms from the query string.

The algorithm uses a three-pass method to determine the largest cluster of hits to determine clip start and end times for a given program. In the first pass, a vector containing the times for each hit is formed. Next, a one-dimensional temporal domain morphological closing operation is performed on the hit vector. Assuming that time is sampled in units of seconds, the size of the structuring element for the closing operation corresponds to the maximum length of time in seconds between hits that can be assigned to a particular candidate clip. We have found that a value of 120 seconds provides reasonable performance for a wide range of content genres. In the final pass, a run-length encoding of the closed hit vector is performed, and the longest run is selected. The tentative clip start time is set to the start time of the sentence containing the first hit of the longest run. The clip end time is set to the end time of the sentence containing the last hit in the longest run. Note that there is no maximum clip length imposed.

It is possible for a program to contain only one instance of the search term, in which case the clip start and end times may be set to correspond to those of the sentence containing the hit. A minimum clip length (of typically 20 seconds) is imposed; it is usually only used in cases where the term occurs only once in a program, and it occurs in a short sentence. If the clip length is less than the minimum, the clip end time is extended to the end time of the next sentence. In other applications it may be desirable to impose a minimum number of sentences constraint instead of a minimum clip duration.

2.5.1 Example

Figure 43.2 represents a timeline of a 30-minute news program (the NBC Nightly News on 6/7/01). The large ticks on the scale on the bottom are in units of 100 seconds. The top row shows occurrences of an example key word, "heart". The second row indicates start times and durations of the sentences. Note that commercial segments have been removed as described above, and sentences or hits are not plotted during commercial segments.

The bar in the second row indicates the extracted clip, in this case a story about heart disease. Notice that the keyword "heart" occurred five times before the story. Most of these are due to "teasers" advertising the upcoming story. The third occurrence of the term is from the mention of the term "heart" in a story about an execution (the first occurrence is from the end of the preceding program which happened to be a local news program, the second is from the beginning of the Nightly News program, and the last two before the clip are just before a commercial break).

Figure 43.2 Timeline plot for a news program.

2.6 TEXT-BASED INFORMATION RETRIEVAL

Text-based information retrieval (IR) is an important part of the overall process of creating personalized multimedia presentations. An in-depth review of IR techniques is beyond the scope of this chapter and further information can be found in the literature [14]. This section will describe how some IR methods could be used to improve the overall accuracy and correctness of the video selections.

Many IR systems attempt to find the best document matches given a single query and a large set of documents. This method will work with reasonable success in this system. We assume that a text transcript of the broadcast is available and that the query will return not just a set of documents, but also locations of the hits within the documents. Each document will relate to a single program. Thus the text processing will return a set of video programs that are candidates for the video segmentation and the additional processing required to create a set of personalized video clips.

In the simplest view of text processing the individual query hits on a document would be analyzed within each program to determine the best candidate areas for a video clip that would satisfy the query. Additional processing on the video and audio would then create the exact bounds of the video clip before it was presented to the user of the system.

A more advanced view would look at the entire profile of the user and use that as input to algorithms for query expansion or topic detection. In addition, suitable user interfaces (e.g., see Section 3.2) could provide the system with excellent information for relevance feedback in making future choices for this topic. The system could log the actual choices that the user makes in viewing the videos and incorporate information from those choices into the algorithms for making future choices of video clips. The view of integrated solutions is the correct one for the future of multimedia information management and there is much work to do in integrating the standard text-based tools with tools for handling multimedia.

2.6.1 Query Expansion

Query expansion should lead to less variation in the responses to similar but different queries [15]. For example, if a query was "home run", then the set of video clips found as a response should be similar to those found for the query "homer". This is only true if the query expansion is clever enough to group these terms in the sports category. Otherwise the query "homer" would be likely to return video clips from a documentary on the artist Winslow Homer.

This context is not available in a system that makes requests with no history or logging but is readily available in a system that not only includes a profile but also maintains and uses logs of previous queries. The system should be set up to be adaptive over time and all the relevant information should be stored with the user's individual profile.

2.6.2 Text Summarization and Keyword Extraction

These techniques, described elsewhere [6,12,20] help the system in determining appropriate clip selection by narrowing the text of a multimedia event to its essential and important components. These techniques can be used in isolation or in combination with relevance feedback to improve overall clip selection and generation.

2.6.3 Relevance Feedback

This goes beyond the realm of text processing but is worth mentioning in the standard IR context. This application lends itself to relevance feedback. Since we maintain profile information on each user we can maintain logs of which videos were viewed for how long and which videos were considered important enough to archive or email to others. Using this log information we can provide relevance feedback to the overall system on the effectiveness of certain searches and we can adapt the system over time to provide more answers similar to the clips that were viewed by the user and reject those that are similar to the clips that were not viewed [28,29].

3. PERSONALIZED VIDEO

We now describe a multimedia personalization service that automatically extracts multimedia content segments (i.e., clips), based on individual preferences (key terms/words, content sources, etc.) that the user identifies in a profile stored in the service platform. The service can alternatively be thought of as an electronic clipping service for streaming media that allows content to be captured, analyzed, and stored on a platform. A centralized server records the video streams of appropriate interest to the target audiences (broadcast television, training videos, presentations, executive meetings, etc.) as they are being broadcast. Additionally, it can include content that has been pre-recorded. The content is preprocessed as described in Section 2.3. User profiles are individually created data structures (stored in XML format) containing the key terms or words of interest to the user. The profiles are compared against the content database using information retrieval techniques and if a match is found, multimodal story segmentation algorithms determine the appropriate length of the video clip as described above. The service then provides the video clips matching the user profile via streaming – eliminating long downloads (assuming that there is sufficient bandwidth for streaming). The user may then view/play these segments. A key feature of the service is to automatically aggregate the clips of diverse sources, providing a multimedia experience that revolves around the user's provided profile. This means that users have direct access to their specific items of interest. The provided segments are dynamically created, thereby the user has recent information including possibly content from live broadcasts earlier in the day. The profiles make searching personalized content easier as the user does not need to retype search strings every time the service is used. Additionally, the service supports e-mailing and archiving (book marking) clips.

3.1   SYSTEM ARCHITECTURE

The architecture will be described from the hardware and software point of view. As we describe each component of the architecture, we will describe the function of each component as well as the hardware and software system details.

3.1.1   Hardware Architecture

The Hardware Architecture is shown in Figure 43.3. The Client (A) requests hypertext markup language (HTML) pages related to the service from the Web/Index Server (C). The retrieved HTML pages contain JavaScript which are responsible for most of the interaction between the Client and the Web/Index Server. From the user profile, the JavaScript builds queries to the Web/Index Server to determine clip information for clips that match the search terms in the profile. After the Web/Index Server has returned the clip information to the Client, the user can navigate the clips by selecting them. The Video Acquisition Server (D) has already captured and recorded past broadcast TV programming. These recorded video files are shipped to the Video Server (B) from which they will be streamed. When the Client selects a clip, the video is streamed from the Video Server.

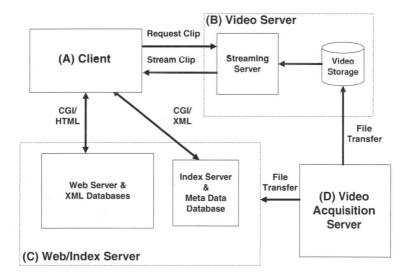

Figure 43.3 Hardware architecture.

3.1.2   Software Architecture

The Software Architecture is shown in Figure 43.4. The architecture employs well-defined interface protocols and XML schema for data storage to maximize interoperability and extensibility. Various JavaScript libraries are used in the Client to access the XML databases, read XML query responses from the Server, and stream the video to the Player Web page. The Client makes extensive use of the JavaScript XML Document Object Model (DOM) Dynamic HTML (DHTML) and of Cascading Style Sheets (CSS). In addition to JavaScript on the Client side, Perl on the Server side is also used to access the XML databases.

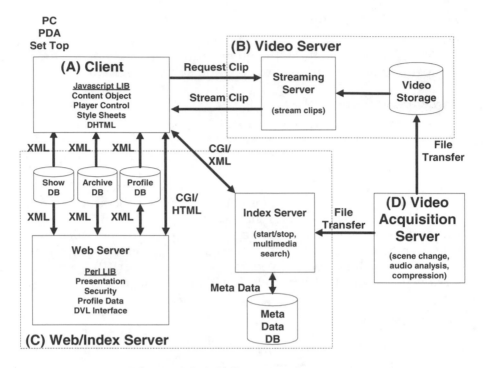

Figure 43.4 Software architecture.

The profile database contains the search topics and associated search terms. This includes restrictions on shows (for example, the search may be limited to certain broadcast TV programs). The show database contains the list of potential shows to be used in search queries. The archive database is simply the clips that the user has saved.

The Web Server handles the client requests for HTML pages related to the service. The Perl common gateway interface (CGI) scripts that the client navigates in order to perform the functions of the service deal with login/registration related pages, home page, profile related pages, archive related pages, and player pages. The player script is launched in a separate Player Web page. The Streaming Server will stream the video from the Video Storage to the Player Web page.

The Index Server handles query requests from the Client (how many clips, which shows have clips, etc.) and requests for clip content (metadata that describes clip content including pointers to video content). The Index Server finds the video clips by searching the previously indexed metadata content associated with the video content. The Index Server also determines the start and stop times of the clips. The Index Server generates XML responses to the queries, which are parsed by the Client JavaScript.

The Video Acquisition Server also performs various other functions including the content preprocessing, multimodal story segmentation, and compression. The Meta Data is shipped to the Meta Data database on the Index Server. Thumbnails are also included as part of the Meta Data. The recorded video files are shipped to the Video Storage.

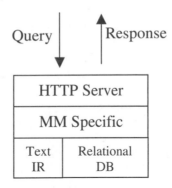

Figure 43.5 Multimedia database.

The Web/Index Server (Figure 43.3) includes a Web (HTTP) interface, a relational database function (e.g., for selecting content from a particular broadcaster), and a full text information retrieval component (Text IR) as shown in Figure 43.5. The lower layer provides the text IR and relational database functions. Each acquisition also has properties such as the program title or date that can be queried against. The MM Specific layer provides the synchronization between multimedia elements (such as thumbnail images and media streams) and the text and other metadata elements. It also includes modules for XML generation and supports dynamically generating clip information. The responses from the MM DB are in standard XML syntax, with application-specific schema as shown in Figure 43.6.

```
<?xml version="1.0" ?>
  <clip start="58.475" duration="268.719">
    <program id="nn">
        <title>Nightly News</title>
        <owner>nbc</owner>
        <date>Tuesday, May 29, 2001 18:30 GMT</date>
        <path>/nbc-data/nn/2001/05/29</path>
    </program>
    <keyword>bush</keyword>
    <keywordtimes>
        58.475, 140.974, 214.915, 220.704, 265.849
    </keywordtimes>
    <img  src="/cgi-bin/upi?/nbc-data/nn/2001/05/29+0012"/>
    <text>Power play -- President Bush goes head-to-head with
California's governor over the energy crunch.</text>
  </clip>
```

Figure 43.6 XML representing a clip.

3.2 USER INTERFACE

Our goal is not merely to create topical video clips, but rather to create personalized video presentations. As such, the user interface is of critical importance, since it is the user who will ultimately be the judge of the quality and performance of the system. The user interface must be designed to gracefully accommodate the errors in story segmentation which any automated system will inevitability produce. Even if the story segmentation is flawless, the determination of which stories are of interest to the user is largely subjective and therefore impossible to fully automate. Further, we cannot know a priori which stories the user is already familiar with (possibly having heard or read about from other sources,) and therefore wishes to skip over. It is often the case that there will be several versions of a given news event covered by several different news organizations. Should we filter the clip list to include only one version, perhaps the one from the user's favorite broadcaster, or if the topic is of sufficient interest, should we offer multiple versions? Obviously, to deal with these issues, tools for browsing the stories will be a key component of the user interface. It is important to balance the need for flexibility and functionality of the interface against the added complexity that having such control brings.

We have chosen to have the system create a clip play list which is used as a baseline for the personalized video presentation. From this the user is given simple controls to navigate among the clips. We move some of the complexity of clip selection to an initial configuration phase. For example, when the user sets up their profile, they choose which content sources will be appropriate for a given topic. This makes the initial setup more complex, but establishes more accurate clip selection, thus simplifying the user interaction when viewing the personalized presentations. The approach of using a baseline clip play list is of even greater importance when we consider viewing this multimedia content on client devices with limited user interaction capabilities (described in detail in Section 3.4.) In these cases the "user interface" may consist of one or two buttons on a TV remote control device. We might have one button for skipping to the next clip within a set of clips about a particular topic and another for skipping to the next topic.

There are fundamental decisions to make in developing the player interface for a web browser:

- Separate player page vs. integrated player page: If we choose a separate player page then we must make sure that there are sufficient controls on the player page to prevent the end user from having to go back and forth often. If we choose an integrated player page then we must make sure that the set of controls on the single page does not overwhelm the customer. (If the player is on a digital set-top box whose display is the TV, then this is a significant issue.)

- Which of the three standard players (MS Media Player, Real, QuickTime) to use: The choice of the player not only specifies the available features and application programmer's interface (API) but also dictates the streaming protocol and servers used since the available players are largely not interoperable.

- Embed the player into a web page or launch it via mime-type mapping: If we do not embed the player then we will have only the controls that are available on the player. If it is embedded, we may decide which controls to expose to the end user, and we may add customized controls.

- Support for additional features:
    o Personal Clip Archive: Maintain an extracted set of video files or a set of pointers to the full video.

    o   E-mail clips: Raises additional security issues and has long-term video storage implications.

    o   Peer-to-peer sharing of clips, lists, profiles, etc.

We chose to develop a browser-based player with the software architecture of the overall system as shown in Figure 43.7. We form a *clip play list* data structure as the result of the data exchange between the Client and the multimedia database (MM DB). This is a list of program video stream file names with clip start times, clip durations, and metadata such as the program title and date. These data are converted from XML format and maintained in JavaScript variables in the client. Playback is done entirely by streaming. During playback, we use JavaScript to implement control functionality, to iterate through the clip play list items, and to display clip metadata such as clip time remaining, etc. JavaScript on the client also generates HTML from the clip metadata for display of the thumbnails and text and for the archival and email features of the player.

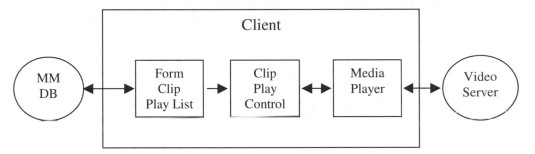

Figure 43.7 Client functional modules.

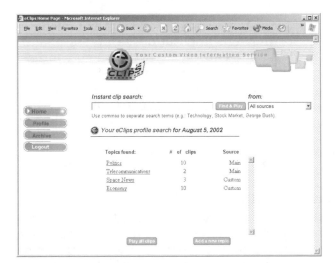

Figure 43.8 Individualized home page for desktop clients.

Figure 43.9 Player page for desktop clients.

The end user interface for the desktop client is depicted in Figure 43.8 and Figure 43.9. The users are assumed to have registered with the service and to have set up a profile containing topics of interest and search terms for those topics. After authentication (login) the individualized home page for the user is displayed (see Figure 43.8) which indicates the number of clips available for each of the topics. The user can select a topic, or may choose to play all topics. The player (see Figure 43.9) displays the video and allows navigation among the clips and topics through intuitive controls.

3.3   PARAMETERS

There are many possible variations of the system parameters that would make the system suitable for different applications. One successful application uses North American broadcast and cable television content sources. That instance of the system allows users to search the most recent 14 days of broadcasts from 23 video program sources which corresponds to about 300 hours of content. Actually the archive maintains over 10,000 hours of indexed video content going back a period of several years, but we only return clips from content less than two weeks old for this application. Also, the user interface displays a maximum of 10 clips per topic. The media encoding parameters are shown in the "Desktop Client" column of Table 43.1.

3.4   ALTERNATIVE CLIENTS

In addition to the desktop client version described above, other user interface solutions were developed for handheld Personal Digital Assistants (PDA), a WebPad that was conceived as a home video control device, and for the telephone. In this section we will describe the challenges of implementing a multimedia service on each of these devices, some of which have very limited capabilities.

3.4.1 Handheld Client

A personalized automated news clipping service for handheld devices can take many forms, depending on the networking capabilities and performance of the device.

The three main networking possibilities for handheld devices are:

1. **No networking:** All content is stored and accessed locally on the device.
2. **Low bit-rate:** bit-rates of up to 56Kbps, e.g., CDPD @19.2Kbps
3. **High bit-rate:** bit-rates over 56Kbps, e.g., IEEE 802.11b @ 11Mbps

The display and computational capabilities of the device limits its ability to render multimedia content and can be classified as follows:

A) Text only
B) Text and images
C) Text, images, and audio
D) Text, images, audio, and video

We implemented option D for both the local content and the high bit-rate scenarios (options 1 and 3). See Table 1 in the Appendix for information detailing the differences between the desktop and the handheld implementations. Some of the key limitations for handheld devices include:

- **Storage:** limits the total number of minutes of playback time for non-networked applications
- **Processing power:** imposes limitations on video decoding and possibly on client player control functionality (e.g., JavaScript may not be supported)
- **Screen Size and Input Modality:** restricts graphical user interface options, and lack of keyboard makes keyword searches difficult.

The limited capabilities of handheld devices require major architectural changes. The goal of these changes is to offload much of the computation from the client to the server. We assume that the user interface for creating and modifying the user profile database is done using a desktop PC with keyboard and mouse. However, since the user interface is based on HTML and HTTP, it is possible (although cumbersome) to modify the profile from a handheld device.

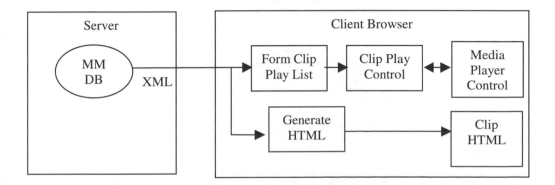

Figure 43.10 Client-server architecture for desktop clients.

The desktop client architecture (see Figure 43.10) makes extensive use of the JavaScript Extensible Markup Language Document Object Model (XML DOM), Dynamic HTML (DHTML), and of CSS. The server is responsible for fielding queries and generating XML representations in response. For the Player Page, the client uses the Microsoft Media Player ActiveX control to play the video stream and generates HTML for the clip thumbnail and text. The user has considerable user interface control over the video content (skipping selected clips, selecting topics, etc.).

For handheld clients, much of the complexity must be moved from the client to the server (see Figure 43.11). The new architecture makes no use of JavaScript, DHTML, or CSS on the client. The clip HTML generation has been moved to the server, and HTML (with no use of CSS) rather than XML is sent to the client. As the server assembles this page, it also creates a clip playlist file in Active Stream Redirector (ASX) format. The "Play" button is linked to this dynamically created file and launches the media player using a Multipurpose Internet Mail Extension (MIME) type mapping. The GUI for navigating clips is limited to the Media Player capabilities. Figure 43.12 and Figure 43.13 represent the user interface.

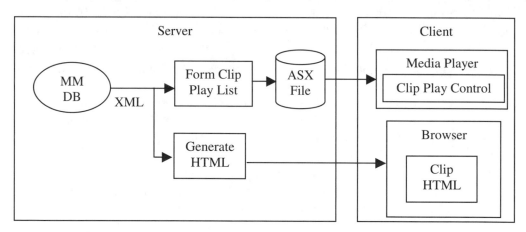

Figure 43.11 Client-server architecture for handheld clients.

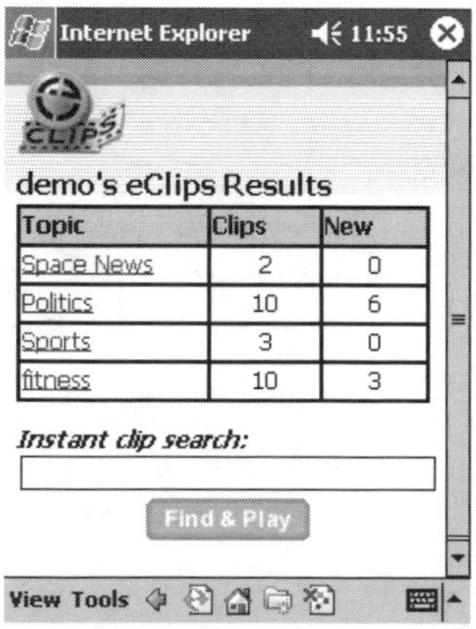

Figure 43.12 Individualized home page for handheld clients.

Figure 43.13 Clip play list display for handheld clients.

3.4.2 WebPad Video Controller

We have also implemented the service using a WebPad (or tablet PC) Video Controller and a custom standalone Video Decoder appliance in order to play the video content with full broadcast television quality on standard TV monitors. It is contrasted with the PC implementation in Table 43.1. The WebPad (essentially a pen-based Windows PC) uses a browser such as MS Internet Explorer like the standard Client PC but the video is not streamed to the tablet. Instead of seeing the video stream (MPEG-4) via the Media Player ActiveX control on the WebPad, the video stream is sent to a special-purpose *video endpoint* that decodes the MPEG-2 video before sending the video directly to the television monitor in analog form. The video screen part of the Player Page on the WebPad is replaced with a still frame graphic (the thumbnail) and pause/volume controls. The other standard playback controls (skip back a clip, stop, play, skip forward a clip, skip topic, play beyond end of clip) work as before. The navigation among clips remains the same (click on thumbnails, play selected clips, etc.) but the video stream is played on the TV instead of on the WebPad. Note that hybrid solutions are possible in which video may be displayed on the tablet, perhaps for preview purposes.

The WebPad implementation is achieved by removing the Media Player ActiveX control from the client and adding a JavaScript object that mimics the Player object. The client causes the video stream to be redirected from the Video Server to the video endpoint. The video endpoint is a small embedded Linux computer with a 100BaseT IP network interface and hardware MPEG-2 decoder. It includes an HTTP Server with custom software. When the user interacts with the Player Page controls on the WebPad, CGI scripts on the video endpoint enable the controls (stop, play, pause, increase/decrease volume, etc.) This implementation supports a range of high quality video from digital cable quality through full DVD quality (4-10Mbps MPEG-2.)

3.4.3 Telephone Client

Lastly, we have implemented the service over the phone by using a front-end VoiceXML (VXML) Gateway (see Figure 43.14.) The client device in this case is any standard touch-tone telephone. During the registration process (using the PC) a new user creates a login and a 4 digit identifier which is needed when using the telephone interface to the service. The audio content is extracted from the video content, resampled to 8 KHz, and quantized to 8 bit μ-law as part of the acquisition process as required by the VXML platform. Similar to the handheld client architecture, the server is responsible for fielding queries and generating XML representations in response. However, instead of generating HTML, here we generate the VXML for the call session. Thus, the VXML is dynamically created using knowledge of the user's profile. After the user calls into the service, Dual-Tone Multi-Frequency (DTMF) input or speech input can be used to navigate the clips. The user can navigate the topics in the profile or the clips within a topic. In one instance of the service, the DTMF interface requires the user to enter "#" to skip to the next clip or "##" to skip to the next topic. The speech interface requires the user to speak the topic name in order to jump to a topic, or the user can say "next clip" or "next topic". At any time, the user can ask for clip details (enter "5" or say "clip details"), which consists of the show name, show date, and the duration of the clip. Because all the segmentation work has already been done during the processing stage, it is relatively straightforward to play clips using a telephone interface. Enhancements to the telephone interface could allow the user to hear more than the clip (play content before the start point or play content beyond the end point).

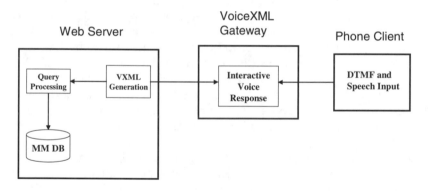

Figure 43.14 Architecture for Telephone Clients.

4. CONCLUSIONS

We have presented several multimodal content processing methods for story segmentation. Some of these algorithms are optimized for content from specific genres, while others are general-purpose and have wider applicability. We have shown how the capability of automatically segmenting multimedia content into semantically meaningful units can enable novel applications. We presented one such application involving multimedia personalization based on user profiles. The techniques presented simplify the user interaction and permit implementations on a wide range of client devices with various user interface and networking capabilities.

Acknowledgements
Ken Huber and Bob Markowitz conceptualized several new services made possible by this technology and provided support for the prototype development. Gary Zamchick is largely responsible for the user interface design. Michael Zalot and Antoinette Lee contributed to the design realization. Larry Ruedisueli is responsible for the video endpoint.

5. APPENDIX

Table 43.1 Parameters for different client devices.

Item	Desktop Client	Handheld Client	WebPad & Video Endpoint
Video Parameters	320x240, 30Hz 512Kbps MPEG-4	224x168, 15Hz 128Kbps MS Video V8	640x480, 30Hz 6Mbps MPEG-2
Audio Parameters	16KHz Stereo 32Kbps MPEG Layer 3	16KHz mono 16Kbps MS Audio V8	44.1KHz stereo 256Kbps MPEG Layer 2
Media Transport	UDP Streaming	UDP Streaming and Local Storage	HTTP Streaming
Graphics	1024x768, 24bit	320x240, 12bit	800x600, 24bit
Keyboard	Yes	No	No
Mouse	Yes	No	Optional
Stylus	No	Yes	Yes
JavaScript	Yes	Limited	Yes
DHTML	Yes	No	Yes
Style Sheets	Yes	No	Yes
Operating System	Windows PC	WinCE 3.0	Windows 98
Web Browser	IE 5.0+	Pocket IE	IE 5.0+
CPU	400 MHz Intel	206 MHz RISC	266 MHz Cyrix
Memory	128 MB RAM	64 MB RAM 16 MB ROM	30 MB RAM
Disk Memory	25 GB	2 GB PC card (Optional)	6 GB

REFERENCES

[1]   J. S. Boreczky and L. A. Rowe, Comparison of Video Shot Boundary Detection Techniques, Storage and Retrieval for Still Image and Video Databases IV, Proceedings of the SPIE 2664, January 1996, pp. 170-179.

[2]   R. Chellappa, C. L. Wilson, and S. Sirohey, Human and machine recognition of faces: A suervey, Proceedings of the IEEE, vol. 83, no. 5, May 1995, pp. 705-741.

[3]   M. Christel, S. Stevens, H. Wactlar, Informedia Digital Video Library, Proceedings of the ACM Multimedia '94 Conference. New York: ACM, October 1994, pp. 480-48.

[4]   B. Croft, What do People Want from Information Retrieval? D-Lib Magazine, Nov 1995.

[5]   M. Flickner, H. Sawhney, W. Niblack, J. Ashley, Q. Huang, B. Dom, M. Gorkani, J. Hafner, D. Lee, D. Petkovic, D. Steele and P. Yanker, Query by Image and Video Content: The QBIC System, IEEE Computer 28(9), September 1995, pp. 23-32.

[6]   Y. Freund and R. E. Schapire, Experiments with a new boosting algorithm, Machine Learning: Proceedings of the Thirteenth International Conference, 1996, p. 148.

[7]   D. Gibbon, Generating Hypermedia Documents from Transcriptions of Television Programs Using Parallel Text Alignment, Handbook of Internet and Multimedia Systems and Applications, Edited by Borko Furht, CRC Press, December 1998.

[8]   U. Gargi, R. Kasturi, S. H. Strayer, Performance Characterization of Video-Shot-Change Detection Methods, IEEE Transaction on Circuits and Systems for Video Technology, Vol. 10, No. 1, February 2000.

[9]   A. G. Hauptmann, and M. J. Witbrock, Story Segmentation and Detection of Commercials in Broadcast News Video, ADL-98 Advances in Digital Libraries, Santa Barbara, CA, April 22-24, 1998.

[10]   A. G. Hauptmann, R. Jin, N. Papernick, D. Ng, Y. Qi, R. Houghton, and S. Thornton, Video Retrieval with the Informedia Digital Video Library System, NIST Special Publication 500-250: The Tenth Text Retrieval Conference, 2001, p. 94.

[11]   Q. Huang, Z. Liu, A. Rosenberg, D. Gibbon, B. Shahraray, Automated Generation of News Content Hierarchy by Integrating Audio, Video, and Text Information, 1999 IEEE International Conference On Acoustics, Speech, and Signal Processing, Phoenix, Arizona, Volume: 6, March 15-19, 1999, pp. 3025-3028.

[12]   A. Hulth, J. Karlgren, A. Jonsson, H. Boström and L. Asker, Automatic Keyword Extraction Using Domain Knowledge, Proceedings of Second International Conference on Computational Linguistics and Intelligent Text Processing, LNCS 2004, Springer, 2001.

[13]   A. K. Jain and R. C. Dubes, Algorithms for Clustering Data, Prentice Hall, 1988.

[14]   K. S. Jones, and P. Willet, Readings in Information Retrieval, Morgan Kaufmann, July 1997.

[15]   W. Ligget and C. Buckley, Query Expansion Seen Through Return Order of Relevant Documents, NIST Special Publication 500-249: The Ninth Text Retrieval Conference (TREC 9), November 2000.

[16]   Z. Liu and Q. Huang, Detecting News Reporting Using Audio/Visual Information, ICIP-1999, Kobe, Japan, October 24-28, 1999.

[17]   Z. Liu and Q. Huang, Adaptive Anchor Detection Using On-Line Trained Audio/Visual Model, Proceedings of the SPIE, San Jose, CA, January 2000.

[18]   Z. Liu and Q. Huang, Content-based Indexing and Retrieval-by-Example in Audio, ICME-2000, New York, NY, July 30 – August 2, 2000.

[19]   M. Maybury, A. Merlino, and J. Rayson, Segmentation, Content, Extraction, and Visualization of Broadcast News Video using Multistream Analysis, AAAI Spring Symposium. Stanford, CA, 1997.

[20]   K. R. McKeown and D. R. Radev, Generating summaries of multiple news articles, Advances in Automatic Text Summarization, Edited by I. Mani and M. Maybury, MIT Press, 1999.

[21]   A. Nagasaka and Y. Tanaka. Automatic Video Indexing and Full-Motion Search For Object Appearances, Proceedings of the Second Working Conference on Visual Databases Systems, September 1991, pp. 113-127.

[22]   M. F. Porter, An algorithm for suffix stripping, Program, 14(3), pp. 130-137, 1980.

[23]   B. Shahraray, Scene Change Detection and Content-based Sampling of Video Sequences, Digital Video Compression: Algorithms and Technologies 1995, Robert J. Sarfanek and Arturo A. Rodriguez Editors, Proc. SPIE 2419, February 1995.

[24]   B. Shahraray and D. Gibbon, Automated authoring of hypermedia documents of video programs, Proceedings of the Third International. Conference on Multimedia (ACM Multimedia'95), San Francisco, CA, Nov. 1995, pp. 401-409.

[25]   M. Siegler, U. Jain, B. Raj, R. Stern, Automatic Segmentation, Classification and Clustering of Broadcast News Audio, Proceedings of the DARPA Speech Recognition Workshop, Chantilly, VA February 1997, pp. 97-99.

[26]   A. Singal, J. Choi, D. Hindle, and F. Pereira. AT&T at TREC-6: SDR track, Proceedings of the Sixth Text Retrieval Conference (TREC-6), pages 227--232, 1998.

[27]   T. Strzalkowski, Robust Text Processing in Automated Information Retrieval, Proceedings of the Fourth Conference on Applied Natural Language Processing, Stuttgart, Germany, October 1994.

[28]   L. Taycher, M. La Cascia, and S. Sclaroff, Image Digestion and Relevance Feedback in the ImageRover WWW Search Engine, Proceedings of the International Conference on Visual Information, San Diego, December 1997.

[29]   P. Wu and B.S. Manjunath Adaptive Nearest Neighbor Search for Relevance Feedback in Large Image Databases, Proceedings of the ACM International Multimedia Conference (MM '01), Ottawa, Canada, October 2001.

[30]   H. Zhang, S. Smoliar, J. Wu, Content-Based Video Browsing Tools, Proceedings of the 1995 SPIE Conference on Multimedia Computing and Networking, San Jose, February 1995.

44

SEGMENTING STORIES IN NEWS VIDEO

Lekha Chaisorn, Tat-Seng Chua, and Chin-Hui Lee
School of Computing
National University of Singapore
3 Science Drive 2, Singapore 117543
{lekhacha, chuats, chl}@comp.nus.edu.sg

1. INTRODUCTION

The rapid advances in computing, multimedia, and networking technologies have results in the production and distribution of large amount of multimedia data, in particular digital video. To effectively manage these sources of videos, it is necessary to organize them in a way that facilitates user access and supports personalization. For example, when we are viewing news on the Internet, we may want to view just the video segments of our interest like sports or only those segments when the Prime Minister is giving a speech.

Research on segmenting an input video into shots, and using these shots as the basis for video organization, is well established [3, 15]. A shot represents a contiguous sequence of visually similar frames. It, however, does not usually convey any coherent semantics to the users. As users remember video contents in terms of events or stories but not in terms of changes in visual appearances as in shots, it is necessary to organize video contents in terms of small, single-story units that represent the conceptual chunks in users' memory. These units can then be organized hierarchically to facilitate browsing by the users. For a specific domain such as the news, these video units can further be classified according to their semantics such as meeting, sport, etc.

This work aims at developing a system to automatically segment and classify news video into semantic units using a learning-based approach. It is well known that the learning-based approaches are sensitive to feature selection and often suffers from data sparseness problems due to difficulties in obtaining annotated data for training. One approach to tackle the data sparseness problem is to perform the analysis at multiple levels as is done successfully in natural language processing (NLP) research [7]. For example, in NLP, it has been found to be effective to perform the part-of-speech tagging at the word level, before the phrase or sentence analysis at the higher level. Thus in this research, we propose a two-level, multi-modal framework to tackle the news story boundary detection problem. The video is analyzed at the shot and story unit (or scene) levels using a variety of features. At the shot level, we use a set of low-level and high-level features to model the contents of each shot. We employ a Decision

Tree to classify the video shots into one of the 13 pre-defined categories. At the story level, we perform HMM (Hidden Markov Models) analysis [19] to identify news story boundaries. To focus our research, we adopt the news video domain, as such video is usually more structured and has clearly defined story units.

Briefly, the content of this paper is organized as follows. *Section 2* describes related research and *Section 3* discusses the design of the multi-modal two-level classification framework. *Section 4* presents the details of shot level classification and *Section 5* discusses the details of story/scene segmentation. *Section 6* discusses the experiment results, and *Section 7* contains our conclusion and discussion of future work.

2. RELATED WORKS

Story segmentation and video classification are hot topics of research for many years now and much interesting research has been done. Because of the difficulty and often subjective nature of video classification, most early works examined only certain aspects of video classification and story segmentation in a structured domain such as sports or news.

Ide et al. [12] used videotext, motion, and face as the features to tackle the problem of news video classification. They first segmented the video into shots and used multiple techniques including clustering to classify each shot into one of the five classes of: Speech/report, Anchor, Walking, Gathering, and Computer Graphics categories. Their classification technique seems effective for this restricted class of problems. Zhou et al. [22] examined the classification of basketball videos into a set of restricted categories of Left-court, Middle-court, Right-court, and Closed-up. They considered only motion, color and edges as the features and employed a rule-based approach to classify each video shot (represented using a key frame). Chen and Wong [4] also used a rule-based approach to classify news video into six classes of news, weather, reporting, commercials, basketball, and football. They used the feature set of motion, color, text caption, and cut rate in the analysis.

Another category of techniques incorporated information within and between video segments to determine class transition boundaries using mostly the HMM approaches. Eickeler et al. [9] considered 6 features, deriving from the colour histogram and motion variations across the frames, and employed HMM to classify the video sequence into the classes of Studio Speaker, Report, Weather Forecast, Begin, End, and Editing Effect. Huang et al. [11] employed audio, colour, and motion as the features and classified the TV program (an input news video) into one of the categories of news report, weather forecast, commercial, basketball game, and football game. Alatan et al. [1] aimed to detect dialog and its transitions in fiction entertainment type videos. They modelled the shots using the features of audio (music/silence/speech), face and location changed, and used HMM to locate the transition boundary between the classes of Establishing, Dialogue, Transition, and Non-dialogue. Greiff et al. [10] used only text from transcript as the feature and employed the HMM framework to model the word sequence. Each word was labelled with a state number from 1 to 251, and a story boundary was located at a word produced from state 1 for the maximum likelihood state sequence.

In summary, most reported works considered only a limited set of classes and features, and provided only partial, intermediate solutions to the general video organization problem. In our work, we want to consider all essential categories of shots and scenes to cover potentially all types of news video. A major difference between our approach and existing works is that we perform the story segmentation analysis at two levels, similar to the approach successfully employed in NLP research. Furthermore, we aim to organize video at shot and story levels to facilitate user access.

3. THE MULTI-MODAL TWO-LEVEL FRAMEWORK

3.1 STRUCTURE OF NEWS

Most news videos have rather similar and well-defined structures. Figure 44.1 illustrates the structure of a typical news video. It typically begins with several Introduction/highlight shots that give a brief introduction of the upcoming news items to be reported. The main body of news contains a series of stories organized in terms of different geographical interests (such as international, regional, and local) and in broad categories of social political, business, sports, entertainment and weather. Each news story normally begins and ends with anchorperson shots and several in-between live-reporting shots. Most news ends with reports on sports, finance, and weather. In a typical half an hour news, there will be several periods of commercials, covering both commercial product and self-advertisement by the broadcasting station.

Although the ordering of news items may differ slightly from broadcast station to station, they all have similar structure and news categories. In order to project the identity of a broadcast station, the visual contents of each news category, like the anchorperson shots, finance and weather reporting etc., tends to be highly similar within a station, but differs from that of other broadcast stations. Hence, it is possible to adopt a learning-based approach to train the system to recognize the contents of each category within each broadcast station.

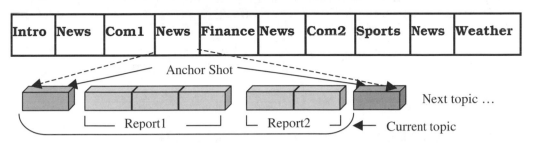

Intro	News	Com1	News	Finance	News	Com2	Sports	News	Weather

Figure 44.1 The structure of local news video under study

3.2 THE DESIGN OF NEWS CLASSIFICATION AND SEGMENTATION SYSTEM

To tackle the problem effectively, we must address three basic issues. First, we need to identify the suitable units to perform the analysis. Second, we need to extract an appropriate set of features to model and distinguish different categories. Third, we need to use a good technique to perform the classification and identify the boundaries between stories. To achieve these, we adopt the following strategies as shown in Figure 44.2.

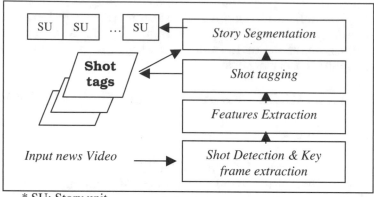

* SU: Story unit

Figure 44.2 Overall system components

a) We first segment input video into shots and extract the representative key frame for each shot using mature techniques.

b) We extract a suitable set of features to model the contents of shots. The features include low level visual and temporal features, and high-level features like faces. We select only those features that can be automatically extracted in order to automate the entire classification process.

c) We employ a learning-based approach that uses multi-modal features to classify the shots into the set of well-defined subcategories. This is known as the shot tagging process.

d) Finally, given a sequence of shots in respective subcategories, we use a combination of shot categories, content, and temporal features to identify story boundaries using the HMM technique. This process is referred to as the story segmentation process as shown in Figure 44.2.

The detailed design of the system is illustrated in Figure 44.3. Further details of the system components are discussed in Section 4 and Section 5.

4. THE CLASSIFICATION OF VIDEO SHOTS

4.1 SHOT SEGMENTATION AND KEY FRAME EXTRACTION

The first step in news video analysis is to segment the input news video into shots. We employ the multi-resolution analysis technique developed in our lab [15] that can effectively locate both abrupt and gradual transition boundaries effectively. We adopt a "greedy" strategy to over-segment the video in order to minimize the chance of missing the shot and hence the story boundaries. This strategy is reasonable as most of the falsely detected shots will be merged into story units in subsequent analysis.

After the video is segmented, there are several ways in which the contents of each shot can be modelled. We can model the contents of the shot: (a) using a representative key frame; (b) as feature trajectories; or, (c) using a combination of both. In this research, we adopt the hybrid approach as a compromise to achieve both efficiency and effectiveness. Most visual content features will be extracted from the key frame while motion and audio features will be extracted from the temporal contents of the shots.

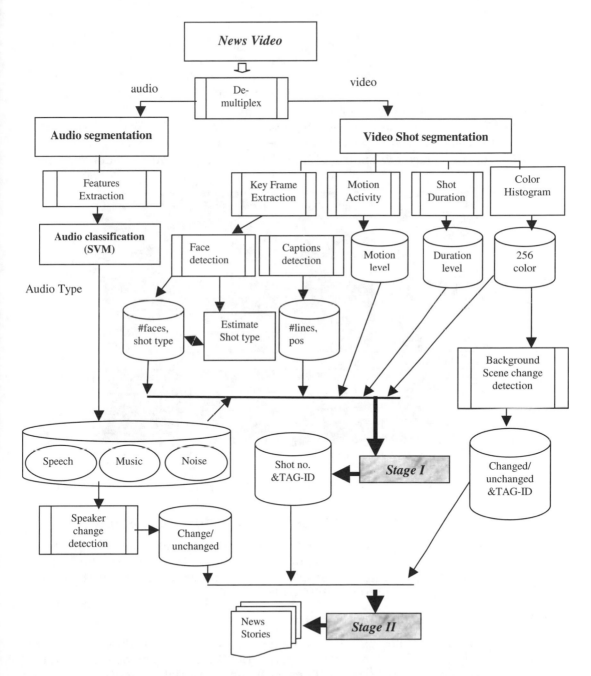

Figure 44.3 The detailed design system components

4.2 SELECTION OF SHOT CATEGORIES

We studied the set of categories employed in related works, and the structures of news video in general and local news in particular. The categories must be meaningful so that the category tag assigned to each shot is reflective of its content and facilitates the subsequent stage of segmenting and classifying news stories. The granularity of the choice of categories is important. Here we want the categories selected to reflect the major structure and concepts in news presentation. Thus, while it is reasonable to have sports and weather categories, it is too fine-grain to consider subcategories of sports like basketball or football. These subcategories are not important in understanding the overall news structure. Based on this consideration, we arrive at the following set of shot categories: *Intro/Highlight, Anchor, 2Anchor, Meeting/Gathering, Speech/Interview, Live-reporting, Still-image, Sports, Text-scene, Special, Finance, Weather,* and *Commercial.* Figure 44.4 shows a typical example in each category.

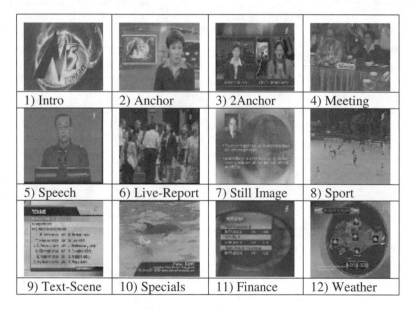

Figure 44.4 Examples of the predefined categories and example shots

4.3 CHOICE AND EXTRACTION OF FEATURES FOR SHOT CLASSIFICATION

For a learning base classification system to work effectively, it is imperative to identify a suitable set of features based on in-depth understanding of the domain. In addition, we aim to derive a comprehensive set of features that can be extracted automatically from MPEG video.

4.3.1 Low-level Visual Content Feature

Colour Histogram: Colour histogram models the visual composition of the shot. It is particularly useful to resolve two scenarios in shot classification. First, it can be used to identify those shot types with similar visual contents such as the weather and finance reporting. Second, it can be used to model the changes in background between successive shots, which provides important clues to determining a possible change in shot category or story. Here, we represent the content of key frame using a 256-colour histogram.

4.3.2 Temporal Features

Background scene change: Following the discussions on Colour histogram, we include the background scene change feature to measure the difference between the Colour histogram of the current and previous shots. We employ a higher threshold than that for shot segmentation to detect background scene change. It is represented by 'c' if there is a change and 'u' otherwise.

Audio type: This feature is very important especially for Sport and Intro/Highlight shots. For Sport shots, its audio track includes both commentary and background noise, and for Intro/Highlight shots, the narrative is accompanied by background music. We adopt an algorithm similar to that discussed in Lu et al. [17] to classify audio into the broad categories of speech, music, noise, speech and noise, speech and music, and silence.

Speaker change: Similar to background scene change feature, this feature measures whether there is a change of speaker between the current and previous shot. It takes the value of 'u' for no change, and 'c' if there is a change. The latter condition also applies to shots that do not contain speech but when there is a change from the previous speech to non-speech shot or vice versa. The detection of non-speech shot from speech shot can be done by detecting the shot's audio type as described earlier.

Motion activity: For MPEG video, there is a direct encoding of motion vectors, which can be used to indicate the level of motion activities within the shot. We usually see high level of motion in sports and certain live reporting shots such as the rioting scenes. Thus, we classify the motion into *low* (like in an Anchor-person shot where only the head region has some movements), *medium* (such as those shots with people walking), *high* (like in sports), or *no* motion (for still frame shots and Text-scene shots).

Shot duration: For Anchorperson or Interview type of shots, the duration tends to range from 20 to 50 seconds. For other types of shots, such as the Live-reporting or Sports, the duration tends to be much shorter, ranging from a few seconds to about 10 seconds. The duration is thus an important feature to model the rhythm of the shots. We set the shot duration to *short* (if it is less than 10 seconds), *medium* (if it is between 10 to 20 seconds), and *long* (for shot greater than 20 seconds in duration).

4.3.3 High-level Object-based Features

Face: Human activities are one of most important aspects of news videos, and many such activities can be deduced from the presence of faces. Many techniques have been proposed to detect faces in an image or video. In our study, we adopt the algorithm developed in [6] to detect mostly frontal faces in the key frame of each shot. We extract in each shot the number of faces detected as well as their sizes. The size of the face is used to estimate the shot types.

Shot type: We use the camera focal distance to model the shot type, which include *closed-up*, *medium*-distance or *long*-distance shot etc. Here, we simply use the size of the detected face to estimate the shot type.

Videotext: Videotext is another type of object that appears frequently in news video and can be used to determine video semantics. We employ the algorithm developed in [21] to detect videotexts. For each shot, we simply determine the number of lines of text that appear in the key frame.

Centralized Videotext: We often need to differentiate between two types of shots containing videotexts, the normal shot where the videotexts appear at the top or bottom of a shot to annotate its contents, the text-scene shot where only a sequence of texts is displayed to summarize an event, such as the results of a soccer game. A text-scene shot typically contains multiple lines of centralized text, which is different from normal shots. This feature takes the value "true" for centralized text and "false" otherwise.

Figure 44.5 presents a view of a news video in our approach.

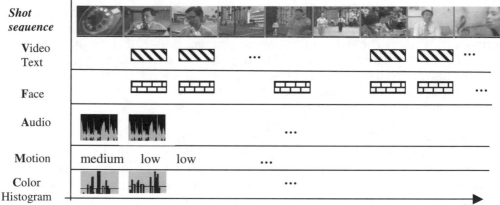

Figure 44.5 The model of a news video   *Time*

4.4 SHOT REPRESENTATION

After all features are extracted, we represent the contents of each shot using a *colour histogram vector* and a *feature vector*. The histogram vector is used to match the content of a shot with the representative shot of certain categories, while the feature vector is used by Decision Tree to categorize the shots into one of the remaining categories. The feature vector of a shot is of the form:

$$S_i = (a, m, d, f, s, t, c) \qquad (44.1)$$

where:

- a the class of audio, $a \in \{t=\text{speech}, m=\text{music}, s=\text{silence}, n=\text{noise}, tn = \text{speech} + \text{noise}, tm= \text{speech} + \text{music}, mn=\text{music}+\text{noise}\}$
- m the motion activity level, $m \in \{l=\text{low}, m=\text{medium}, h=\text{high}\}$
- d the shot duration, $d \in \{s=\text{short}, m=\text{medium}, l=\text{long}\}$
- f the number of faces, $f \geq 0$
- s the shot type, $s \in \{c= \text{closed-up}, m=\text{medium}, l=\text{long}, u=\text{unknown}\}$
- t the number of lines of text in the scene, $t \geq 0$
- c set to "true" if the videotexts present are centralized, $c \in \{t=\text{true}, f=\text{false}\}$

For example, the feature vector of an Anchorperson shot may be (t, l, l, 1, c, 2, f). Note that at this stage we did not include the background scene change and speaker change features in the feature set. These two features are not important for shot classification and will be included in detecting story boundary using HMM.

4.5 CLASSIFICATION OF VIDEO SHOTS

In shot classification process, we first remove the commercials before performing the classification of the remaining shots.

In most countries, it is mandatory to air several black frames preceding or proceeding a block of commercials. However, this is not always the case in many countries, like in Singapore. Our studies have shown that commercial boundary can normally be characterized by the presence of black frames, still frames, and/or audio silence [14]. We thus employ a heuristic approach to identify the presence of commercials and detect the beginning and ending of the commercials blocks. Our tests on six news videos (180 minutes) obtained from the MediaCorp of Singapore demonstrate that we are able to achieve a higher detection accuracy of over 97%.

We break the classification of remaining shots into two sub-tasks. We first identify the shot types that have very similar visual features. Examples of these shot types include the Weather and Finance reports. For these shot types, we simply extract the representative histogram of the respective categories and employ the histogram-matching algorithm developed in [5] to compute the shot-category similarity that takes into consideration the perceptually similar colours. We employ a high threshold of 0.8 to determine whether a given shot belongs to the Weather or Finance category.

For the rest of the shots, we employ Decision Tree (DT), in particular C4.5 algorithm, to perform the classification in a learning-based approach. Decision tree is one of the most widely used techniques in machine learning. The technique has the advantages that it is robust to noisy data, capable of learning disjunctive expression, and the training data may contain missing or unknown values [2, 18].

The Decision Tree approach has been successfully employed in many multi-class classification problems [8, 22]. We thus select the Decision Tree for our shot classification problem.

5. SCENE/STORY BOUNDARY DETECTION

After the shots have been classified into one of the pre-defined categories, we employ HMMs to detect scene/story boundaries. We use the shot sequencing information, and examine both the tagged categories and appropriate features of the shots to perform the analysis. This is similar to the idea of part-of-speech (POS) tagging problem in NLP that uses a combination of POS tags and lexical information to perform the analysis.

HMM is a powerful statistical tool first successfully utilized in speech recognition research. HMM contains a finite set of *states*, each of which is associated with a probability distribution. Transitions among the states are governed by a set of probabilities called *transition probabilities.* In a particular state, an outcome or *observation* can be generated according to the associated probability distribution. We can express the HMM parameters as the following. Each HMM is modelled with a set $\lambda\lambda = (\pi, A, B)$. Here, $\pi = (\pi_1,..., \pi_N)$ is the initial state probability; and N is the number of states represented by $Q = \{1, 2, ...,N\}$. A =

{a$_{ij}$} is the state transition probability matrix where $1 \leq i \leq N$ and $1 \leq j \leq N$; and a$_{ij}$ is the probability of moving from state i to state j. B = {b$_{jk}$} is the observation probability distribution matrix with $1 \leq j \leq N$ and $1 \leq k \leq M$; b$_{jk}$ is the emission of symbol k at state j; and M is the number of symbols. Finally, V is the set of symbols (feature vectors), with V = {$v_1, v_2, v_3......v_M$}. Given a HMM $\lambda\lambda$ and the observation sequence $O = (O_1, O_2, ...O_T)$, $P(O|\lambda)$ is the probability that HMM $\lambda\lambda$ produces the observation sequence O. $P(O|\lambda)$ can be computed by:

$$P(O / \lambda) = \sum_{allQ} P(O / Q, \lambda) P(Q / \lambda) \qquad (44.2)$$

Further details on HMM can be found in [18].

In our approach, we represent each shot by: (a) its tagged category; (b) scene/location change (c= change, u = unchanged); and, (c) speaker change (c = change, u = unchanged). We use the tag id as defined in Figure 44.3 to denote the category of each shot. The commercial category is not used here, so there are 12 categories. Each shot i is thus represented by a feature vector given by:

$$S_i = (t_i \ p_i \ c_i) \qquad (44.3)$$

where t_i is the tag id of shot i; p_i is the speaker change indicator (c or u); and c_i is the scene change indicator (c or u). Thus, each output symbol is represented by 1 of the 12 possible categories of shots, 1 out of 2 possible scene change feature, and 1 out of 2 possible speaker change feature. This gives a total of 12x2x2 = 48 distinct vectors for modeling using the HMM framework.

In our priliminary experiments, we employ the ergodic HMM framework. We perform the experiments by varying the number of states from 4 to 9 to evaluate the results. As we have a small training data set, our initial test indicates that the number of state equals to 4 gives the best result. Figure 44.6 illustrates an ergodic HMM with 4 states in our approach. When 4 states are used, we need to estimate b$_{jk}$ for 4x48 = 192 probabilities.

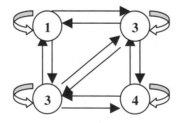

Figure 44.6 The ergodic HMM with 4 hidden states

6. TESTING AND RESULTS

This section discusses the experimental setups and results of the shot-level classification and scene/story boundary detection.

6.1 TRAINING AND TEST DATA

We use two days of news video (one from May 2001, the other from June 2001) obtained from the MediaCorp of Singapore to test the performance of our system. Each day of news video is half an hour in duration. One day is used for training, and the other for testing. In order to eliminate indexing errors, we manually index all the features of the shots segmented using the multi-resolution analysis algorithm [14]. After the removal of commercials, the training data set contains 200 shots and testing data set contains 183 shots. The numbers of story/scene boundaries are respectively 39 and 40 for the training and test data sets.

In Information Retrieval, there are several methods to measure the performance of the systems. One method is to use precision (P) and recall (R) values. The formulas are expressed as below:

$$P = NC \,/(NC +FP) \qquad\qquad (44.4)$$
$$R = NC \,/(NC + FN) \qquad\qquad (44.5)$$

where NC the number of correct boundaries detected
FN the number of False Negatives (missed)
FP the number of False Positives (not a boundary but is detected as a boundary)

By giving equal weights to precision and recall, we can derive an F_1 value to measure the overall system performance as:

$$F_1 = 2*R*P/(R+P) \qquad\qquad (44.6)$$

6.2 SHOT LEVEL CLASSIFICATION

6.2.1 Results of Shot-level Classification

The results of shot-level classification using the Decision Tree are presented in Table 44.1. The diagonal entries in Table 44.1 show the number of shots correctly classified into the respective category, while the off-diagonal entries show those wrongly classified. It can be seen that the largest classification error occurs in the Anchor category where a large number of shots are misclassified as Speech. This is because their contents are quite similar, and we probably need additional features like background or speaker change, and the context of neighboring shots to differentiate them.

Overall, our initial results indicate that we could achieve a classification accuracy of over 95%.

Table 44.1 The classification results from the Decision Tree

Classified as->	a	b	c	d	e	f	g	h	i	j
a) Intro/highlight	26					1				
b) Anchor		16					4			
c) 2anchor			2							
d) Gathering				13						
e) Still image					1					
f) Live-reporting						82		1		
g) Speech		1					11			
h) Sport						1		8		
i) Text-scene									6	
j) Special										5

6.2.2 Effectiveness of the Features Selected

In order to ascertain the effectiveness of the set of features selected, we perform separate experiments by using different number of features. As face is found to be the most important feature, we use the face as the first feature to be given to the system. With the face feature alone, the system returns an accuracy of only 59.6%. If we include the audio feature, the accuracy increases rapidly to 78.2%. However, this accuracy is still far below the accuracy that we could achieve by using all the features. When we successively add in the rest of features in the order of shot type, motion, videotext, text centralization, and shot duration, the performance of the system improves steadily and eventually reaches the accuracy of 95.10%. The analysis indicates that all the features are essential in shot classification. Figure 44.7 shows the summary of the analysis.

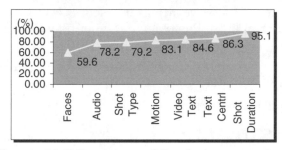

Figure 44.7 Summary of the features analysis

Figure 44.8 gives the interface of the shot classification system and shows the output category of each shot in an input news video.

6.3 SCENE/STORY BOUNDARY DETECTION

6.3.1 Results of Scene/Story Boundary Detection

We set up three experiments (Tests I, II, and III) for scene/ story boundary detection. As explained in Section 5, our experiments indicate that the number of states equal to 4 gives the best results. Thus, we set the number of states to 4 in these three HMM tests.

For Test I, we assume that all the shots are correctly tagged. We perform the HMM to locate the story boundaries and we could achieve a F_1 value of 93.7%. This experiment demonstrates that HMM is effective in news story boundary detection.

Test II is similar to Test I except that we perform the HMM analysis on the set of shots tagged using the earlier shot classification stage with about 5% tagging error. The test shows that we are able to achieve an F_1 measure of 89.7%.

The results of both tests are detailed in Table 44.2.

In Test III, we want to verify whether it is necessary to perform the two-level analysis in order to achieve the desired level of performance. We perform HMM analysis on the set of shots with their original feature set but without the category information. We vary the number of features used from the full feature set to only a few essential features. The best result we could achieve is only 37.6% in F_1 value. This test shows that although in theory a single stage analysis should perform the best, in practice, because of data sparseness, the 2-level analysis is superior.

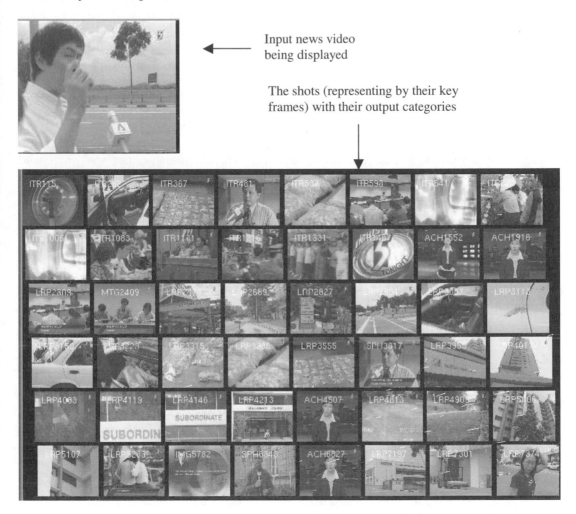

Input news video being displayed

The shots (representing by their key frames) with their output categories

**ITR: Intro/Highlight, MTG: meeting, ACH: Anchor, LRP: Live reporting, SPH: Speech, IMG: still Image.

Figure 44.8 The output category for each shot in an input news video sequence

Table 44.2 The results of HMM analysis in tests I & II

Test	NB	NC	FN	FP	R (%)	P (%)	F_1 (%)
I	40	37	3	2	94.9	92.5	93.7
II	40	35	5	3	87.5	92.1	**89.7**

** NB: the total number of correct boundaries.

6.3.2 Effectiveness of the Features Selected for HMM Analysis

In order to evaluate the importance of each feature used in Test II, we perform another set of experiments using only the individual feature one at a time, and by adding the second and third feature to the Tag-ID feature. The results are listed in Table 44.3.

Table 44.3 Results of using combination of features used in Test II

Feature	NS	NC	FN	FP	R	P	F₁
Tag	6	35	5	6	87.5	85.4	**86.4**
Sp	6	35	5	93	87.5	27.3	41.7
Sc	5	26	14	90	65.0	22.4	33.3
Tag +Sp	6	37	3	7	92.5	84.1	**88.9**
Tag+Sp+Sc	4	35	5	3	87.5	92.1	**89.7**

Table 44.3 indicates that by using only the Tag-ID feature, the system could achieve an F_1 measure of 86.4%. On the other hand, the use of the second and the third feature alone return low F_1 measures of 41.7 and 33.3 respectively. However, by combining the last two features with the Tag-ID feature, the system's F_1 performance improves gradually from 86.4% (with Tag-ID as the only feature) to 88.9% (Tag-ID +Sp), and reaches 89.7% when all the three features are included (Tag-ID +Sp +Sc). The analysis indicates that the first feature (Tag-ID) is the most important feature for scene/story boundary detection. It further confirms that shot classification facilitates the detection of news boundaries, and therefore our two-level approach is effective.

6.3.3 Analysis of HMM Results

Here we analyse how the HMM framework detects the story boundaries. Figure 44.9 lists two examples of the output state sequences resulting from the HMM analysis with 4 states. Figure 44.9 indicates that *State 4* signals the transition from current news topic to the next. In most of the cases, state 4 corresponds to Anchor shot (see the first example in Figure 44.9). However, there are exceptions as indicated in the second example, where state 4 represents the transition from Weather news to Special news. Thus, by detecting *State 4*, we can locate the boundary of the current news topic.

I₁:	*1cc*	*1uu*	*1cu*	....	*1uu*	*2cc*	*4cc*	*4uu*	*6uu*	*6uu* ...
O₁:	*3*	*3*	*3*	...	*3*	*4*	*1*	*1*	*1*	*1* ...

I₂:	*6uc*	*6uu*	*6uu* ...	*6uu*	*10cu*	*10uu*	*10uu*	*10uu* ...
O₂:	*1*	*1*	*1*	*1*	*4*	*3*	*3*	*3* ...

Note: I_i means observation sequence i
O_i means output state sequence i

Figure 44.9 Two examples of the observation sequences and their output state sequences

7. CONCLUSION AND FUTURE WORK

We have developed a two-level framework that can automatically segment an input news video into story units. Given an input video stream, the system performs the analysis at two levels. The first is shot classification, which classifies the video shots into one of 13 pre-defined categories using a combination of low-level, temporal, and high-level features. The second level builds on the results of the first level and performs the HMM analysis to locate story (or scene) boundary. Our results demonstrate that our two-level framework is effective and we could achieve an F_1 performance of over 89% in scene/story boundary detection. Our detailed analysis also indicates that HMM is effective in identifying dominant features that can be used to locate story transitions.

As our training data is rather sparse, our conclusion is preliminary. Although in theory one level analysis should yield better results, two-level analysis, which requires less training data, has been found to be superior. This conclusion is reinforced in NLP research. Nevertheless, we need to do further tests by using a large set of training data, and by using news from different broadcast stations and countries. We will also incorporate speech to text data obtained from the audio track and use text segmentation technique to help identify story boundaries. We hope to fuse information from multiple sources in order to develop a robust and reliable story boundary detection model for news and other types of video.

Our eventual goal is to convert an input news video into a set of news stories together with their classification. This will bring us a major step towards supporting personalized news video for general users.

Acknowledgment
The authors would like to acknowledge the support of the National Science & Technology Board and the Ministry of Education of Singapore for the provision of a research grant RP3960681 under which this research is carried out. The authors would also like to thank Rudy Setiono, Wee-Kheng Leow, and Gao Sheng for their comments and fruitful discussions on this research, and Chandrashekhara Anantharamu for his help in programming technique.

REFERENCES

[1]   A. Aydin Alatan, Alin N. Akansu, and Wayne Wolf (2001). Multi-modal Dialog Scene Detection using Hidden Markov Models for Content-based Multi-media Indexing. Multimedia Tools and applications, 14, pp 137-151

[2]   L. Breiman, J. H. Friedman, R. Olshen, and C. Stone. Classification and Regression Trees, Chapman & Hall, 1993.

[3]   S.-F. Chang and H. Sundaram (2000). Structural and semantic analysis of video, IEEE International Conference on Multimedia and Expo (II).

[4]   Y. Chen and E. K. Wong (2001). A knowledge-based Approach to Video Content Classification, Proceeding of SPIE Vol. 4315, pp.292-300.

[5]   T.-S. Chua and C. Chu (1998). Colour-based Pseudo-object for image retrieval with relevance feedback. International Conference on Advanced Multimedia Content Processing '98. Osaka, Japan, Nov. 148-162.

[6]   Tat-Seng Chua, Yunlong Zhao and Mohan S. Kankanhalli (2000). An Automated Compressed-Domain Face Detection Method for Video Stratification, Proceedings of Multimedia Modeling (MMM'2000), USA, Nov, World Scientific, pp 333-347.

[7]   Robert Dale, Hermann Moisl, and Harold Somers (2000). Handbook of natural language processing, Imprint New York: Marcel Dekker.

[8]   T. G. Dietterich, and G. Bakiri (1995). Solving Multi-class Learning Problems via Error-Correcting Output Codes, Journal of Artificial Intelligence Research, pp 263-286

[9]   Stefan Eickeler, Andreas Kosmala, Gerhard Rigoll (1997). A New Approach To Content-based Video Indexing Using Hidden Markov Models, IEEE workshop on Image Analysis for Multimedia Interactive Service (WIAMIS), pp 149-154.

[10]   Warren Greiff, Alex Morgan, Randall Fish, Marc Richards, and Amlan Kunda [2001]. Fine Grained Hidden Markov Modeling fro Broadcast-news Story Segmentation. Proceedings of Human Language technology Conference, California, March 2001.

[11]   J. Huang, Z. Liu, Y. Wang (1999). Integration of Multimodal Features for Video Scene Classification Based on HMM, IEEE signal processing Society workshop on Multimedia Signal processing, Denmark, pp 53-58.

[12]   Ichiro Ide, Koji Yamamoto, and Hidehiko Tanaka (1998). Automatic Video Indexing Based on Shot Classification, Conference on Advanced Multimedia Content Processing (AMCP'98), Osaka, Japan. S. Nishio, F. Kishino (eds), Lecture Notes in Computer Science Vol.1554, pp 87-102.

[13]   Michael I. Jordan (1998) (Eds). Learning in Graphical Models, MIT Press.

[14]   Chun-Keat Koh and Tat-Seng Chua (2000). Detection and Segmentation of Commercials in News Video, Technical report, The School of computing, National University of Singapore.

[15]   Yi Lin, Mohan S Kanhanhalli, and Tat-Seng Chua (2000). Temporal Multi-resolution Analysis for Video Segmentationtion, Proceedings of SPIE (Storage and Retrieval for Media Databases)., San Jose, USA. Jan 2000, Vol 3972, pp 494-505.

[16]   Zhu Liu , Jingcheng Huang, and Yao Wang (1998). Classification of TV Programs Based on Audio Information using Hidden Markov Models, IEEE Signal Processing Society, Workshop on Multimedia Signal Processing, Los Angeles, pp 27-31.

[17]   Lie Lu, Stan Z. Li and Hong-Jiang Zhang (2001). Content-based Audio Segmentation using Support Vector Machine, IEEE International Conference on Multimedia and Expo (ICME 2001), Japan, pp 956-959.

[18]   J. R. Quinlan (1986). Induction of Decision Trees. Machine Learning vol. 1, pp. 81-106.

[19]   L. Rabiner and B. Juang (1993). Fundamentals of Speech Recognition, Prentice-Hall.

[20]   Hong-Jiang Zhang, A. Kankanhalli and S.W. Smoliar (1993). Automatic Partitioning of Full-motion Video, Multimedia Systems, 1(1), pp 10-28.

[21]   Yi Zhang and Tat-Seng Chua (2000). Detection of Text Captions in Compressed domain Video. Proceedings of ACM Multimedia'2000 Workshops (Multimedia Information Retrieval), California, USA. Nov, pp 201-204.

[22]   WenSheng Zhou, Asha Vellaikal, and C–C Jay Kuo (2000). Rule-based Classification System for basketball video indexing, Proceedings of ACM Multimedia'2000 Workshops (Multimedia Information Retrieval), California, USA. Nov, pp 213-216.

45

THE VIDEO SCOUT SYSTEM: CONTENT-BASED ANALYSIS AND RETRIEVAL FOR PERSONAL VIDEO RECORDERS

Nevenka Dimitrova, Radu Jasinschi, Lalitha Agnihotri, John Zimmerman, Thomas McGee, and Dongge Li
Philips Research
345 Scarborough Road
Briarcliff Manor, NY 10598
(Nevenka.Dimitrova, Radu.Jasinschi,
Lalitha.Agnihotri)@philips.com

1. INTRODUCTION

For many years there has been a vision of the future television where users can "watch what they want, when they want." This vision first emerged through the concept of video-on-demand (VOD), where users could get streaming TV programs and movies on demand. While this approach was expected to be available today, it can currently only be found in hotels and to a limited degree as Pay-Per-View movies available on cable and satellite. Recently, however, a second approach has been in the form of hard disk recorders (HDR) available today from TiVo and ReplayTV. The current models of these devices record 30-140 hours of programs, allowing users to truly watch what they want when they want. Users navigate electronic program guides where they select programs to be recorded. In addition, TiVo "personalizes" the TV experience. Users rate shows and then the TiVo recorders automatically record TV programs that they infer the users will like. Hard disk recorders offer users tremendous flexibility to time-shift and they have pushed the idea of personalization into the living room. However, they are limited because they have no tools for personalizing information at a sub-program level.

Content-based video analysis and retrieval has been an active area of research for the last decade. As we have progressed through the low level algorithms for visual analysis and used them for video indexing at first [Nagasaka], the unfolding possibilities of combining the results from audio, visual, and transcript analysis were considered far fetched. However, the InforMedia project started first by combining visual processing with keyword spotting [Hauptman]. Recently there have been multiple reports of combining different modalities for indexing topical events in multimedia presentations, detecting highlights in baseball TV programs, and others. We have been building a system called Video

Scout, which monitors TV channels for content selection and recording based on multimodal integration of video, audio, and transcript processing. The Video Scout advances existing reported systems in using a comprehensive set of visual characteristics including cuts, videotext, faces, audio characteristics such as noise, silence, music, and speech, and transcript characteristics such as keywords, word histograms, and categories. The system architecture allows for integrated processing of all the different features within a probabilistic framework.

We designed Video Scout so that the users can make content requests in their user profiles and Scout begins recording TV programs. In addition, Scout actually "watches" the TV programs it records and personalizes program segments based on the users' profiles. Scout analyses the visual, audio, and transcript data in order to segment and index the programs. When accessing full programs, users see a high-level overview as well as topic-specific starting points. For example: users can quickly find and playback Dolly Parton's musical performance within an episode of *Late Show with David Letterman*. In addition, users can access video segments organized by topic. For example: users can quickly find all of the segments on Philips Electronics that Scout has recorded from various financial news programs.

2. RELATED WORK

Most research in the area of content based indexing and retrieval of multimedia information focuses on searching large video archives. These systems are similar to ours in that they use features from the visual, audio, and/or transcript data to segment and index video content. However, these systems differ in (1) their scale (they deal with large video archives such as the 600,000 hours of news owned by the BBC) and (2) in their need for precision since they are intended as professional applications. Our research focuses on devices for consumers' homes. We assume a device with limited channel tuning, storage, and processing power in order to be affordable. We also focus on content-based retrieval that consumers would want in their homes such as automatic personalization of content retrieval based on user profiles.

The following are complementary approaches to multimodal processing of visual, audio, and transcript information in video analysis. Rui et al. present an approach that uses low-level audio features to detect excited speech and hits in a baseball game. They employ a probabilistic framework for automatic "highlight" extraction [1]. Syeda-Mahmood et al. present event detection in multimedia presentations from teaching and training videos [2]. The foils (slides) are detected using visual analysis. Their system searches the audio track for phrases that appear as screen text. They employ a probabilistic model that exploits the co-occurrence of visual and audio events.

Another approach based on the observation that semantic concepts in videos interact and appear in context was proposed by Naphade [3]. To model this contextual interaction explicitly, a probabilistic graphical network of multijects or a multinet was proposed. Using probabilistic models for multijects, "rocks", "sky", "snow", "water-body" and "forestry/greenery" and using a factor graph as the multinet, they built a framework for semantic indexing.

Reported systems for content-based access to images and video include Query-By-Image-and Video-Content (QBIC) [4], VisualGrep [5], DVL of AT&T, InforMedia [6], VideoQ [7], MoCA [8], Vibe [9] and CONIVAS [10]. In particular, the InforMedia, MoCA, and VideoQ systems are more related to Video Scouting. The InforMedia project is a digital video library system containing methods to create a short synopsis of each video primarily based on speech recognition, natural language understanding, and caption text. The MoCA project is designed to provide content-based access to a movie database. Besides segmenting movies into salient shots and generating an abstract of the movie, the system detects and recognizes title credits and performs audio analysis. The VideoQ system classifies videos using compressed domain analysis consisting of three modules: parsing, visualization, and authoring.

3. ANALYSIS FOR CONTENT CLASSIFICATION

TV programs fall largely into two categories: narrative and non-narrative. Movies, dramatic series, situational comedies, soap operas, etc., fall into the narrative category. These TV programs all focus on telling a single story or a piece of a story over an entire episode. Narrative programs are often defined by their characters, themes, environments, and unfolding story arcs. Talk shows, news, financial programs, sporting events, etc. fall into the non-narrative category. They often have no single story line but do contain clearly recognizable structure (format). For example: local news programs are often made up of many stories told one at a time. These stories are grouped into thematic segments such as breaking news, local news, sports news, and weather.

3.1 NON-NARRATIVE PROGRAMS

The main focus of our work has been the segmentation of non-narrative programs, which are more easily addressed with current content-based retrieval technology. So far, most segmentation and retrieval advances have been made for news programs as well as some advances for talk shows and sports. The main challenges have been accurate story classification and segmentation (boundary detection).

3.1.1 Segment Classification

In general, program metadata (if available) is useful to determine if programs are either financial news or talk shows. Metadata listing the current program genre is not always available or is sometime inaccurate. Therefore we test each segment in order to classify it properly. In addition, a news program can contain, celebrity, financial, sports, weather, politics and other types of segments. Visual, audio, and transcript data are used together to resolve conflicts and make the right inferences. For example: a segment from a financial news program often contains faces and videotext onscreen at the same time. Also, a segment that has a lot of financial keywords might be financial news; however, if there is background noise such as audience clapping and laughter, it may be a talk show with jokes about the stock market rise or fall.

3.1.2 Story Segmentation

Story segmentation depends on the program genre. For example, news programs typically consist of one story after another grouped thematically: sports stories, weather stories, local stories, national stories; figure skating broadcasts usually show one competitor at a time; talk shows almost always consist of host segments followed by guest segments. The challenge is to correctly identify the boundaries of individual story segments within the broadcast.

We chose to use transcript cues to provide coarse boundary indicators. In addition, genre specific visual and audio cues are used to refine these boundaries. For our prototype we analyzed financial news and talk shows using this process. In the case of talk shows, we identify when the guest is introduced by searching for introduction cues such as: "my next guest…". We then search for the guest's actual appearance on the screen by looking entry cues such as: "please welcome…." We analyze the transcript data that falls between the introduction cue and the entry cue using a categorizer to find the main topic that will be discussed such as the release of a new movie or a new album.

3.2 NARRATIVE PROGRAMS

Segmentation of narrative programs normally implies understanding the high-level semantics in order to identify the underlying structure, the conflict and the resolution [11]. This is a difficult task given the type of processing power we expect users to have in their living rooms. In addition, our user tests on different segmentation concepts revealed that users did not want narrative content cut into pieces unless the segmentation either produced a preview for the entire program or removed all TV commercials. Since creating an accurate preview seemed to be beyond Scout's semantic abilities, we decided instead to focus on a high-level overview for narrative content, clearly identifying program segments and commercial segments. In addition, Scout can provide users with some additional segmentations that may have value to a small number of users. For example: using face and voice detection, we can find all shots that contain a specific actor; using music detection, we can find musical numbers within a musical; examining the colors on screen, we can find baseball game scenes within a movie about baseball.

4. OVERALL SYSTEM DESIGN

Video Scout consists of an archiving module and a retrieval module (see Figure 45.1). Scout takes input from the encoded video, the electronic program guide (EPG) metadata, and the user profile in order to derive high-level information. This process involves the following:

1. Archiving Module

   - Pre-selects programs to record based on matches between the EPG metadata and the user profile.
   - De-multiplexes and decodes TV program.
   - Unimodal Analysis Engine: Analyzes the individual audio, visual, and transcript streams.
   - Multimodal Bayesian Engine: Classifies segments based on integration of the visual, audio, and transcript streams.

2. Retrieval Module

    -   Personalizer: Matches requests in user profile with indexed content.
    -   User interface: Allows users to access video segments organized by either TV program or individual segment topic.
3. User profile: Collects data from the user interface and passes it to the archiving module.

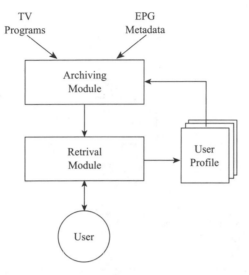

Figure 45.1 System Diagram.

4.1 ARCHIVING MODULE

An overview of the system architecture used for archiving is given in Figure 45.2. The system consists of a host PC with a TriMedia board (a specialized media processor). We should note that the archiving could be performed during program recording at the set-top box end or at the service provider end. In the second case, content descriptions can be encoded in XML in a proprietary format or using MPEG-7 for later use in retrieval. As shown in Figure 45.2, the program for analysis is selected based on the EPG data and the user's interests. Selected fields of the EPG data are matched to the user profile. When a match occurs, the relevant program information is written to a program text file. The MPEG-2 Video related to the one of selected program is then retrieved and decoded in software. During the decoding process, the individual visual, audio, and transcript data are separated out.

The Unimodal Analysis Engine (UAE) reads the separated visual and audio streams from the MPEG-2 video. In addition, it creates the transcript stream by extracting the closed caption data. From the visual stream, the UAE generates probabilities for cuts and performs, videotext, and face detection. From the audio stream, the UAE segments the audio signal and generates probabilities for seven audio categories: silence, speech, music, noise, and speech with background music, speech with speech, and speech with noise. From the transcript stream, the UAE looks for the single, double and triple arrows in text and generates probabilities indicating the category of a segment (e.g., "economy", "weather"). In addition, the UAE produces a descriptive summary for each program. These probabilities and the summary are then stored for each program. The Multimodal Bayesian Engine combines the input from the unimodal analysis engine and delivers high-level inferences to the retrieval module.

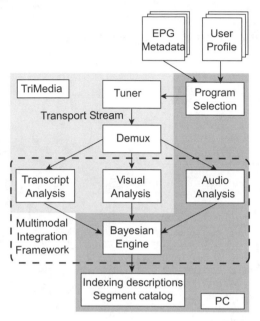

Figure 45.2 Block diagram of the system architecture displaying the PC and TriMedia sides. Items within the dashed line comprise the Multimodal Integration Framework.

4.2 RETRIEVAL MODULE

An overview of the retrieval module is shown in Figure 45.3. The retrieval module contains the Personalizer performing the personalized segment selection, and the User Interface to enable the user to interact with the system. The Personalizer looks for matches between the indexed segments and the user profile. When a match is found, the segment is flagged as "not watched". The user interface allows users to enter their profiles. These profiles use a magnet metaphor-attracting the content users request. In addition, the user interface allows users to access and playback whole programs or video segments organized by topic.

4.3 USER PROFILE

The user profile contains implicit and explicit user requests for content. These requests can be for whole TV programs as well as for topic-based video segments. Implicit requests are based on inferences from viewing history. The profile contains program titles, genres, actors, descriptions, etc. In addition, the profile contains specific topics users are interested in. For example: A profile might contain request for financial news about specific companies. In this case, Scout would store individual news stories and not whole programs. The profile performs the function of a database query in two ways. First, the Analysis Engine uses the profile to determine which programs to record. Second, the

Personalizer uses the profile to select specific segments from the entire indexed and recorded content.

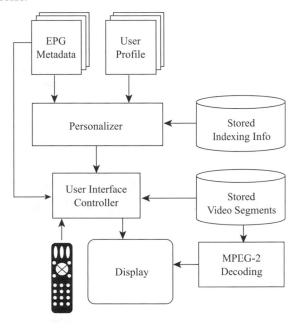

Figure 45.3 Overview of Video Scout retrieval application.

5. MULTIMODAL INTEGRATION BAYESIAN ENGINE

The Multimodal Integration Framework (MIF) is the most vital part of the archiving module. The MIF (Figure 45.4) consists of the Unimodal Analysis Engine and the Multimodal Bayesian Engine from the archiving module. This three-layered framework separates the visual, audio, and transcript streams; and it describes feature extraction at a low, mid, and high-level. In the low-level layer, features are extracted from each of the three domains. At the mid-level layer, identifiable objects and segmentation features are extracted. At the high-level layer, video information is obtained through the integration of mid-level features across the different domains. The Unimodal Analysis Engine extracts features from the low-level layer and produces probabilities for the mid-level features. The Multimodal Bayesian Engine takes the mid-level features and their probabilities and combines them with context information to generate features at the high-level layer.

The low-level layer describes signal-processing parameters as shown in Figure 45.4. In the current implementation, the visual features include color, edge, and shape; the audio features consist of twenty audio parameters [12] derived from average energy, mel-frequency cepstral coefficients, and delta spectrum magnitude; and the transcript features are given by the close-captioning (CC) text. All the features are typically extracted via signal processing operations. They result in high granularity and low abstraction information [13]. As one example, a high granularity information is pixel and/or frame based; the low abstraction character of these features is due to the fact that they correspond to low-level signal processing parameters. Within each domain, information can be combined. For example, in the visual domain, color, edge, and shape information can be combined to define whole image regions associated to parts or whole 2-D/3-D objects.

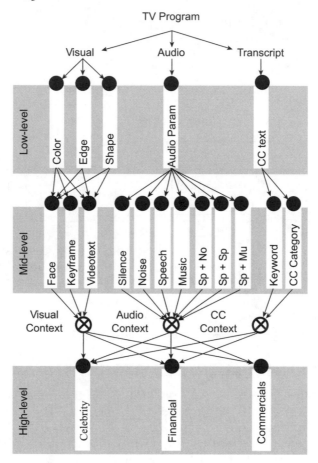

Figure 45.4 Three-layered Multimodal Integration Framework.

The mid-level features are associated with whole frames or collections of frames and whole image regions. In order to achieve this, information is combined within and/or across each of the three domains. The mid-level features in the current implementation are: (i) visual: keyframes (first frame of a "new" shot), faces, and videotext, (ii) audio: silence, noise, speech, music, speech plus noise, speech plus speech, and speech plus music, and (iii) transcript: keywords and twenty predefined transcript categories. We will describe in more detail mid-level feature extraction in section 5.1.

The high-level features describe semantic video information obtained through the integration of mid-level features across the different domains. In the current implementation, Scout classifies segments as either part of a talk show, financial news, or a commercial.

The information obtained through the first two layers represents video content information. By this we mean "objects", such as, pixels, frames, time intervals, image regions, video shots, 3-D surfaces, faces, audio sound, melodies, words, or sentences, etc. In opposition to this, we define video context information. Video context information is multimodal; it is defined for the audio, visual, and/or transcript parts of the video. Video context information denotes the circumstance, situation, underlying structure of the information being processed. The role of context is to circumscribe or disambiguate video information in order to achieve a precise segmentation, indexing, and classification. This is analogous to the role of linguistic context information [14] [15] in the interpretation of sentences according to their meanings.

Contextual information in multimedia processing circumscribes the application domain and therefore reduces the number of possible interpretations. For example: in order to classify a TV program between news, sports, or talk show we can use audio, visual, and or transcript context information. The majority of news programs are characterized by indoor scenes of head/shoulder images with superimposed videotext and speech (or speech with background music). Sports programs are characterized by lots of camera panning/zooming, outdoor scenes, and speech (announcer) with background noise (public cheering, car engine noise, etc.). This circumstantial information about the underlying structure of TV programs is not evident when using strict video content information in terms of "objects". However, taken together with the video content information, accuracy of classification tasks can be increased. In Figure 45.4, we indicate the *combination* of video content and context information with the $\otimes$ symbol.

Using the MIF we can perform two crucial tasks needed for the retrieval of segments: classification and segmentation. The high-level inferences introduced at the high-level layer of the MIF correspond to the first task of topic classification. In our benchmarking test we classified financial, celebrity, and TV commercial segments. The MIF produces probability information associated with each of the segments. The second task of segmentation process determines boundaries according to mid-level features. The initial segment boundaries are determined based on the transcript information. These segments are often too short to make sense for users since many times they may mean a change of speaker, which does not necessarily mean change of topic. MIF uses the output of the visual, audio, and transcript engines to accurately determine the boundary of the segments. These segments are further merged if consecutive segments have the same category. The output of the classification, segmentation, and associated indexing information is used in the retrieval module. We will describe multimodal integration in section 5.2.

5.1 UNIMODAL ANALYSIS ENGINE

In the visual domain, the Unimodal Analysis Engine (UAE) searches for boundaries between sequential I-frames using the DCT information from MPEG-2 encoded video stream [16]. The UAE then outputs both the uncompressed keyframes and a list detailing each keyframe's probability and its location in the TV program. The keyframe's probability reflects the relative confidence that it represents the beginning of a new shot. The UAE examines these uncompressed keyframes for videotext using an edge-based method [17] and examines the keyframes for faces [18]. It then updates the keyframe list indicating which keyframes appear to have faces and/or videotext. We will describe in more detail the visual analysis in section 5.1.1.

In the audio domain, the UAE extracts low-level acoustic features such as short-time energy, mel-frequency cepstral coefficients, and delta spectrum magnitude coefficients. These were derived after testing a comprehensive set of audio features including bandwidth, pitch, linear prediction coding coefficients, and zero-crossings. These features are extracted from the audio stream (PCM .wav files sampled at 44.1kHz) frame by frame along the time axis using a sliding window of 20ms with no overlapping. The audio stream is then divided into small, homogeneous signal segments based on the detection of onsets and offsets. Finally, a multidimensional Gaussian maximum *a posteriori* estimator is adopted to classify each audio segment into one of the seven audio categories: speech, noise, music, silence, speech plus noise, speech plus music, and speech plus speech. We will describe in more detail audio analysis in section 5.1.2.

The transcript stream is extracted from the closed caption (CC) data found in the MPEG-2 user data field. The UAE generates a timestamp for each line of text from the frame information. These timestamps allow alignment of the extracted CC data with the stored video. In addition, the UAE looks for the single, double, and triple arrows in the closed captions text to identify events such as a change in topic or speaker. We will describe in more detail transcript extraction and analysis in section 5.1.3.

5.1.1 Visual Analysis Engine

Visual analysis is one of the basic and most frequently used ways to extract useful information from video. We analyze the visual signal in order to find important temporal information such as shot cuts and visual objects such as faces and videotext. The extracted features are used later on for multimodal video classification.

Keyframe Extraction

Consecutive video frames are compared to find the abrupt scene changes (hard cuts) or soft transitions (dissolve, fade-in and out). We have experimented with several methods for cut detection. The method that performed the best uses the number of changed macroblocks to measure differences between consecutive frames. We have experimented with 2 hours of U.S. TV broadcast video and video segments from the MPEG-7 evaluation content set. The experiments on the subset of MPEG-7 test set show that we can obtain 66% precision and 80% recall.

Text Detection

The goal of videotext detection is to find regions in video frames, which correspond to text overlay (superimposed text), i.e., the anchor name and scene text (e.g., "hotel" street signs, etc.). We developed a method for text detection that uses edge information in the input images to detect character regions and text regions [17]. The method has 85% recall and precision for CIF resolution.

The first step of our videotext detection method is to separate the color information that will be processed for detecting text: R frame of an RGB image or Y component for an image in the YUV format. The frame is enhanced and edge detection is performed. An edge filtering is performed next in order to eliminate frames with too many edges. These edges are then merged to form connected components. If the connected components satisfy size and area restrictions, they are accepted for the next level of processing. If connected components lie within row and column thresholds of other connected components, they are merged to form text boxes. The text boxes are extracted from the original image and local threshold is performed to obtain text as black on white. This is then passed onto an OCR to generate text transcripts.

Figure 45.5 shows examples of edge images and extracted text areas. The text detection can be further used to string together text detected across multiple frames to generate the text pattern: scrolling, fading, flying, static, etc. Scrolling text could possibly mean ending credits and flying text could mean that the text is a part of a commercial.

Figure 45.5 Edge image and extracted image areas.

Text analysis is related to detecting and characterizing the superimposed text on video frames. We can apply OCR on the detected regions, which results in a transcript of the superimposed text on the screen. This can be used for video annotation, indexing, semantic video analysis, and search. For example, the origin of a broadcast is indicated by a graphic station logo in the right-hand top or bottom of the screen. Such station logo, if present, can be automatically recognized and used as annotation. From the transcript, a look-up database of important TV names, public figures, can be created. These names are associated with topics and categories, e.g., Bill Clinton is associated with "president of the U.S.A." and "politics". We will link names to faces if there is a single face detected in an image. Naming can be solved by using a "name" text region under the face that has certain characteristics of text length and height. Names can also be inferred from discourse analysis (e.g., in news, the anchors are passing the token to each other: "and Jim now back to you in New York".) Anchor/correspondent names and locations in a news program are often displayed on the screen and can be recognized by extracting the text showing in the bottom one-third of the video frame. Further, in music videos, musician names and music group names, and in talk shows, talk show hosts and guests, and other TV personalities are also introduced and identified in a similar fashion. So, by detecting the text box and recognizing the text, the video can be indexed based on a TV personality or a location. This information can then be used for retrieving news clips based on proper names or locations. For example, in the case of the personal profile indicating a preference to news read by a particular newscaster, information obtained using superimposed text analysis can help in detecting and tagging the news for later retrieval. Sports programs can be indexed by extracting the scores and team or player names.

Face Detection

We used the face detection method described in [18]. The system employs a feature-based top-down scheme. It consists of the following four stages:

1) Skin-tone region extraction. Through manually labeling the skin-tone pixels of a large set of color images, a distribution graph of skin-tone pixels in YIQ color coordinate is generated. A half ellipse model is used to simulate the distribution and filter skin-tone pixels of a given color image.

2) Pre-processing. Morphological operations are applied to skin-tone regions to smooth each region, break narrow isthmuses, and remove thin protrusions and small isolated regions.

3) Face candidate selection. Shape analysis is applied to each region and those with elliptical shapes are accepted as candidates. Iterative partition process based on k-means clustering is applied to rejected regions to decompose them into smaller convex regions and see if more candidates can be found.

4) Decision making. Possible facial features are extracted and their spatial configuration is checked to decide if the candidate is truly a face.

5.1.2 AUDIO ANALYSIS ENGINE

We have observed that the audio domain analysis is crucial for multimodal analysis and classification of video segments. The background audio information defines the context of the program. For example a segment in news program is distinguished from a comedy program depicting a newscaster (e.g., Saturday Night Live) with the fact that there is background laughter in the comedy program. We perform audio feature extraction, pause detection segmentation, and classification.

Audio Feature Extraction

In the audio domain, the UAE extracts low-level acoustic features such as short-time energy, mel-frequency cepstral coefficients, and delta spectrum magnitude coefficients. These were derived after testing a comprehensive set of audio features including bandwidth, pitch, linear prediction coding coefficients, and zero-crossings [12]. These features are extracted from the audio stream (PCM .wav files sampled at 44.1kHz) frame by frame along the time axis using a sliding window of 20ms with no overlaps.

Audio Segmentation and Classification

Audio segmentation and classification stage follows the feature extraction stage. The first step is that of pause detection. This eliminates the silence segments and further processing is performed only on non-silent segments. The audio stream is divided into small, homogeneous signal segments based on the detection of onsets and offsets. Finally, a multidimensional Gaussian maximum *a posteriori* estimator is adopted to classify each audio segment into one of remaining six audio categories: speech, noise, music, speech plus noise, speech plus music, and speech plus speech.

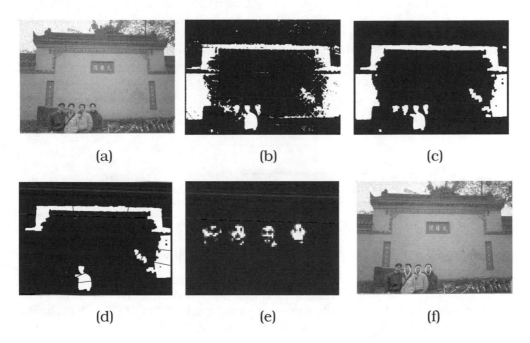

(a)   (b)   (c)

(d)   (e)   (f)

Figure 45.6 a) Original image, b) Skin tone detection, c) Pre-processing result, d) Candidate selection, e) Decision making, f) Final detection.

5.1.3 Transcript Processing and Analysis

Transcript processing begins with the extraction of closed captions from the video stream. Once the transcript is available, the transcript is analyzed to obtain coarse segmentation and category information for each segment.

Transcript Extraction

Transcripts can be generated using multiple methods including speech to text conversion, closed captions (CC), or third party program transcription. In the United States, FCC regulations mandate insertion of closed captions in the broadcast. In the analog domain, CCs are inserted in the Vertical Blanking Interval. In the digital domain there are different standards for encoding closed captions. For this paper we used CC from digital broadcast. The transcripts are generated using the CC from either High Definition Television (HDTV) broadcasts or Standard Definition (SD) program streams. For the HDTV broadcasts we extract the CC from MPEG-2 transport streams using the ATSC Standard EIA-608 format on a Philips TriMedia TM1100. The transport stream is demultiplexed and the program stream is selected for decoding. During the decoding process the decoding chip delivers the CC packets to the TriMedia processor. The CC packets are intercepted before reaching the processor. Each packet contains 32 bits of data. This 32-bit word must be parsed into four bytes of information: data type, valid/invalid bit, CC byte 1, and CC byte 2. The EIA-608 packet structure is depicted in Figure 45.7 with the ASCII values shown. If the valid/invalid bit is zero then the third and fourth bytes contain CC characters in EIA-608 format; if the valid/invalid bit is one then these bytes contain the EIA-708 wrapper information. The third and fourth bytes contain the CC data in 7-parity so that to extract the characters, the most significant bit must be cleared before extracting the data. Once the characters are extracted, all the control characters must be identified to find the carriage returns, line feeds, or spaces in order to build words from the characters. The control information is also used to insert the same information into the transcript. A time stamp generated from the frame information is inserted into the transcript

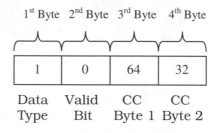

for each line of text.

Figure 45.7 EIA-608 Packet Structure with the ASCII values shown.

In our experiments, standard analog US broadcasts are encoded to create the SD MPEG-2 program streams. While encoding, the CC data found on line 21 of these broadcasts are inserted into the user-data field of the MPEG-2 program stream. The current encoding process uses an OptiVison VSTOR-150. This encoder inserts the CC data into the user-data field in a proprietary format rather than the ATSC EIA-608 standard. Consequently, the TriMedia does not recognize the packets and the CC extraction must be done completely in software. First the MPEG-2 header information is read to locate the user-data field. Then the next 32 bits are extracted. To get the first byte of CC data the first 24 bits are "ANDed" with the ASCII value 127 while the full 32 bits are "ANDed" with this value for the second byte of data. This extraction process is shown in Figure 45.8a and 8b for the first (CC1) and second (CC2) characters, respectively. Next the words must be assembled and the timestamps generated to create the transcript. This process is similar to that used in the EIA-608 case; however, a different set of control characters is used and these are identified and handled accordingly.

The result of either of the above processes is a complete time-stamped program transcript. The time-stamps are used to align the CC data with the related portion of the program. The transcripts also contain all the standard ASCII printable characters (ASCII value 32-127) that are not used as control characters. This means that the transcript then contains the standard single (">"), double (">>"), and triple arrows (">>>") that can be used to identify specific portions of the transcript, for example the change in topic or speaker. Figure 45.9 shows an example of an extracted closed caption text. The first column is the time stamp in milliseconds relative to the beginning of the program and the second is the closed captions extracted from the input video.

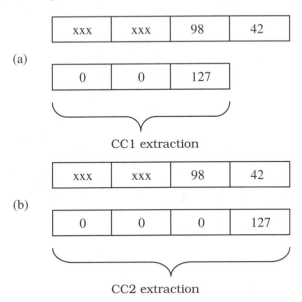

(a)

CC1 extraction

(b)

CC2 extraction

Figure 45.8 CC1 and CC2 extraction from the proprietary format.

Transcript Analysis

Figure 45.10 displays an overview of the transcript analysis. The UAE begins by extracting a high-level table of contents (summary) using known cues for the program's genre. The knowledge database and the temporal database embody the domain knowledge and include the temporal sequence of the cues [19]. For example: when analyzing a ski race, the system attempts to create segments for each racer. The phrases "now at the gate" or "contestant number" can be searched for in the transcript to index when a contestant is starting. Scout also employs a database of categories with the associated keywords/key-phrases for different topics. This database helps find the main topic for a given segment. Finally, the non-stop words (generally nouns and verbs) in each segment are indexed.

```
29863 an earlier decision and voted
31264 to require background checks
32567 to buy guns at gun shows.
34970 But did the senate really make
36141 the gun laws tougher?
38205 No one seems to agree on that.
40549 linda douglass is
41549 on capitol hill.
43613 >> Reporter: The senate debate
44815  dissolved into a frenzy
```

Figure 45.9 Sample extracted closed caption text.

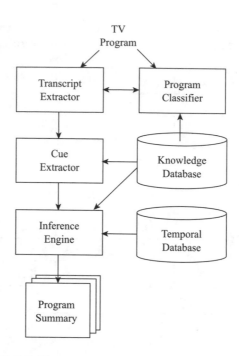

Figure 45.10 Story segmentation and summarization.

5.2 MULTIMODAL BAYESIAN ENGINE

The Bayesian Engine (BE), a probabilistic framework, performs multimodal integration [13]. We chose to use a probabilistic framework for the following reasons: (i) allows for the precise handling of certainty/uncertainty, (ii) describes a general method for the integration of information across modalities, and (iii) has the power to perform recursive updating of information. The probabilistic framework we use is a combination of Bayesian networks [20] with hierarchical priors [21].

Bayesian networks are directed acyclical graphs (DAG) in which the nodes correspond to (stochastic) variables. The arcs describe a direct causal relationship between the linked variables. The strength of these links is given by conditional probability distributions (cpds). More formally, let the set $\Omega \equiv \{x_1, \cdots, x_N\}$ of N variables define a directed acyclic graph (DAG). For each variable $x_i, i = 1, \cdots, N$, there exists a sub-set of variables of Ω, Π_{x_i}, the parent's set of x_i, describing the predecessors of x_i in the DAG, that is, the predecessors of x_i in the DAG, such that

$$P(x_i \mid \Pi_{x_i}) = P(x_i \mid x_1, \cdots, x_{i-1}), \tag{45.1}$$

where $P(\cdot \mid \cdot)$ is a strictly positive cpd. Now, the chain rule [23] tells us that the joint probability density function (jpdf) $P(x_1, \cdots, x_N)$ is decomposed as:

$$P(x_1, \cdots, x_N) \quad P(x_N \mid x_{N-1}, \cdots, x_1) \quad \cdots \times \quad P(x_2 \mid x_1) \times P(x_1). \tag{45.2}$$

According to this, the parent set Π_{x_i} has the property that x_i and $\{x_1, \cdots, x_N\} \backslash \Pi_{x_i}$ are conditionally independent given Π_{x_i}.

In Figure 45.4 we showed a conceptual realization of the MIF. As explained at the beginning of section 5, the MIF is divided into three layers: low, mid, and high-level. These layers are made up of nodes (shown in filled circles) and arrows connecting the nodes. Each node and arrow is associated with a probability distribution. Each node's probability indicates how likely a given property representing that node might occur. For example, for the keyframe node, the probability associated with it represents the likelihood a given frame is a cut. The probability for each arrow indicates how likely two nodes (one from each layer) are related. Taken together, the nodes and arrows constitute a directed acyclic graph (DAG). When combined with the probability distributions, the DAG describes a Bayesian network.

The low-level layer encompasses low-level attributes for the visual (color, edges, shape), audio (20 different signal processing parameters), and transcript (CC text) domains, one for each node. In the process of extracting these attributes, probabilities are generated for each node. The mid-level layer encompasses nodes representing mid-level information in the visual (faces, keyframes, video text), audio (7 mid-level audio categories), and transcript (20 mid-level CC categories) domains. These nodes are causally related to nodes in the low-level layer by the arrows. Finally, the high-level layer represents the outcome of high-level (semantic) inferences about TV program topics.

In the actual implementation of MIF, we used the mid- and high-level layers to perform high-level inferences. This means we compute and use probabilities only for these two layers. The nodes in the mid-level layer represent video content information, while the nodes of the high-level layer represent the results of inferences.

Figure 45.4 shows a $\otimes$ symbol between the mid- and high-level layers. This indicates where we have the input of multimodal information. For each multimodal context information we associate a probability. The multimodal context information generates circumstantial structural information about the video. This is modeled via hierarchical priors [21] which is a probabilistic method of integrating this circumstantial information with that generated by Bayesian networks. For example, a TV talk show contains a lot of speech, noise (clap, laughter, etc.) and background music (with or without speech), faces, and a very structured story line – typically, each talk show has a host and 2 or 3 guests.

In Video Scout, we discover Context Patterns by processing multiple TV programs. For example, by examining the mid-level audio features of several Financial News programs, Scout learned that Financial News has a high probability of Speech and Speech plus Music. In the visual domain, Scout looks for context patterns associated with keyframe rate (cut rate) and the co-occurrence of faces and videotext. In the audio domain Scout looks for context patterns based on the relative probability of six of the seven mid-level audio features. The silence category was not used. In the textual domain the context information is given by the different categories which are related to talk show and financial news programs.

Initial segmentation and indexing is done using closed caption data to divide the video into program and commercial segments. Next the closed captions of the program segments are analyzed for single, double, and triple arrows. Double arrows indicate a speaker change. The system marks text between successive double arrows with a start and end time in order to use it as an atomic *closed captions unit.* Scout uses these units as the indexing building blocks. In order to determine a segment's context (whether it is financial news or a talk show) Scout computes two joint probabilities. These are defined as:

$$\text{p-FIN-TOPIC} = \text{p-VTEXT} * \text{p-KWORDS} * \text{p-FACE} * \qquad\qquad (45.3)$$
$$\text{p-AUDIO-FIN} * \text{p-CC-FIN} * \text{p-FACETEXT-FIN}$$

$$\text{p-TALK-TOPIC} = \text{p-VTEXT} * \text{p-KWORDS} * \text{p-FACE} * \qquad\qquad (45.4)$$
$$\text{p-AUDIO-TALK} * \text{p-CC- TALK} * \text{p-FACETEXT- TALK}$$

The audio probabilities p-AUDIO-FIN for financial news and p-AUDIO-TALK for talk shows are created by the combination of different individual audio category probabilities. The closed captions probabilities p-CC-FIN for financial news and p-CC-TALK for talk shows are chosen as the largest probability out of the list of twenty probabilities. The face and videotext probabilities p-FACETEXT-FIN and p-FACETEXT-TALK are obtained by comparing the face and videotext probabilities p-FACE and p-TEXT which determine, for each individual closed caption unit, the probability of face and text occurrence. One heuristic used builds on the fact that talk shows are dominated by faces while financial news has both faces and text. The high-level indexing is done on each closed captions unit by computing in a new pair of probabilities: p-FIN-TOPIC and p-TALK-TOPIC. The highest value dictates the classification of the segment as either financial news or talk show.

6. SCOUT RETRIEVAL APPLICATION

The retrieval module consists of the Personalizer and the user interface. The Personalizer matches indexed content from MIF with specific requests in the user profile. The user interface allows users to visualize and playback the video segments.

6.1 PERSONALIZER

The Personalizer uses the user profile as a query against the stored, indexed content. The user interface represents these requests using a "magnet" metaphor, allowing users to "attract" different categories within the themes financial and celebrity. For TV financial news requests, the Personalizer searches the transcript and reconciles segment boundaries using the description given by the Bayesian Engine. When a segment has a financial category that appears more than n times (n>= 2), it is indexed under that category name. For example: if a financial news segment mentions Microsoft more than two times, that segment is indexed as "Microsoft" under the corresponding magnet. For celebrities, the Personalizer looks for matches between the summarization information produced by the Unimodal Analysis Engine and users' celebrity magnets. These matches are then stored and invoked during user interaction.

6.2 USER INTERFACE

We divided the user interface into two sections called Program Guide and TV Magnets. The Program Guide allows users to interact with whole TV programs. The TV Magnets section offers users access to their profiles and access to video clips organized by topic. Users navigate the interface on a TV screen using a remote control.

6.2.1 Program Guide

The Program Guide is the first screen users see when they initiate Video Scout (Figure 45.11). Our guide differs from guides found in satellite, cable, and printed TV guides in two ways. First, we display live and stored content in a single interface. Second, we changed the orientation of the time/channel grid to support the metaphor of "raining content." This rotation also helps explain the relationship between broadcast and stored content. Live programs fall across the current time boundary—the dark line in the center—to become stored programs at the bottom. When users select a stored show they see summarized program segments separated by commercial breaks. Users can also access a more detailed view of the segmented program with longer summaries of each segment.

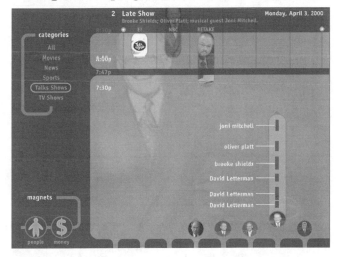

Figure 45.11 Program Guide screen showing a stored talk show's program segments. Each program segment has been summarized with host name and guest names.

6.2.2 TV Magnets

The TV Magnets section displays video clips organized by topic. Figure 45.12 shows the Financial TV Magnet screen. As users navigate up and down the list of financial topics, stored clips matched by the Personalizer appear on the right. Figure 45.12 shows four clips that have been attracted on the topic "Dow" within three programs. Faces pulled from the video stream help to indicate the shows that clips come from. We chose to display the video clip in relation to the whole program for three reasons. First, this display is consistent with the segmentations users see in the program guide section. Second, the length of the clip in relation to the whole show offers a quick preview as to how much detail is covered in the clip. Third, topics that appear at the beginning or appear several times within a program have been given a higher priority by the broadcaster and this information may be important to users who select one clip over another.

The TV Magnets section also provides users with tools for refining the content that gets recorded. Users can add new topics to the list on the left. In addition, they can add filters that limit the amount of content attracted by a specific topic. For example: if users have Microsoft as a topic, they can add the term "game", limiting the content stored on Microsoft to news stories with the word game in them to obtain the latest X-Box stories.

7. EXPERIMENTAL RESULTS

The individual algorithms for unimodal analysis have been benchmarked and the results have previously been published [12, 17, 18, 22]. For Scout, we benchmarked the Bayesian Engine to investigate how the system would perform on financial and celebrity segments. We automatically segmented and indexed seven TV programs: Marketwatch, Wall Street Week (WSW), and Wall Street Journal Report (WSJR) as well as the one-hour talk shows hosted by Jay Leno and David Letterman. The total video analyzed was about six hours. Each of the seven TV programs was classified as being either a financial news program or a talk show. Initially, each segment was classified as either a program segment or a commercial segment.

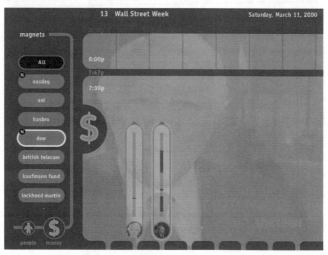

Figure 45.12 TV Magnet showing Financial Magnet results.

Program segments were subsequently divided into smaller, topic-based segments based mainly on the transcript. The Bayesian Engine performed a bi-partite inference between financial news and talk shows on these sub-program units. Visual and audio information from the mid-level layer was also used in addition to the transcript. Next, a post-processing of the resulting inferences for each segment was performed by merging small segments into larger segments. In this inference process, Bayesian Engine combined video content and context information. This turned out to be of great importance for the final outcome of the inferences. In particular, the role of audio context information turned out to be especially important.

There were a total of 84 program segments from the four financial news programs and a total of 168 program segments for talk shows. This gave us a total of 252 segments to classify. When using all multimedia cues (audio, visual, and transcript), we get the following results: (i) total precision of 94.1% and recall of 85.7% for celebrity segments, and (ii) total precision of 81.1% and a recall of 86.9% for financial segments.

8. CONCLUSIONS

The arrival of personal TV products into consumers market has created an opportunity to experiment with and deploy content-based retrieval research into advanced products. In this paper we describe a multimodal content-based approach to personal TV recorders. Our system employs a three-layer, multimodal integration framework to analyze, characterize, and retrieve whole programs and video segments based on user profiles. This framework has been applied to narrative and non-narrative programs, although in the current implementation we focus on non-narrative content. We believe that our research enables consumers to have a completely new level of access to television programs. The access is enhanced through personalization of video content at the sub-program level using multimodal integration.

We can apply personalized content based access to video in both television of the future scenarios mentioned in the introduction of the paper: VOD and HDR. In the VOD scenario, the archiving can be performed at the service provider end and the descriptions sent down to the set-top box. In the HDR scenario the archiving can be also performed in the set-top box as we have shown with our current research and implementation.

Video Scout is currently implemented on a PC and a Philips TriCodec board. We use a 600 MHz PC with a PIII processor and 256 MB of RAM running WindowsNT. The TriCodec board has a 100 MHz TriMedia TM1000 processor and 4MB of RAM. On this low-end platform, the visual, audio, and transcript analyses take less than the length of the TV programs to analyze and extract the features while the Bayesian engine takes less than one minute per one hour of TV program to give the segmentation and classification information. The retrieval application runs on the PC. However this is not computationally intensive and therefore can migrate onto the TriMedia. Users can access Scout using a Philips Pronto programmable remote control.

In the future we plan to explore (i) the use of different retrieval features, (ii) multimodal integration for narrative content, and (iii) delivery of a full-fledged system capable of responding to users' personalization needs. Also, we plan to fully integrate our face learning and person identification methods into the current system.

REFERENCES

[1]   Y. Rui, A. Gupta, and A. Acero, Automatically Extracting Highlights for TV Baseball Programs, presented at ACM Multimedia, Marina Del Rey, 2000.

[2]   T. Syeda-Mahmood and S. Srinivasan, Detecting Topical Events in Digital Video, presented at ACM Multimedia, Marina Del Rey, 2000.

[3]   M. Naphade and T. Huang, A Probabilistic Framework for Semantic Indexing and Retrieval in Video, presented at IEEE International Conference on Multimedia and Expo, New York, 2000.

[4]   W. Niblack, X. Zhu, J. L. Hafner, T. Breuel, D. Ponceleon, D. Petkovic, M. Flickner, E. Upfal, S. I. Nin, S. Sull, B. Dom, B.-L. Yeo, S. Srinivasan, D. Zivkovic, and M. Penner, Updates to the QBIC System, SPIE- Storage and Retrieval for Image and Video Databases VI, vol. 3312, pp. 150-161, 1998.

[5]   R. Jain, Visual Information Retrieval, Communications of the ACM, vol. 40, pp. 71-79, 1997.

[6]   A. Hauptmann and M. Smith, Text, speech, and vision for video segmentation: The Informedia project, presented at AAAI Fall 1995 Symposium on Computational Models for Integrating Language and Vision, 1995.

[7]   S.-F. Chang, W. Chen, H. J. Meng, H. Sundaram, and D. Zhong, VideoQ: An Automated Content Based Video Search System Using Visual Cues, presented at ACM Multimedia, 1997.

[8]   S. Pfeiffer, R. Lienhart, S. Fischer, and W. Effelsberg, Abstracting Digital Movies Automatically, Journal on Visual Communications and Image Representation, vol. 7, pp. 345-353, 1996.

[9]   J.-Y. Chen, C. Taskiran, A. Albiol, E. J. Delp, and C. A. Bouman, Vibe: A Compressed Video Database Structured for Active Browsing and Search, Purdue University 1999.

[10] M. Abdel-Mottaleb, N. Dimitrova, R. Desai, and J. Martino, CONIVAS: CONtent-based Image and Video Access System, presented at ACM Multimedia, Boston, 1996.

[11] H. Sundaram and S.-F. Chang, Determining Computable Scenes in Films and their Structures using Audio-Visual Memory Models, presented at ACM Multimedia, Marina Del Rey, 2000.

[12] D. Li, I. K. Sethi, N. Dimitrova, and T. McGee, Classification of General Audio Data for Content-Based Retrieval, Pattern Recognition Letters, 2001.

[13] R. Jasinschi, N. Dimitrova, T. McGee, L. Agnihotri, and J. Zimmerman, Video Scouting: an Architecture and System for the Integration of Multimedia Information in Personal TV Applications, presented at ICASSP, Salt lake City, 2001.

[14] J. McCarthy, Notes on Formalizing Context, presented at IJCAI, 1993.

[15] R. V. Guha, Contexts: A Formalization and Some Applications: Stanford PhD Thesis, http://www-formal.stanford.edu/, 1991.

[16] N. Dimitrova, T. McGee, L. Agnihotri, S. Dagtas, and R. Jasinschi, On Selective Video Content Analysis and Filtering, presented at SPIE Conference on Image and Video Databases, San Jose, 2000.

[17] N. Dimitrova, L. Agnihotri, C. Dorai, and R. Bolle, MPEG-7 VideoText Description Scheme for Superimposed Text in Images and Video, Signal Processing: Image Communication Journal, pp. 137-155, 2000.

[18] G. Wei and I. Sethi, Omni-Face Detection for Video/Image Content Description, presented at Intl. Workshop on Multimedia Information Retrieval, in conjunction with ACM Multimedia, (MIR2000), 2000.

[19] L. Agnihotri, K. Devara, T. McGee, and N. Dimitrova, Summarization of Video Programs Based on Closed Captioning, presented at SPIE Storage and Retrieval in Media Databases, San Jose, 2001.

[20] J. Pearl, Probabilistic Reasoning in Intelligent Systems: Networks of Plausible Inference: Morgan Kaufmann Publishers, Inc., 1988.

[21] J. O. Berger, Statistical Decision Theory and Bayesian Analysis. New York: Springer Verlag, 1980.

[22] N. Dimitrova, T. McGee, and H. Elenbaas, Video Keyframe Extraction and Filtering: A Keyframe is not a Keyframe to Everyone, presented at ACM Conference on Information and Knowledge Management, Las Vegas, 1997.

Section IX

PANEL OF EXPERTS: THE FUTURE OF VIDEO DATABASES

In this chapter, world-renowned experts answer fundamental questions about the state of the art and future research directions in video databases and related topics.

The following four intriguing questions were answered by each panelist.

1. *What do you see as the future trends in multimedia research and practice?*

2. *What do you see as major present and future applications of video databases?*

3. *What research areas in multimedia would you suggest to new graduate students entering this field?*

4. *What kind of work are you doing now and what is your current major challenge in your research?*

EXPERT PANEL PARTICIPANTS
(alphabetically)

Dr. Al Bovik received the Ph.D. degree in Electrical and Computer Engineering in 1984 from the University of Illinois, Urbana-Champaign. He is currently the Robert Parker Centennial Endowed Professor in the Department of Electrical and Computer Engineering at the University of Texas at Austin, where he is the Director of the Laboratory for Image and Video Engineering (LIVE) in the Center for Perceptual Systems. Al's main interests are in combining the sciences of digital multimedia processing and computational aspects of biological perception.

Al Bovik's mission and professional statement
I would really like to know, and be able to predict (at least statistically), why people look where they do! If I knew this, then I think I would be able to change

the way people think about Multimedia Processing. I have been *technically* inspired since early on by such thinkers as Arthur C. Clarke, Isaac Asimov, Carl Sagan, in general science; by my advisors in graduate school, Dave Munson and Tom Huang for their enthusiasm and storehouse of knowledge, guidance, and intuition; by my many peer colleagues in the field, such as Ed Delp and Rama Chellappa, for their professionalism, knowledge, and humor; but most of all, by the many brilliant students that I have been lucky enough to work with. Although I love my work, I escape from it regularly through my practices of enjoying my wife Golda and daughter Dhivya (one year old at this writing), my practice of yoga, my hobbies of deep-sky astronomy and close-up magic, and by near-daily dips and mile-long swims in Austin's wonderful Barton Springs Pool.

Dr. Alberto Del Bimbo graduated with Honors in Electronic Engineering at the University of Florence, Italy, in 1977. From graduation to 1988, he worked at IBM Italia SpA. In 1988, he was appointed as Associate Professor of Computer Engineering at the University of Florence. He was then appointed, in 1994, as Professor at the University of Brescia and, shortly thereafter 1995, returned to the University of Florence as Professor of Computer Engineering, Florence, in charge of Research and Innovation Transfer.

Prof. Del Bimbo's scientific interests and activities have dealt with the subject Image Technology and Multimedia. Particularly, they address content-based retrieval from 2D and 3D image databases, automatic semantic annotation and retrieval by content from video databases, and advanced man-machine interaction based on computer vision technology.

He authored over 180 publications and was the guest editor of many special issues on distinguished journals. He authored the monograph "Visual Information Retrieval" edited by Morgan Kaufman Publishers Inc., San Francisco, in 1999. He was the General Chairman of the 9th IAPR International Conference on Image Analysis and Processing, Florence 1997 and the General Chairman of the 6th IEEE International Conference on Multimedia Computing and Systems, Florence 1999.

Dr. Nevenka Dimitrova is a Principal Member Research Staff at Philips Research. She obtained her Ph.D. (1995) and MS (1991) in Computer Science from Arizona State University (USA), and BS (1984) in Mathematics and Computer Science from University of Kiril and Metodij, Skopje, Macedonia. Her main research passion is in the areas of multimedia content management, digital television, content synthesis, video content navigation and retrieval, and content understanding.

Dr. Dimitrova's mission and professional statement
I believe that the advancement of multimedia information systems can help improve quality of life (and survival). My professional inspiration is drawn from

colleagues on this panel, particularly Prof. Ramesh Jain. I believe that our inspiration for moving from Hume-multimedia processing to Kant-multimedia processing should come from philosophy and psychology but the research should be firmly grounded on formal mathematical basis.

Dr. Shahram Ghandeharizadeh received his Ph.D. degree in computer science from the University of Wisconsin, Madison, in 1990. Since then, he has been on the faculty at the University of Southern California. In 1992, Dr. Ghandeharizadeh received the National Science Foundation Young Investigator's Award for his research on the physical design of parallel database systems. In 1995, he received an award from the School of Engineering at USC in recognition of his research activities.

During 1994 to 1997, he led a team of graduate students to build Mitra, a scalable video-on-demand system that embodied several of his design concepts. This software prototype functioned on a cluster of PCs and served as a platform for several of his students' dissertation topics. In 1997, Matsushita Information Technology Laboratory, MITL, purchased a license of Mitra for research and development purposes.

Dr. Ghandeharizadeh's primary research interests are in the design and implementation of multimedia storage managers, and parallel database management systems. He has served on the organizing committees of numerous conferences and was the general co-chair ACM-Multimedia 2000. His activities are supported by several grants from the National Science Foundation, Department of Defense, Microsoft, BMC Software, and Hewlett-Packard. He is the director of the Database Laboratory at USC.

Dr. David Gibbon is currently a Technology Consultant in the Voice Enabled Services Research Laboratory at AT&T Labs - Research. He received the M.S. degree in electronic engineering from Monmouth University (New Jersey) in 1990 and he joined Bell Laboratories in 1985. His research interests include multimedia processing for searching and browsing of video databases and real-time video processing for communications applications.

David and his colleagues are engaged in an ongoing project aimed at creating the technologies for automated and content-based indexing of multimedia information for intelligent, selective, and efficient retrieval and browsing. Such multimedia sources of information include video programs, TV broadcasts, video conferences, and spoken documents. Lawrence Rabiner is one of the people who inspired David professionally. His unwavering demonstration of his work ethic, pragmatism, positive attitude, and research excellence serve as models to strive to emulate. David is a member of the IEEE, SPIE IS&T, and the ACM and serves on the Editorial Board for the Journal of Multimedia Tools and Applications. He has twenty-eight patent filings and holds six patents in the areas of multimedia indexing, streaming, and video analysis.

Dr. Forouzan Golshani is a Professor of Computer Science and Engineering at Arizona State University and the director of its Multimedia Information Systems Laboratory. Prior to his current position, he was with the Department of Computing at the Imperial College, London, England, until 1984. His areas of expertise include multimedia computing and communication, particularly video content extraction and representation, and multimedia information integration.

Forouzan is the co-founder of Corporate Enhancement Group (1995) and Roz Software Systems (1997), which was funded and became Active Image Recognition, Inc. in 2002. He has worked as consultant with Motorola, Intel Corp., Bull Worldwide Information Systems, iPhysicianNet, Honeywell, McDonnell Douglas Helicopter Company, and Sperry. He has successfully patented nine inventions and others are in the patent pipeline. Forouzan's more than 150 technical articles have been published in books, journals, and conference proceedings.

He is very actively involved in numerous professional services, mostly in IEEE Computer Society, including chair of the IEEE Technical Committee on Computer Languages and IEEE Distinguished Speaker program. Among his conference activities are: chairing of six international conferences, program chair for five others, and a member of program committee for over 70 conferences. Forouzan is currently the Editor-in-Chief of IEEE Multimedia and has served on its editorial board since its inception. He wrote the Review column and the News column at different times. He also serves on the editorial boards of a number of other journals. He received a PhD in Computer Science from the University of Warwick in England in 1982.

Dr. Thomas S. Huang is William L. Everitt Distinguished Professor of Electrical and Computer Engineering, University of Illinois at Urbana-Champaign. He received Sc.D. from Massachusetts Institute of Technology in 1963. His research interests include image processing, computer vision, pattern recognition, machine learning, multimodal (esp. visual and audio) human computer interaction, and multimedia (esp. images and video) databases.

Dr. Ramesh Jain is Rhesa Farmer Distinguished Chair in Embedded Experiential Systems, and Georgia Research Alliance Eminent Scholar, School of Electrical and Computer Engineering and College of Computing, Georgia Institute of Technology, Atlanta, GA 30332-0250. He received his Ph.D. from Indian Institute of Technology, Kharagpur in 1975.

Dr. Ramesh's mission and professional statement

My professional goal is to address some challenging real problems and develop systems that are useful. People that inspired me professionally include among others Prof. Azriel Rosenfeld and Prof. Hans-Helmut Nagel.

I enjoy taking research out of my lab to practice, particularly to develop products. Working with some bright research students to develop products and build companies has been a rewarding experience. I am also trying to develop devices to take computing to masses – to people in third world countries who are illiterates. It is time that we try to bridge the digital divide by developing 'folk computing'. Folk computers will be multimedia systems that will use text only when essential – exactly opposite of what happens today.

Dr. Dragutin Petkovic obtained his Ph.D. at UC Irvine, in the area of biomedical image processing. He spent over 15 years at IBM Almaden Research Center as a scientist and in various management roles. His projects ranged from use of computer vision for inspection, to multimedia databases and content based retrieval for image and video. Dr. Petkovic received numerous IBM awards for his work and is one of the founders of content-based retrieval area and IBM's QBIC project.

Dr. Petkovic's mission and professional statement

My passion is for multimedia systems, which are very easy to use by non-technical users, offer automatic indexing and data summarization, and which exhibit some form of intelligence. My inspiration comes from technical, scientific, and business leaders who are successful but at the same time compassionate and people oriented.

Dr. Rosalind Picard is Associate Professor of Media Arts and Science, MIT Media Lab and Co-director MIT Things That Think Consortium. She received her Doctorate from MIT in Electrical Engineering and Computer Science in 1991.

Dr. Picard's mission and professional statement

The mission statement for my research group can be found at http://affect.media.mit.edu. Basically, our research focuses on creating personal computational systems with the ability to sense, recognize, and understand human emotions, together with the skills to respond in an intelligent, sensitive, and respectful manner toward the user. We are also developing computers that aid in communicating human emotions, computers that assist and support people in development of their skills of social-emotional intelligence, and computers that "have" emotional mechanisms, as well as the intelligence and ethics to appropriately manage, express, and otherwise utilize such mechanisms. Embracing the latter goal is perhaps the most controversial, but it is based on a variety of scientific findings from neuroscience and other fields, which suggest that emotion plays a crucial role in enabling a resource-limited system to adapt intelligently to complex and unpredictable situations.

In short, we think mechanisms of emotion will be needed to build human-centered machines, which are able to respond intelligently and sensitively to the complex and unpredictable situations common to human-computer interaction. Affective computing research aims to make fundamental improvements in the ability of computers to serve people, including reducing the frustration that is prevalent in human-computer interaction.

With respect to personal challenges, my greatest challenge (and delight) is raising my sons. People that inspired me professionally are Fran Dubner, Anil Jain, Nicholas Negroponte, Alan Oppenheim, John Peatman, Alex Pentland, Ron Schafer, and many other teachers, colleagues, and friends.

Expert Panel Discussion

1. What do you see as the future trends in multimedia research and practice?

Rosalind Picard

With respect to long-term needs and challenges, I see a demand for:

(a) Development of better pattern recognition, machine learning, and machine understanding techniques, especially for enabling semantic and conversational inquiries about content, and (b) Interfaces that recognize and respond to user state during the interaction (detecting, for example, if the user is displeased, pleased, interested, or bored, and responding intelligently to such).

Neither of these is entirely new, but I would like to see a change in their emphasis: from being centered around cool new algorithms and techniques, to being centered around what people are good at. How can multimedia systems enable users to not only feel successful at the task, but also to enjoy the experience of interacting with the system? One way of thinking about this is to build the system in a way that shows respect for what people like to do naturally, vs. expecting people to become trained in doing things the system's way. If a person assumes something is common sense, the system should be capable of understanding this. If the person communicates to the system that it is not functioning right or it is confusing, then the system should acknowledge this, and even be able to take the initiative to offer the user an alternative. If the person communicates that one approach is enjoyable, and another is annoying, then the system could consider adapting to make the experience more enjoyable for that user. If the user communicates in a way that is natural, that another person would understand, but the system still doesn't get it, then we haven't succeeded yet. Clearly we have a long way to go, since today's computers really "don't get it" with respect to understanding most of what we communicate to them.

Ramesh Jain
Ten year ago, people had to plan their budget to create a multimedia environment for acquiring, processing, and presenting audio, video, and related information. Now computers below $1,000 are all equipped with powerful multimedia environment. Mobile phones are coming equipped with video cameras. The trend to make everything multimedia exists and will continue.

This will give rise to some new interesting applications. I believe that one of the very first will be that e-mail will become more and more audio and video. This will be used by people all over the world by people who do not know English, or who don't even know how to read and write in their own languages. Text mail will remain important, but audio and video mail will become popular and someday overtake text in its utility. This has serious implications for multimedia researchers because this will force people to think beyond the traditional notion of everything being a page of text. Storage, retrieval, and presentation of true multimedia will emerge. Current thinking about audio and video being an appendix to text will not take us too far.

Multimedia research is not keeping pace with the technology, however. In research, multimedia has to become multimedia. We all know that 3D is not the same as 3 X 1d. Current multimedia systems rarely address multimedia issues. Look at all the conferences and books; they usually have sections on images, audio, video, and text. Each section is more or less independent. This is definitely not multimedia. Multimedia must address more and more Gestalt of the situation.

Multimedia research must address how the semantics of the situation emerges out of individual data sources independent of the medium represented by the source. This issue of semantics is critical to all the applications of multimedia.

Alberto Del Bimbo
High speed networking will emphasize multimedia applications on the Internet that include 3D digital data and video. It will also enhance interactivity and enrich the type of combinations of media employed. The range of future applications is definitely broad. Applications that will interest large public will be, among the others, interactive TV, multimedia database search, permanent connectivity to the net. Niche applications will include new modalities of man-machine interaction, based on gestures, emotions, speech, and captured motion. We expect that these applications will drive most of research on Multimedia in

the next years. It will develop on different lines: Operating System Support - including network and service management, database management, quality of service; Network Architectures; Multimedia Databases – including digital library organization and indexing, search engines, content based retrieval; Multimedia Standards; Man-machine Interaction – including new terminals for multimedia information access and presentation, multimodal interactivity, simultaneous presence in virtual and augmented environments.

Nevenka Dimitrova

The trend in multimedia research so far was to use existing methods in computer vision and audio analysis and databases. However, in the future, multimedia research will have to break new frontiers and extend the parent fields. First of all, we need to rely on context and memory. Context is the larger environmental knowledge that includes the laws of biology and physics and common sense. In philosophical terms so far, we have been using what I call the "Hume" model of signal processing where the only things that exist in the present frame are real, and we should transcend to the "Kant" model where there is a representation which accounts for common sense knowledge and assumptions about the expected behavior of the entities that are sought for. Memory is an important aiding factor in analysis with longer term goals. In this respect our methods have severe anterograde amnesia and we just keep a very localized information about the current computations. In detection of dissolves, we keep a buffer of frames. In computing scenes we keep a few minutes worth of data. However, in multimedia processing we need to keep more information for longer periods of time, such as full programs, episodes, and genres.

The practice will have to include wide applications that use this technology in support of "normal" activities of the users: their everyday life, work, and entertainment. In all three categories they could be served by storage, communications, and productivity activities.

Thomas Huang

In the past 10 years, researchers have been working on various components of Video Databases, such as shot segmentation and key frame extraction, genre classification, and retrieval based on sketching. In the future, we shall see more work in integrating these components, and in searching for important and meaningful applications.

Al Bovik

I was quite the reader of science fiction when I was in college. Among the most famous works that I liked very much is, of course, Isaac Asimov's Foundation Trilogy. While I was in school, Asimov published a fourth volume in the series entitled *Foundation's Edge,* which wasn't as good as the original trilogy, but part of it still sticks in my mind. What has this to do with future trends in Multimedia? Well, Asimov's work, like much of good science fiction, contains good speculative science writing. One of the things that I remember from *Foundation's Edge* (and little else, actually) is a chapter where the hero of the moment (whose exotic name I forget) is piloting a small starship. The way in which he interacted with the shipboard computer is what I remember most: it was almost a *merging* of the human with the computer at all levels of sensory input: visual, tactile, aural, olfactory, etc. My recollection (25 years later; please don't write to correct me on the details!) is that the hero and computer communicated efficiently, synergistically, and bidirectionally. The impression was given that the spaceship deeply analyzed the operator's sensory and manual capacities and, using this information, delivered maximal sensory information to the operator, which, simply stated, gave the operator the sensation of being part of the ship itself, with awareness of all of its sensory data, control over its mechanics, and complete capability for interaction with every aspect of the ship and its environment. This makes things sound both very mysterious and technical, whereas the writer intended to convey that human and machine were *experiencing* one another, at all levels involving sensory apparatus. Perhaps an efficient way to encapsulate this is to say that the fictional multimedia engineers that had designed the systems on this spaceship were, generally, masters of human-computer interface design, and, specifically, what I would call *Multimedia Ergonomics.*

So, this is my answer regarding which future trends will eventually dominate multimedia research and practice: *Multimedia Ergonomics*! I am sure that there are many important answers to this question that are more immediate and accessible. However, since many of the most relevant to near-term practice will likely be addressed by other authors writing in this space, I have decided to try to take a longer view and be speculative.

So, again, *Multimedia Ergonomics.* What do I mean by this? Well, first, what is meant by *ergonomics*? Most of us think in terms of back-supporting office chairs and comfy keyboard and mouse supports. Dictionary.com gives us this: "The applied science of equipment design, as for the workplace, intended to maximize productivity by reducing operator fatigue and discomfort." Well, I would propose to take the definition and application further, meaning that ergonomics for multimedia should be *closed-loop, active, sensor-driven,* and *configurable in software.* All of these leading to a more relaxed, productive, data-intensive, and yes, comfy multimedia experience.

Certainly, the ultimate multimedia system could deliver an all-enveloping, sense-surround audio-visual-tactile-olfactory-etc. experience; indeed, much has been accomplished in these directions already: visit your local IMAX theatre, advanced video game, or museum. But regardless of the degree of *immersion* in the multimedia experience, I propose that future emphasis be placed on truly closing the loop; rather than just surrounding the participant with an onslaught of sensory data, multimedia systems should familiarize themselves with the human(s) involved, by measuring the parameters, positions, and responses of their sensory organs; by measuring the complete three-dimensional physical form, position, movements, and perhaps even state of health and fitness of the participant(s); by providing exceptionally rich and absorbable sensory data across the range of sensory capacity, and means for delivering it to the full range of sensory organs. In other words, completely merge the human and machine at the interface level, but in as noninvasive a way as possible.

Since all this is highly speculative, I cannot say how to best do all of these things. However, I do observe that much work has been done in this direction already (with enormously more to do): methods for visual eyetracking, head-tracking, position sensing, body and movement measurement, human-machine interfacing, assessing human sensory response to data; immersive audio-visual displays, and even olfactory sensing and synthesis are all topics that are currently quite active, and are represented by dedicated conferences, meetings, and so on. However, most of these modes of interactive sensing still have a long way to go. For example, visual eyetracking, or the means for determining the gaze direction of the human participant(s), has become mostly non-invasive, but still requires fairly expensive apparatus, a cooperative subject, and a tradeoff between calibration time taken and accuracy in each session. Visual algorithms that use visual eyetracking data are still few and far between. Methods for assessing human sensory response to data of various types have advanced rapidly, owing primarily to the efforts of sensory psychologists. Our own contribution to this book details our current inability to even predict whether visual data will be considered "of good quality" to a human observer. Advanced methods for human-machine interfacing are high priorities with researchers and funding agencies, but my desktop computing environment doesn't look much different than it did ten years ago.

So, how to approach these problems and make progress? I think that the answer lies in looking within, meaning to understand better the sensory functioning and sensory processing capacities of human beings. Sensory psychologists are, of course, energetically engaged in this effort already, but rarely with an eye towards the engineering applications, and specifically, towards implications for Multimedia Ergonomics that I view as critical for advancing this art.

For my part, I believe that the answer is for Multimedia Engineers to work closely with sensory psychologists, for their common ground, I have discovered, is actually quite wide when recognized. I plan to continue working with my colleagues in both fields as I have been doing for some time now. Come to think of it, I think I may soon return to Asimov's *Foundation Series* for a reread and some new ideas.

Forouzan Golshani

Let me point out two areas, each with a completely different emphasis as trends that we should follow. One is the general area of "security, biometrics, and information assurance" which is now receiving considerably more attention, and the other is media technologies and arts. I elaborate on each one separately.

Whereas the premise that digital information is no longer limited to the traditional forms of numbers and text is well accepted, why is it that multimedia information is not specifically addressed in such areas as information assurance, security and protection? With the ubiquitous nature of computing and the Internet, the manner in which information is managed, shared and preserved has changed drastically. Such issues as intellectual property (IP) protection, security, privacy rights, confidentiality, authorization, access control, authentication, integrity, non-repudiation, revocation and recovery are at the heart of many tasks that computer users have to face in different degrees. These issues significantly impact both the technology used for information processing (namely hardware, software and the network) and the people who perform a task related to handling of information (i.e., information specialists.) Thus, the major elements are <u>information</u>, <u>technology</u> and <u>people</u>, with *information* being the prime element. This departure from system orientation and emphasis on information itself implies that information must be considered in its most general form, which would encompass multimedia, multi-modal, and multidimensional information for which the traditional methods may not be sufficient. In addition, information fusion is at the heart of the matter. Take, for example, the general security area and one its most essential elements, biometrics. Whether it is face images of 2D or 3D, voice samples, retina scans or iris scans, the multimedia area plays a major role in this area. Similarly, information protection, say intellectual property preservation, clearly overlaps significantly with our field. Examples range from watermarking of objects to content-based comparison of digital documents for identifying significant overlap. Despite these obvious examples, the current discussions on information assurance still center around securing the "information system" with a clear emphasis on protection against unauthorized access and protection from malicious modifications, both in storage and in transit. Shouldn't parameters that our community hold important, say, quality of service, be part of the general area of assurance?

With respect to media and the arts, I would like to focus on a discontinuum that currently exists between multimedia technologies and multimedia content. Arts are mostly about experiences. Many (including Ramesh Jain, a leading multimedia pioneer) have argued for inserting yet another layer, namely "experience" to the traditional hierarchy of "data, information, knowledge, wisdom, ...", where data and information are used to bring about "experiences" which may lead to knowledge. For example, a movie at the data level would be identified by pixels, colors, transitions, etc., and at information level by recognized scenes and characters. All these, however, must convey a certain experience that the artist intended to communicate. Looking at the state of multimedia technology, it seems that sensory data processing, transmission and fusion, data abstraction and indexing, and feedback control, provide new paradigms to artists for engaging their audience in experiences that were not possible before. But how easy are these tools for usage by an artist who is not a programmer? In other words, the tool builders of multimedia community have certainly mastered data and information processing, but the leap to creating new experiences have been left to those artists who are technically astute and have been able to master the digital multimedia technology. Speaking with a friend, who is an internationally recognized composer, about the gap between multimedia tools and artistic contents, he commented that in order to bridge the gap, a phenomenon similar to what happened in the nineteenth century Europe is needed in the digital age. He explained that in those days, piano was the center of all social activities — from homes to schools and other public places — and families and friends gathered around their piano at night, to enjoy music by the best composers of the time. Those composers knew the instruments well and communicated directly with the instrument builders. On the other hand, the instrument builders knew the work of the composers well and built instruments that could make their art better. For example, by knowing the intension of the composer in using "fortissimo", the piano builders perfected certain features that enabled such variations in sound. He concluded that such a dialog between today's tool builders and artists would be of utmost importance. Well... in the 21st century, the computer has become the centerpiece of activities in homes, schools, and many public places. The question is, do we have master artists who can effectively use computers in creating new experiences that were not possible before? Are the existing tools and techniques adequate for artists who would want to make use of the digital technology in their work, or are they limiting factors when it comes to communicating experiences? Some believe that digital arts in general are not ready to claim a leadership role in the aesthetic and emotional development of artistic endeavors across the society. Should the two groups – multimedia technologists and the artists – begin true collaboration in a manner similar to that of the 19th century leading to the creation of piano, we may begin to see a transition... a transition away from media content that uses forms with origins in pre-digital age!

Dragutin Petkovic

Basic technologies and market trends that will significantly help the development of multimedia information systems but which will be mainly addressed by industry and market forces are as follows:

- The volume of information is rapidly increasing, both in size and richness. Media data is growing much faster than other data. At the same time abilities, time and patience of potential creators, editors, and users of such information are staying constant, and the cost of human labor is not going down.
- Cost/performance of all hardware is advancing rapidly: CPU, storage, memory and even network infrastructure can be assumed to be free and not an obstacle.
- Capture and delivery devices as well as sensors are showing rapid advances as well. They range from digital cameras, multimedia cell phones, handheld devices, PCs, to small wireless cameras, and every year they are smaller, more powerful and cheaper.
- Basic multimedia software (operating systems, compression, media players and tools) is reaching maturity.
- Software interface standards are also progressing well and are largely being resolved by international standard bodies and by cooperation of academia, Government and industry. For example: MPEG 4 and MPEG 7 (related to video metadata organization), XML, industry specific metadata standards, WWW services and various wireless standards promise to enable easy connectivity between various building blocks in terms of SW compatibility.
- Media, gaming and entertainment applications are part of our daily lives. New forms of entertainment (Interactive TV, real video on demand) are receiving increased attention by major media corporations
- Integration of media devices, PCs, and WWW at home (or anywhere) offers yet another new frontier for PC, software, and electronic manufacturers.

However, there are some fundamental challenges inhibiting wider proliferation of multimedia information systems that must be addressed by research community:

- Successful multimedia indexing, cross-indexing, summarizing and visualization are today done only by humans. This is a slow and expensive process, and it is feasible only for a few applications where the cost of this process can be amortized over large number of paying customers, such as major entertainment, gaming, and media applications.
- On the usage side, people have only limited amount of time to devote to usage of multimedia information systems, while at the same time the volume of data is growing rapidly. Users are impatient, want easy to use, robust and intuitive systems, expect timely data and fairly accurate indexing, and want the ability to navigate and quickly browse multimedia, rather than use simple linear play (except for entertainment).

David Gibbon

Over the past few years, there has been much emphasis on algorithms and systems for managing structured video such as broadcast television news. While this was a logical starting point for algorithm development, we must now focus on less structured media. Media processing algorithms such as speech recognition and scene change detection have been proven to work well enough on clean, structured data, but are known to have poorer performance in the presence of noise or lack of structure. Professionally produced media amounts to only the tip of the iceberg — if we include video teleconferencing, voice communications, video surveillance, etc. then the bulk of media has very little structure that is useful for indexing in a database system. This latter class of media represents a potentially rich application space.

Accuracy of the individual media processing algorithms is steadily improving through research as well as through microprocessor advances. Multimodal processing algorithms will benefit from the improvements in these constituent algorithms and they will improve intrinsically through intelligent application of domain specific knowledge. The infrastructure for many advanced multimedia systems is available today and will be widespread in a few years. This comprises broadband networking to consumers, terminals (PCs) capable of high quality video decoding and encoding, affordable mass storage devices suitable for media applications, and, importantly, media players (some even with limited media management capability) bundled with operating systems. All of these trends favor the diffusion of multimedia and motivate multimedia research.

Shahram Ghandeharizadeh

Generally speaking, a multimedia system employs human senses to communicate information. Video communication, for example, employs human sight. Its essence is to display 30 pictures a second to fool the human visual perception to observe motion. It is typically accompanied by sound to stimulate our auditory sense. This communication might be either real-time or delayed. An example of a real-time communication is a video teleconference involving multiple people communicating at the same time. This form of communication is almost always interactive. Delayed communication refers to a pre-recorded message. While this form of communication is not necessarily interactive, it is both common place and popular with applications in entertainment, education, news-dissemination, etc. Movies, news clips, and television shows are one example of delayed communication.

Humans have five senses: in addition to our visual and auditory senses, we use our touch, taste, and smell senses. Assuming that all these five senses can be exercised by a system, it may produce a virtual world that is immersive. Touch along with human motor skills is starting to receive attention from the multimedia community. Its applications include entertainment, education, scientific applications, etc. As an example, consider students in a dental program. Their education entails training to obtain sufficient dexterity to perform a procedure such as filling a cavity. This training is typically as follows: a student observes an instructor and then tries to repeat the instructor's movements. An alternative to this is to employ a haptic glove equipped with sensors that record the movement of the instructor when performing a procedure. Next, the student wears the glove and the glove controls the movement of the student's hand to provide him or her with the dexterity of the instructor. The glove may operate in a passive mode where it records the student while performing the operation and then gives feedback as to what the student might have done incorrectly when performing the procedure. A challenge of this research area is to develop mechanical devices that do not result in injuries while training students, data manipulation techniques that compensate for different hand sizes, etc.

Taste and smell are two senses that are at the frontiers of multimedia research. At the time of this writing, it is unclear how one would manipulate these two senses. An intersection of Computer Science, Biomedical Engineering, and Neuroscience may shape the future of this area. This intersection currently offers devices such as Cochlear implants for the hearing impaired individuals, organ transplants such as human hands with nerve re-attachments, and devices that feed data directly into the visual cortex of blind people to provide them with some sight. Given advances in computer hardware technology, it is not far fetched to envision future MEMS devices that interface with our neurons to stimulate our senses in support of taste and sounds.

In passing note that video required approximately a century of research and development, e.g., Thomas Edison authored the first camcorder patent. Maybe in the 21st century, we will be able to develop true multimedia systems that exercise all our five senses.

2. What do you see as major present and future applications of video databases?

Nevenka Dimitrova

With the storage capacity doubling every year, current consumer storage devices should reach the terabyte range by 2005. That means that the professional range will be reaching exabytes. We should look at the generators of the content, not only the 3500 movies that Hollywood and its equivalents around the world produce every year, but also all the camera devices in surveillance, mobile communication, live event streaming, conferencing and personal home video archives.

In creating video databases we travel the round trip: from brains to bits and back. In film production, it first starts with an idea expressed in a script, and then production and capture of this idea into bits. Accessing this information in a video database requires enabling to travel from the bits back to consumption and playback. The applications are in enabling to travel this path from bits back to brains in the enterprise, home environment and accessing public information.

Forouzan Golshani

The major application areas for video databases continue to be education, law enforcement (surveillance, access control, criminal investigations), and entertainment. Here I am including such areas as medicine within the general category of education, and entertainment is broadly used to include sports as well as various forms of arts. Some other applications include: sales and marketing support, advertising, manufacturing and industrial monitoring. Another interesting application area, not tapped to its full potential, is rehabilitation.

The biggest obstacle facing the video database area is a lack of sound business model by which an enterprise can maintain its ownership of the intellectual property and have a reasonable return on its investment. Clearly, we know how to do this on traditional media, e.g., cinema and video rental outlets. Ideally, soon we will have comparable business models for the digital era.

Thomas Huang
Current real-world useful video databases, where the retrieval modes are flexible, and based on a combination of keywords and visual and audio contents, are nonexistent. We hope that such flexible systems will in the future find applications in: sport events, broadcasting news, documentaries, education and training, home videos, and above all biomedicine.

Shahram Ghandeharizadeh
Video-on-demand is an application of video databases that is long overdue. This application faces challenges that are not necessarily technical. These include how to enforce copyright laws, pricing techniques and frameworks that benefit all parties, etc. There are some signs of progress in these directions. As an example, Time Warner is offering four channels of video-on-demand with its digital cable set-top box in Southern California; TiVo enables one household to email an episode of their favorite show to another household, etc.

Another immediate application is to provide storage and bandwidth for households to store their in-home video libraries on remote servers. This service would empower users to quickly search and discover their recorded video clips. A user may re-arrange video segments similar to shuffling pictures in a picture album. A user may organize different video clips into different albums where a sequence might appear in multiple albums. Such a digital repository must provide privacy of content (authentication, and perhaps encryption of content), fast response times, and high throughputs.

Ramesh Jain
Corporate video is one area that I feel has not received as much attention from researchers, as it should. Corporate, training, and educational videos are ready for modernization. Unfortunately at this time tools to manage these are lacking.

What the field needs is a 'videoshop' that will allow people to quickly enhance and manipulate their videos. Powerful editing environment for presenting videos will also help. But the most important need is for a simple database that will allow storing and accessing videos and photographs from cameras onto our systems easily. Current research is addressing problems that may be useful about 10 years from now, and is ignoring important problems to make video databases a reality today.

David Gibbon
There are several major application areas for video databases. We must first look to the entertainment industry, if we are searching for commercially viable applications of these technologies. Video-on-demand systems have been studied and developed for a long time but we have yet to see widespread availability of these systems due to economic constraints. For several years pundits have been predicting that VoD services were just around the corner. Given this history, it

would be unwise to predict this yet again. However, it is clear that storage, compression, and delivery technologies are on a steady trend which reduces the overall cost of operation of these systems. One would conclude that the economics should eventually turn favorable. Another driving force is that consumer expectations in this area are being raised by the availability of advanced PVRs (digital video recorders with personalization.)

A second major application area for video databases is in the video production process. There is a slow but steady trend to automate the production process through the use of digital asset management systems. There is clearly a need for more research here, since the bulk of the video footage is "raw" field material with little structure. In fact, the production process can be thought of as one of creating structure. Video producers will also continue to enhance the quality and availability of the archives of the results of their work to extract value and possibly to repurpose it, but the issues in this case are more of engineering and standardization rather than signal processing research.

We are also beginning to see significant application-specific video databases for academic and pedagogical purposes. Two examples of these are the Survivors of the Shoah Visual History Foundation's visual histories archive, and the Museum of Television and Radio. This trend is encouraging to and should be supported by video database researchers. One challenge here is to balance the dichotomy of the need for customization with the desire for standardization and interoperability.

Finally, it is interesting to consider peer-to-peer file sharing systems as the antithesis of video on demand systems in several respects (video quality, structure of metadata, homogeneity of content, rights management, centralization, economic model, terminal device, etc.). These systems can be thought of as loosely organized distributed video databases. Methods developed to manage this data to promote ease of use and to add features should be applicable to other application areas as well.

Dragutin Petkovic

Some of the compelling possibilities for multimedia information systems are numerous, such as:

- Entertainment and gaming over TV, cable, wireless, or WWW including future radically novel forms that are now in early research stage.
- Training, education, and visual communication where presentations, important talks, meetings and training sessions are captured, indexed, cross-indexed, and summarized automatically and in real or near real time.
- Surveillance and security where large number of video and other sensors deliver and process huge amount of information which has to be adequately analyzed in real time for alerts and threats (including predicting them) and presented so that people can digest it effectively.
- Monitoring and business or government intelligence where vast resources on the WWW and on file systems are constantly monitored in search and prediction of threats (in real or near real time).
- Multimedia information that can be quickly found and delivered to the user anywhere (medical, bio-hazard, troubleshooting, etc.).
- Military applications involving unmanned vehicles with variety of sensors and remote controls.

- Consumer and end-user created applications using home entertainment, WWW and other systems.

We are also certain that in a few years from now we might see some totally new applications that we never even thought about today.

Alberto Del Bimbo

Video database applications will allow faster search of information units ranging from a few seconds of video to entire movies (up to 2 hours). The use of this information goes from the composition of news services by the broadcast industry to private entertainments. Private entertainment applications will probably be available soon to the masses and require fast and effective access to large quantities of information. For enterprise applications the focus will be on the possibility of automatic annotation of video streams. We must distinguish between live logging (real-time automatic annotation for short-term reuse) and posterity logging (automatic annotation for storage and successive reuse) applications. In the near future, it is possible that live logging will be much more interesting than posterity logging for many enterprises and, certainly, for the broadcast industry. Live logging will reduce the time-to-broadcast and, differently from posterity logging, has little impact on the company organization.

Al Bovik

For this question I will be somewhat less speculative! But, perhaps, equally nonspecific. My reason for this is that I think that *visual* databases (not just video) are going to become quite ubiquitous. In some forms, they already are. The Internet is a gigantic, completely heterogeneous visual (and other data forms) database. Most home PCs either have simple resident visual databases (files of photos people they've taken or have sent) or similar databases at commercial websites. I know that I do, and so do other members of my family. Nearly every scientific enterprise today involves imaging in one form or another, and this usually means cataloging huge amounts of data. This is especially true in such visual sciences as the remote sensing, medical, geophysical, astronomical/cosmological, and many other fields. And of course this is all obvious and what is driving the field of visual database design. In my view, the two largest problems for the farther future are, in order of challengingness: (1) visual display and communication of the data, for I assume that the data is not only to be computer-analyzed, but also viewed, by all types of people, of all ages and educational levels, with all types of computing, network, and display capabilities (implying complete scalability of all design); answers for this will, in my opinion, be found along the lines in my answer to the first question. (2) Capability for content-based visual search, meaning without using side or indexing information which, ultimately, cannot be relied upon. The answer to this second question is difficult, but again, I think the clues that are to be found reside within: humans are still better at general visual search than any algorithm promises to be in the near future, viz., ask a human to find unindexed photos containing "elephants," then ask an algorithm … yet I believe that the processes for search that exist in humans are primarily computational and logical and can be emulated and improved on, with vastly greater speed. Yet, current visual search algorithms do not even utilize such simple and efficient mechanisms as visual fixation and foveation. I believe also that the answers to be found in this direction will be found by working with experts in visual psychology.

3. What research areas in multimedia would you suggest to new graduate students entering this field?

Ramesh Jain

I would suggest the following areas:

How to bring domain semantics, identify all other sources of knowledge and use audio and video together to build a system? I would deemphasize the role of image processing and would emphasize closed captions, keyword spotting, and similar techniques when possible. In fact, I would make information the core, and the medium just the medium.

The assimilation of information from disparate sources to provide a unified representation and model for indexing, presentation, and other uses of 'real' multimedia data — audio, video, text, graphics, and other sensors is another interesting problem that I would like to explore.

Another very interesting research issue at the core of multimedia systems is how to deal with space and time in the context of information systems.

Rosalind Picard

I'd suggest learning about pattern recognition, computer vision and audition, machine learning, human-centered interface design, affective interfaces, human social interaction and perception, and common sense learning systems. These are areas that are likely to have a big impact not only on multimedia systems, but in the design of a variety of intelligent systems that people will be interacting with increasingly.

With respect to pattern recognition and machine learning there is particular need for research in improving feature extraction, and in combining multiple models and modalities, with the criterion of providing results that are meaningful to most people. I am happy to see a new emphasis on research in common sense learning, especially in grounding abstract machine representations in experiences that people share. The latter involves learning "common sensory information" — a kind of learned common sense — in a way that helps a machine see which features matter most to people.

Al Bovik

I would encourage them to seek opportunities for cross disciplinary research in Multimedia Processing and in Computational Psychology and Perception.

David Gibbon

In addition to being a rewarding field to pursue, multimedia is also very broad. If a student has a background and interest in signal processing, then such topics as audio segmentation (detecting acoustic discontinuities for indexing purposes) or video processing for foreground/background segmentation may be appropriate. On the other hand, if data structures and systems are of interest, then students can work to benefit the community by developing indexing and storage algorithms for generic labeled interval data, or for searching phonetic lattices output from speech recognizers. Finally, for those with a communications bent, there is clearly room for improvement in techniques for dealing with loss, congestion, and wireless impairments in multimedia communications systems.

Alberto Del Bimbo

Pattern recognition, speech analysis, computer vision, and their applications are important research areas for Multimedia. They support new man-machine interaction with multimedia information, based on gesture or motion capture or affective computing. In video databases, they can provide the means for automatic annotation of video streams, and, supported by knowledge models, the understanding of video contents.

Forouzan Golshani

As stated above, our field has been mostly about technology development. I would encourage students entering this area to consider the end results much more seriously than before. Engineering and computer science curricula are dominated by the question "HOW": how to solve a problem. We must think about the "WHY" question too! By working in interdisciplinary teams that include other students specializing in arts, bioengineering, or sports, engineering students immediately become aware of the problems that the end user faces. The best scenario will be to have the creativity from the domain (arts), and the methodology from science. An added benefit is that the entire team will be exposed to "problem solving in the large", since interdisciplinary projects bring out some real engineering issues. These issues are not generally encountered when working on a small scale project.

Thomas Huang

Important research areas include:

- Video understanding, based on visual and audio contents as well as closed captions or even transcripts if available. Audio scene analysis is an important subtopic.
- Interplay of video processing and pattern recognition techniques with data structure and management issues, when the database is HUGE.
- Flexible user interface. More generally, how best to combine human intelligence and machine intelligence to create a good environment for video data annotation, retrieval, and exploration.

Dragutin Petkovic

Given the above trends and challenges, we suggest the following four basic areas that research community (in collaboration with industry and Government) must work on in order to fulfill the promise of multimedia information systems:

- <u>Automated Real-Time Indexing, Cross-indexing, Summarization, and Information Integration</u>: While we can capture, store, move, and render "data bits" we still lack the ability to get the right meaning out of them, especially in case of video, sound and images. Issues of automated and real-time indexing, cross indexing of heterogeneous data items, information integration and summarization are largely unresolved. Real time or near real time performance are also critical since in many cases the value of information is highest at the time close to its creation. One particular building block that offers great promise but requires more work is robust speech recognition that would adapt to many domain specific vocabularies (e.g. computer science, medical), noise, speaker accents etc.

- Search, Browse, and Visualization: On the receiving side, users have only so much time (and patience) to devote to using such systems. At the same time the volume of information they can or have to assimilate as part of their work or entertainment is rapidly growing. Therefore, the second challenge for research community is to provide solutions for easy search, and automated presentation and visualization of large volume of heterogeneous multimedia information that is optimized for the human users and related delivery devices.

- Making Systems more "Intelligent": We need to make systems that are aware of general context, that can adapt, learn from previous experience, predict problems, correct errors and also have ability to manage themselves to a large extent. Systems should adapt their search and indexing algorithms based on context (i.e. topics of the talk, speaker's accent). In case of very large systems, in times of crisis, resources should be intelligently focused on parts of the data space or servers where the most likely data that is searched for could be found in shortest possible time. Systems should also learn and adapt to the users, "work" with users and present the information in such a way as to optimize information exchange with particular users and their devices. In case of failure or problems in delivery of media data, such systems should be able to adapt, correct and do the best job possible, given the constraints.

- New applications: We need to explore totally new applications, usage and user interfaces for multimedia, benefiting from smaller and cheaper devices, abundant hardware and ability to have all of them interconnected all the time with sufficient bandwidth.

Nevenka Dimitrova
I would delineate the areas into content analysis, feature extraction, representation and indexing, and potential applications.

Multimodal content understanding for all above four areas is the least explored area right now. We have a fair understanding of computer vision technology its applications and limitations. We have also a fair understanding of speech recognition, but to a much less extent of audio scene content analysis and understanding. However, we still have rudimentary approaches to a holistic understanding of video based on audio, visual and text analysis, and synthesis of information.

We need to consolidate the theoretical foundations of multimedia research so that it can be considered on an equal footing with the parenting fields. We need to go beyond the simple application of pattern recognition to all the features that we can extract in compressed and uncompressed domain. We need new pattern recognition techniques that will take into account context and memory. This requires meditating more on the new domain of audio-visual content analysis, on selecting mid-level and high-level features that stem from multimodality, and on scaling up indexing to tera-, peta-, and exa-byte databases. The simplest application is retrieval, however, we need to invest brain cycles on other applications, in user interaction modalities, query languages. An alternative to active retrieval and browsing is content augmentation where the query is given by a user profile and acts as a query against all the abstracted information for personalization and filtering applications in a lean backward mode. We have to expand into applications that serve the purpose of content protection, streaming, adaptation, enhancement, using the same techniques as in multimodal content analysis. In addition, in multiple circumstances storage serves as either the source or the target for these other applications and it makes sense to investigate the synergy. For example, video streaming can use the knowledge of important scenes in order to provide variable bit-rate based on the content; Content adaptation can rely on multimodal analysis to use the best features for transcoding and downscaling higher rates video to be presented on mobile devices with limited resources.

Shahram Ghandeharizadeh

I encourage graduate students to work on challenging, high-risk research topics. Typically, research that investigates an intersection of multiple disciplines is most fruitful. As one example, consider material scientists who investigate light sensitive material that produces electrical current in response to light to stimulate an eye's retina (for visually impaired). A natural hypothesis is how should these devices evolve in order to increase a visually impaired individual's perception. One may start to investigate techniques to make recordings from these devices for subsequent playback. One may manipulate the digital recordings to guide material scientists towards an answer. One may even hypothesize about the design of MEMS-based pre-processors to personalize these devices for each individual. This is one example of an intersection of multiple disciplines to develop a high-risk research agenda. There are many others.

4. What kind of work are you doing now, and what is your current major challenge in your research?

Alberto Del Bimbo

In the field of video databases, we are presently working on automatic annotation of video streams, interpreting video highlights from visual information. The area of application is sports videos. In this research, we distinguish between different sports, model a priori knowledge using finite state machines, recover from the imaged view the real player position in the playground, and distinguish significant highlights – like shot on goal, corner kick, turn over, in soccer games. The ultimate goal is to obtain this semantic annotation in real-time.

We are also working on 3D image database content-based retrieval – addressing automatic recovery of the 3D shape of objects from a 2D uncalibrated view, 3D feature extraction, local and global similarity measures for 3D object retrieval. Finally, research is also being conducted on new interfaces for advanced multimedia applications, based on gestures or motion capture. In gesture-based interfaces the user points at and selects, with simple hand movements, multimedia information displayed on a large screen. With motion-capture interfaces, the 3D motion of the full human body or of its parts is captured and replicated in a virtual environment, for collaborative actions and for eventual training.

Ramesh Jain

I am trying to develop experiential environments for different applications. Experiential environments allow a user to directly utilize his senses to observe data and information of interest related to an event and to interact with the data based on his interests in the context of that event. In this different data sources, are just that – data sources.

Two major interesting problems here are correct representations to capture application semantics and assimilating data from different sources into that representation.

In addition to these technical challenges, a major socio-technical challenge is to change the current academic research culture where all the motivation is on publishing papers rather than solving problems.

Al Bovik

I am working closely with visual psychologists on the problems of (a) visual quality assessment of images and video, both full-reference and no-reference; (b) methods for foveated visual processing and analysis; (c) visual ergonomics, especially visual eyetracking, with application to visual processing of eyetracked data; and (d) analyzing human low-level visual search using natural scene statistics. We feel that our major challenge is this: "why do humans direct their gaze and attention where they do, and how can we predict it?" If we could answer this, then there are many wonderful new algorithms that we could develop that would change the face of Multimedia Processing.

Shahram Ghandeharizadeh

My research conceptualized a video clip as a stream with real-time requirements. Recently, peer-to-peer networks such as Kazaa have become popular. Once installed on a desktop, that desktop may download audio and video clips that are requested by a user. At the same time, the desktop becomes an active participant in the peer-to-peer network by advertising its content for download by other desktops. I am interested in developing streaming techniques that enable peer-to-peer networks to stream video to a desktop to minimize the observed latency. Obviously, the system must approximate certain bandwidth guarantees in order to ensure a hiccup-free display.

I am also interested in ad hoc networks and techniques to stream video clips from a source node to a destination node. An example is an inexpensive wireless device that one may purchase for a vehicle. These devices monitor the status of different road stretches and communicate with one another to provide drivers with up-to-date (a few seconds old) congestion updates. This network is ad hoc because all vehicles are mobile and the network is always changing. Given the monitors in minivans and other vehicle types, passengers of a vehicle may request a movie that is located on a remote server. The remote server must stream a video clip across this ad-hoc network to deliver the content to a specific vehicle. Strategies to accomplish such a streaming to minimize response time while maximizing the number of simultaneous displays is one of my current research interests.

I am also studying haptic devices and their data streams. Similar to a video clip, a stream produced by a haptic glove is continuous and pertains to the movement of a human joint as a function of time. I am collaborating with several USC colleagues to investigate techniques to process these streams. Our target application is a framework to translate hand-signs to spoken English for the hearing impaired. The idea is as follows. A hearing impaired person wears a haptic glove and performs hand signs pertaining to a specific sign language, say American Sign Language (ASL). A computer processes these streams and translates them into spoken English words to facilitate communication between this individual and a person not familiar with ASL. We employ a hierarchical framework that starts by breaking streams into primitive hand postures. The next layer combines several simultaneous hand postures to form a hand sign. Either a static hand sign or a temporal sequence of these signs may represent an English letter. Letters come together to form words, etc. Two key characteristics of our framework are: a) buffers at each level of the hierarchy are used in support of delayed decisions, and b) use of context to resolve ambiguity. Our framework may maintain multiple hypotheses on the interpretation of incoming streams of data at each layer of the hierarchy. These hypotheses are either refuted or verified either in a top-down or a bottom-up manner. In a bottom-up processing, future sequence of streams generated by the glove select amongst alternative hypotheses. In a top-down processing, context is used to choose between competing hypotheses.

Forouzan Golshani

I am currently working on two somewhat inter-related projects. They are: distributed media and arts, and multimedia information fusion.

The distributed media and arts is an interdisciplinary <u>Media and Arts</u> research and education program, where engineers and domain experts are educated collectively. It focuses on finding new paradigms for creating human-machine experiences, with the goal of addressing societal needs and facilitating knowledge. An example of what we hope to bring about is the ability to go from motion capture and analysis to:

* free movements within shared space
* organizing gestures into a phrase
* communicating an emotion through movement.

Currently on the campus of Arizona State University, we have a highly equipped performance area called the Intelligent Stage. It was created with the aim of enabling artists to interact with their environment and each other. It is equipped with many types of sensors and actuators, Video tracking, Vicon 3D motion capture, and novel navigation and interaction mechanisms that bring the performer and the audience closer to each other.

Multimedia information fusion is about working with the semantics of data and information objects, and going beyond the low level, machine oriented modes of search and retrieval. The forty-year-old tradition of keyword (or textual descriptions) search is completely inadequate, particularly when it produces such a dismal result. The challenge in multimedia information fusion is that we do not have an effective mechanism that can cross the artificial barriers created by media-specific tools. What is needed is an ontological[*] representation of

---

[*] *An ontology is an enriched semantic framework for highlighting relationships between concepts and correlating media-specific features to the application-specific concepts.*

domain knowledge, which maps features to topic-specific themes and maps themes to user-specific concepts. Once we can create domain-specific audio-visual ontologies, analogues to what exists for natural language terms, we will have great stride in fusion and integration of all types of information regardless of what type of media they come in.

Nevenka Dimitrova

My major challenge currently is to figure out how to improve quality of life in the domains of the TV/media access and personal memories preserving activities of people. My dream is to insert non-invasive technology for content access everywhere, anytime supported by multimodal video content analysis, indexing, and applications to empower people to easily access the "fiction" fiction memory and the "private" memories. This non-invasive technology requires breaking new grounds into multimodal content analysis, blending of pattern recognition, signal processing, and knowledge representation. This technology has to be pervasive, part of the environment, the "ambient," and perceived as "naturally intelligent" by the users. This requirement translates into important breakthroughs in scaling down the algorithms into silicon and affordably inserting them into recording, storage and communication devices. So, in the end, personal video databases for consumers will include a diverse, distributed multitude of content "processors" and "storers" at home and away from home.

Thomas Huang

I am practicing what I am preaching. So the problems listed before are the challenging issues we are working on currently.

Rosalind Picard

I direct research in affective computing: computing that relates to, arises from, or deliberately influences emotion. This may be surprising at first – but it actually grows out of my efforts to build better multimedia systems. This work involves several challenges: One challenge is recognizing, in real-time, if a user is interested, bored, confused, frustrated, pleased, or displeased, and figuring out how to respond intelligently to such cues. At the core of this challenge is the problem of representing and recognizing patterns of human behavior from face, voice, posture, gesture, and other direct inputs to the computer.

Another challenge is that of incorporating mechanisms of affect in the computer system in a way that improves its learning and decision-making abilities (it is now known that when people lack such mechanisms, many of their cognitive decision-making abilities become impaired.) The latter appear to be critical for enabling the computer to respond in more intelligent ways.

Human pattern recognition and learning are intimately coupled with internal affective state mechanisms that regulate attention and assign value and significance to various inputs. When you are looking for something, a human companion learns how to help with your search not merely by hearing the label you associate with the retrieved item, but

---

Just think of how many interpretations the word "plant" may have if used as a search parameter! An ontology makes explicit all of the relevant interpretations of a given concept.

also by sensing a lot of other context related to your state: how hurried you are, how serious you are, how playful you are, how pleased you are, and so forth. The label you attach to what you are looking for is only part of the information: a smart companion gathers lots of other context, without burdening you with articulating such things as logical queries. The human companion tries to infer which of the many inputs is likely to be most important; he doesn't demand that you stop what you are doing to rank them.

Affective "human state" characteristics play an important role not only in disambiguating a query, but also in reinforcing what are proper and improper responses to your quest. If your state signifies pleasure with the results, this indicates a positive reinforcement to the companion -- enabling better learning of what works. If you look displeased, then this signals negative reinforcement – and the need to learn how to do better. Such cues are important for any system that is to learn continuously. They are related to work on relevance feedback – but the latter techniques currently only scratch the surface of what is possible.

David Gibbon
In keeping with the trend toward focusing on unstructured multimedia processing, we are interested in the domain of video telecommunications. The goal of this work is to improve the ease of use of and enrich the user experience of video telecommunications systems through intelligent applications of media processing algorithms. One of the main challenges here relates to the media quality. Typical conferencing applications have poor acoustical and lighting conditions in comparison with broadcast television. The quality of the transducers (microphones and cameras) is also subpar. Further, the grammar and fluency of conference participants is far removed from that of typical professionally produced video content.

Our research center is focused on several other areas related to multimedia. There are efforts underway in the areas of multimedia data mining, indexing, and retrieval, wireless delivery of multimedia to a range of terminal devices with various capabilities, and multimedia communications over Internet protocols.

INDEX